2019 Devices for Integrated Circuit (DevIC 2019)

Kalyani, India
23-24 March 2019

IEEE Catalog Number: CFP19K13-POD
ISBN: 978-1-5386-6723-1

**Copyright © 2019 by the Institute of Electrical and Electronics Engineers, Inc.
All Rights Reserved**

Copyright and Reprint Permissions: Abstracting is permitted with credit to the source. Libraries are permitted to photocopy beyond the limit of U.S. copyright law for private use of patrons those articles in this volume that carry a code at the bottom of the first page, provided the per-copy fee indicated in the code is paid through Copyright Clearance Center, 222 Rosewood Drive, Danvers, MA 01923.

For other copying, reprint or republication permission, write to IEEE Copyrights Manager, IEEE Service Center, 445 Hoes Lane, Piscataway, NJ 08854. All rights reserved.

*** This is a print representation of what appears in the IEEE Digital Library. Some format issues inherent in the e-media version may also appear in this print version.

IEEE Catalog Number: CFP19K13-POD
ISBN (Print-On-Demand):   978-1-5386-6723-1
ISBN (Online):   978-1-5386-6722-4

Additional Copies of This Publication Are Available From:

Curran Associates, Inc
57 Morehouse Lane
Red Hook, NY 12571 USA
Phone: (845) 758-0400
Fax: (845) 758-2633
E-mail: curran@proceedings.com
Web: www.proceedings.com

3rd international conference on "2019 Devices for Integrated Circuit (DevIC)"

Table of Contents

Message from the General Chair & Editor		xi

1. **Noise Analysis of 1.0 THz GaN IMPATT Source**

   Sudip Chakraborty, Aritra Acharyya, Arindam Biswas , J. N. Roy   1-4

2. **Forecasting Solar Potential Using Support Vector Regression**

   Subham Shaw , Prakash Marimuthu   5-8

3. **The Quantum Cost, Garbage Outputs and Constant Input Optimized Implementation of 2:4 Decoder Using Peres Gate**

   Heranmoy Maity, Arindam Biswas, Anup Kr. Bhattacharjee , Anita Pal   9-11

4. **A Cost Effective Design and Implementation of Arduino Based Sign Language Interpreter**

   Anirbit Sengupta, Md. Tausif Mallick , Abhijit Das   12-15

5. **Hamming Code Generators using LTEx Module of Quantum-dot Cellular Automata**

   Chiradeep Mukherjee, Saradindu Panda, Asish K Mukhopadhyay , Bansibadan Maji   16-20

6. **Role of Stress/Strain Mapping in Advanced CMOS Process Technology Nodes**

   T. P. Dash, J. Jena, E. Mohapatra, S. Dey, S. Das , C. K. Maiti   21-25

7. **Electromagnetic Bandgap Formation in Two-Dimensional Photonic Crystal Structure with DNG Materials under TE Mode**

   Papri Chakraborty, Ratul Ghosh, Anwesha Adhikary, Arpan Deyasi , Angsuman Sarkar   26-29

8. **Design and performance analysis of all-optical Soliton based 4-bit two's complement generator using Reflective Semiconductor Optical Amplifier**

   Kousik Mukherjee, Kajal Maji , Kajal Maji   30-34

9. **A Proposed High-k Dielectric Based Thin Film Transistor for Next Generation Backplane Display Technology**

   Abhishek Kumar Singh, Ashutosh Kumar Dikshit, Brahmdutta Dixit, V.V. Kharche, Kamal, J.S. Rana, P. Chakrabarti, A. Pandey   35-38

10. **Synthesis of Morphological- Variant ZnO nanostructures**

    Ashutosh Kumar Dikshit, Kamal, Abhishek Kumar Singh, J.S. Rana, Rohit K Singh, Nillohit Mukhrjee,P. Chakrabarti   39-41

11. **Analytical Model of Subthreshold Swing in Triangular-Shaped FinFET**

    Subham Banerjee , Buddhadev Pradhan   42-44

3rd international conference on "2019 Devices for Integrated Circuit (DevIC)"

12. **Extensive analysis of band alignment engineering on the open circuit voltage performance of a GaAs/GaSb hetero structure solar cell**
    Girija Shankar Sahoo , Guru Prasad Mishra   45-48

13. **Investigation of a nanoscale grooved stepped gate MOSFET to explore the self-heating effect**
    Subhashree Bhol, Sikha Mishra, Soumya Mohanty , Guru Prasad Mishra   49-52

14. **An extensive analysis of In0.53Ga0.47As/InP surrounding gate MOSFET to enhance the electrostatic performance using δ-doped technique**
    Soumya Mohanty, Sikha Mishra, Sweta Mohanty , Guru Prasad Mishra   53-57

15. **Performance analysis of optical logic XOR gate using dual-control Tera Hertz**
    Optical Asymmetric Demultiplexer (DCTOAD)
    Kajal Maji , Kousik Mukherjee   58-60

16. **Nonmonotonous Electron Mobility in Double Quantum Well Pseudomorphic High Electron Mobility Transistor Structure**
    Sangita Panda, Sudhakar Das, Arttatran Sahu, Ajit Panda , Trinath Sahu   61-64

17. **Performance Prediction of Stacked Nanowire Transistors in the Presence of Random Discrete Dopants and Metal Gate Granularity**
    Suprava Dey, Eleena Mohapatra, Jhansirani Jena, Sanghamitra Dash, Tara Prasanna Das , Chinmay Kumar Maiti   65-69

18. **NBTI Degradation and Recovery in Nanowire FETs**

    S. Das, T.P. Dash, S. Dey, E. Mohapatra, J. R. Jena , C. K. Maiti   70-74

19. **Performance Comparison of Body Biasing and Coupling Capacitor Sense Amplifier for SRAM**
    Phuntso Chotten , Akho John Richa   75-78

20. **Characterisation of Ultra-Small Pocket Si0.7Ge0.3 Junction-less Tunnel FET with SOI**
    Suman Lata Tripathi, Shekhar Verma , Namrata Dhanda   79-83

21. **Performance Analysis of FinFET device Using Qualitative Approach for Low-Power applications**
    Shekhar Verma, Suman Lata Tripathi , Mohinder Bassi   84-88

22. **Analytical Threshold Voltage Model of Schottky source/drain (Schottky-S/D) double gate-all-around (DGAA) Field-Effect-Transistors (FETs)**
    Arun Kumar, P.S.T.N. Srinivas , Pramod Kumar Tiwari   89-93

23. **Performance and Opportunities of Gate-All-Around Vertically-Stacked Nanowire Transistors at 3nm Technology Nodes**
    Suprava Dey, Tara Prasanna Dash, Eleena Mohapatra, Jhansirani Jena, Sanghamitra Das , Chinmay Kumar Maiti   94-98

24.	Vertically-Stacked Silicon Nanosheet Field Effect Transistors at 3nm Technology Nodes	
	Tara Prasanna Dash, Suprava Dey, Eleena Mohapatra, Sanghamitra Das, Jhansirani Jena , Chinmay Kumar Maiti	99-103
25.	Adaptive JAYA Optimization Technique for Economic Load Dispatch Considering Valve Point Effect	
	Sourav Basak , Swaraj Banerjee	104-107
26.	Role of Body Effect on Threshold Voltage of Strained Si-SixGe1-x MOSFET	
	Swarnav Mukhopadhyay , Arpan Deyasi	108-112
27.	A new algorithm for single line outage estimation	
	Mehebub Alam, Shubhrajyoti Kundu, Siddhartha Sankar Thakur , Sumit Baneerjee	113-117
28.	Impact of trapped interface charges on short channel characteristics of WFE high-K SOI MOSFET	
	Priyanka Saha, Pritha Banerjee, Dinesh Kumar Dash , Subir Kumar Sarkar	118-123
29.	A 1.8V 204.8-μW 12-Bit Fourth Order Active Passive ΣΔ Modulator for Biomedical Applications	
	Arifuddin Sohel, Maliha Naaz, Aayesha Al Khadir , Amena Najeeb	124-127
30.	Comparative Investigation of DSG-MOSFET and Some analysis on its Performance	
	Babita Kumari, Dr. Kaushik Mazumdar , Aniruddha Ghosal	128-130
31.	Effect of Uniaxial Strain on Properties of Blue Phosphorene-CNT Heterojunction	
	Arnab Mukhopadhyay, Amretashis Sengupta , Hafizur Rahaman	131-133
32.	Mobility Modulation in V-shaped Double Quantum Well based HEMT Structure	
	Sangeeta K Palo, Ajit K Panda, Trinath Sahu, Narayan Sahoo, , Tarini C Tripathy	134-136
33.	Up-state and Down-state Capacitance Measurement in RF MEMS One-bit Switch Designed at Microwave Frequency Range	
	Pampa Debnath, Arpan Deyasi , Angsuman Sarkar	137-140
34.	Role of Internet of Things (IoT) in Smart Farming: A Brief Survey	
	Monu Bhagat, Dilip Kumar , Deobrata Kumar	141-145
35.	A Cyber Communication Package in the Application of Grid Tied Solar System	
	Rajib Majumdar, Pritam Gayen, Subhradip Mondal, Archak Sadhukhan, Pradip Das , Indranil Kushary	146-150

36.	Counter Based Low Power, Low Latency Wallace Tree Multiplier Using GDI Technique for On-chip Digital Filter Applications	
	Biswarup Mukherjee , Aniruddha Ghosal	151-155
37.	Effect of AlGaN Back Barrier on InAlN/AlN/GaN E-Mode HEMTs	
	Sarosij Adak, Nisarga Chand, Sanjit Kumar Swain , Angsuman Sarkar	156-160
38.	A Systemic Review on ohe Technological Development in the Field of Phonocardiogram	
	Subhashis Maitra , Deepneha Dutta	161-166
39.	QCA Realization of Reversible Gates Using Layered T Logic Reduction Technique	
	Chiradeep Mukherjee, Saradindu Panda, Asish K Mukhopadhyay , B. Maji	167-171
40.	Set point weighted modified Smith predictor for delay dominated integrating processes	
	Somak Karan , Chanchal Dey	172-176
41.	Gameplay using Reinforcement Learning	
	Dilip Kumar, Ritesh Prasad, Kumar Ritu Raj Singh , Surbhi Singh	177-180
42.	Comparative HRV Analysis of ECG Signal in the Context of Sportsperson under Post-Exercise and Relaxed Condition	
	Prashant Kumar, Ashis Kumar Das , Suman Halder	181-185
43.	Influence of Channel Thickness on Analog and RF Performance Enhancement of an Underlap DG AlGaN/GaN based MOS-HEMT Device	
	Akash Roy, Rajrup Mitra , Atanu Kundu	186-190
44.	Optimisation and Classification of EMG signal using PSO-ANN	
	Virendra Prasad Maurya, Prashant Kumar , Suman Halder	191-195
45.	Design and Implementation of Gabor Filter and SVM based Authentication system using Machine Learning	
	Shalini Singh, Indrajit Das, Md Golam Mohiuddin, Amogh Banerjee , Sonali Gupta	196-201
46.	Pd Doped TiO2 – CuO Mixed Metal Oxide Thin Film Sensor for H2 Sensing Application	
	Bikram Biswas, Anup Dey, Subhashis Roy, Subhashis Roy , Subir kumar Sarkar	202-205
47.	Analytical Modeling and Simulation of Low Power Salient Source Double Gate TFET	
	Bijoy Goswami, Debadipta Basak, Ayan Bhattacharya, Koelgeet Kaur, Sutanni Bhowmick , Subir Kumar Sarkar	206-210

48. **Quasi FGMOS Inverter: A Strategy for low power applications**

Alekhya Yalla , Umakanta Nanda 211-215

49. **Structure and Electronic Properties of TiO$_2$ Nanowires of Different Geometrical Shapes: An Abinitio Study**

Debashish Dash, Chandan Kumar Pandey, Saurabh Chaudhury , Susanta Kumar Tripathy 216-220

50. **Empirical Model of Surface Potential and Simulation Analyses for ZnO TFTs**

Anirudh Aggarwal, Rupam Goswami , Kavindra Kandpal 221-225

51. **Smart ATM Service**

Sayan Hazra 226-230

52. **Comparative Analysis of Underlapped Silicon on Insulator and Underlapped Silicon on Nothing Dielectric and Charge Modulated FET based Biosensors**

Khuraijam Nelson Singh , Pranab Kishore Dutta 231-235

53. **Storing Digital Data in Nucleic Acid Memory with Extended Genetic Alphabet**

Saptarshi Biswas, Subhrapratim Nath, Jamuna Kanta Sing , Subir Kumar Sarkar 236-239

54. **Surface potential based Analytical Modeling of Graded Channel Strained High-k Gate stack Dual-Material Double Gate MOSFET**

Pritha Banerjee, Priyanka Saha, Dinesh Kumar Dash , Subir Kumar Sarkar 240-244

55. **Design and Analysis of High Performance Multiplier Circuit**

Inamul Hussain, Chandan Kumar Pandey , Saurabh Chaudhury 245-247

56. **Production of Few Layer Graphene by Electrochemical Delamination using Alkaline Solution**

Swapan Das, Sunipa Roy , Chandan Kumar Sarkar 248-251

57. **Investigating Photonic Bandgap Width for Metamaterial based PhC Structure from Dispersion Relation**

Sangita Das, Urmi Dey, Soumita De , Arpan Deyasi 252-255

58. **An Analytical Approach of EEG Analysis for Emotion Recognition**

Indronil Mazumder 256-260

59. **A UWB Band-pass Filter with a WLAN Notch based on Multi-mode Resonator Structure for Application in Wireless Communication**

Anirban Neogi, Jyoti Ranjan Panda, Saptarshi Sil, Shibaditya Chakraborty , Anupam Tarafdar 261-263

60.	Boundstates Computation for Double Quantum Well Structure with Pöschl-Teller Potential for MWIR Photodetector Design	
	Avik Chakraborty, Suporna Bhowmick, Debarati Chakraborty, Arpan Deyasi , Angsuman Sarkar	264-268
61.	Analytical Drain Current Model of UTBB SOI MOSFET with lateral dual gates to Suppress Short Channel Effect	
	Arighna Basak , Angsuman Sarkar	269-274
62.	Performance Comparison of 2.5-GHz LC Voltage-Controlled Oscillator for Three Different Technology Nodes	
	Shrabanti Das , Sayan Chatterjee	275-280
63.	Cell thickness optimization of dual junction InGaP/GaAs solar cell against temperature variation	
	Shingmila Hungyo, Khomdram Jolson Singh , Rudra Sankar Dhar	281-285
64.	Metal Grain Granularity Induced Variability in Gate-All-Around Si-Nanowire Transistors at 1nm Technology Node	
	Tara Prasanna Dash, Suprava Dey, Jhansirani Jena, Sanghamitra Das, Eleena Mohapatra , Chinmay Kumar Maiti	286-290
65.	SPICE Parameter Extraction of Tri-Gate FinFETs- An Integrated Approach	
	Tara Prasanna Dash, Sanghamitra Das, Suprava Dey, Eleena Mohapatra, Jhansirani Jena , Chinmay Kumar Maiti	291-294
66.	TCAD Modelling of 30nm Strained-Si/SiGe/Si Channel MOSFET	
	Lalthanpuii Khiangte , Rudra Dhar	295-297
67.	Novel Self-Pipelining Strategy for Efficient Multiplication	
	Rahul Pal, Jayanta Ghosh , Aloke Saha	298-301
68.	Smart Power Theft Detection System	
	Nitin Krishna Mucheli, Umakanta Nanda, Debasish Nayak, P. K. Rout, Sanjit Kumar Swain, Satish Kumar Das , Sudhansu Mohan Biswal	302-305
69.	Analysis of Adiabatic flip-flops for Ultra Low Power Applications	
	Samik Samanta, Rajat Mahapatra , Ashis Kumar Mal	306-309
70.	A Low Power LNA using Current Reused Technique for UWB Application	
	Dhananjaya Tripathy, Debasish Nayak, Sudhansu Mohan Biswal , Sanjit Kumar Swain, Biswajit Baral , Satish Kumar Das	310-313
71.	A Novel Driver less SRAM with Indirect Read for Low Energy Consumption and Read Noise Elimination	
	Debasish Nayak, Umakanta Nanda, Prakash Kumar Rout, Sudhansu Mohan Biswal, Dhananjaya Tripathy, Sanjit Kumar Swain, Biswajit Baral , Satish Kumar Das	314-317

72. **Scholastic Approach towards Economic Digital Mileage Meter with GPS**

Anirban Ghosal, Subham Ghosh, Ayan Saha, Nilanjan Bhattacharjee, Srijan Bhar , Indranath Sarkar
318-321

73. **Comparative Study of High K in Silicon Nano Tube FET for Switching Applications**

Avtar Singh, Chandan Kumar Pandey, Saurabh Chaudhury , Chandan Kumar Sarkar
322-325

74. **High Frequency Performance of AlGaN/GaN HEMTs Fabricated on SiC Substrates**

Eleena Mohapatra, Sanghamitra Das, Tara Prasanna Dash, Suprava Dey, Jhansirani Jena , Chinmay Kumar Maiti
326-330

75. **Noise analysis of Dual Halo Dual Dielectric Triple Material Surrounding Gate MOSFET for RF applications**

Prashant Kumar, Neeraj Gupta, Rashmi Gupta , Amit Kumar Sharma
331-333

76. **Optimization of Electrical Parameters for the Gate Stack Double Gate (GSDG) MOSFET using Simplex-PSO Algorithm**

Dibyendu Chowdhury, Bishnu Prasad De, Kanchan Baran Maji, Sumalya Ghosh, Rajib Kar , Durbadal Mandal
334-336

77. **Analysis of Interface Trap Charges on Dielectric Pocket SOI-TFET**

Chandan Kumar Pandey, Avtar Singh , Saurabh Chaudhury
337-340

78. **Ubiquitous and Emerging Concepts of Sensors**

Manisha Sharma, Monu Bhagat , Dilip Kumar
341-347

79. **Smart Wireless Distribution for Micro Grid System**

Sayan Paramanik, Krishna Sarker, Avijit Chakraborty , Biswajit Mahanty
348-355

80. **Smart Grid Power Quality Improvement Using Modified UPQC**

Sayan Paramanik, Krishna Sarker, Debashis Chatterjee , S.K Goswami
356-360

81. **Survey of Smart Grid Network Using Drone & PTZ Camera**

Sayan Paramanik, Partha Sarathi Sarkar, Koustav Kumar Mondol , com Avijit Chakraborty , Sajib Chakraborty , Krishna Sarker
361-364

82. **Effect of High-K Spacer on the Performance of Gate-Stack Uniformly doped DG-MOSFET**

Satish Kumar Das, Sanjit Kumar Swain, Sudhansu Mohan Biswal, Dedasish Nayak, Umakanta Nanda, Biswajit Baral , Dhananjaya Tripathy
365-369

83. **Bandwidth Increment of Piezoelectric Energy Harvester using Multi-beam Structure**

Sourav Naval, Prasun Kumar Sinha, Nikhil Kumar Das, Ashutosh Anand , Sudip Kundu
370-373

84.	An approach towards the development of refreshable Braille Computer Display Unit	
	Moumita Ghosh, Subham Ghosh, Shivam Sarkar, Kaustav Saha, Anirudha Ray , Biswarup Neogi	374-377
85.	An Improved Edge Detection Method based on Median Filter	
	Preeti Topno , Govind Murmu	378-381
86.	Improvement of Output Power in Piezoelectric Energy Harvester under Magnetic Influence	
	Ashuntosh Anand , Sudip Kundu	382-385
87.	Analysis of Four-Stage Charge Pump Circuit for UHF RFID Tag Design	
	Savio Sengupta, Dipanjan Sen, Subhashis Roy, Sudhabindu Ray , Subir Sarkar	386-389
88.	Analytical modelling and simulation of pseudoresistive circuit techniques for biomedical applications	
	Kulbhushan Sharma, Anisha Pathania , Rajnish Sharma	390-393
89.	Design of an Integrator-Differentiator Block For a Transimpedance Amplifier Using 0.18µm Technology	
	Chander Pratap Singh, Anisha Pathania, Kulbhushan Sharma, Jaya Madan , Rajnish Sharma	394-397
90.	Impact Analysis of Dual Material Double Gate Oxide-Stack Junction-Less MOSFET in RFID Memory Cell Realisation	
	Dipanjan Sen, Savio Jay Sengupta, Subhashis Roy, Sudhabindu Ray , Subir Kumar Sarkar	398-403
91.	Design Approach for Artificial Human Ankle Movement	
	Susmita Das, Dalia Nandidas , Biswarup Neogi	404-407
92.	Improvement of the Gain Accuracy of the Instrumentation Amplifier Using a Very High Gain Operational Amplifier	
	Maitraiyee Konar, Rashmi Sahu , Sudip Kundu	408-412
93.	Trade-off Characteristics of Hysteresis Comparator used in Noisy Systems	
	Aashita Raj, Yaswanth Divvela, Geetanjali Singh , Sudip Kundu	413-417
94.	An Impact of Current &Voltage Harmonics in the Power Quality Issues	
	Tapas Halder	418-423
95.	A Low Power Biopotential Amplifier based on Bulk Driven Quasi Floating Gate Technique	
	Sharma Preeti, Kulbhushan Sharma, H S Jatana , Rajnish Sharma	424-427

96. **Synthesis of PPG Waveform Using PSPICE and Simulink Model**

Ankita Mukherjea , Parshati Chaudhury , Alvin Karkun , Soumalya Ghosh ,   428-432
Subhajit Bhowmick

97. **A Hypothetical Analysis to Study the Variations of Complex Dielectric Permittivity for Detection of Various Stages of Cancer of a Biological Target using Microwave Tomography**

Deborsi Basu , Kabita Purkait   433-440

98. **Radio Frequency Identification based Goal Line Technology for Quick Decision Making in a Football Match**

Samrat Ghosh, Suvam Sasmal, Saptarshi Bhui, Sandipan Dutta, Swarnadeep   441-445
Mukherjee, Arko Majumder , Biswarup Ganguly

99. **Performance Comparison of CMOS and MEMS based Thermal Energy Harvesters using Finite Element Analysis**

Indrajit Sil, Kalyan Biswas , Sagar Mukherjee   446-450

100. **High PSNR based Image Fusion by Weighted Average Brovery Transform Method**

Nidhi Taxak , Sachin Singhal   451-455

101. **A Novel Model for Analyzing Current-Voltage Characterization of SiC MOSFET**

Sabuj Sarkar, Saikat Adhikary , Md. Mostafizur Rahman   456-460

102. **Work-function modulated hetero gate charge plasma TFET to enhance the device performance**

Sasmita Sahoo, Sidhartha Dash , Guru Prasad Mishra   461-464

103. **Design of Mems Based Piezoelectric Energy Harvester for Pacemaker**

Ashuntosh Anand , Sudip Kundu   465-469

104. **Comparative computational study of LD-HfO2 and TiO2 as layered dilectrics in RRAM**

Deependra Chettri, Abinash Thapa, Smita Rai, Pronita Chettri, Chandan   470-473
Kumar Sarkar , Bikash Sharma

105. **Diode & Neutral Point Clamped Five-Level Inverter For the Power Quality Issues**

Rabindranath Das Adhikary, Writwik Balow , Dr. Tapas Halder   474-479

106. **Semiconducting Behaviour of Metal Thin Film composed of Green Synthesized Silver Nanoparticles**

Kaushik Roy, Chandan K. Sarkar , Chandan K. Ghosh   480-483

107. **Performance Enhancement of Non-uniformly Doped Junctionless Transistors by Gate and Dielectric engineering**

Muktasha Maji , Gaurav Saini   484-488

108.	**A Low Power Hardware Implementation of Lifting based Reversible Watermarking for Medical Image**	
	Poulami Jana, Goutam Kumar Maity , Himadri Mandal	489-492

109.	**Performance Analysis of Staggered Heterojunction based SRG TFET biosensor for health IoT application**	
	Sudhansu Mohan Biswal, Sanjit Kumar Swain, Biswajit Baral, Debasish Nayak, Umakanta Nanda, Satish Kumar Das , Dhananjaya Tripthy	493-496

110.	**Hardware Architecture of a Decoder for Fractal Image Compression**	
	Hasanujjaman, U. Biswas , M. K. Naskar	497-500

111.	**Fractal Image Compression of an Atomic Image using Quadtree Decomposition**	
	Hasanujjaman, Arnab Banerjee, Utpal Biswas , Mrinal Kanti Naskar	501-504

112.	**RF/Analog & Linearity performance analysis of a downscaled JL DG MOSFET on GaAs substrate for Analog/mixed signal SOC applications**	
	Biswajit Baral, Sudhansu Mohan Biswal, Sanjit Swain, Satish Kumar Das, Debasish Nayak , Dhananjaya Tripathy	505-509

113.	**Effect of High-K Spacer on the Performance of Non-Uniformly doped DG-MOSFET**	
	Sanjit Kumar Swain, Satish Kumar Das, Sudhansu Mohan Biswal, Sarosij Adak, Umakanta Nanda and Asmit Amlan Sahoo, Debasish Nayak, Biswajit Baral, Dhananjaya Tripathy	510-514

114.	**All-optical Walsh-Hadamard code Generation using MZI**	
	Supriti Samanta, Goutam Kumar Maity and Subhadipta Mukhopadhyay	515-518

Message from General Chair & Editor

It has been a real honor and privilege to serve as the General Chair of the 3rd International Conference on **"2019 Devices for Integrated Circuit (DevIC)"** acronym as **"DevIC 2019"** held on March 23-24, 2019 at Kalyani Government Engineering College, Kalyani, organized by IEEE KGEC Student Branch Chapter in association with Department of ECE, KGEC and technically co-sponsored by IEEE EDS Kolkata Chapter.

The response to the conference's call for papers has been overwhelming and the attendance was equally remarkable. I would like to express my sincere gratitude and thanks to the authors, reviewers, Technical Program Committee members, keynote speakers, organizing committee members, steering teams, patrons, sponsors, delegates, session chairs and student volunteers for their contributions, hard work and support in promoting this notable academic event.

I hope that you will find this proceeding interesting and that you will consider participation in the future "DevIC" conferences.

General Chair **& Editor of the proceedings of**

3rd international conference on "2019 Devices for Integrated Circuit (DevIC)"

Dr. Angsuman Sarkar

Kalyani Government Engineering College

2019 Devices for Integrated Circuit (DevIC)

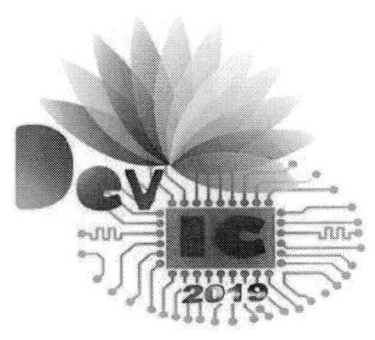

Proceedings of

3rd international conference on

2019 Devices for Integrated Circuit (DevIC)

23-24 March, 2019, Kalyani, Nadia, India

Organized by

IEEE KGEC Student Branch Chapter

In association with

Department of ECE, Kalyani Government Engineering College

Technically co-sponsored by

IEEE EDS Kolkata Chapter

Edited by

Dr. Angsuman Sarkar
Kalyani Government Engineering College

Noise Analysis of 1.0 THz GaN IMPATT Source

Sudip Chakraborty
Department of Electrical Engineering,
NSHM Knowledge Campus,
Arrah, Shibtala, Durgapur, West
Burdwan, West bengal – 713212, India
sudip.0079@gmail.com

Aritra Acharyya
Department of Electronics and
Communication Engineering,
Cooch Behar Govt. Engg. College,
Harinchawra, Ghughumari, Cooch
Behar, West bengal – 736170, India
ari_besu@yahoo.co.in

Arindam Biswas
Centre for Organic Spin-tronics and
Optoelectronics Devices (COSOD) and
Mining Engineering Department,
Kazi Nazrul University,
Asansol, Burdwan,
West bengal – 713340, India
mailarindambiswas@yahoo.co.in

J. N. Roy
Centre for Organic Spin-tronics and
Optoelectronics Devices (COSOD) and
Physics Department,
Kazi Nazrul University,
Asansol, Burdwan,
West bengal – 713340, India
jnroys@yahoo.co.in

Abstract—Noise analysis of 1.0 THz Wz-GaN IMPATT source has been carried out by the authors. A two-dimensional avalanche noise model for IMPATT diodes developed by the authors has been used in the present paper to study the avalanche noise characteristics of the said source. Results reveal that the mean-square noise voltage per unit bandwidth (i.e. noise spectral density) of the source lies in the order of 10^{-16} V^2 s and noise measure remains within the range of 7.3440 – 5.8755 dB due to the variation of bias current within the range of 78.54 – 98.17 mA for a fictitious value of zero series resistance. However, around 3 – 7% increase in noise measure has been obtained by considering earlier calculated series resistance values ranging from 1.5779 – 1.7879 Ω.

Keywords—*Avalanche noise, gallium nitride, noise measure, noise spectral density, terahertz*

I. INTRODUCTION

The THz frequency regime (0.3 – 10.0 THz) has become one of the most attractive frequency regimes among the modern-day researchers due to various fascinating applications and several extraordinary possibilities [1-15]. Among several solid-state THz radiators [16-28], IMPATT oscillators based on Wz-GaN have shown highest potentiality [29]. Only Wz-GaN IMPATT sources can deliver THz power in milli-watts level, while power output of all other available THz semiconductor sources can deliver THz power in the order of few microwatts. Recently the conventional vertical double-drift region (DDR) structure of 1.0 THz Wz-GaN IMPATT diode shown in Fig. 1 has been proposed by Biswas *et al.* [29]. The thickness, radius of all layers as well as doping densities of each of them are already provided in the figure in form of illustrations. Two-dimensional (2-D) large-signal (L-S) simulation of the GaN IMPATT structure exposed the effects of series resistance of the diode on its power delivering capability [29]. It was observed that the series resistance of the diode varied from 1.7879 to 1.5779 Ω due to 78.54 – 98.17 mA change in bias current; the corresponding variations of THz power output and efficiency of the diode are found to be within the ranges 191 – 203 mW and 8.48 – 6.41% respectively.

The THz power output of around 10^2 mW of 1.0 THz GaN source is more than sufficient for all the applications

mentioned earlier. However, one cannot provide a quantitative measure regarding the quality of THz signal generated from GaN IMPATT source without studying the noise performance of it. In this paper, the avalanche noise characteristics of GaN IMPATT source operating at 1.0 THz have been studied. The authors' developed 2-D small-signal noise model [30] is used to carry out the noise analysis in the present work.

II. NOISE SPECTRAL DENSITY

The unsystematic generation of electron-hole pairs as a result of impact ionizations within the multiplication zone of reverse biased DDR structure is the mail cause of its inherent noise. The generation of this kind of noise is directly related to the avalanche multiplication process; as a result it is referred to as avalanche noise. In 2-D small-signal noise simulation technique, two second order differential equations relating real ($e_{nr}(x,x')$, $e_{nr}(y,y')$) as well as imaginary ($e_{ni}(x,x')$, $e_{ni}(y,y')$) parts of noise field ($e_n(x,x') = e_{nr}(x,x') + i\ e_{ni}(x,x')$, $e_n(y,y') = e_{nr}(y,y') + i\ e_{ni}(y,y')$; $i = \sqrt{-1}$) are framed under small-signal condition [30] and those are simultaneously solved under proper boundary conditions at the edge of the active layers in order to calculate the spatial variation of $e_{nr}(x,x')$, $e_{nr}(y,y')$ and $e_{ni}(x,x')$, $e_{ni}(y,y')$ due to the noise sources $\gamma(x')$, and $\gamma(y')$ located at the space points x' and y' respectively. The simultaneous solution of the above-mentioned differential equations requires Runge-Kutta technique [31] of numerical solution of simultaneous differential equations. The magnitude of the resultant noise field can be obtained by using

$$\left| e_n\left((x,x'),(y,y') \right) \right| = \left(\left| e_n(x,x') \right|^2 + \left| e_n(y,y') \right|^2 \right)^{\frac{1}{2}}. \quad (1)$$

The terminal voltage and transfer impedance as a result of a noise source at (x',y') can be obtained as (considering origin $(x = 0, y = 0)$ at left most side of the n-n^+ interface)

$$v_t\left(x',y' \right) = -\int_{y=0}^{y=W_n+W_p} \left(\int_{x=0}^{x=r_0\sqrt{\pi}} \left| e_n\left((x,x'),(y,y') \right) \right| dx \right) dy, \quad (2)$$

$$z_t\left(x',y' \right) = \frac{v_t\left(x',y' \right)}{i_n\left(x',y' \right)}, \quad (3)$$

where $i_n(x',y')$ stands for the average noise current produced in the intervals dx' and dy' as a consequence of the noise sources $\gamma(x')$, and $\gamma(y')$. Therefore, the mean-square noise voltage is calculated from

$$\left\langle v_n^2 \right\rangle = 2\pi\, r_0^2\, q^2\, df \int\limits_{y'=0}^{y'=W_n+W_p} \left(\int\limits_{x'=0}^{x'=r_0\sqrt{\pi}} \left| z_t\left(x',y'\right) \right|^2 \left(\begin{array}{c} \gamma(x')^2 \\ +\gamma(y')^2 \end{array} \right)^{\frac{1}{2}} dx' \right) dy' \qquad (4)$$

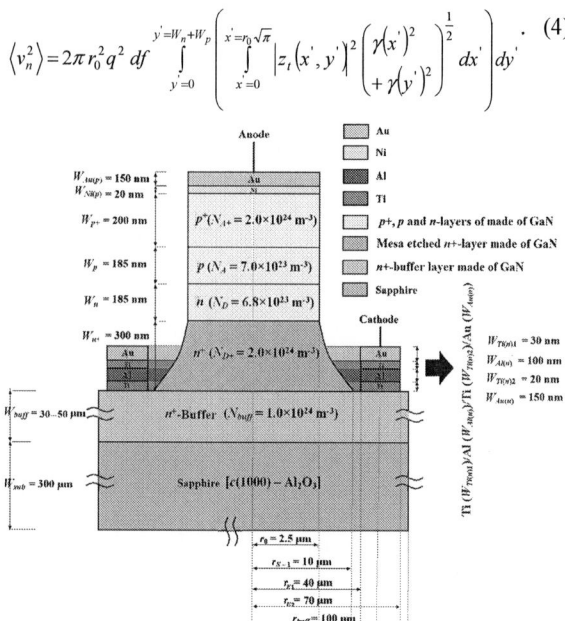

Fig. 1. Illustration of the 1.0-THz GaN DDR IMPATT structure on sapphire substrate grown via metalorganic vapour phase epitaxy (MOVPE) [29].

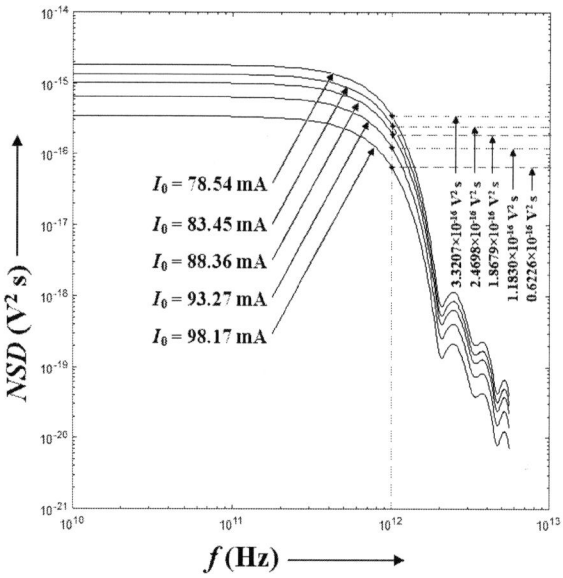

Fig. 2. Mean-square noise voltage per unit bandwidth of the source with frequency.

Therefore its unit is V^2 s. Thus Noise spectral density (*NSD*) can be written as

$$NSD(f) = \frac{\left\langle v_n^2 \right\rangle}{df}. \qquad (5)$$

The *NSD* will be a strong function of frequency since $\left\langle v_n^2 \right\rangle$ is strongly dependent on frequency. The variations of *NSD* of the source with frequency for different values of bias current (I_0) have been illustrated in Fig. 2. It can be seen that the *NSD* of the device remains independence of frequency at lower frequencies (< 100 GHz) and starts to decrease when the frequency increases beyond 100 GHz. Very sharp fall in *NSD* is observed for the frequencies above the operating frequency of the source (i.e. 1.0 THz). It is also noteworthy from Fig. 2 that the *NSD* of the source at any frequency reduces prominently with the enhancement of bias current. It is obvious, because the peak amplitude of the avalanche noise current becomes more insignificant as the DC bias current through the device increases. Therefore, greater bias current always suppresses the avalanche noise in IMPATT devices. The *NSD* values associated with the 1.0 GHz source are found to be 3.3207×10^{-16}, 2.4698×10^{-16}, 1.8679×10^{-16}, 1.1830×10^{-16} and 0.6226×10^{-16} for the bias currents of 78.54, 83.45, 88.36, 93.27 and 98.17 mA respectively.

III. Noise Measure

Noise measure (*NM*) is one of the most important parameters for describing the quantitative assessment of the *NM* of a semiconductor device. The *NM* of IMPATT oscillators can be formulated as [32]

$$M_N(f) = \frac{\left\langle v_n^2 \right\rangle / df}{4 k_B T \left(\left| Z_R(f) \right| - R_S \right)}, \qquad (6)$$

where Boltzmann constant $k_B = 1.38 \times 10^{-23}$ J K^{-1}, absolute temperature *is T*, frequency dependent negative resistance of the diode is $Z_R(f)$ [29, 30], series resistance is R_S.

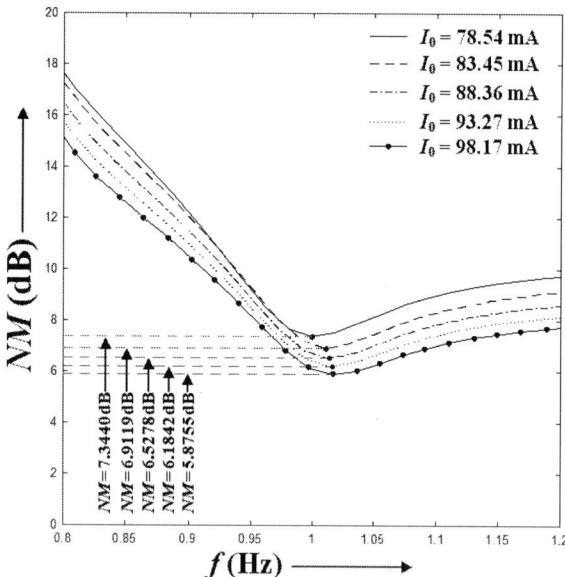

Fig. 3. Variations of noise measure of the source with frequency (assuming $R_S = 0$).

The variations of *NM* (in dB) of the source with frequency have been shown in Fig. 3. These calculations have been carried out for $R_S = 0$. It is observed that the *NM* versus frequency curves attains their minima near the optimum frequency of the source (i.e. around 1.0 THz); it is occurred because negative resistance of the diode (which is at the denominator of equation (6)) becomes maximum near the operating frequency of the source. Also, the decrease of NM with the enhancement of bias current is also observable from Fig. 3. The NM of the device are obtained as 7.3440, 6.9119, 6.5278, 6.1842 and 5.8755 dB for the bias currents of 78.54, 83.45, 88.36, 93.27 and 98.17 mA respectively.

IV. EFFECT OF SERIES RESISTANCE

The influence of R_S on the noise characteristics of 1.0 the source is also investigated here. Variations of noise measure of the source with bias currents for different values of series resistance are illustrated in Fig. 4. The series resistance of the diode as a function of bias current has already been calculated by the authors earlier [29]. The earlier calculated series resistance values, i.e. 1.7879, 1.7370, 1.6834, 1.6303 and 1.5779 Ω for the bias currents of 78.54, 83.45, 88.36, 93.27 and 98.17 mA respectively, are used in this paper for the calculation of NM. It is observed that NM increases from 7.3440 to 7.8429 dB at $I_0 = 78.54$ mA with the increase of R_S from 1.5779 to 1.7879 Ω.

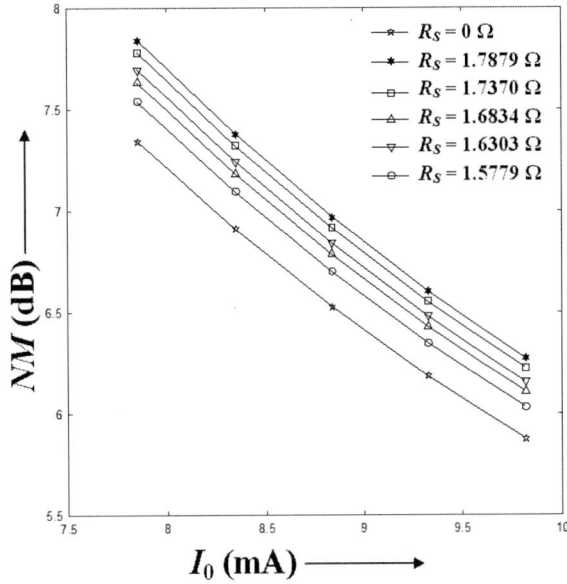

Fig. 4. Variations of noise measure of the source with bias currents for different values of R_S.

V. CONCLUSION

The avalanche noise analysis of 1.0 THz GaN IMPATT source based on Wz-GaN has been presented here. A 2-D small-signal noise simulation technique developed by the authors has been used in the present paper to study the avalanche noise performance of the 1.0 THz GaN IMPATT source. Results show that the mean-square noise voltage per bandwidth of the source lies in the order of 10^{-16} V^2 s and noise measure remains within the range of 7.3440 – 5.8755 dB due to the variation of bias current ranging from 78.54 to 98.17 mA for a fictitious value of zero series resistance. However, around 3 – 7% increase in noise measure has been obtained by considering earlier calculated values of series resistance ranging from 1.5779 – 1.7879 Ω.

The present work is funded by Science and Engineering Research Board (SERB), Government of India, through ECR Award scheme to Dr. Arindam Biswas [grant number ECR/2017/000024/ES].

REFERENCES

[1] P. Martyniuk, J. Antoszewski, M. Martyniuk, L. Faraone, A. Rogalski, "New concepts in infrared photodetector designs," *Appl. Phys. Rev.*, vol. 1, pp. 041102, 2014.

[2] R.M. Woodward, B.E. Cole, V.P.Wallace, R.J. Pye, D.D. Arnone, E.H. Linfield, M. Pepper, "Terahertz pulse imaging in reflection geometry of human skin cancer and skin tissue," *Phys. Med. Biol.*, vol. 47, pp. 3853, 2002.

[3] M. Nagel, P.H. Bolivar, M. Brucherseifer, H. Kurz, A. Bosserhoff, R. Buttner, "Integrated THz technology for label-free genetic diagnostics," *Appl. Phys. Lett.*, vol. 80, no. 1, pp. 154, 2002.

[4] N. Karpowicz, H. Zhong, C. Zhang, K.I Lin, J.S. Hwang, J. Xu, X.C. Zhang, "Compact continuous-wave subterahertz system for inspection applications," *Appl. Phys. Lett.*, vol. 86, no. 5, pp. 054105, 2005.

[5] K. Yamamoto,M. Yamaguchi, F. Miyamaru, M. Tani,M. Hangyo, "Non-invasive inspection of c-4 explosive in mails by terahertz time-domain spectroscopy," *J. Appl. Phys.*, vol. 43, no. 3B, pp. L414, 2004.

[6] K. Kawase, Y. Ogawa, Y. Watanabe, H. Inoue, "Non-destructive terahertz imaging of illicit drugs using spectral fingerprints," *Opt. Express*, vol. 11, no. 20, pp. 2549, 2003.

[7] C. Joerdens, and M. Koch, "Detection of foreign bodies in chocolate with pulsed terahertz spectroscopy," *Opt. Eng.*, vol. 47, no. 3, pp. 037003, 2008.

[8] M. Tonouchi, "Cutting-edge terahertz technology," *Nat. Photonics*, vol. 1, pp. 97, 2007.

[9] P.H. Siegel, "Terahertz technology," *IEEE Trans. Microwave Theory Tech.*, vol. 50, pp. 910, 2002.

[10] B. S. Williams, "Terahertz quantum-cascade lasers," *Nat. Photonics*, vol. 1, pp. 517, 2007.

[11] F. Schuster, D. Coquillat, H.Videlier, M. Sakowicz, F. Teppe, L.Dussopt, B.Giffard, T. Skotnicki, W. Knap, "Broadband terahertz imaging with highly sensitive silicon CMOS detectors," *Opt. Express*, vol. 19, pp. 7827, 2011.

[12] Xinghan Cai, et al. "Sensitive room-temperature terahertz detection via the photothermoelectric effect in graphene," *Nature Nanotechnology*, vol. 9, pp. 814, 2014.

[13] L. Liu, J. L. Hesler, H. Xu, A.W. Lichtenberger, R. M.Weikle, "A broadband quasi-optical terahertz detector utilizing a zero bias Schottky diode," *IEEE Microwave Wireless Compon. Lett.*, vol. 20, pp. 504, 2010.

[14] G. C. Trichopoulos, H. L. Mosbacker, D. Burdette, K. Sertel, "A broadband focal plane array camera for real-time THz imaging applications," *IEEE Trans. Antennas Propag.*, vol. 61, pp. 1733, 2013.

[15] C.M.Watts, D. Shrekenhamer, J. Montoya, G. Lipworth, J.Hunt, T. Sleasman, S. Krishna, D. R. Smith, W. J. Padilla, "Terahertz compressive imaging with metamaterial spatial light modulators," *Nat. Photonics*, vol. 8, pp. 605, 2014.

[16] A. Tiwari, H. Satoh,M. Aoki,M. Takeda, N. Hiromoto, H. Inokawa, "Fabrication and analytical modeling of integrated heater and thermistor for antenna-coupled bolometers," *Sensors and Actuators A: Physical*, vol. 222, pp. 160, 2015.

[17] Banerjee, H. Satoh, A. Tiwari, C. Apriono, E. T. Rahardjo, N. Hiromoto and H. Inokawa, Width dependence of platinum and titanium thermistor characteristics for application in room-temperature antennacoupled terahertz microbolometer, *Jpn. J. Appl. Phys.*, pp. 56, pp. 04CC07, 2017.

[18] A. Banerjee, H. Satoh, Y. Sharma, N. Hiromoto, H. Inokawa Characterization of platinum and titanium thermistors for terahertz antenna-coupled bolometer applications," *Sensors and Actuators: A Physical*, vol. 273, pp. 49-57, 2018.

[19] A. Banerjee, H. Satoh, D. Elamaran, Y. Sharma, N. Hiromoto, and H. Inokawa, Optimization of narrow width effect on titanium thermistor in uncooled antenna-coupled terahertz microbolometer, *Jpn. J. Appl. Phys.*, vol. 57, no. 4, pp. 1–7, 2018.

[20] J. Ward, E. Schlecht, G. Chattopadhyay, A. Maestrini, J. Gill, F. Maiwald, H. Javadi, and I. Mehdi, "Capability of THz sources based on Schottky diode frequency multiplier chains," *IEEE MTT-S Digest.*, pp. 1587-1590, 2004.

[21] S. Heyminck, R. Güsten, U. Graf, J. Stutzki, P. Hartogh, H. W. Hübers, O. Ricken, and B. Klein, "GREAT: ready for early science

aboard SOFIA," *Proc. 20th Intl. Symp. Space THz Techn., Charlottesville, VA.*, pp. 315-317, 2009.

[22] T. W. Crowe, J. L. Hesler, S. A. Retzloff, C. Pouzou, and G. S. Schoenthal, "Solid state LO sources for greater than 2 THz," *2011 ISSTT Digest, 22nd Symposium on Space Terahertz Technology, Tucson Arizona, USA*, 2011.

[23] T. W. Crowe, J. L. Hesler, S. A. Retzloff, C. Pouzou, and J. L. Hester, "Multiplier based sources for frequencies above 2 THz," *36th International Conference on Infrared, Millimeter and terahertz Sources (IRMMW-THz)*, pp. 1, 2011.

[24] A Maestrini, I Mehdi, JV Siles, J Ward, R Lin, B Thomas, C Lee, J Gill, G Chattopadhyay, E Schlecht, J. Pearson, and P Siegel, "First demonstration of a tunable electronic source in the 2.5 to 2.7 THz range," *IEEE Trans. Terahertz Science Techn.*, vol. 3, 2012.

[25] S. Kitagawa, M. Mizuno, S. Saito, K. Ogino, S. Suzuki, and M. Asada, "Frequency-tunable resonant tunneling diode terahertz oscillators applied to absorbance measurement," *Japanese Journal of Applied Physics*, vol. 56, pp. 058002–1–3, 2017.

[26] B. S. Williams, "Terahertz quantum-cascade lasers," *Nature Photonics*, vol. 1, pp. 617-626, 2007.

[27] R. Lai, X. Mei, W. Deal, W. Yoshida, Y. Kim, P. Liu, J. Lee, J. Uyeda, V. Radisic, M. Lange, T. Gaier, L. Samoska, and A. Fung, "Sub 50 nm InP HEMT device with fmax greater than 1 THz," *in Proc. IEEE Int. Electron Devices Meeting*, 2007, pp. 609–611.

[28] W. Deal, X. Mei, V. Radisic, K. Leong, S. Sarkozy, B. Gorospe, J. Lee, P. Liu, W. Yoshida, J. Zhou, M. Lange, J. Uyeda, and R. Lai, "Demonstration of a 0.48 THz amplifier module using InP HEMT transistors," *IEEE Microw. Wireless Compon. Lett.*, vol. 20, no. 5, pp. 289–291, May 2010.

[29] A. Biswas, S. Sinha, A. Acharyya, A. Banerjee, S. Pal, H. Satoh and H. Inokawa, "1.0 THz GaN IMPATT Source: Effect of Parasitic Series Resistance," *Journal of Infrared, Millimeter and Terahertz Waves*, vol. 39, issue 10, pp. 954-974, 2018.

[30] P. Banerjee, A. Acharyya, A. Biswas, A. K. Bhattacharjee and H. Inokawa, "Noise Performance of Magnetic Field Tunable Avalanche Transit Time Source," *International Journal of Electronics and Communication Engineering*, vol. 12, no. 10, pp. 718-728, 2018.

[31] S. K. Roy, J. P. Banerjee, and S. P. Pati, "A Computer analysis of the distribution of high frequency negative resistance in the depletion layer of IMPATT Diodes," *in Proc. 4th Conf. on Num. Anal. of Semiconductor Devices (NASECODE IV) (Dublin), Ireland*, pp. 494–500, 1985.

[32] H. A. Haus, H. Statz, and R. A. Pucel, "Optimum noise measure of IMPATT diode," *IEEE Trans. on MTT*, vol. 19, pp. 801-813, 1971.

Forecasting Solar Potential Using Support Vector Regression

Subham Shaw
Department of Electrical and Electronics Engineeering
NIT Nagaland
Dimapur, India
subhamshawww@gmail.com

M. Prakash
Department of Electrical and Electronics Engineeering
NIT Nagaland
Dimapur, India
prakash@nitnagaland.ac.in

Abstract— **Solar energy is one of the most commonly used renewable energy resources. To obtain reliable output from solar energy, prediction of solar radiation is necessary. In this paper, a solar radiation prediction model has been developed for New Alipore, Kolkata. Easily available meteorological parameters like temperature, pressure and humidity have been utilized as inputs, to build the prediction model. Two years data (2011-2012) have been used to develop the Support Vector Regression (SVR) based solar radiation prediction model. The results obtained from the prediction model have been validated with the help of statistical metrics, Root-Mean-Square Error (RMSE) and Coefficient of Determination (R^2). The results signifies that the performance of the developed model is better in comparison with the models existing in the literature.**

Keywords— Support Vector Regression, Solar Radiation Prediction.

I. INTRODUCTION

Renewable energy is the future of the energy market as non-renewable energies are nearly extinct and have major negative impacts on the environment. F. Antonanzas-Torres et al [1] observed that solar radiation is rarely recorded in isolated rural areas across the countries. Soft computing technique like SVR has been deployed for predicting global solar radiation. Moreover genetic algorithm has been used for tuning the SVR parameters for better efficiency. To validate the model, 14 different meteorological stations data have been used. Shahaboddin Shamshirband et al [2] has found that diffused solar radiation is an elementary parameter which has extreme demand in various applications. An SVM and wavelet transform coupled model is developed for the forecasting of daily horizontal diffuse solar radiation. In this model cloudiness index is correlated with clearness index. The results indicated that the model gives precise result than other models, especially from third degree empirical model.

Kuo-Ping Lin et al [3] explored that for better scheduling and improved balanced area control performance, renewable energy is important. In the study the solar radiation prediction has been investigated with help of hybrid model which comprises of SVR and evolutionary decomposition least square technique. Here also genetic algorithm is also used for selection of best input parameters for the SVR model. The results signifies that the model gives superior forecasting accuracy in comparison to the empirical models. Makbul A.M. Ramli et al [4] has performed a case study, for a remote location in Saudi Arabia to explore the performance of SVM and artificial neural network (ANN) for solar radiation prediction. In the study the prediction of solar radiation is done for a particular tilt angle of solar PV panel. Around 360 days of data is used for the model, averaged from each 5-minute data. After analyzing the results it is seen that SVM has low value of RMSE which shows that the system is robust and has the capability to minimize the errors while computing. The

system is faster in prediction than ANN in case of solar radiation prediction for tilted surfaces.

Zibo Dong et al [5] used a hybrid model based on Self-Organizing Maps (SOM), SVR and Particle Swarm Optimization (PSO) technique to forecast hourly solar irradiance. Here SOM is used for partitioning the whole input space into multiple disjointed regions with different characteristic information on the correlation between input and output. SVR is used to model this disjointed regions to recognize the characteristic correlation. While PSO is used for parameter selection for SVR. Han Seung Jang et al [6] observed that due to increase in load demand and huge increase in the number of large scale photovoltaic plant, the amount of solar energy penetration into the main grid escalates. Depending on climatic conditions this penetration brings heavy fluctuations which can hamper the grid. So prediction of solar energy is very needful. In the study motion vectors of clouds are taken from satellite images and utilizing them a large number of input and output dataset is configured for the learning of SVM model. The proposed model can predict both future amount of clouds and solar radiation simultaneously in the range of 15-300 minutes.

Kasra Mohammadi et al [7] investigated the SVR methodology by predicting horizontal global solar radiation based upon sunshine hours and maximum possible sunshine hours as input parameters. In this paper two models of SVR were employed; one with RBF kernel and other with Poly kernel. The results shows that SVR-RBF model gives higher accuracy in prediction of horizontal global solar radiation than the poly model. S. Belaid et al [8] explored that, for sizing and control of solar energy devices and for better power management prior knowledge of solar radiation is very important. In the study, one-step ahead or one-day ahead solar radiation prediction is done using SVM algorithm. The kernel function used in the computation is RBF. A comparative study of the model is done with neural network based model and other empirical models where the accuracy of SVM is found better than others and it is also seen that SVM based models require few simple parameters to get good accuracy.

ABBREVIATIONS

SVR	-	Support Vector Regression
SVM	-	Support Vector Machine
PSO	-	Particle Swarm optimization
RBF	-	Radial Basis Function
ANN	-	Artificial Neural Network
SOM	-	Self-Organizing Maps

2019 Devices for Integrated Circuit (DevIC), 23-24 March, 2019, Kalyani, India

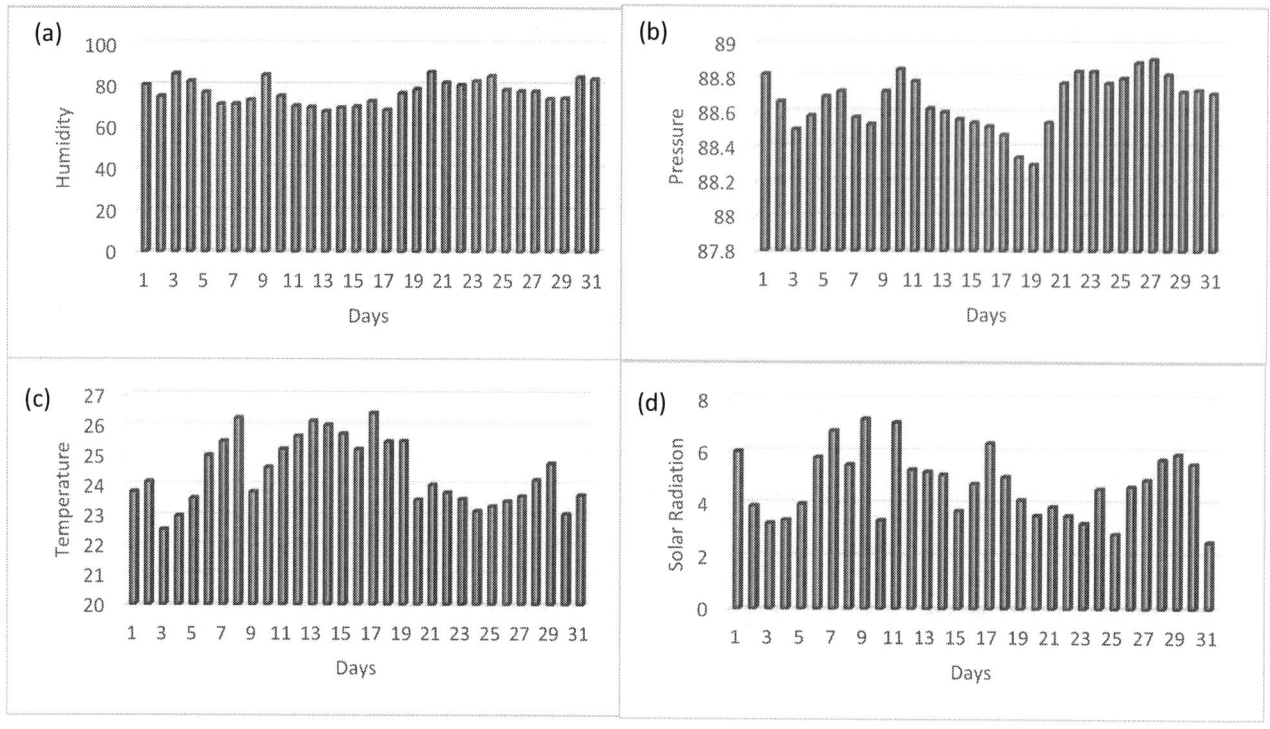

Fig 1 shows distribution of (a) Humidity (b) Pressure (c) Temperature (d) Daily global solar radiation

II. DATASET

The daily mean values of temperature, pressure and humidity has been considered for the period of two years (2011 and 2012). With the help of this three input parameters the dependent variable viz. daily global solar radiation, is estimated. The whole data is divided into two parts; training data and testing data. The data is splitted in such manner that 70% of the data are referred as training data and the remaining 30% are referred as testing data. The Fig. 1 shows the bar chart of all the three input variables and the output solar radiation.

III. SOLAR RADIATION PREDICTION MODEL

SVR has been used as short-term forecasting technique for prediction of daily mean solar radiation value. Fig (2) shows the flowchart for the steps involved in the modelling. Here, dependent and independent variables are taken and with the help of SVR the model is trained. After training the model test data are used to predict the output.

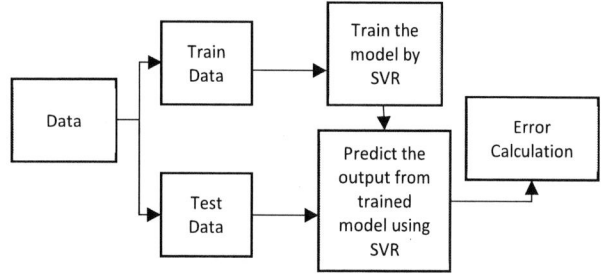

Fig 2 Proposed Model

A. SVR APPLICATION

SVR is another form of SVM. On one hand SVM is used for classification purpose, it just classifies linear data, while on another side SVR is used to find correlation by the mapping the nonlinear data. SVR forms a hyperplane between the variables and then by mapping the distance between the data points and hyperplane is calculated. The property of forming the hyperplane is gained from SVM. It forms the hyperplane between the variables in such a manner that the hyperplane remains at equidistance from the closest data points of both the variables. With the help of this hyperplane it forms a path, within which SVR tries to fit as many data points as possible without violating the marginal limits. The width of this path is defined with the help of the parameter ε. The data points which fall under this path has good forecasting accuracy while the data points which fall far away from this path has less prediction accuracy. The data points which are present over the boundary line of the path or closer to it act as support vectors. On the basis of this support vectors only the data correlation matrix is formed.

In case of SVM the formation of hyperplane becomes easy as the dataset remains linear in nature but in SVR mainly nonlinear dataset is used for which forming the hyperplane becomes difficult. Here lies the main property of SVR which makes it different from SVM, the kernel function. The kernel function converts the dimension of the dataset to a higher dimension and then it forms the hyperplane. After the

978-1-5386-6723-1/19 $31.00 © 2019 IEEE

formation of the hyperplane the kernel function convert back the dimension of the data points to its original dimension retaining the hyperplane with them. Thus the dimension in which the input given to the model gets the output in the same dimension.

B. SVR PARAMETERS

There are three main parameters taken into consideration in SVR model. They are the kernel function, parameter C and parameter ε. The type of kernel function to be assumed depends on the variation of the data points in the dataset. The nature of the data points in the dataset helps the observer to decide the type of kernel function to be used. There are mainly three types of kernel function; viz. linear, polynomial, RBF functions. In this paper, RBF kernel function has been used. The formula for computation of RBF kernel is given as Equation (1)

$$k(x_i, x_j) = exp^{\left(\Sigma_k \theta \,|x_i - x_j|^2\right)} \qquad (1)$$

$$\text{Where } \theta = \left(-\frac{1}{2\sigma^2}\right)$$

Since (x_i, y_i) is the training dataset pairs, where $i = 1,\dots n$. Here σ is the kernel parameter.

The simple method to compute the C and ε values is based on previous experiences of using the SVR techniques. If the user has experienced in prior about the computation of this two parameter then it is very easy to estimate their values depending upon the dataset. But this becomes difficult for a new user to estimate this values. Cherkassky et al. [10] has given mathematical formula to get the two parameters

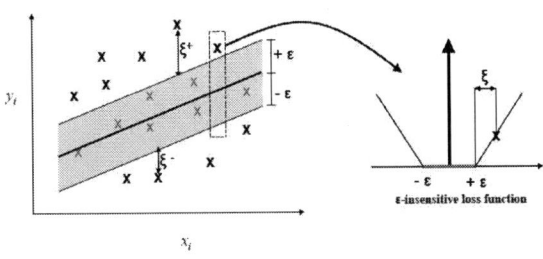

Fig 3 Boundary margin and loss setting for the data points that lie outside the path [9]

For calculating C parameter

$$C = max\left(|\bar{y} + 3\sigma_y|, |y - 3\sigma_y|\right) \qquad (2)$$

Where $\bar{y}$ and σ_y are mean and standard deviation of the y values of the training data

For computing ε parameter

$$\varepsilon = 3\sigma \sqrt{\frac{\ln n}{n}} \qquad (3)$$

After computation of these two parameters, their values can be varied to get the optimal value required to predict the output of the particular data. Computing SVR parameters and choice of kernel function are very important as the model to be developed for forecasting the output mainly depends on these two user defined parameters.

IV. PERFORMANCE EVALUATION OF THE MODEL

The statistical measures that are used to evaluate the estimated output of the model are R^2 and RMSE. The mathematical expression for both the evaluation method are represented in equation (4) & (5) respectively [11].

$$R^2 = \frac{\left[\Sigma(Q_i - \bar{Q}_i)^2 \cdot (P_i - \bar{P}_i)\right]^2}{\Sigma(Q_i - \bar{Q}_i) \cdot \Sigma(P_i - \bar{P}_i)} \qquad (4)$$

$$RMSE = \sqrt{\frac{\Sigma(Q_i - \bar{P}_i)^2}{n}} \qquad (5)$$

Where P_i and Q_i are experimental and predicted values while $\bar{P}_i$ and $\bar{Q}_i$ are the mean values of P_i and Q_i. Here n is total number of observations.

V. RESULTS

At first the data was divided into training and testing data. 70% of the data were taken as training data and the remaining 30% of the data were taken as testing data. Then the model is trained with the training data. The model was analyzed for estimation of daily solar radiation based on three input parameters viz, humidity, temperature and pressure. RBF kernel was the best choice for the data used in the model. As explained RBF kernel requires two user defined input parameters; C and ε. The values of these two parameters assumed for study are given in the table I.

Table I   USER DEFINED PARAMETERS

Kernel function	C	ε
RBF	8.7318	0.3944

After applying the above parameters to the model the predicted output is analyzed with the help of R^2 and RMSE. Table II shows the statistical metrics for both training data and testing data.

Table II   STATISTICAL METRICS

	R^2	RMSE
Training set	0.7773	1.0349
Testing set	0.8492	1.1420

The plot of error values is shown in fig 4. The comparison of prediction accuracy of the developed model and the other models is given in table III.

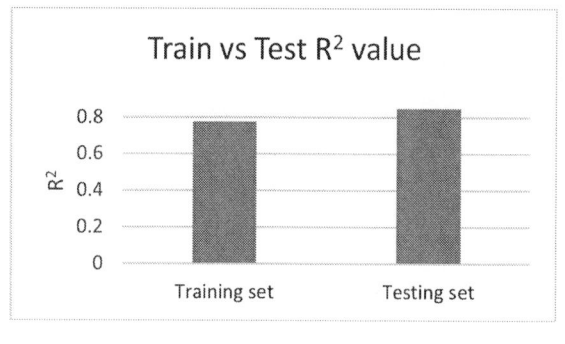

(a)

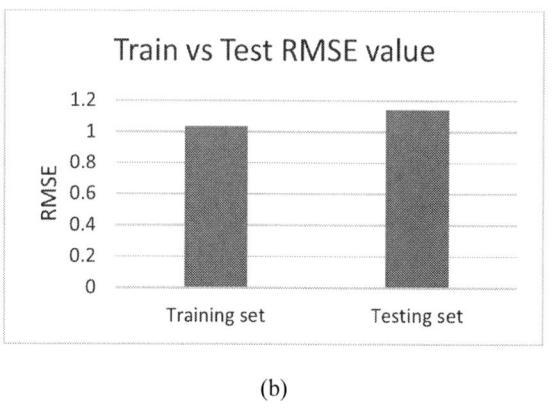

(b)

Fig 4 plotting of the (a) R^2 and (b) RMSE for both training and testing set

Table III COMPARISION BETWEEN SVR-RBF MODEL WITH OTHER MODELS

Reference	Model type	Input Parameters	R^2
Ramedani et al. (2014) [12]	ANN	7	0.799
	ANFIS	7	0.801
	SVR-RBF	7	0.790
Present study	SVR-RBF	3	0.849

VI. CONCLUSION

In the present study the accuracy of SVR method for prediction of solar radiation was illustrated. The model uses temperature, pressure, humidity as the input parameters. Its statistical characteristics shows R^2 of 0.7773 and RMSE of 1.0349 for training data while R^2 of 0.8492 and RMSE of 1.1420 for testing data. From the results, it is concluded that the prediction model for solar radiation using SVR with RBF kernel is found more suitable for the study area (New Alipore, Kolkata) considered.

REFERENCES

[1] F. Antonanzas-Torres, R. Urraca, J. Antonanzas, J. Fernandez-Ceniceros, F.J. Martinez-de-Pison, "Generation of daily global solar irradiation with support vector machines for regression", Energy Conversion and Management, Volume 96, 2015, Pages 277-286, ISSN 0196-8904

[2] Shamshirband Shahaboddin, Mohammadi Kasra, Khorasanizadeh Hossein, Por Yee, Lee Malrey, Petkovic Dalibor, Zalnezhad Erfan, "Estimating the diffuse solar radiation using a coupled support vector machine-wavelet transform model", Renewable and Sustainable Energy Reviews, Volume 56, 2015, Pages 428-435, ISSN 1364-0321

[3] Kuo-Ping Lin, Ping-Feng Pai, "Solar power output forecasting using evolutionary seasonal decomposition least-square support vector regression", Journal of Cleaner Production, Volume 134, Part B, 2016, Pages 456-462, ISSN 0959-6526

[4] Makbul A.M. Ramli, Ssennoga Twaha, Yusuf A. Al-Turki, "Investigating the performance of support vector machine and artificial neural networks in predicting solar radiation on a tilted surface: Saudi Arabia case study", Energy Conversion and Management, Volume 105, 2015, Pages 442-452, ISSN 0196-8904

[5] Zibo Dong, Dazhi Yang, Thomas Reindl, Wilfred M. Walsh, "A novel hybrid approach based on self-organizing maps, support vector regression and particle swarm optimization to forecast solar irradiance", Energy, Volume 82, 2015, Pages 570-577, ISSN 0360-5442

[6] H. S. Jang, K. Y. Bae, H. Park and D. K. Sung, "Solar Power Prediction Based on Satellite Images and Support Vector Machine," in IEEE Transactions on Sustainable Energy, vol. 7, 2016, no. 3, pp. 1255-1263

[7] Kasra Mohammadi, Shahaboddin Shamshirband, Mohammad Hossein Anisi, Khubaib Amjad Alam, Dalibor Petković, "Support vector regression based prediction of global solar radiation on a horizontal surface", Energy Conversion and Management, Volume 91, 2015, Pages 433-441, ISSN 0196-8904

[8] S. Belaid, A. Mellit, "Prediction of daily and mean monthly global solar radiation using support vector machine in an arid climate", Energy Conversion and Management, Volume 118, 2016, Pages 105-118, ISSN 0196-8904

[9] A. J. Smola, B. Sch, and B. Schölkopf, "A Tutorial on Support Vector Regression," Stat. Comput., vol. 14, no. 3, pp. 199–222, 2004.

[10] Cherkassky, V., and Ma, Y., (2004). Practical selection of SVM parameters and noise estimation for SVM regression, Neural Networks, Vol. 17, No. 1: 113 - 126.

[11] Willmott CJ, Matsuura K. Advantages of the mean absolute error (MAE) over the root mean square error (RMSE) in assessing average model performance. Clim Res 2005; 30(1):79.

[12] Zeynab Ramedani, Mahmoud Omid, Alireza Keyhani, Shahaboddin Shamshirband, Benyamin Khoshnevisan, "Potential of radial basis function based support vector regression for global solar radiation prediction", Renewable and Sustainable Energy Reviews, Volume 39, 2014, Pages 1005-1011, ISSN 1364-0321

The Quantum Cost, Garbage Outputs and Constant Input Optimized Implementation of 2:4 Decoder Using Peres Gate

Heranmoy Maity
ECE
NSHM Knowledge Campus Durgapur
Durgapur, India
heranmoy@gmail.com

Arindam Biswas
School of Mines & Metallurgy
Kazi Nazrul University
Asansol, India
mailarindambiswas@yahoo.co.in

Anup Kr. Bhattacharjee
ECE
NIT Durgapur
Durgapur, India
akbece12@yahoo.com

Anita Pal
Mathematics
NIT Durgapur
Durgapur, India
anita.buie@gmail.com

Abstract—This paper presents the quantum cost (QC), garbage output (GO) and constant input (CI) optimized 2:4 decoder using reversible logic gate. The proposed reversible circuit is design using existing reversible logic gate, such as reversible NOT gate, Feynman gate and Peres gate. The proposed decoder can be design using one Peres gate, one NOT gate and four Feynman gate. The QC, GO and CI of the proposed 2:4 decoder is 9, 0 and 2. Which is better w.r.t previously reported work. This is the first reversible 2:4 decoder circuit, which has '0' garbage output. The improvement % of QC, GO and CI are 0 – 74.28 %, 100 % and 33.33 – 75 %.

Keywords—*Quantum computing, quantum cost, constant input, garbage output*

I. INTRODUCTION

The very crucial parameters of the reversible logic design and implementation are the quantum cost, garbage output, constant input and delay [1-6]. Hence, the most important objective in reversible logic design and implementation is to minimize the quantum cost (QC), garbage outputs (GO) constant input (CI) and delay (D). The different logic circuits such as adders, subtractors, multipliers, comparator and different code converter are the important circuit component for a reversible computing system.

The many researchers have proposed several reversible combinational logic circuits [7-15]. However, the design of reversible 2:4 decoders has not been sufficiently developed and all above circuits can be design using basic reversible logic gate. The reversible 2:4 decoder circuit reported in [12-18].

This paper presents the QC, GO and CI optimized 2:4 decoder circuit using reversible logic gate. The proposed 2:4 circuit is very important because of its applications. We can design different combinational/sequential circuit using decoder. So, optimized 2:4 decoder circuit can be used to design different optimized reversible logic circuit.

II. EASIC CONCEPTS

The reversible logic function is an equal number of input and outputs logic functions, the reversible gate having same no. of inputs and outputs logic gate, which can be represented as R×R, where R is the number of input and output. If input vector is $(I_P1, I_P2,......I_PR)$ then the output vector is represented by $(O_P1, O_P2,......O_PR)$. Fig. 1 shows the R×R reversible logic gate.

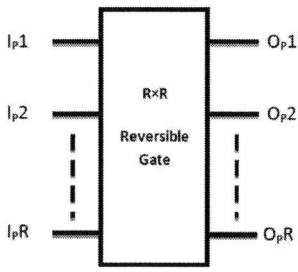

Fig. 1. Basic R×R reversible logic gate.

There are several reversible gates for example reversible NOT gate [1-2], Feynman gate(FG)[5], Peres gate(PG) [6], Fredkin gate[10], Toffoli gate[10], etc. but in this paper only reversible NOT gate, FG and PG is used. Fig. 2 shows the reversible NOT gate with QC 1. Terminal A and P is the input and output terminal, where P=A'. Fig. 3 and Fig. 4 shows the basic block diagram of 2×2 Feynman gate and its quantum representation with QC 1. Fig. 5 and Fig. 6 represents the PG with its quantum representation. The QC of PG is 4.

Fig. 2. Reversible NOT gate with QC 1

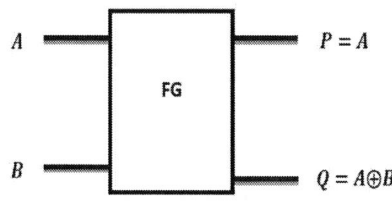

Fig. 3. Block diagram of 2×2 Feynman gate

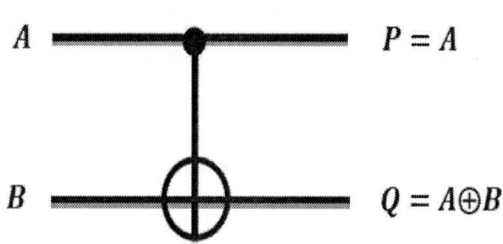

Fig. 4. Quantum representation with QC 1

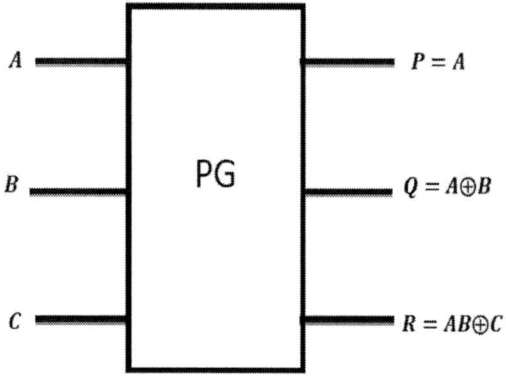

Fig. 5. PG

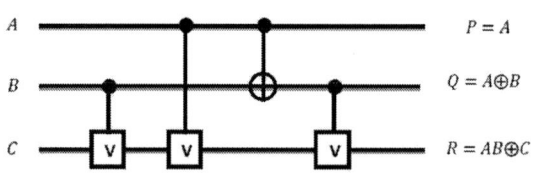

Fig. 6. PG with its quantum representation

III. PROPOSED WORK

In this section we have describe the proposed QC, GO and Delay optimized 2:4 decoder using existing reversible logic gate. The Fig. 7 and Fig. 8 shows the proposed reversible 2:4 decoder circuit using Peres gate, Feynman gate and NOT gate with QC 8 and zero garbage output.

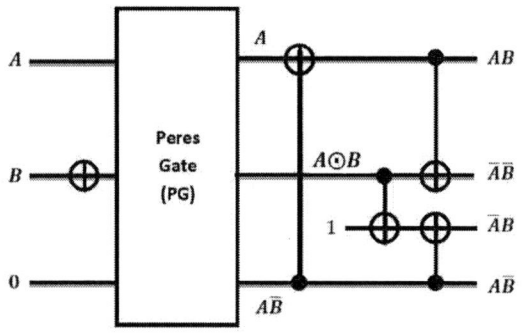

Fig. 7. Reversible 2:4 decoder circuit

Fig. 7 shows that if C input of PG is '0', then $Q=A \odot B$ and $R=AB'=D2$. We know that $A \oplus AB'=AB=D3$, $AB \oplus (A \odot B)=A'B'=D0$ and $(A \odot B) \oplus AB'=A'B=D1$.

The input terminal of decoder circuit is A, B and four output terminals are $D0=A'B'$, $D1=A'B$, $D2=AB'$ and $D3=AB$. The QC, GO, CI of the proposed circuit are 9, 0 and 2. The improvement % of QC, GO, CI is $0-74.28$ %, 100 % and $33.33-75$ % which is shown in table 1.

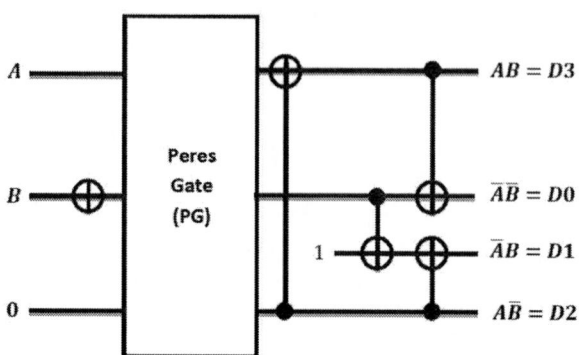

Fig. 8. Reversible 2:4 decoder circuit using Peres gate, Feynman gate and NOT gate

Table 1. Comparison result of 2:4 decoder using PG, FG and NOT.

	QC	GO	CI
Proposed design	9	0	2
Ref [9]	9	1	3
Ref [12]	9	1	3
Ref [13]	11	1	3
Ref [14]	12	2	4
Ref [15]	35	3	8
Ref [16]	15	2	-
Ref [17]	35	3	4
Ref [18]	-	1	3
Improvement %	$0-74.28$	100	33.33 - 75

IV. CONCLUSIONS

The authors have proposed the QC, GO and CI optimized 2:4 decoder using Peres gate, Feynman gate and NOT gate. The QC, GO and CI of the proposed circuit is 9, 0 and 2. The improvement of QC, GO and CI are correspondingly $0-74.28$ %, 100 % and $33.33-75$ %. In future we can design reversible memory like ROM using proposed circuit. So, proposed 2:4 decoder has very useful application in quantum computing.

REFERENCES

[1] M. Haghparast and A. Bolhassani, "Optimized parity preserving quantum reversible full adder/subtractor", *Journal of Circuits, Systems, and Computers*, 14 (2016) 1650019(12 Pages), doi: 10.1142/S0219749916500192

[2] H. Maity, A. K. Barik, A. Biswas, A. K. Bhattacharjee & A. Pal, "Design of Quantum Cost, Garbage Output and Delay Optimized BCD to Excess-3 and 2's Complement Code Converter", *Journal of Circuits, Systems, and Computers*, 27, (2018), 1850184 (11 pages).

[3] C.H. Bennett, "Logical Reversibility of Computation", *IBM J.Research and Development*, 17 (1973) 525-532, doi: 10.1147/rd.176.0525

[4] M. Haghparast, M. Mohammadi, K. Navi and M. Eshghi, "Optimized reversible multiplier circuit", *Journal of Circuits, Systems, and Computers*, 18 (2009) 311-323, doi: 10.1142/S0218126609005083

[5] R. Feynman, "Quantum Mechanical Computers", *Optics News*, (1985) 11–20, doi:10.1364/ON.11.2.000011

[6] A. Peres, "Reversible Logic and Quantum Computers", *Phys. Rev.*, (1985) 3266-3276, doi:10.1103/PhysRevA.32.3266

[7] H. Maity, A. Biswas, A. K. Bhattacharjee & A. Pal, "Quantum cost optimized design of 4-bit reversible universal shift register using reduced number of logic gate", *International Journal of Quantum Information*, 16, (2018) 1850016 (8 pages).

[8] H. Maity, A. Biswas, A. K. Bhattacharjee and A. Pal, Design of Quantum Cost Efficient 4-Bit Reversible Universal Shift Register, *in Proc. 2nd IEEE int. Conf. on Device for Integrated Circuits (DevIC 2017), Kalyani, India,* March, 2017, pp. 44 – 47

[9] N. K. Misra, B. Sen, S. Wairya and B. Bhoi "Testable Novel Parity-Preserving Reversible Gate and Low-Cost Quantum Decoder Design in 1D Molecular-QCA", *Journal of Circuits, Systems, and Computers*, 26(9), (2017), 1750145 (26 pages).

[10] E. Fredkin, T. Toffoli, "Conservative Logic",*International Journal of Theor. Physics*, 21 (1982) 219-253, doi: 10.1007/BF01857727

[11] H. Maity, A. Biswas, A. Pal and A. K. Bhattacharjee, "Design of BCD to Excess-3 Code Converter Circuit with Optimized Quantum Cost, Garbage Output and Constant Input Using Reversible Gate", International Journal of Quantum Information, 16, (2018) 1850061 (5 pages).

[12] M. R. Rahman, Cost efficient fault tolerant decoder in reversible logic synthesis, Int. J. Comput. Appl. 108 (2014) 7–12

[13] L. Jamal, M. M. Alam and H. M. H. Babu, "An efficient approach to design a reversible control unit of a processor", Sustain. Comput. Inf. Syst. 3 (2013) 286–294.

[14] M. Shamsujjoha, H. M. Hasan Babu and L. Jamal, "Design of a compact reversible fault tolerant field programmable gate array: A novel approach in reversible logic synthesis", Microelectron. J. 44 (2013) 519–537.

[15] M. Shamsujjoha and H. M. Hasan Babu, "A low power fault tolerant reversible decoder using MOS transistors", Proc. 26th IEEE Int. Conf. VLSI Design and the 12th Int. Conf. Embedded Systems (2013), pp. 368–373

[16] S. Noor Mahammad and K. Veezhinathan, "Constructing online testable circuits using reversible logic", IEEE Trans. Instrum. Meas. 59 (2010) 101–109.

[17] M. Perkowski, M. Lukac, P. Kerntopf, M. Pivtoraiko, M. Folgheraiter, Y. W. Choi, J.-W. Kim, D. Lee, W. Hwangbo and H. Kim, "A hierarchical approach to computer-aided design of quantum circuits", Paper 228, Electrical and Computer Engineering Faculty Publications and Presentations, Portland State University (2003).

[18] N. K. Misra, S. Wairya and V. K. Singh, "Evolution of structure of some binary group based n-bit comparator, n-to-2n decoder by reversible technique", Int. J. VLSI Des. Commun. Syst. 5 (2015) 9–30.

A Cost Effective Design and Implementation of Arduino Based Sign Language Interpreter

Anirbit Sengupta
Dept. of Electronics and Communication Engineering
Camellia Institute of Engineering & Technology
Budbud, West Bengal
anirbit87sengupta@gmail.com

Md. Tausif Mallick
Dept. of A.K.C.S.I.T
University of Calcutta
Kolkata, West Bengal
tausif.realmadrid@gmail.com

Abhijit Das
Dept. of Information Technology
RCC Institute of Information Technology
Kolkata, West Bengal
ayideep@yahoo.co.in

Abstract—**People suffering from listening impairment and voice disability usually make use of different sign of symbols and languages for their communication purpose. Sign languages are generally dependent on hand-driven gesticulations with various motions explicit to that particular language by which these people communicate. In sign language, gesticulation is basically specific movements of our hands with an explicit form build out of them. Current research focuses on converting the hand gesticulations based on electronic devices. The devices will basically convert the sign language into its speech form to build the communication gap among the voiceless societies with the normal people. Here cloth driven gloves are being used which are Bluetooth-enabled. The glove is tailored with one accelerometer and five flexible sensors. The sensor placement is length ways of each of the fingers with the thumb. This work will facilitate the silent people to make different hand gesticulations wearing this gloves, and these intern will be transformed into respective speeches to the normal peoples recognition. In this case flexible sensors has a key role to play. Also the resistance value change generating from the extent of curvature of the sensors in combination with accelerometer value of slant position of hand to the land surface is measured. This acquired data is further managed by microcontroller module and can be transmitted to any smart-phone user via Bluetooth connectivity. A developed application can further be used to transform the data into text liable to the hand shape detected and will produce a voice signal.**

Index Terms—**Flex Sensors, Accelerometer, Bluetooth, Sign Language, Microcontroller, Android.**

I. Introduction

Communication between human being is not only through speech or voice however such differently abled people who try to communicate to the normal world find it difficult. Hence they use sign language to communicate [1]. Without using voice or speech as communication medium with one another there are several process of communication like hand shape, body and arm movements and facial expression this modes of communication is termed as Sign Language. Mute community for example deaf and dumb people uses sign language as the medium of communication. As it is very tough for normal people to recognize and analyze sign language. Thus the barrier between normal people and mute community is created which can be broken through the study of sign language. All words that are used by normal people can be derived through sign language.The present work also guides different signs of letters to construct words which are correctly unavailable in

sign language dictionary.Sentence construction can be done using sign letters and does wing sign of word id faster. The current research work uses "Bahasa Isyarat Malaysia (BIM) gestures.Fig.1 shows the sign language gesticulations. The gap

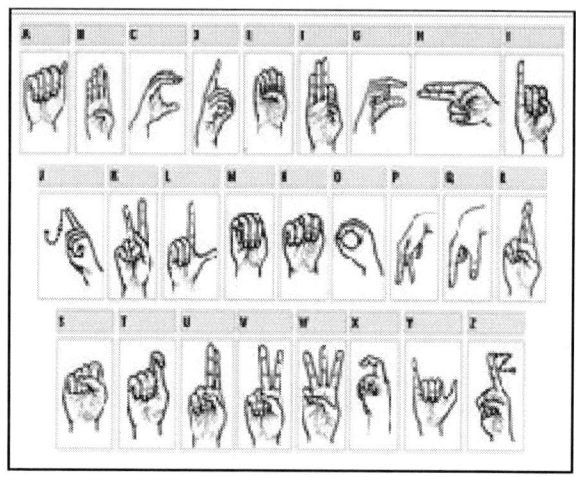

Fig. 1. Sign Language Gestures

between the sign language and verbal language is bridged through introduction of finger spelling which is nothing but the representation of numerical system through hands Finger spelling acts as the master for learning sign language. This phenomenon motivated as to concentrate on finger, gestured word recognition. Around million of the entire population belongs to the mute community as per the new global estimates of WHO [2].Low and middle income countries are the mostly affected with hearing disability due to ear infection as per the reports of WHO. Hearing loss is also caused due to measles and mumps. Researchers are working for the past decades in developing system capable of translating gesture into speech. This aims to bridge the gap between mute community and normal people. Vision approach is one of the approaches taken by the researchers to recognize and apply sign languages or finger spelling.

II. Working Model

The methodology for the construction and the development of the work can be divided into two section shown in Fig.

978-1-5386-6723-1/19 $31.00 © 2019 IEEE

2.These two section shows the hardware design and flow of the process along with the recognition system.The development of the software and communication protocols to be followed are part of the process.

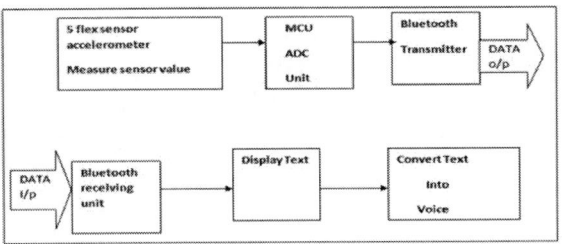

Fig. 2. Sign Language Gestures

III. HARDWARE COMPONENTS AND THEORETICAL BACKGROUND

A. Microcontroller

The microcontroller act as the decision processing device for our research work. The microcontroller used is ATMEGA32 which is embedded with other component in the board named as Adriuno leornardo. It is an 8bit microcontroller having a flash type memory of 32KB and EEPROM of 1KB.It comprises of three flexible timer and counter with compare mode. The microcontroller has an inbuilt programmable watchdog timer with internal oscillator. The device works within a voltage range of 1.8 to 5.5 Volt.The throughput of the device is 1MIPS per MHZ. It has 20 digital input and output pins [3].

B. Flex Sensor

To capture gesticulations of finger movements, mostly flex sensors has been used. The bending of the sensor causes the change in resistance in the sensor. Bending is directly proportional to the change of resistance. The increase in bending causes increase of the re-sistance in the sensor. The resistance range of the flex sensor is approximately in the range of 10K to 40K ohms.Fig. 4 shows the basic circuit connection of the flex sensor. A voltage divider circuit is interfaced with the sensor to provide bridge between microcontroller and the sensor during interfacing.
Fig. 4. Give the basic circuit diagram for the flex sensor connection. Flex sensor acts as interface with a microcontroller when it is connected to the voltage divider circuit.

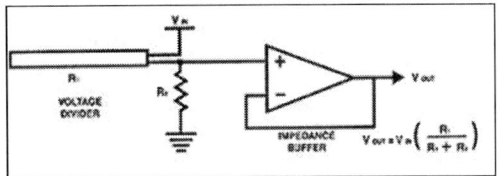

Fig. 3. The basic circuit for the Flex Sensor connection

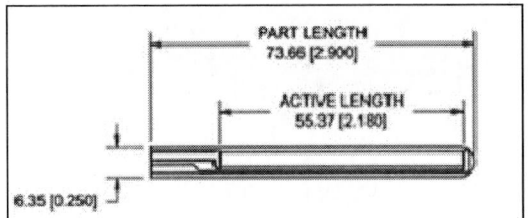

Fig. 4. Flex Sensor

C. Accelerometer

Static acceleration due to gravity is an important parameter in our research work which need to be measured for which we are using an accelerometer shown in fig. 5. Static acceleration is produced when the gloves is titled against the earth and there by calculating the angle we can find static acceleration. To measure the angle at which the gloves is tilted the accelerometer is placed at the middle of the gloves. The voltage range at which the accelerometer works ranges from 1.6 to 3.6 Volt. The X, Y and Z axis is sensed by the accelerometer using a single structure. Depending upon the angle of the tilt, the output voltage of the accelerometer changes [4] [5].

D. Bluetooth

Data transmission from microcontroller to smart phone is done via Bluetooth communication protocol. For the betterment of the quality and for error free communication the Bluebee Bluetooth module is used, which has an on board antennae for better establishment of the wireless communication, this can work with various Bluetooth adopter and phone devices as it performs like a transparent serial port. During the pairing with the smart phone it becomes a Serial Com Port and only provides SPP (Serial Port Profile).The blue tooth has an input voltage of 3.3Volt and 50mA.The operating frequency is within a range of 2.4 to 2.48GHZ and operates of a distance of 20M to 30Min free space. The dimension of the module is 32*24*9mm which makes it more unique to place in on the top of the gloves. The Fig. 6 shows the connection of the Bluetooth device with Arduino board.

E. Bending Variables of Flex Sensors

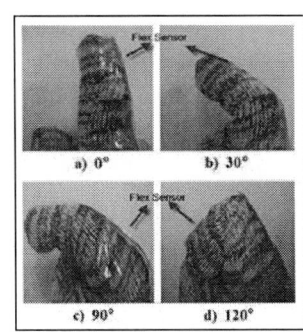

Fig. 5. Bending Activity at different angle of Finger[6]

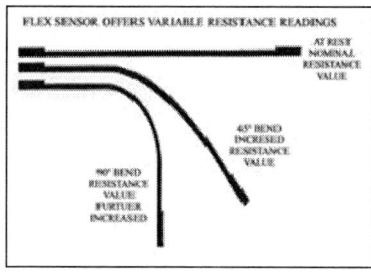

Fig. 6. Bending variables[7]

F. Arduino Software

The Arduino IDE has been used for the programming language processing. As Arduino Id is a java based cross platform application, this is beneficial for the writing process. The software is designed in such a manner that it is very user friendly programming plat forms the new comers and artist.The important features of the Arduino software are syntax highlighting, brace matching and automatic indentation. It is capable of compiling and uploading the hex file just by one click. Practically, there is no requirement to use the command like interface or file editor to run programs. The program can be written in C++ as there is a C++ library in the Arduino IDE called wiring which makes common input and output operation much easier.

- Setup () This function runs at the beginning of the program and does all the initializations.
- Loop () This function is called repetitively till the board power goes off.

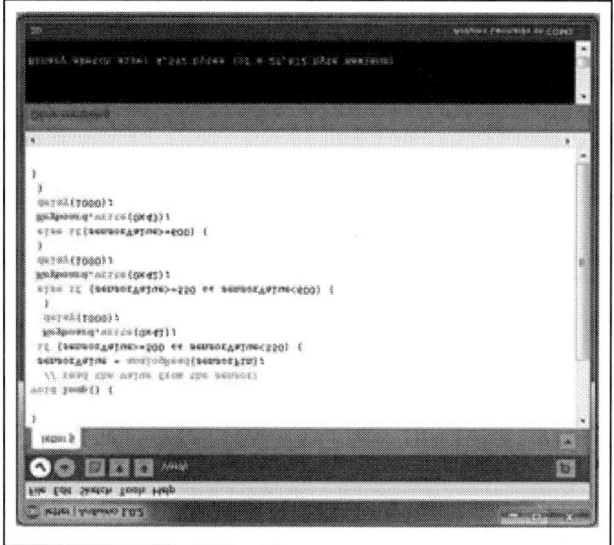

Fig. 7. Arduino Software

G. GUI Software Developments

Text to speech application for our smart phone was developed using a freeware software MIT app inventor software, to convert the text received from the microcontroller to the Smartphone via blue tooth. Search Bluetooth button, connect button, play button and clear button are the four button of the application developed using MIT app inventor having specific function. For searching nearby Bluetooth device we use the search Bluetooth button w display use the connect button to confirm the pairing of the devices. The play button converts the text into speech and for deleting the text we use the clear button.

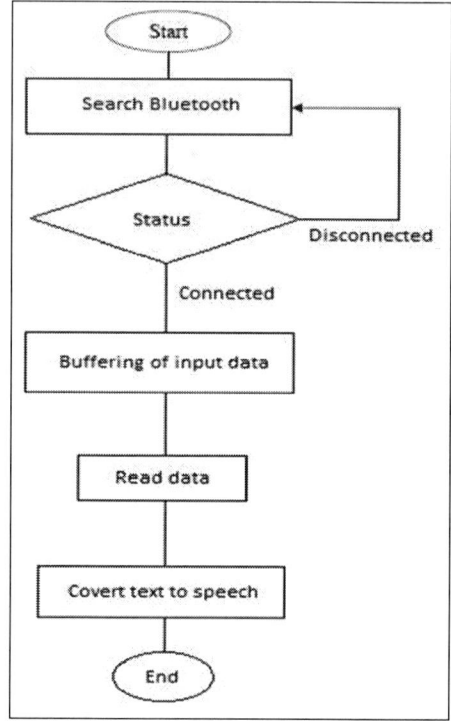

Fig. 8. GUI Flow Chart

IV. EXPERIMENT SETUP AND DATA ANALYSIS

The smart gloves was tested by using hand gesticulations A to Z. Few letters like M,N,O,R,S,T,V,X cannot be demonstrated as these letters have gesticulation similarities with other letters. All the sensors used here recognize the hand gesticulation of the user, transforms it to text and voice with the help of the smart phone application. The hand gesticulation was performed five times for each gestures to gather and calculate the accuracy level of the smart gloves. From this test, the hand gesture was performed 5 times each to collect the accuracy of the smart glove. Fig. 9 depicts the accuracy deposit for each such hand gesticulations.

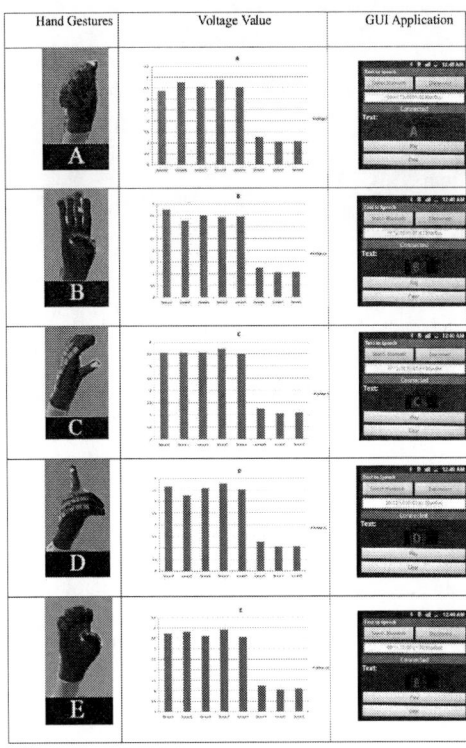

Fig. 9. Result Set

V. CONCLUSION

In this current work a significant amount of accuracy has been achieved for finger spelling recognition using smart gloves. The smart gloves approach proposal is meant to be prototype to check the feasibility of recognizing sign language. The output found was very satisfying. We have used free ware software for design of the application and the design of the system code in Arduino platform, and the former in MIT application inventor platform. The use of Bluetooth as communication protocol has given a great enhancement in the design compared to the other wireless communication protocol like ZigBee or RF link[8] [9]. The future of this technology will evolve greatly if different languages [9] can be converted in sign form and then with the help of these gloves we can convert it. The future proposal for the communication protocol that can be used is Wi-Fi or GSM for long distant communication.

REFERENCES

[1] Chouhan, Tushar, et al.Smart glove with gesture recognition ability for the hearing and speech impaired. 2014 IEEE Global Humanitarian Technology Conference-South Asia Satellite (GHTC-SAS). 2014.

[2] World Health Organization. (WHO), World Health Organization Media Centre Fact Sheet [Online], http://www.who.int/mediacentre/factsheets/fs300/en/index.html. Last Accessed on June 20, 2013.

[3] Arduino.cc Inc. Italy, http://arduino.cc/en/Main/ArduinoBoardLeonardo Last Accessed on December 12, 2012.

[4] Praveen, Nikhita, Naveen Karanth, and M. S. Megha. Sign language interpreter using a smart glove. Advances in Electronics, Computers and Communications (ICAECC), 2014 International Conference on. IEEE, 2014.

[5] Dimension Engineering Inc. http://www.dimensionengineering.com/info/ accelerometers last Accessed on December 12, 2012.

[6] Arduino.cc Inc. Italy, http://arduino.cc/en/Main/ArduinoBoardLeonardo Last Accessed on December 12, 2012.

[7] Bhaskaran, K. Abhijith, et al.Smart gloves for hand gesture recognition: Sign language to speech conversion system. Robotics and Automation for Humanitarian Applications (RAHA), 2016 International Conference on. IEEE, 2016.

[8] Choudhary, Tanay, Saurabh Kulkarni, and Pradyumna Reddy. A Braille-based mobile communication and translation glove for deaf-blind people. Pervasive Computing (ICPC), 2015 International Conference on. IEEE, 2015.

[9] Abdulla, Dalal, et al.Design and implementation of a sign-to-speech/text system for deaf and dumb people. Electronic Devices, Systems and Applications (ICEDSA), 2016 5th International Conference on. IEEE, 2016.

Hand Gesture	Recognized Wrongly As	Accuracy (%)
A		100
B		100
C	E,D	60
D	C	80
E	C	80
F		100
G	L,Q	60
H		100
I		100
J		100
K	P	80
L	G,Q	60
P	K	80
Q	G,L	60
U		100
W		100
Y		100
Z		100

Table1: Accuracy Results

2019 Devices for Integrated Circuit (DevIC), 23-24 March, 2019, Kalyani, India

Hamming Code Generators using LTEx Module of Quantum-dot Cellular Automata

Chiradeep Mukherjee [1, 2]
[1]Department of ECE
NIT Durgapur, West Bengal, India
[2]Department of ECE
UEM Jaipur, Rajasthan, India
chiradeep.mukherjee@uem.edu.in,
chiradeep.1234321@gmail.com

Saradindu Panda[2]
[2]Department of ECE
NIT-JIS Agarpara, Kolkata
India
saradindupanda@gmail.com

Bansibadan Maji [1]
[1]Department of ECE
NIT Durgapur, West Bengal, India
bmajiecenit@yahoo.co.in

Asish Kumar Mukhopadhyay[3]
[3]Professor
Department of ECE,
NIT-JIS Agarpara, Kolkata
India
askm55@gmail.com

Abstract—**Quantum-dot Cellular Automata (QCA) explores a novel paradigm to escape from the notion of traditional semiconductor transistor that has dominated processor design from its inception. The realization of Nano communication network by using QCA becomes promising in the arena of application, and such models of QCA networks are expected to be more reliable in terms of error. Hamming code plays an important role in error detection and correction of communication circuits. This work extends the idea of layered T (LT) gate to develop an** *algorithm* **to model generic QCA Hamming code generator. As a faithful instantiation, even and odd (7,4) Hamming code generators are designed by using QCADesigner simulator to demonstrate the functionality and testability of proposed** *algorithm* **based methodology.**

Keywords—*Layered T Gate, LTEx Module, Hamming Code Generators, O-Cost, Quantum-dot Cellular Automata*

I. INTRODUCTION

The Quantum-dot Cellular Automata (QCA) is an emerging nanotechnology that is capable enough to deploy the apparent positions of an individual electron. Instead of traditional voltage and current flow, the QCA architecture carries information from a specific point inside the circuit to another location by using Coulombic interactions of electrons [1]. The problem of leakage current has been avoided in this new paradigm of nanotechnology specifically when the dimensions are shrunk to the level of few nanometers, and the operation of the circuit became much faster with the speed of hundreds of Giga Hertz. The Heat dissipation & Power consumption are trivial in the Quantum Cellular Automata which avoids the requirements of the heat sink in the electronic circuitry [2]. The QCA designs of several elemental components of the central processing unit (CPU) like adders, multiplexers, registers, counters, memory, etc. are developed to claim the new possibilities to realize nano processor-based computer design [3]. Researchers realized that the ability of detection and correction of errors should be present in nano communication circuits for the further evolution of nanocomputers. Hamming code uses such an improved technique to locate and correct single-bit errors. This work focuses on developing an algorithm to realize generic and efficient" QCA Hamming code generators that can be used in future Nano communication networks [4-5].

Four types of QCA technology occurs. They are metal-dot QCA, magnetic QCA, semiconductor QCA, and atomistic QCA. The metal dot and semiconductor QCA operate near cryogenic temperature, but magnetic QCA and atomistic QCA functions at room temperature [6-8]. In this work, the logical quantum-dot based nano-architectures are generated by using manufacturing technology independent systematic algorithms. Therefore the solutions are general, not architecture specific.

The rest of the paper is organized as follows. Section 2 demonstrates the background of QCA specifically about the QCA clocking and various QCA logic gates. Section 3 discusses the even and odd Hamming codes. An algorithm is proposed in this section to explore the guideline to realize QCA even and odd Hamming code generators. Section 4 provides the outputs of the proposed layouts and compares the design summary with the existing designs. This section also contains the conclusion.

II. BACKGROUND OF QCA

A. QCA cell and QCA clock

The basic unit of QCA is a square structured cell that has four quantum-dots located at its four corners. This square cell is known as QCA cell as demonstrated in figure 1(a). Two excess electrons in a QCA cell accommodate themselves at any two out of four corners to maintain the furthest distance between them. Such orientation of two electrons in a QCA cell are shown in figure 1(b) and (c).

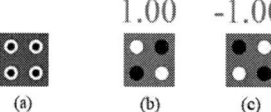

Figure 1. (a) QCA Cell, QCA Cell with Polarization with (b) P=+1 and (c) P=-1

The orientation of electrons as mentioned in figure 1(b), denotes polarization, P=+1, and that of figure 1(c) represents polarization, P=-1. These polarizations, i.e., P=+1 and -1 are equivalent to "logic"1 and "logic 0" of Boolean digital circuits. Electrons can change their locations between the corners of quantum dots of a QCA cell through quantum mechanical tunneling. However, the electrons cannot tunnel between the dots of neighboring cells due to the presence of high potential barriers. An array-like arrangement of QCA cells in a row or column constructs QCA devices, i.e., logic gates and interconnecting wires. The first logic available in QCA is majority voter and inverters. Figure 2(a) and (c) respectively present the QCA cell arrangements of majority gate and inverters. As presented in figure 2(a), the majority voter takes three inputs A, B, and C, to decide the output as the majority of these three input logic. The equation (1) best

978-1-5386-6723-1/19 $31.00 © 2019 IEEE

describes the working of the majority voter as follows:

$$Z=AB+BC+CA \quad\quad\quad (1)$$

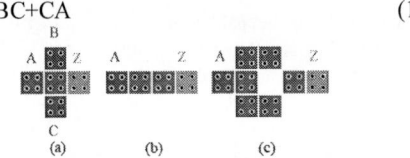

Figure 2. (a) Majority Voter, (b) Binary Wire and (c) Conventional Inverter

Figure 2(c) represents QCA inverter that takes input A and generates $\overline{A}$ as output Z. The arrangement of QCA cells to form interconnecting wire is shown in figure 2(b). The QCA wire can be used to carry either intermediary inputs/ outputs to the next level of a circuit or to provide the information to the QCA gates. The entire assemblage of QCA cells is sequentially groped by using QCA clocks. For proper information flow through the circuit, four phases of QCA clocks are assigned in order as follows: switch, hold, release, relax, switch and so on [9].

B. QCA Logic reduction techniques

Apart from the majority logic reduction technique, nowadays And-Or-Invert (AOI), Nand-Nor Invert (NNI), layered T (LT) logic reduction techniques become significant in post majority design era of QCA [10]. The wired AOI, wireless AOI, NNI and LT gates have been successfully tested and considered in the realization of standard functions. The LT gate provides the scope to build multilayer universal NAND/NOR-based functions with better logic reduction capabilities in [11]. The QCA layout with layer-wise cell arrangements of LT gate is demonstrated in figure 3. The LT gate needs 5 QCA cells. It can also be seen from figure 4 that the first 4 (out of 5) QCA cells are placed "normally" in layer 1. But the remaining cell is polarized as either '+1' or '-1' and placed at layer 2. The layer two has developed 34.5 nanometers (nm) above layer 1.

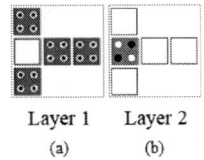

Layer 1   Layer 2
(a)   (b)

Figure 3. Layer-wise cell arrangements of layered T gate, (a) arrangements of 4 cells in layer 1, (b) placement of 1 cell in layer 2

If the polarization of the upper layer cell is fixed at P=+1, then LT gate operates as NAND gate. However, the gate serves as NOR function when the top layer cell is set at P=-1. The logic symbol of LT NAND and NOR gates along with their QCA schematic is shown in figure 4(a) and (b) respectively.

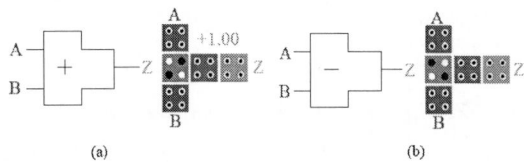

Figure 4. Logic symbol and QCA schematic of (a) Layered T NAND gate, (b) Layered T NOR gate

Equation (2) represents the characteristic equation of layered T NAND and NOR gate as follows below:

$$Z = L_T^+(A, B) = \overline{AB}, \; Z = L_T^-(A, B) = \overline{A + B} \quad (2)$$

Table 1 shows the logic symbols, layered T equations and QCA schematics of AND, OR, Ex-OR and Ex-NOR gates. In the next section, LT Ex-OR (LTEx) gate will be used to realize generic Hamming code generators.

III. HAMMING CODE GENERATORS AND ITS QCA REALIZATION

A. Definition of Hamming Code Generators

The commonly used error-detection and correction code in random access memory (RAM) is called a Hamming code. In this code, k parity bits are added to n-bit information to n-bit information to form (n+k) bits of information with the length of 2^k-1 length. So the number of bits for useful data becomes, $n=2^k-k-1$. The variable 'k' is any random integer [5]. To exemplify the operation of even Hamming code generator, the 4-bit data is chosen as 1011 as an example. The parity bits are included as follows:

Bit Positions	MSB*						LSB**
	D7	D6	D5	P4	D3	P2	P1
	1	0	1	P4	1	P2	P1

*MSB=Most Significant Bit, **LSB=Least Significant Bit

The three parity bits P1 through P4 are positioned in 2^0, 2^1 and 2^2 positions respectively. These parity bits are calculated as below:

P1=D3⊕D5⊕D7=Ex-OR of bit position(3,5,7)=1⊕1⊕1=1
P2=D3⊕D6⊕D7=Ex-OR of bit position(3,6,7)=1⊕0⊕1=0
P4=D5⊕D6⊕D7=Ex-OR of bit position(5,6,7)=1⊕0⊕1=0

TABLE I.   LAYERED T IMPLEMENTATIONS OF AND, OR, EX-OR AND EX-NOR GATES

Layered T function as	Layered T equation	Layered T Symbol	Layered T Schematic
AND	$L_{TA}^+(A, B) = AB$		
OR	$L_{TO}^-(A, B) = A + B$		
Ex-OR	$LTEx(A, B) = A \oplus B$		
Ex-NOR	$LTExn(A, B) = A \odot B$		

978-1-5386-6723-1/19 $31.00 © 2019 IEEE

Thus, the 7 bit coded word becomes 1010101. In this example, a total number of 1's is maintained at even number. An encoder which generates even Hamming Code is called even Hamming Code generator. Likewise, if the encoder generator totals an odd number of 1s, then it is termed odd Hamming Code generator.

B. QCA Realization of Hamming Code Generators

The even and odd Hamming code generators are realized by utilizing LTEx module. The *algorithm 1* to design a generic even or odd LTEx Hamming code generator is given in *algorithm one* as follows.

Algorithm 1:
Step 1: Number the bits from MSB to LSB starting from n^{th} position,
Step 2: Specify the bit numbers in binary,
Step 3: Designate the positions of "powers of two" as parity bits and calculate them in accordance with the following:
  if: even Hamming code generation
  i) P(2^0)=LTEx(bit 3, bit 5, bit 7...(check 1, skip 1, check 1, skip 1...) up to $(n+k)^{th}$ bit)
  ii) P(2^1)=LTEx(bit 3, bit 6, bit 7, bit 10, bit 11,...(check 2, skip 2, check 2, skip 2,...) up to $(n+k)^{th}$ bit)
  iii) P(2^2)=LTEx(bit 5, bit 6, bit 7, bit 12, bit 13, bit 14, bit 15...(check 4, skip 4, check 4, skip 4, ...) up to $(n+k)^{th}$ bit)
  .

  .

  iv) P(2^k)=LTEx(bit 2^k+1, bit 2^k+2,...,bit $2*2^k$+1...(check 2^k, skip 2^k, check 2^k, skip 2^k, ...) up to $(n+k)^{th}$ bit)
  else: odd Hamming code generation
  i) P(2^0)=LTExn(bit 3, bit 5, bit 7...(check 1, skip 1, check 1, skip 1...) up to $(n+k)^{th}$ bit)
  ii) P(2^1)=LTExn(bit 3, bit 6, bit 7, bit 10, bit 11,...(check 2, skip 2, check 2, skip 2,...) up to $(n+k)^{th}$ bit)
  iii) P(2^2)=LTExn(bit 5, bit 6, bit 7, bit 12, bit 13, bit 14, bit 15...(check 4, skip 4, check 4, skip 4, ...) up to $(n+k)^{th}$ bit)
  .

  .

  iv) P(2^k)=LTExn(bit 2^k+1, bit 2^k+2,...,bit $2*2^k$+1...(check 2^k, skip 2^k, check 2^k, skip 2^k, ...) up to $(n+k)^{th}$ bit)
Step 4: Specify the remaining positions as data bits and generate n+k coded word.

From *algorithm 1*, it can be shown that 3 parity bits (i.e. k) are added with 4 bits of data (i.e. n=4=2^k-k-1) to form 7 bit (n+k=2^k-k-1+k=2^k-1) Hamming code. If data bits are taken as 1011, then [D7, D6, D5, D3] are indexed to comply the *step 1* and *2* of *algorithm 1* to set D7=1, D6=0, D5=1 and D3=1. Then the parity bits are positioned at 2^0, 2^1 and 2^2. According to the *step 3* of the proposed algorithm, the calculations of the parity bits are given as follows: P1=LTEx(D3,D5,D7)=LTEx(1,1,1)=1, P2= LTEx(D3,D6,D7)=LTEx(1,0,1)=0 and P4=LTEx(D5,D6,D7)=LTEx(1,1,1)=1. Finally the even (7,4) Hamming code becomes D7=1, D6=0, D5=1, P4=1, D3=1, P2=0 and P1=1. The QCA layout of (7,4) even Hamming code generator is demonstrated in figure 5. The proposed layout needs 4 LTEx gates, consumes 0.87 μm² area with 571 cells to generate the expected output as D7_Out=1, D6_Out=0, D5_Out=1, P4=1, D3_Out=1, P2=0 and P1=1 after 3 clock cycles. Since the design and type of crossover to generate a specific layout depends on the manufacturer, this work

considers the "number of quantum cells in effective area of gate" required to construct LTEx modules. This notion fixes the effective O-Cost (number of quantum cells used in a layout) [12] of proposed even Hamming code generator at 208. Similarly, the QCA realization of odd (7,4) Hamming code generator is illustrated in figure 6. The odd counterpart of figure 6 takes 0.9 μm² area with 578 cells to get the result as D7_Out=1, D6_Out=0, D5_Out=1, P4=1, D3_Out=1, P2=1 and P1=0 after 3 clock cycles.

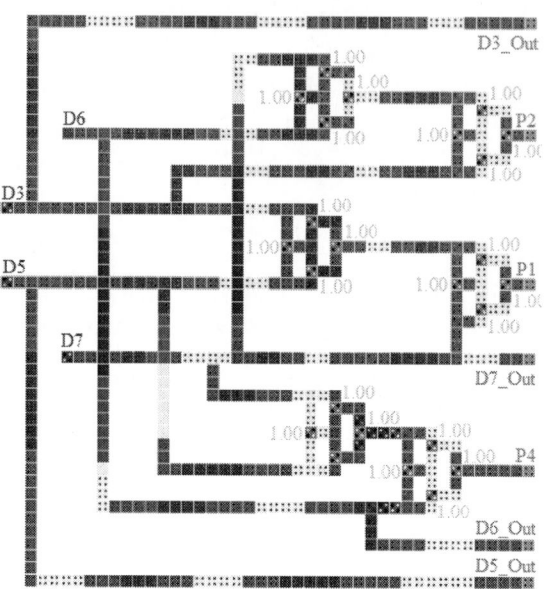

Figure 5. QCA layout of even (7,4) Hamming code generator

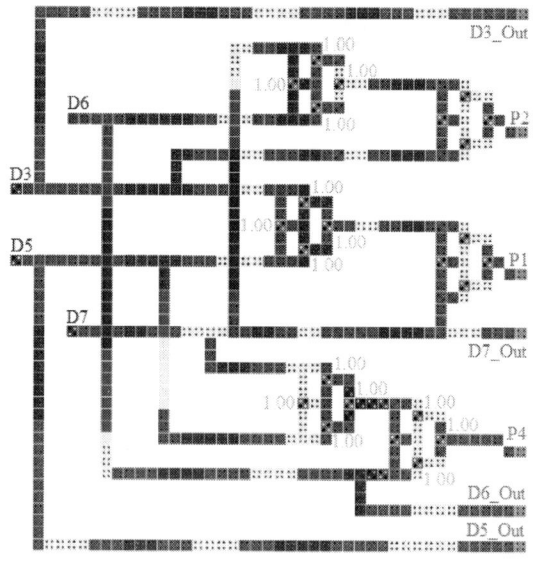

Figure 6. QCA layout of odd (7,4) Hamming code generator

The outputs of the proposed QCA even and odd (7,4) Hamming code generators are shown in figure 7 (a) and (b) respectively. The design summary of even (7,4) Hamming code generator which has been designed by using 3-input MV, AOI, NNI, and LT logic reduction technique is reported in Table II. The ratio of two quantities, i.e., 2^k-k-1 to 2^k-1 is known as *rate of Hamming code*. Figure 8 and nine respectively illustrate the variations of effective O-Cost and number of gate concerning the increasing *rate of Hamming code generators*.

TABLE II.   DESIGN SUMMARY OF (7,4) EVEN HAMMING CODE GENERATORS BY USING MV, AOI, NNI, AND LT LOGIC REDUCTION METHOD

Sl. No.	Parity Bits (k)	Data Bits (2^k-k-1)	3-input MV		AOI			NNI		LT	
			#gate	O-Cost	#gate	O-Cost*	O-Cost**	#gate	O-Cost	#gate	O-Cost
1	2	3	20	256	16	112	480	16	272	16	104
2	3	7	30	384	24	168	720	24	408	24	156
3	4	15	40	512	32	224	960	32	544	32	208
4	5	31	50	640	40	280	1200	40	680	40	260
5	6	63	60	768	48	336	1440	48	816	48	312
6	7	127	70	896	56	392	1680	56	952	56	364
7	8	255	80	1024	64	448	1920	64	1088	64	416

*wireless AOI gate, **wired AOI gate, Here O-Cost represents "useful" quantum cells used to build the logic gates. The quantum cells for interconnecting wires will be added with the number mentioned in O-Cost columns of table II.

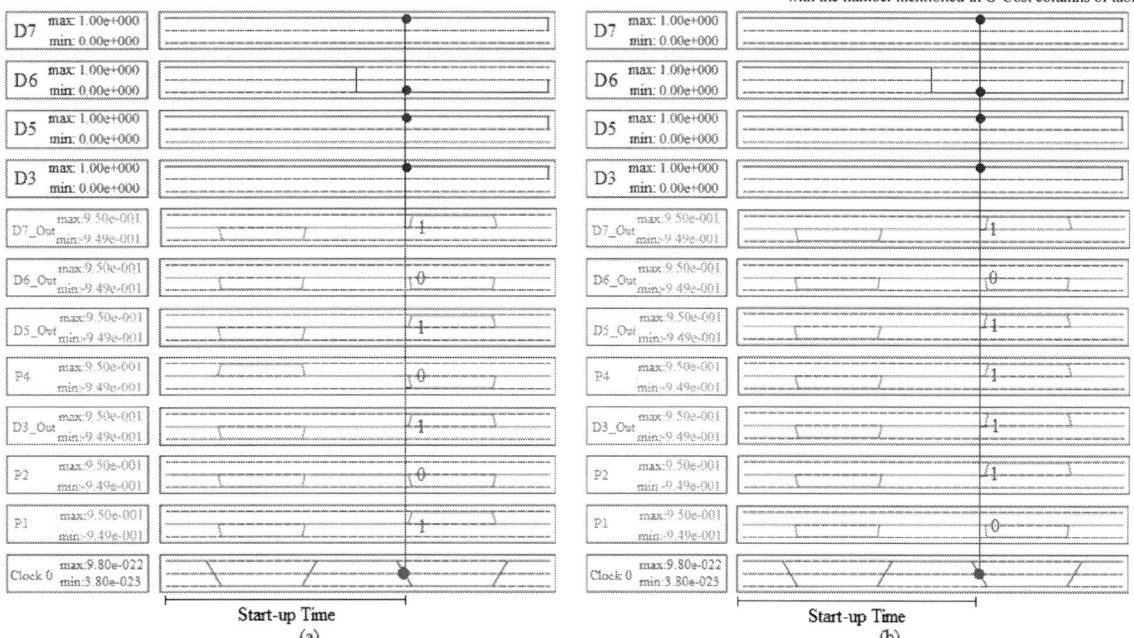

Figure 7. Outputs of (7,4) Hamming Code generator for (a) even and (b) odd counterpart

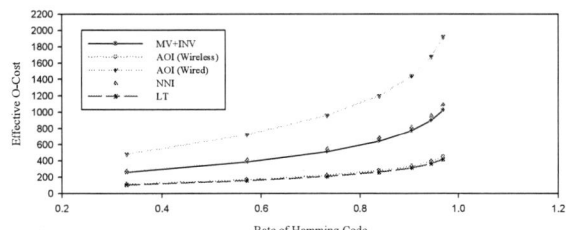

Figure 8. A plot of effective O-Cost versus rate of Hamming code generators

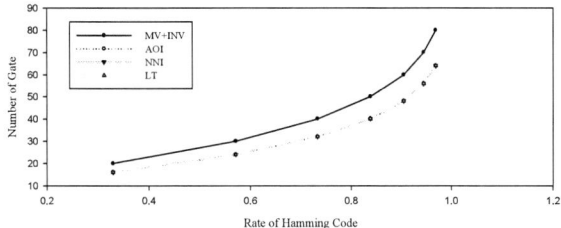

Figure 9. A plot of Number of the gate for various logic reduction techniques versus rate of Hamming code generators

IV. RESULT DISCUSSION AND CONCLUSION

The simulation results of (7,4) even and odd Hamming code generator are demonstrated respectively in figure 7(a) and (b). The QCA simulator, i.e., QCADesigner [16] with the parameters specified in [11], is used to test and verify the proposed QCA layouts. The input vector, [D7=1, D6=0, D5=1, D3=1] is given to even Hamming code generator layout of figure 5. The output [D7_Out, D6_Out, D5_Out, P4, D3_Out, P2, P1] is obtained as [1010101] clearly showing the compliance of parity equations specified in section III. Similarly, the odd counterpart is tested with the same dataset. The output, i.e. [D7_Out, D6_Out, D5_Out, P4, D3_Out, P2, P1] is obtained as [1011110]. The blue color spots of figures 7(a) and (b) denote the input logic values in the outputs of the proposed layouts.

This work offered the characterization of LT gate with a simulation-based presentation. Testing of LT gate in various gate-level schematics are addressed, and appropriate LT equations are presented in table I. The existing works on Hamming code generators in [5, 13-15], do not follow specified guidelines. Even the current layouts do not obey QCA design rules. *Algorithm 1* provides a proper direction to create even and odd generic Hamming code generators by using LTEx modules. The QCA layouts of even (7,4) Hamming code generator is compared with the 3-input Majority voter (3 input MV), AOI and NNI design counterparts. The detailed analysis of design parameters is mentioned in table II. It shows that the LT designs of even Hamming code generators require 20% less number of gates on an average than their MV counterparts. However, the gate requirements for LT, AOI and NNI designs remain the same. The even LT Hamming code generators need 59.38%, 7.15%, and 61.77% less O-Cost than 3-input MV, AOI (wireless) and NNI counterparts respectively. From the detailed analysis of figure 8 and 9, it can be said that the Hamming code generators with high rates consume significantly less effective O-Cost and gate requirements than the MV, AOI and NNI gate based designs. Unlike conventional MV, AOI, and NNI even Hamming code generators, the LT designs operate quite favorably in terms of layout design parameters. So it can be concluded that the LT Hamming code generator designs can be used to produce robust Nano communication networks.

ACKNOWLEDGEMENT

Authors thank anonymous reviewers for their insightful comments and suggestions. They express special thanks to Prof. Debdatta Banerjee for her literary contributions that help authors in organizing the article.

REFERENCES

[1] C. S. Lent and P. D. Tougaw, "A device architecture for computing with quantum dots," in Proceedings of the IEEE, vol. 85, no. 4, pp. 541-557, April 1997. DOI: 10.1109/5.573740

[2] International Technology Roadmap for Semiconductor Report 2015 (ITRS Report 2015), Available at http://www.itrs2.net/itrs-reports.html (accessed on 12th October, 2018).

[3] G. Toth and C. Lent, "Quantum Computing with Quantum-Dot Cellular Automata". Physical Review A, vol. 63, no. 5, April 2001, DOI: 10.1103/PhysRevA.63.052315.

[4] R. W. Hamming, "Error Detection and Error Correcting Codes" The Bell System Technical Journal, vol. 29, no. 2, April 1950.

[5] A. Ahmadpour, A. Ahadpour Sha and M. Ziabari, "A novel formulation of Hamming Code," 6th International Conference on Electrical Engineering/Electronics, Computer, Telecommunications and Information Technology, Pattaya, Chonburi, 2009, pp. 808-811. DOI: 10.1109/ECTICON.2009.5137169.

[6] C. Lent, B. Isaksen and M. Lieberman, "Molecular Quantum-Dot Cellular Automata". Journal of the American Chemical Society, vol. 125, no. 4, pp. 1056-1063, January 2003, DOI: 10.1021/ja026856g

[7] R. Cowburn and M. Welland, "Room Temperature Magnetic Quantum Cellular Automata". Science, vol. 287, no. 5457, pp. 1466-1468, February 2000, DOI:10.1126/science.287.5457.1466.

[8] G. A. Dilabio, R. A. Wolkow, J. L. Pitters, G. Piva. "Atomistic Quantum Dots", May 2015, US Patent: US2015/0060771-A1.

[9] C. S. Lent, P. D. Tougaw and W. Porod, "Quantum cellular automata: the physics of computing with arrays of quantum dot molecules," Proceedings Workshop on Physics and Computation. Phys. Comp. '94, Dallas, TX, USA, 1994, pp. 5-13, DOI: 10.1109/PHYCMP.1994.363705.

[10] C. Mukherjee, S. Panda, A. K. Mukhopadhyay, B. Maji, "Synthesis of standard functions and generic Ex-OR module using layered T gate", International Journal of High Performance Systems Architecture, Vol.7 No.2, pp.70–86, September 2017, DOI: 10.1504/IJHPSA.2017.087164.

[11] C. Mukherjee, S. Panda, A. K. Mukhopadhyay, B. Maji, "QCA Gray Code Converter Circuits Using LTEx Methodology", International Journal of Theoretical Physics, Vol.57, no. 7, pp 2068–2092, July 2018, DOI: https://doi.org/10.1007/s10773-018-3732-4.

[12] C. Mukherjee, S. S. Roy, S. Panda and B. Maji, "T-Gate: Concept of partial polarization in Quantum Dot Cellular Automata," 2016 20th International Symposium on VLSI Design and Test (VDAT), Guwahati, 2016, pp. 1-6, DOI: 10.1109/ISVDAT.2016.8064844.

[13] M. Sukanya and R. Ganeshan, "Nano Design of Communication Parts With Quantum Dot Cellular Automata", International Journal of Engineering Science and Computing, vol. 6, no. 5, May 2016, DOI:10.4010/2016.1266.

[14] M. Ziabari, A. M. Kassai, A. Ziabari, S. E. Maklavani, "Designing a Hamming Coder/Decoder Using QCAs", Journal of Applied Sciences, vol. 8, no. 14, pp. 2569-2576, May 2008, ISSN: 2569-2576.

[15] D. S. Silva, L. H. B. Sardinha, M. A. M. Vieira, L. F. M. Vieira, O. P. Vilela Neto, "Robust Serial Nanocommunication With QCA", IEEE Transactions on Nanotechnology, vol. 14, no. 3, May 2015, DOI: 10.1109/TNANO.2015.2407696.

[16] QCADesigner Tool Version 2.0.3. Available at hyperlink: http://www.mina.ubc.ca/qcadesigner_downloads

2019 Devices for Integrated Circuit (DevIC), 23-24 March, 2019, Kalyani, India

Role of Stress/Strain Mapping in Advanced CMOS Process Technology Nodes

T. P. Dash[1*], J. Jena[2], E. Mohapatra[3], S. Dey[4], S. Das[5] and C. K. Maiti[6]
Department of Electronics and Communication Engineering,
Siksha 'O' Anusandhan (Deemed to be University), Bhubaneswar, Odisha, India-751030
[1*]taradash@soa.ac.in; [2]jhansiranijena@soa.ac.in; [3]eleenamohapatra@soa.ac.in; [4]supravadey@soa.ac.in; [5]sanghamitradas@soa.ac.in; [6]ckmaiti@soa.ac.in.

Abstract— **Multiple-gate MOSFETs have emerged as potential candidates for the future device generations considering the continuous increase in performance requirements. Therefore, a great demand to control strain/stress and their variation in MOSFETs has recently emerged. In this work, biaxial and uniaxial strain techniques are implemented in the device channel for both p- and n-type MOSFETs. Stress/strain mapping in strained-Si and SiGe channel trapezoidal tri-gate FinFET devices are studied through three-dimensional (3D) numerical simulation, with particular focus on enhancement of drain current. Following the strain/stress profiles simulated, the piezoresistive changes are implemented in the simulator to describe the strain effects on device operation.**

Keywords – Strained-Si, Strained-SiGe, stress/strain map, FinFET, Technology computer aided design (TCAD).

I. INTRODUCTION

Stress and strain engineering have so far been use in Si CMOS technology as a technology booster to enhance the carriers transport via band structure modulation[1,2]. Since 22nm technology nodes, the scaling of Metal-Oxide-Semiconductor field effect transistors (MOSFETs) alone are not sufficient to enhance performance of integrated circuits[3]. Beyond the 22 nm node, the planar architecture in bulk silicon has been discarded in favor of non-planar 3D structures with fully depleted channels. Introduction of stress in the nonplanar devices further enhances the device performance essential for future CMOS technology[4,5]. FinFET structures have shown great potential for both the digital and analog applications and there are also some works reported dealing with experimental measurements on strain mapping in FinFETs. Utilization of wafer level, biaxially strain in conjunction with local strain techniques leads to significant mobility enhancement in both n- and p-channel devices. Such an approach will allow the separate optimization of p- and n-channel device performances by locally controlling the stress level [6]. However, no results on simulation of stress/strain mapping data are available addressing the role of strain map in device design.

Straining of silicon fundamentally changes the mechanical, electrical (band structure and mobility), and chemical (diffusion and activation) properties. The introduction of strain causes effective mass change and band splitting and consequently lower scattering. Both of these effects improve the carrier mobility. There are several methods to introduce strain, such as substrate induced biaxial strain and process-introduced uniaxial strain. For successful stress/strain engineering in semiconductor device structures, one must consider all the contributions of the stress field including those not commonly considered for stress analysis, such as, the intrinsic stress. It is necessary to visualize the three-dimensional (3D) stress profiles to understand the device operation. The stress components identify the areas of compressive and tensile stress and the stress field could be used to understand the device performance, and to study the possible relative improvements in drain current.

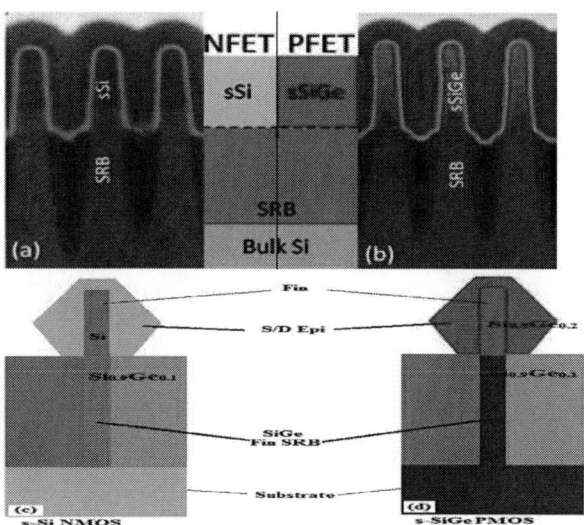

Fig.1. Schematic description of dual-stressed channel materials on strain-relaxed buffer (SRB) and TEM of (a) tensile strained Si FIN, (b) compressive strained SiGe FIN on a common SRB [Source: Chip Works], (c) simulated strained-Si NMOS (Si on $Si_{0.9}Ge_{0.1}$), and (d) simulated strained-SiGe PFET ($Si_{0.8}Ge_{0.2}$ on $Si_{0.9}Ge_{0.1}$) FinFETs, respectively.

This work is motivated from the reported device in [5] shown in Figs. 1(a) and (b) and focuses on the use of channel strain mapping technique for stress/strain tuning in strained tri-gate FinFETs (see Figs 1(c) and (d)) with epitaxial source/drain SiGe for different channel shapes. Keeping the biaxial stress intact and adding etch-stop liners that increase the tensile strain in the channel length direction is studied to show the enhanced electron mobility in n-MOS transistors. High hole mobility enhancement in p-MOS devices is achieved with a tensile stress in channel width direction and a compressive stress in the channel length direction. The main device analog parameter, such as, the drain current is studied in detail.

II. TCAD CALIBRATION

3D Technology CAD (TCAD) process simulation of stress evolution provides valuable insights for technology development and stress management. Studies on stress engineering, performance and reliability trade-off are essential for design and technology explorations. It also requires calibration of model parameters as the predictive range of any process modeling capability can vary significantly between process modules. In this aspect, the transfer characteristics of a 20-nm channel length tri-gate FET with <110> orientation has been calibrated with the experimental data reported in Reference [7]. Such calibration is performed by incorporating the exact device geometry and materials along with other relevant device parameters as reported in the experimental work. The comparative plots shown in Fig. 2 indicate a good agreement between the experimental data and theoretical simulation values.

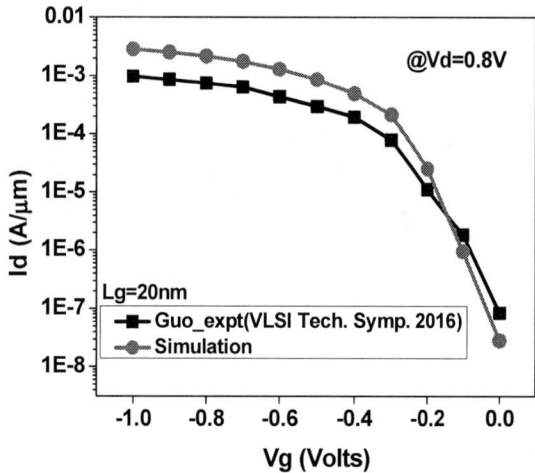

Fig 2. The simulated transfer characteristics has been calibrated with the experimental data taken from Reference [7].

III. SIMULATION ENVIRONMENT

Realistic 7nm NMOS and PMOS FinFET devices are constructed as shown in Figs. 1(c) and (d), with a physical gate length of 14 nm and <110>oriented channels. The analytical doping profile used in simulation is shown in Fig. 3. The technological parameters considered in simulations are shown in Table I.

TABLE I. TECHNOLOGICAL PARAMETERS CONSIDERED IN SIMULATION.

Design parameters	7nm Node	Units
Fin Height	30	nm
Fin Width	5	nm
Fin SRB Height	50	nm
Oxide Thickness	0.5	nm
High-k (HfO$_2$) Thickness	1.5	nm
Gate Length	14	nm
Spacer Length	7	nm
Epi Length	14	nm
Epi Top Width	7	nm
Epi Middle Width	25	nm
Epi Bottom Width	14	nm
Substrate Height	350	nm
Bulk Height	50	nm

Well Doping	1e20	cm^{-3}
Epi Doping	2e20	cm^{-3}
Substrate Doping	1e16	cm^{-3}

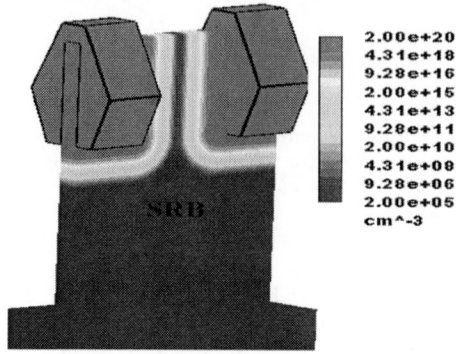

Fig. 3. Net doping profile in NFET.

The effect of mechanical stress can be modeled using the theory of elasticity by relating stress with strain. The relationship between them was first time developed by Hook. According to Hook, the stress tensor and strain tensor are linearly related over a certain range of deformation for Hookean elastic solid. Generally this linear relationship between these tensors can be expressed as:

$$\sigma_{ij} = C_{ijkl}\varepsilon_{kl} \qquad (1)$$

Where, σ_{ij} is stress tensor, ε_{kl} is strain tensor and C_{ijkl} represents the 4th order elastic stiffness tensor with 81 (= 3^4) components. Whereas, C_{ijkl}, can be limited to a tensor of 36 components by considering the symmetries involved for both the strain and stress tensors under equilibrium. The off-diagonal components are equal to one-half of the shear strain. As the off-diagonal strain components are converted to shear strains, so for simplicity the notations used for the strain and stress tensors can be expressed as vectors following the contracted notations:

$$\begin{bmatrix} \varepsilon_{xx} & 2\varepsilon_{xy} & 2\varepsilon_{xz} \\ 2\varepsilon_{yx} & \varepsilon_{yy} & 2\varepsilon_{yz} \\ 2\varepsilon_{zx} & 2\varepsilon_{zy} & \varepsilon_{zz} \end{bmatrix} = \begin{bmatrix} \varepsilon_{xx} & \gamma_{xy} & \gamma_{xz} \\ \gamma_{yx} & \varepsilon_{yy} & \gamma_{yz} \\ \gamma_{zx} & \gamma_{zy} & \varepsilon_{zz} \end{bmatrix} \quad (2)$$

Where γ is the notation used for shear strain. In mechanical engineering, Young's modulus, Y and Poisson ratio, v are commonly used (for homogeneous and isotropic material) to relate strain. These values of compliance coefficient based on Y and v considered in simulation for Si and Ge are shown in Table II.

TABLE II. THE ELASTIC COMPLIANCE COEFFICIENTS C$_{ij}$ [GPA], VALUES FOR SI AND GE.

C$_{ij}$	Si	Ge
C_{11}	165.64	128.7
C_{12}	63.94	47.7
C_{44}	79.51	66.7

2019 Devices for Integrated Circuit (DevIC), 23-24 March, 2019, Kalyani, India

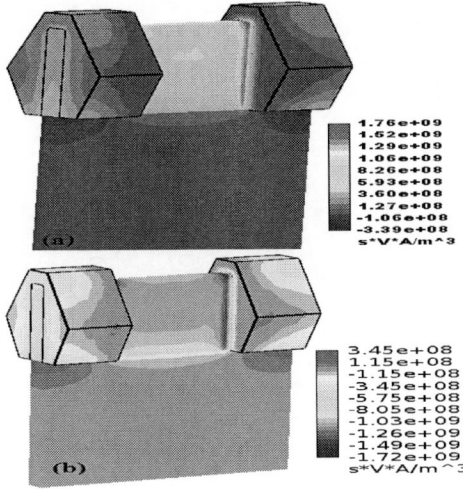

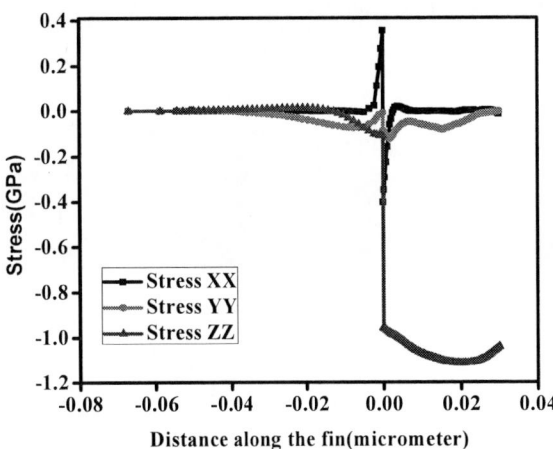

Fig. 4. (a) Axial stress (σ_{zz}) (left) in NFET and (b) axial stress (σ_{zz}) (right) in PFET. The stress is generated due to lattice mismatch between the fin and S/D stressor.

Fig. 6. Vertical stress profile in the device center for a PFET device with $Si_{0.8}Ge_{0.2}$ channel grown on $Si_{0.9}Ge_{0.1}$ SRB.

There are three stress components Stress XX, Stress YY, and Stress ZZ along x, y and z directions, respectively. The axial stress component (Stress ZZ) profile for both NFET and PFET have been shown in Fig. 4. The channel region of fin shows tensile ('+' sign) and compressive stress ('-'sign) for NFET and PFET respectively. Stress values of the order of 1.6 GPa have been reported for similar device with $Si_{0.75}Ge_{0.25}$ SRB [5]. The maximum stress obtained in our simulation is approximately $\pm$ 1.7 GPa which can be observed from the stress map shown in Figs. 4(a) and (b). The stress value ranges from +1.76 GPa to -0.34 GPa for NFET and +0.34 GPa to -1.72 GPa for PFET. The uniaxial stress is applied along the channel length, as presented in the figure, with the maximum value of 1.2 GPa at source/drain interfaces and decays in the direction of the center of the channel, reaching the level of 1 GPa.

Transformation of biaxial strain to uniaxial strain occurs when Si/SiGe film is etched into stripes. Etching of biaxially strained-SiGe films to produce uniaxial strain (for strained-Si) in devices has opened the avenue for various non-planner multi-gate device structures below 45nm technology node currently in use [6, 8]. The stress developed in the channel is uniaxial in nature which has been verified by taking vertical cut at the mid of the fin. Vertical profile of all three stress components are plotted in Figs. 5 and 6 for NFET and PFET, respectively. The stress values along x and y directions are ±405 MPa and ±99 MPa whereas the stress ZZ maintains the maximum value ±1 GPa. Hence, it is confirmed that the uniaxial stress dominates the channel and provides the platform for design of stress-enhanced CMOS. It is important to note that stress ZZ is slightly higher in case of strained-SiGe PFET, compared to strained-Si NFET.

IV. RESULTS AND DISCUSSIONS

Strained-Si channel ultra-thin multi-gate MOSFET structures have been considered in CMOS integration in post 22nm technology nodes. An accurate modeling of such devices relies on the modeling of the sub-band structures of Si and Ge [9]. In this work, a two-band k·p model is used for the device simulation. A six-band k·p model is used with SiGe composition dependent parameters for the PFET devices. The material composition is included in the parameters for various scattering models, which included phonons (acoustic, optical intra and inter-valley), roughness, charged impurities and alloy disorder. The sub-band Boltzmann Transport Equation (SBTE) was solved in the active regions of the device (channel part of the fin), which is then fitted to the density-gradient (DG) simulation of the entire device using an effective mobility [10] in each iteration step. Figs. 7 and 8 show the solution of the SBTE, in terms of the electron and hole spectrum at on-state for NFET and PFET, respectively. The potential variation, carrier concentration and average

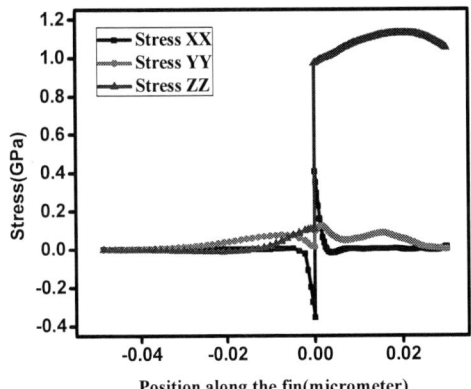

Fig. 5. Vertical stress profile in the device center for a NFET device with Si channel grown on $Si_{0.9}Ge_{0.1}$ SRB.

978-1-5386-6723-1/19 $31.00 © 2019 IEEE   23

carrier velocity are shown in (a), (b) and (c) parts in Figs. 7 and 8, respectively for both the type of devices.

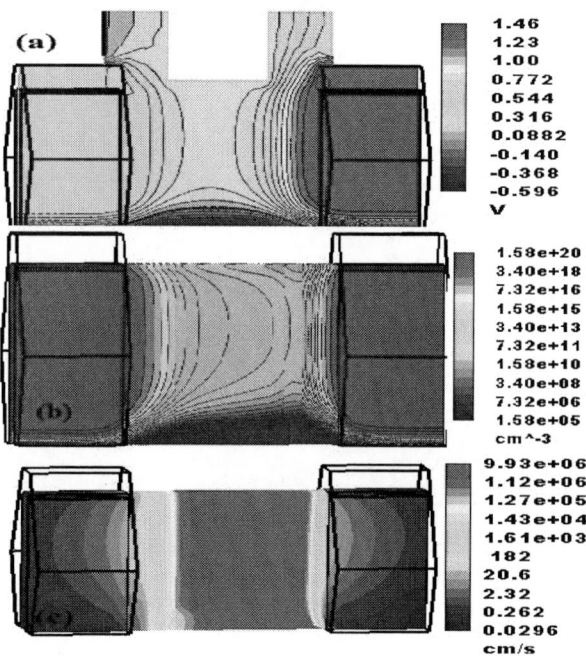

Fig. 7. Cut plane though the NFET device along the fin showing electrostatic potential (a), electron concentration (b), and average electron velocity (c) at $V_{GS} = V_{DS} = 0.7V$.

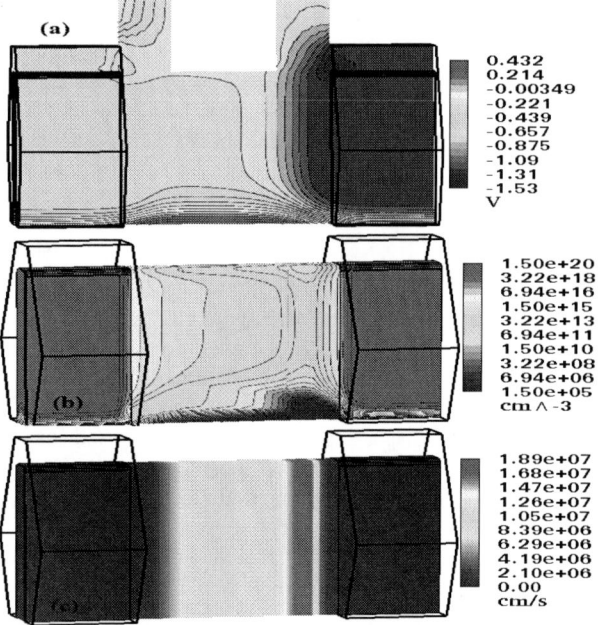

Fig. 8. Cut plane though the PFET device along the fin showing electrostatic potential (a), hole concentration (b), and average hole velocity (c) at $V_{GS} = V_{DS} = -0.7V$.

Piezoresistive approach has been used mostly as it provides the relation of mobility under strained and unstrained conditions. The carrier mobility enhancement tensor, expanded up to second order in stress, is given by [11]:

$$\Delta\mu_{ij} = \frac{\mu_{ij}}{\mu_0} \approx \sum_{k=1}^{3}\sum_{l=1}^{3} \Pi_{ijkl}\sigma_{kl} \qquad (3)$$

where σ_{kl} is a component of the stress tensor. Π_{ijkl} is a component of the first-order electron/hole piezoconductance tensor. Due to symmetry, the mobility for isotropic unstrained cubic crystal is obtained from the piezoconductance tensor Equation (3) can be written as:

$$\Delta\mu_{ij} = \frac{\mu_{ij}}{\mu_0} \approx \sum_{k=1}^{3} \Pi_{jk}\sigma_k \qquad (4)$$

The piezoconductance are related to the components of the piezoresistance tensors. Effective mobility is linked with drift velocity, which corresponds to the average speed of the charge carriers. However, in the context of ultra-short channel transistors in a high-field regime, Drift-Diffusion (DD) simulations must have currents consistent with those resulting from advanced transport models. In this aspect, Density Gradient (DG) model is implemented which has been derived from the method of moments applied to the Wigner equation [12].

In this work, MINIMOS-NT tool [11] has been used in simulation to extract device performance parameters, such as, the drain current as shown in Figs. 9(a) for PFET and 9(b) for NFET, respectively. The enhancement in drain current is found to be 31% in case of strained-SiGe enhanced-PFET (EPMOS) compared to unstressed condition (PMOS) shown in red and black color in the figure. In case of strained-Si enhanced- NFET (ENMOS) and unstressed condition (NMOS), the enhancement factor found to be 13%. The enhancement in EPMOS and ENMOS drain current is due to improvement in mobility caused by uniaxial stress introduced as described by equation (4). However, the percentage improvement is higher in PFETs than in NFETs. This is due to slightly higher stress observed in case of strained-SiGe PFETs compared to strained-Si channel NFET devices as shown in Figs. 5 and 6. DIBL for the devices are found to be 25 and 47 mV/V for PMOS and NMOS, respectively. Subthreshold slope for the devices are found to be 68 and 67 mV/decade for PMOS and NMOS, respectively.

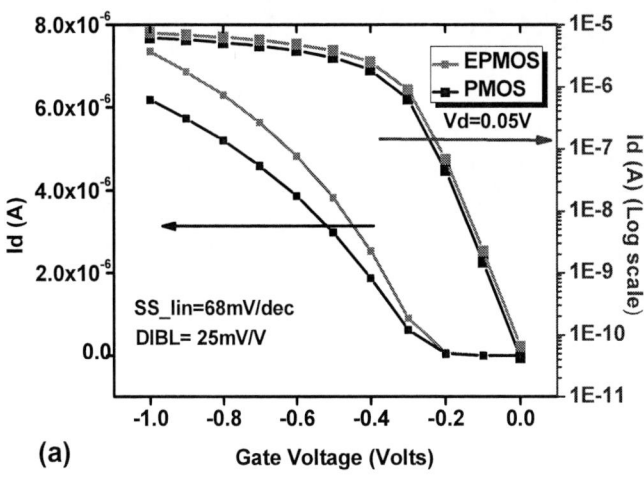

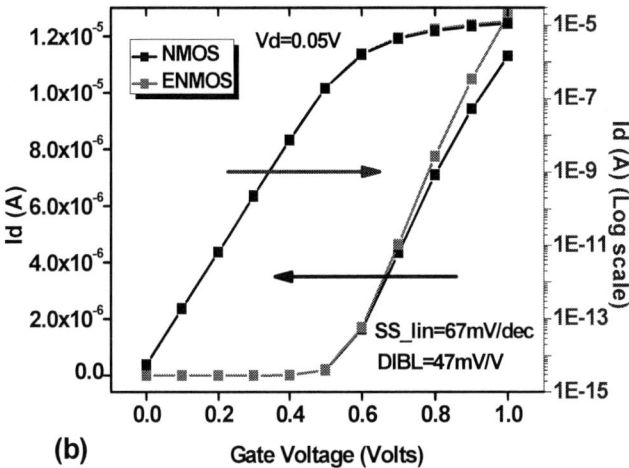

Fig. 9. Transfer characteristic for PFET (a) and NFET (b) at $|V\text{DS}|= 0.05\text{V}$ for devices with $Si_{0.9}Ge_{0.1}$ SRB (red) and no stress (blue). The enhanced devices show a 31% (PFET) and 13% (NFET) increase in on-current compared to their respective unstressed devices.

V. CONCLUSION

Using 3D TCAD simulations, we generated the stress/strain maps in different regions of a state-of-the-art (7N technology node) tri-gate FinFETs with SiGe source/drain stressors. Using the stress/strain maps, in this study, we show the importance of the knowledge of distribution of the biaxial and uniaxial strain states in the devices for the performance enhancement studies of device parameters in advanced CMOS technology nodes. It is shown that channel stress/strain engineering and their control are critical for technology development as strain state fluctuation is a statistical source of variability in nanoscale FinFETs.

REFERENCES

[1] G. A. Armstrong and C. K. Maiti, "TCAD for Si, SiGe and GaAs Integrated Circuits (The Institution of Engineering and Technology (IET), UK, 2008.

[2] R. K. Nanda et al., "Beyond silicon: Strained-SiGe channel FinFETs," *2015 International Conference on Man and Machine Interfacing (MAMI)*, pp. 1-4, 2015.

[3] K. J. Kuhn, "Considerations for Ultimate CMOS Scaling," in *IEEE Transactions on Electron Devices*, vol. 59, no. 7, pp. 1813-1828, 2012.

[4] C. Auth et al., "A 22nm high performance and low-power CMOS technology featuring fully-depleted tri-gate transistors, self-aligned contacts and high density MIM capacitors," *2012 Symposium on VLSI Technology (VLSIT)*, pp. 131-132, 2012.

[5] R. Xie et al., "7nm FinFET Technology Featuring EUV Patterning and Dual Strained High Mobility Channels," *in IEEE International Electron Devices Meeting (IEDM)*, pp. 2.7.1-2.7.4., 2016.

[6] D. Lu et al., "Silicon Germanium FinFET Device Physics, Process Integration and Modeling Considerations," *ECS Transaction*, vol. 64, pp. 337-345, 2014.

[7] D. Guo et al., "FINFET technology featuring high mobility SiGe channel for 10nm and beyond," *2016 IEEE Symposium on VLSI Technology (VLSIT)*, pp. 1-2, 2016.

[8] J. G. Fiorenza, J. S. Park and A. Lochtefeld, "Detailed Simulation Study of a Reverse Embedded-SiGe Strained-Silicon MOSFET," *IEEE Transactions on Electron Devices*, vol. 55, pp. 640-648, 2008.

[9] O. Baumgartner et al., "VSP–a quantum-electronic simulation framework," *Journal of Computational Electronics*, vol. 12, pp. 701–721, 2013.

[10] Z. Stanojevic et al., "Physical modeling – A new paradigm in device simulation," in *IEEE International Electron Devices Meeting (IEDM)*, pp. 5.1.1–5.1.4, 2015.

[11] Minimos-NT User Manual 2017.

[12] A. Wettstein, *Quantum Effects in MOS Devices*. Konstanz: Hartung-Gorre Verlag, 2000.

Electromagnetic Bandgap Formation in Two-Dimensional Photonic Crystal Structure with DNG Materials under TE Mode

Papri Chakraborty
Dept of Electronic Science
A.P.C College
New Barrackpore, India
papri.chakraborty1996@gmail.com

Ratul Ghosh
Dept of Electronic Science
A.P.C College
New Barrackpore, India
ratulghosh777@gmail.com

Anwesha Adhikary
Dept of Electronic Science
A.P.C College
New Barrackpore, India
piu.anwesha@gmail.com

Arpan Deyasi
Dept. of Electronics and Communication Engineering
RCC Institute of Information Technology
Kolkata, India
deyasi_arpan@yahoo.co.in

Angsuman Sarkar
Dept. of Electronics and Communication Engineering
Kalyani Govt Engg College
Kalyani, India
angsumansarkar@ieee.org

Abstract—**Bandgap width within first Brillouin zone in metamaterial-based two-dimensional photonic crystal structure is analytically computed using plane wave expansion method. A wider range of negative refractive index is considered for simulation purpose within physically feasible limit whereas rectangular geometrical shape is taken into account for the analysis with TE mode of propagation. Artificial materials in presence of air holes are considered to achieve negative index, and results are compared with that obtained for conventional SiO₂-air material system with equivalent dimensional configuration. Coordinates of two peak points, 'Γ' point and 'X' point are noted for all the DNG materials which give the indication of blueshift of the valence region with change of negative refractive index. Simulated findings speak in favor of multiple forbidden regions for some specific material systems which can be utilized to design photonic multi-channel filter in beyond THz region.**

Keywords—*DNG material, Electromagnetic bandgap, TE mode, Brillouin zone, Forbidden region, Peak points*

I. Introduction

Design of photonic multi-channel filter is the subject of research in the last decade [1-4] due to its unique property of restricting electromagnetic wave in some spectra, and allowing other regions. This property is exhibited due to the formation of electromagnetic bandgap [5], and that is also dependent on the polarization of the propagating wave [6-7]. Experimental realization of photonic crystal fiber is possible now-a-days [8] for new generation optical communication system [9-10] replacing conventional counterpart; henceforth, photonic crystal (PhC) based devices with their novel properties become a part of all-optical integrated circuit [11-12] which can be used for quantum information processing [13].

Though several literatures are already available in PhC based device design, but negative refractive index materials are rarely used [14] as far the knowledge of the authors.

Investigation of fundamental physical characteristics of optical metamaterials is one of the les highlighted arena, and the present paper primarily focuses in that sector. Since the features of these classified materials is primarily depended on the extent and nature of electromagnetic bandgaps, and that is also critically function of propagating mode, so the authors investigate photonic bandgap within first Brillouin zone for rectangular shape under TE mode of propagation. Though different methods of computation are already published, but authors choose plane wave expansion method due to already established supremacy [15] over other techniques. Widths of lowest two bandgaps are measured for different material compositions, and coordinates of peak points in the equivalent valence band are noted. Quasi forbidden regions are also reported for a few particular systems.

II. Mathematical Formulation

Wave propagation in any arbitrary lossy medium is governed by Maxwell's equation

$$\vec{\nabla} \times \frac{1}{\mu(\vec{r})\varepsilon(\vec{r})} \vec{\nabla} \times \vec{B}(\vec{r}) = \frac{\omega^2}{\mu(\vec{r})C^2} \vec{B}(\vec{r})$$

(1)

where symbols have usual significances. When the electromagnetic wave is passed through any periodic structure, then magnetic field component may be considered as a combination of large number of plane waves, may be expressed in the following form

$$\vec{B}(\vec{r}) = \sum_{\vec{L}_i, \lambda} b_{L_i, \lambda} e^{i(\vec{k} + \vec{L}_i) \cdot \vec{r}} \hat{p}_\lambda$$

(2)

where the positive integer 'λ' helps to compositely define the unit axes ($\hat{p}_\lambda$) perpendicular to the propagation direction, $\vec{L}$ denotes reciprocal lattice vector space, '$b_{L,\lambda}$' is signifies the coefficient of the magnetic field component along $\hat{p}_\lambda$.

Permittivity of the material under consideration depends on the lattice vector, may be expressed in the form

$$\varepsilon(\vec{L}) = \frac{1}{V} \oiiint_\Omega \varepsilon(r) \exp(-i\vec{L}.\vec{r}) \tag{3}$$

Substituting Eq. (2) and Eq. (3) in Eq. (1) gives

$$|k+L||k+L'|\varepsilon^{-1}(L-L') \times$$
$$\sum_{G'} \begin{bmatrix} \hat{p}_2\hat{p}'_2 & -\hat{p}_2\hat{p}'_1 \\ -\hat{p}_1\hat{p}'_2 & \hat{p}_1\hat{p}'_1 \end{bmatrix} \begin{bmatrix} b_1' \\ b_2' \end{bmatrix} = \frac{\omega^2}{C^2}\begin{bmatrix} b_1 \\ b_2 \end{bmatrix} \tag{4}$$

where the column matrix $\begin{bmatrix} b_1' \\ b_2' \end{bmatrix}$ gives physically valid

solutions. Eigenvalues for TE mode can be obtained once we will solve eq. (4) with appropriate boundary conditions

$$|k+G||k+G'|\varepsilon^{-1}(\vec{G}-\vec{G}') \times$$
$$\sum_{G'} (\hat{e}_1 \bullet \hat{e}'_1) b_2(\vec{G}') = \frac{\omega^2}{C^2} b_2(\vec{G}) \tag{5}$$

III. RESULTS AND DISCUSSIONS

Using Eq. (5), results are computed and plotted for different DNG materials. In Fig 1, three materials are considered for electromagnetic bandgap computation. From the plots, it is found out that for some cases, multiple bandgaps are observed [Fig 1c], whereas for the other cases, only single bandgap is noticed, and that is inside the equivalent conduction band; whereas between valence and conduction region, only quasi bandgaps are formed [Fig 1a and Fig 1b]. In these cases, second photonic bandgap exist at higher region of conduction band. Width of the bandgaps also varied with refractive index, and corresponding midband frequency ranges.

From the variation, we have computed width of the lowest two bandgaps [wherever applicable], and peak coordinates of the equivalent valence band. Results are represented in tabular form. Results for positive refractive index materials are also evaluated. for all the dataset, air is set as reference material.

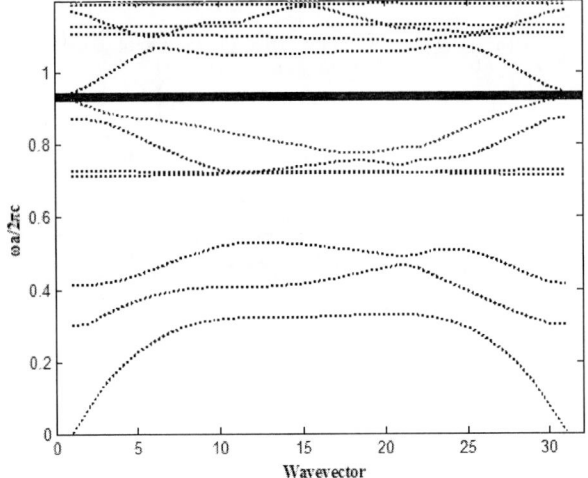

Fig 1a: Photonic bandgap for DNG material with r.i = -25 for TE mode

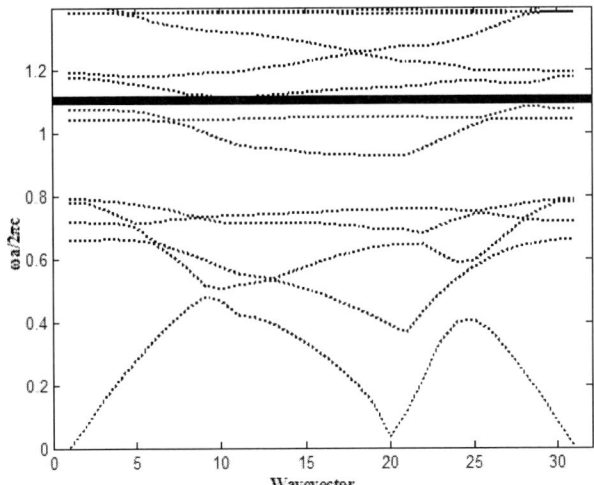

Fig 1b: Photonic bandgap for DNG material with r.i = -8 for TE mode

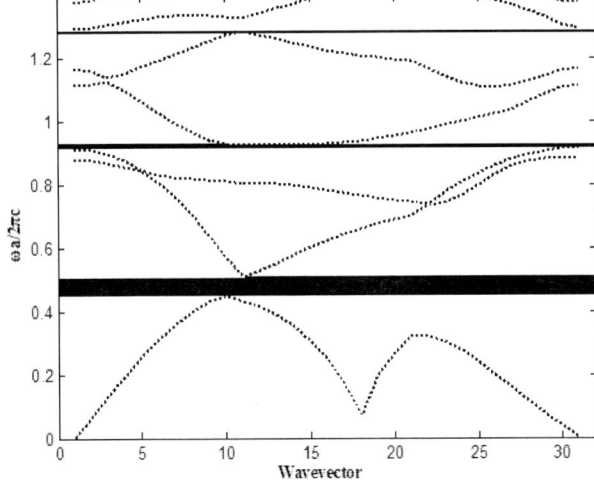

Fig 1c: Photonic bandgap for DNG material with r.i = -0.3 for TE mode

TABLE I. PHOTONIC BANDGAP VARIATION WITH REFRACTIVE INDEX

Refractive index of material	Width of first photonic bandgap	Width of second photonic bandgap
-30	0.011	0.0231
-27	0.0047	0.0048
-24	0.0051	0.0026
-21	0.0045	0.0908
-18	0.0047	0.0086
-15	0.0129	0.0327
-12	0.0057	0.0037
-9	0.0479	0.0093
-6	0.0021	0.0065
-3	0.1231	0.126
3	0.0699	---
6	0.094	0.043
9	0.1282	0.032
12	0.1608	0.017
15	0.1537	---

TABLE II. PEAK POINT COORDINATES OF EQUIVALENT VALENCE BAND

Refractive index of material	Coordinates of first peak point	Coordinates of second peak point
-30	(10, 0.279)	(21, 0.2918)
-27	(11, 0.3026)	(21, 0.3133)
-24	(11, 0.3303)	(21, 0.3391)
-21	(5, 0.06089)	(26, 0.2204)
-18	(3, 0.08495)	(22, 0.3168)
-15	(10, 0.34)	(22, 0.3613)
-12	(4, 0.2874)	(28, 0.2406)
-9	(8, 0.441)	(25, 0.3913)
-6	(11, 0.4641)	(29, 0.1118)
-3	(6, 0.2843)	(26, 0.3111)
3	(11, 0.4898)	(21, 0.5576)
6	(11, 0.418)	(21, 0.4672)
9	(11, 0.3714)	(22, 0.4021)
12	(10, 0.3287)	(22, 0.3644)
15	(10, 0.3043)	(22, 0.3355)

TABLE III. MIDBAND FREQUENCY VARIATION CORRESPONDING TO BANDGAPS

Refractive index of material	Width of first midband	Width of second midband
-30	0.3056	0.67665
-27	0.21495	0.6405
-24	0.32185	0.6319
-21	0.36975	0.6362
-18	0.27435	0.5035
-15	0.23605	0.95695
-12	0.48245	0.89365
-9	0.57055	0.82095
-6	0.80305	0.96625
-3	0.94545	1.19
3	1.01005	---
6	0.9244	1.6915
9	0.8766	1.569
12	0.8424	1.3345
15	0.81655	---

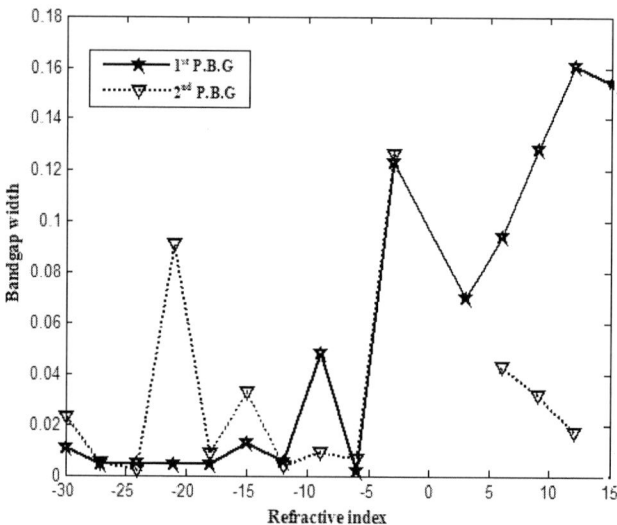

Fig 2: Variations of the lowest two photonic bandgaps with refractive index

Fig 2 shows the variation of bandgap widths with refractive indices, whereas in Fig 3, midband range is plotted. From the plot, it is observed that lowest bandgap width remains almost invariant with higher negative index materials, but starts increasing rapidly as the gap of indices between the constituent materials decreases. Opposite nature is observed for 2nd bandgap width. Again for positive index materials, bandgap width becomes very large, which indicates for larger stopband. Corresponding midband frequency is plotted in Fig 3. It is found form the graph that the midband range becomes monotonically increasing function with increase of refractive index for lowermost bandgap, whereas it again decreases for the second bandgap for conventional materials.

Fig 4 depicted the approximate width of first Brillouin zone. It is measured from the data shown in Table-II. From the figure, it is seen that the zone width remains very high for metamaterials, whereas it becomes low when conventional materials are considered. Thus the information about band structure for DNG material can easily be found from the plot, and hence become very useful for further analysis.

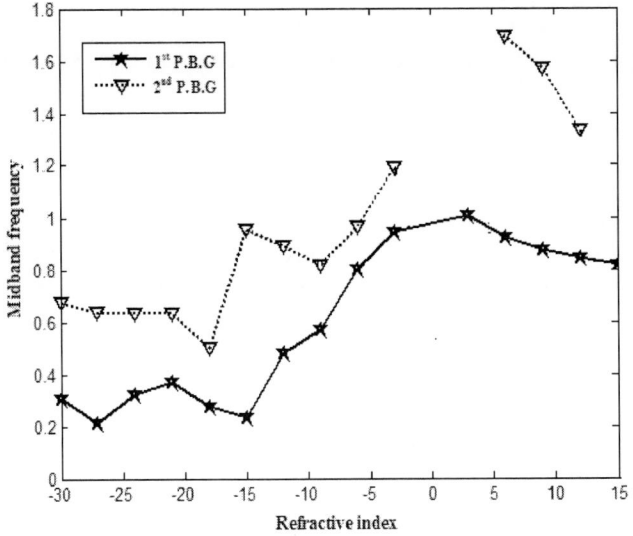

Fig 3: Variations of the lowest two midband frequencies with refractive index

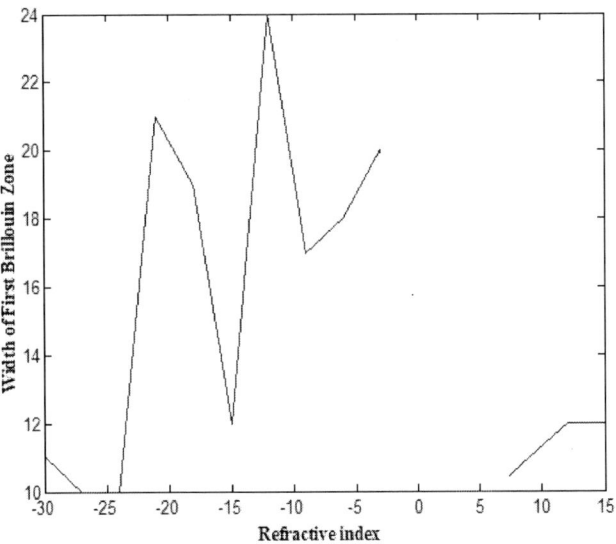

Fig 4: Variation of first brillouin zone width with refractive index

IV. CONCLUSION

Lowest two electromagnetic bandgap widths, width of Brillouin zone, and midband range of the bandgaps are computed for the metamaterial-based 2D PhC structures, and result are compared with that obtained for conventional materials. Rectangular geometry is considered for the analysis, and in some cases, evidence for quasi-bandgap formation is

noted due to cross-over between two consecutive bands. TE mode of propagation is considered, where multi-channel filter design is supported.

REFERENCES

[1] S. Kim, I. Park, H. Lim, C. S. Kee, "Highly Efficient Photonic Crystal-based Multichannel Drop Filters of Three-Port System with Reflection Feedback", *Optics Express*, vol. 12, pp. 5518-5525, 2004

[2] H. Pezeshki, V. Ahmadi, "Nanoscale Effects on Multichannel Add/Drop Filter based on 2-D Photonic Crystal Ring-Resonator Heterostructure", *Journal of Theoretical and Applied Physics*, vol. 6(12), pp. 1-6, 2012

[3] K. P. Sreejith, V. Mathew, "Multichannel Filtering Properties of a One Dimensional Photonic Crystal Composed of Semiconductor Photonic Quantum Well Defect", *Silicon*, vol. 10(6), pp. 2895-2900, 2018

[4] H. Azarshab, A. Gharaati, "Modeling of Multichannel Filter using Defective Nano Photonic Crystal with Thue-Morse Structure", *Progress in Electromagnetics Research Letters*, vol. 71, pp. 61-67, 2017

[5] E. Yablonovitch, "Photonic band-gap structures", *Journal of Optical Society of America B*, vol. 10, pp. 283-295, 1993

[6] S. Foteinopoulou, A. Rosenberg, M. M. Sigalas, C. M. Soukoulis, "In-and Out-of-Plane Propagation of Electromagnetic Waves in Low Index Contrast Two Dimensional Photonic Crystals", *Journal of Applied Physics*, vol. 89(2), pp. 824-830, 2001

[7] H. Hardhienata, A. I. Aziz, D. Rahmawati, H. Alatas, "Transmission Characteristics of a 1D Photonic Crystal Sandwiched by Two Graphene Layers", *11th International Symposium on Modern Optics and Its Applications*, IOP Conference Series: Journal of Physics, vol. 1057, p. 012003, 2018

[8] X. Xu, N. Song, Z. Zhang, W. Cai, F. Gao, "High Precision Photonic Crystal Fiber Optic Gyroscope for Space Application", *Asia Communications and Photonics Conference, OSA Technical Digest*, paper Su1A.3, 2017

[9] Pawel Marć, Natalia Przybysz, Angelika Molska, and Leszek R. Jaroszewicz, "Photonic Crystal Fiber Transducers for an Optical Fiber Multilevel Temperature Threshold Sensor", *Journal of Lightwave Technology*, vol. 36, pp. 898-903, 2018

[10] R. Ahmad, M. Komanec, D. Suslov, S. Zvanovec, "Modified Octagonal Photonic Crystal Fiber for Residual Dispersion Compensation over Telecommunication Bands", *Radioengineering*, vol. 27(1), pp. 10-15, 2018

[11] B. R. Singh, S. Rawal, "Photonic-Crystal based All-Optical NOT Logic Gate", *Journal of Optical Society of America B*, vol. 32, pp. 2260-2263, 2015

[12] S. Khosravi, M. Zavvari, "Design and Analysis of Integrated All-Optical 2×4 Decoder based on 2D Photonic Crystals", *Photonic Network Communications*, 1, 2018

[13] P. Lodahl, "Photonic Quantum-Information Processing with Quantum Dots in Photonic Crystals", *CLEO: 2015, OSA Technical Digest*, paper FW1A.3, 2015

[14] E.o D. Tommasi, A. C. D. Luca, S. Cabrini, I. Rendina, S. Romano, V. Mocella, "Plasmon-like Surface States in Negative Refractive Index Photonic Crystals", *Applied Physics Letters*, vol. 102, p. 081113, 2013

[15] R. A.Norton, R.Scheichl, "Planewave Expansion Methods for Photonic Crystal Fibres", *Applied Numerical Mathematics*, vol. 63, pp. 88-104, 2013

Design and performance analysis of all-optical Soliton based 4-bit two's complement generator using Reflective Semiconductor Optical Amplifier

Kousik Mukherjee
Physics Department
B B College, Asansol, India
klmukherjee003@gmail.com

Kajal Maji
Physics Department
B B College, Asansol, India
kajalmaji200@gmail.com

Ashif Raja
Physics Department
B B College, Asansol
ashifphysics@gmail.com

Abstract- **All Optical Two's complement generator for four bit binary numbers is proposed and analyzed using reflective semiconductor optical amplifier (RSOA). We use Soliton bits to implement the device and therefore, find application in long distance communication systems. The performance is analyzed in terms of input-output bit patterns, quality factor, pseudo eye diagram, relative eye opening and amplified spontaneous emission.**

Key Words- *All Optical Logic, Two's complement; Semiconductor Optical Amplifier; Soliton Pulse; Quality Factor*

I INTRODUCTION

Optical computation and logic implementation are very interesting topics for researchers for the last few decades and more and more people are enriching this fast growing field. Digital signal processing in all optical manners has the advantages of both ultrafast speed of operation and easy fabrication technique. Different all optical logic gates, functional devices, switches, MUX-DEMUX, Flip-Flops have been proposed and implemented during last few years exploiting different mechanisms such as intensity encoding, polarization encoding, frequency encoding [1-7]. The two's complement of a number is the negative number representation of the number itself. This representation technique has the advantage that if we can generate the two's complement of a number we can use an adder for subtraction purpose also. Therefore, we do not need a Subtractor separately. This allows a digital computer to use same circuitry to add and subtract. This makes the hardware simple, cost effective and low power consuming for integrated circuits. In last few years few all optical two's complement generator have been proposed [8-10]. Reflective Semiconductor Optical Amplifier (RSOA) is a versatile gain medium with improvement over conventional single pass Semiconductor Optical Amplifiers (SOA)[11] and finds application in passive optical networks(PON). Soliton pulses are very important for long distance communication due to its particle and long distance transmission is possible without much attenuation and degradation [12]. In this communication for the first time soliton pulses along with RSOA is used to design all optical two's complement generator. The performance of the proposed device is analyzed by numerical simulation for practical feasibility and efficiency of the scheme.

II. WORKING PRINCIPLE AND BASIC BUILDING BLOCK

The two's complement generator utilizes RSOA based cross gain modulation (XGM) optical switching. When there is no pump signal(high power), the RSOA gain is high(unsaturated gain,G_0) and another signal(low power probe) gets amplified and comes out. This corresponds to high state of the optical switch. When the control signal is present, the gain of the RSOA becomes saturated and low. Therefore, the output at the probe wavelength will also be low, which corresponds to low state of the optical switch. In the figure 1, below the block diagram of the RSOA based optical switch is shown and its gain profile. The time dependent gain profile is numerically simulated according to the equations,

$$G(t) = R \exp(2h(t)) \qquad (1)$$

Where h(t) is given by[6]

$$h(t) = \ln\left[1 - \left(1 - \frac{1}{G_0}\right)\exp\left(-\frac{E_{cp}(t)}{E_{sat}}\right)\right] \qquad (2)$$

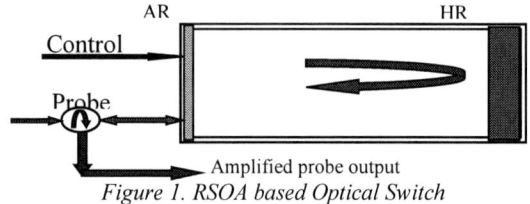

Figure 1. RSOA based Optical Switch

In equation(2), E_{cp} is the control pulse energy, E_{sat} is the saturation energy of the RSOA. The variation of RSAO gain with E_{cp}/E_{sat} is shown in the figure 2 for different unsaturated gain. From figure 2, it is clear that gain saturates faster for higher gains. For better switching higher values of unsaturated gain is better and in this communication 20 dB gain is considered at control pulse energy of 20fJ.

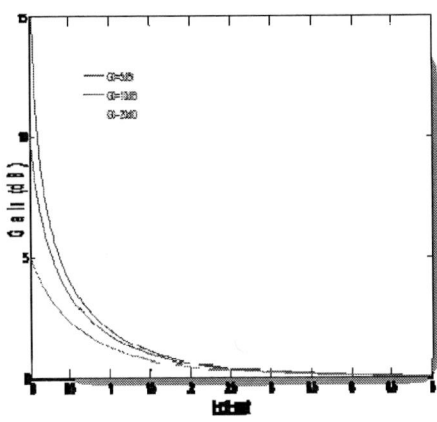

Figure 2. Gain saturation in RSOA

Table 1. Truth table for two's complement generator

Binary				Two's complement			
B_3	B_2	B_1	B_0	T_3	T_2	T_1	T_0
0	0	0	0	0	0	0	0
0	0	0	1	1	1	1	1
0	0	1	0	1	1	1	0
0	0	1	1	1	1	0	1
0	1	0	0	1	1	0	0
0	1	0	1	1	0	1	1
0	1	1	0	1	0	1	0
0	1	1	1	1	0	0	1
1	0	0	0	1	0	0	0
1	0	0	1	0	1	1	1
1	0	1	0	0	1	1	0
1	0	1	1	0	1	0	1
1	1	0	0	0	1	0	0
1	1	0	1	0	0	1	1
1	1	1	0	0	0	1	0
1	1	1	1	0	0	0	1

The truth table for binary to two's complement generator for four-bit binary numbers is shown in the table1. From the truth table it was found that the bits of the two's complements($T_3T_2T_1T_0$)can be expressed in terms of binary bits($B_3B_2B_1B_0$) as, $T_0= B_0$, $T_1=B_0$ **Ex-OR** B_1, $T_2 = B_2$ **Ex-OR** ($B_1 +B_0$), and $T_3= B_3$ **Ex-OR** ($B_0 +B_1 +B_2$).

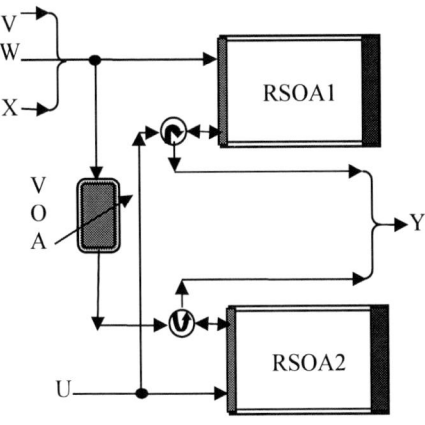

Figure 3. Basic building Block

The basic building block of the two's complement generator is an RSOA based logic device having four inputs, U,V,W and X, and a single output Y = U **Ex-OR** (V+W+X) as shown in the figure 3. Two RSOAs, RSOA1 and RSOA2 are used as switching elements. The inputs U

and (V,W,X) are selected at different wavelengths 1552nm and 1555 nm respectively. The data signals to the RSOAs are selected from the control inputs and a variable optical attenuator (VOA) is used which permit a fixed maximum value of optical power as data signal to the RSOA2. The RSOA 1 receives the same maximum data power as RSOA2. The circulators C_1, C_2 properly routes the probe signals(data) and amplified probe output.

The operation of the building block of figure 3 is as follows. When all the inputs are low, i.e. '0', there will be no output. Now if U is high and the rest(V,W,X all) are low('0'), RSOA1 receives data signal(probe). Since V,W, and X are all low, there will be no gain saturation in it. Therefore, data signal in the RSOA1will be amplified and comes out as high('1') output. Now if U is high and at least one of the three inputs V, W, X, is high, then both the RSOAs experiences gain saturation, and both receives data signal also. But due to gain saturation the final output is low('0'). Now if U is low, i.e. '0', and at least one of the other three inputs are high, then there is gain saturation in RSOA1, but not in RSOA2. This results in high output. The operation is summarized in the table 2 below. From the table 2, using Karnaugh map, it is found that, the output of the basic building block is given by = X **Ex-OR** (U+V+W). Therefore, using this fundamental block we can generate two's complement of any four bit number.

If we choose U =B_0, V =B_1, W =B_2, and X =B_3, then the output Y = B_3 **Ex-OR** ($B_0 +B_1 +B_2$) = T_3.

If we choose X= B_2, U =B_1, V =B_0, W =0, the output will be Y = B_2 **Ex-OR**(B_0+B_1) = T_2.

If we select X =B_0, U =B_1, V=W=0, then output Y= B_0 **Ex-OR** B_1= T_1, and finally T_0 = B_0 is achieved by selecting X=B_0, and U=V=W=0. It should be noted that B_0 directly gives T_0 but for that, some additional circuit is necessary at the output to maintain same intensity levels of the output. This may degrade the performance of the

2019 Devices for Integrated Circuit (DevIC), 23-24 March, 2019, Kalyani, India

device. Using this block, figure 4 gives the design of the two's complement generator.

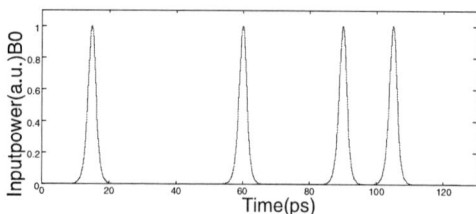

(a)

Table 2. Truth table for the basic building block(fig.3)

U	V	W	X	Y
0	0	0	0	0
0	0	0	1	1
0	0	1	0	1
0	0	1	1	1
0	1	0	0	1
0	1	0	1	1
0	1	1	0	1
0	1	1	1	1
1	0	0	0	1
1	0	0	1	0
1	0	1	0	0
1	0	1	1	0
1	1	0	0	0
1	1	0	1	0
1	1	1	0	0
1	1	1	1	0

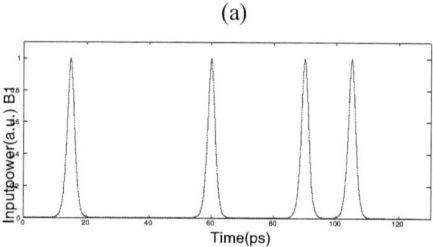

(b)

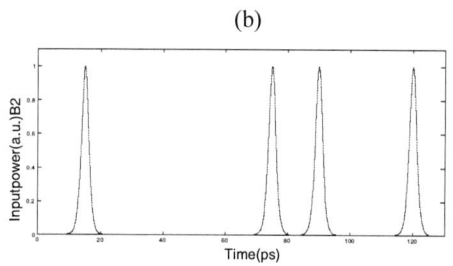

(c)

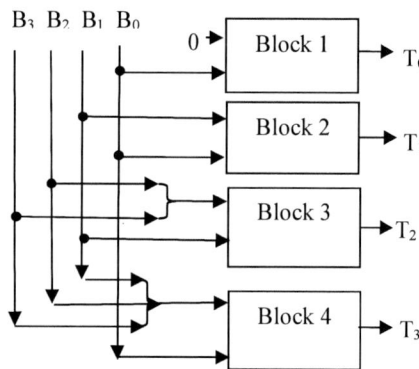

Figure 4. Two's complement Generator

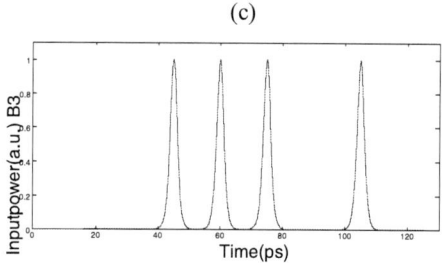

(d)

Figure 5.Input Bit pattern (a) B_0(b) B_1(c) B_2(d) B_3

III. NUMERICAL SIMULATION RESULTS AND DISCUSSIONS

We have considered soliton pulses as input signals and data signals for the operation of the RSOAs used in this communication. They have the $sech^2$ form as shown in the figure 5(a) to 5(d). All the inputs B_0, B_1, B_2 and B_3 are taken as soliton pulse trains. The peak values of the soliton pulse train corresponds to 50fJ for this simulation and full width half maximum to be 1.5 ps. Figure 6 shows

978-1-5386-6723-1/19 $31.00 © 2019 IEEE

the input and simulated output bit patterns of the two's complement generator.

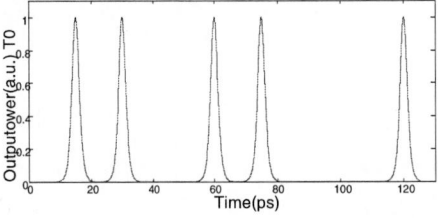

(a)

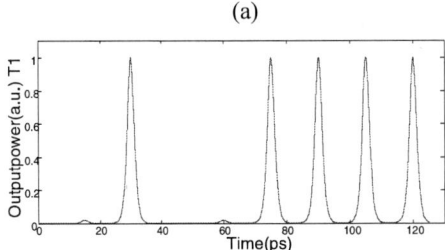

(b)

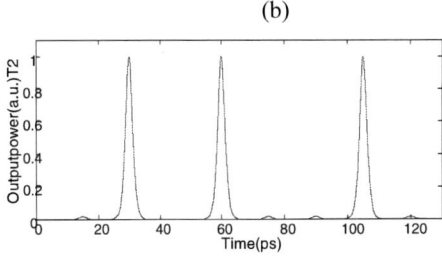

(c)

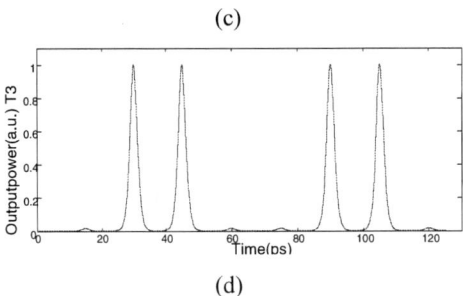

(d)

Figure 6.Output bit patterns(a) T_0 (b) T_1 (c) T_2 (d) T_3

The effect of amplified spontaneous noise(ASE) is also considered for performance evaluation. The ASE noise in RSOA is given by [12,13]$P_{ASE} = 2N_{SP}(G-1)h\upsilon B_0$, where N_{SP} is the amplified spontaneous noise parameter depends on the degree of population inversion, and is unity for ideal case, G is the single pass gain of the active medium, h is Planck's constant, υ, the frequency and B_0 is the bandwidth. This noise power is numerically added to consider the effect of ASE on the performance. Figure 7

gives the wavelength dependence of noise power for different single pass RSOA gains. The effect is large for large RSOA gains and falls of with wavelength. For low gains, the effect is negligible.

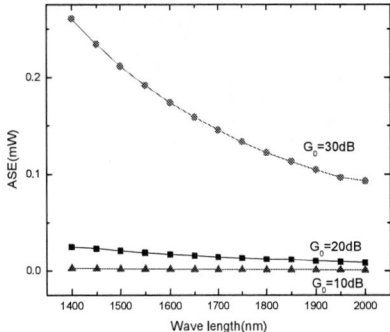

Figure 7. Amplified Spontaneous Noise

Figure 8(a) and (b) shows the pseudo eye diagram for two different currents 215 mA and 185 mA respectively. The eye diagram shows large opening about 98 percent for I = 215 mA and for 82% for I=185 mA. That is for larger biasing current, power difference between high and low states is also larger. Therefore, for lower biasing current effect of noise is significant than for larger current in the eye diagram. The large eye opening also reflects the efficiency of the two's complement generator. The high quality factor ~78 implies negligible bit error rate (BER) and hence error-less transmission is possible. Figure.9 shows variations of relative eye opening with control pulse energy for different biasing current. For larger gains quality factor decreases with control pulse energy faster compared to lower gains. The quality factor of the proposed Two's complement generator is plotted against control pulse energy for different biasing current in the figure 10. For larger biasing current quality factor decreases faster than lower biasing current due to larger effect of ASE noise.

IV.CONCLUSIONS

A high Q(~78) factor for the two's complement generator is found using RSOA for biasing current 215mA. Numerical simulation shows practical feasibility of the device. The ASE noise is included in our simulation and has small effect for low single pass gain. The device shows large eye opening and for larger gain.

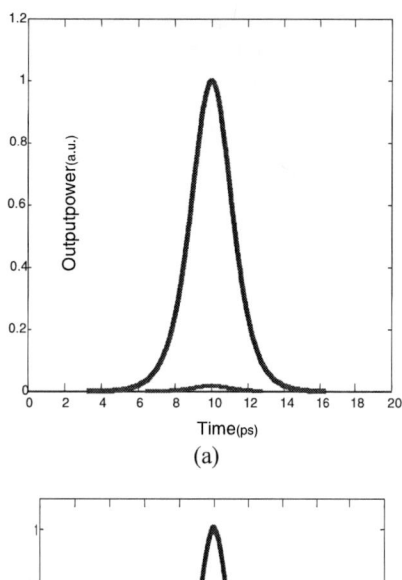

(a)

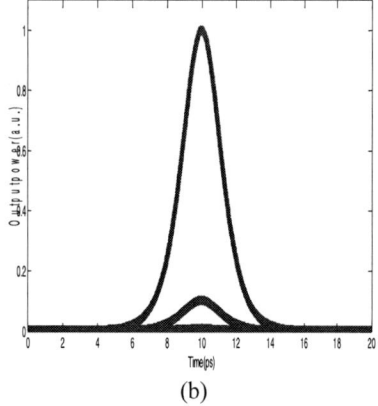

(b)

Fig 8. pseudo-eye-diagram for Biasing current
(a) I=215mA,(b) I=185 mA

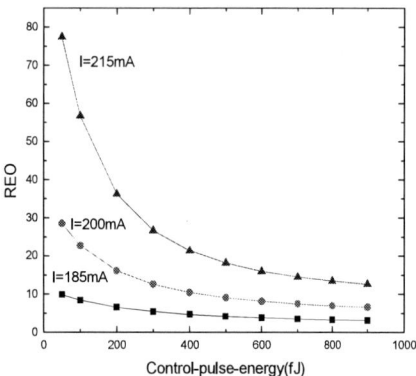

Figure 9. Variations of relative eye opening(REO) with
control pulse energy

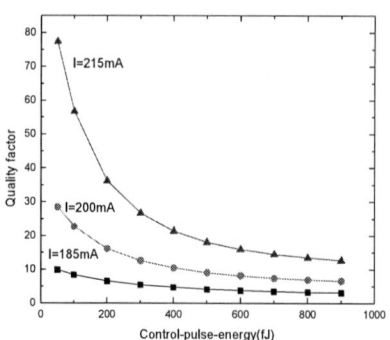

Figure 10. Variation of quality factor Q with control pulse energy

REFERENCES

[1] A. Kotb, K E. Zoiros, C. Guo, All optical XOR, NOR and NAND logic functions with parallel Semiconductor Optical Amplifier based Mach Zehnder Interferometer modules, Optics and Laser Technolgy, 108, pp 426, (2018).

[2] Amer Kotb, Theoretical analysis of soliton NOR gate with semiconductor optical amplifier-assisted Mach–Zehnder interferometer Opt Quant Electron (2017) 49:180 DOI 10.1007/s11082-017-1017-

[3] K. Mukherjee, Frequency encoded optical four bit adder/ Subtractor control input using semiconductor optical amplifier, Optik, 125,20,6183(2014).

[4] E. Dimitriadou, K. E. Zoiros, All optical XOR gate using single quantum dot SOA and optical filter Journ, Of Lightwave Technology, 31, 3813,(2013).

[5] A.E. Willner, S. Khaleghi, M. R. Chitgarah, O. L. Yilmaz, All Optical Signal Processing, Journ, Of Lightwave Technology, 32, 660,(2014).

[6] T. Chattoipadhyay, "All-optical clocked delay flip- flop using a single terahertz optical asymmetric demultiplexer - based switch: a theoretical study Applied Optics, Vol. 49, No. 28(2010)

[7] K. Maji, K. Mukherjee, A. Raja, "Analysis of Tera Hertz Optical Asymmetric Demultiplexer(TOAD) based Optical Switch using soliton pulse", Presented in IEEE EDCON 2018, Kolkata

[8] T. chattopadhyay, D. K. Gayen, All-Optical two's complement conversion scheme without binary addition, IET Optoelectronics, 11,1,1-7(2016) doi: 10.1049/iet-opt.2015.0087

[9] A. Bhattacharya, D. K. Gayen, T. Chattopadhyay, " 4-bit all optical binary to two's complement converter", *2012 International Conference on Communications, Devices and Intelligent Systems (CODIS)*,2012, doi: 978-1-4673-4700-6/12/$31.00 c_2012 IEEE

[10] R. Katti, S. Prince, All optical binary to two's complement conversion based on microring resonator, International Conference on Fibre Optics and Photonics 2016 ,Kanpur India,4–8 December 2016,ISBN: 978-1-943580-22-4

[11] L. Q. Guo, M. J. Connelly, "A novel approach to all optical wavelength conversion by utilizing a reflective semiconductor optical amplifier in a co propagation scheme", Optics communications, 281,4470(2008).

[12] K. Mukherjee, K. Maji, A. Raja, "Frequency encoded all optical universal logic gates using tera hertz optical asymmetric demultiplexer", IJPOT, 4,3,1-7(2018)

[13] K. Komatsu, G. Hosoya, H. Yashima, "All-optical logic NOR gate using a single quantum dot SOA assisted an optical filter", Opt. Quant. Electronics,50,131,pp 1-18(2018).

A Proposed High-k Dielectric Based Thin Film Transistor for Next Generation Backplane Display Technology

Abhishek Kumar Singh[1]*, Ashutosh Kumar Dikshit[2], Brahmdutta Dixit[3], V.V. Kharche[4], Kamal[$], J.S. Rana[5]
P. Chakrabarti[6], A. Pandey[7]

Dept. of Electronics Engineering, Indian Institute of Technology (Banaras Hindu University), Varanasi – 221005, India
$ Department of ECE, Motilal Nehru National Institute of Technology Allahabad, UP 211004, India
3 Department of ECE , School of Engineering &Technology, Mizoram University, Aizawl, India.

[1]*aksingh.rs.ece16@iitbhu.ac.in, [2]ashutoshdikshit.rs.ece17@itbhu.ac.in, [3]brahmdutta.iist@gmail.com
[4]kvivek143143143@gmail.com, [$]kamaleced@gmail.com, [5]jogendrasrana@gmail.com, [6] pchakrabarti.ece@itbhu.ac.in
[7]amrit.ece@iitbhu.ac.in

Abstract– **A Zinc oxide (ZnO)-based thin film transistor (TFT) has been proposed to be built using a high-k $La_xTa_{1-x}O_y$ as an insulator to enhance their effectuation for backplane display technology. The device has been scrutinized on Silvaco ATLAS™ 2D simulator to examine the switching performance of the device for possible application in display driver circuits. To improvise the working of the TFT for the targeted application the atomic compositions of La and Ta in $La_xTa_{1-x}O_y$ have been converted to attune the dielectric constant of the insulator. The study reveals that the high-k insulator-based ZnO TFT can be modified to obtain high on/off current ratio of the order of 108 to ensure the high-speed operation with low sub-threshold swing 0.49 V/dec which is desirable for low power applications. It has been demonstrated that high-k dielectric insulator-based ZnO TFT has great potential for the development of cost-effective large-area display systems.**

Keywords—High-k dielectric insulator, TFT, ZnO, Display Driver, $La_xTa_{1-x}O_y$

I. Introduction

ZnO thin film transistors have been studied since many decades given their potential applications as sensors, detectors, and other optoelectronics devices [1]-[3]. One of the major areas of application of TFTs is in the driver circuits of active-matrix LED display systems. For applications in display drivers, the TFTs need to have a high value of the on-off current ratio. However, conventional ZnO based TFTs has very poor on-off ratio making them unfit for driver applications. A variety of ZnO related materials such as InGaZnO (IGZO), GaZnO (GZO), InZnO (IZO), etc. have been tried by the researchers in an attempt to improve the current on-off ratio [4]. However, the cost of these TFTs becomes unacceptably high as compared to ZnO based TFTs. Therefore, low-cost ZnO - TFTs still need to be investigated to match the performance with IGZO or other doped ZnO based TFTs. The key issue for ZnO TFTs is the proper choice of the insulator for gate. Among the different dielectric-materials such as SiO_2, $(Pb, Zr)TiO_3$, HfO_2, ZnMgO

and Y_2O_3 high dielectric constant $Bi_{1.5}Zn_{1.0}Nb_{1.5}O_7$ (BZN) has been reported to <4V voltage which is pretty low in TFTs with ZnO filling in as the electron semiconductor channel [5]. The performance TFT is highly dependent on the trait of the interface of channel and dielectric material. In general, high-k dielectric materials exhibit lower bandgap energies as compared to conventional dielectrics like SiO_2 and Al_2O_3 and therefore are likely to increase the risk of enhancing the gate leakage current because of direct tunneling across the insulation layer, by Schottky emission or Pool-Frenkel effect [6]. Among various possible high-k dielectrics, $La_xTa_{1-x}O_y$ exhibits the potential to overcome the problem of leakage current and thereby helps in scaling the size of the transistor. By using the high dielectric $La_xTa_{1-x}O_y$, it is, therefore, possible to achieve a high value of capacitance per unit without compromising the insulator- layer of thickness. Ta_2O_5 has high- k value (~26) with low bandgap of order 4.4 eV which offers the scarcity of high concentration of oxygen resulting in charge trapping [7][8], On the other side, La_2O_3 exhibits a very large band gap (~6eV) and hygroscopic in nature causing instability caused by formation of hydroxide [9]. In this work, we have demonstrated the possible use of $La_xTa_{1-x}O_y$ as an insulator in a ZnO based TFT to achieve high-performance ratings for deployment in display drivers. The performance ratings of different ZnO based TFTs employing various types of insulator or gate dielectric reported till date are summarized in Table I. In order to make ZnO based TFTs suitable for display driver applications, it is necessary to achieve higher on/off current ratio of about $10^7 – 10^8$ so as to compete with expensive alloy based ZnO TFTs such as IGZO, GZO TFTs. In Table, I the performances of ZnO based TFTs reported by having been summarized with conventional insulator dielectric such as SiO_2, Al_2O_3, SiN_x, HfO_2/Ta_2O_5. From Table I it is apparent that the researchers are still struggling in order to high on-off current ratio of the order of 108 with thick gate insulator. In this work, we have made a modest attempt to demonstrate the importance of high-k dielectric in improving the on-off current ration and low threshold voltage operation of ZnO based TFTs for future display technology applications.

2019 Devices for Integrated Circuit (DevIC), 23-24 March, 2019, Kalyani, India

TABLE I. PERFORMANCE OF DIFFERENT DIELECTRIC BASED TFTS

S.No.	Gate Dielectric	On-Off Ratio	Threshold V_T	S.S (V/dec)	Field-effect mobility (cm^2/V-Sec)	Thickness (Oxide/S.C) (nm/nm)	W/L ratio	Refer.	Deposition Method of ZnO
I.	SiO$_2$	2.43×10^6	-12.5 to 14.7	1.21 to 24.1	8.82×10^{-3} to 6.11×10^{-3}	60/100	80	[10]	Atomic Layer Deposition
II.	SiO$_x$/SiN$_x$	2.7×10^7	----	---	5.2	40/110	50/20	[11]	RF magnetron Sputtering
III.	SiNx	10^7	6	---	0.67	200/300	500/10	[12]	Sol Gel
IV.	Al$_2$O$_3$	4.2×10^6	~2.6	0.190	19-21	45/15	40/20	[13]	Pulsed Laser Deposition
V.	HfO$_2$/Al$_2$O$_3$	~10^7	0.5	0.14	~100	15/5	15/5	[14]	DC Magnetron Sputtering
VI.	HfO$_2$/Ta$_2$O$_5$	~10^6	---	0.5	1.3	100/250 &100/40	50/20	[15]	RF Sputtering
VII.	Al$_2$O$_3$/Ta$_2$O$_5$	~10^7-10^8	-0.3-2.1	0.38-2.5	0.3-2.5	25/(100+2+4+8)	500/100	[16]	Atomic Layer Deposition

II. DEVICE STRUCTURE AND SIMULATION

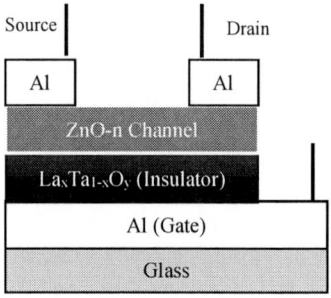

Fig. 1. Schematic of bottom gate ZnO TFT

It is understood from previous reports that TFTs with bottom gate structures are superior to their top-gate counterparts because of easy fabrication steps involved in the former case and high-performance ratings and improved electrical characteristics as compared to the later. ZnO based low-cost aluminum gated, La$_x$Ta$_{1-x}$O$_y$ dielectric structure as shown in Fig.1. The simulated proposed structure has been carried out and optimized using ATLAS^{TM} tool from Silvaco International, Singapore. In this manuscript, various important parameters of the polycrystalline ZnO have been taken into account [18]-[19] as indicated in Table II. The 2D simulation of La$_x$Ta$_{1-x}$O$_y$ thin film with different percentage content of Lanthanum (La) is used in the virtual design the gate dielectric for best performance of the FET to incorporate the advantages of both Ta$_2$O$_5$ and La$_2$O$_3$ while on other side curbing their disadvantages mentioned previously. In the present work, we have simulated the device structure for four different compositions of La$_x$Ta$_{1-x}$O$_y$ by varying the content of La percentage such 22.7%, 56.2%, 76.4%, and 88.3%. The corresponding values of the dielectric constant of La$_x$Ta$_{1-x}$O$_y$ have been listed in Table III. For the simulation of different La content based La$_x$Ta$_{1-x}$O$_y$ film, parameters have been taken from ref [20]. The thickness of the ZnO layer and dielectric layer have been considered to be 150nm and 300nm respectively. The width and length of the channel are taken to be 50µm and 10µm respectively. The thickness of source, drain and gate contacts is considered as 10nm each. Simulation was carried out for achieving high performance of display technology keeping in a view of low cost and feasible technology cut the cost, low cost aluminum material was used as gate dielectric source and drain contacts due abundance availability. For 2D simulation drift-diffusion SRH recombination, mobility and impact ionization models have been used. To determine carrier concentration Fermi Dirac Distribution was used in simulation. Performance analysis were performed at 300 K.

TABLE II. PARAMETERS OF ZNO MATERIAL FOR 2-D DEVICE SIMULATION

Material constant for ZnO	Values
Bandgap E$_G$ (300K)	3.367 eV
Electron Effective mass	0.19m$_o$
Hole Effective mass	1.21m$_o$
Dielectric	8.5
Mobility	150 cm^2/V-s
Electron affinity	~04.3 eV
N_C	2.9348×10^{24}/m^3
N_V	1.1436×10^{25}/m^3
SRH life time of electrons	4.18×10^{-6} s
SRH life time of holes	6.6×10^{-6} s
Trap density	~10^{23} m^3

The drain current when $V_{DS} \geq V_{GS}$-V_T i.e. in saturation region is given by [21]

$$I_{DS} = \frac{1}{2} \mu_n \, C_{ox} \left(\frac{W}{L}\right)(V_{GS} - V_T)^2 \qquad (1)$$

where C_{ox} is the capacitance per unit area of the oxide layer, W is the width of the channel, L is length of the channel, μ_n mobility of the charge carriers of channel. The source voltage is kept at 0 V for simulation.

The C_{ox} can be written as

$$C_{ox} = \frac{\epsilon_0 \epsilon_r}{t} \qquad (2)$$

where ϵ_0 permittivity of free space is, ϵ_r is relative permittivity and t is the thickness of the dielectric layer.

From equation (1) and (2), it can be observed that the value of drain source current is directly proportional of ϵ_r for fixed thickness of the dielectric layer.

The sub-threshold swing is given by [22],

$$SS = \left(\frac{d\log(I_{DS})}{dV_{GS}} | max\right)^{-1} V/dec \qquad (3)$$

This parameter describes the required V_{GS} to increase I_{DS} by one-decade (V/decade). It is desirable to have a low value of sub-threshold voltage for high speed with low power consumption.

978-1-5386-6723-1/19 $31.00 © 2019 IEEE

The density of interface & deep bulk trap per unit area and per unit energy, N_{tt} can be extracted from sub-threshold swing [23]. Assuming N_{tt} independency on trap energy, we may write

$$N_{tt} = \frac{c_{ox}}{q}\left(\frac{SS}{\frac{kT}{q}\ln 10} - 1\right) \qquad (4)$$

where k the Boltzmann's constant, q is electron charge and T the temperature in Kelvin.

III. RESULTS AND DISCUSSIONS

Figs. 2(a)-(f) show the transfer characteristics (I_D-V_G) as well as log I_D-V_G plots of the simulated ZnO TFT device structure for different dielectric constant values of 9.28, 10.6, 10.9, 11.2, 12.6 and 14.8 obtained by tailoring the composition of La and Ta. The structure is simulated with channel width of 50μm and channel length of 10μm. The gate voltage V_G is varied from -2.5 V to 15 V keeping drain voltage constant at 5 V. It was found that current at V_G=15 V is of order 10^{-3} A and current at V_G= -2.5 V is of order 10^{-11} A. For ZnO based TFT with dielectric constant of 12.6, the drain current at V_G=-0.8V is 1.38×10^{-11}A (OFF State).

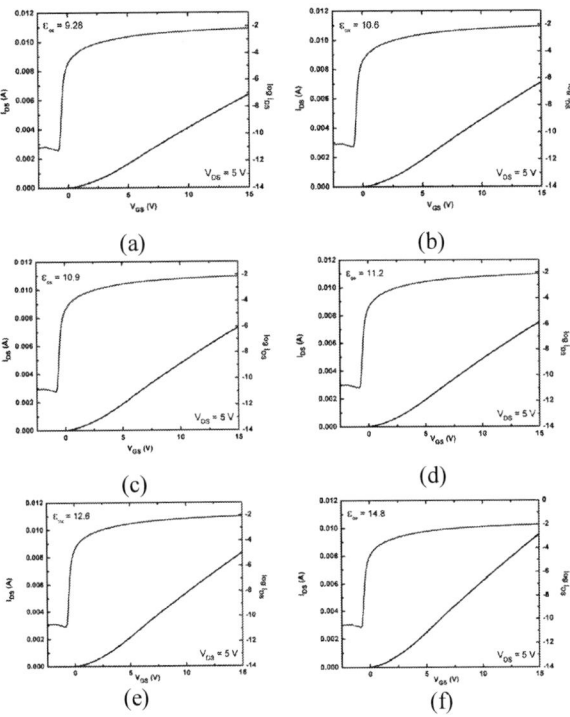

Fig. 2 Transfer characteristics (I_D-V_G) of ZnO based TFT with dielectric constant (a)9.28 (b)10.5 (c)10.9 (d)11.2(e)12.6 (f) 14.8

The drain current at V_G=5V is 1.6×10^{-3}A (ON State) and hence on-off ratio is obtained as 1.16×10^8. Similarly, results were obtained for other ZnO based TFTs with dielectric constant of 9.28, 10.6, 10.9, 11.2 and 14.8 as shown in Table III. From the results shown in Fig. 2, it can be observed that as value of ϵ_r

Increases the value of drain current also increases. In this study, we have demonstrated that it is possible to improve the performance of TFTs based on Ta_2O_5 and La_2O_3 dielectrics by making use of $La_xTa_{1-x}O_y$ as the dielectric in a TFT wherein the advantages of both the dielectrics can be best utilized while undesirable properties can be suppressed. For example, Ta_2O_5 has high- k value (~26), but due to the small bandgap of ~4.4eV which leads to charge trapping due to the high concentration of oxygen. Another side , La2O3 has large bandgap which is very unstable in air. However, $La_xTa_{1-x}O_y$ offers a much better alternative dielectric that outperforms TFTs based on other dielectrics.

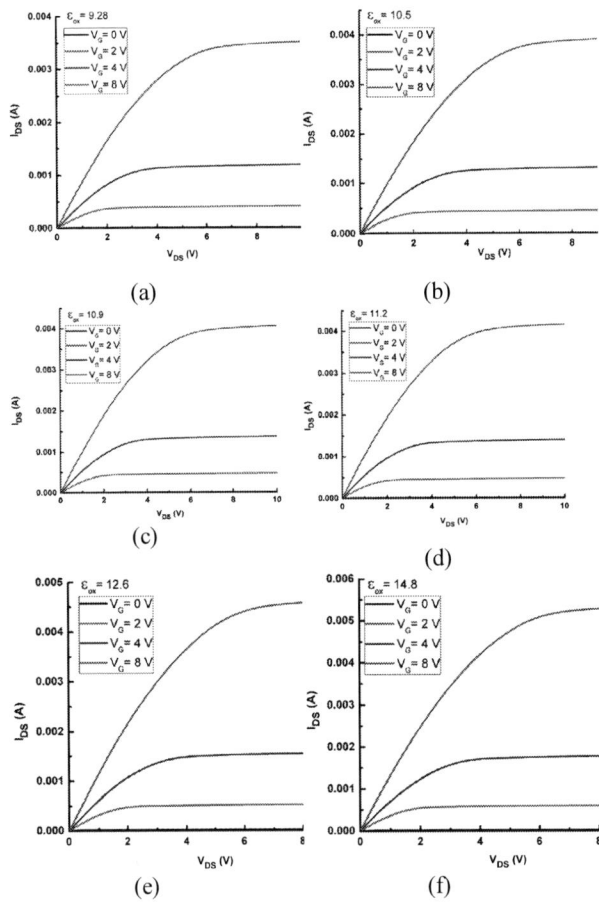

Fig. 3. Output (I_D-V_D) characteristics of ZnO based TFT with dielectric constant (a) 9.28 (b)10.6 (c)10.9 (d)11.2 (e)12.6 (f)14.8

Figs 3(a)-(f) show output characteristics of simulated ZnO TFT device structure for different dielectric constant values of 9.28, 10.6, 10.9, 11.2, 12.6 and 14.8 obtained by tailoring the composition of La and Ta. It can be easily observed that the channel current increases with increase in drain voltage until the field from the drain cancels the gate field.

TABLE III. DEVICE PARAMETERS FOR DIFFERENT COMPOSITION OF GATE DIELECTRIC

x	1	0.883	0.764	0.562	0.227	0
ϵ_r	9.28	10.5	10.9	11.2	12.6	14.8
$I_{off}(10^{-12} A)$	6.02	8.48	9.41	10.15	13.8	20.8
$I_{on}/I_{off}(\times 10^{8})$	1.05	1.08	1.09	1.12	1.16	1.22
$SS(V/dec)$	0.520	0.509	0.503	0.498	0.490	0.478
$V_T(V)$	-0.659	-0.664	-0.665	-0.667	-0.674	-0.688
$C_{ox}(nF/cm^2)$	26.55	30.975	32.155	33.04	37.17	43.66
$N_{tt}(\times 10^{12}/cm^3)$	1.4	1.47	1.512	1.53	1.695	1.936

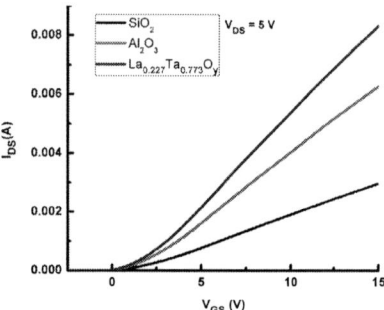

Fig. 4 Comparison between transfer (I_D-V_G) characteristics of ZnO based TFT with dielectric layer as SiO₂ (ϵ_r=3.9), Al₂O₃ (ϵ_r=9.1) and La $_{0.227}$Ta $_{0.773}$O$_y$.

The encouraging results in Table III and in Fig.4 reveal that by varying the content of La in La$_x$Ta$_{1-x}$O$_y$, it is possible to design a ZnO - TFT with a much better on-off ratio, sub-threshold swing is quite low along with low threshold voltage as compared to conventional ZnO based TFTs using Al₂O₃ and SiO₂ as gate dielectrics. The results obtained from simulation studies of ZnO TFT based on a high dielectric-k insulator are presented in Table III. The sub-threshold swing for the dielectric constant of 12.6 is obtained as 490 mV/decade.

IV. CONCLUSION

The simulation of ZnO-TFT with La$_x$Ta$_{1-x}$O$_y$ insulator has been presented in this paper. Different compositions of La and Ta have been used for designing the proper dielectric material for best performance. The study revealed that an exciting on-off ratio of 1.16×10^8 could be achieved by using La$_{0.227}$Ta$_{0.773}$O$_y$ as the gate insulator. Also it is observed that as the La -content decreases, the value of dielectric constant increases which results in the lower value of sub-threshold swing. With higher value of the on to off ratio.The proposed ZnO-TFT device is expected to be used as an alternative of existing thin film transistors using derivatives of ZnO for high frequency applications.

REFERENCES

[1] C. I. Kuan, H. C. Lin and P. W. Li, "Improving the Performance of ZnO Thin-Film Transistors with ZnO Source/Drain Contacts," *in IEEE Transactions on Electron Devices*, vol. 64, no. 7, pp. 2849-2853, July 2017.

[2] S. Jon et al., "Nanometer-scale oxide thin film transistor with potential for high-density image sensor applications", *ACS Appl. Mater. Interfaces*, vol. 3, no. 1, pp. 1-6, 2011.

[3] J. YKwon, J. K. Jeong, "Recent progress in high performance and reliable n-type transition metal oxide-based thin film transistors", *Semicond. Sci. Technol.*, vol. 30, no. 2, pp. 024002, 2015.

[4] C. Shih and A. Chin, "Remarkably High Mobility Thin-Film Transistor on Flexible Substrate by Novel Passivation Material", *Scientific Reports*, vol. 7, no. 1, 2017.

[5] U. Ozgur, D. Hofstetter and H. Morkoc, "ZnO Devices and Applications: A Review of Current Status and Future Prospects," in Proceedings of the IEEE, vol. 98, no. 7, pp. 1255-1268, July 2010.

[6] J Robertson, "Interfaces and defects of high-k oxides on silicon. Solid State Electron. ". *Solid State Electronics*, Vol 49, no. 3, pp. 283-293, 2005.

[7] R. Ramprasad, "First principles study of oxygen vacancy defects in tantalum pentoxide", *Journal of Applied Physics*, vol. 94, no. 9, pp. 5609-5612, 2003.

[8] H. Sawada and K. Kawakami, "Electronic structure of oxygen vacancy in Ta₂O₅", *Journal of Applied Physics*, vol. 86, no. 2, pp. 956-959, 1999.

[9] Y. Zhao, M. Toyama, K. Kita, K. Kyuno and A. Toriumi, "Moisture-absorption-induced permittivity deterioration and surface roughness enhancement of lanthanum oxide films on silicon", *Applied Physics Letters*, vol. 88, no. 7, p. 072904, 2006.

[10] S. Lim, S. Kwon and H. Kim, "ZnO thin films prepared by atomic layer deposition and rf sputtering as an active layer for thin film transistor", *Thin Solid Films*, vol. 516, no. 7, pp. 1523-1528, 2008.

[11] T. Hirao et al., "Bottom-Gate Zinc Oxide Thin-Film Transistors (ZnO TFTs) for AM-LCDs," *in IEEE Transactions on Electron Devices*, vol. 55, no. 11, pp. 3136-3142, Nov. 2008.

[12] H. Cheng, C. Chen and C. Tsay, "Transparent ZnO thin film transistor fabricated by sol-gel and chemical bath deposition combination method", *Applied Physics Letters*, vol. 90, no. 1, p. 012113, 2007.

[13] R. A. Chapman, R. A. Rodriguez-Davila, I. Mejia and M. Quevedo-Lopez, "Nanocrystalline ZnO TFTs Using 15-nm Thick Al2O3 Gate Insulator: Experiment and Simulation," *in IEEE Transactions on Electron Devices*, vol. 63, no. 10, pp. 3936-3943, Oct. 2016.

[14] Y. Song, R. Xu, J. He, S. Siontas, A. Zaslavsky and D. C. Paine, "Top-Gated Indium–Zinc–Oxide Thin-Film Transistors With In Situ Al₂O₃/HfO₂ Gate Oxide," *in IEEE Electron Device Letters*, vol. 35, no. 12, pp. 1251-1253, Dec. 2014.

[15] H. J. H. Chen, B. B. L. Yeh, H. C. Pan and J. S. Chen, "ZnO transparent thin-film transistors with HfO₂/Ta₂O5 stacking gate dielectrics," *in Electronics Letters*, vol. 44, no. 3, pp. 186-187, January 31 2008.

[16] F. Alshammari, P. Nayak, Z. Wang and H. Alshareef, "Enhanced ZnO Thin-Film Transistor Performance Using Bilayer Gate Dielectrics", *ACS Applied Materials & Interfaces*, vol. 8, no. 35, pp. 22751-22755, 2016.

[17] X. Zhang, J. Zhang, W. Zhang and X. Hou, "Fabrication and comparative study of top-gate and bottom-gate ZnO–TFTs with various insulator layers", *Journal of Materials Science: Materials in Electronics*, vol. 21, no. 7, pp. 671-675, 2009.

[18] A. K. Singh, V. V. Kharche and P. Chakrabarti, "Performance Optimization of ZnO based Thin Film Transistor for Future Generation Display Technology," 2017 14th IEEE India Council International Conference (INDICON), Roorkee, 2017, pp. 1-5.

[19] C. Periasamy, P. Chakrabarti, "Effect of Annealing on the Characteristics of Nanocrystalline ZnO Thin Films", *Sci. Adv. Mate.* vol. 3, pp. 73-79 2011.

[20] C. Y. Han, W. M. Tang, C. H. Leung, C. M. Che, and P. T. Lai, "High mobility pentacene thin-film transistor by using Lx Ta(1−x)Oy as gate dielectric," Organic Electronics, vol. 15, no. 10, pp. 2499–2504, Oct. 2014.

[21] A. Correia, P. Barquinha and J. Goes, "Thin-Film Transistors", Springer Briefs in Electrical and Computer Engineering, pp. 5-15, 2015

[22] J.-H. Lee, S.-T.Wu, D.N. Liu, Introduction To Flat Panel Displays (Wiley,West Sussex, 2008),p. 280.

[23] W.L. Kalb, B. Batlogg, "Calculating the trap density of states in organic field-effect transistors from experiment: a comparison of different methods," *Phys. Rev. B* vol. 81, no. 035327, 2010.

Synthesis of Morphological- Variant ZnO nanostructures

Ashutosh Kumar Dikshit[1*], Kamal[2], Abhishek Kumar Singh[1], J.S. Rana[1], Rohit K Singh[3], Nillohit Mukhrjee[4], P. Chakrabarti[1,4]

[1]Dept. of Electronics Engineering, Indian Institute of Technology (Banaras Hindu University), Varanasi – 221005, India
[2]Department of ECE, Motilal Nehru National Institute of Technology Allahabad, UP 211004, India
[3]Dept. of Metallurgical Engineering, Indian Institute of Technology (Banaras Hindu University), Varanasi – 221005, India
[4]Center of Excellence for Green Energy and Sensor Systems, IIEST Shibpur, Kolkata-711103, India

ashutoshdikshit.rs.ece17@itbhu.ac.in, kamal24041996@gmail.com, aksingh.rs.ece16@iitbhu.ac.in,
jogendrasrana@gmail.com, rohit.ksingh.met14@itbhu.ac.in, nillohit.mukherjee@gmail.com, pchakrabarti.ece@itbhu.ac.in

Abstract– Nanostructures of zinc oxide have great potential in optoelectronics as well as in sensing applications. Here, we report the controlled and systematic growth of ZnO nanostructure by hydrothermal and vapor-liquid-solid (VLS) techniques, which can be used in various electronics and optoelectronic devices. The effect of various growth parameters (growth temperature, annealing time, the thickness of seed layer, to name a few) on the morphology of ZnO nanostructures has been studied. Ultra-long nanowires of length more than 300μm have been synthesized with aspect ratio around 10.

Keywords— Zinc oxide, Nanorod, Nanowire, Hydrothermal, Vapor-Liquid-Solid (VLS).

INTRODUCTION

The advancement in the field of nanotechnology has felicitated the increased use of zinc oxide (ZnO) nanostructures. ZnO 1-D nanostructures are being widely used due to their enormous potential in optoelectronic and electronic devices, catalysis, biomedical and sensing applications. The wide bandgap of ~3.37 eV (for bulk) and large excitonic binding energy of 60 meV at room temperature make ZnO an ideal candidate [1-2] for the blue-ultraviolet (UV) optical devices [3]. With the help of hydrothermal and vapor-liquid-solid (VLS) methods, a variety of nanostructures of ZnO can be created easily. The advantages of these methods are: ease of synthesis, low-cost, eco-friendly, high yielding, minimum loss of materials, to name a few. The beauty of these techniques is that, these nanostructures can be grown on any substrate such as transparent conducting oxide coated (TCO) glasses, plain glass, silicon, plastic, fabric etc. Especially, hydrothermal synthesis of ZnO is a low temperature technique so the synthesis of nanostructures on flexible/polymer substrates is possible; which can be easily integrated with flexible organic electronics.

In this text, we report the synthesis of various nanostructures of ZnO grown by hydrothermal and VLS mechanism. The key idea is to change the growth parameters to obtain different morphologies of the nanostructures such as nanorods, nanowires, nano-urchins, nanobelts.

EXPERIMENTAL DETAIL

For the fabrication of ZnO nanostructures, p-Si wafer (100) of thickness 400 μm and resistivity of 5-10 Ωcm was used. Firstly, a p-Si wafer was cut into five pieces with 0.5 cm × 2.0 cm dimension. Then the wafers were thoroughly cleaned with soap solution and then they were rinsed in trichloroethylene (TCE) vapor for 10 minutes followed by rinsing in 1-propanol and deionised (DI) water for 5 and 2 minutes, respectively. The next step is piranha cleaning, in which sulphuric acid and hydrogen peroxide in the ratio of 3:2 were used and the wafer was kept in the solution for 15 min. After this process, the substrates were gently washed with cold DI water followed by dipping it into the solution of $HF:H_2O$ (1:10) to remove the native oxide layer. Finally, the substrates were again rinsed in deionised (DI) water for a minute and air dried. All chemicals were commercially available analytical grade regent and were used without any further purification.

A. HYDROTHERMAL METHOD

Pure and Al-doped ZnO nanorods were fabricated on the Si substrates by the two-step hydrothermal process. At first, ZnO seed layers were prepared by using 0.01M and 0.05M solutions of zinc acetate dihydrate $[Zn(CH_3COO)_2.2H_2O]$ in 2-methoxyethanol $(C_3H_8O_2)$ and it was stirred at 60°C for an hour [7-8]. Further, monoethanolamine (MEA, C_2H_7NO) was added until the solution becomes transparent, and then the solution was further stirred for an hour. Here, MEA acts as a stabilizer for ZnO particles/seed layer. After aging the solution for 24 hours, 100 μL of it was spin coated on the cleaned p-Si substrates at 3000 rpm for 30 seconds.

In the second step, 0.1 M aqueous solution of zinc nitrate hexahydrate $[Zn(NO_3)_2.6H_2O]$, hexamethylenetetramine [HMT; $C_6H_{12}N_4$] (0.1M) and aluminium chloride hexahydrate (for doping at. 2% of Zn) were used [5-6]. Now, the previously prepared substrates with the seed layers (developed using both concentration) were put vertically into the vials and they were kept in an oven. The temperature of the oven was different for different set of samples (viz. 90°C, 100°C, 110°C) and the time for which the heat treatment is provided is also varied (2h, 4h).

The samples were then cooled and washed with DI water and dried in N₂ atmosphere. An attempt was also made to grow ZnO hydrothermally by taking a thin layer of radio frequency (RF) sputtered gold (Au, 10 nm) and ZnO (300 nm) as the seed layer, instead of chemically synthesized ZnO seed layer.

B. Vapor-liquid-solid method

In this technique [4], a tube furnace equipped with inlet and outlet for Argon (Ar) gas was used. A source of zinc i.e. zinc pellets and substrate i.e. p-Si were kept at 1-2 cm apart in this furnace. The distance was pre-optimized. The temperature of the furnace was kept at 700°C. In this experiment, an argon gas was used as carrier gas, the flow of which was controlled by MFC (Mass Flow Controller). To maintain constant vapor pressure (50 mTorr) inside the tube, a vacuum pump was used.

C. Characterization

Morphology of the prepared ZnO nanostructures was studied with the help of a field emission scanning electron microscope (FESEM; CARL ZEISS MA15/18). In this paper, only the optimized results have been provided.

RESULT AND DISCUSSION

A. Hydrothermal Method

Fig.1 depicts the SEM micrographs of ZnO grown by varying the molar concentration of ZnO seed layer on p-Si substrate. Here, growth time was fixed for 2 hrs. The morphological variation in the nanostructures has been observed. Various parameters such as diameter, shape and orientation of nanostructures were found to change. Nanostructures grown on the seed layer of thickness 300 nm, which was deposited by RF sputtering at Ar : O₂ ratio of 5:1, had a rod instead of nanorods, its average length was 750 µm and the rods were not grown in a well-defined pattern as shown in Fig.1(a).

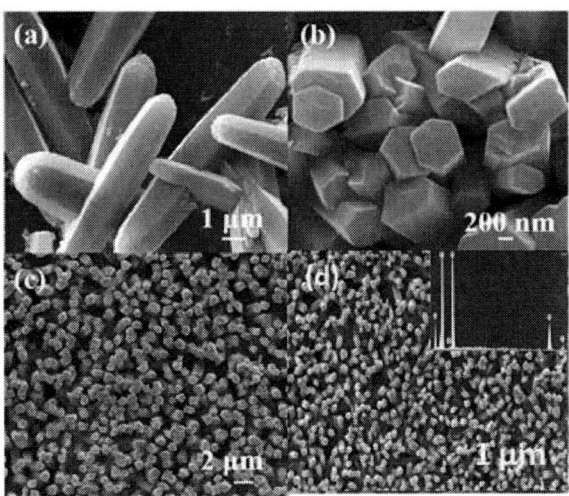

Fig. 1. SEM micrographs of nanorods grown by hydrothermal method: (a) with RF sputtered ZnO seed layer (b) seed layer obtained from 0.01 M precursor (c) seed layer grown by gold

(d) seed layer obtained from 0.05 M precursor used for Al:ZnO growth (inset: EDX pattern).

The nanorods which were grown on the single seed layer by spin coating method by using a precursor of 0.05 M concentration, have a mean diameter of 750 nm, but even these nanorods were not aligned properly and exhibited a contorted polygonal shape.

In Fig.1(b) the concentration of the precursor for the seed layer was 0.01M resulting symmetric growth of the nanorods which is nearly perpendicular to the surface with a polygonal geometry at the end of the nanorod, giving it a tapered shape. Now, instead of using ZnO as a seed layer, thin gold layer of 10 nm was also used. From the Fig.1(c) it is clear that here we get well aligned nanorods in dense form and at the top they exhibit hexagonal structure. The diameter of the nanorod was around 200 nm. Most of the nanorods exhibited a good hexagonal shape, which indicates good crystal growth along the [0 0 0 1] direction. Fig. 2(d) depicts the plane view of the nanorods, which is different from all the figures shown above it. For growing seed layer, we have used 0.05M precursor and coated with the help of spin coater and after that heated at 300°C, this step was repeated at least for four times and finally the samples were heated at 500°C in furnace for an hour in air. On this seed layer, Al doped ZnO were grown using the technique as reported in experimental section.

B. Vapour-Liquid-Solid method

In VLS technique, the aspect ratio is significantly top higher, as compared to the hydrothermal process. The aspect ratio is of the order of 10 and the reason behind such high aspect ratio is the thermodynamic as well as kinetic control over the shape of the nanorods.

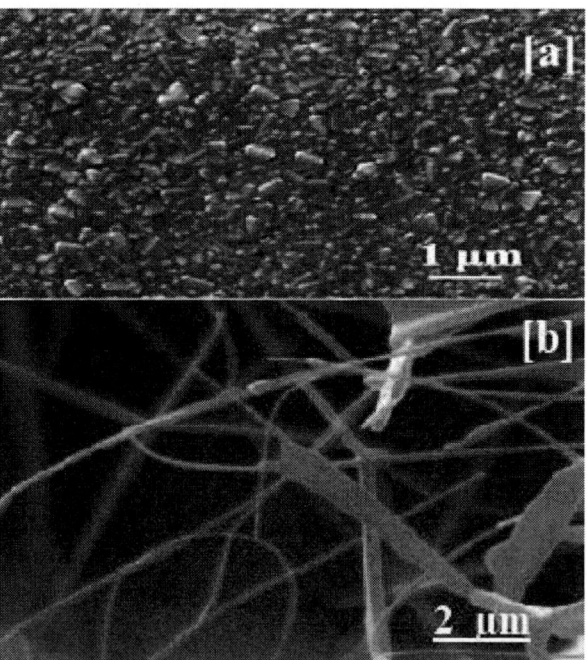

Fig 2: SEM image of nanorods grown on Si substrate with (a) without vapour flow (b) with vapour flow.

The factors which influence the growth of nanowires are temperature, growth duration and most importantly; the vapour flux. By this technique ZnO nanorods were grown continuously without any length limitation. Fig.2 shows the micrograph of un-synthesized nanorods; when vacuum pump was not connected at the end.

Fig.3: SEM micrograph of nano-urchin. The inset clearly shows the shape and justifies the name.

After connecting the vacuum pump at the end, vapor flow was maintained and a good aspect ratio for the nanorods was obtained. The diameter and length of the rod were found to be around 50 nm & 50 μm respectively as shown in Fig. 2(b). Here the tube temperature was fixed the 700°C. As the temperature of the furnace was increased, the diameter of the rod decreased to a large extent, slowly transforming to nanowires. The synthesis of nanorods was carried out in presence of argon and oxygen mixture. The gas and the vapour flux play a crucial role in the growth process of such nanostructures. Replacing argon gas by a mixture of O2 and N2 in the ratio of 2:1 yields an urchin like nanostructure: nano-urchin as depicted in Fig. 3.These structures have a large surface area to volume ratio and this property is quite useful in applications such as gas sensing and catalysis. The large surface area provides sites for molecular interaction.

CONCLUSIONS

ZnO and Al:ZnO nanostructures with different morphologies were successfully synthesized via a two-step hydrothermal method on p-Si substrates where a seed layer was grown by different techniques viz. spin coating and RF sputtering. It was observed that the seed layer plays a crucial role in the growth and alignment of the nanorods. For proper orientation of the nanorods, crystallization of the nucleus is important. Random orientation of nanorods was observed when the seed layer was made using RF sputtering due to the unfavorable oxygen partial pressure. Also, since the samples were not annealed, the nucleation process didn't occur properly. On the contrary, when the seed layer was grown by spin coating method and there after annealed properly, the crystallization of nuclei takes place and the quality of nanorods/nanowires is enhanced whereas, in VLS technique, the vapor phase decides the growth and morphology of the nanorods.

REFERENCES

[1] L.K.V. Vugt, S. Ru€hle, D. Vanmaekelbergh, Nano Lett. 6 (2006) 2707–2711.

[2] C. Periasamy, P. Chakrabarti, Sci. Adv. Mater. 3 (2011) 73–79.

[3] P. N. Mbuyisa, O. M. Ndwandwe, et.al "Controlled growth of zinc oxide nanorods synthesised by the hydrothermal method," Thin Solid Films, vol. 578, pp. 7–10, 2015.

[4] R. P. Sugavaneshwar et.al, "Uninterrupted and reusable source for the controlled growth of nanowires," Sci. Rep., vol. 3, pp. 1–7, 2013.

[5] T. Jannanea et.al, "Sol-gel Aluminum-doped ZnO thin films: Synthesis and characterization," J. Mater. Environ. Sci., vol. 8, no. 1, pp. 160–168, 2017.

[6] V. Balaprakash et.al "Preparation and characterization of aluminum doped zinc oxide (AZO) nanorods," Sādhanā, vol. 43, no. 6, p. 86, 2018.

[7] B. Liu et.al, "Hydrothermal synthesis of ZnOnanorods in the diameter regime of 50 nm," J. Am. Chem. Soc., vol. 125, no. 15, pp. 4430–4431, 2003.

[8] K. H. Tam et.al "Defects in ZnOnanorods by a hydrothermal method," J. Phys. Chem. B, vol. 110, no. 42, pp. 20865–20871, 2006.

2019 Devices for Integrated Circuit (DevIC), 23-24 March, 2019, Kalyani, India

Analytical Model of Subthreshold Swing in Triangular-Shaped FinFET

Subham Banerjee
Electronics & Communication
Engineering
Techno India University
Kolkata, India
subhamb741@gmail.com

Buddhadev Pradhan
Electronics & Communication
Engineering
Techno India University
Kolkata, India
bpjnity@gmail.com

Abstract— **FinFET, a quasi-planar device has gained tremendous importance in the semiconductor industry because it has the ability to suppress all the short-channel effects. The silicon fin of FinFET can be of different shapes-rectangular, trapezoidal, triangular, convex, concave etc. In this paper, we have discussed subthreshold swing, threshold voltage of different fin structures and drain current based on angle of inclination, fin width and applied gate voltages. We also include change of channel capacitance of triangular fin based on different fin width. Some novel analytical formula expression for subthreshold swing has been discussed. In MATLAB, we have plotted the variation of different graphs based on different parameters of FinFETs and obtain their respective results to prove that our dimensions are superior to other values.**

Keywords—angle of inclination, channel capacitance, different fin shapes, different parameters of FinFET, subthreshold swing.

I. INTRODUCTION

Moore's law is an evolutionary trend which has led to the revolution in electronics. It is just an observation that the number of components embedded in a semiconductor circuit doubled each year [1]. Transistor structure has evolved, starting from vacuum tube to 3-Dimensional structure, FinFET. One of the major challenges faced in the coming years is MOSFET scaling. Some of the challenges and it's causes are described below. Oxide thickness cannot be scaled below 2.3 nm due to increase of leakage current in that range. Junction depth cannot be reduced below 30 nm due to reduction of resistance in that range. Threshold voltage cannot be reduced below 0.25 V since leakage increases in that range. Channel length and Gate length cannot be reduced below 0.06 μm and 0.1 μm respectively due to increase of leakage current in that range [2-3]. FinFETs are easy to fabricate and offer higher drive current. It is known as quasi-planar device because current conduction in the plane of the wafer but the channel protrudes out of the wafer plane. For triple gate transistor the width of the transistor is $2(H_{fin}) + W_{fin}$, where H_{fin} is height of fin and W_{fin} is width of fin [4]. FinFET has better channel controllability and multi-fin FinFET has an aspect ratio of 2. Aspect ratio is defined as fin height/fin width. Therefore current drive of FinFET increases as fin height is twice that of fin width [5].

Subthreshold swing is inverse of subthreshold slope. FinFETs, indeed have small subthreshold swing. FinFET with rectangular-shaped fin is an ideal fin. But practically, it takes the shape of a trapezium. The top fin width can be

further modulated to give triangular FinFET. In this paper, we have studied the effect on subthreshold swing of the triangular-shaped FinFET. The small the value of subthreshold swing, the steeper is the slope which results in low leakage and high switching speed. The higher the value of subthreshold swing, smaller is the slope, which results in high leakage and the device becomes slow. It is better to have low values of subthreshold swing.

Design and modelling is discussed in section II. Analytical solution is provided in section III. In section IV, result and discussion is being given.

II. DESIGN AND MODELLING

Here, 20 nm triangular-shaped FinFET is being used. At this technology node subthreshold swing is low and leakage current is very low for triangular-shaped FinFET. At a doping concentration of the order of 1e17, triangular FinFET has lowest value of subthreshold swing and low leakage current. The following schematic diagram represents an independent-gate triangular-shaped SOI FinFET. The source and the drain are located perpendicular to the plane of the paper. In the following structure, dimensions are as follows: bottom fin width is 15 nm, top fin width is 1 nm, height of fin is 1 nm, doping concentration is 5e17 /cm³ and gate voltage is assumed to be 0.75 V. The aforesaid dimensions are used to reduce the short-channel effects.

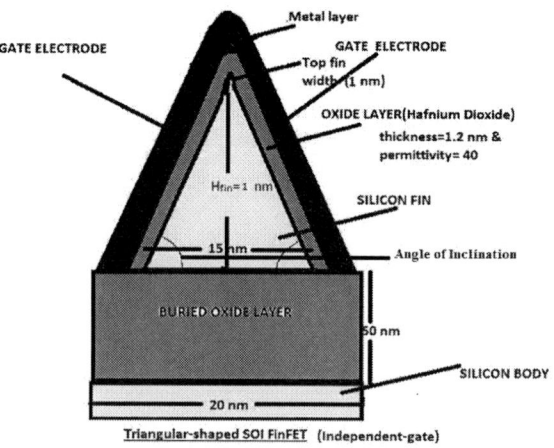

Fig. 1. Schematic Diagram of Triangular-Shaped SOI FinFET.

978-1-5386-6723-1/19 $31.00 © 2019 IEEE

H_{fin} is the fin height, α is velocity saturation index (usually 1 for short channel devices), I_{DO} is drain current at $V_{GS}=V_{DS}=V_{DD}$, V_{DD} is the power supply voltage, V_{GS} is the gate-to-source voltage, V_{DS} is the drain-to-source voltage, V_{TH} is the threshold voltage, a_0, a_1, p_0, p_1, β are the parameters used in [6] and I_{ON} is the drain current. V_{FB}, v_T, N_{si}, n_i, C_g, C_{ch}, Q_d are the flat-band voltage, thermal voltage, doping concentration, intrinsic carrier concentration, gate-oxide capacitance per unit length, channel capacitance per unit length and depletion charge per unit length.

III. ANALYTICAL SOLUTION

The total saturation current through the fin is

$$I_{ON} = \beta \int_0^{H_{fin}} (p_0 + p_1 y \cot\Phi)^\alpha \, dy \, [6] \qquad \dots\dots (1)$$

Putting $p_0=V_{GS}- a_0 + a_1 t_0$ & $p_1=2a_1$ [6] in equation (1) and then solving,

$$I_{ON} = \left[\frac{\beta}{\{2a_1(\alpha+1)\cot\Phi\}}\right] \left[(V_{GS} - V_{TH})^{\alpha+1}\right.$$
$$\left. - (V_{GS} - V_{TH} - 2a_1 H_{fin}\cot\Phi)^{\alpha+1}\right]$$
$$\dots\dots (2)$$

where $\beta = \frac{I_{D0}}{(V_{DD}-V_{TH})^\alpha}$

Subthreshold swing is given by [7]

$$ss = \frac{dV_{GS}}{d(log_{10}(I_{ON}))}$$

$$ss = \frac{(ln10)[(V_{GS}-V_{TH})^{\alpha+1}-(V_{GS}-V_{TH}-2a_1 H_{fin}\cot\Phi)^{\alpha+1}]}{(\alpha+1)[(V_{GS}-V_{TH})^\alpha-(V_{GS}-V_{TH}-2a_1 H_{fin}\cot\Phi)^\alpha]}$$
$$\dots\dots\dots (3)$$

From equation (3), we can observe that subthreshold swing is depended threshold voltage of the device. The threshold voltage, in turn is depended on top fin width of the device. The threshold voltage is given by [8]

$$V_{TH} = V_{FB} - \frac{Q_d}{C_g} + 2v_T ln\frac{N_{si}}{n_i} - v_T \ln\left[\frac{C_{ch}}{C_g}\left(1 - e^{\frac{Q_d}{v_T C_{ch}}}\right)\right]$$
$$\dots\dots\dots (4)$$

Where $Q_d=-qN_{si}H_{fin}W_{fin}$, q is electronic charge.

$$W_{fin} = W_{fin,top} + \frac{\lambda}{\lambda+1}\left(W_{fin,bottom} - W_{fin,top}\right) \dots\dots (5)$$

$$\lambda = \frac{2W_{fin,bottom} + W_{fin,top}}{2W_{fin,top} + W_{fin,bottom}}$$

The gate oxide capacitance per unit length is given by [8]

$$C_g = \frac{1}{W_{eff.}}\left[\frac{3.02*\frac{3\varepsilon_{ox}}{2}}{\ln\left(1+\frac{3t_{ox}}{2H_{fin}}\right)} - \frac{\frac{5\varepsilon_{ox}}{4}}{\ln\left(1+\frac{5t_{ox}}{4H_{fin}}\right)} + \frac{\frac{5\varepsilon_{ox}}{4}}{\ln\left(1+\frac{5t_{ox}}{4W_{fin}}\right)}\right] \dots (6)$$

The channel capacitance per unit length is given by [8]

$$C_{ch} = \frac{1}{W_{eff.}}\left[\frac{W_{fin}\varepsilon_{si}}{H_{fin}} + \frac{2H_{fin}\varepsilon_{si}}{W_{fin}}\right] \qquad \dots\dots (7)$$

The effective width of the device is given by
$W_{eff.} = 2 * H_{fin} + W_{fin}$

IV. RESULT AND DISCUSSION

Equation (3) leads to the plot of subthreshold swing with angle of inclination in MATLAB.

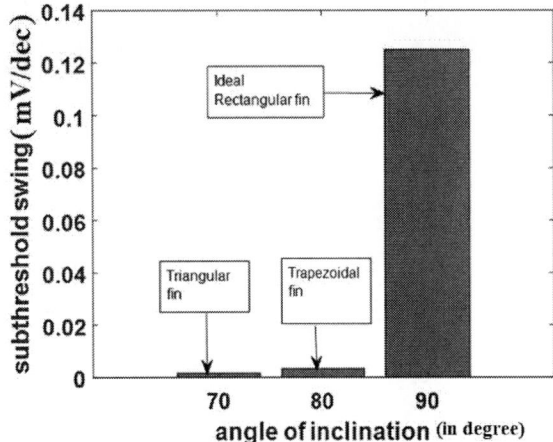

Fig. 2. Variation of subthreshold swing with angle of inclination

From Fig. 2 we can observe that subthreshold swing for triangular FinFET is the least and that for rectangular FinFET is maximum. For triangular FinFET, the drain current is 6%-9% less than that of rectangular FinFET [9]. Therefore to maintain I_{ON}/I_{OFF} ratio, leakage current of triangular FinFET is less than that of rectangular FinFET.

Among the two shapes of FinFET, used in laboratory, triangular FinFET has lower value of subthreshold swing than the trapezoidal FinFET. Thus, triangular FinFET has low leakage and high switching speed, making it suitable for SRAM cell design.

From equation (2), we get the variation of saturation drain current with gate voltage for triangular-shaped FinFET and drain current variation for three types of FinFET. The curves are shown in Fig. 3 and Fig. 4. respectively.

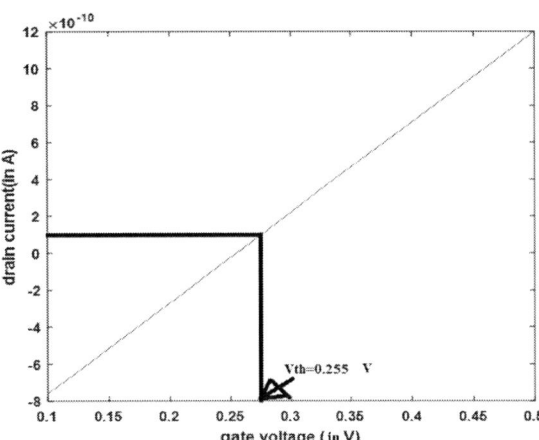

Fig. 3. Variation of drain current with gate voltage

In Fig. 3, the value of threshold voltage for triangular-shaped FinFET is 0.255 V. It is clear that when gate voltage is below threshold voltage, drain current is coming negative. It is due to diffusion current due to holes.

In Fig.4 shows that the drain current variation for triangular, trapezoidal and rectangular FinFET.

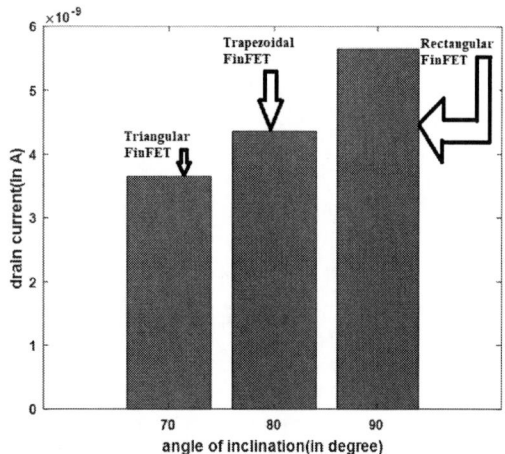

Fig. 4. Variation of drain current for triangular, trapezoidal and rectangular FinFET

From Fig. 4, it is clear that drain current is least for triangular-shaped FinFET. Also, we know that leakage current is least for triangular-shaped FinFET. Thus it does not violate the I_{ON}/I_{OFF} ratio rule.

In Fig. 5 shows the variation of threshold voltage with fin width. The unit of threshold voltage is V and that of fin width is nm.

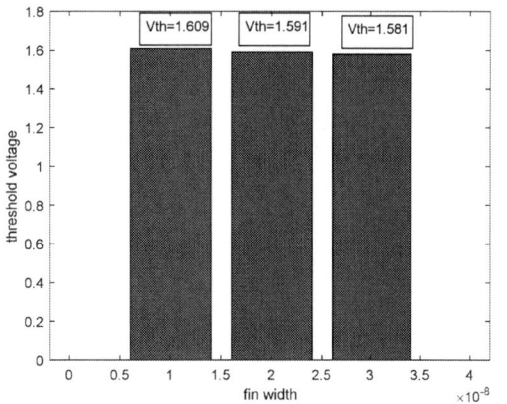

Fig. 5. Variation of threshold voltage with fin width

From Fig. 5, we can observe that the threshold voltage decreases with increase in fin width for any shape of FinFET with the help of equation (4). In our analytical modeling, the value of fin width is 10.04 nm. We can observe that if we increase the value of fin width above that value, the threshold voltage decreases. Hence, subthreshold leakage current increases.

Fig. 6 shows the variation of channel capacitance with fin width. The unit of channel capacitance is in nF.

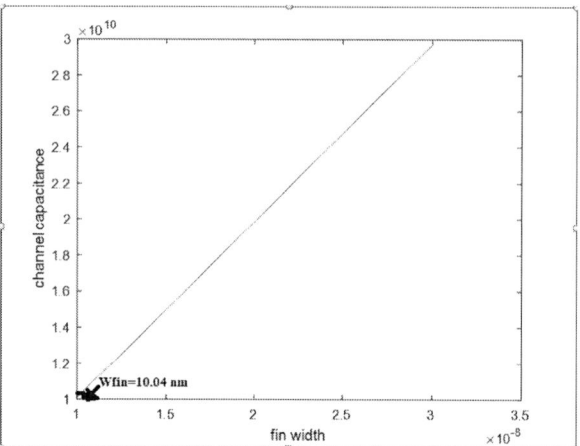

Fig. 6. Variation of channel capacitance with fin width.

From Fig. 6, we observe that at fin width 10.04 nm the channel capacitance is very small, which implies that delay is very small. Above curve is obtained with the help of equation (7).

Thus, we can conclude that our aforesaid dimensions are beneficial over other values.

ACKNOWLEDGMENT

The work is supported by Techno India University.

REFERENCES

[1] C.A. Mack, "Fifty years of Moore's law" IEEE transactions on semiconductor manufacturing, vol. 24, no. 2, may 2011, pp.202.

[2] S. Thompson, "MOS Scaling: Transistor Challenges for the21st Century," Intel Technology Journal Q3' 98, pp. 1.

[3] C. Hu,FinFET Modeling for IC simulation and Design, 2015, pp.2.

[4] V. Hariharan, "FinFETs", EEs801 seminar report, Electrical Engineering Dept., I.I.T. Bombay, 28thApril, 2005, pp.06-09.

[5] Hui-Wen Cheng, "Electrical characteristics dependence on the channel fin aspect ratio of multi-fin field effect transistors," in Proceedings of Semiconductor science and technology in IOP Publishing ltd., Vol. 24, pp.1, 27 October2009.

[6] R. Rao, "Accurate modeling and analysis of currents in Trapezoidal FinFET devices," 2007 IEEE International SOI Conference Proceedings, pp.47.

[7] S. Shukla, "Comparative Simulation Analysis of process parameter variations in 20 nm Triangular FinFET.," pp. 6, Volume 2017.

[8] F. Afrin, "Statistical analysis of leakage current of trapezoidal FinFETs," 2015 IEEE international WIE conference on Electrical and computer engineering(WIECON-ECE), pp.407-408.

[9] K. Wu, "Performance advantage and energy saving of Triangular-Shaped FinFETs," 2013 IEEE, pp.143.

Extensive analysis of band alignment engineering on the open circuit voltage performance of a GaAs/GaSb hetero structure solar cell

Girija Shankar Sahoo
Device Simulation Lab., Department of Electronics and Communication Engineering
S 'O'A Deemed to be University
Bhubaneswar, India
girijashankarsahoo@soa.ac.in

Guru Prasad Mishra
Department of Electronics and Telecommunication Engineering
National Institute of Technology Raipur
Raipur, India
gpscmishra.etc@nitrr.ac.in

Abstract—Maximum use of solar spectrum is possible through the selection of suitable band gap material in a solar cell. Single junction hetero structure solar cell provides a better opportunity by opting different materials with different band gap in the preparation of such device. But the problem arises with the large lattice mismatch and band discontinuity among these materials, which drastically reduces the open circuit voltage (V_{oc}) as well as the fill factor of the cell. In this work, band alignment engineering has been introduced for such type of problems. The device is simulated and verified using Silvaco TCAD suite. An extensive study is carried out in terms of SRH, radiative recombination and its effect on V_{oc} of the cell for different band offset values. It is found that the reduction in band discontinuity improves the V_{oc} of the device.

Keywords—Band alignment, Hetero structure, SRH recombination, Radiative recombination, Solar cell

I. INTRODUCTION

Earth abundant Si is a less superior material for photovoltaic (PV) applications as it suffers from many type of disadvantages such as; lower than optimum band gap that can harness highest efficiency (From Shockley and Queisser thermo-dynamical limit graph), low optical absorption coefficient, highly sensitive towards radiation damage, lack of incompatible material that can form a good class of hetero junction, etc. [1-3]. III-V semiconductor materials have the ability to overcome these problems and allow the cell to show good performance [2]. III-V semiconductors are more sensitive towards lower energy proton radiation damage compared with Si, but they can be easily removed by covering the solar cell with a glass plate. When a hetero structure is formed of III-V materials with different band gaps, a discontinuity between valence band (VB) maxima or conduction band (CB) minima is noticed [4]. This difference in CB and VB creates a barrier to the electrical transport across their interface. As a result the device performance is critically governed by the conduction band offsets (CBO's), valence band offset (VBO's), interface quality, such as roughness and interface defects. Depending on the alignment of energy band, heterostructures are categorised into different classes such as; straddling gap (Type-I), staggered gap (Type-II), broken gap (Type-III) and more [4-5]. With this new degree of freedom in case of present trend of solar cell design, it is important to understand the physics of heterostructure devices [6]. The generated offsets in the material band edge of a hetero structure forms a well in the CB and a pedestal in VB. Both the well and the pedestal produces confined energy states, which helps in PV applications [6-7]. So the discontinuities in CB and VB are of great interest as they purely depend on the physical property of the material and provide facts about the device operation and opportunities to boost the device performance [8]. A small positive discontinuity in CB earns a high efficiency by considerably reducing the recombination at surface of solar cell, resulting in higher open circuit voltage (V_{oc}) and Fill Factor (FF) [9]. An additional rise in CB discontinuities imposes an energy barrier against the photogenerated electrons under forward bias [9]. In addition to this the charge recombination at the interface is a strong function of CBO's. When the CBO forms a cliff structure in the emitter and base interface, the interface recombination becomes dominant and degrades the V_{oc} of the cell. When a spike structure is formed at the hetero interface the recombination is reduced drastically and enhances the V_{oc} of the device [10]. The V_{oc} of a solar cell under illumination is basically calculated by estimating the effective electronic bandgap. It is the difference between the lowest unoccupied molecular orbital (LUMO) of the acceptor and the highest occupied molecular orbital (HOMO) of the donor. The actual V_{oc} is somehow less than this calculated value due to carrier recombination at the electrodes and incomplete expansion of the quasi Fermi levels [11]. As a result it is strongly understood that the lower conversion efficiency in a hetero interface device is essentially the result of poor band alignment along with unsuitable energy gap [12]. So at the time of design, the role of CBO must be considered carefully, as it plays an important role in carrier transport phenomena and recombination mechanism in the trapping states [12]. To overcome such issues in case of a P-N hetero-junction, many techniques are used, such as; insertion of i-layer, insertion of electron or hole blocking layer, fabrication of some regular 3-D patterning of absorbing medium, and insertion of ultra-thin dielectric buffer layers [12].

From the above study it is found that the efficiency (E_{ff}) of PV is dependent on V_{oc}, FF, short circuit current density (J_{sc}). So to improve the V_{oc}, we must change the CBO, VBO and reduce the interface recombination [13]. In this study we have changed the CBO and VBO of the hetero-junction device by the help of band alignment engineering with the Silvaco TCAD suite and have studied its effect on recombination and V_{oc} of the cell along with photogeneration rate.

II. CARRIER DYNAMICS OF THE PROPOSED MODEL

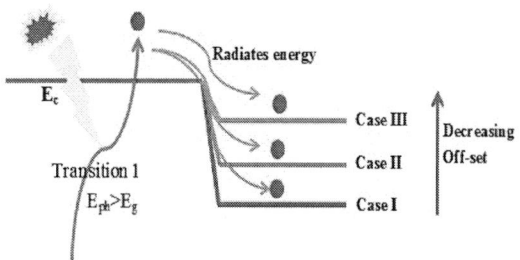

Fig. 1. Pictorial representation of carrier movement in case of heterostructure solar cell

Radiated photons of the sun irradiate the surface of the solar cell. If that photon is having an energy > band gap (E_g) of that active material used in the cell, it excites an electron from the valence band (VB) to conduction band (CB) (Transition 1 of Fig. 1). Now the electrons from the CB are collected as photocurrents. Let us consider a situation, when the photons having excess amount of energy ($>2E_g$, i.e. shorter wavelength area of the spectrum). Then that excites the electron to a position very far from the conduction band. Now the excited electron radiates this extra amount of energy to reach the CB and get collected as photo current. This process gives rise to a phenomenon called radiative recombination, which ultimately reduces the open circuit voltage (V_{oc}) of the cell. It adversely affect the fill factor (FF) and efficiency (E_{ff}) of the cell. Fig. 1 explains such a situation. To get rid of this situation, in this study the band alignment engineering concept is proposed, which reduces the band off-set between the CB of two active materials. This is shown in Fig. 1 by considering three case studies. From Fig. 1 it clear that the amount of radiated energy are in the order of case I> case II> case III. So, in the case III the amount of energy loss is very small and the off-set between the CB of active material is very low. This preserves the V_{oc} of the cell. So by using this concept, the hot carriers can be utilized fully to enhance the efficiency by avoiding the complex situations in carrier selective solar cell.

III. DEVICE MODELING AND SIMULATION

Different layers	SJ GaAs	Hetero Structure (HS) GaAs/GaSb
Window	$In_{0.49}Ga_{0.51}P$	$In_{0.49}Ga_{0.51}P$
Emitter	GaAs	GaAs
Base	GaAs	GaSb
BSF	$In_{0.5}(Al_{0.7}Ga_{0.3})_{0.5}P$	GaAs
Substrate	GaAs	GaAs

Fig. 2. Schematic of the proposed model along with SJ GaAs solar cell

Fig. 2 represents the schematic of a SJ GaAs cell along with a GaAs/GaSb heterostructure solar cell. Important physical design constraints of the cell like doping and thickness of each layer are considered from several well recognized published literatures [14-21]. Proposed design under investigation is simulated using Silvaco ATLAS TCAD suite by using the AM1.5G spectrum provided by ASTM [22]. In the simulation process of optical device like solar cell, optical intensity at each mesh point plays an important role to calculate different essential parameters like efficiency (E_{ff}), fill factor (FF), short circuit current density (J_{sc}) and open circuit voltage (V_{oc}). It is done by using the ray tracing method. After finding optical intensity, the simulator solves the carrier continuity and Poison's model at each point along with other physical models such as Shockley read hall (SRH) recombination model, OPTR, Auger and Radiative recombination model to calculate the J_{sc}. From the J_{sc} curve, all other figure of merits are estimated.

IV. RESULTS AND DISCUSSIONS

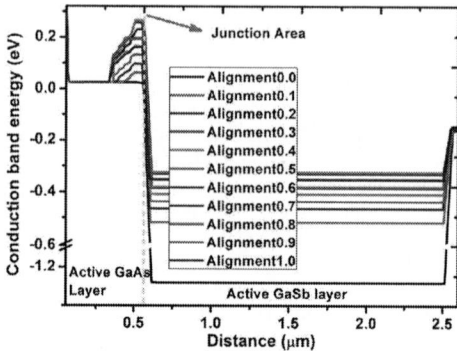

Fig. 3. Band offset adjustment through alignment engineering

Fig. 3 shows only the conduction band of GaAs/GaSb active layers to avoid the complexity of whole band analysis of the cell. It concludes that the conduction band offset can be adjusted between GaAs and GaSb layers by using band alignment engineering. Though we want to adjust the band of GaSb layer, but it is found that the conduction band edge of GaAs layer are also get affected. This little change in band edge (barrier height problem, see Fig. 3) of GaAs is tolerable as the excited electron can tunnel through it. So the expansion of conduction band of GaSb towards the conduction band of GaAs results in a drastic degradation in radiative as well as SRH recombination of the cell (See Fig. 4 and 5). This recombination resulted due to large lattice mismatch between GaAs (5.65 Å) and GaSb (6.09 Å). This strong lattice mismatch can also lead to formation of defects like threading dislocation (TD) or misfit dislocation (MD) [18]. This dislocation propagates into the active region of the cell and plays a significant role in the reduction of the carrier lifetime [18]. On the other hand, it can be inferred that this TD and MD gives rise to a large number of recombination centers in the cell that minimizes the performance of the PV cell by adversely hampering both the V_{oc} and J_{sc} [18]. However, we are thankful towards the technologies such as interfacial misfit dislocation (IMF), dislocation filtering layer technique and RIBER COMPACT 21 DZ solid−source molecular beam epitaxy (MBE), which are help full in fabricating such large lattice mismatch hetero structures [18].

Total recombination in solar cell can be represented by following relation.

$$R_{TOT} = R_{SRH} + R_{RAD} + R_{AUG} + R_{S,I,GB}$$

where, $R_{S,I,GB}$ stands for surface, interface and grain boundary recombination.

In this study we are focusing on radiative and SRH recombination as this two are the primary causes behind the voltage degradation in a cell.

In Fig. 4 and 5 we have only considered four situations to avoid the haziness of the graphs. It can be seen from Fig. 4 that there is a higher rate radiative recombination in the hetero GaAs/GaSb cell (alignment=0). This is due to the strong lattice mismatch between GaAs and GaSb. However, the band is adjusted in such a way that it reduces the radiative recombination in the order of few decades and becomes an alternative choice in case of designing hetero structure solar cell, which can reduce the fabrication complexity a little. A similar kind of result can be inferred from Fig. 5 also. However, in GaSb active area it shows a little high SRH recombination for '0' alignment cases. This drastic reduction in recombination is due to the spike change in band offset at the hetero interface of GaAs/GaSb structure. This has already been explained in the introduction section.

Now the ultimate focus goes towards the V_{oc} issue in case of hetero structure solar cell. This is shown in Fig. 6. It is clearly visible from Fig.6 that, with increasing band alignment the cell shows an exponential growth in V_{oc} and reaches a maximum of 1.42 eV. It is the result of such large degradation in both radiative as well as SRH recombination in the cell.

Optical properties of the cell, which plays important roles in determining the performance of the device is under investigation stage. However, photogeneration rate which is one of the important optical properties is discussed here briefly.

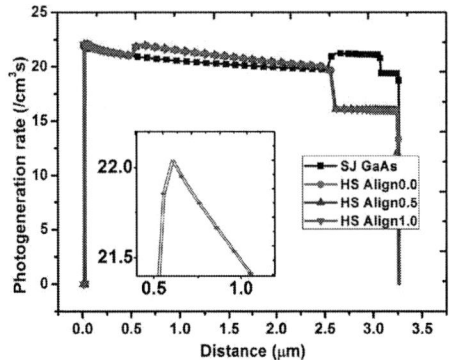

Fig. 7. Photogeneration rate in the device under different cases

There is a sharp increment in photogeneration rate in the GaSb region due to its higher absorption coefficient. The details regarding absorption coefficient can be studied from literature [18]. But there is a sharp decrement in the BSF region due to use of GaAs as BSF instead of AlInGaP. However, this higher photogeneration may help in achieving higher photo current. The inset view of Fig. 7 states that with the change in band off-set the photogeneration also changes.

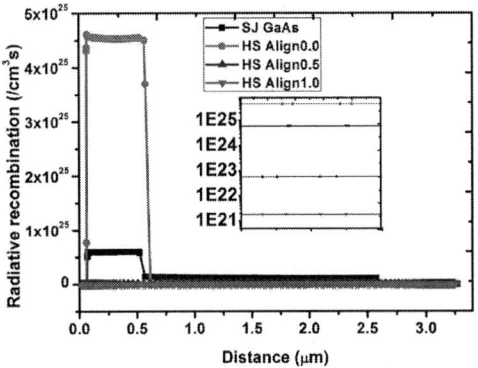

Fig. 4. Comparison of radiative recombination generated in the cell for different conditions

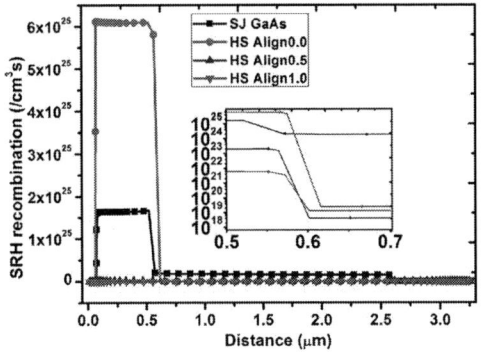

Fig. 5. Comparison of SRH recombination in the cell for various conditions

V. CONCLUSION

This study suggests that insertion of i-layer, electron or hole blocking layer, ultra-thin dielectric buffer layer and fabrication of some regular 3-D patterning of absorbing medium can be avoided in a device that are used to solve the high recombination phenomena and reduced V_{oc} issue. This will reduce the fabrication complexity as well as material cost of the device.

REFERENCES

[1] W. Shockley, H. J. Queisser, "Detailed balance limit of efficiency of PN junction solar cells," Journal of Applied Physics, vol. 32, no. 3, pp. 510-519, 1961.

[2] L. W. James, "III-V Compound Heterojunction solar cells," in: International Electron Devices Meetings, Washington, DC, USA, 1-3 Dec, 1975, pp. 87-90.

[3] V. M. Andreev, "Heterostructure Solar Cells," Semiconductors, vol. 33, no. 9, pp. 942-945, 1999.

[4] B. Roul, M. Kumar, M. K. Rajpalke, T. N. Bhat, S. B. Krupanidhi, "Binary group III-Nitride based heterostructures: band offsets and transport properties," Journal of Physics D: Applied Physics, vol. 48, no. 42, pp. 423001-1-21, 2015.

[5] S. Wang, "Fundamentals of Semiconductors: Theory and Device Physics," Prentice Hall Series in Electrical and Computer Engineering.

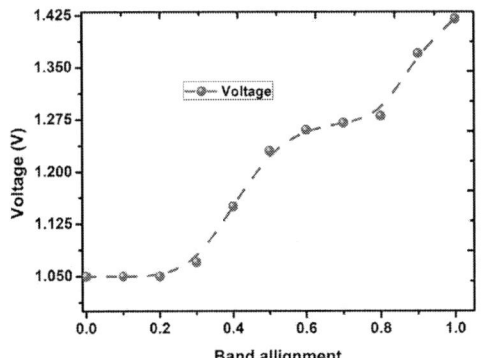

Fig. 6. Variation in voltage with respect to band adjustment

[6] S. J. Chua and A. Ramam, "Schottky contacts and conduction band offsets in bandgap engineered InGaAlAs/ InP heterostructures," in: Proceedings of IEEE International Conference on Semiconductor Electronics, Penang, Malaysia, 26-28 Nov., 1996, pp. 73-77.

[7] A. Panchak, A. Luque, A. Vlasov, V. Andreev, and A. Marti, "The effects of band offsets in quantum dots," Solar Energy Materials and Solar Cells, vol. 145, pp. 180-184, 2016.

[8] A. Cheriet, M. Mebarki, P. Christol, H. Ait-Kaci, "Energy band diagram and band offsets of $Ga_{0.6}Al_{0.4}As_{0.034}Sb_{0.966}(p)/Ga_{0.6}Al_{0.4}As_{0.034}Sb_{0.966}(n)/InAs_{0.9}Sb_{0.1}$ double interface determined by capacitance voltage measurement," Materials Science in Semiconductor Processing, vol. 66, pp. 50-55, 2017.

[9] Y. Chen, X. Tan, S. Peng, C. Xin, A. E. Delahoy, K. K. Chin, and C. Zhang, "The influence of conduction band offset on CdTe Solar Cells," Journal of Electronic Materials, vol. 47, no. 2, pp. 1201-1207, 2018.

[10] C. Ding, Y. Zhang, F. Liu, Y. Kitabatake, S. Hayase, T. Toyoda, K. Yoshino, T. Minemoto, K. Katayama, Q. Shen, "Effect of the conduction band offset on interfacial recombination behaviour of the planar perovskite solar cells," Nano Energy, vol. 53, pp. 17-26, 2018.

[11] D. C. Olson, S. E. Shaheen, M. S. White, W. J. Mitchell, M. F. A. M. Hest, R. T. Collins, and D. S. Ginley, "Band-Offset Engineering for Enhanced Open-Circuit Voltage in Polymer–Oxide Hybrid Solar Cells," Adv. Funct. Mater., vol. 17, no. 2, pp. 264–269, 2007.

[12] K. Javaid, W. Wu, J. Wang, J. Fang, H. Zhang, J. Gao, F. Zhuge, L. Liang, and H. Cao, "Band offset engineering in ZnSnN2-based heterojunction for low-cost solar cells," ACS Photonics, vol. 5, no. 6, pp. 2094-2099, 2018.

[13] S. A. Al Kuhaimi, "Conduction and valence band offsets of CdS/CdTe solar cells," Energy, vol. 25, pp. 731-739, 2000.

[14] G. S. Sahoo, P. P. Nayak, and G. P. Mishra, "An ARC less InGaP/GaAs DJ solar cell with hetero tunnel junction," Superlattices Microstructure, vol. 95, pp. 115–127, 2016.

[15] G. S. Sahoo and G. P. Mishra, "Effective use of spectrum by an ARC less dual junction solar cell to achieve higher efficiency: A simulation study," Superlattices Microstructure, vol. 109, pp. 794–804, 2017.

[16] G. S. Sahoo and G. P. Mishra, "Use of ratchet band in a quantum dot embedded intermediate band solar cell to enrich the photo response," Materials Letter, vol. 218, pp. 139–141, 2018.

[17] G. S. Sahoo and G. P. Mishra, "Use of InGaAs/GaSb Quantum Ratchet in p-i-n GaAs Solar Cell for Voltage Preservation and Higher Conversion Efficiency," IEEE Transactions on Electron Devices, vol. 66, pp. 153-159, 2019.

[18] G. S. Sahoo and G. P. Mishra, "Efficient Use of Low-Bandgap GaAs/GaSb to Convert More than 50% of Solar Radiation into Electrical Energy: A Numerical Approach," Journal of Electronics materials, vol. 48, pp. 56-570, 2019.

[19] G. S. Sahoo and G.P. Mishra, "Design and modeling of an efficient metamorphic dual-junction InGaP/GaAs solar cell," Opt. Quant. Electron., vol. 48, no. 9, pp. 420-1-16, 2016.

[20] G. S. Sahoo and G. P. Mishra, "Effect of wideband gap tunnel diode and thickness of the window layer on the performance of a dual junction solar cell," Procedia Technology, vol. 25, pp. 684-691, 2016.

[21] G. S. Sahoo and G. P. Mishra, "Design and modeling of an SJ infrared solar cell approaching upper limit of theoretical efficiency,"

International Journal of Modern Physics B, vol. 32, no. 2, pp. 1850014-1-15, 2018.

[22] SILVACO Data Systems Inc, Silvaco ATLAS User's Manual, 2010.

Investigation of a nanoscale grooved stepped gate MOSFET to explore the self-heating effect

S. Bhol[1], S. Mishra[2], S. S Mohanty[2], G.P. Mishra[3]*

[1]Device Simulation Lab, Dept. of Electronics and Instrumentation Engg.
Siksha 'O' Anusandhan University, Bhubaneswar, 751030, India.
[2]Device Simulation Lab, Dept. of Electronics and Communication Engg.
Siksha 'O' Anusandhan University, Bhubaneswar, 751030, India.
[3]Dept. of Electronics & Telecommunication Engg, National Institute of Technology Raipur,
Raipur, Chhattisgarh- 492010, India
gpscmishra.etc@nitrr.ac.in

Abstract—**This work reports a new grooved gate silicon-on-insulator (GG-SOI) MOSFET with multi-layered (SiO$_2$/Si$_3$N$_4$/SiO$_2$) buried insulator structure to reduce self-heating effect (SHE). The proposed model is simulated using the Sentaurus TCAD simulator. As the thermal conductivity of SiO$_2$/Si$_3$N$_4$/SiO$_2$ buried insulator is higher than SiO$_2$ buried layer, this innovative grooved gate SOI model is able to reduce the self-heating effect of the conventional GG-SOI MOSFET. Thus appropriate for high temperature solicitations. Performance comparison has been done between the multi-layer-buried recessed channel SOI and conventional SiO$_2$ based GG-SOI MOSFET. The presented structure has well controlled the device temperature against the low thermal conductivity of conventional GG-SOI MOSFET. Further step gate concept is used for the improvement of analog performance and short channel effects (SCEs). Simulation results reveal the enhanced performance manifested by the proposed structure in terms of increased drain current, reduced device temperature and increased electron mobility.**

Keywords— *Recessed channel, multi-layered buried oxide, Negative junction depth (NJD), self-heating effect.*

I. INTRODUCTION

The reduced device dimension of the MOS structure confronted numerous provocations, such as hot carrier effects (HCEs), short channel effects (SCEs), which causes degradation of device performance [1-3]. SOI technique is the most prominent method for reduction of short channel effects (SCEs). SOI MOSFETs provide an advantage for high speed applications, because of the low parasitic capacitance. For the future expertise, prominence of the SOI MOSFET is so high that various alterations are dedicated to its performance exploration [4-8]. SCEs are mostly created due to the penetration of the drain side barrier towards the source, promoting barrier lowering. Thus recessed gate concept in SOI-MOSFET alleviates SCEs and HCEs [9-17]. Though the SOI MOSFET promises a valuable device with extremely stimulated productivity but this technology suffers from a challenging situation termed as a self-heating effect (SHE), which arises due to the presence of low thermal conductivity thin buried SiO$_2$ insulating layer [18]. Hence, heat produced in SOI MOSFET causes a considerable temperature

intensification in the device body. The temperature hike results in a reduction in drive current and leads to additional severe reliability difficulties in terms of boosted impact ionization [19]. Further, SHE will be more serious for the SOI structures with reduced nanoscale dimensions. Different structures have been explored by researchers to suppress the self-heating effect (SHE) [20–24]. But this critical issue of the SHE has not analyzed in the nanoscale groove gate SOI structures. Therefore, a new structure is investigated to overwhelm the harmful thermal effect in groove gate SOI MOSFET.

This work proposes a new groove gate SOI structure to overwhelm the self-heating consequence, where the electrostatic performance of groove gate SOI MOSFET is simulated with a multilayer SiO$_2$/Si$_3$N$_4$/SiO$_2$ buried oxide. By doing this, multi-layered buried groove gate SOI MOSFET can replace the conventional GG-SOI MOSFET with almost similar transfer characteristics and improved device temperature. Further, to get enhanced device performance a stepped gate concept is incorporated in multi-layered buried GG-SOI MOSFET. Excellent electrostatic gate control is reached due to the smaller oxide thickness near the source end, which helps to enhance the mobility in the channel region. Again larger oxide thickness at drain end gives lower on-resistance [25-27].

Thus the present work designates a standardized exploration of analog performance of multi-layered (ML) buried grooved steeped gate (GSG) SOI MOSFET on Sentaurus TCAD platform [28].

II. DEVICE STRUCTURE AND SIMULATION

A schematic translational outlook of the proposed device is shown in fig.1.The structure consists of a 30nm length metal gate, whose work function is 4.77eV. Gate region in proposed ML buried GSG-SOI MOSFET is split into three equal regions, where the oxide thickness is gradually increased towards the drain side with a step of 1nm. The thickness of multilayer SiO$_2$/Si$_3$N$_4$/SiO$_2$ buried oxide (t$_b$) consists of a 50nm Si$_3$N$_4$ layer sandwiched between 25nm SiO$_2$ layer.

Structure simulation has been executed on Sentaurus TCAD simulator. Hydrodynamics (HD) model is considered, as the

978-1-5386-6723-1/19 $31.00 © 2019 IEEE

drift-diffusion model fails to clarify the speed overshoot and charge carrier dissemination. This model provides better precision for the non-local transport phenomena. Further, Lombardi CVT mobility model has been used to get perfect results. The simulation has been performed by using SRH (Shockley-Red-Hall), low field mobility (CONMOB), Auger recombination models, parallel electric-field-dependent (FLDMOB),Shockley-Read-Hall (SRH) models to provide a better device performance.

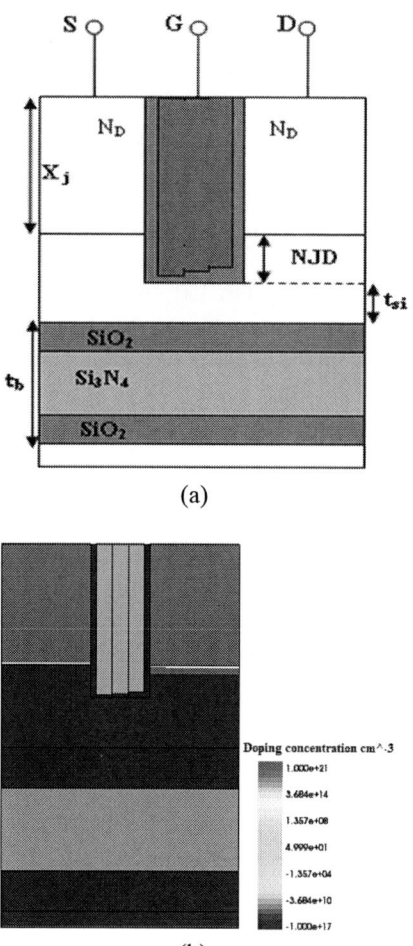

(a)

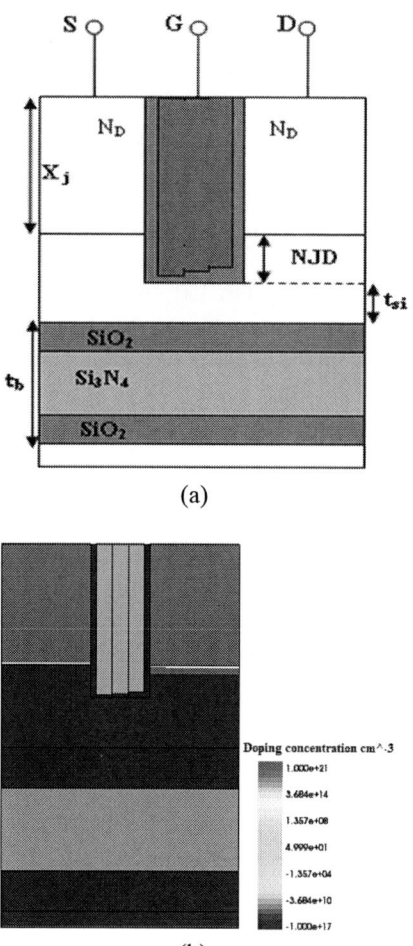

(b)

Fig.1 Schematic structure of multi-layered buried grooved stepped gate (ML buried GSG) SOI MOSFET with device dimension: BOX thickness t_b=100nm, junction depth X_j = 48nm, NJD= 20nm, oxide thickness t_{ox}= 2nm at the source side and Source/Drain doping N_D=10^{21} cm^{-3}.

RESULTS AND DISCUSSION

The input/output characteristics and corresponding conductance for all the devices are shown in fig. 2 and 3. From the transfer characteristics, conventional GG-SOI MOSFET and ML buried GG-SOI structure have 0analogous input characteristics with the threshold voltage of 0.638V and 0.645V respectively. So SiO$_2$/Si$_3$N$_4$/SiO$_2$ insulator layer can substitute the Si dioxide buried layer without affecting the

electrical performance of the device. Further, the improvement in device parameters are reflected in the proposed structure by implementing the stepped gate in ML buried GG-SOI structure. With the presence of the groove gate and lesser oxide thickness at the source side enhances the charge carrier density along the channel and lowers the threshold voltage which is prominently reflected in fig.2. From the transfer characteristics, it is clear that the ML buried GSG-SOI MOSFET provides 44% enhancement in on current and 22% improvement of peak transconductance as compared to ML buried GG-SOI MOSFET. Thus with higher transconductance, the proposed structure can predict a better cut-off frequency.

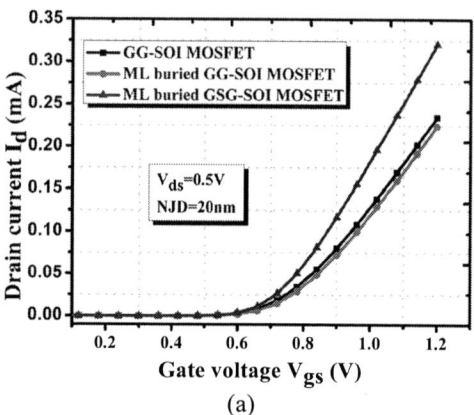

(a)

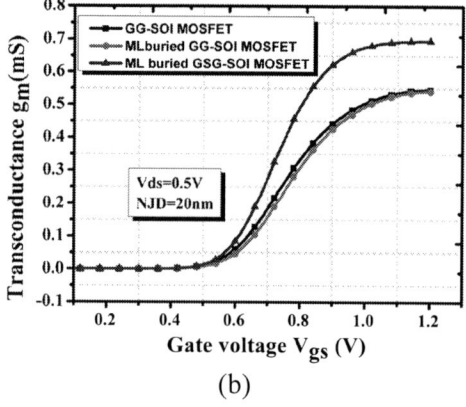

(b)

Fig.2. Simulated (a) Transfer characteristics and (b) Transconductance for GG-SOI MOSFET, ML buried GG-SOI MOSFET and ML buried GSG- SOI MOSFET

Similarly, due to the smaller potential barrier at the drain side, the current density along the channel for a stepped gate MOSFET is more. Thus ML buried GSG-SOI MOSFET with a saturated drain current of 0.26mA provides an enhancement of 36% in output current as compared to ML buried GG-SOI MOSFET. Further, the proposed structure gives an improved drain conductance which is reflected in fig.3.

2019 Devices for Integrated Circuit (DevIC), 23-24 March, 2019, Kalyani, India

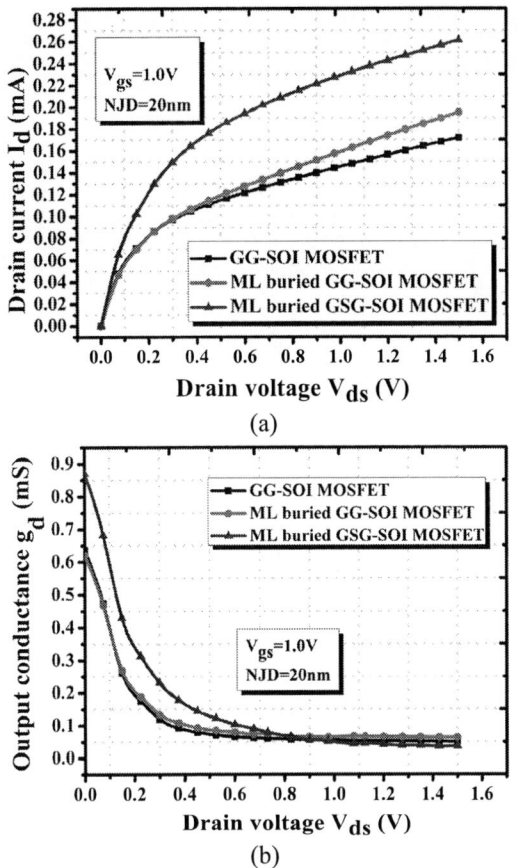

(a)

(b)

Fig.3. Simulated (a) Output drain characteristics and (b) Drain Conductance for GG-SOI MOSFET, ML buried GG-SOI MOSFET and ML buried GSG-SOI MOSFET.

Fig. 4 indicates the dependence of device temperature with the drain voltage for all simulated grooved gate SOI MOSFETs. In Fig.4 (a) GG-SOI device with $SiO_2/Si_3N_4/SiO_2$ buried layer can overwhelm the self-heating effect effectively as compared to the Si dioxide based buried GG-SOI MOSFET. Further, due to the presence of the stepped gate, a slight improvement in device temperature is reflected in the proposed structure, because of reduced hot carrier effect, which is clearly reflected in Fig. 4b.

The mobility temperature dependent formula is given by [29-31]

$$\mu_{eff} = \mu_{eff,0} \, (T/T_0)^{-2}$$

where μ_{eff} = Effective mobility,
$\mu_{eff,0}$ =The effective mobility at ambient temperature,
T= The average device temperature and
T_0 = The ambient temperature.

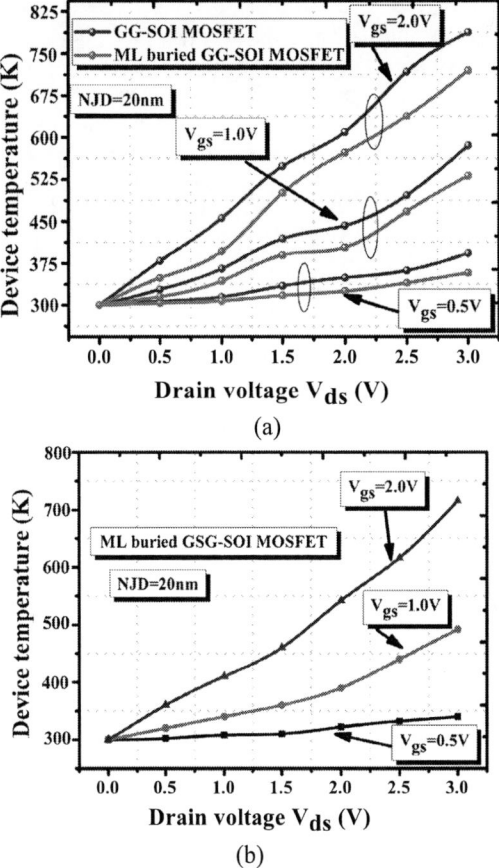

(a)

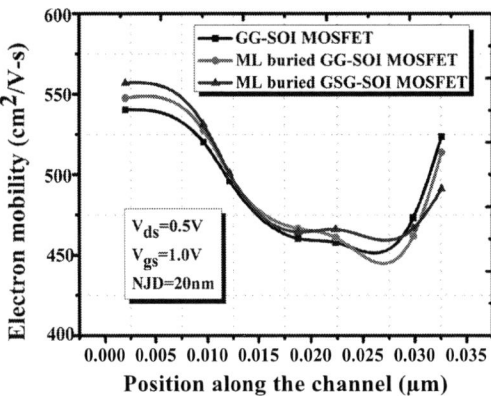

(b)

Fig. 4. Variation of device temperature with drain voltage V_{ds} at different gate voltage V_{gs}.

The mobility equation states, SHE enhances the device temperature thus carrier scattering, which leads to mobility degradation.

Fig.5. Effective electron mobility along the channel length of GG-SOI MOSFET, ML buried GG-SOI MOSFET and ML buried GSG- SOI MOSFET.

For more explication, in Fig. 5 the electron mobility has been demonstrated for all the presented structure at biases V_{ds} =

978-1-5386-6723-1/19 $31.00 © 2019 IEEE   51

0.5V and V_{gs} = 1 V. So with a better device temperature, the multi-layered buried GSG-SOI structure has improved electron mobility and consequently, the device current is enhanced in the proposed model as compared to other GG-SOI structures. Further, with higher source side and lower drain side electron mobility of proposed structure interprets more carrier density along the channel with reduced hot carrier effects.

CONCLUSION

This work presents the performance analysis of ML buried GSG-SOI structure. From the investigation, it is concluded that with similar transfer characteristics $SiO_2/Si_3N_4/SiO_2$ buried layer GG-SOI MOSFET can substitute the Si dioxide buried insulator without affecting the electrical features of the device. Further, performance enhancements are reflected in the proposed ML buried GSG-SOI MOSFET. With stepped gate the presented model exhibits a superior device temperature, threshold voltage and driving current in contrast to buried Si dioxide based GG-SOI MOSFETs, therefore proving it's adequacy for high temperature application.

REFERENCES

[1] Y. Taur, D. A. Buchanan, W. Chen, D. J. Frank, K. E. Ismail, S.H. Lo, A. George, S. Halasz, R.G. Vishwanathan, H. Jen, C. Wann, S. J. Wind and H. S. Wong, "CMOS scaling in Nanometer regime," proceedings of IEEE, Vol.85(4), pp.486-504, 1997.

[2] Seong-Dong Kim, Cheol-Min Park, Jason C. S. Woo, "Advanced Model and Analysis of Series Resistance for CMOS Scaling Into Nanometer Regime—Part II: Quantitative Analysis," IEEE Tran. on Electron Devices, Vol. 49(3),pp.526–532, 2002.

[3] D. J. Frank, R. H. Dennard, E. Nowak, P. M. Solomon, Y. Taur, and H.-S. P. W. H.-S. P.Wong, "Device scaling limits of Si MOSFETs and their application dependencies," Proc.IEEE, vol. 89 (3), pp. 259–288, 2001.

[4] K. Suzuki and S. Pidin, "Short channel single gate SOI MOSFET model," IEEE Trans. Electron Devices, Vol.50, pp.1297–1305,2003.

[5] G. P. Katti, N. Das Gupta and A. Das Gupta, "Threshold Voltage Model for Mesa-Isolated Small Geometry Fully Depleted SOI MOSFETs Based on Analytical Solution of 3-D Poisson's Equation,"IEEE Tran. On Electron Devices, Vol.51, pp.1169-1177,2004.

[6] M. K. Anvarifard and Ali A. Orouji, "Voltage difference engineering in SOI MOSFETs: A novel side gate device with improved electrical performance," Materials Science in Semiconductor Processing, Volume 16, Issue 6, pp. 1672-1678 ,2013.

[7] Ali A. Orouji and M. K. Anvarifard, " Novel Reduced Body Charge Technique in Reliable Nanoscale SOI MOSFETs for Suppressing the Kink Effect," Superlattices and Microstructures,Vol.72,pp.111-125, 2014.

[8] Mohammad K.Anvarifard, Successfully Controlled Potential Distribution in a Novel High-Voltage and High-Frequency SOI MESFET, IEEE Trans. Dev.Mater. Reliab., Vol.16 , pp.631-637,2016.

[9] J.Y Seo,K.J. Lee,Y.S. Kim ,S.Y.Lee, S.J.Hwang ,C.K.Yoon, "Reliability for recessed channel structure n-MOSFET,"Microelectronics Reliability, Vol.45,pp.1317–1320, 2005.

[10] R. Chaujar, R. Kaur, M. Saxena, M. Gupta and R.S.Gupta, " TCAD Assessment of Gate Electrode Workfunction Engineered Recessed Channel (GEWE-RC) MOSFET and Its Multi-layered Gate Architecture—Part I: Hot-Carrier-Reliability Evaluation," IEEE Trans. Electron Devices, Vol. 55, no. 10, pp. 2602–2613 ,2008.

[11] Priyanka Malik, R.S.Gupta, Rishu Chaujar and Mridula Gupta, "AC analysis of nanoscale GME-TRC MOSFET for microwave and RF applications," Microelectronics Reliability, Vol. 52, pp.151–158,2012.

[12] Ajay Kumar, Neha Gupta, Rishu Chaujar," TCAD RF performance investigation of Transparent Gate Recessed Channel MOSFET," Microelectronics Journal,Vol.49,pp.36–42,2016.

[13] S. Mishra, A.S Lenka, S.S Mohanty,U. Bhanja, G.P. Mishra, "Effect of RRC on SOI MOSFET to Improve the SCEs," Devices for Integrated Circuit (DevIC), IEEE, pp. 536-540, 2017.

[14] S. Mishra,U. Bhanja,G.P.Mishra,"Variation of source gate workfunction on the performance of dual material gate rectangular recessed channel SOI-MOSFET,"Int J Numer Model,Vol.32(1), e2487,2019.

[15] M.Singh, S.Mishra, S.S. Mohanty, G.P. Mishra, " Performance analysis of SOI MOSFET with rectangular recessed channel," Adv. Nat. Sci.: Nanosci. Nanotechnol,Vol.7 ,015010 (8pp) 2016.

[16] Annada Shanker Lenka, Sikha Mishra, Satyaranjan Mishra, Urmila Bhanja, Guru prasad Mishra,"An extensive investigation of work function modulated trapezoidal recessed channel MOSFET,"Superlattices and Microstructures, Vol.43, pp.878-888,2017.

[17] Sikha Mishra, Urmila Bhanja and Guru Prasad Mishra "Impact of structural parameters on DC performance of recessed channel SOI-MOSFET," Int.J.Nanoparticles (Inderscience) (Accepted).

[18] A.K. Goel, T.H. Tan," High-temperature and self-heating effects in fully depleted SOI MOSFETs," Microelectronics Journal,Vol.37,pp.963–975,2006.

[19] Zheng Xuan Zhang, Qing Lin, Ming Zhu ,"Cheng Lu Lin. A new structure of SOI MOSFET for reducing self-heating effect,"Ceramics International,Vol. 30, pp.1289–1293,2004

[20] Mohammad K. Anvarifard, Ali A. Orouji, "Improvement of self-heating effect in a novel nanoscale SOI MOSFET with undoped region: A comprehensive investigation on DC and AC operations," Superlattices and Microstructures,Vol.60 , pp.561–579,2013.

[21] Ali A. Orouji , Sara Heydari , Morteza Fathipour ," Double step buried oxide (DSBO) SOI-MOSFET: A proposed structure for improving self-heating effects," Physica E ,Vol.41 , pp.1665–1668,2009.

[22] M. Rahimian, Ali A. Orouji, "A novel nanoscale MOSFET with modified buried layer for improving of AC performance and self-heating effect," Mater. Sci. Semicond. Process,Vol. 15 ,pp. 445–454, 2012.

[23] Majid Ghaffari, Ali A. Orouji, "A novel nanoscale SOI MOSFET by embedding undoped region for improving self-heating effect," Superlattices and Microstructures,Vol.118 , pp.61–78,2018.

[24] S E Jamali Mahabadi, Ali A Orouji, P Keshavarzi,Hamid Amini Moghadam,"A new partial SOI-LDMOSFET with a modified buried oxide layer for improving self-heating and breakdown voltage," Semicond. Sci. Technol.Vol. 26, 095005(12pp),2011.

[25] M. J. Kumar and A. Bansal, "Improving the breakdown voltage, ON resistance and gate-charge of InGaAs LDMOS power transistors, "Semicond. Sci. Technol., vol. 27, no. 10, pp. 105030-37, 2012.

[26] D. G. Lin, S. L. Tu, Y. C. See, and P. Tam, "A novel LDMOS structure with a step gate oxide," in IEDM Tech. Dig, pp. 963–966,1995.

[27] M. Jagadesh Kumar and Radhakrishnan Sithanandam, "Extended-p+ Stepped Gate LDMOS for Improved Performance, "IEEE Trans. Electron Device, Vol. 57(7),pp.1719-1724,2010.

[28] Sentaurus Device User Guide, Synopsys, Inc., Mountain View, USA, (2013).

[29] S.M. Sze, K.K. Ng, Physics of Semiconductor Devices, 3rd ed., John Wiley & Sons, New Jersey, 2007.

[30] B.M. Tenbroek, M.S.L. Lee, R. Redman-White, R. John, T. Bunyan, M.J. Uren, "Self- heating effects in SOI MOSFETs and their measurement by small signal conductance techniques," IEEE Trans. Electron Dev., Vol.43 pp.2240–2248,1996.

[31] Sajad A. Loan, S. Qureshi, S. Sundar Kumar Iyer, "A Novel Partial-Ground-Plane-Based MOSFET on Selective Buried Oxide: 2-D Simulation Study," IEEE Trans.Electron devices, Vol. 57, pp.671-680, 2010.

An extensive analysis of $In_{0.53}Ga_{0.47}As/InP$ surrounding gate MOSFET to enhance the electrostatic performance using δ-doped technique

S. S. Mohanty[1], S. Mishra[1], S. Mohanty[1], G.P. Mishra[2]*

[1]*Device Simulation Lab, Dept. of Electronics and Communication Engg.*
Siksha 'O' Anusandhan University, Bhubaneswar, 751030, India.
[2]*Dept. of Electronics & Telecommunication Engg, National Institute of Technology Raipur,*
Raipur, Chhattisgarh- 492010, India
gpscmishra.etc@nitrr.ac.in

Abstract—Surrounding gate (SG) heterostructure metal oxide semiconductor field effect transistor (HMOSFET) has been embraced for generating the future device, which could limit the working range and decrease the static standby power dissipation. This paper presents an investigation of source δ-doped $In_{0.53}Ga_{0.47}$ As /InP based SGHMOSFET to enhance the device performance. As the channel is encompassed by the all-around gate with the δ-doped region in the source end, there is a better electrostatic control around the HMOSFET, which is obvious through the smaller DIBL and SS as compared to conventional SGHMOSFET. The proposed work deals with a detailed simulation based study of the device performance parameters such as the surface potential, electric field, electron mobility, On resistance, threshold voltage and drain current. The simulated results are compared with conventional SGHMOSFET. It has been revealed that δ-doped SGHMOSFET gives a superior insusceptibility to short channel effects (SCEs) when compared with conventional SGHMOSFET.

Keywords— *δ-doped, SGHMOSFET, $In_{0.53}Ga_{0.47}As/InP$, On resistance.*

I. INTRODUCTION

As indicated by ITRS [1], continual downsizing of the conventional MOSFET will rapidly hit its limit forced by short channel impacts and extreme channeling of the gate dielectric. So as to stretch out MOSFET performance in sub 20nm regime, different nanoscale sub-atomic materials and devices are being extensively examined [2-3]. Apart from the above, the non-planar device structures have been introduced due to their great degree of scaling.

Surrounding gate (SG) MOSFET [4-5] is one of the non-planar challenging structures for the development of CMOS innovation as it allows an exceptional mechanism over the charge carrier in the channel [6-8]. SGMOSFET structures reflect enhanced Off state performance, such as better DIBL and SS because of their greater output current as compared to the conventional planner structures [9].

On the other hand, to improve the device drive current capacity, the group III-V based compound materials are used in the channel area [10]. In the recent past, there are various device structures are being developed using compound semiconductors by various researchers [11-13]. As heterostructures are formed with the significantly flexible III-V semiconductors, it offers high drive current as well as

greater electron transport efficiency [14]. Utilization of InP/InGaAs/InP heterostructure in the channel area and high-k dielectric in the gate area helps in achieving high On current along with the minimal leakage [15]. Use of high k material ultimately reduces the equivalent dielectric thickness in the gate region [16]. In the recent past, several silicon-based channel SGMOSFET is extensively studied in order to enhance the device performance [17–19]. One of the emergent charge based on–planer InP/InGaAs/InP surrounding gate heterostructure MOSFETs have been introduced due to the larger gate control mechanism along the channel [20]. The δ-doped epitaxial structure can be created by using different mechanisms such as Metal Organic Vapor Phase Epitaxy (MOVPE) and Molecular Beam Epitaxy (MBE) [21-22]. The δ-region can be made by the industry manufacturing process, which shows excessive field affect adaptability, sensible doping reproducibility and appealing soundness [23-24].

In this work, a SGHMOSFET has been designed with a δ-doped region inserted in the source side. The δ-doped region is formed exactly 12nm away from the channel interface in order to obtain the optimum results. There are various analog performance metrics like electron velocity, electron temperature, surface potential, electric field, On current, peak transconductance and output conductance are estimated by using 2D Sentaurus TCAD device simulator. From the simulation, it reveals that as the channel is incorporated by the multifaceted gate with the δ-doped region in the source end, there exists a superior electrostatic channel control. It helps to enhance the device performance as compared to conventional SGHMOSFET.

II. DEVICE STRUCTURE AND SIMULATION

Fig.1. (a) and (b) indicates the 2D cross-sectional view of δ-doped SG $In_{0.53}Ga_{0.47}As/InP$ hetero MOSFET and a 3D simulated view of SGMOSFET respectively. In the proposed structure the work function of the gate material is considered as 5.01eV. The channel region of SGHMOSFET consists of a narrow band $In_{0.53}Ga_{0.47}As$ layer (3nm) sandwiched between two wideband InP layers with a thickness of 3nm each. The channel is confined near the heterostructure interface because InGaAs is lattice matched with the InP layer, which minimizes the traps at the interface. The length of source/drain is of 20 nm each and doped with an n-type impurity concentration of

10^{20} cm^{-3}. The δ-doped (n++) layer, thickness of 2nm with n-type impurity and concentration of 10^{21} cm^{-3} is inserted in the source region. The layer is placed exactly 12 nm away from the channel interface to get optimum result. The channel length is considered as 60nm. The InGaAs and InP layers of the hetero body are kept undoped to minimize the mobility degradation. Here the equivalent oxide thickness (EOT) of HfO$_2$ is considered with a thickness of 1.2nm. Using 3D Sentaurus TCAD device simulator a comparative valuation has been carried out between conventional and δ-doped SGHMOSFET. In the proposed SGHMOSFET, hydrodynamic model and carrier transport model are considered as the drift-diffusion model fails to clarify the speed overshoot and charge carrier dissemination. Though this model provides accuracy using SRH (Shockley-Red-Hall), but low field mobility (CONMOB), Auger recombination, parallel electric-field-dependent (FLDMOB), Shockley-Read-Hall (SRH) models are required to realize better device performance. The device dimensions, doping profile and the physical properties of the wideband and narrowband materials are listed in Table-1 and Table-2 respectively.

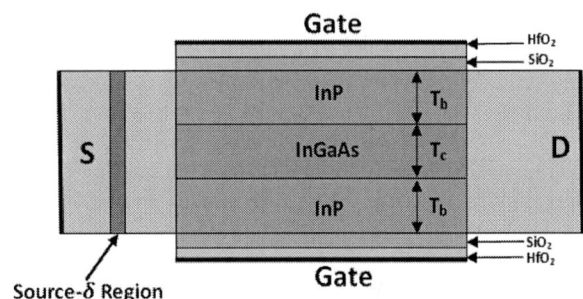

(a)

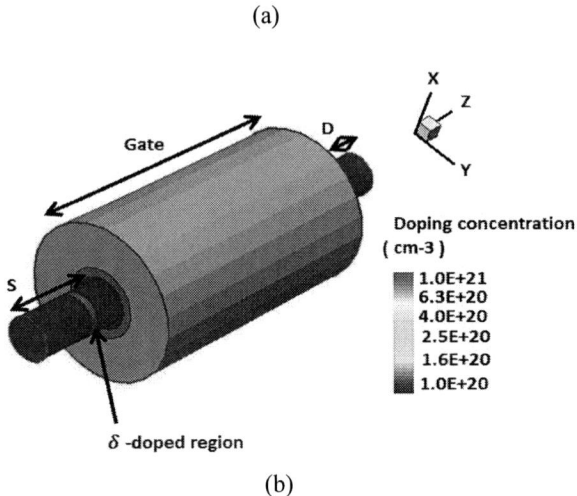

(b)

Fig.1. (a) 2D cross-sectional (b) 3D simulated view of the δ-doped SGHMOSFET

RESULTS AND DISCUSSION

A. Electron characteristics

To improve the charge density along the channel, the electron mobility at the source end must be larger than the drain end. The insertion of the δ-doped layer at the source region enhances the electron transport efficiency, hence increases the device drive current ability. Fig. 2 (a) and (b) indicates the graph between electron mobility and electron velocity with respect to position along the channel (L) for V_{gs}=1.0V and V_{ds}=0.5V respectively. From the figure it is seen that, δ-doped SGHMOSFET shows better charge transport efficiency and high On current ability as compared to the conventional SGHMOSFET, because of existence of additional intensity region at the source end. Fig. 2(b) demonstrates the reduced electron velocity of the δ-doped SGHMOSFET at the drain edge. It shows that the proposed structure is less articulate to hot carrier effects.

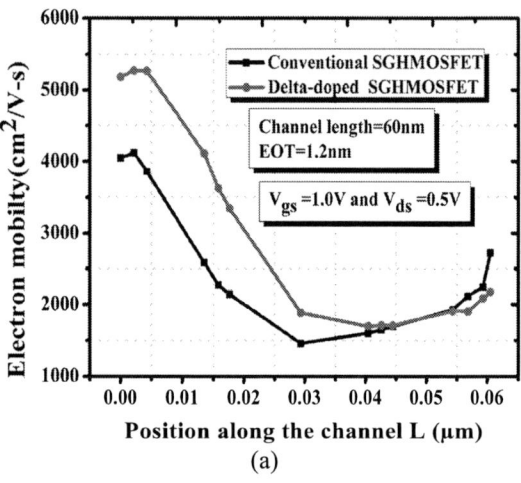

(a)

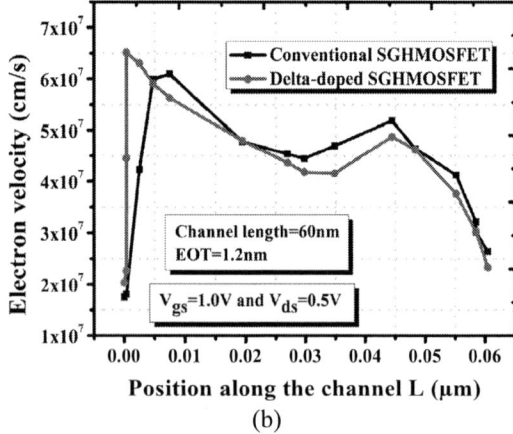

(b)

Fig.2. (a) electron mobility (b) electron velocity versus position along the channel for conventional and δ-doped SGHMOSFET

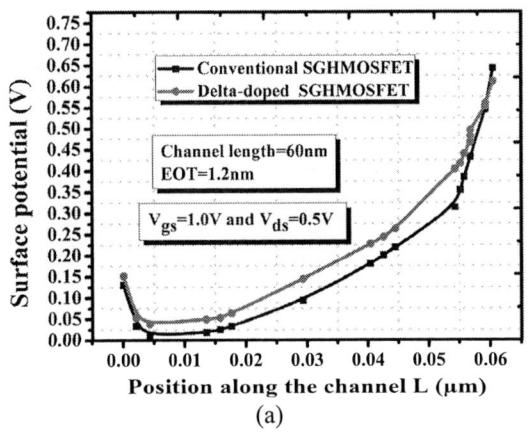

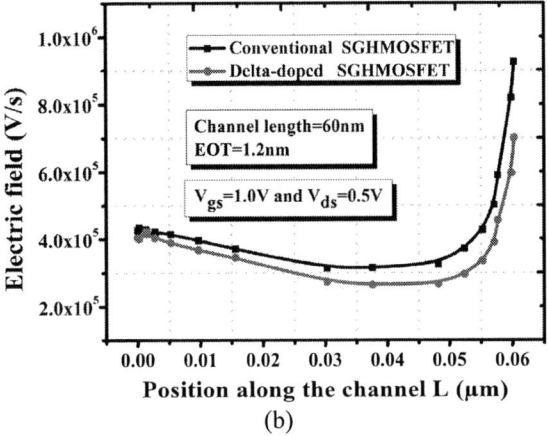

Fig.3. (a) electrostatic potential (b) electric field versus position along the channel for conventional and δ-doped SGHMOSFET

Fig. 3 (a) and (b) shows the surface potential and electric field with respect to position along the channel length (L) for V_{gs} =1.0V and V_{ds} =0.5V respectively. From the graph, it is clearly seen that the proposed structure gives better electrostatic control along the source and channel due to the additional doping, which causes the least electrostatic potential towards the source end. Similarly, the low peak electric field at the drain end offers improved drain induced barrier lowering (DIBL) and sub-threshold slope (SS).

B. Drain characteristics

To investigate the analog performance, the input and output characteristics of both the structures have been studied. Fig. 4(a) and (b) shows the variation of drain current (I_d) and transconductance (g_m) with respect to gate voltage for the proposed and conventional SGHMOSFET with V_{ds}=0.5V. With the incorporation of the delta-doped region in the source end enhances the carrier density along the channel region, which ultimately improves the drive current. Improved drive current gives rise to higher transconductance as compared to conventional SGMOSHFET.

Fig. 5(a) and (b) shows the variation of drain current and output conductance (g_{ds}) with respect to drain voltage for the proposed and conventional DGSGHMOSFET. Here the effect of V_{ds} on drain current of the proposed model is analyzed with a finite slope, which can be clearly reflected in the saturation region. This arises due to the negligible channel length modulation. Thus with the high-intensity region at the source side of the proposed model achieves higher drain current and lower output conductance in the subthreshold region as compared to conventional SGHMOSFET. This indicates that the device is more suitable for high power and high switching operation.

TABLE 1. Dimensions of the device and doping concentration

Device Parameters	δ-doped SGHMOSFET
Source/Drain region extensions	20nm
Source δ -doped region	2nm
Channel length	60 nm
Narrowband layer thickness (T_c)	3nm
Each wideband layer thickness (T_b)	3nm
Source/Drain doping	$1*10^{20}$cm^{-3}
Delta doping region	$1*10^{21}$ cm^{-3}

TABLE 2. Dimensions of the structure with a doping profile

Parameters	In$_{0.53}$Ga$_{0.47}$As	InP
Lattice-matched constant (A^0)	5.8688	5.8688
Permittivity(ϵ_r)	13.94	12.50
V_{sat} (cm/sec)	$2.5*10^7$	$1*10^7$
E_G (eV)	0.75	1.35
μ_n (cm^2/V-sec)	5400	2000
μ_p (cm^2/V-sec)	240	100

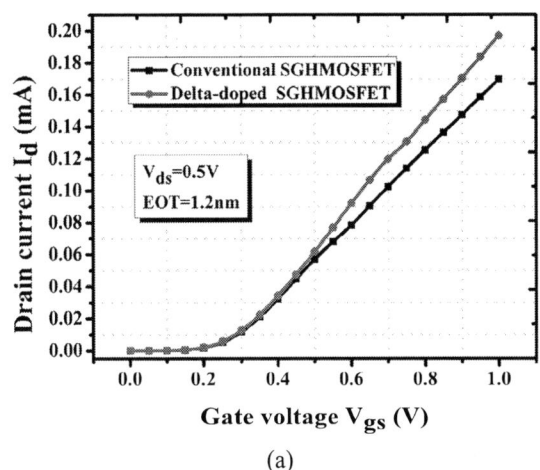

(a)

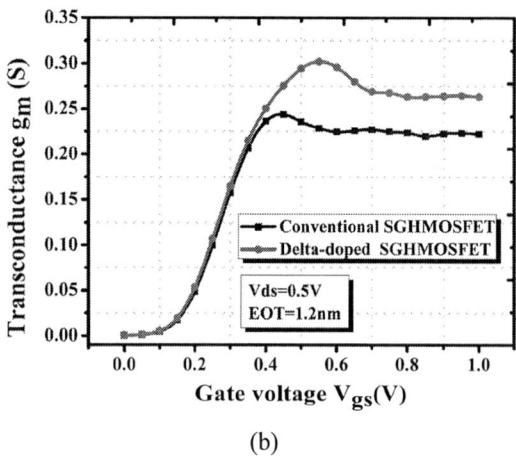

(b)

Fig.4. (a) transfer characteristics (b) transconductance versus gate voltage for conventional and δ-doped SGHMOSFET

.

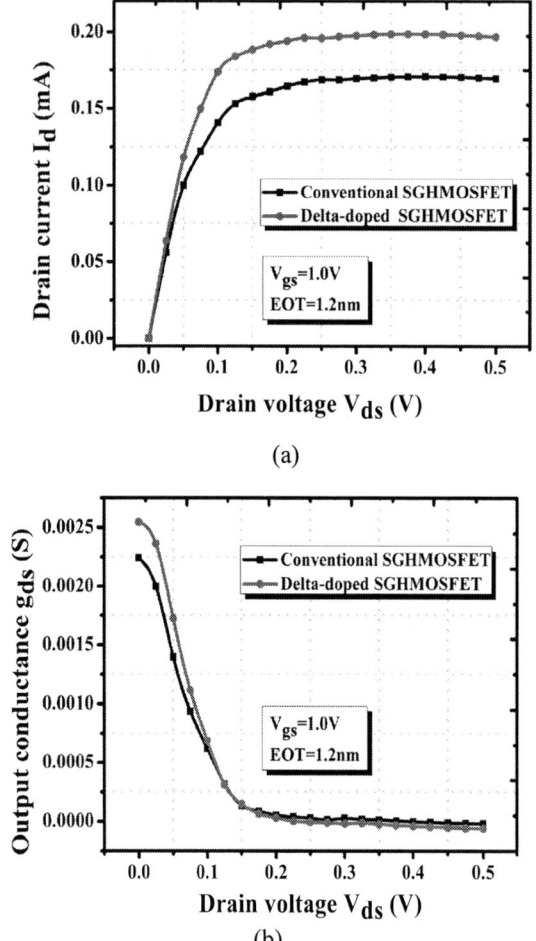

(a)

(b)

Fig.5. (a) output characteristics (b) output conductance versus drain voltage for conventional and δ-doped SGHMOSFET

CONCLUSION

In the proposed work, the introduction of the δ-doped layer at the source end of SGHMOSFET is studied with 3D Sentaurus TCAD device simulator. It is revealed that an additional δ-doped layer in the source end enhances the device performance metrics in terms of electron mobility, electron velocity, electric field, On current and transconductance. The δ-doped SGHMOSFET provides superior carrier mobility and electron velocity. An amazing On current of 0.2mA is achieved with proposed structure, which significantly exhibits the improvements of transconductance and intrinsic gain. Moreover, the shifting of minimum surface potential towards the source side of the proposed structure contributes to the improvement of subthreshold regime performance parameters like DIBL and SS.

REFERENCES

[1] ITRS, International Technology Roadmap for Semiconductors, 2007, Available at http://www.itrs.org

[2] J.M. Seminario, P.A. Derosa, L.E. Cordova, and B.H. Bozard, "A molecular device operating at terahertz frequencies: theoretical simulations," IEEE Trans. Nanotechnol, vol.3, pp. 215–218, 2004.

[3] Y. Tao, J. He, X. Zhang, T.Y. Man, and M. Chan, "Full-band quantum transport based simulation for carbon nanotube field effect transistor from chirality to device performance," Mol. Simul,vol.34.pp.73–80, 2008.

[4] C.P. Auth and J.D. Plummer, "Scaling theory for cylindrical, fully depleted, surrounding-gate MOSFETs," IEEE Electron. Dev. Lett. , vol.18, pp. 74–76, 1997.

[5] S.H. Oh, D. Monore, and J.M. Hergenrother, " Analytical description of short-channel effects in fully-depleted double-gate and cylindrical, surrounding-gate MOSFETs," IEEE Electron. Dev. Lett., vol. 21, pp. 445–447, 2000.

[6] F.L. Yang, D.H. Lee, H.Y. Chen, C.Y. Chang, S.D. Liu, C.C. Huang, T.X. Chung, H.W. Chen, C.C. Huang, Y.H. Liu et al., "5 nm-Gate nanowires FinFET, "VLSI Symp. Tech. Dig., vol. 5, pp. 196–197, 2004.

[7] N. Singh, A. Agarwal, L.K. Bera, T.Y. Liow, R. Yang, S.C. Rustagi, C.H. Tung, R. Kumar, G.Q. Lo, N. Balasubramanian, and D.-L. Kwong, "High-performance fully depleted silicon nanowire (diameter # 5 nm) gate-all-around CMOS devices," IEEE Electron. Dev. Lett., vol. 27, pp. 383–386, 2006.

[8] B. Yang, K. Buddharaju, S. Teo, N. Sing, G. Lo, and D. Kwong, "Vertical silicon–nanowire formation and gate-all-around MOSFETs," IEEE Electron. Dev. Lett., vol. 29, pp. 791–794, 2008.

[9] J.P. Colinge, "Multiple-gate SOI MOSFETs," Solid State Electron., vol.48, pp. 897–905. 2004.

[10] S. Oktyabrsky, and D.Y. Peide, "Fundamentals of III–V Semiconductor MOSFETs," Springer, New York, 2010.

[11] A.Sarkar, A. K. Das, Swapnadip, and C. K. Sarkar, "Effect of gate engineering in double-gate MOSFETs for analog/RF applications," Microelectronics Journal, vol 43, pp. 873–882, 2012.

[12] R. Chau, S. Datta, and A. Majumdar, "Opportunities and Challenges of III-V Nanoelectronics for Future High-Speed, Low-Power Logic Applications Compound Semiconductor Integrated Circuit Symposium," Palm Springs, CA, 2005.

[13] S. K. Pati, H . Perdeshi , C.K.Sarkar, "Impact of gate length and barrier thickness on the performance of InP/InGaAs Based Double Gate Metal-Oxide- Semiconductor Heterostructure Field- Effect Transistor (DG MOSHFET)," Superlattices and Microstructures, vol.55, pp.124001-6, 2013.

[14] M. Alomari, F. Medjdoub, J. F., Carlinet ,"InAlN/GaN MOSHEMT with self-aligned thermally generated oxide recess," IEEE Electron Device Lett, vol.30, pp.1131, 2009.

[15] C.Robert, B. Doyle, and S. Datta, " Integrated nanoelectronics for the future," Nat Mater, vol.6, no 11, pp.810–20, 2007.

[16] A. Sarkar and R.Jana, "The influence of gate underlap on analog and RF performance of III–V heterostructure double gate MOSFET, "Superlattices and Microstructures, vol. 73, pp.256–267, 2014.

[17] A. Pal, A. Sarkar, "Analytical study of Dual Material Surrounding Gate MOSFET to suppress short-channel effects (SCEs), "Engineering Science and Technology, an International Journal, vol.17, pp.205-212, 2014.

[18] N. Gupta, J. B. Patel, and A. K. Raghav, "A Study of Traditional and Surrounding Gate MOSFET using TCAD Simulations," Indian Journal of Science and Technology, vol 9, DOI: 10.17485/ijst/2016/v9i47/101747, October 2016.

[19] S. K. Sharma, B. Raj, and M. Khosla, "A Gaussian approach for the analytical subthreshold current model of cylindrical nanowire FET with quantum mechanical effects," MicroelectronicsJournal, vol.53, pp.65–72, 2016.

[20] E. G. Marin, F. G. Ruiz, I. M. Tienda-Luna, A. Godoy, P. Sánchez-Moreno et al. , "Analytic potential and charge model for III-V surrounding gate metal-oxide-semiconductor-field-effect transistors,"J. Appl. Phys. 112, 084512 (2012); doi: 10.1063/1.4759275.

[21] E. F. Schubert, "Delta doping of III-V compound semiconductors: Fundamentals and device applications." J. Vac. Sci. Technol. A, vol. 8, pp.2980-2996, 1990.

[22] B. Sciana, D. Radziewicz, B. Paszkiewicz, M. Tlaczala, M. Utko, P. Sitarek, G. Sek, J. Misiewicz, R. Kinder, J. Kovac, R. Srnanek, "MOVPE technology and characterization of silicon >-doped GaAs and AlxGa1-xAs," Thin Solid Films, vol. 412: pp.55-59, 2002.

[23] P. Kim, K. M Lee, E. W Lee, Y. Jo, D. H Kim, H. J Kim, K. Y Yang, H. Son, H. C Choi, "A delta-doped amorphous silicon thin-film transistor with high mobility and stability," Journal of the Korean Physical Society, vol. 61, pp.1835-1839, 2012.

[24] Schubert EF. Delta doping of semiconductors, Springer 1996.

2019 Devices for Integrated Circuit (DevIC), 23-24 March, 2019, Kalyani, India

Performance analysis of optical logic XOR gate using dual-control Tera Hertz Optical Asymmetric Demultiplexer (DCTOAD)

K . Maji
Post Graduate Department of Physics
B.B College Asansol
Asansol, India
kajalmaji200@gmail.com

K . Mukherjee
Post Graduate Department of Physics
B.B College Asansol
Asansol, India
klmukherjee003@gmail.com

Abstract— **In this paper we have proposed performance of optical logic XOR gate using dual control Tera Hertz Optical Asymmetric Demultiplexer (DCTOAD). For the first time DCTOAD based XOR gate with soliton pulse is proposed and analyzed in terms of eye diagram and quality factor. Extinction Ratio (ER), Contrast Ratio(CR) and relative eye opening are also calculated. High Q factor implies bit error free operation.**

Keywords—XOR gate, Dual-control TOAD, Q value, Gain saturation, cross gain modulation, Interferometric structure.

I. INTRODUCTION

Semiconductor Optical Amplifier (SOA) based switching has attracted a large number of researchers in optical computation and signal processing again as an alternative technique in last few years [1-6]. And owing to this in last few years SOA based logic processors have been proposed using varieties of switching mechanism four wave mixing, non linear polarization rotation, cross gain modulation etc[1-9]. X-OR gate is an important logic gate finds application in many areas of computation and communication systems [8,9]. TOAD based optical switching is an important candidate in this category due to its high speed, reasonable noise power, ease of integration etc [4]. Most of the TOAD based logic uses single SOA in the fiber loop[1-5,7,10].In this communication, we have analyzed double SOA based TOAD or dual control TOAD (DCTOAD) and its switching performance by characterizing gain and output powers and pseudo eye diagram of the proposed X-OR gate(controlled NOT gate). Extinction ratio and Contrast ratio, relative eye opening are also calculated. Moreover, use of soliton pulses as control results output as soliton pulses, which may be useful for long distance communication network. This DCTOAD based switch will also find applications in future complex logic processors and networks.

II. THEORETICAL ANALYSIS:

The XOR gate consists of a loop mirror with two identical SOAs(both having offset from the loop midpoint by an amount Δ) biased with same current, two control signal input points(A and B), a 2x2 coupler, and proper circulator, filters to introduce data signal and to extract output. The switch works in such a way that the output signals can be switched from port-1 to port-2 by the application of control signals. The data signal after entering the loop, is divided into two counter propagating components, Pcc and Pc. When the both controls are absent both counters - clockwise (Pcc) and clockwise (Pc) components of data signal P_{data} experience the same unsaturated gain in each SOA. In this situation, there is no phase difference between these two counter propagating components. Therefore, the data signal comes out to the port 2 since there is constructive interference at port 2. Now if any of the control enters into the DCTOAD, one of the SOA gain becomes saturated, and therefore these two counters - clockwise (Pcc) and clockwise (Pc) components experience different gains and a phase shifts, which can be adjusted to give a phase difference π between them. When they interfere at the coupler, data signal comes out of the port 1 as constructive interference happens there. When both the control enter into the dual control-TOAD both of these component experience same unsaturated gains and phase shift resulting zero phase difference between them and data will exit from port 2 again. It is interesting to note that, both the data signal components experience unsaturated SOA gain G_u at least one and maximum two times depending on the conditions of control signal's presence and absence. This enhances the performance of the switching and the logic gate will be clear from the following sections.

The DCTOAD output power at port 1 can be expressed by [2]

$$Y_{XOR} = \frac{G_u P_{data}}{4}\left(G_1 + G_2 - 2\sqrt{G_1 G_2}\,Cos\,\theta\right) \qquad (1)$$

Where G_1 and G_2 are the power gains of the left and right SOA and are functions of time. G_u is the unsaturated gain of the SOAs and θ is the phase difference between counter - clockwise (Pcc) and clockwise (Pc) components when they interfere.

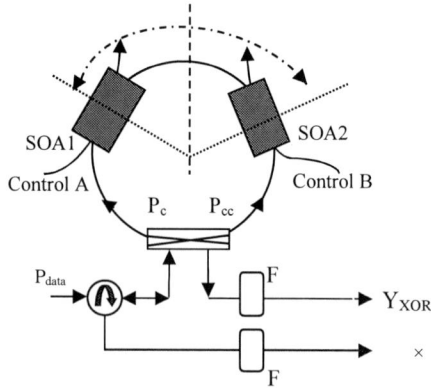

Fig1. Dual-control TOAD based XOR gate

we consider P_{cp} as the power of the soliton pulse [3,4] given by

978-1-5386-6723-1/19 $31.00 © 2019 IEEE

$$P_{cp}(t) = \sum_{n=1}^{n=N} a_{nA,B} P_{soli} \sec h^2\left(1.763\frac{(t-n\chi)}{\tau_{fwhm}}\right) \quad (2)$$

Where $P_{soli} = \left(\dfrac{1.763}{2\pi}\right)^2 \dfrac{A_{eff}\lambda^3 D}{n_2 c \tau_{fwhm}^2}$ is the soliton peak

power. D is the dispersion constant. n_2 is the nonlinear coefficient. λ and c are the wavelength and velocity of light. A_{eff} is the fiber effective area. τ_{fwhm} is the full width half maximum.

$E_{cp}(t \to \infty) = P_{soli} \times \tau_{fwhm} = E_c$ is the total control pulse energy.

Operation principle of XOR gate: For X-OR gate port 1 of the DCTOAD is used.

1. When the both controls A=0 and B=0 only data signal enter into the dual control-TOAD, then the two components Pcc and Pc experience a zero phase difference but high gain. So the data will exit in the port 2 and there is no output power present in the port 1 i.e. Y_{XOR}=0. So this gives '0' logic state
2. When control A=1and B=0 then two components Pc and Pcc experience different gains (G_1 = low as left SOA becomes gain saturated and high as $G_2=G_u$, since right SOA does not receive control signal) and the phase shift adjusted to give a phase difference $\theta=\pi$. When they recombine at the coupler, the data will exit in the port 1 i.e. Y_{XOR}=1. So this gives '1' logic state.
3. When control A=0 and B=1 then the two components Pc and Pcc experience different gains(G_2 = low as right SOA gain saturates and $G_1=G_u$, since left SOA does not receive control signal) and phase shift, so that phase difference $\theta=\pi$ again. In this case also, the data will exit in the port 1 i.e. Y_{XOR}=1. So this gives '1' logic state.
4. When the both controls A=1 and B=1, then the two components Pc and Pcc experience phase shift and a phase difference of $\theta=\pi$. In this case, both the SOAs experience gain saturation.Therefore, the output from the SOAs is low, and when they interfere at the coupler with same frequency. Output power $Y_{XOR}\approx 0$. I.e. low. So this gives '0' logic state.

From the above discussion we have seen that in port 1, the output power generates a bit sequence 0 1 1 0 giving XOR operation.

III.RESULT AND DISCUSSION:

We have used control signal pulse and data signal pulse as a soliton pulse. Figure 3 shows the control signal A, control signal B and the output bit patterns of DCTOAD based XOR gate. The output quality factor or Q value is defined as the ratio of the difference between average output power($P^{1,0}_{mean}$) of '1' and '0' to the sum of the standard deviations($sd_{1,0}$) of the output power '1' and '0'[2].

$$Q = \frac{P^1_{mean} - P^0_{mean}}{sd^1 + sd^0} \quad (3)$$

The output extinction ratio is defined as the ratio of minimum values of peak power of '1' to the maximum values of peak power of '0'[5].

$$E.R. = \frac{\text{Minimum values of peak power of 1}}{\text{Maximum values of peak power of 0}} \quad (4)$$

The output contrast ratio is defined as the ratio of average values of peak power of '1' to the average values of peak power of '0'.[2,5]

$$C.R. = \frac{\text{Average values of peak power of 1}}{\text{Average values of peak power of 0}} \quad (5)$$

The relative eye-opening(REOP) is defined as the ratio of difference between the minimum and maximum output power at '1' and '0' to the minimum output power at '1'[2].

$$REOP = \frac{P^1_{min} - P^0_{max}}{P^1_{min}} \quad (6)$$

Table1. SOA parameters used in simulation

Symbol	Parameters	Value
c	Velocity of light	3×10^8 m/s
Γ	Confinement factor of SOA	0.48
α_N	Differential gain	3.3×10^{-20} m^2
w	Width of the SOA	1.5 μm
d	Depth of the SOA	250 nm
λ	Wave length of light	1550 nm
Gu	Unsaturated single-pass amplifier gain	15dB
E_c	Control pulse energy	50 fJ
τ_{fwhm}	Full width half maximum of control pulse	1.5 ps
n_2	Nonlinear coefficient	2.6×10^{-20} m^2/w
D	Dispersion constant	1 ps/(nm-km)
A_{eff}	Fiber effective area.	5×10^{-13} m^2

Fig2 shows the output pseudo [6] eye diagram of the XOR gate of the dual control TOAD. This shows a clear eye opening with relative eye opening of 97% and is indicator of high Q factor and error less transmission.

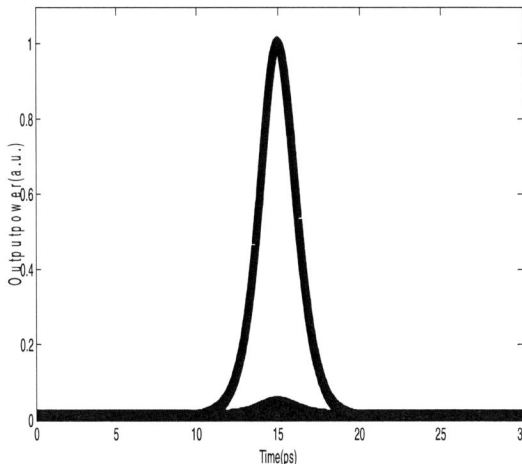

Fig2. Output pseudo-eye-diagram for XOR operation

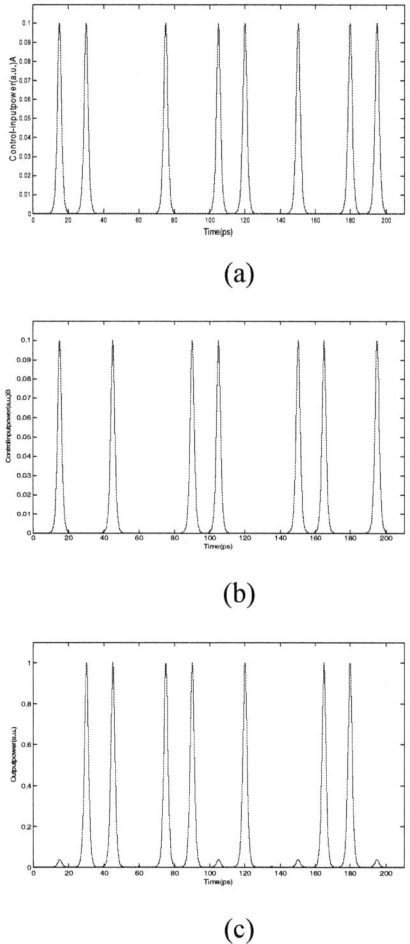

(a)

(b)

(c)

Fig.3 (a) Input power A (b) Input power B (c) Output power
Of the XOR operation of the dual control TOAD

After optimization and using the SOAs parameters unsaturated Gain Gu=15 dB and control pulse energy Ec= 50fJ we found E.R.C.R., Q value and relative eye opening. The corresponding bit error rate (BER) is negligible.

E.R	C.R.	Q value	REOP
13.76 dB	16.77 dB	16.68 dB	97%

CONCLUSIONS

Performance of Optical logic XOR gate using dual-control Tera Hertz Optical Asymmetric Demultiplexer analyzed and characterized for the first time. It is clear from the simulation that the Control pulse energy should be kept at 50fJ for best ER and CR, which are found to be 13.76 dB and 16.77dB respectively. X-OR gate is an important logic finds varieties of application in parity generating/checking, packet switching, comparison, header processing, adder/Subtractor, cryptography etc. In future DCTOAD has potential to become an important candidate for all optical logic and signal processing.

REFERENCES

[1] A.K Mondal, "Full-Optical TOAD based Walsh-Hadamard code generation", Optical and Quantum Electronics, 49,9,(2017)DOI 10.1007/s11082-017-1130-4.

[2] T. Chattoipadhyay, "All-optical clocked delay flip-flop using single terahertz optical asymmetric demultiplexer –based switch: a theoretical study Applied Optics, Vol. 49,No. 28(2010)

[3] K.Maji, K.Mukherjee, A.Raja "Frequency Encoded all-Optical Universal Logic Gates using Terahertz Optical Asymmetric Demultiplexer".IJPOT,Vol. 4,Iss. 3, pp: 1-7, Sep(2018)

[4] K.Maji,K.Mukherjee, A.Raja "Analysis of Tera Hertz Optical Asymmetric Demultiplexer(TOAD)based Optical Switch using soliton pulse" presented in 1st International conference "2018 IEEE Electron Device Kolkata(2018IEEE EDKCON)" Nov 24-25.

[5] J.N.Roy, D.K.Gayen, "Integrated all-optical logic and arithmetic operations with the help of TOAD based interferometer device – alternative approach", Appl. Opt. 46(22), 5304-5310, (2007).

[6] H. Le Minh, Member, IEEE, Z. Ghassemlooy, Senior Member, IEEE,and Wai Pang Ng,Member, IEEE, "Characterization and Performance Analysis of a TOAD Switch. Employing a Dual Control Pulse Scheme in High-speed OTDM Demultiplexer ", IEEE COMMUNICATIONS LETTERS, VOL. 12, NO. 4, APRIL 2008

[7] Saha,S, Manna, A, Bandyopadhyay,p, Rahaman, H, "All optical design of hybrid adder circuit using terahertz optical asymmetric demultiplexer", 2018 International Symposium on Devices, Circuits and Systems (ISDCS), 29-31 March 2018 ,DOI: 10.1109/ISDCS.2018.8379631

[8] K. Mukherjee, Frequency encoded optical four bit adder/ Subtractor control input using semiconductor optical amplifier, Optik, 125,20,6183(2014).

[9] Dimitriadou, K. E. Zoiros, All optical XOR gate using single quantum dot SOA and optical filter Journal, Of Lightwave Technology, 31, 3813,(2013)

[10] K.E. Zoiros, G. Papadopoulos, T. Houbavlis, G.T. Kanellos, "Theoretical analysis and performance investigation of ultrafast all-optical Boolean XOR gate with semiconductor optical amplifier-assisted Sagnac interferometer", Optics Communications 258 (2006) 114- 134

Nonmonotonous Electron Mobility in Double Quantum Well Pseudomorphic High Electron Mobility Transistor Structure

Sangita R. Panda, Sudhakar Das, Arttatran Sahu, Ajit Kumar Panda and Trinath Sahu

Department of Electronics and Communication Engineering
National Institute of Science and Technology
Palur Hills, Berhampur-761 008, Odisha, India
email: tsahu_bu@rediffmail.com

Abstract— Non-monotonic mobility μ of electrons is obtained in a pseudomorphic GaAs/In$_x$Ga$_{1-x}$As high electron mobility transistor having a double quantum well structure with asymmetric doping concentrations in the outer barriers. A dip in μ occurs at the resonance of subband states because of the shift of the subband wave functions, which affects the subband mobilities limited by interface roughness scattering. The dip in μ amplifies with increase in the asymmetry of the width of wells.

Keywords—GaAs/InGaAs p-HEMT structures, asymmetric double quantum well, multi-subband electron mobility, inter-subband scattering

I. INTRODUCTION

The InGaAs based pseudomorphic high electron mobility transistors (p-HEMTs) have been extensively used in microwave and millimeter range devices [1, 2]. The p-HEMTs support high power gain, low noise and exceptional cut-off frequency and also exhibit excellent logic characteristics [2-6]. The p-HEMTs with delta-doped channels show good performance in high speed digital applications [6-8]. The outcome of a device is based upon the enhanced 2D-electron density and mobility μ. Several efforts have been made for the study of μ adopting pseudomorphic InGaAs quantum wells by altering the well widths and doping concentrations [8-10]. In a multi-subband occupied system, μ substantially depends on the inter-subband effects [11-19].

A double quantum well (DQW) structure demonstrates the effects of quantum size and tunnelling coupling [17-18]. When the structure parameters and doping concentrations of the DQW are symmetric, the potentials of the wells are also symmetric. The energy levels of both the wells of the DQW display resonance. The electron subband wave functions has the equal extension into the wells. The lowest level wave function ψ_0 extends symmetrically and the first excited subband wave function ψ_1 extends antisymmetrically. However, a minor change in the potential destroys the resonance. The subband wave functions lie in individual wells separately [15]. Therefore, change in structure parameters displaces the subband wave functions between the wells. This affects the scattering potentials and hence the mobility.

In this paper, we attempt to study the doping concentration dependence of μ in a GaAs/InGaAs strained DQW around the resonance of the subband states. The doping concentration ND1 lies in the substrate side barrier and ND2 lies towards the surface. We keep the sum of ND1 and ND2 unchanged. We analyze μ by varying ND1. We find that when the well widths ww1 and ww2 of the wells are same the resonance occur at ND1 = ND2 which causes lowering of μ. When the well widths ww1 ≠ ww2, the system is asymmetric at ND1 = ND2 and therefore, the system behaves as potentially symmetric at around another point ND1 ≠ ND2. The subband wave functions in the wells change significantly around resonance. The subband energy levels also exhibit anti-crossing near resonance. Accordingly, the scattering potentials show discontinuity causing a sudden dip in μ. We take alloy disorder scattering (ADS), ionized impurity scattering (IIS) and interface roughness scattering (IRS) and find that the IRS mostly causes the dip in μ. The nonlinearity of μ is more when there are unequal well widths. Our results can be used for the improvement of the performance of DQW based p-HEMT devices.

II. THEORY

We consider GaAs/In$_x$Ga$_{1-x}$As pseudomorphic DQW structure. There is delta-doping in the outer barriers of width dw in the structure. ND1 is doping concentrations in the substrate side and ND2 towards the surface. sw is the spacer width. ww1 and ww2 are well widths and bw is the width of the barrier in the centre (Fig. 1 (a)).

The impurity distribution N$_D$ (z) along z-axis is:

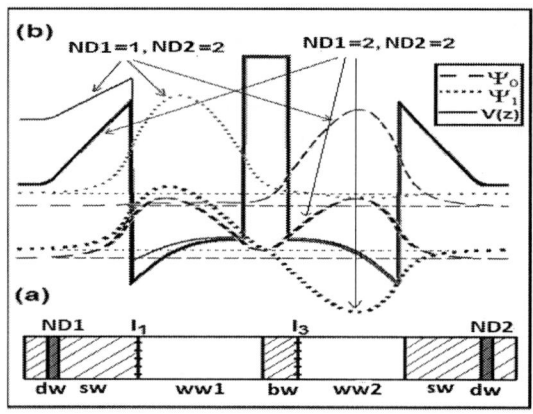

Fig. 1. (a): Schematic diagram of the DQW structure. (b): The potential structure, subband energy levels and wave functions for (i) ND1 = 1 × 10^{18} cm^{-3}, ND2 = 2 × 10^{18} cm^{-3}, and (ii) ND1 = ND2 = 2 × 10^{18} cm^{-3} taking ww1 = ww2 =150 Å, bw = 60 Å, sw = 100 Å and dw = 20 Å.

$$N_D(z) = \begin{cases} ND1 & -(dw+sw+ww1+bw/2)<z<-(sw+ww1+bw/2) \\ ND2 & (bw/2+ww1+sw)<z<(bw/2+ww1+sw+dw) \\ 0 & \text{Otherwise} \end{cases}$$

(1)

At absolute zero K, the 2D-electron density N_s and the Fermi energy E_F are related [12-14]. We got the energy levels E_n and wave functions ψ_n basing upon the selfconsistent solution of the Poissons's and Schrödinger equations. In Fig. 1(b), we show ψ_0 and ψ_1 for a symmetric as well as asymmetric DQW potential structure. For symmetric case both ψ_0 and ψ_1 equally penetrate into the wells. For asymmetric DQW profile, ψ_0 lies in one well while ψ_1 lies in another well.

For a multi-subband occupied structure, the subband relaxation time τ_n is attained through the Boltzmann transport equation [12, 14]. For the occupation of two lowest subbands (n = 0, 1), τ_0 and τ_1 can be expressed through the intra-subband, and inter-subband scattering rate matrix elements (SRME) [14]. The SRME can be written using screened scattering potential $V_{nm}^{eff}(q)$ obtained through random phase approximation (RPA) [14].

The screened IIS, IRS and ADS potentials are expressed as:

$$\left|V_{nm}^{IIS}(q)\right|^2 = \frac{4\pi^2 e^4}{\varepsilon_0^2 q^2}\left[ND1\int_{-(dw+sw+ww1+bw/2)}^{-(sw+ww1+bw/2)} dz_i \left|\sum_{n'm'}\varepsilon_{nm,n'm'}^{-1}(q)S_{n'm'}(q,z_i)\right|^2\right.$$

$$\left.+ND2\int_{(bw/2+sw+ww2)}^{(dw+bw/2+sw+ww2)} dz_i \left|\sum_{n'm'}\varepsilon_{nm,n'm'}^{-1}(q)S_{n'm'}(q,z_i)\right|^2\right]$$

(2)

$$\left|V_{nm}^{IRS}(q)\right|^2 = V_b^2\,\pi\Lambda^2\,\Delta^2\,e^{-q^2\Lambda^2/4}\left|\left.\sum_{n'm'}\psi_{n'}^*(z)\psi_{m'}(z)\right|_{z=z_1}\varepsilon_{nm,n'm'}^{-1}(q)\right|^2$$

(3)

$$\left|V_{nm}^{ADS}(q)\right|^2 = \left[a^3(\delta V)^2 x(1-x)/4\right]\times\int dz\left|\sum_{n'm'}\psi_{n'}(z)\psi_{m'}(z)\varepsilon_{nm,n'm'}^{-1}(q)\right|^2$$

(4)

where

$$S_{n'm'}(q,z_i) = \int_{-\infty}^{\infty} dz\,\psi_{n'}(z)\psi_{m'}(z)e^{-q|z-z_i|}$$

(5)

x, 'a', and δV are the alloy fraction, lattice constant, and the alloy disorder scattering potential respectively. Δ and Λ are the parameters of IRS. The substrate side interfaces (I1 and I3) are rough surfaces [11]. The subband mobility and transport lifetime are related as $\mu_n(E_F) = (e/m)\tau_n(E_F)$. The mobility for any scattering mechanism (I = IIS, IRS ,ADS) is obtained using $\mu^I = \Sigma_n\, n_n\mu_n^I/\Sigma_n\, n_n$. The total mobility μ is determined by adopting Matthiessen's relation.

III. RESULTS AND DISCUSSION

We calculate the effect of asymmetric doping concentration on the electron mobility of GaAs/In$_x$Ga$_{1-x}$As pseudomorphic DQW structure. For x = 0.2, the potential barrier height including strain V_0 = 140 meV [19]. The effective mass of the electron m = 0.62m$_0$ [19]. The parameters, Δ = 2.83 Å, Λ = 100 Å [19] and δV = 530 meV [19]. The structure parameters sw = 100 Å, and dw = 20 Å.

In Fig. 2, we plot the total mobility μ as functions of ND1 for ww1 = ww2 = 150 Å and ww1 = 100 Å, ww2 = 200 Å taking ND1 + ND2 = 3 × 10^{18} cm^{-3}. The barrier width bw = 40 Å. We vary ND1 from 1 to 2 × 10^{18} cm^{-3}. For these structure parameters, double-subband is occupied all through the range of ND1. We show the mobilities μ^{IIS}, μ^{ADS} and μ^{IRS} in the inset of the figure. We show that μ is due to both IIS and ADS because of the alloy channel. There is almost no change in μ^{IIS} and μ^{ADS} with increase in ND1. Whereas, there is a remarkable change in μ^{IIS}, which causes a dip near resonance through inter-subband effects. In case of ww1 = ww2 = 150 Å, the system is symmetric at ND1 = ND2 = 1.5 × 10^{18} cm^{-3}. Whereas, for ww1 = 100 Å and ww2 = 200 Å, the asymmetry in the potential due to unequal well widths is compensated by varying the doping concentrations. The potential symmetry is achieved around ND1 = 1.9 × 10^{18} cm^{-3} leading to the resonance of subband states. The magnitude of the dip in μ increases considerably.

In Fig. 3, we plot μ_0^{IRS} and μ_1^{IRS} for ww1 = ww2 = 150 Å. There occurs a substantial change in μ_0^{IRS} and μ_1^{IRS} around ND1 = 1.5 × 10^{18} cm^{-3} because of the symmetric potential. The inset figure shows the anti-crossing of the energy levels E_0 and E_1 near that point. At ND1 = ND2, both ψ_0 and ψ_1 almost equally extend into the wells. However, for ND1 < 1.5 × 10^{18} cm^{-3} the wave functions ψ_1 almost lies in the substrate side well while the wave functions ψ_0 lies in the surface side well. The interface towards the substrate being rough, we have $\mu_1^{IRS} < \mu_0^{IRS}$. However, once ND1 > 1.5×10^{18} cm^{-3}, the wave functions swap between the wells leading to $\mu_0^{IRS} < \mu_1^{IRS}$. Such change in

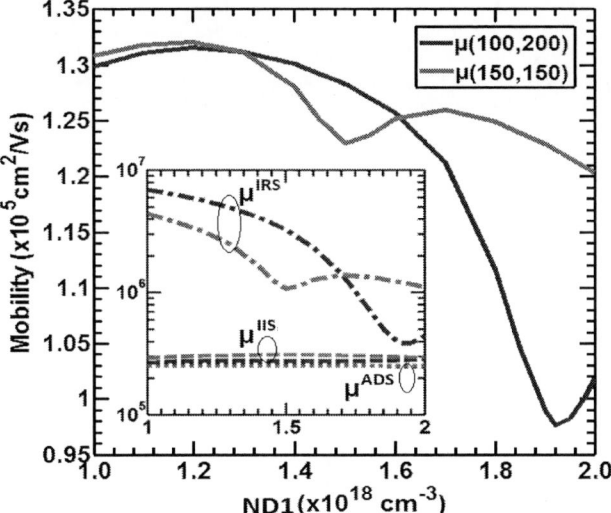

Fig. 2. Mobility μ versus ND1 for ND1 + ND2 = 3×10^{18} cm^{-3} and bw = 40 Å for (ww1 = ww2 = 150 Å) and (ww1 = 100 Å and ww2 = 200 Å) in red and blue colour lines, respectively. μ (100,200) represents mobility for well widths ww1=100 Å and ww2=200 Å and μ(150,150) means mobility for ww1=ww2=150 Å. The inset figure shows the mobilities μ^{IIS}, μ^{ADS} and μ^{IRS} for above mentioned parameters

.

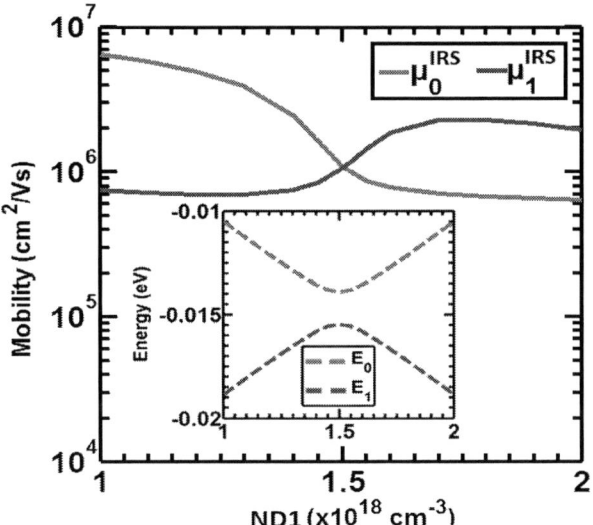

Fig. 3. Mobilities μ_0^{IRS} and μ_1^{IRS} versus ND1 taking ND1 + ND2 = 3×10^{18} cm^{-3} and bw = 40 Å for ww1 = ww2= 150 Å. Inset figure shows the energy eigen values E$_0$ and E$_1$ for the above parameters.

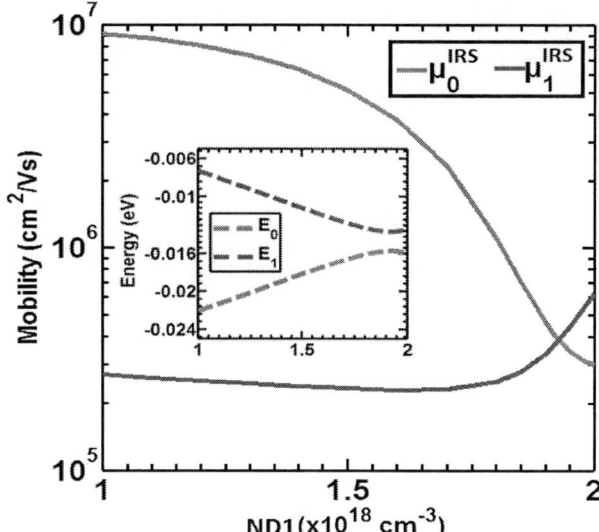

Fig. 4. Mobilities μ_0^{IRS} and μ_1^{IRS} versus ND1 taking ND1 + ND2 = 3×10^{18} cm^{-3} and bw = 40 Å for ww1 = 100 Å and ww2 = 200 Å. Inset figure shows the energy eigen values E$_0$ and E$_1$ for the above parameters.

The subband mobilities at the point of symmetry of potential causes a dip in μ^{IRS} and hence a dip in μ.

In Fig. 4, we describe μ_0^{IRS} and μ_1^{IRS} for ww1 = 100 Å and ww2 = 200 Å. In the inset figure, we have shown the anti-crossing of the corresponding energy levels. We note that there is considerable change in μ_0^{IRS} compared to μ_1^{IRS} due to the sharp variation of ψ_0 and ψ_1 near resonance leading to larger drop in μ, compared to the case of equal well widths.

IV. CONCLUSION

This work analyzes the influence of asymmetric doping concentrations on the non-monotonic electron mobility μ in a GaAs/InGaAs pseudomorphic double quantum well (DQW) high electron mobility transistor structure with delta doped outer barriers. We analyze μ as a function of asymmetry in doping concentration and show that a dip in μ occurs near the resonance of subband states because of the shifting of subband wave functions between the wells. The dip in μ is larger when the differences in the widths of the wells are more. The results of our work can be used to study the effect of fine tuning of μ near resonance for the improvement of the performance of p-HEMT devices.

REFERENCES

[1] S. J. Mahon, A. Dadello, A. P. Fattorini, A. Bessemoulin, and J. T. Harvey, "35 dBm, 35 GHz power amplifier MMICs using 6-inch GaAs pHEMT commercial technology," IEEE MTT-S Int Microwave Symp. Dig., pp. 855–858, 2008.

[2] S. J. Cho, C. Wang, and N. Y. Kim, "High power density AlGaAs/InGaAs/GaAs PHEMTs using an optimized manufacturing

process for Ka-band applications," Microelectron. Eng., vol. 113, pp.11–19, 2014.

[3] Z. Wang, J. Mou, W. Yu, and X. Lv, "Design of double delta-doped $Al_{0.22}Ga_{0.78}As/In_{0.22}Ga_{0.78}As$ pseudomorphic HEMTs," Appl. Mech. Mater. vol. 229-231, pp. 2007–2009, 2012.

[4] K. Kalna, and A. Asenov, "Role of multiple delta doping in PHEMTs scaled to sub-100 nm dimensions," Solid State Electron., vol. 48, pp. 1223–1232, 2014.

[5] D. H. Kim, J. A. del Alamo, J. H. Lee, and K. S. Seo, "Logic suitability of 50-nm $In_{0.7}Ga_{0.3}As$ HEMTs for beyond-CMOS applications," IEEE Trans. Electron Devices, vol. 54, pp. 2606–2613, 2007.

[6] J. Lin, T. W. Kim, D. A. Antoniadis, and J. A. del Alamo, "A self-aligned InGaAs quantum-well metal-oxide-semiconductor field-effect transistor fabricated through a lift-off-free front-end process," Appl. Phys. Express, vol. 5, pp. 064002, 2012.

[7] D. H. Kim, B. Brar, and J. A. del Alamo, "f_T = 688 GHz and f_{max} = 800 GHz in Lg = 40 nm $In_{0.7}Ga_{0.3}As$ MHEMTs with g_{m_max} > 2.7 mS/μm.," IEDM Tech. Dig., pp. 319–322, 2011.

[8] C. Y. Lee, H. P. Shiao, K. C. Kuo, H. Y. Wu, and W. H. Lin, "Mobility and charge density tuning in double δ-doped pseudomorphic high-electron-mobility transistors grown by metal organic chemical vapor deposition," J. Vac. Sci. Technol. B, vol. 24, pp. 2597–2600, 2006.

[9] K. Y. Chu, S. Y. Cheng, M. H. Chiang, Y. J. Liu, C. C. Huang, T. Y. Chen, C. S. Hsu, W. C. Liu, W. Y. Cheng, and B. C. Lin, "Comprehensive study of InGaP/InGaAs/GaAs dual channel pseudomorphic high electron mobility transistors," Solid State Electron., vol. 72, pp. 22–28, 2012.

[10] D. Y. Protasov, and K. S. Zhuravlev, "The influence of impurity profiles on mobility of two-dimensional electron gas in AlGaAs/InGaAs/GaAs heterostructures modulation-doped by donors and acceptors," Solid State Electron., vol. 129, pp. 66–72, 2017.

[11] D. Y. Protasov, and K. S. Zhuravlev, "The influence of impurity profiles on mobility of two-dimensional electron gas in AlGaAs/InGaAs/GaAs heterostructures modulation-doped by donors and acceptors," Solid State Electron., vol. 129, pp. 66–72, 2017.

[12] T. Sahu, and K. A. Shore, "Multi-interface roughness effects on electron mobility in a $Ga_{0.5}In_{0.5}P/GaAs$ multisubband coupled quantum well structure," Semicond. Sci. Technol., vol. 24, pp. 095021, Aug. 2009.

[13] P.K.Subudhi, S. Palo and T. Sahu, "Effect of strain on multisubband electron transport in GaAs/InGaAs coupled quantum well structures", Superlattices and Microstructures, vol. 51, pp.430-442 , 2012.

[14] Narayan Sahoo, Ajit K Panda and Trinath Sahu, "Enhancement of multisubband electron mobility in square-parabolic asymmetric double quantum well structure", Superlattices and Microstructures, vol. 105, pp-11-21, 2017.

[15] S. Das, R. K. Nayak, A. K. Panda, and T. Sahu, "Multisubband electron mobility in asymmetric GaAs/AlGaAs quantum well structures," Superlattices Microstruct., vol. 66, pp. 39-47, Feb. 2014.

[16] S. Das, R.K. Nayak, T. Sahu, and A.K.Panda, ""Enhancement of electron mobility in asymmetric coupled quantum well structures," J. Appl. Phys., vol. 115, p. 073701, Feb. 2014.

[17] S. Das, M. Mohapatra, R. K. Nayak, A. K. Panda, and T. Sahu, "Improved two-dimensional electron mobility in asymmetric barrier delta-doped GaAs/AlGaAs modulation-doped field-effect transistor structures," Jpn. J. Appl. Phys., vol. 56, pp. 034001, 2017.

[18] Sujaul Chowdhury and Md. Jabed Iqbal, "Nanostructure physics of coupled quantum well : parametric variation of energy spectrum", American Academic Press (2015)

[19] Meryleen Mohapatra, Arttatran Sahu, Sangita R. Panda, Sudhakar Das, Trinath Sahu, and Ajit K. Panda, "Nonlinear electron transport in GaAs/InGaAs asymmetric double-quantum-well pseudomorphic high-electron-mobility transistor structure," Jpn. J. Appl. Phys., vol. 56, p. 064101, May 2017.

978-1-5386-6723-1/19 $31.00 © 2019 IEEE

Performance Prediction of Stacked Nanowire Transistors in the Presence of Random Discrete Dopants and Metal Gate Granularity

S. Dey[1*], E.Mohapatra[2], J.Jena[3], S. Das[4], T. P. Dash[5] and C. K. Maiti [6]

Department of Electronics and Communication Engineering,
Siksha 'O' Anusandhan (Deemed to be University),
Bhubaneswar, Odisha, India--751030

[1*]supravadey@soa.ac.in; [2]eleenamohapatra@soa.ac.in; [3]jhansiranijena@soa.ac.in; [4]sanghamitradas@ soa.ac.in; [5]taradash@soa.ac.in; [6]ckmaiti@soa.ac.in.

Abstract–Gate-all-around nanowire field effect transistors (GAA-NW-FETs) in a horizontal configuration is now being considered as a strong candidate to extend today's CMOS technology to its ultimate scaling limits. In this paper, full 3-D device simulations are performed to study the effect of random discrete dopants (RDD) and metal gate granularity (MGG) on the performance of a 10nm channel length vertically stacked silicon nanowire FETs. The impact of metal grain crystallographic orientation on the gate work function and presence of discrete dopants on transistor threshold voltage is reported. The discrete dopants have been distributed randomly in the source/drain and channel regions of the device. Due to the small dimensions of the transistor a quantum transport formalism has been deployed in simulation. Our results show the magnitude and importance of RDD and MGG and the need for process optimization to minimize device parameter variations in sub-10nm technology nodes.

Keywords- Random discrete dopants, metal grain granularity, Schrodinger-Poisson, Variability

I. INTRODUCTION

Variations in integrated circuits are basically the deviations from the intended performance due to device parameter variations and are of serious concern for circuit designers [1].Usually, the variations in devices are caused by various physical factors which cause a permanent variation in device parameters. Variations are generally caused by the lack of exact process control during the fabrication and are statistical in nature.

Process variations has impact on the device structure and thus, they alter the circuit electrical performance. The main process variation sources can be outlined as follows: Random Dopants Fluctuations (RDF), Metal Gate Granularity (MGG), Channel Length Variations, Gate Oxide Thickness Variations, Line Edge Roughness (LER) and Channel Width Variations. LER refers to the roughness introduced on the channel edge during the gate patterning. Transistor channel width (W) will have variations as well due to the lithography limitations. The channel width variations will contribute to V_{th} variations due to the Narrow-Width Effects (NWEs) which is, however, minimal compared to the impact due to gate length (L) variation since width is typically 3-4 times larger than L.

With the shrinking of device size, the reliability and performance of Si-based MOSFETs are of serious concern. With CMOS technology scaling, the number of dopant impurities in the channel depletion layer decreases, especially with nanoscale devices. The average number of dopants in a $10x 5x 10\ nm^3$ volume with $10^{20}\ cm^{-3}$ doping concentration is equal to 50. This value is approximately the number of dopants surrounding the channel within a distance of around two Fermi wave lengths of a device with $10x5\ nm^2$ cross section. The atomicity of the dopants in the channel does not allow a constant concentration of dopants to appear across the channel rather as a random discrete dopant. Thus, it is very unlikely to have two neighbouring transistors with the same number and placement of dopants. This random number and arbitrary placement of the dopants cause uncertainty in the transistor threshold voltage, V_{th}. The statistical distribution of V_{th} due to RDF is found to follow a normal distribution [2]. The standard deviation of V_{th} distribution due to RDF is modeled as [3]:

$$\sigma_{V_{th}} = \sqrt[4]{4q^3\varepsilon_{Si}N_a\varphi_F}\frac{T_{ox}}{\varepsilon_{ox}}\frac{1}{\sqrt{2W*L}} \qquad (1)$$

where q represents the electron charge, ε_{ox} and ε_{Si} are the dielectric constants of the gate oxide and silicon respectively, N_a represents the channel dopant concentration, φ_F represents the difference between intrinsic level and Fermi level, T_{ox} represents the gate oxide thickness, and L and W represent the channel length and width, respectively. From equation (1) it is clearly observed that $\sigma_{V_{th}}$ is varying inversely proportional to the square root of the active device area. Thus, sizing up the transistors can help mitigating these variations, which is one of the most commonly used techniques in analog circuit design to decrease transistors mismatch.

In addition to that the variation in transistor, channel length has a direct effect on various transistor electrical parameters; however, the most impacted parameter is the transistor threshold voltage, V_{th} which can be modeled as [2]:

$$V_{th} \approx V_{tho} - (\xi + \eta V_{DS})exp\left(\frac{L}{L_{to}}\right) \qquad (2)$$

where ξ is the charge sharing coefficient, V_{th0} is the long channel threshold voltage, η is the DIBL coefficient and L_{t0} is the characteristic length. Thus, a slight variation in L will introduce huge variation in V_{th}. In order to limit those, NWFET which has better gate controllability than other multigate structures are preferred. Nanowire FETs can be implemented in a lateral or a vertical configuration. The fluctuation in the characteristics induced by various random dopants in gate all around Si nanowire (GAA NW) have been reported by several authors [4].

Technology CAD (TCAD) tools are widely used to model the electrical characteristics of microelectronic components. This is mainly due to the fact that the development of TCAD technologies can reduce the cost and time of development and also ensure the choice of technology. The modeling of devices using TCAD often describes either the distribution of carriers or transport of carriers inside a given device by solving the Poisson equation coupled to a Density-Gradient model (DG) and also to the continuity equation (with these models of associated transport). As such, simulations are first compared to available experimental data for calibration as shown in Fig. 1.The simulations are performed using MINIMOS-NT [5]. The use and calibration of available TCAD models to efficiently model the device shows a good match between the simulation and experiment.

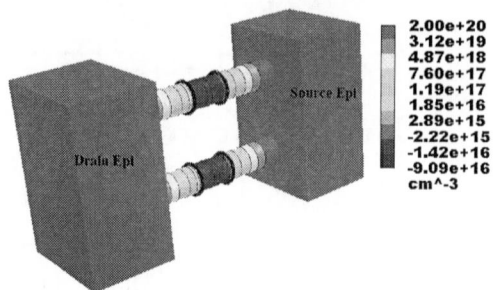

Fig. 2. Schematic of the simulated structure with net concentration. .

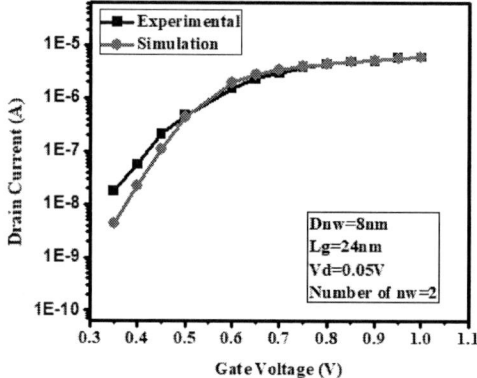

Fig. 1. Comparison of simulation and experimental I_D-V_G characteristics at a V_D of 0.05 V for a 24nm 2NW structure [6].

The paper is organized as follows: The simulation environment is given in section II. The discrete dopants effects in channel and SD extension are discussed in section III. In addition, the metal grain orientation effects on the performance of NW structures are presented in the same section. The conclusion is given in section IV.

II. SIMULATION ENVIRONMENT

The structure of the simulated device is given in Fig.2. The geometry of the simulated device is provided in Table 1.Each nanowire has gate length of 10nm. The diameter of each nanowire is 5nm.The channel doping and the source/drain doping are $1 \times 10^{17} cm^{-3}$ and $2 \times 10^{20} cm^{-3}$, respectively. The structure of the simulated device and the potential distribution at a particular gate and drain voltage are presented in Figs . 2 and 3 respectively.

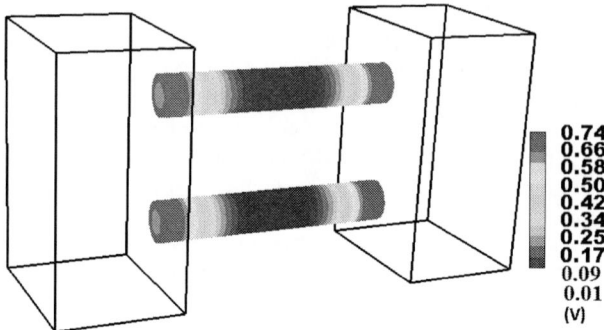

Fig. 3. Potential distribution in 2 nanowire structure at V_G=0V and V_D=0.05V.

The potential distribution in the channel is very less as the no. of carriers are less in the interface due to quantum confinement. Here for the simulation purpose the quantum correction method has been taken into account. In addition to that the bandgap narrowing model, discrete dopant model and metal gate granularity model are also used in the simulation. The electron current density are presented in equation (3) is modeled in [5].

$$J_n = q \cdot \mu_n \cdot n \left(grad \left(\frac{\varepsilon_C}{q} - \psi - \gamma_n \right) + \frac{k_B \cdot T_L}{q} \cdot \frac{N_{C,0}}{n} \cdot \right.$$
$$\left. grad \left(\frac{n}{N_{C,0}} \right) \right) \qquad (3)$$

Where ε_C is the position-dependent band edge energies ,ψ is the electrostatic potential and T_L is the local lattice temperature. The quantum correction potential for electron are modeled as in (4).

$$\gamma_n = \frac{\hbar}{12 \cdot \lambda_n \cdot m_0} \cdot div \; grad \frac{\psi + \gamma_n - \varepsilon_C/q}{k_B \cdot T_L} \qquad (4)$$

Where $\Box$ is the reduced Planck's constant, ψ is the electrostatic potential, ε_c is the local band edge energy for electrons, k_B presents the Boltzmann constant and T_L is the local lattice temperature. The change in bandgap energy is given by [7]

978-1-5386-6723-1/19 $31.00 © 2019 IEEE

$$\Delta \varepsilon_g = \varepsilon_{ref.} \cdot \left(ln\left(\frac{c_I}{N_{ref}}\right) + \sqrt{ln^2\left(\frac{c_I}{N_{ref}}\right) + \frac{1}{2}} \right) \qquad (5)$$

The physical based model contains no free parameters except the ratio $\Delta\varepsilon_c / \Delta\varepsilon_g$. So, the change in conduction and valency band edges are given by

$$\Delta\varepsilon_C = \left(\frac{\Delta\varepsilon_C}{\Delta\varepsilon_g}\right) \cdot \Delta\varepsilon_g \qquad (6)$$

$$\Delta\varepsilon_V = \Delta\varepsilon_g - \Delta\varepsilon_C \qquad (7)$$

In addition to this the discrete dopants are modeled by randomly generating dopant locations at Si lattice sites with a probability determined by the donor doping concentration in the particular region of the device. Discrete dopants are generated randomly distributed in the simulation domain or a given segment. For each grid point the number of dopants is determined using a Poisson distribution. The position within the control volume of each grid point follows a uniform distribution. Metal Gate Granularity (MGG) occurs due to random grain orientation which leads to variation ineffective work function of high-k metal gate stack in nanoscale devices. The random grain orientation leads to variation in potential distribution. As a result of which the drain current varies. In simulation, we use the Poisson-Voronoi model to simulate the polycrystalline grain structure of materials. For the present study of variability, we have generated different 4.43 and 4.33 work-function patterns with two different average grain sizes (1nm,10nm).The geometry of the simulated devices are presented in Table 1.

TABLE 1. DEVICE GEOMETRY AND OTHER PARAMETERS USED IN THE SIMULATION.

Parameters for the simulation	Value
Channel Diameter	5nm
Channel length	10nm
Source/Drain Doping	$2 \times 10^{20} cm^{-3}$
Channel Doping	$10^{17} cm^{-3}$
SiO2 thickness	1nm
HfO2 thickness	1.2nm
Number of nanowire	2

The simulation results of the devices are presented in section III.

III. RESULTS AND DISCUSSION

The variation is induced in the same type of devices due to discrete dopants in different regions. Here the discrete dopants in the source drain extension region are shown in Fig.4 (a) and the effect of discrete dopants on the potential distribution is shown in Fig.4(b).It can be observed that in the presence of discrete dopants the potential is enhanced in the channel.

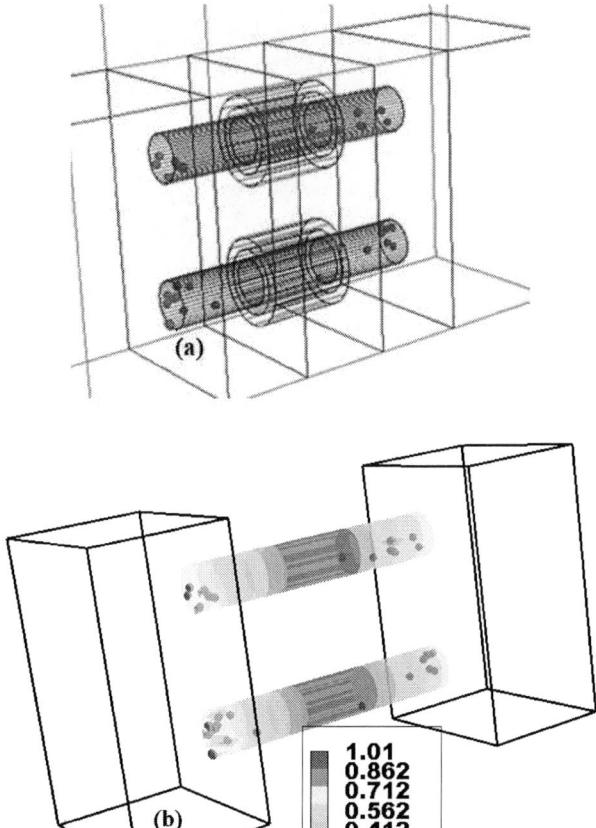

Fig. 4. (a) Distribution of discrete dopants in S/D extension region for a particular configuration and (b) Potential distribution in 2 nanowire structure at V_G=0V with discrete dopants in S/D extension region. Donor distribution for one configurations of random dopants. Donors are randomly distributed in a 10nm region from the channel edges towards the source and drain, respectively.

As the discrete dopants trap majority carriers and reduce the mobile electron concentration, a sharp Coulomb potential well is created resulting in threshold voltage fluctuation. The effects of discrete dopants in the channel is presented in Fig. 5. The source /drain doping is higher than channel doping. As the channel is undoped (less doped),it may happen that the dopants may diffuse from source /drain region. Here the effect of one dopant is shown in Fig .5. Since the number of discrete dopants are less in the channel region, so the fluctuation in the threshold voltage is less which is clearly mentioned in Table 2. The V_{th} and I_{off} value without discrete dopants are found to be 496.96 mV and 0.00038 pA , respectively. The I_{on} and SS value without discrete dopant is found to be 6.842 µA and 71.039 mV/dec , respectively.The effects of discrete dopants on the electrical characteristic are shown in Fig.6.

978-1-5386-6723-1/19 $31.00 © 2019 IEEE

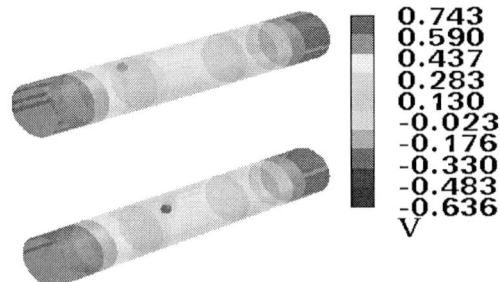

0.743
0.590
0.437
0.283
0.130
-0.023
-0.176
-0.330
-0.483
-0.636
V

Fig. 5. 3D Potential distribution in 2 nanowire structure at V_G=0V with discrete dopants in the channel. Only one discrete dopant is observed in the channel of each nanowire.

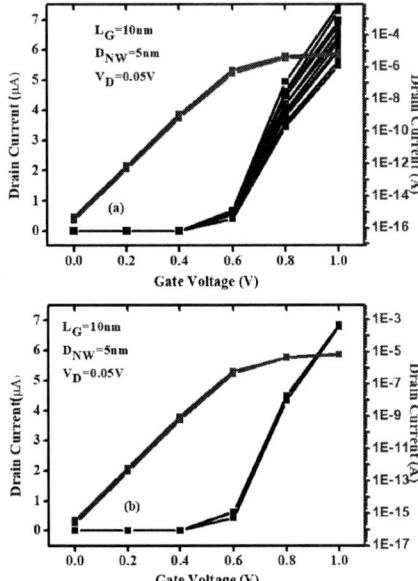

Fig. 6. I_D-V_G plot for (a) discrete dopants in S/D extension, (b) discrete dopants in channel (linear and logarithmic scale) from DG simulations of 10 devices with random dopants at V_D=50mV.

From the above figure it can be seen that in presence of discrete dopants the drain current is getting reduced due to scattering effect. Since more number of discrete dopants are present in SD extension region, so the variation in drain current is more in SD extension region than in channel region. The standard deviation of V_{th}, I_{off}, I_{on}, SS are presented in Table 2 which is shown below for ten different configurations.

TABLE 2. VARIATION OF DEVICE PARAMETERS FOR TEN DIFFERENT CONFIGURATIONS OF DISCRETE DOPANTS IN THE CHANNEL AND SOURCE DRAIN EXTENSION REGIONS.

Discrete Dopants (location)	σV_{th} (mV)	σI_{off} (pA)	σI_{on}(micro Amp)	σSS (mV/ dec)
Channel	4.15	4.88664E-5	1.508 E-2	0.351 58
SD Extension	4.39	4.7491E-5	6.615 E-2	0.506 24

In a metal gate, different grain orientation will occur with different probabilities. As a result of which fluctuation in threshold voltage [8] occurs which leads to variation in drain current, off current and SS. Here the average grain size is taken as 1nm and 10nm. The I_D-V_G plot for different grain size is provided in Fig. 7. Larger grain sizes exhibit higher standard deviations of V_{th} which shown in Fig.7.

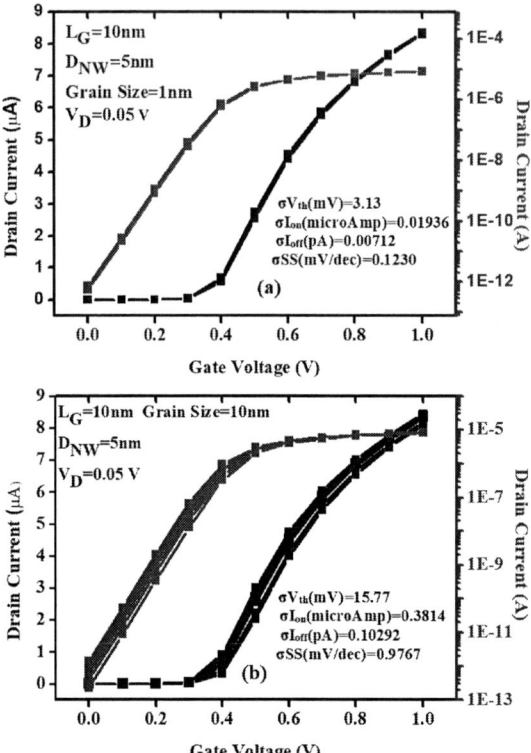

Fig. 7. I_D-V_G plot for different grain size for ten different configurations used in simulation with two different grain size.

IV. CONCLUSIONS

In this work, we have studied the metal grain and discrete dopant induced nanowire transistor device parameter variation in a 10nm gate length vertically stacked 2wire NW transistors using quantum corrected 3-D DG simulations. We have used Poisson-Voronoi model to simulate the polycrystalline grain structure of materials for local work function variation. The deviation in V_{th} is larger for the bigger size of grains of 10 nm. It is shown that when the grain size is comparable to the gate size, the distribution of V_{th} becomes wider, the variation in on current, off current, SS are more. Furthermore, as the device size decreases, our approach for analysis of the effects on the device performance due to discrete dopants, grain shape and size distribution has demonstrated the advantages of TCAD for variability studies.

REFERENCES

[1] S. Borkar, T. Karnik, S. Narendra, J. Tschanz, A. Keshavarzi, and V. De, "Parameter variations and impact on circuits and micro-architecture," Proc. of the IEEE Design Automation Conference (DAC'03), pp. 338-342, 2003.

[2] Y. Taur and T. H. Ning, "Fundamentals of modern VLSI devices," New York, NY, USA: Cambridge University Press, 1998.

[3] T. Mizuno, J. Okumtura, and A. Toriumi, "Experimental study of threshold voltage fluctuation due to statistical variation of channel dopant number in MOSFETs," IEEE Trans. on Electron Dev., vol. 41, pp. 2216-2221, November 1994.

[4] W. Sung and Y. Li, "DC/AC/RF Characteristic Fluctuations Induced by Various Random Discrete Dopants of Gate-All-Around Silicon Nanowire n-MOSFETs," in IEEE Transactions on Electron Devices, vol. 65, no. 6, pp. 2638-2646, June 2018.

[5] MINIMOS-NT User's Manual, 2017.

[6] F. M. Bufler, R. Ritzenthaler, H. Mertens, G. Eneman, A. Mocuta, and N. Horiguchi, "Performance Comparison of n–Type Si Nanowires, Nanosheets, and FinFETs by MC Device Simulation," in IEEE Electron Device Letters, vol. 39, no. 11, pp. 1628-1631, 2018.

[7] J. Slotboom and H. de Graaff, "Measurements of Bandgap Narrowing in Si Bipolar Transistors," Solid-State Electron., vol. 19, pp. 857–862, 1976.

[8] K. Nayak, S. Agarwal, M. Bajaj, P. J. Oldiges, K. V. R. M. Murali and V. R. Rao, "Metal-Gate Granularity-Induced Threshold Voltage Variability and Mismatch in Si Gate-All-Around Nanowire n-MOSFETs," in IEEE Transactions on Electron Devices, vol. 61, no. 11, pp. 3892-3895, Nov. 2014.

2019 Devices for Integrated Circuit (DevIC), 23-24 March, 2019, Kalyani, India

NBTI Degradation and Recovery in Nanowire FETs

S. Das[1*], T.P. Dash[2], S. Dey[3], E. Mohapatra[4], J. R. Jena[5] and C. K. Maiti[6]

Department of Electronics and Communication Engineering,
Siksha 'O' Anusandhan (Deemed to be University), Bhubaneswar, Odisha, India--751030
[1*]sanghamitradas@soa.ac.in; [2]taradash@soa.ac.in; [3]supravadey@soa.ac.in, [4]eleenamohapatra@soa.ac.in; [5]jhansiranijena@soa.ac.in;
[6]ckmaiti@soa.ac.in.

Abstract—Following the downscaling roadmap for planar MOSFETs, non-planar (3-D) multiple-gate architectures are becoming essential for ultimate scaling of CMOS devices. Negative bias temperature instability (NBTI) is one of the key device reliability issues which exhibit some different features at nanoscale. In this work, the NBTI reliability issues of p-channel gate-all-around silicon nanowire transistors (SNWTs) have been investigated. When stressed, NBTI behavior in SNWTs show fast initial degradation, quick degradation saturation and then a special recovery behavior.

Keywords - Nanowire, NBTI, nonradiative multiphonon model, time dependent defect spectroscopy, trapped hole concentration, stress, recovery.

I. INTRODUCTION

Nanowire FETs and Gate-all-around (GAA) FET architectures are considered to be the ultimate refinement of the planar MOSFETs to address the scaling challenges below 22nm technology nodes [1, 2]. The nanowire transistors show excellent immunity to short channel effects (SCE) due to good electrostatic control over the channel as well as high I_{ON}/I_{OFF} ratio. However achieving a good reliability of these 3D devices is still very challenging. In particular, many studies have reported that Bias Temperature Instability (BTI) is strongly enhanced in narrow transistors like FinFET or nanowire [3]. Particularly NBTI is a critical concern, which occurs in negatively biased ($V_{GS} < 0$) p-type Si nanowire FETs at an higher temperatures due to generation of interface traps at the channel/oxide interface. As a result, the threshold voltage increases with time with decrease in the ON current [4]. Another interesting property of NBTI is the recovery of threshold voltage when the negative bias stress is removed i.e. the threshold voltage of the nanowire shifts towards the value before stress. Hence analysis of the NBTI characteristics of p-channel nanowires is essential for its application in high speed circuits [5].

From the various studies conducted so far, it is clear that bias temperature instability is mainly caused by two types of defects: (i) defects present inside the oxide and (ii) defects present at the interface [6]. These oxide and interface defects can trap the channel carriers during the device operation and degrade the performance. In this work, for the first time we use the four-state nonradiative multiphonon (NMP) model for degradation studies in advanced silicon nanowire devices.

II. TCAD CALIBRATION

The simulated NBTI degradation characteristics have been presented along with available experimental data for long channel nanowires. The experimental p-type nanowires are of length 400-750 nm [7]. As the simulation of long channel nanowires are not feasible, hence p-type nanowires were simulated with a gate length of 30 nm. As such the simulated results are higher (in magnitude) than the experimental results for the same stress voltage as the effect of NBTI is more severe with device downscaling.

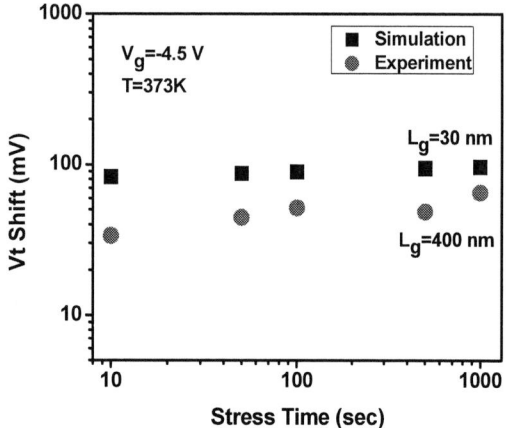

Figure 1. Threshold voltage shift vs. stress time for simulated and experimental p-type nanowires.

III. DEVICE STRUCTURE AND SIMULATION ENVIRONMENT

A. Device Structure

A p-type nanowire FET has been simulated with the following design parameters. The two dimensional view of the transverse slice of the 3D nanowire is presented in figure 2(a) which shows one half of the device. The doping concentrations in the source/drain and channel region are shown in Figure 2(b).

978-1-5386-6723-1/19 $31.00 © 2019 IEEE

TABLE 1. DEVICE GEOMETRY USED IN THE SIMULATION.

Parameters of Nanowires	Value
Channel Diameter	4nm
Channel Length	30nm
Channel Doping	10^{18} cm^{-3}
Source Drain Doping	10^{20} cm^{-3}
Oxide thickness	1nm

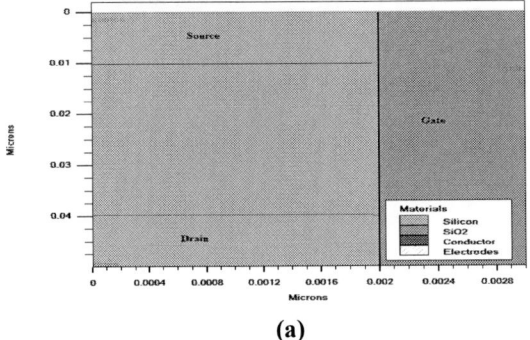

(a)

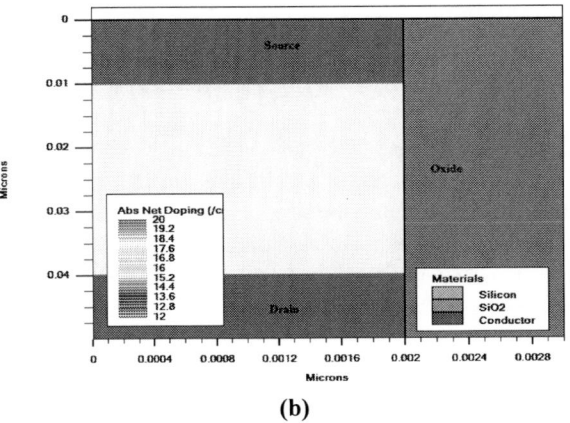

(b)

Figure 2. (a) Two dimensional (2D) view of the transverse slice of the 3D nanowire device, (b) The doping concentrations in the source/drain and channel region of the nanowire.

B. Simulation Environment:

When the devices are scaled down to nanometer range, quantum effects play a vital role while determining the device characteristics. Hence for the nanowire device characteristic simulation, these effects have to be taken into consideration. In this work, the semiclassical drift diffusion mode space model including quantum corrections has been used to describe the carrier transport mechanism in nanowires. The model describes quantum effects in transverse direction and also includes the drift diffusion models for mobility, recombination, impact ionization, and band-to-band tunneling. In transverse direction, the solution is given by Schrodinger equation along with one dimensional transport equations in each sub-band. This 1D sub band transport,

characterized by sub-band index v and effective mass index b is described as [8]

$$\frac{1}{q}\frac{\partial J_{vb}}{\partial x} = R_{vb} - G_{vb} \tag{1}$$

$$J_{vb} = q\mu_{vb}n_{vb}\frac{\partial E_{vb}}{\partial x} - qD_{vb}\frac{\partial n_{vb}}{\partial x} \tag{2}$$

where $J_{vb}(x)$ and $n_{vb}(x)$ are the sub-band current sub-band carrier density respectively and $E_{vb}(x)$ represents the eigen energy in transverse direction for the sub bands.

The diffusion coefficient D_{vb} is related to mobility as:

$$D_{vb} = \mu_{vb}\left(\frac{\partial n_{vb}}{\partial E_{vb}}\right)^{-1} \tag{3}$$

the relationship between sub-band carrier densities (cm^{-3}), quasi-fermi levels and eigen energies for 1D confinement in y direction is given as:

$$n_{vb} = 2\frac{k_BT}{A\pi h^2}\sum_v \sqrt{m_x^{vb}m_z^{vb}}\ln\left[1 + exp\left(-\frac{E_{vb}-E_{F,vb}}{k_BT}\right)\right] \tag{4}$$

In order to model the reliability degradation characteristics of the p-channel silicon nanowire due to NBTI, the most popular four state nonradiative multiphonon (NMP) model has been used [9]. The NMP model can accurately describe the hole capture and emission process in the oxide traps as observed from the time dependent defect spectroscopy (TDDS) experiments.

The state diagram of the four state NMP model is shown below:

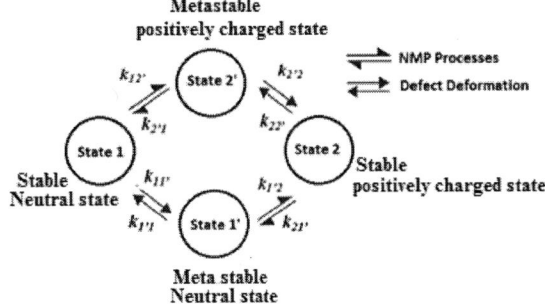

Figure 3. State diagram of the NMP model showing the transition between the four states.

The non radiative multiphonon transition rates are given as [10]:

$$k_{12'} = \sigma v_{th}pe^{-\beta\varepsilon_{12'}}, \tag{5}$$

$$k_{2'1} = \sigma v_{th}pe^{-\beta\varepsilon_{12'}}e^{-\beta(E_{TF}-\varepsilon_{T2'})}, \tag{6}$$

$$k_{1'2} = \sigma v_{th}pe^{-\beta\varepsilon_{1'2}}, \tag{7}$$

$$k_{21'} = \sigma v_{th}pe^{-\beta\varepsilon_{1'2}}e^{-\beta E_{T'F}}. \tag{8}$$

Where p denotes the hole concentration in channel, v_{th} and σ represent their thermal velocity and capture cross-section, respectively.

The behavior of the interface traps which are responsible for the permanent BTI degradation can be described using the double-well model [9, 11] as represented in figure 4.

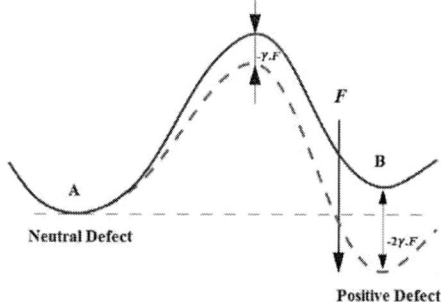

Figure 4. Configuration coordinate diagram of the interface reaction.

The forward and reverse rates of the interface reaction are given by:

$$K_{AB} = v\, e^{-(E_A - \gamma.F)/K_B T} \tag{9}$$

$$K_{BA} = v\, e^{-(E_A - E_B + \gamma.F)/K_B T} \tag{10}$$

Here v: attempt frequency
    E_A: barrier height.
    γ : thermal activation energy coefficient

IV. RESULTS AND DISCUSSIONS

The NBTI degradation results for the p-type silicon nanowire FETs are presented in this section. Figure 5(a) shows the cross-sectional view of the nanowire with traps present inside the oxide and at the oxide/channel interface. When the device is stressed with a constant gate bias, more number of traps are generated and the trap concentration increases with stress time. These traps are majorly responsible for the reliability degradation of the device. Figure 5(b) shows the trapped hole concentration inside a particular section of the gate oxide for three conditions i.e. before stress, after stressing the device for 100s, and after relaxation (after 1000s), respectively. The device was stressed with a gate bias of -2.0 V.

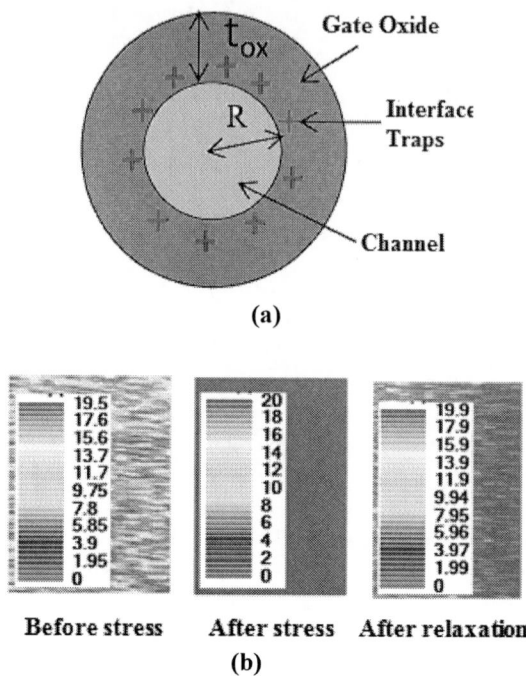

Before stress   After stress   After relaxation

(b)

Figure 5. (a) Cross-section of the nanowire with interface traps, and (b) trapped hole concentration inside the gate oxide of the nanowire; before stress, after stressing the device for 100s, and after relaxation (after 1000s).

It has been observed that the trapped hole concentration inside the oxide increases significantly after stressing the device. Before applying the stress bias, the concentration of positively charged traps (by capturing a hole) was in the order of 10^{15} in most parts of the gate oxide. The concentration has increased to 10^{20} after stressing the device for 100s. This shows that almost all of the traps have captured holes and are in the positively charged state. The concentration increases with the increase in the stress time. After removing the stress bias, the device is relaxed and the traps tend to move back to the neutral state by the emission of holes. However, complete recovery of the trap states is not possible, hence some of the traps permanently remain in the positively charged condition even after relaxation. This is also evident from Figure 5(b) which shows that the trapped hole concentration decreases after relaxing the device for 1000s, but it is higher than that before stressing the device.

Due to these holes trapped inside the gate oxide, threshold voltage of the nanowire shifts to higher value and the mobility also degrades which leads to lower drain current. This is proved from the device transfer characteristics (Figure 6) which is shown in three bias conditions.

2019 Devices for Integrated Circuit (DevIC), 23-24 March, 2019, Kalyani, India

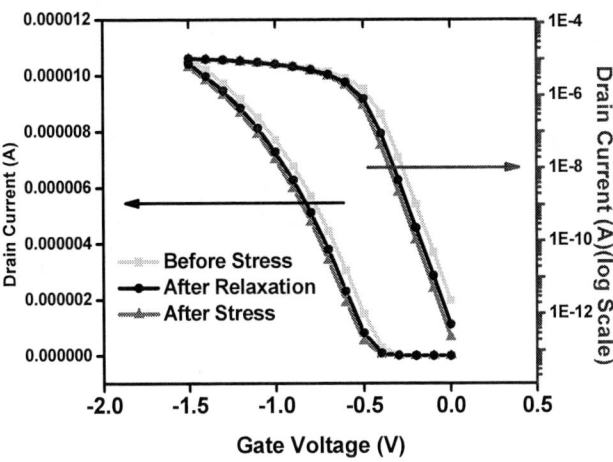

Figure 6. Drain current degradation and recovery due to NBTI in p-type nanowire FET.

The threshold voltage (Vt) of the device is observed to be -0.4 V before applying the stress bias. After stressing the device for 100 seconds, the threshold voltage shifts to -0.474 V due to increased trap concentrations. Then, after relaxing the device for 1000 seconds, the threshold voltage shifts back to -0.443 V. Hence after relaxation, the Vt shift reduces to 43 mV from 74 mV (after stress). This 43mV shift in threshold voltage is attributed to the permanent component of BTI. Also, it has been reported that reliability degradation in nanowire FETs is more in comparison to other nonplanar structures like FinFETs and the planar MOSFETs [5]. This is due to the cylindrical structure of the nanowire which leads to enhanced electric field and larger rate of trap generation inside the gate oxide. The threshold voltage degradation depending on electric field (E_{ox}), stress time (t) and temperature (T) has been modeled as [5]:

$$\Delta V_T = A e^{\beta E_{ox}} e^{-E_a/kT} t^n \tag{11}$$

Where A and β are constants. E_a represents the activation energy and k is the Boltzmann constant. Figure 7 shows the shift in threshold voltage of the nanowire with stress time under two different stress bias conditions.

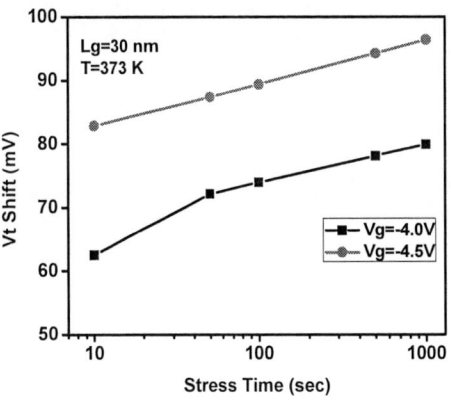

Figure 7. Threshold voltage shift of the nanowire with stress time.

With increase in the stress bias, the threshold voltage shift increases due to higher electric field which leads to higher rate of trap generation. Temperature plays a major role in the NBTI degradation along with stress time and electric field. The influence of temperature on the threshold voltage shift is shown in Figure 8. The Vt shift increases with increase in temperature which is evident from the relationship shown in equation (11).

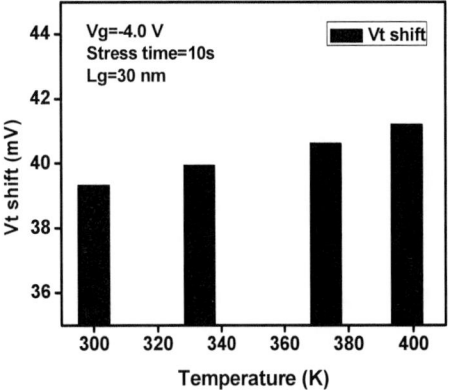

Figure 8. Effect of temperature on the threshold voltage shift of the nanowire.

V. CONCLUSION

NBTI behavior of the gate-all-around (GAA) Si nanowire transistors (SNWT), which is considered as one of the promising candidates for ultimate scaling is discussed. Our simulation results show, NBTI behavior in SNWTs with a fast initial degradation, quick degradation saturation and recovery behavior. It was observed that the NBTI degradation effect in nanowire FETs and in conventional planar MOSFETs is different and is also a serious concern for ultra-scaled devices. The degradation in nanowires tends to saturate quickly due to the small interface area of the SNWT. Also, the device degrades faster than planar transistors in the initial stage of the stressing. This can be attributed to the enhanced electric field resulting from the cylindrical structure which is responsible for faster trap generation.

References

[1] H.-C. Chin *et al.*, "III-V multiple-gate field-effect transistors with highmobility in 0.7Ga0.3As channel and epi-controlled retrograde-doped fin," *IEEE Electron Device Lett.*, vol. 32, no. 2, pp. 146–148, Feb. 2011.

[2] M. Karner, O. Baumgartner, Z. Stanojevi´c, F. Schanovsky, G. Strof, C. Kernstock, HW. Karner, G. Rzepa, T. Grasser, "Vertically Stacked Nanowire MOSFETs for Sub-10 nm Nodes: Advanced Topography, Device, Variability, and Reliability Simulations," in IEDM16, pp.762-765, 2016.

[3] X. Garros, A. Laurent, S. Barraud, J. Lacord, O. Faynot, G. Ghibaudo and G. Reimbold, "New insight on the geometry dependence of BTI in 3D technologies based on experiments and modelling," in 2017 Symposium on VLSI Technology Digest of Technical Papers, pp.T134-T135, 2017.

[4] D. K. Schroder, "Negative bias temperature instability: What do we understand?" *Microelectron. Rel.*, vol. 47, no. 6, pp. 841–852, Jun. 2007.

[5] Om Prakash, S. Beniwal, S. Maheshwaram, A. Bulusu, N. Singh, and S. K. Manhas, "Compact NBTI Reliability Modeling in Si Nanowire MOSFETs and Effect in Circuits," IEEE Transactions On Device And Materials Reliability, vol. 17, pp.404-413, 2017.

[6] T. Grasser, Editor, Bias Temperature Instability for Devices and Circuits, Springer Science+Business Media, New York, 2014.

[7] N. Singh, A. Agarwal, L. K. Bera, T. Y. Liow, R. Yang, S. C. Rustagi, C. H. Tung, R. Kumar, G. Q. Lo, N. Balasubramanian, and D.-L. Kwong, "High-performance fully depleted silicon nanowire (diameter ≤5 nm) gate-all-around CMOS devices," *IEEE Electron Device Lett.*, vol. 27, no. 5, pp. 383–386, May 2006.

[8] ATLAS User's Manual, 2018.

[9] T. Grasser, Stochastic Charge Trapping in Oxides: From Random Telegraph Noise to Bias Temperature Instabilities, Microelectron. Reliab., vol. 52, pp. 39 -70, 2012.

[10] A. Stesmans, "Dissociation Kinetics of Hydrogen-Passivated P_b Defects at the (111) Si/SiO$_2$ Interface", Physical Review B, Vol. 61, pp.8393–8403, 2000.

Performance Comparison of Body Biasing and Coupling Capacitor Sense Amplifier for SRAM

Phuntso Chotten, Akho John Richa

Department of Electronics and Communication Engineering
North Eastern Regional Institute of Science and Technology
Nirjuli, India
anuphuntso96@gmail.com, akhojohn@gmail.com

Abstract—**A comparative study of a voltage mode sense amplifier (VMSA) using different techniques to understand their performance and applicability has been presented in this paper. The circuits to be studied are VMSA using a coupling capacitor and body biasing along with the conventional circuit. Various performance measuring parameters of the sense amplifiers (SA) have been examined and analyzed. It has been realized that both the techniques heighten the performance of the sense amplifier in comparison to the conventional method, but their mutual comparison reveals that the coupling capacitor has better performance than the body biasing technique. Detail analysis of the circuits in term of sensitivity, power consumption and delay has been made for a better understanding of their performances. The circuits in the paper have been simulated by using Tanner EDA tool of 0.25 um technology.**

Index Terms—**SRAM, voltage mode sense amplifier, body biasing, capacitor coupling, pull down network.**

I. INTRODUCTION

In digital designing, data storage is one critical part which occupies a large area. In today's microprocessors with high-performance, half of the transistors belonging to the caches memories and might increase more in the future. There is a need to create on-chip caches which are faster for the high-speed processors. The caches memories are made up of small blocks of static random access memories (SRAM) [1]. They are created by using a multitude of small blocks of SRAMs. Thus, the SRAM performance factors such as speed, area, power dissipation, etc. strongly affect the whole system overall performance. The cache covers most of the area on the chip and is regarded as the source of the majority of the power consumption of the whole system. The SRAM cell structures are fixed by the technology and enhancement of the performance can be achieved by either changing the technology or modification of the peripheral circuits.

The sense amplifier (SA) is considered as a critical peripheral circuit of the SRAM. It quickly amplifies the differential voltage between the 2-bit lines (BL) and the data lines, which can be then read at the output. It can enhance the performance of memory, reduces the power consumption during the read operation, and also provides an improvement in yield for SRAM [2-4]. Inside the memory cell, every column of it contains one SA. So, in memory chips, thousands of such identical SAs are present. Due to the demand for increasing

memory capacity along with low power consumption and high speed, a new operating environment for SA has been developed. The SA can be operated in three fashion namely voltage, current, and charge [5-8]. The voltage mode SA (VMSA) can sense small differential signal but at the cost of more considerable delay because it depends on the bit line capacitance charging and discharging whereas the other SAs provide less delay due to the non-dependence on the bit-line capacitance [9]. The static power consumption of current SA (CSA) is significant due to the flow of biased current before the SA is active. In charge transfer SA, an increment/decrement in the transistor biasing from its optimal value changes the bit-line value from its actual one and thus, the differential voltage is also change leading to a different value in the output. Though VMSA has a large delay, it also has high sensitivity and small area. The delay of the VMSA can be improved by using techniques such as capacitor coupling [10] and body biasing [11].

In this work, a study on the performance comparison of VMSA by using the concept of body biasing and capacitor coupling techniques are presented by using simulation done through the use of Tanner. Utilization of body biasing technique decreases the threshold voltage of the pull-down transistor, thereby enhancing the driving capability of the pull-down network (PDN) of the SA. The driving capability of the PDN can also be increased by the capacitor coupling technique by providing a negative voltage at the SA virtual ground [10].

The organization of the remaining sections of this paper is presented as follows. Section-II discusses the working principle of VMSA by using conventional, body biasing and capacitor coupling techniques. Section-III discusses the simulation results and the comparison of the performance parameters. Finally, the conclusion is made in section-IV.

II. WORKING PRINCIPLE

A. Conventional VMSA

A conventional VMSA is shown in Fig.1 consist of pairs of a cross-coupled inverter formed by transistor P6-N1 and P7-N2 with positive feedback connection to amplify the differential voltage develop at internal nodes (X and Xb) by input BLs to full swing at outputs. The memory cell comprises two inputs that are BL and BLB connected at the column bit lines. The pre-charge circuit is formed by transistors P3-P5 is used to

2019 Devices for Integrated Circuit (DevIC), 23-24 March, 2019, Kalyani, India

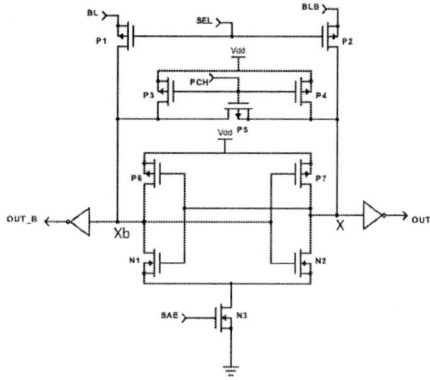

Fig. 1. Schematic of Conventional VMSA

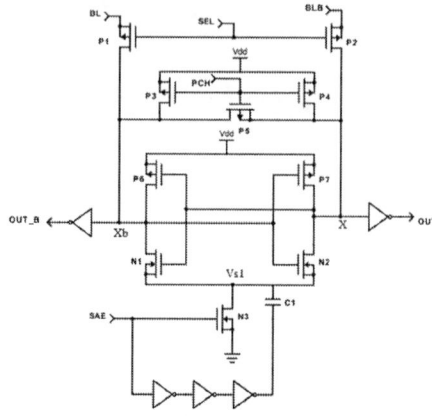

Fig. 2. Schematic of VMSA using Coupling Capacitor

charge the internal nodes through the BLs before the read operation. The connection of memory cell with the SA is made by transistors P1 and P2 while N3 is used to enable the SA. The two output inverters are used to set apart the internal nodes from the external loads.

It operates in two phases: pre-charging and evaluation phase. In the pre-charging phase, and through the BLs the internal nodes are pre-charged to high by giving PCH signal low. During the evaluation phase, the connection of memory cell to the SA is made by keeping SEL low. The data stored in the cell makes one of the internal node low and one high. If we assume storage of 1 in the memory cell, then the BL will be high and the BLB low. Once the differential voltage is created at the BLs, the SA gets enable by pulling SAE (sense amplifier enable) high. This type of amplifier makes use of differential discharge to sense the differential voltage and then convert them to full swing at the outputs.

Due to variation in the process the SRAM, cell current decreases leading to the slower discharge of BLs. SAE path is faster when compared to the generation of differential voltage leading to the more substantial requirement of differential voltage. This also creates less differential voltage at the internal nodes making the driving capability of NMOS transistors low, thereby, increasing the reaction time of the SA.

B. VMSA Using Coupling Capacitor

The modification of conventional SA shown in Fig. 2 is done by adding a coupling capacitor between the node of V_{s1} and SAE which enhances the V_{s1}, the virtual ground of the SA. The chief benefits of using this approach are the generation of negative V_{s1} voltage and quick integration with the available SA circuit. At the node of V_{s1}, a negative coupling should be made to gain a substantial dip and must be done when the node reaches the possible minimum level in which an odd number of inverter chains is added. The inverter chains insert a delay as the SAE signal entering one terminal of the coupling capacitor node and increase the coupling voltage. Further, if the SAE signal is high which will keep one end of capacitor low, then, it will lead to the steep ramp of the capacitor coupled with the

V_{s1} node which provides extra drive to the NMOS transistors, i.e., the gate to source voltage increases and decreases the SA reaction time. Thus, this technique compensates the loss of differential voltage caused by variations in the device or due to the mismatch between the timing of the SAE signal.

C. VMSA Using Body Biasing

The conventional SA can also be modified as the circuit shown in Fig. 3 by using a body bias voltage through the control circuitry to control the performance of the SRAM cell. Applying body biasing to reduce delay and improving stability is a challenging task at the nanometer level. If it is implemented at the gate level, there is area overhead, and at the architecture level, there is a power overhead problem. If it is applied at the word level (row style based) of the SRAM architecture, it helps in reducing delay and power dissipation. The advantage of using the controller is that by forwarding body biasing the PDN in active mode, it leads to speed up of the read as well as the write operation.

A potential driver is formed by P9 and N5 transistors that divide the supply voltage into the half. Then the BLs are ON so that the BLs voltages couples with the cross-coupled SA at the same time BB is switched is ON. Initially, the BB remains high and switch ON the N4 transistor and shorting of the substrate and the source of N1 and N2 is done. As soon as the SAE is turned ON, the source voltage is pulled down, then a negative voltage arises at the source, thereby, lowering the threshold voltage. Thus, the read operation gets speeds up.

III. RESULTS AND DISCUSSION

The SAs have been simulated by using Tanner EDA tool of 0.25 um technology. Various performance parameters such as sensitivity, power consumption, and delay of the SAs have been studied to understand their performance and their applicability. Fig. 4 is the transient response of conventional VMSA with a supply voltage of 2.5 V which shows a reading of 1 in the output of the memory; to read 0, the OUT terminal will go low when the SAE is enabled. Transient response of other SAs has not been included due to their similar nature.

978-1-5386-6723-1/19 $31.00 © 2019 IEEE

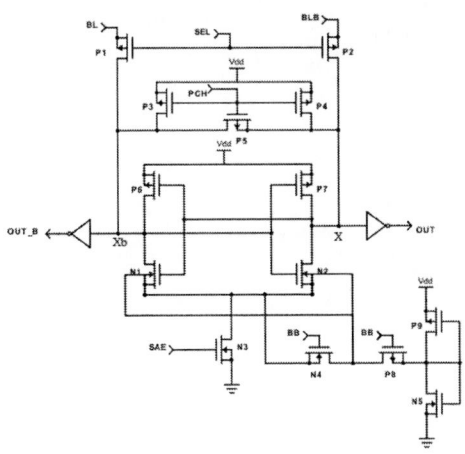

Fig. 3. Schematic of VMSA using body biasing

TABLE I
SENSITIVITY COMPARISON FOR VARIOUS VMSAS

S.No.	V_{dd} (V)	ΔBL_{min} (V)		
		Conv.	Coupling Capacitor	Body Biasing
1	2.5	1.5	1.3	1.4
2	3	3	1.5	1.8
3	3.5	3.5	1.7	2
4	4	4	2.2	2.4

A. Sensitivity

Sensitivity is the minimum differential BLs (ΔBL_{min}) voltage that can be sensed by the SA. Lower the value of ΔBL_{min} better is the sensitivity as it can detect lower differential in the input. The analysis proves that VMSA using coupling capacitor has the maximum sensitivity when compared with the other two VMSAs as it can detect the lowest BLs differential voltage of 1.3 V when the supply is 2.5 V. Sensitivity in terms of ΔBL_{min} for various input voltages have been given in Table I. It also shows that when supply increases, sensitivity also increases as the difference between the supply voltage and the ΔBL_{min} increases.

B. Power Consumption

The power consumption is evaluated for all the VMSA using different methods and is presented in Table II. It shows that VMSA using capacitor consumes less power when compared with the body biasing during the operation of reading 1. It is also found that with the increase in supply, power consumption also increases which can be explained according to the following equation

$$P = C_L V_{dd}^2 f \qquad (1)$$

where, P is the dynamic power consumption, C_L is the load capacitance, V_{dd} is the supply voltage and f is the switching frequency.

TABLE II
POWER CONSUMPTION COMPARISON FOR VARIOUS VMSAS

S.No.	V_{dd} (V)	Power consumption of VMSAs (uW)		
		Conv.	Coupling Capacitor	Body Biasing
1	2.5	503.26	346. 54	414.81
2	4	1841.86	1179.30	1664.00

TABLE III
DELAY COMPARISON FOR VARIOUS VMSAS

S.No.	V_{dd} (V)	Delay of VMSAs (s)		
		Conv.	Coupling Capacitor	Body Biasing
1	2.5	5.45×10^{-7}	2.45×10^{-9}	4.097×10^{-8}
2	4	5.43×10^{-10}	3.39×10^{-12}	7.63×10^{-11}

C. Delay

The delay was calculated for all the three VMSAs for different V_{dd} as shown in Table III. It was found that VMSA using body biasing provides more delay when compared with the other two techniques. It has been seen that with the increase in supply voltage, the delay decreases for all the circuits.

IV. CONCLUSION

In this paper, VMSA using the conventional, coupling capacitor and body biasing techniques have been compared to study their performance. It was found from the above analysis that sensitivity and power consumption increases when the supply voltage increases for all VMSAs while the delay got reduced with the increase of supply voltage. The reduction in the delay in case of the coupling capacitor and body biasing techniques can be attributed to the rise in the driving capability of the PDN when compared with conventional VMSA. It is also found that the VMSA using body biasing enhances the speed of memory when compared with the conventional but at the cost of a large increase in power consumption. Performance comparison of all the three configuration reveals that coupling capacitor technique has the best performance among them. For a supply voltage of 4 V, the coupling capacitor was found to be more sensitive by a margin of 0.2 V than body biasing and 1.8 V than the conventional. The power consumption was found to be lesser by 484.7 uW and 662.56 uW than body biasing and conventional techniques respectively for the same supply voltage. The delay was also found to be lesser by a factor of 1/10 and 1/100 than body biasing and conventional techniques respectively for the same supply voltage. So, it can be concluded that both the techniques are improved version of the conventional VMSA but, the coupling capacitor is indeed the better performing VMSA.

REFERENCES

[1] J. Rabaey, A. Chandrakasan and B. Nikolic, Digital integrated circuits, 2nd ed. Upper Saddle River, New Jersey: Prentice Hall International, 2003.

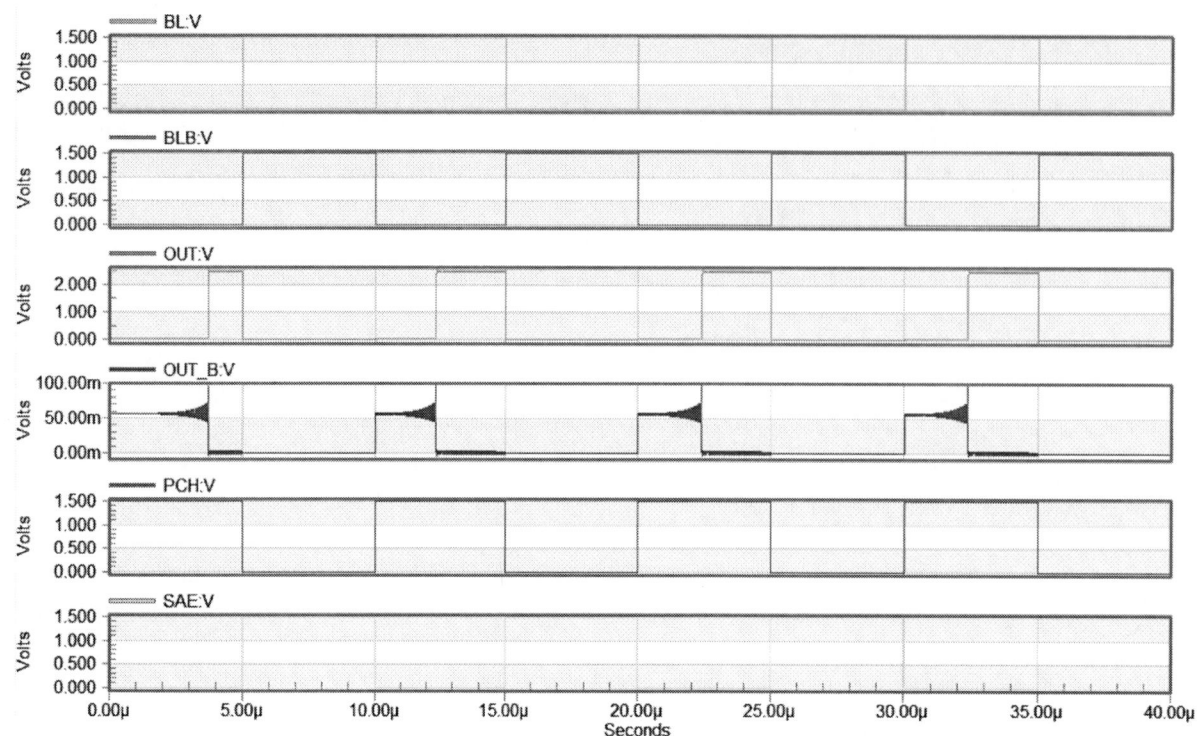

Fig. 4. Transient analysis of Conventional VMSA

[2] B. Liu, J. Cai, J. Yuan, and Y. Hei, "A Low-Voltage SRAM Sense Amplifier With Offset Cancelling Using Digitized Multiple Body Biasing," *IEEE Transactions on Circuits and Systems II: Express Briefs*, vol. 64, no. 4, pp. 442–446, Apr. 2017.

[3] K. W. Mai et al., "Low-power SRAM design using half-swing pulse-mode techniques," *IEEE Journal of Solid-State Circuits*, vol. 33, no. 11, pp. 1659–1671, Nov. 1998.

[4] T. N. Blalock and R. C. Jaeger, "A high-speed clamped bit-line current-mode sense amplifier," *IEEE Journal of Solid-State Circuits*, vol. 26, no. 4, pp. 542–548, Apr. 1991.

[5] L. Heller, D. Spampinato, and Ying Yao, "High-sensitivity charge-transfer sense amplifier," *IEEE International Solid-State Circuits Conference. Digest of Technical Papers*, pp. 112–113, Philadelphia, USA, 1975.

[6] E. Seevinck, P. J. van Beers, and H. Ontrop, "Current-mode techniques for high-speed VLSI circuits with application to current sense amplifier for CMOS SRAM's," *IEEE Journal of Solid-State Circuits*, vol. 26, no. 4, pp. 525–536, Apr. 1991.

[7] P. Murugeswari, G. Anusha, P. Venkateshwarlu, M. Bhaskar, and B. Venkataramani, "A wide band voltage mode sense amplifier receiver for high speed interconnects," *TENCON, IEEE Region 10 Conference*, pp. 1–5,Hyderabad, India, 2008.

[8] M. Sharifkhani, E. Rahiminejad, S. M. Jahinuzzaman, and M. Sachdev, "A Compact Hybrid Current/Voltage Sense Amplifier With Offset Cancellation for High-Speed SRAMs," *IEEE Transactions on Very Large Scale Integration (VLSI) Systems*, vol. 19, no. 5, pp. 883–894, May 2011.

[9] Kamal Kant Joshi, and Jyoti Kedia,"Performance Analysis of Sense Amplifiers", *International Journal of Advanced Engineering Research and Technology (IJAERT*, vol. 3, pp. 210-214, June, 2015.

[10] A. Pullareddy and G. Sreenivasulu, "Design and Implementation of High Speed Sense Amplifier for SRAM," *WSEAS Transaction on Electronics*, vol. 9, pp. 112-120, 2018.

[11] R. Lorenzo and S. Chaudhury, "Optimal Body Bias to Control Stability, Leakage and Speed in SRAM Cell," *Journal of Circuits, Systems and Computers*, vol. 25, no. 08, p. 1650096, Aug. 2016.

Characterisation of Ultra-Small Pocket Si$_{0.7}$Ge$_{0.3}$ Junction-less Tunnel FET with SOI

Suman Lata Tripathi
Lovely Professional University
Jalandhar, India
suman.21067@lpu.co.in

Shekhar Verma
Lovely Professional University
Jalandhar, India
shekhar.14572@lpu.co.in

Namrata Dhanda
Amity University, Lucknow,
India
ndhanda510@gmail.com

Abstract--- **The Tunnel FET has substantial potential to overcome limitations imposed due the scaling in low voltage region because of its steep subthreshold slope relative to its corresponding junction based MOSFET counterpart. Use of Si$_{0.7}$Ge$_{0.3}$ material as pocket region (0.5nm) enhances band-to-band tunneling by decreasing tunneling distance. The proposed pocket ultra small pocket Si$_{0.7}$Ge$_{0.3}$ Junction-less TFET(JLTFET) exploits advantage junction-less regions and p-type pocket region to improve device performance in subthreshold region showing improvement in subthreshold slope and better ON and OFF-state drain current ratio as compared to other similar TFET structures. Proposed ultra small pocket JLTFET shows high value I$_{ON}$/I$_{OFF}$ ratio and good subthreshold behaviour even with 2nm gate length and body thickness 0.5nm. All the designs proposed for n JLTFETs are designed on visual TCAD 2D/3D device simulator.**

Keyword: Subthreshold slope, Junction-less Tunnel FET, band-to-band tunneling, Narrow band gap material etc.

I. INTRODUCTION

Down scaling of conventional MOS transistors imposes thermal limits and undesirable ON/OFF characteristics. Multi-gate MOSFETs including double gate and triple gate MOSFET, FinFET and Gate-All-Around MOSFET structures on Bulk silicon or SOI (buried oxide region) are further examined by the researchers to optimise the device performance under subthreshold conditions[1-4]. The drain current versus gate voltage characteristics mainly depends on ON-state and OFF-state current which further decide the subthreshold performance of transistors. Some of the researches brought into the notice that TFET have a lot of potential to obtain good agreement between down scaling of device with subthreshold performance under limit and lesser dependencies on thermal limits[5]. Hetero-junction vertical p and n TFET structures were proposed with very small subthreshold slope, low threshold voltages

and off current up to the value of 10^{-15}A/um which is the important design consideration while down scaling of device [6]. Inclusion of narrow band gap material such as SiGe, Ge and InAs with silicon were utilised to enhance band to band tunneling in TFET with better on and off current ratio [7]. Pocket regions under the gate included with the channel reduces tunneling distance and increases band to band tunnel effect resulting in more steep subthreshold characteristics[8]. InAs interface with GaSb was explored to design 10nm pTFET to get 60mV/decade subthreshold slope for several orders of current in comparison to its performance with FinFET structure [9]. Longer channel region with L and T shaped TFET implemented with Si-Ge pocket region to support band bending resulting to more BTBT tunneling [10-11]. UTFET was proposed with delta region (5e19cm^{-3}) under hetro-junction SiGe pocket (doping 1e20cm^{-3})with enlarged tunneling area and low leakage current [12] with reduced miller capacitance[13] suitable for switching of digital circuit. ZS-TFET further increases on state current and lower subthreshold slope as compared to other vertical TFET by suppressing lateral parasitic as well as providing more tunneling area and improved analog/RF performance[14]. Dual Material Gate TFET designed with metal gate of high work function overcome the problem of drain induced barrier lowering and improves on current with suitable high-K dielectric material under gate contact below 20nm channel length and Vdd=0.5V[15]. Mostly the tunnel FET designed shows considerable on current with longer channel length which indicates the compromise in terms of dimensions. Inclusion of junction-less source-channel and channel-drain region controls the parasitic resistance and improves on current [16-17]. Buried oxide region (SOI) under the channel region substantially improves subthreshold performance [18]. Junction-less tunnel FET shows increase in both ON and OFF current that leads to low I$_{ON}$/I$_{OFF}$ratio which a major drawback of junction-less transistors [19]. In this paper an ultra small pocket $Si_{0.7}Ge_{0.3}$nJLTFET is proposed and performance has been optimised in terms of gate contact material/oxide region to obtain better ON/OFF performance as well as good value of ON current. A comparison made between proposed pocket JLTFET and Junctions based Pocket Tunnel FET similar in structure and dimensions. The proposed design of nJLTFET can be a part of future high performance memory design of DRAM [20] cell and SRAM [21] cell.

II. DEVICE DESIGN AND STRUCTURES

2-D visual TCAD (Cogent) has been used to implement the new nJLTFET. The source/drain doping is kept at $10^{20}cm^{-3}$. The channel doping is kept at low value 10^{16} cm^{-3}. The channel doping is kept low to improve carrier mobility in this region. Fig. 1 indicates a new 2D nJLTFET structure with body thickness 0.5nm, buried oxide region with 3nm and p-type substrate region of 2nm. Thickness of oxide region under gate gate is kept 0.1nm. A metal gate with high value of work function Pt(5.7eV) used as gate contact material with gate length 2nm.Oxide regionmade with High dielectric constant material HfO_2(25) used replacing SiO_2(3.9)under gate and performance compared. To increase band to band tunneling property a pocket region of $Si_{0.7}Ge_{0.3}$ of 0.5nm thickness is included near the source region under influence of gate. Fig. 2 shows 3D structure of pocket JLTFET indicating device material region view.

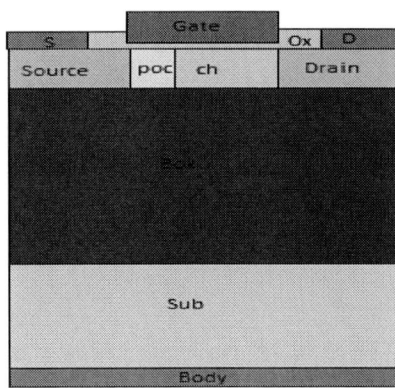

Fig. 1 2D structure of nJLTFET with SiGe pocket region

III. DEVICE SIMULATIONS AND PERFORMANCE

Firstly simulation performed for conventional junction based Tunnel FET then performance compared with pocket tunnel FET. Then proposed pocket JLTFET is

simulated in linear (V_{ds}=0.1V) and saturation region (V_{ds}=1V) with varying gate voltage.

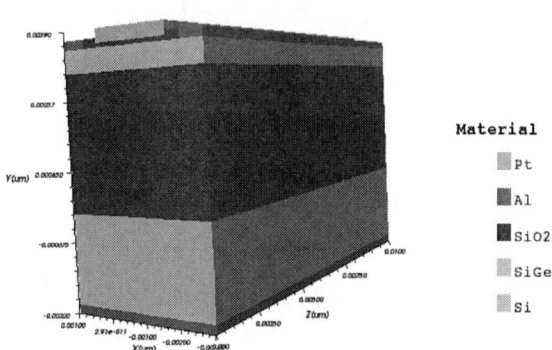

Fig.2 3D simulation view of nJLTFET Device material view

Fig. 3 shows sharp I_d versus V_{gs} characteristics of Junction-less TFET with or without pocket region. $Si_{0.7}Ge_{0.3}$ Pocket region introduced reduces off state current by maintaining on current and shows sharp subthreshold characteristic which further supports scaling limits. Id versus V_{gs} characteristics (Fig. 4) of JLTFET(with or without pocket region) shows almost similar behaviour in linear (V_{ds}=0.1V) and saturation region (V_{ds}=1V). Comparison of JLTFET with different gate and oxide material have been done in Fig. 5 which shows advantage of Pt/HfO2 interface as metal-gate/oxide region over other metal gate/oxide interface. Fig. 6 shows that Pocket JLTFET has higher I_{ON}/I_{OFF}current ratio and sharp subthreshold slope with V_{gs} variations in range of 0-0.5V(V_{ds}=0.1V).

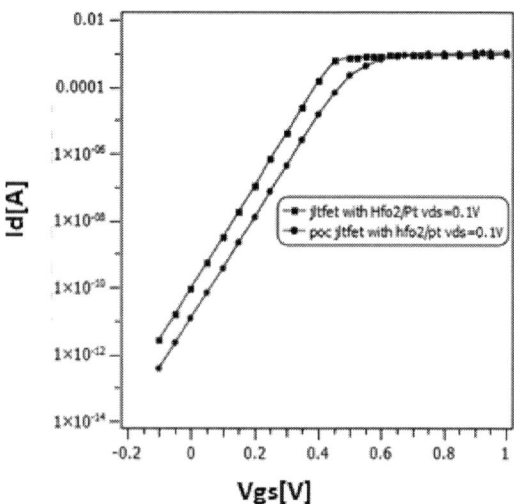

Fig. 3 I_d vs V_{gs} of pocket JLTFET (V_{ds}=0.1V)

2019 Devices for Integrated Circuit (DevIC), 23-24 March, 2019, Kalyani, India

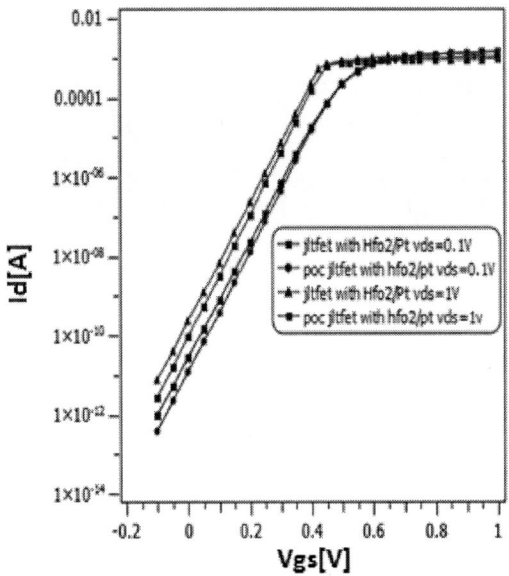

Fig. 4 I_d vs V_{gs} of pocket JLTFET in linear(Vds=0.1V) and saturation region(Vds=1V)

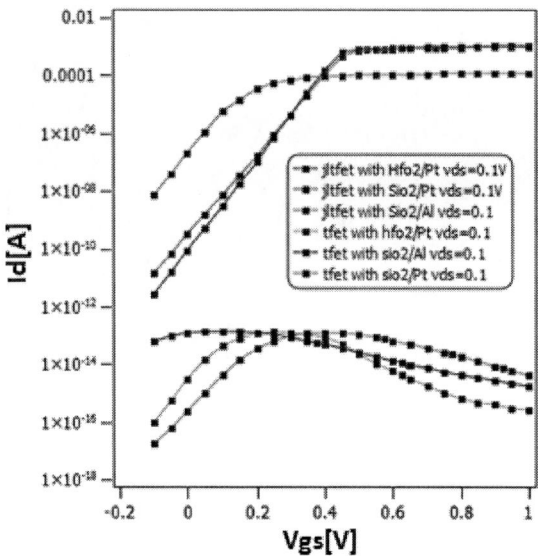

Fig. 5 I_d vs V_{gs} of pocket JLTFETin comparison to junction based TFET(V_{ds}=0.1V)

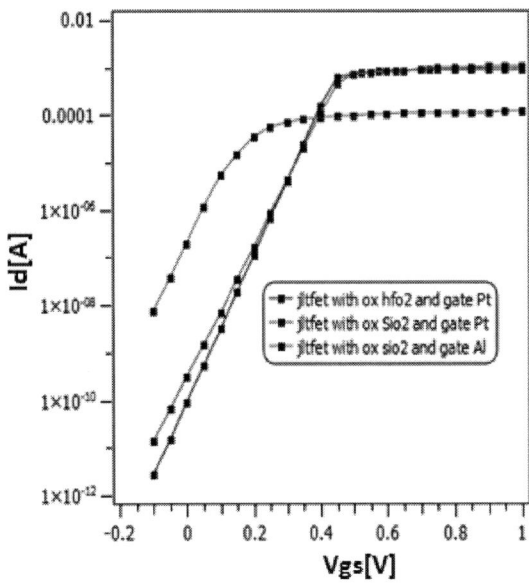

Fig. 5 I_d vs V_{gs} of pocket JLTFET with varying gate contact and oxide material (V_{ds}=0.1V)

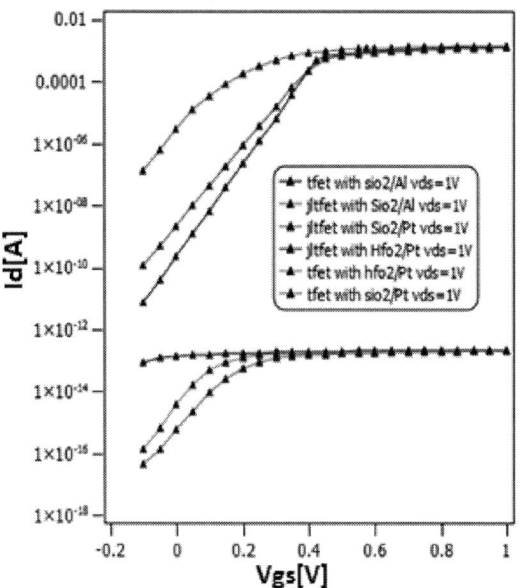

Fig. 6 I_d vs V_{gs} of pocket JLTFET in comparison to junction based TFET (V_{ds}=1V)

JLTFET with or without pocket region shows good performance in linear and saturation region but comparatively JLTFET with SiGe pocket region has better I_{ON}/I_{OFF} current ratio of the order of 2.64×10^8 as well as high value of subthreshold slope(124.25 mV/decade) even with ultra-small dimensions (L_g=2nm, Body thickness=0.5nm).

Similar characteristics of pocket JLTTFET obtained in saturation region (V_{ds}=1V) comparing other TFET structures(Fig. 7). Pocket JLTFET shows ideal I_d versus V_{ds} characteristics with variations in Vgs in range of 0-1V (Fig. 8).

978-1-5386-6723-1/19 $31.00 © 2019 IEEE   81

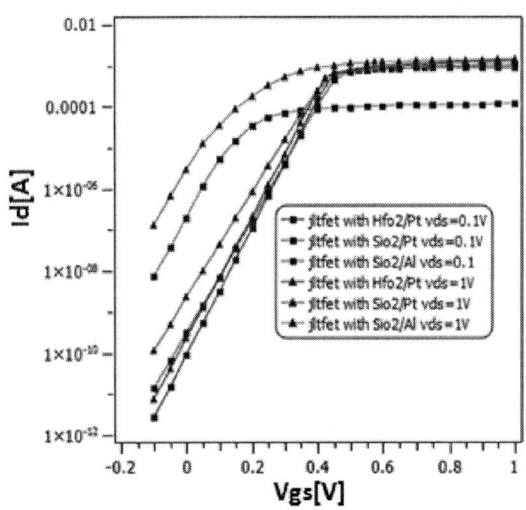

Fig. 7 Id vs Vgs of pocket JLTFET in linear and saturation region

Such ultra-small JLTFET has potential improve on chip performance, low power dissipation and lesser on chip area for low voltage operations (0.5V).

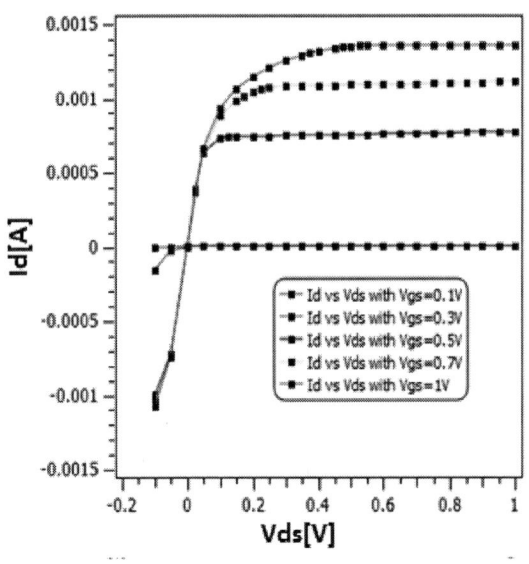

Fig. 8 I_d vs V_{ds} of pocket JLTFET with varying V_{gs} (0-1V)

V. CONCLUSIONS

Proposed ultra small pocket $Si_{0.7}Ge_{0.3}$ JLTFET exhibit high value of I_{ON}/I_{OFF} current ratio (Table1) above than 10^8 in comparison junction based Tunnel FET with similar dimensions with or without pocket regions.

Although the dimensions are very small even then the device subthreshold performance is under limit with tremendously sharp subthreshold slope of 76mV/Decade and DIBL have a value of 23mV/V. This support the further the scaling trend with advancement of technology incorporating future applications.

Table1: Performance comparison between TFET and JLTFET with or without Pocket region

Device type	I_{OFF}	I_{ON}	I_{ON}/I_{OFF}
TFET with pocket$Si_{0.7}Ge_{0.3}$(HfO2/Pt)	8.75E-17	4.06E-14	4.64E2
TFET with Pocket$Si_{0.7}Ge_{0.3}$(SiO2/Pt)	4.48E-17	4.96E-14	1.11E3
JLTFET with Pocket$Si_{0.7}Ge_{0.3}$	2.65E-12	7.00E-04	2.64E8
JLTFET without Pocket	3.51E-13	7.65E-05	2.17E8

REFERENCES

[1] C. R. Manoj, Meenakshi Nagpal, etal. "Device Design and Optimization Considerations for Bulk FinFETs" IEEE Transactions on Electron Devices, 55, 2008

[2] J. P. Colinge, "The new generation of SOI MOSFETs," Rom. J. Inf. Sci. Technol, 11, 3–15, 2008.

[4] Hui Zhao, Yee-Chia Yeo, etal.. "Analysis of the Effects of Fringing Electric Field on FinFET Device Performance and Structural Optimization Using 3-D Simulation" IEEE transactions on Electronics Devices, 55, 2008.

[3] Suman Lata Tripathi, Sanjeet Kumar SinhaPriti Gupta "Design of Triple material Junctionless CG MOSFET" IEEE conference ICICS, 2018

[5] Thomas Nirschl, Peng-Fei Wang, etal. "The tunneling field effect transistor (TFET): the temperature dependence, the simulation model, and its application" IEEE conference 2004

[6] Yasin Khatami, and Kaustav Banerjee "Steep Subthreshold Slope n- and p-Type Tunnel-FET Devices for Low-Power and Energy-Efficient Digital Circuits" IEEE transactions on electron devices, 56, 2009

[7] U. E. Avci et al., "Understanding the feasibility of scaled III–V TFET for logic by bridging atomistic simulations and experimental results,"in Proc. VLSI Technol. (VLSIT) Symp., Honolulu, HI, USA, 183–184, 2012

[8] Uygar E. Avci, Daniel H. Morris, and Ian A. Young "Tunnel Field-Effect Transistors: Prospects and Challenges" Journal of Electron Device society, IEEE, 2014

[9] Ankit Sharma, A. Arun Goud, and Kaushik Roy, "P-channel Tunneling Field Effect Transistor (TFET): Sub-10nm Technology Enablement by GaSb-InAs with Doped Source Underlap" in Device research conference (DRC) IEEE, 2015

[10] Zhaonian Yang "Tunnel Field-Effect Transistor Withand L-Shaped Gate"IEEE electron device letters, 37, 2016

[11] Wei Li, Hongxia Liu, Shulong Wang, Shupeng Chen and Zhaonian Yang "Design of High Performance

Si/SiGeHeterojunction Tunneling FETs with a T-Shaped Gate" Nanoscale Research Letters, 2017

[12]Wei Li, Hongxia Liu, Shulong Wang, and Shupeng Chen "Reduced Miller Capacitance in U-Shaped Channel Tunneling FET by Introducing Heterogeneous Gate Dielectric" IEEE Electron Device Letters, 38, 2017

[13] Wei Wang, Peng-Fei Wang etal."Design of U-Shape Channel Tunnel FETs WithSiGe Source Regions" IEEE transactions on electron devices, 61, 2014.

[14] RouzbehMolaeiImenabadi , Mehdi Saremi , and William G. Vandenberghe, "A Novel PNPN-Like Z-Shaped Tunnel Field-Effect Transistor With Improved AmbipolarBehavior and RF Performance"IEEE transactions on electron devices, 64, 2017

[15]SnehSaurabh and M. Jagadesh Kumar "Novel Attributes of a Dual Material Gate Nanoscale Tunnel Field-Effect Transistor" IEEE Transactions on Electron Devices, 58, 1023–1029, 2012.

[16]S. Sahay and M. J. Kumar, "Controlling L-BTBT and volume depletionin nanowire JLFETs using core–shell architecture," IEEE Trans. Electron Devices, 63, 3790–3794, 2016.

[17] M. J. Kumar and S. Sahay, "Controlling BTBT-induced parasitic BJTaction in junctionless FETs using a hybrid channel," *IEEE Trans.Electron Devices*, 63, 3350–3353, 2016.

[18] S. Sahay and M. J. Kumar, "Realizing efficient volume depletion in SOI junctionless FETs," *IEEE* J. Electron Devices Soc., 4, 2016

[19]AvinashLahgere, MamidalaJagadesh Kumar "A Tunnel Dielectric-Based Junctionless Transistor With Reduced Parasitic BJT Action" IEEE Transactions on Electron Devices, 2017

[20] Li et al The Programming Optimization of Capacitorless 1T DRAM Based on the Dual-Gate TFET, Nanoscale Research Letter 12, 524, 2017

[21] Ahmad et al Robust TFET SRAM cell for ultra-low power IOT applications, AEU - International Journal of Electronics and Communications, Elsevier, 89, 70-76, 2018

2019 Devices for Integrated Circuit (DevIC), 23-24 March, 2019, Kalyani, India

Performance Analysis of FinFET device Using Qualitative Approach for Low-Power applications

Shekhar Verma
Electronics & Communication
Enginnering
Lovely Professional University
Phagwara, India
shekharverma05@gmail.com

Suman Lata Tripathi
Electronics & Communication
Enginnering
Lovely Professional University
Phagwara, India
suman.21076@lpu.co.in

Mohinder Bassi
Electronics & Communication
Enginnering
Lovely Professional University
Phagwara, India
mohinderbassi@gmail.com

Abstract—The introduction of Field Effect Transistor (FinFET) Technology played a leading contender in today microelectronics. FinFET structure allows to scale the device at sub-nanometer. Short channel effects can be suppressed by formation of ultra-thin fin in FinFET device. In this paper we compared the performance of the 20nm FinFET device by using different dielectric materials. We have considered only n-channel FinFET device. Simulation carried on the electron mobility, potential distribution, energy band of hole and electron, on-off current ratio (Ion/Ioff) and power dissipation of device with respect to the applied gate voltage. Mobility enhancement and higher current ratio (Ion/Ioff) is observed in proposed FinFET device having high k-dielectric material at lower voltage. This designed can be useful for low power applications due to low power dissipation. In high k-dielectric material, 1.41% improvement is observed in potential voltage with respect to low k- dielectric material when V_{gs} at low voltage and 0.98% improvement is observed when V_{gs} at high voltage. In high k- dielectric material 15% hike is observed in the energy conduction band as compared to low k-dielectric material when V_{gs} at low voltage and 14% hike is observed when V_{gs} at high voltage.

Keywords— FinFET, mobility, high- K dielectric material, electrical characteristics, threshold voltage, Short channel effect, Complementary Metal-Oxide-Semiconductors(CMOS).

I. INTRODUCTION

To achieve the high performance of semiconductor devices is a challenging aspect in now days, as we scale down the size of the transistor size. If we continuing the scaling down of the MOSFET in the nanometer regime that will be challenging aspect due to increases in the leakage current (Ioff) [1-3]. The drain potential of MOSFET in narrow channel length, influenced the electrostatics of the channel and consequently, the gate is unable to off the channel in the shut-mode of operation. The use of higher- k dielectric materials and thinner gate oxides helps to overcome this problem by increasing the gate-channel capacitance. [4-6].

FinFET are considered as most promising devices to minimize the short channel effects and leakage and it acts as an alternative for replacing the CMOS devices below 22nm [7-8]. For providing the good control for electrical, it provided by the wrapped-around gate structure that leads to decreasing in the short channel effects and subthreshold leakage current.

FinFET device has many advantages as compared to the planner structure of MOSFET device in terms of short channel effects, for reducing the subthreshold swing and small threshold voltage roll-off [8-9]. In industries they used the non-uniform shape similar to trapezoidal [10].In this paper we presented the evaluation of on-off current ratio

(Ion/Ioff), electron mobility, subthreshold voltage and energy band level of electron and hole on 20-nm FinFET device. All the simulation carried on the Visual TCAD by cogenda. This paper divided into three section, sections-1 as an introduction and section-2, shows the earlier work on FinFETs and Section-3 explain the simulation setup and structure design along with the results discussion. In section -4 we presents the conclusion.

II. LITERATURE REVIEW

For the scaling down of the oxide thickness and gate length, can be attained by changing the device structure of MOSFET. For the structure of 10-30nm channel length, double-gate MOSFET has been considered as a most promising device. Performance of the FinFET device generally effected by the oxide thickness, height of the fin (H_{fin}) , fin width (Wf_{fin}), gate length (L_{gate}). For getting the better performance of the devices, these parameter can be vary. The thickness of the fins in FinFET devices need to be kept minimum as compared to (1/3) channel length to reduce the short channel effects [11]. High k-dielectric material gate stack used, for reducing the gate oxide leakage current. [12]. Gate stack of high- k dielectric material improved the gate to channel capacitive coupling with degrading in gate oxide layer. For developing the compact model for FinFETs is very challenging task due smaller size with multi-dimensional structure. [14]. for reducing the SCEs effect in FinFET can be attain by increasing the height of the fin. This lead to better subthreshold slope and significant DIBL [14].

III. SIMULATION SETUP & STRUCTURE DESIGN

A. Structure Design Parameter and material Composition

The 20-nm channel length of FinFET with rectangular fins shape, designed on Cogenda's GDS2 Mesh 3D[15]. In Table-1, dimensions of the design are mentioned. High k-dielectric material Hafnium Oxide is used as a gate oxide and silicon is used for substrate and fin. Pictorial view of 3D rectangular FinFET as shown in figure 1.a , 1.b and 1.c.

Figure 1.a) Active region of rectangular fin shaped of low k-dielectric material

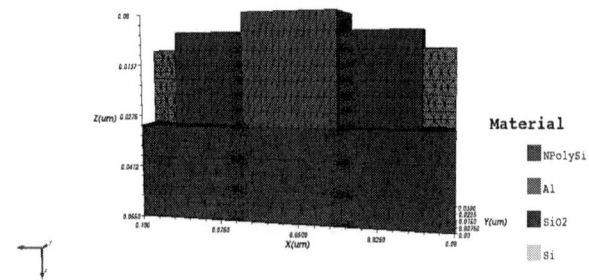

978-1-5386-6723-1/19 $31.00 © 2019 IEEE

Figure 1.b) Internal View of Active region

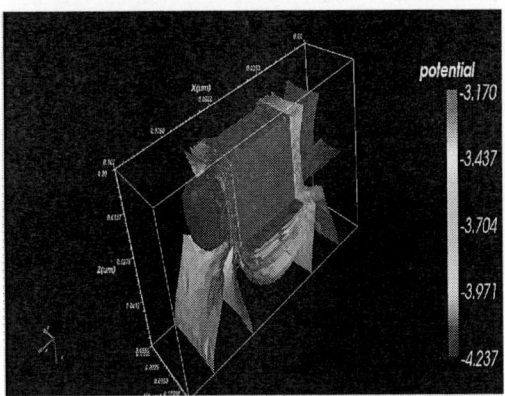

Figure 1.c) Active region of rectangular fin with high k-dielectric material

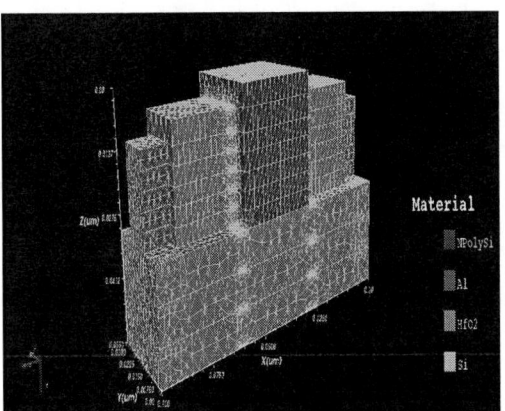

TABLE 1: Dimension of the Proposed FinFET

S.No	FinFET Parameter	Value
1	Fin Width (nm)	2
2	Oxide thickness (nm)	15
3	Fin Height (nm)	10 to 40
4	Doping Concentration (/cm^3)	10^{15} to 10^{18}

B. Simulation Setup

In this paper, we simulated rectangular FinFET of 20 nm gate length on Cogenda's Visual TCAD by using high-k dielectric material of HFO2[16]. In this work we applied the step voltage of 1V across the V_{gs} and constant voltage of 1 V across the V_{ds}. For calculating the driving force of electron and holes, quantum potential included in density gradient model of FinFETs [17, 18]. Quantum corrections of the hole and electrons included in simulation as per below equations 1 [20].

$$\nabla_n = -((\hbar^2 \gamma n)/(6qm_n^*)) * ((\nabla^2 \sqrt{n})/(\sqrt{n})) \quad (1)$$

$$\nabla_p = -((\hbar^2 \gamma n)/(6qm_p^*)) * ((\nabla^2 \sqrt{p})/(\sqrt{p})) \quad (2)$$

Where mn* and mp* are the electron and hole effective mass, n and p are the electron and hole concentration in conduction band and valance band, q is electron charge and h is Planck's constant.

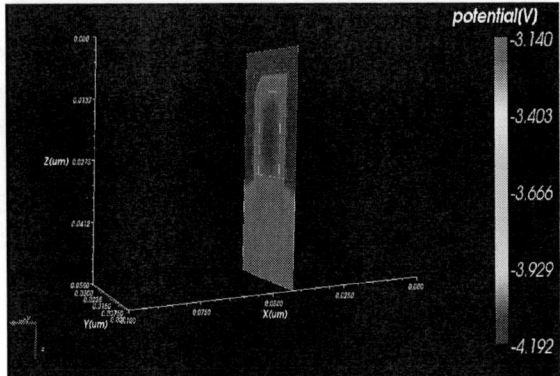

Figure 2: Potential structure of FinFET Design with high-k Material at high voltage

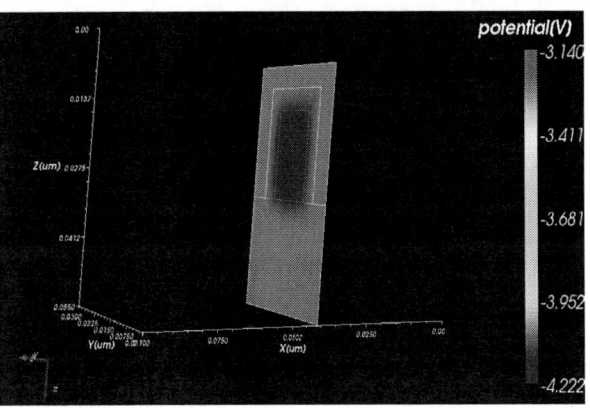

2019 Devices for Integrated Circuit (DevIC), 23-24 March, 2019, Kalyani, India

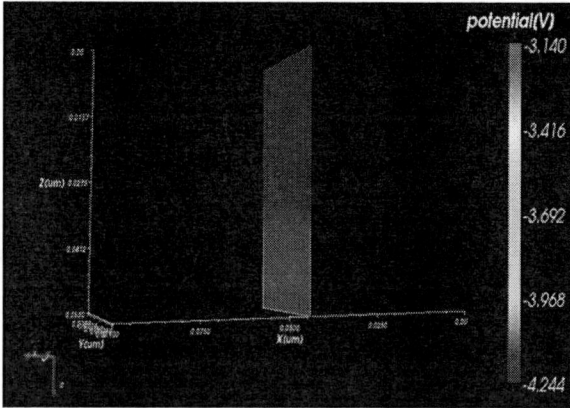

Figure 3: Potential structure of FinFET Design with high-k Material at Low Voltage

Figure 4: Potential structure of FinFET Design with Low-K Material at High Voltage

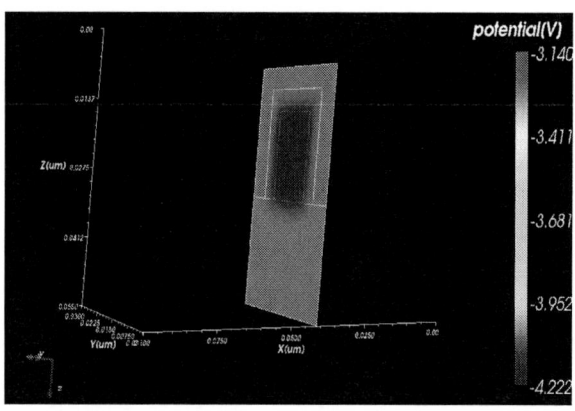

Figure 5: Potential structure of FinFET Design with Low-K Material at Low Voltage

C. Results and Discussion

In simulations results, variation of the potential distribution as shown in figure 2 and figure 4 respect to the Distance (Cut Line with respect to X axis – 0.05nm) at 1 V across the gate source voltage for low k-dielectric material and as well as high k-dielectric material of FinFET device. It has been observed in case of high k-dielectric material, potential distribution as shown in figure 6 is good as compared to the low k-dielectric material due to . At low voltage, potential distribution of low k-dielectric material high as compared to the high k-dielectric material as per

figure 3 and figure 5 & 6. When device in on state then in low k-dielectric material, more leakage current will there due to high voltage difference along the x-axis. Same case will be happen device in off mode due to high potential difference in low k-dielectric material. For high – k material and low- k material, analyzed the energy band level of electron and hole at low voltage and high voltage.

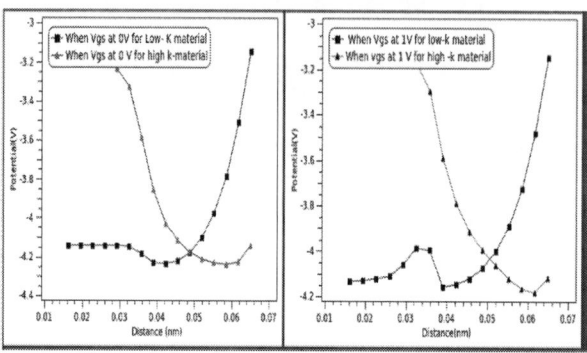

Figure 6: Combine Potential structure of FinFET

As per figure 7, it has been observed that energy band level of electron in case of high k- dielectric material close to 1 but in case of low k- dielectric material its 0.886 at high voltage. In case of higher k-dielectric material based designed of FinFET, conduct faster as compared to the low k-dielectric material. This designed more reliable for those applications where required fast switching.

As per Figure 8, drain current of the high-k material linearly vary with respect to the gate source voltage. But in case of low-K material, drain current not vary linearly. When gate voltage at 0V, leakage current (Ioff) flowing in the FinFET device. High k-dielectric material based FinFET designed more immune to the subthreshold leakage current as compared to low k- dielectric material.

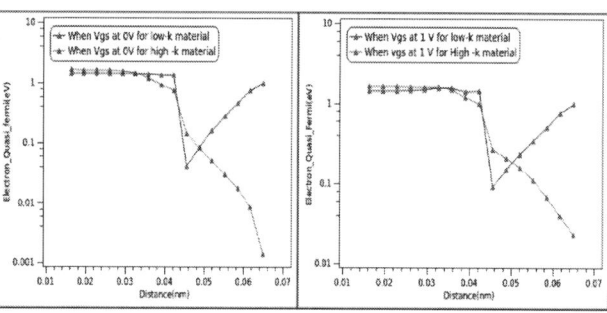

Figure 7: Combine Electron Fermi level of FinFET

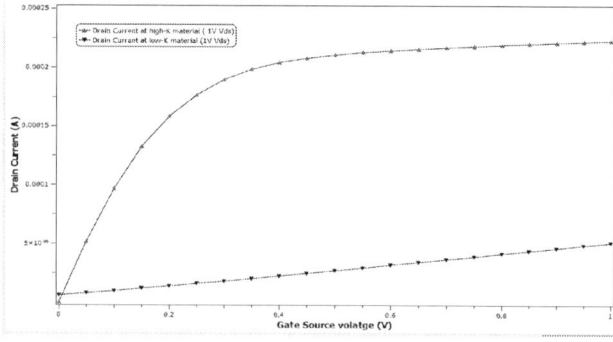

978-1-5386-6723-1/19 $31.00 © 2019 IEEE   86

Figure 8: Drain Current Vs Gate Voltage for different dielectric material

Figure 9: Power Vs. Gate Voltage for different dielectric material

In figure 9, compared the power dissipation for both the material. In case of High-K dielectric material, power is less as compared to the low-K dielectric material. Thus, threshold voltage is less affected by variation in applied voltage and the use of HFO2 as a gate dielectric scale the oxide thickness without increasing current

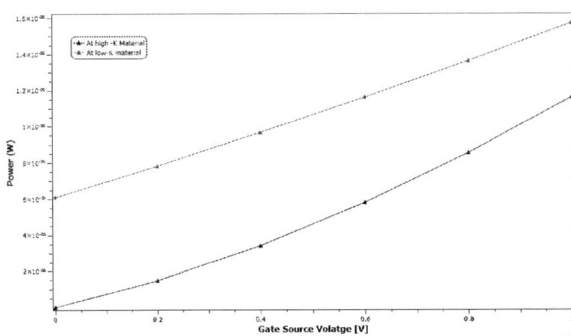

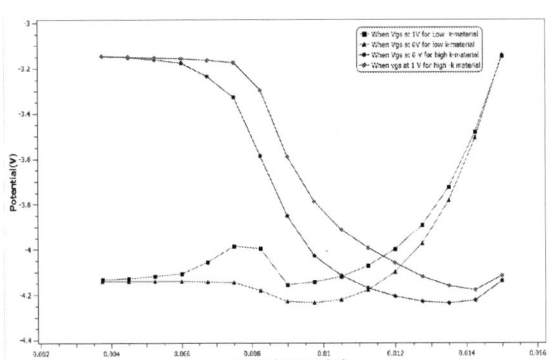

Figure 10: Combine Potential structure of FinFET along Y-Axis

IV. CONCLUSION

From the simulations results, it has been observed that variation of process parameter of FinFET has considerable impact on the performance parameter of the FinFET. In high k-dielectric material, 1.41% improvement is observed in potential voltage with respect to low k- dielectric material when Vgs at low voltage and 0.98% improvement is observed when Vgs at high voltage. In high k- dielectric material 15% hike is observed in the energy conduction band as compared to low k-dielectric material when Vgs at low voltage and 14% hike is observed when Vgs at high voltage. Due to high-K dielectric material (HFO2), gate control over channel is improved and it leads to better potential distribution and Fermi level of electron and holes for FinFET

device. Leakage current is minimum in high-k material due to lower off current and in return, it improve the subthreshold leakage current of the devices and reduces the short channel effect. Analytical modeling may be considered as a part of prospective work in future.

REFERENCES

[1] A. Arasteh, M. Hossein Moaiyeri, M. Taheri, K. Navi, and N. Bagherzadeh, "An energy and area efficient 4:2 compressor based on FinFETs," Integr. VLSI J., vol. 60, no. June 2017, pp. 224–231, 2018.

[2] F. Jazaeri, C. M. Zhang, A. Pezzotta, and C. Enz, "Charge-Based Modeling of Radiation Damage in Symmetric Double-Gate MOSFETs," IEEE J. Electron Devices Soc., vol. XX, no. XX, pp. 1–9, 2017.

[3] Deepika Singh, Sunil Pandey, Kaushal Nigam, Dheeraj Sharma, Dharmendra Singh and Pravin Kondekar, "A Charge Plasma- Based Dielectric-Modulated Junctionless TFET for Biosensor Label-Free Detection.", IEEE Transactions On Electron Devices, vol.64, no.1, pages 271-278, Jan 2017.

[4] Rajashree Das, Rupam Goswami and Srimanta Baishya "Tri-gate Hetrojunction SOI e-FinFets." Superlattices and Microstructure(2016), vol 19, no. 03, pages 51-61 Mar 2016

[5] Chuyang Hong, Jun Zhou, Jiasheng Huang, Rui Wang, Wenlong Bai, James B. Kuo,."A General and Transformable Model Platform Emerging Multi-Gate MOSFETs." IEEE Electronic Device Letters , vol. 11, no. 12, pages921–924, Mar 2017.

[6] Hooman Farkhani,Ali Peiravi and Jens Madsen Kargaar, "Comparative Study of FinFETs versus 22nm Bulk CMOS technologies:SRAM Design Perspective." System-on-Chip Conference (SOCC), 27th IEEE International, vol. 5, no. 10, pages 449-454, Nov 2016.

[7] M. P. King, X. Wu, M. Eller, S. Samavedam, M. R. Shaneyfelt, A. I. Silva, B. L. Draper, W. C. Rice, T.Meisenheimer, J. A. Felix, K. J. Shetler E. X. Zhang, T. D. Haeffner, D. R. Ball, M. L. Alles, J. S. Kauppila, L. W. Massengill, "Analysis of TID Process, Geometry, and Bias Condition Dependence in 14-nm FinFETs andImplications for RF and SRAM Performance." IEEE Transactions on Nuclear Science, vol. 64, no. 1, pages 285-292, Jan 2017.

[8] Brad D. Gaynor and Soha Hassoun, "Fin Shape Impact on FinFET Leakage With Application to Multithreshold and Ultralow-Leakage FinFET Design." IEEE Transactions On Electron Device, vol. 61, no. 8 pages 2738-2745 Aug 2014.

[9] Yao-jen lee, guang-li luo, fu-ju hou, min-cheng chen, chih-chao yang, chang-hong shen, wen-fa wu, jia-min shieh, and wen-kuan yeh, "Ge GAA FETs and TMD FinFETs for the Applications Beyond Si—A Review." vol. 4 no. 5 pages 286-294 Sept 2016.

[10] Youngtaek Lee, Sand Changhwan Shin, "Impact of Equivalent Oxide Thickness on Threshold Voltage Variation Induced by Work-Function Variation in Multigate Devices." IEEE Transactions on Electron Devices, vol 64, no 5, pages 2452-2456 Mar 2017.

[11] A. M. Bughio, S. Donati Guerrieri,and G. Ghione, "Multi-gate FinFET Mixer Variability assessment through physics-based simulation." IEEE Electron Device Letters vol. 38, no. 8 pages-1004-1007 Aug 2017.

[12] Venkata P. Yanambaka, Saraju P. Mohanty Elias Kougianos, Dhruva Ghai, and Garima Ghai, "Process Variation Analysis and Optimization of a FinFET based VCO." IEEE Transactions on Semiconductor Manufacturing, vol.30, no.2 pages-126-134, Feb 2017.

[13] Keunwoo Kim and J. G. Fossum, "Double-Gate CMOS: Symmetrical- Versus Asymmetrical-Gate Devices" IEEE Trans. Electron Devices, vol. 48, pp. 294–299, Feb 2001

[14] A. Wang, and A. Chandrakasan, "A 180-mVSubthreshold FET Processor Using a Minimum Energy Design Methodology," IEEE J. Solid-State Circuits, vol. 40, no. 1, pp. 310-319, Jan. 2005.

[15] http://www.cogenda.com/article/Gds2Mesh.

[16] R. Rooyackers, E. Augendre, B. Degroote, N. Collaert, A. Nackaerts, A. Dixit, T. Vandeweyer, B. Pawlak, M. Ercken, E. Kunnen, G. Dilliway, F. Leys, R. Loo, M. Jurczak, and S. Biesemans. Doubling or quadrupling mugfet fin integration scheme with higher pattern fidelity, lower cd vari- ation and higher layout efficiency. In Electron Devices Meeting, pages 1–4, December 2006.

[17] Xin Sun, V. Moroz, N. Damrongplasit, Changhwan Shin, and Tsu-Jae King Liu. Variation study of the planar ground-plane bulk mosfet, soi finfet and trigate bulk mosfet designs. IEEE Transactions on Electron Devices,58(10):3294–3299, October 2011.

[18] Brian Swahn and Soha Hassoun. Gate sizing: finfets vs 32nm bulk mosfets. In ACM/IEEE Design Automation Conference (DAC), pages 528–531, 2006.

[19] M.J.H. van Dal, N. Collaert, G. Doombos, G. Vellianitis, G. Curatola, B.J.Pawlak, R. Duffy, C. Jonville, B. Degroote, E. Altamirano, E. Kunnen, M. Demand, S. Beckx, T. Vandeweyer, C. Delvaux, F. Leys, A. Hikavyy, R. Rooyackers, M. Kaiser, R.G.R. Weemaes, S. Biesemans, M. Jurczak, K. Anil, L. Witters, and R.J.P. Lander. Highly manufacturable finfets with sub-10nm fin width and high aspect ratio fabricated with immersion lithography. In IEEE Symposium on VLSI Technology, pages 110–111, June 2007.

[20] Alf J. van der Poorten. Some problems of recurrent interest. Technical Report 81-0037, School of Mathematics and Physics, Macquarie Univer- sity, August 1981.lications," IEEE Trans. Electron Devices, vol. 51 (9), pp. 1468-1474, 2004.

[21] Gu, J. Keane, S. Sapatnekar, and C. Kim. Width quantization aware finfet circuit design. In IEEE Custom Integrated Circuits Conference (CICC), pages 337 –340, September 2006.

[22] [B. Muddu S. V. Nakagawa S Gupta, P. Kahng. Modeling edge place- ment error distribution in standard cell library. In The International Sociaty For Optical Engineering, pages 6156–57, 2006.J. Van der Geer, J.A.J. Hanraads, R.A. Lupton, J. Sci. Commun. 163 (2000) 51–59.

[23] L. Beloni Devi, Kundan Singh, A. Srivastava. "Impact of Substrate Bias and Dielectrics on the Performance parameters of Symmetric Lateral Bipolar Transistor on SiGe-OI for Mixed signal applications", MicroelectronicsJournal, 2018.

[24] Ankit Gaurav, Sandeep S. Gill, Navneet Kaur, Munish Rattan. "Density gradient quantum corrections based performance optimization of triangular TG bulk FinFETs using ANN and GA", 2016 20th International Symposium onVLSI Design and Test (VDAT), 2016

Analytical Threshold Voltage Model of Schottky-source/drain (Schottky-S/D) double gate-all-around (DGAA) Field-Effect-Transistors (FETs)

Arun Kumar[1], P.S.T.N. Srinivas[1], and Pramod Kumar Tiwari[1]

[1]Department of Electrical Engineering Indian Institute of Technology Patna, Bihta, Bihar, 801106, India

Email: arun.pee16@iitp.ac.in, srinu.teja3509@gmail.com, pktiwari@iitp.ac.in

Abstract—This paper reports an analytical model of threshold voltage for the Schottky-source/drain (Schottky–S/D) double gate-all-around (DGAA) FETs. In order to develop the threshold voltage model, the method of parabolic potential approximation is used to solve the 3-D Poisson's equation to formulate the surface potentials and threshold voltage expressions. The proposed model of threshold voltage is based on a modified definition of the threshold voltage. The influence on threshold voltage due to variation in device physical parameters like core and oxide thickness, channel thickness, and source-drain metal work function variation, and drain voltages have been investigated. The proposed analytical model results have been compared and verified against the results obtained from 3-D ATLAS device simulator from SILVACO.

Keywords—Schottky-source/drain; DGAA FETs; threshold voltage; device modeling

I. INTRODUCTION

Several innovative multi-gate MOSFET architectures such as Trigate FETs, quadruple-gate (QG) FETs and gate-all-around (GAA) FETs with improved immunity against short-channel effects (SCEs) have been proposed for sustaining the MOSFET scaling for next generation computational efficient and high-speed ULSI applications [1-7]. Recently, double gate-all-around (DGAA) MOSFET structures are being pursued rigorously [8-9] in the quest of both excellent SCEs immunity and superior current driving capability in a single device. In DGAA MOSFETs, a tube-like silicon channel is enclosed by outer and inner gates that are known as shell and core gates, respectively [8-15]. This unique device structure with double gate provides better electrostatic control over entire channel region. However, from the fabrication perspective, the fabrication of heavily doped source-drain regions may require a high-temperature process, which is one of the significant demerits of conventional MOSFETs [16]. Further, the formation of abrupt p-n junctions (channel-source and channel-drain) is challenging in an undoped/lightly doped short-channel non-conventional MOSFETs because of the lateral spreading of implanted impurities. Consequent to the above limitations, The Schottky-source/drain (Schottky-S/D) barrier based MOSFET structures can be used in place of silicon source/drain based MOSFETs [16-18]. The significant advantages of Schottky-S/D structure include low source-drain sheet resistance, low channel doping, low parasitic

bipolar effect [16-18]. N-channel Schottky-S/D MOSFET was fabricated and tested by Mochizuki *et .al.*[19].They observed that for long channel Schottky-S/D MOSFETs the electrical characteristics were almost similar to conventional MOSFETs of the same channel length [19]. Wang *et al.* [20] depicted that the short-channel effects can be controlled in nano-scale devices by replacing conventional source-drain by Schottky-S/D. Gate-all-around (GAA) MOSFETs with Schottky-S/D show extended scalability in the nanometer regime. Thus, considering the advantages of unique double gate-all-around structure and Schottky-source/drain contacts in a single structure, Schottky-S/D double gate-all-around (DGAA) FETs could be a probable choice for ULSI applications [21-24]. Therefore, in this work, an analytical model of threshold voltage is derived for Schottky-S/D DGAA FETs. The method of parabolic potential approximation is used to solve the 3-D Poisson's equation to formulate the surface potentials and threshold voltage model of the device. The results attained from the proposed analytical model have been compared and validated against the results obtained from commercially available ATLAS™ TCAD [25] simulation data.

II. DEVICE STRUCTURE

Figure 1 demonstrates the simulated 3-D device schematic of Schottky-S/D DGAA MOSFET. The device gate length is denoted as L, as depicted in Fig.1. The core radius inner/outer gate oxide thickness and the channel thicknesses are t_c, t_{ox} and t_{si}, respectively. Source and drain regions with high doping profile are replaced by metallic contacts having a work function $\varphi_{S/D} = 4.1 eV$. The channel region is uniformly and lightly doped and, the doping density is $N_a = 10^{15} cm^{-3}$. The electronic charge is represented by q. Supply voltage V_{gs} is provided to both of the inner and outer gates and φ_G represents the gate work function of the device. The permittivity value of silicon is considered as $\epsilon_{si} = 11.8 \times 8.854 \times 10^{-14} F/cm$ and for SiO_2 is $\epsilon_{ox} = 3.97 \times 8.854 \times 10^{-14} F/cm$. A two-dimensional (2D) cross-sectional picture of the device structure is depicted in Fig. 2. To analyze the device structure, the cylindrical co-ordinate system is used with an origin (0,0,0) at the center of source contact, as shown in Fig.1.

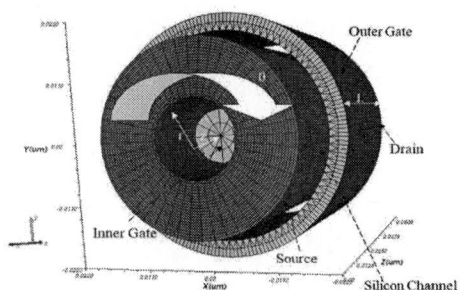

Fig.1: 3-D simulated structure of DGAA FET

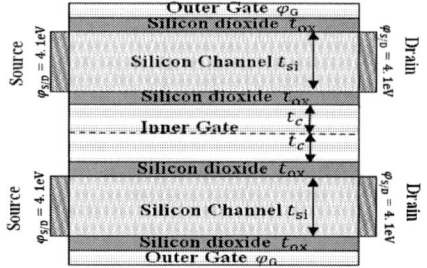

Fig.2: 2-D cross-section picture of the device.

III. MODEL DERIVATION

A. Electrostatic potential distribution formulation

The channel electrostatic potential distribution $\phi(r,\theta,z)$ of Schottky-S/D DGAA MOSFET has been formulated by solving 3-D Poisson's equation in the channel region as follows

$$\frac{1}{r}\frac{d}{dr}\left[r\frac{d}{dr}\phi(r,\theta,z)\right] + \frac{1}{r^2}\frac{d^2}{d\theta^2}\phi(r,\theta,z) + \frac{d^2}{dz^2}\phi(r,\theta,z) = \frac{qN_a}{\epsilon_{si}} \quad (1)$$

potential $\phi(r,\theta,z)$ does not vary along θ owing to an angular symmetry in that direction, so Eq. (1) can be modified as:

$$\frac{1}{r}\frac{d}{dr}\left[r\frac{d}{dr}\phi(r,z)\right] + \frac{d^2}{dz^2}\phi(r,z) = \frac{qN_a}{\epsilon_{si}} \quad (2)$$

The electrostatic potential variation in the radial direction can be given assuming parabolic potential as

$$\phi(r,z) = C_1(z) + C_2(z)r + C_3(z)r^2 \quad (3)$$

where, C_1, C_2 and C_3 are z dependent coefficients. Let ϕ_{s1} and ϕ_{s2} represent the surface potentials at inner and outer Si-SiO$_2$ interfaces as

$$\phi_{s1}(z) = \phi(r = r_1, z) \quad (4)$$

$$\phi_{s2}(z) = \phi(r = r_2, z) \quad (5)$$

where, $r_1 = t_c + t_{ox}$ and $r_2 = t_c + t_{ox} + t_{si}$
Channel potential $\phi(r,z)$ can be specified at $z = 0$ and L as follows

$$\phi(r, z = 0) = V_{bi,S(D)} \quad (6)$$

$$\phi(r, z = L) = V_{bi,S(D)} + V_{ds} \quad (7)$$

where, V_{ds} is an applied drain to source voltage, $V_{bi,S(D)}$ is the built-in potential developed at source-drain boundaries. $V_{bi,S(D)}$ can be evaluated as [22]

$$V_{bi,S(D)} = \chi_{si} + \frac{E_g}{2q} + V_T \ln\left(\frac{N_a}{n_i}\right) - \varphi_{S/D} \quad (8)$$

where, χ_{si} and E_g are electron affinity and the energy band gap of Silicon, V_T is the thermal voltage. It may be observed that in the subthreshold region of device operation the source/drain barrier lowering owing to image force may be neglected. Intrinsic carrier concentration of silicon is represented as n_i. We have already developed the channel and surfaces potential of double gate-all-around (DGAA) FETs in our previous work [12]. We can adopt those equations for predicting the electrostatic channel potential of Schottky-S/D DGAA MOSFET with modified boundary conditions of Eqs. (6) and (7). The following equations of inner and outer surface potentials ϕ_{s1}, ϕ_{s2}, respectively can be written as

$$\phi_{s1}(z) = k_1 e^{\sqrt{\alpha_1}z} + k_2 e^{-\sqrt{\alpha_1}z} - \frac{\beta_1}{\alpha_1} \quad (9)$$

$$\phi_{s2}(z) = k_3 e^{\sqrt{\alpha_2}z} + k_4 e^{-\sqrt{\alpha_2}z} - \frac{\beta_2}{\alpha_2} \quad (10)$$

where, k_1, k_2, k_3 and k_4 have been evaluated using the boundary conditions mentioned in Eqs. (6) and (7), given as follows

$$k_1 = \frac{-(PS_1)e^{-\sqrt{\alpha_1}z} + (PD_1)}{2\sinh\sqrt{\alpha_1}L} \quad (11)$$

$$k_2 = \frac{(PS_1)e^{\sqrt{\alpha_1}z} - (PD_1)}{2\sinh\sqrt{\alpha_1}L} \quad (12)$$

$$k_3 = \frac{-(PS_2)e^{-\sqrt{\alpha_2}z} + (PD_2)}{2\sinh\sqrt{\alpha_2}L} \quad (13)$$

$$k_4 = \frac{(PS_2)e^{\sqrt{\alpha_2}z} - (PD_2)}{2\sinh\sqrt{\alpha_2}L} \quad (14)$$

where,

$$PS_1 = \left(V_{bi,S(D)} - \sigma_1\right) \quad (15)$$
$$PD_1 = \left(V_{ds} + V_{bi,S(D)} - \sigma_1\right) \quad (16)$$

$$PS_2 = \left(V_{bi,S(D)} - \sigma_2\right) \quad (17)$$

$$PD_2 = \left(V_{ds} + V_{bi,S(D)} - \sigma_2\right) \quad (18)$$

$$\sigma_1 = -\frac{\beta_1}{\alpha_1} \quad (19)$$

$$\sigma_2 = -\frac{\beta_2}{\alpha_2} \quad (20)$$

where the values of β_1, β_2 and α_1, α_2 are calculated in Ref. [12].
The channel potential $\phi(r,z)$ is also determined as [12]

$$\phi(r,z) = \phi_{s1}(z)\left[1 - \frac{\epsilon_{ox}}{\epsilon_{si}}(A - Br + Cr^2)\right] + \frac{\epsilon_{ox}}{\epsilon_{si}}(V'')(m - nr + pr^2) \quad (21)$$

$V'' = V_{gs} - V_{fb}$, V_{fb} is flat band voltage; V_{gs} is the applied gate to source voltage. The constants mentioned in Eq. 21 are already evaluated in our previous work [12]. The locus of minimum surface potential; z_{min}, in channel length direction can be extracted by taking the differentiation of the surface potential equation and equating this with zero. So the position of minimum channel potential for inner and outer Si-SiO₂ interfaces can be given as

$$z_{min1} = \frac{ln\left(\frac{k_2}{k_1}\right)}{2\sqrt{\alpha_1}} \tag{22}$$

$$z_{min2} = \frac{ln\left(\frac{k_4}{k_3}\right)}{2\sqrt{\alpha_2}} \tag{23}$$

Utilizing Eq. (22) in Eq. (9) and Eq. (23) in Eq. (10) the minimum potentials at inner and outer Si-SiO₂ interfaces can be evaluated as follows

$$\phi_{s1,min} = 2\sqrt{k_1 k_2} + \sigma_1 \tag{24}$$

$$\phi_{s2,min} = 2\sqrt{k_3 k_4} + \sigma_2 \tag{25}$$

B. Threshold voltage formulation

For an extremely short channel length device, the conventional definition of the threshold voltage ($\phi_{s\,min}|_{V_{gs}=V_{th}} = 2\phi_f$; where $\phi_f = V_T ln(N_A/n_i)$) is not appropriate. For a short channel length undoped/lightly doped device, the surface potential is assumed to be pinned at $2\phi_f + 6V_T$, where V_T is the thermal voltage, as $6V_T$ becomes comparable to $2\phi_f$ [26]. Thus, the threshold voltage for a short channel length device is described as that particular gate voltage value at which the minimum of surface potential value reaches to $2\phi_f + 6V_T$ i.e. $\phi_{smin}|_{v_{gs}=V_{th}} = 2\phi_f + 6V_T$ .As the present device is inherently asymmetric, two different threshold voltages for inner and outer Si-SiO₂ interfaces can be determined. Let V_{th1} and V_{th2} are the threshold voltages of inner and outer Si-SiO₂ interfaces, therefore

$$\phi_{s1\,min}|_{v_{gs}=V_{th1}} = 2\phi_f + 6V_T \tag{26}$$

$$\phi_{s2\,min}|_{v_{gs}=V_{th2}} = 2\phi_f + 6V_T \tag{27}$$

Solving Eq. (26), gives the following expression of the threshold voltage V_{th1}

$$V_{th1} = V_{fb} + \left[\frac{-b+\sqrt{b^2-4ac}}{2a}\right] \tag{28}$$

where,
$$a = b_1 b_2 - b_3^2 \tag{29}$$

$$b = b_1 b_4 + b_2 b_5 - 2b_6 b_3 \tag{30}$$

$$c = b_4 b_5 - 2b_6^2 \tag{31}$$

$$b_1 = \frac{b_3}{sinh\sqrt{\alpha_1}L}\left(1 - e^{-\sqrt{\alpha_1}z}\right) \tag{32}$$

$$b_2 = \frac{b_3}{sinh\sqrt{\alpha_1}L}\left(e^{-\sqrt{\alpha_1}z} - 1\right) \tag{33}$$

$$b_3 = \frac{Z_1}{\alpha_1 X_1} \tag{34}$$

$$b_4 = \frac{\left(e^{\sqrt{\alpha_1}z}-1\right)V_{bi,S(D)}+\xi\left(1-e^{\sqrt{\alpha_1}z}\right)-V_{ds}}{sinh\sqrt{\alpha_1}L} \tag{35}$$

$$b_5 = \frac{\left(1-e^{-\sqrt{\alpha_1}z}\right)V_{bi,S(D)}+\xi\left(e^{-\sqrt{\alpha_1}z}-1\right)+V_{ds}}{sinh\sqrt{\alpha_1}L} \tag{36}$$

$$b_6 = 2\phi_f + 6V_T - \xi \tag{37}$$

$$\xi = -\frac{qN_a}{\alpha_1 X_1 \epsilon_{si}} \tag{38}$$

Similarly, the threshold voltage for the outer Si-SiO₂ interface can be evaluated by solving Eq. (27)

$$V_{th2} = V_{fb} + \left[\frac{-y+\sqrt{y^2-4xz}}{2x}\right] \tag{39}$$

where,

$$x = d_1 d_2 - d_3^2 \tag{40}$$

$$y = d_1 d_4 + d_2 d_5 - 2d_6 d_3 \tag{41}$$

$$z = d_4 d_5 - 2d_6^2 \tag{42}$$

$$d_1 = \frac{d_3}{sinh\sqrt{\alpha_2}L}\left(1 - e^{-\sqrt{\alpha_2}z}\right) \tag{43}$$

$$d_2 = \frac{d_3}{sinh\sqrt{\alpha_2}L}\left(e^{-\sqrt{\alpha_2}z} - 1\right) \tag{44}$$

$$d_3 = \frac{Z_2}{\alpha_2 X_2} \tag{45}$$

$$d_4 = \frac{\left(e^{\sqrt{\alpha_2}z}-1\right)V_{bi,S(D)}+\xi_1\left(1-e^{\sqrt{\alpha_2}z}\right)-V_{ds}}{sinh\sqrt{\alpha_2}L} \tag{46}$$

$$d_5 = \frac{\left(1-e^{-\sqrt{\alpha_2}z}\right)V_{bi,S(D)}+\xi_1\left(e^{-\sqrt{\alpha_2}z}-1\right)+V_{ds}}{sinh\sqrt{\alpha_2}L} \tag{47}$$

$$d_6 = 2\phi_f + 6V_T - \xi_1 \tag{48}$$

$$\xi_1 = -\frac{qN_a}{\alpha_2 X_2 \epsilon_{si}} \tag{49}$$

where, the constants used in Eqs. (34) and (45) have already been presented in [12]. Now, as we have defined two different threshold voltages V_{th1} and V_{th2} for inner and outer surfaces of the channel, respectively, we can define the device threshold voltage, V_{th} , as that one which is smaller than the other one

$$V_{th} = \begin{cases} V_{th1}, & if \ V_{th1} < V_{th2} \\ V_{th2}, & if \ V_{th2} < V_{th1} \end{cases} \tag{50}$$

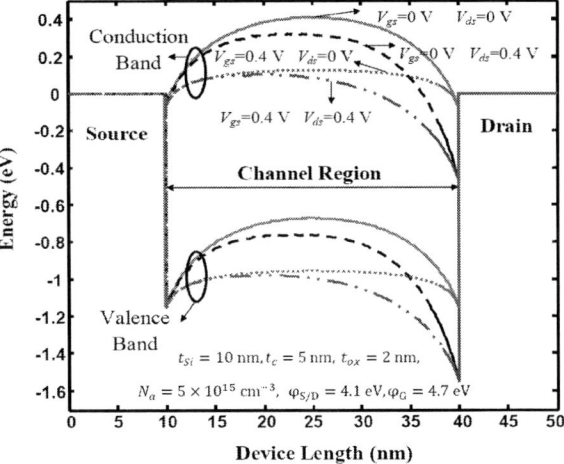

Fig.3: Energy band diagram along device length with gate to source (V_gs) and drain to source voltage (V_ds) variations.

IV. RESULTS AND DISCUSSION

In this part, we have discussed and validated the results of the developed model against the results from 3-D numerical device simulator. ATLAS, a 3D device simulator from SILVACO has been employed to simulate the device structure and obtain the simulation results. Drift-Diffusion carrier transport model, Philips unified mobility model, Lombardi high field mobility model and S-R-H model for recombination are activated during simulation [25]. In this complete analysis of Schottky-S/D DGAA MOSFET, the device channel length is limited up to 20nm to restrict ballistic transport of the carriers. In order to obtain the threshold voltage value, the constant current method is adopted for both models as well as for simulation [12].

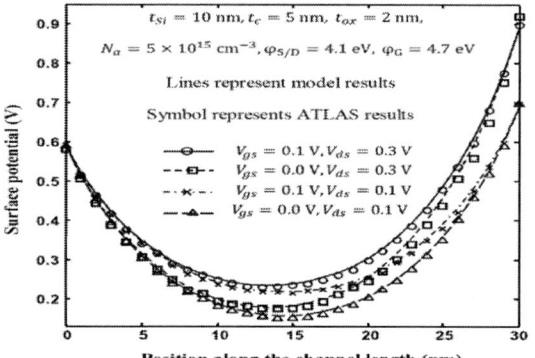

Fig.4: Surface potential variation against channel length for different values of gate to source voltage (V_{gs}) and drain to source bias voltage (V_{ds}).

Figure 3 depicts the valence energy bands and conduction bands variation along length of channel for a different combination of V_{gs} and V_{ds}. A negative shift in the channel conduction and valance bands is found with an increase in the gate bias voltage. This may be attributed to the barrier lowering effect in the silicon channel under the impact of the vertical electric field. As seen from the fig. 3, with the increase in gate bias voltage, the height of the energy barrier at the interface of source/channel decreases thereby enabling the flow of the carriers from source to channel region. Figure 4 depicts the inner surface potential variation along the length of channel region for distinct combinations of V_{gs} and V_{ds}. The magnitude of minimum surface potential increases with an increase in the applied gate voltage (V_{gs}). The height of the source/channel energy barrier and the Schottky barrier thickness is found to be decreased with an increase in the V_{gs} value.

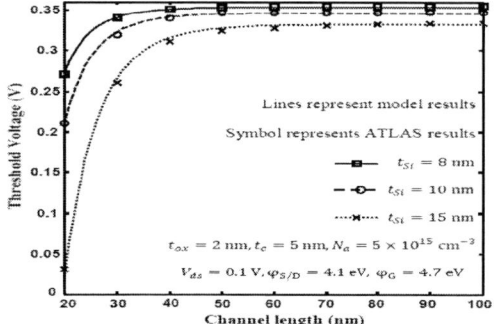

Fig.5: Threshold voltage variation against channel length for multiple values of channel thickness (t_{si}).

Figure 5 presents the threshold voltage (V_{th}) variation against channel length for three distinct channel thickness (t_{si}) values. A positive increment in value of threshold voltage is observed if channel thickness is decreased for a given channel length. This may be partially imputed to better gate and channel regions electrostatic coupling. A device structure with thinner channel thickness offers better short channel effects immunity and this may be partially due to strong electrostatics coupling between gate and channel region and reduced charge carriers shared by the drain region. Figure 6 demonstrates the threshold voltage variation against the channel length variation considering gate oxide thickness (t_{ox}) as a variable parameter. The device demonstrates excellent immunity against threshold voltage roll-off provided the device is designed with an ultrathin gate oxide.

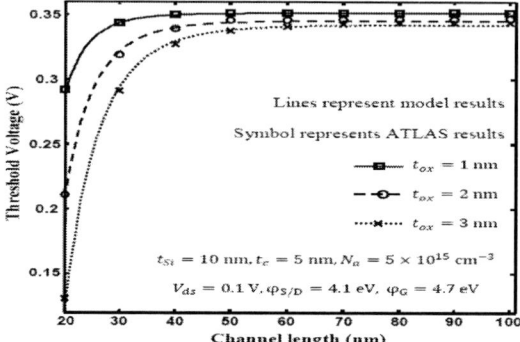

Fig.6: Threshold voltage variation against channel length for multiple values of oxide region thickness (t_{ox}).

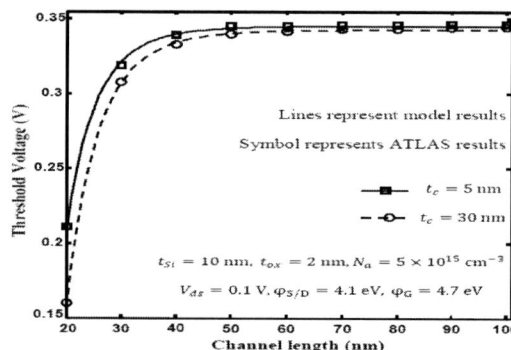

Fig.7: Threshold voltage variation against channel length for multiple values of core channel thickness (t_c).

Figure 7 presents the variation in threshold voltage with respect to channel length for different core radius (t_c). As observed from figure, the device threshold voltage can be varied up to some extent by changing the core radius of the device. However, compared to the change in the core radius (from 5nm to 30nm), the change in the threshold voltage (less than 10mV at the channel length of 30nm) is not significant. This confirms the fact that by increasing the core radius while keeping the tube thickness constant, the drive current can be enhanced without compromising much with the short-channel induced effects.

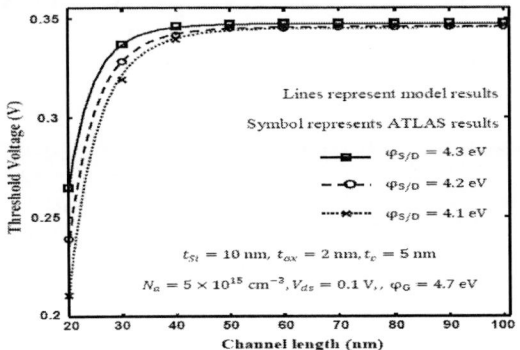

Fig.8: Threshold voltage variation against channel length for different values of source-drain metal work-function ($\varphi_{S/D}$).

In Fig. 8, the change in threshold voltage V_{th} is plotted against the channel length where the work-function of source-drain ($\varphi_{S/D}$) is a variable parameter. A positive increment in the threshold voltage value is noticed with an increase in the metal work-function of source and drain. This may be attributed to the rise in the height of the energy barrier for the electron at the interfaces of source-drain with the channel region.

V. CONCLUSION

In this work, an analytical model of threshold voltage has been demonstrated for the Schottky-S/D DGAA MOSFETs. The method of parabolic potential approximation is used to solve the 3-D Poisson's equation to formulate the surface potential and threshold voltage model of the device. A modified definition of threshold voltage is used to formulate the threshold voltage of the device. The results extracted from the analytical model have been compared and validated against the results obtained from commercially available ATLAS™ TCAD [25] simulation data.

REFERENCES

[1]. International technology roadmap for semiconductors (ITRS). [Online].Available:http://www.semiconductors.org/clientuploads/Research_Technology/ITRS/2015/0_2015%20ITRS%202.0%20Executive%20Report%20(1).pdf

[2]. K. K. Young, "Short-channel effect in fully-depleted SOI MOSFETs", *IEEE Trans. Electron Devices*, vol. 36, pp. 339, 1989.

[3]. J. P. Colinge, "Multiple-gate SOI MOSFETs," *Solid-State Electron.*, vol. 48, no. 6, pp. 897-905, June 2004.

[4]. W. Lu, P. Xie, C. M. Lieber, "Nanowire transistor performance limits and applications", *IEEE Trans. Electron Devices*, vol. 55, no. 11, pp. 2859-2876, Nov. 2008.

[5]. J. P. Colinge, "The new generation of SOI MOSFETs," *Romanian J. Inf. Sci. Technol.*, vol. 11, no. 1, pp. 3–15, Nov. 2008.

[6]. J. Song, B. Yu, Y. Yuan, Y. Taur, "A review on compact modeling of multiple-gate MOSFETs", *IEEE Trans. Circuits Syst.*, vol. 56, no. 8, pp. 1858-1869, Aug. 2009.

[7]. Y. B. Kim, "Challenges for nanoscale MOSFETs and emerging nanoelectronics," *Trans. Elect. Electron. Mater.*, vol. 11, no. 3, pp. 93–105, Jun. 2010.

[8]. H. M. Fahad, and M. M. Hussain, "Are nanotube architectures more advantageous than nanowire architectures for field effect transistors?," *Nature Sci. Rep.*, vol.2, no. 475, pp. 1-7, June 2012.

[9]. D. Tekleab, H. H. Tran, J. W. Sleight, D. Chidambarrao, Silicon nanotube MOSFET, U.S. Patent US20120217468 A1, Aug.30 (2012).

[10]. A. Kumar and P. K. Tiwari, "An Explicit Unified Drain Current Model for Silicon-Nanotube-Based Ultrathin Double Gate-All-Around mosfets," *IEEE Trans. Nanotech.*, vol. 17, no. 6, pp. 1224-1234, Nov. 2018.

[11]. H. M. Fahad, C. E. Smith, J. P. Rojas, and M. M. Hussain, "Silicon nanotube field effect transistor with core-shell gate stacks for enhanced high-performance operation and area scaling benefits." *Nano Lett.*, vol.11, no. 10, pp. 4393-4399, Sep. 2011.

[12]. P. K. Tiwari, V. R. Samoju, T. Sunkara, S. Dubey, and S. Jit,"Analytical modeling of threshold voltage for symmetrical silicon nano-tube fieldeffect-transistors (Si-NT FETs)," J. Comput. Electron., vol. 15, no. 2, pp. 516–524, Apr. 2016.

[13]. A. Kumar, S. Bhushan, P. K. Tiwari, "A Threshold voltage model of silicon-nanotube based ultra-thin double gate-all-around (DGAA) MOSFETs incorporating quantum confinement effects", *IEEE Trans. Nanotech.*, vol. 16, no. 5, pp. 868-875, Sep. 2017.

[14]. A. Kumar, S. Bhushan, P. K. Tiwari, "Analytical modeling of subthreshold characteristics of ultra-thin double gate-all-around (DGAA) MOSFETs incorporating quantum confinement effects," *Superlattices Microstruct*, vol. 105, pp. 567-578, Sept. 2017.

[15]. A. Kumar, S. Bhushan, and P. K. Tiwari "Drain current modeling of double gate-all-around (DGAA) MOSFETs", *IET CIRC DEVICE SYST*, 10.1049/iet-cds.2018.5201

[16]. J. M. Larson, J. P. Snyder, "Overview and status of metal S/D Schottky-barrier MOSFET technology", *IEEE Trans. Electron Devices*, vol. 53, no. 5, pp. 1048-1058, May 2006.

[17]. M. Balaguer, B. Iñíguez, and J. B. Roldán, "An analytical compact model for Schottky-barrier double gate MOSFETs," Solid-State Electron., vol. 64, no. 1, pp. 78–84, 2011.

[18]. S. Zhu, J. Chen, M.-F. Li, S. J. Lee, J. Singh, C. X. Zhu, A. Du, C. H. Tung, A. Chin, D. L. Kwong, "N-type Schottky barrier S/D MOSFET using Ytterbium silicide", *IEEE Electron Device Lett.*, vol. 25, no. 8, pp. 565-567, Aug. 2004.

[19]. T. Mochizuki, K. D. Wise, "An n-channel MOSFET with Schottky source and drain", *IEEE Electron Device Lett.*, vol. EDL-5, no. 4, pp. 108-111, Apr. 1984.

[20]. C. Wang, J. P. Snyder, J. R. Tucker, "Sub-40 nm PtSi Schottky S/D metal-oxide-semiconductor field-effect transistors", *Appl. Phys. Lett.*, vol. 74, no. 8, pp. 1174-1176, Feb. 1999.

[21]. G. Zhu, X. Zhou, T. S. Lee, L. K. Ang, G. H. See, S. Lin, Y.-K. Chin, K. L. Pey, "A compact model for undoped silicon-nanowire MOSFETs with Schottky-barrier source/drain", *IEEE Trans. Electron Devices*, vol. 56, no. 5, pp. 1100-1109, May 2009..

[22]. M. Kumar, S. Haldar, M. Gupta, R.S.Gupta, "Ambipolarity reduction in DMG asymmetric vacuum dielectric schottky barrier GAA MOSFET to improve hot carrier reliability", *Superlattices Microstruct*, vol. 111, pp. 10-22, Nov. 2017.

[23]. C. Ho, Y. Chang, "Evaluation of Schottky barrier source/drain contact on gate-all-around polycrystalline silicon nanowire MOSFET", *Mat. Sci. Sem. Proc.*, vol. 61, pp. 150-155, 2017.

[24]. E. J. Tan, K. L. Pey, N. Singh, G. Q. Lo, D. Z. Chi, Y. K. Chin, K. M. Hoe, G. Cui, P. S. Lee, "Demonstration of Schottky barrier NMOS transistors with erbium silicided source/drain and silicon nanowire channel", *IEEE Electron Device Lett.*, vol. 29, no. 10, pp. 1167-1170, Oct. 2008.

[25]. ATLAS Users Manual, 3D-Device Simulator, Silvaco International, Santa Clara, CA, USA (2013).

[26]. Y. Tsividis, Operational Modeling of the MOS Transistor, New York, NY, USA:McGraw-Hill, 1999.

2019 Devices for Integrated Circuit (DevIC), 23-24 March, 2019, Kalyani, India

Performance and Opportunities of Gate-All-Around Vertically-Stacked Nanowire Transistors at 3nm Technology Nodes

S. Dey[1*], T. P. Dash[2], E.Mohapatra[3], J.Jena[4], S. Das[5], and C. K. Maiti [6]

Department of Electronics and Communication Engineering,
Siksha 'O' Anusandhan (Deemed to be University),
Bhubaneswar, Odisha, India--751030
[1*]supravadey@soa.ac.in; [2] taradash@soa.ac.in;[3]eleenamohapatra@soa.ac.in;[4] jhansiranijena@soa.ac.in;[5]sanghamitradas@ soa.ac.in;
[6]ckmaiti@soa.ac.in.

Abstract— **Gate-all-around (GAA) cylindrical channel Si nanowire field effect transistor (NW-FET) devices have the potential to replace FinFETs in future technology nodes because of their better channel electrostatics control. In this work, 3-D TCAD simulations are performed for the first time to evaluate the potential of NW-FETs at extreme scaling limits of 3nm gate length. The performance of n-type silicon nanowire transistors is benchmarked using predictive TCAD device simulation.**

Keywords– Vertically-stacked Nanowire FETs, quantum confinement, technology computer aided design (TCAD), density-gradient (DG) model.

I. INTRODUCTION

Since early 2000s, simply reducing dimensions has not been enough to follow Moore's Law [1]. New processes, architectures and materials have come into extend the quest for doubling computing power every two years. It's about More-than-Moore. Previously managed by the ITRS, the specification roadmaps for logic circuits have been set (since 2016) by the International Roadmap for Devices Semiconductors (IRDS). In addition to reducing the dimensions of the components, the successive recommendations of the ITRS (then the IRDS) show a decrease in supply voltages. One solution is to increase the on-current so that we can get the same drive at a lower voltage. The other option is to find a way to get the subthreshold slope below 60mV/decade using alternative channel materials [2].

The miniaturization of transistors is a major concern of component manufacturers since the production of the first integrated circuits. Reduced production costs on a large scale (more transistors on a single wafer), increased working frequency (shorter transistors), and reduced power consumption (with the reduction of supply voltages) are important requirements for circuit designers. However, the miniaturization of the transistors leads to short channel effects.

For down scaled devices, the gate oxide thickness has to be reduced leading to a high leakage current by tunnel effect (quantum tunneling effect). By continuing to use SiO_2 as gate insulation, there is a risk of increasing the tunneling leakage current. To counter this effect, microelectronics industry is now using so-called materials: the high-k gate dielectrics.

Better performances of multiple-gate transistors (MUGFETs) can in part be explained by volume inversion rather than surface inversion. The channel forming in the heart of the film rather than at the film-oxide interface, the carriers are no longer disturbed by defects interface and their mobility and slope below the threshold are increased. The MUGFETs offer the best electrostatic control of the channel for different configurations, such as, Gate-All-Around (GAA). The Nanowire Field-Effect Transistors (NW-FETs) are promising with a great future in the field of microelectronics because of the cylindrical shape of the channel, which offers homogeneous control throughout the device. IRDS considers nanowires for the ideal transistor [3]. Like FinFETs, several nanowires are used in parallel for the formation of the channel. GAA NW-FETs consist of several pairs of nanowires in silicon, each pair corresponding to two horizontal nanowires suspended one above. However, drain currents are relatively low compared to the conventional transistors.

The surrounding gate devices offer tighter capacitive coupling to the device channel region from all directions and hence better control over short channel effects. However, the problem with nanowire devices is in obtaining a high drive current from a single narrow cylindrical Si channel. Quantum confinement is stronger in a cylindrical structure, since the confinement comes from all directions.

In manufacturing, silicon nanowire is made by epitaxial growth on a SiGe layer on silicon (the substrate) following a FinFET technology entirely in Si to reduce the number of manufacturing steps (and thus the cost of production). The Si film is covered with a layer of SiGe, and then a new Si layer is grown (to make the second nanowire). Spacer SiGe is etched to transform the Si/SiGe/Si/SiGe/Si stack by a stack Si/empty/Si/empty/Si. The two nanowires are then covered by the insulator and gate is formed, thus forming a pair of superimposed nanowires. However, if the lithographic steps of etching SiGe are not carried out, giving rise to FinFET with a stack of Si and SiGe leading to superlattice FinFETs (super-FF) [4]. Compared to the ideal case (the etching of the spacers does not affect the silicon nanowires), superlattice FinFETs offer a 25% higher current. As a result, superlattice FinFET offers the possibility of higher current compared to NW-FET GAA, currently available in integrated circuits.

978-1-5386-6723-1/19 $31.00 © 2019 IEEE

As it takes over ten years for a technology to go from the research laboratory to volume production, it is essential to use predictive Technology CAD (TCAD) tools for performance evaluation at the beginning. As the sub-5nm technology must fit within the CMOS manufacturing and design infrastructure, device options and trade-offs are now to use nanowire technology. As predicted, 5nm technology is not going to be ready for production until 2025 and it will be some sort of FinFET (possibly gate-all-around silicon nanowire or similar type of devices), it is time to search for advanced device structures such as multiple stacked wires (or perhaps vertical nanowires). Stacking vertically NW channels could be the ultimate of conventional CMOS scaling).

The paper is organized as follows: After describing the vertically-stacked multi-channel device structure in Section II, we use an experimental benchmark (sub-10nm) NW-FET for technology TCAD calibration with the reported experimental data. Advanced simulation models used in this work are described in the simulation environment (Section III). The objectives of the simulation runs are to determine various dc performance metrics of vertically-stacked horizontal multiple nanowire channel transistors. In Section IV, we present the electrical characterization results obtained on NW-FETs.

II. DEVICE STRUCTURE

Nanowire FETs can be implemented in a lateral or a vertical configuration. Devices in the lateral configuration use conventional 2D layouts, and hence their scaling into advanced nodes will eventually hit physical limits, similarly to FinFETs. However, the diminishing cross-section to obtain the better short-channel effect immunity is traded off against the drain current enhancement. To obtain larger drive current for a given layout (nanowire diameter) with a high immunity against the short-channel effects, the vertical integration of nanowire is effective. In this structure, the current density can be proportionally increased to the stacking level (three levels as shown in Fig. 1) without the layout surface area penalty.

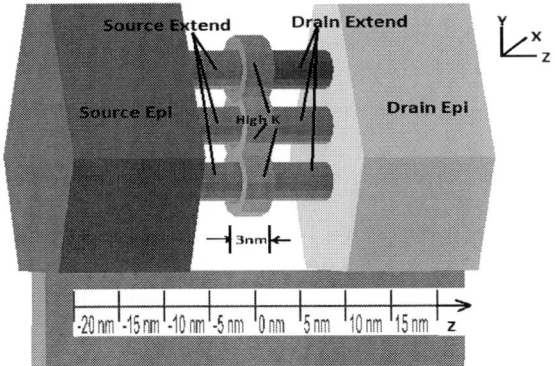

Fig. 1. Vertically stacked multi nanowire (3) nanowire device schematic with a gate length as 3nm.

TABLE I. DEVICE GEOMETRY AND OTHER PARAMETERS USED IN SIMULATION.

Parameters	Value	Units
Number of wire(s)	1/2/3	-
Nanowire Diameter	5	nm
Channel Length	3	nm
High-k (HfO$_2$) Thickness	1.5	nm
Oxide (SiO$_2$) Thickness	0.5	nm
Epi Length	14	nm
Epi Top Width	8	nm
Epi Middle Width	25	nm
Epi Bottom Width	14	nm
Substrate Height	350	nm
Channel Doping	10^{16}	cm^{-3}
S/D Doping	10^{20}	cm^{-3}

To understand the challenges facing advanced device design, one must first understand the design parameters available to a device designer and their significance. Instead of going through an expensive and time-consuming fabrication process, computer simulations can be used to predict the electrical characteristics of a device design quickly and cheaply. TCAD can also be used for reducing design costs, improving device design productivity, and obtaining better device and technology designs. However, accurate TCAD simulations and modeling of physical devices depend critically on calibrated physical models and proper input data.

An ideal TCAD simulation tool should be able to precisely predict changes in process and device geometry providing reliable results for ultimate technology nodes. For these reasons, advanced device modeling strategies are required to account for various aspects, such as quantum effects, mobility models, channel material properties, complex device architectures etc. As such, in this work, simulations are first compared to available experimental data [3] for TCAD calibration as shown in Fig. 2. Simulations are performed with MINIMOS-NT [5]. The detail of the experimental device are as follows: n–type twin–nanowire device with gate length L_G = 24nm, nanowire diameter D_{nw}= 8nm and spacer thickness dspacer = 11nm. The gate oxide consists of 0.8nm interfacial oxide and 1.8 nm HfO$_2$. The channel orientation is in <110> direction. The use and calibration of available TCAD models to efficiently model the nanowire device shows a good match between the simulation and experiment.

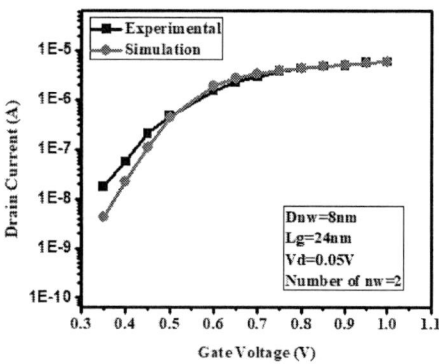

Fig. 2. Comparison of simulation and experimental I_D-V_G characteristics at a V_D = 0.05 V for a 24nm twin nanowire structure [experimental data after Reference 3]

III. SIMULATION ENVIRONMENT

In order to improve the accuracy of the Drift-Diffusion equations (DD) on complex device structures some quantum corrections cannot be ignored anymore. The Density Gradient (DG) model was developed by observing how the gradient of density in the electron gas, in addition to its density, impacts on its energy. The macroscopic origin of the DG model may be found from the work reported by Ancona and Iafrate [6, 7]. They developed a macroscopic description of transport of electrons in a semiconductor in which the equation of state for the electron gas was generalized to include a dependence on the gradient of the density.

The DG model can be seen, in its simplest form, as a direct enhancement of the DD model. By generalizing the equation of state of the electron gas to include density-gradient dependences, the standard description can be extended to describe much of the quantum-mechanical behavior exhibited by strong inversion layers. Quantum effects, which affect mainly the spatial distribution of the inversion charge in the channel and the threshold voltage, are included. The DG scheme activates the equations needed for the density gradient extension. Density-Gradient model incorporated in the MINIMOS-NT simulation tool makes use of the phenomenology method, which is the most common approach in quantum mechanical analysis. In simulation one needs to solve Eqs. 1 and 2:

$$J_n = q \cdot \mu_n \cdot n \left(grad \left(\frac{\varepsilon_C}{q} - \psi - \gamma_n \right) + \frac{k_B \cdot T}{q} \cdot \frac{N_{C,0}}{n} \cdot grad \left(\frac{n}{N_{C,0}} \right) \right) \quad (1)$$

$$J_p = q \cdot \mu_p \cdot p \cdot \left(grad \left(\frac{\varepsilon_V}{q} - \psi - \gamma_p \right) - \frac{k_B \cdot T}{q} \cdot \frac{N_{V,0}}{p} \cdot grad \left(\frac{p}{N_{V,0}} \right) \right) \quad (2)$$

The quantities γ_n and γ_p are defined as the quantum corrected potentials, and this quantities are the summary of a more complex quantum analysis, which, according to a microscopic theory, involves solving the Wigner's distribution equation by the method of moments [8].

$$\gamma_n = \frac{\hbar}{12 \cdot \lambda_n \cdot m_0} \cdot div \ grad \ \frac{\psi + \gamma_n - \varepsilon_C/q}{k_B \cdot T} \quad (3)$$

$$\gamma_p = \frac{\hbar}{12 \cdot \lambda_p \cdot m_0} \cdot div \ grad \ \frac{\psi + \gamma_p - \varepsilon_V/q}{k_B \cdot T} \quad (4)$$

The above set of Eqs. (1-4) are solved by the simulation tool. The DG model takes the quantum electron evanescence into its macroscopic theory as a prime purpose. The physical principles of the DG model are classical, and consists of the conservation of charge and momentum, meaning that the conservation of the total electron or hole charge in a volume increases in time only due to entry through its surface via generation process inside the volume. The most important material response functions of the DG model are the equations of state that describe the electron and hole gases as follows:

$$J_n = -q \cdot \mu_n \cdot n \cdot \nabla(\phi + \phi_{DGn}) + q \cdot D_n \cdot \nabla(n) \quad (5)$$
$$J_p = -q \cdot \mu_p \cdot p \cdot \nabla(\phi + \phi_{DGp}) - q \cdot D_p \cdot \nabla(p) \quad (6)$$

Where D_n and D_p are the diffusivity terms of electrons and holes, respectively, and they account for position-dependent band edges and effective masses. The quantities ϕ_{DGn} and ϕ_{DGp} are called the generalized quantum potentials and given as:

$$\phi_{DGn} = 2b_n \left(\frac{\Delta\sqrt{n}}{n} \right) \quad , \quad b_n = \frac{h^2}{12 \cdot q \cdot m_e^*} \quad (7)$$

$$\phi_{DGp} = -2b_p \left(\frac{\Delta\sqrt{p}}{p} \right), \quad b_p = \frac{h^2}{12 \cdot q \cdot m_p^*} \quad (8)$$

Where b_n and b_p parameters are material response functions. Eqs. 5 and 6 are the quantum drift-diffusion current equations, but these are the result of the inclusion of the quantum potentials ϕ_{DGn} and ϕ_{DGp} in the two generalized current equations as follows [9]:

$$J_n = -q \cdot \mu_n \cdot n \cdot \nabla(\phi) + q \cdot D_n \cdot \nabla(n) - 2q \cdot \mu_n \cdot b_n \cdot n \cdot \nabla \left(\frac{\Delta\sqrt{n}}{\sqrt{n}} \right) \quad (9)$$

$$J_p = -q \cdot \mu_p \cdot p \cdot \nabla(\phi) - q \cdot D_p \cdot \nabla(p) + 2q \cdot \mu_p \cdot b_p \cdot p \cdot \nabla \left(\frac{\Delta\sqrt{p}}{p} \right) \quad (10)$$

Eqs. 9 and 10 may be simplified mathematically to get to Eqs. 5 and 6. The third quantum corrected term is incorporated as a quantum potential in the first term and the latter can be treated as current driven by an effective electric field [10].

The simulated device is a three-dimensional structure with gate-all-around silicon channel and its cut plane is shown in Fig. 3. Device simulations are performed using MINIMOS-NT device simulator environment. The device cross section is 3nm (cylindrical) while its effective channel length is also 3nm. The channel region is assumed to be p doped ($N_A = 10^{16}$ cm^{-3}) and the source and drain extension regions are heavily n-doped ($N_D = 10^{20}$ cm^{-3}) with abrupt doping profiles. A metallic gate with the mid-gap work function of 4.7 eV is assumed and the gate oxide is 2 nm thick.

IV. RESULTS AND DISCUSSIONS

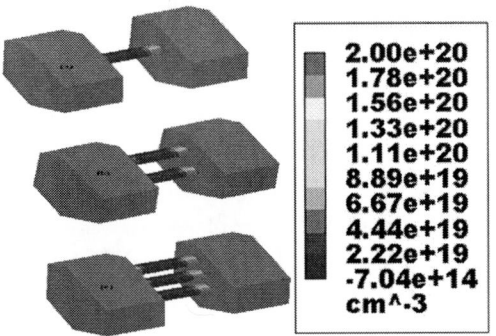

Fig. 3. Schematic of vertically stacked device structure with (a) one nanowire, (b) two nanowires, and (c) three nanowires. Net doping concentrations are also shown.

Schematic of vertically stacked nanowire device structure with (a) one nanowire, (b) two nanowires, and (c) three nanowires along with the net doping concentrations is shown in Fig. 3. The potential distributions in one, two and three nanowires at $V_G=0$ and $V_D=0.05V$ are shown in Fig. 4.

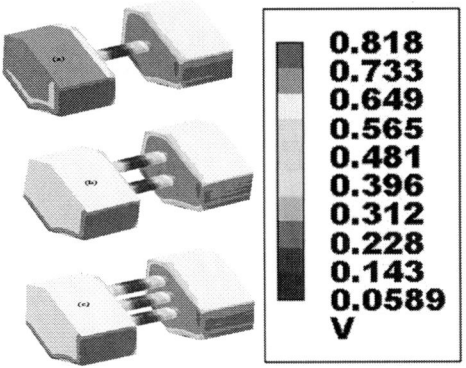

Fig. 4. Potential distribution in one, two and three nanowires in the device at $V_G=0$ and $V_D=0.05V$.

The transfer and output characteristics of the three (1 nanowire to 3 nanowires present in the channel) devices were obtained using the calibrated density gradient (DG) model. The effect of using the DG model on drain current is shown

in Fig. 5. We studied the scaling behavior of NW-FETs by varying the number of nanowires in the channel of the device. Fig. 5 shows the variation in transfer characteristics when the number of nanowires is increased from one to three. The cross section of the nanowires was 5nmx5nm. Fig. 5 shows that the drain current in the three nanowire device is more than two times that of one nanowire FET at $V_D=0.05V$. When the drain voltage in increased to 0.7V, the I_D-V_G plot of stacked nanowire shows that the drain current is also 3x higher.

The output characteristics I_D-V_D curves of stacked (1-3) nanowire FETs at $V_G=0.7V$ are shown in Fig. 6. A high saturation current of the stacked NW-FET is due to the fact that it has more electron conducting channels, higher mobility, and lower S/D parasitic resistance than the single NW-FET. Other performance metrics, such as I_{off}, I_{on}, V_{th}, DIBL and SS extracted from the transfer and output characteristics are shown in Fig. 7. It may be noted that both the I_{on} and I_{off} currents increase as the number of nanowires are increased in the channel although the Ion/Ioff ratio remains almost same at $3x10^5$ (at $V_D = 0.7V$). Variation in subthreshold slope is found to be almost same (80 mv/dec) even when the number of nanowires is increased to 3.

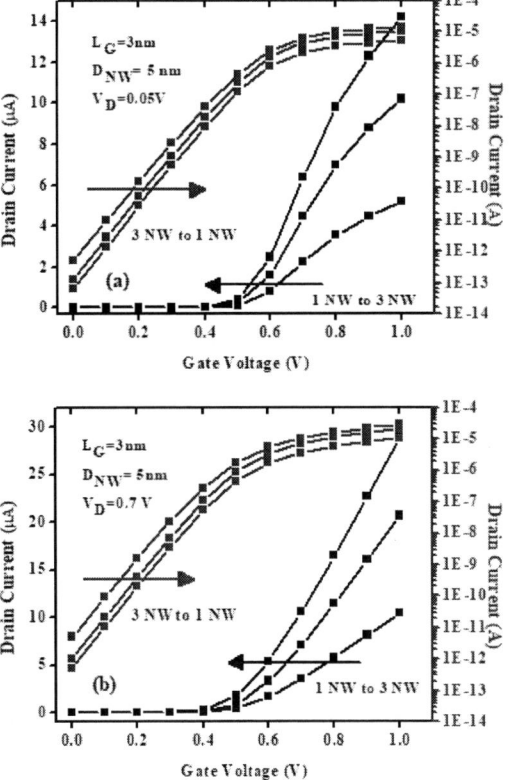

Fig. 5. (a) I_D-V_G plot of one, two and three vertically stacked nanowire devices at $V_D = 0.05V$, and (b) at $V_D=0.7$ V.

2019 Devices for Integrated Circuit (DevIC), 23-24 March, 2019, Kalyani, India

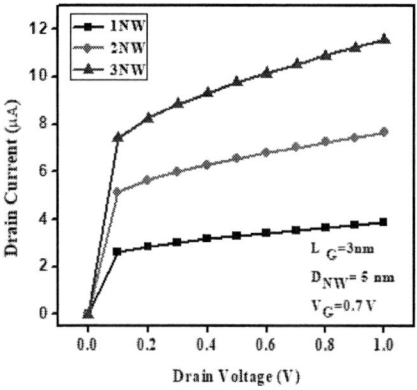

Fig. 6. Comparison of I_D-V_D characteristics of one, two and three nanowire devices at $V_G = 0.7V$.

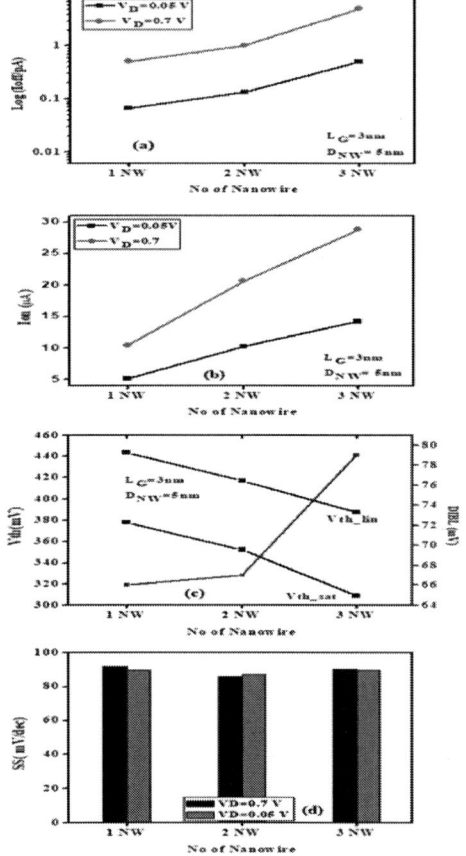

Fig. 7. Variation of Ioff, Ion, Vth, DIBL, SS vs. number of nanowires are shown in (a), (b), (c), and (d), respectively.

V. CONCLUSION

Using physics-based path-finding 3-D TCAD simulations and device optimization at 3nm technology node, we are able to point out critical improvements possible for the vertically stacked NW-FETs to surpass current FinFET technology. Using an advanced simulation framework, we calibrated a recent sub-10 nm technology demonstration [3] based on stacked nanowire transistors. We have performed a quantum simulation of Silicon Nanowire Field-Effect Transistors by using the DG Quantum Potential model to include quantum confinement effects in the simulations. We have shown that stacked nanowires offer versatile design options for performance and power management essential for More-than-Moore applications at 3nm technology node and beyond.

REFERENCES

[1] G. E. Moore, "Cramming more components onto integrated circuits", Electronics, p. 38, 1965.

[2] IRDS, Executive Summary. Rapport technique, International Roadmap for Devices and Systems, 2017.

[3] F. M. Bufler, R. Ritzenthaler, H. Mertens, G. Eneman, A. Mocuta, and N. Horiguchi, "Performance Comparison of n–Type Si Nanowires, Nanosheets, and FinFETs by MC Device Simulation," in IEEE Electron Device Letters, vol. 39, no. 11, pp. 1628-1631, 2018.

[4] G. Hellings, H. Mertens, A. Subirats, E. Simoen, T. Schram, L.-A. Ragnarsson, M. Simicic, S.-H. Chen, B. Parvais, D. Boudier, B. Cretu, J. Machillot, V. Pena, S. Sun, N. Yoshida, N. Kim, A. Mocuta, D. Linten, and N. Horiguchi, "Si/SiGe superlattice I/O finFETs in a vertically-stacked Gate-All-Around horizontal Nanowire Technology," in 2018 Symposium on VLSI Technology Digest of Technical Papers, pp. 85-86, 2018.

[5] MINIMOS-NT User's manual, 2018.

[6] M. G. Ancona, "Density-gradient theory: a macroscopic approach to quantum confinement and tunneling in semiconductor devices," Journal of computational electronics, vol. 10 pp. 65-97, 2011.

[7] M. G. Ancona, G. J. Iafrate, "Quantum correction to the equation of state of an electron gas in a semiconductor," Phys. Rev. B 39, p. 9536, 1989.

[8] A. Wettstein, "Quantum Effects in MOS Devices". Konstanz: Hartung-Gorre Verlag, 2000.

[9] M. G. Ancona and H. F. Tiersten, "Density-Gradient Theory of Electron Transport in Semiconductors," Naval Research Lab Washington DC, 1989.

[10] R.-C. Chen and J.-L. Liu, "A quantum corrected energy-transport model for nanoscale semiconductor devices," Journal of Computational Physics, vol. 204, p.131-156, 2005.

978-1-5386-6723-1/19 $31.00 © 2019 IEEE

2019 Devices for Integrated Circuit (DevIC), 23-24 March, 2019, Kalyani, India

Vertically-Stacked Silicon Nanosheet Field Effect Transistors at 3nm Technology Nodes

T.P. Dash [1*], S.Dey [2], E. Mohapatra [3], S. Das [4] J. Jena [5] and C. K. Maiti [6]

Department of Electronics and Communication Engineering,
Siksha 'O' Anusandhan (Deemed to be University), Bhubaneswar, Odisha, India-751030

[1*]taradash@soa.ac.in; [2]supravadey@soa.ac.in; [3]eleenamohapatra@soa.ac.in; [4] sanghamitradas@soa.ac.in; [5]jhansiranijena@soa.ac.in; [6]ckmaiti@soa.ac.in.

Abstract— **Feasibility of vertically-stacked silicon nanosheet FETs (SNS-FETs) for extreme scaling at 3nm technology node are investigated for the first time as one of the possible solutions to continue to enhance the performances of the CMOS technology. With the end of happy scaling era, change of device architecture has raised integration complexity along with several sort channel effects, mobility degradation, variability and quantum tunneling leakage. These are the major challenges as device dimensions are scaled for ultimate scaling below 7nm technology nodes. Towards low power and high speed (More-than-Moore applications), nanowires and nanosheet transistors are being proposed. Today, the question of FinFET downscaling is still open and more than ever alternatives to CMOS transistors, such as, vertically-stacked SNS-FETs are showing their potential to surpass the FinFETs. In this work, we use 3-D predictive simulations to study the performance potential of SNS-FETs at 3nm technology node.**

Keywords – Nanosheet FETs (NS-FETs), quantum confinement, technology computer aided design (TCAD), drift diffusion, density-gradient.

I. INTRODUCTION

Nanosheets, whether they are one, two or three-dimensional, have been studied in detail from both the theoretical and experimental points of view. This is not only because of the effects related to the strong surface-to-volume ratio, but also because of the new scale-specific properties in nanoscale. These specific properties stem mainly from the modification of the electronic structure of the conduction and valence bands of matter at the nanoscale. As the integrated circuit technology scaling continues, the vertically stacked horizontal nanosheet FET has emerged as a potential successor to the FinFET [1]. However, stacking horizontal nanosheets on top of each other puts a hard limit to the oxide thickness needed for the transistors. The objective of this paper is to explain how advanced simulations can be used to predict vertically stacked silicon nanosheet field effect transistor (SNS-FET) performance. Two dimensional Schrodinger Poisson simulations using the tool MINIMOS NT can be used for successfully predicting the effect of carrier confinement in silicon nanosheets at 3nm technology nodes.

II. TCAD CALIBRATION

Short channel gate all around (GAA) nanosheet FETs [2] (with $L_g < 10$ nm) have been fabricated using various compatible CMOS top-down processes [3]. These devices showed excellent immunity to short-channel effects. Computer-aided design (CAD) is universally accepted for design and development of ICs. As the semiconductor manufacturing becoming increasingly complex, prediction of device performance with changing process and device geometry at ultimate technology nodes using TCAD has become essential. The use of available TCAD models and calibration provide a way to model the device efficiently. As such, simulations are first compared to available experimental data for calibration as shown in Fig. 1. It shows a good match between the simulation and experiment. Simulations are performed using MINIMOS-NT [4].

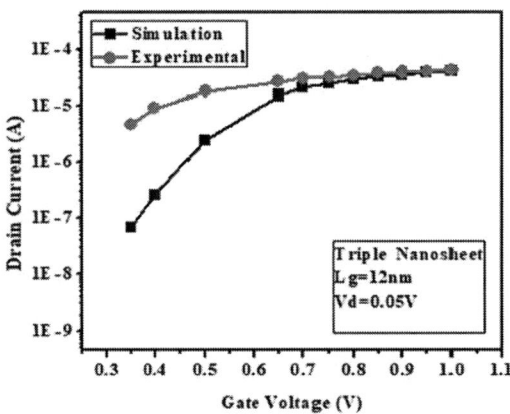

Fig. 1. Comparison of simulation and experimental ID-VG characteristics at a Vd of 0.05 V for a 12nm gate length 3NS structure [Experimental data after Reference 2].

III. DEVICE STRUCTURE

The device structure with three nanosheet and 3nm gate length has been shown in Fig 2 and followed for simulation. The detailed technological parameters considered in simulation has been listed in Table I.

978-1-5386-6723-1/19 $31.00 © 2019 IEEE

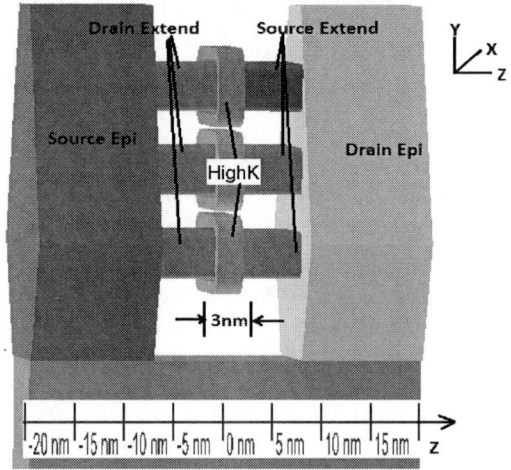

Fig. 2. Vertically stacked multi (3) nanosheet device schematic with a gate length of 3nm used in simulation.

TABLE I. TECHNOLOGICAL PARAMETERS CONSIDERED IN SIMULATION.

Parameters	Value	Units
Number of sheets	1/2/3	-
Nanosheet Height	6	nm
Nanosheet Width	12	nm
Channel Length	3	nm
High-k (HfO$_2$) Thickness	1.5	nm
Oxide (SiO$_2$) Thickness	0.5	nm
Epi Length	14	nm
Epi Top Width	8	nm
Epi Middle Width	25	nm
Epi Bottom Width	14	nm
Substrate Height	350	nm
Channel Doping	1e16	cm^{-3}
S/D Doping	1e20	cm^{-3}

IV. SIMULATION ENVIRONMENT

In a GAA Si nanosheet FET (SNS-FET), the total current in the channel is decided by the number of nanosheets present in the channel. In general, as one single nanosheet carries a smaller current, nanosheet transistors are used as multi-channel structure to increase the drive current [5]. For an ideal rectangular nanosheet (without volume inversion and corner effect etc.), the current per nanosheet depend on the area of the nanosheet i.e., the nanosheet width, (W_{NS}) and height, (H_{NS}), the top and bottom surface mobility, (μ_{top}) and (μ_{bot}) and the surface mobility on the side, (μ_{side}). The on-current in GAA SNS-FET with multi-finger are given by

$$I_{ON}^{nanowwire} = n \times (2W_{NS}\mu_{top} + 2H_{NS}\mu_{side}) \frac{C_{ox}}{L_{eff}} \frac{(V_{DD} - V_T)^{\alpha}}{2m}$$

where n represents the number of the nanosheets and α ($1 < \alpha < 2$) is the degree of velocity saturation in short-channel devices. With a nanosheet pitch of (W_{pitch}), the current per unit width is given by

$$\frac{I_{ON}^{nanowire}}{W_{pitch}} = \frac{I_{ON}^{planar}}{W} \cdot \frac{2W_{NS}\mu_{top} + 2H_{NS}\mu_{side}}{\mu_{top}W_{pitch}}$$

where $I_{ON}^{nanowire}$ is the current in the single-gate, planar device occupying the same area as the multi-nanosheet device.

The threshold voltage roll-off for a nanosheet transistors can be expressed as [6]:

$$SCE \equiv |\Delta V_T| = 0.64 \frac{\varepsilon_s}{\varepsilon_{ox}} EI \times \Phi_D$$

where

$$EI = \left(1 + \frac{x_j^2}{L_{eff}^2}\right) \frac{t_{ox}}{L_{eff}} \frac{W_{dm}}{L_{eff}}$$

is the electrostatic integrity, ε_s and ε_{ox} are the semiconductor and oxide permittivity, respectively.

The subthreshold slope for nanosheet transistors can written as a function of the surface potential, Ψ_s,

$$SS = \frac{kT}{q}\ln(10) \cdot \left(\frac{\partial V_G}{\partial \Psi_s}\right) = \frac{kT}{q}\ln(10) \cdot \left(1 + \frac{1}{C_{ox}}\frac{\partial Q_B}{\partial \Psi_s}\right)$$

The subthreshold swing (SS) is affected by the barrier potential lowering in short-channel devices which as a function of gate length can be expressed as:

$$SS = \frac{kT}{q}\ln(10) \cdot \left[1 + \frac{1}{C_{ox}}\frac{\partial Q_B}{\partial \psi_s} + \frac{\varepsilon_s}{\varepsilon_{ox}}\frac{t_{ox}}{L_{eff}}\frac{x_j}{L_{eff}}\left(1 + \frac{3}{4}\frac{W_{dm}}{L_{eff}}\right)\sqrt{1 + 2\frac{V_{DS}}{\phi_D}}\right]$$

The DIBL can be expressed as:

$$DIBL = 0.80\frac{\varepsilon_s}{\varepsilon_{ox}}EI \times V_{DS}$$

Short channel gate all around (GAA) SNS-FETs (with L$_g$ < 10 nm) have been fabricated recently using CMOS process [2]. To evaluate the benefit of the multi nanosheet transistors, in this work, the current of GAA SNS-FET are calculated with several nanosheet dimensions.

V. RESULTS AND DISCUSSIONS

Device structure used in simulation along with the potential distribution in one, two and three nanosheets at V$_G$=0 and V$_D$=0.05V are shown in Fig. 3. The simulated nanosheets are 3D GAA structures with a p-type doping concentration of 10^{16} cm^{-3} in the channel region and n-type doping of 10^{20} cm^{-3} in the source and drain regions. Work function of the metal gate is assumed to be 4.7 eV with a gate oxide thickness of 2 nm.

2019 Devices for Integrated Circuit (DevIC), 23-24 March, 2019, Kalyani, India

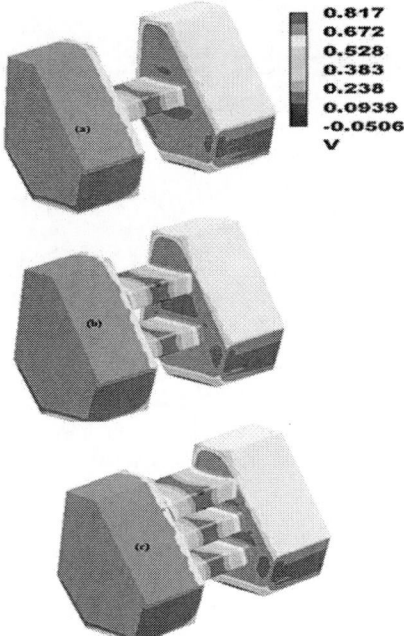

Fig. 3. Potential distribution in one, two and three nanosheets transistors at VG=0 and VD=0.05V.

Quantum confinement becomes important in nano-scale device due to the thin EOT. There are two frequently used methods for including quantum correction in classical DD device simulation are the Density Gradient (DG) approach and the Effective Potential approach. The density gradient model provides a description of transport in terms of macroscopic quantities, e.g., densities of the particles and their currents. In this aspect, it is similar to the classical hydrodynamic transport and drift diffusion model but in addition it includes contributions that account for certain aspects of the quantum nature of the particles. In DG model, one uses thermodynamic considerations introducing a lower order dependency on the density gradient into the internal energy per particle and reach an equation of the Schrodinger type for a static one-dimensional system. The derivations of the DG model are based on quantum statistical mechanics, i.e., the microscopic transport is described by either the Wigner-Boltzmann equation [10] or its Fourier transformed counterpart, the von Neumann equation which governs the evolution of the density matrix [11]. Applying the method of moments yields the classical transport models with additional quantum corrections which are then called quantum hydrodynamic solution [12].

Figure 4 shows the device transfer characteristics using the DG model while the number of nanosheets are varied from one to three. The cross section of the nanosheet was 6nmx12nm. Fig. 4 shows that the drain current in three nanosheet device is approximately three times that of one nanosheet FET at V_D=0.05V. When the drain voltage is increased to 0.7V, the I_D-V_G plot of multiple stacked nanosheet shows that the drain current is about two times higher compared to single sheet. This can be attributed to the

decreased source/drain parasitic resistance (RSD) of the stacked multiple NS in comparison with that of the single (SNS-FET). The RSD value of the stacked NS is nearly 8.7 times lower compared to the FinFET due to larger cross-sectional area of stacked NS-FET for conducting current at S/D.

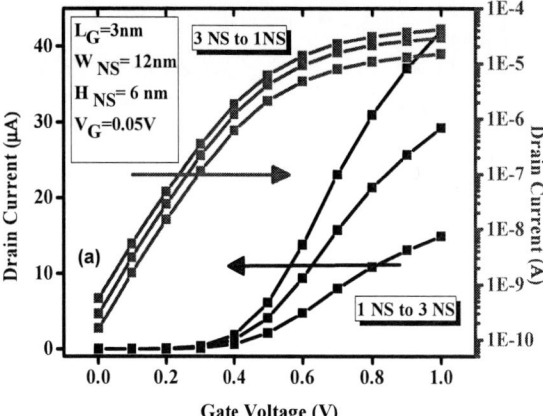

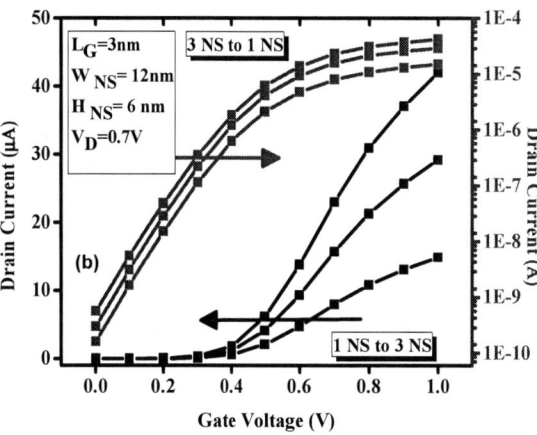

Fig. 4. (a) I_D-V_G plot of stacked (1-3 sheets) nanosheet transistor at V_D=0.05V. Linear scale I_D–V_G plot shows that the drain current in three nanosheet device is approximately three times that of one nanosheet FET, and (b) I_D-V_G plot of stacked nanosheet at V_D=0.7V. The linear scale plot of I_D-V_G shows that the drain current is about two times higher compared case (a).

The output characteristics I_D-V_D curves of stacked (1-3) nanosheet FETs at V_G=0.7V are shown in Fig. 5. The high saturation current of the stacked SNS-FET can be attributed to more no. of electron conducting layers, higher mobility, and lower S/D parasitic resistance compared to the Single SNS-FET. Other performance metrics, such as Ioff, Ion, Vth, DIBL and SS extracted from the transfer and output characteristics are shown in Fig. 6.

978-1-5386-6723-1/19 $31.00 © 2019 IEEE

2019 Devices for Integrated Circuit (DevIC), 23-24 March, 2019, Kalyani, India

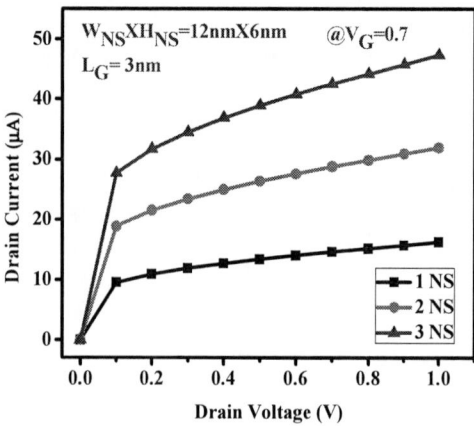

Fig. 5. The output characteristics I_D-V_D curves of stacked (1-3) nanosheet at VG=0.7V.

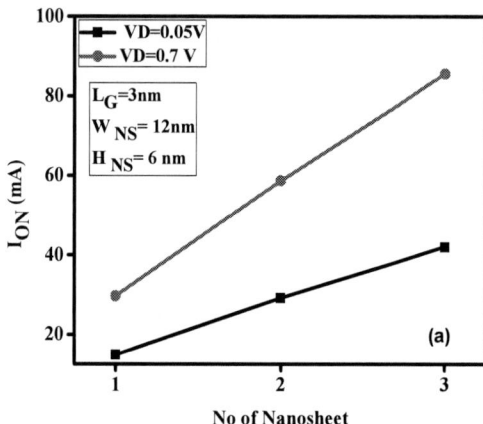

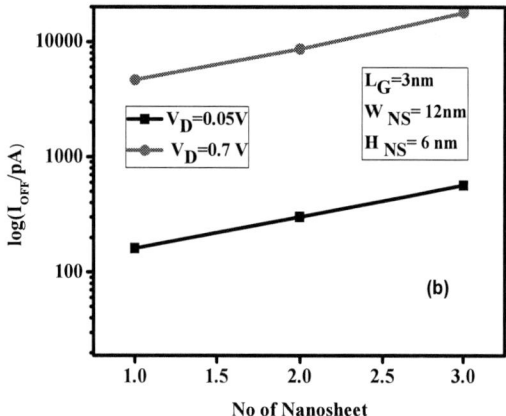

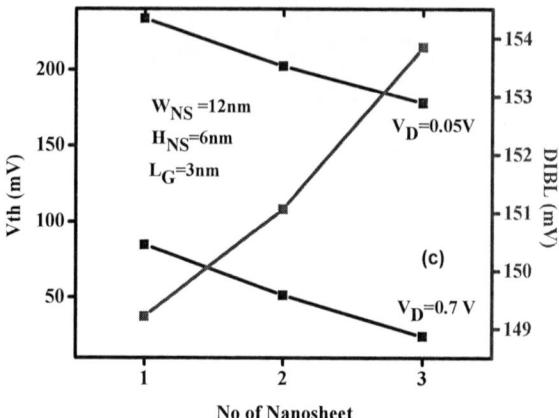

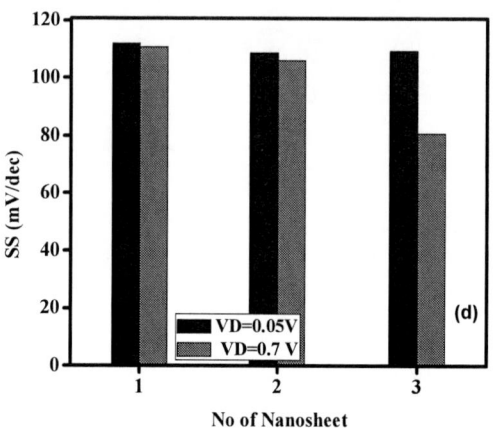

Fig. 6. Plot of Ion, Ioff, Vth, DIBL and SS vs. no of nanosheet are shown in (a), (b), (c) and (d), respectively.

VI. CONCLUSION

The simulation study of silicon Nanosheet FETs (SNS-FETs) is presented in this work. The effect of quantum confinement in SNS-FETs has been included in simulations by using the density gradient model. The models presented in this work feature a projection for 3nm technology nodes based on vertically stacked multiple nanosheet field effect transistors. Simulations show a clear advantage for the nanosheet FETs in terms of switching frequency and power consumption.

REFERENCES

[1] IRDS: Executive Summary, Rapport technique, International Roadmap for Devices and Systems, 2017.

[2] N. Loubet, T. Hook, P. Montanini, C.-W. Yeung, S. Kanakasabapathy, M. Guillorn, T. Yamashita, J. Zhang, X. Miao, J. Wang, A. Young, R. Chao, M. Kang, Z. Liu, S. Fan, B. Hamieh, S. Sieg, Y. Mignot, W. Xu, S.-C. Seo, J. Yoo, S. Mochizuki, M. Sankarapandian, O. Kwon, A. Carr, A. Greene, Y. Park, J. Frougier, R. Galatage, R. Bao, J. Shearer, R. Conti, H. Song, D. Lee, D. Kong, Y. Xu, A. Arceo, Z. Bi, P. Xu, R. Muthinti, J. Li, R. Wong, D. Brown, P. Oldiges, R. Robison, J. Arnold, N. Felix, S. Skordas, J. Gaudiello, T. Standaert, H. Jagannathan, D. Corliss, M.-H. Na, A. Knorr, T. Wu, D. Gupta, S. Lian, R. Divakaruni,

T. Gow, C. Labelle, S. Lee, V. Paruchuri, H. Bu, and M. Khare, "Stacked Nanosheet Gate-All-Around Transistor to Enable Scaling Beyond FinFET", in 2017 Symposium on VLSI Technology pp. T230-T231, 2017.

[3]   T. Skotnicki, C. F.-Beranger, C. Gallon, F. Boeuf, S. Monfray, F. Payet, A. Pouydebasque, M. Szczap, A. Farcy, F. Arnaud, S. Clerc, M. Sellier, A. Cathignol, J.-P. Schoellkopf, E. Perea, R. Ferrant, and H. Mingam, "Innovative Materials, Devices, and CMOS Technologies for Low-Power Mobile Multimedia," IEEE Trans. Electron Devices, vol. 55, no. 1, pp. 96–130, 2008.

[4]   MINIMOS-NT User's Manual, 2017.

[5]   J. P. Colinge, "Multi-gate SOI MOSFETs," Microelectronic Engineering, vol. 84, pp. 2071–2076, 2007.

[6]   T. Skotnicki, G. Merckel, and T. Pedron, "The voltage-doping transformation: A new approach to the modeling of MOSFET short-channel effects," IEEE Electron Device Lett., vol. 9, no. 3, pp. 109–112, 1988.

[7]   S. D. Suk, M. Li, Y. Y. Yeoh, K. H. Yeo, K. H. Cho, I. K. Ku, H. Cho, W. J. Jang, D.-W. Kim, D. Park, and W.-S. Lee, "Investigation of nanowire size dependency on TSNWFET," in IEDM Tech. Dig., pp. 891–894, 2007.

[8]   M. Saitoh, Y. Nakabayashi, H. Itokawa1, M. Murano, I. Mizushima, K. Uchida, and T. Numata, "Short-Channel Performance and Mobility Analysis of <110>- and <100>-Oriented Tri-Gate Nanowire MOSFETs with Raised Source/Drain Extensions," in VLSI Tech. Dig., pp. 169–170, 2010.

[9]   J. Chen, T. Saraya, K. Miyaji, K. Shimizu, and T. Hiramoto, "Experimental Study of Mobility in [110]- and [100]-Directed Multiple Silicon Nanowire GAA MOSFETs on (100) SOI," in VLSI Tech. Dig., pp. 32–33, 2008.

[10] G. J. Iafrate, H. L. Grubin, and D. K Ferry, "Utilization of Quantum Distribution Functions for Ultra-Submicron Device Transport", Journal de Physique., vol. 42 (Colloq. 7), 10, 307–312, 1981.

[11] E. Wigner, "On the Quantum Correction For Thermodynamic Equilibrium", Phys. Rev., vol. 40, pp. 749–759, 1932.

[12] D. K. Ferry and J.-R. Zhou, "Form of the quantum potential for use in hydrodynamic equations for semiconductor device modeling", Phys. Rev. B., vol. 48, pp. 7944–7950, 1993.

ADAPTIVE JAYA OPTIMIZATION TECHNIQUE FOR ECONOMIC LOAD DISPATCH CONSIDERING VALVE POINT EFFECT

Sourav Basak
Electrical and Electronics Engineering
National Institute of Technology Nagaland
Dimapur, INDIA
souravkr.basak@yahoo.com

Swaraj Banerjee
Electrical and Electronics Engineering
National Institute of Technology Nagaland
Dimapur, INDIA
srjbanerjee@gmail.com

Abstract—**ELD with the help of Jaya Optimization Technique has been proved as one of the better options. Along with the cost curves or cost coefficients valve point effect has also very important impact in ELD. So in this paper, for ELD with valve point effect has been optimized with the help of Adaptive Jaya Optimization Technique. 13 thermal unit has been taken to get the optimal power output and optimal cost, which has been compared to the other optimization techniques' results. Advantage of this proposed optimization technique has been shown in the current work.**

Keywords—Economic dispatch, Jaya algorithm, valve point effect.

I. INTRODUCTION

This Reliable rate of electricity has a great impact in power system economics. Reducing the cost of electricity is the main goal of power system economics. To depreciate the electricity cost, it's very important to depreciate the power generation. Optimization problem of ELD is a procedure of serving all the demand constantly at the minimum cost while satisfying equality and inequality constraints of units committed by UC. Particle Swarm Optimization, Global minimization, Genetic Algorithm, Classical Evolutionary Programming Fast Evolutionary Programming or FEP, Modified FEP, or Lambda iteration methods are there to solve the ELD problem. These techniques are having limitations, like execution speed, executions of repeated stages, require common controlling parameters i.e. population size, number of generations etc. Now it's very important to solve ELD most cost effective way. Jaya optimization technique [1, 2] has been proved as good and acceptable optimization technique for ELD as it has significant role in ELD. Here proposed Adaptive Jaya Optimization Technique makes out better output for ELD problems. It's having strong potential to solve the optimization of ELD problem with constrains.

The generating modules are having various valve-system turbine, which has huge effects on the fuel cost function. This is a non-linear function because of the ripple effect for opening of valve here. So in this scenario it's very essential to consider this valve point effect while optimize the generation as different valve-system turbine makes effects on the cost function. Without considering this effects ELD problem is incomplete and inappropriate. So in this present work this effect has been considered and solved 13 unit system ELD problem with the help of proposed Adaptive Jaya Optimization Technique.

In section II, formulation of ELD has been described. In section III, Jaya Optimization Technique has been discussed. Section IV stands for the description of implementation. In section V, dataset and result has been shown and discussed. Conclusion has been discussed in section VI.

II. FORMULATION OF ELD PROBLEM

This Economic Load Dispatch is the problem of cutting down of the power generation cost in cost-effective way, also fulfill the load demand as well as the equality, inequality constrains at the same time.

A. Cost function of thermal units

The Cost equation of a generator is a quadratic equation. Here this is the objective function as the function of cost, so we have to minimize this cost.

$$C_j = (a_j P_j^2 + b_j P_j + c_j) \tag{1}$$

$$C_{Total} = C_1 + C_2 + \ldots + C_n$$

So, $C_{Total} = \sum_{j=1}^{n} (a_j P_j^2 + b_j P_j + c_j) \tag{2}$

a_j, b_j and c_j are cost equation coefficients or the coefficient of objective function for j^{th} unit. Generation cost of jth unit is C_j. Here total cost of generation is coming, $C_1 + C_2 + \ldots + C_n = C_{Total}$ [3, 4].

As the generating modules are consists of various valve-system turbine, which has huge effects on the cost function. So here to consider this effect, its mathematical presentation has been stated as, by adding the abs value of sinusoidal function with a quadratic function [1, 4] of the cost function.

$$C_j = (a_j P_j^2 + b_j P_j + c_j) + |e_j \times \sin(f_j(P_{j,min} - P_j))| \tag{3}$$

$$C_{Total} = \sum_{j=1}^{n} (a_j P_j^2 + b_j P_j + c_j) + |e_j \times \sin(f_j(P_{j,min} - P_j))| \tag{4}$$

Where, e_j and f_j are the coefficients of jth generating module for the valve-point effects [1, 4, 5].

B. Constrains

$$\sum_{j=1}^{n} P_j - P_L - P_D = 0 \tag{5}$$

$$P_j^{min} \le P_j \le P_j^{max} \tag{6}$$

Here, P_j is the power shared by generator j (in MW), P_L is the total transmission loss, P_D is the load demand, P_j^{min} and P_j^{max} are the the minimum and maximum limit of jth unit [1, 3, 4].

978-1-5386-6723-1/19 $31.00 © 2019 IEEE

III. JAYA OPTIMIZATION TECHNIQUE

A. Algorithm and Flowchart

In Jaya Algorithm [1, 2, 6], Let f(x) is objective function here, so it has to be minimized. According to this optimization technique, for i^{th} iteration, design variables are 'm' (i.e. j = 1, 2,...,m) in number and candidate solutions are n (i.e. k = 1, 2,...,n) in number. Now among these candidate solutions, here best candidate will make out the best f(x) value, which solution is the best, that is f(x)$_{best}$. Worst candidate among them will make out the worst solution which is the worst value of f(x) i.e. f(x)$_{worst}$ (say). If $X_{i,j,k}$ is the solution of jth design variable, k^{th} candidate solution during i^{th} iteration, this value may be the possible solution and finally this value has been updated as per 5,

$$X'_{j,k,i} = X_{j,k,i} + r_{1,j,i} \times (X_{j,best,i} - |X_{j,k,i}|) - r_{2,j,i} \times (X_{j,worst,i} - |X_{j,k,i}|) \tag{7}$$

$X_{j,best,i}$ and $X_{j,worst,i}$ are the best and worst candidate solution here respectively for j^{th} design variable at i^{th} iteration. $X'_{j,k,i}$ is the new value, which is the revised value of $X_{j,k,i}$. Two numbers which lies between 0 to 1 are taken randomly specified in the equation as $r_{1,j,i}$ and $r_{2,j,i}$ for the jth design variable at i^{th} iteration.

The term "$r_{1,j,i} \times (X_{j,best,i} - |X_{j,k,i}|)$" is here for representing the affinity of solution so by adding this term it's possible to move on to a better solution and the term, which indicates the tendency to avoid the solution, to get the optimal solution is "$r_{1,j,i} \times (X_{j,best,i} - |X_{j,k,i}|)$". So by subtracting that term it's possible to avoid the worse solution.

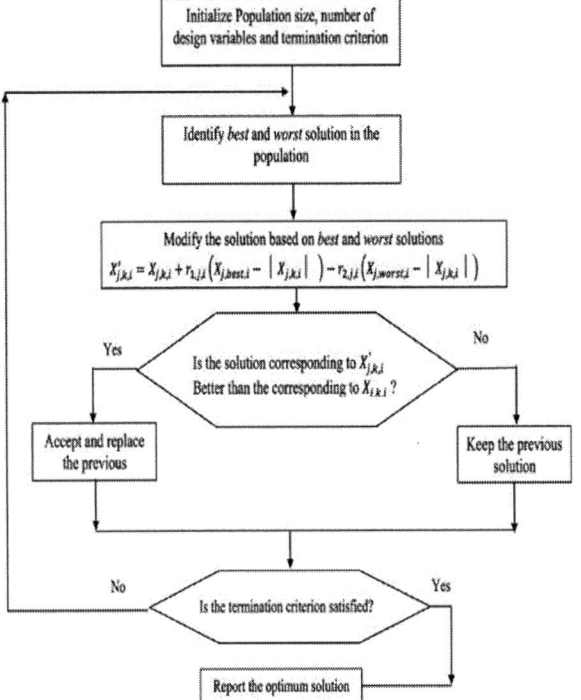

Fig. 1.   Flow Chart of Jaya Optimization Technique [1, 2, 6].

In this Jaya Optimization Technique first step is the initialization part, as the population, objective function, number of variables (design). In this state it can identify which is the best and which is the worst solution among the entire candidate. After that the solution has updated on the basis of best and worst solution according to 5. If the solution is better, then the updated solution has been accepted and if not better then it has been avoided and previous solution has been kept. If all the termination criterion satisfied then it will print the optimal solution or else this process will repeat from the solution identifying step.

B. Mathematical Illustration of JA

To make this optimization technique easier and simpler, two generator problem has been taken here. Max and min power generation limits has been given as 120- 600 for the unit P$_1$ and 50 – 400 is for the unit P$_2$. Now total demand, P$_D$ without considering transmission losses has been given as 750 MW. Being a minimization problem having constrains, this problem of ELD has to satisfy all the constraints given here. Here two design variables are there as P$_1$ and P$_2$. If there are 7 numbers candidate solutions and here for the best solution 5th candidate has been assumed while for the worst solution 1st candidate has been assumed, then there is the step to get the more optimum output for two upper mentioned units. Objective function has to be reduced to make the cost of generation cheaper on the basis of coefficients of generators. If the best solution is 5th candidate and the worst solution is the 1st candidate in this case then the value of $X'_{1,2,1}$ and $X'_{2,2,1}$ has to be updated as below. Assume, the numbers taken randomly are r_1=0.35, r_2=0.76 for P$_1$ and r_1=0.58, r_2=0.28 for P$_2$. The new values has been calculated as per 5. During first iteration, here the updated value of second candidate will be,

$$X'_{1,2,1} = X_{1,2,1} + r_{1,1,1} \times (X_{1,5,1} - |X_{1,2,1}|) - r_{2,1,1} \times (X_{1,1,1} - |X_{1,2,1}|)$$
$$= 470 + 0.35 \times (550 - |470|) - 0.76 \times (500 - |470|)$$
$$= 475.2$$

$$X'_{2,2,1} = X_{2,2,1} + r_{1,2,1} \times (X_{2,5,1} - |X_{2,2,1}|) - r_{2,2,1} \times (X_{2,1,1} - |X_{2,2,1}|)$$
$$= 220 + 0.58 \times (200 - |220|) - 0.12 \times (250 - |220|)$$
$$= 214$$

Here, during the 1st iteration the updated value of the 2nd candidate has been considered like that.

Here Adaptive Jaya Optimization Technique is the proposed method to solve the constrained ELD problem. Here it helps to vary slightly more with the tendency and affinity to get the better optimal solution. The Adaptive Jaya Optimization Technique has been shown below.

$$X'_{j,k,i} = X_{j,k,i} + r_{1,j,i} \times (X_{j,best,i} - |X_{j,k,i}|) - r_{2,j,i} \times (X_{j,worst,i} - |X_{j,k,i}|) + r_{3,j,i} - r_{4,j,i} \tag{8}$$

During i^{th} iteration, for j^{th} candidate solution the numbers taken randomly are $r_{3,j,i}$ and $r_{4,j,i}$.

IV. IMPLEMENTATION

The goal of the optimization of ELD problems is to schedule the optimal power generation for each unit. Which units are already committed and can serve the entire demand at lower operating cost and also satisfy the equality, inequality constraints.

At first generating units or the design variables are 'm' (i.e. j = 1, 2,…,m) in number which are having the lower and the upper limit (mentioned in problem) of each unit, that has been initialized. Then candidate solutions are 'n' (i.e. k = 1, 2,…,n) in number has also been defined. Mentioned thermal generating units have to serve the demanded power in the next step. Random generation has been formed between lower and upper limit of each thermal unit in matrices form where the length of row is j and length of column is k. Next step is to calculate mismatch by subtracting the total demand from summation value of generation for each candidate solution. Then that mismatch has been shared among the committed units. Here it has also been considered that if a unit is giving less than the lower limit, then that unit output will be equal to the lower limit and if any generating unit is giving more than upper limit, then that unit output will be equal to the upper limit. This step has been done here for certain iterations. Like this, the generation has been updated here. Then the cost has been calculated with the help of 4.

Now the next step is the cost calculation step as sort out the minimum cost and then cut down the generating cost while there are no mismatch between generation and demand and also satisfying the equality, inequality constrains of the problem mentioned here. This minimization has been done according to 8. And then it will give the optimum output.

Implementation ELD problem here, has been performed by the software, Matlab 2014a, on a 2.2-GHz AMD, E2 processor with 4 GB RAM.

V. DATASET AND RESULT

A. Dataset

Here min and max limit has been mentioned in Table 1 for each generator and a_j, b_j, c_j, e_j and f_j are cost function coefficients which has been mentioned in Table I.

TABLE I.   LOWER, UPPER BOUND AND COST COEFFICIENTS OF UNIT.

Unit	P^{min} (MW)	P^{max} (MW)	a	b	c	e	f
1	0.00	680.00	0.000280	8.100	550.0	300.0	0.0350
2	0.00	360.00	0.000560	8.100	309.0	200.0	0.0420
3	0.00	360.00	0.000560	8.100	307.0	200.0	0.0420
4	60.00	180.00	0.003240	7.740	240.0	150.0	0.0630
5	60.00	180.00	0.003240	7.740	240.0	150.0	0.0630
6	60.00	180.00	0.003240	7.740	240.0	150.0	0.0630
7	60.00	180.00	0.003240	7.740	240.0	150.0	0.0630
8	60.00	180.00	0.003240	7.740	240.0	150.0	0.0630
9	60.00	180.00	0.003240	7.740	240.0	150.0	0.0630
10	40.00	120.00	0.002840	8.600	126.0	100.0	0.0840
11	40.00	120.00	0.002840	8.600	126.0	100.0	0.0840
12	55.00	120.00	0.002840	8.600	126.0	100.0	0.0840
13	55.00	120.00	0.002840	8.600	126.0	100.0	0.0840

Now with respect to dataset, the output of Adaptive Jaya Optimization Technique has been compared with Classical evolutionary programming (CEP) [4], Fast evolutionary programming (FEP) [4], Fast evolutionary programming (MFEP) [4], Improved Fast evolutionary programming IFEP

[4], Jaya Optimization Technique, and the results has been shown below for the demand of 1800 MW.

TABLE II.   COST OF DIFFERENT EVOLUTION METHODS.

Evolution methods	Mean Time (sec)	Best time (sec.)	Mean cost($)	Max cost($)	Min cost($)
CEP	294.96	293.41	18190.3	18404	18048.2
FEP	168.11	166.43	18200.8	18453.8	18018
MFEP	317.12	315.98	18192	18416.9	18028.1
IFEP	157.43	156.81	18127.1	18267.4	17994.1
JOT	92.56	90.02	18118.3	18222.4	18008.5
AJOT	**62.52**	**19.16**	**18115.6**	**18238.8**	**17998.2**

Here the optimum output of each unit has been given bellow for the minimum cost,

The Optimal Fuel Cost (in $/hr.) = 17998.1985

The Optimal Generation of :
Generator-1(in MW/hr.) = 538.620751
Generator-2(in MW/hr.) = 224.809090
Generator-3(in MW/hr.) = 299.272865
Generator-4(in MW/hr.) = 60.000000
Generator-5(in MW/hr.) = 159.716085
Generator-6(in MW/hr.) = 60.000000
Generator-7(in MW/hr.) = 60.000000
Generator-8(in MW/hr.) = 109.912867
Generator-9(in MW/hr.) = 60.000000
Generator-10(in MW/hr.) = 40.000000
Generator-11(in MW/hr.) = 40.277126
Generator-12(in MW/hr.) = 55.000000
Generator-13(in MW/hr.) = 92.391216

Gen1+Gen2+Gen3+Gen4+Gen5+Gen6+Gen7+Gen8+Gen9+Gen10+Gen11+Gen12+Gen13 = 1800.0000.

No Violation of Equality Constraint.
Total Number of Inequality Constraint Violation = 0
Number of Iterations Required for Convergence = 425
The Total Simulation Time (in Seconds) = 19.1578

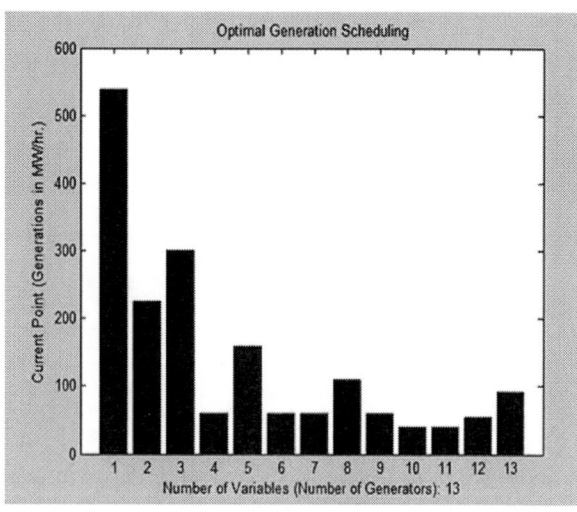

Fig.2.   Optimal Generation Scheduling.

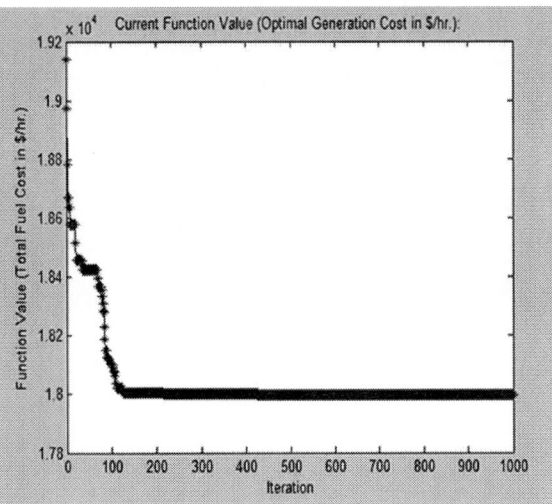

Fig. 3.   Function value v/s iteration.

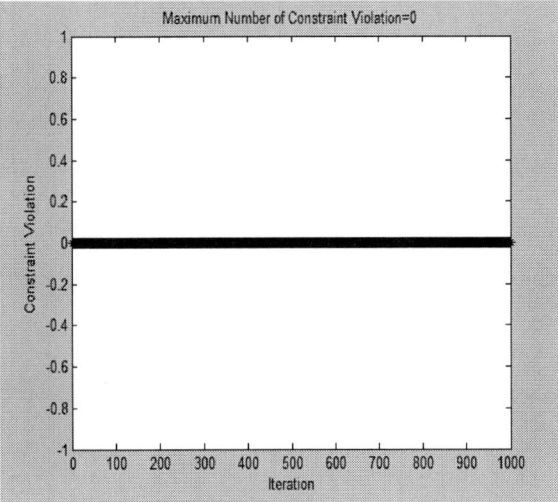

Fig. 4.   Violation v/s iteration.

The corresponding graphs of the optimal generation also has been mentioned above.

VI. CONCLUSION

ELD with the thermal units with the help of Adaptive Jaya Optimization Technique has been successfully done in the Matlab 2014a software. This paper is for implementing the Adaptive Jaya Optimization Technique to cut down the generation cost of ELD in power system economics. 13 thermal units system ELD problem has been implemented here and the optimal result has been shown above.

Here the generation cost of units after implementation of ELD with the Adaptive Jaya Optimization Technique is coming less than the other optimization techniques, so it will definitely has an great impact on minimizing the electricity cost. Here it has also consumed less time than the other techniques. The number of iterations required for convergence the cost function is also less here. It's also faster optimization technique according to the comparison done here.

REFERENCE

[1]   Swaraj Banerjee, Dipu Sarkar, "Comparative Analysis of Jaya Optimization Algorithm for Economic Dispatch Solution", IJRASET, Volume 5 Issue VIII, August 2017.

[2]   Venkata Rao R, "Jaya: A simple and new optimization algorithm for solving constrained and unconstrained optimization problems", International Journal of Industrial Engineering Computations, Volume 7 Issue 1 pp. 19-34, 2016.

[3]   Zwe-Lee Gaing, "Particle Swarm Optimization to Solving the Economic Dispatch Considering the Generator Constraints", IEEE Transactions on power systems, vol. 18, no. 3, August 2003.

[4]   N. Sinha, R. Chakrabarti and P. K. Chattopadhyay, "Evolutionary programming techniques for economic load dispatch," in IEEE Transactions on Evolutionary Computation, vol. 7, no. 1, pp. 83-94, Feb. 2003.

[5]   Al-Sumait, J. S., A. K. Al-Othman, and J. K. Sykulski. "Application of pattern search method to power system valve-point economic load dispatch." International Journal of Electrical Power & Energy Systems 29.10 (2007): 720-730.

[6]   M. Bhoye, M. H. Pandya, S. Valvi, I. N. Trivedi, P. Jangir and S. A. Parmar, "An emission constraint Economic Load Dispatch problem solution with Microgrid using JAYA algorithm," 2016 International Conference on Energy Efficient Technologies for Sustainability (ICEETS), Nagercoil, 2016, pp. 497-502.

2019 Devices for Integrated Circuit (DevIC), 23-24 March, 2019, Kalyani, India

Role of Body Effect on Threshold Voltage of Strained Si-Si$_x$Ge$_{1-x}$ MOSFET

Swarnav Mukhopadhyay
Dept. of Electronics and Telecommunication Engineering
Jadavpur University
Kolkata, India
swarnavhome@gmail.com

Arpan Deyasi
Dept. of Electronics and Communication Engineering
RCC Institute of Information Technology
Kolkata, India
deyasi_arpan@yahoo.co.in

Abstract—Role of body effect coefficient on threshold voltage of strained Si/SiGe MOSFET is analytically investigated. Effect of dielectric thickness, doping concentration and dielectric material are computed on the threshold condition both in presence and absence of body effect. Simulation findings reveal that introduction of strained material in otherwise ideal structure enhances carrier mobility which, in turn, reduces threshold voltage. Sharp peak is observed when body effect is taken into account for a particular heterostructure composition due to enhancement of tunneling probability which decrease of barrier potential, and thus carrier flow is augmented. A few results for n-channel MOSFET are also represented to further justify the importance of novelty of the paper. Result also suggests that higher doping or thicker dielectric region leads to depletion mode of operation. An optimized design criterion is evaluated for the minimum threshold under inversion condition.

Keywords—Body effect, Band bending, Inversion condition, Strained device, Threshold voltage, Structural parameters, Material composition

I. INTRODUCTION

Strained Silicon devices are proposed in recent past [1-2] in order to achieve less transport time of carriers and thus by increasing the overall mobility of the device [3]. When silicon is grown on a SiGe substrate, the silicon atoms are not properly aligned with SiGe atoms in spite of lattice mismatch, thereby creating stress at Si/SiGe junction. Literatures are published to prove supremacy of strained Si devices over conventional MOS devices [4-5] in terms of higher speed of operation, less power dissipation. Threshold voltage is one of the important parameter which governs its electrical characteristics, and its indirect method of estimation is proposed from capacitance-voltage profile [6]. Dual channel CMOS architecture is proposed [7] with variable material composition for performance improvement, and its effect on gate leakage [8] and reliability is measured. Proposed technology also substantiates the reduction of self-heating effect [8].

Fabrication techniques including oxidation kinetics, nature of surface roughness greatly controls the device performance [9], which is reflected from transconductance measurement. Double heterostructures are also proposed with relaxed Si/SiGe layer, and corresponding mobility enhancement is reported [10]. Ultrathin body is recently mentioned to improve current [11] as well as better gate control. Different material systems are proposed in this context including SOI feature. Transfer characteristics are also obtained to measure the impact of strain [12] for different material and structural parameters. Buried channel FET is designed [13] for very good quality RF performance

even at extremely low temperature. But in these articles, body effect is not considered over transistor characteristic.

In the present work, analytical modeling is done incorporating the body effect to calculate the threshold voltage, and different material system and structural parameters are considered. The motivation behind the idea is that with change of potential barrier height at the hetero-interface, threshold voltage linearly increases, which does not validate with the actual observations under practical situation. But incorporation of body effect clearly shows the occurrence of maximum threshold where change of potential barrier at either sides will reduce the threshold. This will help to identify the operating point of the device when body bias is provided. Moreover, role of dielectric is also studied. The change of slope for threshold in presence of body effect is drastically modified for different structural parameters, and the effect is graphically represented. Result is quite interesting from the point of view of reducing the barrier height at certain combination of design parameters.

II. DESIGN OF DEVICE

We have considered the strained Si-SiGe pMOSFET according to the Fig 1 [14].

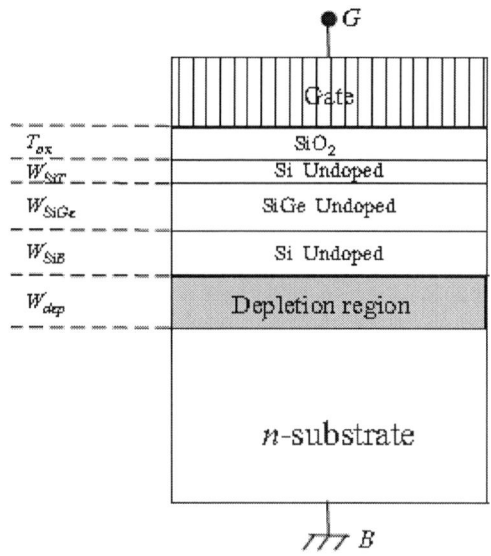

Fig. 1. Schematic structure of strained Si-SiGe nMOSFET

where W_{SiT}, W_{SiGe}, W_{SiB} are the corresponding width of the regions. We have also assumed that inversion takes place only at hetero-interface.

Surface potential at Si-SiO₂ interface is given by

$$\phi_{s1} = -2V_t \ln\left(\frac{N_D}{n_i}\right) \tag{1}$$

and at Si-SiGe interface

$$\phi_{s2} = -2V_t \ln\left(\frac{N_D}{n_i}\right) + \Delta E_V \tag{2}$$

where ΔEv is valence band discontinuity, other symbols have customary significances.

When we apply negative gate voltage the holes accumulate in the surface of Si-SiO₂ and the Si-SiGe interface. But due to less strain the mobility of holes in the Si-SiGe interface is greater than Si-SiO₂ surface. So the inversion layer is created mostly in Si-SiGe interface, mathematically written as $|\Phi s2| < |\Phi s1|$. As we apply the gate voltage the substrate is depleted and a depletion layer is created. The width of the depletion region is given by [14]

$$W_d = -W_{SiB} - \frac{\varepsilon_{Si}}{\varepsilon_{SiGe}} W_{SiGe}$$

$$+ \sqrt{\frac{2\varepsilon_{Si}}{qN_D}(-\phi_{s2}) + \left(W_{SiB} + \frac{\varepsilon_{Si}}{\varepsilon_{SiGe}} W_{SiGe}\right)^2} \tag{3}$$

Threshold voltage is a function of depletion width as

$$\phi_{th} = \phi_{s2} - qN_D W_d \left(\frac{t_{ox}}{\varepsilon_{ox}} + \frac{W_{SiT}}{\varepsilon_{Si}}\right) \tag{4}$$

where we have excluded flatband factor.

If we consider the body effect coefficient, then Eq. (4) is modified in the following form

$$\phi_{th} = -\phi_{SB} - \phi_{s2} - qN_D W_d \left(\frac{t_{ox}}{\varepsilon_{ox}} + \frac{W_{SiT}}{\varepsilon_{Si}}\right) \tag{5}$$

III. RESULTS AND DISCUSSIONS

Using Eq. (4) and Eq. (5), threshold voltage is computed and plotted in absence and presence of body effect respectively. Result is shown in Fig 2. From the plot, it is seen that in absence of body effect [14], threshold voltage monotonically increases with increasing mole fraction. In this context, it may be mentioned that we have considered the mole fraction of Si is within 0.4 so that the material composition can vary from $Si_{0.1}Ge_{0.9}$ to $Si_{0.4}Ge_{0.6}$. Our computation results are compared with [14], and shows excellent agreement when body effect is absent.

When body effect is considered, a sharp peak is observed in the profile for a particular value of mole fraction keeping all other parameters unchanged. For the present set of parameters, it is noted that the maximum value is obtained at $x = 0.28$. This is due to the maximum band bending, which occurs as a result of increased mobility of holes in the Si-Ge interface. At a particular design consideration, maximum majority carriers are accumulated at the surface of Si and Si-Ge interface under inversion condition, which causes maximum lowering/bending of the band. That reduces the opposing potential barrier, and hence justifies threshold potential reduction.

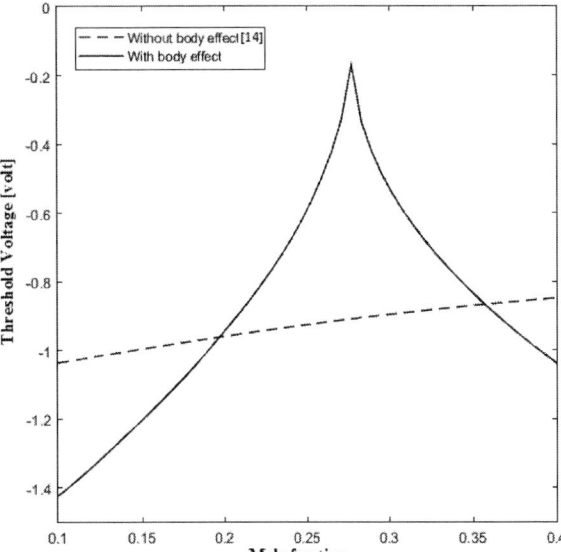

Fig. 2. Threshold voltage variation in presence and absence of body effect with material composition

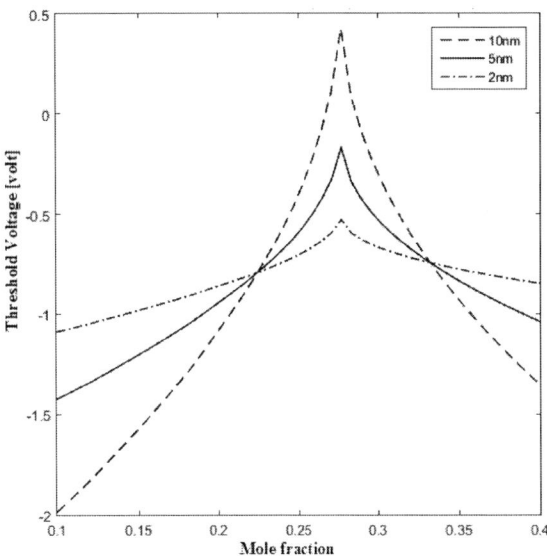

Fig. 3. Threshold voltage variation for different dielectric thickness in presence of body effect with material composition

Fig 3 shows the variation for different dielectric thickness considering SiO₂ as material. It is observed from the figure that as we moves towards nano-dimension, threshold voltage remains almost constant w.r.t material system, whereas sharp change is seen for larger dielectric thickness. This is due to the fact that higher oxide layer width forces more band bending in the semiconductor layer causes greater fluctuation in threshold. Larger band bending also changes the mode of operation, and device starts to work in depletion mode. this causes positive threshold. Similar type of variation is also observed for different dielectric materials also, as represented in Fig 4. As electric permittivity increases, effect of gate voltage on the inversion layer decreases, causes less change in threshold voltage. This is reflected when we make a comparative study between SiO_2 and HfO_2.

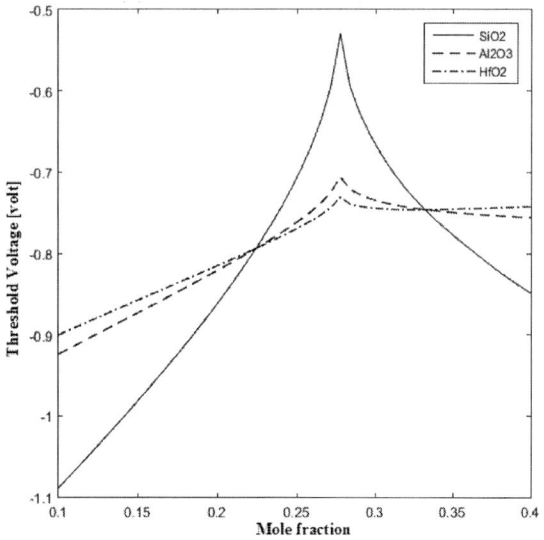

Fig. 4. Threshold voltage variation for different dielectric materials in presence of body effect with material composition

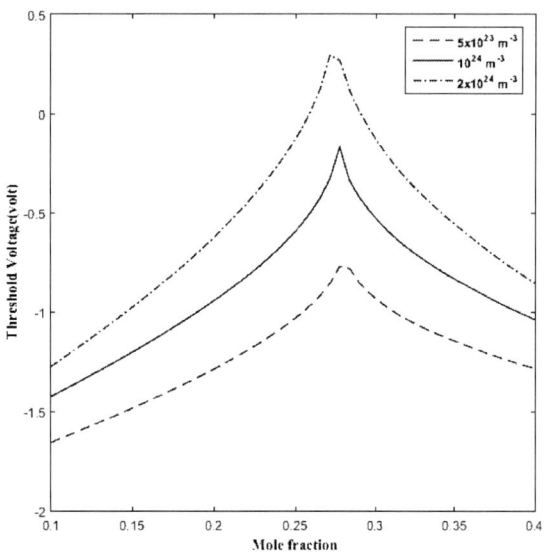

Fig. 5. Threshold voltage variation for different doping concentrationsin presence of body effect with material composition

Computation is also carried out for different doping concentrations. In this case, higher doping is considered for simulation as we are considering short-channel device, which leads to positive threshold voltage. Higher doping

concentration reduces threshold as presence of more majority carriers at inversion region helps the carrier flow in the channel. A closer inspection of Fig 5 reveals the fact that for higher doping, sharp peak is modified in flat top. This suggests that with increase of doping, higher magnitude of threshold voltage can be obtained for a wider range of material composition. This may be due to the fact that very high doping increases the carrier flow inside the channel, and the threshold condition is inverted. This is reflected in Fig 5, which reveals that the p-channel MOSFET operates in depletion mode instead if enhancement mode. So a channel is created flow.

Similar variations are also plotted in absence of body effect. This is shown in Fig 6. The threshold voltage gets increased [negative scale] if we increase the doping concentration in the bulk. We can get the maximum V_{th} by choosing the optimum value from the analytical curves. As we increase the mole fraction the threshold voltage gets reduced. In this case we have taken the oxide thickness as 10nm.

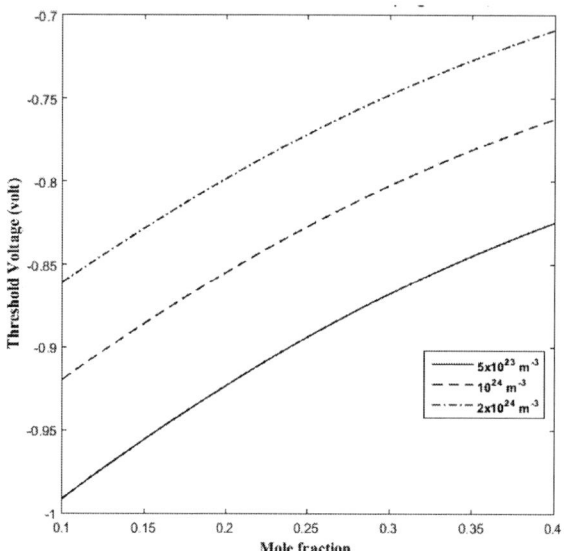

Fig. 6. Threshold voltage variation for different doping concentrationsin absence of body effect with material composition

The peak points for different sets of structural parameters and material systems are represented in tabular forms. Table I gives the threshold voltages for different dielectric materials [t_{ox} = 5 nm, N_D = 10^{24} m^{-3}], whereas table II and table III give the data for different oxide thickness.

TABLE I.   MINIMUM THRESHOLD VOLTAGE FOR DIFFERENT DIELECTRIC MATERIALS

Dielectric	Threshold voltage
SiO_2	−0.17
Al_2O_3	−0.61
HfO_2	−0.674

TABLE II.   MINIMUM THRESHOLD VOLTAGE FOR DIFFERENT OXIDE THICKNESS

Oxide thickness [nm]	Threshold voltage	
	SiO_2	HfO_2

2	-0.53	-0.73
3	-0.41	-0.71
4	-0.29	-0.69
4.5	-0.23	-0.68
5	-0.17	-0.674
5.5	-0.11	-0.66
6	-0.05	-0.655
7	0.07	-0.635
8	0.19	-0.615
9	0.3	-0.6
10	0.43	-0.58

TABLE III.   MINIMUM THRESHOLD VOLTAGE FOR DIFFERENT DOPING CONCENTRATIONS

Doping concentration [m⁻³]	Threshold voltage	
	SiO₂	HfO₂
5×10^{23}	-0.52	-0.7
6×10^{23}	-0.45	-0.695
7×10^{23}	-0.38	-0.69
8×10^{23}	-0.3	-0.685
8.5×10^{23}	-0.25	-0.68
9×10^{23}	-0.2	-0.675
9.5×10^{23}	-0.18	-0.676
10^{24}	-0.17	-0.674
1.5×10^{24}	0.08	-0.653
1.75×10^{24}	0.23	-0.635
2×10^{24}	0.43	-0.612

We have also investigated the magnitude of body effect on threshold voltage. Result is displayed in Fig 7 for three different heterostructure compositions. It has been observed from the plot that threshold voltage increases with increase of negative body bias, and attains a high value at a particular value of the voltage. Further increase of body bias reduces threshold. It has also been noted that higher negative bias also makes negligible effect of mole fraction, and this function is indistinguishable around the maximum threshold.

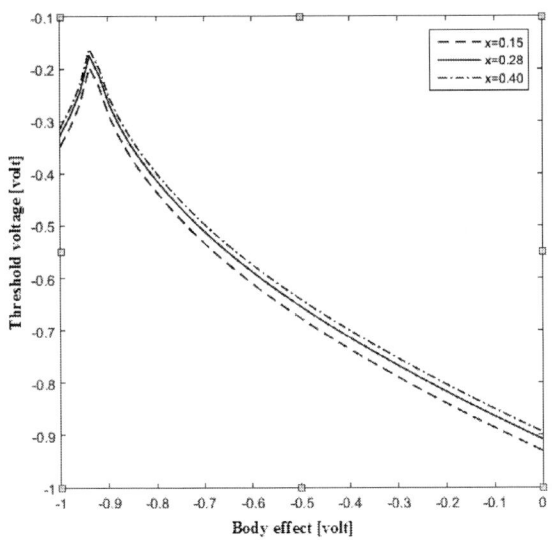

Fig. 7. Threshold voltage variation for different body bias for three different hetero-compositions

Similar computation is also performed for n-channel device, both in presence and absence of body effect. It is found that in presence of body effect, threshold voltage

decreases for a particular composition. When substrate voltage is made negative w.r.t source voltage, then the junction formed between source and channel gets reverse biased. This causes a large increase of electrons/negatively charged ions inside the channel. The overall effect is to reduction of threshold voltage.   But change of barrier potential causes a restriction of current flow which enhances the threshold.

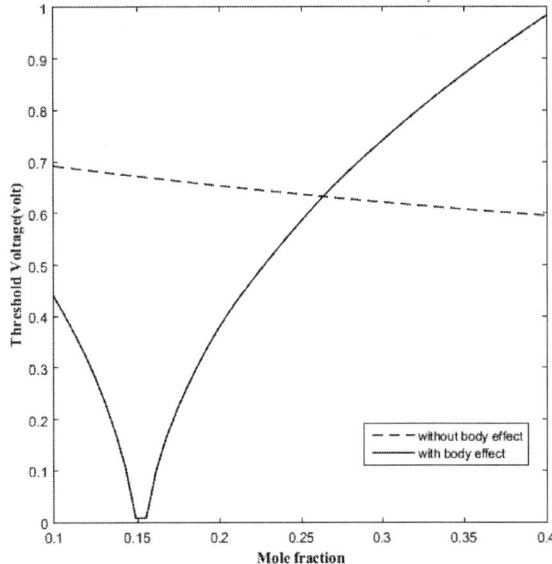

Fig. 8. Thrshold voltage with material compositon for n-channel device in presence and absence of body effect

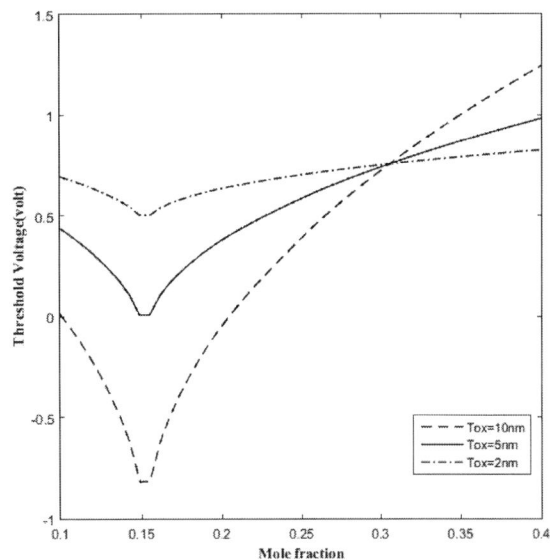

Fig. 9. Threshold voltage with different dielectric thcikness for n-channel device in presence of body effect

Higher increase of dielectric thickness also makes change of threshold voltage, as the device works as enhancement type. This causes negative value of V_{th}.

IV. CONCLUSION

Role of body effect is analytically investigated on threshold voltage of strained MOSFET, and variations due to dielectric materials, its thickness and doping concentrations at bulk are also studied. Lowest threshold is obtained for a particular design consideration, and that will help to use the device at nano-dimensions. Change of operational modes is also reflected due to higher doping or larger depletion, and result is justified for both types (p and n-channel) of MOSFET. Results are also compared with the ideal biasing configuration, and strain effect is studied over a wide range of practical interest.

REFERENCES

[1] K. Rim, J. L. Hoyt, J. F. Gibbons, "Fabrication and Analysis of Deep Submicron Strained-Si n-MOSFET's", *IEEE Transactions on Electron Devices*, vol. 47, pp. 1406-1415, 2000

[2] R. Ohba, T. Mizuno, "Nonstationary Electron/Hole Transport in Sub-0.1 /spl mu/m MOS Devices: Correlation with Mobility and Low-power CMOS Application", *IEEE Transactions on Electron Devices*, vol. 48, pp. 338-343, 2001

[3] A. Lochefeld, D. A. Antoniadis, "Investigating the Relationship between Electron Mobility and Velocity in Deeply Scaled NMOS via Mechanical Stress", *IEEE Electron Device Letters*, vol. 22(12), pp. 5791-593, 2001

[4] T. Pešić-Brđanin, B. L. Dokić, "Strained Silicon Layer in CMOS Tech9nology", *Electronics*, vol. 18(2), pp. 63-69, 2001

[5] K. Rim, "Strained Si Surface Channel MOSFETs for High-performance CMOS Technology", *IEEE International Solid-State Circuits Conference*, 2001

[6] S. Chattopadhyay, K. S. K. Kwa, S. H. Olsen, L. S. Driscoll, A. G. O'Neill, "C–V Characterization of Strained Si/SiGe Multiple Heterojunction Capacitors as a Tool for Heterojunction MOSFET Channel Design", *Semiconductor Science and Technology*, vol. 18(8), pp. 738-744, 2003

[7] S. H. Olsen, K. S. K. Kwa, L. S. Driscoll, S. Chattopadhyay, A. G. O'Neill, "Design, Fabrication and Characterisation of Strained Si/SiGe MOS Transistors", *IEE Proceedings - Circuits, Devices and Systems*, vol. 151(5), pp. 431-437, 2004

[8] S. H. Olsen, L. Yan, R. Agaiby, E. Escobedo-Cousin, A. G. O'Neill, P. E. Hellström, M. Östling, K. Lyutovich, E. Kasper, C. Claeys, E. H. C. Parker, "Strained Si/SiGe MOS Technology: Improving Gate Dielectric Integrity", *Microelectronic Engineering*, vol. 86(3), pp. 218-223, 2009

[9] S. H. Olsen, A. G. O'Neill, D. JNorris, A. G. Cullis, S. J. Bull, S. Chattopadhyay, K .S. K. Kwa, L.S. Driscoll, A. M. Waite, Y. T. Tang, A. G. R. Evans, "Thermal Oxidation of Strained Si/SiGe: Impact of Surface Morphology and Effect on MOS Devices", *Materials Science and Engineering: B*, vol. 109(1-3), pp. 78-84, 2004

[10] M. L. Lee, E. A. Fitzgerald, M. T. Bulsara, M. T. Currie, A. Lochtefeld, "Strained Si, SiGe, and Ge Channels for High-mobility Metal-Oxide-Semiconductor Field-Effect Transistors", *Journal of Applied Physics*, vol. 97, p. 011101, 2005

[11] I. Aberg, C. N. Chleirigh, J. L. Hoyt, "Ultrathin-body Strained-Si and SiGe Heterostructure-on-Insulator MOSFETs", *IEEE Transactions on Electron Devices*, vol. 53(5), pp. 1021-1029, 2006

[12] H. Batwani, M. Gaur, M. J. Kumar, "Analytical Drain Current Model for Nanoscale Strained-Si/SiGe MOSFETs", *The International Journal for Computation and Mathematics in Electrical and Electronic Engineering*, vol. 28(2), pp. 353-371, 2009

[13] K. Fobelets, J. E. Velazquez-Perez, T. Hackbarth, "Study of MOS-gated Strained-Si Buried Channel Field Effect Transistors", *IETE Journal of Research*, vol. 53(3), pp. 253-262, 2014

[14] K. Iniewski, S. Voinigescu, J. Atcha, C. A. T. Salama, "Analytical Modeling of Threshold Voltages in p-channel Si/SiGe/Si MOS Structures", *Solid State Electronics*, vol. 36(5), pp. 775-783, 1993

A new algorithm for single line outage estimation

Mehebub Alam[1], Shubhrajyoti Kundu[2],Siddhartha Sankar Thakur[3]
Department of Electrical Engineering
NIT Durgapur
Durgapur, India
mehebubjgec1990@gmail.com[1],
sk.17ee1102@phd.nitdgp.ac.in[2],
sst@ee.nitdgp.ac.in[3]

Sumit Banerjee
Department of Electrical Engineering
B.C Roy Engineering College
Durgapur, India
sumit_9999@rediffmail.com

Abstract—This paper presents a novel algorithm for estimation of line outages using phasor angle measurements. In this algorithm, phasor angles obtained from load flow simulation for various outage cases and thereby simulated bus power mismatches (SBPM) are to be stored. On occurrence of actual outage, bus power mismatches are to be computed using PMU provided post outage as well as pre outage phasors .Thereafter comparison between simulated bus power mismatches and computed bus power mismatches (CBPM) is done through least square norm minimization approach to find out the actual outage case.Moreover, random Gaussian noise with zero mean and standard deviation from 1% to 5% is introduced in the proposed model to check the feasibility in real power system.Performance of the algorithm is tested on IEEE 5bus, 14 bus and 30 bus system. The simulation results show the efficiency and viability of the proposed algorithm.

Keywords—Line outage; Phasormeasurement unit(PMU); blackout; power systems

I. INTRODUCTION

Nowadays reliability and security of power systems has become an important area of research to the power system planners as well as academicians. To ensure the reliability and security of power systems it is essential to provide the real time information of power systems to the power systems operators so that any abnormal situation can be detected accurately as early as possible.One of the important area of situational awareness is line outage estimation(LOE).

It is to be noted that the line outage is uncertain in nature and can happens at any instant of time due to various reasons like breaker failures ,tree falling, lightning etc.Various blackout events have occurred during the past decades which has given a thrust to theresearchers to develop suitable algorithm for line outage identification to avoid such blackouts.An investigation was carried out on the August 14, 2003 blackout by North American Electric Reliability Council (NERC) to find out the root causes of the outage. After investigations it has been found that the lack of situational awareness and deficiencies of real time monitoring of system conditions were the major reasons [1]-[2]. So,it is utmost important to develop a new suitable algorithm for real -time monitoring and identifying transmission line outages. Phasor Measurement Units (PMUs) are now becoming popular day by day due to their special features like capability of providing

real-time measurements of voltage phasors and current phasors for monitoring of power systems [3].So, PMUs are placed at various buses of power systems to maintain the network observability. PMUs are having a very vital role for detecting line outages. When line outages occur phasor angles are altered and system power flow is also disturbed. Overloading due to single line outage may lead to cascading tripping which further may cause complete system blackout [4]. Major benefits of PMUs over traditional supervisory control and data acquisition (SCADA) system is illustrated in [5] considering features like resolution, observation capability, coverage area etc.

Several works [6]-[12] have been carried out related with line outage over the past few years.In [6], single line outage detection is presented by comparing variation of voltage angles using PMU. In this paper optimization technique based on normalized angle differences is presented to detect the outage event. This work is then further extended to detect double line outage detection in [7]. In reference [8], authors suggestedmixed integer programming based optimization technique to identify line single outages. A new approach using compressive sensing based method with the help of least absolute shrinkage and selection operator (LASSO) is presented in [9].The proposed LASSO based method has the potential to detect multiple line outages but performance is dependent on the fixed prior parameter like outage probability and perturbed noise.The work in [10] presents a sparse over complete representation for multiple line outage detection.In this paper DC power flow model is reformulated as a sparse over complete expansion and testing was done with affordable complexity.In [11],the various issues of PMU uncertainty and its effect on the line outage identification is discussed by adopting a multi-hypothesis test and a general Bayesian approach .A least square based method is presented for line outage detection by comparing simulated phasor angles with measured phasor angles from PMU in [12]. Single as well as double line outage detection has been validated through results considering IEEE 5 bus,14 bus and 30 bus system in [12].Impact of line outages is discussed in paper [13] with reference to change of voltage phasors. In [13], it is shown that due to line outages both voltage magnitude of buses and angles are altered in comparison with base case.

II. LINE OUTAGE ESTIMATION MODEL

The proposed model is basically based on comparison of simulated bus power mismatches (SBPM) with computed bus power mismatches (CBPM) .In modern era storage system has been advanced a lot. Utilizing this modern storage devices it is very easy to store the simulated angles for different outage cases. Now the bus power mismatches to be computed using susceptance matrix of full network. In this proposed algorithm, DC power flow model is used employing the *bus susceptance matrix of full network*.Let consider a power system network with N number of buses and L number of lines. Popular Fast Decoupled Load Flow Method is used for simulation purpose.

According to DC power flow method, power flowing (P_{ij}) from i bus to j bus can be expressed as,

$$P_{ij} = \frac{1}{x_{ij}}(\theta_i - \theta_j) \qquad (1)$$

The real power injections vector (p) and voltage phasor anglesvector (θ) can be written as,

$$P = [p_1, p_2, \dots, p_N]^T \qquad (2)$$

$$\theta = [\theta_1, \theta_2, \dots, \theta_N]^T \qquad (3)$$

In general power injection vector (P) can be expressed as,

$$P = B\theta \qquad (4)$$

For line l ($l = 1,2 \dots \dots L$) that connects between buses j and i, incidence vector is denoted by $\boldsymbol{m_l}$ .From this definition bus line incidence matrix (NXL-order) can be expressed as,

$$M = [m_1, m_2 \dots m_L] \qquad (5)$$

Elements of M can be expressed as,

$$M_{kl} = \begin{cases} 1; if\ k = j \\ -1; if\ k = i \\ 0; otherwise \end{cases} \qquad (6)$$

Susceptance matrix (B) can also be written as,

$$B = MD_xM^T \qquad (7)$$

Where D_x is the diagonal matrix with $\frac{1}{x_l}$ as its $l - th$ diagonal element.

Let post outage power injection using susceptance matrix (B) *of full network* can be written as,

$$P' = B\theta' \qquad (8)$$

Where θ' is the post outage voltage phasor angle vector
Similarly, post outage power can be written as,

$$P = B\theta \qquad (9)$$

Here, θ is the pre outage voltage phasor angle vector
Now, difference between post outage and pre outage bus power injection be expressed as,

$$\Delta P = P' - P = B(\theta' - \theta) \qquad (10)$$

Let simulated voltage phasor angles for various outage cases can be expressed as,

$$\theta_s^i = [\theta_s^1, \theta_s^2, \dots, \theta_s^n] \qquad (11)$$

Where, $i = 1,2, \dots, n; n$ is the total number of outage cases.
For a particular outage casesimulated angles can be written as,

$$\theta_s = [\theta_{s,1}, \theta_{s,2}, \dots, \theta_{s,N}]^T \qquad (12)$$

θ_s is a N dimensional vector.

Measured voltage angles from PMU after actual outage be denoted by,

$$\theta_m = [\theta_{m,1}, \theta_{m,2}, \dots, \theta_{m,N}]^T \qquad (13)$$

Now simulated bus power injection mismatch for various outage cases can be written as,

$$\Delta P_s^i = [\Delta P_s^1, \Delta P_s^2, \dots, \Delta P_s^n] \qquad (14)$$

Similarly, for a particular simulated outage case bus power mismatch can be written as,

$$\Delta P_s = B(\theta_s - \theta) \qquad (15)$$

Computed bus power mismatches using PMU measurements after actual outage occurs be expressed as

$$\Delta P_m = B(\theta_m - \theta) \qquad (16)$$

Now difference of power mismatch vector between measured value and computed value for a particular outage cases can be written as,

$$\Delta P_m - \Delta P_s = B(\theta_m - \theta - \theta_s + \theta) = B(\theta_m - \theta_s) \qquad (17)$$

However, the PMU measurements are subject to error due to noise addition. Hence a Gaussian noise vector (τ) is introduced with zero mean and standard deviation sigma (σ).

In ideal condition simulated bus power mismatches for one outage cases will be closely matched with the computed bus power mismatches using PMU. Hence, line outage estimation problem can be written as,

$$\boldsymbol{F(i; \Delta P_m, \Delta P_s^i) = \left\| \Delta P_m - \Delta P_s^i + \tau \right\|_2^2} \qquad \mathbf{(18)}$$

Therefore, the line outage can be detected easily by solving,

$$\boldsymbol{i = min\left(F(i; \Delta P_m, \Delta P_s^i) \right)} \qquad \mathbf{(19)}$$

As the proposed model estimates the line outages by comparing bus power injection mismatch, so the model is named as comparison of bus power injection mismatch (CBPIM) model.

Now define, outage vector, $s_o = [s_{0,1}, s_{0,2}, \dots, s_{0,L}]^T$ is zero one binary vector whose elements, $s_{0,l} = 1$ if $l \in \mathcal{L}_0$ and $s_{0,l} = 0$, $otherwise$. $\mathcal{L}_0$ is the set of outage lines. $\mathcal{L}$ is the set of lines.

III. PROPOSED ALGORITHM

Proposed line outage estimation algorithm is discussed below:

1. Input system data, noise etc.
2. Store the voltage phasor for various outage cases obtained through load flow study
3. Retrieve voltage phasor measurement after actual outage happens
4. Compute difference between measured bus power mismatches (ΔP_m) and vector of simulated bus power mismatches for various outage cases (ΔP_s^i) using (17)
5. Calculate objective function value using (18) for possible outage cases
7. Find minimum least square normalized value and the corresponding index
8. Obtain the line outage associated with the index
9. Display results

IV. RESULTS AND DISCUSSIONS

The proposed line outage estimation algorithm has been tested on standard IEEE 5-bus, IEEE14-bus and IEEE 30-bus system. The logical program was run with MATLAB 7.10.0 (R2013a) in Intel core -i3 processor (2.4 GHz) with 4 GB RAM. Noise standard deviation (SD) was varied from 1%-5%.

A. IEEE 5 bus system

For IEEE 5 bus system seven different single line outage cases are possible .Let consider outage of line no. 6 with noise standard deviation (SD) 2%.The obtained norm value is presented in figure 1.In this case, minimum norm value obtained is 0.0469 which corresponds to outage of line no. 6 .Hence line outage vector will be, $s_o = [0\ 0\ 0\ 0\ 0\ 1\ 0]^T$.So line outage is detected successfully and is depicted in figure 2.

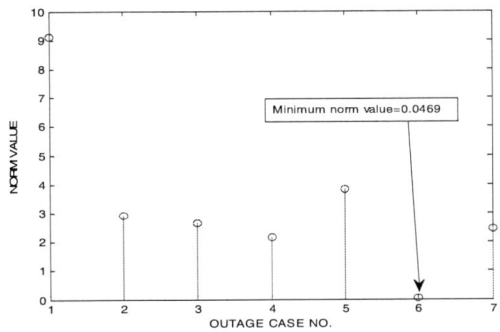

Fig.1. Norm value for different outage cases of IEEE 5 bus system

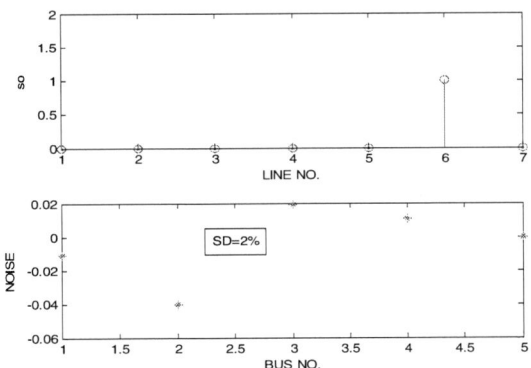

Fig.2. Outage of line no. 6 with noise SD of 2%

The obtained norm values for outage of line no. 3 with noise SD 1% is depicted in figure 3 which shows minimum value corresponds to outage of line no. 3. For this case line outage vector will be, $s_o = [0\ 0\ 1\ 0\ 0\ 0\ 0]^T$. The simulation results for this case is shown in figure 4 which clearly demonstrates the outage of line no. 3.

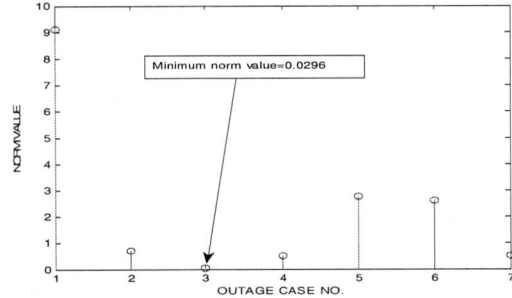

Fig.3. Norm value for different outage cases of IEEE 5 bus system (SD=1%)

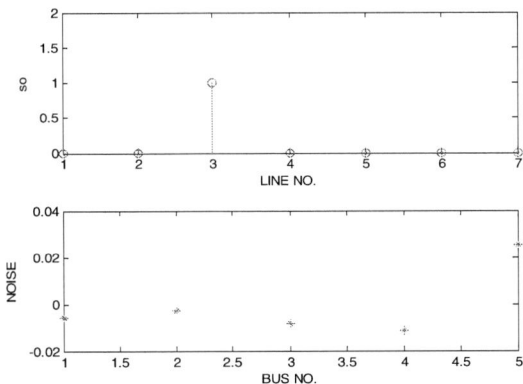

Fig.4. Line No. 3 outage with noise SD of 1%

B. IEEE 14 bus system

IEEE 14 bus system is having 20 lines so twenty different single line outage cases are possible. Here for example, we take lineno 13 outage with noise SD of 4%. The norm values for various single line outage cases are presented in figure 5.Minimum norm value obtained is 0.1476 which refers to outage of line no.13.The simulation results with 4% noise SD is presented in figure 6.

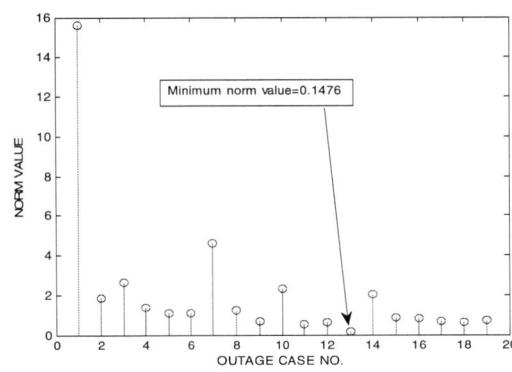

Fig.5. Norm value for different outage cases of IEEE 14 bus system(SD=4%)

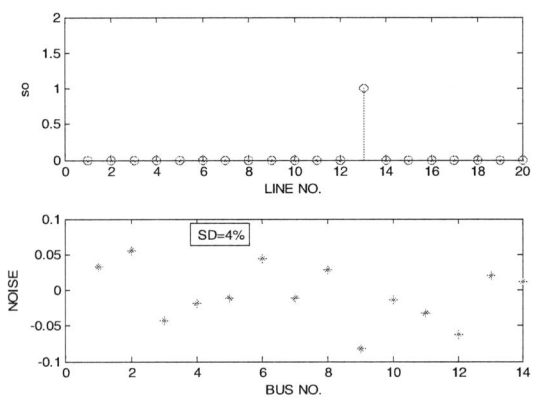

Fig.6. Line No. 13 outage with noise SD of 4%

Similarly, the norm values obtained considering noise SD 5% is presented in figure 7 which shows that minimum norm value for outage case 10.So, outage occurs at line no. 10 which is accurate. The outage index and associated noise variation with SD of 5% for IEEE 14 bus system is shown in figure 8.

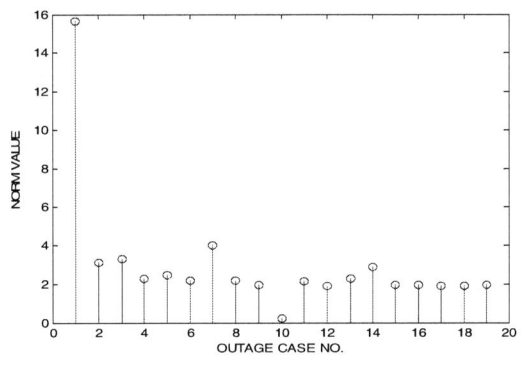

Fig.7. Norm value for different outage cases of IEEE 14 bus system (SD=5%)

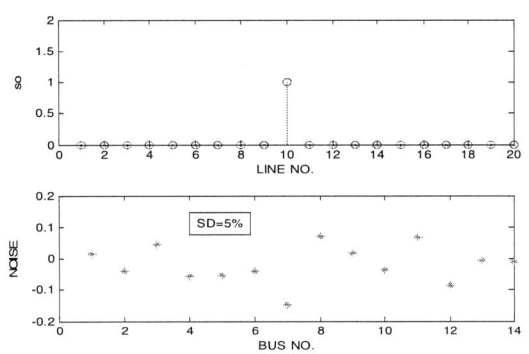

Fig.8. Line No. 10 outage with noise SD of 5%

C. IEEE 30 bus system

For example outage we consider outage of line no. 12 with noise SD 5%.The obtained norm values are shown in figure 9.This figure clearly implies that the minimum norm value obtained for outage case no. 12 which refers to outage of line no. 12.The estimated line outage is shown in figure 10 with associated noise SD of 5%.

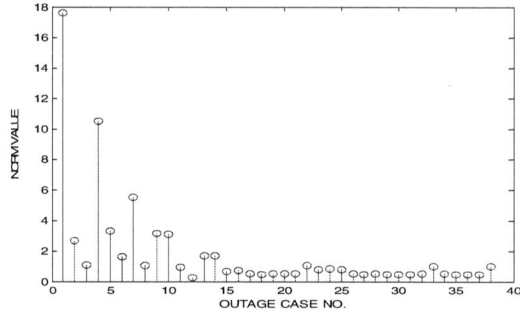

Fig.9. Norm value for different outage cases of IEEE 30bus system (SD=5%)

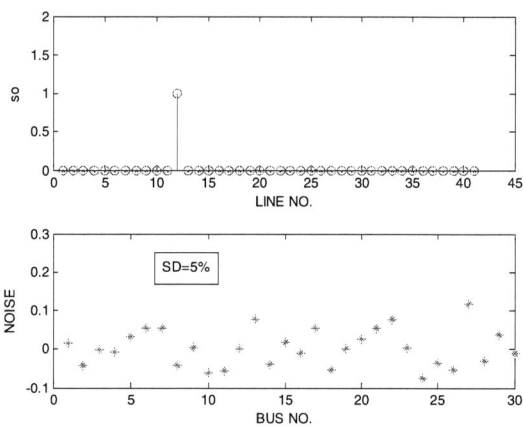

Fig.10. Line No. 12 outage with noise SD of 5%

We have checked all possible single line outage cases through simulations and line outage estimation obtained accurately.However it is to be mentioned that for few outage cases load flow solution do not converge and these cases can not be identified as post outage phaor angles could not be obtained.These unstsble and non convergence cases are referred as critical cases [12] which are not considered and beyond the scope of the paper.This assumption aslo mentioned in litearature [6].The critical cases of single line outage estimation for various test systems is given in table 1.
Estimation accuracy of the proposed algorithm for single line outage estimation obained 100%.Estimation accuracy is the ratio of number of succesfull trials to the total number of

trials. While computing estimation accuracy critical cases are excluded from total number of trials.Estimation accuracy of single line outage estimation for various test systems are shown table 2.

TABLE I.   CRITICAL CASES OF SINGLE LINE OUTAGE

Test system	Line outage no.
IEEE 5 bus	NIL
IEEE 14 bus	14
IEEE 30 bus	13,16,34

TABLE II.   ESTIMATION ACCURACY OF SINGLE LINE OUTAGE

Test system	Total lines	Critical case nos.	Estimation accuracy
IEEE 5 bus	7	NIL	7/7=100%
IEEE 14 bus	20	1	19/19=100%
IEEE 30 bus	41	3	38/38=100%

V. CONCLUSION

A completely new algorithm for line outage estimation is presented in this paper through comparison of simulated bus power mismatches with computed bus power mismatches. Few test results of standard IEEE 5 bus, 14 bus, and 30 bus system are presented. To reflect the measurement errors random noises with zero mean and standard deviation from 1%-5% is considered during the simulation. This infers the robustness of the proposed algorithm under practical scenario. The proposed algorithm is capable to detect the line outage accurately using network topology except the critical cases where load flow solution do not converges. So, this proposed algorithm will be helpful for estimating line outages and monitoring the power systems so that reliable and secure operation can be ensured. Identification of double or more line outages can be future scope of work considering large network systems.

REFERENCES

[1]. "U.S.–Canada power system outage task force, final report on the August 14th blackout in the United States and Canada," Dept. Energy, Apr. 2004 [Online]. Available: https://reports.energy.gov/

[2]. G. Andersson *et al.*, "Causes of the 2003 major grid blackouts in North America and Europe, and recommended means to improve system dynamic performance," in *IEEE Transactions on Power Systems*, vol. 20, no. 4, pp. 1922-1928, Nov. 2005.

[3]. R. F. Nuqui and A. G. Phadke, "Phasor measurement unit placement techniques for complete and incomplete observability," *IEEE Transactions on Power Delivery*, vol. 20, no. 4, pp. 2381-2388, Oct. 2005.

[4]. K. Yamashita, S.-K. Joo, J. Li, P. Zhang, and C.-C. Liu, "Analysis, control, and economic impact assessment of major blackout events," *European Transactions on Electrical Power*, vol. 18, pp. 854–871, 2008.

[5]. B. Singh, N. Sharma, A. Tiwari, K. Verma, and S. Singh, "Applications of phasor measurement units (PMUs) in electric power system networks incorporated with FACTS controllers," *International Journal of Engineering, Science and Technology*, vol. 3, no. 3, pp. 64–82, 2011.

[6]. J. E. Tate and T. J. Overbye, "Line Outage Detection Using Phasor Angle Measurements," *IEEE Transactions on Power Systems*, vol. 23, no. 4, pp. 1644-1652, Nov. 2008.

[7]. J. E. Tate and T. J. Overbye, "Double line outage detection using phasor angle measurements," *2009 IEEE Power & Energy Society General Meeting*, Calgary, AB, 2009, pp. 1-5.

[8]. R. Emami and A. Abur, "Tracking changes in the external network model," *North American Power Symposium 2010*, Arlington, TX, 2010, pp. 1-6.

[9]. JinpingHao, R. J. Piechocki, D. Kaleshi, WoonHauChing and Zhong Fan, "Smart grid health monitoring via dynamic compressive sensing," *IEEE PES ISGT Europe 2013*, Lyngby, 2013, pp. 1-5.

[10]. H. Zhu and G. B. Giannakis, "Sparse Overcomplete Representations for Efficient Identification of Power Line Outages," *IEEE Transactions on Power Systems*, vol. 27, no. 4, pp. 2215-2224, Nov. 2012.

[11]. C. Chen, J. Wang and H. Zhu, "Effects of Phasor Measurement Uncertainty on Power Line Outage Detection," *IEEE Journal of Selected Topics in Signal Processing*, vol. 8, no. 6, pp. 1127-1139, Dec. 2014.

[12]. M. Alam, B. Mishra and S. S. Thakur, "A New Approach of Multiple Line Outage Identification Using Phasor Measurement Unit (PMU) with Bad Data," *2018 International Conference on Current Trends towards Converging Technologies (ICCTCT)*, Coimbatore, 2018, pp. 1-6.

[13]. M. Alam, B. Mishra and S. S. Thakur, "Assessment of the Impact of Line Outage in Modern Power System," *2018 International Conference on Current Trends towards Converging Technologies (ICCTCT)*, Coimbatore,2018, pp.1-6.

Impact of trapped interface charges on short channel characteristics of WFE high-K SOI MOSFET

Priyanka Saha
Department of Electronics and Telecommunication Engineering
Jadavpur University
Kolkata-700032, India
priorient06@gmail.com

Pritha Banerjee
Department of Electronics and Telecommunication Engineering
Jadavpur University
Kolkata-700032, India
prithaedu7@gmail.com

Dinesh Kumar Dash
Department of Electronics and Telecommunication Engineering
Jadavpur University
Kolkata-700032, India
dineshdash123@gmail.com

Subir Kumar Sarkar
Department of Electronics and Telecommunication Engineering
Jadavpur University
Kolkata-700032, India
su_sircir@yahoo.co.in

Abstract: **In the present exertion, we have developed a surface potential and threshold voltage model for linearly graded work function engineered high-K Silicon On Insulator MOSFET including interface charges trapped at both front/channel and buried oxide/channel interfaces. The analytical model is derived utilizing 2D Poisson's equation and criteria of minimum surface potential is further employed for exploring device performance in terms of potential profile, electric field distribution and threshold voltage decay. The interface trapped charges modulate the short channel characteristics of the device structure which is investigated thoroughly in this research endeavor. The same is compared ignoring the trapped charge to grab its effect on the device performance. All the analytical results show excellent concordance when matched with data of ATLAS device simulator thus validating the precision of our model.**

Keywords: *Work function engineering, silicon on insulator, high-k gate oxide, trapped interface charges, threshold voltage roll off*

I.   INTRODUCTION

Relentless shrinking of MOSFET devices in nanometer regime to meet the scaling demands of microelectronics industries poses serious threats as Very Large Scale Integration (VLSI) technology progresses. The major bottlenecks associated with scaling include several short channel effects e.g. threshold voltage roll off (TVRO), hot carrier effect (HCE), drain induced barrier lowering (DIBL), subthreshold slope (SS) degradation, etc [1]. Contemporary literature studies report several innovative or modified MOSFET structures including

Silicon-On-Insulator (SOI) MOSFET, Silicon-On-Nothing (SON) MOSFET, Germanium On Insulator (GeOI) MOSFET and multi gate topology to curb the limitations associated with scaling [2-5]. A few more alternatives to conquer the aforementioned bottlenecks encompass the gate work function engineering concept, hetero gate dielectric engineering, graded channel, strained silicon channel [6-9], etc.

One of the recently emerging gate material engineering scheme is linearly graded work function engineered (WFE) gate electrode. It consists of binary metal alloy (K_xL_{1-x}) with work function laterally varying from source (100% K) to drain end (100% L). This continuous mole fraction variation reduces asymmetry in surface potential and tunes the vertical electric field subsequently controlling the DIBL effect [10]. Further improvisation are conducted as published in [11, 12] where conventional silicon dioxide gate dielectric is replaced by high-k gate stack to reduce gate leakage current and simultaneous enhancement in carrier mobility. Several analytical models are reported [13, 14] that predict the short channel characteristics of SOI devices and examine the effectiveness of the devices in suppressing the short channel effects (SCEs). However, the reported models do not address the effect interface charges trapped at front oxide/channel and buried oxide/channel interfaces. Accumulation of these localized charges at the oxide interface greatly affects the flat band voltage thus impacting the overall device performance especially when it is operating in ultra low dimension.

In this work, analytical surface potential and threshold voltage model is developed by incorporating the linearly graded work function engineering concept in a high-K gate stack SOI MOSFET. This model takes into account the charges trapped at oxide interfaces to investigate trapped charge based device response. Comparative analysis is also presented with the same device configuration neglecting the trapped interface charges to establish the significance of such charges on device characteristics.

II. PROPOSED STRUCTURE

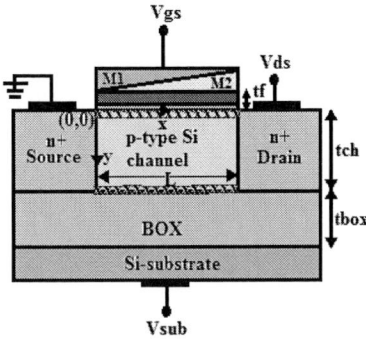

Fig.1. *Schematic view of WFE high-K SOI MOSFET with interface trapped charges*

Cross sectional view of WFE high-K SOI MOSFET with interface trapped charges is shown in figure 1. The length (L) and thickness of the channel (t_{ch}) are taken along x and y direction respectively. $N_{s/d}$ is considered as the doping concentration of source/drain. Effective Oxide Thickness (EOT) of high-k gate stack replacing conventional silicon dioxide (SiO_2) gate oxide is given by [11]:

$$t_f = t_2 + \frac{\varepsilon_2}{\varepsilon_1} t_1 \qquad (1)$$

Here, t_2 and t_1 present the thickness of interfacial SiO_2 layer and HfO_2 (high-k gate oxide considered here) with relative permittivity of ε_2 and ε_1 respectively. ε_{ch} is the relative permittivity of silicon (Si) channel doped with N_{ch} dopants. The thickness of the buried oxide layer (BOX) is denoted by t_{box}. The existence of trapped charges with density of N_f =10^{12}/cm^2 at the interface of Si/oxide is shown in the figure in the form of crossed marks.

The gate electrode consists of binary metal alloy with linearly graded work function from source to drain end. The effective work function ($\phi_m(x)$) of such linearly graded binary metal alloy considered is expressed by [10]:

$$\phi_m(x) = \frac{x}{L}\phi_2 + (1 - \frac{x}{L})\phi_1 \qquad (2)$$

Here, ϕ_1 and ϕ_2 represent the constituent work functions of the metals M1 and M2 at source (x=0) and drain end (x=L) respectively.

V_{gs} and V_{ds} represent applied gate/source bias and drain bias respectively. V_{sub} is the back substrate bias.

III. ANALYTICAL MODELING

The presence of trapped interface charges and linearly graded work function engineered gate electrode modulates the flat band voltage of the device which in turn is reflected in device characteristics.

2D Poisson's equation for determining the electrostatic potential of WFE high-K SOI MOSFET with interface trapped charges can be expressed as [15]:

$$\frac{d^2\phi(x,y)}{dx^2} + \frac{d^2\phi(x,y)}{dy^2} = \frac{qN_{ch}}{\varepsilon_{ch}} \qquad (3)$$

Here, $\phi(x,y)$ is 2D potential distribution in the silicon channel of present structure.

Young's Parabolic Approximation used to solve the channel potential is given by [16]:

$$\phi(x,y) = \phi_{fs}(x) + p_1(x)y + p_2(x)y^2 \qquad (4)$$

Here, $\phi_{fs}(x)$ denotes the front surface potential.

$p_1(x)$ and $p_2(x)$ are arbitrary constants depending on x. Similarity, back surface potential (at y=t_{ch}) is given by :

$$\phi_{bs}(x) = \phi_{fs}(x) + p_1(x)t_{ch} + p_2(x)t_{ch}^2 \qquad (5)$$

The boundary conditions utilized for solving Poisson's equation are given below:

I.   Electric flux being continuous at the interface of front high-k gate dielectric and silicon channel presents :

$$\left.\frac{d\phi(x,y)}{dy}\right|_{y=0} = \frac{C_{ox}}{\varepsilon_{ch}}(\phi_{fs}(x) - V_{gs}') \qquad (6)$$

978-1-5386-6723-1/19 $31.00 © 2019 IEEE   119

II. Likewise, electric flux at the buried oxide layer is continuous with the channel interface which can be represented by :

$$\frac{d\phi(x,y)}{dy}\bigg|_{y=t_{ch}} = \frac{C_{box}}{\varepsilon_{ch}}(V_{sub}^{'} - \phi_{bs}(x)) \quad (7)$$

Here,

$$V_{gs}^{'} = V_{gs} - (V_{f,fb}(x) - \frac{qN_f}{C_{ox}}) = V_{gs} - (\phi_m(x) - \phi_{ch}) + \frac{qN_f}{C_{ox}} \quad (8)$$

$$V_{sub}^{'} = V_{sub} - V_{b,fb} \quad (9)$$

$V_{f,fb}(x)$ and $V_{b,fb}$ are the front and back channel interface flat band voltages respectively.

And, front gate oxide and buried oxide capacitance are denoted as:

$$C_{ox} = \frac{\varepsilon_2}{t_f}, C_{box} = \frac{\varepsilon_{box}}{t_{box}}$$

Using I and II, we have derived $p_1(x)$ and $p_2(x)$ as:

$$a_1(x) = \frac{C_{ox}}{\varepsilon_{ch}}(\phi_f(x) - V_{gs}^{'}) \quad (10)$$

$$a_2(x) = \frac{\frac{C_{box}}{\varepsilon_{ch}}(V_{sub} - V_{b,fb} - \phi_s(x)) - \frac{C_{fox}}{\varepsilon_{ch}}(\phi_s(x) - V_{gs}^{'})(1 + \frac{C_{box}}{C_{ch}})}{t_{ch}(2 + \frac{C_{box}}{C_{ch}})} \quad (11)$$

With $C_{ch} = \frac{\varepsilon_{ch}}{t_{ch}}$

Now, substituting (10) and (11) in (3), applying parabolic approximation, we obtain:

$$\frac{d^2\phi_s(x)}{dx^2} - (D_1 + D_2)\phi_s(x) - D_2\phi_m(x) + D_1V_{bg} + D_2V_{fg} = \frac{qN_{ch}}{\varepsilon_{ch}} \quad (12)$$

Here,

$$D_1 = \frac{\frac{2C_{box}}{\varepsilon_{ch}}}{t_{si}(2 + \frac{C_{box}}{C_{ch}})}, D_2 = \frac{\frac{2Cf}{\varepsilon_{ch}}(1 + \frac{C_{box}}{C_{ch}})}{t_{si}(2 + \frac{C_{box}}{C_{ch}})}$$

$$V_{fg} = V_{gs} + \phi_{ch} + \frac{qN_f}{C_{ox}} \quad (13)$$

$$V_{bg} = V_{sub} + V_t \log(\frac{N_{ch}}{N_i}) + \frac{qN_f}{C_{box}} \quad (14)$$

(12) is now solved to obtain the expression of front surface potential of the present device structure and is given by :

$$\phi_{fs}(x) = P\exp(\lambda x) + Q\exp(-\lambda x) + \frac{D_2(\phi_1 - \phi_2)x}{\lambda^2 L} - \frac{N}{\lambda^2} \quad (15)$$

With $N = \frac{qN_{ch}}{\varepsilon_{ch}} - D_1V_{bg} - D_2V_{fg} + D_2\phi_1$ and

$$\lambda = \sqrt{D_1 + D_2}$$

Coefficients P and Q are now calculated by considering the source/channel and drain/channel interface potentials described as:

$$\phi(0,y) = \phi_s(0) = V_{bi} \quad (16)$$

$$\phi(L,y) = \phi_s(L) = V_{bi} + V_{DS} \quad (17)$$

$$P = \frac{(V_{bi} + V_{ds}) - (V_{bi}\exp(-\lambda L)) - ((D_2(\phi_1 - \phi_2)) + N(1 - \exp(-\lambda L)))/\lambda^2}{2\sinh(\lambda L)} \quad (18)$$

$$Q = \frac{(V_{bi}\exp(\lambda L)) - (V_{bi} + V_{ds}) + ((D_2(\phi_1 - \phi_2)) - N(1 - \exp(-\lambda L)))/\lambda^2}{2\sinh(\lambda L)} \quad (19)$$

The minimum surface potential position x_{min} is obtained by first order derivate of (15) and equating to zero [17]:

$$x_{min} = \frac{1}{2\lambda}\log(\frac{B\lambda^3 L}{D_2(\phi_1 - \phi_2)}) \quad (20)$$

Substituting x_{min} in (15), an expression of minimum front surface potential is calculated as:

$$\phi_{s,min} = P\sqrt{\frac{Q\lambda^3 L}{D_2(\phi_{m1} - \phi_{m2})}} + Q\sqrt{\frac{D_2(\phi_1 - \phi_2)}{Q\lambda^3 L}} + \frac{D_2(\phi_1 - \phi_2)}{2\lambda^3 L}\log(\frac{Q\lambda^3 L}{D_2(\phi_1 - \phi_2)}) - \frac{N}{\lambda^2} \quad (21)$$

Finally, by equating (21) to $2\phi_f$ (strong inversion condition) and solving for V$_{gs}$, analytical expression of threshold voltage is derived for the present structure.

IV. RESULTS AND DISCUSSIONS

The evaluative performance of WFE high-K SOI MOSFET is examined here with focus on the impact of trapped charges at oxide/channel interface. The consequence of such trapped charges on short channel behavior of WFE high-K SOI MOSFET is demonstrated based on analytical model based study which is further compared with ATLAS simulated data [18] for validation of the accuracy of the derived model. Modeling and simulation have been carried out by considering the device parameters enlisted in Table 1.

Figure 2 depicts the potential distribution of WFE high-K SOI MOSFET with potential minima being shifted towards the source side whereas it is located at the central of the channel in Single material Gate SOI MOSFET counterpart. This favorable position of potential minima screens the channel potential from drain-bias fluctuation and offers higher alleviation to DIBL effect. The figure further captures the impact of trapped charges existing at the silicon/oxide interface on surface potential profile of the device. Presence of positive/negative trapped charges deceases/increases the minimum source/channel potential barrier in comparison to Nf=0. This in turn effects the interface trap charge induced threshold voltage degradation (ITTVD) of the device.

In figure 3, gate to source bias and drain bias variation is examined on device surface potential profile with Nf=10^{12}/cm^2. As V_{ds} varies, surface potential is affected only at the drain end leaving major portion of the channel unaltered which manifests the upgraded immunity of the device to SCE in nanometer regime. Again, gate controllability of the device is revealed by increment in V_{gs} which leads to gradual variation in potential barrier height as depicted in the figure.

The electric field distribution against the channel position of WFE high-K SOI MOSFET is compared with Single material Gate SOI MOSFET counterpart in figure 4. At the drain side, the electric field peak is minimum for positive interface trapped charges for both V_{ds}=0.5V and 0.75V as compared to Nf=0. This minimization of electric field peak at the drain side can be inferred as reduced hot carrier effect which is

TABLE 1. DEVICE PARAMETERS

Parameters	t1	t2	tsi	L
Value	2nm	1nm	10nm	60nm
Parameters	N_{ch}	$N_{s/d}$	N_f	ϕ_1, ϕ_2
Value	10^{22}/m^3	10^{26}/m^3	10^{16}/m^2	5.3 eV, 4.4 eV

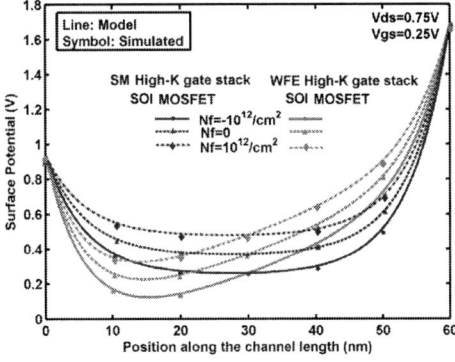

Fig. 2. *Impact of interface trapped charges on surface potential against the position along the channel length of WFE and SM high-K SOI MOSFET*

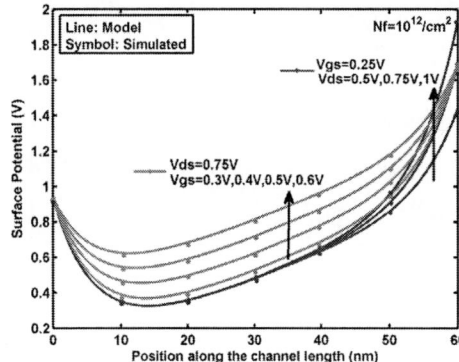

Fig. 3. *Impact of V_{gs} and V_{ds} variations on surface potential against the position along the channel length of WFE high-K SOI MOSFET with Nf=10^{12}/cm^2*

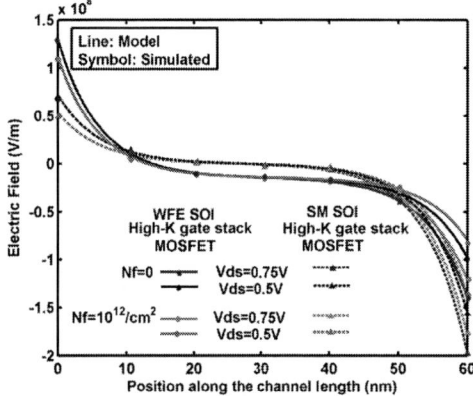

Fig. 4. *Impact V_{ds} variations on electric field distribution against the position along the channel length of WFE and SM high-K SOI MOSFET with Nf=0 and Nf=10^{12}/cm^2*

further lowered in WFE high-K SOI MOSFET in comparison to Single material Gate SOI MOSFET equivalent.

Figure 5 demonstrates the threshold voltage characteristic of WFE high-K SOI MOSFET against the channel length for Nf=0 and Nf=10^{12}/cm^2. Results reveal that threshold voltage is higher for Nf=0 in comparison to positive interface trapped charge. This is attributed to the dependency of threshold voltage on position of surface potential minima. As previously discussed, potential minima is lower for Nf=0 than Nf=10^{12}/cm^2 which in turn raises the value of threshold voltage of the fresh device in comparison to device with charges trapped at the oxide interface. Impact of interface trapped charges is again reflected in threshold voltage degradation in the same figure.

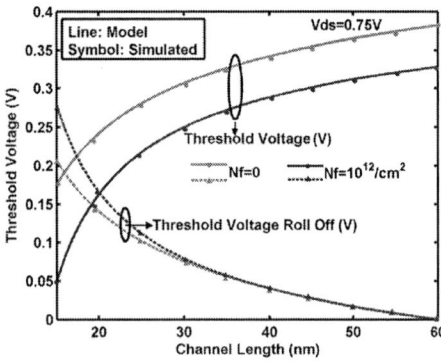

Fig. 5. *Impact of interface trapped charges on threshold voltage against the position along the channel length of WFE high-K SOI MOSFET with Nf=0 and Nf=10^12/cm²*

It portrays that the device experiences higher threshold voltage decay due to the presence of interface trapped charges than the fresh device making the device prone to SCE.

V.   CONCLUSION

This work highlights the impact of trapped interface charges on short channel characteristics of WFE high-K SOI MOSFET. Presence of trapped charges alters the overall device performance which is well established in the results obtained. ITTVD becomes more pronounced and degrades the device functional efficiency when trapped charges are included in modeling. Thus, incorporation of the effects of trapped charges explores the short channel characteristics of the device efficiently which is validated by subsequent comparison with ATLAS simulated data.

ACKNOWLEDGMENT

Priyanka Saha thankfully acknowledges this publication as an outcome of the R&D work undertaken project under the Visvesvaraya PhD Scheme of Ministry of Electronics & Information Technology, Government of India, being implemented by Digital India Corporation.

REFERENCES

[1] Fabio D'Agostino, Daniele Quercia, "Introduction to VLSI design (EECS 467), Short-Channel Effects in MOSFETs", December 11th, 2000.

[2] Sanjoy Deb, N.B.Singh, Nurul Islam and Subir Kumar Sarkar, "Work Function Engineering With Linearly Graded Binary Metal Alloy Gate Electrode for Short Channel SOI MOSFET", IEEE Transaction on Nanotechnology ,Volume: 11 ,Issue: 3 Page(s): 472 – 478, May 2012

[3] Pritha Banerjee ,Priyanka Saha and Subir Kumar Sarkar, "Analytical Modelling and Performance Analysis of Gate Engineered Tri-gate SON MOSFET", IET Circuits, Devices & Systems, DOI: 10.1049/iet-cds.2017.0473, March 2018

[4] Chandrima Mondal and Abhijit Biswas, "2-D Compact Model for Drain Current of Fully Depleted Nanoscale GeOI MOSFETs for Improved Analog Circuit Design",IEEE Transactions On Electron Devices, Vol. 60, No. 8, August 2013

[5] Verhulst, A.S., Sorée, B., Leonelli, D., Vandenberghe, W.G., Groeseneken, G.: Modeling the single-gate, double-gate, and gateall- around tunnel field-effect transistor. J. Appl. Phys. 107, 024518 (2010)

[6] Priyanka Saha, Saheli Sarkhel and Subir Kumar Sarkar," Compact 3D Modeling and Performance Analysis of Dual Material Tri-Gate Tunnel Field Effect Transistor",   DOI :10.1080/02564602.2018.1428503, IETE Technical Review , Taylor & Francis, January 2018

[7] B. Padmanaban , R. Ramesh , D. Nirmal , S. Sathiyamoorthy,"Numerical modeling of triple material gate stack gate all-around (TMGSGAA) MOSFET considering quantum mechanical effects",Superlattices and Microstructures, Vol.82,pp-40–54,Feb 2015

[8] Ekta Goel, Sanjay Kumar, Kunal Singh, Balraj Singh, Mirgender Kumar, and Satyabrata Jit,"2-D Analytical Modeling of Threshold Voltage for Graded-Channel Dual-Material Double-Gate MOSFETs" , IEEE Transactions On Electron Devices, Vol. 63, No. 3, MARCH 2016

[9] Kumar, M.J., Venkataraman, V., Nawal, S.: A simple analytical threshold voltage model of nanoscale single-layer fully depleted strained-silicon-on-insulator MOSFETs. IEEE Trans. Electron Devices 53, 2500–2506 (2006)

[10] Bibhas Manna, Saheli Sarkhel, N.Islam, S.Sarkar, Subir Kumar Sarkar, "Spatial Composition Grading of Binary Metal Alloy Gate Electrode for Short-Channel SOI/SON MOSFET Application", IEEE Transaction on Electron Device, Vol-59, Issue-12, Pp-3280-3287, December 2012.

[11] Manoj Saxena , Subhasis Haldar, Mridula Gupta , R.S. Gupta , "Modeling and simulation of asymmetric gate stack (ASYMGAS)-MOSFET", Solid-State Electronics, Vol. 47,pp- 2131–2134, May 2003

[12] Pritha Banerjee , Priyanka Saha and Subir Kumar Sarkar,"3-D Analytical Modeling and comprehensive analysis of SCEs of a high-K gate stack dual-material tri-gate Silicon-On-Insulator MOSFET with dual-material bottom gate", IEEE Calcutta Conference, CALCON 2017, Lalit Great Eastern, Dalhousie Square, Kolkata-69, 2-3rd, December 2017

[13] M. Jagadesh Kumar and Anurag Chaudhry, "Two-Dimensional Analytical Modeling of Fully Depleted DMG SOI MOSFET and Evidence for Diminished SCEs", IEEE Transactions On Electron Devices, Vol. 51, No. 4, APRIL 2004

[14] F. Lime, R. Ritzenthaler, M. Ricoma, F. Martinez, F. Pascal, E. Miranda, O. Faynot, and B. Iniguez, "A physical compact DC drain current model for long-channel undoped ultra-thin body (UTB) SOI and asymmetric double-gate (DG) MOSFETs with independent gate operation," Solid- State Electron., vol. 57, no. 1, pp. 61–66, Mar. 2011.

[15] T.K Chiang, M.-L. Chen, " A new two-dimensional analytical model for short-channel symmetrical dual-material double-gate metal-oxide-semiconductor field effect transistors", Japanese Journal of Applied Physics, 46(6A), 3283–3290 (2007)

[16] K.K.Young, "Short-channel effect in fully depleted SOI MOSFETs", IEEE Trans. Electron Dev. 36(2), 399–402 (1989)

[17] Priyanka Saha, Saheli Sarkhel and Subir Kumar Sarkar, "Compact 2D threshold voltage modeling and performance analysis of ternary metal alloy work-function-engineered double-gate MOSFET", in Journal of Computational Electronics, Springer, Vol.16.No.3,pp-648-657,June 2017

[18] ATLAS User Manual: Silvaco International. Santa Clara, CA (2015)

A 1.8V 204.8-μW 12-Bit Fourth Order Active Passive ∑Δ Modulator for Biomedical Applications

Arifuddin Sohel
Professor and Head, ECED
MJCET
Hyderabad, India
e-mail: arif.sohel@gmail.com

Maliha Naaz
Assistant Professor, ECED
MJCET
Hyderabad, India
e-mail: maliiha.naaz@gmail.com

Aayesha al khadir
Student, M.E. (D.S.)
MJCET
Hyderabad, India
e-mail: aayeshaalkhadir@gmail.com

Amena Najeeb
Student, M.Tech. (E.S.)
NSAKCET
Hyderabad, India
e-mail: amena.najeeb123@gmail.com

Abstract — This paper presents a 1.8V fourth-order 12.64 bits active-passive delta-sigma modulator. Passive integrators reduce power consumption which employs a switched capacitor circuit that operates from 1V to 1.8V supply voltage. The modulator is implemented in a 0.18μm CMOS process and achieves 77.85 dB SNR within 500 Hz at a sampling frequency of 256 kHz and consumes 204.8μW from a 1.8V supply. The Sigma-Delta modulator becomes the appropriate choice for high resolution as well as low frequency applications due to its highly linear property derived from a linear single-bit quantizer and oversampling technique.

Keywords - *Switch capacitor, Active-Passive SDM, OTA*

I. INTRODUCTION

The medical implants like pacemakers pose a requirement of having an A/D converter that consume low power with different resolutions and speed in the initial measuring stages [1],[2]. Notably, the modulators with low power have become popular where measurements of high resolution are required.

One of the most preferred A/D conversion technique to achieve high resolution is Sigma-Delta modulation. The bio-potential signals consist of critical information that allow the doctors to judge the health of a person. The various bio-electrical signal frequency varies in the range of DC up to 10 KHz with amplitude range from μV scale to mV scale [6].

The use of techniques like oversampling and noise shaping by a delta-sigma modulator facilitates in providing a high Signal-to-Noise Ratio as the oversampling technique samples the signal at a frequency much greater than the

Nyquist frequency; in turn reducing the signal-band quantization noise. Further, the noise is pushed out of the signal band by noise-shaping technique. Thereby making it more preferred choice for the A/D interface system.

Section II highlights the problems encountered while designing a low-power design. The requirements of OTA specifications, passive integrator, switch capacitor circuit and a quantizer will be discussed. The results are discussed in Section III, followed by the conclusion.

II. ARCHITECTURAL DESIGN

Delta-sigma modulators (∑Δ) based on switched-capacitors (SC) are ideal for low-power medium conversion due to its insensitivity towards clock jitter and process [17].

Switched capacitors:

A switched capacitor circuit is as presented in Figure 1. For low voltage design, the most common technique is to make use of the charge pumps that can generate voltages greater than supply voltage in order to drive the switches [18]. The alternate method could be the implementation of switched operational amplifier instead of charge pumps [4].

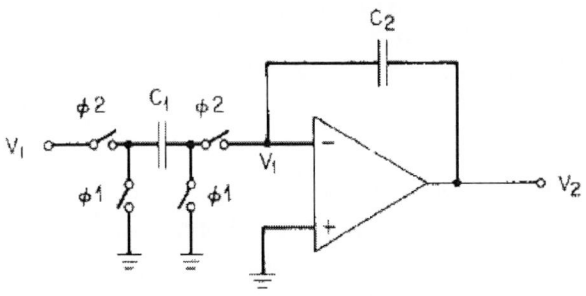

Figure 1: *A switch capacitor circuit*

The schematic of a switched capacitor circuit is shown in Figure 2.

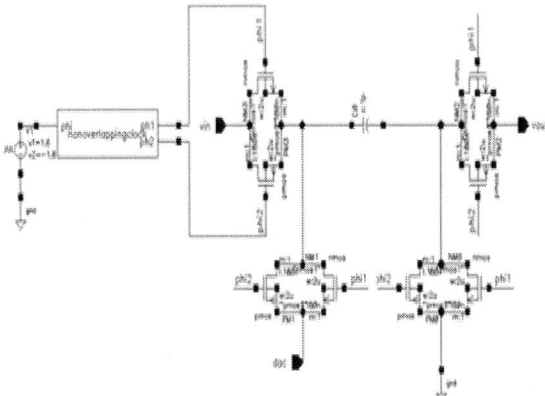

Figure 2: *Schematic of a switched capacitor*

Operational Transconductance Amplifier (OTA):

The active Δ-Σ modulators are designed using the conventional distributed architecture employing OTAs. These amplifiers make an essential block of the Δ-Σ ADC; they devour most of the energy [13], [16].

The factors that determine better functioning of a low power amplifier include gain bandwidth, efficiency, low operational voltage, voltage swing, dc gain, etc. As a result, the single-stage OTA is used. Since multi-stage OTA needs frequency compensation, which is implemented using large capacitors; thus, consuming high power.

The schematic of OTA is shown in Figure 3 using the specifications described in Table 1, and its layout is shown in Figure 4. The minimum area is 0.012515mm².

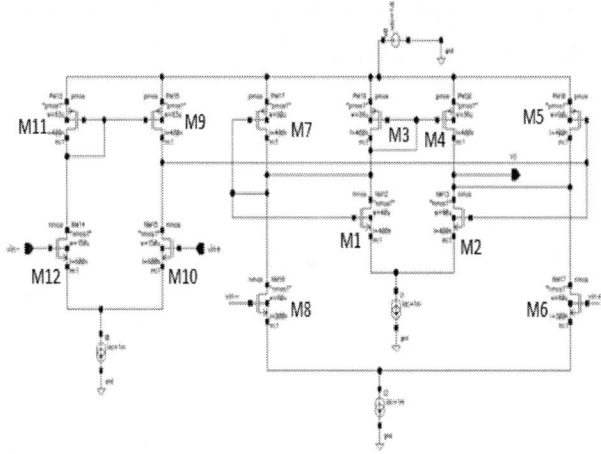

Figure 3: *Schematic of Proposed Amplifier*

Parameter	Values
V_{dd}	1.8V
Unity Gain Bandwidth	30MHz
Phase margin	≥60º
Slew Rate	20 V/us

Table 1: *Specifications of Amplifier*

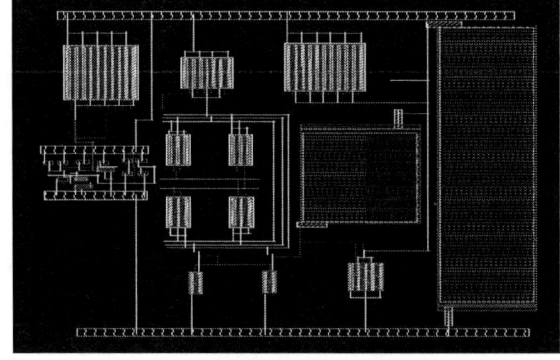

Figure 4: *Layout of OTA*

The minimum layout area is 0.012515mm².

4th Order Active-Passive Σ-Δ Modulator:

The modulator design with low-power and low-voltage modulator designs have made advancements using full feedforward architecture [13], [14]. This architecture presents a drawback for the passive modulators implemented using traditional cascade-of-integrators feedback topology [19]. It results in a simpler the comparator design eliminating the use of the preamp stage before the comparator [20].

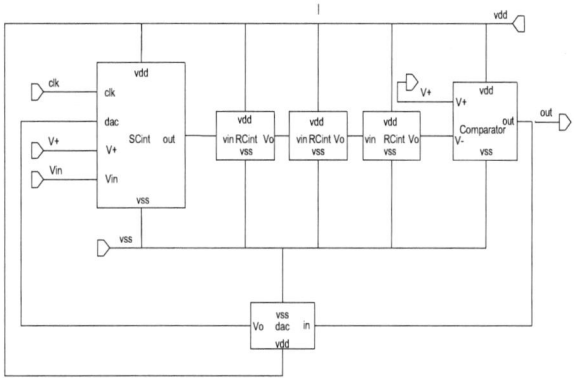

Figure 5: *Block Diagram of fourth order modulator*

Single-loop fourth order modulator topology is shown in Figure 5. At the first stage, an active integrator is used, while the following stages comprise of three SC passive filters.

Passive Filter Design:

The second, third and fourth stages employ an integrator, which is a passive filter, shown in Figure 6. Here the switches used are the NMOS switches.

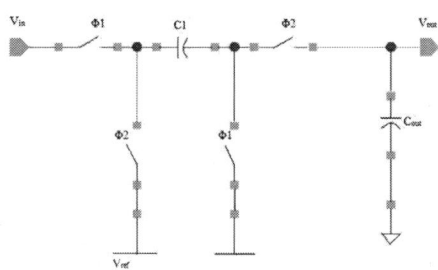

Figure 6: *Passive Integrator*

Single-bit Quantizer:

A single bit quantizer is composed of an SR latch preceded by a regenerative comparator [10]. The previously proposed designs of the passive ADCs [8], [9], [11], [12] make use of the Cascade-of- Integrators Feedback (CIFB) architecture. The challenging task here is the design of comparator that results in signal suppression. Pre-amp stages are used before the comparator circuit, as a power consuming solution, is essential for detecting the weak signal.

III. MEASUREMENT RESULTS

The implementation of the modulator is carried out in a 0.18μm standard CMOS process and the layout is presented in Figure 7. The measured SNR is 77.85 dB with 204.8μW power consumption and the active number of bits is 12.64 bits. Figure 8 highlights the output spectrum.

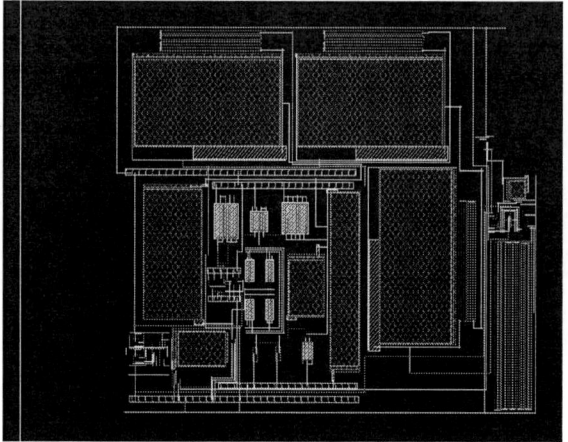

Figure 7: *Layout of the modulator*

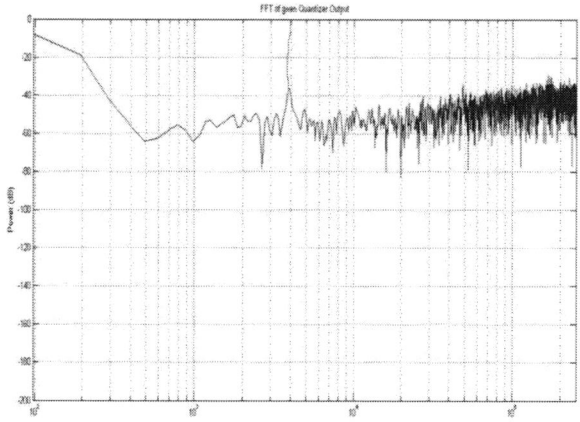

Figure 8: *Output spectrum*

IV. CONCLUSION

A high-resolution low-power fourth-order active-passive Σ-Δ modulator has been designed in a 0.18μm CMOS process that consumes 204.8μW of power and has achieved SNR of 77dB. The design implementation is done efficiently to reduce the power consumption by implementing active integrator in the first stage and passive integrator in the second, third and fourth stages.

Reference	[3]	[5]	[7]	This work
Technology	0.25μm	0.65μm	0.18μm	0.18μm
ENOB	14 bits	-	8.9 bits	12.64 bits
Power	15mW	400μW	24.5 mW	204.8 μW
Signal Bandwidth	1.1 MHz	500Hz	5 MHz	500 Hz
Supply voltage	2.5V	0.7V	1.8V	1.8V
SNDR	100dB	80.3dB	55.33dB	-
OSR	64	-	-	256
Sampling Frequency	150 MHz	256 KHz	320 MHz	256 KHz
SNR	86dB	84dB	-	77.85 dB

Table 2: *Performance Comparison of the modulator*

V. REFERENCES

[1] L. S. Y. Wong, S. Hossain, A. Ta, J. Edvinsson, D. H. Rivas, and H. Nääs, "A very low-power CMOS mixed-signal IC for implantable pacemaker applications," IEEE J. Solid-State Circuits, vol. 39, no. 12, pp. 2446–2456, Dec. 2004.

[2] A. Gerosa, A. Maniero, and A. Neviani, "A fully integrated two-channel A/D interface for the acquisition of cardiac signals in implantable pacemakers," IEEE J. Solid-State Circuits, vol. 39, no.7, pp. 1083–1093, Jul. 2004.

[3] Tongyu Song and Shouli Yan "A Low Power 1.1MHz CMOS Continuous-Time Delta-Sigma Units Modulator with Active-Passive Loop-filters," IEEE ISCAS 2006.

[4] V. Peluso, P. Vancorenland, A.M Marques, M.S.J. Steyaert, and W. Sansen, "A 900-mv low-power ∑Δ A/D converter with 77dB dynamic range," IEEE J. Solid State Circuits, vol. 33, no.12, pp.1887-1897, Dec.1998.

[5] Ali Fazli Yeknami and Atila Alvandpour, "A 0.7 V 400-nW Fourth-Order Active-Passive Delta Sigma Modulator with One Active Stage", 21st International Conference on Very Large-Scale Integration.

[6] Chih-Han Hsu, Kea-Tiong Tang, "A 1V Low Power Second-Order Delta-Sigma Modulator for Biomedical Signal Application," IEEE 2013.

[7] Wen-Cheng Lai, Jhin-Fang Huang, Wei-Chih Chen and Fan-Tsai Kao. "A Continuous-Time Low-Pass Sigma-Delta ADC Chip Design for LTE Communication Application and Bio-Signal Acquisitions."

[8] F. Chen and B. Leung, "A 0.25-mW Low-Pass Passive Sigma-Delta Modulator with Built-In Mixer for a 10-MHz IF Input," IEEE J. Solid-State Circuits, vol. 32, no. 6, pp. 774-782, June 1997.

[9] F. Chen, B. Bakkaloglu, and S. Ramaswamy, "Design and Analysis of a CMOS Passive ΣΔ ADC for Low Power RF Transceivers," J. of Analog Integr Circ Sig Process, Vol. 59, Issue 2, pp. 129-141, 2009.

[10] A. Fazli Yeknami and A. Alvandpour, "A 2.1 μW 76 dB SNDR DT ΔΣ Modulator for Medical Implant Devices," Proceedings of IEEE Norchip Conference, Nov. 2012, pp. 1-4.

[11] A. Fazli Yeknami and A. Alvandpour, "A 0.5-V 250-nW 65-dB SNDR Passive ΣΔ Modulator for Medical Implant Devices," IEEE Int. Symposium on Circuits and Systems (ISCAS), May 2013, pp. 1-4.

[12] R. Yousry, E. Hegazi, and H. F. Ragai, "A 3rd-Order 9-Bit 10-MHz CMOS ΔΣ Modulator with One Active Stage," IEEE Trans. on Circuits Syst. I, vol. 55, no. 9, pp. 2469-2482, Oct. 2008.

[13] J. Roh, S. Byun, Y. Choi, H. Roh, Y. G. Kim, and J. K. Kwon, "A 0.9-V 60-μW 1-bit Fourth-Order Delta-Sigma Modulator with 83-dB Dynamic Range," IEEE J. Solid-State Circuits, vol. 43, no. 2, pp. 361-370, Feb. 2008.

[14] J. Zhang et al., "A 0.6-V 82-dB 28.6-μW Continuous-Time Audio Delta- Sigma Modulator," IEEE J. Solid-State Circuits, vol. 46, no. 10, pp. 2326– 2335, Oct. 2011.

[15] R. Schreier and G. C. Temes, IEEE Press "Understanding Delta-Sigma Data Converters," A John Wiley and Sons publication, 2005.

[16] L. Yao, M. Steyaert, and W. Sansen, "A 1-V 140-μW 88-dB Audio Sigma-Delta Modulator in 90-nm CMOS," *IEEE J. Solid-State Circuits*, vol. 39, no. 11, pp. 1809-1818, Nov. 2004.

[17] S. R. Norsworthy, R. Schreier, and G. C. Temes, Delta-Sigma Data Converters: Theory, Design, and Simulation. New York, NY, USA: Wiley, 1997, pp. 168–169.

[18] S. Rabii and B. A. Wooley, The Design of Low-Voltage, Low-Power Sigma-Delta Modulators. Norwell, MA: KAP,1999.

[19] Ali Yeknami, Atila Alvandpour. Low-Power Low-Voltage ΔΣ Modulator Using Switched-Capacitor Passive Filters.

[20] Alex Orailoglu; H. Fatih Ugurdag; Luís Miguel Silveira; Martin Margala; Ricardo Reis. 21th IFIP/IEEE International Conference on Very Large-Scale Integration - System on a Chip (VLSISoC), Oct 2013, Istanbul, Turkey.

Comparative Investigation of DSG-MOSFET and Some analysis on its Performance

Babita Kumari
Department of electronics engineering,
Indian institute of technology(ISM),
Dhanbad-826004,India
gudbabi123@gmail.com

Kaushik Mazumdar
Department of electronics engineering,
Indian institute of technology(ISM)
Dhanbad-826004, India
kaushik@iitism.ac.in

Aniruddha Ghosal
Institute of Radio physics and
Electronics, University of Calcutta,
Kol-700009, India
Aghosal2008@gmail.com

Abstract— AlGaN/GaN-based DSG-MOSFET is upcoming model of MOSFET. In this paper, comparative study of DSG-MOSFET V_s normal MOSFET and analyzed the performance of DSG-MOSFET. The main features of it are there is negotiable barrier between source and drain so current moves more freely and current is more than normal MOSFET.As if current is more than normal MOSFET then crack length is minimum in DSG-MOSFET.

Keywords—MOSFET, Griffith's equation, MOSFET, Nanocrack formation

I. INTRODUCTION

In this modern era of electronics, the technological advancement leads to the invention of new devices. The detection of new material with different properties is in demand, so as to design new devices with better performance. The silicon based devices ruled the semiconductor industries because of its simplicity, availability and low cost. But as there was increase in the demand of high output power, low power consumption, higher bandwidth, and higher efficiency, use of silicon based devices decreased in market because silicon was unable to fulfill those demands. Due to these limitations in silicon based devices, some novel semiconductor materials were investigated. The semiconductor such as AlGaN, GaN, InAlN, AlGaAs, SiGe, SiC and group III – V, III-N have been included to change the existing silicon based technology [3].

MOSFET is being used in industry since many years. The operation of the MOSFET is different from FET and as a result it gives better performance than MOSFET. Some electrons moving from the n-type region to the crystal lattice and some are nearer to the hetero-junction. These electrons form a layer that consists of one thick layer, which is recognized as a two dimensional electrongas(2DEG). The electrons can move freely within these regions because no other donor electrons or other impurities collide with the electron, and the mobility of the electrons in 2DEG is very high.A bias applied to the gate known as the schottky barrier diode is used to modulate the number of electrons in the 2DEG channel, which controls the device's conductivity.This can be compared to more traditional FET types where the gate bias changes the width of the channel.

Now days, GaN/AlGaN is on demand as it has many features such as high efficiency and high voltage operation, shorter gate lengths that addresses higher frequency operation, lower loss in high power electronics applications and low energy consumption. The key property of GaN is its Wide range of energy gaps of 3.4 eV [1] that is a major reason for its usage in high frequency and high power transistors. The ordinary transistors fails to operate at high frequencies, so as to operate MOSFET at higher frequencies, GaN is used. In Power electronics applications AlGaN/GaN based MOSFET are the most implementable devices.

A. *New Features in GaN as Compared to Si /Ge*

- There is a hexagonal lattice system with little symmetry of crystals.
- The performance of the device depends mainly on the effect of polarization.
- Lower intrinsic concentration as it have wide band gap.

There is a hexagonal lattice system in which few crystal symmetry exists. There is a lattice mismatch of AlGaN on GaN. After application of high voltage they experiences high tensile strain which stores some elastic energy in these crystal at rest. If the stored energy exceeds decisive value, mechanical twist will be there in the crystal. Then the defects present in the DSG-MOSFET will change the device uniqueness. Device performance mainly depends on polarization effect.

AlGaN / GaN is very good piezoelectric materials, so if we apply high voltage in AlGaN / GaN based HEMT there will be great stress. Large electrical field appears at the transistor's gate edge under high voltage operation, which causes high mechanical stress in it.

B. Lower intrinsic concentration as it have wide band gap:- The layers of GaN, AlGaN, and AlN exhibit a wide band gap when compared to silicon material.

II. DESIGN FORMATION OF DSG-MOSFET PROCESS AND THERE ANALYSIS

DSG-MOSFET is having AlGaN/GaN heterojunction epitaxial wafer. On the AlGaN surface, there is a GATE electrode, schottky electrode or a metal insulator-semiconductor (MIS) electrode. There is laser-lift-off or etching process which can remove the substrate and which make GaN vertical schottky barrier diode [3].After removing substrate, the source and the drain are on the GaN surface and the gate is on another side below Fig. 1 depicts the DSG-MOSFET .

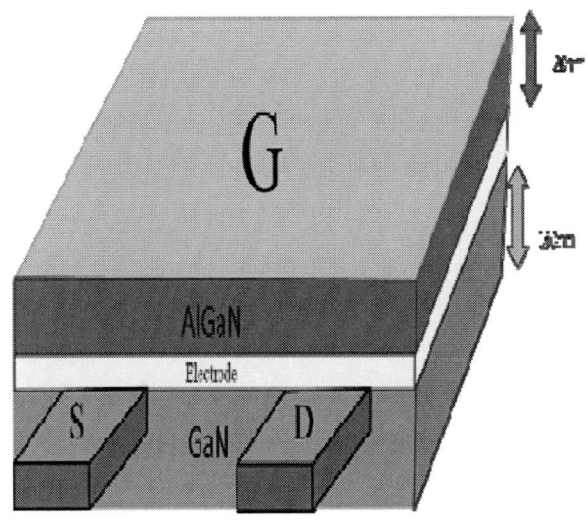

Fig. 1: Model of DSG-MOSFET

In the model, the Green and Blue layers are GaN and AlGaN, with thicknesses 180 nm and 20 nm respectively. On the layer, there is a electrode which is gate electrode. The source electrode and drain electrode are the purple squares in the GaN layer.

III. RESULTS AND DISCUSSION

In this theoretical experiment under voltage variation how the is stress is varying in the DSG-MOSFET in respect to normal MOSFET .Stress under traditional MOSFET and DSG-MOSFET both are having the nanocracks as but in the DSG-MOSFET source side, drain are opposite side with respect to gate so there is a channel formation between source and drain and there is no barrier in between them so current passes more than traditional MOSFET. The characteristics showing drain current vs. drain voltage (I_{ds} Vs. V_{ds}) and tensile stress vs. crack length are illustrated below

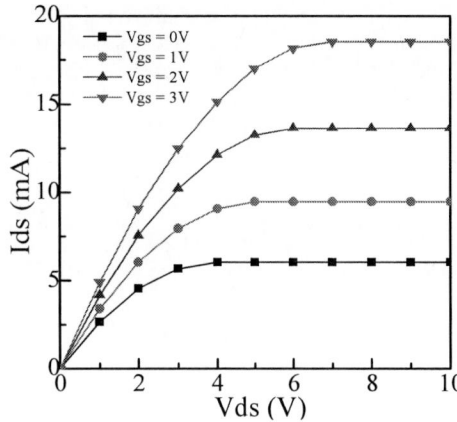

Fig. 2: I_d -V_{ds} curve for DSG-MOSFET

Figure. 2 indicate the characteristics of output (I_d -V_{ds}) of a DSG-MOSFET, plotted between drain current and drain-to-source voltages for various. The Threshold voltage was kept constant at V_T = -4V, drain to source voltage sweeps from 0 to 10V, Vgs is taken from 0V to 3V and the step was of 1V.

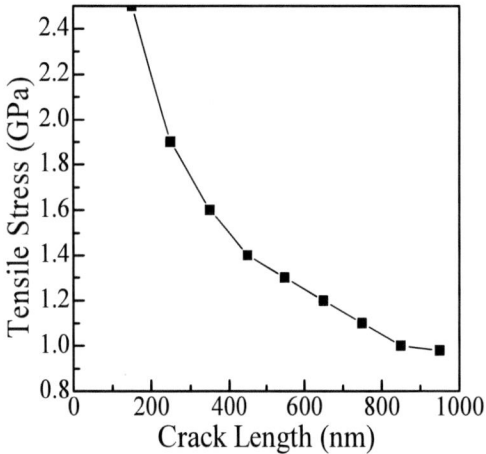

Fig. 3: Plot for Tensile stress Vs created crack length in DSG-MOSFET

The Tensile stress and crack length is calculated by Griffith's Equation and their plot is shown in Fig. 3. From the above figure it can be seen that, DSG-MOSFET has less tensile stress than that of traditional MOSFET. When there is no crack in DSG-MOSFET, the stress is maximum, but when the crack started generating the surface energy released and the value of tensile stress a decreased. And hence crack length increased and exponential curve is obtained.

IV. CONCLUSION

The observation on DSG-MOSFET is a structure that can generate greater current than traditional MOSFET. The result of the comparison shows the crack length in DSG-MOSFET is minimum than traditional.

Theoretically study how crystallographic defects affect electrical properties of DSG-MOSFET & AlGaN/GaN MOSFET crack results in parasite resistance but in DSG-MOSFET & AlGaN/GaN MOSFET.

Crack results in parasitary resistance but in DSG-MOSFET it is less than traditional MOSFET.

V. ACKNOWLEDGEMENT

This research investigation has been supported by DST having project NO. DST(SERB)/(192)/2018-19/568/ECE.

VI. REFERENCES

[1] Kaushik Mazumdar, Sanam Kala, Nanocrack Formation Due to Inverse Piezoelectric Effect in AlGaN/GaN MOSFET, Superlattices and Microstructures, S0749-6036(17)33049-5

[2] Zhe Cheng, Yun Zhang, A NEW STURCTURE AlGaN/GaN MOSFET, 978-1-5090-0176-7/15/$31.00 ©2015 IEEE

[3] J.P. Jones, E. Heller, D. Dorsey, S. Graham, Transient stress characterization of AlGaN/GaN MOSFETs due to electrical and thermal effects, Microelectron. Reliability 55 (12) (2015) 2634–2639.

[4] O. Ambacher, J. Smart, J.R. Shealy, N.G.Wiemann, K. Chu,M. Murphy,W.J. Schaff, L.F. Eastman, R. Dmitrov, L. Wittmer, M. Stutzmann, W. Rieger, J. Hilsenbeck, Two dimensional electron gas induced by spontaneous and piezoelectric polarization charges in N-face and Ga-face AlGaN/GaN heterostructures, J. Appl. Phys. 85(1999)3222–323

[5] Abdelmalek Douaraa, Bouaza Djelloulib, Abdelaziz Rabehic, Abderrezzak Zianec and Nabil Belkadic,

[6] I-V Characteristics Model For AlGaN/GaN MOSFETs Using Tcad-Silvaco, AlGaN/GaN Heterojunction Field-Effect Transistors," Journal of Applied Physics, Vol. 106, 2009, pp. 1-7.

2019 Devices for Integrated Circuit (DevIC), 23-24 March, 2019, Kalyani, India

Effect of Uniaxial Strain on Properties of Blue Phosphorene-CNT Heterojunction

Arnab Mukhopadhyay[1], Amretashis Sengupta[2], Member, IEEE, Hafizur Rahaman[2], Member, IEEE

[1]*C.V. Raman College of Engineering, Bhubaneswar, Odisha, 752054, India*

[2]*School of VLSI Technology, Indian Institute of Engineering Science and Technology, Shibpur, Howrah, 711103, India*

E-mail: arnabm.electinstru@gmail.com

Abstract—**We investigate the effects of uniaxial tensile and compressive strain on the material and transport properties of semiconducting Carbon Nanotube and the device properties of Blue-Phosphorene-CNT heterojunction devices. We see that the material and transport properties of the semiconducting CNT can be tuned through the application of uniaxial strain. The device properties of the heterojunction can also be modulated by strain. The variation in the current is significant in tensile strain zone.**

Keywords—Blue Phosphorene, CNT, DFTB, Strain

I. INTRODUCTION (*HEADING 1*)

Carbon nanotube is an element of interest due to their good carrier mobility and tunable optical/electronic properties. [1,2] Their unique electronic structure make them appropriate for molecular nanoscale devices. [3]

Phosphorene, the two-dimensional structure of black phosphorus has attracted research interest for its interesting properties such as thickness dependent band gap, mechanical flexibility, structural stability etc. for its honeycomb like orthorhombic structure. [4] The thermal stability of two-dimensional blue phosphorene is nearly similar to the monolayer black phosphorene. [5]

Nanotube heterojunctions have created huge research interests due to their unique shapes and properties. [6] Due to the recent advancement in fabrication technology in nanoscale dimensions, synthesis of nanotube heterojunctions are being performed nowadays to explore interesting physics and various applications. In this work, we have evaluated the effects of various percentage of uniaxial strain on the material as well as transport properties of Blue Phosphorene-Carbon Nanotube (CNT) based heterojunction devices.

II. METHODOLOGY

First, we consider (6,5) Carbon nanotube which is semiconducting according to its chirality, to evaluate the effects of uniaxial tensile and compressive strain on the material properties like Density of States (DOS), Optical Spectrum, Total Energy and corresponding effects on the transport property such as Transmission Spectrum in QuantumWise Atomistic Tool Kit (ATK). [7] Then we evaluate the effects of similar uniaxial strain on Device Density of States, Transmission Spectrum and Current-Voltage (I-V) Characteristics of the Blue-Phosphorene-CNT based heterojunction devices in ATK.

We perform semi Slater Koster based semi empirical DFTB calculations with 1×1×11 Monkhorst-Pack K-grid to evaluate the Density of States (DOS), Optical Spectrum, Total Energy and Transmission Spectrum in ATK. [7-9] We use DFTB-NEGF formalism incorporated in ATK to evaluate the transport properties of the Blue-Phosphorene-CNT heterojunction devices. The DFTB-NEGF method is used due to its speed and computational efficiency in case of nanotube and nanoribbon structures which contain large number of atoms. [10,11] To investigate the stability of the heterojunction device, we calculate the formation energy of the device [12]

$$E_F = E_{Tot}^{BP-CNT} - E_{Tot}^{BP} - E_{Tot}^{CNT} \qquad (1)$$

Where, E_{Tot}^{BP-CNT}, E_{Tot}^{BP}, E_{Tot}^{CNT} represent the total energies of Blue-Phosphorene-CNT heterojunction device, Blue-Phosphorene and CNT respectively.

III. RESULTS AND DISCUSSIONS

Fig.1. shows the structure of (6, 5) Carbon Nanotube. The length and diameter of the Nanotube are 4nm and 0.75nm respectively. The main motivation behind choosing (6,5) CNT is its low interface energy with substrate.[13]

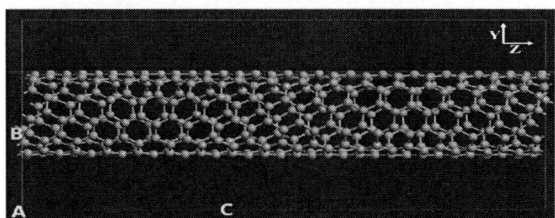

Fig.1. The structure of the (6, 5) Carbon Nanotube.

In Fig.2. we have shown the Density of States (DOS) of the CNT for different percentage of uniaxial strain. In case of unstrained CNT, we see that there are more than 200 states near -5eV energy region. The DOS profile gradually decreases and become smoother with increase in the compressive strain. The number of available states at 0eV energy with increase in percentage of compressive strain. Whereas, the DOS peaks maintain nearly same value in case of tensile strain. But the number of states at 0eV gradually increases with the increase in the percentage of tensile strain.

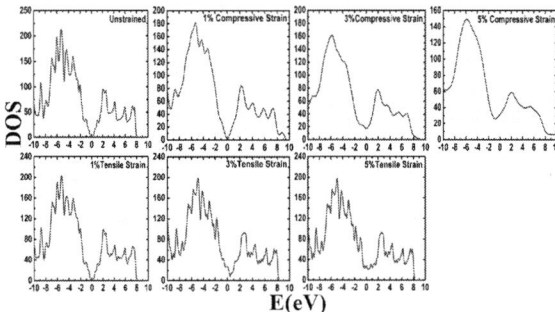

978-1-5386-6723-1/19 $31.00 © 2019 IEEE

Fig.2. The Density of States (DOS) of (6, 5) CNT for various uniaxial strain.

Fig.3. shows the Transmission spectrum of CNT for different percentage of strain. We see that the window around the 0eV energy, where the transmission is 0, gradually become smaller with increase in both compressive and tensile strain. We observe that there is a transition from semiconductor to metal from 3% compressive and tensile strain. This phenomenon is more dominant in compressive strain zone.

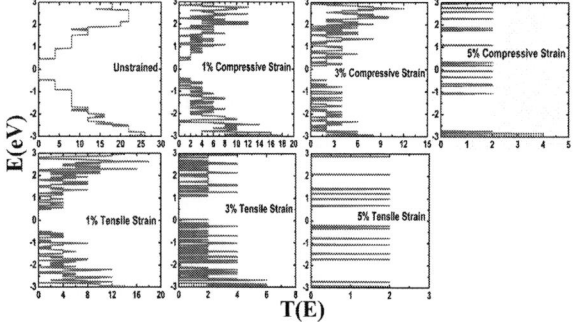

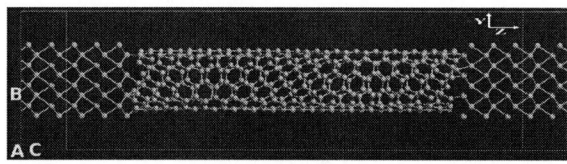

Fig.3. The Transmission Spectrum of (6, 5) CNT for various uniaxial strain.

In Fig.4. we have shown the schematic structure of the Blue-Phosphorene – CNT heterojunction device. We consider four-layer Blue-Phosphorene as the electrodes. The length of each electrode is 0.56nm.

Fig.4. The side-view of the schematic structure of the Blue-Phosphorene – CNT heterojunction device.

Fig.5. shows the Device DOS characteristics of the device for various strain. In case of 1% compressive strain, the states near 0eV are very high in number. In case of 1% tensile strain, there are significant number of available states as well near 0eV.

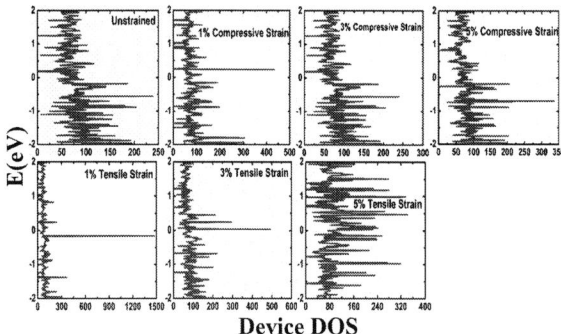

Fig.5. The schematic structure of the Blue-Phosphorene – CNT heterojunction device.

Fig.6. shows the Current-Voltage characteristics of the device for different percentage of strain. The obtained current is higher in case of compressive strain than that tensile strain with an exception in case of 5% compressive strain. The current has highest value for 5% tensile strain. Interestingly, a saturation region starts to appear in the I-V characteristics from 3% tensile strain. Similar pattern of

output current is observed in the experimental result for CNT-MoS$_2$ heterojunction based devices as shown by Jariwala et. al.[14] From the calculated values of formation energy for the Blue-Phosphorene-CNT heterojunction devices, we can say that the device structures are stable because of the negative value of formation energy.

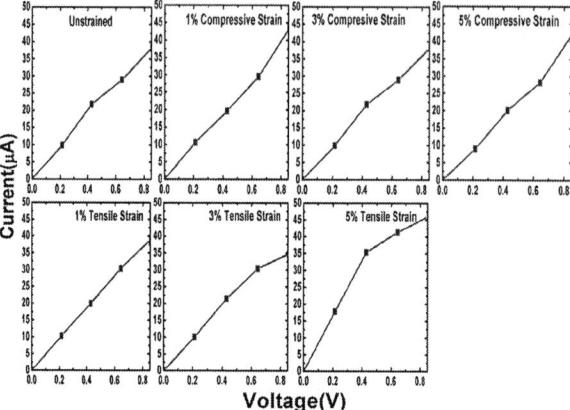

Fig.6. The I-V Characteristics of the heterojunction device for different percentage of uniaxial strain.

IV. CONCLUSION

Analyzing the results obtained, we can conclude that we can tune the material as well as transport properties of (6,5) semiconducting CNT. In addition to that, we can significantly modulate the device properties of the Blue-Phosphorene-CNT based heterojunction devices particularly in the tensile strain zone.

ACKNOWLEDGMENT

The work is supported by DST, Govt. of India under the INSPIRE Faculty Scheme (Grant No.: DST/INSPIRE/04/2013/000108).

REFERENCES

[1] Yi Jia et. al., "Achieving High Efficiency Silicon-Carbon Nanotube Heterojunction Solar Cells by Acid Doping," Nano Lett. 2011, 11, 1901–1905.

[2] W. Yan *et al.*, "CNT-graphene heterostructures: First-principle study of electrical and thermal conductions," *2017 18th International Conference on Electronic Packaging Technology (ICEPT)*, Harbin, 2017, pp. 1319-1322.

[3] R. Martel et. al., "Single- and multi-wall carbon nanotube field-effect transistors," Appl. Phys. Lett. 73, 2447 (1998).

[4] Yan Li and Fei Ma, "Size and strain tunable band alignment of black–blue phosphorene lateral heterostructures," Phys. Chem. Chem. Phys., 2017, 19, 12466-12472.

[5] Zhu, Z.; Tomanek, D. Semiconducting Layered Blue Phosphorus: A Computational Study. Phys. Rev. Lett. 2014, 112, 176802.

[6] Liu Hongxia et al., "Electronic structures of an (8, 0) boron nitride/carbon nanotube heterojunction," J. Semicond. 2010, 31, pp. 013001-013003.

[7] *QuantumWise simulator*, Atomistix ToolKit (ATK), [Online]. Available: http://www.quantumwise.com/.

[8] J. C. Slater and G. F. Koster, Phys. Rev. 94, 1498 (1954).

[9] H. J. Monkhorst and J. D. Pack, Phys. Rev. B. 13, 5188 (1976).

[10] E. Erdogan, I. H. Popov, A. N. Enyashin, and G. Seifert, Eur. Phys. J. B 85, 33 (2012).

[11] See http://www.quantumwise.com/documents/manuals/latest/Reference Manual/index.html/chap.atomicdata.html for more information about ATK Reference Manual.

[12] Liao, J., Sa, B., Zhou, J., Ahuja, R. & Sun, Z. Design of high efficiency visible-light photocatalysts for water splitting: MoS2/AlN (GaN) heterostructures. *J. Phys. Chem. C* **118,** 17594–17599 (2014).

[13] V. I. Artyukhov, E. S. Penev, B. I. Yakobson, "Why nanotubes grow chiral," Nature Communication, 5, 4892 (2014).

[14] Deep Jariwalaa, Vinod K. Sangwana, Chung-Chiang Wua, Pradyumna L. Prabhumirashia, Michael L. Geiera, Tobin J. Marksa,b,1, Lincoln J. Lauhona, and Mark C. Hersam, "Gate-tunable carbon nanotube–MoS2 heterojunction p-n diode," PNAS, 110, 45, 18076–18080 (2013).

Mobility Modulation in V-shaped Double Quantum Well based HEMT Structure

S. K Palo, A. K Panda and T. Sahu
Department of ECE
National Institute of Science and Technology
Berhampur-761 008, Odisha, India
E-mail: tsahu_bu@rediffmail.com

N. Sahoo*
Department of Electronic Science,
Berhampur University,
Berhampur-760 007, Odisha, India
*Email: narayansahoo.cvrp@gmail.com

T. C. Tripathy
Department of ECE,
Roland Institute of Technology,
Berhampur, Odisha, India
Email: tarinitripathy@yahoo.com

Abstract—In the present work, modulation of low temperature mobility μ is studied theoretically with the application of electric field F_e in a double quantum well HEMT structure whose channel is craved from $Al_xGa_{1-x}As$ having V-shaped potential. We show that there is an unusual rise in μ at the transition field where the change in subband occupancy occurs, unlike that of the conventional square quantum well systems.

Keywords—*V-shaped quantum well, Electric field, Scattering matrix.*

I. INTRODUCTION

In the last few decades, low dimensional structures like quantum wells, wires, and dots with non-square potential profiles have attracted much attention due to their specific shape which can manipulate the electronic subbands of the system [1]. From these, interests are focused on the development of the devices based on non-square V-shaped quantum well (VQW) structures [2]. The VQWs can be fabricated from the $Al_xGa_{1-x}As$ by linearly grading the alloy concentration x and hence have higher alloy scattering than that of the conventional square quantum well (SQW). In addition to this here the confinement potential is higher as compared to SQW structures leading to interesting two-dimensional electron gas (2-DEG) properties. The systems designed with double quantum well give rise to engrossing results on electron mobility due to the effects of tunnelling coupling and quantum size effect [2-5].

In this work, we analyse the mobility modulation with the variation of gate electric field F_e in a double VQW. We consider mobility up to two occupied subbands. We show that mobility μ rises near the transition of the occupation of subbands, i.e., from single to double subband occupancy, unlike that of a conventional SQW structure, where a drop in μ is normally seen. Under the occupation of two subbands, μ shows the behaviour of oscillation through the ionised impurity scattering. As the width of the well and middle barrier broadens the magnitude of oscillation increases. Our results of oscillatory mobility will be helpful for the improvement of optical properties of non-square quantum well based new devices.

II. THEORETICAL FRAMEWORK

We propose an $Al_xGa_{1-x}As$ double VQW based high electron mobility transistor (HEMT) system as shown in Fig.1. The middle barrier has a width b. The two symmetric wells have well widths W_w. Carriers are introduced by doping Si in the side barriers at a distance s from their respective well-barrier edges. N_d is the doping concentration, and d is the doping width.

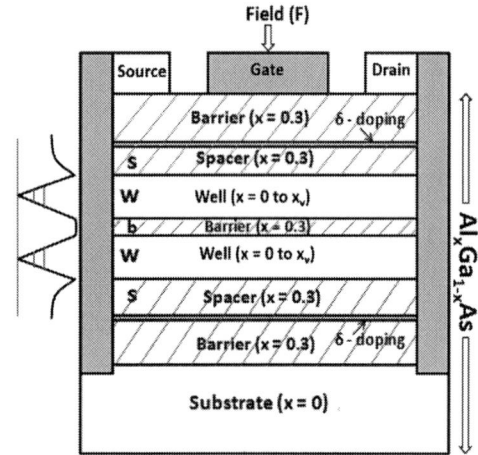

Fig.1 Structural layout of $Al_xGa_{1-x}As$ V-shaped double quantum well based HEMT system.

The electrons diffuse from the outside barriers into the wells leading to band bending. The energy levels E_k and wave functions $\psi_k\ (z)$ are calculated from the transmission probability across the potential of the structure. The potential $V(z)$ contains the potential due to the Coulomb interaction, potential due to the V-shaped quantum well structure and external electric field F_e. We adopt multi-step potential approximation to calculate E_k and $\psi_k(z)$ [6]. The multisubband electron transport life time τ_k is derived–from the Boltzmann transport equation [7, 8]:

$$\sum_{l=0}^{J} C_{kl}\tau_l = 1, \tag{1}$$

J is occupied levels. k and l are subband indices. Under occupation of two subbands k, $l = 0$ and 1. The transport matrix elements of C_{kl} can be written as: $C_{00} = P_{00} + Q_{01}$, $C_{11} = P_{11} + Q_{10}$, $C_{01} = (-k_{F1}/k_{F0})R_{01}$ and $C_{10} = (-k_{F0}/k_{F1})R_{10}$. When only the lowered subband is occupied, then $k = l = 0$, $C_{00} = P_{00} = 1/\tau_0$. Here P_{kk} is intrasubband while Q_{kl} and R_{kl} are the intersubband scattering rate matrix elements, which can be rewritten in terms of $V^{eff}_{kl}(q_{kl})$ [7]. The above screening of the scattering potentials is derived by utilising static dielectric response function formalism within the random phase approximation [7]. We derive the mobility by adopting ionised impurity (*Imp*-) and alloy disorder (*Al*-) scatterings. At T = 0° K, the subband mobility μ_k can be written as μ_k $(E_F) = (e/m)$ $\tau_k(E_F)$, where τ is the subband transport life time. The total mobility μ is calculated as $1/\mu = 1/\mu^{Imp} + 1/\mu^{Al}$ [7].

III. RESULTS AND ANALYSIS

The impact of electric field F_e on mobility μ in a coupled V-shaped quantum well (VQW) system is studied. The structural design factors are: $s = 60$ Å, $N_d = 1 \times 10^{18}$ cm^{-3}, $d = 20$ Å. The alloy concentration x inside the well varies from 0 to 0.3 from the centre to the edges of the well respectively. The barrier height at the well edge as well as at *GaAs/ $Al_{0.3}Ga_{0.7}As$* interface is 228 meV [9]. The applied field F_e is towards the negative z-axis (i.e., substrate side of the structure).

Fig. 2 represents μ^{Imp}, μ^{Al} and μ (10^4 cm^2/Vs) by varying F_e (kV/cm) for $W_w = 200$ Å and $b = 60$ Å. The mobility μ is governed by both μ^{Imp} and μ^{Al}. Initially, for $F_e = 0$ kV/cm double subbands are occupied which continues up to $F_e = 4.5$ kV/cm then only the lowest subband is occupied. At this transition point, μ^{Imp} decreases but μ^{Al} increases due to intersubband effects. For single subband range, μ remains almost constant, but in the double subband range, it fluctuates nonlinearly. Although μ^{Imp} controlled this fluctuation, μ^{Al} governs the overall magnitude of μ.

The behavior of μ^{Imp} and μ^{Al} can be explained through their subband mobilities. In Fig. 3, we represent μ_0^{Imp}, μ_1^{Imp}, μ_0^{Al}, and μ_1^{Al} with F_e. The elements of the scattering matrix for *Imp*-scattering are shown in its inset figure to analyse the nonlinear fluctuation of μ. The opposite trends of μ_0^{Al} and μ_1^{Al}, during the double subband occupancy, make it almost flat. From the main figure, it is clear that the nonlinear behaviour of μ^{Imp} is mainly due to μ_0^{Imp}. The intersubband factors Q_{01}^{Imp} and R_{01}^{Imp} are negligible as compared to that of the intrasubband factors P_{00}^{Imp} and P_{11}^{Imp}.

In Fig. 4, we show μ vs F_e by considering $W_w = 200$ and 250 Å with other parameters remain unchanged. As W_w decreases, the nonlinearity in μ decreases but the range of two-subband occupancy increases. The decrease in W_w reduces mobility μ due to μ^{Imp} and μ^{Al}. It is engrossing to show that there is an improvement of μ at the alteration point of one- subband to two-subband occupancy, unlike that of the SQW structure where a drop in μ is usually seen. This anomalous increase in μ is due to the higher potential

confinement in VQW and dominance of *Al*-scattering as compared to that of *Imp*-scattering. The enhancement of μ increases with an increase in well width, and it is maximum for $W_w = 250$ Å.

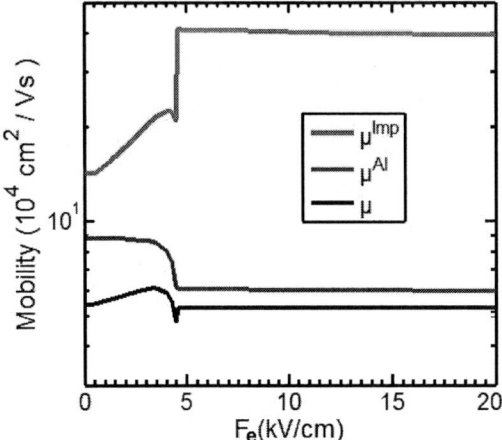

Fig. 2 μ^{Imp}, μ^{Al} and μ as function of F_e of V-shaped double quantum well structure for $W_w = 200$ Å. $s = 60$ Å, $b = 60$ Å, $d = 20$ Å, $x_V = 0.3$, $N_d = 1 \times 10^{18}$ cm^{-3}.

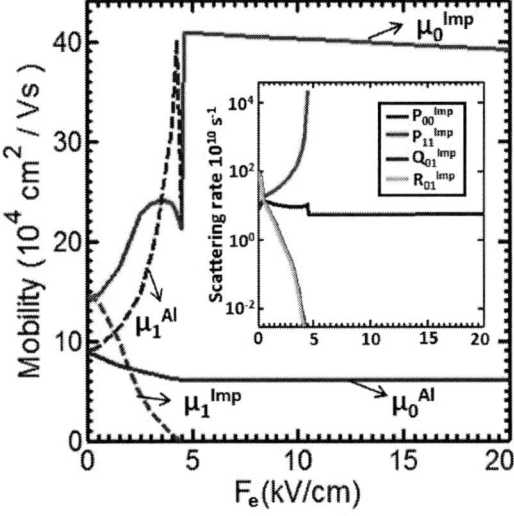

Fig. 3 μ_0^{Imp}, μ_1^{Imp}, μ_0^{Al} and μ_1^{Al} vs. F_e of coupled VQW structure with the structure parameters to that of Fig.1. The inset figure shows the corresponding P_{00}^{Imp}, P_{11}^{Imp}, Q_{01}^{Imp}, and R_{01}^{Imp}.

In Fig. 5, we plot μ vs. F_e by considering different middle barrier width. Here we take $b = 20$, and 60 Å for $W_w = 200$ Å case and other parameters remain unchanged. The inset figure shows their corresponding subbabd Fermi energy values. The DQW structure is coupled through the middle barrier b. The coupling strength has an essential effect on μ. As b increases, the coupling strength decreases, thereby making the two wells behave independently. Higher oscillatory μ is obtained for $b = 60$ Å, but the range of F_e for which double subband occupancy occurs decreases as compared to $b = 20$ Å. The variation in mobilities for $b = 20$ and 60 Å is because of difference in the subband Fermi energy values (inset figure).

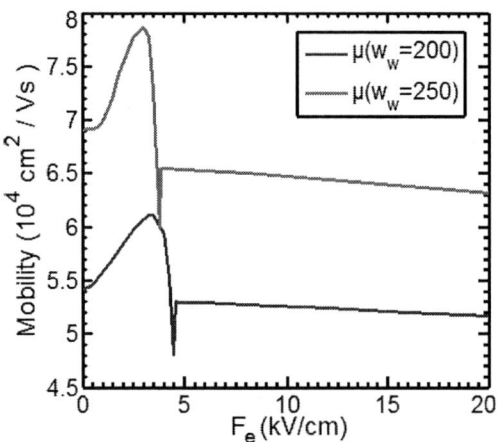

Fig. 4 μ vs F_e of double VQW structure for $W_w = 200$ and 250 Å, with $s = 60$ Å, $b = 60$ Å, $d = 20$ Å, $x_V = 0.3$, $N_d = 1 \times 10^{18}$ cm^{-3}.

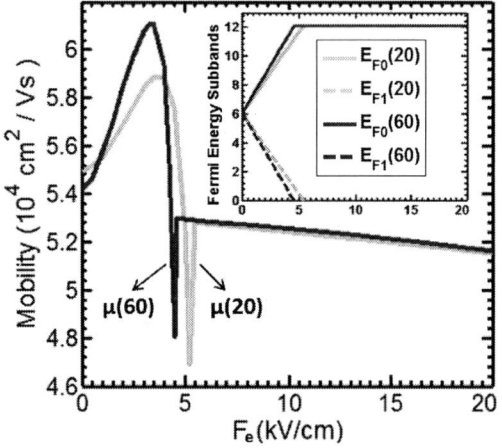

Fig. 5 μ vs F_e of coupled VQW structure for $b = 20$ and 60 Å by taking $W_w = 200$ Å. $s = 60$ Å, $d = 20$ Å, $x_V = 0.3$, $N_d = 1 \times 10^{18}$ cm^{-3}. Inset figure: shows the Fermi Energy Subbands for $b = 20$ and 60 Å.

IV. CONCLUSIONS

The main focus of this work is to study the mobility modulation in the channel of coupled V-shaped quantum well (VQW) HEMT system through gate electric field F_e. Here, the increase of μ is revealed at the field F_e where single to double subband occupied. The rise in μ increases with the increase in well width W_w as well as barrier width b. Our proposed structure will help to analyse the fundamental physics on the effect of non-square shaped potential for the development of emerging devices.

ACKNOWLEDGEMENT

S.P. acknowledges thanks for the financial funding from SERB, Govt. of India through an N-PDF (File No - PDF/2016/002779).

REFERENCES

[1] E. Ozturk, "Non-linear intersubband transitions in different shaped quantum wells under intense laser fields," Superlattices Microsturct., vol. 82, pp. 303-312, June 2015.

[2] U. Yesilgul, F. Ungan, E. Kasapoglu, H. Sari, and I. Sokmen, "The linear and nonlinear intersubband optical absorption coefficients and refractive index changes in a V-shaped quantum well under the applied electric and magnetic fields," Superlattices Microsturct., vol. 50, pp. 400-410, August 2011.

[3] T. Ando, A. B. Fowler, and F. Stern, "Electronic properties of two dimensional systems," Rev. Mod. Phys., vol. 54, pp. 437-672, April 1982.

[4] E. H. Kim, Y. H. Shin, Y. Kim, and S. J. Noh, "Nonlinear optical transitions of GaAs/AlGaAs asymmetric double-well structures," Appl. Phys. Lett., Vol. 89, pp. 032114, July 2006.

[5] N. C. Mamani, G. M. Gusev, O. E. Raichev, T. E. Lamas, and A. K. Bakarov, "Nonlinear transport and oscillating magnetoresistance in double quantum wells," Phys. Rev. B, vol. 80, pp-075308, August 2009.

[6] T. Sahu and N. Sahu, "Oscillating electron mobility in GaAs/Al$_x$Ga$_{1-x}$As double quantum well structure under applied electric field," Superlattices Microsturct., vol. 77, pp. 162-170, January 2011.

[7] N. Sahoo, and T. Sahu, "Mobility modulation in inverted delta doped coupled double quantum well structures," Physica B, vol. 498, pp. 49-54, June 2016.

[8] N. Sahoo, A. K. Panda, and T. Sahu, "Enhancement of electron mobility in square-parabolic asymmetric double quantum well structure," Sperlattices Microstruct., vol. 105, pp. 11-21, May 2017.

[9] S. K. Palo, T. Sahu, and A. K. Panda, "Effect of non-square structure potential on the multisubband electron mobility in double quantum well structure," Physica B, vol. 545, pp. 62-68, June 2018.

Up-state and Down-state Capacitance Measurement in RF MEMS One-bit Switch Designed at Microwave Frequency Range

Pampa Debnath
Dept of Electronics and Comm Engg
RCC Institute of Information Technology
Kolkata, India
poonam.4feb@gmail.com

Arpan Deyasi
Dept of Electronics and Comm Engg
RCC Institute of Information Technology
Kolkata, India
deyasi_arpan@yahoo.co.in

Angsuman Sarkar
Dept of Electronics and Comm Engg
Kalyani Govt Engg College
Kalyani, India
angsumansarkar@ieee.org

Abstract—**Insertion loss, isolation factor and return loss of one-bit RF MEMS switch designed at higher microwave frequency ranges is numerically measured for computation of up-state and down-state capacitance. SiO_2 is the material considered for design purpose and simulation is performed over the entire microwave frequency range in order to investigate the position of maximum loss (peak point). Overlap cross-sectional area is varied over the possible fabrication range, and losses are measured for both actuated as well as unactuated states of the device over the varying overlap region. Both up and down state capacitances are measured which are higher with increase of active area. Return loss of -50 dB is observed for unactuated state whereas it becomes very low (~ -7.5 dB) for actuated device. Also for down state capacitance measurement, isolation increases upto -40 dB. Results are very useful for phase-shifter design at microwave spectrum.**

Keywords—*RF MEMS, Up-state capacitance, Down-state capacitance, Insertion loss, Isolation, Return loss, One-bit switch*

I. INTRODUCTION

Low loss RF MEMS switches are the subject of research due to their capability of lower loss, higher quality factor [1], lower power consumption, higher tunable characteristics [2] which gives the edge compared with semiconductor-based passive component communication system designed at higher frequency. Among different RF MEMS components, switches are analyzed extensively in the last decade due to immense possibility in several commercial low and medium power applications [3], precisely due to its operating ability at very low actuation voltage [4-5]. It offers the flexibility form integration point-of-view over different substrates [6], with major drawback of lower switching speed (2 – 40 µs) compared with solid-state FET/HEMT/PIN diode based switches [7]. Inspite of that, its candidature is already applicable in satellite, ATE, anti-collision RADAR, telecom infrastructure [8], consumer electronics.

Programmable pattern control X-band antenna is proposed a few years back [9] using RF MEMS switch with improved performance, where 587° and 952° swing are achieved for discrete and continuous slot length tuning. Capacitive shunt switch [10] is also designed at X-band for high power applications with very low insertion loss. This model is further advanced by Mafinejad and co-workers [11] for wideband applications, with high amount of return loss. Design is experimentally realizable with lower cost due to the simplified processing methodology. Later, multi-band switches are proposed [12] without reduction of down-state response. This structure is applicable for reconfigurable RF systems, and the structure is made of cantilever type. Semiconductor based RF MEMS switch is recently projected [13] in the entire microwave frequency range with lower insertion and higher isolation loss. High-K dielectric is later used [14] which makes less overlap area, and thus switch cross-section is significantly reduced. Introduction of HfO_2 also enhances bandwidth twice compared to that obtained for conventional material. MMIC technology is also proposed for greater power handling capability and enhanced reliability [15]. In this case, GaAs material is used for fabrication compatibility with existing MMIC process, and design is made for Ka-band. Research is also carried out to make it integrable with CMOS transreceivers [16], and corresponding loss parameters are computed. Inductive switch is also designed for enhancement of Q-factor [17], which effective utilized for tuning over narrow frequency region.

Effect of design geometries and cross-sectional area over RF and mechanical performances are investigated [18] for capacitive shunt switches. Multi-frequency operation on the same wafer is possible by replacing the conventional DOM composition by MIM structure [19], where just by varying the length of floating metal, resonant frequency can be tuned. Very recently, this switch is used for frequency-reconfigurable antenna use [20]. In the present paper, return loss, isolation factor and insertion loss of one-bit RF MEMS switch is calculated over the entire microwave spectrum, and both up-state and down-state capacitances are measured. Effect of overlap cross-sectional area over the both the capacitances are studied, and result is compared with the published literatures. Result is very significant for phased-array design.

II. DESIGN OF ONE-BIT RF MEMS SWITCH

Fig 1 shows the layout of the switch before and after defining layer parameters and ports. IE3D software is used for the design purpose, where the established design methodology is followed. At this point, care is taken about the Bragg frequency and optimum actuation voltage. SiO_2 material is considered for the device. In this context, it may be mentioned that electro-mechanical properties of the switch will critically depend on the material used for fabrication.

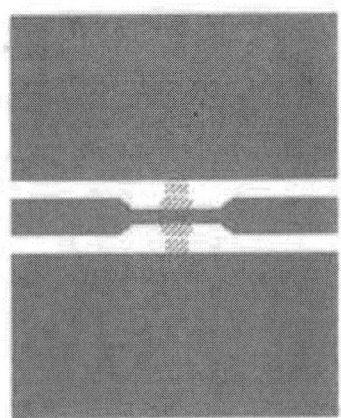

Fig 1a: L-Edit layout for switch with overlap area 140×70 μm^2 and switch length 352 μm

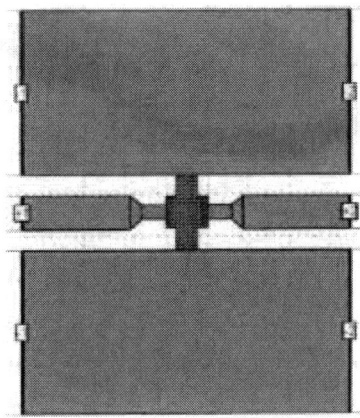

Fig 1b: Improved layout of the switch with overlap area 140×70 μm^2 and switch length 352 μm using IE3D after defining layer parameters and ports

Similar procedure is also adopted for other overlap cross-sectional area. Based on this design, up-state and down-state capacitances are calculated keeping air-gap and dielectric thickness constant.

III. RESULTS AND DISCUSSIONS

Based on the design mentioned in section-II, return loss is obtained for both unactuated as well as actuated states. Results are shown in Fig 2. From Fig 2a, it is found that with increase in overlap area, peak of the return loss profile, i.e., resonance frequency shifts towards lower frequency range, and magnitude also enhanced. This is obtained for unactuated state. Table-I represent s the peak value of return loss with resonant frequency. But when simulation is performed for actuated state, monotonically increasing profile is seen over the frequency spectrum. In this case, higher return loss is seen for

increased overlap area. At very high frequency (30 – 40 GHz), return loss becomes almost independent on the overlap area.

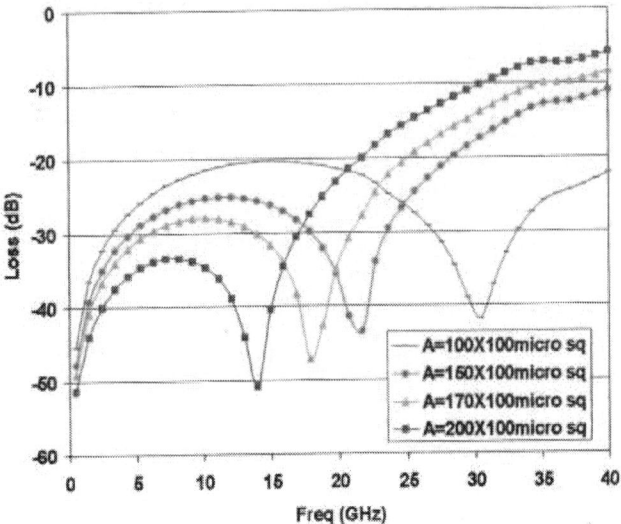

Fig 2a: Return loss in unactuated state with varying overlap area

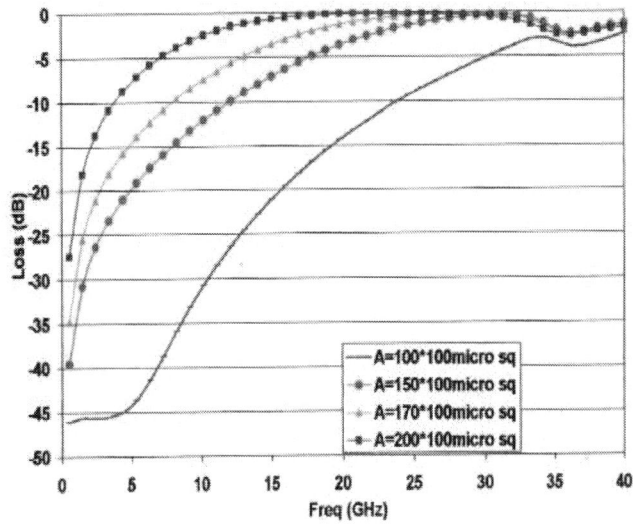

Fig 2b: Return loss in actuated state with varying overlap area

TABLE I.   MAXIMUM RETURN LOSS WITH RESONANT FREQUENCY

Overlap area [μm^2]	Maximum return loss [dB]	Resonant frequency [GHz]
100×100	-43	30
150×100	-44	22
170×100	-47	17.5
200×100	-50	14.5

Next we have computed the insertion loss in unactuated state, and graphically represented in Fig 3. From the plot, it is seen that at lower microwave range, insertion loss is insignificant, and almost independent with the overlap area, whereas after 20 GHz, its magnitude rapidly increases. Higher overlap cross-section leads to higher insertion loss.

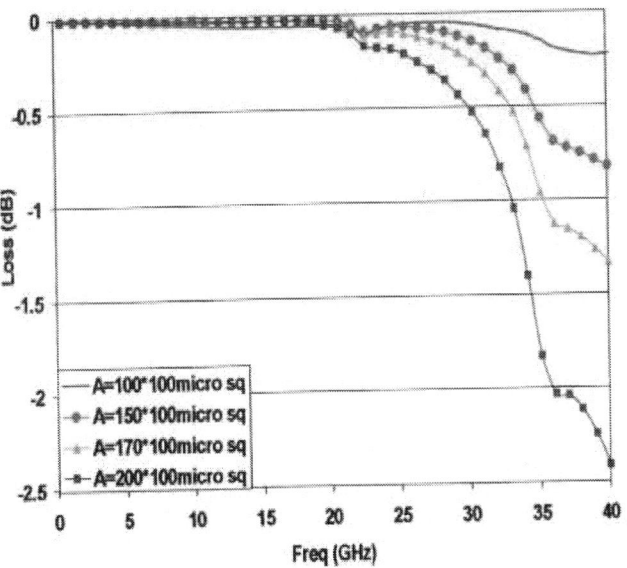

Fig 3: Insertion loss in unactuated state with varying overlap area

We also have computed the isolation factor under actuated condition, as evident from Fig 4. Like in Fig 2b, in this case, we also found that higher overlap cross-section increases isolation, and also makes a redshift of the peak position.

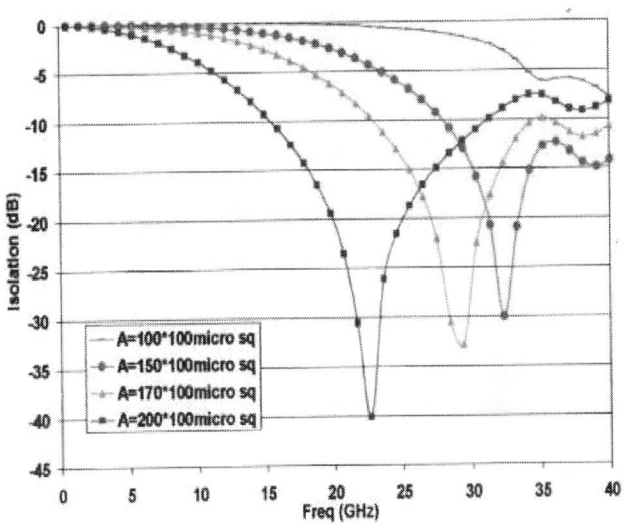

Fig 4: Isolation loss in actuated state with varying overlap area

Effect of return and insertion loss on up-state capacitance can be analyzed from Fig 2a and Fig 3. As already shown, magnitude of insertion loss is similar for all C_{up} values upto 20 GHz. Since devices with larger capacitances have higher reflection coefficients, thus reflection power can be reduced

either by decreasing overlap area or by increasing gap height. But reduction is reflected voltage by tuning gap height increases actuation voltage. Hence optimization is required.

From Fig 3, it has been shown that higher capacitance switch is suitable for lower frequency range only because of small reflection coefficient. But smaller capacitance switch is suitable for higher frequency range owing to lower insertion loss and higher return loss. Table-II gives the dataset for up-capacitance for various overlap area.

TABLE II. UP-CAPACITANCE FOR VARIOUS OVERLAP AREA

Overlap area [μm²]	Air gap [μm]	Dielectric thickness [μm]	Dielectric constant	Up-capacitance [fF]
100×100	3	1	3.97	27.19
150×100	3	1	3.97	40.78
170×100	3	1	3.97	46.22
200×100	3	1	3.97	54.37

The effect of down-capacitance on the isolation characteristics and resonant frequency is analyzed from Fig 2b and Fig 4. Switches with larger capacitance show better isolation at lower frequency range. With increase of active overlap area, down capacitance also increases which further reduces resonant frequency. Table-III gives the dataset for down-capacitance for various overlap area.

TABLE III. DOWN-CAPACITANCE FOR VARIOUS OVERLAP AREA

Overlap area [μm²]	Air gap [μm]	Dielectric thickness [μm]	Dielectric constant	Down-capacitance [fF]
100×100	3	1	3.97	320
150×100	3	1	3.97	520
170×100	3	1	3.97	590
200×100	3	1	3.97	690

In table-IV, we have shown comparative analysis of various loss parameters with the data obtained in published literatures. From the comparative study, we find that our result is more conclusive, and better compared to the other simulated findings.

IV. CONCLUSION

Up-state and down-state capacitances are calculated for various overlap area for one-bit switch, and corresponding losses are much better for both actuated and unactuated states compared to the results published in various literatures. Tailoring of the capacitances are also discussed with the consideration of overlap area, and result shows extremely low up-state capacitance compared to down-state value. Simulated findings can be extended for distributed phased-array design.

TABLE IV.   COMPARATIVE STUDY OF DIFFERENT LOSSES

Lit.	Dielectric material	Overlap area [μm²]	Return loss [dB] unactuated	Return loss [dB] actuated	Isolation [dB]	Insertion loss [dB]
[12]	---	150×20	< 36.8 at 25 GHz	---	[i] 48.8 at 4.5 GHz [ii] 54.56 at 9.7 GHz	<0.1 at 25 GHz
[13]	AlN [ε = 9.5]	---	<16	---	20	0.44
[14]	HfO₂ [ε = 25]				52.7 at 9 GHz	0.06 at 9 GHz
[10]	---	---	---	---	30 at 10 GHz	0.1 at 10 GHz
[20]	---	320×120	---	30 over 30 GHz	26 over 30 GHz	0.1 over 30 GHz
[11]	---	---	20 at 4-27 GHz	---	[i] 10 at 4-8 GHz [ii] 18 at 18-27 GHz	---
[16]	---	---	---	---	85 upto 25 GHz	0.08 upto 25 GHz
[15]	---	---	---	---	20.4 over 27-40 GHz	0.27 over 27-40 GHz
Present work	SiO₂ [ε = 3.97]	200×100	50 at 14.5 GHz	-7.5 at 5 GHz	40 at 22.5 GHz	0.06 at 15 GHz

REFERENCES

[1] H. Jaafar, K. S. Beh, N. A. M. Yunus, W. Z. W. Hasan, S. Shafie, O. Sidek, "A Comprehensive Study on RF MEMS Switch", Microsystem Technologies, vol. 20(12), 2014

[2] Z. Liu, L. Li, "RF MEMS Switch and Its Applications", ECS Transactions, vol. 44(1), pp. 1361-1366, 2012

[3] K. S. Beh, H. Jaafar, N. A. M. Yunus, S. Shafie, "Design and Simulation of Low Voltage RF MEMS Series Switch Array", Proceeding of IEEE Circuits and Systems, 2013

[4] M. Manivannan, R. J. Daniel, K. Sumanga, "Low Actuation Voltage RF MEMS Switch using Varying Section Composite Fixed-Fixed Beam", IInternational Journal of Microwave Science and Technology, vol. 2014, A.Id: 862649, 2014

[5] A. K. Sharma, N. Gupta, "Investigation of Actuation Voltage for Non-Uniform Serpentine Flexure Design of RF-MEMS Switch", Microsystem Technologies, vol. 20, pp. 413-418, 2014

[6] H.S. Newman, "RF MEMS Switches and Applications", 40th Annual IEEE International Reliability Physics Symposium, 2002

[7] A. Saxena, V. K. Agrawal, "Comparative Study of Perforated RF MEMS Switch", Procedia Computer Science, vol. 57, pp. 139 – 145, 2015

[8] A. K. Sharma, N. Gupta, "An Improved Design of MEMS Switch for Radio Frequency Applications", International Journal of Applied Electromagnetics and Mechanics, vol. 47(1), pp. 11-19, 2015

[9] Y. Radi, S. Nikmehr, A. Pourziad, "A Novel Bandwidth Enhancement Technique for X-Band RF MEMS Actuated Reconfigurable Reflectarray", Progress in Electromagnetics Research, vol. 111, pp. 179-196, 2011

[10] A. Ziaei,; S. Bansropun, P. Martins, M. Le Baillif, "Fast High Power Capacitive RF-MEMS Switch for X-Band Applications", 45th European Solid State Device Research Conference, 2015

[11] Y. Mafinejad, A. Kouzani, K. Mafinezhad, R. Hosseinnezhad, "Low Insertion Loss and High Isolation Capacitive RF MEMS Switch with Low Pull-in Voltage", The International Journal of Advanced Manufacturing Technology, vol. 93(1-4), pp. 661-670, 2017

[12] M. Angira, K. J. Rangra, "A Novel Design for Low Insertion Loss, Multi-band RF-MEMS Switch with Low Pull-in Voltage", Engineering Science and Technology, an International Journal, vol. 19(1), pp. 171-177, 2016

[13] T. L. Narayana, K. G. Sravani, K. S. Rao, "Design and Analysis of CPW based Shunt Capacitive RF MEMS Switch", Cogent Engineering, vol. 4(1), A. Id: 1363356, 2017

[14] M. Angira, "High Performance Capacitive RF-MEMS Switch based on HfO₂ Dielectric", Transactions on Electrical and Electronic Materials, pp. 1-8, 2018

[15] C. Chu, X. Liao, H. Yan, "Ka-band RF MEMS Capacitive Switch with Low Loss, High Isolation, Long-Term Reliability and High Power Handling based on GaAs MMIC Technology", IET Microwaves, Antennas & Propagation, vol. 11(6), pp. 942-948, 2017

[16] A. Bojesomo, N. Saeed, I. M. Elfadel, "A Multiband RF MEMS Switch with Low Insertion Loss and CMOS-compatible Pull-in ,Voltage", Symposium on Design, Test, Integration & Packaging of MEMS and MOEMS, 2018

[17] E. S. Shajahan, M. S. Bhat, "Fabrication and Characterisation of RF MEMS Capacitive Switches tuned for X and Ku Bands", International Journal of Mechatronics and Automation, vol. 6(2/3), pp. 143-149, 2018

[18] E. A. Savin, K. A. Chadin, R. V. Kirtaev, "Design and Manufacturing of X-band RF MEMS Switches", Microsystem Technologies, vol. 24(6), pp. 2783-2788, 2018

[19] S. Gopalakrishnan, A. DasGupta, D. R. Nair, "Novel RF MEMS Capacitive Switches with Design Flexibility for Multi-frequency Operation", Journal of Micromechanics and Microengineering, vol. 27(9), p. 095013, 2017

[20] Y. Xu, Y. Tian, B. Zhang, J. Duan, L. Yan, "A Novel RF MEMS Switch on Frequency Reconfigurable Antenna Application", Microsystem Technologies, vol. 24(9), pp. 3833-3841, 2018

2019 Devices for Integrated Circuit (DevIC), 23-24 March, 2019, Kalyani, India

Role of Internet of Things (IoT) in Smart Farming: A Brief Survey

Monu Bhagat
CSE Department
NIT Jamshedpur
Jamshedpur, India
2018rscs002@nitjsr.ac.in

Deobrata Kumar
CSE Department
NIT Jamshedpur
Jamshedpur, India
2018rscs004@nitjsr.ac.in

Dilip Kumar
CSE Department
NIT Jamshedpur
Jamshedpur, India
dilip.cse@nitjsr.ac.in

Abstract—**Presently there is huge improvement in technologies. Lots of tools and techniques are available in the agriculture sector. Internet of Things plays very important role to improve productivity, efficiency, global market. It also reduces human intervention, cost and time which are major factors in agriculture. The Internet of Things (IoT) can be defined as a system which interrelate computing devices, objects, machines (like mechanical and digital), living beings. The IoT components are provided with unique identifiers and have the ability to transfer data over a network without requiring human-to-human or human-to-computer interaction. So, in order to increase productivity, IoT works in synergy with agriculture to get smart farming. Role of IoT, different tools, hardware and software used in smart farming is discussed in this paper.**

Index Terms—**Smart Farming, IoT, sensors, productivity.**

I. INTRODUCTION

For enhancing agricultural productivity, it is needed to understand and forecast crop performance under different conditions of soil, irrigation, fertilization and environment. Agricultural productivity can be improved by investigating which crop type has produced the greatest yield under similar conditions of soil, climate, irrigation and fertilization. To see the growth of the crop under changing real-world conditions like soil quality, environmental conditions etc. The typical studies of crop involves phenotyping to understand the key factors like optimum temperature, soil humidity, pH levels of soil, the rate of Nitrogen depletion etc. that affects its growth. Introduction of IoT technologies in agriculture can be cost effective and increase the scale of such studies by collecting different related data. The time series data can be gathered from sensor networks, spatial data from imaging sensors, and human observations can be recorded via mobile smart phone applications. There are also some IoT devices that can be used to capture the pH levels of soils and the rate of Nitrogen depletion as time-series data which are helpful for the researchers in their research purpose. Smart farming can be done using smart phones and IoT devices.In order to monitor agricultural sector, farmer can get information regarding soil, environmental, irrigation conditions data by using IoT devices and smart phones.

II. SMART AGRICULTURE USING IOT

For the Economical growth of any Country having farming as primary resource, the agriculture is its main backbone.

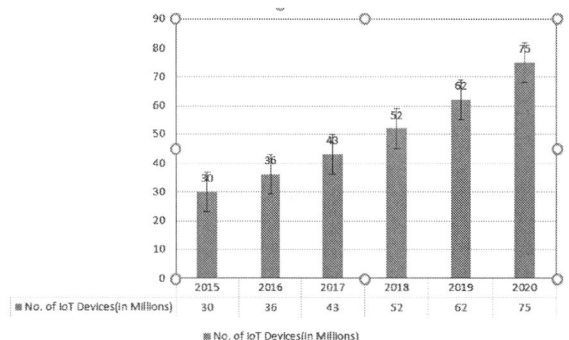

Fig. 1. Graph showing Estimated No.of IoT Devices till year 2020.

Climate change is the most important barrier in tradition farming. Climate change effect includes heavy rainfall, intense storm, less rainfall, heat waves, low or high temperature, humidity etc. Due to these problems farm productivity is effected at a large extent. These effects can alter seasonal life cycle of plants. Hence an innovative technology is needed to forecast and monitor such calamity so that farm productivity can be increased to a great extent. Internet of Things provides different type of sensors like soil moisture sensor, temperature sensor, heat wave sensors, actuators, networking devices, hardware and software platform that can collectively gather data related to different condition like environment, soil, fertilization, irrigation etc. Thus IoT can help in smart agriculture. The report generated by Garner says that there will be approximately 30 percentage rise in number the of connected devices in 2016 as compared to 2015. Further, he manifests that this number will reach to 26 billion by 2020 [5]

III. LITERATURE SURVEY

In this section a literature survey of some published research paper on application of Internet of Things(IoT) in agriculture is presented. IoT devices brought significant change in agriculture. Smart Farming is not possible without IoT devices. A lots of work has been published on Smart farming, some of which are discussed below in table.

978-1-5386-6723-1/19 $31.00 © 2019 IEEE

TABLE I
TABLE SHOWING SOME RELATED WORK OF IoT IN THE FIELD OF AGRICULTURE

Sl.No.	Author(s), Title of Publication & year	Tools/Technology Discussed	Problem Discussed	Solution
1	Nurzaman Ahmed et al., Internet of Things (IoT) for Smart Precision Agriculture and Farming in Rural Areas [1], 2018	Fog Computing, WiLD network, ifogsim, 6LoWPAN, Cooja, Contiki etc.	Existing techniques for smart farming are not suitable for long range coverage,low latency and high throughput.	Introduction of fog computing and WiLD network in existing wireless sensor network will cover long range with lesser delay. Crossed layer based MAC and routing solution will improve delay and throughput.
2	Manlio Bacco et al., IEEE 802.15.4 Air-Ground UAV Communications in Smart Farming Scenarios [2], 2018	UAVs, IEEE 802.15.4, ZigBee.	Precision agriculture in rural area.	Vehicle that are able exchange data with ground sensors are low lost and easy to deploy. These vehicles can be used for monitoring and controlling smart farming. It can be used to achieve precision farming.
3	Inkyu Sa et al., weedNet: Dense Semantic Weed Classification Using Multispectral Images and MAV for Smart Farming [17], 2018	Micro Aerial Vehicle(MAV), SegNet, embedded GPU (Jetson TX2), CNN, FCNNs, weedNet.	Challenges in autonomous crop management which is necessary for yield and crop health.	Dense semantic weed classification technique along with multispectral images that are collected by a micro aerial vehicle is used. Also neural network and SegNet are used to classify datasets
4	Deepak Vasisht et al., FarmBeats: An IoT Plateform for Data-Driven Agriculture [22], 2017.	FarmBeats, UAV, sensors, Drones, cloud.	Difficulties in agricultural data collection which decreases productivity.	Data driven techniques can increase agricultural yield and decrease cost of production. An end to end IoT platform called FarmBeats is introduced that provides smooth data collection from different devices like sensors, camera and drones.
5	Prem Prakash Jayaraman, Internet Of Things Platform for Smart Farming [10], 2016	SmartFarmNet, an IoT based platform that integrate IoT device such as sensrs,camera, weather stations etc.	Slow collection of Crop performance data.	SmartFarmNet automate collection of environmentat soil, fertilization and irrigation data in cloud that automately corelate such data and filter out invalid data from the perspective of assessing crop performance faster.
6	M.Stoces et al., Internet of Thing in Agriculture-Selected Approach [19], 2016	FarmBot, GSM, LTE, Bluetooth, Wifi,Sigfox, LoRaWAN, IEEE P802 etc.	Fractured nature of platforms and communication protocols which leads to incompatibility issues between various IoT devices and between different area where IoT is implement.	For purpose of IoT development a specialized infrastructures has to be established.
7	Nikesh Gondchawar et al., IoT based Smart Agriculture [7], 2016	sensors, wifi, ZigBee modules, camera, actuators, raspberry pi.	Smart Agriculture by modernizing the current traditional methods of agriculture for development of agriculture country.	Smart GPS based remote controlled robot perform tasks like weeding, spraying, moisture sensing, bird and animal scaring, keeping vigilance. Smart irrigation with smart control and intelligent decision making based accurate real time field data.
8	Jaiganesh.S , Gunaseclan.K et al. IoT Agriculture to improve Food and farming technology [9], 2017	Cloud computing, BigData, IoT 3G, NFC, LTE, GSM, RFID, ZigBee etc.	Improving food and farming technology.	Explained application of IoT Cloud computing, BigData, sensors, agriculture virtualized information database to improve food and farming technology.
9	Vinayak N. Malavade et al., Role of IoT in Agriculture [13], 2016	IoT sensors, RFID chips, EPC, PH level etc.	Role of IoT in agriculture that leads to smart farming.	IoT can be used for water management, soil management, crop monitoring, control of insectisides & pesticides.
10	N. Suma et al., IoT Based Smart Agriculture Monitoring system [21], 2017	PIC microcontroller, GSM module soil moisture sensor, temperature sensor, PIR sensor, proteus 8 simulator.	Environmental factor monitoring is not the complete solution to increase the yield of crops. An automated system is needed to monitor other factors.	integrated system is configured using sensors, microcontroller, software simulation tools to overcome mentioned difficulties

IV. ROLE OF IoT IN SMART AGRICULTURE

IoT has a wide range of applications in different domains like Smart Heathcare, Smart City etc. IoT plays very important role in making agriculture smart. Some of its vital role are discussed below:

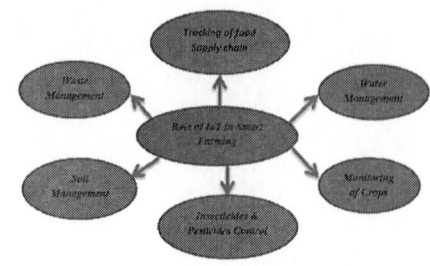

Fig. 2. Diagram showing role of IoT in Smart Farming.

A. In Soil Management

Soil management means measuring different parameter of soil like its pH level, moisture content etc. By using IoT sensors these parameters can be easily measured. Accordingly farmers can take actions like fertilization, irrigation etc. Soil management helps in choosing perfect breed in plant. It also helps in detecting need of fertilizer in soil. It requires network of low latency for immediate action.

B. In Water Management

Water plays significant role in production of crop. Using different IoT sensors water management can be done very efficiently. This save water from wastage and increase crop productivity. It helps in detecting water born disease in pond and agricultural area. It also need low latency of network for immediate action.

C. In Monitoring Crops

Crop monitoring means identifying and monitoring health of crop. IoT sensors and RFID chips can identify the diseases present in plants and crops. The reader can get these information from RFID tags and shares across the internet. These information are assessed by the farmer remotely and accordingly actions are taken. This will protect crops from diseases.

D. Insecticides nd Pesticides Control

There are some automated IoT pest control systems that are developed by some company like Spensa of West Lafayette. These automated systems can be monitored remotely from android phones.

E. In Waste Management

IoT provides a waste management solution. Intelligent wastebins can be made using IoT sensor. This can be used to read, collect and transmit waste related data over the communication network. Using some intelligent and optimized algorithms waste management can be done.

F. In Tracking Supply Food Chain

IoT provides ability to automatically track supply food chain. This tracking of food chain be done by collecting, transmitting, and publishing data centrally to track, monitor and manage information about shipments moving throughout the supply food chain.

V. SMART FARMING APPS

It is used to keep track of the data of wireless sensors. These data are used for predictive analysis. A few IoT agriculture apps are discussed below:-

A. Phenonet Project (OpenIot)

It is used to assist plant breeder to test the condition of different types of wheat by measuring the humidity, air temperature and soil temperature. This helps farmers to predict the harvest time, irrigation time and nutrients needed for the growth of plants.

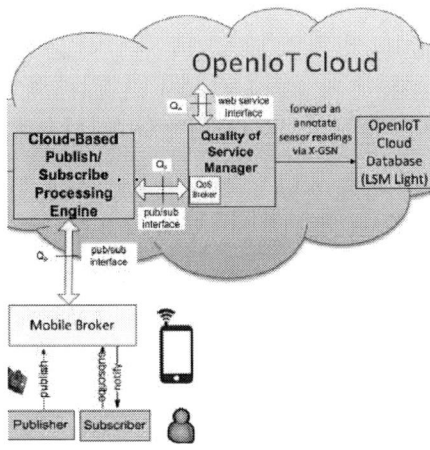

Fig. 3. Diagram showing structure of OpenIoT.

B. FarmBot

It utilizes freely available hardware and software in IoT in agriculture. It focuses on open source automated precision farming machine for humanity.

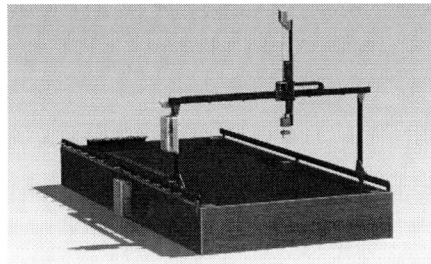

Fig. 4. Diagram of FarmBot.

C. SmartFarmNet

It provides real time response to queries that are made over time-series data streams from sensors. It uses a combination of non SQL (NoSQL) and semantic data stores. This combination manage user data, sensor data, aggregated data and caching of commonly used information.

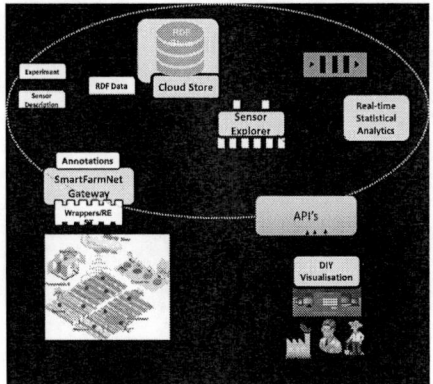

Fig. 5. Diagram showing structure of SmartFarmNet.

D. CLAAS Equipment

It produces equipment that run on autopilot. Autopilot provides advice on increasing crop flow and on reducing the losses. It allow the program to optimize the equipment automatically.

Fig. 6. Diagram of CLAAS Equipment.

E. Precision Hawks UAV sensor platform

It is introduced by Precision Hawk. The unmanned Aerial Vehicle (UAV) performs a series of land related tasks which were done by the labors previously. Surveying and mapping of the land are its main tasks.

Fig. 7. Diagram of Precision Hawks UAV sensor platform.

VI. HARDWARE AND SOFTWARE TOOLS USED IN SMART FARMING

In this section different hardware and software tools used in smart farming is discussed based on the survey of application of IoT in agriculture. It would not be possible to make agriculture smart without using these tools.

TABLE II

TABLE SHOWING SOME IoT HARDWARE AND THEIR APPLICATIONS IN AGRICULTURE

Sl.No.	Name of Hardware	Applications
1	PIC Microcontroller [14]	It is used in security, remote sensors, home appliances and industrial automations. An EEPROM is also featured which is used to store the information permanently.
2	Raspberry Pi [4]	It is credit card sized computer. It is used to do small computing and networking operations. It provides internet access to connect the automation system with remote location controlling device.
3	ZigBee Module [20]	It is used for achieving wireless communication between different node. Its power consumption and cost is very low. It is usually used to establish wireless local area network.
4	GSM module [8]	It can accept any GSM network operator SIM and it can act just like a mobile phone with its own unique phone number. It can use RS-232 protocol which can be easily connected to the controller.
5	DHT11 [3]	It is a basic, low cost digital temperature & humidity sensor.
6	LM35 [12]	It is precision IC temperature sensor. It is very low cost sensor. Its output voltage is directly proportional to temperature.
7	PIR Sensor [18]	It detect infrared radiation emitted or reflected by an object. It is used to detect the movement of people, animals or other objects.
8	Soil Moisture Sensor [6]	It is a sensor that senses the moisture content of the soil. When soil is dry , the current will no pass through it and output is said to be maximum otherwise zero.

TABLE III

TABLE SHOWING SOME IoT SOFTWARE USED IN AGRICULTURE

Sl.No.	Software Used	Application
1	Contiki [1]	It is an operating System. It is used in low power IoT devices.
2	Cooja [1]	To simulate Contiki network Cooja is preferred i.e. It is Contiki network simulator.
3	Proteus 8 Simulator [23]	It is a simulator software which is used to test programs and embedded design for electronics before actual hardware testing.
4	Raspbian OS [15]	It provides the basic set of program and utilities for Raspberry Pi. It is free and open source operating system which is Debian based and optimized for Raspberry Pi.
5	Dip Trace [16]	It is EDA/CAD software for creating schematic diagrams and printed circuit boards.
6	AVR Studio Version 4 [11]	To perform desired operations on microcontroller, it is needed which performs write, build, compile and debug the embedded c program codes in microcontroller.
7	Sina Prog [16]	It is Hex downloader application with AVR Dude and Fuse Bit Calculator. It is used to download code and to set fuse bits of all AVR based microcontroller.
8	iFogSim [1]	It is a simulator that is used to model Fog environment and Internet of Things. It can measure latency, energy consumption, network congestion and cost.
9	SegNet [17]	It is used in computer vision and Robotics. It is a deep convolution encoder-decoder architecture used for multiclass pixelwise segmentation.
10	FarmBeats [22]	It uses low-cost sensors, drones, machine learning and vision algorithm to solve the problem of data driven farming.

VII. DIFFERENT WIRELESS SENSOR NETWORK (WSN) USED IN SMART FARMING

Wireless Sensor Network is main backbone of Smart Farming. It consists of distributed devices, sensors, routers and coordinator. It has many applications in IoT like weather monitoring, indoor air quality monitoring, soil moisture monitoring, surveillance system, smart grid etc. In this section some WSN used in Smart Farming is discussed.

A. 6LoWPAN

It is low power Wireless Personal Area Network over IPv6. It brings IP protocol to the low-power devices having limited processing capacity. It works in network layer of IoT. It works with the 802.15.4 link layer protocol.

B. Wifi

It is IEEE 802.11 standard . It is collection of wireless local area network and communication standard. It works in link layer of IoT devices.

C. WiLD

WiLD network possess long range connection capability with high data rate. This network is capable to process huge amount of data. It is used to connect rural area from remote locations where internet connectivity is difficult.

D. IEEE 802.15.4(LR-WPAN)

It is also Known as low-rate wireless personal area networks(LR-WPANs). It is IEEE standard IoT protocol that works in Link layer of IoT. It forms the basis of specifications for high level communication protocols like ZigBee. It provides low cost and low speed communication for power constrained devices.

TABLE IV

TABLE SHOWING SOME CHARACTERISTICS OF WSN USED IN SMART FARMING

Characteristics	6LoWPAN	Wifi	WiLD	IEEE 802.15.4 (LR-WPAN)
Data rate	250 Kb/s	1 Mb/s to 6.75 Gb/s	comparable to wifi	40 Kb/s to 250 Kb/s
Frequecy	2.4 GHz	2.4 GHz to 5 GHz	comparable to wifi	868/915/2450 MHz
Range	PAN	LAN	longer than Wifi range	PAN
Power consumption	lowest	high	low	low
Cost	low	affordable	low	low
Processing capability	limited	high	hugh	low
IoT protocol layer	network	link	-	link

VIII. CHALLENGES OF SMART FARMING

In this section different challenges in implementing Smart Farming has been discussed. It is not an easy task to collect real time data using sensors in agricultural field. There are also other challenges in Smart Farming like huge volume of agricultural data needs massive storage system, interoperability issues between different networking devices and sensors, inefficiency of existing IoT platform. These challenges are listed below.

A. IoT hardware and Software implementation challenges

Smart farming generates huge data at different sensors involved in it. To store large volume agricultural data and processing these data need good software and hardware.

978-1-5386-6723-1/19 $31.00 © 2019 IEEE

B. Organisational challenges and their interoperability issues

Agricultural data is collected over different sensors and different software and hardware are needed to construct smart farming model. It is difficult to integrate heterogeneous data coming from different sensors (like moisture sensor, soil sensors, temperature sensor etc.) IoT system in Smart farming.

C. Networking Challenges

Communication between different sensors nodes are challenging task in agriculture. Sensors perform massive computation for which they consume energy but battery power of sensors are limited. So efficient energy management is needed in network.

D. Inefficient storage for massive agricultural data

From Smart Farming huge volume of data generates from different IoT devices(like sensors, camera, weather stations etc.). For storing such a large volume of data, a great repository is needed. Generally database is used to store these information but it is not sufficient to handle such a huge data.

E. Inefficiency of existing IoT platform

Real-time data ingestion, visualization of massive volume of sensor data and their quick analysis is not supported by existing IoT platform.

CONCLUSION AND FUTURE WORK

This paper presents the role of Internet of Thing in Smart Farming. Now a days farming can, With the introduction of IoT, an revolutionary change has been remarked in agricultural sector. IoT plays a vital role in huge production agriculture by monitoring different factors that are responsible in the production of agriculture. In this paper a survey Role of Internet of Things (IoT) in Smart farming along with its challenges, different agricultural apps, software and hardware used is explained briefly. From the survey it is revealed that still there are many things that are needed to improve to get good agricultural yield. In Future I will try to implement some new agricultural model that will overcome the challenges that present IoT model for Smart Farming is facing.

REFERENCES

[1] AHMED, N., DE, D., AND HUSSAIN, I. Internet of things (iot) for smart precision agriculture and farming in rural areas. *IEEE Internet of Things Journal 5*, 6 (2018), 4890–4899.

[2] BACCO, M., BERTON, A., GOTTA, A., AND CAVIGLIONE, L. Ieee 802.15. 4 air-ground uav communications in smart farming scenarios. *IEEE Communications Letters 22*, 9 (2018), 1910–1913.

[3] BALAJI, G. N., NANDHINI, V., MITHRA, S., PRIYA, N., AND NAVEENA, R. Iot based smart crop monitoring in farm land. *Imperial Journal of Interdisciplinary Research (IJIR) Vol 4*, 88–92.

[4] BALAMURUGAN, C., AND SATHEESH, R. Development of raspberry pi and iot based monitoring and controlling devices for agriculture. 207–215.

[5] CHASE, J. The evolution of the internet of things. *Texas Instruments* (2013), 1.

[6] EL MARAZKY, M. S. A., MOHAMMAD, F. S., AND AL-GHOBARI, H. M. Evaluation of soil moisture sensors under intelligent irrigation systems for economical crops in arid regions. *American Journal of Agricultural and Biological Sciences 6*, 2 (2011), 287–300.

[7] GONDCHAWAR, N., AND KAWITKAR, R. Iot based smart agriculture. *International Journal of Advanced Research in Computer and Communication Engineering (IJARCCE) 5*, 6 (2016), 177–181.

[8] HEGADE, A. S., JADHAV, S. H., JADHAV, S. A., AND GAIKWAD, N. M. Gsm based automation in agriculture.

[9] JAIGANESH, S., GUNASEELAN, K., AND ELLAPPAN, V. Iot agriculture to improve food and farming technology. In *Emerging Devices and Smart Systems (ICEDSS), 2017 Conference on* (2017), IEEE, pp. 260–266.

[10] JAYARAMAN, P. P., YAVARI, A., GEORGAKOPOULOS, D., MORSHED, A., AND ZASLAVSKY, A. Internet of things platform for smart farming: Experiences and lessons learnt. *Sensors 16*, 11 (2016), 1884.

[11] KUMAR, S. S., MURTHY, R. V., AND TULASI, P. K. Renovate the conventional methods of agriculture using internet of things.

[12] LIU, C., REN, W., ZHANG, B., AND LV, C. The application of soil temperature measurement by lm35 temperature sensors. In *Electronic and Mechanical Engineering and Information Technology (EMEIT), 2011 International Conference on* (2011), vol. 4, IEEE, p. 18251828.

[13] MALAVADE, V. N., AND AKULWAR, P. K. Role of iot in agriculture. *IOSR Journal of Computer Engineering (IOSR-JCE) e-ISSN*, 2278–0661.

[14] MANIMARAN, P., AND ARFATH, D. Y. An intelligent smart irrigation system using wsn and gprs module. *International Journal of Applied Engineering Research 11*, 6 (2016), 3987–3992.

[15] PRABHAVATHI, M., AND KIRANMAI, A. Smart security for agriculture using iot.

[16] RAUT, M. D. R., MASANE, M. G. V., NILESH, M. R. A. N. M., AND VERULKAR, M. Smart farming: Present and future.

[17] SA, I., CHEN, Z., POPOVIĆ, M., KHANNA, R., LIEBISCH, F., NIETO, J., AND SIEGWART, R. weednet: Dense semantic weed classification using multispectral images and mav for smart farming. *IEEE Robotics and Automation Letters 3*, 1 (2018), 588–595.

[18] SHINDE, N., SARAF, R., YADAV, P., AND KULKARNI, S. Farm monitoring and crop disease diagnosis.

[19] STOCES, M., VANEK, J., MASNER, J., AND PAVLIK, J. Internet of things (iot) in agriculture-selected aspects. *Agris on-line Papers in Economics and Informatics 8*, 1 (2016), 83.

[20] SUÁREZ BARÓN, J. C. Application of zigbee technology for monitoring enviromental variables in greenhouses. In *XVIII Congreso Argentino de Ciencias de la Computación* (2013).

[21] SUMA, D. N., SAMSON, S. R., SARANYA, S., SHANMUGAPRIYA, G., AND SUBHASHRI, R. Iot based smart agriculture monitoring system. *International Journal on Recent and Innovation Trends in Computing and Communication 5*, 2 (2017), 177–181.

[22] VASISHT, D., KAPETANOVIC, Z., WON, J., JIN, X., CHANDRA, R., SINHA, S. N., KAPOOR, A., SUDARSHAN, M., AND STRATMAN, S. Farmbeats: An iot platform for data-driven agriculture. In *NSDI* (2017), pp. 515–529.

[23] XIUMEI, X., AND JINFENG, P. The simulation of temperature and humidity control system based on proteus. In *Mechatronic Science, Electric Engineering and Computer (MEC), 2011 International Conference on* (2011), IEEE, pp. 1896–1898.

A Cyber Communication Package in the Application of Grid Tied Solar System

R. Majumdar, P. K. Gayen
Electrical Engineering Department
Kalyani Government Engineering
College
Kalyani, Nadia, West Bengal, India.

S. Mondal, A. Sadhukhan, P. K. Das
Electrical Engineering Department
Kalyani Government Engineering
College
Kalyani, Nadia, West Bengal, India.

I. Kushary
Department of Electrical Engineering
JIS College of Engineering
Kalyani, Nadia, West Bengal, India.

Abstract- **In this paper, development of cyber communication package in the application of grid connected solar system has been presented. Here, implemented communication methodology supports communication process with reduced latency, high security arrangement with various degrees of freedom. Faithful transferring of various electrical data for the purpose of measurement, monitoring and controlling actions depend on the bidirectional communication strategy. Thus, real-time communication of data through cyber network has been emphasized in this paper. The C# language based coding is done to develop the communication program. The notable features of proposed communication process are reduction of latency during data exchange by usage of advanced encryption standard (AES) algorithm, tightening of cyber security arrangement by implementing secured socket layer (SSL) and Rivest, Shamir and Adleman (RSA) algorithms. Various real-time experiments using internet connected computers have been done to verify the usability of the proposed communication concept along with its notable features in the application.**

Keywords- Solar system, cyber security, latency reduction, advanced encryption standard (AES), Rivest, Shamir and Adleman (RSA), secured socket layer (SSL).

I. INTRODUCTION

THE penetration level of solar energy conversion system into electrical grid [1] is rapidly increasing throughout the world. The increased trend of usage of this renewable energy source for generation of electrical energy draws special attention of researchers for technological development in its working area. Nowadays, the energy management system (EMS) [2] or supervisory control and data acquisition (SCADA) system [3] for grid connected solar system depends on the correctness of bidirectional data flow i.e. sensed electrical data from measuring point to EMS/SCADA server and decision/control command from EMS/SCADA server to device/plant via intermediate computer(s). In addition to this, protection action of various devices in this application also relies on the communication issues. Thus, various data from client computers are finally processed by the server or main computer situated in central processing / control centre for making any decision on the operation of solar system. In this process, suitable communication link is required to transfer data of various electrical quantities. Various types of communication links are mentioned in published paper [4]. In

this context, importance of cyber network is reported in [4]. The extensive usage of cyber network in connection with distributed generation based power system is also cited in [4]. Here, the integrated system is exposed to open market via cyber network. This fact obviously leads to exploration of power market in respect of new possibilities, but there is risk of vulnerabilities. Thus, chance of cyber-attack or cyber-threat may occur and evidences of such attack are also stated in some references [5]. So cyber security is one of the primary requirements for the grid connected solar system like other control mechanisms/operational requirements. Different issues of cyber security in power communication network are elaborately described in papers [5]. The standards of smart grid communication and its security measure are cited and discussed in [6].

Communication latency [7] is the other performance parameter to be dealt in practice. Heavy security arrangement can increase communication delay which introduces delay in control and management decision i.e. system performance can be affected. Therefore, suitable solution has to be thought for reduction of conflict between cyber security and communication latency.

From the above discussions, it can be stated that the development and testing/verification of the communication concept in renewable power application such as grid connected solar system is of utmost importance. In this paper, three-fold cyber communication arrangement by combining three potent algorithms, namely secured socket layer (SSL), advanced encryption standard (AES) and Rivest, Shamir and Adleman (RSA) algorithms [12] are proposed and applied in the work. Here, the SSL technique is used for generating authenticated digital signature for a particular server-client pair at the time of installation of developed client-server program package. The RSA and AES algorithm in the package is used to provide excellent communication security with reduced latency. Besides, some additional security measures like token based verification and data format checking have been incorporated. Thus, in the proposed concept, selective usage of three algorithms (SSL, AES and RSA algorithms) in different stages of communication process maintains satisfactory communication latency without compromising cyber security i.e. smart communication is established

978-1-5386-6723-1/19 $31.00 © 2019 IEEE

between client and server.

For validating proposed communication methodology, graphical model of grid connected solar system with its control logics are framed using SIMULINK software. The sensed real-time electrical data at various parts of the model (during run time of the model) are transmitted to the server via client. In reverse direction, any real-time command from server is sent to the client computer for controlling the operation of SIMULINK software based graphical model. This experiment of bidirectional communication between client and server via internet is implemented by writing various codes using C#. The developed program also blocks unauthorized client from the main server. Therefore, development of smart cyber communication i.e. optimum performance in respect of cyber security and latency is targeted in the work presented in this paper. It can be mentioned here that large power network data can be handled (sensing and controlling) with secured communication in real-time using the developed program. Real-time testing of any communication logic associated with any power network can also be done even before their real world implementation.

The overall test system studied in this paper with its schematic is discussed in section II. Section III presents realization of proposed smart cyber communication with its features. In section IV, experimentation and associated outcome of the work are presented for validation of the proposed concept. This paper concludes in section V.

II. DESCRIPTION OF TEST SYSTEM

The schematic diagram of test system set-up considered in this paper is presented in Fig. 1. Various computers required for the communication test are connected via internet as shown in Fig. 1(a). The model of physical system (grid connected solar system) with its control loops and various sensing points are built using MATLAB-SIMULINK software and are loaded in one of those computers (marked as physical system model in Fig. 1(a)). The developed C# language based communication programs are loaded in main server and client computers. The schematic diagram of the model of physical system and related closed loop control logics are shown in Fig. 1(b). In the model, two different values of load are connected at point of common coupling (PCC) to simulate change of load condition. In Fig. 1(b), two control loops, namely perturb and observe (P & O) based maximum power point tracking (MPPT) logic and inverter's current control loop are designed. The proposed communication concept is justified using the test system. Here, the structure of the test system is equivalent to physical scenario for validating the communication concept.

III. DEVELOPMENT OF PROPOSED CYBER COMMUNICATION FOR GRID CONNECTED SOLAR SYSTEM

The codes for transmission control protocol (TCP) based communication are developed using C# language. The reason is C# language and its IDE (integrated development environment) provide an excellent interface for developing

program using .NET framework in a user friendly manner that also facilitates graphical user interface for the program. The implemented steps for communication process are presented in Fig. 2 with demarcation of server and client side activity and also, with mentioning of related algorithm. The communication process is developed as follows:

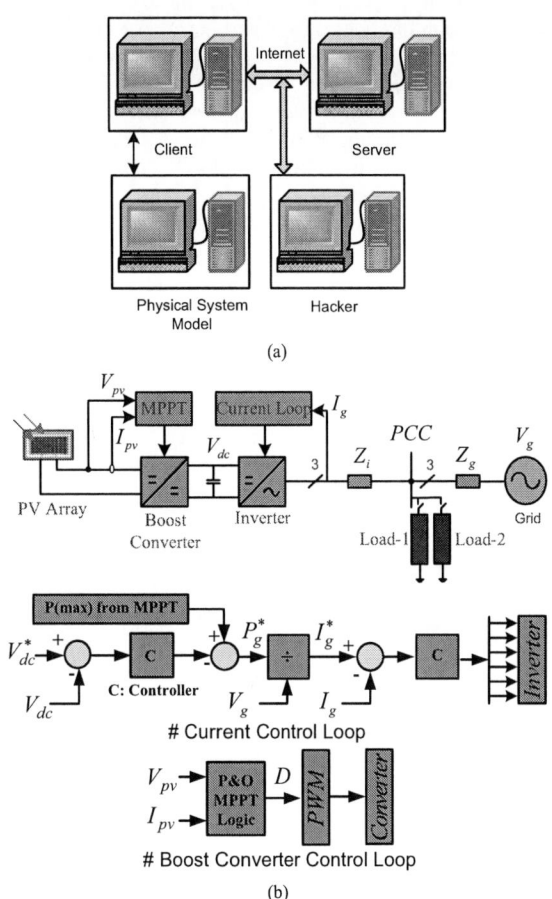

(a)

(b)

Fig. 1. Schematic diagram of (a) test system, (b) grid connected solar system and two control loops.

A. Developed Communication Process

The total communication process comprises of three successive stages as described below,

1) Primary Credential Exchange with Server by Client

On the very first connection set up, basic credentials of aspiring client are sent to main server. Those credentials include a set of two constituents – (a) connection token, (b) public key. In the developed program, connection token is a string of alphanumeric character having length of 12 digits. Connection token is required for maiden client else session token received in last session is used for the purpose of identifying intended client. Connection token is generated using such a logic that uniquely identifies client machine. Public key is used to send server generated secret key to the client on validation of other credentials of the same. Two credentials with digital signature reach to the server for verifications.

2) Validation of Client and Sharing of Generated Secret Key & New Session Token by Server

After the arrival of primary credentials, the server checks digital signature. If it is found valid then, checking of token is initiated by server. In the next step, if token is validated by server then, the client is declared as valid client by server. After validation of client, server will generate secret key using AES algorithm. A fresh session token is also generated and sent for using while setting up connection for the next time. The secret key is shared by server and client. The secret key is encrypted by client-supplied public key. This encrypted secret key and newly generated token is sent to the client with digital signature. Then, the client decrypt the secret key and token with own private key and finally gets them.

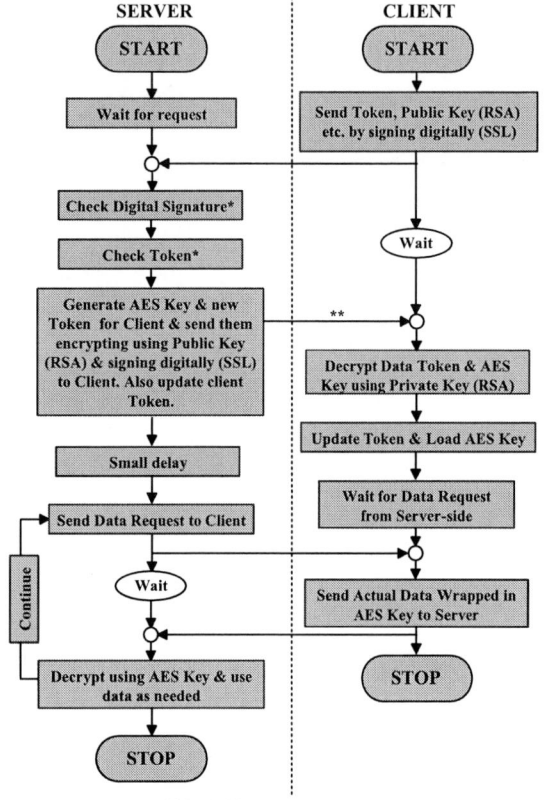

*Proceed to next step if & only if checking succeeds, else discards client request.
**In case of any communication failure, server disconnects client & client try to reconnect.

Fig. 2. Steps of proposed communication process.

It can be mentioned here that server discards any client request for connection on failure of validity of either digital signature or token.

3) Data transfer between Client and Server

Next step onwards, safe data transfer takes place with optimum speed with the help of secret key.

In the described communication process, the role and features of various algorithms are outlined as follows:

B. Provision of Secured Socket Layer (SSL)

In the above mechanism, digital signature is generated by SSL logic for a definite server-client pair at the installation stage of the developed package and the digital signature is designed by manufacturer/designer of the program. The SSL ensures secured data transmission between main server and client at the very beginning of communication commencement. Here, the server and client(s) are verified via respective digital signature embedded in them to authenticate each other for a particular client-server combination. In the implemented logic, 4096-bits long SSL signature is used which is considered to be strong enough to withstand any practical cyber attack of present time. It is note-worthy that 128-bits long SSL signature is widely used in online bank transaction.

C. Usage of Rivest, Shamir, and Adleman (RSA) Algorithm

The public-private key pair is generated by client using RSA algorithm. The RSA algorithm is used at the initial stage of communication commencement to share token and AES secret key, which is used for rest of the communication duration. 4096-bits long RSA key is used for providing higher security level. But RSA algorithm is relatively slower than AES algorithm and hence, it is not used for entire communication period to reduce latency.

D. Usage of Advanced Encryption Standard(AES) Algorithm

The role of AES algorithm is to encrypt or decrypt the transmitted data with its generated secret key. The AES secret of length 4096-bits is used for transmitting the actual data in encrypted manner through the cyber communication lines in the work. This provides improved communication speed without compromising security standard.

E. Combined Usage of SSL-RSA-AES Algorithm

The SSL is static kind of security arrangement i.e. fixed and only used at the very beginning of communication. Thus, there may be chance of information leakage due to its invariableness. On the other hand, both RSA and AES have dynamic security key i.e. these keys are changed during every session of communication. They are strong due to variable pattern of key from session to session. The AES key may be considered relatively inferior than RSA as same secret key is used for either encryption or decryption purpose at a particular session. On the other hand, RSA key (public key) for encryption is different from RSA key (private key) for decryption purpose. The RSA private-public key pair is generated in the client side and the public key is shared with the server. In return, server generates the AES secret key and shares it with client by using the public key provided by client. Now, the retrieval of the encrypted secret key only can be done by private key which is in the hand of client only. Here, the private key can be treated as confidential key to client and it is not shared with the server or any other computer in the network. Thus, secrecy of secret key (AES) is guaranteed in the transferring process. The exchange of data by AES logic is comparatively faster than that of the RSA logic. Thus, AES logic is used for transferring of data. The comparative performance is presented in the result section.

Thus, the notable features of developed communication package are summarized as,

i. Support of bidirectional communication,
ii. Usage of three-fold algorithms, namely SSL, AES and RSA algorithms for maintaining secured cyber communication,
iii. Reduction of latency without hampering security by selective usage of above said algorithms,
iv. Availability of graphical user interface (GUI) makes usage of the program attractive and user friendly,
v. Multithreading capability makes the program faster and more efficient for better communication with large power network,
vi. Validity and testing is possible with the modeling of physical system in MATLAB-SIMULINK software which is recognized as widely used software in academic and research purpose. The proposed concept can be implemented for any other power network application.

IV. RESULTS

The proposed communication concept is validated in the laboratory using three computers with internet connected client-server as shown in Fig. 3. Fig. 3 shows that left most computer loads the physical model using MATLAB-SIMULINK software, the middle computer (client) collects sensed data and transfers to the right most computer (server). Any command by server is communicated to the physical model in reverse way. Various codes are written using C# language to support proposed bidirectional communication process in encrypted form for grid connected solar system. Overall study is presented as follows:

A. Validation of Bidirectional Communication

The run-time responses of various quantities in SIMULINK platform are shown in Fig. 4. In initial stage, the load at PCC is taken as 120 kW (power factor = 1 i.e. unity). Here, total load is shared as 100 kW and 20 kW by solar system and grid respectively. The solar system is operated at temperature of 25ºC and solar irradiation level of 1000W/m^2 and the corresponding generated maximum power is 100 kW by the action of MPPT controller. In the next step, command on change of load is initiated from the server side and is sent through internet link and client computer for activating the total load of 70 kW in SIMULINK model instead of 120 kW load at simulation time (t) = 0.75 sec. After load change, it is observed that total load of 70 kW is supplied from the solar system alone and rest of the solar generation (100-70=30 kW) is injected in to the grid. Before and after load change conditions, various sensed data of solar system, grid and loads (total 15 variables) are transferred via internet. These data are shown in Fig. 5 which is displayed on server monitor. These displayed data match with the parameters values of dynamic responses of the SIMULINK model. From this part of the study, the bidirectional communication capability of data has been validated via cyber network with proper integrity.

B. Testing of Cyber Security

Failure of the communication process due to improper security reason is tested here in respect of unauthorized digital signature, wrong token and file format. The wrong digital signature has been intentionally put to test and at that time, hacker fails to connect to the server with the indication of failure message on the hacker computer as shown in Fig. 6. In similar way, intentionally altered token is used at the time of communication to justify the failure of connection. The corresponding observations are shown in Fig. 7. The acceptable file format for the server is .xml. Incorrectly formatted data from any client leads to immediate disconnection of that client with failure message as shown in Fig. 8.

From various observations of this part of study, it is apparent that tight security arrangement has been established in the proposed concept.

Fig. 3. Experimental set-up.

(a) Different power responses

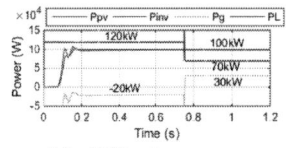

(b) Output voltage and current waveforms of solar cell

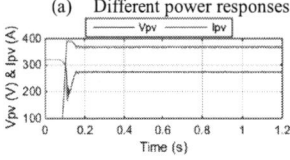

(c) Grid side voltage and current responses

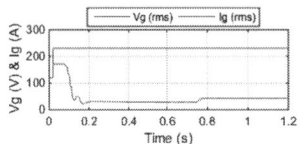

(d) Load-1 voltage and current

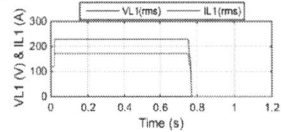

(e) Load-2 voltage and current

Fig. 4. Various dynamic responses in SIMULINK platform.

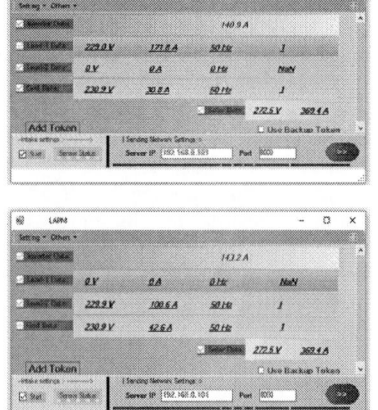

Fig. 5. Display boxes of sensed data in server monitor before (upper) and after (lower) load changing condition using developed program package.

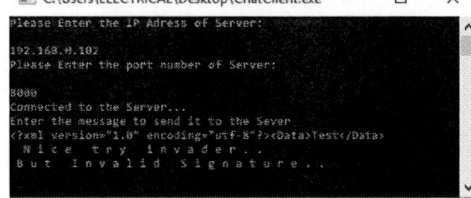

Fig. 6. Failure message of invalid signature on hacker computer

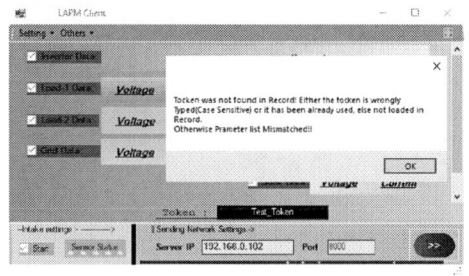

(a) Message on hacker computer

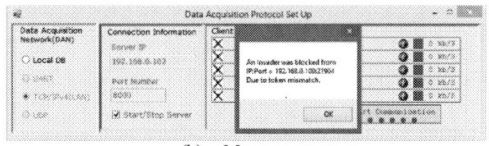

(b) Message on sever

Fig. 7. Failure messages of wrong token using developed program.

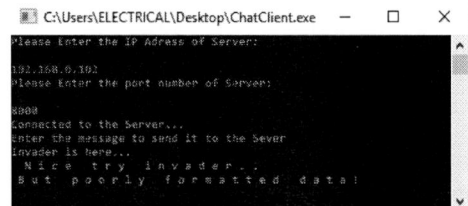

Fig. 8. Failure message of improper file format on hacker.

C. Testing of Latency

The communication using AES and RSA algorithm of size 4096-bits is done separately. The encryption and decryption time taken by AES and RSA algorithms are compared. The respective time is displayed in server and client computer which is shown in Fig. 9. Thus, latency reduction by employing AES algorithm is demonstrated in this part of study.

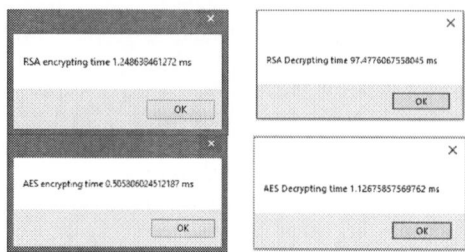

Fig. 9. Message of latency using different algorithms.

V. CONCLUSION

In this paper, the C# language based communication package using three-fold security arrangement is developed to facilitate bidirectional cyber communication for grid connected solar system. The transmitting of sensed data from client computer to server and any command from server to the client system are established. In the communication process, latency is reduced by usage of AES algorithm without compromising security level. Here, cyber security arrangement is strengthened by employing SSL and RSA logics during connection of client to the server. The coding using C# language and its IDE makes the developed program user friendly, cost effective, faster and graphical interface ability. The designed communication program can be tested with the real-time model using MATLAB-SIMULINK software before its practical application to the controlled renewable energy conversion system.

REFERENCES

[1] Y. Wu, J. Lin, and H. Lin, "Standards and Guidelines for Grid-Connected Photovoltaic Generation Systems: A Review and Comparison," in *IEEE Trans. Ind. Appl.*, vol. 53, no. 4, pp. 3205-3216, July/August 2017.

[2] N. Mahmud, A. Zahedi, and A. Mahmud, "A cooperative operation of novel PV inverter control scheme and storage energy management system based on ANFIS for voltage regulation of grid-tied PV system," *IEEE Trans. Ind. Informat.*, vol. 13, no. 5, pp. 2657-2668, Oct. 2017.

[3] M. Liserre, T. Sauter, and Y. J. Hung, "Future energy systems-Integrating renewable energy sources into the smart power grid through industrial electronics," *IEEE Trans. Ind. Electron.*, vol. 4, no. 1, pp. 18-37, Mar. 2010.

[4] M. Kuzlu, M. Pipattanasomporn, and S. Rahman, "Communication network requirements for major smart grid applications in HAN, NAN and WAN," *Comput. Netw.*, vol. 67, pp. 74-88, Jul. 2014.

[5] E. Bou-Harb, C. Fachkha, M. Pourzandi, M. Debbabi, and C. Assi, "Communication security for smart grid distribution networks," *IEEE Commun.Mag.*, vol. 51, no. 1, pp. 42-49, Jan. 2013.

[6] V. C. Gungor, D. Sahin, T. Koçak, S. Ergüt, C. Buccella, C. Cecati, and G. P. Hancke, "Smart grid technologies: Communication technologies and standards," *IEEE Trans. Ind. Inf.*, vol. 7, no. 4, pp. 529-539, Nov.2011.

[7] P. Kansal and A. Bose, ''Bandwidth and latency requirements for smart transmission grid applications,'' *IEEE Trans. Smart Grid*, vol. 3, no. 3, pp. 1344-1352, Sep. 2012.

2019 Devices for Integrated Circuit (DevIC), 23-24 March, 2019, Kalyani, India

Counter Based Low Power, Low Latency Wallace Tree Multiplier Using GDI Technique for On-chip Digital Filter Applications

Biswarup Mukherjee
Dept. of Electronics and Communication Engineering
Neotia Institute of Technology, Management And Science
Diamond Harbour Road, India
biswarup80@gmail.com

Aniruddha Ghosal
Institute of Radio Physics and Electronics
University of Calcutta
Kolkata, India
aghosal2008@gmail.com

Abstract— **This paper represents a new design of a low power, low latency Wallace tree multiplier. Wallace Tree algorithm is one of the most commonly used operations in modern days DSP applications as it can provide a fast and area efficient strategy for higher operand multiplication. For higher bits of multiplications the addition operation of partial products includes greater delay and complexity. In this present communication a number of techniques are applied in the partial products addition circuitry to optimize the area delay and speed of the Wallace multiplier. Proposed Design is synthesized for 4x4 bit multiplication using standard CAD tool design compiler in 250nm process technology. Simulation results show that the proposed multiplier design has the best power and delay results as compared to other available multipliers.**

Keywords—counter, full adder, GDI technique, partial products, Wallace tree multiplier

I. INTRODUCTION

In modern days with the exclusive extension in mobile computing and portable multimedia applications low power, compact and high speed digital signal processing (DSP) systems has been increasing demand for niche devices[1]. A major task for DSP system is being handled by multipliers. Not only in DSP chips but also in many public-key cryptosystems such as Elliptic Curve Cryptography (ECC) and RSA, high operand multiplications are widely used. Therefore multiplication of higher operands has been one of the topics of interest for researchers, all the time. Multiplication is just a process which deals with lot of additions. A binary multiplication of NxN bit can be represented pictorially as follows,-

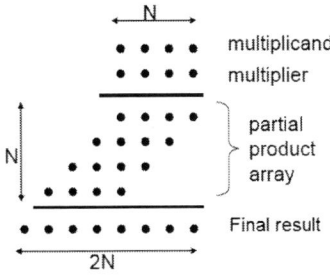

Fig. 1. Simple NxN bit multiplication

In an NxN bit binary multiplication, each bit in the multiplier is multiplied with N bit multiplicand. The results are called partial products. If the multiplier bit is a 1, then the corresponding product is simply an appropriately shifted copy of the multiplicand. If the multiplier bit is a zero, then

the product rows are also containing N nos. of zero[2]. In such way when all Partial products are generated then the corresponding columns are accumulated from top to bottom on an N-bit adder. For an N bit adder the delay of producing the output become significant as the no of transition increases [3]. Therefore in order to make a low latency multiplier circuit for higher bit of operands the accumulation of partial products must be done with modified techniques. One of such faster accumulation is performed in Booth multiplication techniques by reducing the no. of partial products with a recoding technique [4]. But such operation includes additional encoder circuitry and increases area burden. In order to make it area efficient serial multipliers or array multipliers can be used. Unfortunately, such designs suffer with poor latency. A solution to such problem can be minimized by Wallace tree algorithm. In 2010 Ron S. Waters & Earl E. Swartzlander introduced a novel architecture of modified Wallace tree architecture [5]. A high speed counter based Wallace architecture was introduced by Shahzad Asif, Yinan Kong in 2015 [6]. Many types of counters design are also communicated recently [7][8].

The main objective of the design is to achieve a novel ASIC design of Wallace tree multiplier circuit which would be efficient in power, area and delay with respect to other state of art designs. The next sections are categorized as follows. The detailed design of the proposed multiplier has been discussed in the following section. CAD tool based simulation and comparison with existing designs in terms of power, delay and area (transistor count) has been discussed later. Finally we conclude with the merits and demerits of proposed design.

II. PROPOSED WALLACE TREE MULTIPLIER DESIGN

From the above literature survey, it has been found that there are many types of multiplier architectures available for digital designs. Among all, Wallace tree can show low latency and high speed response. Such multiplication is done by the following three steps:

- For a NxN bit multiplication N^2 partial products are generated and each partial product carries a weight which depends on the bit position.
- In the next stage, the numbers of partial products are reduced by using counters/ compressors, full adders and half adders. This is done by combining the vertical partial products of same weight. The resulting sum carries the same weight but the

978-1-5386-6723-1/19 $31.00 © 2019 IEEE
151

2019 Devices for Integrated Circuit (DevIC), 23-24 March, 2019, Kalyani, India

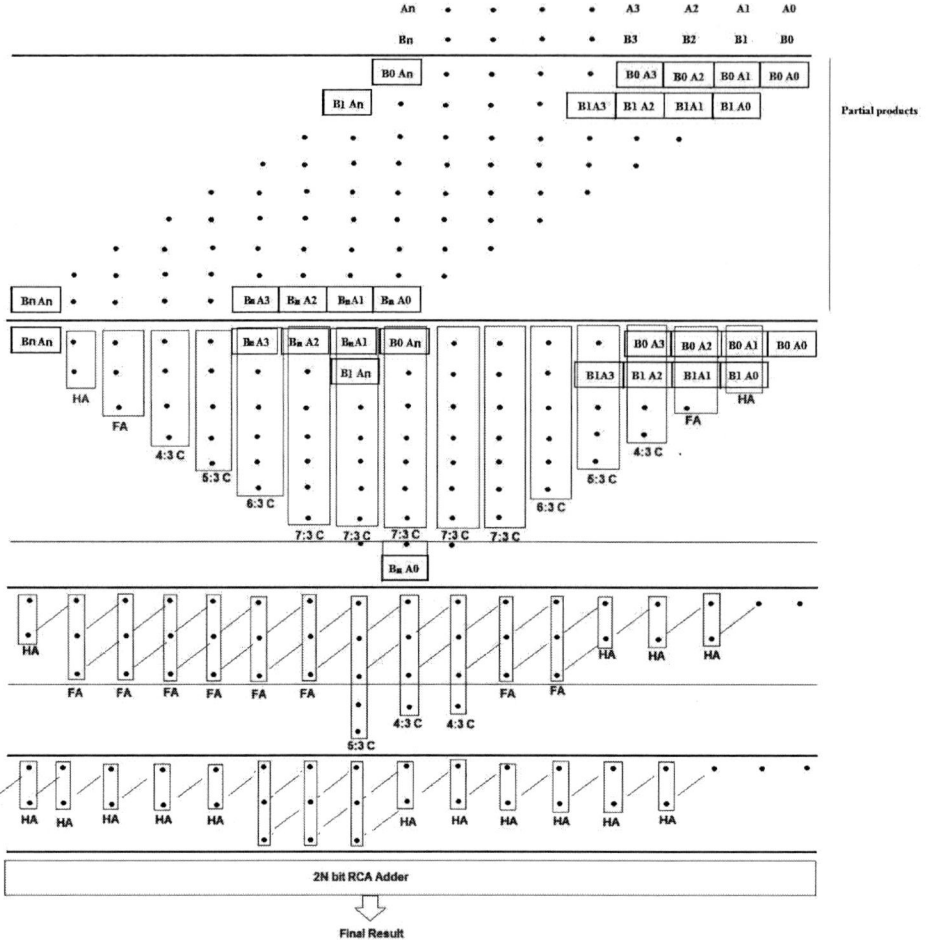

Fig. 2. Generalized architecture of modified structure of NxN Wallace tree Multiplier

generated carry out has higher weight. This process continues upto that level when there are only two layers of partial products.

- The final two layers are added using RCA adder to get the result.

The proposed multiplier uses the Wallace tree structure with modified architecture. For reducing the area, power and delay the GDI technique has been introduced in the building blocks. Fig. 2 shows general architecture for NxN bit Wallace tree architecture. The architecture consists of three major parts. Namely,- 1) Partial product generator, 2) Counter based partial product accumulation and 3) RCA adder for final result generation.

A. Partial Product Generator

First, the multiplicand each bit is multiplied with multiplier bits . The result is a partial products array. Simple AND operation can perform this job. For NxN bit multiplication operation N^2 nos. of AND gates are required. It may increase the area overhead. In order to make this stage area and power efficient a GDI based AND gates are introduced. Gate Diffusion Input (GDI) is a low power technique by which various Boolean functions can be performed by only 2 nos. of MOS transistors [9]. Figure 3

shows transistorized schematic of a partial product generator using GDI MUX based AND gate [2].

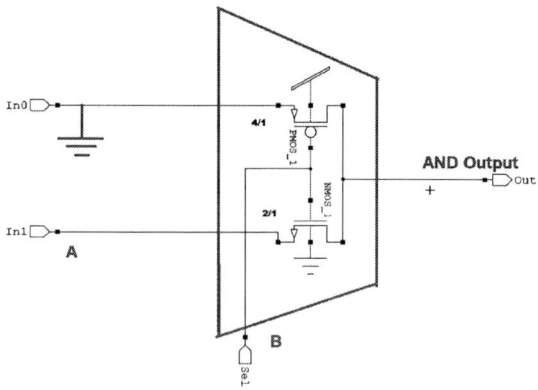

Fig. 3: GDI MUX based AND gate

B. Counter Based Partial Product Accumulation

The reduction of partial products can be done by counters or compressors and full adders, half adders. For more than three partial products having same weight counter or compressor circuits are preferred instead of full adder series. An (m,p) counter takes an m bit input bits from a column and produces a p bits output which detects the number of input 1 bits. That means it counts the number of input bits set to 1.Where as A (m:p) compressor is any circuit that takes in m equal weight input sum and carry in bits and produces p bit count along with additional carry bit. In our proposed design counter based reduction technique has been applied as it reduces the no of reduction stages than compressor based technique. Conventional high speed counter designs can be found from [10].

In the proposed design the counters have been implemented using GDI gates. A transistorized schematic of GDI AND gate has been shown in fig. 3. To design different types of counter circuit we also require designing GDI OR gate and XOR gate. Two transistor based GDI OR gate and three transistors based XOR gate schematics have been shown in fig. 4 & 5 respectively.

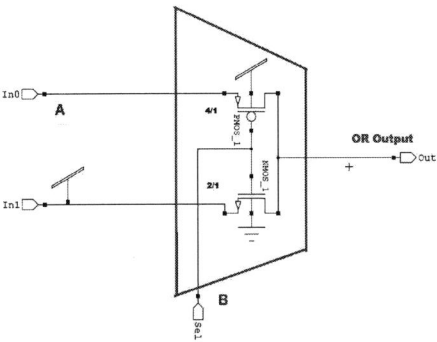

Fig. 4. GDI MUX based OR gate

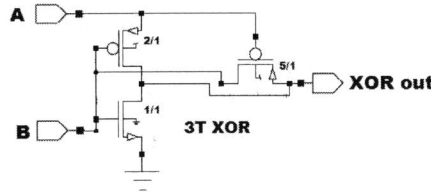

Fig. 5. Three transistors based XOR gate schematic

For accumulation of the partial products of same weights 7:3, 6:3, 5:3 and 4:3 counters are used along with full adders and half adders.

1) 7:3 Counter

A 7:3 counter can take seven input bits (x0, x1, x2, x3, x4, x5, x6) and produces three outputs (Sum, Cout1, Cout2) as follows,-

$$
\left.\begin{aligned}
Sum &= \left[(x_0 \oplus x_1) \oplus (x_2 \oplus x_3)\right] \oplus \left[(x_4 \oplus x_5) \oplus x_6\right] \\
C_{out1} &= (m_1 \oplus m_2) \oplus m_3 \\
C_{out2} &= (m_1 . m_2) + ((m_1 \oplus m_2) . m_3)
\end{aligned}\right\} \quad (1)
$$

where

$$m_1 = x_0.x_1 + x_2.x_3 + ((x_0 + x_1).(x_2 + x_3))$$
$$m_2 = \left[((x_4 + x_5).x_6 + x_4.x_5)\right]$$
$$m_3 = \left[x_0 x_1 x_2 x_3 + ((x_0 \oplus x_1) \oplus (x_2 \oplus x_3))\right].\left[(x_4 \oplus x_5) \oplus x_6\right]$$

The GDI gates based schematics of 7:3 counter is shown in fig. 6. The total no. transistors required to design the counter

is (8 XOR gates x 3T)+(8 AND gates x 2T)+ (8 OR gates x 2T) equal to 56.

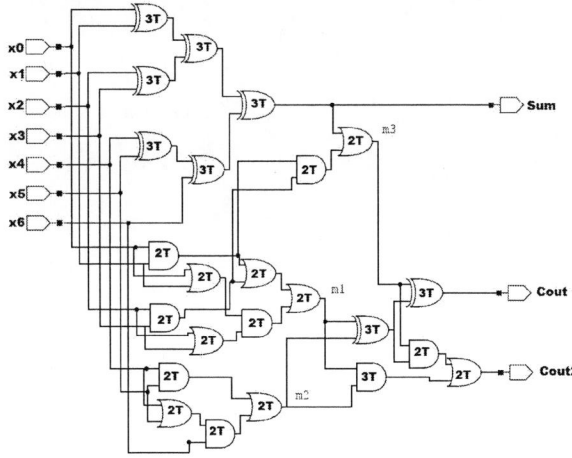

Fig. 6. GDI AND (2T), GDI OR(2T) and XOR (3T) gate based 7:3 counter schematic

2) 6:3 Counter

A 6:3 counter can take seven input bits (x0, x1, x2, x3, x4, x5) and produces three outputs (Sum, Cout1, Cout2) as follows,-

$$
\left.\begin{aligned}
Sum &= p_0 \oplus p_1 \oplus p_2 \\
C_{out1} &= (p_0.p_1 \oplus p_0.p_2 \oplus p_1.p_2) \oplus (g_0 \oplus g_1 \oplus g_2) \\
C_{out2} &= (g_0.g_1 + g_0.g_2 + g_1.g_2) + ((p_0.p_1).g_2) \\
&\quad + ((p_0.p_2).g_1) + ((p_1.p_2).g_0)
\end{aligned}\right\} \quad (2)
$$

where,

$$
\left.\begin{aligned}
p_0 &= x_0 \oplus x_1, & p_1 &= x_2 \oplus x_3, & p_2 &= x_4 \oplus x_5 \\
g_0 &= x_0.x_1, & g_1 &= x_2.x_3, & g_2 &= x_4.x_5
\end{aligned}\right\} \quad (3)
$$

The GDI gates based schematics of 6:3 counter is shown in figure 7. The total no. transistors required to design the counter is (10 XOR gates x 3T)+(12 AND gates x 2T)+ (5 OR gates x 2T) equal to 64.

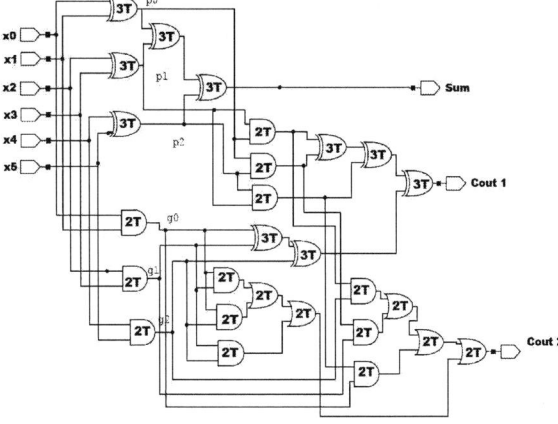

Fig. 7. GDI AND (2T), GDI OR(2T) and XOR (3T) gate based 6:3 counter schematic

3) 5:3 Counter

A 5:3 counter can take seven input bits (x0, x1, x2, x3, x4) and produces three outputs (Sum, Cout1, Cout2) as follows,-

$$Sum = p0 \oplus p1 \oplus x4$$
$$C_{out1} = (g0 \oplus g1 \oplus h0) \oplus (h1 \oplus h2) \oplus ((p0.p1) \oplus h3)$$
$$C_{out2} = (g0 . g1 + g0 . h2) + (g0 . h3) + (g1 . h0 + g1 . h1)$$
(4)

where

$$h0 = x0.x4 , h1 = x1.x4 , h2 = x2.x4 , h3 = x3.x4$$

The GDI gates based schematics of 5:3 counter is shown in figure 8. The total no. transistors required to design the counter is (10 XOR gates x 3T)+(12 AND gates x 2T)+ (4 OR gates x 2T) equal to 62.

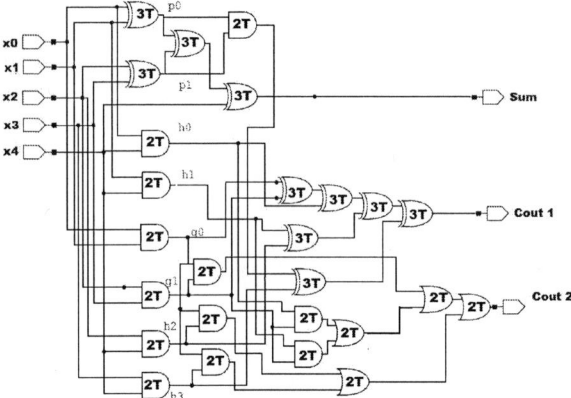

Fig. 8. GDI AND (2T), GDI OR(2T) and XOR (3T) gate based 5:3 counter schematic

4) 4:3 Counter

A 4:3 counter can take seven input bits (x0, x1, x2, x3) and produces three outputs (Sum, Cout1, Cout2) as follows,-

$$Sum = p0 \oplus p1$$
$$C_{out1} = (p0.p1) + (g0 \oplus g1)$$
$$C_{out2} = g0 . g1$$
(5)

The GDI gates based schematics of 4:3 counter is shown in figure 9. The total no. transistors required to design the counter is (4 XOR gates x 3T)+(4 AND gates x 2T)+ (1 OR gates x 2T) equal to 22.

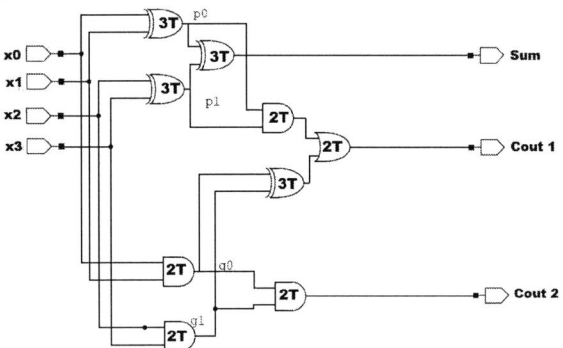

Fig. 9. GDI AND (2T), GDI OR(2T) and XOR (3T) gate based 4:3 counter schematic

5) 8T Full Adders & 5T Half Adders

In order to accumulate three partial products of same weight full adder can be used. In the proposed design full adders have been implemented using two nos. of 3T XOR gates and a GDI MUX. Similarly, half adders are used to accumulate two nos. of partial products of same weight. In the proposed design half adders are implemented using a 3T XOR and 2T GDI MUX [2].

C. Final 2N bit RCA ADDER Stage

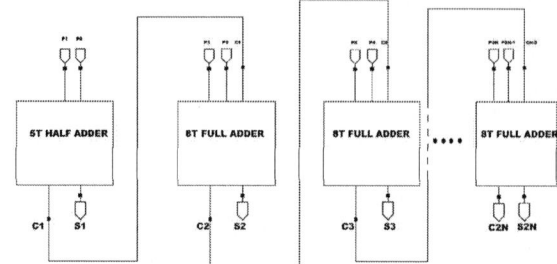

Fig. 10. Final stage 2N bit RCA adder schematic

When the partial products of same weight are reduced into two rows the final addition is done by RCA adder. The proposed RCA adder consists of (2N-1) nos. of eight transistors based full adders and five transistors based half adder circuit. The outputs of the RCA adder is (2N+1) bits long produces the final multiplied output.

III. SIMULATION RESULTS AND ANALYSIS

The proposed design is implemented for 4x4 bit multiplication example. A generalized schematic of 4x4 bit Wallace tree multiplier circuit is shown in fig. 11. The multiplicand is taken as $(1101)_2$ and the multiplier is $(1011)_2$. The simulation environment is set with standard EDA simulator using generic 250nm technology. Table I shows the input specifications for functional test of the design.

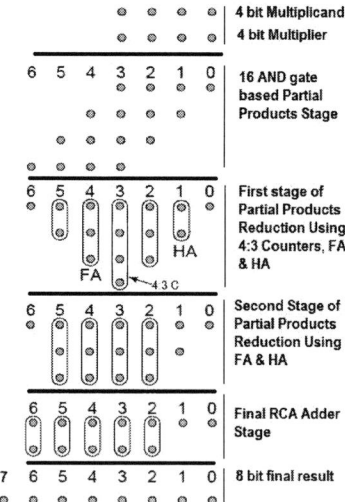

Fig. 11. 4x4 bit Wallace tree multiplier schematic

TABLE I. INPUT SPECIFICATION FOR SIMULATION

Input Type	Bit
Multiplicand	1101
Multiplier	1011
Zero value	0 v
One value	2.5 v
Bit duration	10ns
Rise and fall time	1ns

The simulation results are compared with existing designs of multipliers along with our proposed counter based GDI Wallace tree multiplier. The comparisons are made in terms of timing and power analysis and transistors counts. The proposed 4x4 counter based GDI Wallace tree multiplier consists of 16 AND gates for partial product generation. The partial product reduction is done by one 4:3 GDI counter, 6 nos. of 8 transistorized full adders and 7 nos. of 5 transistorized half adders in two stages. The circuit shows 5% improvement in latency and 43% improvement in power consumption with respect to conventional Booth multiplier. Table II shows detailed simulation results of proposed design along with other state of art designs.

TABLE II: DETAILED SIMULATION RESULT

	4x4 Array Multiplier	4x4 Vedic Multiplier	4x4 Wallace tree multiplier	4x4 Booth multiplier	Proposed Design simulated for 4x4 multiplier
Power (mw)	56.94	76.35	55.73	17.34	9.87
Delay (ns)	1.99	1.75	1.67	1.39	1.32
Area (Trs. count)	544	304	384	226	137

IV. CONCLUSION & FUTURE SCOPE

Analyzing the simulation results, we can conclude that the proposed counter based GDI Wallace tree multiplier shows better results in low power consumption, area concern and delay performance. The improvement will be more for higher no of bits. The Wallace tree structure can be utilized for modified Booth Wallace Tree multiplier circuit for further improvement in signed bit multiplications. Such on-chip multipliers can be implemented in different portable micro-systems and MEMS processor units.

REFERENCES

[1] Maskell, D.L.: "Design of efficient multiplierless FIR filters", IET Circuits Device Syst., 2007, 1, (2), pp. 175–180

[2] B.Mukherjee,B.Roy, A.Biswas, A. Ghosal, "Design of a Low Power 4x4 Multiplier Based on Five Transistor (5-T) Half Adder, Eight Transistor (8-T) Full Adder & Two Transistor (2-T) AND Gate" IEEE conference C3IT, pp: 1-5, 2015

[3] Manan Mewada, Mazad Zaveri, and Anurag Lakhlani,"Estimating the Maximum Propagation Delay of 4-bit Ripple Carry Adder Using Reduced Input Transitions" VDAT 2017, CCIS 711, pp. 15–23, 2017.

[4] Manjunath , Venama Harikiran , Kopparapu Manikanta , S Sivanantham , K Sivasankaran, "Design and implementation of 16×16 modified booth multiplier"IEEE International Conference on Green Engineering and Technologies (IC-GET), Nov. 2015

[5] Ron S. Waters, Earl E. Swartzlander, "A Reduced Complexity Wallace Multiplier Reduction", IEEE Transaction on Computers, Vol. 59, No. 8,PP- 1134-1137, August 2010

[6] Shahzad Asif , Yinan Kong, "Design of an algorithmic Wallace multiplier using high speed counters", Proceedings of IEEE International Conference on Computer Engineering & Systems (ICCES), PP: 133 - 138,Cairo, Egypt, 2015

[7] Sreehari Veeramachaneni, Lingamneni Avinash, M. Kirthi Krishna, M.B. Srinivas; "Novel Architectures for Efficient (m, n) Parallel Counters"; Proceedings of ACM Great Lakes Symposium on VLSI ;PP: 188-191; Stresa - Lago Maggiore, Italy, March 11-13, 2007

[8] Christopher Fritz , Adly T. Fam;"Fast Binary Counters Based on Symmetric Stacking"; IEEE Transactions on Very Large Scale Integration (VLSI) Systems; Volume: 25 , Issue: 10; PP: 2971 - 2975; 2017

[9] ArkadiyMorgenshtein, Alexander Fish, and Israel A. Wagner, "Gate-Diffusion Input (GDI): A Power-Efficient Method for Digital Combinatorial Circuits", IEEE Transaction on VLSI Systems, Vol. 10, No. 5,pp. 566-581, October 2002

[10] Shahzad Asif, Yinan Kong, "Design of an Algorithmic Wallace Multiplier using High Speed Counters", Proceedings of Tenth International Conference on Computer Engineering & Systems (ICCES), Egypt, 2015

[11] Christopher Fritz , Adly T. Fam; "Fast Binary Counters Based on Symmetric Stacking"; IEEE Transactions on Very Large Scale Integration (VLSI) Systems; Volume: 25, Issue: 10; pp: 2971 - 2975, 2017.

2019 Devices for Integrated Circuit (DevIC), 23-24 March, 2019, Kalyani, India

Effect of AlGaN Back Barrier on InAlN/AlN/GaN E-Mode HEMTs

Sarosij Adak
Electronics & Communication Engineering
Brainware University
Kolkata, India
sarosijadak@gmail.com

Nisarga Chand
Electronics & Communication Engineering
Adamas University
Kolkata, India
nisargaece@gmail.com

Sanjit Kumar Swain
Electronics & Communication Engineering
Silicon Institute of Technology
Bhubaneswar, India
sanjit.swain@gmail.com

Angsuman Sarkar
Electronics & Communication Engineering
Kalyani Government Engineering College
Kalyani, India
angsumansarkar@gmail.com

Abstract— **This paper reports the effect of AlGaN back barrier on the performance of lattice matched $In_{0.17}Al_{0.83}N$/AlN/GaN Recess Gate E HEMT Device. The use of AlGaN back barrier on this device relaxes the GaN channel, which in turn limits the SCEs. Moreover reduced the leakage current through gate (Ig) and simultaneously improves carrier confinement and off state breakdown voltage. The numerical modeling are carried out with the help of 2D Sentaurus TCAD simulator using Hydrodynamic model, which is standardized with respect to already published fabricated results. Different performance parameters are studied using the simulations and a wide comparison was done with and without considering AlGaN back barrier (BB). Addition of AlGaN BB has added benefits in performance parameters w.r.t without BB i.e. threshold voltage raised to 0.93 volt with respect to 0.75 volt, drop in DIBL from 100mv/V to 36mv/V and substantial reduction in gate leakage current. These results reveal that use of AlGaN BB in such devices can be an alternative solution for high power and high frequency switching applications.**

Keywords— InAlN/GaN Heterostructure; AlGaN Back barrier; Gate-Recess; E-Mode HEMT; SCE.

I. INTRODUCTION

In current years Nitride based HEMTs have emerged as a potential candidate for high-speed and high-power applications and more ever the uniqueness of $In_{0.17}Al_{0.83}N$/GaN proposed by Kuzmik has recently became an attractive alternative w.r.t usually employed AlGaN/GaN HEMTs [1-6]. The InAlN/GaN has numerous structural benefits over AlGaN/GaN system i.e. lattice matched with 17% of indium in $In_{0.17}Al_{0.83}N$ can be grown and the resulting structure can less vulnerable to relaxation related degradation and is excepted to have better stability and reliability [7,8]. The huge polarization break existing between the interface of InAlN and GaN resulted in better carrier confinement and induces higher 2DEG in the channel which overcomes the limitations of SCEs [1] and resulted in an improvement of microwave performances by

scaling down the gate length and InAlN barrier thickness [9,10]. Recent updated performance of InAlN/GaN has reported that a current density of 2.5 A/mm for Lg of 100 nm can be achieve with 6.9-nm barrier thickness [8]. The recessed gate normally off InAlN/GaN HEMT structure proposed by R.Wang reported the transconductance (g_m) in the order of 800ms/mm can be achieved [11]. The InAlN/GaN structure can be exploited to have f_T of 205 GHz with a gate length of 55nm as proposed by Sun et al [4] and also provide good thermal stability, delivering very high power and operating frequency over a wide range of temperature up to 1000^0C in harsh environment [7]. Very high f_T of 210 GHz is obtained by using dielectric free passivation process (DFP), which is reported by R.Wang et al. [11]. In spite of these competitive results, SCEs effects can't be overlooked as gate lengths continued to scale down to achieve the high frequency performance. To overcome these issues of scaling without altering the performance, the introduction of back barriers can be an ideal alternative. In latest years few groups have studied the effect of BB layer to improve the characteristics of AlGaN/GaN and InAlN/GaN structures [12-19]. The effect AlGaN back barrier structures have successfully deployed by many groups [14, 17, 18], which mainly focus on the normally ON operation. Henceforth, it is important to considered improved enhanced mode operation to have E/D IC applications as well as for digital switching. To the best of our information the p GaN back barrier suggested by Adak et.al in Gate-Recessed E-mode HEMT device [18] can be a better substitute for improved E/D circuit applications with (f_T) 123GHz and threshold voltage of 0.4 volt. In the proposed work for the first time we have implemented the AlGaN BB in Gate-Recessed E-mode HEMT device which can be a substitution to p-GaN BB considering same Lg. The simulation result reflects that with AlGaN back barrier the, V_{th} increased to 0.93volt and improved gate leakage current as compare to without back barrier. Further TCAD simulations are carried out to discover the performance of

978-1-5386-6723-1/19 $31.00 © 2019 IEEE   156

the proposed device w.r.t transconductance, Ig, DIBL, subthreshold slope (SS), I_{on}/I_{off} ratio respectively. Simulation model is calibrated by comparing with previously published result [11]. Therefore, the proposed device considered to be a better substitute for E/D circuit applications.

II. DEVICE DESCRIPTION

Fig. 1 and Fig. 2 shows the cross sectional view of InAlN/AlN/GaN Recess Gate E HEMTs; thicknesses of InAlN and AlN are 4.8nm and 1nm respectively. Former one having GaN buffer while the later one is having 0.4μm AlGaN back barrier (t_{bb}) with 30nm UID GaN channel (T_c). The said devices are having same L_g 150 nm, insulated with 140nm and 200nm SiN prior to and after gate definition respectively. The gate prepared of Pt/Au material and W_g of 2×50μm and the recess to Aluminium nitride spacer layer surface. Source and drain made of Titanium material and R_c of 0.6 Ω-mm incorporated in numerical modeling [11]. The doping concentrations of source/drain regions are 5E20 cm^3. GaN having thinner bandgap layer underneath the InAlN wide bandgap front barrier layer bounds channel material at hetero interface. InAlN front barrier offers a heavy carrier confinement in the QW at hetero boundary and insertion of Aluminium nitride spacer increases 2DEG and mobility [20, 21]. In Fig. 3 we have standardized our recommended device model with previously published fabricated result having L_g of 150nm [11].

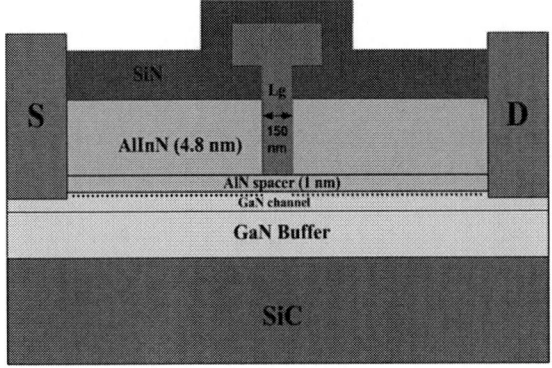

Fig. 1 Structure of InAlN/AlN/GaN E HEMT device without AlGaN BB.

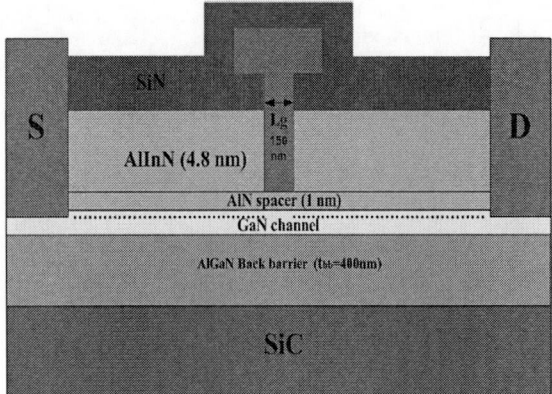

Fig. 2 Structure of InAlN/AlN/GaN E HEMT structure with AlGaN BB.

III. NUMERICAL MODEL CALIBRATION AND EXPERIMENTAL COMPARISON

The physical properties of GaN and $In_{0.17}Al_{0.83}N$ are listed in Table 1. 2D numerical simulations are done by using Synopsys TCAD [22]. The simulations are completed using the Hydrodynamic (HD) mobility model. Other important physical effects are taken into considerations for example band gap narrowing, variable effective mass, doping dependent mobility at high electric fields and spontaneous polarizations respectively.

The numerically simulated transfer characteristic (Fig.3) of the device (Fig.1) evident great matching with the given fabricated result, hence authenticating our carrier transport model as well as other typical parameters. As the model is authenticated, then extensive numerical modeling of InAlN/AlN/GaN Recess Gate E HEMT device with AlGaN BB have been performed.

Table1. Physical Properties of $In_{0.17}Al_{0.83}N$ and GaN [19]

Material	GaN	$In_{0.17}Al_{0.83}N$
Eg (eV)	3.4	4.70
CBO (eV)	0.31	-
VBO (eV)	0.39	-
ε0	9.5	11.7
Lattice Constant (A)	3.186	3.190
μ_e (cm^2/Vs)	1160	1540
μ_h (cm^2/Vs)	22	82

2019 Devices for Integrated Circuit (DevIC), 23-24 March, 2019, Kalyani, India

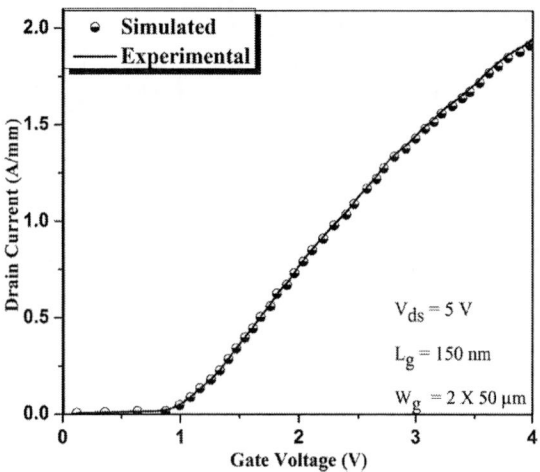

Fig 3. Fabricated (solid line) and numerically modeled (symbol) I_d vs V_g for InAlN/AlN/GaN E HEMT without AlGaN BB after regulating the simulation model parameters, presenting a great matching with fabricated [11] and numerically modeled data.

IV. RESULTS AND DISCUSSION

Fig. 4 shows the I_d vs V_g characteristics of the InAlN/AlN/GaN Recess Gate E HEMT (Fig. 2) having 400 nm thickness AlGaN BB. The maximum drain current is found to be 1.62 A/μm at V_{ds}= 5V which is less than 1.9 A/μm resulted from the device by considering without back barrier (Fig 1) and it is also evident from the experimental result [11]. The device with the AlGaN back barrier shows a reduction in drain current due to the decrease in charge density at the interface of AlInN and GaN and this results in increase in the threshold voltage from 0.75V for device without BB to 0.93 V with AlGaN BB. Greater positive V_{th} is greatly acceptable for power switching [23]. Further reduction in the Lg of the discussed device can be an attractive feature for E/D mode circuit applications due to improved characteristics [24].

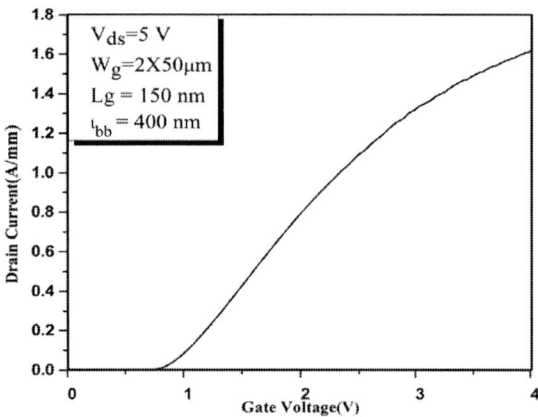

Fig 4. I_d vs V_g characteristics curve of InAlN/AlN/GaN Recess Gate E HEMT with AlGaN BB.

Fig. 5 represents the variation of g_m with V_{gs} for the device with and without AlGaN back barrier. As a result of the greater access resistance and sheet resistance, the device having AlGaN BB demonstrate the transconductance of 761 mS/mm, reduced by 5% as compare to without BB (800 mS/mm).

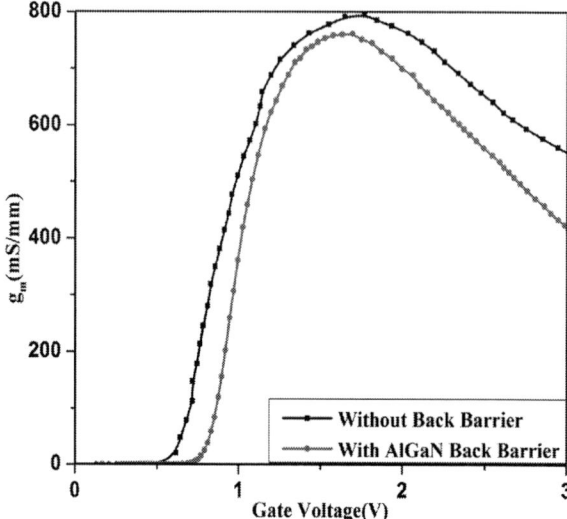

Fig 5. g_m vs V_{gs} for InAlN/AlN/GaN Recess Gate E HEMT devices with and without AlGaN BB.

To enhancement the performance of the Gallium nitride based HEMTs we have to decrease the leakage current through gate. Gate leakage current is determined by three components i.e. TE, PF and FN tunneling. The Poole-Frenkel (PF) and Fowler Nordheim (FN) tunneling are considered at room temperature and reverse bias condition [25]. Since simulations are done considering room temperature and reverse bias condition, hence we have taken PF and FN component of gate leakage current. After adding of AlGaN BB in InAlN/AlN/GaN Gate-Recessed E-Mode device the Ig leakage current is substantially decreased by more than 10 times as compare to the without BB as indicated in Fig 6. Reduced gate leakage current (Ig) is further beneficial to get higher I_{on}/I_{off} ratio, smaller standby power dissipation and higher breakdown voltage for the AlGaN BB device.

978-1-5386-6723-1/19 $31.00 © 2019 IEEE   158

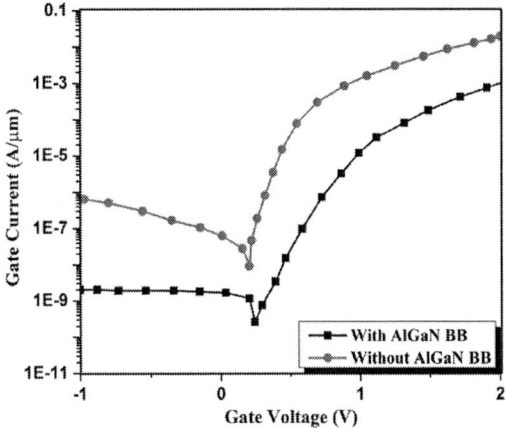

Fig6. Ig current of $Al_{0.83}In_{0.17}N/AlN/GaN$ Recess Gate E HEMT with and without AlGaN BB, having L_g = 150nm, W_g= 2×50 μm, t_{bb}= 400 nm.

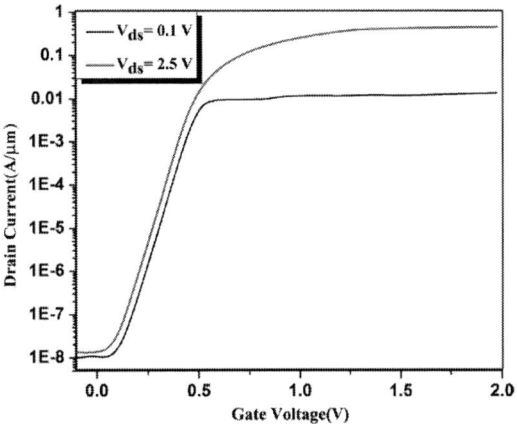

Fig 7. Subthreshold characteristics of an AlInN/AlN/GaN Gate-Recessed Enhancement- Mode HEMT with AlGaN back barrier layer, (DIBL=36 mV/V, I_{on}/I_{off} ratio>$3x10^7$ and SS =64.475 mV/dec).

From Fig. 7 we can easily measure the impact of AlGaN back barrier on different type of short channel effects(SCE) parameters such as drain induced barrier lowering (DIBL), subthreshold slope (SS) and I_{on}/I_{off} ratio of AlInN/AlN/GaN Gate-Recessed E -Mode HEMT device. It is desirable to have a steep subthreshold slope for switching off the transistor rapidly and low DIBL for electrostatic integrity of HEMTs. From the transfer characteristics of the AlInN/AlN/GaN Gate-Recessed Enhancement-Mode HEMT device with AlGaN back barrier layer at V_{ds} = 0.1 and V_{ds} = 2.5 V the DIBL is measured, which is found to be 36mV/V. The obtained value is considerably lower as compared to 44 mV/V in case p-GaN back barrier and 84mV/V without back barrier respectively [19,10]. Moreover SS of 64.475 mV/decade is observed as compared to the 100 mV/decade stated for the device without back barrier [10]. High I_{on}/I_{off} ratio greater than $3x10^7$ is obtained for the proposed device due to the lower gate leakage

current. Therefore, the above decreased SCEs indicate an outstanding gate control device.

V. CONCLUSION

In this paper we have studied the advantages of AlGaN BB on the characteristics of InAlN/AlN/GaN Recess Gate E HEMT. This BB numerical modeling shows in a large reduction of gate leakage current, lowering the DIBL value, significant reduction in SS and better I_{on}/I_{off} ratio as compared to the previously published results for the device without back barrier. These enhanced performances are very attractive for high power switching application as well as E/D mode circuits. Moreover, the optimization of this BB and gate length will be left for future work.

ACKNOWLEDGMENT

S. Adak and S.K. Swain would like to express their thanks to Brainware University, Barasat, Kolkata 700125 and Silicon Institute of Technology, Bhubaneswar 751024 respectively for their valuable support and cooperation for carrying out this research work. Authors would also like to thank Prof. C. K. Sarkar of Jadavpur University, Kolkata, for his valuable technical support.

REFERENCES

[1] Kuzmík, J., 2001. Power electronics on InAlN/(In) GaN: Prospect for a record performance. IEEE Electron Device Letters, 22(11), pp.510-512.

[2] Medjdoub, F., Carlin, J.F., Gonschorek, M., Feltin, E., Py, M.A., Ducatteau, D., Gaquiere, C., Grandjean, N. and Kohn, E., 2006, December. Can InAlN/GaN be an alternative to high power/high temperature AlGaN/GaN devices?. In Electron Devices Meeting, 2006. IEDM'06. International (pp. 1-4).

[3] Kuzmík, J., 2002. InAlN/(In) GaN high electron mobility transistors: some aspects of the quantum well heterostructure proposal. Semiconductor Science and Technology, 17(6), p.540.

[4] Sun, H., Alt, A.R., Benedickter, H., Feltin, E., Carlin, J.F., Gonschorek, M., Grandjean, N.R. and Bolognesi, C.R., 2010. 205-GHz (Al, In) N/GaN HEMTs. IEEE Electron Device Letters, 31(9), pp.957-959.

[5] Adak, S., Swain, S.K., Rahaman, H. and Sarkar, C.K., 2016. Impact of gate engineering in enhancement mode n++ GaN/InAlN/AlN/GaN HEMTs. Superlattices and Microstructures, 100, pp.306-314.

[6] Adak, S., Swain, S.K., Raj, G., Rahaman, H. and Sarkar, C.K., 2016, March. Performance analysis of gate material engineering in enhancement mode n++ GaN/InAlN/AlN/GaN HEMTs. In Devices, Circuits and Systems (ICDCS), 2016 3rd International Conference on (pp. 89-92). IEEE.

[7] Kuzmik, J., Pozzovivo, G., Ostermaier, C., Strasser, G., Pogany, D., Gornik, E., Carlin, J.F., Gonschorek, M., Feltin, E. and Grandjean, N., 2009. Analysis of degradation mechanisms in lattice-matched InAlN/GaN high-electron-mobility transistors. Journal of Applied Physics, 106(12), p.124503.

[8] Wang, H., Chung, J.W., Gao, X., Guo, S. and Palacios, T., 2010. Al2O3 passivated InAlN/GaN HEMTs on SiC substrate with record current density and transconductance. physica status solidi c, 7(10), pp.2440-2444.

[9] Lee, D.S., Chung, J.W., Wang, H., Gao, X., Guo, S., Fay, P. and Palacios, T., 2011. 245-GHz InAlN/GaN HEMTs with oxygen plasma treatment. IEEE Electron Device Letters, 32(6), pp.755-757.

[10] Pardeshi, H., Raj, G., Pati, S., Mohankumar, N. and Sarkar, C.K., 2013. Influence of barrier thickness on AlInN/GaN underlap DG

MOSFET device performance. Superlattices and Microstructures, 60, pp.47-59.

[11] Wang, R., Saunier, P., Xing, X., Lian, C., Gao, X., Guo, S., Snider, G., Fay, P., Jena, D. and Xing, H., 2010. Gate-recessed enhancement-mode InAlN/AlN/GaN HEMTs with 1.9-A/mm drain current density and 800-mS/mm transconductance. IEEE Electron Device Letters, 31(12), pp.1383-1385.

[12] Rennesson, S., Damilano, B., Vennegues, P., Chenot, S. and Cordier, Y., 2013. AlGaN/GaN HEMTs with an InGaN back - barrier grown by ammonia - assisted molecular beam epitaxy. physica status solidi (a), 210(3), pp.480-483.

[13] Dickerson, J.R., Ravindran, V., Moudakir, T., Gautier, S., Voss, P.L. and Ougazzaden, A., 2012. A study of BGaN back-barriers for AlGaN/GaN HEMTs. The European Physical Journal-Applied Physics, 60(3).

[14] Lee, D.S., Gao, X., Guo, S. and Palacios, T., 2011. InAlN/GaN HEMTs with AlGaN back barriers. IEEE Electron Device Letters, 32(5), pp.617-619.

[15] Lee, D.S., Gao, X., Guo, S., Kopp, D., Fay, P. and Palacios, T., 2011. 300-GHz InAlN/GaN HEMTs with InGaN back barrier. IEEE Electron Device Letters, 32(11), pp.1525-1527.

[16] Swain, S.K., Adak, S., Pati, S.K. and Sarkar, C.K., 2016. Impact of InGaN back barrier layer on performance of AlInN/AlN/GaN MOS-HEMTs. Superlattices and Microstructures, 97, pp.258-267.

[17] Lee, H.S., Piedra, D., Sun, M., Gao, X., Guo, S. and Palacios, T., 2012. 3000-V 4.3-mΩ • cm2 InAlN/GaN MOSHEMTs With AlGaN Back Barrier. IEEE Electron Device Letters, 33(7), pp.982-984.

[18] Bo, L., Zhihong, F., Shaobo, D., Xiongwen, Z., Guodong, G., Yuangang, W., Peng, X., Zezhao, H. and Shujun, C., 2013. An extrinsic fmax> 100 GHz InAlN/GaN HEMT with AlGaN back barrier. Journal of Semiconductors, 34(4), p.044006.

[19] Adak, S., Sarkar, A., Swain, S., Pardeshi, H., Pati, S.K. and Sarkar, C.K., 2014. High performance AlInN/AlN/GaN p-GaN back barrier gate-recessed enhancement-mode HEMT. Superlattices and Microstructures, 75, pp.347-357.

[20] Gonschorek, M., Carlin, J.F., Feltin, E., Py, M.A. and Grandjean, N., 2006. High electron mobility lattice-matched AlInN/GaN field-effect transistor heterostructure. Applied physics letters, 89(6), p.062106.

[21] Xie, J., Ni, X., Wu, M., Leach, J.H., Özgür, Ü. and Morkoç, H., 2007. High electron mobility in nearly lattice-matched AlInN/AlN/GaN heterostructure field effect transistors. Applied Physics Letters, 91(13), p.132116.

[22] Synopsys. TCAD Sentaurus device user's manual VG-2012.06.

[23] Kachi, T., 2007, October. GaN power device for automotive applications. In Proc. IEEE Compound Semicond. IC Symp. Tech. Dig 2007 (pp. 13–16). IEEE.

[24] Schuette, M.L., Ketterson, A., Song, B., Beam, E., Chou, T.M., Pilla, M., Tserng, H.Q., Gao, X., Guo, S., Fay, P.J. and Xing, H.G., 2013. Gate-recessed integrated E/D GaN HEMT technology with f T/f max> 300 GHz. IEEE Electron Device Letters, 34(6), pp.741-743.

[25] Turuvekere, S., Karumuri, N., Rahman, A.A., Bhattacharya, A., DasGupta, A. and DasGupta, N., 2013. Gate leakage mechanisms in AlGaN/GaN and AlInN/GaN HEMTs: comparison and modeling. IEEE Transactions on Electron Devices, 60(10), pp.3157-3165.

A SYSTEMIC REVIEW ON THE TECHNOLOGICAL DEVELOPMENT IN THE FIELD OF PHONOCARDIOGRAM

Subhashis Maitra
Department of Electronics and Communication
Kalyani Government Engineering College
Kalyani, Nadia, West Bengal, India
maitrasubhashis3@gmail.com

Deepneha Dutta
Department of Medical Electronics
Central Calcutta Polytechnic
Kolkata, West Bengal, India
deepneha_90@yahoo.com

Abstract- Medical advancement in the field of computation and technology has made the process of continuous surveillance of an individual's health much easier and pocket friendly. Three of the well approved tests by medical personnel to assess the cardiac health of an individual are the Holter Monitoring, Phonocardiogram and the 12 lead Electrocardiograph. In urban India, due to busy lifestyle, people suffer from lack of time, whereas, in rural India there is a huge lack of awareness regarding health and also there are a very few number of health centers to guide the rural people with proper medical assistance. This has lead to an unexpected rise of cardiovascular diseases, leading to death. This paper provides a comprehensive review on the various articles published, concentrating on the key factors revolving around the tele-monitoring of an individual's cardiac condition by designing a cardiac abnormality detector using the electronic stethoscope.

Keywords- Holter Monitoring, Phonocardiogram, Electrocardiograph, tele-monitoring, electronic stethoscope.

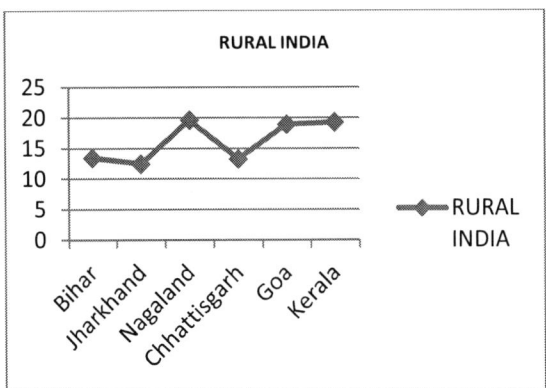

Fig.1. a. Percentage scale vs. state wise evaluation corresponding to average 10 year risk of cardiovascular disease in rural areas.

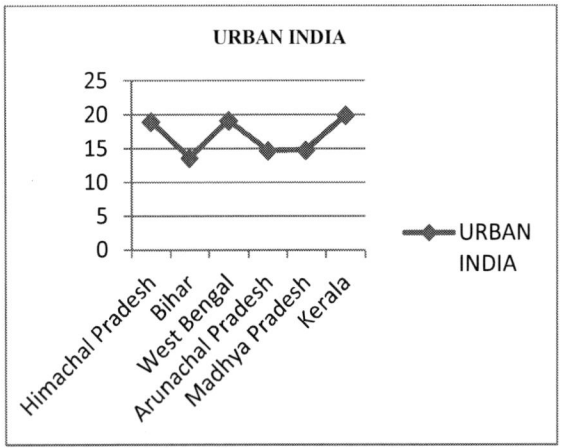

Fig.1.b. Percentage scale vs. state wise evaluation corresponding to average 10 year risk of cardiovascular disease in urban areas.

I. INTRODUCTION

Cardiac diseases are the main cause of fatality in maximum countries of the world. According to the Global Burden of Disease study, in India, death rate due to heart disease is 272 per 100000 population, which is way higher than the global average of 235 per 100000 population. According to study, in 1990, cardiovascular diseases saw a mortality rate of 23.2million in India, which took an abrupt surge of 59%, leading to 37 million in 2010 and has taken an upward slope thereafter. Three in every ten deaths recorded is due to cardiac health problems. The Valvular Heart Disease (VHD) is one of the typical heart diseases as recorded. This disease is caused due to deposition on the heart valves, leading to unsynchronized movement of the valves. So, diagnosis of the cardiac diseases plays a vital role in increasing the longevity of one's life. One of the most trusted tools for heart disease detection is the Electrocardiograph (E.C.G). This is a non-invasive, easy to use device. But the main drawback of this device is that it lacks the ability to detect any abnormality or defect in the structure or functioning of the natural as well as artificial heart valves of an individual. However, in primary healthcare centers, the physicians still rely on the auscultation process to detect the cardiac health problems, due to either absence of the E.C.G machines or lack of proper knowledge to operate those machines. On top of that, in rural India, the primary health centers are run by untrained and inexperienced physicians, who even lack the ability to perform auscultation method and detect any abnormality. As a result, there is an urgent need for a user- friendly, low cost machine, which can automatically detect the feeble heart sounds and conclude in an automated way. The main drawback lies on the sensing of the heart sound signals, due to its complex and non-stationary nature. However latest developments in the technological field has given much importance in this field. The development of electronic stethoscope has mainly concentrated on the design of highly sensitive sensor, as well as proper processing and transmission of the obtained data to the nearest computer for further analysis. The analysis of frequency deviation in the various obtained wavelets helps in establishing a relation between the obtained heart sound signal and the related heart defects. Heart sounds are produced by the normal beating of the heart muscles and the flow of blood through the cardiac chambers. In a normal healthy adult, two main heart sounds can be heard.

The first heart sound (HS1) is produced during the closure of the bicuspid and tricuspid valves, while the second heart sound (HS2) is heard during the closure of the semi lunar valves. In case of cardiac abnormality, the third (HS3) and the fourth (HS4) heart sounds can be heard in between HS1 and HS2, along with murmurs. HS3 is heard when there is a sudden drop in flow of blood, while moving from left auricle to left ventricle. This sound is pretty normal in children and in adults, residing in the age group till 40. After the age of 40, the fourth heart sound correlates to excessive volume of blood in the ventricles. The fourth heart sound is heard due to the vibrations caused by the heart valves, their

surrounding structures and the walls of the ventricles and this sound is also considered as one of the symptoms of heart failure. The frequency of HS1 is lower than that of HS2, but the duration of HS1 is longer than that of HS2. The HS3 occurs from 0.1 to 0.2 seconds after HS2, while HS4 occurs from 0.07 to 0.1 s before HS1. In addition to these heart sounds, heart murmurs can be heard, which are characterized according to their time of occurrence, such as, systolic, diastolic, continuous. In case of mitral stenosis and tricuspid stenosis, the closing of these valves produces a high- pitched snap noise, called diastolic sound. Another abnormal heart sound, termed as the ejection sound or the systolic sound can be heard when there is sudden stoppage of the semi lunar valves during its functioning in early period of systole. Similarly, there are different heart murmurs, which can be heard during abnormal functioning of the heart.

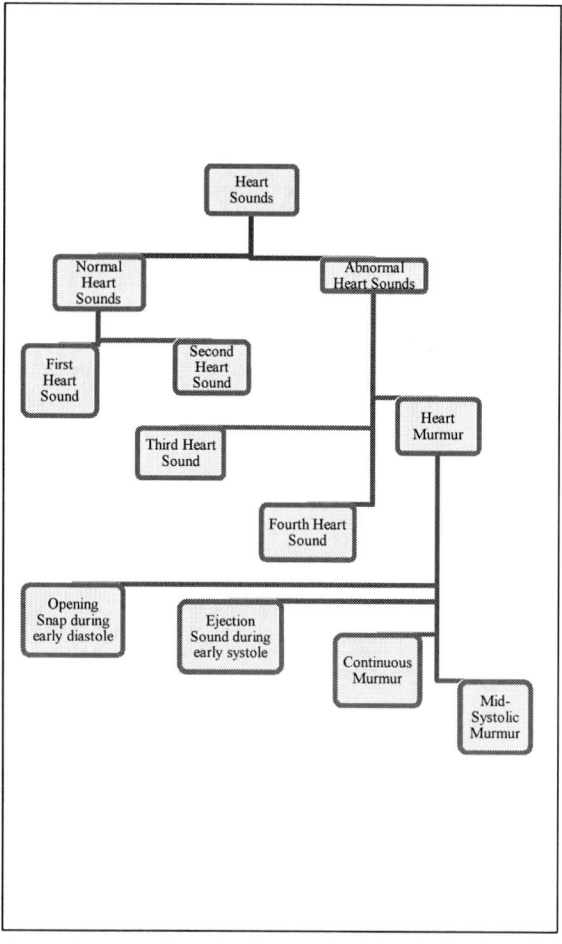

Fig.2. Chart showing the different heart sounds.

II. SYSTEM OVERVIEW

In the computer based cardiac abnormality detector, there are three main modules. Firstly, there is the signal sensing module, inclusive of the electronic stethoscope, next is the pre-processor module and thirdly lies the signal processing module. The signal acquisition module senses the desired heart sound signals with the help of a suitable electronic stethoscope, which is then transformed digitally and further sent to the pre-processing module for filtration and segmentation. The signal is filtered to overcome the sound interferences and thereby passed on to the signal processing module to extract the features and further classify them. At the end, the processed data helps the medical personnel to conclude the medical condition of an individual. There are many technical experts, who have worked in the field of phonocardiogram to detect valvular structure and their subsequent abnormalities. Some of the details of their work are being reviewed below.

The research done by Abdelghani Djebbari and Fethi Bereksi-Reguig, concluded by the development of an algorithm[2], which helped in the detection of any kind of split behavior in the aortic or pulmonary heart valves during the second heart sound.

The paper presented by Ignacio Foche-Perez and team suggested the development of an electronic stethoscope[3], performing in real time. The data collected by the stethoscope can be collected in two ways. It can be retrieved by a microcontroller and further processed to transfer the processed signal to the nearest computer and the received signal can also be heard by the ear piece after subsequent amplification.

The paper published by Yi-Li Tseng, Pin-Yu Ko` and Fu-Shan Jaw suggested the detection of the third and the fourth heart sounds[4]. These heart sounds are very feeble in nature and are mere heart murmurs. These murmurs are very difficult to detect and on top of that these sounds hold vital information about the cardiac well being of an individual. The application of a specific transform theory by Hilbert and Huang helped in the detection of the third and the fourth heart sounds.

Another paper on the acquisition of phonocardiography signal digitally by Mohamad Ali

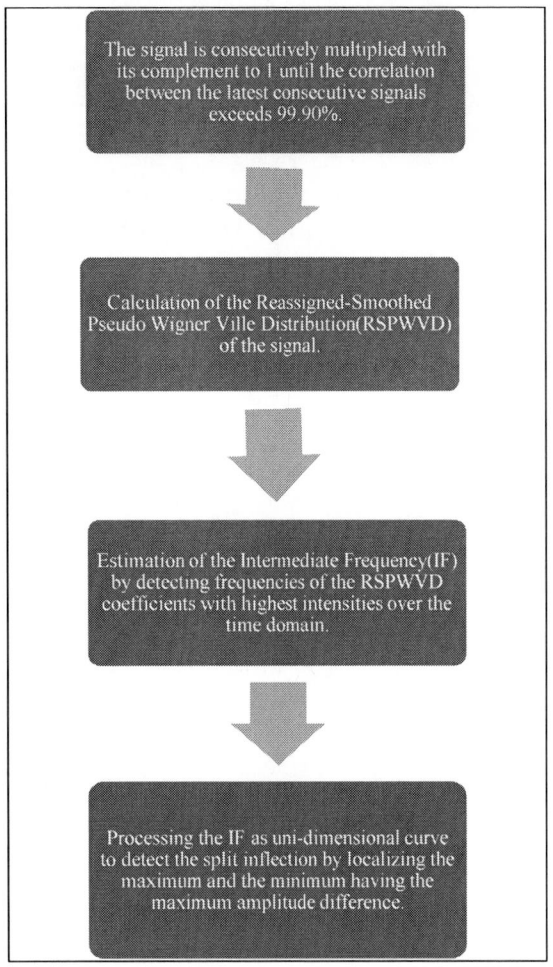

Fig.3. Algorithm for split behavior detection in aortic or pulmonary heart valves during second heart sound.

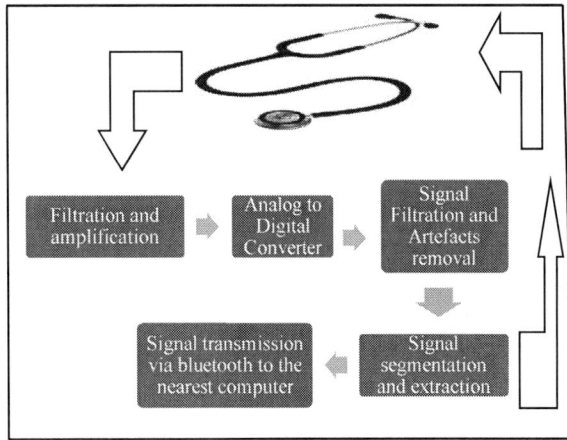

Fig.4. Real time signal transmission with the help of electronic stethoscope.

Akbari and team prompted the application of heart sound electrodes on the anterior region of the chest wall of a group of adults and infants. The received signals were transferred to the nearest computer by interfacing it through Universal Serial Bus. The obtained data is then retrieved using the Gold Wave software[5].

The work by Zifu Xiao and team concentrated mainly on the measurement and evaluation of the cardiac reserve [8] of an individual by calculating the change in amplitude of the first heart sound, while different workloads were applied on him. Similarly the change in value of the second heart sound can also be calculated and thus the change in ratio of heart sounds with that of change in work load can be calculated.

An article by T.A. Aliev and team presented a paper on analyzing the cardiac proceedings of an individual by the process of auscultation and thereby obtaining the heart sound data with the help of laptop or smart phone[9]. The obtained data is thereby compared to the reference ones and the cardiac condition is ascertained.

Another advanced research by Xuan Zhang, Louis-Gilles Durand and others suggested the retrieval of the noisy and feeble heart sound signals with the help of Matching Pursuit method[16].

IV. DISCUSSION

The research paper as discussed above suggests an algorithm for the detection of valvular split. But the real time implementation of this paper is of big concern. After the recording of an individual's heart sound, the recording needs to be further processed and analyzed to come to any kind of conclusion. This will require a lot of time consumption. Any kind of medical emergency cannot be dealt with immediately. As a result, our work mainly concentrates on real time recording, as well as analyzing the cardiac disorder of the patient by detecting the change in time-frequency response of the recorded phonocardiogram by comparing the pre-set and obtained data.

The second illustration is in a close approximation with my work. But the paper proposes transfer of the cardiac data from the chest piece to the remote computer via Bluetooth. This limits the transfer

distance to a maximum of 30 meter. So, our work suggests the transfer of the data via radio frequency signal. As a result, the device can be utilized at any rural area, where the basic medical camp is at a far off distance. This radio frequency data can be received by any mobile device and thereby the subsequent helpful suggestions can be received in no time.

The third one deals mainly with the detection of S3 and S4 heart sounds. Whereas, the importance of detection and analysis of S1 and S2 heart sounds is also a way too much. The S1 and S2 sounds detect the actual scenario of the patient's heart. Also, the paper suggests the receiving of the heart sounds and thereby further processing to be done in the laboratory. Our work deals with real-time receiving and further analysis of the heart sounds and simultaneously displaying the cardiac condition. Further this will help in giving equal importance to each and every heart sound of an individual.

The fourth one is an advanced analysis on heart murmurs. The paper mainly dealt with colour spectroscopic images of the different individuals and from the obtained image the signals with the maximum frequency were given importance and were denoted as S1 and S2. While the remaining were characterized as heart murmurs. The change in frequency response of the different hearts sounds were observed and analyzed. The main challenge with this work is it needs a computer for the utilization of the Gold wave software. This way, it becomes completely impossible for an individual to get an on-spot report of his cardiac condition as well as the idea of 24 hours surveillance of the individual is completely discarded. This situation is easily resolved by our research work, as the phonocardiograph machine is hand-held and user-friendly and gives the spot-on analytical report of any individual.

The fifth paper mainly deals with the change in cardiac contractibility of an individual with the application of varying loads. The amount of load to be applied is assigned at the beginning and applied accordingly. But this paper does not evaluate the sudden change in load, which can occur due to abrupt changes in daily life. Our research work mainly deals with the real-time recording and analysis of the different changes in heart conditions in the 24 hours time span of an individual. Sudden changes in cardiac conditions are therefore easily evaluated from the changes in time-frequency response of the significant heart sounds.

The sixth one deals with acquiring of the heart sound signals and transferring them to the computer for further analysis. The obtained data is thereby compared with the reference data and the change is analyzed. But the paper lacks real-time analysis and any deviation in cardiac condition from the stored data fails to be detected. Our work concentrates only on the acquired data and neglects any kind of reference signal.

The main objective of our work is the acquisition and processing of the heart sound signals, collected from any subject in real time and thereby transmission of the signals via R.F waves to any mobile device or F.M radio set for proper assessment of subject's heart condition. Our designed phonocardiogram is handy and equipped with low-cost components. The fact that the physiological heart signals generated are of very small amplitudes, filtering noise and generation of a noise free signal is a big challenge. There is also a need for data analysis to establish the accuracy of data, acquired by our device, when compared with standard instruments available. The selection of the range of audio signal transmission is also a challenge, as per selection of the telemetry procedure.

REFERENCES

1. Shuang Leng, Ru San Tan, Kevin Tshun Chuan Chai, Chao Wang, Dhanjoo Ghista and Liang Zhong,(2015) The electronic stethoscope, p. 1-37

2. Abdelghani Djebbari and Fethi Bereksi-Reguig, (2013) Detection of the valvular split within the second heart sound using the reassigned smoothed pseudo Wigner–Ville distribution, p. 1-21

3. Ignacio Foche-Perez, Rodolfo Ramirez-Payba, German Hirigoyen-Emparanza, Fernando Balducci-Gonzalez, Francisco-Javier Simo-Reigadas, Joaquin Seoane-Pascual, Jaime Corral-Peñafiel and Andres Martinez-Fernandez1, (2012) An open real-time tele-stethoscopy system, p. 1-17

4. Yi-Li Tseng, Pin-Yu Ko and Fu-Shan Jaw, (2012) Detection of the third and fourth heart sounds using Hilbert-Huang transform p. 1-13

5. Mohammad Ali Akbari, Kamran Hassani1, John D Doyle, Mahdi Navidbakhsh, Maryam Sangargir, Kourosh Bajelani and Zahra Sadat Ahmadi, (2011) Digital subtraction phonocardiography (DSP) applied to the detection and characterization of heart murmurs, p. 1-14

6. Reza Ramezani Sarbandi, John D Doyle, Mahdi Navidbakhsh, Kamran Hassani and Hassan Torabiyan,(2011) A color spectrographic phonocardiography (CSP) applied to the detection and characterization of heart murmurs: preliminary results, p. 1-10

7. Vladimir Kudriavtsev, Vladimir Polyshchuk and Douglas L Roy,(2007) Heart energy signature spectrogram for cardiovascular diagnosis, p. 1-22

8. Shouzhong Xiao, Xingming Guo, Xiaobo Sun and Zifu Xiao, (2002) A relative value method for measuring and evaluating cardiac reserve, p. 1-5

9. T.A. Aliev, N.E. Rzayeva, U.E. Sattarova, (2016) Robust correlation technology for online monitoring of changes in the state of the heart by means of laptops and smart phones, p. 1-8

10. Azra Saeidi, Farshad Almasganj, (2016)3D heart sound source localization via combinational subspace methods for long-term heart monitoring, p.1-10

11. S.P. Reddy, L.Y. Shyu, B.E. Hurwitz, J.H. Nagel, N. Schneidennan,(1988) Improved reliability of impedance cardiography by new signal processing techniques, In: IEEE Engineering in Medicine and Biology Society 10th Annual International Conference —0043, p. 1-2

12. Hayrettin KÖymen, Ahmet Baykal, Ziya I der, (1988) comparative time domain modeling of natural heart valve and mechanical heart valve sounds, In: IEEE Engineering in Medicine and Biology Society 10th Annual International Conference —0177, p. 1-2

13.Miguel A. Peiia C, RamCln Gonzhlez C, Tomb Aljaxna C, Salvador Carrasco S, Rocio Ortiz P, Vargas CG, Valencia G,(1995) Comparison of abdominal ECG and phonocardiography for instantaneous fetal heart rate detection, p. 1-2

14. M. Cozic, L.-G. Durand, R. Guardo, (1995) Development of a cardiac acoustic mapping system, p. 1-2

15. Sakari Lukkarinen, Pekka Korhonen, Anna Angerla, Annit-Leena Nloponen, Kari Sikio,Raimo Sepponen,(1996) Multimedia personal computer based phonocardiography, In: 18th Annual International Conference of the IEEE Engineering in Medicine and Biology Society, Amsterdam, p. 1-2

16. Xuan Zhang', Louis-Gilles Durand, Lotfi Senhadji, Howard C. Lee, Jean-Louis Coatrieux, (1996)Application of the matching pursuit method for the analysis and synthesis of the phonocardiogram, In: 18th Annual International Conference of the IEEE Engineering in Medicine and Biology Society, Amsterdam, p. 1-2

17. Hideaki Shino, Hisashi Yoshida, Kazuo Yana Kensuke Harada, Jiro Sudoh, and Eishi Harasawa, (1996)Detection and classification of systolic murmur for phonocardiogram screening, In: 18th Annual International Conference of the IEEE Engineering in Medicine and Biology Society, Amsterdam, p. 1-2

18. Xuan Zhang, Louis-Gilles Durand, Lotfi Senhadji, Howard C. Lee, and Jean-Louis Coatrieux,(1998) Analysis-synthesis of the phonocardiogram based on the matching pursuit method, In: IEEE transactions on Biomedical Engineering, vol. 45, no. 8, August 1998, p. 1-10

19. Xuan Zhang, Louis-Gilles Durand, Lotfi Senhadji, Member, Howard C. Lee, and Jean-Louis Coatrieux, (1998)Time-frequency scaling transformation of the phonocardiogram based of the matching pursuit method, In: IEEE transactions on Biomedical Engineering, vol. 45, no. 8, August 1998, p. 1-8

20. Zhenyu Guo, Chris Moulder, Louis-Gilles Durand and Murray Loew, (1998)Development of a virtual instrument for data acquisition and analysis of the phonocardiogram, In: Proceedings of the 20th Annual International Conference of the IEEE

Engineering in Medicine and Biology Society, vol. 20, no. I, 1998, p. 1-4

21. Dhanjoo N. Ghista, (1997) Noninvasive diagnosis of diseased heart wall and valve, based on their dynamics modeling, echo & phonocardiography, In: Proceedings - 19th International Conference - IEEE/EMBS Oct. 30 - Nov. 2, 1997 Chicago, IL. USA, p. 1-4

22. Herkole Sava and Louis-Gilles Durand,(1997) Automatic detection of cardiac cycle Based on an adaptive time-frequency analysis of the phonocardiogram, In: Proceedings - 19th International Conference - IEEE/EMBS Oct. 30 - Nov. 2, 1997 Chicago, IL. USA, p. 1-4

2019 Devices for Integrated Circuit (DevIC), 23-24 March, 2019, Kalyani, India

QCA Realization of Reversible Gates Using Layered T Logic Reduction Technique

Chiradeep Mukherjee [1, 2]
[1]Department of ECE
NIT Durgapur, West Bengal, India
[2]Department of ECE
UEM Jaipur, Rajasthan, India
chiradeep.mukherjee@uem.edu.in,
chiradeep.1234321@gmail.com

Saradindu Panda[2]
[2]Department of ECE
NIT-JIS Agarpara, Kolkata
India
saradindupanda@gmail.com

Bansibadan Maji [1]
[1]Department of ECE
NIT Durgapur, West Bengal, India
bmajiecenit@yahoo.co.in

Asish Kumar Mukhopadhyay[3]
[3]Academic Advisor
KEI, Kolkata
West Bengal, India
askm55@gmail.com

Abstract— **Quantum-dot cellular automata (QCA) becomes a promising model of computation as it possesses extreme-high packing density, ultra-high speed and low power dissipation for various nanoscale computing architectures. In this work, QCA based designs of Feynman, Toffoli, Fredkin and Peres gates are presented. These elementary gates are realized by utilizing layered T logic reduction technique. The QCA designs are evaluated in terms of QCA design metrics like the number of quantum cells, area, and delay. The analysis shows significant improvements over existing models in terms of QCA design metrics. As a result, the proposed layered T based QCA layouts of elementary reversible gates become an excellent candidate for developing multilevel reversible circuits.**

Keywords—Quantum-dot Cellular Automata, Reversible Computation, Layered T gate, Reversible gate

I. INTRODUCTION

The main feature of a processor-based computer system is information; it's storage, processing, and transfer. Conventional and reversible computations of information are definite ways to build a high-speed processor with cheap storage. The traditional design of a computer by using a micro-electronics device that associate information digitization has reached its saturation limit to validate Moore's law [1]. Here, the saturation indicates the size of minimum feature size, specifically to the continuous scaling of its length, breadth, and height.

Moreover, typical architecture design by using conventional AND, OR, NOT gate, implies the loss of kTln2 joules per bit, where k is Boltzmann constant, and T is operating temperature in Kelvin. At this pivotal point, arises the need for a reversible computer, where the information is preserved. Like conventional digital systems, standard and universal gates exist in reversible computation, known as a reversible gate. The reversible gates have a one-to-one relationship between inputs and outputs, popularly known as objective functions [2]. There are several ways of implementing reversible computers; it may be a flow of electrons or complex relationships of interconnecting wires. Quantum dot Cellular Automata (QCA) explores another exciting way to realize reversible gates to form all the elemental components of the reversible central processing unit (CPU) [3]. Many prominent developments of QCA in following fields force the researchers to explore the new possibilities of area-delay optimized reversible gates.

The QCA which becomes a prominent alternative amongst its other competitors like carbon nanotubes (CNT), resonant tunneling devices, single electron transistors (SET), etc. has ultra-small feature size, fast information processing speed, and high device density. Extreme small feature size is handled by quantum mechanical tunneling, hence the problems of CMOS technology do not arise here. There are four types of QCA technology: metal-dot QCA, magnetic QCA, semiconductor QCA, and atomistic QCA. The metal dot and semiconductor QCA are operable in cryogenic temperature, but magnetic QCA and atomistic QCA can function at room temperature [4]. In QCA, the arrangements of electrons within quantum dots expose the way information flows. To realize reversible gates by using QCA the existing works have serious drawbacks like non-optimal design layouts and delays. This work mainly emphases on scheming important reversible gates to bridge the gap amongst QCA technology and reversible computation. Feynman gate, Toffoli gate, Fredkin gate, and Peres gate are the examples of such reversible gate.

The authors divide their papers into four sections to help manufacturers and researchers to understand their work in a lucid way. These four sections are i) Introduction, which briefly explains the motivation of the research, ii) Overview of QCA, which provide a brief concepts and underlying mechanism of layer T gate (LT) of QCA, iii) Reversible gates and their LT logic realizations, which describes the LT logic implementations of standard reversible gates and iv) Result Discussion and Conclusion, which analyses the output and concludes the work.

II. OVERVIEW OF QCA

A. QCA cell, clock, and crossover

In QCA the information propagates in the circuit using coulombic interactions of electrons instead of traditional voltage or current flow. The basic unit of QCA is a square structured cell that has four quantum-dots located at its corners. This square cell is known as QCA cell. Two excess electrons in a QCA cell accommodate themselves at any two of four corners to maintain the furthest distance between them. The orientations of these two electrons in four corners of a QCA cell are shown in figure 1.

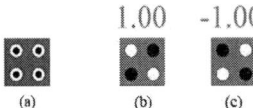

Fig. 1. (a) QCA Cell, QCA Cell with Polarization with (b) P=+1 and (c) P=-1

The orientation of electrons as mentioned in figure 1(a), denotes polarization, P=+1, whereas the orientation of figure 1(b) represents polarization, P=-1. These polarizations, i.e., P=+1 and -1 are equivalent to "logic"1 and "logic"0 of a standard digital circuit. Electrons can change their orientations

978-1-5386-6723-1/19 $31.00 © 2019 IEEE   167

(or locations) between the quantum dots of a QCA cell through quantum mechanical tunneling. However, electrons cannot tunnel between the dots of neighboring cells due to the presence of high potential barriers. An array-like arrangement in a row or columns forms elemental QCA devices, i.e., QCA logic gates and interconnecting wires. The primary logic available in QCA is majority voter and inverters. Figure 2(a) and (c) represent the QCA cell arrangements of majority gate and inverters. As presented in figure 2(a), majority voter takes three inputs, i.e., A, B, C, and makes the decision in favor of the majority of these three input logic and finally generates the output. The equation, (1) best describes the working of majority voter.

$$Z = AB + BC + CA \qquad (1)$$

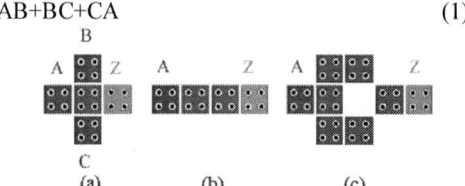

Fig. 2. (a) Majority Voter, (b) Binary Wire and (c) Conventional Inverter

Figure 2(c) represents QCA inverter that takes input A and generates $\overline{A}$ as output Z. The arrangement of cells to form QCA interconnecting wire is shown in figure 2(b). The QCA wire can be used either to carry intermediary inputs/outputs to the next level of the circuit or to provide the information to the QCA gates. The entire assemblage of QCA cells is sequentially groped by using QCA clocks. Four phases of QCA clocks are assigned to ensure proper information flow through the circuit. These phases are given as follows: switch, hold, release, relax, switch and so on. The information transfer from a specific point of the circuit to another point in QCA occurs by either coplanar or multilayer cross over. While the coplanar crossover used the regular and rotated QCA cells in one layer, the multilayer crossover use more than one layer of the cells. Though the multilayer cross over is difficult to fabricate, it is broadly used due to several reasons, as specified in [5].

B. Layered T logic Reduction Technique

Although the existing logic reduction techniques of QCA can be easily adapted to realize AND, OR and standard functions, they suffer from the disadvantage that universal NAND or NOR function cannot be achieved by using these gates. Additionally, these gates are not preferable in the optimal design of various CPU elements. The layered T (*LT*) gate provides the scope to build universal NAND/NOR based functions and definitely with better logic reduction capabilities in [4-7]. The logic schematic and QCA layout of *LT* gate are illustrated in figure 3. This is five-cell multilayer gate. From figure 3(a), it is seen that four cells are situated in layer 1 and the remaining one has placed 34.5 nm above layer one, i.e., a new layer, layer 2. Figure 3(b) shows layer 2 of *LT* gate. It is vital to note that layer-2 cell has been polarized with '+1' or '-1' polarization.

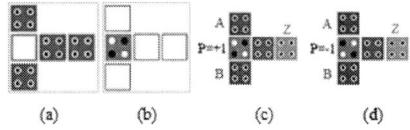

Fig. 3. (a) Arrangements of 4 cells in layer 1, (b) placements of 1 cell in layer 2, (c) *LT* gate operating as NAND gate and (c) *LT* gate operating as NOR gate

If the upper layer cell is fabricated with P=+1, then the output, Z of the gate becomes $\overline{AB}$. But the gate produces $\overline{A+B}$ when the same cell is fabricated with P=-1. The logic function of *LT* gate is given in equation (2) as follows:

$$L_T^+(A,B) = \overline{AB} \text{ when P=+1,}$$
$$L_T^-(A,B) = \overline{A+B} \text{ when P=-1} \qquad (2)$$

Figure 3(c) and (d) respectively show the QCA layouts of *LT* NAND and NOR gates. Apart from universal NAND and NOR functions, the *LT* gate followed by an inverter can produce AND and OR functions respectively. If two cells are diagonally placed with five cell structure (as demonstrated in figure 3), then the entire cell arrangement produces AND or OR function. If the polarization of the upper layer cell is fixed at P=+1, then it generates logic AND output. However, if the same is fixed at P=-1, then it operates as a logic OR function. Equation (3) demonstrate AND or OR functions.

$$L_{TA}^+(A,B) = AB \text{ when P=+1,}$$
$$L_{TO}^-(A,B) = A+B \text{ when P=-1} \qquad (3)$$

$L_{TA}^+()$ And $L_{TO}^-()$ represent logic AND and logic OR operation respectively. The QCA cell arrangements for *LT* AND and *LT* OR gate are shown in figure 4(a) and (b) respectively.

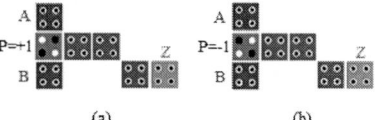

Fig. 4. LT gate operating as (a) AND gate and (b) OR gate

It is depicted in [5] that arranged with proper clocking cycle the four *LT* NAND gates produce exclusive OR (Ex-OR) operation. The Ex-OR function by using *LT* gate is represented by the abbreviation *LTEx()*. Like AND and OR functions, followed by two diagonally placed QCA cells the *LTEx* gate produce exclusive-NOR operation (EX-NOR). The Ex-NOR function by using the *LT* gate is represented by *LTExn()*. Equation (4) best describes *LTEx()* and *LTExn()*.

$$LTEx(A,B) = A \oplus B,$$
$$LTExn(A,B) = A \odot B \qquad (4)$$

The cell arrangements of *LTEx* gate and *LTExn* gate are shown respectively in figure 5(a) and (b).

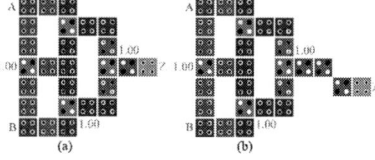

Fig. 5. Layered T gate operating as (a) Ex-OR gate, i.e., LTEx gate and (b) Ex-NOR gate, i.e., LTExn gate

The various symbols (or gate level diagram) of *LT* NAND, NOR, Ex-OR, Ex-NOR, AND, OR gates which are used in this work to design reversible gates, is demonstrated in figure 6(a)-(f).

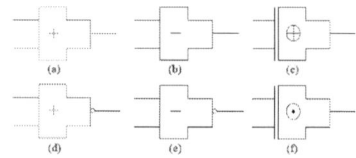

Fig. 6. Symbols of (a) *LT* NAND gate, (b) *LT* NOR gate, (c) *LT* Ex-OR gate, (d) *LT* AND gate, (e) *LT* OR gate and (f) *LT* Ex-NOR gate

978-1-5386-6723-1/19 $31.00 © 2019 IEEE

III. Reversible Gates and their *LT* Logic Realization

It can be noted from the previous sections that the *LT* gate can be constructed to develop the irreversible functions like NAND, NOR, AND, OR, Ex-OR and Ex-NOR operations. However, this work focuses on constructing well known reversible gates [8] by using "*LT logic reduction technique*" to develop a guideline to generate their area-delay efficient QCA counterparts. In this section, the QCA implementations of Feynman gate, Toffoli gate, Fredkin gate, and Peres gate are presented.

A. Feynman Gate

The prototypical reversible logic gate is Feynman gate. This gate constitutes two lines, known as the control line and target line, respectively. The gate-level diagram and QCA layout of Feynman gate are respectively shown in figure 7(a) and (b). If the control line is set to 0, then the target line forwards the input directly. If the control line is reset to 1, then the target line gets flipped. As the operation of Feynman gate resembles the classical Ex-OR gate, so the gate may be summarized by using the equation (5).

$$P = A, Q = A \oplus B \qquad (5)$$

The LT format for the equation (6) can be obtained as:

$$P = A, Q = LTEx(A, B) \qquad (6)$$

The *LT* schematic of Feynman gate, as mentioned in figure 7(b) requires one *LTEx* module to generate proper output. This *LTEx* module performs basic Ex-OR operation. The QCA layout of *LT* realization of Feynman gate consumes 30 quantum cells and consumes 29346 nm² area. It merely takes three clock zones to process the output, making the delay as 0.75.

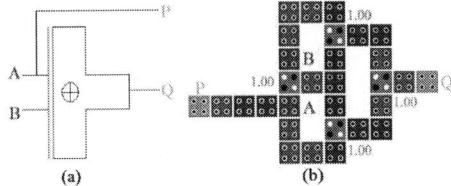

Fig. 7. (a) Gate-level logic diagram and (b) QCA layout of LT Feynman Gate

B. Toffoli Gate

Figure 8(a) and (b) respectively show *LT* schematic and QCA layout of the Toffoli gate. Like Feynman gate, it satisfies bijective property between inputs, i.e., A, B and C, and outputs P, Q and R. The first two outputs, P and Q copy two inputs, A and B respectively and the remaining output, R controls the inversion of input signal C by utilizing A and B. The equation (7) justifies the expression of Toffoli gate as follows:

$$P = A, Q = B, R = AB \oplus C \qquad (7)$$

The *LT* format of equation (6) is given as follows:

$$P = A, Q = B, R = LTEx(L_{TA}^+(A, B), C) \qquad (8)$$

The QCA layout of Fredkin gate needs 37 quantum cells, consumes 60358 nm² area to evaluate the output after 1.25 clock cycle.

C. Fredkin Gate

The 3*3 Fredkin gate has three inputs, A, B, C, and outputs, P, Q, R, as shown in figure 9. The control line A, i.e.,

input A is mapped directly to the target line P. If A = 0, then no swap is performed and other control lines, i.e., B and C map to Q and R respectively.

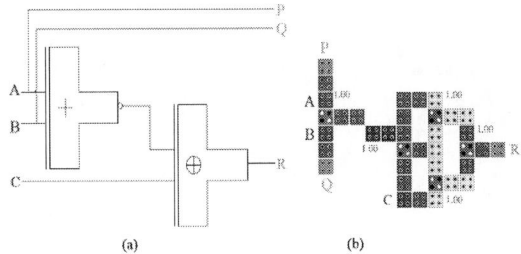

Fig. 8. (a) Gate-level logic diagram and (b) QCA layout of LT Toffoli Gate

Otherwise, if A=1, then the logic at two target lines are swapped, i.e., Q and R are swapped in such a manner that Q maps to C and R maps to B. It is easy to see that this circuit is reversible, i.e., "undoes" itself when running backward.

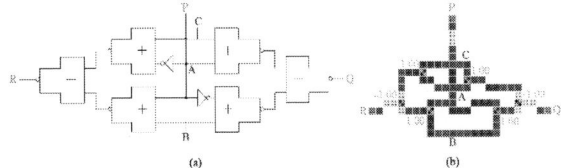

Fig. 9. (a) Gate-level logic diagram and (b) QCA layout of LT Fredkin Gate

The characteristic equation and its *LT* counterpart are given respectively in equation (9) and (10) as follows:

$$P = A, Q = \overline{A}B + AC, R = AB + \overline{A}C \qquad (9)$$
$$P = A, Q = L_{TO}^-\left(L_{TA}^+(\overline{A}, B), L_{TA}^+(A\,C)\right), R = L_{TO}^-\left(L_{TA}^+(A, B), L_{TA}^+(\overline{A}, C)\right) \qquad (10)$$

The QCA layout of Fredkin gate needs 84 quantum cells, consumes 158364 nm² area to evaluate the output after 1.25 clock cycle.

D. Peres Gate

The 3*3 reversible circuit with inputs, A, B, C, and outputs P, Q, R is shown in figure 10. One of three target lines, i.e., output Q follows B when control line, A becomes O, otherwise, Q=B when A becomes 1. Such a functionality leads toward exclusive-OR operation between two inputs A and B to evaluate Q. The first target line, P amongst three outputs directly follows control line, A., On the other hand, the target line, R is entirely determined by the product, AB. If AB becomes O, then R follows the control line C, otherwise. R when AB=1, produces C.

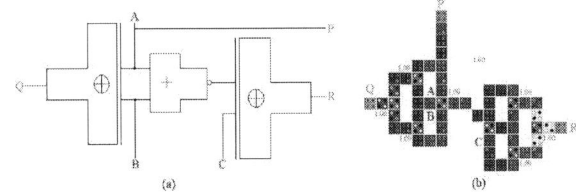

Fig. 10. (a) Gate-level logic diagram and (b) QCA layout of LT Peres Gate

To configure R regarding AB and C, one exclusive OR function between AB and C is highly needed. The characteristic equation and its layered T counterpart are given in equation (11) and (12) respectively.

$$P = A, Q = A \oplus B, R = AB \oplus C \quad (11)$$

$$P = A, Q = LTEx(A, B), R = LTEx(L_{TA}^+(A, B), C) \quad (12)$$

The equations (11) and (12) strongly demand one *LT* AND gate and one *LTEx* gate to couple four literals (A, B) and (AB, C) respectively. The *LT* schematic and QCA layout of Peres gate are respectively shown in figure 10(a) and (b). The QCA layout of Peres gate needs 59 quantum cells, consumes 87204 nm^2 area to evaluate the output after 0.75 clock cycle.

E. Outputs of proposed gates

The QCA layouts of proposed *LT* reversible gates, as mentioned in figure 7-10, are simulated in the renowned computer-aided design tool, QCADesigner 2.0.3. The bistable approximation engine of QCADesigner is used to verify the operability of the QCA layouts. Figure 11(a), (b), (c) and (d) demonstrate the outputs of *LT* Feynman gate, *LT* Toffoli gate, *LT* Fredkin gate, and *LT* Peres gate respectively. The red colored boxes in figure 11(a)-(d) represent the valid outputs following the equations (5)-(12).

IV. RESULT DISCUSSION AND CONCLUSION

QCA devices are promising for nano-scale computation. In this work, the concept of *LT* gate is utilized to design, build and characterize the AND, OR, NAND, NOR, Ex-OR and Ex-NOR functions. The *LT* gate forms the "*LT*" logic reduction technique, i.e., a universal set of all elementary logic gates that can be used to realize irreversible and reversible functions.

As Feynman, Toffoli, Fredkin and Peres gates are used to develop multi-control reversible circuits, so this work extends

the idea of *LT* logic reduction technique to realize these reversible gates in QCA. The functionality of the proposed *LT* counterparts have been tested with the parameters specified in [5] and confirm their outputs in figure 11. The QCA design metrics like number of quantum cells, area in micrometer^2 (μm^2) and delay of *LT* layouts of figure 7-10 are compared to existing QCA-based reversible gates in Table I. It is observed that the proposed *LT* counterpart of Feynman gate (*LT* Feynman) requires 23.07% less quantum cells than the Feynman gate design mentioned in [21]. Although both the designs consume the same area, *LT* Feynman becomes 25% faster than Reshi Feynman design specified in [21]. In another case, Das Feynman design, as mentioned in [10], have 0.50 delay but it consumes 54 quantum cells and 0.038 μm^2 area. So *LT* Feynman design needs less cost than Das Feynman counterpart. *LT* Toffoli design requires ~35% fewer quantum cells than Abdullah-Al-Shafi design, as mentioned in [11]. Since both the models need the same area, the cost to realize *LT* Toffoli design is less than Abdullah-Al-Shafi Toffoli counterpart. *LT* Peres gate needs 32.91% fewer quantum cells than the existing Peres gate design of [22]. All the Peres gate, i.e., the current Peres designs and *LT* Peres gate consume 0.08 μm^2 area, but *LT* counterpart becomes 25% faster than the existing Peres gates, as specified in [21-22]. *LT* Fredkin gate requires ~13% fewer quantum cells and becomes ~28% faster than the Fredkin counterpart, as mentioned in [11]. It can be observed that the Fredkin design mention in [13] consumes lesser area than *LT* Fredkin design. On the contrary, *LT* Fredkin gate requires 16% fewer quantum cells and generates outputs 50% faster.

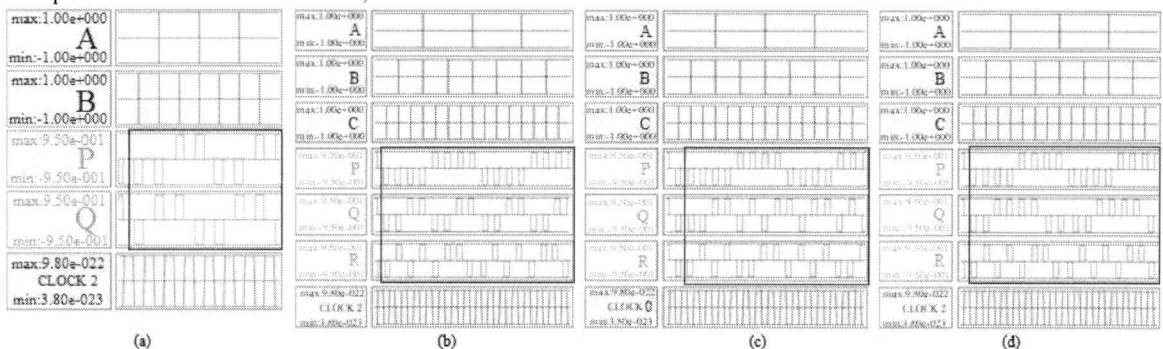

Fig. 11. Outputs of (a) *LT* Feynman gate, (b) *LT* Toffoli gate, (c) *LT* Fredkin gate, and (d) *LT* Peres gate

TABLE I DESIGN SUMMARY OF PROPOSED *LT* GATES WITH THE EXISTING DESIGNS [9-22]

Sl. No.	Gate	Proposed	#cell	Area in μm^2	Delay	Sl. No.	Gate	Proposed	#cell	Area in μm^2	Delay
1	Feynman	In [9]	75	0.08	1.25	2	Toffoli	In [14]	101	0.043	1.25
								In [11]	57	0.06	0.75
		In [10]	54	0.038	0.50			In [15]	48	0.067	1
		In [11]	53	0.07	0.75			In [16]	44	0.081	1.25
		In [12]	43	0.038	0.75			In [19]	108	0.14	1
		In [13]	34	0.036	0.75			In [20]	155	0.227	3
		In [19]	62	0.11	1			*LT* Toffoli	37	0.06	1.25
		In [20]	102	0.13	2.25	3	Peres	In [11]	99	0.08	1
		In [21]	39	0.029	1			In [22]	87	0.08	1
		LT Feynman	30	0.029	0.75			*LT* Peres	59	0.08	0.75
4	Fredkin	In [9]	187	0.19	2						
		In [11]	97	0.15	1.75						
		In [13]	100	0.092	3						
		LT Fredkin	84	0.15	1.25						

So it can be concluded that the proposed reversible gate layouts exploiting *LT* logic reduction technique can be used as basic building blocks for developing reversible benchmark functions.

ACKNOWLEDGEMENT

Authors thank anonymous reviewers for their insightful comments and suggestions. They express special thanks to Prof. Debdatta Banerjee for her literary contributions that help authors in organizing the article.

REFERENCES

[1] International Technology Roadmap for Semiconductor Report 2015 (ITRS Report 2015), Available at http://www.itrs2.net/itrs-reports.html (accessed on 12th January, 2019)

[2] C. P. Williams, Explorations in Quantum Computing, vol. XXII, Springer, 2011, ISBN: 978-1-84628-886-9

[3] C. S. Lent, P. D. Tougaw and W. Porod, "Quantum cellular automata: the physics of computing with arrays of quantum dot molecules," Proceedings Workshop on Physics and Computation. Phys. Comp. '94, Dallas, TX, USA, 1994, pp. 5-13, DOI: 10.1109/PHYCMP.1994.363705

[4] C. Mukherjee, S. Panda, A. K. Mukhopadhyay, B. Maji, "Synthesis of standard functions and generic Ex-OR module using layered T gate", International Journal of High Performance Systems Architecture, vol. 7, no. 2, September 2017, pp. 70-86, DOI: 10.1504/IJHPSA.2017.087164.

[5] C. Mukherjee, S. Panda, A. K. Mukhopadhyay, B. Maji, "QCA Gray Code Converter Circuits Using LTEx Methodology", International Journal of Theoretical Physics, vol. 57, no. 7, July 2018, pp. 2068-2092, DOI: https://doi.org/10.1007/s10773-018-3732-4

[6] C. Mukherjee, A. S. Sukla, S. S. Basu, R. Chakrabarty, A. Khan and D. De, "Layered T full adder using Quantum-dot Cellular Automata," *2015 IEEE International Conference on Electronics, Computing and Communication Technologies (CONECCT)*, Bangalore, 2015, pp. 1-6. DOI: 10.1109/CONECCT.2015.7383867

[7] C. Mukherjee, S. Panda, A. K. Mukhopadhyay, B. Maji, "Generic parity generators design using LTEx methodology: A quantum-dot cellular automata based approach", International Journal of Nano Dimensions, vol. 9, no. 3, Summer 2018, pp. 215-227

[8] M. A. Nielson and I. L. Chuang, Quantum Computation and Quantum Information, edition 10, Cambridge University Press, 2010, ISBN 978-1-107-00217-3

[9] P. Biswas, N. Gupta, N. Patidar, "Basic Reversible Logic Gates and It's QCA Implementation", International Journal of Engineering Research and Applications, vol. 4, no. 6, pp.12-16, 2014.

[10] J. C. Das and D. De, "Reversible Binary to Grey and Grey to Binary Code Converter using QCA" IETE Journal of Research, vol. 61, no. 3, pp. 223–229, May 2015, DOI: https://doi.org/10.1080/03772063.2015.1018845

[11] M. Abdullah-Al-Shafi, M. Shifatul, and A. N. Bahar, "A Review on Reversible Logic Gates and its QCA Implementation," International Journal of Computer Applications, vol. 128, no. 2, pp. 27–34, Oct. 2015

[12] J. C. Das and D. De, "Quantum-dot cellular automata based reversible low power parity generator and parity checker design for nanocommunication," Frontiers Inf Technol Electronic Eng, vol. 17, no. 3, pp. 224–236, Mar. 2016

[13] A. N. Bahar, S. Waheed, and M. A. Habib, "A novel presentation of reversible logic gate in Quantum-dot Cellular Automata (QCA)," in 2014 International Conference on Electrical Engineering and Information Communication Technology (ICEEICT), 2014, pp. 1–6.

[14] M. Rolih, "Analysis of possible logical reversible gate realization in ternary quantum-dot cellular automata," engd, Univerza v Ljubljani, 2013

[15] A. N. Bahar, M. A. Habib, and N. K. Biswas, "A Novel Presentation of Toffoli Gate in Quantum-dot Cellular Automata (QCA)," International Journal of Computer Applications, vol. 82, no. 10, pp. 1–4, Nov. 2013

[16] B. Cvetkovska, I. Kostadinovska, and J. Danek,"Implementing the Toffoli gate in Quantum-dot Cellular Automata", Seminar project at University of Ljubljana in the winter semester of the academic, (2013) 1-12.

[17] P. K. Biswas, A. N. Bahar, M. A. Habib, M. Abdullah-Al-Shafi, "Efficient Design of Feynman and Toffoli Gate in Quantum dot Cellular Automata (QCA) with Energy Dissipation Analysis", Nanoscience and Nanotechnology, vol. 7, no.2, pp.27-33, 2017, DOI: 10.5923/j.nn.20170702.01

[18] M. Sarvaghad-Moghaddam and A. A. Orouji, "A new design and simulation of reversible gates in quantum-dot cellular automata technology", March 2018, Accessed from arXiv:1803.11017 on 12th January 2018

[19] D. Kunalan, C. L. Cheong, C. F. Chau, A. B. Ghazali, "Design of a 4-bit Adder using Reversible Logic in Quantum-Dot Cellular Automata (QCA)", 2014 IEEE International Conference on Semiconductor Electronics (ICSE2014), 2014, DOI:10.1109/SMELEC.2014.6920795

[20] J. F. Chaves, D. S. Silva, V. V. Camargos and O. P. V. Neto, "Towards reversible QCA computers: Reversible gates and ALU," 2015 IEEE 6th Latin American Symposium on Circuits & Systems (LASCAS), Montevideo, 2015, pp. 1-4.DOI: 10.1109/LASCAS.2015.7250458

[21] J. I. Reshi and M. T. Banday, "Efficient Design of Reversible Code Converters Using Quantum Dot Cellular Automata", Journal of Nano and Electronic Physics, vol. 8, no. 2, Article ID:02042, June 2016

[22] J. I. Reshi and M. T. Banday, "Realization of Peres Gate as Universal Structure using Quantum Dot Cellular Automata", Journal of Nanoscience and Technology, vol. 2, no. 2, pp.115-118, June 2016

Set point weighted modified Smith predictor for delay dominated integrating processes

Somak Karan
Dept. of Applied Electronics & Instrumentation Engineering
Haldia Institute of Technology, Haldia
West Bengal, India
Somakkaran91@gmail.com

Chanchal Dey
Instrumentation Engineering, Dept. of Applied Physics
University of Calcutta, Kolkata
West Bengal, India
cdaphy@caluniv.ac.in

Abstract — **A process having any pole at origin is said to be integrating in nature. In practice, it is difficult to obtain the desired output from such integrating processes with significant time delay. Pure integrating processes have inherent non self-regulating feature and hence if they are disturbed from their equilibrium condition, process output continuously fluctuates over a considerable period of time. In practice a number of industrial processes like combustion chamber, distillation column, chemical reactors are well-known integrating processes with considerable time delay. Smith predictor based control technique is an established methodology for controlling such processes with considerable dead time. But, this technique fails to perform satisfactorily for pure integrating processes with considerable time delay. Modified Smith predictor method by Majhi and Atherton may be considered to be a good alternative. However, its performance is not found to be quite satisfactory due to undesired overshoot and sluggish recovery. In addition, complexity lies with the tuning of three controllers involved in multi-loop structure. To overcome this limitation, a simple tuning methodology is proposed here for modified Smith predictor with two controllers only. Efficacy of the proposed mechanism is substantiated through performance evaluation of pure integrating processes with significant time delay in comparison with well-known modified Smith predictor tuning reported by Majhi and Atherton.**

Keywords— **Integrating process; large dead time; Smith predictor; modified Smith predictor**

I. Introduction

A good number of industrial processes like chemical reactor, combustion chamber, distillation column etc. are found to be integrating in nature and majority of them contain considerable dead time [1]. Controlling of such processes with non-self-regulating feature and large dead time is truly a difficult task for process engineers [2]. Conventional feedback control methodology [3] completely fails in such cases and hence Smith predictor based tuning [4] is the most widely accepted scheme. But, in presence of process model uncertainty and undesired disturbances, conventional Smith technique [4] fails to perform satisfactorily. Hence, to ascertain desired process response, numbers of research findings have been reported [5-13] over the last three decades towards extension and modification of the conventional Smith predictor technique. Among these reported methodologies,

modified Smith predictor technique by Majhi and Atherton [12] is quite popular. However, for purely integrating processes modified Smith predictor technique by Majhi and Atherton [12] fails to eliminate the overshoot during set point tracking and load recovery is not up to the mark. Hence, there is a scope for performance enhancement of modified Smith predictor by incorporating a superior and simpler tuning scheme with lesser number of controllers.

Here, we suggest an easy but enhanced tuning scheme for modified Smith predictor [12] based multi-loop controller. Only two controllers are involved (instead of three as with [12]) in the proposed designing. A proportional-integral (PI) controller is present in the feed-forward path and a proportional controller (P) is introduced in feedback path of the modified Smith predictor structure. To restrict the process overshoot during set point tracking, set point weighting [13] mechanism is incorporated with PI controller. P controller present in the feedback path helps for faster recovery subsequent to load changes. Hence, we can expect that during close-loop operation an overall performance enhancement may be obtained under transient and steady state operating phases. Efficacy of the proposed scheme is verified through simulation study with well-known integrating process models with time delay (IPTD). Considerable amount of perturbation is introduced to the process parameters during simulation study to apprehend the model uncertainty of process plants. Superiority of the proposed methodology is substantiated through graphical responses during set point tracking and load rejection phases along with quantitative estimation of performance indices – integral error (IAE, ISE), integral time multiplied error (ITAE, ITSE) criterion [14]. In addition to improved close-loop response substantial robustness is also observed for our proposed methodology in presence of considerable uncertainty in process model parameters.

II. Modified Smith Predictor [12]

Well accepted modified Smith predictor technique reported by Majhi and Atherton [12] is an enhancement of the conventional Smith predictor [4] towards achieving improved close-loop response. Especially, for purely integrating processes with considerable dead time, conventional Smith predictor technique [4] fails during both the transient and steady state operational phases. Structure of the modified Smith predictor [12] is depicted in Fig. 1 which contains three controllers – $G_{C1}(s)$ is in the feed-forward path, $G_{C2}(s)$ and $G_{C3}(s)$ are in the feedback path. Plant dynamics is realized by the model $G_m(s)e^{-\theta_m s}$ where $e^{-\theta_m s}$ represents the estimated process dead time. The actual plant model $G(s)e^{-\theta s}$ is present in the forward path of the control loop as shown in Fig. 1.

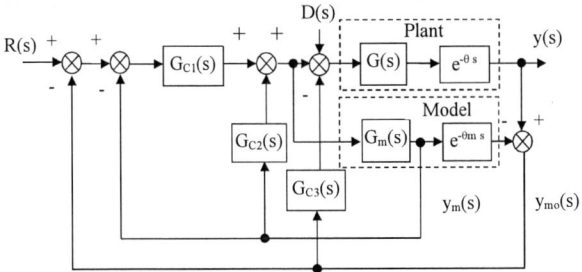

Fig. 1. Modified Smith predictor [12] based close-loop structure.

$D(s)$ represents the disturbance signal, introduced in the process input. Here, in Fig. 1, $G_{C1}(s)$ is a proportional-integral (PI) controller targeted towards satisfactory set point response, $G_{C2}(s)$ is a proportional (P) controller to restrict the process overshoot, and $G_{C3}(s)$ is also a proportional (P) controller to minimize the effect of undesired disturbances.

Relation of the process output due to set point change and load variation are given by Eqn. (1) and Eqn. (2) respectively

$$\frac{Y(s)}{R(s)} = \frac{G.G_{C1}e^{-\theta_m s}}{1+G_m(G_{C1}+G_{C2})}, \tag{1}$$

$$\frac{Y(s)}{D(s)} = \frac{Ge^{-\theta_m s}}{1+G_m(G_{C1}+G_{C2})} \cdot \frac{1+G_m.\left(G_{C1}+G_{C2}-G_{C1}G_{C3}e^{-\theta_m s}\right)}{1+G_{C3}G(s)e^{-\theta_m s}}. \tag{2}$$

From Eqn. (1), it is found that the denominator expression doesn't contain any delay term, and controllers $G_{C1}(s)$ and $G_{C2}(s)$ are responsible for determining the nature of set point response. Moreover, Eqn. (2) which is related to load response, all three controllers $G_{C1}(s)$, $G_{C2}(s)$ and $G_{C3}(s)$ are responsible for load recovery and delay term is present in both the numerator and denominator of Eqn. (2).

Here, we consider pure integrating processes with dead time i.e., IPDT in nature as given by

$$G(s) = \frac{K_m}{s} e^{-\theta_m.s}. \tag{3}$$

Here, K_m is the open loop gain and θ_m is the time delay. Expression of the controllers $G_{C1}(s)$, $G_{C2}(s)$, and $G_{C3}(s)$ are given by :

$$G_{C1}(s) = K_C \left(1 + \frac{1}{T_i s}\right), \tag{4}$$

$$G_{C2}(s) = K_f, \tag{5}$$

$$G_{C3}(s) = K_d. \tag{6}$$

In the expression of PI controller $G_{C1}(s)$ is given by Eqn. (4), K_C is the proportional gain and T_i is the integral time. Expressions for the two P controllers $G_{C2}(s)$ and $G_{C3}(s)$ are given by Eqn. (5) and Eqn. (6) respectively. All the three controllers $G_{C1}(s)$, $G_{C2}(s)$, and $G_{C3}(s)$ are tuned based on the guideline as reported in [12]. But, these tuning parameters fail to provide satisfactory close-loop response for IPDT processes. To overcome this limitation, we have proposed an alternative simpler tuning mechanism of the modified Smith predictor with two controllers only as shown in Fig. 2.

III. PROPOSED TUNING

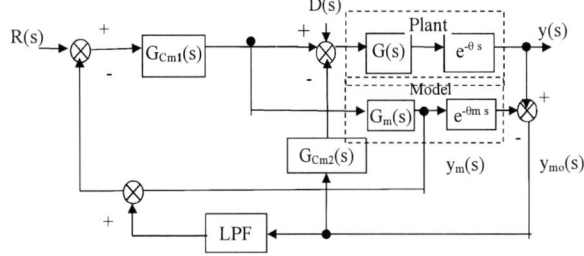

Fig. 2. Proposed modified Smith predictor structure.

To obtain improved set point tracking performance for IPDT processes, set point weighting (weighting factor b<1) scheme [13] is incorporated with $G_{Cm1}(s)$ which is a PI controller. P controller $G_{Cm2}(s)$ present in the feedback path helps to reduce oscillation during load recovery. Setting for $G_{Cm2}(s)$ is obtained from Routh stability analysis [15]. In the proposed scheme, a first-order low pass filter (LPF) is introduced in the outermost feedback path to eliminate the fluctuations present in process output. Therefore, an improved servo response as well as enhanced load regulation may be achieved for our proposed modified Smith predictor.

The most important feature of the proposed methodology (Fig. 2) is that only two controllers are employed instead of three controllers involved in modified Smith predictor [12] designing (Fig. 1). Expression of $G_{Cm1}(s)$, $G_{Cm2}(s)$ and the low pass filter (LPF) of our proposed modified Smith predictor structure are given by

$$G_{Cm1}(s) = K_c \left[\{b.R(s) - Y(s)\} + \frac{1}{T_i s}\right], \tag{7}$$

$$G_{Cm2}(s) = \gamma K_c, \tag{8}$$

$$LPF = \frac{1}{(0.01\theta_m)s+1}. \tag{9}$$

In Eqn. (7), K_C is the proportional gain, T_i is the integral time and b is the weighting factor for improved set point tracking. In the expression of the P controller as given by Eqn. (8) an additional tuning parameters γ is introduced in the controller relation. Selection for the value γ is very crucial towards achieving the desired close-loop response.

Using Mason's gain formula, transfer function related to set point change and load disturbance for the proposed modified Smith predictor due to set point change and load disturbance can be defined by Eqn. (10) and Eqn. (11) respectively.

$$\frac{Y(s)}{R(s)} = \frac{G_{Cm1}.G_m e^{-\theta_m s}}{1+G_{Cm1}.G_m}. \tag{10}$$

$$\frac{Y(s)}{D(s)} =. \frac{(1+G_{Cm1}G_m-G_{Cm1}G_m e^{-\theta_m s}).LPF.G_m.e^{-\theta_m s}}{(1+G_{Cm1}.G_m)(1+G_{Cm2}G_m.e^{-\theta_m s})}. \tag{11}$$

Based on Routh's stability criterion [15], the additional tuning parameter γ present in the expression of $G_{Cm2}(s)$ can be given by

$$\gamma = \frac{0.525}{\theta_m K_m K_c}. \tag{12}$$

Performance of the proposed modified Smith predictor is compared with the conventional [4] and well-known modified Smith predictor tuning by Majhi and Atherton [12]. To have a quantitative estimation, integral error indices IAE and ISE as well as time integral error indices ITAE and ITSE [14] are calculated for two well-known integrating IPDT models reported in [12]. Based on the close-loop responses and calculated values of performance indices it is found that the proposed tuning methodology offers significant amount of performance enhancement compared to the conventional [4] as well as reputed modified Smith predictor technique [12].

IV.   RESULTS

In this section, simulation results are reported during close-loop operation for two well-known IPDT models under set point change and load variation. Robustness of the conventional [4], modified [12], and the proposed Smith predictor based controllers are also evaluated with perturbation in dead time and open-loop gain of the process model. A unit step change is introduced at time t = 0 and once the process reaches the steady state, pulse like load disturbances D(s) is incorporated in the process input.

A. Model-I

We consider a very popular integrating process model with dead time as given by the Eqn. (13)

$$G(s) = \frac{1}{s} e^{-5s}. \tag{13}$$

This model has a time delay of 5 sec with unity open-loop gain i.e. $\theta = 5\ sec$, $K = 1$. Tuning parameters for the controllers $G_{C1}(s)$, $G_{C2}(s)$ and $G_{C3}(s)$ of Fig. 1 are calculated as per the guideline provided in [12]. For our proposed scheme, $G_{Cm1}(s)$ is tuned with $K_c = 1$ and $T_i = 0.1\ sec$ as given by Eqn. (14) and the set point weighting factor b is considered to be 0.7. The expression for $G_{Cm2}(s)$ is depicted by Eqn. (15) and the relation for the LPF is given by Eqn. (16).

$$G_{Cm1}(s) = \frac{0.1s+1}{0.1s}, \tag{14}$$

$$G_{Cm2}(s) = 0.105, \tag{15}$$

$$LPF = \frac{1}{0.05s+1}. \tag{16}$$

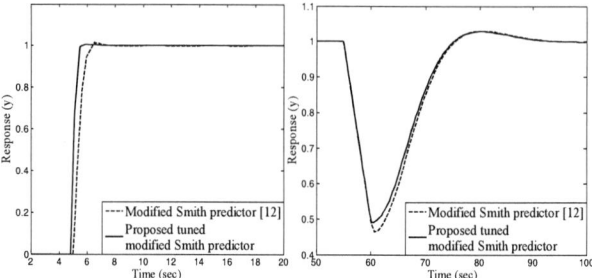

Fig. 3. Set point and load responses of the proposed modified Smith predictor and modified Smith predictor reported by Majhi and Atherton [12] for nominal Model-I as given by Eqn. (13).

Responses during set point tracking and load rejection phases are shown in Fig. 3 where solid line represents the response of the proposed modified Smith predictor and dashed line for the well-known modified Smith predictor by Majhi and Atherton [12]. From the responses as depicted in Fig. 3, it is clearly found that the proposed scheme offers an overall improved performance compared to modified Smith predictor technique. Quantitative performance indices – IAE, ISE, ITAE, and ITSE also substantiate the superiority of the proposed scheme as listed in Table I.

Table I. Performance indices for Model-I

Control technique	IAE	ITAE	ISE	ITSE
Smith predictor [4]	Produces large overshoot and fails to recover under load change			
Modified Smith predictor [12]	0.87	27.47	0.28	0.88
Proposed tuned modified Smith predictor	0.22	6.58	0.06	0.05

To verify the robustness of the proposed scheme against uncertainty of the process model, a considerable amount of perturbation (+20%) is introduced in process parameters (i.e.

open-loop process gain and dead time) of Eqn. (13) and the resulting perturbed model is given by Eqn. (17)

$$G'(s) = \frac{1.2}{s}e^{-6s}. \tag{17}$$

Responses of this perturbed model during set point change and load rejection phases are depicted in Fig. 4 which also justify the improved and robust feature of the proposed scheme in comparison with modified Smith predictor tuning by Majhi and Atherton [12].

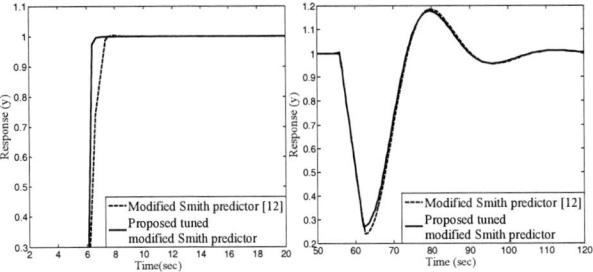

Fig. 4. Set point and load responses of the proposed modified Smith predictor and modified Smith predictor reported by Majhi and Atherton [12] for Model-I with +20% perturbation as given by Eqn. (17).

B. Model-II

We consider another well-known IPDT model as given by Eqn. (18) for performance evaluation

$$G(s) = \frac{0.2}{s}e^{-7.4s}. \tag{18}$$

Delay present in this process model is 7.4 sec and the open-loop gain is 0.2, i.e. $\theta = 7.4\ sec$ and $K = 0.2$. Tuning parameters for $G_{C1}(s)$, $G_{C2}(s)$ and $G_{C3}(s)$ are obtained from the guideline provided in [12]. In case of our proposed controller, tuning parameters for PI controller $G_{Cm1}(s)$ is given by $K_c = 1$ and $T_i = 0.1$ sec given by Eqn. (19). Expressions for P controller $G_{Cm2}(s)$ and LPF are given by the Eqn. (20) and Eqn. (21) respectively. For $G_{Cm1}(s)$ the value of set point weighting factor b is considered to be 0.6.

$$G_{Cm1} = \frac{0.1s+1}{0.1s}. \tag{19}$$

$$G_{Cm2} = 1.48. \tag{20}$$

$$LPF = \frac{1}{0.07s+1}. \tag{21}$$

Close-loop responses for the process model as given by Eqn. (18) i.e. Model-II are depicted in Fig. 5 and the related performance indices are listed in Table. II.

Responses during set point tracking and load rejection phases for the proposed modified Smith predictor and well-known modified Smith predictor tuning scheme [12] are shown in Fig. 5.

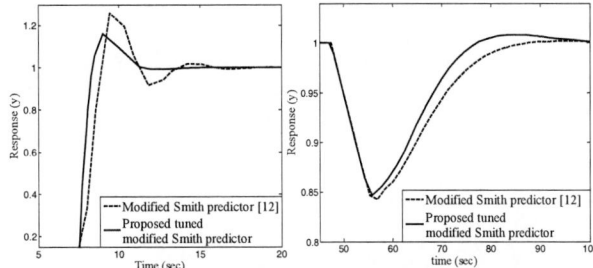

Fig. 5 Set point and load responses of the proposed modified Smith predictor and modified Smith predictor reported by Majhi and Atherton [12] for nominal Model-II as given by Eqn. (18).

Table. II. Performance indices for nominal Model- II

Control technique	IAE	ITAE	ISE	ITSE
Smith predictor [4]	Large overshoot and fails to recover load change			
Modified Smith predictor [12]	1.50	10.44	0.65	0.53
Proposed tuned modified Smith predictor	0.68	2.54	0.25	0.11

Here, solid line represents the responses of our proposed scheme and the dashed line for modified Smith predictor by Majhi and Atherton [12]. From the responses as depicted in Fig. 5 and listed performance indices of Table II, it is quite clear that our proposed technique offers enhanced load recovery compared to conventional [4] and modified Smith predictor [12]. In contrast, set point tracking performance is not satisfactory due to moderate overshoot (but smaller than [12]). Similar to the previous case, performance evaluation is also made on the perturbed (+10%) model as given by Eqn. (22).

$$G'(s) = \frac{0.22}{s}e^{-8.14s}. \tag{22}$$

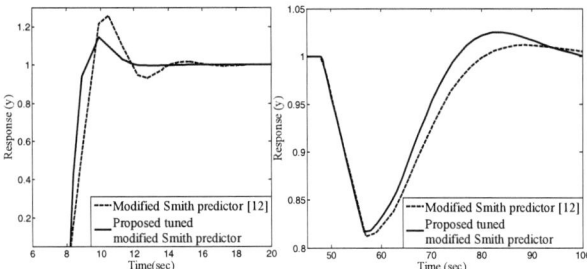

Fig. 6 Set point and load responses of the proposed modified Smith predictor and modified Smith predictor reported by Majhi and Atherton [12] for Model-II with +10% perturbation as given by Eqn. (22).

Considerable improvement is also observed during set point tracking and load recovery of the perturbed model compared to the setting reported by Majhi and Atherton [12].

V. CONCLUSION

An improved designing of modified Smith predictor is reported here for pure integrating processes with significant time delay. Superiority of the proposed scheme is substantiated through performance comparison with well-known modified Smith predictor setting by Majhi and Atherton [12]. One of the important feature of the proposed controller is that only two controllers are employed for the designing of the proposed modified Smith predictor. In addition, a clear guideline is provided for the tuning of the two controllers involved in the proposed designing. Due to incorporation of a low pass filter in the outer most feedback path it helps to reduce undesired fluctuations present in the process output. Close-loop performance analysis of the two reputed pure integrating models with time delay during set point change and load variation clearly substantiate the superiority of the proposed mechanism. In addition, improved robustness of the proposed tuning is also verified by incorporating considerable perturbation in the process model parameters.

References

[1] A. O. Dwyer, "The estimation and compensation of process with time delay," Ph. D. Thesis, Dublin city University, 1996.

[2] J. E. Normey-Richo and E. F. Camacho, "Control of dead-time processes," London: Springer-Verlag, 2007.

[3] P. Harriott, "Chemical reactor design," New York: Marcel Dekker, 2003.

[4] O. J. M. Smith, "A controller to overcome dead time," ISA Transactions, vol. 6, no. 2, pp. 28-33, 1959.

[5] K. Watanabe and M. Ito, "A process-model control for linear systems with delay," IEEE Transactions on Automatic Control, vol. 26, no. 6, pp. 1261-1269, 1981.

[6] K. J. Aström, C. C. Hang, and B. C. Lim, "A new Smith predictor for controlling a process with an integrator and long dead time," IEEE Transactions on Automatic Control, vol. 39, no. 2, pp. 343-345, 1994.

[7] I. Kaya and D. P. Atherton, "A new PI-PD Smith predictor for control of processes with long dead time," In Proc. 14th IFAC World Congress, 1999.

[8] S. Majhi and D. P. Atherton, " A new Smith predictor and controller for unstable and integrating processes with time delay," In Proc. IEEE conf. on Decision and Control, 2000.

[9] S. E. Hamamci and A. Ucar, "A robust model-based control for uncertain systems," Transactions of the Institute of Measurement and Control, vol. 24, no. 5, pp. 431-445, 2002.

[10] T. Liu, Y. Z. Cai, D. Y. Gu, and W. D. Zhang, "New modified Smith predictor scheme for integrating and unstable processes with time delay," IEE Proc.- Control Thery and Applications, vol. 152, no. 2, pp. 238-246, 2005.

[11] M. R. Matušek and A. D. Micić, "A modified Smith predictor for controlling a process with an integrator and long dead-time," IEEE Transactions on Automatic Control, vol. 41, no. 8, pp. 1199-1203, 1996.

[12] S. Majhi and D.P. Atherton, "Obtaining controller parameters for a new Smith predictor using auto-tuning," Automatica, vol.36,no. 11, pp. 1651-1658, 2000.

[13] A.S. Rao, V.S.R. Rao and M. Chidambaram, "Set pointweighted modified Smith predictor for integrating and double integrating processes with time delay," ISA Trans., vol. 46, pp. 59-71, 2007.

[14] C. Bissel, "Control Engineering," London: CRC Press, 1994.

[15] E. J. Routh, "A treatise on the stability of a given state of motion: Particularly steady motion," Macmillan, 1877.

Gameplay using Reinforcement Learning

Dr. Dilip Kumar*, Ritesh Prasad†, Kumar Ritu Raj Singh‡ and Surbhi Singh§
Department of Computer Science and Engineering
National Institute of Technology, Jamshedpur
* Email : dilip.cse@nitjsr.ac.in
† Email : ritesh.prasad289@gmail.com
‡ Email : skritu13@gmail.com
§ Email : surbhi3649@gmail.com

Abstract—**In this article we present a study of different Reinforcement Learning algorithms and how these algorithms can be used to train agents which can achieve human level performance in different games and control tasks.**
Index Terms—**Reinforcement Learning, Gameplay**

I. INTRODUCTION

Machine Learning is broadly divided into three categories - i) Supervised Learning , ii) Unsupervised Learning and iii) Reinforcement Learning. In supervised learning, models are trained on features and associated labels; and then they are required to predict the labels given a set of features. In unsupervised learning, the targets are not given and models are trained to infer similarities or dissimilarities among the data points. In reinforcement learning agents learn to perform certain tasks by interacting with environment and receiving feedback for the actions performed.

Reinforcement Learning algorithms closely resemble the working of a child's brain. Just as a child learns to perform a task by taking actions and getting feedback for those actions; similarly most reinforcement learning algorithms work on the principle of action and rewards. This ability of reinforcement learning algorithms to handle uncertainty so well has led to its rapid growth in popularity in the last decade.

TD-Gammon [6] was probably the first computer program to achieve human level performance in the game of backgammon. Recently, agents powered by reinforcement learning and deep learning have received media attention as they defeated the world champions in the game of Go and DOTA. *AlphaGo* [5] developed by Google's DeepMind is the world's first computer program to defeat the Go world champion. Similarly OpenAI's 1v1 bot defeated top DOTA 2 players of the world. Apart from gameplay, reinforcement learning is being used in various other fields like robotics, self-driving cars, traffic control systems [1], personalized recommendation systems [9], advertising [2] etc.

II. REINFORCEMENT LEARNING

A. State

The current situation of the environment is known as the state. It can be represented numerically as a measure of different environmental components.

B. Action

The moves which an agent takes is known as the action. This is how the agent interacts with the environment. An action transitions the state of the environment from one to another.

C. Reward

Feedback for an action is given in the form of rewards. The type of reward depends on the kind of environment. When the agent takes an action the state transitions from one to another and agent receives a reward for that action. Here reward may not mean something positive always. If the state transitions to something undesirable(like our player dies), then the reward returned is comparatively low or negative.

Idea of reward maximization: All of the reinforcement learning algorithms, depend on the central idea of maximizing rewards. The agents are trained to perform actions which maximizes the expected cumulative reward.

$$C_t = R_{t+1} + \gamma R_{t+2} + \gamma^2 R_{t+3} + ...$$

Where, C_t is the expected reward at time step t. R_T is the reward collected at time step T. γ is the discount rate, whose value lies between $[0, 1)$.

While playing games, it maybe possible that an action taken at the current time step results into a significant reward at a much later time step. However, such later reward is not guaranteed and is less probable than a reward expected in the next time step. Hence in the above equation, a discounted cumulative reward is considered.

D. Policy

Policy can be thought of as a function, which gives the information about the action to be taken at a particular state. A deterministic policy tells the exact action to be taken by the agent, where as a stochastic policy gives probabilities of different possible actions. Policy is just a function of state and is usually represented by π.

E. Value Function

Value function depends on the current state and the policy which the agent follows. It gives information about how valuable a state s is for an agent to be in, if it follows a particular policy π. Value function for a state s_t is the expected total reward an agent can get starting from state s_t.

$$V^{\pi}(s_t) = E[\sum_{i=1}^{T} \gamma^{i-1} R_{t+i} | s_t]$$

III. Algorithms

A. Behavioural Cloning

In this approach we ask an expert to play the game/perform the task for us. At every instance we record the state of the game and the action which the expert takes. This is done continuously for several episodes. At the end of it, we have a dataset consisting of $\{state_i, action_i\}$ pairs. And now we simply train our agent using supervised learning techniques, to predict the correct action for an input state. Hence our agent just tries to clone the behaviour of an expert.

B. Q-Learning

A Q-function gives the total expected reward an agent can get, if it is in a state s_t and takes an action a_t governed by a policy π [7].

$$Q^{\pi}(s_t, a_t) = E[\sum_{i=1}^{T} \gamma^{i-1} R_{t+i} | s_t, a_t]$$

The Q-function helps us to evaluate how good is it to take an action a_t if the current state is s_t. And hence for an optimal policy π^*, for a given state s_t, the action which gives higher expected reward, has higher Q-value.

In Q-learning, a Q-table is constructed which is a matrix having number of rows equal to the number of possible states the game can have; and the number of columns is equal to the number of actions the agent can take. At the end of training, the Q-table is filled up in such a way that, if for an agent at state s_i, it is best to take the action a_j, then among all values in the ith row of Q-table, the one in the jth column would be maximum.

The Q-Learning algorithm is as follows:

1) Initialize the Q-table arbitrarily with small values for all state-action pairs. Initialize ϵ.
2) Until convergence or the learning is stopped, repeat the following steps:
   - Get a random variable ϵ' between 0 and 1. If ϵ' is less than ϵ, then choose a random action, else choose the action guided by the Q-table.
   - In the current state s, perform the action a and transition to the new state s', getting a reward r.
   - Update the Q value as:

   $$Q(s, a) \leftarrow (1-\alpha)Q(s, a) + \alpha Q'(s, a)$$

   Where,

   $$Q'(s, a) = r + \gamma max_{a'} Q(s', a')$$

   - Gradually decrease ϵ.

In the above algorithm, a variable ϵ is introduced to handle the exploration-exploitation trade-off. In the beginning when Q-table is randomly initialized, it is better for the agent to explore and try out different actions. However as the number of iteration increases, the agent gets more confident with the Q-values and then it is better to exploit the environment, i.e., take the action suggested by the Q-table.

C. REINFORCE Algorithm

REINFORCE is a policy gradient algorithm, which directly tries to optimize the policy function π. Our policy can be defined as a function of parameters θ, and can also be written as π_θ. In most of the cases we want our agent to follow a particular trajectory τ which maximizes the reward. A trajectory τ can be represented as a chain of state action pairs [8].

$$\tau = (s_1, a_1, s_2, a_2, s_3, a_3, ..., s_n, a_n)$$

The total reward of a trajectory τ can be represented as $r(\tau)$.

$$r(\tau) = \sum_t r(s_t, a_t)$$

We need to train our policy in such a way that $r(\tau)$ is maximized. Hence our Reinforcement Learning objective is to maximize the total expected reward following a policy π. Total expected reward for a trajectory τ is the product of probability of that trajectory following policy π and the corresponding reward. Mathematically, we want to maximize,

$$J(\theta) = E_{\tau \sim \pi_\theta(\tau)}[r(\tau)]$$

More formally, we want to find optimal parameters θ^* for our policy π such that $J(\theta)$ is maximized.

$$\theta^* = arg\ max_\theta J(\theta)$$

This optimal set of parameter θ^* can be obtained by tried and tested method of gradient ascent.

$$\theta \leftarrow \theta + \alpha \nabla_\theta J(\theta)$$

The complete REINFORCE algorithm can be written as :
- Repeat until convergence :
  1) Run the policy $\pi_\theta(a_i|s_i)$ to sample trajectories $\{\tau^1, \tau^2 ... \tau^N\}$
  2) Calculate gradient of our objective as:

  $$\nabla_\theta J(\theta) \approx \frac{1}{N} \sum_{i=1}^{N} (\sum_{t=1}^{T} \nabla_\theta log \pi_\theta(a_t^i|s_t^i))(\sum_{t=1}^{T} r(s_t^i, a_t^i))$$

  3) Update θ as:

  $$\theta \leftarrow \theta + \alpha \nabla_\theta J(\theta)$$

In the above algorithm, we basically try to shift our policy in the direction where total expected reward is higher and opposite to the direction where total expected reward is less. However, the above implementation of REINFORCE has high variance and low convergence. This can be tackled by introducing causalities and baselines.

Causalities: This simply work on the fact that the policy at later time can not effect the reward at current time. Hence, $\nabla_\theta J(\theta)$ can be written as,

$$\nabla_\theta J(\theta) \approx \frac{1}{N} \sum_{i=1}^{N} (\sum_{t=1}^{T} \nabla_\theta log\pi_\theta(a_t^i|s_t^i))(\sum_{t'=t}^{T} r(s_{t'}^i, a_{t'}^i))$$

Baselines: This reduces variance without adding any bias.

$$\nabla_\theta J(\theta) \approx \frac{1}{N} \sum_{i=1}^{N} (\sum_{t=1}^{T} \nabla_\theta log\pi_\theta(a_t^i|s_t^i))(\sum_{t'=t}^{T} r(s_{t'}^i, a_{t'}^i) - b)$$

The b in the above equation is called a baseline. Usually the average reward over the sample trajectories is used as the baseline, however any function which stays unbiased with the expectation can be chosen as a baseline.

$$b = \frac{1}{N} \sum_{i=1}^{N} r(\tau)$$

D. Deep Q-Learning

Deep Q-Learning is similar to Q-learning, only in this instead of a Q-table, we use a deep neural network model to approximate the Q-values [4]. When the frame of game or stack of frames is used as the state of the game, naturally the state space becomes very high, and in that case constructing and updating a Q-table becomes inefficient. Hence in such cases deep neural network can replace the Q-table, which takes the state as input and gives action Q-values for different actions.

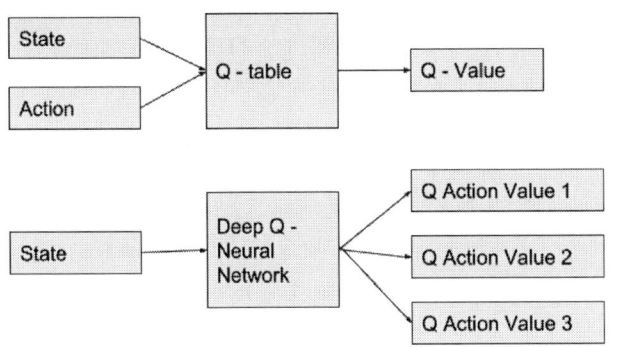

Fig. 1. Comparison of Q-Learning and Deep Q-Learning

In most of the games, the state can be represented by a frame of the game, however a single frame can not capture motion of the objects in the game. One way of tackling this is to stack up a few continuous frames of the game and use it as an input for our neural network.

In Deep Q-Learning, a technique called experience replay [3] is used, in which we store the experiences of our agent as tuple $e_t = (s_t, a_t, r_{t+1}, s_{t+1})$ in a dataset D. We apply Q-learning updates to a sample of experiences of fixed length, $e \sim D$, drawn at random from D.

The deep Q-learning algorithm with experience replay can be written as:

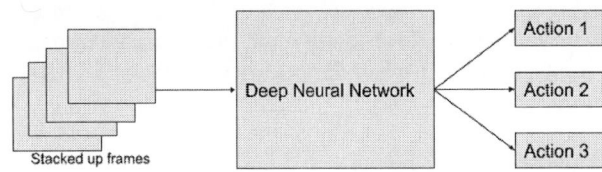

Fig. 2. Using stacked up continuous frames of the game as input

1) Initialize the weights θ for our neural network. Also initialize $\epsilon = 1$
2) Initialize the replay memory D to capacity N.
3) While learning is not stopped
   - Get a random variable ϵ' between 0 and 1. If $\epsilon' < \epsilon$ select a random action a_t, else select $a_t = argmax_a Q(s_t, a)$.
   - In the current state s_t, perform the action a_t and transition to a new state s_{t+1}, receiving a reward r_{t+1}.
   - Store the transition as $e_t = (s_t, a_t, r_{t+1}, s_{t+1})$ in our replay memory D.
   - Sample a random minibatch of transitions from our replay memory D.
   - Set $\widehat{Q} = r_{t+1}$ if the current episode ends at s_{t+1}, else set $\widehat{Q} = r_{t+1} + \gamma max_{a_{t+1}} Q(s_{t+1}, a_{t+1})$.
   - Make a gradient descent step with loss : $(\widehat{Q} - Q(s,a))^2$
   - Gradually decrease ϵ.

IV. EXAMPLES OF GAMEPLAY

There can be numerous types of gaming or control environments, each of which has different input state space and action space. The gaming environment is represented by an emulator with which our agent can easily interact. *OpenAI's* gym consist of different environments which allow us to train and test our RL agents easily. Examples of such environments are :

A. Cartpole classic control game

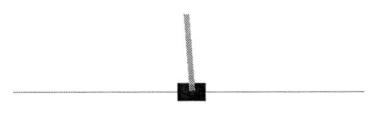

Fig. 3. CartPole

In the above game a pole is attached by an un-actuated joint to a cart which moves along a frictionless rope. We need to keep the pole upright for as long as possible in the center, by applying leftward or rightward force to the cart. The episode ends when the cart moves away more than 2.4 unit from the centre of the rope or when the pole is more than 15 degrees from the vertical.

State: Position and velocity of the cart, angle of the pole and pole velocity at the tip.

Actions: Pushing the cart to left or right. It is represented numerically as 1 or 0.

Reward: +1 for every step taken.

B. Making a humanoid robot run in a fast physics simulator

Fig. 4. Humanoid robot

The above is a physics simulated environment, in which we need to make a 3-Dimensional bipedal robot walk as fast as possible, without falling over.

State: A 376-dimensional vector representing the location, angle and velocity of various joints.

Action: A 17-dimensional vector representing the movement of different joints.

Reward: Number of steps the robot can take.

C. Breakout atari game

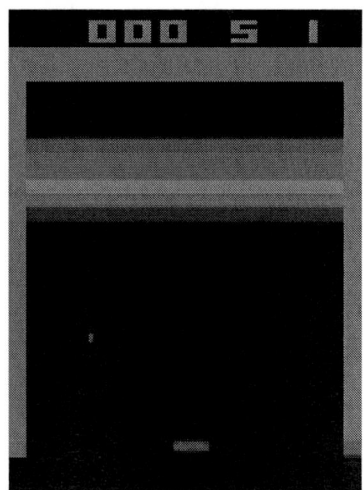

Fig. 5. Breakout

It is an Atari 2600 game, in which we need to break the brick and maximize the score.

State: A frame of the game, represented by an array of shape (210,160,3).

Action: One of the 18 possible moves like moving the paddle left, right or fire etc. .

Reward: The score in the game.

V. CONCLUSION

In this paper we discussed how reinforcement learning techniques can be used to train agents which can play games and perform other control tasks. Gameplay is the most widely used domain for evaluating reinforcement learning algorithms and observing them in action.

REFERENCES

[1] Itamar Arel, Cong Liu, T Urbanik, and AG Kohls. Reinforcement learning-based multi-agent system for network traffic signal control. *IET Intelligent Transport Systems*, 4(2):128–135, 2010.

[2] Junqi Jin, Chengru Song, Han Li, Kun Gai, Jun Wang, and Weinan Zhang. Real-time bidding with multi-agent reinforcement learning in display advertising. *arXiv preprint arXiv:1802.09756*, 2018.

[3] Long-Ji Lin. Reinforcement learning for robots using neural networks. Technical report, Carnegie-Mellon Univ Pittsburgh PA School of Computer Science, 1993.

[4] Volodymyr Mnih, Koray Kavukcuoglu, David Silver, Alex Graves, Ioannis Antonoglou, Daan Wierstra, and Martin Riedmiller. Playing atari with deep reinforcement learning. *arXiv preprint arXiv:1312.5602*, 2013.

[5] David Silver, Aja Huang, Chris J Maddison, Arthur Guez, Laurent Sifre, George Van Den Driessche, Julian Schrittwieser, Ioannis Antonoglou, Veda Panneershelvam, Marc Lanctot, et al. Mastering the game of go with deep neural networks and tree search. *nature*, 529(7587):484, 2016.

[6] Gerald Tesauro. Temporal difference learning and td-gammon. *Communications of the ACM*, 38(3):58–68, 1995.

[7] Christopher John Cornish Hellaby Watkins. *Learning from delayed rewards*. PhD thesis, King's College, Cambridge, 1989.

[8] Ronald J Williams. Simple statistical gradient-following algorithms for connectionist reinforcement learning. *Machine learning*, 8(3-4):229–256, 1992.

[9] Guanjie Zheng, Fuzheng Zhang, Zihan Zheng, Yang Xiang, Nicholas Jing Yuan, Xing Xie, and Zhenhui Li. Drn: A deep reinforcement learning framework for news recommendation. In *Proceedings of the 2018 World Wide Web Conference on World Wide Web*, pages 167–176. International World Wide Web Conferences Steering Committee, 2018.

2019 Devices for Integrated Circuit (DevIC), 23-24 March, 2019, Kalyani, India

Comparative HRV Analysis of ECG Signal in the Context of Sportsperson under Post-Exercise and Relaxed Condition

Prashant Kumar
Electrical Engineering Department
National Institute of Technology
Durgapur, India
raja.prashant89@gmail.com

Ashis Kumar Das
Electrical Engineering Department
National Institute of Technology
Durgapur, India
Faculty of Technology
Uttar Banga Krishi Viswavidyalaya
Coochbehar, India
ashiskd11@gmail.com

Dr. Suman Halder
Electrical Engineering Department
National Institute of Technology
Durgapur, India
sum_hal@yahoo.co.in

Abstract— Heart rate variability (HRV) analysis of electrocardiogram (ECG) signal is used to examine the health condition of the sportsperson. For the present work, a total of 22 numbers of data were taken from eleven participants in two different physical conditions who are under regular sporting activity. One set of data were taken just after playing badminton and 400-600 m run and another set of ECG data were taken in the normal i.e., relaxed condition. Acquired ECG signal is pre-processed for removal of powerline frequency, baseline wander (BW) removal, removal of low-frequency noise. The smoothening filter is also used for removal of high-frequency noise for smoothing of the wave. The filtered signals are used for HRV analysis. HRV is associated with average heart rate (HR) tachycardia and bradycardia. Main HRV parameters like mean heart rate, mean RR-interval, SDNN, RMSSD, NN50, pNN50 are analyzed in two different conditions namely post exercise and relaxed condition. In addition to HRV, the complexity analysis of the signal is also being done in the contest of sample entropy, approximate entropy etc.

Keywords— Electrocardiogram (ECG), Heart rate variability (HRV), RR-interval, SDNN, RMSSD, ApEn.

I. INTRODUCTION

An electrocardiogram (ECG) is an electrical signal generated from the heart. It is formed through a common cardiac cycle consists of contractions and relaxations of the human heart muscles [1]. ECG is the quasi-periodic signal which frequency content change over time. The frequent change in frequency change is the main event of interest for heart-related study. The ECG is an efficient non-invasive mechanism for various biomedical applications such as measurement of the heart rate, heartbeats rhythm monitoring, heart abnormalities diagnosis, emotion recognition and biometric identification.

ECG signals can be acquired by affixing the Ag-AgCl electrodes on specific locations such as surfaces of the chest, arms, and legs of the human body [2]. The electrodes detect electrical impulses coming from different directions within the heart. ECG gives the rate at which the heart is beating along with its rhythms whether steady or irregular [3].

ECG signal pattern of the same individual may change with time and different physical condition [4]. Schematic ECG wave showed in Fig. 1, comprises of the P wave, PQ segment, QRS wave, ST segment and T waves and RR interval. It defines the features as time intervals (millisecond) and amplitudes (millivolts). P wave occurs

due to atrial depolarization, PQ segment is an isoelectric line denoting AV delay. QRS complex due to ventricular depolarization which may sometimes affect by conduction disorder i.e. bundle block branch, ST segment is an isoelectric line harmonious with PQ segment and T wave due to ventricular repolarization [5, 6]. Exercise leads to higher heart rate and it decreases again once the body gets relaxed. On the other hand, an increase in HRV is manifested in the exercise condition. Heart rate complexity has also a reduction in its value after exercise and attaining some rest. The decrement of cardiovascular HRV indices is independent of relaxed condition but related to post-exercise [7].

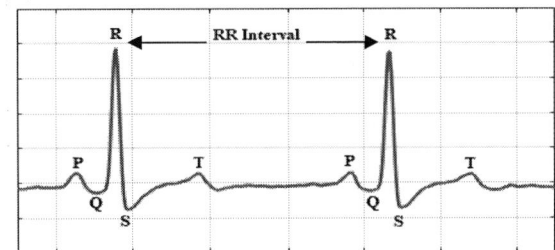

Fig. 1. Schematic ECG wave

The remaining parts of this paper are organised as material and methods, data processing, results and conclusion in sections II, III, IV, and V respectively.

II. MATERIAL AND METHODS

A. Participants

Eleven participants (seven male and four female) age group 25-30 year have voluntarily participated in the study. Each participant is selected as they used to perform some sporting activity in the morning. Before the experiment, all participants were informed about the research and they filled the questionnaire form regarding this study. All Participants have been instructed to abstain consumption of alcohol a day before and refrain from smoking before the ECG data acquisition. A participant's undersigned written consent form has been taken from each participant as an approval of the use of their ECG data for the different study.

B. Data Acquisition

Participants' data is taken in a sitting posture on the chair as shown in Fig. 2.

978-1-5386-6723-1/19 $31.00 © 2019 IEEE   181

For the study, Biopac MP45 biomedical signal acquisition device is used for data acquisition. Biopac MP45 is a two-channel computer interfaceable data acquisition device. Only one channel is used for ECG signal acquisition in Lead II configuration as it is recommended to use Lead II configuration because of its highest R-waves peak [8]. Lead II configuration of ECG signal acquisition are as follows: Positive terminal is connected to Left Leg (LL), negative terminal is connected to Right Arm (RA) and the ground (GND) is connected to Right Leg (RL). For this study, ECG data is acquired from participants who have performed outdoor sporting activities like playing badminton and after than a run of 400-600 m. Another set of ECG data of the same participants is taken in relax condition sitting in the lab. Total twenty-two ECG data, two from each of eleven participants of 5-minute duration, were acquired as part of this study.

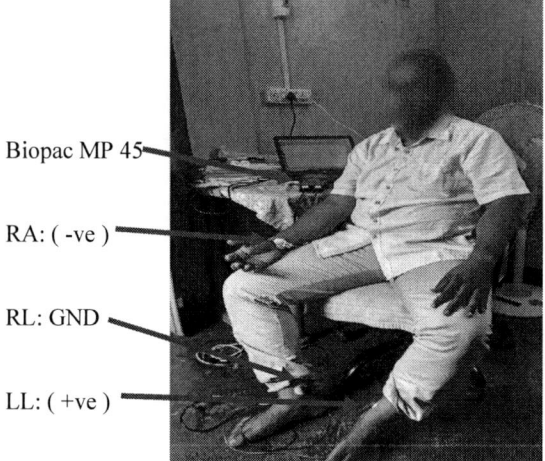

Fig. 2. ECG Data Acquisition from Biopac MP 45

III. DATA PROCESSING

The useful frequency range of the ECG signal is 0.5–45 Hz [9]. The frequencies other than these are unwanted for processing and treated as noise to the ECG signal as it deteriorates the prognostic and diagnostic of HR & HRV assessment. Thus, removal of BW and power line interference (PLI) artifacts from ECG signals is the first step and considered as data preprocessing.

A. Removal of power Line Interference

Cables carrying ECG signal experiences electromagnetic interference to power frequency (50Hz) by power supply lines. PLI is treated as high frequency according to the useful frequency range of the ECG signal. And this can be eliminated by designing a notch filter based on Fast Fourier Transform (FFT) methods. Design parameters for the notch filter are the location and angle of zeros in a unit circle which incorporates transfer function (T.F.). Parameters of the designed notch filter have been written below.

Zeros on unit circle at an angle,

$$\omega_0 = \pm 2\pi \frac{f_0}{f_s} \qquad (1)$$

Zeros locations,

$$(z_1, z_2) = (\cos \omega_0 \pm j \sin \omega_0) \qquad (2)$$

T.F. of the notch filter,

$$T.F = (1 - Z^{-1} z_1)(1 - Z^{-1} z_2) \qquad (3)$$

The poles are so chosen that there should be a sharp transition from pass band to notch frequency and notch frequency to pass band again.

B. Removal of Baseline Wander

Baseline Wander is mainly due to respiration and muscles movement and it is considered as very low-frequency noise.

Wavelet-based ECG noise removal methods are used for this study as suggested in [10]. Tenth level wavelet decomposition of the signal is used which gives noise. Subtracting it with the preprocessed signal by notch filter gives the filtered signal.

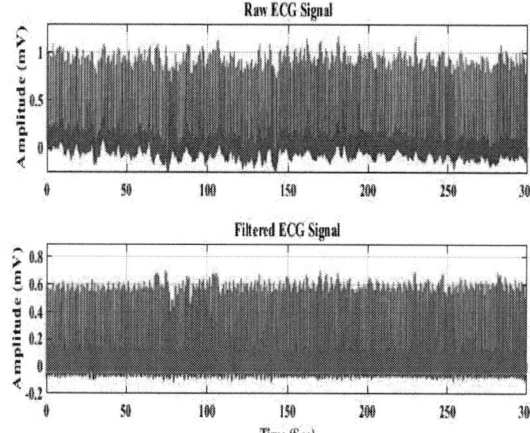

Fig. 3. Plot of Raw ECG Signal and Filtered ECG Signal of Relax Data

Using an average filter, glitches are removed to increase the performance of peak detection of the ECG signal. The raw ECG signal and the filtered ECG signal is shown in Fig. 3.

C. R-Peak Detection

R-peak is the peak magnitude of the QRS complex having positive magnitudes. The QRS complex mainly contains three deflections in ECG waveforms. Depolarization of right and left ventricles is being reflected by the QRS complex which is the prominent features of human ECG signal. For the R-peak detection, wavelet decomposition method of the signal is used as wavelets decomposed signals into time-varying components and separates signal components into different frequency bands which results in easier analysis and estimation of the signal with sparser data.

The 'symN' wavelets are also known as Daubechies' least-asymmetric wavelets have been used for QRS complex detection where N is the vanishing moments. Due to resemblance to the QRS complex, Sym4 wavelet is the best-suited choice for QRS detection. Comparison of Sym4 wavelet and QRS complex is shown in Fig. 4.

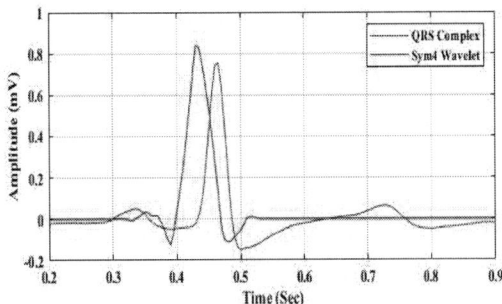

Fig. 4. Comparison of Sym4 Wavelet and QRS Complex

Maximal overlap discrete wavelet transforms (MODWT) is used to enhance the R peaks of the ECG signal as it handles arbitrary sample sizes. The absolute value of the squared signal from the wavelet coefficient is employed to identify peak and location of R wave using the peak finding algorithm. Filtered signal, wavelet reconstructed signal and filtered squared signals for the particular period are shown in Fig. 5.

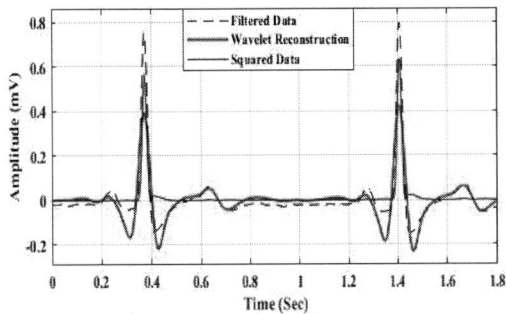

Fig. 5. Wavelet Reconstructed R Peak Location Detection

RR interval is calculated by the successive difference of location of R peak for HRV analysis.

D. HRV Analysis

HRV is the assessment of cardiovascular mechanism which provides the reliable, non-invasive technique to monitor parasympathetic activity. HRV is associated with average heart rate (HR) which means it is the difference in a time interval of heartbeat *i.e* analysis of beat-to-beat fluctuations in heart rate. In other ways, it is the variation of the time interval between R-peaks.

Both linear and nonlinear HRV analysis has performed. Linear HRV analysis performs time-domain and frequency domain analysis.

Time domain analysis is the statistical analysis based on beat-to-beat or NN beats and is termed as

RR (msec) - Mean of all NN interval,
HR - Mean heart rate,
SDNN (msec) - Standard deviation of NN interval in which average NN interval is taken for a short time,
RMSSD (msec) - Root mean square of successive different NN interval,
NN50 - No. of pairs of NN that is more than 50 msec,
pNN50 - The percentage of no. of pairs of RR that is more than 50 [8, 11].

Frequency domain analysis of the HRV uses power spectral density (PSD) curve which measures area/energy. The PSD is divided into separate frequency band as very low frequency (VLF) with dubious physiologic significance usually less than 0.04 Hz, low frequency (LF) which is derived from short term regulations of blood pressure ranging between 0.04-0.15 Hz and high frequency (HF) band ranging 0.15-0.50 Hz which reflects respiratory influences respiratory sinus arrhythmia [12,13].

Heart rate is one of the complex systems in humans. Control system for HRV is nonlinear because of its high complexity and the interactions between the physiological subsystem are nonlinear. The nonlinear indices are fractal measures, entropy measures, symbolic dynamics measures and Poincare plot. Poincare plot is a scattered plot of each RR interval concerning the previous RR interval. The short-term (SD1) and long-term variability (SD2) by ellipse fitting techniques from Poincare plot can be expressed as equation 4 and 5.

$$SD1 = \frac{1}{\sqrt{2}} SDSD \qquad (4)$$

$$SD2 = \sqrt{2SDRR^2 - \frac{1}{2}SDSD^2} \qquad (5)$$

Where, SDRR is the standard deviation of RR intervals and SDSD is the standard deviation of the successive differences of the RR interval.

Nonlinear HRV also gives information about HRV complexity such as approximate entropy, sample entropy and Detrended Fluctuation Analysis (DFA). Approximate Entropy (ApEn) is negative of the average natural algorithm and hence a smaller value of approximate entropy is desired as it is regular and predictable [8,13]. DFA is spectral ratios of LF/(HF+LF) and VLF/(LF+VLF) regarding frequency-weight.

HRV analysis was performed on ECG signal time series by using MATLAB and various results are validated by Kubios HRV Standard 3.1 demo software.

IV. RESULTS

All parameters of HRV analysis for each participant just after outdoor sporting activities and in a relaxed state are calculated. For the simplicity, heart rate stair plot, Poincare plot of both post-exercise and relaxed data are shown only for one participant. Fig. 6 and Fig. 7 shows heart rate stair plot in both post-exercise and relaxed condition.

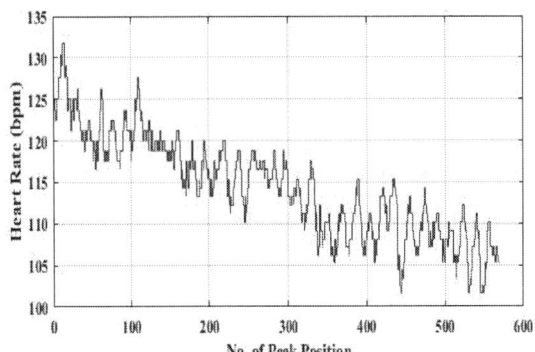

Fig. 6. Stair Plot of Heart Rate of Participant I Post Exercise

Fig. 6 shows a continuous decline in heart rate as long as ECG data is being acquired and it is finally coming down to resting position but the heart rate variation is very minimum in resting condition as shown in Fig 7.

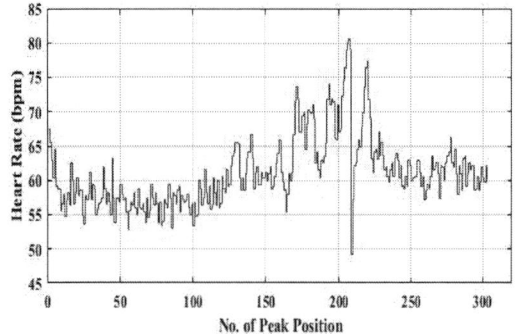

Fig. 7. Stair Plot of Heart Rate of Participant I Relax Condition

Fig. 8 and Fig. 9 show the Poincare plot in both post exercise and relaxed condition which is the plot of the successive difference of RR interval. From the Poincare plot, it can be seen that a decline in RR interval in post-exercise condition as mentioned in the literature.

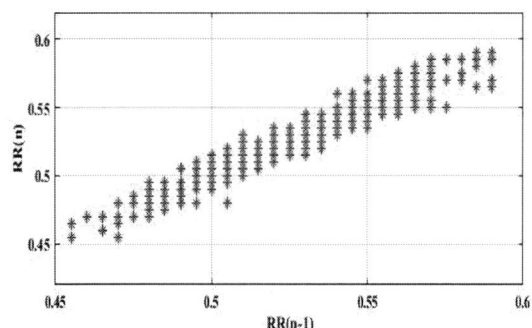

Fig. 8. Poincare plot of Participant I Post Exercise

The short-term variability (SD1) and long-term variability (SD2) results from the Poincare plot are shown in Table II. There is an increase in the ratio of long-term variability (SD2) and short-term variability (SD1) is seen.

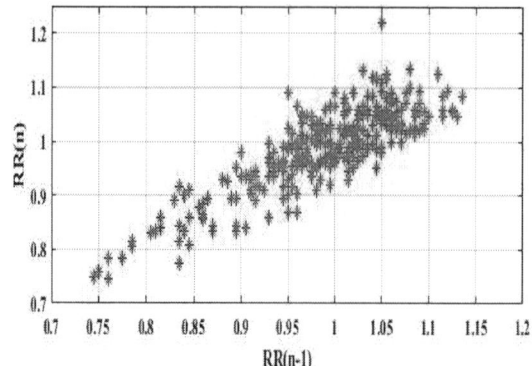

Fig. 9. Poincare plot of Participant I Relax Condition

Fig. 10 shown are frequency domain analysis for the post-exercise of participants I. Plots data were drawn using HRV analyser Kubios HRV Standard 3.1 software.

Frequency-Domain Results(FFT spectrum)

Variable	Units	VLF	LF	HF
Frequency band	(Hz)	0.00-0.04	0.04-0.15	0.15-0.40
Peak frequency	(Hz)	0.033	0.127	0.163
Power	(ms²)	53	1126	534
Power	(log)	3.971	7.027	6.281
Power	(%)	3.09	65.69	31.18
Power	(n.u.)		67.79	32.17
Total power	(ms²)	1714		
Total Power	(log)	7.447		
LF/HF ratio		2.107		
EDR	(Hz)	-		

Fig. 10. FFT spectrum of Participant I Exercise Period

Different time-domain HRV parameters of all eleven participants in both post-exercise and relaxed condition are listed in Table I. It can be analysed that there is decrease in mean RR interval time, SDNN, RMSSD, HRV triangular index and TINN (triangular interpolation of NN interval) for exercise data in corresponds to relax data along with an increase in heart rate (beats/min).

TABLE 1. TIME DOMAIN HRV PARAMETER

Parameter	Unit	Participants	I	II	III	IV	V	VI	VII	VIII	IX	X	XI
Mean RR	msec	Exercise	581.07	488.94	479.66	558.26	565.04	559.79	577.05	580.08	657.73	633.13	526.14
		Relaxed	938.9	793.74	808.71	707.73	698.61	683.45	1008.5	785.04	852.53	830.82	987.52
STD RR (SDNN)	msec	Exercise	47.226	9.0016	8.2051	13.182	20.568	29.19	12.283	25.651	27.368	40.322	10.989
		Relaxed	78.332	40.844	34.611	36.824	28.747	29.88	51.899	30.921	45.516	36.926	46.368
Mean HR	Beats/min	Exercise	103.26	122.72	125.09	107.48	106.19	107.18	103.98	103.43	91.223	94.767	114.04
		Relaxed	63.905	75.592	74.192	84.779	85.884	87.79	59.496	76.429	70.379	72.218	60.758
RMSSD	msec	Exercise	44.384	6.0648	7.157	6.8021	13.197	34.341	5.9954	14.563	15.107	33.061	7.4176
		Relaxed	82.499	42.459	40.031	32.554	28.846	26.148	55.448	25.825	32.58	29.398	49.537
NN50	-	Exercise	25	0	2	0	4	3	0	4	1	10	0
		Relaxed	146	96	83	23	32	27	104	17	43	24	71
pNN50	%	Exercise	4.8638	0	0.32154	0	0.75758	0.56285	0	0.7767	0.22026	2.1186	0
		Relaxed	46.057	25.6	22.554	5.4632	7.4766	6.1785	35.135	4.4737	12.321	6.7039	23
HRV Triangular Index	-	Exercise	7.0548	2.8419	2.6511	4.4958	4.7232	4.7257	3.4145	5.3196	6.8939	9.2745	3.4072
		Relaxed	11.778	10.743	10.25	8.7917	8.58	9.7333	11.423	7.0556	12.069	9.9972	12.12
TINN	msec	Exercise	315	54	55	66	128	385	72	153	141	183	54
		Relaxed	372	210	172	232	139	155	311	190	208	364	342

TABLE II. NON-LINEAR HRV PARAMETER

Parameter	Unit	Participants	I	II	III	IV	V	VI	VII	VIII	IX	X	XI
SD1	msec	Exercise	31.415	4.292	5.0649	4.8144	9.3408	24.306	4.2456	10.308	10.694	23.409	5.24496
		Relaxed	58.429	30.064	28.345	23.047	20.421	18.51	39.277	183286	23.071	20.817	35.086
SD2	msec	Exercise	58.994	11.995	10.447	18.023	27.552	33.401	16.86	34.771	37.211	49.338	14.634
		Relaxed	94.227	49.223	39.812	46.764	35.183	38.002	61.286	39.644	60.066	47.898	55.403
SD2/SD1	-	Exercise	1.879	2.7947	2.0627	3.7436	2.9464	1.3742	3.9711	3.3733	3.4795	2.1076	2.7877
		Relaxed	1.6127	1.6373	1.4045	2.0291	1.7228	2.053	1.5604	2.168	2.6035	2.3009	1.579
ApEn	-	Exercise	0.8343	1.2407	1.2559	1.1894	1.1331	0.9390	1.1196	1.1006	1.1449	1.1516	1.3324
		Relaxed	1.0953	1.2332	1.1116	1.2383	1.2305	1.2352	1.1455	1.254	1.1094	1.1903	1.1558
SampEn	-	Exercise	0.7632	1.3648	1.3979	1.3728	1.2783	0.9123	1.2139	1.1532	1.4248	1.1693	1.6649
		Relaxed	1.7524	1.9541	1.6597	1.5839	1.7903	1.7645	1.8745	1.8081	1.6424	1.8051	1.8296
DFA, Alpha1	-	Exercise	1.1788	1.2493	1.2055	1.7226	1.3974	0.76616	1.8298	1.5219	1.5773	1.3	1.4873
		Relaxed	1.1463	0.9871	0.7801	1.1246	1.054	1.2589	1.162	1.2274	1.3714	1.2289	1.0232
DFA, Alpha2	-	Exercise	0.2244	0.6935	0.8047	0.3273	0.3561	0.4951	0.4317	0.3853	0.4364	0.4959	0.4196
		Relaxed	0.1156	0.3463	0.2837	0.4531	0.2907	0.2691	0.1614	0.4627	0.2269	0.3158	0.2726

HRV complexity analysis of R-R interval is computed for all eleven participants in post-exercise and relaxed condition are shown in Table II.

From the above Table II, it can be commented that there is an increase in values of approximate entropy in some post-exercise cases showing random and uncertain time series but according to literature the lower value of approximate entropy indicates regular and predictive data set but since in some cases ApEn value is increased so no conclusion can be made from approximate entropy alone. In the case of sample entropy complexity is reduced with a decrement in its value. Here, sample entropy value is decreasing in post-exercise data in comparison with relaxed data, which means complexity reduces in the first case. Thus, the study gets validated while considering non-linear HRV parameters.

V. CONCLUSION

Time-domain and frequency-domain parameters of HRV analysis provide crucial information about heart responses. The non-linear parameters of the HRV analysis are also useful regarding the dynamics of the foreboding condition of the heart. Differences between HRV parameters in the ECG signal is correlated with the changes in HR. In this paper, different HRV parameters are analyzed for two different physical conditions, post-exercise and relaxed state. A significant decrement in RR interval for exercise condition and increment in heart rate is observed. RR interval, SDNN, RMSSD, NN50, pNN50 values are observed which shows the decreased values in post-exercise in comparison with the relaxed condition. In addition to this, a decrease in HRV complexity parameter (SampEn) is observed. This study manifested the tool for monitoring the cardiac activity sportspersons with a regular check-up and also for monitoring the health condition of cardiac patients.

REFERENCES

[1] Alberdi, A. Aztiria and A. Basarab, "Towards an automatic early stress recognition system for office environments based on multimodal measurements: A review", Journal of Biomedical Informatics, vol. 59, pp. 49–75, February, 2016.

[2] M. Abo-Zahhad, "ECG signal compression using discrete wavelet transform—Theory and applications", Discrete Wavelet Transform, pp. 143–168, April, 2011.

[3] Mohamed Hammad, Asmaa Maher, Kuanquan Wang, Feng Jiang and Moussa Amrani, "Detection of abnormal heart conditions based on

characteristics of ECG signals", Measurement, vol 125, pp. 634-644, September, 2018.

[4] W. Jiang and S.G. Kong, "Block-Based neural networks for personalized ECG signal classification", IEEE Transactions Neural Network, vol 18(6), pp. 1750–1761, November, 2007.

[5] I. Odinaka, L. Po-Hsiang, A.D. Kaplan, J.A. O'Sullivan, E.J. Sirevaag and J.W. Rohrbaugh, "ECG biometric recognition: A comparative analysis", IEEE Transactions Information Forensics Security, vol 7(6), pp. 1812–1823, December, 2012.

[6] Inderbir Kaur, Rajni Rajni and Anupma Marwaha, "ECG signal analysis and arrhythmia detection using wavelet transform", Journal of The Institution of Engineers (India), vol 9(4), pp.499-507, December, 2016.

[7] Javorka M, Zila I, Balhárek T and Javorka K., "Heart rate recovery after exercise: relations to heart rate variability and complexity", Brazilian Journal of Medical and Biological Research, vol 35(8), pp. 991-1000, August, 2002.

[8] G. Vega-Martínez, C. Toledo-Peral, C. Alvarado-Serrano, L. Leija Salas, O. G. Aztati-Aguilar and A. de Vizcaya-Ruiz, "SDNN index of heart rate variability as an indicator of change in rats exposed to fine particles: study of the impact of air pollution in Mexico City", International Conference on Electrical Engineering, Computing Science and Automatic Control, pp. 1-4, Campeche, September-October, 2014.

[9] Rishi Raj Sharma and Ram Bilas Pachori, "Baseline wander and power line interference removal from ECG signals using eigenvalue decomposition", Biomedical Signal Processing and Control, vol 45, pp. 33-49, August 2018.

[10] S.K. Yadav, R. Sinha and P.K. Bora, "Electrocardiogram signal denoising using non-local wavelet transform domain filtering", IET Signal Processing, vol. 9, pp. 88–96, December, 2015.

[11] Jakub S. Gąsior, Jerzy Sacha, Piotr J. Jeleń, Jakub Zieliński and Jacek Przybylski "Heart rate and respiratory rate influence on heart rate variability repeatability: Effects of the correction for the prevailing heart rate", Frontiers in Physiology, vol 7, pp. 1-11, August, 2016.

[12] Ahsan Habib Khandoker, Chandan Karmakar, Michael Brennan, Marimuthu Palaniswami and Andreas Voss, "Poincaré plot methods for heart rate variability analysis", Springer Publication, pp. 10-11, 2003.

[13] K. K Tripathi, "Respiration and heart rate variability: A review with special reference to its application in aerospace medicine", Indian Journal of Aerospace Medicine, vol 48(1), pp. 64-75, 2004.

Influence of Channel Thickness on Analog and RF Performance Enhancement of an Underlap DG AlGaN/GaN based MOS-HEMT Device

Akash Roy
Department of Electronics and Communication Engineering
Heritage Institute of Technology
Kolkata, India
akash_roy@ieee.org

Rajrup Mitra
Department of Electronics and Communication Engineering
Heritage Institute of Technology
Kolkata, India
rajrupmitra.94@gmail.com

Atanu Kundu
Department of Electronics and Communication Engineering
Heritage Institute of Technology
Kolkata, India
kundu.atanu@gmail.com

Abstract—This paper elucidates a comprehensive, illustrative and qualitative study on the Analog and RF performance of an Underlapped Double-Gate (U-DG) AlGaN/GaN heterojunction-based MOS-HEMT device with Hafnium-based high-k dielectric gate material. This paper presents the effect of GaN channel thickness variation on the drain current (I_d), the transconductance (g_m), output resistance (R_O), the intrinsic gain ($g_m R_o$), transconductance generation factor (g_m/I_d) and the RF FOMs- total intrinsic gate capacitance (C_{gg}) and cut-off frequency (f_T). These hetero-structured devices show superior performance as power transistors due to its enhanced efficiency, cost-effectiveness, reliability and controllability over silicon based conventional DG-MOS and HEMT transistors.

Index Terms— *GaN/AlGaN, Underlapped Dual Gate (U-DG) MOS-HEMT, Symmetric Underlap, Channel Thickness variation, High-k Dielectric*

I. INTRODUCTION

With the growing demand for technology, channel lengths of FETs are being scaled down, for better performances. One way of increasing performance is by modulating the channel length and width of the device. But this would increase DIBL and other Short Channel Effects (SCE)[1]. It has also been noted that conventional MOSFET behaves like a resistor in deep sub-micron level [2]. Demand for faster devices to replace present MOSFET technology is growing. AlGaN/GaN heterojunction-based HEMT devices were first demonstrated over a decade ago[3] and consequently, a great deal of work has been carried out on these devices to maximize their overall device performance[4-6]. MOS-HEMT imbibes gate efficiency of MOS-HEMT devices and the gate sensitivity of MOS transistors [7].

Research shows that GaN-based MOS-HEMT devices perform better than other subsequent materials [8]. AlGaN/GaN heterojunction-based MOS-HEMTs have been widely studied in the recent years because of their promising performance in the RF domain as well as in fast switching applications [9-10]. Better electric field effects are seen in the channel region for high-k materials such as Hafnium Oxide (HfO_2) and thus, HfO2 has been used as the gate oxide material for our study [11].

Double gate MOS-HEMTs [12] have superior Analog and RF characteristics than a single gate MOS-HEMTs [13], due to better channel utilization. Hence double gate MOS-HEMTs are used keeping the channel lengths greater than or equivalent to that of MOSFETs, thereby reducing the Short Channel Effects (SCE) [14]. Underlapped structures are used in MOS-HEMT structures because it helps to reduce parasitic capacitances and improve RF performances [15]. As the underlap length increases, the effect of DIBL reduces [16], thus making the device more gate sensitive and further reducing fringing losses.

In this paper using an Underlapped Double-Gate (U-DG) AlGaN/GaN heterojunction-based MOS-HEMT structure, firstly an optimal range of channel widths is attained where the best (I_{on}/I_{off}) ratio is achieved; then a detailed comparison of drain current (I_d), the transconductance (g_m), output resistance (R_o), intrinsic gain ($g_m R_o$), transconductance generation factor (g_m/I_d) has been carried out on this optimal range. RF parameters are extracted from the device using the Non-Quasi Static (NQS) model of the device [17]. RF performance is done by extracting total intrinsic gate capacitance (C_{gg}) and thereby the cut off frequency for unity gain (f_T).

II. DEVICE STRUCTURE

The cross-sectional view of a U-DG AlGaN/GaN MOS-HEMT with source and drain underlap is depicted in Fig. 1. The device dimensions have been taken in consideration is gate length (L_g), of 200 nm, oxide thickness (oxo), of 10 nm, source/drain length of 200 nm according to the ITRS 2008 for RF and Mixed Signal application [18]. A hetero-structured layer of AlGaN/GaN is grown on the oxide layer with AlGaN thickness (d) of 18 nm and GaN thickness (Buf) has been varied from 140 nm to 180 nm to study the Analog and RF performance. The optimized underlap length taken on both sides of the source/drain electrode is 200 nm.

The strained interface between the AlGaN and GaN crystal layer forms on the GaN interface which is abundant with highly mobile electrons all along the channel. The gate electrodes deplete a region below the gate contacts on both the side of the device otherwise, the device behavior is similar to the n-channel enhancement mode MOSFETs.

Here Hafnium Oxide (HfO_2) has been used as the dielectric gate material because it increases the gate capacitance thereby leading to a better device performance.

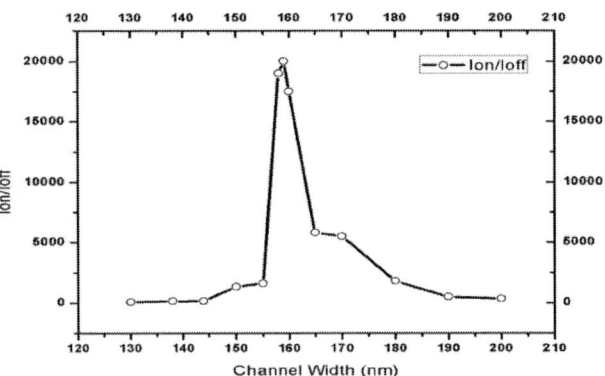

Fig. 2: Variation in I_{on}/I_{off} ratio with varying channel widths for a U-DG AlGaN/GaN MOS-HEMT device.

The drain electrode is applied with a voltage, V_{ds} of 2.0 V and the variable gate to source voltage is varied from -3.0 V to 1 V for simulation of the Analog performance of this U-DG AlGaN/GaN MOS-HEMT. The simulation and analysis of RF performance is conducted by giving an applied frequency of 10 GHz, with V_{ds} taken as 2.0 V and varying V_{gs} from -3.0 V to 2.75 V.

The following section describes a comparative study on the analog and RF performance of the buffer-engineered U-DG AlGaN/GaN MOS-HEMT devices.

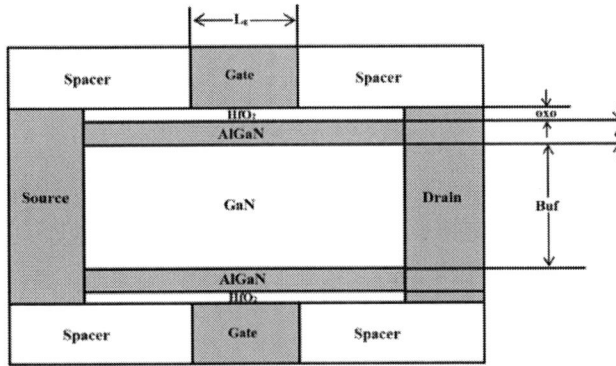

Fig. 1: Cross section of a U-DG AlGaN/GaN MOS-HEMT with source and drain underlap.

III.   SIMULATION PROCEDURE

TCAD device simulator has been used to simulate and analyse all the electrical behaviours of the device [19]. The Albrecht Model is incorporated to simulate low field mobility calibration in the device and other physical models used in device simulation are Shockley-Read-Hall (SRH) recombination model, Fermi Dirac statistics model for much correct device processing. Polarization model has been incorporated for GaN-based devices and Newton is used as the numerical model here as the device computation has strongly coupled system of equations and quadratic convergence. Molybdenum has been used as the gate material and the mole fraction, x of Al in Al$_x$Ga$_{1-x}$N is considered as 0.3. The underlap regions are used to minimize the effect of overlap parasitic capacitance on the gate electrodes.

For achieving the optimized device performance, an array of (I_{on}/I_{off}) ratio is studied for a varying range of channel thickness, where I_{on} is the current obtained at $V_{gs}=V_{ds}=V_{DD}$ and superior gate control is possible to be achieved at better (I_{on}/I_{off}) ratio. The optimal range of channel thickness as shown in Fig. 2 is secured between 140 nm to 180 nm, which has been used in this paper to study the Analog and RF FOMs of this device.

IV.   ANALOG AND RF PERFORMANCE ANALYSIS

This section analyses the impact of varying channel thickness on the Analog parameters, i,e., I_d, g_m, g_m/I_d, R_O, g_mR_O and the RF figure-of-merits- total intrinsic gate capacitance (C_{gg}), cut-off frequency (f_T).

It is evident from Fig. 3 that the conduction band energy decreases along length of the channel and is lowered with increasing channel thickness. This observation confirms an increase in average electron velocity along the channel with increase in channel thickness, which leads to increase in drain current (I_d).

It is evident from Fig. 4 that as the channel thickness reduces, the barrier across drain to source also reduces leading to reduction in gate control over the device. The reduction in channel thickness leads to interference of the vertical electric fields formed due to the double gate structure.

It is observed from Fig. 3 and Fig. 4 that while the device changes from OFF state to ON state, the barrier across the channel drops the most for the device having higher channel thickness which makes it easier for the electrons to flow from source to drain. Fig. 5 shows the (I_d/V_{gs}) curve for three different values of channel thickness in the linear scale. Steeper slopes are obtained for devices with increased channel thickness, so it is prominent that the device with 180 nm channel thickness has better gate control. It is also evident that the device with 180 nm channel thickness has higher values of ON current.

978-1-5386-6723-1/19 $31.00 © 2019 IEEE

2019 Devices for Integrated Circuit (DevIC), 23-24 March, 2019, Kalyani, India

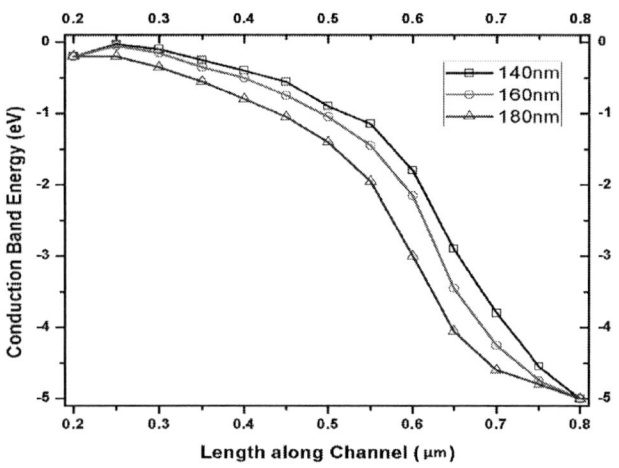

Fig. 3: Conduction band energy along the channel of a 200 nm U-DG AlGaN/GaN MOS-HEMT device of varying channel thickness, at V_{ds} = 5.0 V for an applied voltage at gate V_{gs}= 1.0V, when the device is in ON state.

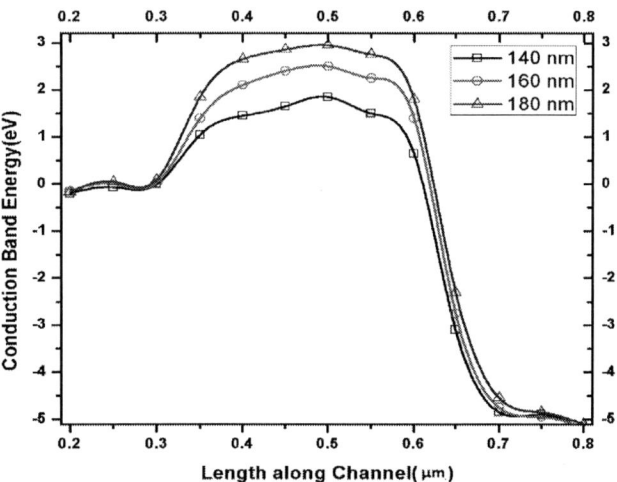

Fig. 4 Conduction band energy along the channel of a 200 nm U-DG AlGaN/GaN MOS-HEMT device of varying channel thickness, at V_{ds} = 5.0 V for an applied voltage at gate V_{gs}= -5.0V, when the device is in OFF state.

Fig. 6 illustrates the variation of the drain current for three different values of channel thickness in (I_d/V_{ds}) curve. It is evident from the result that highest value of drain current is achieved at 180 nm of channel thickness. It is observed from the conduction band diagrams of Fig. 3 that the device with 180 nm channel thickness has the lowest conduction band energy in the On state, which directly implies of it having the highest drain current among the three devices, for any specific value of V_{ds}.

In simplified terms, the transconductance (g_m) is quantified as the variation in drain current with change in gate control voltage while keeping the drain to source voltage constant. Hence better the g_m, better is the gate sensitivity of the device. It is discerned from Fig. 7, that the device with channel thickness of 180 nm has the highest peak value of transconductance which means for a small change in gate bias voltage, this device has an enhanced drain current which

justifies the fact that the device of 180 nm channel thickness is a better gate sensitive device for a constant drain to source voltage V_{ds} of 2.0 V.

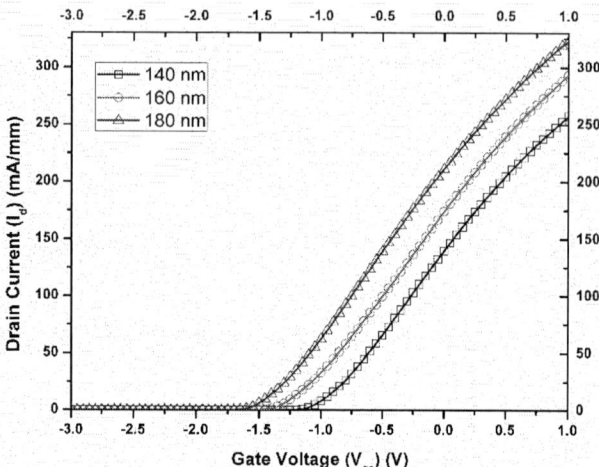

Fig. 5: Variation of I_d in linear scale as a function of V_{gs}, at V_{ds}= 1.0 V, for 200 nm U-DG AlGaN/GaN MOS-HEMT device with different values of Buf.

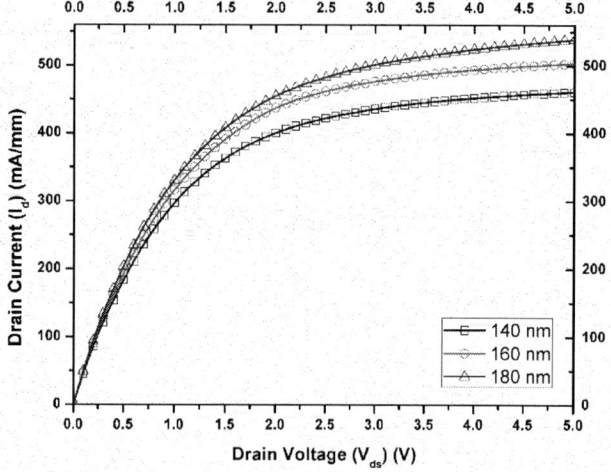

Fig. 6: Variation of I_d in linear scale as a function of V_{ds} at V_{gs}= 1.0 V, for 200 nm U-DG AlGaN/GaN MOS-HEMT device with different values of Buf.

Transconductance generation factor [TGF] (g_m/I_d) is a significant device parameter in analog circuit designing as it denotes the gain available per units of power dissipated. From Fig. 8, it is noticeable that, the ratio (g_m/I_d)is maximum in deep sub-threshold region, and as the drain current increases with increasing gate voltage as depicted from Fig. 5; this factor degrades severely while approaching the super-threshold region. Fig. 8 also highlights that device with 140 nm channel thickness, has higher TGF hence making the device more efficient.

Higher output resistance or on-resistance (R_O) is an essential device parameter for switching applications like output driver and switching power supply. Fig. 6, exemplifies that with decreasing channel thickness, the slope of (I_d/V_{ds}) decreases, hence shifting the early voltage to a much negative value of

978-1-5386-6723-1/19 $31.00 © 2019 IEEE   188

V_{ds} needed for channel pinch-off. As this pinch-off point shifts to the left for decreasing channel thickness, this results in a much lower drain conductance ($1/R_O$), making the R_O of the device to increase with decrement in channel thickness as shown in Fig. 9.

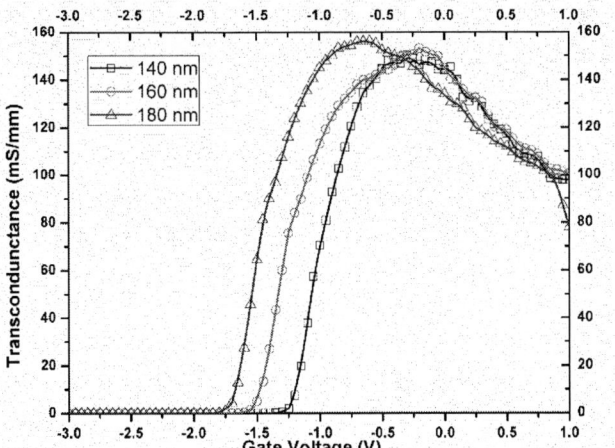

Fig. 7: Variation of g_m as a function of V_{gs} for a 200 nm U-DG MOS-HEMT device with different values of *Buf* at V_{ds} =2.0 V.

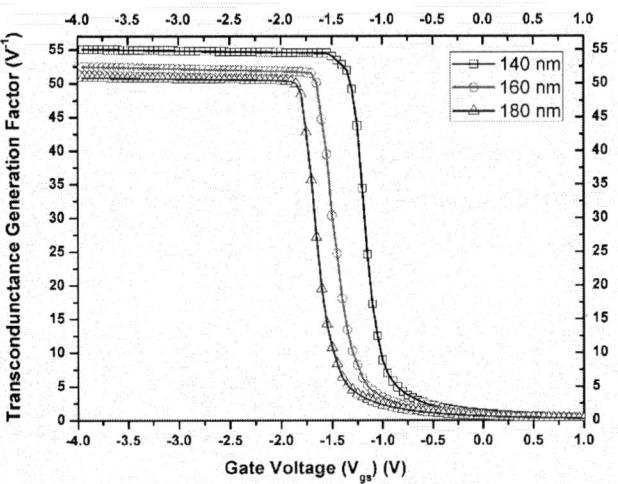

Fig. 8: Variation of g_m/I_{ds} as a function of V_{gs} for a 200 nm U-DG AlGaN/GaN MOS-HEMT device with different values of *Buf* at V_{ds} =2.0 V.

Intrinsic gain $(g_m R_O)$ in dB scale of the device has the higher peak value with decrease in channel thickness as shown in Fig. 10. As discussed earlier, with decreasing channel thickness, there is an increase in the output resistance (R_O). Thus a device with thinner channel exhibits better gain than that of a thicker channel device.

To examine the RF performance of this device, necessary device parameters have been extracted using small signal equivalent circuit with Non-Quasi-Static modelling for attaining model accuracy in higher frequency [17]. Y-parameter study of the AC small signal model is conducted to extract the RF parameters.

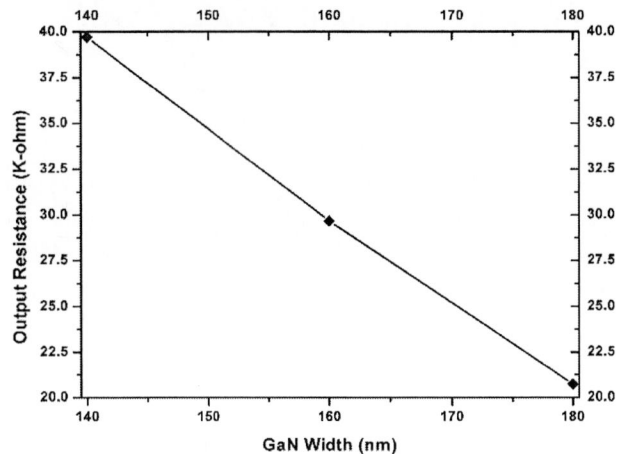

Fig. 9: Output resistance R_0 curve as a linear function of channel thickness at V_{gs}= 1.0 V, for a 200 nm U-DG AlGaN/GaN MOS-HEMT device.

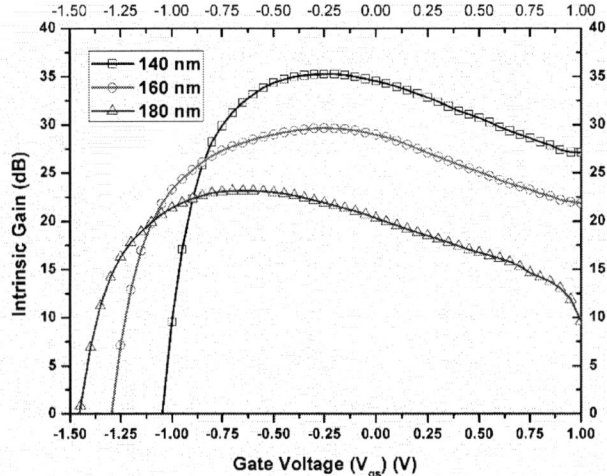

Fig. 10: Variation of $g_m R_o$ as a function of V_g for a 200 nm U-DG AlGaN/GaN MOS-HEMT device with different values of *Buf* at V_{ds} =2.0 V.

The total intrinsic gate capacitance (C_{gg}) is the summation of intrinsic gate-to-source/drain capacitances, i.e., (C_{gg} =C_{gs}+ C_{gd}). Fig. 11 illustrates the change in (C_{gg}) with respect to varying gate voltage, while a constant drain voltage (V_{ds}) is fed to the device. There is a steady increase in the (C_{gg})in from the sub-threshold region to super-threshold region with increasing gate voltage, due to the increment in the fringing field lines emerging from Gate ends [20].

From Fig. 12, it is observed that the cut off frequency (f_T) of a U-DG AlGaN/GaN MOS-HEMT device for a channel thickness of 180 nm is maximum in sub-threshold regime and is vice-versa in the super-threshold region. Hence it is inferred from the graph that increase in (f_T) in the sub-threshold region follows the same trend as (C_{gg}) as depicted in Fig. 11, with channel thickness of 180 nm having higher values; but in the super-threshold region, (f_T) is largely dominated by (g_m), following the same curve pattern as that of Fig. 7.

2019 Devices for Integrated Circuit (DevIC), 23-24 March, 2019, Kalyani, India

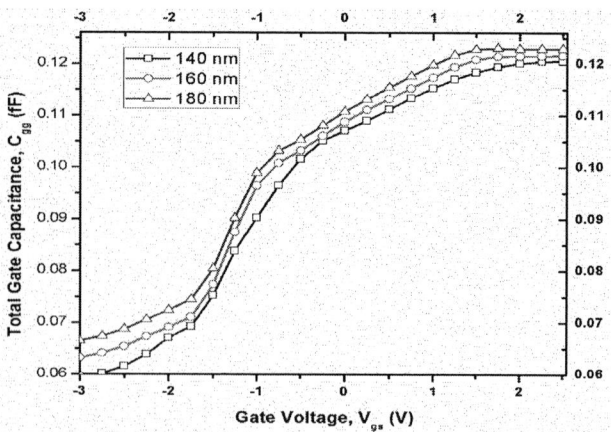

Fig. 11: Variation of C_{gg} as a function of V_{gs} for a 200 nm U-DG AlGaN/GaN MOS-HEMT device with varying channel thickness at $V_{ds} = 2.0 V$

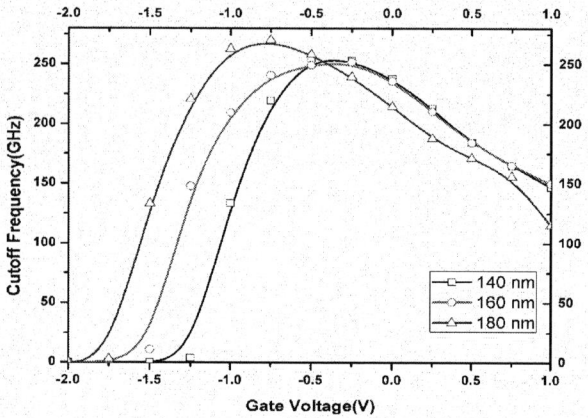

Fig. 12: Variation of f_T as a function of V_{gs} for a 200 nm U-DG AlGaN/GaN MOS-HEMT device with varying channel thickness at $V_{ds} = 2.0 V$

V.   CONCLUSION

This paper demonstrates a comparative analysis of the Analog and RF performance of a U-DG AlGaN/GaN MOS-HEMT device with varying channel thickness. From the (I_{on}/I_{off}) ratio, the channel thickness of 140nm, 160 nm and 180nm were taken to analyze different parameters such as (I_d/V_{ds}) curve, (I_d/V_{gs}) curve, transconductance (g_m) ,output resistance (R_O),TGF (g_m/I_{ds}), intrinsic gain (g_mR_O), total intrinsic gate capacitance (C_{gg}) and cut-off frequency (f_T). From the study, it is evident that increasing the channel thickness of the device there is an increase of 15% and 27% in the drain current of (I_d/V_{ds}) curve at (V_{ds}) = 2.0 V and (I_d/V_{gs}) curve at (V_{gs}) = 0.9 V respectively, 4% in the transconductance peak value ($g_{m,max}$), 6% in the cut-off frequency (f_T) and a better gate control as realized from the ($Cond_{on}$) and ($Cond_{off}$) curves. However, higher output resistance (R_O) and around 20 dB increase in the maximum intrinsic gain achieved, is spotted for the device having channel thickness of 140 nm.

VI.   ACKNOWLEDGMENT

The authors would like to thank the IEEE EDS Center of Excellence, Heritage Institute of Technology

for providing laboratory facilities.

VII.   REFERENCES

[1] Y. Awano, M. Kosugi, K. Kosemura, T. Mimura and M. Abe, "Short-channel effects in subquarter-micrometer-gate MOS-HEMTs: simulation and experiment," in IEEE Transactions on Electron Devices, vol. 36, no. 10, pp. 2260-2266, Oct. 1989.

[2] J. R. Brew\, ti. K. Ng, and R. ti. Watts, .'The submicrometer silicon MOSFET." in Submicron /~ire,qru/ed Circuit.\. R. K. Watts. Ed. New York: Wiley. 1989. ch. 1.

[3] M. A. Khan, J. M. Van Hove, J. N. Kuznia, and D. T. Olsen, "High electron mobility GaN–AlGaNheterostructures grown by LPMOCVD," Appl. Phys. Lett., vol. 58, no. 21, pp. 2408–2410, May 1991.

[4] U. K. Mishra, P. Parikh, and Y. F. Wu, "AlGaN/GaNMOS-HEMTs—An overview of device operation and applications," Proc. IEEE, vol. 90, no. 6, pp. 1022–1031, Jun. 2002.

[5] Y. F. Wu, A. Saxler, M. Moore, R. P. Smith, S. Sheppard, P. M. Chavarkar, T. Wisleder, U. K. Mishra, and P. Parikh, "30-W/mm GaNMOS-HEMTs by field plate optimization," IEEE Electron Device Lett., vol. 25, no. 3, pp. 117–119, Mar. 2004.

[6] Y. F. Wu, M. Moore, T. Wisleder, P. M. Chavarkar, U. K. Mishra, and P. Parikh, "High-gain microwave GaNMOS-HEMTs with source-terminated field-plates," in IEDM Tech. Dig., Dec. 2004, pp. 1078–1079.

[7] Medjdoub, F., N. Sarazin, M. Tordjman, M. Magis, M. A. di Forte-Poisson, M. Knez, E. Delos, C. Gaquiere, S. L. Delage, and E. Kohn. "Characteristics of al/sub 2/o/sub 3//allnn/ganmosMOS-HEMT." Electronics letters 43, no. 12 (2007): 691-692.

[8] J.H. Edgar, S. Stride, I. Akasaki, H. Amano, C. Wetzel (Eds.), Gallium Nitride and Related Materials, INSPEC, London, 1999.

[9] R. P. Smith, S. Sheppard, Y.-F. Wu, S. Heikman, S. Wood, W. Pribble, et al., "AlGaN/GaN-on-SiCMOS-HEMT technology status," in Proc. IEEE Compound Semicond. Integr. Circuits Symp., Oct. 2008, pp. 1–4.

[10] A. M. H. Kwan, X. Liu, and K. J. Chen, "Integrated gate-protected MOS-HEMTs and mixed-signal functional blocks for GaN smart power ICs," in Proc. IEEE IEDM, Dec. 2012, pp. 7.3.1–7.3.4.

[11] B. Cheng, M. Cao, R. Rao, A. Inani, V.V. Paul, W.M. Greene, J.M.C. Stork, Y. Zhiping, P.M. Zeitzoff, J.C.S. Woo, The impact of High-K gate dielectrics and metal gate electrodes on sub-100nm MOSFETs, IEEE TED 46 (7) (1999) 1537-1544.

[12] J. Wei et al., "Enhancement-mode GaN double-channel MOS-HEMT with low on-resistance and robust gate recess," 2015 IEEE International Electron Devices Meeting (IEDM), Washington, DC, 2015, pp. 9.4.1-9.4.4. doi: 10.1109/IEDM.2015.7409662

[13] N. Wichmann, I. Duszynski, S. Bollaert, J. Mateos, X. Wallart and A. Cappy, "100nm InAlAs/InGaAs double-gate MOS-HEMT using transferred substrate," IEDM Technical Digest. IEEE International Electron Devices Meeting, 2004., San Francisco, CA, 2004, pp. 1023-1026.

[14] Breitschadel O, Kley L, Grabeldinger H, Hsieh JT, Kuhn B, Scholz F, et al. Shortchannel effects in AlGaN/GaNMOS-HEMTs. Mater SciEng B 2001;82(1–3):238–40.

[15] H. Pardeshi, Analog/RF Performance of AlInN/GaN Underlap DG MOSMOS-HEMT, Superlattices and Microstructures (2015), doi: 10.1016/j.spmi.2015.10.009.

[16] B.C. Paul, A. Bansal, K. Roy, Underlap DGMOS for digital-subthreshold operation, IEEE Trans. Electron Devices 53 (4) (Apr. 2006) 910e913.

[17] I.M. Kang, H. Shin, Non-quasi-static small-signal modeling and analytical parameter extraction of SOI FinFETs, IEEE Trans. Nanotechnol. 5 (3) (May 2006) 205-210.

[18] International Technology Roadmap for Semiconductor, 2008.

[19] Sentaurus TCAD Manuals, Synopsys Inc., Mountain View, CA 94043, USA. Release C-2009.06.

[20] Koley, Kalyan, Arka Dutta, BinitSyamal, Samar K. Saha, and Chandan Kumar Sarkar. "Subthreshold analog/RF performance enhancement of underlap DG FETs with high-k spacer for low power applications." IEEE Transactions on Electron Devices 60, no. 1 (2013): 63-69.

Optimisation and Classification of EMG signal using PSO-ANN

Virendra Prasad Maurya
Electrical Engineering Department
National Institute of Technology
Durgapur, India
vir61191@gmail.com

Prashant Kumar
Electrical Engineering Department
National Institute of Technology
Durgapur, India
raja.prashant89@gmail.com

Dr. Suman Halder
Electrical Engineering Department
National Institute of Technology
Durgapur, India
sum_hal@yahoo.co.in

Abstract— Qualitative feature extraction from Electromyogram (EMG) signal has become necessary to assess the fitness of human being. Till date, various analysis tools have been employed to examine the EMG signal. Here the authors are endeavored to apply PSO-ANN based optimisation and two classification tools, namely KNN (nearest neighbor) and SVM (support vector machine) to extract features from EMG signal. EMG signal represents the signal generated by neuron from the brain, which is transmitted through the spinal cord into the body to which part is guided by the brain. The EMG signal is computed by Biopac MP45 Biomedical measurement device which is further divided into five-second segments for each activity. Unwanted EMG signal is regarded as noise and is filtered by an appropriate filter to improve the signal to noise ratio. Fourteen different time-domain and frequency domain features have been extracted for different hand movement (Weight lifting Up, Weight lifting Down, movement of Hand Gripper). Both hands are utilized for acquisition of EMG for hand grip movement. Classifier Model is used in classifying the optimised features and calculation of sensitivity, selectivity and precision of those features. From results it is evident that better accuracy is achieved for classifier KNN with respect to SVM.

Keywords—Electromyogram (EMG), Particle Swarm Optimisation (PSO), K-Nearest Neighbor (KNN), Support Vector Machine (SVM)

I. INTRODUCTION

The brain controls all bodily function. It transmits the communication signal to the spinal cord with the help of neurons. The action of any body part depends on action potential propagation through the neurons and that action takes place if the potential is more than the threshold [1]. The signal from the anterior horns of the spinal cord is transmitted in the form of motor neuron which propagates to the muscle cell through the nerve. In this way the electrical signal is generated by the spinal cord in membrane, these electrical signals converted into chemical energy in nerve ending which stimulates the neuromuscular junction and enters the nerve motor end plate in the form of electrical signal. That electrical signal is the EMG signal.

In this paper, the EMG signal is utilised in extracting the feature of the different hand movement. This signal can be employed for the detection of the disease like myasthenia gravis, amyotrophic lateral sclerosis [2], cerebral palsy. Other wields of the acquired EMG signals are soft six-finger employment in chronic stroke peasant as a robotic finger [3], gesture recognition [4]. Gesture recognition supports in communicating within a noisy environment also fruitful for deaf-people. EMGs is estimated by the non-invasive method, by providing electrodes at the specific location on the body. In this experiment, three electrodes were applied at a fixed position. The acquired signal is in nonlinear form so different filter has been used for improving the signal characteristics such as butterworth filter. Data is obtained at the sampling frequency (Fs) of 2000 Hz, and band pass filter with the aid by butterworth filter having bandwidth 5 Hz to 500 Hz. Feature extraction can be accomplished in three domains which are mainly time, frequency and time-frequency domain. Average amplitude change, approximate entropy, difference absolute standard deviation value, integrated EMG, kurtosis, log detector, mean absolute value, root mean square, sample entropy, simple square integral, variance, wavelength, skewness are the major time-domain features while mean frequency, mean power, spectral moment, total power are frequency-domain features and in time-frequency form wavelet transform are considered for feature extraction.

For classification of EMG signal, data is optimised. There are various optimisation methods namely particle swarm optimisation (PSO), genetic algorithm (GA), Fuzzy analysis and hybrid techniques. Extracted features are classified into different classes with the use of classifiers like KNN (K-Nearest Neighbor), support vector machine (SVM), best first, linear forward selection. Fig. 1 shows the EMG signal of weight lifting in which waveform showing in time duration between 4.5-7 sec is weight-lifting up EMG signal, that of between 10-12.5 sec is for weight-lifting down and the rest is the resting position with the weight either at up position or down position.

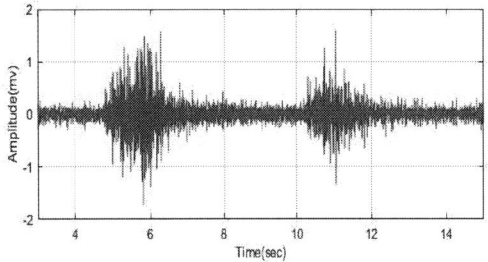

Fig. 1. EMG due to weight lifting

II. MATERIAL AND METHODS

A. Participants

Total twenty-one number of interested participants have taken part in this study, from which four are female and seventeen are male. All the basic information about the study is provided to the contributor and a written consent form has been filled by each participant for this study.

B. Data Acquisition

Participant data is taken in the standing position as shown in Fig. 2.

EMG is obtained with the help of Biopac MP45. Biopac MP45 consists of two channel's which is computer interfaceable. For this EMG signal study, only one channel is employed. Three electrode positions are mentioned in Fig. 2, which is coupled to the channel through the lead wire. White lead near the elbow as a negative terminal electrode, the red electrode as a positive terminal and ground electrode in black color near the wrist.

Total twenty-one contributor's data are extracted of 65-sec duration, which included dumble-up, dumble-down with resting period included in between them. Another set of data were accumulated with Hand Gripper of the same period.

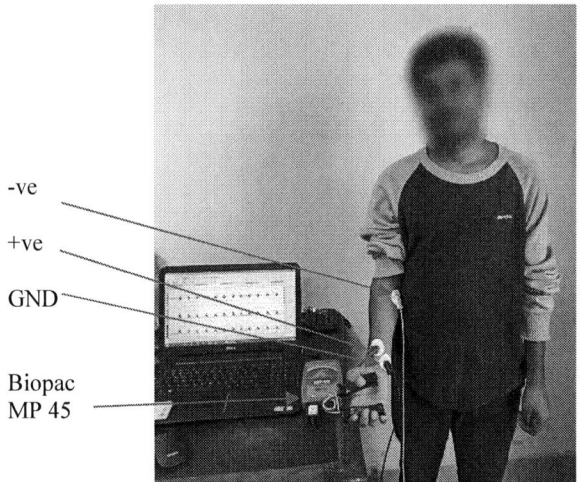

Fig. 2. Position of Electrode for EMG Data Acquisition

III. DATA PROCESSING

A. Removal of Power Line Frequency

The power line frequency in India is 50 Hz which act as interference to the original signal, so notch butterworth filter is employed to remove the particular frequency. Butterworth filter is an analog filter. Equation (1) represents the butterworth filter of Nth order.

$$\left| B_a(\Omega) \right|^2 = \frac{1}{\left(1 + \left(\dfrac{\Omega}{\Omega_c} \right)^{2N} \right)} \qquad (1)$$

Where, B_a shows the magnitude of the butterworth at any particular frequency of desire order. Ω manifests the particular frequency at which analysis has to be done. Ω_c Cut-off frequency and it is the frequency where magnitude 3 db attenuation occurs. N is the order of the filter.

For expulsion of the power line frequency

$$\Omega_0 = \pm 2\pi \frac{f_0}{f_s} \qquad (2)$$

Where, f_o is the power line frequency and f_s is sampling frequency.

B. Removal of Noise using Butterworth Filter

For an ideal low pass filter, the slope at its cut-off frequency should be 90° which is practically not possible. An increase in the order will improve low pass response but it will

lead to higher cost of filter design. So, order is selected considering economic concern also and there is a trade-off between SNR and order of the filter.

$$B_p = \frac{\Omega_c^N}{\Pi_{k=1}^{N/2}(s^2 + b_k \Omega_c s + \Omega_c^2)} ; \qquad \text{N=Even} \qquad (3)$$

$$B_p = \frac{\Omega_c^N}{(s + \Omega_c)\Pi_{k=1}^{(N-1)/2}(s^2 + b_k \Omega_c s + \Omega_c^2)} ; \quad \text{N= Odd} \quad (4)$$

$$b_k = 2\sin\left(\frac{(2k-1)\pi}{2N} \right) \qquad (5)$$

When N is even, all poles are complex conjugate, no real pole but, when N is odd, there must be one real pole and other poles are complex as seen in (3) and (4).

Fig. 3 shown are the EMG filtered signal after power line frequency dismissal, low pass and band pass filtering. Zoomed view of the waveform is also shown to clear envision.

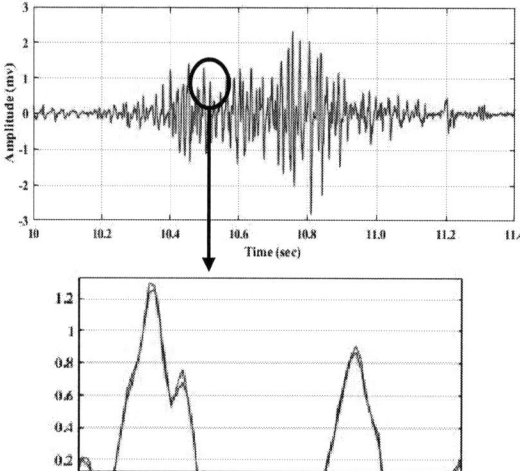

Fig. 3. EMG filtered Signal (Normal and Zoomed View)

C. Feature Extraction

After applying the filter on raw data, the signal to noise ratio improved, and then the feature is extracted mainly in time domain and frequency domain [5,6].

- Time Domain

  The amplitude of any signal contained the energy of the signal, which is helpful in transmitting the signal from one place to another place. The average change in amplitude represents the variation in energy. In given data amplitude varies from -2 to 2 mV. So average aptitude changes in EMG signal represented by the plotted wave. Approximate Entropy represents the disorder in given data set in time series form. The difference absolute standard deviation value notify that how much absolute value of data deviated from its mean value. Integrated EMG signal gives the rectified absolute value of EMG data. Kurtosis grand a bundle of data across the center of the distribution. Skewness assigns the symmetry or lack of symmetry in data. For perfect symmetry, the value of skewness is zero.

- Frequency domain

Mean frequency shows that the mean of the data which is handy in obtaining deviation in data from the mean value. Median frequency assigns the middle of the data set.

All features parameters used for this study are shown in TABLE I.

TABLE I. Feature Extraction Methods [6]

Sr. No	Features	Mathematical Definition
1	Average Amplitude Change	$AAC = \dfrac{1}{M} \sum\limits_{m=1}^{M-1} \lvert S_{m+1} - S_m \rvert$
2	Approximate Entropy	$ApEn(p,q,M) = \left[\Psi^p(q) - \Psi^{p+1}(q) \right]$ $Where, \Psi^p(q) = (M-p+1)^{-1} \sum\limits_{m=1}^{M-p+1} \log C_m^p(q)$
3	Difference Absolute Standard Deviation Value	$DASDV = \dfrac{1}{M-1} \sqrt{\sum\limits_{m=1}^{M-1} \left(s_{m+1} - s_m \right)^2}$
4	Integrated EMG	$IEMG = \sum\limits_{m=1}^{M-1} \lvert s_m \rvert$
5	Kurtosis	$Kurt = \left(\dfrac{1}{m} \sum\limits_{m=1}^{M} \left[\dfrac{s_m - s}{6} \right]^4 \right) - 3$
6	Log detector	$LOG = e^{\frac{1}{M} \sum\limits_{m=1}^{M} \log \lvert s_m \rvert}$
7	Mean absolute value	$MAV = \dfrac{1}{M} \sum\limits_{m=1}^{M} \lvert s_m \rvert$
8	Root Mean Square	$RMS = \sqrt{\dfrac{1}{M} \sum\limits_{m=1}^{M} \lvert s_m \rvert^2}$
9	Simple Square Integral	$SSI = \sum\limits_{m=1}^{M} \lvert s_m \rvert^2$
10	Variance	$VAR = \dfrac{1}{M-1} \sum\limits_{m=1}^{M} \lvert s_m \rvert^2$
11	Skewness	$Skew = \left(\dfrac{1}{M} \sum\limits_{m=1}^{M} \left[\dfrac{s_m - s}{6} \right]^3 \right)$
12	Mean frequency	$MNP = \left(\sum\limits_{k=1}^{P} f_k x_k \right) \Big/ \sum\limits_{k=1}^{P} x_k$
13	Mean Power	$MNP = \left(\sum\limits_{k=1}^{P} x_k \right) \Big/ P$
14	Total power	$TTP = \sum\limits_{k=1}^{P} x_k$

D. Optimisation Technique

For optimising the data, there are several techniques, for example, Particle swarm optimisation (PSO), Genetic algorithm, Fuzzy analysis, Differential Evolutionary. In this paper, the PSO technique is wielded for optimising the data.

In the PSO algorithm, some defined function like weight and some constant value in velocity function change iteratively according to best optimisation.

Fitness function selection for any depends on the fitness value, in this paper standard deviation is used as a fitness function. More than one solution is called population in PSO, by increasing the population convergence iteration decrease but increase the computational time. In general constant factors C1 and C2 in velocity function are in the range of 0 to 4, but in the practical case, it is between 1 to 4, and weight value in that function in the range of 0.4 to 0.9 [7]. For this study, both C1, C2 and w values are taken as 2 and 0.4 respectively.

E. Classification

It alleviates in the best way to select the data from the raw data. For this study, KNN (K-Nearest Neighbor), SVM (Support Vector Machine) are used as the classifier.

K-NN is instance-based learning, here k is variable and select as required. K=h, means unselected 'h' point which is the least distance from the select point. Euclidean distance and Manhattan distance are mainly used for distance measurement. Euclidean distance method evaluates the distance using (6). Manhattan distance use the real vector sum of their absolute distance and is given in (7).

$$D_i = \sum_{k=1}^{h} \sqrt{(x_i - x_k)^2 + (y_i - y_k)^2} \qquad (6)$$

Where, D_i is distance measure from the chosen point to the other point of the taken domain. (x_i, y_i) The chosen point and (x_k, y_k) is another point in the taken domain.

$$D_i = \lvert x_i - x_k \rvert + \lvert y_i - y_k \rvert \qquad (7)$$

SVM: In Support Vector Machine learning classifier hyperplane best splits the data, As for as possible from support vector. Hyperplane divides the data set into a particular group of data set, which help in separating the data in the same domain from the raw data [8]. Tuning parameters in SVM are Kernel, Regularisation, Gamma and Margin.

IV. CLASSIFIER MODEL

Confusion matrix is used to describe the performance of the classifier model. It is a matrix which represents the relation between Actual and Predictive event. It is also known as Error Matrix.

TN (True Negative): It describes that prediction matches the actual negative event.

FN (False Negative): Prediction is contradicted the actual event means Actual negative event shows the Positive Prediction.

FP (False positive): Actual event happens but prediction shows opposite the actual event.

TP (True Positive): Prediction Match with the Actual Event.

TABLE. II CLASSIFIER MODEL

	Predicted No	Predicted Yes
Actual No	TN	FN
Actual Yes	FP	TP

TABLE II is the classifier model of confusion matrix from which specificity, precision and accuracy can be calculated as indicated in (8), (9) and (10) respectively.

$$Specificity = \frac{TN}{TN + FP} \qquad (8)$$

$$\mathrm{Pr}\,ecision = \frac{TP}{TP + FP} \qquad (9)$$

$$Accurecy = \frac{TP + TN}{TP + TN + FP + FN} \qquad (10)$$

V. RESULTS

In this study, total twenty-one participants EMG data are considered and the total fourteen number of features are taken both in time and frequency domain are extracted after the removal of various noises. These features are optimized using PSO and six optimised features are the consequence for further classification. Two classifier KNN and SVM are used for calculation of accuracy, sensitivity, specificity etc. The result for the accuracy of the above-mentioned classifier is 99.7% & 99.6% respectively.

Fig. 4 and Fig. 5 are the confusion matrix of KNN classifier using Classification Learner app in MATLAB 2018a. This gives overall sensitivity and specificity of greater than 97 % using the KNN classifier.
The nomenclature used in the plots are aberrated as

LHC: Left Hand Clip
LHD: Left Hand Down
LHU: Left Hand Up
RHC: Right Hand Clip
RHD: Right Hand Down
RHU: Right Hand Up

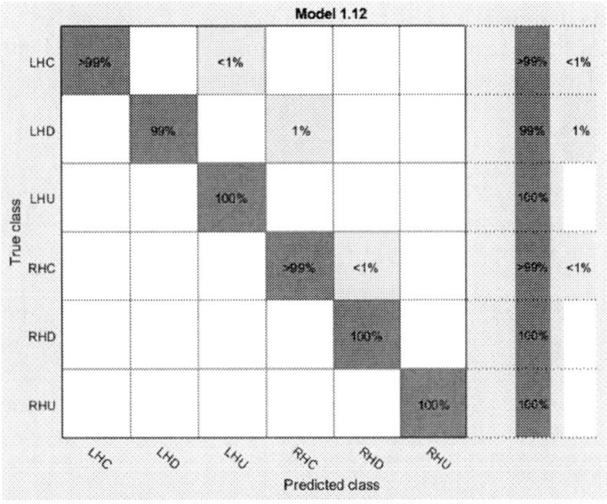

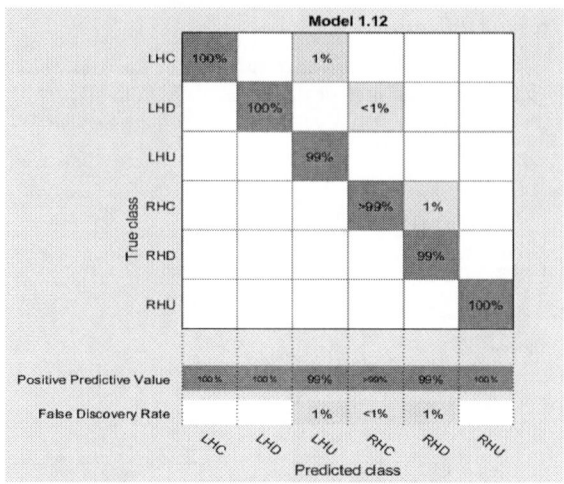

Fig. 5. Confusion Matrix using KNN for PPV, FDR

In the same manner, confusion matrix for SVM classifier has been plotted in Fig. 6 and Fig. 7. This gives overall sensitivity and specificity of 97 % and specificity of 98% using the KNN classifier.

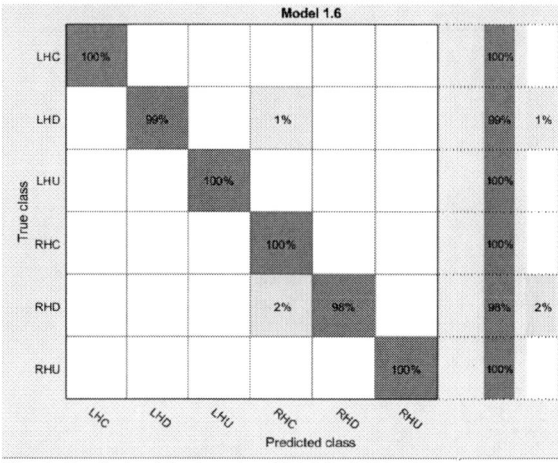

Fig. 6. Confusion Matrix using SVM for TPR, FNR

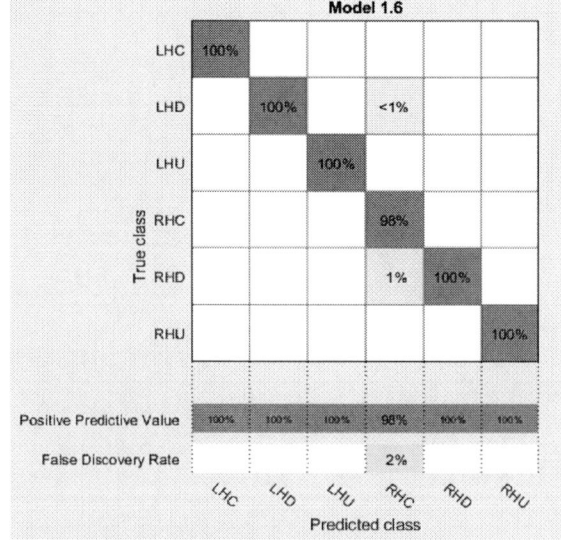

Fig. 4. Confusion Matrix using KNN for TPR, FNR

Fig. 7. Confusion Matrix using SVM for PPV, FDR

978-1-5386-6723-1/19 $31.00 © 2019 IEEE

Fig. 8 and Fig. 9 are the scatter plot of two different classifiers namely KNN and SVM respectively of each class.

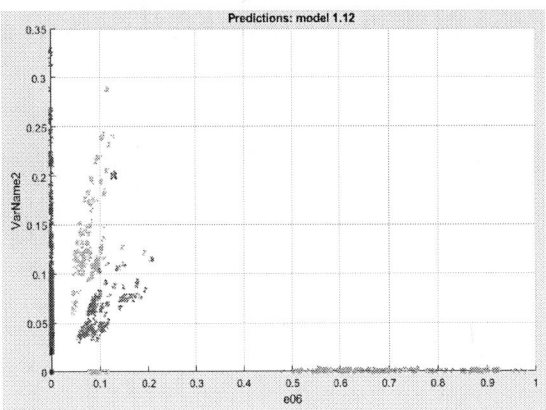

Fig. 8. Scatter Plot using KNN Classifier

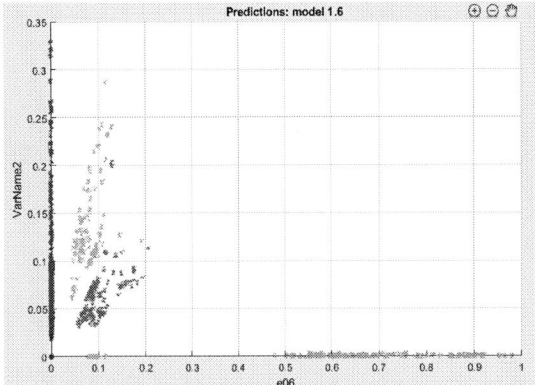

Fig. 9. Scatter Plot using SVM Classifier

For the simplicity, ROC plot of the KNN classifier only is shown in Fig. 10 which shows the area under the curve of 1.00 that is the probability of the model ranks random positive example is highest among the random negative example and an indication of proper classification result.

Fig. 10. Region of Convergence Plot using KNN Classifier

TABLE III COMPARISON OF ACCURACY IN DIFFERENT ALGORITHM

Algorithm	Method	Accuracy (%)
[6]	MEDA (std dev)	91.8
Proposed PSO-KNN	Butterworth (std dev)	99.7

In TABLE III, different comparison shown in which proposed PSO-KNN with butterworh is provided more accuracy as compare to Minimum entropy deconvolution analysis [6].

VI. CONCLUSION AND FUTURE SCOPE

In this paper, the EMG signal is considered and the feature is extracted considering time-domain and frequency-domain. For rejection of various noise in EMG signal, butterworth filter is used for removing the power line frequency and to improve the signal to noise ratio. In the process of feature extraction, fourteen features are used which are aberrated as ACC, ApEn, DASD, IEMG, Kurt, LOG, MAV, SSI, RMS, VAR, SKEW, MNF, MNP, TTP. For optimisation, PSO is used for a better feature. Six optimised features have been selected among fourteen features. Classifier Model is used further for classifying the optimise data into the different class. The accuracy of this KNN method is 99.7% and that of SVM technique is around 99.6%. Adaptive filter can be used for removal of power line frequency in the future. Along with that a combined time-frequency domain, wavelet transformation can also be used for feature extraction with another hybrid classifier.

REFERENCES

[1] Daniel W. Stashuk, "EMG signal decomposition: how can it be accomplished and used", Journal of Electromyography and Kinesiology, vol.11, pp. 151–173, 2001.

[2] M chen,X. Zhang And P.Zhou, "A novel validation approach for high-density surface EMG decomposition in motor neuron disease". IEEE Transaction on Neural Systems And Rehabilitaion Engineering, vol. 26(6), pp.1161-1168, June, 2018.

[3] Hussain,G.Salvietti,G. Spagnoletti And D. Prattichizzo, "The Soft-Sixth finger: A noble EMG controlled robotic extra-finger for grasp compensation in chronic strike patients". IEEE Robotics And Automation Letters, vol. 1(2), pp 1000-1006, July, 2016.

[4] A. Moin et.al, "An EMG gesture recognition system with flexible high-density sensors and brain-inspired high-dimensional classifier", IEEE International Symposium on Circuit And Systems, pp.1-5, Florence, April, 2018.

[5] Doulah, A. B. M. S. U., Ahmad, M. O., Fattah, S. A. and Zhu, "DCT domain feature extraction scheme based on motor unit action potential of EMG signal for neuromuscular disease classification". Healthcare Technology Letters, vol. 1(1), pp.26–31., 2014.

[6] Powar, O. S., Chemmangat, K. and Figarado, S, "A novel pre-processing procedure for enhanced feature extraction and characterization of electromyogram signals", Biomedical Signal Processing and Control, vol. 42, pp. 277–286, 2018.

[7] Hongqiang LiEmail, Xiuli FengLu Cao, Enbang LiHuan and Liang Xuelong Chen, "A new ECG signal classification based on WPD and ApEn feature extraction", Circuits, Systems, and Signal Processing, vol. 35(1), pp 339–352, January, 2016.

[8] Abdulhamit Subasi, "Classification of EMG signals using PSO optimized SVM for diagnosis of neuromuscular disorders", Computers in Biology and Medicine, vol. 43(1), pp. 576-586, June, 2013.

Design and Implementation of Gabor Filter and SVM based Authentication system using Machine Learning

Shalini Singh
Dept. of Information Technology
Meghnad Saha Institute of Technology
Kolkata, India
shalini70396@gmail.com

Amogh Banerjee
Dept. of Information Technology
Meghnad Saha Institute of Technology
Kolkata, India
amoghbanerjee20@gmail.com

Indrajit Das
Dept. of Information Technology
Meghnad Saha Institute of Technology
Kolkata, India
indrajitdas1979@hotmail.com

Md Golam Mohiuddin
Dept. of Information Technology
Meghnad Saha Institute of Technology
Kolkata, India
mdgolammohiuddin892@gmail.com

Sonali Gupta
Dept. of Information Technology
Meghnad Saha Institute of Technology
Kolkata, India
sonaliguptag2@gmail.com

Abstract— **The most vital requirement in today's world is to overcome the different types of attacks. Human behavioral and physiological features in biometrics have the largest scope as a solution for security issues. However, the existing biometric systems such as faces, iris, palm, voice or fingerprints are highly complex in terms of time or space or both, and thus are not suitable in high security. So the design and implementation of finger-vein authentication method is proposed in this paper. This system is implemented using a combination of image processing and machine learning algorithm. Lacunae, fractal dimension and gabor filter are the algorithms used for feature extraction and the classification of the extracted feature is done using the Support Vector Machine. The accuracy of classification algorithm for One-Versus-One and One-Versus-All is 98.75 % and 97.92 % and the execution time is 0.168 Seconds and 0.187 Seconds respectively. At the end the comparative analysis between different classification algorithm and previous research work related to Finger Vein Authentication System using Machine learning is provided.**

Keywords—Biometric Authentication System, Finger Vein Detection, CLAHE, Gabor Filter, Support Vector Machine, Fractal Dimension, Lacunae

I. INTRODUCTION

Nowadays the security of personal information is of high importance. The Conventional biometrics personal authentication system, such as faces, iris, palm, voice or fingerprints have several drawbacks [1] i.e. the features can be mimicked because of physical appearance. Thus, are vulnerable to damage or change with time.

Finger print authentication has several drawbacks such as it can easily be ripped off by using a fake finger. If there are some injuries or dirts present in finger it is often unrecognized. Since Iris recognition depends upon the brightness of the light during the biometric capture process it makes recognition less precise [1].The light condition, distance from the camera, facial expressions etc creates problem at the time of Facial recognition. Another biometric is voice recognition that is likewise utilized in some places. Yet, the fundamental issue with it is that spoofing can be done using recorded voice. The above mentioned physiological characteristics and behavioral pattern have certain limitations. Hence, a more secure biometric

authentication system is needed. Finger vein pattern recognition method is a solution to it.

Vein is a collection of blood vessels which transfers blood towards the heart. The main advantage of using finger vein patterns in security purpose is that it lies under the skin and is unique. Unlike existing biometric methods, for example, faces, iris, voice, palm and fingerprints, it is impossible to replicate. Finger-vein features show a few other astounding advantages and these include:

- It gives great distinction between people as it includes identical twins.
- They are static in nature.
- Finger vein patterns cannot be easily forged, observed, damaged, or obscured like others.
- It is gaining importance as the capturing gadget is comparatively little and recognition accuracy is generally high.
- The images can be caught noninvasively with contactless sensors for convenience and cleanliness of the user.
- Other fingers can be used as an alternative for authentication, if something unexpected happens to one finger of person [2].

Along with the advantages there are some disadvantages associated with it which needs to be eliminated for achieving higher performance in real world.

- In acquisition process, the position of the back of the finger and the camera is very close, so the optical blurring on the captured image occurs, deteriorating the quality of the image [3].
- The images may appear extremely dark or bright depending on the lighting at the capturing device [4].
- Under the poor guidance of the finger position, misalignment occurs which further decreases the recognition rate [5].
- Matching process fails if the finger position is not properly aligned.

978-1-5386-6723-1/19 $31.00 © 2019 IEEE

- Provided that the skin layer isn't straightforward and uniform, light scattering happens because of various thicknesses of bones and skin [6].
- The noise associated with it needs to be removed as much as possible [7].

In this paper a low cost and higher accuracy finger vein image acquisition device (developed in house) is provided with the help of Near Infrared Imaging and CCD camera [8]. While capturing image, region of interest is extracted through threshold value, followed by the enhancement of the input image which is done using Contrast Limited Adaptive Histogram Equalization (CLAHE). Then the enhanced image is converted to gray scale and feature is extracted through texture (Fractal Dimension and Lacunae) and edge extraction (Gabor filter) and then the extracted images are learned using multi SVM, finally classification is done using SVM.

The following part of the paper includes literature survey, proposed methodology and conclusion respectively.

II. LITERATURE SURVEY

Over the time, various researchers throughout the world have already been working in the domain of vein pattern identification system and have utilized and imagined a heap of estimation that evaluates the execution of their proposed methodologies and plans. In this segment, we have advanced a few survey of such existing research along with their corresponding accuracy and computational execution time.

Researcher J-D Wu et al. [9] have proposed an individual recognizable system utilizing finger-vein designs with component analysis and neural network methodology. In this paper, the finger-vein designs are caught by a gadget that can transmit near infrared through the finger and record the pattern for confirmation which comprises of a combination of feature extraction using PCA, pattern classification using BP network and ANFIS respectively. To check the impact of the proposed ANFIS in the pattern classification, the BP network is contrasted with the proposed framework. The exploratory outcomes showed the proposed framework utilizing ANFIS has preferred execution over the BP network for personal identification utilizing the finger-vein patterns and its accuracy and execution time is 99% and 4.5 seconds.

Researchers K-Q Wang et al. [10] have advocated that finger vein identification technique is a steady biometrics method that has numerous points of interest when contrasted with different strategies. In this paper, local binary pattern variance (LBPV) has been proposed to overcome the featured rotationally invariant challenges and to describe the nearby complexity data into one-dimensional LBPH. Further global matching technique is utilized to additionally stimulate the matching method and to diminish feature measurements using distance measurement and minimize computational expense. The classification is done by Support vector machine (SVM), which gives different accuracy and execution time i.e. for little, middle and ring finger are (79%, 2.3 seconds), (90%, 2.1 seconds) and (96%, 1.78 seconds) respectively .

Researchers A.N. Hoshyar et al. [11] have advised distinctive techniques incorporating gait, face, fingerprint, iris, retina, voice, DNA, hand vein information while iris, fingerprint, face etc. are considered as conventional ID strategies. Every strategy has its own disadvantages. Standing out from different biometrics, finger vein designs makes the frameworks increasingly secure and recognizable on the grounds that they are covered up inside the body and the circumstances of outside skin can't affect on that. This paper researched about Smart Access Control utilizing Finger Vein authentication and Neural Network. Fourteen finger vein pictures are gathered from people by putting finger under close infrared light, then cropped followed by noise reduction process. After that pattern extraction is done by Morphological Operation and Maximum Curvature Points is done. After extracting the finger vein image features, Neural Network was utilized for training purpose and the classification accuracy is 93% with 2.754 seconds execution time.

Researchers J-D Wu et al. [12] considered SVM strategy for finger-vein Identification system. In this work, LDA and PCA are used for image pre-processing as measurement decreases and features are extracted. For pattern classification, this framework has utilized SVM and ANFIS. Noise reduction is done using PCA whereas LDA is used for holding the main features. These extracted features are then used for classification. The outcome demonstrates a better performance (98%) to the artificial neural network of ANFIS with minimum execution time 0.015 seconds.

Researchers S. Khellat-kihel et al. [13] suggested that biometric identification is advantageous than other authentication systems like password that are subjected to falsification. This paper intends to display a finger vein authentication framework utilizing SVM in view of a supervised training. The system has two preprocessing (Histogram Equalization and Median Filter and the second one is based on Gabor filter) method which are used for better understanding the efficiency of Gabor filter. Texture extraction of veins is done by Gabor filter that catches local orientation. This features vector will be used for learning and classification of the individual. The classification is done by SVM. The accuracy and execution time is 98.75% and 0.00246 seconds respectively.

Researchers S.A. Radzi [14] suggested convolutional neural system for finger-vein authentication system. In ordinary finger-vein authentication system, complex image handling is required to evacuate noise, extract and upgrade the previous image classification performed to accomplish superior accuracy, whereas in CNN its capacity is to extricate features at same time, lessen information dimensionality and classify in one network structure makes it better. What's more, the strategy requires just minimal image preprocessing since the CNN is vigorous to noise and little misalignments of the obtained images. Here a four-layer CNN is fused with convolutional sub sampling architecture and for network training a stochastic diagonal

Levenberg–Marquardt algorithm is used which have resulted in a quicker convergence time (0.1574 sec) and better accuracy (99.38%).

The consecutive section explains the proposed methodology.

III. PROPOSED METHODOLOGY

Finger-vein authentication system can be divided through three methodologies. The primary methodology deals with image processing whereas the second methodology is implemented by applying machine learning system. This paper deals with a combined approach of image processing and machine learning (shown in Fig .1).

A. Image Acquisition

Finger vein patterns are imperceptible to the bare eye and visible by a sensor sensitive to near-infrared light (wave lengths somewhere in the range of 700 and 1000 nanometers) [15-16]. In this paper 850 nanometers NIR are used for taking finger vein images. Near infrared light passes through the tissues of the human body where it is further obstructed by hemoglobin or melanin.

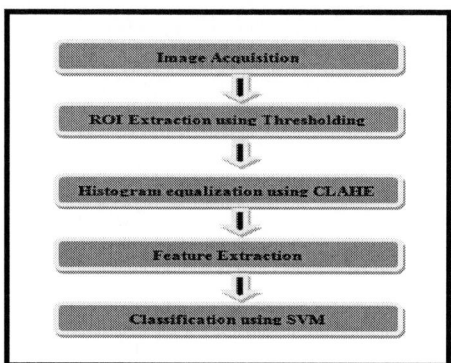

Fig. 1. Proposed Method

As hemoglobin or melanin exists thickly in veins, near infrared light sparkling makes the veins show up as dark lines of shade. There are two strategies utilized for capturing the vein pattern "light reflection" and "light transmission" as shown in Fig. 2 and Fig. 3. In this paper light transmission method is utilized, where the finger is set between the CCD and the NIR light source goes through the finger where it is caught by the CCD.

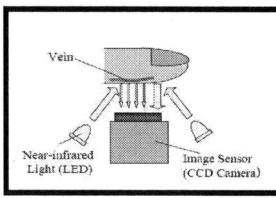
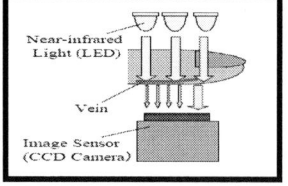

Fig. 2. Light Reflection Method   Fig. 3. Light Transmission Method

B. ROI Extraction using thresholding

The Finger region extraction i.e ROI (Region of Interest) is acquired by thresholding the image. The thresholding is done by multidimensional filtering. The ROI extracted image is acquired by combining the cropped and threshold image [17]. The following procedures are given below:

> **Step1:** The mask of size is 4×20 where height (h)=4 and width(w)=20.
> **Step2:** Calculate the lower half beginning point.
> **Step3:** Initially the mask of size 4×20 =0, upper half of mask = -1 and lower half of mask =1
> **Step 4:** Mask is applied on the image border. Maximum of upper half and Minimum of lower half =1
> **Step5:** Fill the gap between upper half and lower half =0

C. Histogram Equalization using CLAHE

Adaptive Histogram Equalization is needed to equalize the low contrast bright and dark region. One disadvantage of this method is over amplification of the noise in homogeneous region of images. To overcome this Contrast Limited Adaptive Histogram Equalization is used. To limit the amplification CLAHE used cumulative distribution function (CDF). The cut limit rely on the normalization of histogram and the area of neighborhood and it is redistributed equally to whole histogram. This way the redistribution procedure is repeated until the extra value goes to negligible.

D. Feature Extraction

It consists of two parts one is texture extraction another is extraction of edges. Texture extraction is done through fractal Dimension and Lacunae algorithms. Later is done through Gabor filter.

Fractal Dimension

Fractal Dimension is a proportion giving a measurable record of complexity in fractal geometry. It compares pattern changes in detail with the scale at which it's measured. A box counting method is a popular approach for measuring Fractal dimensions. Three dimensionally extended box checking strategies have likewise been proposed for analyzing 3D volumetric information. The self-similarity concept is involved to predict fractal dimensions. Fractal dimension (FD) is expressed by:

$$FD = \log (Nr)/\log (1/r) \quad\dots\dots\dots\dots\dots\dots\dots\dots\dots\dots(1)$$

Here the fractal object is partitioned into N copies of a crude shape, every one scaled up or somewhere around a factor of r. There is a logarithmic connection among N and r. When Nr is computed for different estimations of r, the fractal measurement can be evaluated as the slope of the line. The line is assessed by the log (N) plotted against log (r). Here a software program is implemented that gauges fractal measurements by solving an equation.

Lacunae

Lacunae mean unfilled space or gap. It is used for measuring intensity varying from point to point. A situation might occur where different images may have same value of fractal dimensions; to differentiate such images lacunae algorithm is used [18].

Gabor Filter

Vein Detection is possible through extracting edges. The edges are sharp transition of pixels. Modulating a sinusoidal signal along with a Gaussian, Gabor Filter is determined. Let us take a function f(x, y, θ, φ) defining a Gabor filter. As θ and φ are the spatial frequency in the orientation, Gabor filter is centered at the origin with θ and φ [19]. The Gabor filter can be viewed as:

$$f(x, y, \theta, \phi) = \exp((-x^2 + y^2)/\sigma^2)\exp(2\pi\theta i(x\cos\phi + y\sin\phi)) \dots(2)$$

Here, σ is Standard Deviation which depends on θ.

E. Classification

Many new learning and classification algorithms are invented now a days, most of which are based on the statistical theory of learning. Support Vector Machine is one of them. Here support vector machine is used for better classification. It is a collection of training technique that is aimed to solve discrimination and regression problem. The main aim of the classifier is to calculate the boundary among the groups of points. For data separation a number of linear classifiers are used that are included in SVM [20]. Another objective of SVM is not only to find a separator between the classes, but also to find optimum hyper plane.

Gaussian Kernel is used for changing data space in SVM to avoid non-linearity case shown in Fig. 4. The SVM is nothing but the combination of two stages:

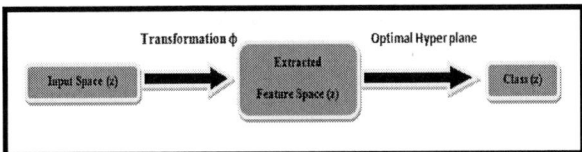

Fig. 4. SVM Space

Training

In this case quadratic program and Lagrange multipliers are used to maximize the margin to get an optimal hyper plane of separation.

Test

Lagrange multipliers that are determined are applied to the decision function of the test examples for determining class. The disadvantage of using SVM was that it can't handle real life problem which consist of multi class. To resolve this issue the only way was to extend the SVM principle for the problems consisting of more than two classes.

Two algorithms one-versus-all and one-versus-one are used and Gaussian is used to support kernel. In one-versus-one method each class is differing from another so N (N-1)/2 decision functions are required but in case of one-versus-all approach one class is differing from all the other classes, so here N decision functions are required.
In the next section, experimental results are well stated.

IV. EXPERIMENTAL RESULT

The experiment was performed using Python IDLE 3.6.4 on Windows 8.1 operating system running on Intel Core i5 with 2.4 GHZ. SQLITE 3 database was utilized for the reason for storing data (like Name and Finger subtleties of

people). Here the finger vein image acquisition system (developed in house) includes a near infrared LED with wavelength 850 nm and an infrared sensitive Charge – Coupled Device (CCD) camera (shown in Fig. 5). The CCD camera was used for its higher sensitivity in the NIR light than CMOS image sensor, which overcome the disadvantages such as optical blurring on the captured image, image deteriorate, misalignment of finger position and light scattering due to varied thickness of bones .

Fig. 5. Finger Vein Detector and NIR Light

Here database is used for training and testing purpose. The database consists of 10 individuals and each individual have 3 other folders containing the images of index, ring and middle finger. Each folder contains 6 images each, so a total of 18 images are stored for every individual. So the total number of test samples is 180.

After Image acquisition Finger region extraction i.e. ROI (Region of Interest) is acquired by thresholding the finger image of 10 individuals as shown in fig. 6. For normalizing the image here Histogram Equalization using CLAHE is used, so that each portion of image can get equal contrast as shown in fig 7. Followed by the feature extraction i.e. Texture extraction and Edge extraction is done. The Texture extraction is done by fractal Dimension and Lacunae algorithms and edge extraction is done with Gabor filter (shown in Fig. 8). Further the Classification is done by SVM. Two approaches are used i.e. one-versus-one and one-versus-all. Both the approaches are applied consisting of training set and test set consists of 10 classes with 18 trials resulting into 180 feature vectors in total. The test set also contains 10 classes along with four tests of each class i.e. venous networks are represented by a total of 80 feature vectors. The input images are taken, which are further compared with all 180 feature vectors by using N (N-1)/2 decision functions and if a match is found it will show a successful recognition. In this proposed methodology the average recognition accuracy for One-Versus-One and One-Versus-All is 98.75 % and 97.92 % and the execution time is 0.168 Seconds and 0.187 Seconds respectively. At the end the comparative analysis between different classification algorithm and different research work related to Finger Vein Authentication using Machine learning with this proposed methodology is provided in fig. 9 and Table. 1.

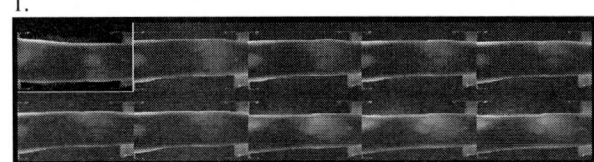

Fig. 6. ROI Extraction using thresholding

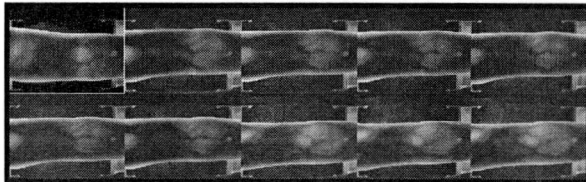

Fig. 7. Histogram Equalization using CLAHE

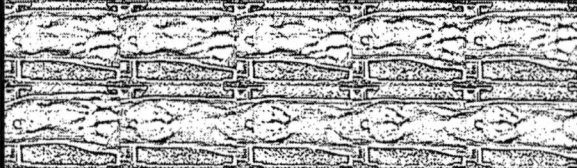

Fig. 8. Feature extraction

From figure 9 and Table 1 there are two algorithm ANFIS (99%) and Convolutional Neural Network (99.38%) whose classification accuracy is better than proposed methodology. But ANFIS has execution time 4.5 sec which is much larger than proposed methodology One-Versus-One (0.168 sec) and One-Versus-All (0.187sec) and CNN is not suitable for non ideal finger-vein cases, but the proposed methodology has no such problem like ANFIS which has larger execution

time (4.5 sec) and CNN which is only suitable for ideal finger vein cases.

Hence, the proposed methodology has better classification accuracy in compare to other algorithms except ANFIS and CNN and also has better advantages in terms of execution time, sensitivity of noise and suitability in non-ideal finger-vein cases.

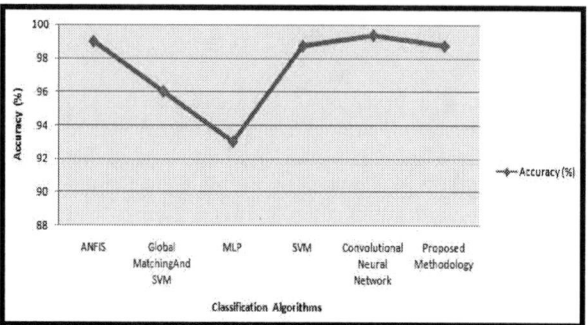

Fig. 9. Comparative analysis of Classification Accuracy

TABLE I. COMPARATIVE ANALYSIS OF DIFFERENT FINGER VEIN METHODOLOGY USING MACHINE LEARNING

	Database	No. of subjects	No. of test samples	Preprocessing	Feature extraction	Classifier	Accuracy (%)	Execution time (s)	Weaknesses
Proposed Methodology	Developed in-house (using CCD camera)	10	180	Thresholding and CLAHE	Fractal Dimension, Lacunae and Gabor Filter	SVM	One-Versus-One 98.75 % and One-Versus-All 97.92 %	One-Versus-One 0.168 and One-Versus-All 0.187	Too little number of subjects
[8]	Developed in-house (using CCD camera)	10	100	ROI extraction and Image resize	PCA	ANFIS	99%	4.5	Too few number of subjects, the execution time is too long.
[9]	Developed in-house (using CCD camera)	10	800	Gaussian matched filter	Local Binary Pattern Variance (LBPV)	Global Matching And SVM	79%, 90% and 96% For little, middle And Ring finger respectively	2.3, 2.1 and 1.78 for little, Middle and ring finger respectively	Too few number of subjects, low accuracy as for middle and little fingers.
[10]	Developed in-house (using digital camera with CCD sensor and IR filter)	7	14	ROI extraction, Median filtering and Histogram equalization	Morphological operation and Maximum curvature points	MLP	93%	2.754	Too few number of subjects, low accuracy
[11]	Developed in-house (using CCD camera)	10	100	ROI extraction and Image resize	PCA and LDA	SVM	98%	0.015	Too few numbers of subjects, SVM is sensitive to noise.
[12]	PKU Finger-vein Database(V2, V4)	20	48	Histogram equalization, Contrast amelioration, Median filtering and Gabor filter	Global thresholding and Gabor filter	SVM	98.75%	0.00246	Too few number of subjects
[13]	Developed in-house (VeCAD-UTM)	81	162	ROI extraction and Image resize	Convolutional Neural Network	Convolutional Neural Network	99.38%	0.1574	Not suitable for non-ideal finger-vein cases.

The following section provides the conclusion of our paper.

V. CONCLUSION

The driven conclusion is that with an increase in globalization, the average person's life has being threatened by the increasing heinous occurrences of attacks of personal information, verification and identity, so there is a great demand of high security. Hence the human physiological features are being used to provide better security solutions. Such existing methods are finger print, iris, palm, voice, face recognition etc, but they all can be easily cheated, copied and dubbed at one level and so better solution is needed for the security requirements which are cost efficient and reliable. Hence, embedded finger-vein recognition technique is approached in this paper. Feature extraction is done using algorithms like Lacunae, fractal dimension, and Gabor filter and the classification of the extracted feature is carried out using Support Vector Machine. The resulting experimental accuracy of recognition for One- Versus- One and One- Versus- All is 98.75% and 97.92% respectively and the resulting time of execution is 0.168 seconds and 0.187 seconds respectively. The comparative study of various classification algorithm and research works related to Finger Vein Authentication using Machine Learning with this proposed methodology is presented at the end. Thus, the proposed methodology is better regarding classification accuracy when contrasted with ANFIS and CNN and also have some more advantages in case of time of execution, sensitivity of noise and suitability in non-ideal finger-vein cases.

REFERENCES

[1] Jian-DaWu, Siou-HuanYe, "Driver identification using finger-vein patterns with Radon transform and neural network." Expert System with Applications, Vol. 36, Issue. 3 , Part. 2, April 2009; Pg. 5793-5799.

[2] Yanagawa T, Aoki S, Ohyama T. Human, "Finger vein images are diverse and its patterns are useful for personal identification." 21st Century COE Program, Development of Dynamic Mathematics with High Functionality; April. 2007. pp. 1-8.

[3] Lee EC, Park KR. , "Image restoration of skin scattering and optical blurring for finger vein recognition." , Optics and Lasers Engineering, 2011, pp. 816-828.

[4] U.D. Podgantwar, U.K. Raut, " Extraction of Finger-vein patterns using Gabor Filter in Finger vein image profiles.", International Journal of Engineering Research and Technology , Vol. 2, Issue. 6, June. 2013, pp. 3294 – 3298.

[5] J . Yang , Y. Shi , " Finger-vein ROI localization and vein ridge enhancement." , Pattern Recognition Letters , Vol. 33, Issue. 12, Sept. 2012, pp. 1569-1579.

[6] N . Miura , A .Nagasaka , T .Miyatake , " Feature extraction of Finger-vein patterns based on repeated line tracking and its application to personal identification." , Machine Vision and Applications , July, 2004, pp. 194-203.

[7] Z . Liu, Y .Yin, H .Wang , S .Song, Q . Li, "Finger vein recognition with manifold learning.", Journal of Network and Computer Applications , 2010, pp. 275-282.

[8] H. G Hong, M. B Lee and K .R Park "Convolutional Neural Network-Based Finger-Vein Recognition Using NIR Image Sensors" Sensors, Volume 17, issue 6 6 June 2017.

[9] J-D Wu, C-T Liu, " Finger-vein pattern identification using principal component analysis and the neural network technique." Expert Systems with Applications. 2011, Vol. 38, Issue. 5, pp. 5423–5427.

[10] K-Q. Wang, A. S. Krisa, X-Q Wu, Q-S Zhao, "Finger vein recognition using LBP variance with global matching." International

Conference on Wavelet Analysis and Pattern Recognition (ICWAPR), July, 2012, pp. 196-200.

[11] A.N. Hoshyar, R .Sulaiman, A.N.Houshyar, "Smart access control with finger vein authentication and neural network." Journal of American Science , 2011, Vol. 7, Issue. 9 , pp. 192–200.

[12] J-D Wu, C-T Liu , " Finger-vein pattern identification using SVM and neural network technique." Expert Systems with Applications. 2011, pp. 14284–14289.

[13] S. Khellat-kihel , R. Abrishambaf, N. Cardoso, J. Monteiro, M. Benyettou, " Finger vein recognition using Gabor filter and Support Vector Machine.", International Image Processing, Applications and Systems Conference (IPAS), 2014, pp.1-6.

[14] S.A. Radzi, M . Khalil-Hani, R. Bakhteri, "Finger-vein biometric identification using Convolutional Neural Network." Turkish Journal of Electrical Engineering and Computer Sciences. July, 2014, pp. 1863 – 1878.

[15] Hitachi, " Finger Vein Authentication- White Paper," Copyright , 2006.

[16] M. Naoto, A. Nagasaka, and M. Takafumi, " Feature extraction of finger-vein patterns based on repeated line tracking and its application to personal identification," Machine Vision and Applications, Springer- Verlag, 2004, pp. 194 -203.

[17] P. Gupta and P. Gupta, "An accurate finger vein based verification system", Digital Signal Processing, Volume 38, Issue C, March 2015, pp. 43 -52

[18] Z. Liu and S. Song, "An Embedded Real Time Finger Vein Recognition System For Mobile Devices", IEEE Transactions on Consumer Electronics, Vol. 58, No. 2, May 2012, pp. 522-527.

[19] A. Kumar and Y. zhou, "Human Identification Using Finger Images", IEEE Transactions on Image Processing, Vol. 21, NO. 4, April 2012,pp.2228-2244.

[20] K.R. Park, " Finger Vein Recognition by Combining Global and Local Features based on SVM", Computing and Informatics, Vol. 30, 2011, pp. 295 -309.

2019 Devices for Integrated Circuit (DevIC), 23-24 March, 2019, Kalyani, India

Pd Doped TiO₂ – CuO Mixed Metal Oxide Thin Film Sensor for H₂ Sensing Application

Bikram Biswas*
Department of E.T.C.E
Jadavpur University
Kolkata, India
*bikram94biswas@gmail.com

Anup Dey
Department of E.T.C.E
Jadavpur University
Kolkata, India
*anupetce@gmail.com

Subhashis Roy
Department fo E.T.C.E
Jadavpur University
Kolkata, India
subhashisaec@gmail.com

Sudhabindu Ray
Department of E.T.C.E
Jadavpur University
Kolkata, India
sudhabin@yahoo.com

Subir Kumar Sarkar
Department of E.T.C.E
Jadavpur University
Kolkata, India
su_sircir@yahoo.co.in

Abstract— **In this present work, high response Pd modified TiO₂-CuO thin film hydrogen sensors have been prepared by using sol-gel process. Prepared thin film was characterized by field emission scanning electron microscope (FESEM). Sensitivity of hydrogen sensors was observed at operating temperature 50⁰C-250⁰C and different gas concentrations (3000 ppm). It has been observed that Pd-doped thin film sensors have better responsivity in comparison with the un-doped thin film sensors.**

Keywords— Sol-gel-process, TiO₂-CuO thin film, Pd-modification, Surface morphology, H₂ sensing.

I. INTRODUCTION

Over the years, the fabrications of low size devices and reliable gas sensors with reduced power consumption have attracted considerable attention [1-3]. In this substance, several studies have been discussed on hydrogen detection due to safety reasons as an illustration, gas detection in industries, gas detection in food industries, environmental gas detection, gas detection in biomedical and space applications. Hydrogen is an odorless, colorless, non-toxic, non-metallic, highly combustible diatomic and explosive gas. So, it is a major challenge to detect gas efficiently.

Over the last few decades, the concentration measurement and detection of hydrogen gas leakage has been done and various approaches proposed to realize hydrogen sensor with effective performance [4-6]. However, the presently used H₂ sensors have some drawbacks like low response time, recovery time and poor longevity. So there is a requirement of high performance H₂ sensor for different industries.

Different metal oxides have been used for sensing material like ZnO, WO₃, CuO, TiO₂, Cr₂O₃, Mn₂O₃, Co₃O₄, NiO, Ta₂O₅, La₂O₃, CeO₂, Nd₂O₃ ,SrO, In₂O₃, V₂O₃, Fe₂O₃, GeO₂, Nb₂O₅, MoO₃, [7]. Now-a-days, mixed metal oxides are used for better sensing application. Some nobel metal are used like (Pd, Pt, Au, Ag) to modified the surface of the metal oxides for better sensitivity[8-10].These metal oxides are useful for the detection of toxic gases like H₂, CO, NO₂, CO₂ etc.

Mixed metal oxide based sensor have been widely used for gas detection. Various processes are available for deposition of TiO₂ -CuO mixed metal thin film, such as sol gel process, RF sputtering, pulse laser deposition (PLD), molecular beam epitaxy (MBE), metal organic chemical vapor deposition (MOCVD) etc. Sol gel process is very

efficient by virtue of being low cost, room temperature operation and simplicity.

In this proposed work, TiO₂-CuO mixed metal oxide thin film has been prepared by the use of sol-gel technique. To enhance the sensitivity of mixed metal oxides we incorporated PdCl₂ solution on the top of thin film sensors. We found that, Pd/ TiO₂-CuO thin film sensors obtained better sensitivity in comparison with the undoped thin film sensors. At operating temperature 150⁰C, proposed Pd-doped thin film sensors gets high sensitivity as compared to the undoped prepared samples.

II. EXPERIMENTAL DETAILS

Sensitivity of the thin film sensor has been measured by the equation 1.

$$S(\%) = \frac{R_n - R_g}{R_n} \times 100 \qquad (1)$$

Where, Rn is the resistance of the sensor in carrier gas (N₂) and Rg is the resistance in the presence of hydrogen gas.

The experimental setup is shown in Fig. 1. and the device structure is shown in Fig. 3 Nitrogen has been used as a carrier gas with a mass flow controller (MFC) which controls the gas flow and hydrogen as the target gas with a mass flow meter (MFM) which controls the gas flow. Hydrogen and nitrogen are mixed in the mixing chamber. The electrical behaviour of the thin film sensors in presence of H₂ and absence of H₂ are measured by using agilent multimeter (U1252A). Temperature of the measurement chamber is controlled by the temperature controller.

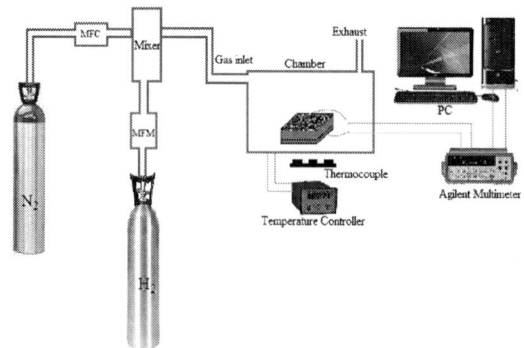

Fig. 1. Experimental Setup.

978-1-5386-6723-1/19 $31.00 © 2019 IEEE

III. FABRICATION STEPS

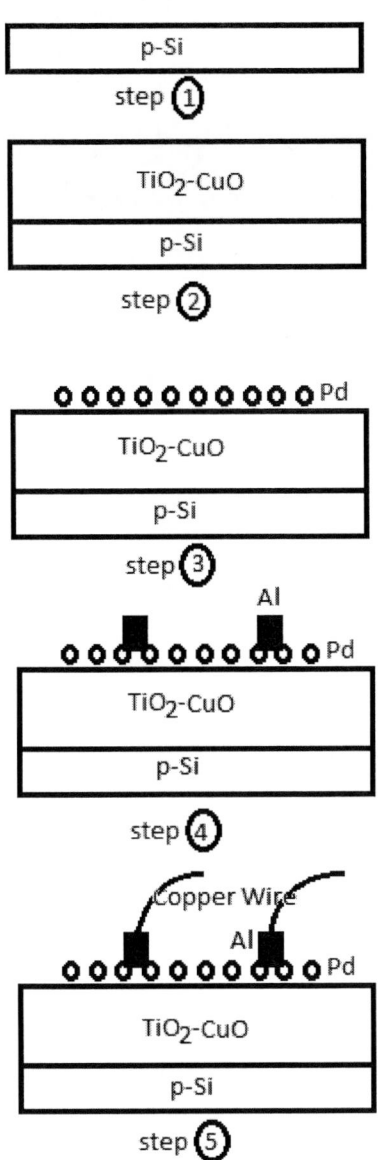

Fig. 2. Fabrication steps of prepared thin film sensor.

Fabrication steps of Pd modified thin film sensor for detection of hydrogen is shown in the Fig. 2 In step (1), p-Si substrate is taken and cleaned. Afetr cleaning in step (2), by using dip-coating meathod, prepared TiO_2-CuO thin film is deposited on the p-Si substrate. In step (3), $PdCl_2$ solution is prepared and deposited on the top of TiO_2-CuO layer by spin coating meathod (1200 rpm and 25 sec). Al metal electrodes is formed in the step (4). In step (5), copper wire is taken for connectivity of the prepared thin film devices.

IV. DEVICE STRUCTURES

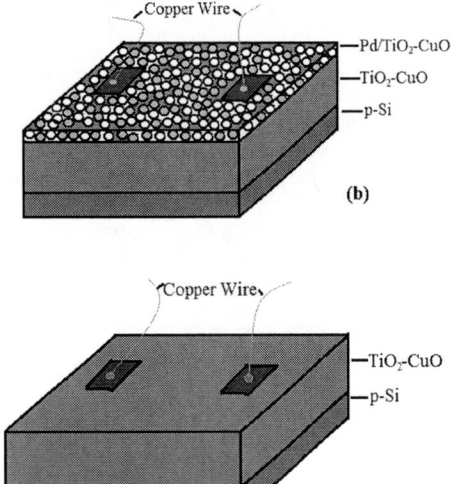

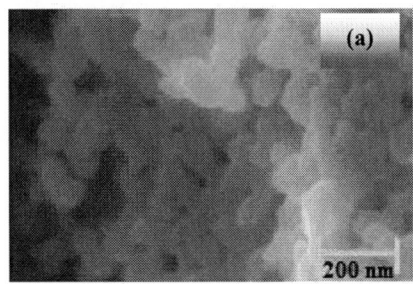

Fig. 3. (a) TiO_2-CuO/p-Si (b)Pd/TiO_2-CuO/p-Si

V. FESEM

Surface morphology (Field Emission Scanning Electron Microscopy) of prepared thin film sensors are shown in the Fig. 4 FESEM suggests that sol-gel grown metal oxides have different polycrystalline size (78 nm-92nm). FESEM characteristics also found that deposited Pd-doped TiO_2-CuO thin have sharper grain size in comparison with the unmodified TiO_2-CuO thin film allowing better adsorption H_2 in the letter case.

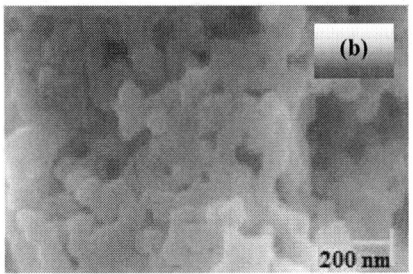

Fig. 4. FESEM image of (a) undoped and (b) Pd-doped TiO_2-CuO thin film sensor.

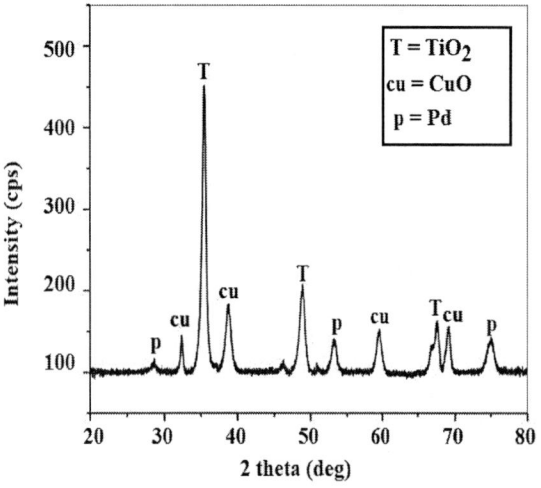

Fig. 5. XRD pattern of Pd modified TiO2-CuO thin film sensor.

Fig. 5 shows the XRD peaks of Pd modified TiO$_2$-CuO thin film sensors at operating temperature 150^{0}C. Prepared sample clearly indicates that deposited nano film is available in the sample which different peaks obtained at different angles (2θ= 29.2^0-75.1^0).

VI. RESULT AND DISCUSSION

The variation of sensitivity w.r.t gas concentration changes is shown in Fig. 6 With the increase in gas concentration the sensitivity also increases and after 1100 ppm of concentration the sensor shows highest sensitivity. After that the sensitivity saturate because hydrogen gas molecules are desorbed by the high conductive Pd/TiO$_2$-CuO surface.

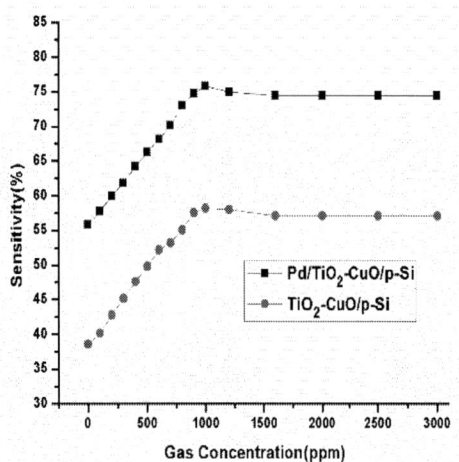

Fig. 6. Sensitivity vs gas concentration

How sensitivity has changed with operating temperature of the sensor is shown in the Fig. 7 It has been observed that the sensor has shown the highest sensitivity at the temperature of 150^{0}C. Pd modified TiO$_2$-CuO sensor has given the highest sensitivity compared to unmodified TiO$_2$-CuO sensor.

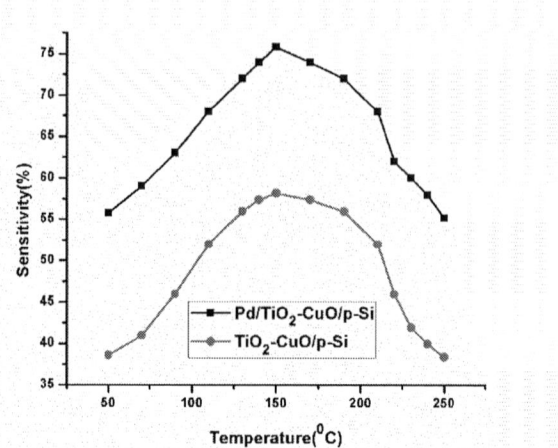

Fig. 7. Sensitivity vs temperature.

VII. CONCLUSION

This work, TiO$_2$-CuO thin film has been prepared by the use of sol-gel synthesis technique. Surface morphology of the prepared sample was characterized by FESEM. The experimental result shows that hydrogen sensors huge affect on prepared thin film TiO$_2$-CuO samples. It also shows that at Pd doped mixed thin film sensors obtained better sensitivity (78%) in comparison to the undoped thin film. In addition, we found that prepared mixed metal oxide thin film sensors in the environmental monitoring had a huge impact towards high sensitivity hydrogen sensing application. The experimental results shows that TiO$_2$-CuO based mixed metal oxide thin film sensors are highly sensitive to variations in H$_2$ concentration and are hence sensible for H$_2$ sensors.

ACKNOWLEDGMENT

The authors thankfully acknowledge the UGC UPE phase II project Ref. No. R-11/183/17 fund for the financial support and also acknowledge IC Design and Fabrication Centre and School of Material Science and Nanotechnology, Jadavpur University for providing infrustructure facilities to conduct the present work.

REFERENCES

[1] A. Hazra, K. Dutta, B Bhowmik and P Bhattacharyya, "Highly repeatable low-ppm ethanol sensing characteristics of p-TiO2-based resistive devices," IEEE Sensor Journal vol-15 pp-408–415, 2015.

[2] A. Dey, B. Kantha and S. K. Sarkar., "Study the Effects of Annealing Temperatures on Sol-Gel derived TiO2 Sensing Element," Computational Science and Engineering, Taylor & Francis Group, pp-247-250, ISBN 978-1-138-02983-5.2016.

[3] A. Dey, B. Kantha and S. K. Sarkar. "Sol–gel grown Pd modified WO3 thin film based methanol sensor and the effect of annealing temperatures,".Microsystem Tech.,vol 23, pp-4195-4201,Sep 2017.

[4] S. Roy, A. Dey, B. Biswas, and S. K. Sarkar, "Analytical Modeling of Voltage to Time Conversion Circuit for Gas Sensing System and Some Experimental Studies," Journal of Nanoelectronics and Optoelectronics, ASP publisher, vol-12, pp. 1-8 , 2017.

[5] O. Lupan, G. Chaic and L. Chow."Novel hydrogen gas sensor based on single ZnO nanorod," Microelectronic Engineering, Elsevier, vol-85, pp-2220-2225, 2008.

[6] V. Aroutiounian, V. Arakelyan, V. Galstyan, K. Martirosyan and Patrick Soukiassian. "Hydrogen Sensor Made of Porous Silicon and

Covered by TiO 2−x or ZnO ⟨ Al ⟩ Thin Film," IEEE Sensors Journal, vol-9, pp-9-12, 2009.

[7] E.Kanazawa, G. Sakai, K. Shimanoe, Y. Kanmura, Y. Teraoka, N. Miura and N. Yamazoe,"Metal Oxide Semiconductor N_2O Sensor for Medical Use," Sens. Actuators B., vol-77, pp-72-77, 2001.

[8] B. Kantha, S. Roy and S. K. Sarkar, "Implementation of Pd modified WO3 thin film gas sensing system with Bulk-CMOS and SOI-CMOS for monitoring leakage of hydrogen gas, Journal of Nanoelectronics and Optoelectronics," ASP publisher, Vol. 10, pp-1-8, 2015.

[9] Z. D. Lin , S. J. Young and S. J. Chang. "Carbon Nanotube Thin Films Functionalized via Loading of Au Nanoclusters for Flexible Gas Sensors Devices," IEEE Transactions on Electron Devices, vol-63, pp-476-480, 2016.

[10] L. Rajan, C. Periasamy and V. Sahula. "Comprehensive Study on Electrical and Hydrogen Gas Sensing Characteristics of Pt/ZnO Nanocrystalline Thin Film-Based Schottky Diodes Grown on n-Si Substrate Using RF Sputtering," IEEE Transactions on Nanotechnology, vol-15, pp-201-208, 2016.

978-1-5386-6723-1/19 $31.00 © 2019 IEEE

Analytical Modeling and Simulation of Low Power Salient Source Double Gate TFET

Bijoy Goswami
ETCE Department
Jadavpur University
Kolkata, India
*bijoy.ete@aec.ac.in

Debadipta Basak
ETCE Department
Jadavpur University
Kolkata, India
db.rini1991@gmail.com

Ayan Bhattacharya
ETCE Department
Jadavpur University
Kolkata, India
*ayanbrkmahs@gmail.com

Koelgeet Kaur
ETCE Department
Jadavpur University
Kolkata, India
koelgeetkaur45@gmail.com

Sutanni Bhowmick
ETCE Department
Jadavpur University
Kolkata, India
sutannibhow94@gmail.com

Subir Kumar Sarkar
ETCE Department
Jadavpur University
Kolkata, India
su_sircir@yahoo.co.in

Abstract— The analytical surface potential model of 22nm salient source Double Gate TFET (SS-DG-TFET) is presented in this paper. The surface potential is analyzed as the performance parameter along with an assessment of improved ON/ OFF current ratio. The variation of tunnel current is examined under same front and back gate bias together with identical oxide thickness. The source region has been extended symmetrically in both directions in order to enhance the conductivity of the channel region and it has been efficiently deployed in the proposed model. The execution of low power functionality and lower sub-threshold slope is also established in this model. The analytical results have been suitably validated using Silvaco, Atlas.

Keywords— *Low power, SS-DG-TFET, sub-threshold slope, Analytical model, Surface potential.*

I. INTRODUCTION

The requirement for high speed, low area, high packing density and low power consumption steer to the advancement of CMOS technology which involves scaling of the MOS devices. The scaling leads to Nano scale realm that results into short channel effects (SCE) and quantum effects. Another characteristic of MOSFET is sub-threshold slope (SS) having the maximum value of 60mV/decade [1]. This can be further reduced by the increase of ON current to OFF current ratio for better performance. TFET [2] is commendatory in this regard that follows Band to Band Tunneling (BTBT) mechanism. The tunneling occurs in highly doped source and intrinsic channel junction [3], [4].

Performance study [5], [6] has been conducted previously along with the development of analytical models for the description of TFET devices [7]-[8]. The deficiency of proper drain current model [9] and in some of the models [10] the influence of depletion region on the drain current is limited to the source side only and ambipolar conduction at the channel drain interface [11],[12] has been overlooked. Previously the tunnel current was determined by numerical integration of carrier generation rate upon depletion region rather than tunneling length [13], [14]. We consider the tunneling process at the source channel interface in the proposed design. The boundary conditions are estimated considering the maximum

and minimum tunneling length at the interfaces for the bias conditions [13].

This work is presented in the following sequence in this paper: In Section II, device structure with optimized parameters are described. In Section III, analytical model of the surface potential is presented and has been verified with the Silvaco simulated outcomes. Comparison of transfer characteristics with conventional double gate TFET has been discussed in Section IV. All the simulations are executed by Silvaco, Atlas.

II. DEVICE STRUCTURE WITH OPTIMIZED PARAMETERS

The 2D cross sectional view of the proposed SS-DG-TFET structure has been depicted in Fig. 1. The structure contains an extended source region which is symmetric on both vertical sides resulting into the vertical dimension of the source to be 28nm whereas the drain region as well as the channel region are restricted to 14nm of height. The channel length is considered to be 22nm for best performance and length of both the source and drain regions are taken as 10nm each. So the total device length is determined to be 42nm. The optimum doping concentration after investigation is found to be 1×10^{21} cm^{-3}, $1 \times 10^{15} cm^{-3}$ and 1×10^{19} cm^{-3} in source, channel and drain regions respectively. It can be clearly identified in Fig.2 where y axis indicates doping concentration in logarithmic scale. The source region is heavily doped P+ type, channel is lightly doped N type and drain is heavily doped N+ type silicon. The oxide region is extended laterally as well as vertically at source side to avoid contact of the gate with source region as the source is also extended and the same technique is adopted at the drain side as well to maintain the symmetric structure. The lateral dimension of extended oxide on both side of gate is 1nm each which results into the total oxide length of 22nm whereas the optimum gate thickness is found to be 4nm with 20nm length. So the vertical dimension of oxide regions are t_{ox} = 1nm under gate and 5nm at the extended parts. This oxide and gate structure is followed in both the upper and lower gates of the dual gate structure for symmetry. The dielectric used for gate

978-1-5386-6723-1/19 $31.00 © 2019 IEEE

oxide is SiO$_2$ with dielectric constant 3.9 and N type polysilicon is used as gate material.

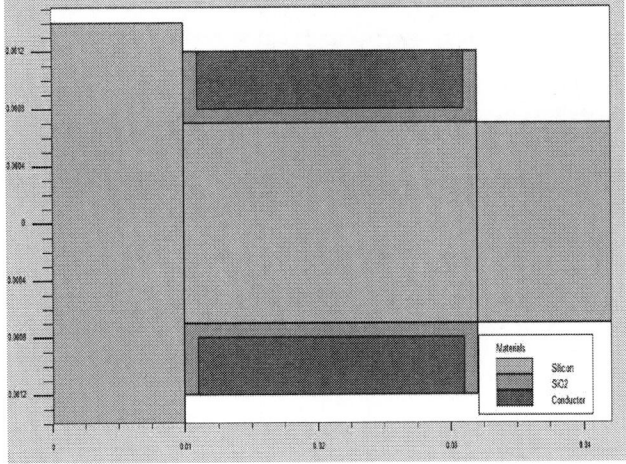

Fig. 1: 2D cross sectional view of salient source DG-TFET

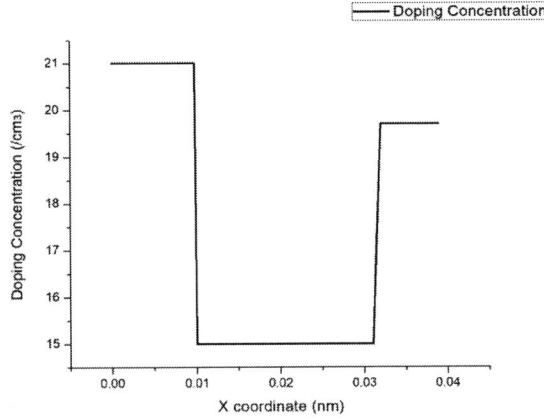

Fig. 2: Doping concentration along the device

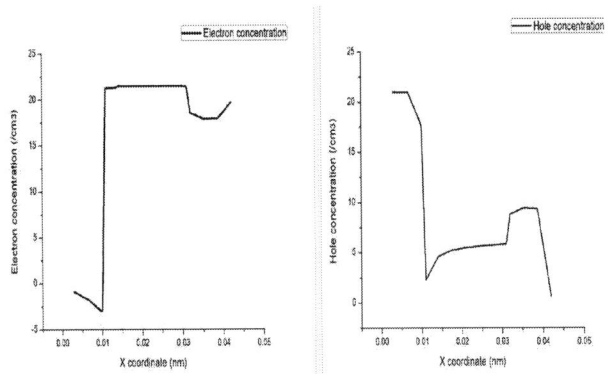

Fig. 3: Electron and hole concentration in SS-DG-TFET

Further we can estimate the behavior of charge carriers (i.e. electrons and holes) in SS-DG-TFET along the length of source, channel and drain region from Fig.3. As source region doping is p-type Si whereas channel and drain consist of n-type Si, hence electron and hole concentration along the length (x-coordinate) of the device is self-explanatory from the figure.

A. Study of gate voltage-drain current characteristic

From the curves of drain current with respect to gate to source voltage it is prominent that the OFF current decreases resulting into lowering of ambipolar conduction with respect to conventional TFET. Thus in low power also the ON to OFF current ratio is in the order of 10^{10} that depicts good performance characteristics of the proposed model. We have investigated the gate current for three different drain to source voltages such as 0.4V, 0.5V and 0.6V. It is evident from Fig. 4 that sub-threshold slope (SS) decreases with the decrease of V$_{ds}$. This phenomenon is striking .It can be concluded that the device works as better TFET in lower drain-to-source voltage. When V$_{ds}$ = 0.4 Volt that is at the lowest voltage among the three compared here, SS is lowest which exhibits the best performance of the proposed model. It indicates that the structure operates as the best in low power.

TABLE I
VALUES OF SUB-THRESHOLD SLOPE CORRESPONDING TO V$_{ds}$

V$_{ds}$(V)	Subthreshold Slope(SS)
0.4 Volt	39 mV/decade
0.5 Volt	48 mV/decade
0.6 Volt	54 mV/decade

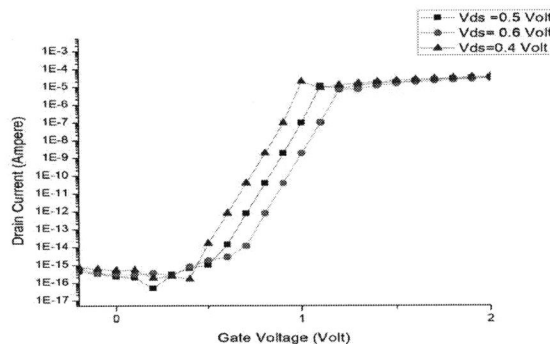

Fig. 4: I$_{ds}$ –V$_{gs}$ characteristic at V$_{ds}$=0.4V, 0.5V and 0.6V

B. Band Diagram Analysis

Fig. 5 depicts the band diagram of the SS-DG-TFET for three different V$_{ds}$ conditions specifically (a) Vds=0.4V, (b) Vds=0.5V and (c) Vds=0.6V. From Fig. 5 it is evident that the tunneling at the source side is almost similar in all the three cases considered here despite of drain to source voltage variation. As the tunnel width is very less so the tunneling effect is predominant at source side. This improved tunneling behavior is the consequence of the extension of source in vertical dimension that provides the piping effect from the source to channel with increased conductivity. For Fig.5(c) V$_{d}$S=0.6V.As the voltage is more so tunneling probability is high but for Fig.5 (a), where V$_{ds}$=0.4V, tunneling is not significantly less compared to Fig.5 (c). If we consider SS then also the characteristics of the device at V$_{ds}$=0.4V is quite appreciable.

2019 Devices for Integrated Circuit (DevIC), 23-24 March, 2019, Kalyani, India

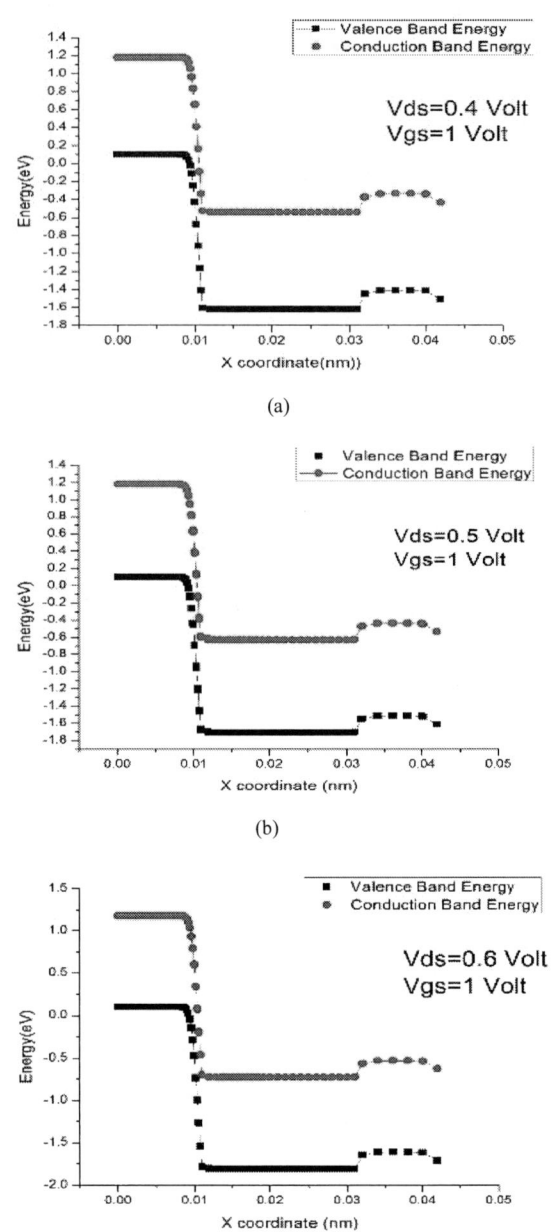

(a)

(b)

(c)

Fig. 5: Band diagram of SS- DG-TFET structure for (a) Vds=0.4V, (b) Vds=0.5V and (c) Vds=0.6V.

C. Electric Field Study

The electric field along the channel length is observed for three different drain to source bias conditions. We have studied that the electric field for V_{ds}= 0.4V, V_{ds}= 0.5V and V_{ds}= 0.6V are almost similar and close to the conventional TFET electric field. So it can be concluded that there is no degradation of electric field profile in lower voltages thus the model is suited for low power application. Fig. 6 shows the electric field along and X and Y direction. As the channel is n-type so the electric field in X direction has a peak in downward direction in source-channel intersection. And for same reason a much smaller peak in opposite direction is achieved in channel-drain intersection.

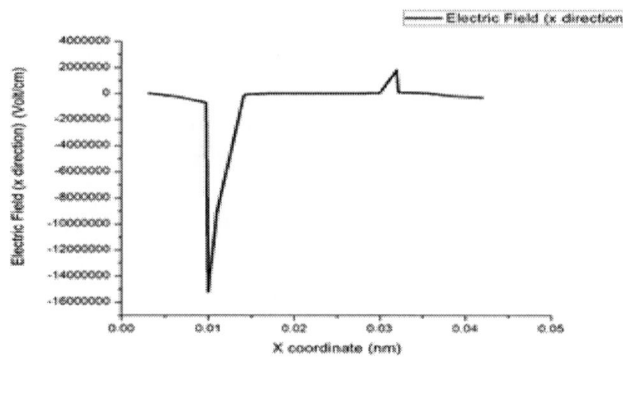

(a)

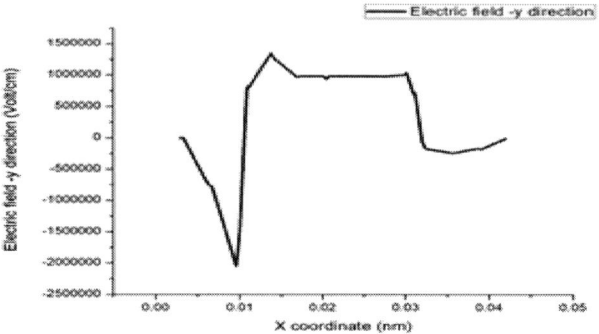

(b)

Fig.6: Electric field along (a) X and (b) Y direction

III. SURFACE POTENTIAL

In this section we will examine analytical modeling as well as simulation based outcome of the surface potential.

A. Analytical Modeling

The analytical surface potential model of the proposed structure is presented in this section. The 2D Poisson's equation is solved for the channel region for calculating the surface potential that is to be verified with the simulation results.

From Fig.1 it is evident that the channel thickness is y ($0 \leq y \leq t_{si}$) and L denotes the channel length that varies along the X direction. The channel region is affected by both the front and back gate bias so the equation is applied to the channel region only. The Poisson equation at the 2D channel region is

$$\frac{\partial^2 \phi(x, y)}{\partial x^2} + \frac{\partial^2 \phi(x, y)}{\partial y^2} = \frac{qN_c}{\varepsilon_{si}} \quad (1)$$

Where $\Phi(x, y)$ is electrostatic potential, q is electronic charge, N_c is the channel concentration and ε_{si} is defined as permittivity of silicon. Now potential profile is considered as Young's parabolic approximation:

$$\phi(x, y) = \phi_s(x) + c_1(x)y + c_2(x)y^2 \quad (2)$$

978-1-5386-6723-1/19 $31.00 © 2019 IEEE

Where $\phi_s(x)$ surface potential, c1 and c2 is are functions of x. The boundary conditions at the source channel and channel drain interface for the electrostatic potential; is defined as described in [13],

$$\left.\frac{d\phi(x,y)}{dy}\right|_{y=0} = \frac{\varepsilon_{ox}}{\varepsilon_{si}}\frac{\phi_s(x)-V_{GS1}}{t_f} \tag{3}$$

$$\left.\frac{d\phi(x,y)}{dy}\right|_{y=t_{si}} = \frac{\varepsilon_{ox}}{\varepsilon_{si}}\frac{V_{GS1}^{'}-\phi_B(x)}{t_B} \tag{4}$$

Where $V_{GS1} = V_{GS}-V_{FB}$, Flat band voltage $V_{FB} = \phi_M - \phi_{Si}$,

$\phi_{Si} = \chi + \dfrac{E_g}{2q} + \phi_B$ and $\phi_B = V_T \ln \dfrac{N_c}{n_i}$.

Differentiating (2) with respect to y we get,

$$\frac{d\phi(x,y)}{dy} = c_1(x) + 2c_2(x)y \tag{5}$$

and applying the boundary condition (3), the values of c1(x) is found to be

$$c_1(x) = \frac{\varepsilon_{ox}}{\varepsilon_{si}}\frac{\phi_s(s)-V_{GS1}}{t_f} \tag{6}$$

Applying the boundary condition (4) and replacing the value of $V_{GS1}^{'}$ with V_{GS1} as are same for both gates and putting the value of y=t$_{si}$ in (2) the value of c$_2$(x) is found to be

$$c_2(x) = \frac{\varepsilon_{ox}}{2t_{si}\varepsilon_{si}}\frac{2V_{GS1}-\phi_B(x)-\phi_s(s)}{t_f} \tag{7}$$

and here in the proposed structure t$_B$=t$_f$.

Where the parameters of the equations are as follows:

χ =electron affinity

E_g = band gap energy

N_c = doping concentration at the channel region

ϕ_{Si} = silicon work function

ϕ_B = built-in potential

ϕ_M = metal work function

To solve $\phi_s(s)$, the values of c$_1$(x) and c$_2$(x) is put into (2) and as $\phi(x,y)|_{y=t_{si}} = \phi_B(x)$ therefore from (2) it is found that

$$\phi_B(x) = \phi_s(x) \tag{8}$$

Now considering (8) and replacing the values of c$_1$(x) and c$_2$(x) from (6) and (7) into (2), the modified expression is found to be

$$\phi(x,y) = \phi_s(x) + \frac{\varepsilon_{ox}}{\varepsilon_{si}}\left(\frac{\phi_s(x)-V_{GS1}}{t_f}\right)y + \frac{\varepsilon_{ox}}{\varepsilon_{si}t_{si}}\left(\frac{V_{GS1}-\phi_s(x)}{t_f}\right)y^2 \tag{9}$$

Differentiating (9) with respect to x and y and replacing in equation (1) the differential equation of the surface potential is found to be

$$\frac{\partial^2\phi_S(x)}{\partial x^2} - \phi_S(x)k_1 = k_2 \tag{10}$$

Where $k_1 = \dfrac{2\varepsilon_{ox}}{t_{si}\varepsilon_{si}t_f}$ \hfill (11)

and $k_2 = -\dfrac{qN_c}{\varepsilon_{si}} - \dfrac{2\varepsilon_{ox}}{t_{si}\varepsilon_{si}t_f}V_{GS1}$ \hfill (12)

Solving the above mentioned second order differential equation the surface potential is found to be

$$\phi_s(x) = A\exp(\sqrt{k_1}\,y) + B\exp(-\sqrt{k_1}\,y) - \delta \tag{13}$$

Where the expression of A, B and δ is as follows [13],

$$A = \frac{k_2\left(e^{\sqrt{k_1}L}-1\right)}{k_1\left(e^{2\sqrt{k_1}L}-1\right)} + \frac{\left[V_{bi}\left(e^{\sqrt{k_1}L}-1\right)+V_{Ds}e^{\sqrt{k_1}L}\right]}{\left(e^{2\sqrt{k_1}L}-1\right)} \tag{14}$$

$$B = \frac{k_2e^{\sqrt{k_1}L}\left(e^{\sqrt{k_1}L}-1\right)}{k_1\left(e^{2\sqrt{k_1}L}-1\right)} + \frac{e^{\sqrt{k_1}L}\left[V_{bi}\left(e^{\sqrt{k_1}L}-1\right)-V_{Ds}\right]}{\left(e^{2\sqrt{k_1}L}-1\right)} \tag{15}$$

$$\delta = \frac{k_2}{k_1} \tag{16}$$

TABLE II
VALUES OF DIFFERENT PARAMETERS

Different parameters	Values
N_c	$10^{15}/cm^3$
E_g	1.1 eV
ε_{ox}	3.9
ε_{si}	11.7
χ	1.39 eV
L	22 nm
V_T	26 mV
V_{DS}	0.4 Volt
n_i	1.5x10^{10}/cm^3
t$_f$	1nm
ϕ_M	4.15 V

The surface potential of the salient source TFET and analytical model has been plotted along the channel length .In Fig7 it is depicted that both the curves are in close proximity with maximum potential of 1V at the channel region. This suggests that surface potential for proposed simulated model and its analytical output are analogous.

978-1-5386-6723-1/19 $31.00 © 2019 IEEE

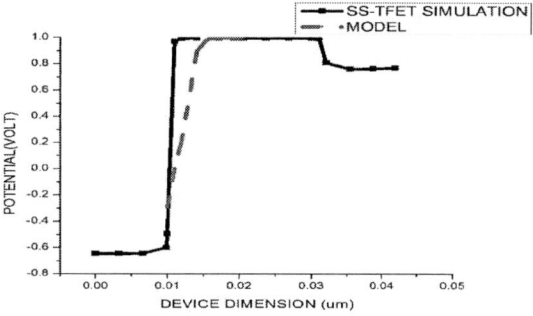

Fig. 7: Surface potential of SS-DG-TFET

B. Simulation Based Study

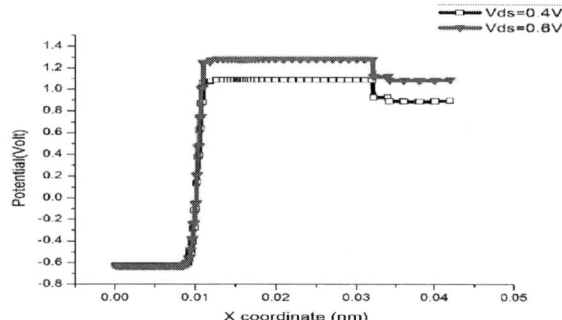

Fig. 8: Comparison of surface potential for V_{ds}=0.4V and V_{ds}=0.6V.

It is evident from the Fig.8 that for greater voltage (V_{ds}=0.6V) the surface potential is intuitively more compared to V_{ds}= 0.4V. For both the cases V_{gs} is considered to be 1V. It can be easily determined that power consumption is more for 0.6V but if we consider the previous analysis, specifically SS, then it establishes the fact that the device is useful enough for low power applications.

IV. RESULTS AND DISCUSSIONS

Further we have also studied that our proposed structure is better in terms of ON current to OFF current ratio than the conventional double gate TFET. Our model has symmetrical extended source that gives a piping effect through channel. Hence more current is flown through channel of SS-DG-TFET

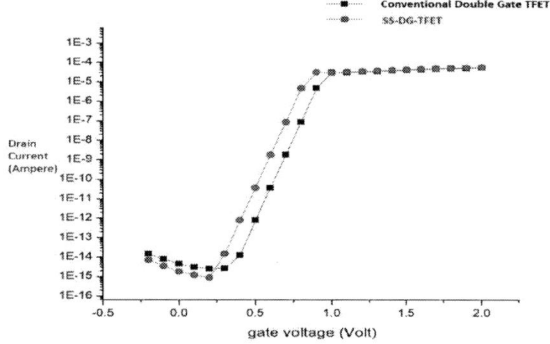

Fig. 9: Comparison of gate-voltage drain-current characteristic of SS-DG-TFET and conventional double gate TFET at Vds=0.4 volt

V. CONCLUSION

The proposed structure of this paper, SS-DG-TFET displays good performance in low power operating conditions. The performance has been characterized in terms of lower SS, increased ON to OFF current ratio, along with improved tunneling at the source end of the channel. The proposed model exhibits comparatively equivalent results with the analytical model in terms of surface potential. The performance parameters are compared for three different V_{ds} conditions and a rigorous examination determines that the device operates well enough in low power.

REFERENCES

[1] W. Y. Choi, B.-G. Park, J. D. Lee, and T.-J. K. Liu, "Tunneling field-effect transistors (TFETs) with subthreshold swing (SS) less than 60 mV/dec", IEEE Electron Device Lett., vol. 28, no. 8, pp. 743–745, Aug. 2007.

[2] N. Patel, A. Ramesha, and S. Mahapatra, "Drive current boosting of n-type tunnel FET with strained SiGe layer at source,"Microelectron. J., vol. 39, no. 12, pp. 1671–1677, Dec. 2008.

[3] K. Boucart and A. M. Ionescu, "Double-gate tunnel FET with high-κ gate dielectric," IEEE Trans. Electron Devices, vol. 54, no. 7, pp. 1725–1733, Jul. 2007.

[4] A. M. Ionescu and H. Riel, "Tunnel field-effect transistors as energy efficient electronic switches, "Nature, vol. 479, no. 7373, pp. 329–337, Nov. 2011..

[5] L. Liu, D. Mohata, and S. Datta, "Scaling length theory of double-gate interband tunnel field-effect transistors," IEEE Trans. Electron Devices, vol. 59, no. 4, pp. 902–908, Apr. 2012.

[6] K. K. Bhuwalka, J. Schulze, and I. Eisele, "A simulation approach to optimize the electrical parameters of a vertical tunnel FET,"IEEE Trans. Electron Devices, vol. 52, no. 7, pp. 1541–1547, Jul. 2005.

[7] K. Boucart and A. M. Ionescu, "A new definition of threshold voltage in tunnel FETs," Solid-State Electron., vol. 52, no. 9, pp. 1318–1323, Sep. 2008.

[8] N. Bagga and S. K. Sarkar, "An Analytical Model for Tunnel Barrier Modulation in Triple Metal Double Gate TFET," in IEEE Transactions on Electron Devices, vol. 62, no. 7, pp. 2136-2142, July 2015.

[9] B. Manna, S. Sarkhel, N. Islam, S. Sarkar and S. K. Sarkar, "Spatial Composition Grading of Binary Metal Alloy Gate Electrode for Short-Channel SOI/SON MOSFET Application," in IEEE Transactions on Electron Devices, vol. 59, no. 12, pp. 3280-3287, Dec. 2012

[10] B. Goswami, D. Bhattachariee, D. Dash, A. Bhattacharya, and S. K. Sarkar, "Demonstration of T-Shaped Channel Tunnel Field-Effect Transistors," 2nd International Conference on Electronics, Materials Engineering & Nano-Technology (IEMENTech), 2018.

[11] C. Anghel, Hraziia, A. Gupta, A. Amara, and A. Vladimirescu, "30-nm tunnel FET with improved performance and reduced ambipolar current," IEEE Trans. Electron Devices, vol. 58, no. 6, pp. 1649–1654, Jun. 2011.

[12] B. Goswami, D. Bhattacharjee, A.Bhattacharya, and S.K.Sarkar "Drain-Doping Engineering and its Influence on Device Output Characteristics and Ambipolar Conduction on a Splitted-Drain TFET Model," Advances in Communication, Devices and Networking. Lecture Notes in Electrical Engineering, vol 537. Springer, Singapore, 2019.

[13] N. Bagga and S. K. Sarkar, "Surface Potential and Drain Current Analytical Model of Gate All Around Triple Metal TFET," IEEE Transactions on Electron Devices, 2017, 64, (2), pp. 606–613.

[14] S. Roy, A. Chatterjee, D. K. Sinha, R. Pirogova and S. Baishya, "2-D Analytical Modeling of Surface Potential and Threshold Voltage for Vertical Super-Thin Body FET," in IEEE Transactions on Electron Devices, vol. 64, no. 5, pp. 2106-2112, May 2017.

Quasi FGMOS Inverter: A Strategy for low power applications

AlekhyaYalla, Umakanta Nanda, *Member, IEEE*

School of Electronics Engineering
VIT-AP University
Amaravathi-522237, India
alekhyayalla@gmail.com, uk_nanda@yahoo.co.in

Abstract—Low voltage operation and low power consumption is of paramount need in integrated circuits which are employed in portable devices. Floating gate MOS (FGMOS) transistor is an analog technique to achieve low power while maintaining performance in various applications such as neural networks, PLL, D/A and A/D converters and memory circuits. This paper deals with various FGMOS techniques with supply voltage V_{DD}=220mV operating in sub-threshold region which is very crucial for lowering power dissipation and achieves higher speed compared to CMOS circuits. The proposed work is validated through inverter circuits using TSMC 0.18µm technology in mentor graphics tool.

Keywords-Multi-input FGMOS,QFGMOS,FGMOS transistors, FGMOSFET,Inverter.

I. INTRODUCTION

Technology scaling and development of portable devices are of high demand and have become the center of research need during last ten years but battery life time did not meet the market requirement where they play crucial role in some biomedical products which have implanted within patients for longer duration of time. Therefore, this challenging research has motivated the designers for development of low voltage/low power electronic systems [1-2].

Multi input FGMOS transistors can operate at low supply voltage levels and consumes low power compared to CMOS devices. They are flexible to process complex functions with minimal circuit complexity. The ability to shift the threshold voltage allows the circuit to operate with supply voltage lower than the V_{TH} while operating circuits up to several hundred MHz range [3].FGMOS transistors allow tuning the circuits in order to make the device operate in intended region [4].

Due to low voltage/low power advantage, FGMOS based designs gained paramount importance in applications on biomedical products like pulse dosimeter, cochlea implants, defibrillator and pacemakers [5], trimming elements[6], analog memory elements in neural networks[7], processors, analog multipliers, D/A converters[8], compressor designs, processors and radio frequency applications.

Due to charge retention characteristics of FGMOSFETs they are used to store data in nonvolatile memories [9]; however the problem still remains to control the initial charge on the floating gate. Fowler-Nordheim tunneling and hot carrier injection process [10] are used but programming using

these methods involves charge leakage problems which require high supply voltages [11].

To overcome these problems Quasi or Pseudo floating gate MOS uses refresh circuitry providing resistance to avoid leakages due to charge traps on the floating gate[12].

In this article we present how FGMOS transistor with low voltage low power provides better performance in comparison to CMOS process. Section II describes about literature study, section III describes modeling of FGMOS transistor. In section IV FGMOS based inverter and Quasi- FGMOS based inverter is presented. Section V demonstrates the simulation and results followed by conclusion in section VI.

II. LITERATURE SURVEY

Due to its special characteristics that every input induces charge on the floating gate and voltage on the floating gate varies linearly with input transistor voltages. Multi-input FGMOS transistor allows its application in mixed signal circuits like phase locked loop [13-14]. Present days, FGMOS based analog nonvolatile memory is used in personal computers. Other attractive feature like tuning process which is used to modify the accumulated charge on floating gate is implemented [15].

Yu and Geiger [16] scaled down V_{th} and injected charge using tunneling process. In order to control charge, Fowler-Nordheim tunneling current must be generated that requires EEPROM technology to force the current which in turn increases chip cost [17].

Rodriguez-Villegas and Barnes [18] provided solution to the accumulated charge without need of post-processing techniques which increases cost. Quasi (or) pseudo floating gate overcomes the initial charge problem by connecting large +/- value resistor within the circuit depending on NMOS (or) PMOS devices. Leakage resistor is kept high to float the gate effectively at low frequency values, so that its operation is not affected [19].

III. MODELLING OF MULTI INPUT FGMOSFET

The schematic representation of Multi-input FGMOSFET is shown in Fig. 1.It consists of 'n' control inputs to the floating gate. Let us assume that total charge on the floating gate be Q_{FG}, Floating voltage be represented as V_{FG} and voltage due to i^{th} control gate be V_{Gi}

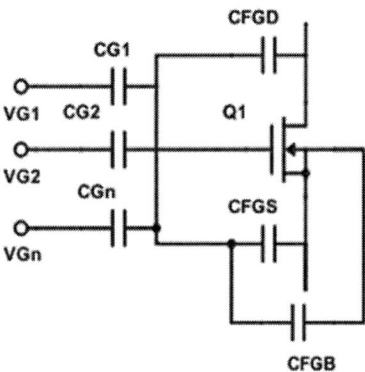

Fig. 1. Schematic representation of n-input Floating gate NMOS Transistor

$$V_{FG} = \frac{Q_{FG} + C_{FGD}V_D + C_{FGS}V_S + C_{FGB}V_B + \sum_{i=1}^{n} C_{Gi}V_{Gi}}{C_{TOTAL}} \quad (1)$$

Where as

$$C_{TOTAL} = C_{FGD} + C_{FGS} + C_{FGB} + \sum_{i=1}^{n} C_{Gi} \quad (2)$$

The parasitic capacitances of floating gate are represented as C_{FGD}, C_{FGS} and C_{FGB}.

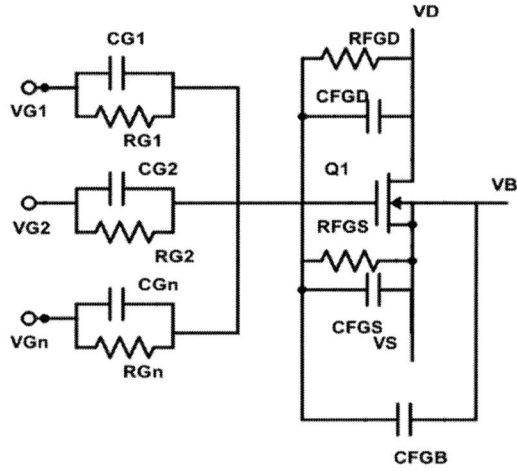

Fig. 2. Simulation model of n-input Floating gate NMOS Transistor with passive components

Fig. 2 represents simulation model in which C_{Gi} to C_{Gn} represent floating gate capacitances at each input, C_{FGD}, C_{FGS} and C_{FGB} represent parasitic capacitances and R_{G1} to R_{Gn} represents resistances which are employed to overcome convergence problem in simulation. The resistor values are chosen large enough so that the current flow through them is very small.

If $R_{G1}C_{G1}=R_{G2}C_{G2}=R_{G3}C_{G3}= \cdots =R_{Gn}C_{Gn}=R_{FGD}C_{FGD}=R_{FGS}C_{FGS}= R_{FGB}C_{FGB}$ then n-input floating gate will not be affected due to resistors. So by writing the nodal equation for the Fig. 2 we get

$$V_{FG}\left[\frac{1}{R_{FGD}}+\frac{1}{R_{FGS}}+\frac{1}{R_{FGB}}+\sum_{i=1}^{n}\frac{1}{R_{Gi}}\right]-\frac{V_D}{R_{FGD}}-\frac{V_S}{R_{FGS}}-\frac{V_B}{R_{FGB}}-\sum_{i=1}^{n}\frac{V_{Gi}}{R_{Gi}}=0 \quad (3)$$

$$V_{FG} = \left[\frac{V_D C_{FGD} + V_S C_{FGS} + V_B C_{FGB} + \sum_{i=1}^{n} C_{Gi}V_{Gi}}{C_{TOTAL}}\right] \quad (4)$$

V_{FG} is a linear weighted sum of multiple input gates. The equations modeling the large signal behavior of the FGMOS can now be obtained by replacing V_{GS} in the equations describing the large signal behavior of the MOS transistor, with the expression describing the voltage between the floating gate and source which can be obtained by referring V_{FG} to the source terminal rather than the bulk.

IV. FGMOS INVERTER AND QFGMOS INVERTER

A. FGMOS Inverter

In order to reduce power dissipation in standard CMOS technology we scale down the transistors but it causes leakage current to increase drastically. By reducing the power supply voltage below the threshold voltage into the sub-threshold region, power consumption is reduced by several orders of magnitude. This reduction in supply voltage also reduces the circuit speed. CMOS technology which is very slow when compared to FGMOS in sub-threshold region is less preferred in many applications. FGMOS can change the threshold voltage which in turn aids in increasing the performance.

In a case study with V_{DD}=220mV, V_{Tn}=372mV and V_{Tp}=-394mV, both standard CMOS inverter(Fig.3) and basic FGMOS inverter(Fig. 4) have been simulated in sub-threshold region, where the inverter using ideal FGMOS provides better results.

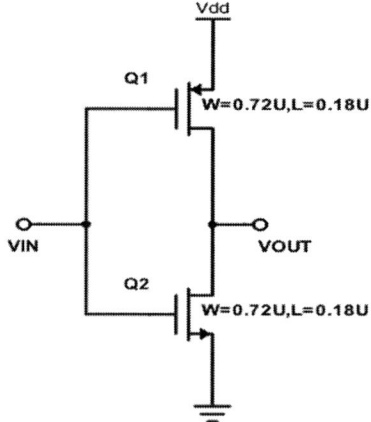

Fig. 3. Standard CMOS inverter schematic circuit

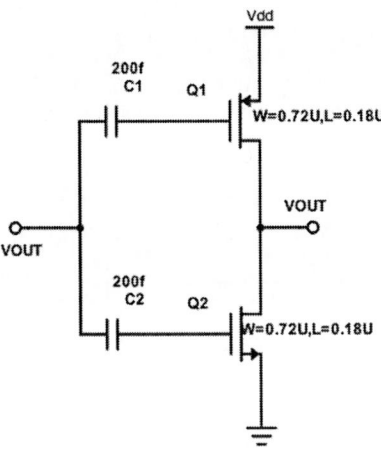

Fig. 4. Basic FGMOS inverter schematic circuit

Fig. 5 shows the voltage transfer characteristics of standard CMOS inverter with supply voltage varying from 0 to 220mV in 20mV.It is observed that inverter does not operate below 25mV even the DC characteristics are correct. This affects the switching frequency of the circuit leading to slow transitions.

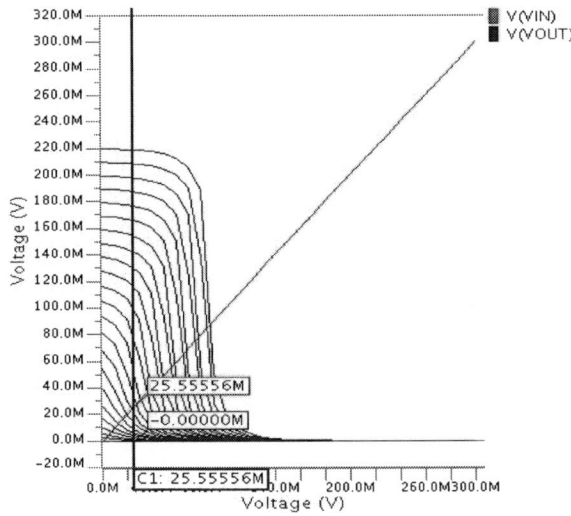

Fig. 5. DC Characteristics of standard CMOS inverter with V_{DD}=220mV

But practically it is not possible to achieve no leakage condition as size of transistors is decreased. To solve the problem of basic FGMOS inverter, refresh circuits having large resistances are incorporated to reduce the charge effects on the circuits. A new technique called Quasi or Pseudo-Floating gate MOS is introduced. Here the gate is semi-floating by connecting resistors or MOS transistors for shifting the threshold voltage to operate in the weak inversion region.

B. Quasi-FGMOS Inverter

As depicted in Fig. 6the Quasi-FGMOS inverter with fixed threshold voltage by connecting the effective input voltage V_{IN} with resistances of 1k ohm in parallel to coupling capacitances $C1$ and $C2$ order of 200fF.

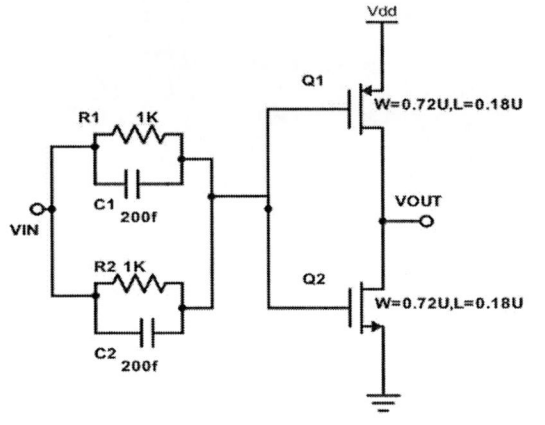

Fig. 6. QFGMOS inverter schematic circuit with resistance at the Floating gate

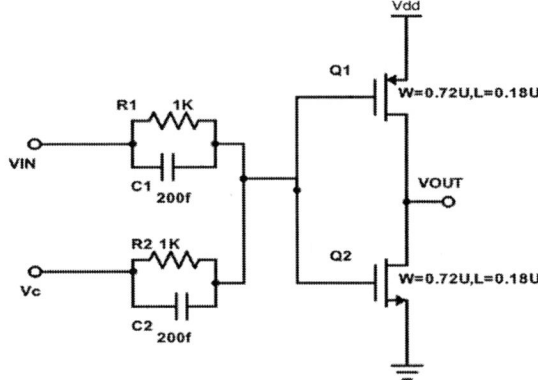

Fig. 7. QFGMOS inverter schematic circuit with varying control voltage V_C at the Floating gate

The change in threshold voltage of QFGMOS inverter can be observed in the DC analysis and power dissipation is observed as shown in Fig. 8

Fig. 9 shows another QFGMOS inverter circuit followed by a MOSFET inverter. In the buffer circuit threshold voltage value is not significant as output is same as the floating gate voltage V_{FG}. By altering the control voltage V_C the threshold voltage of QFGMOS inverter can be programmed.

The performance comparison between QFGMOS inverter and MOSFET inverter is depicted in Fig. 10 by sweeping Vin between 0 to 220mV and changing V_C in steps of 0.2V.The objective of incorporating second inverter to regenerate the same transient response and show that it can produce results in compatible with CMOS buffer.

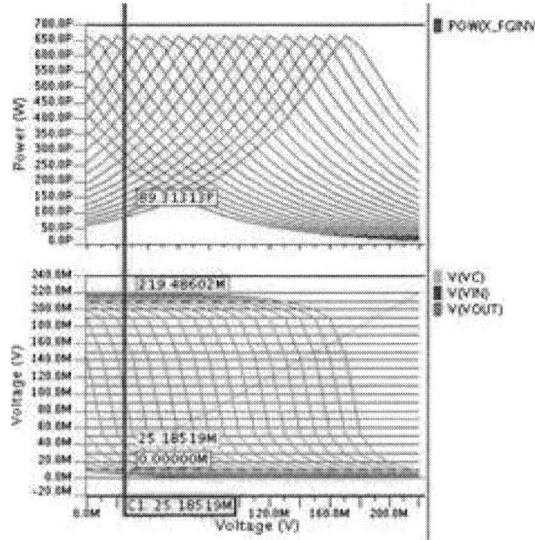

Fig. 8. DC Analysis of QFGMOS inverter by sweeping the supply voltage V_{DD}=220mV

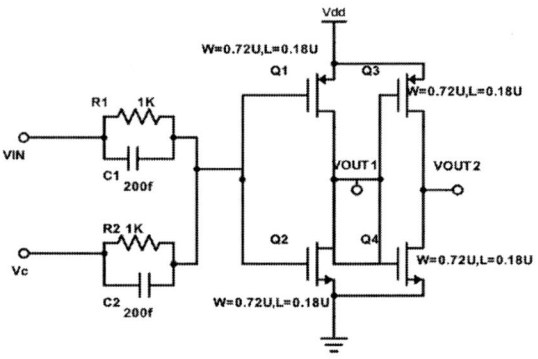

Fig. 9. QFGMOS inverter followed by MOSFET inverter

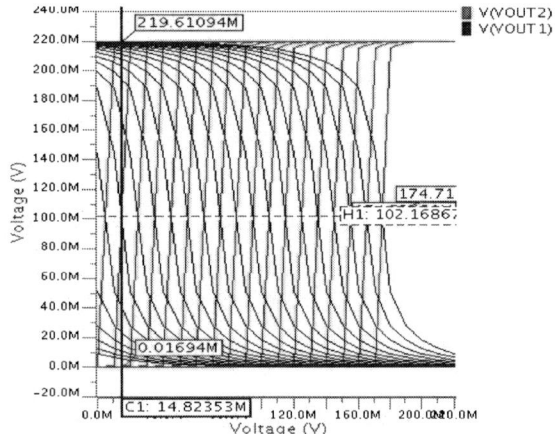

Fig. 10. Input-Output Characteristics of QFGMOS inverter (V_{out1}) followed by MOSFET inverter (V_{out2})

QFGMOS inverter with MOS transistor having large resistance is shown in Fig. 11.QFG PMOS transistor whose DC gate voltage is set to $V_{GMP,DC}$=V_{SS} and NMOS transistor whose DC gate voltage is set to $V_{GMN,DC}$=V_{DD}. It operates properly below sub-threshold region when compared to other FGMOS design techniques.

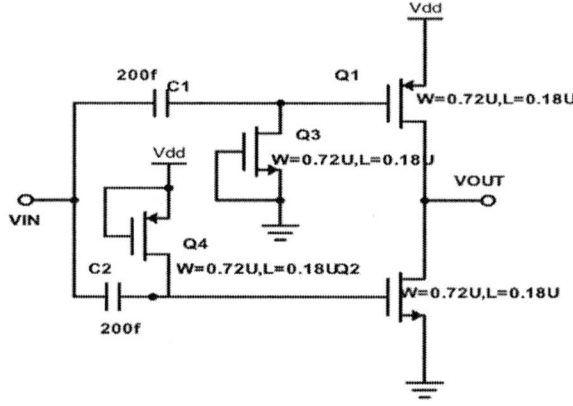

Fig. 11. QFGMOS inverter schematic circuit with MOS transistor at the Floating gate

V. SIMULATION RESULTS

The circuits are simulated in Mentor graphics in 180nm CMOS technology. The simulations carried out here are for supply voltage of 220mV, V_{Tn}=372mV and V_{Tp}=-394mV.In TABLE I, comparison among an ideal CMOS inverter, basic FGMOS inverter and two Quasi FGMOS inverters with circuits (mentioned above) are demonstrated.

The resistances used in the above circuits are in the range of order kΩ which ensures low impact on circuit behavior in the low frequency range. The input terminals to the gate are capacitive coupled to the floating gate are in the order of 200fF.

TABLE I.   DELAY AND POWER DISSIPATION OF CMOS AND FGMOS INVERTER CIRCUITS

TOPOLOGY	DELAY (ns)	POWER DISSIPATION (nW)	POWER DELAY PRODUCT (aJ)
Standard CMOS	2.9746	0.0618	0.18382
Basic FGMOS with fixed and varying V_{FG}	3.0445	0.0618	0.18817
	19.702		1.21779
QFGMOS with fixed and varying V_{FG}	3.3347	0.0618	0.20604
	0.1651		0.01019
QFGMOS with MOS transistor fixed and varying V_{FG}	0.2598	7.2376	1.87453
	0.0862		0.62448

Efficiency of any circuit is measured in terms of power dissipated and performance. Table I explains about delay and power dissipated for simulated circuits mentioned above. Table II presents the comparison of different FGMOS inverter considered in this work. It can clearly be observed that QFGMOS inverter with resistors better but it is practically difficult to fabricate circuits with resistance values. Hence It is essential to use circuits that has MOS transistor at the floating node to design the Quasi FGMOS. Trade-off between increasing circuit area and power delay product is inevitable.

TABLE II.   COMPARISON OF DIFFERENT FGMOS INVERTERS

Parameters	[9]	[7]	[15]	This work
Technology	350nm	500nm	100nm	180nm
Topology	QFGMOS	QFGMOS	FGMOS	FGMOS, QFGMOS
Supply Voltage	400mV	450mV	1.8V	220mV
V_{Tn}, V_{Tp}	120mV -800mV	670mV -960mV	397mV	372mV -394mV
Coupling Capacitance	-	0.2pF	-	200fF
Power	0.14uW	-	-	7.23nW
Delay	3ns	-	-	0.08ns

VI.   CONCLUSION

This work focuses on designing various Multi-input FGMOS based inverters which operate with supply voltage less than threshold voltage and have reduced power dissipation without compromising on speed. Simulation results are included for validating presented techniques. FGMOS circuits allow data processing to be carried out using passive components providing wide operating frequency range. A wide scope of work is extended in developing Multi-input FGMOS inverters employing MOSFETs are resistors. It is observed that the inverter using FGMOS and QFGMOS produces better results with respect to power consumption and speed. This work can further be extended to design analog and digital blocks especially for memory elements in neural networks and bio-medical applications using proposed inverters.

REFERENCES

[1] A.J Lopez-Martin,J. Ramirez-Angulo,R.G Carvajal and L.A costa,"CMOS transconductors with continuous tuning using FGMOS balanced output current scaling,"*IEEE Journal of Solid State Circuits*,pp.1313–1323,2008.

[2] Elisa Spano,Stefano Di Pascoli and Giuseppe Iannaccone,"Low-Power wearable ECG monitoring system for multiple-patient remote monitoring,"*IEEE Sensors Journal*,pp.13,vol. 16,July 2016.

[3] Esther Rodriguez-Villegas,"Low Power and Low Voltage circuit design with the FGMOS transistor,"*IET Circuits,Devices and Systems*,vol. ,pp.2,2006.

[4] F.A.M Rezali,S.F.W.M Hatta and N Soin,"Scaling impact on design performance metric of sub-micron CMOS devices incorporated with halo,"*IEEE Regional Symposium on Micro and Nanoelectronics*,pp.1-4,2015.

[5] A.Paul,K.P Menon,"Digital implementation of hyperbolic calculation unit for electronic cochlea on FPGA,"*IEEE International Conference on Recent Trends in Electronics,Information and Communicatation Technology*,pp.1322-1325,2017.

[6] Weinan Gao,W.M.Snelgrove,"Floating gate charge-sharing:a novel circuit for analog trimming,"*IEEE International Symposium on Circuits and Systems*,vol.4,pp.315-318,1994.

[7] A. L. Martin, D. Orradre, M. P. Garde, P. Sanchis, E. Gubia, G. Perez, D. Astrain, J.R. Angulo, "Energy Harvesting Microsystems Based on the QFG MOS Transistor", *IEEE 15th Int. Conf. on Environment and Electrical Engineering (EEEIC)*,pp. 2035-2039, June 2015.

[8] Fabian Khateb,Tomasz Kulej,Harikrishna Veldandi and Winai Jaikla,"Multiple-input Bulk-driven Quasi-floating-gate MOS transistor for low-voltage low-power integrated circuits,"*AEU International Journal of Electronics and Communications*,vol.18,Dec.2018.

[9] E. Rodriguez-Villegas, J. M. Quintana, M. J. Avedillo, A. Rueda, "High-speed low-power logic gates using floating gates",*IEEE Int. Symposium on Circuits Systems*,vol. 5, pp. 389-392,May 2002.

[10] Ochoa-Padilla,Gomez-Casta,Moreno,"Floating-gateMOScharge programming using pulsed hot-electron injection,"*IEEE 10th International conference on Electrical Engineering, Computing Science and Automatic Control*,pp. 478–481,2013.

[11] F Razaghian,S Bonakdarpour,"Reducing the leakage current and PDP in the Quasi-Floating Gate Circuits,"*IEEE Spring congress on Engineering and Technology(S-CET) Spring Congress*,pp.1-4,2012.

[12] N.Sharma,R.Chandel,"Performance analysis of SRAM cell designed using floating gate MOS,"*IEEE International Conference on Inventive Communication and Computational Technologies*,pp.160-165,2017.

[13] U Nanda,D P Acharya,S K Patra, "A new transmission gate cascode current mirror charge pump for fast locking low noise PLL," *Circuits, Systems, and Signal Processing, Springer*,vol. 33,Issue 9, pp. 2709-2718, 2014.

[14] U Nanda,D P Acharya,"Adaptive PFD Selection Technique for Low Noise and Fast PLL in Multi- standard Radios, *Microelectronics Journal*, vol. 64, pp.92–98,2017.

[15] F.A.S. Musa, D. Nurulain, N. Ahmad, M. Mohamad Isa and Muhammad M. Ramli," Implementation of floating gate MOSFET in inverter for threshold voltage tunability," *International Conference on Applied Photonics and Electronics*,pp.020040-1–020040-6,2017.

[16] C. GYu,R.L Geiger,"Very low voltage operational amplifiers usingFloating Gate MOS transistor,"*IEEE International Symposium on Circuits and Systems*,vol. 2,May3–6,pp.1152–1155, 1993.

[17] Sherif M. Sharroush,"A voltage-controlled ring oscillator based on an FGMOS transistor,"*Microelectronics Journal*,vol.66,pp.167-186,May 2017.

[18] E.Rodriguez-Villegas,H.Barnes,"Solution to the trapped charge inFGMOS transistors,"*IEE Electronic Letters*,vol.39(19),pp.1416–1417, 2003.

[19] E. Mendez-Delgado and G.J. Serrano, "A 300mV low-voltage start-upcircuit for energy harvesting systems,"*International Symposium on Circuits and Systems*,pp.829-832,2011.

Structure and Electronic Properties of TiO$_2$ Nanowires of Different Geometrical Shapes: An Ab-initio Study

Debashish Dash
Dept. of ECE
Madanapalli Institute of
Technology and Science
Madanapalli, India
debashish.dash@ieee.org

Chandan Kumar Pandey
Dept. of EE
National Institute of Technology
Silchar, India
chandankumarpandey@gmail.com

Saurabh Chaudhury
Dept. of EE
National Institute of Technology
Silchar, India
saurabh1971@gmail.com

Susanta Kumar Tripathy
Dept. of ECE
National Institute of Technology
Silchar, India
susanta96@gmail.com

Abstract—**This paper investigates on the structural stability and electronic properties of titanium dioxide (TiO$_2$) nanowires of different shapes using first-principle based density functional approach. Out of linear, ladder, and saw tooth shaped atomic configuration, the ladder shape atomic configuration is energetically most stable. After computation of lattice parameters as well as various mechanical properties of nanowire TiO$_2$, it is seen that highest bulk moduli is obtained for linear TiO$_2$ nanowire which shows the highest mechanical strength for the structure whereas ladder configuration has lowest bulk moduli which shows the lowest mechanical strength for the structure. Analysis of various electronic properties show that different configurations of TiO$_2$ nanowires can have different utility as solid state materials.**

Keywords— Nanostructures; Ab-initio calculations; Stability; Electronic structure; Bulk modulus

I. INTRODUCTION

For the past few decades, several families of nanostructures have been developed which includes quantum dots, Carbon Nano Tubes [1], zero band gap graphene etc., and semiconductor nanowires [2], [3]. Each of the above materials exhibit many interesting properties. The main motivation towards investigating a single material is to study its magnificent properties, although a specific single material with outstanding properties may not constitute a total new technology. Specifically, the potential to create new nanostructures and assembling with a tunable composition, enables radical change in society and future technologies. In other words, in lieu of exploring a single nanomaterial, investigations can be carried out for a systems in which different structure, shape, composition and properties analogous to them can be tuned.

Considering above facts, semiconductor nanowires plays as one of the most powerful quasi single dimensional (1D) platforms available in today's nano arena [4]–[7]. The nanowire is an excellent 1D structure with electrical carriers are restricted in other two directions. Today, nanowires have numerous applications in different research areas like LASERs, sensitive polarized photodetectors, light emitting diodes (LEDs), logic gates, various family of field effect transistors, renewable energy devices like solar cells, single electron storage devices, single hole transistor, lithium- ion storage batteries as well as selected DNA detectors [8]–[10]. As we know, our universe is facing today different problems due to consumption of fossil fuels like, greenhouse effect which causes global warming leading to sudden climatic changes. Thus, solar cells are the best alternative renewable energy sources. For making solar cells different materials are used like, monocrystalline silicon [11], polycrystalline silicon [12], silicon thin film [13], [14], cadmium telluride [15], copper indium gallium selenide [16], gallium nitride [17], gallium arsenide [18] etc. These materials can be used for making solar cells with an efficiency of 14- 19% whereas, newly developed dye sensitized solar cells (DSSCs) have efficiency of 32% under standard test conditions and a two level tandem DSSC embodiment could reach 46% efficiency as per Prof. Michael Graetzel, Ecole Polytechnique Federale de Lausanne (EPFL)

[19]. These advanced solar cells are made from Titanium Dioxide which is known for its non- toxicity, low cost, easy availability, long term stability and its superb photo catalytic nature [20]. TiO_2 has natural and high pressure phases out of which natural TiO_2 is wide band gap semiconductor whereas high pressure TiO_2 is perfect for making renewable panel. Previously, TiO_2 has been enormously investigated as a photo anode for photo electrochemical (PEC) water splitting by Fujishima and Honda [21]. They have found TiO_2 has great photo catalytic activity, proper band edge positions as well as superior photo chemical stability. Many research groups have done the experiments on TiO_2 nanowires by implementing various experimental procedure. Lou et al. [22] reported a large quantity of solvothermal synthesis of natural anatase TiO_2 nanowires, which shows outstanding photo catalytic activity for the degradation of Rhodamine B. TiO_2 is also used to improve the efficiency of various solar cells and it can be used to improve the performance of gas sensors. Despite of all these studies which provides some interesting results of TiO_2, few research groups have studied the transfer properties of 1D TiO_2 nanostructures.

Many groups have performed lot of experiments on TiO_2 synthesis and growth aspect but stability and various electronic properties of different size and shapes has not been checked neither experimentally nor theoretically by any research groups. It's a fact known to all that in nanometer regime, geometrical shape and structure plays an important role for various change in the electronic and other properties compared to normal shapes [23]. It is seen that, many research groups have performed theoretical calculations on different shapes and structures of other materials like, gallium phosphide (GaP) [24], gallium nitride (GaN) [8], gallium arsenide (GaAs) [8], gallium antimonite (GaSb) [8], aluminium phosphide (AlP) [25], aluminium arsenide (AlAs) [25], aluminium antimonite (AlSb) [25] nanowires. But none of the groups worked on proper theoretical investigation focusing on different shapes of the titanium dioxide (TiO_2) nanowires. Tafen et al. [26] investigated four types of small perimeter TiO_2 nanowires such as round, octagonal, hexagonal as well as square shape along the 'z' direction. They have found that, except the square structure, all other structures are stable in nature. But besides this four shapes, there are many more shapes of TiO_2 nanowires and are not investigated yet. It motivates us to consider different shapes and to investigate individual shape's stability and various electronic properties of thin nanowires, containing up to six atoms. The band structure and density of states are thoroughly analyzed for all shapes of nanowires.

II. COMPUTATIONAL METHODS

All the simulations are being carried out within the framework of Density Functional Theory (DFT) [27], [28], which is the basic theory used to find out properties of any material. In this paper, three different shapes of TiO_2 i.e. linear, ladder, and saw tooth are considered. The popular Atomistix Tool Kit (ATK) [29] has been employed for optimization of different geometrical shapes and have been performed under Limited memory Broyden- Fletcher- Goldfarb- Shanno (LBFGS) [30]–[33] approximation. The Generalized Gradient Approximation [34], [35]with Revised Perdew, Burke and Ernzerhof (RevPBE) is used as an exchange correlation potential. ATK is an advanced version of previously available TranSIESTA-C [36], [37] which is basically based on the new molecular technology, models and various algorithms developed in the platform of TranSIESTA. Some part of this tool has been developed using McDAL [38], which employs different localized basis sets, developed earlier in SIESTA [39]. Here, using supercell technique, nanowires are placed in the 'z' direction. The linear, ladder, and saw tooth configurations of TiO_2 nanowires which are considered here for simulation, are shown in Fig. 1 (a – c). The cut- off energy is considered as 100 Ry and the first Brillouin – zone integration is considered with a Monkhorst – Pack scheme using $1 \times 1 \times 30$ k – points. The cut off energy and the number of k – points are continuously changed to find out convergence under the force tolerance having value 0.05 $eV \text{ Å}^{-1}$ for the various values reported here.

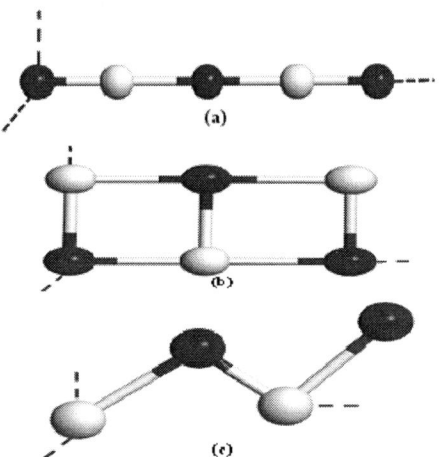

Fig. 1 Atomic configuration of TiO_2 nanowires (a) Linear type (b) Ladder type (c) Saw tooth type
(Blue coloured atoms are 'O' atoms whereas White coloured atoms are 'Ti' atoms.)

III. RESULTS AND DISCUSSIONS

A. Stability Analysis

The stability analysis of titanium dioxide nanowires for different configurations such as, linear, ladder, and saw tooth are considered first, where the equilibrium lattice constant is obtained by minimizing the total energy in a self-consistent manner. Fig. 2 (a – c) shows the plot of total energy expressed in 'eV' vs total volume expressed in 'Ang³' for all the nanowires for different atomic configurations. From the Fig. 2 (a – c), it can be observed that, the ladder shaped nanowire configurations has the lowest energy which implies that it can be the most stable configuration compared to all other configurations. The bond length of all the four structure of nanowires have been computed and listed in Table – 1. As we know that, Bulk and Shear modulus are the parameters for the measurement of hardness of any crystalline solid. The result of simulation as Bulk, Shear, and Young modulus for different structures of TiO_2 nanowires are shown in Table – 1. From the Table -1, it can be observed that, ladder structure has the lowest stiffness whereas, linear structure has the highest stiffness.

Table 1. Column 1 represents the atomic configuration, column 2 represents bond length expressed in Ang. and final column represents various mechanical properties like bulk modulus (B), Shear modulus (G) and Young's modulus (Y)

Structure	Bond Length (Å)	Mechanical Properties		
		B	G	Y
Linear	2.09	27.42	6.95	316.3
Ladder	2.05	-12.69	69.79	3.20
Saw tooth	2.23	22.97	10.84	56.78

B. Electronic Properties

The band diagram of TiO_2 nanowires for different shapes are shown in fig. 3 (a- c) along with other mechanical properties, like Bulk modulus, Shear modulus as well as Young's modulus. All the moduli's for each configuration of nanowire is found out by the 1- D analogy of Murnaghan's Equation of State [40]. Mathematically, it can be given as

$$P(V) = \frac{3B_0}{2}\left[\left(\frac{V_0}{V}\right)^{\frac{7}{3}} - \left(\frac{V_0}{V}\right)^{\frac{5}{3}}\right]\left\{1 + \frac{3}{4}(B_0' - 4)\left[\left(\frac{V_0}{V}\right)^{\frac{2}{3}} - 1\right]\right\}$$

(1)

where,

P= pressure, V_0= reference volume, V=deformed volume, B_0= Bulk modulus, B_0'= derivative of the bulk modulus with respect to pressure.

It can be observed that all nanowire configurations have indirect band gap except saw tooth structure. Further from Fig. 3 (a), it can be observed that linear structure has a narrow band gap of 0.48 eV between Γ to Ƶ point. It can also be observed that, no energy band from conduction band is crossing the Fermi level, which confirms that linear structured nanowire is not metallic in nature. But some of the conduction bands are much closer to Fermi level which means linear nanowire can be a good semiconductor. In Fig. 3(b), the band diagram of ladder type nanowire shows a nominal band gap of 0.028 eV and the transition occurs indirectly from Γ to Ƶ. The band gap mainly depends on the number of atoms taken into consideration and the pseudopotential taken for computation. In ladder structure nanowire, one conduction band is crossing the Fermi level in the band diagram as shown in Fig. 3(b). Thus, the ladder type of nanowire is metallic in nature. Fig. 3(c) shows the band diagram of saw tooth nanowire having a small band gap of 0.094 eV and interestingly transition occurs directly at Ƶ point. As none of the conduction bands are crossing to Fermi level, this structure of nanowire can be treated as semiconductor.

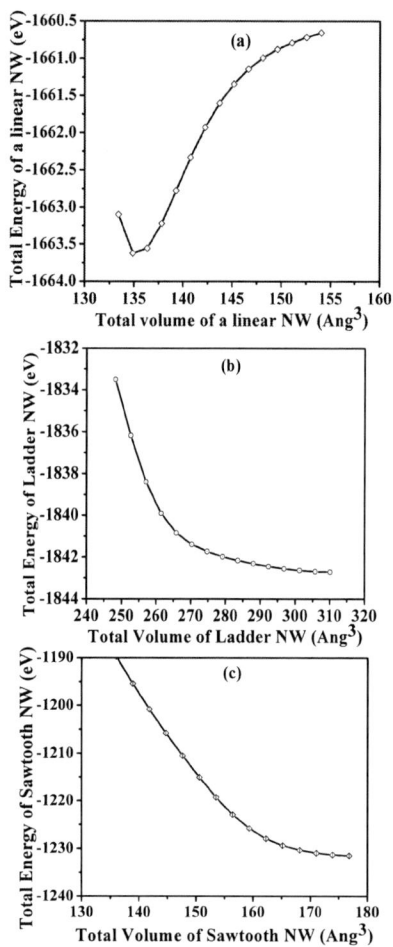

Fig. 2 (a) Total Energy vs total Volume of linear NW (b) total Energy vs total volume of ladder NW (c) total energy vs total volume of saw tooth NW

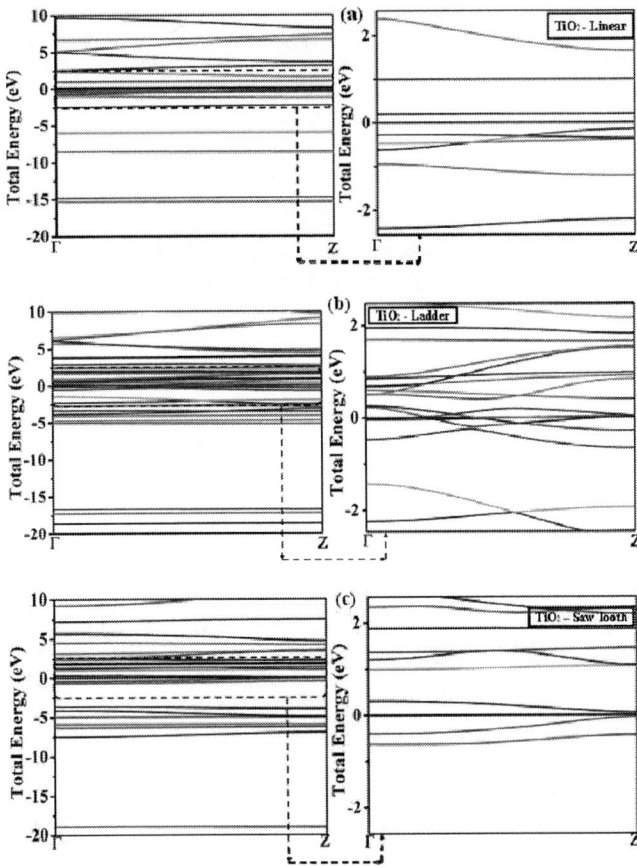

Fig. 3 (a) Band diagram of linear NW configuration (b) band diagram of ladder structured NW (c) band diagram of saw tooth type NW

The density of states (DOS) is another important property to study and analyse the electronic properties of a material. The DOS for each the nanowire configurations is shown in fig. 4 (a- c). It can be seen from Fig. 4 (a) that has a small peak in valence band region and two prominent peaks occurs at the conduction band region. The peak which occurs at $0.6\ eV$ has the highest magnitude and is much closer to Fermi level. It confirms that, good amount of localization of states have occurred near the Fermi level. Similarly, Fig. 4 (b) shows the DOS in ladder structured nanowire. From the figure, it is seen that two peaks occurs in valence band region near to -17.2 and -3.5 eV. But the highest peak occurs at $0.9\ eV$ which lies in the conduction band region. As this peak is much closer to Fermi level, localization of states are very near to Fermi level. Similarly, the DOS of saw tooth can be seen from Fig. 4 (c). Here, two prominent peaks are observed in the valence band region at -19.6 and -5.6 eV respectively. But, in a similar manner like ladder shape nanowires, the highest peak occurs near to 1.5 eV at the conduction band region which shows allowed states are much closer to Fermi level.

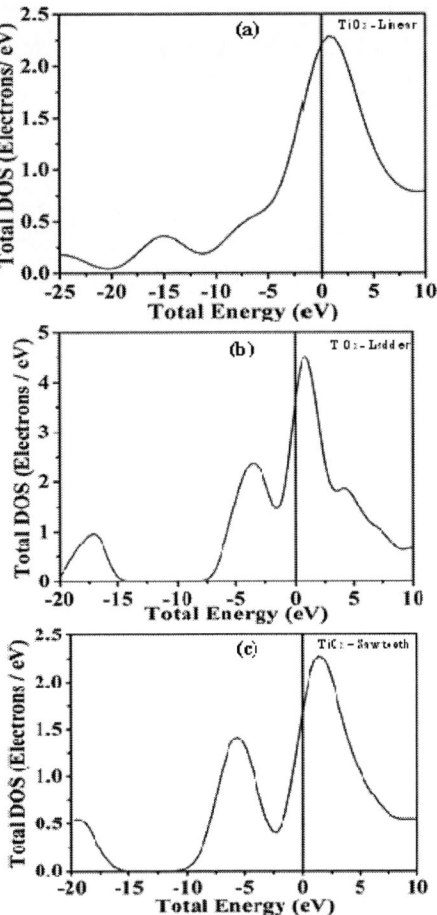

Fig. 4 (a) Density of States diagram for linear TiO_2 NW (b) Density of States diagram for ladder TiO_2 NW (c) Density of States diagram for saw tooth TiO_2 NW.

IV. CONCLUSION

In this paper, a DFT based first- principle method is implemented to investigate the structural stability of TiO_2 nanowires with different atomic configurations. The ladder structure is found to be energetically the most stable configuration among all other configurations. Ladder shaped atomic configuration have very low bulk modulus which proves the softening of the material if dimensions get reduced. Linear, and saw tooth shaped atomic configurations show semiconducting nature whereas, ladder shaped nanowire show metallic nature.

ACKNOWLEDGMENT

We acknowledge National Institute of Technology, Silchar for providing adequate infrastructure for carrying out this research.

REFERENCES

[1]   N. Nouri and S. Ziaei-Rad, "Mechanical property evaluation of carbon nanotube sheets," *Sci. Iran.*, vol. 17, no. 2, 2010.

[2]   S. K. Sinha and S. Chaudhury, "Impact of oxide thickness on gate capacitance - A comprehensive analysis on MOSFET, nanowire FET, and CNTFET devices," *IEEE Trans. Nanotechnol.*,

vol. 12, no. 6, pp. 958–964, 2013.

[3]   C. M. Lieber, "Semiconductor nanowires: A platform for nanoscience and nanotechnology," *MRS Bulletin*, vol. 36. pp. 1052–1063, 2011.

[4]   W. Lu and C. M. Lieber, "Semiconductor nanowires," *J. Phys. D-Applied Phys.*, vol. 39, no. 21, pp. R387–R406, 2006.

[5]   W. Zhang, R. Zhu, X. Liu, B. Liu, and S. Ramakrishna, "Facile construction of nanofibrous ZnO photoelectrode for dye-sensitized solar cell applications," *Appl. Phys. Lett.*, vol. 95, no. 4, pp. 2–5, 2009.

[6]   A. S. Zuruzi, A. Kolmakov, N. C. MacDonald, and M. Moskovits, "Highly sensitive gas sensor based on integrated titania nanosponge arrays," *Appl. Phys. Lett.*, vol. 88, no. 10, pp. 102904–102906, 2006.

[7]   B. O'Regan and M. Grätzel, "A low-cost, high-efficiency solar cell based on dye-sensitized colloidal TiO_2 films," *Nature*, vol. 353, no. 6346, pp. 737–740, 1991.

[8]   A. Srivastava, N. Tyagi, and R. Ahuja, "First-principles study of structural and electronic properties of gallium based nanowires," *Solid State Sci.*, vol. 23, pp. 35–41, 2013.

[9]   J. Wang, M. S. Gudiksen, X. Duan, Y. Cui, and C. M. Lieber, "Highly polarized photoluminescence and photodetection from single indium phosphide nanowires," *Science (80-.).*, vol. 293, no. 5534, pp. 1455–1457, 2001.

[10]   A. I. Yanson, G. Rubio Bollinger, H. E. Van Den Brom, N. Agraït, and J. M. Van Ruitenbeek, "Formation and manipulation of a metallic wire of single gold atoms," *Nature*, vol. 395, no. 6704, pp. 783–785, 1998.

[11]   S. Chander, A. Purohit, A. Sharma, S. P. Nehra, and M. S. Dhaka, "Impact of temperature on performance of series and parallel connected mono-crystalline silicon solar cells," *Energy Reports*, vol. 1, pp. 175–180, 2015.

[12]   C. Becker *et al.*, "Polycrystalline silicon thin-film solar cells: Status and perspectives," *Sol. Energy Mater. Sol. Cells*, vol. 119, pp. 112–123, 2013.

[13]   F. Meillaud *et al.*, "Recent advances and remaining challenges in thin-film silicon photovoltaic technology," *Mater. Today*, vol. 18, no. 7, pp. 378–384, 2015.

[14]   J. Müller, B. Rech, J. Springer, and M. Vanecek, "TCO and light trapping in silicon thin film solar cells," *Sol. Energy*, vol. 77, pp. 917–930, 2004.

[15]   J. D. Poplawsky, "Cadmium telluride solar cells: Record-breaking voltages," *Nat. Energy*, vol. 1, no. 3, p. 16021, 2016.

[16]   J. Ramanujam and U. P. Singh, "Copper indium gallium selenide based solar cells – a review," *Energy Environ. Sci.*, vol. 10, pp. 1306–1319, 2017.

[17] Soumyaranjan Routray, Biraj Shougaijam, T. R. Lenka, "Exploiting Polarization Charges for High-Performance (000-1) Facet GaN/InGaN/GaN Core/Shell/Shell Tringular Nanowire Solar Cell," *IEEE J. Quantum Electron.*, vol. 53, no. 5, pp. 1–8, 2017.

[18]   J. S. Blakemore, "Semiconducting and other major properties of gallium arsenide," *J. Appl. Phys.*, vol. 53, p. R123, 1982.

[19]   M. Grätzel, "Dye-sensitized solar cells," *Journal of Photochemistry and Photobiology C: Photochemistry Reviews*, vol. 4(2). pp. 145–153, 2003.

[20]   D. Dash, C. K. Pandey, S. Chaudhury, and S. K. Tripathy, "Structural, electronic, and mechanical properties of cubic TiO_2: A first-principles study," *Chinese Phys. B*, vol. 27, no. 1, pp. 1–9, 2018.

[21]   A. Fujishima and K. Honda, "Electrochemical photolysis of water at a semiconductor electrode," *Nature*, vol. 238, no. 5358, pp. 37–38, 1972.

[22]   H. Bin Wu, J. S. Chen, H. H. Hng, and X. W. Lou, "Nanostructured metal oxide-based materials as advanced anodes for lithium-ion batteries," *Nanoscale*, vol. 4, no. 8. pp. 2526–2542, 2012.

[23]   Q. Zhang, C. Sun, J. Yan, X. Hu, S. Zhou, and P. Chen, "Perpendicular rutile nanosheets on anatase nano fi bers: Heterostructured TiO_2 nanocomposites via a mild solvothermal method," *Solid State Sci.*, vol. 12, no. 7, pp. 1274–1277, 2010.

[24]   Pankaj Srivastava, Satyendra Singh, "Electronic Properties of GaP nanowires of Different Shapes," *J. Nanosci. Nanotechnol.*, vol. 11, no. 12, pp. 10464–10469, 2011.

[25]   A. Srivastava and N. Tyagi, "Structural and electronic properties of AlX (X=P, As, Sb) nanowires: Ab initio study," *Mater. Chem. Phys.*, vol. 137, no. 1, pp. 103–112, 2012.

[26]   D. Tafen and J. Lewis, "Structure, stability, and electronic properties of thin TiO_2 nanowires," *Phys. Rev. B*, vol. 80, no. 1, pp. 014104–014108, 2009.

[27]   P. Hohenberg and W. Kohn, "The Inhomogeneous Electron Gas," *Phys. Rev.*, vol. 136, no. 3B, p. B864, 1964.

[28]   W. Kohn and L. J. Sham, "Self-consistent equations including exchange and correlation effects," *Phys. Rev.*, vol. 140, p. A1133, 1965.

[29]   D. Dash, S. Chaudhury, and S. K. Tripathy, "First principle investigation of structural and optical properties of cubic titanium dioxide," in *AIP Conference Proceedings*, 2018, vol. 1953.

[30]   C. G. Broyden, "The convergence of a class of double-rank minimization algorithms 1. General considerations," *IMA J. Appl. Math. (Institute Math. Its Appl.*, vol. 6, no. 1, pp. 76–90, 1970.

[31]   R. Fletcher, "A new approach to variable metric algorithms," *Comput. J.*, 1970.

[32]   D. Goldfarb, "A family of variable-metric methods derived by variational means," *Math. Comput.*, vol. 24, pp. 23–26, 1970.

[33]   D. F. Shanno and P. C. Kettler, "Optimal conditioning of quasi-Newton methods," *Math. Comput.*, vol. 24, pp. 657–664, 1970.

[34]   Y. Zhang and W. Yang, "Comment on 'generalized gradient approximation made simple,'" *Physical Review Letters*, vol. 80. p. 890, 1998.

[35]   B. Hammer, L. B. Hansen, and J. K. Nørskov, "Improved adsorption energetics within density-functional theory using revised Perdew-Burke-Ernzerhof functionals," *Phys. Rev. B - Condens. Matter Mater. Phys.*, vol. 59, p. 7413, 1999.

[36]   M. Brandbyge, J. L. Mozos, P. Ordejón, J. Taylor, and K. Stokbro, "Density-functional method for nonequilibrium electron transport," *Phys. Rev. B - Condens. Matter Mater. Phys.*, vol. 65, no. 16, p. 165401, 2002.

[37]   K. Stokbro, J. Taylor, M. Brandbyge, and P. Ordejón, "TranSIESTA: A Spice for Molecular Electronics," in *Annals of the New York Academy of Sciences*, 2003, pp. 212–226.

[38]   J. Taylor, H. Guo, and J. Wang, "Ab initio modeling of quantum transport properties of molecular electronic devices," *Phys. Rev. B*, vol. 63, p. 245407, 2001.

[39]   J. M. Soler *et al.*, "The SIESTA method for ab initio order-N materials simulation," *J. Phys. Condens. Matter*, vol. 14, no. 11, pp. 2745–2779, 2002.

[40]   F. D. Murnaghan, "The Compressibility of Media under Extreme Pressures," *Proc. Natl. Acad. Sci.*, vol. 30, no. 9, pp. 244–247, 1944.

2019 Devices for Integrated Circuit (DevIC), 23-24 March, 2019, Kalyani, India

Empirical Model of Surface Potential and Simulation Analyses for ZnO TFTs

Anirudh Aggarwal
Department of Electrical and
Electronics Engineering
BITS Pilani
Pilani, Rajasthan - 333031
f2015222@pilani.bits-pilani.ac.in

Rupam Goswami
Department of Electrical and
Electronics Engineering
BITS Pilani
Pilani, Rajasthan - 333031
rupam.goswami@pilani.bits-pilani.ac.in

Kavindra Kandpal
Department of Electrical and
Electronics Engineering
BITS Pilani
Pilani, Rajasthan - 333031
kavindra.kandpal@pilani.bits-pilani.ac.in

Abstract— **This paper presents a study on the performance of a bottom gate ZnO thin film transistor (TFT) model through 2-D TCAD device simulations and proposes a surface potential analytical model for the same. The simulation has been calibrated with a fabricated ZnO TFT. The analytical model is found to be in good agreement with the simulated measurements. Through further analyses on TCAD tool, the electrical characteristics of ZnO TFT have been investigated to comprehensively deduce the effect of ZnO active layer thickness, the dielectric material and the drain voltage. To obtain a significant ratio of on and off currents, and a positive gate voltage switching, gate workfunction engineering has been demonstrated.**

Keywords— ZnO, Thin film transistor (TFT), Surface Potential, Two-dimensional (2-D) modelling, Sentaurus TCAD.

I. INTRODUCTION

In the last decade the limitations of smart devices have been overcome via invention, miniaturization and large-scale production of thin film transistors (TFTs) [1]. Among the available active channel layer materials for TFTs, ZnO has become of great importance as an attractive wide band gap semiconductor for realizing a transparent thin film transistor. There are a number of oxide-TFTs which have been proposed so far with the objectives of improving their electrical characteristics. The most notable among them are [2,3,4]. Several experimental reports on ZnO TFTs have demonstrated the surface potential modelling and simulations of device characteristics [5,6]. But there is still a need to comprehensively explore the interplay between the analytical surface potential-based model and the device behaviour. In addition, threshold voltage, subthreshold swing and other such important parameters are difficult to be ambiguously determined for oxide thin film transistors.

This study focusses mainly on defining an analytical gate voltage dependent model for surface potential of the bottom gate ZnO thin film transistors (TFTs) and exploring the effects of varying dielectric material, oxide thickness and the gate voltage of the device through 2D TCAD based simulations. The proposed analytical model for the device describes the behaviour of surface potential along the device channel using a unified equation, without incorporation of the threshold voltage and other uncertain device parameters. The device has been simulated using Synopsys® Sentaurus™. In recent studies, there has been investigation on oxide thin film transistors through application of various oxide materials [7,8]. But in order to have good characteristics at the oxide channel layer and high-k dielectric interface, ZnO channel

layer is deposited atop the dielectric layer in the simulated model. Unlike the non-oxide/high-k dielectric interface, ZnO is expected to have no deteriorating chemical interactions in practical scenario. The bottom gate TFT model was investigated because it is much more convenient to design and fabricate the bottom gate device for practical applications. The device characteristics and performance parameters have been investigated and compared for different dielectric materials namely, SiO_2, Al_2O_3, HfO_2 and Si_3N_4 and varied oxide channel layer thicknesses.

The paper is organized in the following manner. Section II introduces the device structure and simulation parameters. Section III discusses about the proposed surface potential model and inferences. Section IV summarizes various electrostatic effects and characteristics of the ZnO-based TFT model using 2-D device simulation results. Finally, the study is concluded in section V.

II. DEVICE STRUCURE AND SIMULATION SETUP

Figure 1 represents the schematic outlook of a ZnO-based thin film transistor. The current-voltage characteristics of the TFT corresponding to different dielectric materials and applied gate potentials were studied by carrying out 2-D TCAD simulations using Synopsys® Sentaurus™ Device structure editor. ZnO is assumed to have formed a continuous and homogenous active channel layer with a uniform doping of Indium to provide a characteristic n-type behavior to the device. In order to study device characteristics, ZnO layer thickness is varied between $20\ nm$ and $300\ nm$.

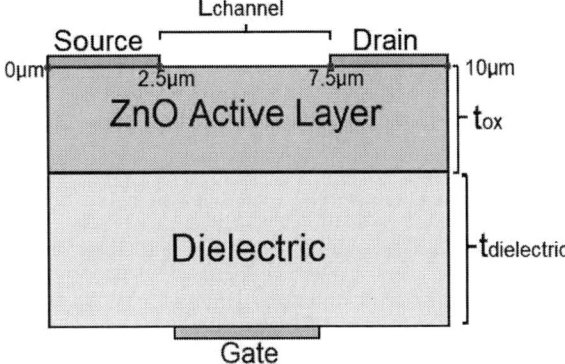

Figure 1. Schematic of a bottom gate ZnO-based Thin Film Transistor (TFT)

978-1-5386-6723-1/19 $31.00 © 2019 IEEE   221

The semiconductor parameters are default for silicon in Sentaurus. For ZnO-based TFT device simulations, some parameters were specified to define ZnO as an active layer material in a thin film transistor structure. For 2-D device simulation, experimentally optimized measurements of the TFT structure were considered in order to obtain better TFT performance for practical applications. The simulation parameters were calibrated by reproducing the existing experimental results reported for oxide thin film transistors [3] as enumerated in Table 1.

Table 1. Device Simulation Parameters

Parameter	Value
Device Depth	$0.45\,\mu m$
Device Length	$10\,\mu m$
Channel length	$5\,\mu m$
ZnO Thickness	$0.1\,\mu m$
ZnO Dielectric Constant	8.12
ZnO Bandgap	3.37 eV
ZnO Electron Affinity	4.29 eV
N-type doping in ZnO Layer	$1x10^{17}\ cm^{-3}$
Si_3N_4 Thickness	$0.35\,\mu m$
Source and drain doping	$1x10^{17}\ cm^{-2}$
Si_3N_4 Dielectric constant	7.55

III. EMPIRICAL MODEL

From the simulations and physics-based nature of the surface potential [9,10], it is clearly seen that the exponential distribution of surface potential for the model can be approximated as a quadratic exponential function of the channel length for modelling of ZnO TFTs. The analytical surface potential expression reads as follows

$$\varphi_S = V_G.e^{(aV_G^2+bV_G)} + cV_D + V_{ref} \qquad (1)$$

where V_G is the applied gate voltage, V_D is the drain voltage (10 V), V_{ref} is the base reference voltage (1.5 V) that has been used across the device structure in Synopsys® Sentaurus™ TCAD making the source voltage as 1.5 V. The variables a, b and c are defined as quadratic and exponential functions of the TFT channel length respectively as

$$a(y) = (\alpha y^2 + \beta y + \gamma) \qquad (2)$$
$$b(y) = (uy^2 + vy + w) \qquad (3)$$
$$c(y) = e^{py^2+qy+r} \qquad (4)$$

where $\alpha, \beta, \gamma, u, v, w, p, q$ and r are constants with y as the variable defining the ZnO TFT channel length in μm.

Table 2. Surface Potential values for varied gate voltages and channel positions

V_G (Volts)	0			5			10		
y (μm)	5	7	7.5	5	7	7.5	5	7	7.5
φ_S (Volts)	2	3.8	9.5	3.67	7	10.74	5	9	11.12

The modelled equations are valid for a portion of the device channel region ranging from 6 to 7.5 μm. The behavior of surface potential curve from 2.5 to 6 μm is linear and is realized by interpolation of points from source end to the mid-point of the channel. In order to find the defined constants, surface potential values for a set of points were noted down using Synopsys® Sentaurus™ Device TCAD simulations for V_G (0,5 and 10 V) and y (5,7 and 7.5 μm). All noted values for the surface potential are for the device surface and have been summarized in table 2.

Now assuming V_G to be zero and referring to the observed surface potential values from the table 2.

$$\varphi_S = c(y)V_D + V_{ref}$$

where $c(y) = e^{py^2+qy+r}$,

for $y = 5$, $\qquad \varphi_S = 10c(5) + 1.5 = 2$

similarly, for $y = 7$, $\qquad \varphi_S = 10c(7) + 1.5 = 3.8$
and for $y = 7.5$, $\qquad \varphi_S = 10c(7.5) + 1.5 = 9.5$
implying $c(5) = 0.05, c(7) = 0.2$ and $c(7.5) = 0.80$

Putting these values in the expression for c(y)
for c(5), $\qquad 0.05 = e^{25p+5p+r}$
$$-3 = 25p + 5q + r$$
similarly, for c(7), $\qquad 0.23 = e^{49p+7p+r}$
$$-1.47 = 49p + 7q + r$$
and for c(7.5) $\qquad 0.80 = e^{56.25p+7.5p+r}$
$$-0.223 = 56.25p + 7.5q + r$$
implying $\quad p = 0.7,\ q = -7.57,\ r = 17.2 \qquad (5)$

Now for constants corresponding to a(y) and b(y)
for $y = 5$,
Case (i): $V_G = 5$, $\qquad \varphi_S = 5.e^{25b+5b} + 2 = 3.67$
Case (ii): $V_G = 10$, $\qquad \varphi_S = 10.e^{100a+10b} + 2 = 5$
$\quad a(5) = 0.019784, \qquad b(5) = -0.31824$

for $y = 7$,
Case (i): $V_G = 5$, $\qquad \varphi_S = 5.e^{25a+5b} + 3.8 = 7$
Case (ii): $V_G = 10$, $\qquad \varphi_S = 10.e^{100a+10b} + 3.8 = 9$
$\quad a(7) = 0.004772, \qquad b(7) = -0.11312$

for $y = 7.5$,
Case (i): $V_G = 5$, $\qquad \varphi_S = 5.e^{25a+5b} + 9.5 = 10.74$
Case (ii): $V_G = 10$, $\qquad \varphi_S = 10.e^{10a+10b} + 9.5 = 11.12$
$\quad a(7.5) = 0.018313, \qquad b(7.5) = -0.37037$

Using the obtained values of a(5), a(7) and a(7.5) in the expression $a(y) = \alpha y^2 + \beta y + \gamma$
$$a(5) = (25\alpha + 5\beta + \gamma) = 0.019784$$
$$a(7) = (49\alpha + 7\beta + \gamma) = 0.004772$$
$$a(7.5) = (56.25\alpha + 7.5\beta + \gamma) = 0.018313$$
$\alpha = 0.0138352, \beta = -0.173528, \gamma = 0.541546 \qquad (6)$

Using the obtained values of b(5), b(7) and b(7.5) in the expression $b(y) = uy^2 + vy + w$
$$b(5) = (25u + 5v + w) = -0.31824$$
$$b(7) = (49u + 7v + w) = -0.11312$$
$$b(7.5) = (56.25u + 7.5v + w) = -0.37037$$
$u = -0.246744, v = 3.06345, w = -9.46688 \qquad (7)$

Finally,

$$\varphi_S(V_G, y) = V_G \cdot e^{(a(y)V_G^2 + b(y)V_G)} + 10c(y) + 1.5 \qquad (8)$$

and from the above obtained values of the constants using (5), (6) and (7), the following is inferred

$$a(y) = 0.0138352y^2 - 0.173528y + 0.541546$$
$$b(y) = -0.246744y^2 + 3.06345y - 9.46688$$

$$c(y) = e^{0.7y^2 - 7.57y + 17.2}$$

The derived analytical model for surface potential of the device along the channel is in very good agreement with the simulated curves for surface potential from Sentaurus TCAD simulations. The suited gate voltage range for the model is -1 to 1 V. The range is justified for the model as the proposed device is based on a high-k dielectric material which brings the threshold voltage of the ZnO-based TFT in the above said ambits [1].

IV. RESULTS AND DISCUSSION

This section presents verification of the surface potential model proposed in Section III, and discusses the effect of various parameters (gate dielectric, ZnO thickness and drain voltage) on the transfer characteristics through TCAD simulations.

A. Empirical Model of Surface Potential

In the current work, the surface potential characteristics of experimental ZnO TFT were investigated by TCAD simulations. The simulations were performed considering a simplified ZnO TFT model with negligible effect of grain boundaries. The surface potential model has been analytically developed for Si_3N_4 as the dielectric material with a layer thickness ($t_{dielectric}$) of $0.35\,\mu m$ and ZnO thickness of $0.1\,\mu m$. The device characteristics were reproduced with very good accuracy using the defined model. It is demonstrated that the surface potential characteristics along the channel of the model appeared to reproduce well the simulated results in Sentaurus TCAD as presented in Fig. 2.

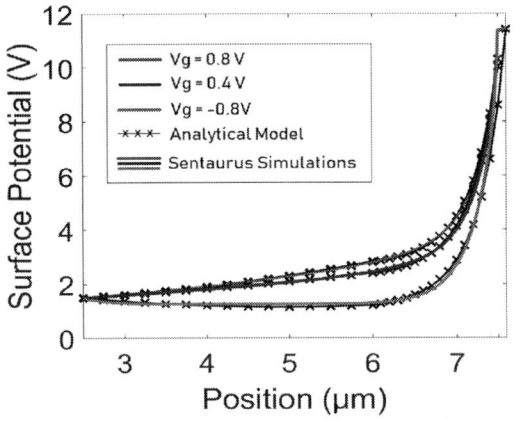

Figure 2. Comparison of modelled and simulated Surface Potential Curves along the device channel

The applied gate voltage was increased in steps of 0.2 V and the analytically modelled equation was used to plot and investigate the surface potential curves. Fig. 3 shows the plots for surface potential along the device active channel. As expected, the surface potential curves shift to higher voltage values as the gate voltage is increased.

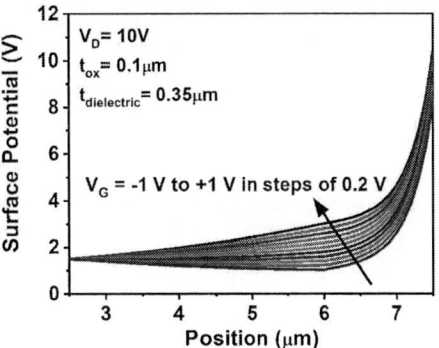

Figure 3. Surface Potential Curves along the device channel for gate voltage varying from -1 to 1 V

B. Dielectric Modulation

The design of thin film transistors is primarily affected by varied gate dielectric materials as the gate dielectric plays a vital role in defining the transistor performance [11,12,13]. 2-D device simulations for different gate dielectric materials namely, SiO_2, Al_2O_3, HfO_2 and Si_3N_4 have been studied to investigate the dependence of thin film transistor properties on the device dielectric material in Fig. 4. A shift in the positive direction in the threshold voltage is observed from the simulation results as the dielectric constant is increased and matches with experimental results shown in [8].

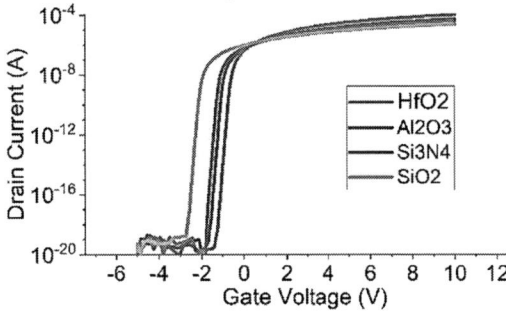

Figure 4. Drain current-Gate voltage characteristics of the n-type ZnO-based TFTs for different dielectric materials

The gradual increase in the threshold voltage for the chosen dielectric materials for the model is because of the variation in the dielectric capacitance. Opposite charges are induced in the channel due to this when gate voltage is applied to the device. Although, there is no consideration of interface effects in the modelled device, this analysis provides sufficient understanding to the concerned designer/researcher to infer about the effect of dielectric material on the device parameters. Subthreshold swing is observed to improve from SiO_2 to Al_2O_3 and there is a steep rise for TFT with HfO_2 because of its permittivity.

The simulated ZnO-based thin film transistor with HfO_2 as gate dielectric is found to have more than 10 times improvement in value of on-state drain current than the TFT with SiO_2 from the comparative study. This makes HfO_2 a preferred candidate for ZnO-based transistors. However, the characteristics of Fig. 4 depict depletion mode operation which results in poor on-off current ratios. To mitigate this problem, one can opt for gate material with higher work function. Fig. 5 shows the effect of using Pt (Φ_m = 5.75 eV) as gate electrode on transfer characteristics which shows a shift towards positive gate voltage which results in enhancement mode operation, thereby, improving on-to-off current ratio.

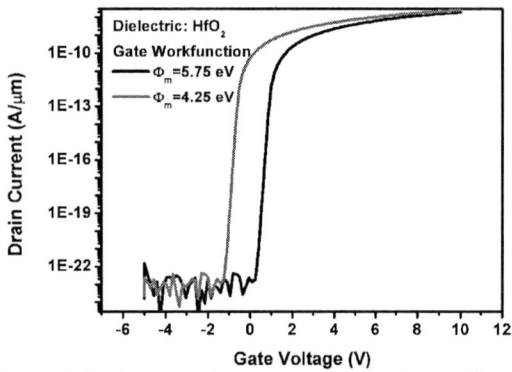

Figure 5. Drain current-Gate voltage characteristics of the n-type ZnO-based TFT showing the effect of gate workfunction engineering

C. ZnO Thickness Modulation

The ZnO active layer thickness (t_{ox}) was modulated to investigate the practicality of controlling the threshold voltage (V_{th}) of the TFT. There was no significant change observed in the subthreshold swing or in $I_{on/off}$ ratio of the device as the value of threshold voltage of the ZnO-based transistor was linearly adjusted by decreasing t_{ox} [14]. Interestingly, there is a systematic shift in the transfer characteristics towards the positive values of gate voltage as the ZnO thickness is decreased from 300 to 20 nm as shown in Fig. 6. The threshold voltage of the Zinc Oxide-based thin film transistors linearly increases. The relatively large

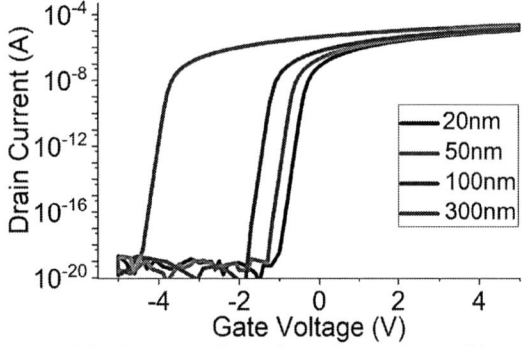

Figure 6. Drain current-Gate voltage characteristics of the n-type ZnO-based TFTs for different oxide layer thicknesses

conduction band density of the ZnO active layer is believed to be the reason associated with the negative threshold voltage values for the n-type oxide thin film transistors [15]. The number of free carriers in the active channel layer is proportional to the thickness. This is the reason why the gate voltage required to turn on the drain current becomes more negative for the TFT with the thicker oxide channel layer. A thin layer of oxide for the device channel is ideally needed as the on-to-off current ratio and mobility are better. But in

practical sense, it is tough to obtain a silicon like single crystalline structure in a very thin ZnO layer. The limit governing how thin the ZnO layer can be made depends on its ability to just support the formation of accumulation region [1].

D. Drain Voltage Dependence

The device was simulated for four different drain voltages as shown in Fig. 7. There was no noticeable shift observed for the curves in the subthreshold exponential region whereas significant increase in the drain current values was seen for increasing values of drain voltage. The subthreshold swing for the devices remained almost the same showing negligible dependence on drain voltage. The simulated results were in accordance with the existing theoretical knowledge.

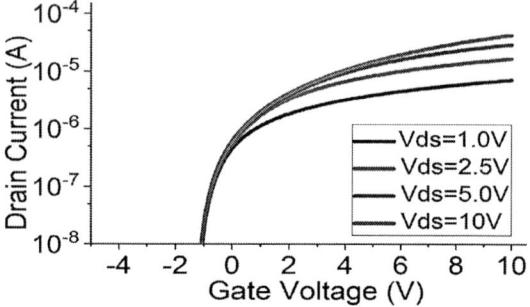

Figure 7. Drain current-Gate voltage characteristics for n-type ZnO-based TFT for different drain voltages

V. CONCLUSION

In this paper, we studied the properties of bottom gate ZnO TFTs and observed that the active channel layer thickness and dielectric material of the device influence the characteristics of ZnO TFTs significantly. The empirical model developed for the surface potential profiles along the TFT channel was found to efficiently reproduce the simulation curves. In results, both the dielectric constant modulation and channel thickness were observed to play a crucial role in determining the mode of operation and threshold voltage of the device.

However, it should be taken into account that the device model that has been considered is simple and does not include several important factors that play a role in an actual environment. Nevertheless, this model presents an efficient description of experimentally observed results. Therefore, it is our belief that this study will encourage further experimental and modelling based explorations of bottom gate ZnO thin film transistors.

REFERENCES

[1] Kandpal, K., & Gupta, N. (2018). Perspective of zinc oxide based thin film transistors: a comprehensive review. *Microelectronics International, Vol. 35, Issue:1*, pp.52-63.

[2] Kandpal, K., & Gupta, N. (2016). Investigations on high-κ dielectrics for low threshold voltage and low leakage zinc oxide thin-film transistor, using material selection methodologies. *Journal of Materials Science: Materials in Electronics*, Volume 27, Issue 6, pp 5972–5981.

[3] Hossain, F. M., Fujioka, H., Ohno, H., Koinuma, H., & Kawasaki, M. (2003). Modeling and simulation of polycrystalline ZnO thin-film transistors. *Journal of Applied Physics, Volume 94, Issue 12,* , pp. 7768-7777.

[4] Lee, S., & Nathan, A. (2016). Conduction Threshold in Accumulation-Mode InGaZnO Thin Film Transistors. *Scientific Reports*, 6, 22567.

[5] Lee, S. J. (2015). Transparent semiconducting oxide technology for touch free interactive flexible displays. *Proc. the IEEE 103, 644.*

[6] Sun, K. G., Nelson, S. F., & Jackson, T. N. (2015). Modeling of Self-Aligned Vertical ZnO Thin-Film Transistors. *IEEE TRANSACTIONS ON ELECTRON DEVICES*, VOL. 62, NO. 6.

[7] Yamada, S., Yokoyama, S., & Koyanagi, M. (1990). Two-Dimensional Device Simulation for Polycrystalline Silicon Thin-Film Transistor. *Japanese Journal of Applied Physics*, Vol.29, No.12, pp. L2388-L2391.

[8] Walker, B., K, A., Pradhan, & Xiao, B. (2015). Low temperature fabrication of high performance ZnO thin film transistors with high-k dielectrics. *Solid State Electronics, Volume 111*, p. 58-61.

[9] Galup Montoro, C., Pahim, V., Rios, R., & Schneider, M. (2005). Comparison of Surface Potential and Charge-based MOSFET Core Models. *TechConnect Briefs*, 13-18.

[10] Gildenblat, G., Chen, H. W.-L., & Gu, X. (2004). SP: An Advanced Surface-Potential-Based Compact MOSFET Model. *IEEE JOURNAL OF SOLID-STATE CIRCUITS* , VOL. 39, NO. 9,.

[11] S.Vallisree, R.Thangavel, & Lenka, T. R. (2017). Comparative Characteristics Study of the Effect of Various Gate Dielectrics on ZnO TFT. *International Conference on Energy, Communication, Data Analytics and Soft Computing (ICECDS)*. IEEE.

[12] Sharma, P. a. (2015). Investigation on material selection for gate dielectric in nanocrystalline silicon (nc-Si) top-gated thin film transistor (TFT) using Ashby's, VIKOR and TOPSIS. *Journal of Material Science: Materials in Electronics, Vol.26, No.12*, 9607-9613.

[13] Moon, Y.-K., Lee, S., & Park, J.-W. (2009). Characteristics of ZnO-based TFT Using La2O3 High-k Dielectrics. *Journal of the Korean Physical Society*, Vol. 55, No. 5,pp. 1906-1909.

[14] Park, J.-S., Jeong, J. K., Mo, Y.-G., Kim, H. D., & Kim, C.-J. (2008). Control of threshold voltage in ZnO-based oxide thin film transistors. *Applied Physics Letters*, 93, 033513.

[15] Chung, J., Lee, J., Kim, H., b, N. J., & Kim, J. (2007). Effect of thickness of ZnO active layer on ZnO-TFT's characteristics. *Thin Solid Films*, 516, pp. 5597-5601.

Smart ATM Service

Sayan Hazra
Department of Electrical Engineering
University Institute Of technology
The University Of Burdwan
Burdwan, India
hazrasayan201442@gmail.com

Abstract— **Automated Teller Machine (ATM) is an electronic telecommunications device, which enables customers to perform banking without the need for direct interaction with bank staff. For this, every account holder must have a unique id card for the individual account having a unique pin. On the absence of this card, whatever be the adverse situation the use of this ATM service is not permitted. So, an Internet Of Things and Computer Vision based Smart ATM service is being proposed here, using Raspberrypi microcontroller based embedded system, where each person will be their own identity, where Fingerprint, Face, OTP verifications are key features for security, which in turn reduces the issue of fraud transactions, fraud ATM cards, hence security issue gets resolved.**

Keywords—ATM, IOT, Computer Vision, Raspberrypi, microcontroller, embedded system

I. INTRODUCTION

ATM service [1] stands as one of the most important facility, while it comes to cashless travel. In ATM service every account holder, have individual unique ATM card, provided from the corresponding bank. Using that ATM cards finite limits of transactions are to be permitted within a finite time interval, from any ATM. On the other hand, digitalization becomes very much easier while, having an ATM card, especially during online money transfer or online marketing facility. During ATM service the ATM card is needed to be inserted into the ATM machine along with the unique password or id for that ATM card, and then if the credentials of that ATM card gets matched with the server data, the transaction gets permitted. According to ATM industry association (ATMIA), there are around 3.5 million ATMs are installed worldwide. But, an issue is the execution of these ATM service is bound to the unique ATM card, if it is not there no permission will be granted, whatever be the adverse situation, ATM service will no more permitted, if it gets lost or is not present at that very moment. On the other hand, as more and more people are being globally digitalized, and as it is being so much used, there are being so many cases of having fraud ATM cards, fraud transactions ever since. In this era of global digitalization, ATM cards are linked with most of the digital wallets, account for achieving easiest transaction facility, along with online banking. So, somehow if the card details including a/c number, password, CVV code is known by some third party, it can be misused in thousands way. So, to get rid of these dreadful issues of security of this service, the efficient way out is to go for biometric security, which is proposed here throughout this model. In this present era of massive technological growth, Internet of things, Computer Vision are becoming cutting-edge technologies. Thus, the association of Internet Of Things(IOT) and Computer Vision with the ATM service makes the ATM service much more smart, advanced and user-friendly, too. Since the last few years some works has been done to reduce those problems.

Transaction from ATM using mobile banking apart from using ATM is proposed in [4], in order to reduce time of transaction, but there might be security problem, if the system is compared to any biometric security. On the other hand, here the mobile is needed each time for ATM service, when this the proposed model deals with only the user, when it comes to account holder transaction, which is more advantageous than this. The research work described Secured pin authentication (SEPIA) as a service for ATM using mobile, and wearable, in [7] does moreover same as described before, in addition to that a new "mcard" is introduced there to avail m-payment facility, which is mobile banking facility using the same card instead of using ATM card. On the other hand, some facilities like checking authentication from any mobile or wearable devices, to prove co-location with cloud based server and to generate a secure pin for banking. It is much advantageous, but not secured and here, the service becomes bound to a card, which is less secured and advantageous too. In the research work [10] a concept of using multipurpose smart SIM card is proposed, which consists of mobile and all the other available facility along with the ATM service. The issue of this proposed model is implementation, because the ATM services can be changes throughout, if it comes to insecurity and need of advancement, but according to proposed model in [10] a set of new SIM card is needed that avails all the facilities of smart cards into one, here all the SIM cards are needed to be changes, which in turn will need to change a whole system, which is not possible at all, as there are more easier ways out like using biometrics, which is being proposed here in this model. In the paper [9], a new way of ATM service is proposed where, it is aimed to connect all the ATMs using IPv6, where a Near-Field Communication (NFC) is proposed, which would communicate via NFC enabled mobile phone with the ATMs, after inserting the ATM. The process is secured but time taking and bounded to have ATM cards and particular mobile having such facilities. In[11] RFID, GSM, GPS based smart ATM card is proposed for security and authentication, which is not cost effective, and main issue is the model is still that card bounded, and still less secured than biometric ones. In [12] fingerprint based biometric authentication is proposed, which is a great move to achieve to goal of improving security of ATM service, thus advantageous. But, no system is error free, and fingerprint checking is one of those efficient ways, but it can also be bypassed. In addition to that, it can be said that, fingerprint authentication is nothing but a sensor based authentication, like the way a blind person tries to execute perception related tasks, though it's efficient, great but incomplete. So, if vision and sensing are added to the system, any security system gets fulfilled. That is the reason, why here two major efficient biometric facilities are put together, which is no more bound to ATM cards while transaction by a/c holders, and much secured as the account holder is the only one gateway of transaction.

978-1-5386-6723-1/19 $31.00 © 2019 IEEE

II. SYSTEM ARCHITECTURE AND IMPLEMENTATION

In this proposed model, no ATM card is needed for transaction, so for security purpose, face-recognition along with the aliveness checkup of face, fingerprint verification, OTP (one time password) verification security checks have been taken into consideration. A brief of face recognition concept will be described here.

A. Face Recognition

In this proposed architecture face-recognition plays an important role. Face-recognition is one of most advanced biometric technology, where a digital image of an individual customer will be taken, and compared to pre-trained system images in a database. The face-recognition algorithm in this proposed system, consists of few major sections-
1. Face-detection and building the dataset, 2. Building a recognition model, 3. Recognize faces in live video feed

All these sections will be explained using the block diagram shown in Fig. 1.

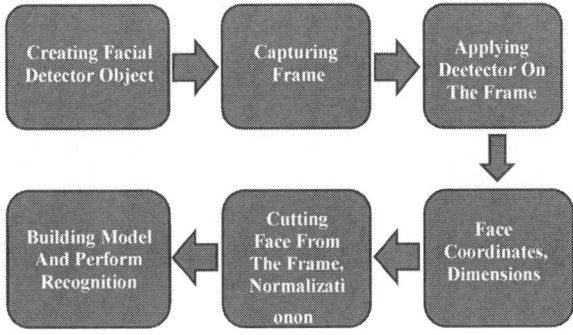

Fig. 1. Block diagram of the workflow of face recognition

1. Face-detection:

In this proposed system, face-detection is done by using Haar-featured based Cascade classifiers, as proposed by Paul Viola and Michole-Jones, in "rapid Object Detection" [2], which is consisted of Haar features extraction: where different square shaped shapes are traversed throughout the images or frames, to analyze the frame. Each of those squares, which can be rotated or extended in all possible ways during traversal, consists of two sections black and white. In this entire process of traversal feature extraction is done by finding difference between overall pixel intensity between black and white regions of the squared shapes, like:
Σ White intensity- Σ Black intensity= Output,
high output is obtained when regions are similar. In haar feature extraction edge, line and center surrounded features are extracted.
In haar feature extraction method, some limitations are observed, so Integral image and Adaboost methods are used. In this **Adaboost** method from linear combination of various weak classifiers, a strong classifiers are obtained from the feature detection process, which is defined by:

$$F(x) = \alpha_1 f_1(x) + \alpha_2 f_2(x) + \ldots + \alpha_n f_n(x),$$

where f(x), F(x) are weak classifier and strong classifier,

α_i is the weight corresponds to i th weak classifier, the bigger the weight, the more relevant the feature will be. Depending upon the value of these weights the classifiers are ranked.

In **Integral image**, each of the pixel intensity is replaced by the sum of all the pixel intensities around it. Then comes the one of most efficient step, called cascading.

Cascading, consists of traversing the input image through various separated classifiers, which actually finds the facial-object from the image, features are distinguished by the classifiers, after traversing through all the classifiers the face is finally detected inside the image or frame.
The demonstration of the proposed model has implemented using **python**, and **OpenCV [3]** computer vision tool for executing image manipulation, facial recognition. Various pre-trained models are there in OpenCV.
"haarcascade_frontalface_alt", is such pre-trained model for face detection, is used in this system for face-detection.
"haarcascade_frontalface_alt", is such pre-trained model for face detection, is used in this system for face-detection. The algorithm of face-detection is represented by the block diagram in Fig.1. According to the block diagram shown in the figure, firstly the **CascadeClassifier,** which creates the classifier is called and thereafter, the Facial detector object is created using the haarcascade frontal face standard face detection model.
Then the image or frames are capured containg faces and, detector is applied on each of the frames. While applying detector objects over the frames, some parameters are passed, which includes

scale-factor: To detect faces inside image, the detector first takes an area and applies classifier to detect faces inside that, if not then it increases or reduces the area by the scale_factor times of the previous area, here scale_factor is taken as 1.2.

min_neighbors: It signifies, no. of neighboring points in the image is to be flagged positive to detect the face.

min_size: this is in the format (width, heights) which signifies the minimum size of the face that can be detected inside the frame.

flag: it means the proerty of the images, which signifies how many features are atleast needed to be satisfied in oder to detect the faces inside the frame.

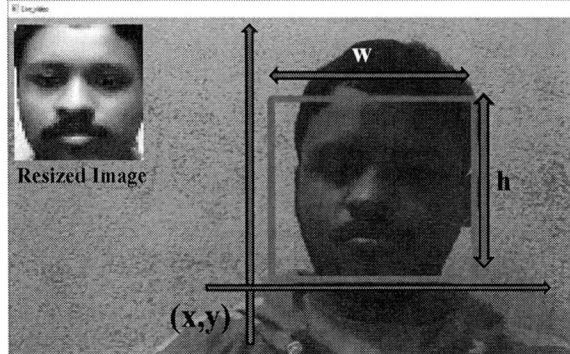

Fig. 2. Demo of the detected face along with the resized face before

After applying the detector, a numpy array is obtained which, consists of (x,y), i.e. the coordinate of the left bottom corner of the rectangular region, which covers the face inside the frame, along with the width(w) and height(h) of the rectangle. After applying the concepts, face is detected and, using the obtained parmeters regarding the position of face inside frame, rectangle is drawn around the faces as demonstrated in the Fig. 2.

2. Building a recognition model:

After the face in a frame gets detected, the face is cut from the frame. The dimension of the region or area which is cut from the frame, that contains the face is

height=h, and
width=w-(2*(w_rm))

Where, **w_rm=absolute value of (0.2*w/2).** So, w_rm is the portion of width deducted from the actual obtained width by the detector, as it can be seen that, some of the width is noisy, so to remove this portion, width gets modulated. After the face is cut, it is first converted into grey-scale image, there after the face is **normalized**, by which, the contrast of the face gets enhanced in the complete greyscale spectrum range, which is between 0 to 255.After the grey scale facial image gets normalized, the face is **resized** to a finite size having standard pixel dimension, here it is (50, 50). The demo of this resized form of a facial image is shown in the Fig. 2. Now, in this way final resized facial data from various consecutive is stored in a particular folder to construct a recognizer model. Workflow of recognization section

Resized dataset→ Pass the recognizer algorithm→ Train the system→ Obtain frame from live video→ apply reconizer→ Check prediction result→ Show the status of recognition

After obtaining the resized facial image set, to build the recognizer model, **Local Binary Pattern Histogram (LBPH)** algorithm, as proposed in **[6]** is used here. In this lbph algorithm, the whole image is not treated at once, it takes a small section of the image, where pixel at the middle becomes the threshold to the neighbors, if the neighboring pixel greater than threshold, then it gets replaced by 1 else replaced by zero, by doing this, a certain local binary pattern on the image gets formed. Then, spatial information are incorporated by lbph. So, this way on applying the lbph over the resized dataset of particular person, face recognizer model of the dataset of that particular person gets built, and trained at the same time. Final step is recognition.

3. Recognize faces in live video feed:

In the previous way, an unknown detected face from a frame is detected, normalized, resized and then gray scale resized image is passed through a predictor, which uses the pre-trained saved recognized model of saved database for checking with the input image, and returns the prediction value, if the prediction value along with the corresponding dataset name with which it's prediction is close, of the model

less than the pre-calibrated threshold, then face got successfully recognized with the corresponding dataset of stored database. In Fig. 3. A demo of such recognized face is shown in python live video window.

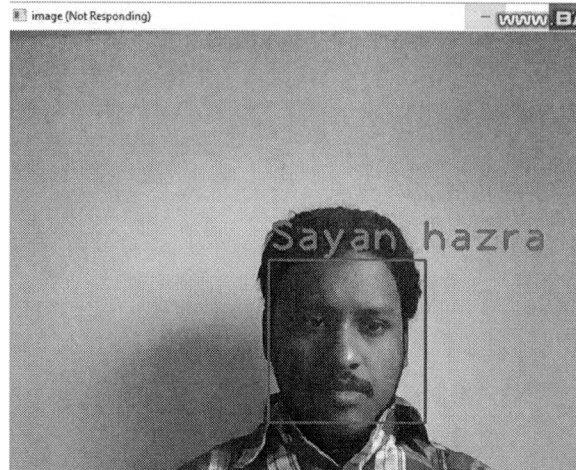

Fig. 3. Demo of the final recognized face from live video

After a face got recognized it is needed to validate to aliveness of that face to distinguish between actual face and image of that face. In order to check aliveness, smile feature is needed to be deteceted on the recognized face, when face got recogized, user will be requested to smile for aliveness checkup.

B. Aliveness Check

After the face gets recognized, to check the aliveness of the face, further smile is detected on the cropped resized facial grey scale image. To **detect smile, haar cascade based classifier,** whose concept was proposed in [5], is constructed using **"haarcascade_smile.xml"**, a pre-trained model, which returns a numpy array consisting location of the smiled portion inside the facial image, if array returns greater than zero, then smile detected on the face and aliveness.
Note: In order to check aliveness of a recognized face, both recognition function and aliveness function should be run in a thread.

C. System Setup

Fig. 4 shows the schematic diagram of the system setup. In this system, **Raspberrypi 3** is used as a microcontroller as shown in the figure. Keypad is connected to enter input from the user during ATM service, fingerprint sensor is used to check the fingerprint of users and operation result will be displayed on the display screen. **Raspberrypi camera module**, which having good light sensitivity, is used to face-recognition. **Global System for Mobile Communication (GSM)** module is used here to send the required message to the user phone number, whether it may be one time password (OTP) or any other confirmation message. **L293d motor driver module** is used to operate the motor setup of the cash transaction slit, using digital output from the analog to digital converter chip depending upon the instruction from Raspberrypi.

2019 Devices for Integrated Circuit (DevIC), 23-24 March, 2019, Kalyani, India

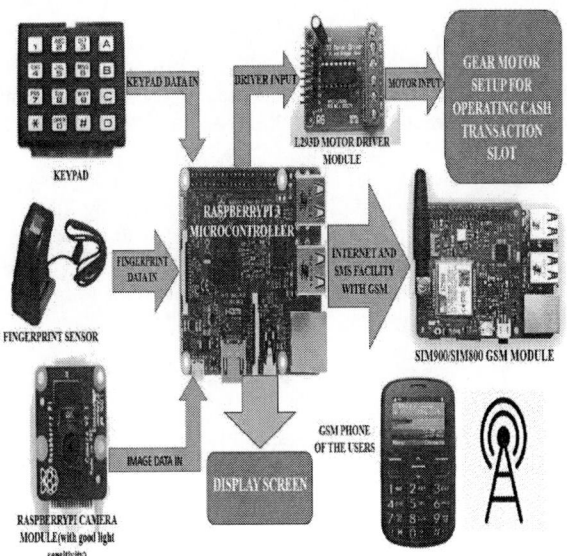

Fig. 4. Schematic of the system setup

D. Banking Process

Banking process is consisted of three categories, these are

1. Banking by account holder: When account holder selects the account holder banking option, then the workflow is explained by the block diagram shown in Fig. 5. If both fingerprint and face got recognized, user can continue the transaction, if anything wrong happens, a message of wrong attempt will be sent to the registerer mobile number of the account holder and the photograph will be stored in the cloud database of bank, if user wants to see the photograph and other details of wrong entry, then details of wrong entry along with photograph can be seen after logging in to the user account.

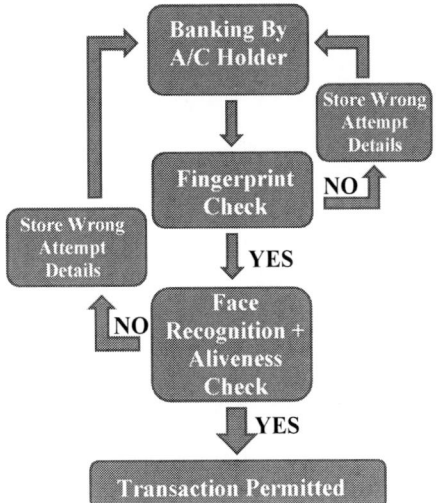

Fig. 5. Block diagram of workflow of a/c holder banking

Fig. 6 shows the demo of account holder banking transaction on the python shell output window.

```
Python 3.6.4 (v3.6.4:d48eceb, Dec 19 2017, 06:54:40) [MSC v.19
Type "copyright", "credits" or "license()" for more informatio
>>>
================ RESTART: D:\7th sem\final seminar\main code.py
WELCOME TO 24 HOURS ATM SERVICE
1.ENTER 1 FOR USER BANKING,
2.ENTER 2 FOR PROVIDING SUB-USER DETAILS,
3.ENTER 3 FOR BANKING BY SUB-USER
1
AS A SUBSTITUTE OF FINGERPRINT WE ARE TAKING ACCOUNT NUMBER:
79801504211234
['ATANU PAL', 98765432112345.0, 'ATUL']
['NABARUN SENGUPTA', 12345678912345.0, 'NABA']
['SAYAN HAZRA', 79801504211234.0, 'Sayan0506']
PROCEED
LOOK AT THE CAMERA FOR FACE RECOGNITION:
face count 1
LBPH faces:->Sayan hazra124.94813873724755 1
YOUR CREDENTIALS ARE
text:'SAYAN HAZRA'
number:79801504211234.0
YOU CAN CONTINUE YOUR TRASACTION
THANKS FOR BANKING!
1.ENTER 1 FOR USER BANKING,
2.ENTER 2 FOR PROVIDING SUB-USER DETAILS,
3.ENTER 3 FOR BANKING BY SUB-USER
```

Fig. 6. Demo of account holder banking on python shell

2. Banking by sub-user: Sub-user is addressed here, to one who is apart from account holder, who needs to conduct transaction from account. Here, to do such

2.1 The first step is sub-user entry, which will be explained using the block diagram as shown in Fig. 6. After logging in into the registered account of the account holder created online using OTP verification, the account holder needed to enter the name of sub-user, their phone number, and the max amount that an individual sub-user can transact. Max, three sub-user entry from the account per day is permitted, and the sub-user details will be valid for only 24 hours, after which these details will be reset to null.

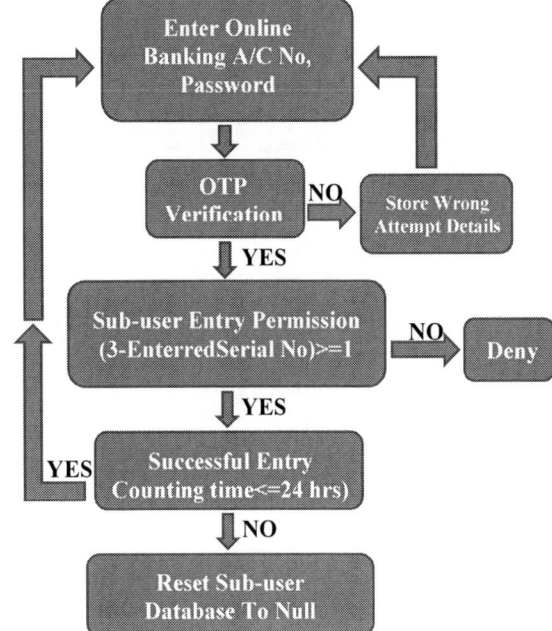

Fig. 6. Block diagram of workflow of sub-user entry by the a/c holder

2.2 Sub-user banking workflow is shown in Fig. 7. During banking by sub-user, sub-user first needed to ask the account number, sub-user details consisting name, phone number and amount to be transacted. If phone number matched with corresponding sub-user details in the database, and transaction amount if less than the calculated transaction limit of that sub-user, a random OTP will be sent to the entered phone number using GSM by Rasppberrypi. Within a finite

978-1-5386-6723-1/19 $31.00 © 2019 IEEE   229

time from then, if the OTP is entered by the sub-user correctly, then the transaction by sub-user is permitted.

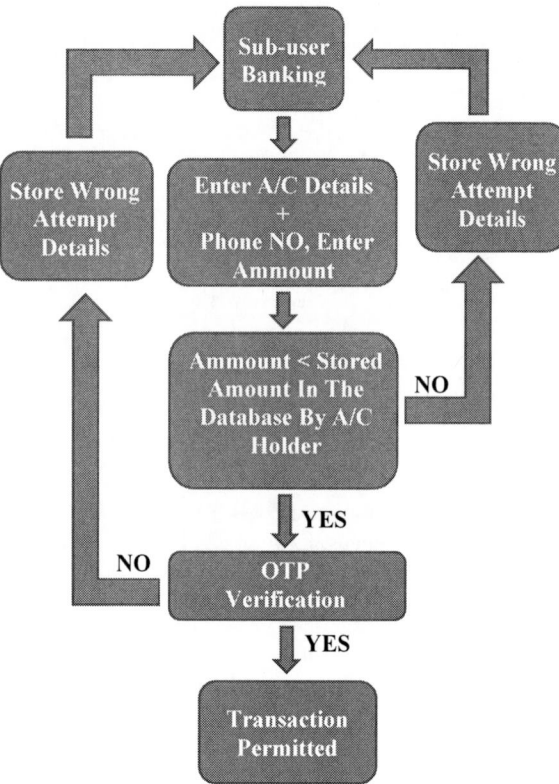

Fig. 7. Block diagram of workflow of sub-user banking

The demo of this sub-user banking along with sub-user details entry in python shell window is shown in Fig. 8.
The Fig. 9 shows the screenshot of OTP sent from the system to the sub-user phone.

```
WELCOME TO 24 HOURS ATM SERVICE
1.ENTER 1 FOR USER BANKING,
2.ENTER 2 FOR PROVIDING SUB-USER DETAILS,
3.ENTER 3 FOR BANKING BY SUB-USER
2
ENTER THE A/C PASSWORD GIVEN FROM THE BANK!
Sayan0506
YOUR CREDENTIALS ARE AS FOLLOWS
A/C HOLDER NAME: text:'SAYAN HAZRA'
A/C NUMBER: number:79801504211234.0
ENTER NUMBER OF SUB-USERS(IT MUST BE WITHIN 3 PER DAY)
2
ENTER SUB-USER NAME
TANDRA HAZRA
ENTER SUB-USER CONTACT NUMBER
7980150421
ENTER MAX AMMOUNT TO BE TRNSACTED
500
ENTER SUB-USER NAME
BIKASH HAZRA
ENTER SUB-USER CONTACT NUMBER
9831535142
ENTER MAX AMMOUNT TO BE TRNSACTED
500
SUB-USER ENTRY SUCCESSFUL
['TANDRA HAZRA', 'number:79801504211234.0', '7980150421', '500']
['BIKASH HAZRA', 'number:79801504211234.0', '9831535142', '500']
THANKS FOR BANKING!
1.ENTER 1 FOR USER BANKING,
2.ENTER 2 FOR PROVIDING SUB-USER DETAILS,
3.ENTER 3 FOR BANKING BY SUB-USER
3
ENTER YOUR NAME IN CAPS!
TANDRA HAZRA
ENTER YOUR REGISTERED PH_NO
7980150421
ENTER A/C NO
79801504211234
HOW MUCH YOU WANT TO TRANSACT!
500
{"return":true,"request_id":"aodmtqvx8413hr6","message":["Message
ENTER THE OTP SENT TO THE GIVEN MOBILE NUMBER
7434
CONTINUE THE TRANSACTION
```

Fig. 8. Demo of the sub-user banking in the python shell output window

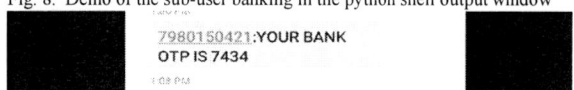

Fig. 9. Screenshot from the sub-user mobile after receiving the OTP

III. CONCLUSION

It can be concluded that, this article provides a solution for card less more secured less time taking user-friendly cash transaction service where face-recognition along with aliveness checkup is used for security purpose followed by fingerprint checkup and for sub-user transaction can also be conducted on the basis of stored information by account holder, where OTP verification has included for validity of sub-users. The demonstration result of working model verifies the proposed smart secured efficient fast ATM service concept.

IV. FUTURE DEVELOPMENT

In this proposed model of ATM service, for security haar-cascade classifier based face recognition along with smile detection has been proposed, in future rather neural network can be incorporated to increase the accuracy of the face recognition system. Again, 3-D biometric face recognition can also be incorporated in future, which will be much more efficient for security checkup. The system should be properly maintained. Some special classifiers or algorithms must be included with the system in order to detect mobbing activity or weapon carried by user in the ATM service room.

REFERENCES

[1] V. Cuervo, "Automated teller machine dispenser of debit cards," U.S. Patent 6,105,009, August 2000

[2] P. Viola and M. Jones, "Rapid Object Detection using a Boosted Cascade of Simple Features," Proc. IEEE Comp. Soc. Conf. USA, vol. 1, pp. 1-1, December 2001

[3] G. Bradski and A. Kaehler, "Learning OpenCV: Computer vision with the OpenCV library," O'Reilly Med. Inc. USA, 2008

[4] N. Bansal and N. Singla, "Cash withdrawl from ATM machine using Mobile banking," Int. Conf. Computational The. Inform. And Communication Tech. (ICCTICT) India, pp. 535-539, March 2016

[5] J. Whitehill, G. Littlewort, I. Fasel, M. Bartlett and J. Movellan, "Toward Practical Smile Detection," IEEE Trans. Pattern Analysis and Intelligence IEEE Comp. Soc., vol. 31, pp. 2106-2111, November 2009

[6] T. Ahonen, A. Hadid and M. Pietikainen, "Face Description with Local Binary Patterns: Application to Face Recognition," IEEE Trans. Pattern Analysis and Machine Intelligence IEEE Comp. Soc., vol. 28, pp. 2037-2041, December 2006

[7] R. Khan, R. hasan, J. Xu, "SEPIA: Secure-PIN-Authentication-as-a-Service for ATM Using Mobile and Wearable Devices," IEEE 3rd Int. Conf. Mobile Cloud Computing, Services, and Engg. , pp. 41-50, March 2015

[8] M. S. Uddin, N. C. Das and A. Barua, "The mCard approach for Bangladesh: A smart phone based Credit/Debit/ATM card," 16th Int. Conf. Computer and Inform. Tech. Bangladesh, pp. 209-212, March 2014

[9] S. Sridharan and K. Malladi, "New generation ATM terminal services," Int. Conf. Computer Communication and Inform. (ICCCI) India, pp. 1-6, January 2016

[10] H. R. F. Najafabadi and M. R. F. Derakhshi, "Multipurpose smart SIM card based on mobile database and location dependent query," 6th Int. Conf. Application Inform. and Communication Tech. (AICT) Georgia, pp. 1-5, October 2012

[11] Nelligani, B. M. Reddy, NV U. reddy and N. Awasti, "Smart ATM security system using FPR, GSM, GPS", Int. Conf. Inventive Computation Tech.(ICICT) India, vol. 3, pp. 1-5, August 2016

[12] Christiawan, B. A. Sahar, A. F. Rahardian, and E. Muchtar, "Fingershield ATM – ATM Security System using Fingerprint Authentication," Int. Symposium Electronics and Smart Devices (ISESD) Indonesia, January 2019.

Comparative Analysis of Underlapped Silicon on Insulator and Underlapped Silicon on Nothing Dielectric and Charge Modulated FET based Biosensors

Khuraijam Nelson Singh, Pranab Kishore Dutta
Department of Electronics and Communication Engineering
North Eastern Regional Institute of Science and Technology
Nirjuli, India
nelsonkhuraijam16@hotmail.com, pkdutta07@gmail.com

Abstract—A biosensor based on silicon on insulator (SOI) MOSFET provides many advantages over the conventional biosensors but still suffers from the inherent problem which exists in SOI structures. Silicon on nothing (SON) MOSFET which is a derivative of SOI MOSFET is an option which has been considered by many as an alternative due to its excellent performances. In this study, underlapped silicon on insulator (USOI) and underlapped silicon on nothing (USON) dielectric and charge modulated FET (DCMFET) has been compared for biosensing application. Modulation of the devices electrical characteristics, namely surface potential, threshold voltage, and sensitivity have been studied to understand the sensing of biomolecules without labeling. Detection of biomolecules is evaluated as a shift of threshold voltage of the devices with the change in biomolecules dielectric constants and charges. The data obtained from the analytical models showed both the devices as highly sensitive; however, USON-DCMFET is found to be the better choice. The analytical models have also been verified by using the data obtained from the 2-D numerical simulations performed using SILVACO ATLAS.

Index Terms—Dielectric modulation, underlap, surface potential, threshold voltage, analytical modeling.

I. Introduction

Interest in the utilisation of field effect transistor (FET) in the sensing of biomolecules has seen immense growth in the recent period owing to its advantages such as high sensitivity, detection without the need of labelling, CMOS fabrication process compatibility, and its ease of miniaturisation [1-6]. Such features of these biosensors have pulled the attention of researchers from various fields in its studies. They can be used in different applications such as drug development, medical diagnosis, food industries, environmental studies and monitoring, the study of biomolecules properties, etc. Since the advent of the first nanogap FET biosensor, there has been no turning back [2]. The nanogap FET biosensor has been fabricated by etching a part of the oxide and the gate metal to create a space where the biomolecules are immobilised. Upon the filling of the space by the biomolecules the effective dielectric constant of the region in the sensing area changes, which in turn changes the threshold voltage of the device. This change in the dielectric of the oxide upon immobilisation of biomolecules is term as dielectric modulation, and its effect of changing the threshold voltage (V_{th}) has been used as a matrix in the recognition of the biomolecules [2-10]. The nanogap dielectric modulated FET (DMFET) has few limitations such as low stability of the structure due to the suspended gate over the nanogap, lower probability of binding of the biomolecules in the nanogap region and high cost [4]. To overcome these problems, underlap DMFET (UDMFET) was developed [5]. In UDMFET, the underlapped region is created by etching a part of the gate material completely and part of the gate oxide in the sensing region. A portion of the gate oxide is still preserved in the sensing region to act as the immobilisation layer and bind the biomolecules with the biosensor.

The development in the field of MOSFET studies has a direct impact on the studies and development of FET based biosensor due to the latter being a derivative of the first. Device dimension reduction has benefits such as higher number of chips per wafer, lower power consumption, lower V_{th}, etc. but also is a source of many unwanted effects collectively known as short channel effects (SCEs) [11-12]. With the ongoing research in nanoscale level, different device structures have been introduced and studied to reduce SCEs to increase performance and reliability of the devices [11-21]. Silicon on insulator (SOI) MOSFET is an unconventional MOSFET structure in which the substrate is insulated from the channel region by an insulating oxide layer which results in enhanced performance and reduction in SCEs [11-15]. It enjoys many benefits such as less junction capacitance, lower power consumption, latch-up free, etc. Instead of the oxide, the channel can also be separated from the substrate by inserting a spacer in place of the oxide [18-21]. These structures are called silicon on nothing (SON) MOSFET. They show very small SCEs, low self-heating, better for low voltage operating, etc. In this study, we present a comparative analysis of underlapped SOI (USOI) and underlapped SON (USON) dielectric and charge modulated FET (DCMFET) as biosensors. The various

2019 Devices for Integrated Circuit (DevIC), 23-24 March, 2019, Kalyani, India

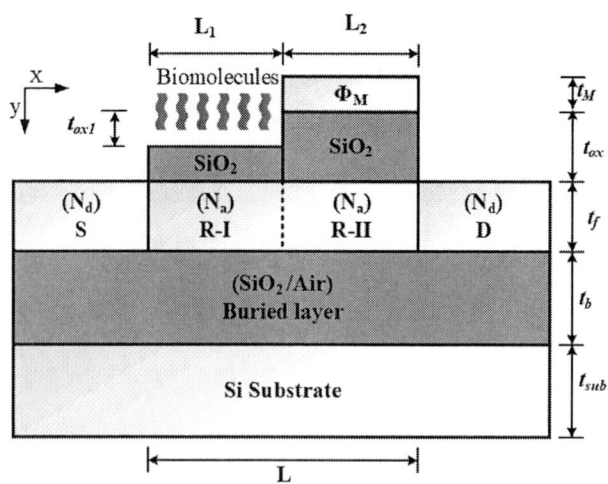

Fig. 1: Schematic structure of USOI/USON DCMFET

electrical parameters of the two devices will be analytically modeled, and analysis will be made based on the results obtained from the analytical models. Experimental study of underlap FET biosensor and SON has already been reported [5,17]. So, fabrication of both USOI-DCMFET and USON-DCMFET based biosensors are practically feasible, and their simple structure is also an additional advantage on top of their inherent advantages. The analytical models are verified by using the data extracted from the 2-D numerical simulations performed by using SILVACO ATLAS.

The remaining portion of the paper is organized as follows. Section-II presents the structures of the USOI/USON DCM-FET devices, section-III present the analytical modeling of the devices, section-IV presents the results and discussion. Finally, the conclusion of the paper is presented in section-V.

II. STRUCTURE OF THE DEVICES

The schematic structures of USOI/USON DCMFET are presented in Fig. 1. The channels of both the devices are separated into two regions, region-I (R-I) and region-II (R-II) due to the gate modification and are doped uniformly with acceptor impurities. The source and drain are doped equally with donor impurities. The gate electrode and part of the gate oxide above R-I have been etched to form an underlap region which will act as a site to immobilize biomolecules. The thin oxide above R-I serves as an adhesive layer between the biomolecules and the biosensor. The remaining part of the gate is kept as it is and it will be used to control the channels of the devices. The layer which insulates the channel and the silicon (Si) substrate can be either (SiO$_2$) or air for USOI-DCMFET and USON-DCMFET respectively. These structural parameters used for the study are presented in Table I.

III. ANALYTICAL MODELING

A. Surface Potential

The surface potential $\phi_i(x,y)$ in the channels of the USOI/USON DCMFET can be solve from the 2-D Poisson's

TABLE I: Structural Parameters for USOI/USON DCMFET

Physical Parameter	Symbol	Value
Total channel length	L	40 nm
Length of R-I	L_1	20 nm
Length of R-II	L_2	20 nm
Length of source or drain	L_{sd}	20nm
Oxide thickness above R-I	$(t_{ox} - t_{ox1})$	1 nm
Oxide thickness above R-II	t_{ox}	7 nm
Thickness of gate electrode	t_M	1 nm
Channel thickness	t_f	10 nm
Buried oxide thickness	t_b	40 nm
Doping of channel	N_a	$1 \times 10^{17} cm^{-3}$
Doping of source/drain	N_{sd}	$1 \times 10^{20} cm^{-3}$
Workfunction of gate electrode	ϕ_M	4.7 V

equation given by

$$\frac{\partial^2 \phi_i(x,y)}{\partial x^2} + \frac{\partial^2 \phi_i(x,y)}{\partial y^2} = \frac{qNa}{\epsilon_{Si}} \quad (1)$$

where, $i = 1$ or 2 for R-I and R-II respectively. N_a is the channel doping concentration, q is the charge of an electron and ϵ_{Si} is the dielectric constant of Si. The surface potentials can be approximated as a parabolic equation as shown in [8]

$$\phi_i(x,y) = K_{i0}(x) + K_{i1}(x)y + K_{i2}(x)y^2 \quad (2)$$

where, K_{i0}, K_{i1}, and K_{i2} can be determined by using the boundary conditions given below.

$$\phi_i(x,0) = \phi_{fs,i}(x) \quad (3)$$

$$\left.\frac{d\phi_i(x,y)}{dy}\right|_{y=0} = \frac{C_i}{\epsilon_{Si}}[\phi_{fs,i}(x) - V'_{gs,i}] \quad (4)$$

$$\left.\frac{d\phi_i(x,y)}{dy}\right|_{y=t_f} = -\frac{C_B}{\epsilon_{Si}}[\phi_b(x) - V'_{sub}] \quad (5)$$

$$\phi_i(x,t_f) = \phi_{bs}(x) \quad (6)$$

where, $\phi_{fs,i}$ is the front surface potential for the ith region, and ϕ_{bs} is the back surface potential. The effective capacitance above R-I is denoted as C_1, capacitance above R-II as C_2 and back capacitance as C_B which are given by

$$C_1 = \frac{C'_{ox}C_f}{C'_{ox} + C_f}, C_2 = \frac{\epsilon_{ox}}{t_{ox}}, C_B = \frac{\epsilon_{ox/air}}{t_b}, C'_{ox} = \frac{\epsilon_{Si}}{(t_{ox} - t_{ox1})}$$

where ϵ_{ox} is the dielectric constant of SiO$_2$, $\epsilon_{ox/air}$ is the dielectric constant of the back insulating layer, and C_f denotes the fringing capacitance and is calculated by using the same method in [18]

$$C_f = \frac{2\epsilon_{bio}}{n\pi L_1} \sinh\left[\cosh^{-1}\left(\frac{t_{ox1} + t_M}{t_{ox1}}\right)\right]$$

where $n = \frac{2}{\pi}\sin^{-1}(1)$. Also, $V'_{gs,i} = V_{gs} - V_{FBf,i}$ and $V'_{sub} = V_{sub} - V_{FBb}$, where V_{gs} and V_{sub} are the gate-source voltage and substrate voltage respectively, $V_{FBf,i}$ is the flatband voltage at the front for i^{th} region and V_{FBb} is the flatband voltage at the back given by

$$V_{FBf,1} = V_{FBf,2} - \frac{qQ_{bio}}{C_1}, V_{FBf,2} = \phi_M - \phi_{Si}$$

978-1-5386-6723-1/19 $31.00 © 2019 IEEE

where, Q_{bio} is the biomolecules surface charge density, ϕ_M is the gate metal workfunction and ϕ_{Si} is Si workfunction. Using the parabolic approximated value of $\phi_i(x, y)$, the Poisson's equation can be simplified as

$$\frac{\partial^2 \phi_{fs,i}(x)}{\partial x^2} - \alpha_i \phi_{fs,i}(x) = \beta_i \quad (7)$$

where,

$$\alpha_i = \frac{2(1 + C_i/C_B + C_i/C_f)}{t_f^2(1 + 2C_f/C_B)} \quad (8)$$

$$\beta_i = \frac{qN_a}{\epsilon_{Si}} - 2V'_{gs,i} \left[\frac{C_i/C_B + C_i/C_f)}{t_f^2(1 + 2C_f/C_B)} \right]$$
$$- 2V'_{sub} \left[\frac{1}{t_f^2(1 + 2C_f/C_B)} \right] \quad (9)$$

The solutions of (7) can be written as

$$\phi_{fs,1}(x) = Ae^{\lambda_1 x} + Be^{-\lambda_1 x} + \sigma_1 \quad (10)$$
$$\phi_{fs,2}(x) = Ce^{\lambda_2(x-L_1)} + De^{-\lambda_2(x-L_1)} + \sigma_2 \quad (11)$$

where, $\lambda_i = \sqrt{\alpha_i}$, and $\sigma_i = -\frac{\beta_i}{\alpha_i}$ for the ith region. The values of A, B, C, and D are solved by using the following boundary conditions.

$$\phi_{fs,1}(0) = V_{bi1} \quad (12)$$
$$\phi_{fs,1}(L_1) = \phi_{fs,2}(L_1) \quad (13)$$
$$\left. \frac{\partial \phi_1(x,y)}{\partial x} \right|_{x=L_1} = \left. \frac{\partial \phi_2(x,y)}{\partial x} \right|_{x=L_1} \quad (14)$$
$$\phi_{fs,2}(L) = V_{bi2} + V_{ds} \quad (15)$$

where, V_{bi1} and V_{bi2} are the source-channel and drain-channel built in potentials and V_{ds} is the voltage at the drain terminal. Using (12)-(14), the value of A, B, C, and D are found as

$$A = \frac{(V_{bi2} + V_{ds} - \sigma_2) - (V_{bi1} - \sigma_1)e^{-\lambda_1 L}}{2\sinh(\lambda_1 L)}$$
$$- \frac{(\sigma_1 - \sigma_2)\cosh(\lambda_1 L_2)}{2\sinh(\lambda_1 L)} \quad (16)$$

$$B = \frac{(V_{bi1} - \sigma_1)e^{\lambda_1 L} - (V_{bi2} + V_{ds} - \sigma_2)}{2\sinh(\lambda_1 L)}$$
$$+ \frac{(\sigma_1 - \sigma_2)\cosh(\lambda_1 L_2)}{2\sinh(\lambda_1 L)} \quad (17)$$

$$C = Ae^{\lambda_1 L_1} - \left(\frac{\sigma_2 - \sigma_1}{2} \right) \quad (18)$$

$$D = Be^{\lambda_2 L_1} - \left(\frac{\sigma_2 - \sigma_1}{2} \right) \quad (19)$$

B. Threshold Voltage Modeling

The threshold voltage V_{th} can be defined as the value of V_{gs} where the minimum value of $\phi_{fs,i}(x)$ equal to twice the value of Fermi potential. As R-I is underlapped, minimum $\phi_{fs,i}(x)$ position should lie in R-II due to the metal workfunction (ϕ_M) in R-II, but electrical characteristics in R-I is a function of the properties of the biomolecules; therefore, we

cannot confirm whether the minimum value of $\phi_{fs,i}(x)$ will lie in R-I or R-II as it was done in [8,11,15,17]. So, there is a necessity to model V_{th} for both the regions. The minimum value of the front surface potential ($\phi_{fs,i,min}$) is calculated by using the position of $\phi_{fs,i,min}$ in (10) and (11). The values of $\phi_{fs,i,min}$ are given by

$$\phi_{fs,1,min} = 2\sqrt{AB} + \sigma_1 \quad (20)$$
$$\phi_{fs,2,min} = 2\sqrt{CD} + \sigma_2 \quad (21)$$

Now, by equating (20) and (21) to $2\phi_F$, where ϕ_F is the Fermi potential, V_{th} can be found as

$$V_{th} = \frac{-k_2 + \sqrt{k_2^2 - 4k_1k_3}}{2k_1} \quad (22)$$

if $\phi_{fs,1,min}$ lie in R-I, and

$$V_{th} = \frac{-k_5 + \sqrt{k_5^2 - 4k_4k_6}}{2k_4} \quad (23)$$

if $\phi_{fs,1,min}$ lie in R-II.

IV. RESULTS AND DISCUSSION

The analysis of USOI-DCMFET and USON-DCMFET as biosensors have been performed based on the 2-D analytical models' data which are verified by using the 2-D numerical simulations results obtained from SILVACO ATLAS. The models which have been used for the simulation are CON-MOB, SRH, AUGER, BGN, and FLDMOB. The appearance of biomolecules is simulated by putting a dielectric (ϵ_{bio}) whose value is larger than 1 in the underlap area. Insertion of air in the underlap area is considered as the absenteeism of biomolecules. The biomolecules charge property is simulated by using a surface charge density of $Q_{bio} = -1 \times 10^{11} cm^{-2}$ to $-3 \times 10^{11} cm^{-2}$.

A. Surface Potential

The variation of the surface potential along the channel of the USOI-DCMFET and USON-DCMFET biosensors are presented in Fig. 2. Filling of the underlapped region with biomolecule(s) increase the effective dielectric constant of the gate oxide in R-I. The change in the surface potential for both USOI-DCMFET and USON-DCMFET along their channels with the shift in the dielectric constant of the biomolecules (ϵ_{bio}) considering a constant charge of the biomolecules (Q_{bio}) is shown in Fig. 2 (a). It is apparent from the figure that as the value of ϵ_{bio} increases, the minimum surface potential value gets lower. This lowering of the minimum surface potential is due to the rise in the fringing field because of the increasing dielectric value, and as a result, it increases the potential separation between the source and the channel. Fig. 2 (b) show the change of the surface potential for both types of DCMFET with the shift of Q_{bio} along their channels keeping $\epsilon_{bio} = 2.5$. Increase in the negative charge of the biomolecules push away the charge carriers responsible for the formation of the channel in the devices, resulting in the lowering of the minimum surface potential.

2019 Devices for Integrated Circuit (DevIC), 23-24 March, 2019, Kalyani, India

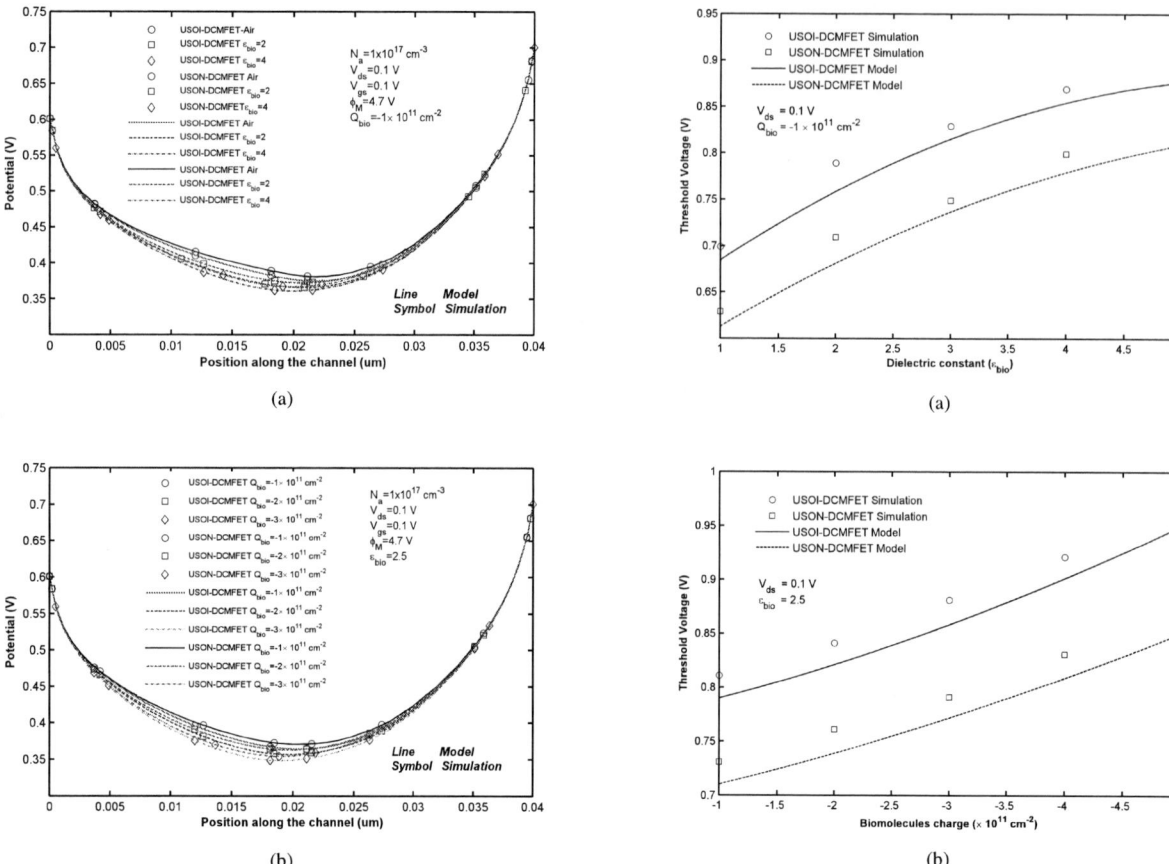

(a)

(b)

Fig. 2: Surface potential variation along the channel of USOI-DCMFET and USON-DCMFET. (a) Effect of different dielectric values of the biomolecules. (b) Effect of different charges of the biomolecules.

(a)

(b)

Fig. 3: Threshold voltage variation of USOI-DCMFET and USON-DCMFET. (a) Effect of different dielectric values of the biomolecules. (b) Effect of different charges of the biomolecules.

Fig. 2 also reveals that USON-DCMFET has lower minimum surface potential value than USOI-DCMFET which rises the need for more gate potential to form a strong inversion layer in the channel. It is also clear that, as the value of ϵ_{bio} and Q_{bio} increases, the minimum surface potential position shifts towards the source, resulting in more uniform electric field in the channel which will improve the mobility of the devices.

B. Threshold Voltage

Often threshold voltage (V_{th}) is considered as a sensing matrix of biomolecules due to its simplicity and applicability. The shift in threshold voltage (ΔV_{th}) when the sensing area is filled with biomolecules in comparison with the V_{th} in the absence of biomolecules is one of the popular methods to detect the biomolecules. Fig. 3 (a) shows the variation of V_{th} with different values of ϵ_{bio} for USOI-DCMFET and USON-DCMFET, keeping Q_{bio} constant. The rise in V_{th} with the increase in dielectric constant of the biomolecules is due to the reduction in the vertical electric field. Fig. 3 (b) shows shifting the value of V_{th} with varying amounts of Q_{bio} values, keeping

ϵ_{bio} as a constant. The graphs show that USOI-DCMFET has a higher V_{th} than USON-DCMFET which can also be inferred from the surface potential graph in Fig. 2 as USOI-DCMFET has lower minimum surface potential value than USON-DCMFET. In Fig. 3 (a), the graph suddenly rises when ϵ_{bio} goes from 1 to 2, this is due to the sudden inclusion of charge and increase in dielectric constant, both of which tend to increase the V_{th}. In Fig. 3, there is an offset of nearly 20 mV in between the model data and simulation data which is quite negligible. This offset arises due to the utilization of different techniques for threshold voltage extractions. The models used the minimum surface potential approach while the simulations used the maximum transconductance method. But, the modeling data follow the simulated data quite closely which shows the validity of the analytical model. Another inference from the graph is that, though V_{th} is higher for USOI-DCMFET, their relative V_{th} change is nearly the same. Thus both device can be used as a biosensor irrespective of the insulating layer, but USON-DMFET is a better choice due to the lower threshold voltage.

978-1-5386-6723-1/19 $31.00 © 2019 IEEE   234

TABLE II: Sensitivity for USOI/USON DCMFET

Biomolecule permittivity (ϵ_{bio})	Sensitivity (ΔV_{th}) (mV)	
	USOI DCMFET	USON DCMFET
2	60	60
3	110	100
4	160	150
5	190	190

C. Sensitivity

Sensitivity of the biosensor can be defined by using the definition provided in [7]. It define sensitivity (ΔV_{th}) as the difference in the threshold voltage with a particular biomolecule and threshold voltage with no biomolecule, i.e,

$$\Delta V_{th} = V_{th}(with\, biomolecule) - V_{th}(with\, air) \quad (24)$$

Table II presents the comparison of the sensitivities of the two devices for different value of biomolecules dielectric values by keeping the charge constant ($Q_{bio} = -1 \times 10^{11} cm^{-2}$). For both the devices the sensitivity, ΔV_{th}= 190 mV when the ϵ_{bio} = 5 which is much larger than the sensitivity provided in [7] which has $\Delta V_{th} \approx 160$ mV and from [22] which has $\Delta V_{th} \approx 60$ mV. Thus, the two structure compared herein also provide high sensitivity even with small size.

V. CONCLUSION

The surface potential behavior and the threshold voltage of USOI-DCMFET and USON-DCMFET have been analyzed with the help of the analytical models and have been verified by using the data obtained from numerical simulations performed by using SILVACO ATLAS. The modulation of surface potential and the threshold voltage of the devices by the dielectric and charges of the biomolecules have been studied to understand their applicability as biosensors. Comparison of USOI-DCMFET and USON-DCMFET reveals that the change in the threshold voltage is nearly the same for both the devices resulting in having almost identical sensitivity, but USON-DCMFET has a lower threshold voltage than USOI-DCMFET. Thus, both device can be used as biosensors, but USON-DCMFET is a better choice among the two as it has a lower threshold voltage which will ensure faster switching, lower dynamic power consumption, and better scaling of the supply voltage while still having nearly the same sensitivity. From the results, it is also clear that consideration of both the properties of biomolecules, i.e., dielectric and charge is necessary while performing their studies.

REFERENCES

[1] P. Bergveld, "The development and application of FET-based biosensors," *Biosensors*, vol. 2, no. 1, pp. 15–33, Jan. 1986.

[2] H. Im, X.J. Huang, B. Gu, and Y.-K. Choi, "A dielectric-modulated field-effect transistor for biosensing," *Nature Nanotechnology*, vol. 2, no. 7, pp. 430–434, Jul. 2007.

[3] B. Gu, T.J. Park, J.-H. Ahn, X.-J. Huang, S. Y. Lee, and Y.-K. Choi, "Nanogap Field-Effect Transistor Biosensors for Electrical Detection of Avian Influenza," *Small*, vol. 5, no. 21, pp. 2407–2412, Nov. 2009.

[4] J.M. Choi, J.W. Han, S.J. Choi, and Y.-K. Choi, "Analytical Modeling of a Nanogap-Embedded FET for Application as a Biosensor," *IEEE Transactions on Electron Devices*, vol. 57, no. 12, pp. 3477–3484, Dec. 2010.

[5] K.W. Lee et al., "An underlap field-effect transistor for electrical detection of influenza," *Applied Physics Letters*, vol. 96, no. 3, p. 033703, Jan. 2010.

[6] Y. Pratap, M. Kumar, S. Kabra, S. Haldar, R. S. Gupta, and M. Gupta, "Analytical modeling of gate-all-around junctionless transistor based biosensors for detection of neutral biomolecule species," *Journal of Computational Electronics*, vol. 17, no. 1, pp. 288–296, Mar. 2018.

[7] A. Chakraborty and A. Sarkar, "Analytical modeling and sensitivity analysis of dielectric-modulated junctionless gate stack surrounding gate MOSFET (JLGSSRG) for application as biosensor," *Journal of Computational Electronics*, vol. 16, no. 3, pp. 556–567, Sep. 2017.

[8] E. Rahman, A. Shadman, I. Ahmed, S. U. Z. Khan, and Q. D. M. Khosru, "A physically based compact I – V model for monolayer TMDC channel MOSFET and DMFET biosensor," *Nanotechnology*, vol. 29, no. 23, p. 235203, Jun. 2018.

[9] P. Dwivedi and A. Kranti, "Dielectric Modulated Biosensor Architecture: Tunneling or Accumulation Based Transistor?," *IEEE Sensors Journal*, vol. 18, no. 8, pp. 3228–3235, Apr. 2018.

[10] B. V. Chandan, K. Nigam, and D. Sharma, "Junctionless based dielectric modulated electrically doped tunnel FET based biosensor for label-free detection," *Micro & Nano Letters*, vol. 13, no. 4, pp. 452–456, Apr. 2018.

[11] K. K. Young, "Short-channel effect in fully depleted SOI MOSFETs," *IEEE Transactions on Electron Devices*, vol. 36, no. 2, pp. 399–402, Feb. 1989.

[12] M. J. Kumar and A. Chaudhry, "Two-Dimensional Analytical Modeling of Fully Depleted DMG SOI MOSFET and Evidence for Diminished SCEs," *IEEE Transactions on Electron Devices*, vol. 51, no. 4, pp. 569–574, Apr. 2004.

[13] V. P. Trivedi and J. G. Fossum, "Nanoscale FD/SOI CMOS: thick or thin BOX?," *IEEE Electron Device Letters*, vol. 26, no. 1, pp. 26–28, Jan. 2005.

[14] Wei-Yuan Lu and Yuan Taur, "On the scaling limit of ultrathin SOI MOSFETs," *IEEE Transactions on Electron Devices*, vol. 53, no. 5, pp. 1137–1141, May 2006.

[15] P. K. Tiwari, S. Dubey, M. Singh, and S. Jit, "A two-dimensional analytical model for threshold voltage of short-channel triple-material double-gate metal-oxide-semiconductor field-effect transistors," *Journal of Applied Physics*, vol. 108, no. 7, p. 074508, Oct. 2010.

[16] A. Chaudhry and M. J. Kumar, "Controlling Short-Channel Effects in Deep-Submicron SOI MOSFETs for Improved Reliability: A Review," *IEEE Transactions on Device and Materials Reliability*, vol. 4, no. 1, pp. 99–109, Mar. 2004.

[17] M. Jurczak et al., "Silicon-on-Nothing (SON)-an innovative process for advanced CMOS," *IEEE Transactions on Electron Devices*, vol. 47, no. 11, pp. 2179–2187, Nov. 2000.

[18] J. Pretet, S. Monfray, S. Cristoloveanu, and T. Skotnicki, "Silicon-on-Nothing MOSFETs: Performance, Short-Channel Effects, and Backgate Coupling," *IEEE Transactions on Electron Devices*, vol. 51, no. 2, pp. 240–245, Feb. 2004.

[19] E. Goel, S. Kumar, K. Singh, B. Singh, M. Kumar, and S. Jit, "2-D Analytical Modeling of Threshold Voltage for Graded-Channel Dual-Material Double-Gate MOSFETs," *IEEE Transactions on Electron Devices*, vol. 63, no. 3, pp. 966–973, Mar. 2016.

[20] P. K. Dutta, B. Manna, and S. K. Sarkar, "Analytical modeling of linearly graded alloy material gate recessed ultra thin body source/drain SON MOSFET," *Superlattices and Microstructures*, vol. 77, pp. 64–75, Jan. 2015.

[21] P. Banerjee and S. K. Sarkar, "3-D Analytical Modeling of Dual-Material Triple-Gate Silicon-on-Nothing MOSFET," *IEEE Transactions on Electron Devices*, vol. 64, no. 2, pp. 368–375, Feb. 2017.

[22] Ajay, R. Narang, M. Saxena, and M. Gupta, "Modeling of gate underlap junctionless double gate MOSFET as bio-sensor," *Materials Science in Semiconductor Processing*, vol. 71, pp. 240–251, Nov. 2017.

[23] *ATLAS User's Manual*, SILVACO Int., Santa Clara, CA, USA, 2015.

2019 Devices for Integrated Circuit (DevIC), 23-24 March, 2019, Kalyani, India

Storing Digital Data in Nucleic Acid Memory with Extended Genetic Alphabet

Saptarshi Biswas
Computer Science Engineering
Meghnad Saha Institute of Technology
Kolkata, India

Subhrapratim Nath
Computer Science Engineering
Jadavpur University
Kolkata, India

Jamuna Kanta Sing
Computer Science Engineering
Jadavpur University
Kolkata, India

Subir Kumar Sarkar
Electronics and Telecommunication Engineering
Jadavpur University
Kolkata, India

Abstract— The rapid improvement of semiconductor technologies have led to a digital revolution globally which accelerated the rate of generation of data exponentially. When researchers are working hard to develop better compression algorithms, they came up with the idea of Nucleic Acid Memory (NAM). Various papers have been studied with their respective merits and demerits. This paper proposes a new scheme of encoding digital data into genetic nucleotide sequence which demonstrates the use of non-standard nucleotides and unnatural base pairs along with standard nucleotide bases. This combination aims in enhancing the efficiency of the encoding scheme much better than the predefined encoding model where the proposed scheme is compared with the existing encoding models.

Keywords—DNA computer, Nucleic Acid Memory, Extended genetic alphabet, Unnatural base pairs, Nonstandard nucleotide

I. Introduction

Nucleic acids are the fundamental genetic components of every living organism. The nucleic acids hold all the information regarding the physiological and biochemical properties of every living being. Nucleic acids are of two types namely, RNA (Ribonucleic acid) and DNA (Deoxyribonucleic acid). The nucleic acids can exist in the single-stranded or double-stranded molecular formation. The double-stranded molecular formation of nucleic acids is more stable than single-stranded formation which makes it less prone to genetic mutations than the single-stranded nucleic acid.

Since the last few decades it is observed that the data production around the globe has increased exponentially and in all probabilities, it is going to gear up the pace in the future. As a result, it is expected that the silicon memory will soon get exhausted in the near future. As an alternative, the nucleic acid memory has been chosen since it has numerous advantages. It has very high data density, high scalability, very high retentivity (> 100 years in the atmosphere), very low latency (< 100 microseconds/bit) as well as it is highly robust [1]. Moreover, the cost of sequencing the DNA memory has declined sufficiently (9×10^{16} bits / USD) which has elevated the interest of the researchers to get inclined towards the Nucleic Acid Memory [1]. Various attempts have been made to expand the use of nucleic acids to store digital data, encoded

in the form of biological data, into the DNA (Deoxyribonucleic acid) since the last century.

Generally, standard DNA consists of a sequence of 4 standard nucleotides. Later four unnatural nucleotides were developed in vitro artificially. These unnatural nucleotides proved to exhibit similar behaviour as that of natural standard nucleotides. This proposed paper combines both the standard as well as nonstandard nucleotides into a single system and assigns a 3-bit binary encoding scheme to each nucleotide which made it possible to achieve 2^{3N} number of nucleotide permutations, where N is the number of nucleotides.

The remaining of the paper is ordered as follows: Section II explains the basic structure of a DNA molecule and introduces with the nonstandard nucleotides. Section III explores the previously designed approaches for encoding digital data into NAM. Section IV describes the proposed methodology of this work. Section V analyses the result of this work with the previously proposed works. Finally, Section VI concludes this work.

II. Preliminaries on DNA Structure

The Deoxyribonucleic Acid (DNA) is known as the genetic memory of any living cell of an organism. The DNA serves as the control unit of all the basic metabolic and physiological behaviour and state of any living cell of an organism. The DNA base pair combinations hold the information of all the proteins that need to be synthesized in a living cell for carrying out the life process. It also stores information about the genetic hierarchical inheritance. A DNA sequence also acts as time and process controller by acting as a genetic clock. The DNA base pair arrangements represent the memory that holds the information of the proteins that are needed in a complete genetic cycle by the clock. The appearance of an organism is also expressed by the information held by its genetic memory.

Any change in the metabolic or physical behaviour of a living organism over a long course of time is known as 'Evolution'. The evolution is the effect of the change in the sequence of the DNA base pairs in the cells of a living organism. This directly represents the change in the genetic data held by the DNA sequence. In a living organism, different cells differ in action and due to their different genetic memory.

978-1-5386-6723-1/19 $31.00 © 2019 IEEE

DNA is a complex molecule composed of nitrogen-containing nucleotide bases (Adenine, Thymine, Cytosine and Guanine) and a phosphate-sugar backbone (deoxyribose sugar). Structurally the molecule takes the shape of a double-stranded helix. It is composed of two chains joined in anti-parallel fashion (3' to 5' strand and 5' to 3' strand). The two antiparallel strands are linked by intermolecular Hydrogen bonds between the base pairs. The nucleotide bases remain bonded with a pentose sugar molecule which in turn binds with the phosphate atom. This forms the basic molecular unit of a DNA molecule. The phosphate links one sugar molecule with another thus forming a bridge. This forms the structural backbone of the DNA molecule as shown in Fig. 1. DNA strands reside in the form of highly condensed chromosomes within the mitochondria and nucleus of a living cell.

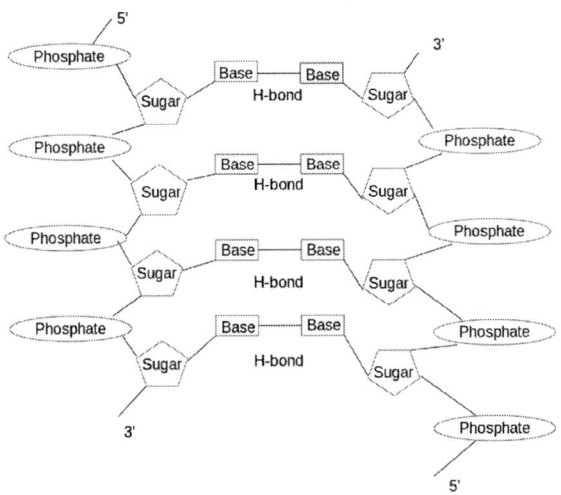

Fig. 1 Basic structure of a DNA molecule

The nucleotide bases are divided into two types, namely the purines and pyrimidines. Purines are classified into Adenine (Mol. Wt. - 135.13 g/mol) and Guanine (Mol. Wt. - 151.13 g/mol) and pyrimidines are classified into Cytosine (Mol. Wt. - 111.1 g/mol) and Thymine (Mol. Wt. - 126.11 g/mol). In a DNA strand Adenine can bind with only Thymine with a double Hydrogen bond (A = T) as shown in Fig. 2 (a) and Cytosine can bind only with Guanine via a triple Hydrogen bond (C ≡ G) as shown in Fig. 2 (b).

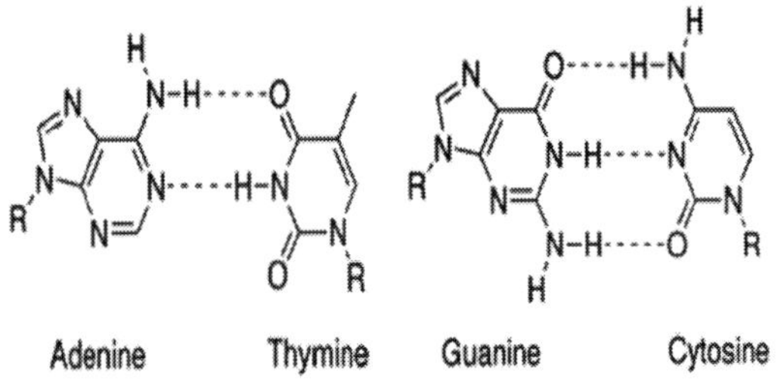

Fig. 2 (a) H-bond between Adenine and Thymine (b) H-bond between Cytosine and Guanine

During the last decade, researchers invented four new nucleotide bases in vitro condition. These artificially synthesized bases are known as unnatural nucleotides [2], [3]. These bases are 7-(2-thienyl) imidazo [4, 5-b] pyridine (Ds) (Mol. Wt. - 201.247 g/mol), 2-nitro-4-propynylpyrrole (Px) (Mol. Wt. - 150.137 g/mol), 5SICS (Mol. Wt. - 289.35 g/mol)

and NaM (Mol. Wt. - 272.30 g/mol) as shown in Fig. 3(a) and Fig. 3(b).

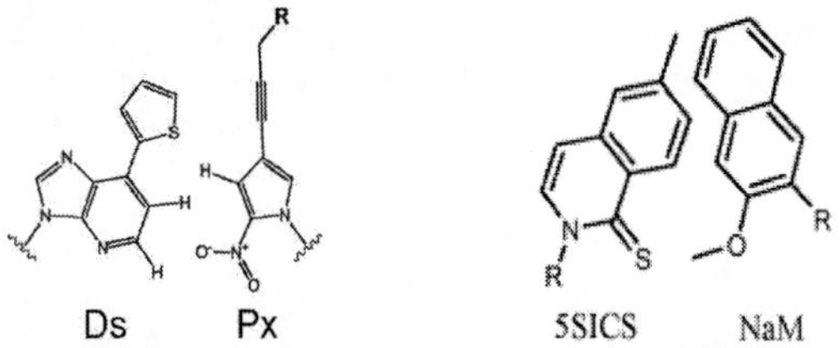

Fig. 3 (a) Molecular structure of Ds and Px base pair [2]
(b) Molecular structure of 5SICS and NaM base pair [3]

It was observed that they exhibit the property of forming base pairs with each other in the same way the natural nucleotides did. The base pairs formed by these bases are known as unnatural base pairs. It was also verified that these unnatural bases behave the same way as the natural bases. They have the capacity to express genetic data and can also be amplified using the PCR (Polymerase Chain Reaction) method of DNA amplification [4].

III. RELATED WORKS WITH NAM

The theory behind the nucleic acid memory began with the concept of encoding the capital English alphabets, digits from 0 to 9 and some punctuation characters with triplet codons of standard nucleotides [5]. The method also required to maintain a memory to recognize the triplet and character pair. G. M. Church et al. in 2012 discovered a new way of encoding the 4 standard nucleotides to digital data [6]. They used A and C to represent 0 and G and T to represent 1. The main drawback of this coding scheme was that one nucleotide corresponds to only one binary data. Although there exist 4 different depictions of genetic data (4 nucleotides) yet only two of them are exploited (as A = C and G = T). Later it was made possible to correlate all the 256 ASCII characters using the two-bit representation of the standard genetic alphabets [7]. It was further modified to exploit the system of extended genetic alphabets [8]. Its limitation was that it was unable to exploit the use of hex-codes rather it was designed to work with an octal representation of digital data. A recent work proposed encoding 16 triplets into hex-codes [9]. It is observed that standard genetic alphabets can generate 64 permutations of triplet codon. Out of which only 16 could be used while the rest of the 48 codons remained unused. Moreover, the repeating sequences and the start codons are not considered for encoding which leads to unused DNA sequence. Another methodology applied for efficient data storage in DNA was DNA fountain [10]. The work introduces the constraint of run length and GC content. The problem was efficiently solved as claimed by the literature and the author achieved an experimental data density of 1.83 bits/nucleotide. Most recent work by W.Song et. al. explains a new technique of solving the problem of run length and GC content where the coding scheme achieved a maximum data density of 1.90 bits/nucleotide [11]. However, these constraints were not applied in the present proposal.

IV. PROPOSED METHOD ON NAM

A. Encoding Scheme

In this proposed work a double-stranded DNA as a memory unit since double-stranded DNA is more robust and less prone to mutation than single-stranded DNA. A 3-bit binary number is assigned to each nucleotide as shown in Table-I.

Since the system is incorporating 3 bits of digital data into a single nucleotide it is possible to represent 8 binary numbers through a single nucleotide i.e. 000-111. This allows the total number of unique arrangements possible with N nucleotides to be 8^N i.e. 2^{3N}

TABLE I

Nucleotide	Binary Number	Nucleotide	Binary Number
A	000	T	111
C	001	G	110
Ds [2]	010	Px [2]	101
5SICS [3]	011	NaM [3]	100

B. Encoding Procedure

- The ASCII value of a character is detected
- Convert it into its corresponding 12-bit binary number (by concatenating a required number of 0s before the original binary ASCII value)
- Starting encoding consecutive 3 bits at a time into a single nucleotide
- This generates a quadruplet consisting of 4 nucleotides representing a 12 bit binary ASCII value (4 nucleotide x 3 bit/nucleotide)

A 12-bit representation of binary data is intentionally chosen to permit the generation of a triplet hex-code. The 12-bit data can be grouped into 3 consecutive hex-codes. Finally, this triplet hex-code unit can be used ubiquitously in any computer system to represent any form of digital data. While decoding the genetic message, the nucleotides are identified and decoded directly to the 12-bit binary data. The encoding scheme described here is demonstrated in Table II.

TABLE II

ASCII Code	Hex-Code	Character	12 Bit Binary Number	Genetic Code
65	#041	A	000001000001	ACAC
66	#042	B	000001000010	ACADs
67	#043	C	000001000011	ACA5SICS
68	#044	D	000001000100	ACANaM
69	#045	E	000001000101	ACAPx
70	#046	F	000001000110	ACAG
71	#047	G	000001000111	ACAT
72	#048	H	000001001000	ACCA
73	#049	I	000001001001	ACCC
74	#04A	J	000001001010	ACCDs
75	#04B	K	000001001011	ACC5SICS
76	#04C	L	000001001100	ACCNaM
77	#04D	M	000001001101	ACCPx
78	#04E	N	000001001110	ACCG
79	#04F	O	000001001111	ACCT
80	#050	P	000001010000	ACDsA
81	#051	Q	000001010001	ACDsC
82	#052	R	000001010010	ACDsDs
83	#053	S	000001010011	ACDs5SICS
84	#054	T	000001010100	ACDsNaM
85	#055	U	000001010101	ACDsPx
86	#056	V	000001010110	ACDsG
87	#057	W	000001010111	ACDsT
88	#058	X	000001011000	AC5SICSA
89	#059	Y	000001011001	AC5SICSC
90	#05A	Z	000001011010	AC5SICSDs

V. RESULT

In contrast with the previously proposed approaches, the result of this work is tabulated in Table III. The theoretical capacity per nucleotide, type of digital data encoded to nucleotides, the theoretical average data density obtained in bytes per gram and the applicability of the encoding scheme on the type of data to be encoded were analyzed and compared with the preexisting encoding models.

TABLE III

Field	[6]	[7]	[8]	[9]	Proposed Encoding Scheme
Theoretical capacity (bits/nucleotide)	1	2	3	1.33	3
Type of encoding used for digital data	Binary	Binary	Octal	Hexadecimal	Binary + Hexadecimal
Average theoretical data density without constraints (bytes/g)	0.332×10^{21}	0.663×10^{21}	NIL	0.441×10^{21}	0.774×10^{21}
Applicability on the type of data to be encoded	Universal	Universal	Character Encoding	Universal	Universal

Analyzing the results from Table III, it can be established that besides maintaining a high average data density the encoded data can be represented with universal binary as well as hexadecimal representation simultaneously. It can also be observed that three bits of data can be represented simultaneously by a single genetic alphabet and the pairing alphabets are assigned with binary numbers such that they are the complement of each other. It was possible to obtain an average theoretical data density of 0.774 Zettabytes/gram of DNA. This design can be effectively used to enhance self-correction during encoding and decoding process. Besides the encoding procedure being straightforward, any form of digital data can be encoded easily using this encoding scheme. Moreover, it doesn't require any explicit use of memory for facilitating the mapping technique, rather the encoding and decoding procedure is direct. This can significantly improve the speed of the encoding and decoding process as a whole.

Error rate defined in Table IV is the Average Mutation Frequency (AMF) which is taken as a parameter for comparison with previous works. To maintain uniformity the AMF is evaluated based on 26 English alphabets only.

978-1-5386-6723-1/19 $31.00 © 2019 IEEE

TABLE IV

Field	[6]	[8]	Proposed Encoding Scheme
Error rate reported (errors/bit)	1.8975×10^{-6}	1.2697×10^{-6}	1.8970×10^{-6}
Type of encoding used for digital data	Binary	Octal	Binary + Hexadecimal
Applicability on the type of data to be encoded	All forms of digital data could be encoded	Only 26 English alphabets could be encoded	All forms of digital data can be encoded

The AMF is calculated by the formula:

AMF = Mutation Frequency x Frequency of English Alphabets x (8 - R) ÷ 7

where R is the degree of redundancy associated with the mapping of genetic data to digital data. Note that the redundancy associated with the proposed work is always 1 since the encoding scheme follows a one-to-one mapping. Due to this reason, the AMF gets reduced to:

AMF = Mutation Frequency x Frequency of English Alphabets

The mutation frequency used is the same as reported in the experimental results of Church. et. al. The experimental result as reported by Church et. al. [6] shows that their coding scheme encoded 658,776 bytes of data in which 10 errors were identified. This leads to an error rate of 1.8975×10^{-6} errors/bit of data. On the other hand, A.Jimenez-Sanchez [8] in his encoding scheme found an average mutation frequency of 1.2697×10^{-6} errors/bit. In the proposed encoding scheme it was found that the average error rate was 1.8970×10^{-6} errors/bit which is close to that of Church et. al for encoding 26 capital English alphabets. However, this encoding scheme not only removes the constraints of octal encoding and its applicability on digital data as mentioned in the Table-III and Table-IV but it is also capable of removing the redundancy. This causes an increase in the data density per nucleotide and ensures effective utilization of the encoding space. Modern processors have the capability to work on raw binary and hexadecimal data. There occurs a disparity if the data is encoded in the octal format whereas a universal processor work with standard hexadecimal representations of digital data.

VI. CONCLUSION

The proposed work demonstrated the use of standard as well as non-standard nucleotides for encoding digital data into NAM. Integration of non-standard nucleotides permitted the natural memory capacity of a single nucleotide to expand from two to three bits per nucleotide.

Besides utilizing the complete permutation set the number of possible permutations for a single nucleotide was also increased. It was also made possible to utilize the hex encoding along with the binary data which are universally accepted by any computing system. Moreover, the self-correcting and anti-parallel nature of double-stranded DNA makes the encoding robust.

The nucleic acid memory proves itself to be an efficient and robust alternative for silicon memory in the near future. Further works need to concentrate on error correction and implementation of more stable nonstandard base pairs to enhance the encoding efficiency. Further, it is also required to focus on the development of efficient electronic equipment for DNA sequencing and nucleotide detection during reading and write operations.

REFERENCES

[1] V. Zhirnov, R. M. Zadegan, G. S. Sandhu, G. M. Church, and W. L. Hughes, 2016. Nucleic acid memory. Nature materials, 15(4), pp. 366-370.

[2] I. Hirao, M. Kimoto and S. Yokoyama. An unnatural base pair system for in vitro replication and transcription. Nucleic acid Symposium Series. 2006, vol. 50, pp. 33-34.

[3] I. Hirao, M. Kimoto and S. Yokoyama. An unnatural hydrophobic base pair system: Site-specific incorporation of nucleotide analogs into DNA and RNA. Nature Methods. 2006, vol. 3(9), pp. 729-735.

[4] D. A. Malyshev, T. Lavergne, P. Ordoukhanian and A. Torkamani. Efficient and sequence independent replication of DNA containing a third base pair establishes a functional six-letter genetic alphabet. Proceedings of the National Academy of Science. 2012, vol. 109, pp. 12005-12010.

[5] C. T. Clelland, V. Risca & C. Bancroft. Hiding messages in DNA microdots. Nature, 1999, 399(6736), pp. 533-534.

[6] G. M. Church, Y. Gao, and S. Kosuri. Next-generation digital information storage in DNA. Science, 2012, 337(6102), pp. 1628-1628.

[7] A. Jiménez-Sánchez. A proposal for a DNA-based computer code. International Inventions Journal Biochemistry and Bioinformatics, 2013, vol. 1, pp. 1-4.

[8] A. Jiménez-Sánchez. DNA computer code based on expanded genetic alphabet. Journal of Computer Science and Technology, 2014, vol. 2(4), pp. 8-20.

[9] K. Suyehira, S. Llewellyn, R. M. Zadegan, W. L. Hughes & T. Andersen. A Coding Scheme for Nucleic Acid Memory (NAM), IEEE Workshop on Microelectronics and Electron Devices (WMED), 2017.

[10] Y. Erlich, D. Zielinski, DNA fountain enables a robust and efficient storage architecture, Science, 2017, 355(6328), 950-954.

[11] W.Song, K.Cai, C.Yuen, Codes with run-length and GC-content constraints for DNA-based data storage, IEEE Communications Letters, 2018, pp. 1-4.

2019 Devices for Integrated Circuit (DevIC), 23-24 March, 2019, Kalyani, India

Surface potential based Analytical Modeling of Graded Channel Strained High-k Gate stack Dual-Material Double Gate MOSFET

Pritha Banerjee
Department of Electronics &Telecommunication Engineering
Jadavpur University
Kolkata, India
prithaedu7@gmail.com

Priyanka Saha
Department of Electronics &Telecommunication Engineering
Jadavpur University
Kolkata, India
priorient06@gmail.com

Dinesh Kumar Dash
Department of Electronics &Telecommunication Engineering
Jadavpur University
Kolkata, India
dineshdash123@gmail.com

Subir Kumar Sarkar
Department of Electronics &Telecommunication Engineering
Jadavpur University
Kolkata, India
sksarkar@etce.jdvu.ac.in

Abstract— **Gate work function engineered MOSFET coupled with the channel and dielectric engineering benefits has always been an important topic owing to their supreme immunity over Short Channel Effects (SCEs). In this paper, the analytical model of a Graded Channel Strained High-k Gate stack Dual-Material Double Gate MOSFET has been presented by solving the Poisson's equation in 2D and following parabolic potential approximation approach. The analytical results have been substantiated by ATLAS simulated data.**

Keywords— **Gate work function engineered MOSFETs, channel engineering, dielectric engineering, Short Channel Effects (SCEs).**

I. INTRODUCTION

The recent advancement in the field of nano electronics is mainly attributed to the development of innovative devices characterized by device scaling. However, miniaturization leads to device performance degradation due to severe SCEs [1] which call for the requirement of novel devices that have appreciable channel control by gate. The enhanced channel control leads to diminished SCEs and to achieve this, a combination of device scaling, implementation of innovative technologies and material property improvement is inevitable. In this context, gate work function engineered devices [2] have been found to perform extremely well. Employing more than one metal like material in the gate electrode results in formation of stepped surface potential profile thereby reducing major SCEs like Drain-induced barrier lowering (DIBL) as well as Hot Carrier Effect (HCE). Engineered channel concepts like Graded channel MOSFETs [3], halo implanted MOSFETs [4] are popular in reducing SCEs. Material property improvement such as Strained MOSFETs [5] is well-known for boosting the carrier mobility and reducing the SCEs.

The device performance can be remarkably amended by skillful implementation of a high-k gate stack configuration [6] boosting its subthreshold characteristics. Therefore, considering these advantages, electrostatic features based performance investigation of Graded Channel Strained High-k gate stack Dual-Material Double Gate MOSFET has been presented along with its ATLAS simulated data.

II. PROPOSED STRUCTURE

The schematic two dimensional representation of the device structure proposed is displayed in Figure 1. The entire channel length is L. The dual material gate is made of materials with work functions φ_{M1} (=4.8eV) with gate length L_1 and φ_{M2} (=4.6eV) with gate length L_2, (L= L_1+ L_2). The HfO$_2$ oxide thickness is t_2 (=2nm) and that of SiO$_2$ is t_1 (=1nm). The channel thickness is t_c (=10nm).

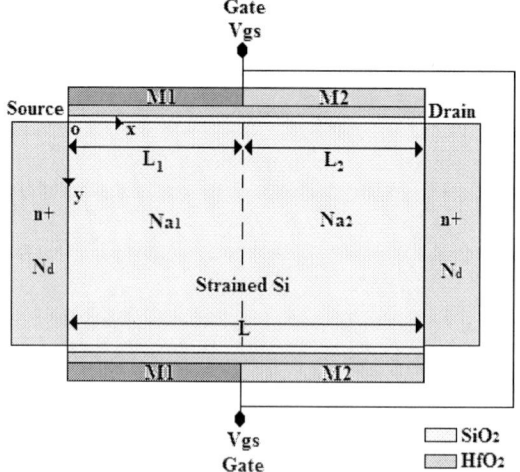

Fig.1 2-D schematic representation of the device under consideration

978-1-5386-6723-1/19 $31.00 © 2019 IEEE   240

The channel is graded with higher concentration Na_1 ($=1\times10^{17}cm^{-3}$) towards the source side and lower concentration Na_2 ($=1\times10^{16}cm^{-3}$) towards the drain side. The channel is a single layer strained silicon channel. The channel is divided into two regions, region I with boundaries $(0 \leq x \leq L_1, 0 \leq y \leq t_c)$ and region II with boundaries $(L_1 \leq x \leq L, 0 \leq y \leq t_c)$.

III. ANALYTICAL MODELING

A. Strain inclusion effects on the various band related parameters:

Strain inclusion in silicon results in alteration of band related parameters. The electron affinity elevates and carrier effective masses and band gap lessen due to strain addition. The strain related effects thus follows [7]:

Increase in electron affinity is

$$\left(\Delta\varepsilon_c\right)_{strainedSi} = 0.57x \tag{1}$$

Decrease in bandgap is denoted as

$$\left(\Delta\varepsilon_g\right)_{strainedSi} = 0.4x \tag{2}$$

$$V_T \ln \frac{N_{V,Si}}{N_{V,strainedSi}}$$

$$= V_T \ln \left(\frac{m^*_{h,Si}}{m^*_{h,strainedSi}}\right)^{3/2} \approx 0.075x \tag{3}$$

Here x is equivalent Ge mole fraction strain.

The front and the back flat band voltage changes can be modeled as:

$$\left(V_{FB}\right)_{strainedSi} = \left(V_{FB}\right)_{Si} + \Delta V_{FB} \tag{4}$$

Where, $\left(V_{FBi}\right)_{Si} = \phi_{Mi} - \phi_{Si}$

Where i =1 and 2 for regions 1 and 2 respectively.

$$\Delta V_{FB} = \frac{-\left(\Delta\varepsilon_c\right)_{strainedSi}}{q} + \frac{\left(\Delta\varepsilon_g\right)_{strainedSi}}{q} - V_T \ln \frac{N_{V,Si}}{N_{V,strainedSi}}$$

The built-in voltage change across the source-body and drain-body junctions can be modeled as:

$$V_{bi,strainedSi} = V_{bi,Si} + \left(\Delta V_{bi}\right)_{strainedSi} \tag{5}$$

where

$$\left(\Delta V_{bi}\right)_{strainedSi} = -\frac{\left(\Delta\varepsilon_g\right)_{strainedSi}}{q} + V_T \ln \frac{N_{V,Si}}{N_{V,strainedSi}}$$

B. Derivation of surface potential

To deduce the surface potential of the proposed device structure, 2-D Poisson's equation is solved which is

$$\frac{\partial^2 \psi_i}{\partial x^2} + \frac{\partial^2 \psi_i}{\partial y^2} = \frac{qN_{ai}}{\varepsilon_{Si}} \tag{6}$$

Considering PPA [8], the potential distribution within silicon film is given by:

$$\psi_i(x,y) = \psi_{Si}(x) + C_{1i}(x)y + C_{2i}(x)y^2 \tag{7}$$

It is necessary to find the coefficients C_{1i} and C_{2i}. The boundary conditions to be followed to evaluate the above cited constants are [9]:

$$\left.\frac{\partial \psi_i(x,y)}{\partial y}\right|_{y=0}$$
$$= \frac{\varepsilon_f}{\varepsilon_{Si}t_{eff}}\left(\psi_i(x,0) - V_{gs} + (V_{FBi})_{strainedSi}\right) \tag{8}$$

$$\left.\frac{\partial \psi_i(x,y)}{\partial y}\right|_{y=t_c}$$
$$= \frac{\varepsilon_b}{\varepsilon_{Si}t_b}\left(V_{gs} - (V_{FBi})_{strainedSi} - \psi_i(x,t_c)\right) \tag{9}$$

Here, the structure being symmetric with similar high-k gate stack in both front and back sides,

$$\varepsilon_f = \varepsilon_b = e_1 \text{ and } t_{eff} = t_b$$

Where, $t_{eff} = t_1 + t_2\left(\frac{e_1}{e_2}\right)$

Here e_1 and e_2 are dielectric constants for SiO_2 and HfO_2 respectively.

The coefficients thus obtained are:

$$C_{1i} = \frac{\varepsilon_f}{\varepsilon_{Si}t_{eff}}\left(\psi_i(x,0) - V_{gs} + (V_{FBi})_{strainedSi}\right)$$

$$C_{2i} = \frac{\left(1 + \frac{C_f}{C_{Si}} + \frac{C_f}{C_b}\right)\left(V_{gs} - (V_{FBi})_{strainedSi} - \psi_{Si}(x)\right)}{t_c^2\left(1 + \frac{2C_{Si}}{C_b}\right)}$$

Here, $C_f = C_b$

Using the expressions of C_{1i} and C_{2i} in equation (8) and (9) and then putting in equation (6) at y=0, we get,

$$\frac{\partial^2\psi_{si}(x)}{\partial x^2} - \alpha\psi_{si}(x) = \beta_i \tag{10}$$

Where

$$\alpha = \frac{2\left(1 + \frac{C_f}{C_{Si}} + \frac{C_f}{C_b}\right)}{t_c^2\left(1 + \frac{2C_{Si}}{C_b}\right)},$$

$$\beta_i = \frac{qNa_i}{\varepsilon_{Si}} - 2\alpha\left(V_{gs} - (V_{FBi})_{strainedSi}\right)$$

The general solutions [10] to equation (10) are:

$$\psi_{s1}(x) = A_1 e^{\eta x} + B_1 e^{-\eta x} - \theta_1 \qquad (11)$$

$$\psi_{s2}(x) = A_2 e^{\eta(x-L_1)} + B_2 e^{-\eta(x-L_1)} - \theta_2 \qquad (12)$$

Where $\eta = \sqrt{\alpha}, \theta_i = \dfrac{\beta_i}{\alpha}$

To obtain expressions for A_1, B_1, A_2 and B_2, the boundary conditions to be regarded are [11]:

$$\psi_{s1}(0) = V_{bi1,strainedSi} \qquad (13)$$

$$\psi_{s2}(L) = V_{bi2,strainedSi} + V_{ds} \qquad (14)$$

$$\psi_{s1}(L_1) = \psi_{s2}(L_1) \qquad (15)$$

$$\left. \frac{d\psi_{s1}(x)}{dx} \right|_{x=L_1} = \left. \frac{d\psi_{s2}(x)}{dx} \right|_{x=L_1} \qquad (16)$$

Subthreshold Swing is given by [12]:

$$SS = 2.3 V_t \left[\frac{d\psi_s(\min)}{dV_{gs}} \right]^{-1}$$

IV. RESULTS AND DISCUSSIONS

This section deals with the discussions of results that have been obtained using analytical modeling and simulation. Here, $L_1:L_2$ ratio is considered as 1:1 in both simulation and modeling. In Fig.2, detailed observation of surface potential distribution of both the figures reveals that there is unique step profile with source-side shifted minimum in our demonstrated structure in contrast to the single material counterpart. Thus reduced DIBl is ensured in our structure. Also, the surface potential slope approaching the drain end is minimized when compared to that in the single material counterpart. This allows a diminished HCE in the device. Therefore, the proposed device structure is more insusceptible to SCEs in contrast to its single material gate equivalent. In Fig.3 as observed from the plot, the increase/ decrease of strain results results in change of source-to-channel barrier height thereby altering the threshold voltage of the device. Hence, tuning of the x-value results in tuning of the threshold voltage. Also, the surface potential course towards the source and drain gets reversed compared to the surface potential minima. This happens since the built-in voltage decreases with strain addition. In Fig.4, the electric field approaching drain end of our demonstrated device structure is reduced ensuring diminished HCE in the device whereas the enhanced electric field in the single material gate equivalent results in increased HCE thereby degrading its performance causing reduced

efficiency in terms of power.

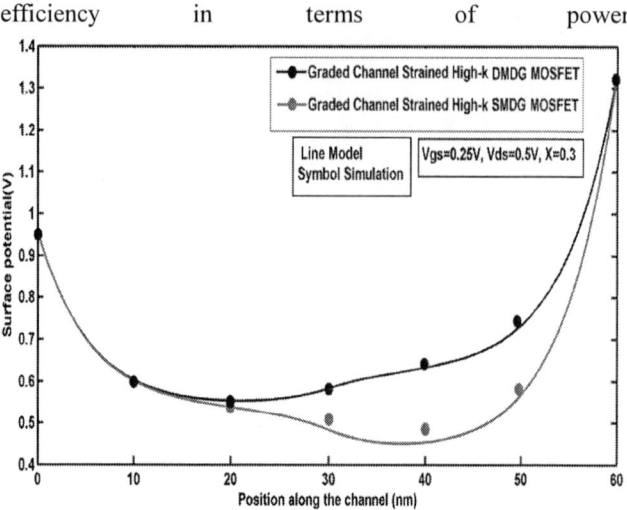

Fig.2 Surface potential distribution vs. channel length for the proposed device structure and its single material equivalent

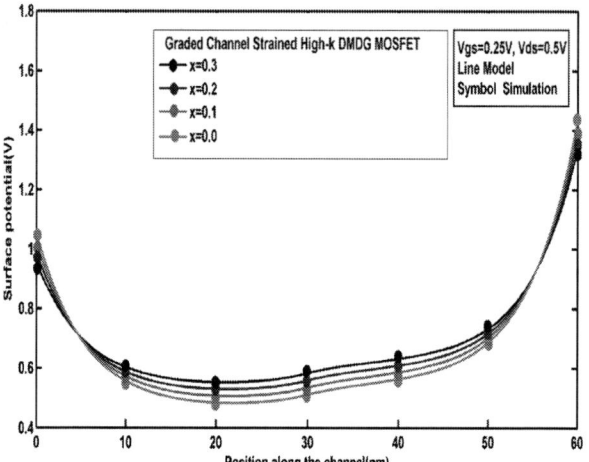

Fig.3 Surface potential distribution vs. channel length for the proposed device structure by varying x

In Fig.5, as the value of strain is enhanced, the electric field approaching drain end gets significantly diminished emanating in suppressed HCE in the device. Therefore, with apt strain value selection, the SCE in the device can be suitably controlled. In Fig.6, the configuration with higher-k gate oxide in our demonstrated device structure offers reduced Subthreshold swing ensuring smaller gate-oxide leakages thereby significantly improving the device performance in terms of power consumption. The analytical results have been validated using simulation results from ATLAS.

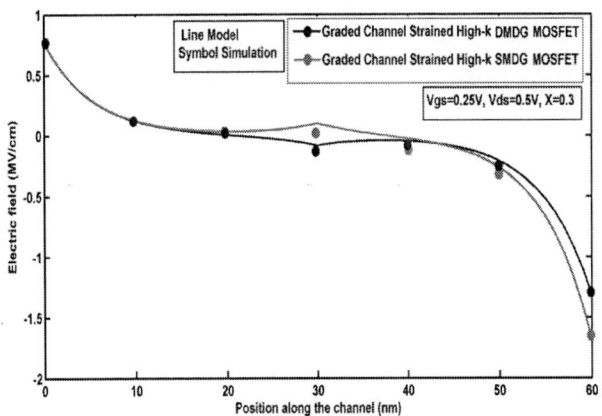

Fig.4 Electric field distribution vs. channel length for the proposed device structure and its single material equivalent

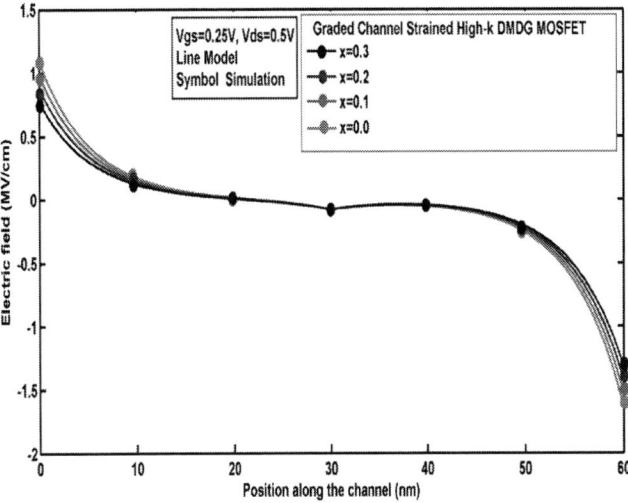

Fig.5 Electric field distribution vs. channel length for the proposed device structure by varying x

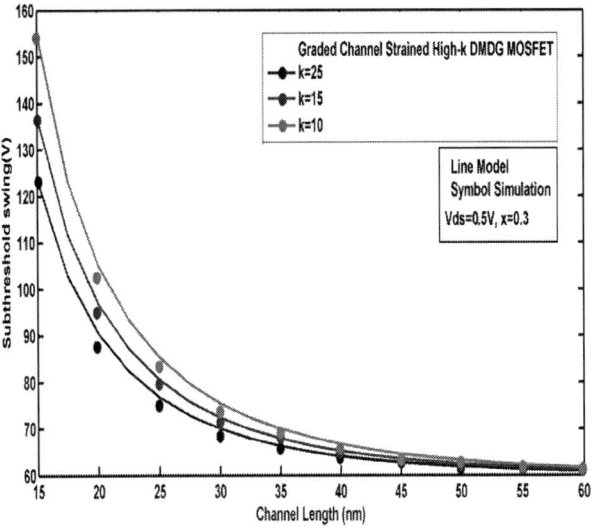

Fig.6 Subthreshold swing distribution vs. channel length for the proposed device structure for different high-k gate oxide

V. CONCLUSION

The current investigation centralizes the analytical model based performance assessment of Graded Channel Strained High-k gate stack Dual-Material Double Gate MOSFET. Detailed studies of device features reveal that the demonstrated structure possesses significant advancement in terms of SCE immunity. Also, apt high-k gate stack configuration selection can effectively suppress SCE like Subthreshold slope degradation by notably ameliorating the subthreshold characteristics in the device. The analytical data are in line with the simulated data confirming that our proposed structure is a proper contender for application in progressive upcoming generation VLSI circuits.

ACKNOWLEDGMENT

Priyanka Saha thankfully acknowledges this publication as an outcome of the R&D work undertaken project under the Visvesvaraya PhD Scheme of Ministry of Electronics & Information Technology, Government of India, being implemented by Digital India Corporation.

REFERENCES

[1] F. D'Agostino and D. Quercia, "Short-channel effects in MOSFETs," in Proc. Introduction VLSI Design (EECS 467), Dec. 2000, pp. 1–15.

[2] Naskar, S., Sarkar, S.K.: Quantum analytical model for inversion charge and threshold voltage of short-channel dual-material double-gate SON MOSFET. IEEE Trans. Electron. Dev., 60(9), 2734–2740 (2013)

[3] H. Kaur, S. Kabra, S. Haldar, and R. S. Gupta, "An analytical drain current model for graded channel cylindrical/surrounding gate MOSFET," Microelectron. J., vol. 38, no. 3, pp. 352–359, 2007

[4] Chetan Gupta , Sagnik Dey, Harshit Agarwal , Ravi Goel, Chenming Hu, and Yogesh Singh Chauhan, "Analysis and Modeling of Temperature and Bias Dependence of Current Mismatch in Halo-Implanted MOSFETs", IEEE TRANSACTIONS ON ELECTRON DEVICES, VOL. 65, NO. 9, SEPTEMBER 2018

[5] Pritha Banerjee and Subir Kumar Sarkar. 3-D analytical modeling of high-k gate stack dual material tri-gate strained silicon-on-nothing MOSFET with dual-material bottom gate for suppressing short channel effects," J. Comput. Electron. , September 2017, Volume 16, Issue 3, pp 631–639

[6] Padmanaban, B., Ramesh, R., Nirmal, D., Sathiyamoorthy, S.: Numerical modeling of triple material gate stack gate all-around (TMGSGAA) MOSFET considering quantum mechanical effects. Superlattices Microstruct. 82, 40–54 (2015)

[7] Sarkhel, S., Manna, B., Sarkar, S.K.: A compact two dimensional analytical modeling of nanoscale fully depleted dual material gate strained SOI/SON MOSFETs for subdued SCEs. J. Low Power Electron. ASP 10(3), 383–391 (2014)

[8] Young, K.K.: Short-channel effect in fully depleted SOI MOSFETs. IEEE Trans. Electron. Dev. 36(2), 399–402 (1989)

[9] Priyanka Saha, Saheli Sarkhel and Subir Kumar Sarkar, "Compact 2D threshold voltage modeling and performance analysis of ternary metal alloy work-function-engineered double-gate MOSFET", Journal of Computational Electronics, Springer, Vol.16.No.3, pp-648-657, June 2017

[10] P. K. Tiwari, S. Dubey, M. Singh, and S. Jit, "A two-dimensional analytical model for threshold voltage of short-channel triple-material double-gate metal-oxide-semiconductor field-effect transistors," J. Appl. Phys., vol. 108, no. 7, pp. 074508-1–074508-8, 2010.

[11] Pritha Banerjee and Subir Kumar Sarkar,"3-D Analytical Modeling of Dual-Material Triple-Gate Silicon-on-Nothing MOSFET", IEEE TRANSACTIONS ON ELECTRON DEVICES, VOL. 64, NO. 2, FEBRUARY 2017

[12] S. Basak, P. Saha, and S. K. Sarkar, "A quasi-two-dimensional analytical threshold voltage model for short-channel junctionless doublegate AQ:6 451 nanoscale SON MOSFET," in Proc. RAECS, 2014, pp. 1–5.

2019 Devices for Integrated Circuit (DevIC), 23-24 March, 2019, Kalyani, India

Design and Analysis of High Performance Multiplier Circuit

Inamul Hussain
Electrical Engineering Department
NIT Silchar
Silchar, India
ihinamul07@gmail.com

Chandan Kumar Pandey
Electrical Engineering Department
NIT Silchar
Silchar, India
saurabh1971@gmail.com

Saurabh Chaudhury
Electrical Engineering Department
NIT Silchar
Silchar, India
saurabh1971@gmail.com

Abstract— **Multiplier is of the most important blocks of many VLSI application. So, it is required to design high performance multiplier to boost up the performance of those circuits and systems. In this work, a high performing Multiplier has been designed by using Wallace tree algorithm. The multiplier has been designed in three modules. Initially partial products are generated followed by partial products processing. Finally, final addition is computed in the 3rd module. Partial products have been generated by using AND gates. Partial products are computed by using Wallace tree logarithm. Final addition has been done by fast adder. The performance of the proposed multiplier circuit is evaluated by using 90nm CMOS technology at the Synopsys tool. The performance of the same is compared conventional multiplier circuits. The performance of the proposed adder has been found to be satisfactory.**

Keywords— Wallace tree, partial products, fast adder, power, delay, EDP, PDP.

I. INTRODUCTION

The multiplier is one of the most prominent modules of VLSI circuits and systems [1-5]. It has many applications such as in DSP processor, image processing, digital filters, communication systems etc. [6-7]. So, the performance of these applications depends on the performance of multiplier in terms of speed and power consumption. So it is important to design multiplier circuits for VLSI applications [8].

Different approaches and architectures are proposed to design a multiplier circuit to improve the performance [9-10]. Wallace tree is one of the oldest and popular techniques used to design multiplier circuit and such multipliers are known as Wallace multiplier [13-15]. The popularity of the architecture of Wallace multiplier is due to the high speed of operation [16].

Wallace multiplier is categorized in conventional Wallace multiplier and Reduced Complexity Wallace multiplier [17-20]. In conventional Wallace multiplier partial products are generated, then the partial products are processed and finally, final addition is performed [21]. The processing of the partial products is done by rearranging the partial products in different stages. The number of rows of partial products in a particular stage is calculated by equation 1 [22-23].

$$T_{(i+1)}=2(T_i/3)+T_i \bmod 3 \qquad (1)$$

Where, T_i is the number of rows in a particular stage and $T0=N=$number of bits.

In this work, a 4x4 Wallace multiplier has been designed and simulated in 90nm CMOS technology by using Synopsys tool. The multiplier is designed in three modules:

partial products generation module, partial products processing module and final addition module. Partial products generation module has been designed by AND gate, partial products are processed by the Wallace tree algorithm and finally, final addition module is designed to get the final products.

The organization of this paper is as follows. Section II explores detail of proposed multiplier. In Section III, performance analysis and discussion are reported. Section IV concludes this paper.

II. PROPOSED MULTIPLIER

The proposed adder has been designed as per the block diagram is shown in Figure 1. From two inputs, partial products are generated. Then the partial products are processed and finally, final addition are done. Thus the proposed adder has three parts: partial product generation, partial product processing, and final addition.

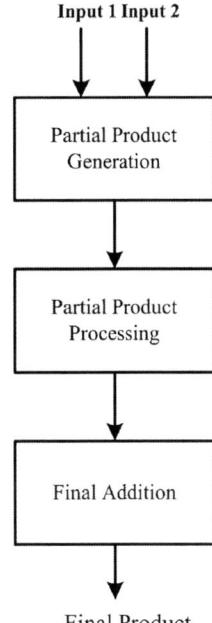

Figure 1 Block diagram of the designed Multiplier

A. Partial product generation module:

The designed multiplier is of 4x4 bits. So, it has two inputs of 4-bits each. Thus a total number of partial products generated is 16. AND gates are used for producing the partial products. Thus the multiplier needs total 16 AND gates and there will be 4 rows of partial products. If the two inputs are

978-1-5386-6723-1/19 $31.00 © 2019 IEEE   245

a3a2a1a0 and b3b2b1b0, their partial products will be as shown in the Figure 2. Thus, for N multiplier, the number of AND gates required will be NxN.

$$a3a2a1a0$$
$$b3b2b1b0$$
$$\overline{}$$
$$p03p02p01p00$$
$$p13p12p11p10$$
$$p23p22p21p20$$
$$p33p32p31p30$$

Figure 2 Partial product generation

B. Partial Product Processing Module:

In this module, the partial products that are created in the previous module are computed. The reduction of partial products is done by the Wallace tree algorithm. In this algorithm, the columns of the partial products are processed in a tree like parallel fashion. For that purpose, the partial products are rearranged in different stages by equation 1 until the last stage contains only two rows of partial products. Each stage, both full adders and half adders are used for addition purpose depending on the bits available. Thus the designed partial product processing module has 4 half adder (HA) and 4 full adder (FA) cells. Where there are two bits in a column half adder is used and when there are 3 bits full adder is used. The half adder and full adder used are reported in [14] and [3] respectively.

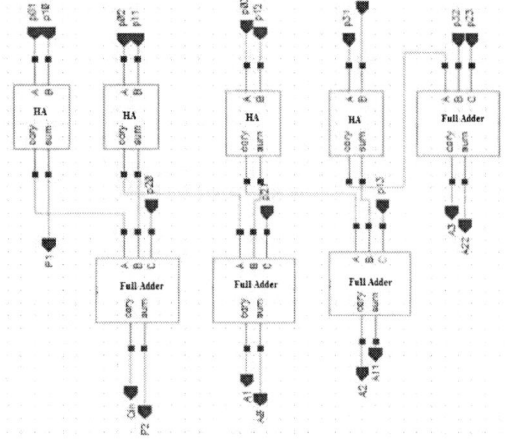

Figure 3 Partial product processing module

C. Final Addition Module:

Partial products are processed by grouping them into different stages by using equation 1 and stopped once the final stage contains only two rows of partial products. These two rows are computed by carry save adder which is fast of its kind.

III. PERFORMANCE ANALYSIS AND DISCUSSION

The multiplier has been designed by cascading the partial product generation module, partial product processing module and final addition module. The simulation has been carried out in the Synopsys tool by using 90nm CMOS technology at room temperature. To simulate the designed multiplier circuit, a test bench as shown in Figure 4 has been prepared to simulate in real time environment. The performance analysis of the multiplier has been done in terms of power, delay, power-delay product, and energy-delay product shown in Table 1. It is observed that the designed multiplier has better performance in terms of PDP and EDP. It is due to the Wallace algorithm that is used to reduce the partial products in tree-like parallel structure. This reduces the computational time. Moreover, by using a fast adder in the final addition make the multiplier more productive in terms of performance.

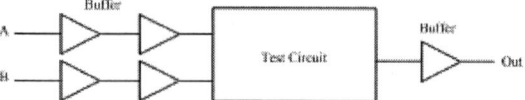

Figure 4. Test bench for test circuit

A. Total Power Calculation

The total power consumed by the circuit has been calculated as the summation of both static and dynamic power. A comparison graph has been shown in the Figure 5. It is observed that the power consumed by the designed multiplier is higher because the designed multiplier requires more number of full adder cells, which can be overcome by improved architecture by reducing FA.

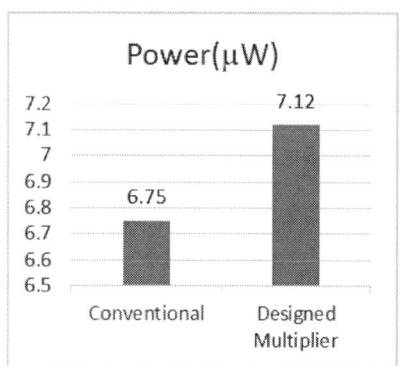

Figure 5 Power comparison

B. Delay Calculation

Delay between inputs and outputs are computed and worst case delay are reported in the Table 1. It is observed that the designed multiplier has better delay as compared to conventional multiplier is due to the Wallace tree algorithm for partial product reduction.

C. PDP and EDP Calculation

To comment on the performance of the multiplier, power delay product (PDP) and energy delay product (EDP) is calculated. In both the cases worst case delay has been considered.

A comparison graph has been shown in the Figure 6, the parameters considered here are delay, PDP and EDP. Though the designed multiplier consume more power, it has less propagation delay that yields less PDP and EDP. The main reason behind the less delay is due to the partial product reduction. Wallace tree algorithm is used for grouping the partial products, whereas for addition high speed half adder and hybrid full adder has been used. Moreover, for final addition carry save adder has been used that reduces the computational time, eventually speed up the operation.

Table 1. SIMULATION RESULTS FOR NAND, NOR, MUX AND XOR

Type	Power(µW)	Delay (ns)	PDP (fJ)	EDP(zJs)
Conventional	6.75	25	168.75	4.22
Designed Multiplier	7.12	20.7	147.38	3.05

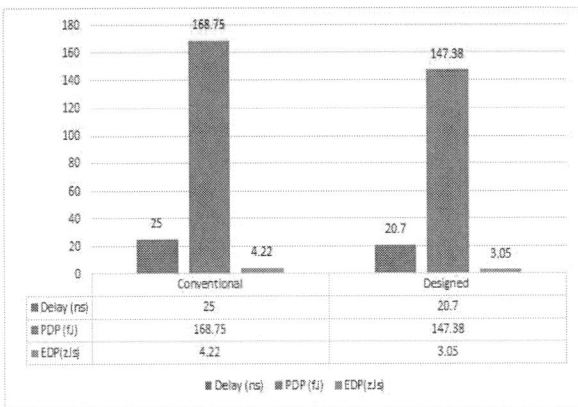

Figure 6 Comparison of delay, PDP and EDP

IV. CONCLUSION

In this work, a new low power multiplier has been designed by using a Wallace tree algorithm. The multiplier has three modules partial product generation module, partial product processing module and final addition. Partial product generation module has been designed by AND gates and it is observed that if the multiplier of N bits, it requires N2 number of AND gates. Partial products generation module has been designed by using Wallace tree algorithm and finally, final addition is done by carry save adder that reduces the computational time. The simulation has been carried out at room temperature in Synopsys tool with 90nm technology. It is found that the proposed multiplier has better power consumption, PDP and EDP. Though the designed multiplier has drawbacks in terms of, it could be overcome by more improved hybrid architecture.

REFERENCES

1. C. S. Wallace. "A suggestion for a fast multiplier." IEEE Transactions on Electronic Computers, vol. EC-13, no. I, pp. 14-17, February 1964.

2. Inamul Hussain, and Manish Kumar: 'Design and Performance Analysis of a 3-2 Compressor by Using Improved Architecture', Journal of Active and Passive Electronic Devices, 2017, 12, (3-4), pp.173-181.

3. S. Goel, A. Kumar, and M. A. Bayoumi: 'Design of Robust, Energy-Efficient Full Adders for Deep-Submicrometer Design Using Hybrid-CMOS Logic Style', IEEE Transactions on Very Large Scale Integration (VLSI) Systems, 2006, 14, (12), pp. 1309-1321.

4. I. Hussain and S Chaudhury, "Performance Comparison of 1-Bit Conventional and Hybrid Full Adder Circuits", In: Bera R., Sarkar S., Chakraborty S. (eds) Advances in Communication, Devices and Networking" Lecture Notes in Electrical Engineering, vol 462, 2018

5. R. S. Waters and E. E. Swartzlander, "A reduced complexity Wallace multiplier reduction," IEEE Transactions on Computers, vol. 59, no. 8, pp. 1134-1137, August 2010.

6. S. P and A. A. Khan, "Comparison of Braun Multiplier and Wallace Multiplier Techniques in VLSI," 2018 4th International Conference on Devices, Circuits and Systems (ICDCS), Coimbatore, 2018, pp. 48-53.

7. J Rabaey, A Chandrakasan, B Nikolic, "Digital Integrated Circuits A Design Perspec tive, 2nd Ed., New Jersey : Prentice- Hall Inc, c2003.

8. Neil H. E. Weste, David Harris, Ayan Banerjee, "CMOS VLSI Design- A Circuit and Systems Perspective", Pearson, 2006,pp.345357.

9. Chou C and Kuo K, Low Power Multiplier with Bypassing and Tree Structures, IEEE Asia Pacific Conference on Circuits and Systems, 2006.

10. Huang Y, Lin J, Sheu M and Sheu C, Low Power Multiplier Designs Based on Improved Column Bypassing Schemes. IEEE Asia Pacific Conference on Circuits and Systems. 2006.

11. Moshnyaga VG and Tamaru K, A comparative study of switching activity, Reduction techniques for design of lowpower multipliers, IEEE International Symposium on Circuits and Systems, 1995, 1560-1563.

12. Ohban J, Moshnyaga VG and Inoue K, Multiplier energy reduction Through bypassing of partial products, Asia Pacific Conf on Circuits and Systems, 2, 2002, 13-17, 2002.

13. Parhami B, Computer Arithmetic, Algorithms and Hardware Designs, Oxford University Press, 2000. Wen M.C, Wang S.J and Lin Y.N, Low-power parallel multiplier with column bypassing, IEE Electronics Letters, 41(10), 2005, 581– 583, 12.

14. Inamul Hussain, and Manish Kumar: "A Fast and Reduced Complexity Wallace Tree Multiplier", Journal of Active and Passive Electronic Devices, vol.12, pp.63-71, 2017.

15. Khan, S., Kakde, S., and Suryawanshi, Y., "VLSI Implementation of Reduced Complexity Wallace multiplier using energy efficient CMOS full adder", IEEE International Confer¬ence on Computational Intelligence and Computing Research, Enathi, pp. 1-4, 2013.

16. Hiremath, Y., "A Novel 8-bit Carry Select Adder using 180nm CMOS Process Technol¬ogy", International Journal of Emerging Engineering Research and Technology,vol. 2, no. 6, pp. 187-194, 2014.

17. S. Asif and Yinan Kong, "Performance analysis of Wallace and radix-4 Booth-Wallace multipliers," 2015 Electronic System Level Synthesis Conference (ESLsyn), San Francisco, CA, 2015, pp. 17-22.

18. I. Hussain, R. K. Sah and M. Kumar "Performance Comparison of Wallace Multiplier Architectures", International Journal of Innovative Research in Science, Engineering and Technology, vol. 4, no. 1, 18729-18734, 2015.

19. G. C. Ram, D. S. Rani, R. Balasaikesava and K. B. Sindhuri, "Design of delay efficient modified 16 bit Wallace multiplier," 2016 IEEE International Conference on Recent Trends in Electronics, Information & Communication Technology (RTEICT), Bangalore, 2016, pp. 1887-1891

20. S. Kakde, S. Khan, P. Dakhole and S. Badwaik, "Design of area and power aware reduced Complexity Wallace Tree multiplier," 2015 International Conference on Pervasive Computing (ICPC), Pune, 2015, pp. 1-6.

21. A. Habibi and P. A. Wintz, "Fast Multipliers," in IEEE Transactions on Computers, vol. C-19, no. 2, pp. 153-157, Feb. 1970.

22. A. A. AlJuffri et al., "ASIC realization and performance evaluation of scalable microprogrammed FIR filters using Wallace tree and Vedic multipliers," 2015 IEEE 15th International Conference on Environment and Electrical Engineering (EEEIC), Rome, 2015, pp. 1995-1998.

23. S. Asif and Y. Kong, "Analysis of different architectures of counter based Wallace multipliers," 2015 Tenth International Conference on Computer Engineering & Systems (ICCES), Cairo, 2015, pp. 139-144.

Production of Few Layer Graphene by Electrochemical Delamination using Alkaline Solution

S.Das
Electronics and Telecommunication Engineering,
Jadavpur University,
Kolkata, India
swapan_1976@yahoo.co.in

S.Roy[*]
Electronics and Communication Engineering,
Guru Nanak Institute of Technology,
Kolkata, India
sunipa_4@yahoo.co.in

C.K. Sarakar
Department of Physics
Bengal Engineering and Science University
Howrah, India
phyhod@yahoo.co.in

Abstract— **few layer graphene produced by an undemanding electrochemical delamination method has been presented here using potassium hydroxide. Copper strip is kept at zero potential and carbon rod is kept at positive or negative potential. The few layer graphene is characterized by XRD, FESEM, Raman and AFM. By employing KOH, larger dimension few layer graphene of 4.5nm thickness with are of 11μm is obtained. Production of few layer graphene was confirmed by RAMAN. These graphene flakes are dissolved in dimethylformamide (DMF) to form a homogeneous solution. After drop casting, a skinny layer of graphene is obtained. After that, annealing is done at 250°C to eliminate some functional groups present.**

Keywords-Electrochemical delamination, KOH, carbon rod, copper strip, graphene, DMF.

I. INTRODUCTION

Graphene having its important intrinsic properties such as fast electron mobility, excellent conductivity [1], high electrocatalytic activity, high surface-to- volume ratio, good mechanical properties, make this substance as a first preference ideally suited to specialized roles and applications for the gas sensor in the near future.

So many researchers focus on the synthesis of graphene by different techniques, such as mechanical exfoliation [2],epitaxial growth on silicon carbide or metal [3], chemical vapor deposition (CVD) [4], thermal exfoliation [5], and chemical reduction of graphene oxide (GO)[6], are there to produce graphene.

The mechanical exfoliation method is measured to be another most cost effective technique to producing the enormous quantity of graphene. Due to the disrupter of electrons in the graphene layer, the conductivity of the graphene sheets decreased.

Monolayer graphene along with its astonishing electronic property by epitaxial growth on silicon carbide (SiC) is beneficial than the other synthesis approach. But the requirement of thickness control and reiterate of large area graphene layer are the central issue of the above technique due to the requirement of excessive temperature and ultrahigh vacuum setting.

CVD (chemical vapour deposition) is the best renowned technique for the production of graphene. But the class of the film in CVD graphene is inferior to mechanically cleaved graphene; subsequently the electronic quality of graphene depends mainly on the presence of flawless grain boundaries. Moreover, the removal of graphene layer from the substrate is not an easy task and if done it doesn't confirms the total elimination of catalytic metal layer present on surface of the base.

In that sense, Hummers' method is the most-popular methods in the delamination of graphene. Chemical reduction has been the most desired one for fabricating graphene oxide (GO) with low price involvement. But the oxidation process damages the honeycomb lattice of graphene and also reduction of graphene oxide engages the high temperature requirement. Further, it often indulges the presence of harmful chemicals (e.g. hydrazine) and thus this method is not an environment friendly method to delaminating graphene.

Here electrochemical delamination technique is suggested for producing graphene to avoid the difficulties attached with the other techniques. It is the simplest and economical approach for synthesis of graphene.

Electrochemical delamination technique can manage better control the thickness of graphene sheet. This work suggests the use of inexpensive copper strip as zero potential electrode and carbon rod as positive and negative potential electrode. Potassium hydroxide (KOH) is used here as electrolyte. This is an easy production of graphene directly onto the substrate including flexible type. The electrochemical delamination leads to better yield, larger sideways extension and no disorder. Production of few layer graphene by electrochemical delamination using alkaline solution is presents here and the X-ray diffraction (XRD), field emission scanning electron microscopy (FESEM), atomic force microscopy (AFM), and RAMAN spectra are done here which give us the surface morphology, crystallographic orientation, and number of layer of graphene.

II. EXPERIMENTAL

A. Electrochemical Delamination of Graphene

Fig. 1 shows a schematic representation of the experimental setup. Electrochemical delamination was done in an electrolysis cell has 250ml borosil beaker , carbon rod (CB154 Makita carbon:15x10x6mm) used as positive or negative potential electrode and 99.9 percent unblended Cu strip of one millimeter thickness used as zero potential electrode.

An optical illustration of electrodes is shown in Fig. 2. This whole electrodes setup was dipped inside the KOH solution maintaining the contact just above the electrolyte facade. The carbon rod anode or cathode and copper strip grounded electrodes were kept at a distance of 3 cm. Present work used Potassium hydroxide (KOH, Merck, Germany; 85%) as electrolyte dissolved in demineralization (DI) water. The wholesome alkaline electrolyte is beneficial of using for delamination that it will not corrode the carbon rod especially at the boundaries. The electrochemical delamination was executed by first putting on DC voltage of +1.5V for 30 sec. Then +6V and -6V were connected on the carbon rod at an interim of 8 sec. The use of these voltages on the carbon rod result is the delamination of graphite through its boundaries. . After applying +6V in carbon rod anode, negatively charged ions enter into the surface layers of carbon. The result is the expansion of the surface layer and delamination of graphene.

The result is the expansion of the surface layer and delamination of graphene.

Similarly, the use of -6V in carbon rod cathode, positively charged ions enter from the electrolyte into the surface layers of carbon rod, followed by expansion of the surface layer and delamination of graphene. The combination of +6V and -6V to the carbon rod indulges continuous expansion and delamination, supporting faster production of graphene. An optical illustration of the electrochemical delamination process after 15 minutes it starts is shown in Fig. 3. After 20 min, it was detected that a significant quantity of graphene flakes were produced. It is worthy to mention that, two types of graphitic flakes were formed; thick and thin.

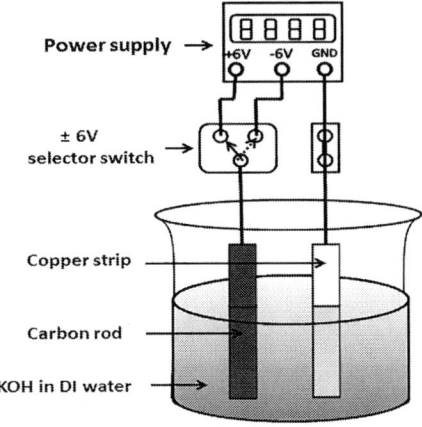

Figure 1. Experimental setup

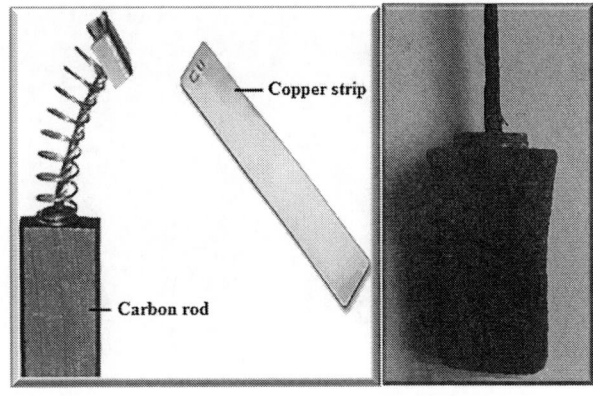

Figure 2. Optical images of (a) carbon rod (15mm x10mm x 6mm) and copper strip (35mm x7mm x1mm) (b) carbon rod after delamination

Thicker one gets accumulated at the bottom and thinner one floats on the top of the alkaline solution. Above mention floating flakes are semitransparent and these are consisting of few layer graphene (FLG).

These flakes are now gathered from the top and the centrifugal rotation was applied to filter it at 5000 rpm. Deionized water was used to washed filtered graphitic flakes several times to remove residual products and then were dehydrated in a desiccator (155mm) for the time period of 48 h in normal atmosphere to have the delaminated graphene powder.

III. CHARACTERIZATION OF GRAPHENE FILM

The graphene film was examined by the X-ray diffraction measurements, Field emission scanning electron microscopy (FESEM), Raman spectroscopy analysis, and Atomic force microscope (AFM) analysis were used to characterize graphene. To study the particle diameter, surface finish, percentage of elements present and crystalline character of the as prepared graphene sheets.

A. Preparation of Graphene Thin Films and Device

Finally, the 1mg graphene powder was mixed with 10ml of DMF (Dimethylformamide) to make a solution and to get a consistent homogeneous mixture of graphene. It was ultrasonicated for a time period of 30 minutes .Fig. 3(b) shows the optical image of dispersed graphene in the DMF solution. Micro-pipet was then used to drop the solution on the SiO_2 layer and was annealed at 250°C for 30 minutes. After annealing thin films were formed. The device was prepared on a SiO2/Si substrate with size was taken 5 mmx5 mm, and two electrodes was taken with gap of 1 mm with size of 1 mmx1 mm. For contact electrode, silver paste was used directly on the samples. Finally thin copper strand was used to take the contact out. Fig.4 shows the pictorial view of the device.

(a)   (b)

Figure 3. Optical images of (a) electrochemical delamination setup after 15 min (b) dispersed graphene in DMF solution

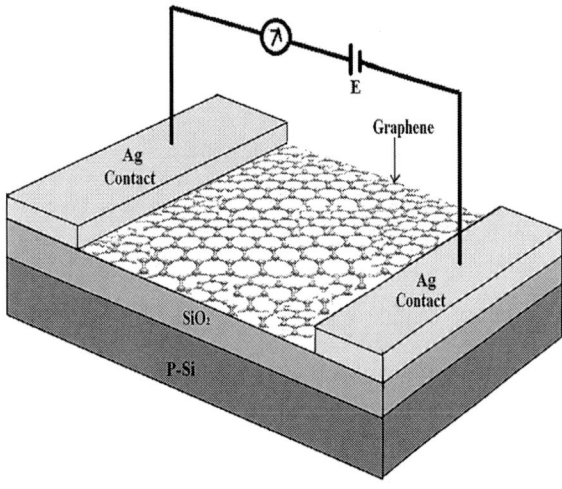

Figure 4. Pictorial view of the device

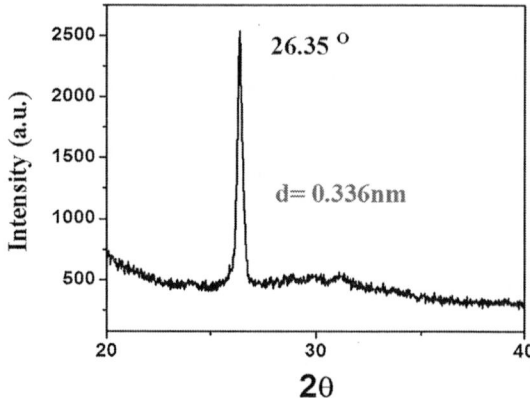

Figure 5. XRD pattern of as prepared graphene

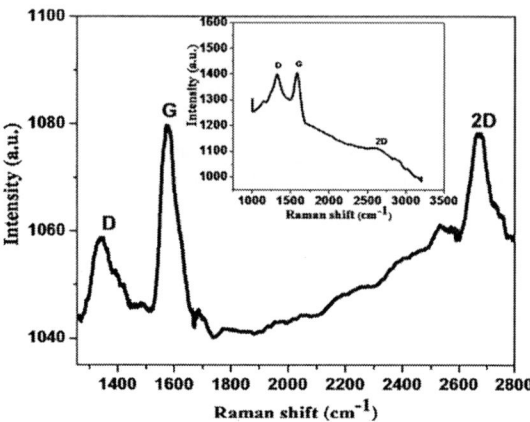

Figure 6. Raman spectra (excited by 512 nm laser) of graphene (inset graphite)

IV. RESULTS AND DISCUSSIONS

A. XRD Analysis

The formation of the electrochemically delaminated graphene was analyzed by X-ray Diffractometer (XRD) and is shown in Fig. 5. The delaminated graphene exhibits a strong and sharp peak located at 26.35°. This is in regular crystal structure. The peak of 26.35 ° corresponding to a basal spacing d002 = 0.336 nm as anticipated.

B. RAMAN Spectroscopy Analysis

Raman spectra investigated mainly to locate the imperfection present and the thickness of the graphene layer. Fig. 6 shows the full Raman spectra (excited by 512 nm laser) of graphene. Two major peaks were observed at 1345 cm^{-1}

and 1581 cm^{-1} correspond to D and G band and complying the research demonstrated so far. D line reveals the inherited structural turmoil of graphene.

2D band as usual, located at 2675 cm^{-1} is found in the spectra validating the fabrication of few sheets graphene. G band is sharp enough. The higher ratio of (I_{2D}/I_G), is higher the extent of sp2 hybridization. In this case (I_{2D}/I_G) is 0.56 which encourages the fabrication of few sheets graphene. As mentioned in the literature [7], p-type doping is confirmed by the blue shift (left) of the 2D peak position, and n- type doping is confirmed by red shift. As per the previous report uncontaminated graphene is confirmed by the position of 2D peak at 2690 cm-1 to 2700 cm-1. Present work reflects the n-type conductivity of graphene layer as the existing place of 2D peak is at 2675 cm-1.

C. FESEM Analysis

A FESEM image of graphene was investigated using field emission scanning electron microscopy. Fig. 7 shows the FESEM images of graphene. From images, it can be concluded that the length

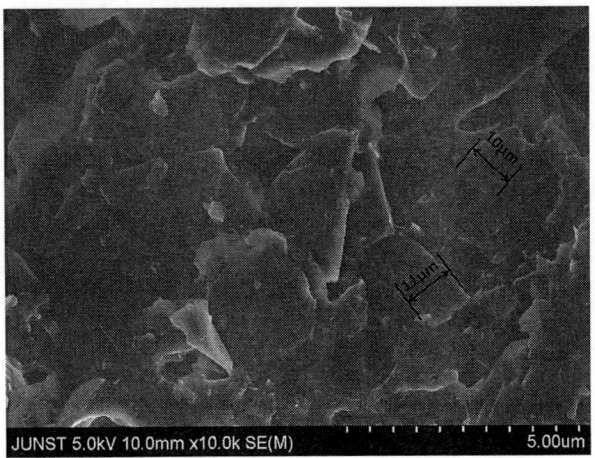

Figure 7. FESEM image of electrochemically delamination graphene

of the graphene sheets is relatively larger and up to ~11μm with regular structure.

D. AFM Analysis

Outer surface roughness of the graphene sheet was typified by atomic force microscope. Fig. 8 shows an AFM image of graphene (4.5 nm thick) drop-filled on SiO_2 layer. Layer numbers of sheets were calculated which reveals that the layer number of electrochemically delaminated graphene ranges from 7 to 9.

E. Sheet Resistance

The electrical testing was done in a small glass cylinder with nonstop pressure of nitrogen (500sccm). The device was kept in the cylinder for 1 hour prior to test starts. Temperature of the chamber was varied from room temperature to 250 °C and sheet resistance was measured accordingly which is shown in Fig. 9.

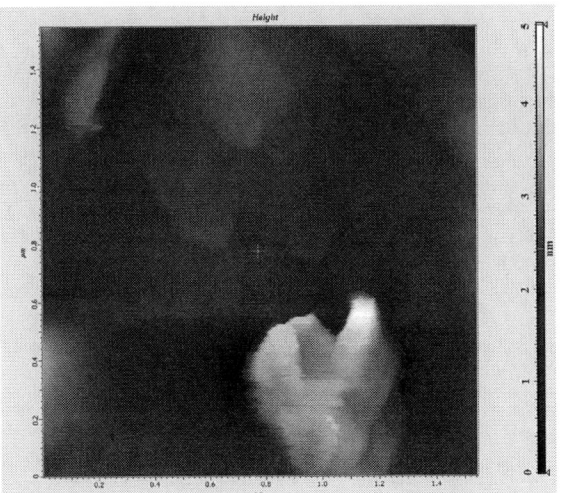

Figure 8. AFM micrograph of electrochemically delaminated graphene

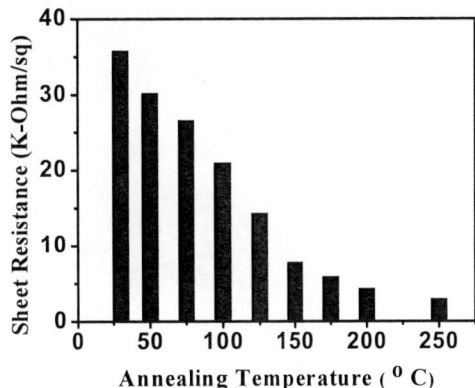

Figure 9. Statistical analysis of graphene sheet resistance with annealing temperature

Enhancement in conductivity has been observed after annealing at 250°C, film quality was also improved, probable cause may be the densification of the film.

V. CONCLUSION

In conclusion, a facile method of producing high quality few layer graphene is demonstrated in this paper by electrochemical delamination of graphite. The delaminated graphene sheets exhibit lateral size up to ~11μm. This work provides an easy approach of producing high-quality scalable graphene with no toxic byproducts.

ACKNOWLEDGMENT

We acknowledge the cooperation of IISER, Nadia, for providing us the laboratory facility.

REFERENCES

[1] S Stankovich, D.A Dikin, G.H.B Dommett, K.M Kohlhaas, E.J Zimney., E.A Stach, R.D Piner., S.T Nguyen,., R.S Ruoff,.,. Nature 442, 282–286. 2006

[2] K S Novoselov, A K Geim, S V Morozov, et al. Electric field effectin atomically thin carbon films. Science, 306: 666–669, 2004

[3] C Berger, Z M Song, X B Li, et al. Electronic confinement and coherence in patterned epitaxial graphene. Science, 312: 1191–1196, 2006

[4] D. Dutta, S.K. Hazra, J. Das, C.K. Sarkar,l And S. Basu, "Temperature- and Hydrogen-Gas-Dependent Reversible Inversion of n-/p-Type Conductivity in CVD Grown multilayer graphene (MLG) Film" Journal of Electronic Materials, DOI: 10.1007/s11664-016-4381-0 _ 2016 The Minerals, Metals & Materials Society

[5] W. Lv, D M. Tang, Y B. He, et al. Low-temperature exfoliated graphenes: Vacuum-promoted exfoliation and electrochemical energy storage. ACS Nano, 3: 3730–3736, 2009

[6] D. Li, M B Muller, S. Gilje, et al. Processable aqueous dispersions of graphene nanosheets. Nat Nanotechnol, 3: 101–105, 2008

[7] A. Das, B. Chakraborty, AK Sood. Raman spectroscopy of graphene on different substrates and influence of defects. *Bull.Mater. Sci.*, 31, 579–584, 2008

2019 Devices for Integrated Circuit (DevIC), 23-24 March, 2019, Kalyani, India

Investigating Photonic Bandgap Width for Metamaterial based PhC Structure from Dispersion Relation

Sangita Das
Department of Electronic Science
A.P.C College
24 PG(N), INDIA
dsangita626@gmail.com

Urmi Dey
Department of Electronic Science
A.P.C College
24 PG(N), INDIA
deyurmi13@gmail.com

Soumita De
Department of Electronic Science
A.P.C College
24 PG(N), INDIA
soumita.de29@gmail.com

Arpan Deyasi
Dept of Electronics and Communication Engineering
RCC Institute of Information Technology
Kolkata, INDIA
deyasi_arpan@yahoo.co.in

Abstract—**Characteristic Lowest photonic bandgap width is analytically computed for double negative index materials from dispersion relation under propagation of TE wave. Three well-established metamaterials are considered for simulation purpose, and two lower photonic bandgaps are computed with varying incidence angle for a few materials. Results are compared with existing bandgap widths obtained for positive refractive indices based materials. Results suggest that negative index materials offer higher bandgap width than existing positive index materials, and hence may be considered as better candidate for optical filter design.**

Keywords— *Photonic bandgap, Metamaterial, Dispersion Relation, TE mode, Incidence angle, Photonic crystal*

I. INTRODUCTION

Research on electromagnetic bandgap structure has been initiated two decades ago after the initial breakthrough of Yablonovitch [1] based on the groundbreaking theoretical work of Loudon [2]. Novel theoretical findings are obtained compared to its optoelectronic counterparts [3-4] which not only establishes its superiority, but also initiates other directions in research for complete photonic circuits [5]. The basic advantage it exhibits is the property of restriction of electromagnetic wave at a certain frequency ranges [6] while allowing others. Thus it can be considered as a novel photonic filter [7-8] which is further explored for designing transmitter [9], receiver [10], sensor [11], photonic crystal fiber [12], waveguide [13], quantum information processing [14]. Effect of defects plays an important role of determining these device's performances [15].

Research on metamaterial is closely related with study of EBG structure due to performance improvement in antennas [16-17]. However, role of these artificial materials are hardly investigated for investigating photonic properties of those structures. Works are published with photonic bandgap

engineering [18-19] where materials with positive refractive indices are considered. Role of refractive indices of constituent materials are recently investigated [20], and single negative materials exhibit zero frequency gap [21].

In the present paper, lowest two photonic bandgaps are investigated for metamaterial based photonic crystal structure from dispersion relation, under TE mode of propagation, for a range of incidence angle varies upto 90°. It may be noted down in this context that the different bandgaps (both complete and quasi) occur in the propagation of electromagnetic wave inside the grating structure due to the angle caused by displacement vector with the direction of propagation. If displacement vector is oriented along the plane of propagation, then results the zero bandgap. If the angle increases but remains as acute angel with very small magnitude, then it causes quasi-bandgap. But if the vectors are exactly perpendicular to each other then only complete bandgap is formed inside first Brillouin zone. For computation purpose, three different metamaterials namely, parallel nanorod, nano-fishnet with elliptical void and nano-fishnet with triangular voids are taken into account; and from normalized frequency profile, bandgaps are estimated. The simulated results become very important for optical device design using metamaterials.

II. MATHEMATICAL FORMULATION

Mathematical formulation for dispersion relation begins with Maxwell's equation [22]

$$\left[\frac{d^2}{dz^2} + k_r^{\ 2} \varepsilon(z) \right] \xi(z) = 0 \tag{1}$$

where the term 'ε' has different values for different constituent materials. The electric field in the periodic structure will follow Bloch theorem

978-1-5386-6723-1/19 $31.00 © 2019 IEEE

$$\xi(z+d) = \exp[-ik_r d]\xi(z) \qquad (2)$$

'd' being the thickness of any layer. Using transfer matrix technique, we can obtain the master equation as [23]

$$\cos(k_r d) = \cos(\phi_1)\cos(\phi_2)$$
$$-\frac{\alpha_1^2 + \alpha_2^2}{2\alpha_1\alpha_2}\sin(\phi_1)\sin(\phi_2) \qquad (3)$$

where φ_1, φ_2, α_1, α_2 are matrix parameters.

For TE mode, we can write

$$TE(f,\theta) = \cos(k_{r1}d_1 f)\cos(k_{r2}d_2 f)$$
$$-\frac{1}{2}\left(\frac{k_{r1}}{k_{r2}} + \frac{k_{r2}}{k_{r1}}\right)\sin(k_{r1}d_1 f)\sin(k_{r2}d_2 f) \qquad (4)$$

III. RESULTS AND DISCUSSIONS

Based on the Eq. (4), we first calculated dispersion relation profile for three metamaterials, and also computed for positive refractive index materials. Parallel nanorod (n=-0.3), nano-fishnet with rectangular void (n=-1), nano-fishnet with elliptical void (n=-4) are the materials considered for simulation, and results are compared with SiO_2 –TiO_2 composition. Fig 1 shows the frequency variation with incidence angles for TE mode of propagation.

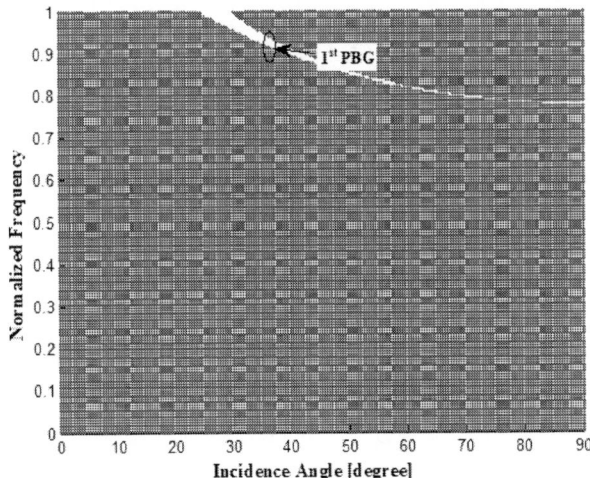

Fig 1a: Normalized frequency variation for parallel nanorod structure under TE mode

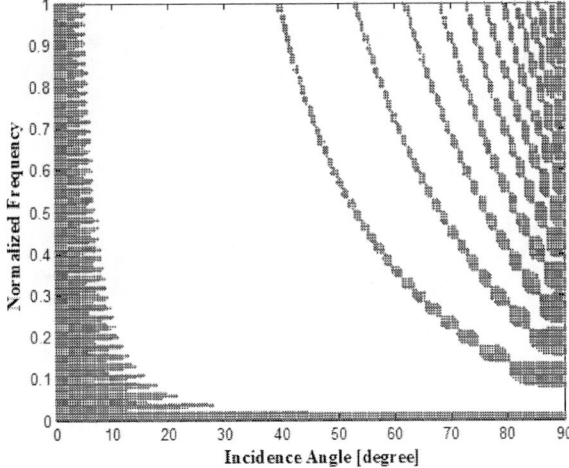

Fig 1b: Normalized frequency variation for nano-fishnet with rectangular void under TE mode

It may be observed from the plot that only one photonic bandgap is observed for parallel nanorod **(Fig 1a)**, whereas multiple bandgaps are observed for other structures. Large stopband is shown for nano-fishnet with rectangular void (Fig 1b). Fig 1c shows the transition between complete and quasi bandgaps. This is due to the fact that with propagation, the direction of displacement vector starts changing w.r.t propagation vector. At a particular value of incident angle, bandgap becomes zero. This is explained in the previous section, as at that point, displacement vector is tonally in-plane with the propagation vector.

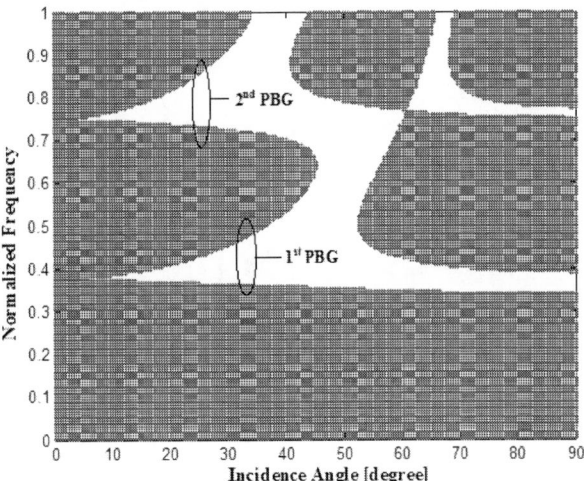

Fig 1c: Normalized frequency variation for nano-fishnet with elliptical void under TE mode

2019 Devices for Integrated Circuit (DevIC), 23-24 March, 2019, Kalyani, India

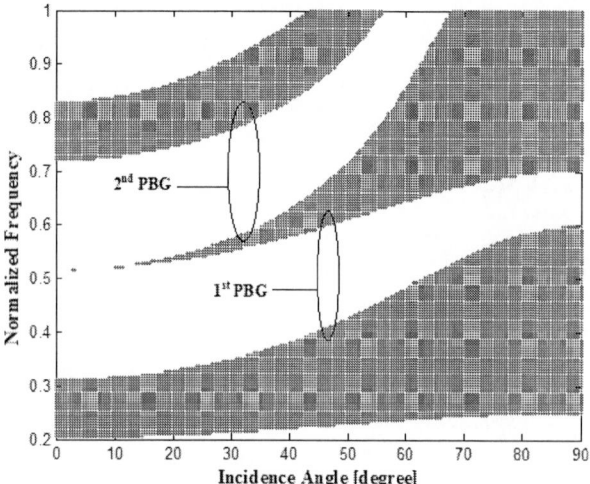

Fig 1d: Normalized frequency variation for SiO_2-TiO_2 structure under TE mode

After computing the normalized frequency response for TE mode, we have analyzed the variation of first photonic bandgap with incidence angle, and result is compared with that obtained from positive index materials [23]. Result is graphically represented in Fig 2. It is seen from the plot that for parallel nanorod, photonic bandgap width is very small, but remains almost constant over a wide range of incidence angle. Random fluctuations are observed for nano-fishnet structures. Careful observation reveals that for elliptical void, photonic bandgap is found at very low incidence angle, whereas for higher angle is required when rectangular void is considered. This is because, for the later, a large stopband is generated at very low frequency, and hence is excluded for bandwidth calculation. This feature is evident from Fig 1b.

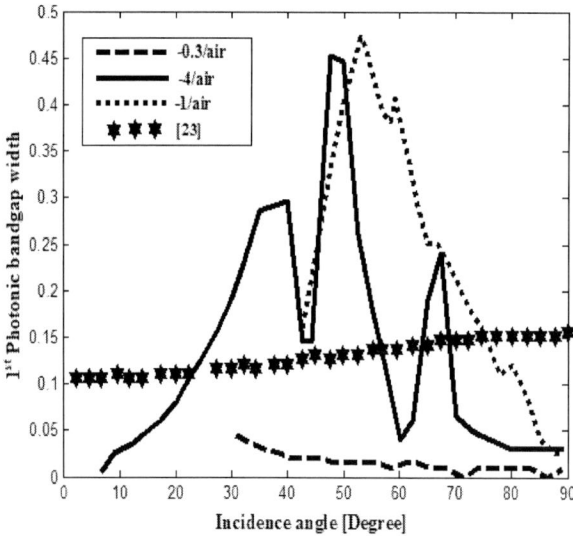

Fig 2: Variation of first photonic bandgap width with incident angle for different metamaterials

Second bandgap is not observable for all the structures. It is evident from Fig 1c that some parts of two bandgaps are joined, and hence at these portions, existence of 2nd PBG is not considered. Based on this assumption, result is again computed, and compared with existing results. Results are shown in Fig 3. Comparative studies are again carried out with the data available in published literature.

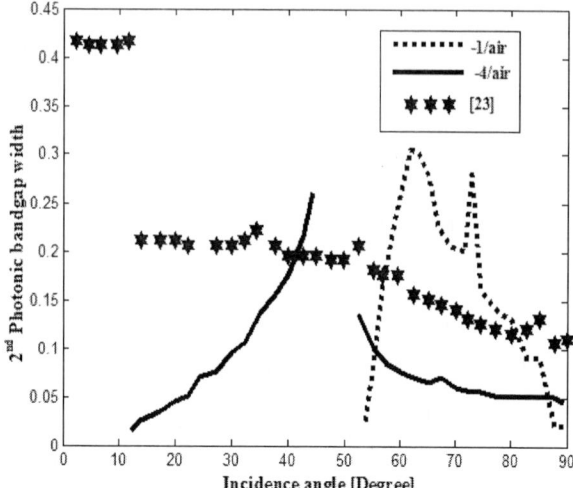

Fig 3: Variation of 2nd photonic bandgap width with incident angle for different metamaterials

The variaitons are represented in tabualr form in table-I and table-II where the comparative studies are also provided. Table-I shows the data for lowermost (first) photonic bandgap, whereas table-II gives the dataset for the next order.

TABLE-I: LOWERMOST PHOTONIC BANDGAP FOR DIFFERENT MATERIALS WITH INCIDENT ANGLE

-0.3/air		-4/air		-1/air		[23]	
Inc . anlge	PBG	Inc . anlge	PBG	Inc . anlge	PBG	Inc . anlge	PBG
30.79	0.0452	6.837	0.005	42.54	0.1608	2.317	0.106
35.76	0.0301	12.26	0.0352	47.51	0.3317	6.837	0.106
38.48	0.0251	14.07	0.0452	49.77	0.402	9.549	0.1111
43.45	0.0201	19.94	0.0804	53.84	0.4623	14.07	0.106
48.42	0.015	24.46	0.1256	57.01	0.392	19.94	0.1111
51.13	0.0151	27.18	0.1558	58.36	0.3819	22.2	0.1111
56.1	0.0151	32.15	0.2311	61.98	0.3367	29.89	0.1161
61.07	0.0151	37.57	0.2915	65.14	0.2513	34.41	0.1161
66.05	0.01	42.54	0.1458	69.21	0.2262	39.83	0.1211
68.76	0.01	44.35	0.1458	71.57	0.191	42.54	0.1261
73.73	0.0101	49.77	0.4472	74.63	0.1608	47.51	0.1261
75.99	0.0101	52.49	0.2613	77.8	0.1105	49.77	0.1312
81.41	0.01	57.46	0.1156	82.77	0.0854	55.2	0.1362
86.38	0	62.43	0.0603	87.74	0.0251	59.72	0.1362
89.1	0.01	65.14	0.1909	89.55	0.0251];	62.43	0.1412

978-1-5386-6723-1/19 $31.00 © 2019 IEEE   254

TABLE-II: LNEXT HIGHER ORDER (2ND) BANDGAP FOR DIFFERENT MATERIALS WITH INCIDENT ANGLE

-1/air		-4/air		[23]	
Inc. anlge	PBG	Inc. anlge	PBG	Inc. anlge	PBG
53.84	0.0251	12.26	0.015	2.317	0.4171
55.2	0.0854	14.07	0.0251	9.549	0.412
57.01	0.1658	17.23	0.0352	17.23	0.2111
58.36	0.2111	22.2	0.0503	27.18	0.206
61.98	0.3065	24.46	0.0704	37.57	0.2061
63.33	0.3015	27.18	0.0754	39.83	0.196
69.21	0.2061	29.89	0.0955	47.51	0.1909
71.57	0.2011	32.15	0.1055	52.49	0.2061
74.63	0.1608	34.86	0.1357	59.72	0.1758
77.8	0.1377	37.57	0.1558	70.11	0.1407
85.03	0.0904	42.54	0.2161	72.37	0.1306
87.74	0.0201	44.35	0.2613	85.03	0.1306

IV. CONCLUSION

Two lower photonic bandgaps are computed for different metamaterial structures under TE mode of propagation with varying incidence angle, and results show that higher bandgap width can be obtained compared to that provided by conventional positive index materials. Results are critically suitable for optical filter design.

References

[1] E. Yablonovitch, "Inhibited Spontaneous Emission in Solid-State Physics and Electronics", Physical Review Letters, vol. 58, pp. 2059-2061, 1987

[2] R. Loudon, "The Propagation of Electromagnetic Energy through an Absorbing Dielectric", Journal of Physics A, vol. 3, pp. 233-245, 1970

[3] L. H. Wang, S. H. Yang, "Nano Photoelectric Material Structures - Photonic Crystals", International Symposium on Next-Generation Electronics, 2013

[4] R. Sakakibara, V. Stelmakh, W. R. Chan, M. Ghebrebrhan, J. D. Joannopoulos, M. Soljacic, I. Celanovic, "Improved Omnidirectional 2D Photonic Crystal Selective Emitter for Thermophotovoltaics", IOP Conference Series: Journal of Physics: vol. 1052, p. 012056, 2018

[5] J. Chovan, F. Uherek, "Photonic Integrated Circuits for Communication Systems", Radioengineering, vol. 27(2), pp. 357-363, 2018

[6] D. Mao, Z. Ouyang, J. C. Wang, "A Photonic-Crystal Polarizer Integrated with the Functions of Narrow Bandpass and Narrow Transmission Angle Filtering", Applied Physics B, vol. 90, pp. 127-131, 2008

[7] S. Saravi, T. Pertsch, F. Setzpfandt, "Photonic Crystal Waveguides as Integrated Sources of Counter propagating Factorizable Photon Pairs",

Conference on Lasers and Electro-Optics, OSA Technical Digest, FTh1G.2, 2018

[8] G. Moloudian, R. Sabbaghi-Nadooshan, M. Hassangholizadeh-Kashtiban, "Design of All-Optical Tunable Filter based on Two-Dimensional Photonic Crystals for WDM (wave division multiplexing) Applications", Journal of the Chinese Institute of Engineers, vol. 39(8), pp. 971-976, 2016

[9] Y. Xu, J. Lin, R. Dubé-Demers, S. LaRochelle, L. Rusch, W. Shi, "A Single-laser Flexible-grid WDM Silicon Photonic Transmitter using Microring Modulators", Optical Fiber Communication Conference, OSA Technical Digest, paper W1I.3, 2018

[10] K. Nozaki, S. Matsuo, T. Fujii, K. Takeda, E. Kuramochi, A. Shinya, M. Notomi, "Forward-biased Photonic Crystal Photodetector towards Amplifier-Free Bias-Free Receiver", Conference on Lasers and Electro-Optics, OSA Technical Digest, paper STh4N.1, 2017

[11] T. W. Lu , C. C. Wu, P. T. Lee, "1D Photonic Crystal Strain Sensors", ACS Photonics, vol. 5(7), pp. 2767-2772, 2018

[12] S. Xing, S. Kharitonov, J. Hu, C. Brès, "Fiber Fuse in GeAsSe Photonic Crystal Fiber and its Impact on Undamaged Segment", Conference on Lasers and Electro-Optics, OSA Technical Digest, paper JTh2A.93, 2018

[13] B. Lang, R. Oulton, D. M Beggs, "Optimized Photonic Crystal Waveguide for Chiral Light-Matter Interactions", Journal of Optics, vol. 19, p. 045001, 2017

[14] J. C. Norman, D. Jung, Y. Wan, J. E. Bowers, "Perspective: The Future of Quantum Dot Photonic Integrated Circuits", APL Photonics, vol. 3, p. 030901, 2018

[15] 15. D. H. Ge, J. H. Zhang, L. Q. Zhang, L. Lu, X. K. Huang, "Effect of Point and Linear Defects on Bandgap Properties in Triangular-Honeycomb Structure Photonic Crystals", IOP Conference Series: Material Science and Engineering, vol. 170, p. 012005, 2017

[16] R. Inum, M. M. Rana, K. N. Shushama, M. A. Quader, "EBG Based Microstrip Patch Antenna for Brain Tumor Detection via Scattering Parameters in Microwave Imaging System", International Journal of Biomedical Imaging, vol. 2018, Article ID 8241438, 2018

[17] T. A. Elwi, "A Slotted Lotus Shaped Microstrip Antenna based an EBG Structure", Journal of Material Sciences & Engineering, vol. 7(2), p. 439, 2018

[18] Y. Karla, R. K. Sinha, "Photonic Bandgap Engineering in 2D Photonic Crystals", PRAMANA - Journal of Physics, vol. 67(6), pp. 1155-1164, 2006

[19] Y. Karla, R. K. Sinha, "Modeling and Design of Complete Photonic Bandgaps in Two-Dimensional Photonic Crystals, PRAMANA - Journal of Physics, vol. 70(1), pp. 153-161, 2008

[20] R. Moukhtari, A. Hocini, D. Khedrouche, "Study of Two-Dimensional Photonic Crystal Microcavities as a Function of Refractive Index, ACTA Physica Polonica A, vol. 129(4), pp. 556-558, 2016

[21] D. W. Yeh, C. J. Wu, "Analysis of Photonic Band Structure in a One-Dimensional Photonic Crystal containing Single Negative Materials, Optics Express, vol. 17(19), pp. 16666-16680, 2009

[22] H. Hitoshi, M. Atsushi, "Dispersion Relation of Electromagnetic Waves in One-Dimensional Plasma Photonic Crystals", Journal of Plasma Fusion Research, vol. 80(2), pp. 89-90, 2004

[23] O. Barkat, "Theoretical Investigation of Transmission and Dispersion Properties of One Dimensional Photonic Crystal", Journal of Electrical and Electronic Engineering, vol. 3(2), pp. 12-18, 2015

An Analytical Approach of EEG Analysis for Emotion Recognition

Indronil Mazumder
*Department of Electronics &
Communication Engineering.*
Brainware University
Barasat, India
nil.indra12@gmail.com

Abstract— **Emotion is the fundamental behavioral attributes of humans. To identify emotional variations from Electroencephalogram signals have currently expanded consideration amid BCI researchers. In this work, emotion recognition from EEG is performed using 21channel EEG acquisition device employing 10-20 method of electrode placement. The experiment being performed on issues of the peer group of 20-25 years of 16 university students (eight females and eight males). Audio-visual stimuli are used for bringing four dissimilar emotions (Happy, Sad, Fear and Relaxed) and corresponding signals are processed for emotion classification. At first EEG signals are filtered using Butterworth 4th order filter which is band limited by 0.5-60 Hz after that smoothened with the help of Surface Laplacian filter. Filtered EEG signals are feature extracted using Power Spectral Density, Wavelet Decomposition, Hjorth Parameter and AR parameter. After that Linear SVM classifier is used. Support Vector Machine classifier generates the best result when used with Wavelet coefficient feature extraction technique (96.81%). The experimental result also shows the diminutive interval EEG can be used for sensing the emotional thought variations effectively. We found that the EEG signals contained adequate information to separate four different emotion classes.**

Keywords — EEG; PSD; Wavelet decomposition; Hjorth parameter; AR parameter; SVM.

I. INTRODUCTION

Human Emotion recognition by using BCI is getting increasingly prevalent day by day. EEG replicates the inner emotional fluctuations related to further orthodox techniques (face recognition, gesticulations, language, etc.). Since, Electroencephalogram signals are unswervingly attached to the scalp and interpretation the beginning variations in mind bustle gives more consistent evidence subject to emotional state deviations. Besides, the emotive taxation thought are highly used in humanoid automaton communications for evolving supporting arrangements for aged persons. Even though small numerals are likened with the exertions being made in the direction of intent-conversion resources, few investigators want to recognize human appliance interaction by emotion thoughtful ability. Usual gears subjected to the analysis emotive state are centered upon the footage and numerical enquiry based on the biological signals from the both CNS ANS. Numerous tactics are testified by various investigators for searching the interrelation among the emotional variation and EEG signals. The emotion taxation by EEG, the interval period subject to given emotive spurs, important part is taken channel number, band of frequencies and the characteristics of numerical feature abstraction procedures and these extracted features shows the most

substantial part. Our research engrossed on identifying emotions from human brain behavior, determined by EEG signals. We have planned a system to examine EEG signals and classify them into FOUR emotional classes HAPPY, SAD, FEAR, RELAXED. To perform this this task, a database is formed with EEG signals. This is done by evaluating Electroencephalogram signals from people that are emotionally stimulated by Audio-Visual clips. This scheme permitted us to clarify our system the correlation among the characteristics of the brain signals and the emotion.

II. EXPERIMENTAL MATERIALS AND METHODOLOGY

A. Data Acquisition and Preprocessing

A virtuous dataset is extremely crucial for designing a smart emotion acknowledgment arrangement. No general dataset is available for this task. For this reason a custom-made data acquirement procedure is considered using video clips to encourage four distinct emotions. Different resources like internet, international standard dataset etc. are accessed for the collection of video clips. An experimental panel study is being directed over 16 university students prior to viewing the clips to the experimental subjects to select the best leading emotive spur to induce exclusive emotion on the issues throughout the data gathering. Eight women and eight men in the peer group of 20-25 years were engaged as subjects of the experimentation. Once agreement forms were filled-up, a normal outline subject to the research exertion and phases of test was given. Collection of EEG is done through using 21channels 10-20 System and filtered with Butterworth 4th order filter band limited by 0.5- 60 Hz and smoothened by using surface Laplacian filter.

B. Featuré Extraction

Feature Extraction Plays vital role in signal investigation. EEG entails of numerous data arguments which is compacted into few constraints for representation, denoted as features. These constraints are characterized the behavior of EEG. So to define and investigative resolve, features are used to indicate the EEG data. For Feature Extraction purpose an algorithm of four basic tools such as Wavelet Decomposition method, Power Spectral Density (PSD), Autoregressive Parameter (AR Parameter) and Hjorth Parameter is used.

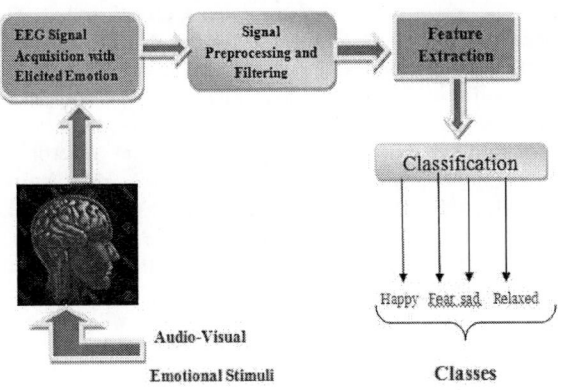

Fig 1 The Overall System

B.1 Wavelet Decomposition Method

It is tough to attract some assumption around the materials of the original data as the convenient info is concealed in the original data. Hence, the extraction of appropriate feature from original EEG data is very important. DSP transformations are useful to the data to get secreted info around data which is willingly unavailable in the original signal. In case of, STFT adopts signal is immobile for small period of time, by noticing the signal over little frame size. However conferring to Heisenberg uncertainty principle, STFT suffers from time frequency resolution problem. Although, WT overwhelms time-frequency resolution problem by MRA. Which investigates the signal at dissimilar frequencies by diverse resolution; it is feasible since maximum normal signals having high frequency elements for small period of time and lesser frequency parts for extended interval. Since EEG is a dynamic signal, it is possible to separate the signal into lower and higher frequency band. Generally high frequency quantities give the substantial info about the materials of the signal. Daubechies wavelet (Db2) which is used for four level segregation. A signal is basically an oscillating function of time or space and it is periodic in nature and wavelets are confined in the signals. They have energy concerted in time or space and useful to study of transient behavior of wave. The wavelets of limited energy are used in WT method. For the wavelet study, the signal to be analyzed is multiplied with a wavelet function and then the transform is computed for each segment is created. In WT, the width of the wavelet function varies with each spectral component. At high frequencies, it provides decent time resolution and reduced frequency resolution, while at low frequencies; the WT gives decent frequency resolution and reduced time resolution.

B.II Power Spectral Density (PSD)

PSD is one of the highest significant features extraction techniques. To start with, we consider x(t) is the voltage across the unit resistance. It refers to $x_2(t)$ is instantaneous power in x(t). The anticipated instantaneous power will be

$$E[x2(t)] = Rxx(0) = 1/2\pi d\omega \quad ....(1)$$

where, $Sxx(j\omega)$ is the CTFT of the autocorrelation function $Rxx(\tau)$. when x(t) is ergodic in correlation, so that time mean and ensemble mean are equal in correlation accumulations, then one also can write the time-mean power in any ensemble value. Since $Rxx(\tau) = Rxx(-\tau)$ and we know $Sxx(j\omega)$ is real and even in ω; a easier notation such as $Pxx(\omega)$ is may be useful, still it should be confined to $Sxx(j\omega)$ to keep away a rapid increase of notational conventions, it is obvious that actually the quantity is the F.T of $Rxx(\tau)$. The integral of overhead equation proposes that we can contemplate, the anticipated power in a frequency band of width $d\omega$ given by $(1/2\pi)Sxx(j\omega)d\omega$. To test this perception further, assume the extraction a band of frequency components of x(t) by flowing x(t) via an perfect band pass filter. Since the way y(t) is obtain from x(t), the presume power in the output y(t) can be understood as the anticipated power that x(t) has in the selected pass band. Considering this, we obtain

$$S_{yy}(j\omega) = |H(j\omega)|2S_{xx}(j\omega)..............(2)$$

We see that this presume power can be computed as

$$E\{y_2(t)\} = Ryy(0) = \frac{1}{2\Pi}\int_{-\infty}^{\infty}Syy(j\omega)d\omega = \frac{1}{2\Pi}\int_{PassBand}^{\infty}Syy(j\omega)d\omega...(3)$$

B.III Hjorth Parameter

Hjorth parameter, in the present context has been used to figure the quadratic mean and the leading frequency of EEG signals on each side of brain. Hjorth parameters were initially established for numerous EEG analyses. Let us assume the spectral moment of order zero and order two.

$$m_2 = \int_{-\Pi}^{\Pi}\omega^2 S(\omega)d\omega = \frac{1}{T}\int_{t-T}^{t}\left(\frac{df}{dt}\right)^2 dt........(5)$$

Here $s(\omega)$ is the power density spectrum and f(t) the EEG signal.

Initial two Hjorth parameters are

$$Activity: h_0 = m_0$$

$$Mobility: h_1 = \sqrt{\frac{m_2}{m_0}}$$

h_0 denotes the square of the quadratic mean and h_1 imitates the prevailing frequency. The distinct forms of these quantities, $h_0(k)$ and $h_1(k)$ at sampled time k. It is noted that Hjorth has also used the fourth-order spectral moment m_4 to define a measure of the bandwidth of the signal, called complexity.

$$Complexity: h_2 = \sqrt{\frac{m_4}{m_2} - \frac{m_2}{m_0}}$$

B.IV Autoregressive Parameter

Autoregressive Parameter of spectral approximation is a technique for modeling signals. AR models the Encephalogram signal as the output random signal of a linear time invariant filter, where the input is white noise with an average of zero and a fixed variance of $\sigma 2$. The goal of the AR method is to get the filter coefficients, since it is presumed that unlike thinking actions will generate different filter coefficients. The filter coefficients will be utilized as the features of the signal. Autoregressive Parameter accepts that the transfer function of the filter will only have poles in the denominator. The number of poles in the denominator relates to the order of the autoregressive model. The guess of an all-pole filter forms the filter coefficients calculation easier since it is only essential to solve linear equation. Autoregressive parameter is one of the best frequently used techniques to extract frequency related EEG features. The pth order AR model that defines an EEG signal $y_k(t)$ at channel (electrode) k as:

$$Y_k(t) = a_1, k y(t-1) + a_2, k y(t-2) + \ldots + a_p, k y(t-p) + E(t) \ldots \ldots (6)$$

Here, a_i, k signifies the ith order AR parameter modeling the Encephalogram signal at channel k and E(t) is white noise with zero mean and determinate variance. There is a straight communication among the AR parameters and the autocorrelation function and this communications can be inverted to define the parameters from the autocorrelation function using the Yule-Walker equations. It is projected AR parameters for each EEG channel that is utilized for this training using least-squares approximation. We have considered parameters for each single trial of overlapping windows. Four EEG channels were used namely f_3; f_4; f_z; P_z. These channels related to electrodes which are possible to deliver instructive depths about brain activities.

C. Classification of Emotions

The goal of the classification phase in a BCI method is to recognize user's intentions on the foundation of a feature vector that characterizes the brain action provided by the feature step. Any classification algorithm can be used to achieve this goal. Classification algorithms use the features extracted as self-governing variables to describe boundaries among the dissimilar targets in feature space. The widely used and most popular classifiers are Linear Discriminant analysis (LDA), Quadratic Discriminant Analysis (QDA), KNN, PNN, SVM etc. In this work we use SVM only to compare the usefulness of dissimilar features for emotion recognition.

C.I Support Vector Machine (SVM)

Most of the accessible supervised classification methods are based on traditional information, which can offer supreme results when sample size is tending to infinity. However, only finite samples can be acquired in practice. SVM a powerful method developed from statistical learning and has made considerable achievement in some field. In machine learning, support vector machines are supervised learning models with linked learning algorithms that examine data and identify designs, used for classification and regression study. Classifying samples are the main fragments in machine learning and SVM is the best classifier because it utilizes a special method called 'Kernel trick' to transform the data and based on these transformations SVM catches an optimal boundary among the possible outputs. The thought of SVM is to generate a hyperplane in amid data sets to specify which class it goes to. The task is to train the machine to recognize structure from sample and mapping with the accurate class label, for greatest output, the hyperplane has major distance to the adjacent training data points of any class. We can classify linearly distinguishable and non-linearly distinguishable sample by SVM. It maps input vector to an upper dimensional space where a maximal separating hyperplane is created. Two parallel hyperplanes are made on each side of the hyperplane that splits up the data. The unraveling hyperplane is the hyperplane that make the most of the distance between the two parallel hyperplanes. A supposition is made that the larger the margin or distance between these parallel hyperplanes the better the generalization error of the classifier will be. Let us assume $\{(x_1,y_1),(x_2,y_2),(x_3,y_3),\ldots,(x_n, y_n)\}$.where y_n=1/-1, a constant signifying the class to which x_n fits. Where, n = sample no. Every x_n remains p-dimensional real vector. Scaling is vital to sentinel in contrast to variable by means of superior alteration. To see the Training data, through dividing hyperplane, this takes

$$w. x + b = 0 \ldots \ldots \ldots \ldots (7)$$

Taken into consideration that 'b' is scalar and 'w' denotes p-dimensional Vector. Vector 'w' represents the normal to splitting hyperplane. The margin can be increased by the offset parameter 'b'. If 'b' is absent the hyperplane is compiled to pass via origin and the solution will be restricted. So, with the interest to maximize the margin, we are interested in SVM and the parallel hyperplanes. Parallel hyperplanes can be explained by

$$w.x + b = 1 \ldots \ldots \ldots .. (8)$$
$$w.x + b = -1 \ldots \ldots \ldots .. (9)$$

If the training data are linearly distinguishable, hyperplanes can be so selected that no points should lie among them and then distance is maximized. According to geometry, the separation among the hyperplanes is $2 / \mid w \mid$. So $\mid w \mid$ is minimized.

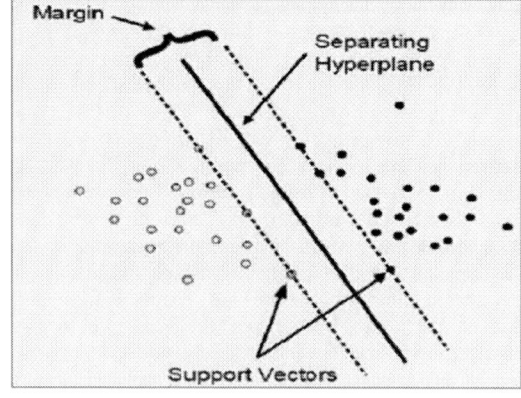

Fig 2 Linear Separable data

If samples are detached by a nonlinear area shown in below fig: 3

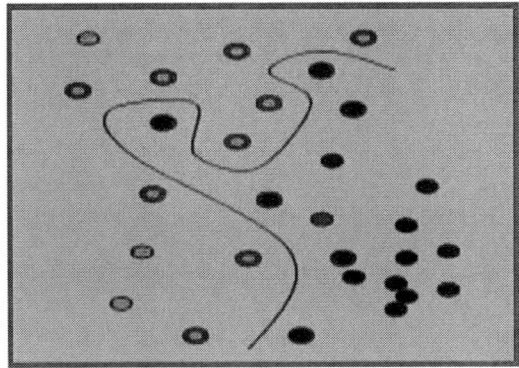

Fig 3 Nonlinear Separable data

In this case SVM systems with dissimilar kernels are there to compare for four unlike human emotions. As SVM is a binary classifier i.e. it can classify only between two groups of classes we used one compared to one tactics is followed in this paper to classify four classes. This is shown in the below fig: 4 the kernel function can convert the samples into an upper dimensional Space to make it feasible to accomplish the segregation.

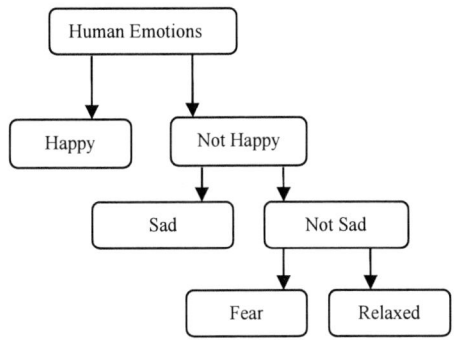

Fig 4 The SVM Tree

III. Experimental Results

Table 1						
CLASSIFICATION ACCURACY (%) OF SVM CLASSIFIER WITH DIFFERENT FEATURE EXTRACTION METHOD						
Feature Extraction Method	Dimension	Different types of Emotions				Average (%)
		Happy	Sad	Fear	Relaxed	
PSD	69	92.25	91.33	92.11	95.33	92.75
Hjorth Parameter	03	88.75	86.33	82.53	90.50	87.02
AR Parameter	10	86.25	83.84	84.50	85.65	85.06
Wavelet Coefficient	128	99.50	94.45	95.20	98.10	96.81

In the said research, Encephalograms are composed by 21 channels 10-20 System in the age cluster of 20-25 years for determination of distinct emotions. Video clips are utilized on behalf of convincing four diverse human emotions (Happy, Sad Fear, and Relaxed) are preprocessed and sampled for emotion classification. So, there are a overall number of 460 EEG epochs of four distinct emotions. Besides, every epoch is framed to hook on 5 second duration to raise the feature vector size for effective emotion acknowledgment and overlying of mounts with 50% duration. Amount of samples in every epoch is subjected to the time duration of audiovisual spurs. A precise attention is maintained for determining the greatest emotive prevailing stimulus by pilot panel study and thus authorizes the uniqueness of this EEG dataset.

IV. CONCLUSION

We used four feature extraction techniques in this work to reduce the dimension and extract relevant feature vectors from the data set. Then only SVM classifier is used to classify the feature vectors. The Classification accuracy for the emotions HAPPY, SAD, FEAR and RELAXED with separate feature extraction methods and the average classification accuracy is shown in above result table. According to the table we can conclude that Support Vector Machine (SVM) Classifier produces the best result when it is fed to Wavelet coefficient feature extraction method (96.81%). PSD is the second best (92.75) followed by Hjorth Parameter (87.02) and AR Parameter (85.06).

REFERENCES

[1] Patil, Anita, Ashish Panat, and Supriya Ambadas Ragade. "Classification of human emotions from electroencephalogram using support vector machine." In Information Processing (ICIP), 2015 International Conference on, pp. 404-408. IEEE, 2015.

[2] Murugappan, M. and Murugappan, S., 2013, March. Human emotion recognition through short time Electroencephalogram (EEG) signals using Fast Fourier Transform (FFT). In Signal Processing and its Applications (CSPA), 2013 IEEE 9th International Colloquium on (pp. 289-294). IEEE.

[3] Graimann, B., Allison, B. and Pfurtscheller, G., 2009. Brain–computer interfaces: A gentle introduction. In Brain-Computer Interfaces (pp. 1-27). Springer, Berlin, Heidelberg.

[4] Vaid, S., Singh, P. and Kaur, C., 2015. Classification of human emotions using multiwavelet transform based features and random forest technique. Indian Journal of Science and Technology, 8(28).

[5] Vidaurre, C., Krämer, N., Blankertz, B. and Schlögl, A., 2009. Time domain parameters as a feature for EEG-based brain–computer interfaces. Neural Networks, 22(9), pp.1313-1319.

[6] Levine, S.P., Huggins, J.E., BeMent, S.L., Kushwaha, R.K., Schuh, L.A., Rohde, M.M., Passaro, E.A., Ross, D.A., Elisevich, K.V. and Smith, B.J., 2000. A direct brain interface based on event-related potentials. IEEE Transactions on Rehabilitation Engineering, 8(2), pp.180-185.

[7] Farwell, L.A. and Donchin, E., 1988. Talking off the top of your head: toward a mental prosthesis utilizing event-related brain potentials. Electroencephalography and clinical Neurophysiology, 70(6), pp.510-523.

[8] Sitaram, R., Caria, A., Veit, R., Gaber, T., Rota, G., Kuebler, A. and Birbaumer, N., 2007. FMRI brain-computer interface: a tool for neuroscientific research and treatment. Computational intelligence and neuroscience, 2007, p.1.

[9] Garrett, D., Peterson, D.A., Anderson, C.W. and Thaut, M.H., 2003. Comparison of linear, nonlinear, and feature selection methods for EEG signal classification. IEEE Transactions on neural systems and rehabilitation engineering, 11(2), pp.141-144.

[10] Vidaurre, C., Krämer, N., Blankertz, B. and Schlögl, A., 2009. Time domain parameters as a feature for EEG-based brain–computer interfaces. Neural Networks, 22(9), pp.1313-1319.

[11] Cincotti, F., Mattia, D., Aloise, F., Bufalari, S., Schalk, G., Oriolo, G., Cherubini, A., Marciani, M.G. and Babiloni, F., 2008. Non-invasive brain–computer interface system: towards its application as assistive technology. Brain research bulletin, 75(6), pp.796-803.

[12] Neuper, C. and Pfurtscheller, G., 1999. Motor imagery and ERD, in Event-Related Desynchronization. Handbook of Electroencephalography and Clinical Neurophysiology. Elsevier, Amsterdam303–325.

2019 Devices for Integrated Circuit (DevIC), 23-24 March, 2019, Kalyani, India

A UWB Band-pass Filter with a WLAN Notch based on Multi-mode Resonator Structure for Application in Wireless Communication

Anirban Neogi[1,2], Jyoti Ranjan Panda[1], Saptarshi Sil[2], Shibaditya Chakraborty[2], Anupam Tarafdar[2]

[1]School of Electronics Engineering, Kalinga Institute of Industrial Technology (Deemed to be University)
Bhubaneswar-751024, Odisha, India
[2]Supreme Knowledge Foundation Group of Institution, Hooghly-721139, West Bengal, India
E-Mail:- anirban07.neogi@gmail.com, jyotiranjan.pandafet@kiit.ac.in

Abstract— **A simple band-pass filter is designed in Ultra-Wide band range which can create a sharp notch at 5.5 GHz, which is the major region of communication in wireless systems. The resonator structure is used for the UWB band-pass filter is multi-mode resonator (MMR) whose bandwidth extends from 3 GHz to 10 GHz. In the MMR structure a spur line is introduced to form a sharp narrow notched band at 5.5 GHz. The return loss within the pass-band of UWB, is higher than 10 dB.**

Index Terms— *Band-pass filter, UWB, spur-line, multi-mode resonator (MMR)*

I. INTRODUCTION

THE range 3.1 GHz to 10.6 GHz [2]is known as ultra-wideband (UWB) range. This band has the potential in the development of various modern communication systems. The applications of UWB filters have got huge attention in both the industry and academia. In the earlier days UWB filters generating transmission zeros of two numbers that is below and the above range of the pass-band. The insertion loss and overall size is increased, this is the main disadvantage in these filters.

In this paper, a band-pass filter with notch at 5.5 GHz in the UWB frequency range is proposed by embedding a spur-line within MMR structure. It is explained that with the two parallel coupled lines, constituted MMR's first two resonant mode can be used on both the sides as the input and output port [3-6]. After that a spur-line structure is created at the center of the MMR structure to generate a notch at 5.5 GHz band in the UWB spectrum to eliminate the potential interference in the pass-band region.

II. ULTRA WIDEBAND (UWB) NOTCH FILTER DESIGN AND RESULTS

A. UWB Band pass Filter Design and Results

The design of UWB band-pass filter (filter 1) is made and simulated with FR4_epoxy with height 1.6 mm and having the dielectric constant ε_r=4.4 with the help of HFSS. The central structure is known as the multi-mode resonator (MMR) as visible in the Fig 1. The parameters for the optimized designs

978-1-5386-6722-4/19/$31.00 ©2019 IEEE

of the filter are W_1= 3 mm, L_1=2 mm, W_2=0.1 mm, L_2=7 mm, W_3=0.1 mm, L_3=1.05 mm, W_4=0.9 mm, L_4=0.35 mm, L_5=6.3 mm and L6=8.34 mm. The whole structure is in resemblance with the stepped impedance resonator (SIR). The substrate width is 27.04 mm. The SIR is designed for the effective enlargement the stop-band in the upper frequency range above the pass-band of dominance of the band-pass filter. In the design of the filter the first two resonances are used to obtain the pass-band of the wide dominance. The resonant frequencies of the first and second order provide the demarcation of the cut-off frequencies of the lower and upper region of the wide pass band. An ultra-wide pass band results with good insertion loss $|S_{21}|$ and transmission coefficient $|S_{11}|$. Simulated transmission coefficient $|S_{11}|$ in dB and insertion loss $|S_{21}|$ in dB of the filter is shown in the Fig 2. The extension of the pass-band is from 2.9 GHz to 10.1 GHz, in which the entire pass-band of the UWB spectrum (3.1 GHz to 10.6 GHz) is almost covered. Within the pass-band, there are three transmission poles at f_1=3.6 GHz, f_2=7.05 GHz, f_3=9.1 GHz. It is clear that the first three transmission poles signify the three resonant modes (f_1, f_2 and f_3) of the stepped impedance multimode resonator [12, 15].

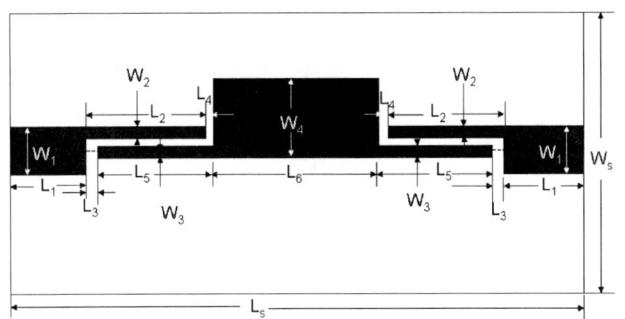

Fig1. Geometry and configuration of UWB band-pass filter (filter 1). L_S=27.04mm, W_S=10mm.

B. Analysis of Multimode Resonator (MMR)

Fig 3 (a) shows the structure of the multi-mode resonator (MMR). It consists of low impedance line section (characteristic impedance Z_1) in the middle and on the both

978-1-5386-6723-1/19 $31.00 © 2019 IEEE   261

sides there are high impedance sections (characteristic impedance Z_2) of the line. $2\theta_1$ and θ_2 are the electrical lengths of the low and high impedance lines respectively. The physical structure of the MMR is same as that of the SIR [14]. The larger frequency gap between the resonant modes of the first and the second order is the main aim of the SIR [14]. Generally, in the UWB band-pass filters based on MMR technology, first three resonant modes are responsible for generating an ultra-wide pass-band with the help of frequency dispersive coupled lines [22].

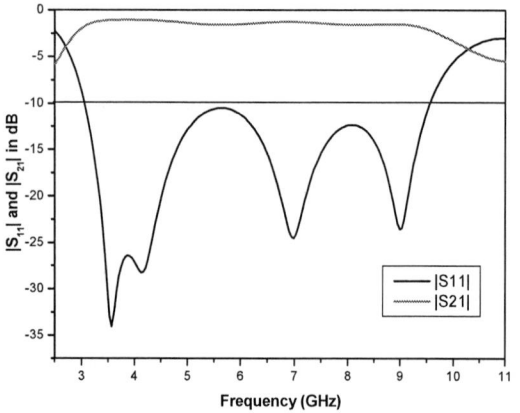

Fig. 2. Simulated insertion loss ($|S_{21}|$) in dB and transmission coefficient ($|S_{11}|$) in dB of UWB band pass filter (filter 1).

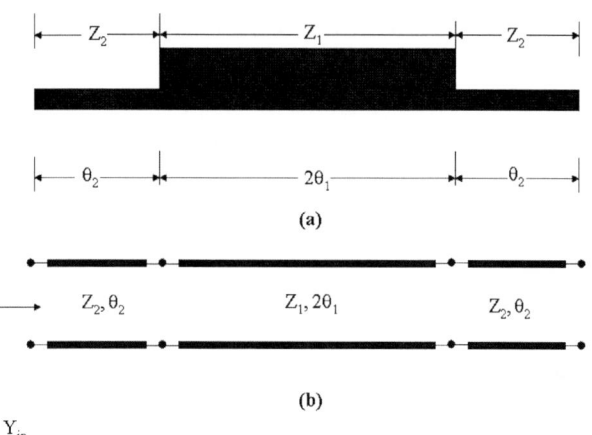

Fig3. (a) Geometry of the open-circuited stepped-impedance multimode resonator (MMR) (b) Equivalent transmission line model.

Fig 3 (b) shows the equivalent transmission line model for the MMR structure, where the two end of the MMR is open circuited. The input admittance Y_{in} at the left end, can be obtained in the following way [22]

$$Y_{in}=jY_2\frac{2(\tan\theta_1+\tan\theta_2)(R-\tan\theta_1\tan\theta_2)}{R(1-\tan^2\theta_1)(1-\tan^2\theta_2)-2(1+R^2)\tan\theta_1\tan\theta_2} \quad (1)$$

Where $R=Z_2/Z_1$ is the impedance ratio of high- and low-impedance line sections. At the resonances,

$$Y_{in}=0 \quad (2)$$

From (2), a set of resonance frequencies (f_x, f_y and f_z) can be evaluated from θ_1 and θ_2. In the case when $\theta_2=2\theta_1=\theta$, then

$$\theta(f_x)=\tan^{-1}\sqrt{\frac{R}{R+2}}$$
$$\theta(f_y)=\tan^{-1}\sqrt{\frac{R+2}{R}} \quad (3)$$
$$\theta(f_z)=\frac{\pi}{2}$$

Therefore

$$\frac{f_y}{f_x}=\frac{\theta(f_y)}{\theta(f_x)}=\frac{\tan^{-1}\sqrt{\frac{R+2}{R}}}{\tan^{-1}\sqrt{\frac{R}{R+2}}}$$
$$\frac{f_z}{f_x}=\frac{\theta(f_z)}{\theta(f_x)}=\frac{\pi}{2\tan^{-1}\sqrt{\frac{R}{R+2}}} \quad (4)$$

D. UWB Band-pass Notch Filter Design and Results

The configuration of an UWB band- pass notch filter (filter 2) whose structure and characteristics are same as the previous one is shown in the Fig 4 (a). A spur-line is embedded in the central low impedance section of the MMR and its geometry and configuration is depicted in Fig. 4 (b). W_1= 3 mm, L_1=2 mm, W_2=0.1 mm, L_2=7 mm, W_3=0.1 mm, L_3=1.05 mm, W_4=0.9 mm, L_4=0.35 mm, L_5=6.3 mm and $L6$=8.34 mm, W_6=L_7=0.1 mm, $W5$=0.65mm and L_8=7.9mm. Simulated $|S_{21}|$ and $|S_{11}|$ in dB are shown in Fig 4. It is visible that there is a deep downward notch is created having the center notch frequency at 5.5 GHz at -22 dB insertion loss. In this way, by embedding a spur-line in the (MMR) based UWB band-pass filter, a notch is created, [23] which eliminates the potential interference of UWB system with the wireless communication system.

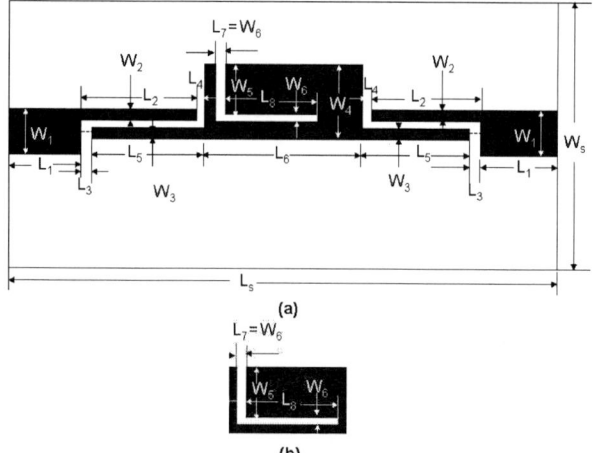

Fig4. (a) Geometry and configuration of UWB band-pass notch filter (filter 2) and (b) Geometry and of spur-line.

Fig. 6 is showing the variation of Group delay in the ultra wide band range. From the graph it is clear that the group delay has the peak up to 0.40 ns. There is a sharp disturbance at 5.5 GHz where the group delay is deeply reduced to -1.2 ns.

2019 Devices for Integrated Circuit (DevIC), 23-24 March, 2019, Kalyani, India

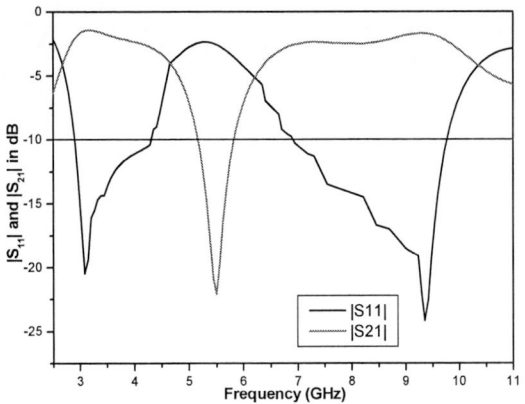

Fig5. Simulated insertion loss ($|S_{21}|$) in dB and transmission coefficient ($|S_{11}|$) in dB of UWB band-pass notch filter (filter 2).

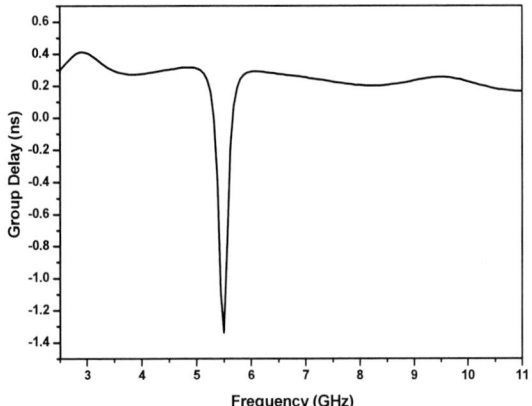

Fig6. Simulated group delay of the UWB band-pass notch filter (filter2)

III. CONCLUSION

In this paper we propose a planar UWB band-pass notched filter based on the multimode resonator (MMR) technique. A spur-line is introduced on the low impedance section of the MMR which is responsible for creating a narrow deep notch in the UWB pass-band at 5.5 GHz with 22 dB insertion loss. Since FR4 epoxy is used as the substrate this will make the design economical. Length of the substrate is limited to 27.04 mm. Size compactness is the important feature of the proposed UWB notch band-pass filter and it provides good frequency response in the entire UWB spectrum. This make the proposed UWB notched filter is attractive for use in the UWB system for purpose of mitigation of potential interference with the narrowband systems such as WLAN.

REFERENCES

[1] First Report and Order, "Revision of Part 15 of the Commission's Rule Regarding Ultra-Wideband Transmission systems FCC 02-48," Federal Communication Commission, 2002.

[2] H. Ishida and K. Araki, "Design and analysis of UWB band-pass filters with ring filter," *in IEEE MTT-S Int. Dig.*, vol. 3, pp. 13071310, Jun. 2004.

[3] C. –L. Hsu, F. –C. Hsu and J. –T, Kuo, "Microstrip band-pass filters for ultra-wideband (UWB) wireless communication, "*in IEEE MTT-S Int. Dig.*, pp. 679-682, Jun. 2005.

[4] W. Menzel, M. S. R Tito and L. Zhu, "Low-loss ultra-wideband (UWB) filters using suspended strip line," *in Proc. 2005 Asia-Pacific Microwave Conf.*, vol. 4, pp. 2148-2151, Dec. 2005.

[5] R. Gomez-Garcia and J.I Alonso, "Systematic method for the exact synthesis of ultra-wideband filtering responses using high pass and low pass sections," *IEEE Trans. Microwave Theory Tech.*, vol. 54, no. 10, pp. 3751-3764, Oct. 2006.

[6] C. –W. Tang and M. –G. Chen, " A microstrip ultra-wideband band pass filter with cascaded broadband band pass and band stop filters," *IEEE Trans. Microwave Theory Tech.*, vol. 55, no. 11, pp. 2412-2418, Nov. 2007.

[7] W. T-. Wong, Y. –S. Lin, C. –H. Wong and C. Chen, "Highly selective microstrip band pass filters for ultra-wideband (UWB) applications," *in Proc. Asia-Pacific Microwave Conf.*, vol. 5, pp. 2850-2853, Dec. 2005.

[8] R. Gomez-Garcia, J. Bonache and F. Martin, "Application of electromagnetic bans gaps to the design of ultra-wideband band-pass filters with out-of-band performance," *IEEE Trans. Microwave Theory Tech.*, vol. 54, no. 12, pp. 4136-4140, Dec. 2006.

[9] J. –S. Hong and K. Li, "Recent development of ultra-wideband (UWB) filters," *In Proc. IEEE Microwave. Antenna Propagation and EMC Technologies Int. Symp.*, pp. 442-445, Aug. 2007.

[10] L. Zhu, H. Bu, and K. Wu, "Aperture compensation technique for innovative design of ultra-broadband microstrip band pass filter," *in IEEE MTT-S Int. Dig.*, vol. 1, 2000, pp. 315-318.

[11] W. Menzel, L. Zhu, K. Wu, and F. Bogelsack, "on the design of novel compact broad-band planar filters," *IEEE Trans. Microw. Theory Tech.*, vol. 51, no. 2, pp. 364-370, Feb. 2003.

[12] L. Zhu, W. Menzel, K. Wu, and F. Boegelsack, "Theoretical characterization and experimental verification of a novel compact broadband microstrip band-pass filter," *in Proc. Asia-Pacific Microwave Conf.*, pp. 625-628, Dec. 2001.

[13] *HFSS* version 13, Ansys, USA.

[14] M. Makimoto and S.Yamashita, "Band pass filters using parallel coupled strip line stepped impedance resonators," *IEEE Trans Microw. Theory Tech.*, vol. 28, no. 12, pp. 1413-1417, Dec. 1980.

[15] L. H. Hsieh and K. Chang, "Compact, low insertion-loss, sharp rejection, and wide-band microstrip band pass filters," *IEEE Trans. Microw. Theory Tech.*, vol. 51, no. 4, pp. 1241-1246, Feb. 2003.

[16] J. T. Kuo and E. Shih, "Microstrip stepped impedance resonator band pass filter with an extended optimal rejection bandwidth," *IEEE Trans. Microwave Theory Tech.*, vol. 51, no. 5, pp. 1554–1559, May 2003.

[17] S. F. Chang, Y. H. Jeng, and J. L. Chen, "Dual-band step-impedance band pass filter for multimode wireless LANs," *Electron. Lett.*, vol. 40, no. 1, pp. 38–39, Jan. 2004.

[18] C.-F. Chen, T.-Y. Huang and R.-B. Wu, "Design of dual- and triple pass band filters using alternately cascaded multiband resonators," *IEEE Trans. Microwave Theory Tech.*, vol. 54, no. 9, pp. 3550–3558, Sept. 2006.

[19] A. F. Sheta, J. P. Coupez, G. Tanne, S. Toutain, and J. P. Blot, "Miniature microstrip stepped impedance resonator band pass filter and diplexers for mobile communications," *in IEEE MTT-S Int. Dig.* pp. 607–610, June 1996.

[20] R. N. Bates, "Design of microstrip spur line band stop filters," *IEE J. Microw., Optics, Acoustics*, vol. 1, no. 6, pp. 209-214, Nov. 1977.

[21] C. Nguyen and K. Chang, "on the analysis and design of spur line band stop filters," *IEEE Trans. Microw. Theory Tech.*, vol. 33, no. 12, pp. 1416-1421, Dec. 1985.

[22] J. R. Panda, A. Neogi, S. K. Dash and S. P. Kar, "A 5/6 GHz UWB Band pass Notch Filter Based on MMR Structure with a Spur line for Application in Wireless Communication System", in Proc. *IEEE International Conference on Applied Electromagnetics, Signal Processing and Communication, (IEEE AESPC-2018)*, Bhubaneswar, India, 22nd-24th October 2018.

[23] L. -Qi, L. -C-. Hong, Z. Wei, and X. –W. Feng "Novel compact UBW band pass filter with notched band" *Microwave Opt. Technol. Lett.*, vol. 52, no.2, pp.280-283, Jul. 2010.

978-1-5386-6723-1/19 $31.00 © 2019 IEEE

Boundstates Computation for Double Quantum Well Structure with Pöschl-Teller Potential for MWIR Photodetector Design

Avik Chakraborty
Dept of Electronics and Comm Engg
Jalpaiguri Govt Engg College
Jalpaiguri, India
chakraborty.avik.ece@gmail.com

Suporna Bhowmick
Dept of Electronics and Comm Engg
RCC Institute of Information Technology
Kolkata, India
bhowmicksuporna@gmail.com

Debarati Chakraborty
Dept of Electronics and Comm Engg
RCC Institute of Information Technology
Kolkata, India
chakrabortydebarati6991@gmail.com

Arpan Deyasi
Dept. of Electronics and Communication Engineering
RCC Institute of Information Technology
Kolkata, India
deyasi_arpan@yahoo.co.in

Angsuman Sarkar
Dept. of Electronics and Communication Engineering
Kalyani Govt Engg College
Kalyani, India
angsumansarkar@ieee.org

Abstract— **Lowest two quantum states of a double-well-triple-barrier structure are numerically analyzed for Pöschl-Teller potential profile. Kane-type band nonparabolicity of first order is considered for rational replication purpose, and effective mass mismatch between well and barrier layers are included in the computation following the BenDaniel Duke boundary conditions. Transmission coefficient is plotted as a function of input energy flux, and peaks are identified for eigenstate determination in presence and absence of external electric field. Results are compared with that obtained for ideal rectangular geometry, and better quasi-peak characteristics speaks in favor of the proposed structure for photodetector application, where tailoring of subband energy can be achieved by means of structural parameter variations.**

Keywords—Pöschl-Teller potential, Double quantum well, Eigenstate, Band nonparabolicity, Transmission coefficient, Propagation matrix method

I. INTRODUCTION

Quantum well photodetector design is not a very new subject of research as evident from various published literatures [1-4], and several complex mathematical models for potential configuration are already tried to match well with experimentally available data [5-6]. Performance of a quantum photodetector precisely depends on the width of quasi-peak, as larger quasi-peak makes possible of flatter absorption coefficient characteristics owing to existence of closer energy bands available for transition. Henceforth, shaper peak is always a subject of research, leads to higher quantum-confined structures [7-8] with improved characteristics at the cost of higher fabrication complexity and price.

Among various potential geometries, triangular [9-10], parabolic [11-12] and Gaussian [13-14] are researched most due to their usefulness in predicting electronic and optical properties of low-dimension devices, as compared with earlier works on rectangular profile [15-17]. Different numerical techniques are also considered for accurate calculation, as complex potentials are not easily solvable via analytical methods. Among different techniques, finite element method [18-19] and finite difference method [20-21] are considered most due to their exactness, but at the same time, these methods exhibit disadvantages in terms of memory consumption and time of computation. Transfer matrix technique is only applicable for rectangular structure [22], though consumes minimum time. As an intermediate solution, propagation matrix method [23] draws attention of workers due to lower time of estimation along with possibility of fitting most of the potential structures. Henceforth, present authors considered it as suitable to tackle the problem.

Ricardo *et. al.* [24] applied transfer matrix method to solve Pöschl-Teller potential with very low error for design of photodetector at MWIR-LWIR bands. Later canonical transformation method is applied on the potential [25]. Ikot and others [26] used Klein-Gordon equation to solve various hypergeometric potentials including Pöschl-Teller, and bound state solutions are obtained. Nikiforov–Uvarov method is also tried for wavefunction determination [27], and later generated scatterings are interpreted [28] for device applications. Concept of position-dependent effective mass is very recently introduced [29] with a pure mathematical variation. But very little works are carried out so far on the optical properties of quantum structures based on Pöschl-Teller potential, as far the knowledge of authors.

In the present work, eigenstates of double quantum well structure using Pöschl-Teller potential is computed by propagation matrix technique. Position-dependent effective mass is considered following BenDaniel Duke condition, and conduction band nonparabolicity is assumed. Field is applied within practical limit, and transmission characteristics are investigated based on resonant tunneling phenomenon. Results are compared with already published literatures, and also with rectangular profile.

II. MATHEMATICAL FORMULATION

Electron motion in any quantum-confined region can always be expressed by Schrödinger equation as

$$-\frac{\hbar^2}{2m^*}\frac{d^2\psi(z)}{dz^2}+V(z)\psi(z)=E(z)\psi(z) \qquad (1)$$

where symbols have usual meanings. Introducing position-dependent effective mass concept at the barrier and well regions

$$-\frac{\hbar^2}{2}\frac{d}{dz}\left[\frac{1}{m(z)^*}\frac{d\psi(z)}{dz}\right]+V(z)\psi(z)=E(z)\psi(z) \qquad (2)$$

$V(z)$ is the potential of the structure, and for Pöschl-Teller configuration, it may be represented as

$$V(z)=-\frac{\hbar^2}{2m^*}\alpha^2\frac{\lambda(\lambda-1)}{\cosh^2(\alpha z)} \qquad (3)$$

where 'λ' is termed as depth parameter [independent of well/barrier], and 'α' is called width parameter. The wave functions in regions p and $p+1$ are

$$\psi_p = A_p \exp[ik_p z]+B_p \exp[-ik_p z] \qquad (4.1)$$

$$\psi_{p+1}=C_{p+1}\exp[ik_{p+1}z]+D_p\exp[-ik_{p+1}z] \qquad (4.2)$$

where A and C are coefficients for the wave function moving left to right in regions 'p' and 'p+1' respectively; B and D are the corresponding right-to-left propagating-wave coefficients. Introducing phase information between potential steps separated by distance L_p, we get

$$\psi A_p \exp[ik_p L_p]=\psi C_p \qquad (5.1)$$

$$\psi B_p \exp[-ik_p L_p]=\psi D_p \qquad (5.2)$$

This can be formulated as-

$$\begin{bmatrix} A_p \\ B_p \end{bmatrix}=I\begin{bmatrix} C_{p+1} \\ D_{p+1} \end{bmatrix} \qquad (6)$$

where

$$I=\begin{bmatrix} \exp[-ik_p L_p] & 0 \\ 0 & \exp[ik_p L_p] \end{bmatrix} \qquad (7)$$

Thus, composite propagation matrix for p^{th} region can be written as:

$$M_p = \frac{1}{2}\begin{bmatrix} \left(1+\dfrac{m_p}{m_{p+1}}\dfrac{k_{p+1}}{k_p}\right)\exp[-ik_p L_p] & \left(1-\dfrac{m_p}{m_{p+1}}\dfrac{k_{p+1}}{k_p}\right)\exp[-ik_p L_p] \\ \left(1-\dfrac{m_p}{m_{p+1}}\dfrac{k_{p+1}}{k_p}\right)\exp[ik_p L_p] & \left(1+\dfrac{m_p}{m_{p+1}}\dfrac{k_{p+1}}{k_p}\right)\exp[ik_p L_p] \end{bmatrix} \qquad (8)$$

For Kane-type band nonparabolicity of first order, we write

$$E(k)=\frac{\hbar^2}{2m^*}k^2(1-\gamma k^2) \qquad (9)$$

where γ is the band nonparabolicity coefficient of first order. Finally transmission coefficient may be formulated in the form

$$T(E)=\frac{1}{(M_{11})^2} \qquad (10)$$

III. RESULTS AND DISCUSSIONS

Using Eq. (10), transmission coefficient is computed and plotted as a function of input energy flux for Pöschl-Teller potential, and compared with rectangular potential well. From Fig 1, it is seen that though rectangular geometry provides eigen-peak at lower energy values, but Pöschl-Teller potential generates sharper peak. A closer introspection reveals that magnitude of quasi-peak is higher for rectangular potential, and increases for higher energy states. Compared to that, sharper peaks are observed for the present potential configuration. This speaks in favor of sharper absorption coefficient profile, one key feature for photodetector application.

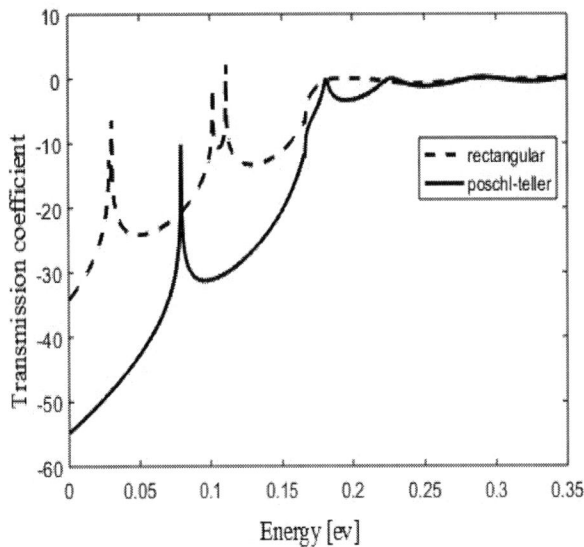

Fig 1: Comparative study of transmission coefficient for Pöschl-Teller potential with rectangular potential

Fig 2 shows the biasing effect on Pöschl-Teller configuration. It is found that with increase of applied bias, energy peaks appear at lower values, which is quite expected owing to reduction of quantum confinement. But the interesting feature noted in this case is the non-existence of quasi-peak even at higher eigenstates, which exist in rectangular profile.

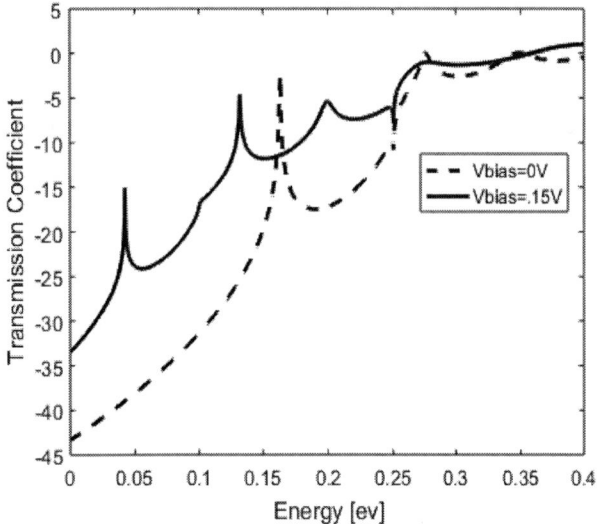

Fig 2: Transmission coefficient as a function of energy in presence and absence of bias

Effect of varying dimensions are computed and plotted in Fig 3. Dimensions are considered within fabrication limit, and well width, contact barrier width as well as middle barrier width is modified to observe the change in magnitude and positions of eigen-peaks. Fig 3a shows that for varying well width, position of lowermost eigenstate remains unchanged, whereas higher energy states make a redshift. This is evident in Fig 3a, which speaks in favor of tuning the intersubband transition.

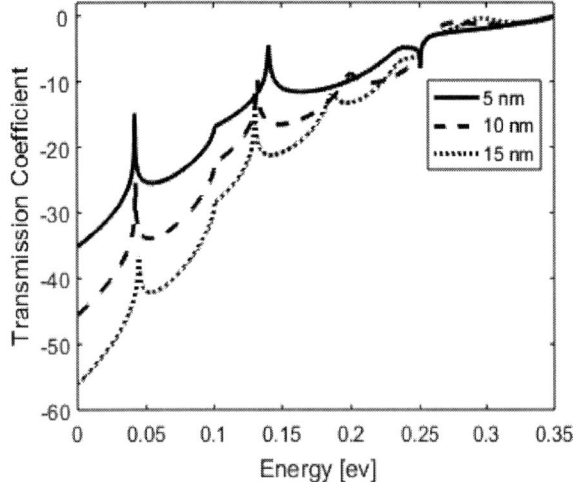

Fig 3a: Transmission coefficient as a function of energy for three different well widths in presence of bias

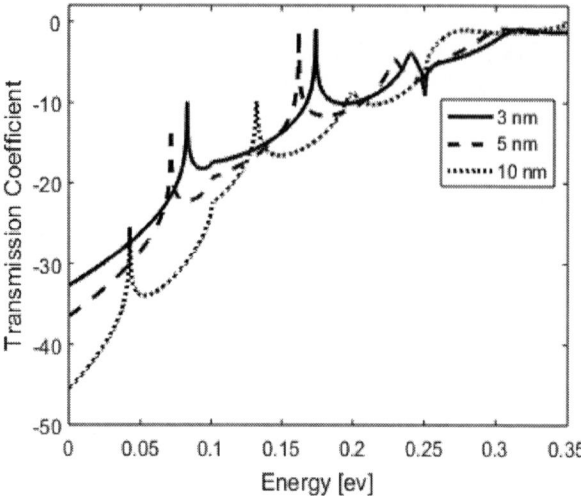

Fig 3b: Transmission coefficient as a function of energy for three different contact barrier widths in presence of bias

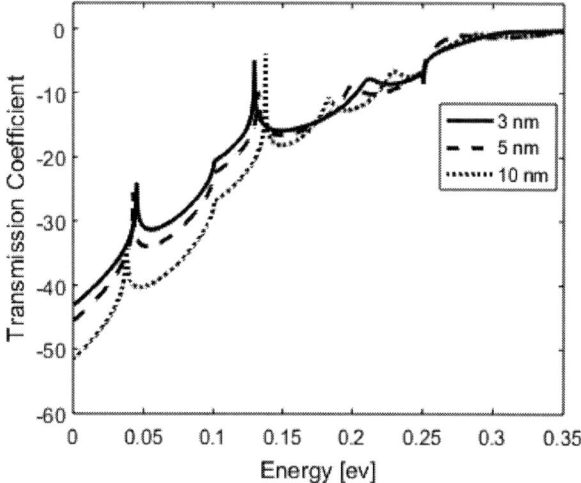

Fig 3c: Transmission coefficient as a function of energy for three different middle barrier widths in presence of bias

From Fig 3b, it is observed that the magnitude of intersubband transition remains almost independent on contact barrier dimensions, whereas middle barrier width can greatly influence it. This is plotted in Fig 3c. Larger middle barrier width increases the subband energy, which speaks for photodetector operating at lower wavelength region.

Fig 4 shows the variations for different material compositions. Lowering the Al content in barrier material reduces the quantum confinement which, in turn, decreases the potential barrier. But the intraband energy remains almost constant. Result is also plotted for type-II system, where $x = 0.5$ is considered. In this context, it may be mentioned that we have considered symmetric DQW structure, i.e., barrier material composition in all the layers are exactly same.

978-1-5386-6723-1/19 $31.00 © 2019 IEEE   266

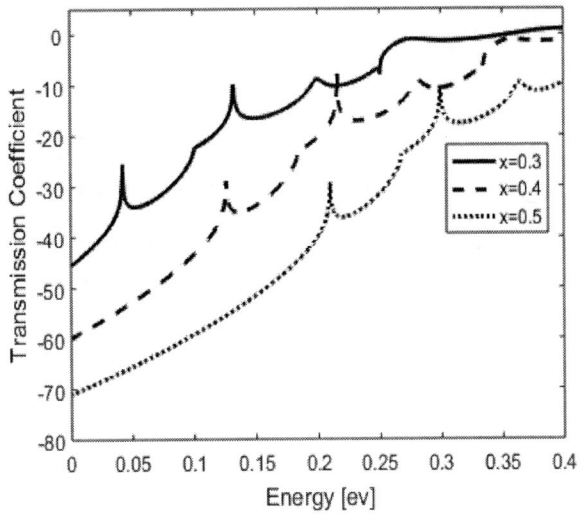

Fig 4: Transmission coefficient as a function of energy for three different material compositions in presence of bias

Fig 5 and Fig 6 show the variations for 'depth' and 'width' parameters respectively. From the plot, it is found that that higher the depth, higher the shift of eigenstates at lower energy values. This also increases the number of energy-peaks, and thus become suitable for optical applications. Similar situation can be observed for larger width parameter also.

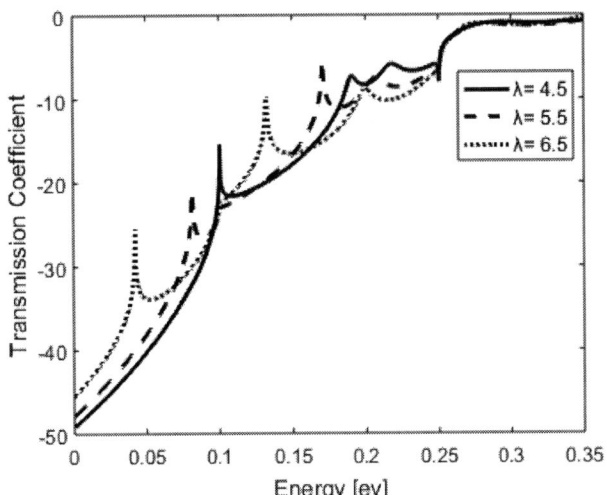

Fig 5: Transmission coefficient as a function of energy for three different depth parameters in presence of bias

Results obtained for dimensional variations and depth and width parameters tuning are represented in tabular form for two lowermost energy states. Data are shown n table – I and table – II respectively. Corresponding intersubband transition energy is also estimated.

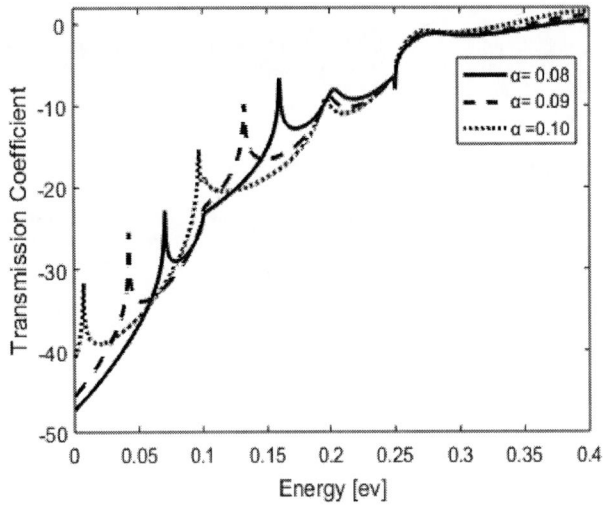

Fig 6: Transmission coefficient as a function of energy for three different width parameters in presence of bias

TABLE I. LOWERMOST TWO EIGENSTATES FOR VARIATION IN WELL DIMENSIONS

Name of parameter	Dimension [nm]	E_1 [eV]	E_2 [eV]	ΔE_{21} [eV]
Well width	5	0.04203	0.14	0.09797
	10	0.04253	0.132	0.08947
	15	0.04453	0.13	0.08547
Contact barrier width	3	0.08303	0.174	0.09097
	5	0.07153	0.162	0.09047
	10	0.04253	0.1325	0.08997
Middle barrier width	3	0.04503	0.1295	0.08447
	5	0.04253	0.1325	0.08997
	10	0.03753	0.1375	0.09997

TABLE II. LOWERMOST TWO EIGENSTATES FOR VARIATION IN WIDTH AND DEPTH PARAMETERS

Name of parameter	Dimension [nm]	E_1 [eV]	E_2 [eV]	ΔE_{21} [eV]
Width parameter	0.08	0.07053	0.1595	0.08897
	0.09	0.04253	0.132	0.08947
	0.1	0.007031	0.09703	0.08999
Depth parameter	4.5	0.01005	0.1905	0.18045
	5.5	0.08203	0.1705	0.08847
	6.5	0.04253	0.1325	0.08997

IV. CONCLUSION

Lowest intersubband energy is computed for a double quantum well structure using Pöschl-Teller potential for different structural, material and characteristic parameter variations. Result is significant compared to the DQW structures in the respect that the peaks appear in transmission coefficient characteristic curve have no quasi-peak, which speaks in favor of very low spreading of energy bands. Results are significant for photodetector design at MWIR range.

References

[1] K. K. Choi, J. Sun, K. A. Olver, R. Fu, "Parameter Study of Resonator-Quantum Well Infrared Photodetectors", *IEEE Journal of Quantum Electronics*, vol. 53(5), p. 4000410, 2017

[2] S. D. Gunapala, S. V. Bandara, J. K. Liu, J. M. Mumolo, B. Rafol, D. Z. Ting, A. Soibel, C. Hill, "Quantum Well Infrared Photodetector Technology and Applications", *IEEE Journal of Selected Topics in Quantum Electronics*, vol. 20(6), p. 3802312, 2014

[3] A. Rogalski, "Quantum Well Photoconductors in Infrared Detector Technology", *Journal of Applied Physics*, vol. 93(8), pp. 4355-4390, 2003

[4] M. K. Das, R. K. Lal, "Modeling of Quantum Well Infrared Photo Detector for Long Wavelength Infrared Detection", *IETE Journal of Research*, vol. 63(5), pp. 719-727, 2017

[5] V. Apalkov, G. Ariyawansa, A. G. U. Perera, M. Buchanan, Z. R. Wasilewski, H. C. Liu, "Polarization Sensitivity of Quantum Well Infrared Photodetector Coupled to a Metallic Diffraction Grid", *IEEE Journal of Quantum Electronics*, vol. 46(6), pp. 877-883, 2010

[6] C. H. Tsai, G. E. Chang, "GeSn/Ge Quantum Well Photodetectors for Short-Wave Infrared Photodetection: Experiments and Modeling", *Proceeding of SPIE: Optical Sensors*, vol. 10231, p. 102310J, 2017

[7] A. Nasr, "Theoretical Characteristics of Quantum Wire Infrared Photodetectors under Illumination Conditions", *Journal of Optical Communications*, vol. 30(3), pp. 126-132, 2009

[8] P. Martyniuk, A. Rogalski, "Quantum-Dot Infrared Photodetectors: Status and Outlook", *Progress in Quantum Electronics*, vol. 32(3), pp. 89-120, 2008

[9] N. Verma, Parveen, J. Jogi, "Effect of Variation in Channel Thickness on Eigenenergies of Double Triangular Quantum Well in Double Gate InAlAs/InGaAs HEMT", *TENCON* 2014

[10] S. Rabbaa, J. Stiens, "Validation of a Triangular Quantum Well Model for GaN-based HEMTs used in pH and Dipole Moment Sensing", *Journal of Physics D: Applied Physics*, vol. 45(47), p. 475101, 2012

[11] A. Keshavarz, N. Zamani, H. Nadgaran, "Optical Gain in Double Semi-Parabolic Quantum Well Laser Typical of AlGaAs/GaAs", *Numerical Simulation of Optoelectronic Devices*, 2014

[12] N. Sahoo, S. K. Palo, A. K. Panda, T. Sahu, "Effect of Parabolic Potential on Improvement of Electron Mobility in Hybrid Double Quantum Well Structure", *AIP Conference Proceedings*, vol. 2005(1), p. 070003, 2018

[13] D. Sarkar, A. Deyasi, "Oscillator Strength of Gaussian Double Quantum Well for Intersubband Transition", *Springer Proceedings in Physics: Advances in Optical Science and Engineering*, chapter 53, pp. 433-438, 2016

[14] D. Sarkar, A. Deyasi, "Calculating Absorption Coefficient of Gaussian Double Quantum Well Structure with Band Nonparabolicity for Photodetector in Microwave Spectra", *CRC Press: Foundations and Frontiers in Computer, Communication and Electrical Engineering*, chapter 47, pp. 225-229, 2016

[15] O. Donmez, A. Erol1, M. C. Arikan, H. Makhloufi, A. Arnoult, C. Fontaine, "Optical Properties of GaBiAs Single Quantum Well Structures Grown by MBE", *Semiconductor Science and Technology*, vol. 30(9), p. 094016, 2015

[16] M. Y. Vinnichenko, R. M. Balagula1, I. S. Makhov, A. A. Shumilov, D. A. Firsov, L. E. Vorobjev, "Optical Properties of GaAs/AlGaAs Double Quantum Wells in Lateral Electric Field", *Journal of Physics: Conference Series*, vol. 741(1), p. 012149, 2016

[17] G. Vashisht, V. K. Dixit, S. Haldar, T. K. Sharma, "Effect of Disorders on the Optical Properties of Excitons in InAsP/InP Quantum Wells Investigated by Magneto-photoluminescence Spectroscopy", *Journal of Optical Society of America B*, vol. 35, pp. 2405-2411, 2018

[18] K. Nakamura, A. Shimizu, M. Koshiba, K. Hayata, "Finite-Element Analysis of Quantum Wells of Arbitrary Semiconductors with Arbitrary Potential Profiles", *IEEE Journal of Quantum Electronics*, vol. 25(5), pp. 889-895, 1989

[19] K. Hayata, M. Koshiba, K. Nakamura, A. Shimizu, "Eigenstate Calculation of Quantum Well Structures using Finite Elements", *Electronics Letters*, vol. 24(10), pp. 614-616, 1988

[20] I. W. Sudiarta, "Non-Standard Finite-Difference Time-Domain Method for Solving the Schrödinger Equation", *Pramana – Journal of Physics*, vol. 91(52), 2018

[21] B. Lô, S. B. Gueye, "Numerical Verification of Transition's Energies of Excitons in Quantum Well of ZnO with the Finite Difference Method", *Journal of Modern Physics*, vol. 7, pp. 329-334, 2016

[22] W. Li, "Generalized Free Wave Transfer Matrix Method for Solving the Schrödinger Equation With an Arbitrary Potential Profile", *IEEE Journal of Quantum Electronics*, vol. 46(6), pp. 970-975, 2010

[23] J. G. S. Demers, R. Maciejko, "Propagation Matrix Formalism and Efficient Linear Potential Solution to Schrödinger's Equation", *Journal of Applied Physics*, vol. 90(12), pp. 6120-6129, 2002

[24] R. A. T. Santos, F. D. P. Alves, C. G. R. Taranti, L. A. Faria, A. A. Quivy, "Quantum Well Infrared Photodetector Design using Transfer Matrix Method", *SBMO/IEEE MTT-S International Microwave and Optoelectronics Conference*, 2009

[25] J. Morales, J. García-Martínez, J. García-Ravelo, J. J. Peña, "Exactly Solvable Schrödinger Equation with Hypergeometric Wavefunctions", *Journal of Applied Mathematics and Physics*, vol. 3, pp. 1454-1471, 2015

[26] A. N. Ikot, H. P. Obong, I. O. Owate, M. C. Onyeaju, H. Hassanabadi, "Scattering State of Klein-Gordon Particles by Parameter Hyperbolic Poschl-Teller Potential", *Advances in High Energy Physics*, vol. 2015, p. 632603, 2015

[27] S. Meyur, "Bound State Energy Level for Three Solvable Potentials", *Bulgerian Journal of Physics*, vol. 38, pp. 347-356, 2011

[28] D. Çevik, M. Gadella, S. Kuru, J. Negro, "Resonances and Antibound States for the Pöschl–Teller Potential: Ladder Operators and SUSY Partners", *Physics Letters A*, vol. 380, pp. 1600–1609, 2016

[29] H. R. Christiansen, M. S. Cunha, "Solutions to Position-dependent Mass Quantum Mechanics for a New Class of Hyperbolic Potentials", *Journal of Mathematical Physics*, vol. 54, p. 122108, 2013

2019 Devices for Integrated Circuit (DevIC), 23-24 March, 2019, Kalyani, India

Analytical Drain Current Model of UTBB SOI MOSFET with lateral dual gates to Suppress Short Channel Effect

Arighna Basak
Department of Electronics & Communication Engineering
Brainware University
Barasat, West Bengal, India
arighnabsk060891@gmail.com

Angsuman Sarkar
Department of Electronics & Communication Engineering
Kalyani Government Engineering College, Kalyani
Nadia, West Bengal, India
angsumansarkar@ieee.org

Abstract— In this paper, we present a 2D analytical modeling of UTBB SOI MOSFET by introducing a gap in the gate for which this new structure behaves like a dual gate MOSFET and compared the result with TCAD simulation. A 2D Poisson's equation is used for solving surface potential profile, electric field distribution, threshold voltage, DIBL and drain current of UTBB SOI MOSFET structure through parabolic approximation method. A comparative study for increasing negative voltage on control gate of this structure has been carried out. Here we observe surface potential profile, electric field distributions, threshold voltage, DIBL and drain current through applying negative voltage on the right gate of the proposed structure. Result reveals that this structure have higher efficacy to reduce short channel effect (SCE) due to the existence of step change in the surface potential distribution and for increasing negative control gate voltage this structure provides better performance for suppression short channel effect.

Keywords— *Analytical Modeling; SOI MOSFET; Lateral dual gates; Short Channel Effects;*

I. INTRODUCTION

Now a days for downscaling MOSFET at range of below 20 nm the channel length is decreasing so the performance of the device degrades and arises short channel effects which is also a major building blocks [1]. SCE includes threshold voltage roll off, DIBL and sub threshold swings degradations. To overcome these situations many nonconventional MOSFET structure introduces [2-3] such as double gate MOSFET [4-9], triple gate MOSFET [10], FINFET [11], and surrounding gate MOSFET [12] but since for more than 25 years in the sub nanometer regime the UTBB SOI MOSFET is an interesting architecture for downscaling MOSFET structure and suppression of short channel effect [13-14] and it also used in many low power applications, power devices etc. [15]. This UTBB SOI MOSFET is used to suppress the short channel effect (SCE) by applying channel/gate engineering techniques, thickness modulation of the buried oxide layer etc. [16-20] which are reported already.

Wei et al. [21] at 2017 analyzed a new structure of UTBB SOI MOSFET with dual gates in lateral direction for DIBL suppression where they introduced a gap between the gates. But the boundary conditions have not analytically derived yet now so less attention has been showed for analytical model of this new structure. So in this paper we analytically model this new structure where we also placed a gap between the gates so the structure behaves like a dual gates and performs better for suppression of short channel effect.

In the sub-threshold region for drain voltage the barrier of the potential near the channel and the drain is dropped so to shield the drain voltage a gap in introduced between the gates. Moreover we also analytically calculate the total drain current of this structure. We also observed surface potential, electric field distribution, threshold voltage for different conditions. The effect of applying negative gate voltage on the control gate is also explained for suppressing of short channel effect.

II. MODEL DERIVATION

A. Device Structure

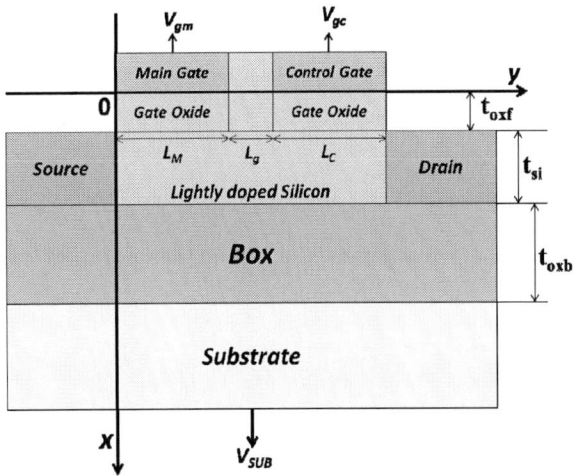

Fig 1: Structure of UTBB SOI MOSFET with lateral dual gates

The cross sectional figure of the proposed structure is shown in the fig 1 for analytical calculation and simulation result. Poly-silicon material of workfunction 5.5 ev is applied on two gates, the main gate and control gate. A gap filled with SiO_2 is introduced between the gates for which the two gates are obtained. The length of the two gates is equal. The length of the gap is 2 nm. SiO_2 is used as gate oxide material. V_{GM} and V_{GC} are voltage applied on the main gate and control gate respectively. Here negative voltage of the control gate can play an important role for suppressing short channel effect. L_M, L_g, L_C are length of main gate, gap and the control gate respectively. V_{DS} & V_{SUB} are the drain voltage and the substrate voltage respectively which applied on the structure. The typical parameters are shown in the table 1. The structure divided into three regions, region I is

978-1-5386-6723-1/19 $31.00 © 2019 IEEE   269

from length of 0 to L_M, region II is from L_M to (L_M+L_g) and region III is from (L_M+L_g) to $(L_M+L_g+L_C)$.

Table 1: The characteristic parameters of UTBB SOI MOSFET through lateral dual gates

Parameter	Value
Source/drain doping	1×10^{20} cm^{-3}
Doping concentration of silicon film N_A	1×10^{15} cm^{-3}
Thickness of gate oxide	1 nm
Thickness of Silicon film t_{si}	5 nm
Thickness of Buried Oxide (BOX)	2 nm
Gate work functions	5.5 ev
Total channel length	52 nm

B. Model development

The 2-D analytical explanation of Poisson's equation for the distribution of potential into the three regions of the channel can be expressed as

$$\frac{\partial^2\phi(x,y)}{\partial x^2}+\frac{\partial^2\phi(x,y)}{\partial y^2}=\frac{qN_A}{\varepsilon_{si}} \qquad (1)$$

Where N_A is the concentration of doping in the channel region, q is electronic charge, ε_{si} is relative permittivity of the silicon, $\phi(x,y)$ is 2D solution of potential distribution in the area of the channel.

From Young [22] observation the distribution of electrostatic potential in the x-dependence can be determined by solving the simple parabolic approximation and can be expressed as

$$\phi_i(x,y)=\phi_{si}(x)+C_{i1}(x)y+C_{i2}(x)y^2 \qquad (2)$$

where i=1, 2, & 3 for region of I, II, & III respectively. $C_{i1}(x)$ and $C_{i2}(x)$ are solved by the boundary condition. Because of different workfunctions the surface potential are different for different region. These values of coefficient are calculated through solving the following boundary conditions

1. The electric field in the front gate is continuous for four regions:

$$\frac{\partial\phi_1(x,y)}{\partial y}\Big|_{y=0}=\frac{\varepsilon_{ox}}{\varepsilon_{si}}\frac{\phi_{s1}(x)-\left(V_{gm}-V_{fb1}\right)}{t_{oxf}} \qquad (3)$$

$$\frac{\partial\phi_2(x,y)}{\partial y}\Big|_{y=0}=\frac{\varepsilon_{ox}}{\varepsilon_{si}}\frac{\phi_{s2}(x)}{t_{oxf}} \qquad (4)$$

$$\frac{\partial\phi_3(x,y)}{\partial y}\Big|_{y=0}=\frac{\varepsilon_{ox}}{\varepsilon_{si}}\frac{\phi_{s3}(x)-\left(V_{gc}-V_{fb2}\right)}{t_{oxf}} \qquad (5)$$

Where t_{oxf} is the front gate thickness, ε_{ox} & ε_{si} are the permittivity of SiO$_2$ and silicon. V_{GM} is voltage of main gate and V_{GC} is voltage of control gate. V_{fb1} & V_{fb2}, are flat band voltage for the region I & III respectively.

2. Also the electric field in the back gate is continuous for four regions:

$$\frac{\partial\phi_1(x,y)}{\partial y}\Big|_{y=t_{si}}=\frac{\varepsilon_{ox}}{\varepsilon_{si}}\frac{V_{SUB}-\phi_{b1}(x)}{t_{oxb}} \qquad (6)$$

$$\frac{\partial\phi_2(x,y)}{\partial y}\Big|_{y=t_{si}}=\frac{\varepsilon_{ox}}{\varepsilon_{si}}\frac{-\phi_{b2}(x)}{t_{oxb}} \qquad (7)$$

$$\frac{\partial\phi_3(x,y)}{\partial y}\Big|_{y=t_{si}}=\frac{\varepsilon_{ox}}{\varepsilon_{si}}\frac{V_{SUB}-\phi_{b3}(x)}{t_{oxb}} \qquad (8)$$

Where t_{oxb} is the thickness of the buried oxide, $\phi_{b1}(x)$, $\phi_{b2}(x)$ & $\phi_{b3}(x)$ are the potential distribution of the backside oxide-silicon interface for three regions.

3. At the junction of the main gate and gap the surface potential distribution can be expressed as

$$\phi_1\left(L_M,0\right)=\phi_2\left(L_M,0\right) \qquad (9)$$

At the junction of gap and control gate, the surface potential is continuous and can be expressed as

$$\phi_2\left(L_M+L_g,0\right)=\phi_3\left(L_M+L_g,0\right) \qquad (10)$$

4. At the interfaces of the three regions the electric field distributions are also continuous and can be stated as

$$\frac{d\phi_1(x,y)}{dx}\Big|_{x=L_M}=\frac{d\phi_2(x,y)}{dx}\Big|_{x=L_M} \qquad (11)$$

$$\frac{d\phi_2(x,y)}{dx}\Big|_{x=L_M+L_g}=\frac{d\phi_3(x,y)}{dx}\Big|_{x=L_M+L_g} \qquad (12)$$

5. At the junction of source side and channel region,

$$\phi_1\left(0,0\right)=\phi_{s1}\left(0\right)=V_{bi} \qquad (13)$$

where V_{bi} is equivalent to the built in potential

6. At the junction of control gate and drain,

$$\phi_{s4}\left(L_M+L_g+L_C,0\right)=V_{DS} \qquad (14)$$

where V_{DS} is equal to drain voltage.

Substitute the boundary conditions equations (3)-(8) and then in (1) we find the value of all coefficients.

Finally the distributions of the potential can be expressed as

$$\frac{d^2\phi_{si}(x)}{dx^2}-\alpha^2\phi_{si}(x)=\beta_i \qquad (15)$$

where i=1, 2, & 3 for region I, II, & III respectively

$$\alpha^2 = \frac{2(1 + \frac{C_{of}}{C_{si}} + \frac{C_{of}}{C_{ob}})}{t_{si}^2(1 + \frac{2C_{si}}{C_{ob}})}$$

$$\beta_1 = \frac{qN_A}{\varepsilon_{si}} - \frac{2V_{SUB}}{t_{si}^2(1 + \frac{2C_{si}}{C_{ob}})} - \frac{2(V_{gm} - V_{fb1})(\frac{C_{of}}{C_{ob}} + \frac{C_{ob}}{C_{si}})}{t_{si}^2(1 + \frac{2C_{si}}{C_{ob}})}$$

$$\beta_2 = \frac{qN_A}{\varepsilon_{si}} - \frac{2V_{SUB}}{t_{si}^2(1 + \frac{2C_{si}}{C_{ob}})}$$

$$\beta_3 = \frac{qN_A}{\varepsilon_{si}} - \frac{2V_{SUB}}{t_{si}^2(1 + \frac{2C_{si}}{C_{ob}})} - \frac{2(V_{gc} - V_{fb2})(\frac{C_{of}}{C_{ob}} + \frac{C_{ob}}{C_{si}})}{t_{si}^2(1 + \frac{2C_{si}}{C_{ob}})}$$

After solving the equations we obtain the solution of the potential in the three regions and can be expressed as

$$\phi_{s1}(x) = Ae^{\alpha x} + Be^{-\alpha x} - \frac{\beta_1}{\alpha^2} \tag{16}$$

for $0 \le x \le L_M$

$$\phi_{s2}(x) = Ce^{\alpha x} + De^{-\alpha x} - \frac{\beta_2}{\alpha^2} \tag{17}$$

for $L_M \le x \le L_M + L_g$

$$\phi_{s3}(x) = Ee^{\alpha x} + Fe^{-\alpha x} - \frac{\beta_3}{\alpha^2} \tag{18}$$

for $(L_M + L_g) \le x \le (L_M + L_g + L_C)$

Solving all equations we get the values of the coefficient A, B, C, D, E & F.

The distribution of electric field is given by solving the differentiation of the potential in the four regions and can be expressed as

$$E_1 = \frac{d\phi_{s1}(x)}{dx} = A\alpha e^{\alpha x} - B\alpha e^{-\alpha x} \tag{19}$$

$$E_2 = \frac{d\phi_{s2}(x)}{dx} = C\alpha e^{\alpha x} - D\alpha e^{-\alpha x} \tag{20}$$

$$E_3 = \frac{d\phi_{s3}(x)}{dx} = E\alpha e^{\alpha x} - F\alpha e^{-\alpha x} \tag{21}$$

for $(L_M + L_g) \le x \le (L_M + L_g + L_C)$ \hfill (22)

C. Threshold Voltage Model

In this arrangement the solution of the minimum value of potential is always placed at main gate [23]. So the minimum value of surface potential can be found by

$\frac{d\phi_{s1}(x)}{dx} = 0$. So we get

$$x_{min} = \frac{1}{2\alpha} \ln\left(\frac{B}{A}\right) \tag{23}$$

The value of minimum potential is determined through substituting x_{min} in the equation (16) we get

$$\phi_{s,min} = 2\sqrt{AB} - \frac{\beta_1}{\alpha^2} \tag{24}$$

The term threshold voltage is defined as the minimum value of gate voltage to create a conducting path between the drain terminal and the source terminal. It is used for measuring SCE as a parameter. It is equal to the twice the value of the potential of bulk for inducing conducting channel on the surface of this structure.

So the threshold voltage is calculated from the equation $\phi_{s,min} = 2\phi_F$

where ϕ_F is the difference between intrinsic and extrinsic Fermi levels of the region of bulk
Therefore we get

$$V_{th} = \frac{-u_2 \pm \sqrt{u_2^2 - 4u_1 u_3}}{2u_1} \tag{25}$$

D. Drain Current Model

The drain current in the linear and saturation regions of the channel is determined by the equation [24-29]

$$I_{D,lin} = \left((V_{gm} - V_{th})V_{DS} - \frac{1}{2}V_{DS}^2\right) \times \left(\frac{2W\mu C_{of}}{(L_{eff} - l_d - \frac{V_{DS}}{E_C})} + \lambda_a \frac{2WC_{of}}{(L_{eff} - l_d)^2}\right) \tag{26}$$

$$I_{d,sat} = \left((V_{gm} - V_{th})V_{Dsat} - \frac{1}{2}V_{Dsat}^2\right) \times \left(\frac{2W\mu C_{of}}{(L_{eff} - l_d - \frac{V_{DS}}{E_C})} + \lambda_a \frac{2WC_{of}}{(L_{eff} - l_d)^2}\right) \tag{27}$$

where W is width of the device, μ is the electron mobility, L_{eff} is the effective length of channel, l_d is channel length modulation [37,40], λ_a is the velocity overshoot effect [38] and V_{Dsat} is the saturation voltage which is given by

$$V_{Dsat} = \frac{V_{gm} - V_{th}}{1 + \frac{V_{gm} - V_{th}}{L_{eff} E_C}}$$

where $E_C = \frac{2v_{sat}}{\mu}$, v_{sat} is saturation velocity of electron [24].

On behalf of impact ionization effect and parasitic BJT effect the total drain current in saturation region can be written as

2019 Devices for Integrated Circuit (DevIC), 23-24 March, 2019, Kalyani, India

$$I_{D,sat} = GI_{d,sat} + HI_{CBO}$$

where G, H, I_{CBO} are determined from [29].

III. RESULT & DISCUSSION

For verification of analytical model of the structure, the graphs were plotted using MATLAB and compared the results with TCAD simulation [30].

Fig 2 shows the surface potential variation of UTBB SOI MOSFET structure with length of the channel for different negative control gate voltage which is obtained from the model and the stimulation result. From the figure it is observed that a step change in the potential is obtained at the junction of the gap and main gate in the structure. This step change is responsible to increase carrier velocity and carrier transport efficiency. From the figure it also observed that the minimum surface potential is obtained for $V_{gc} = -1.5$ V.

negative control gate voltage exhibits a good threshold voltage roll off.

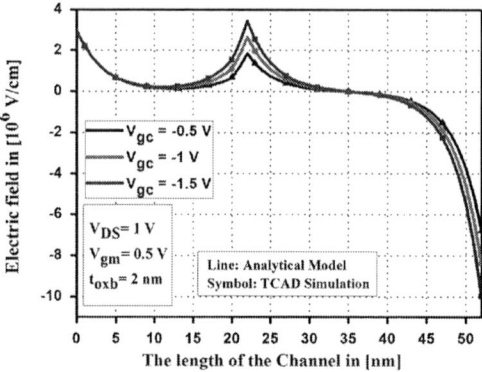

Fig 3: Variation of the electric filed distribution of the structure with length of the channel for different negative control gate voltage

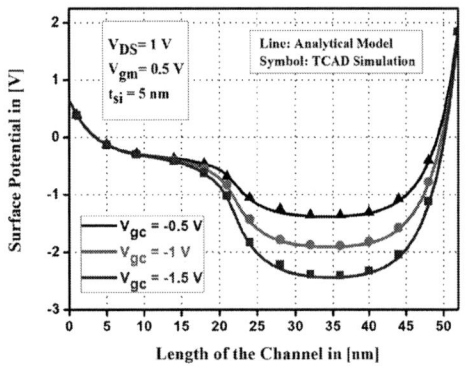

Fig 2: Surface potential variation of the structure through length of the channel for different negative control gate voltage

Fig 3 shows electric field variation of the structure with channel length for different negative control gate voltage which are obtained from the model and the stimulation result. From the figure it is observed that a step is changed in the distribution of electric field at the junction of main gate and gap and also a low electric field at the drain side influences the suppression of hot carrier effect. From the observation it is clear that the hot carrier effect is eliminated for increasing of negative control gate voltage.

Fig 4 shows the threshold voltage variation of this structure with respect to length of the channel for different negative control gate voltage which are obtained from the model and the stimulation result. From the figure it is observed that when the gate length is reduced then the threshold voltage is also reduced. This effect is known as threshold voltage roll off. To suppress this effect, apply negative gate control voltage for this structure so SCE is reduced. So from this figure it is observed that when negative control gate voltage increases, the threshold voltage is increased that's why

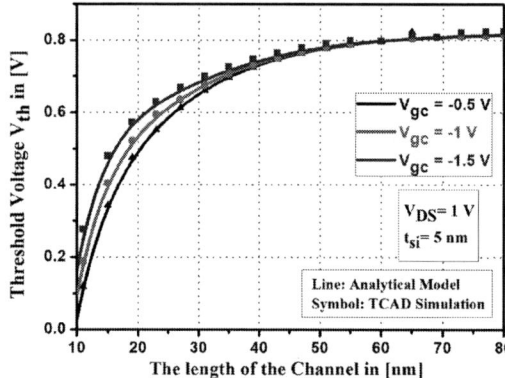

Fig 4: Threshold voltage variation with length of the channel for different negative control gate voltage

Fig. 5 shows DIBL variation with respect to length of channel for UTBB SOI MOSFET for increasing negative gate voltage. DIBL is defined as the ratio of a small change in V_{th} to a small change in V_{DS}.

$$DIBL = \frac{\Delta V_{th}}{\Delta V_{DS}}$$

From the figure it is observed that the structure shows the better performance of variation of DIBL for increasing negative gate voltage due to increase screening of the threshold voltage region caused by step in the potential profile.

Fig 6 shows drain current variation of this structure with main gate voltage for different negative control gate voltage which is obtained from the model and the stimulation result. From fig 6 it is observed that decreasing V_{gc} to negative values is advantageous for suppressing the off-state current and improving ON state current.

978-1-5386-6723-1/19 $31.00 © 2019 IEEE

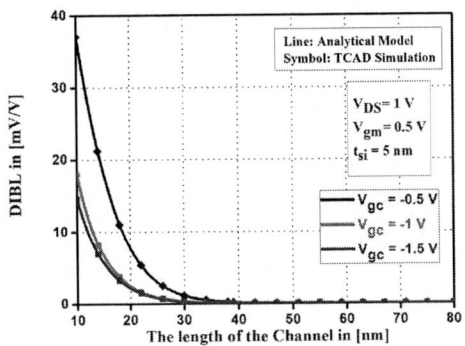

Fig 5: DIBL variation with length of the channel of this structure for different negative control gate voltage

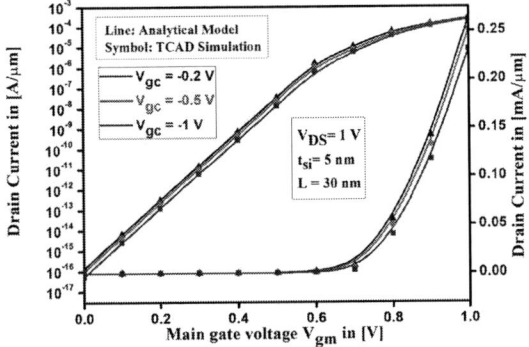

Fig 6: Drain current variation with main gate voltage V_{gm} for different negative control gate voltage at $V_{DS} = 1V$

IV. CONCLUSION

In this paper we solve a 2D analytical model of drain current of the device structure by solving Poisson's equation for suppressing SCE and compared the results with TCAD simulation. Here we observed surface potential, electric field distribution, threshold voltage variation, drain current for this structure. Here we also observed that this structure has better ability to reduce SCE (short channel effect) by reducing threshold voltage roll off, improved ON current and suppressing off-set current for increasing negative gate control voltage. This model suggests that negative gate voltage is important parameters for reducing SCE.

REFERENCES

[1] J. Y Taur, DA Buchanan, W Chen, DJ Frank, KE Ismail, S-H Lo, et al. "CMOS scaling into the nanometer regime," Proc IEEE ,vol. 85 no. 4, pp. 486-504, 1997.
[2] X. Zhou, W. Long, "A novel hetero-material gate (HMG) MOSFET for deep submicron ULSI technology," IEEE Trans. Elec. Dev. 45, pp.2546-2548, 1998.
[3] TK Chiang, "A new two-dimensional subthreshold behavior model for the short-channel asymmetrical dual-material double-gate (ADMDG) MOSFET," Microelectronics Reliability 49, pp. 693-698, 2009.

[4] B. Cousin, M. Reyboz, O. Rozeau, M.A. Jaud, T. Ernst, J. Jomaah, "A unified shortchannel compact model for cylindrical surrounding-gate MOSFET," Solid State Electronics, vol. 56, pp 40-46, 2011.
[5] RK Sharma, R Gupta, M Gupta and RS Gupta, "Dual Material Double-Gate SOI n-MOSFET: Gate Misalignment Analysis", IEEE transaction on Electron Devices, vol. 56, no. 6 pp. 1284- 1291, 2009.
[6] PK Tiwari, S Dubey, K Singh and S jit, "Analytical modeling of subthreshold current and subthreshold swing of short-channel triple-material double-gate (TM-DG) MOSFETs" Superlattices Microstruct, vol. 51, pp. 715-724, 2012.
[7] RK Sharma, R Gupta, M Gupta and RS Gupta, "Dynamic performance of graded channel DG FD SOI n-MOSFETs for minimizing the gate misalignment effect", Microelectronics Reliability, vol. 49, pp. 699-706, 2009.
[8] NC Roy, A Gupta and S rai, "Analytical surface potential modeling and simulation of junctionless double gate (JLDG) MOSFET for ultra low-power analog/RF circuits", Microelectronics Journal, vol. 46, pp. 916-922, 2015.
[9] V Shrivastava, A Kumar, C. Sahu and J. Singh, "Temperature Sensitivity analysis of dopingless charge-plasma transistor", Solid State Electronics, vol. 117, pp. 94-99, 2016.
[10] M.C. Lemme, T. Mollenhauer, W. Henschel, T. Wahlbrink, M. Baus, O. Winkler, R. Granzner, F. Schwierz, B. Spangenberg, H. Kurz, "Sub-threshold behavior of triple-gate MOSFETs on SOI material sub-threshold behavior of triple-gate MOSFETs on SOI material," Solid State Electronics, vol. 48, pp.529-534, 2004.
[11] D. Hisamoto, L. Wen-Chin, J. Kedzierski, H. Takeuchi, K. Asano, C. Kuo, E. Anderson, K. Tsu-Jae, J. Bokor, H. Chenming," FinFET-a self-aligned double-gate MOSFET scalable to 20 nm," IEEE Trans. Electron. Device, vol. 47, pp.2320-2325, 2000.
[12] B. Cousin, M. Reyboz, O. Rozeau, M.A. Jaud, T. Ernst, J. Jomaah, "A unified shortchannel compact model for cylindrical surrounding-gate MOSFET," Solid State Electronics, vol. 56, pp 40-46, 2011.
[13] K. K. Young, "Short-channel effect in fully depleted SOI MOSFETs". IEEE Trans Electron Devices, vol. 36, pp. 399–402, 1989.
[14] L. Grenouillet, Q. Liu, R. Wacquez et al, "UTBB FDSOI scaling enablers for the 10 nm node", IEEE SOI-3D-subthreshold microelectronics technology unified conference (S3S), pp 1–2, 2013.
[15] M. Stephane, S. Thomas, "UTBB FDSOI: evolution and opportunities" Solid-State Electron, vol. 125, pp. 63–72, 2016.
[16] T. A. Karatsori, A. Tsormpatzoglou, C. G. Theodorous, Analytical compact model for lightly-doped nanoscale ultrathinbody and box SOI MOSFETs with back-gate control", IEEE Trans Electron Devices, vol. 62, page: 3117–3124, 2015.
[17] N.A. Srivastava, V. K. Mishra, R. K. Chauhan, "Analytical modelling of surface potential of modified source FD-SOI MOSFET", International conference on emerging trends in communication technologies (ETCT), pp 1–4, 2016.
[18] T. Yamada, S. Abe, Y. Nakajima, "Quantitative extraction of electric flux in the buried-oxide layer and investigation of its effects on MOSFET characteristics", IEEE Trans Electron Devices, vol. 60, pp. 3996–4001, 2013.
[19] T. Yamada, Y. Nakajima, T. Hanajiri, "Suppression of drain-induced barrier lowering in silicon-on-insulator MOSFETs through source/drain engineering for low-operating-power system-on-chip applications", IEEE Trans Electron Devices, vol. 60, pp. 260–267, 2013.
[20] T. Yamada, Y. Nakajima, T. Hanajiri, Corrections to ''suppression of drain-induced barrier lowering in silicon-on-insulator MOSFETs through source/drain engineering for lowoperating- power system-on chip applications''. IEEE Trans Electron Devices, vol. 60, pp. 4281–4283, 2013.
[21] S. Wei, G. Zhang, Z. Shao, L. Geng, C. Yang, "Analysis of a high-performance ultra-thin body ultra-thin box silicon-on-insulator MOSFET with the lateral dual-gates: featuring the suppression of the DIBL", Microsystem technologies, vol. 24, pp. 3949-3956, 2018.
[22] K.K. Young, "Short-channel effects in fully depleted SOI MOSFET's," IEEE Trans. Electron. Device, vol. 36, pp.399-402, 1989.
[23] I. Polishchuk, P. Ranade, K.T. Jae, H. Chenming, "Dual work function metal gate CMOS transistors by Ni-Ti interdiffusion," IEEE Electron. Dev. Lett, vol. 23, pp. 200-202, 2020.
[24] K.Suzuki, and T.Sugii, "Analytical models for n+-p+ double gate SOI MOSFET's," IEEE Trans. Electron. Device, vol. 42, no. 11, pp. 1940-1948, Nov 1995.
[25] N. D. Arora, R. Rios, C. L. Huang, and K. Raol, "PCIM: A physically based continuous short-channel IGFET model for circuit simulation," IEEE Trans. Electron Devices, vol. 41, no. 6, pp. 988–997, Jun. 1994.

[26] J. B. Roldan, F. Gamiz, J. A. Lopez-Villanueva, and J. E. Carceller, "Modeling effects of electron velocity overshoot in a MOSFET," IEEE Trans. Electron Devices, vol. 44, no. 5, pp. 841–846, May 1997.

[27] J. B. Roldan, F. Gamiz, J. A. Lopez-Villanueva, P. Cartujo, and J. E. Car-celler, "A model for the drain current of deep submicrometer MOSFET's including electron-velocity overshoot," IEEE Trans. Electron Devices, vol. 45, no. 10, pp. 2249–2251, Oct. 1998.

[28] G. V. Reddy and M. J. Kumar, "A new dual material double gate (DMDG) nanoscale SOI MOSFET: Two dimensional analytical modelling and simulation," IEEE Trans. Electron Devices, vol. 4, no. 2, pp. 260–268, Mar. 2005.

[29] Y. G. Chen, J. B. Kuo, Z. Yu, and R. W. Dutton, "An analytical drain current model for short-channel fully-depleted ultrathin siliconon- insulator nMOS devices," Solid State Electron., vol. 38, no. 12, pp. 2051–2057, Dec. 1995.

[30] ATLAS 2D DEVICE Simulator, SILVACO Int., Santa Clara, C, 2012.

Performance Comparison of 2.5-GHz LC Voltage-Controlled Oscillator for Three Different Technology Nodes

Shrabanti Das
Electronics and Telecommunication Engineering
Jadavpure University
Kolkata, India
shrabanti.08@gmail.com

Sayan Chatterjee
Electronics and Telecommunication Engineering
Jadavpure University
Kolkata, India
sayan1234@gmail.com

Abstract— **This paper has been written to compare the performances of LC Voltage Controlled Oscillator (VCO) at 2.5 GHz to see the effects of device performances with different technology node on the same while keeping the circuits identical. Accordingly, as a case study, three instances of LC-VCO have been designed in Cadence Spectre using GPDK 45nm, 90nm, and 180nm technology and three different power supply levels of 0.8 V, 1 V, and 1.2 V respectively. ObservedPhase Noise (PN) for 45nm, 90nm, and 180nm technology are -120.8 dBc/Hz, -124.64 dBc/Hz and -126.12 dBc/Hz @ 1MHz offset respectively. Output power, Figure of Merit (FOM) and K$_{VCO}$ also have been measured subsequently. The best possible FOM obtained are -181.7dB, -184.9dB and -186.2dB for 45nm, 90nm and 180nm technology node respectively.**

Keywords— CMOS, PLL, VCO, Phase noise, Jitter, FOM

I. INTRODUCTION

The requirement for additional accessible channels in mobile communication applications has been increased due to the modern exponential growth in the wireless communication system. This requirement has forced further severe constraints on the phase noise of the Local Oscillators (LO). The overall performance of PLL could be measured by the pure impulse signal power in the frequency domain generate by a VCO. If it fails then it is said that VCO is suffering from Phase noise issue [1]. In the digital field, phase noise is known as jitter. The timing margins are directly influenced by Clock jitter and therefore limit the performance of a system [2]. Therefore Phase and frequency variations have become the matter of numerous studies [3]-[4]. Though numerous forms have been built up for different kinds of oscillators, each of these kinds builds limiting hypothesis applicable merely to a restricted group of oscillators. The majority of these forms have been designed using the concept of a linear time-invariant (LTI) system and undergo through the electrical noise sources, like device noise, which turn into the phase noise. Generally, a VCO is used to produce the LO frequency which is mixed in the Mixer along with RF frequency to generate the Inter-mediate Frequency (IF) in a receiver. The most desired conditions for voltage-controlled oscillator (VCO) circuits are full integration of tank circuit, low-phase noise, and high current efficiency [5-6]. In [7] a VCO for 5GHz wireless applications has been designed using 0.35μm CMOS technology which provides -90dBc/Hz at 100 KHz and for a

0.7 - 2.7V control voltage series a 200MHz or 4.3% tuning range is presented. A fully integrated 2-GHz LC-tank VCO has been designed for very low Phase Noise and to minimize Flicker Noise upconversion [8]. The design of low-phase-noise LC-oscillators is mostly depended on the design of a high-quality inductor. A Phase Noise of -125.1 dBc/Hz at 600-kHz offset and -138 dBc/Hz at 3 MHz can be achieved by means of an in-house-developed integrated inductor simulator-optimizer [8]. Further, in [9] a new optimization technique for VCO has been demonstrated by modifying the design of conventional LC-VCO, leading to a 2.4 GHz fully integrated, LC-VCO. And the design of the circuit has been implemented with 0.35μm MOSFET technology. The values of phase-noise have been measured are -121, -117, and -115 dBc/Hz at 600-kHz offset from 1.91, 2.03, and 2.60-GHz carrier frequencies, correspondingly and also 26% of tuning range has been achieved. To minimize Phase Noise two different techniques has been taken in [10] one is inductive degeneration and the other is capacitive filtering. Both of these techniques avoid low-frequency tail current noise from being altered into phase noise. The major phase noise diminution (up to 6–7 dB at 3-MHz offset frequency from the carrier) has been realized through inductive degeneration. However, capacitive filtering furthermore considerably decreases the phase noise at high offset frequencies and may, therefore, turn into a convincing alternate to inductive degeneration. A VCO with a very low phase noise and a wide tuning range is one of the most vital and critical building blocks in a Phase-Locked Loop (PLL) of the RF receiver and broadband wireless system [13]. When designing a VCO, the spectral purity (i.e. the Phase Noise should be as low as possible) is the most crucial part. A VCO with a wide tuning range can be attained by using a single- varactor device with sharp C-V characteristics i.e. large Cmax /Cmin [11]. On the other hand, the employment of single-varactor can result in an enormously high tuning sensitivity K$_{VCO}$ [12], which causes poor Phase Noise and spurious behavior. To decrease such sensitivity within the noise created by the Phase-locked Loops (PLL) building blocks, very low VCO gain is thus very much advantageous. In general, the low gain as well as a wide tuning range can be obtained by using a capacitor bank or inductances redesigned with Field Effect Transistor (FET) switches [14]. Using a capacitive division technique swing and phase-noise enhanced dynamic-threshold VCO has been presented in [15]. The anticipated VCO gives

Phase Noise of −122.0 dBc/Hz @ 1MHz offset frequency. In [16] a novel low Phase Noise and a wide tuning range VCO has been offered with 180nm CMOS technology with an Integrated Passive Device (IPD) transformer which gives a high-quality (Q) factors (Q > 30). The high-quality IPD transformer has been employed to reduce the phase noise of the VCO and, therefore, leads to a superior FOM. For 0.8 V supply voltage, the VCO provides phase noise of −124.2 dBc/Hz at a 1-MHz offset from the 6.26-GHz. For wide frequency tuning range and low phase noise application a millimeter-wave (mm-wave) dual-mode VCO topology with switchable coupled VCO-cores has been introduced in [17]. VCO shows very small Phase Noise from -87.5 dBc/Hz to -93.5 dBc/Hz @ 1 MHz offset over the total tuning range. A low voltage and low Phase Noise LC-VCO IC with simplified noise filtering with 180nm CMOS technology have been reported in [18] which give a Phase Noise of -119.4 dBc/Hz @ 1 MHz offset frequency from a 2.25 GHz carrier frequency at power supply voltage of only 0.5 V.

The design of three LC VCO at 2.5 GHz has been reported in this paper. In the present work, conventional CMOS Cross-Coupled LC VCO has modified with three different technology nodes i.e. 45nm, 90nm, 180nm which are shown in the Fig. 1, Fig. 6 and Fig. 11 respectively to compare the parameters of VCO at 2.5 GHz. The details of design parameters for a 2.5 GHz output frequency have been listed in Table I. The VCO topologies used in this experiment be made up of a pair of Cross-Coupled PMOS and NMOS with a tail current source and one differential LC tank that uses two typical NMOS transistor as varactors. The VCO circuits have been designed on the said nodes CMOS technology accompanied by inductor quality factors (Q) of around 8 at 2.5 GHz. This topology permits a very low voltage operation and the tuning range of the VCO can be changed by changing the value of LC tank as the oscillation frequency be based on the value of L and C according to equation number (1).

$$f_{osc} = \frac{1}{2\Pi}\left(\sqrt{LC_{TOT}}\right) \qquad (1)$$

The frequency of oscillation is controlled by the tank circuit. Hence, the tank is a natural place to locate the tuning varactor and the varactor C(V) is usually placed in shunt with a fixed capacitance C for tuning. However, the oscillation frequency varies linearly with applied VCO control voltage. This can be depicted by the equation number 2.

$$f_{osc} = f_0 + K_{vco}V_{CTL} \qquad (2)$$

A small change in f_{osc} with respect to control voltage V_{CTL}, i.e. $df_{osc}/dV_{CTL} = K_{vco}$ is frequently referred to as the gain of VCO or VCO sensitivity and explains how much the oscillator frequency varies per volt of applied control voltage.
The mostly used MOSFET varactor in CMOS applications is the accumulation mode varactor as it has the highest capacitance per unit area, better Q, and wider voltage tuning range than inversion/depletion MOSFET or PN junction varactors [19]. Here in this paper the gate terminals of the MOS varactors are connected to the tank to minimize

capacitive loading by the parasitic capacitances so as to not overload the tank circuit. The control voltage is applied to the drain source connection as shown in Fig. 1, Fig. 6 and Fig. 11.

II. CIRCUIT TOPOLOGY

A. VCO circuit in 180nm technology :
The simulation of a CMOS cross coupled LC-VCO has been done in Cadence Spectre, using GPDK 180 nm technology is shown Fig. 1. For this circuit 1.2V power supply has been used.

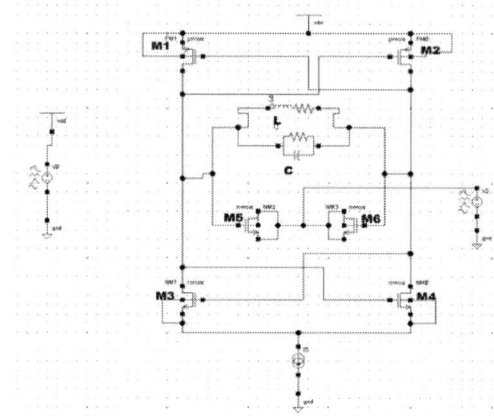

Fig. 1 : Schematic view of CMOS Cross coupled LC VCO in 180nm technology

Simulating the VCO circuit of Fig. 1 with 1.2 V power supply and 2mA bias, the obtained Phase Noise is -126.13 dBc/Hz @ 1MHz which is shown in Fig. 3. The VCO is tunable from 2.5413 to 2.5421 GHz and consumes 7.329 dBm powers. the plot of VCO tuning range and output power has been shown in Fig. 4 and Fig. 5 respectively. The obtained K_{VCO} for this circuit is 993.93 KHz/V.

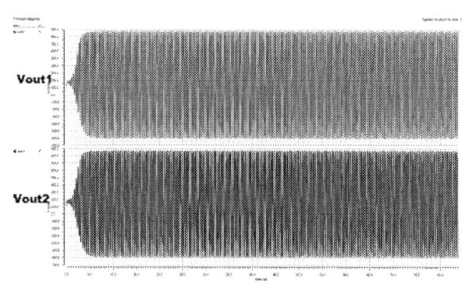

Time (ns.)
Fig. 2 : Plot of VCO transient output

978-1-5386-6723-1/19 $31.00 © 2019 IEEE

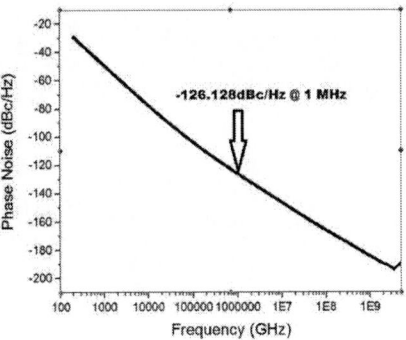

Fig. 3 : Plot of VCO Phase Noise vs Frequency

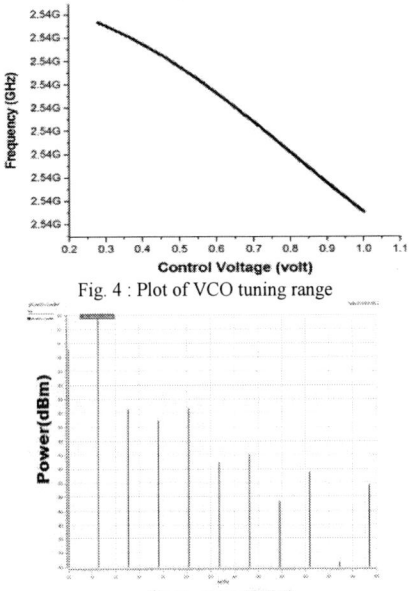

Fig. 4 : Plot of VCO tuning range

Fig. 5 : Frequency Spectrum at vtune

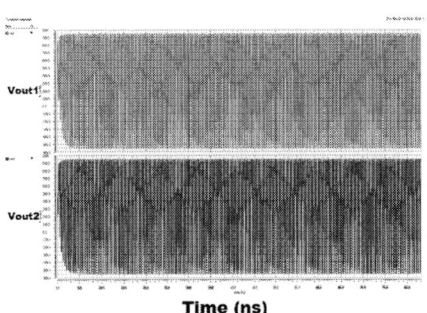

Fig. 6 : Schematic view of CMOS Cross coupled LC VCO in 90nm technology

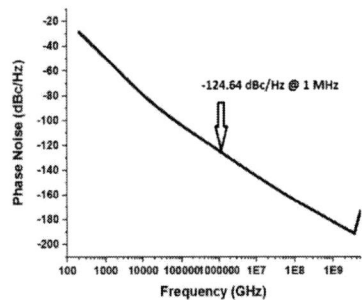

Fig. 7 : Plot of VCO transient output

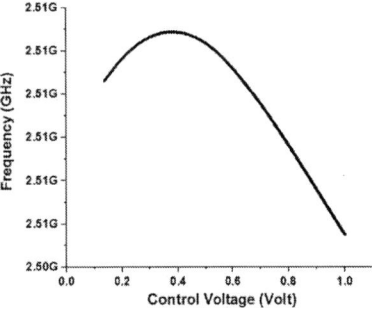

Fig. 8 : Plot of VCO Phase Noise vs Frequency

B. VCO circuit in 90nm technology :

Next a CMOS Cross Coupled LC VCO of Fig. 1 has been modified and simulated in Cadence Spectre, using GPDK 90 nm technology is shown Fig. 6 with 1V power supply. Simulating the VCO circuit of 90nm technology with 1V power supply voltage and 2mA bias current, the obtained Phase Noise is -124.64 dBc/Hz @ 1MHz as shown in the Fig. 8. The simulated VCO is tunable for range of 2.5057 to 2.5089 GHz. The whole circuit consumes 7.23 dBm powers. From the simulated results it can be noticed that the Phase Noise is higher for VCO designed in 90 nm technology than the VCO designed with 180 nm technology

Fig. 9 : Plot of VCO tuning range

The obtained K_{VCO} of this circuit is 3.59 MHz/V which is higher than the value obtained from simulating the VCO circuit with 180nm technology node. The plot of tuning range and frequency spectrum has been shown in Fig. 9 and Fig.10respectively.

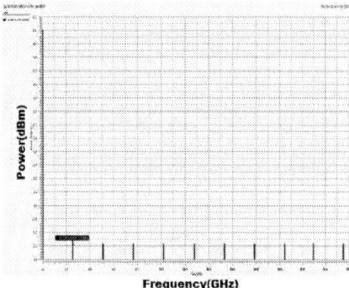

Fig. 10 : Frequency Spectrum at vtune

C. VCO circuit in 45nm technology :

The CMOS Cross Coupled topology for VCO of Fig. 1 and Fig. 6 has been changed a bit to compare VCO parameters, with respect to MOSFET technology node and power supply. For that the next circuit of LC VCO has been designed and simulation is done with Cadence Spectre, using GPDK 45 nm technology and 0.8 V power supply where MOSEFT channel length has been kept constant for cross coupled CMOS as well as in MOS varactor at 45 nm. However the width of the MOSFET's has been changed according to the need for the oscillation to start. The above mentioned LC VCO circuit has been shown in the Fig. 11.

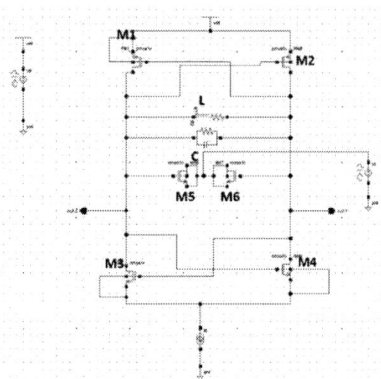

Fig. 11 : Schematic view of CMOS Cross coupled LC VCO in 45nm technology

Circuit of Fig. 11 have been simulated with 0.8V power supply and 2 mA bias current. Transient response of the VCO circuit has been shown in Fig. 12. The simulated results are as follows - the obtained Phase Noise is -120.8 dBc/Hz @ 1MHz, the K_{VCO} is 7 MHz/V and output power of the whole circuit is about 7.44 dBm which are shown in Fig. 13, Fig. 14 and Fig. 15 respectively.

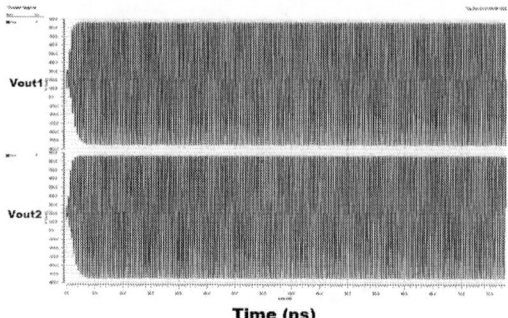

Fig. 12 : Plot of VCO transient output

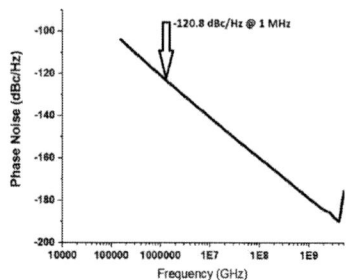

Fig. 13 : Plot of VCO Phase Noise vs Frequency

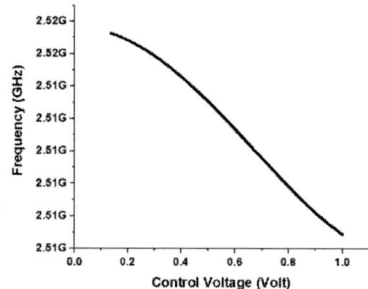

Fig. 14 : Plot of VCO tuning range

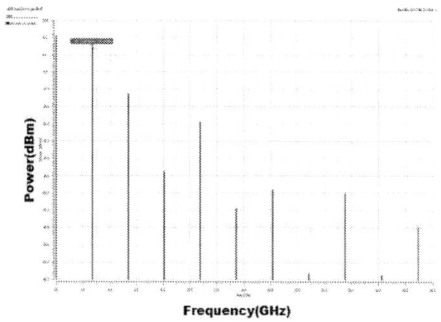

Fig. 15 : Frequency Spectrum at vtune

Figure of Merit (FOM) calculation

To calculate the general performance of a VCO, the Figure of Merit (FOM) has been taken into account, which is given by the following relation [1]:

$$FOM = L\{f_{offset}\} - 20\log\left(\frac{f_0}{f_{offset}}\right) + 10\log\left(\frac{P_{DC}}{1mW}\right) \quad (5)$$

Where f_0 is the oscillation frequency, $L\{f_{offset}\}$ is the Phase Noise at a certain frequency offset (f_{offset}), and P_{DC} is the power dissipation. The FOM mainly depends on the value of Phase Noise for a particular offset frequency. Using this above relation for 1 MHz offset frequency, the obtained FOM for the circuits of Fig. 1, Fig. 6 and Fig. 11 are -181.7, -184.9 and -186.2 respectively.

LC-VCO circuits of the said nodes have been simulated using 180nm, 90nm and 45 nm CMOS technology to generate 2.5 GHz oscillation frequency. These circuits having varactors implemented by MOSFETs with same technology node as used for CMOS cross coupling. For the three circuits the actual circuit topology has not been changed and W/L ratio of the core also has been kept constant. Only the used technology node of the transistors has been changed to observe the VCO performance. Even the values of Inductor and capacitor for the three circuits with three different technologies have been kept constant. From the above three discussions it has been observed that simultaneous increase in technology nodes, the Phase Noise of the LCVCO has been improved. Also from the results it can be noticed that the circuit with lowest technology node consumes more power than other two circuits with higher channel length transistors. Table I shows the proposed circuit parameters related technology, the simulated results and similar prior results from [8] [9] [15] [16] .

TABLE. I

	This work			[8]	[9]	[15]	[16]
Technology	CMOS 45 nm	CMOS 90 nm	CMOS 180 nm	0.65μm BiCMOS	0.35μm BiCMOS	CMOS 0.11μm	CMOS 180nm + IPD
f_0 (GHz)	2.5 GHz	2.5 GHz	2.5 GHz	2.0 GHz	2.4 GHz	2.31 GHz	2.26 GHz
V_{dd} (V)	0.8 V	1.0 V	1.2 V	1.8 V	2.5 V	0.38/0.46 V	0.8 V
L (H)	2.7 nH	2.7 nH	2.7 nH	1.64 nH	-	-	-
Phase Noise (dBc/Hz)	-120.8 @ 1 MHz	-124.6 @ 1 MHz	-126.1 @ 1 MHz	-125.1 @ 600 KHz	-115 @ 600 KHz	-122 @ 1 MHz	-124.2 @ 1 MHz
Output power	7.44 (dBm)	7.39 (dBm)	7.32 (dBm)	34.2 mW	-	-2.49 (dBm)	9.2 mW
FOM	-181.7	-184.9	-186.2	-	-	-192.7	-190.5

III. CONCLUTION

In this paper three CMOS cross coupled LC VCO has been implemented for 2.5 GHz oscillation frequency with three different technology nodes in Cadence Spectre and using three different power supplies of 0.8 V, 1 V and 1.2 V. Fig. 1 showing the LCVCO circuit with 180nm technology which has a Phase Noise of -126.8 dBc/Hz whereas the circuits in Fig. 6 and Fig. 11 have phase noise such as -120.8 dBc/Hz and -126.64dBc/Hz for 90nm and 45nm respectively. So from the above results it has been noticed that with increasing channel length of transistors the Phase Noise also gets better. Even the FOM of Fig. 1 is better than the other two circuits. For 90 nm technology, a power spectrum has been noticed at lower frequency than 2.5 GHz. That problem

cannot be minimized by this circuit topology. So it could be a future research work where this higher power spectrum in the side lobs can be reduce or cancelled out by using different circuit topology with same technology node.

IV. ACKNOWLEDGMENT

The authors of this paper would like to thank Integrated Circuit Centre of Electronics Telecommunication Engineering Department of Jadavpur University for allowing them to use the Laboratory.Also the authors are grateful to Special Manpower Development Program phase-III (funded by Deity, under Ministry of Communication & IT, Government of India) for giving them with the necessitated Laboratory conveniences and softwares like cadence and synopsys.

REFERENCES

[1] Behzad Razavi, "RF Microelectronics" 2nd Edition, Prentice Hall PTR.

[2] E. J. Baghdady, R. N. Lincoln, and B. D. Nelin, "Short-term frequency stability: Characterization, theory, and measurement," *Proc. IEEE*, vol. 53, pp. 704–722, Jul. 1965.

[3] J. Rutman, "Characterization of phase and frequency instabilities in precision frequency sources; Fifteen years of progress," *Proc. IEEE*, vol. 66, pp. 1048–1174, Sept. 1978.

[4] J. Craninckx and M. Steyaert, "Low-noise voltage controlled oscillators using enhanced LC-tanks," *IEEE Trans. Circuits Syst.–II*, vol. 42, pp. 794–904, Dec. 1995.

[5] Craninckx, J., M. Steyaert, "A 1.8GHZ low-phase-noise spird-LC CMOS VCO," *Digest of Technical Papers Symposium on VLSI circuits*, pp. 30-31, Jun. 1996.

[6] Razavi, B., "A 1.8 GHzVoltage-Controlled Oscillator, " *ISSCC Digest of Technical Papers*, pp. 388-389, Feb., 1997.

[7] Peter Kinget, Bell Labs., Lucent Technologies, Murray Hill, NJ, "A Fully Integrated 2.7V 0.35pm CMOS VCO for 5GHz Wireless Applications" 1998 *IEEE International Solid-State Circuits Conference*.

[8] Bram De Muer, M. Borremans, M. Steyaert, and G. Li Puma "A 2-GHz Low-Phase-Noise Integrated LC-VCO Set with Flicker-Noise Upconversion Minimization" *IEEE Journal of Solid-State Circuits*, vol. 35, no. 7, Jul. 2000.

[9] Donhee Ham, and Ali Hajimiri, "Concepts and Methods in Optimization of Integrated LC VCOs" *IEEE Journal of Solid-State Circuits*, vol. 36, no. 6, Jun. 2001.

[10] Pietro Andreani, and Henrik Sjöland, "Tail Current Noise Suppression in RF CMOS VCOs" *IEEE Journal of Solid-State Circuits*, vol. 37, no. 3, Mar. 2002.

[11] Jung, B., and Harjani, R.: 'A 20 GHz VCO with 5 GHz tuning range in 0.25 mm SiGe BiCMOS'. ISSCC Dig. Tech. Pprs, San Francisco, Feb. 2004, pp. 178–179.

[12] Ali Hajimiri, and Thomas H. Lee, "A General Theory of Phase Noise in Electrical Oscillators" *IEEE Journal of Solid-State Circuits*, vol. 33, NO. 2, Feb. 1998.

[13] S.A. Osmany, F. Herzel, J.C. Scheytt, K. Schmalz and W. Winkler "Integrated 22 GHz low-phase-noise VCO with digital tuning in SiGe BiCMOS technology" *Electronics Letters* 1st Jan. 2009 Vol. 45 No. 1

[14] Kral, A., Behbahani, F., and Abidi, A.A.: 'RF-CMOS oscillator with switched tuning'. *Custom IC Conf.*, Santa Clara, CA, USA, 1998, pp. 555–558.

[15] Chang-Hun Lee, Changwoo Lim, and Tae-Yeoul Yun, "Swing and Phase-Noise Enhanced VCO with Capacitive-Division Dynamic-Threshold MOS" *IEEE Microwave and Wireless components letters*, vol. 26, no. 3, Mar. 2016

[16] Sen Wang, and Po-Hung Chen, "A Low-Phase-Noise and Wide-Tuning-Range CMOS/IPD_Transformer-Based VCO with High FOMT of −206.8 dBc/Hz", *IEEE Transactions on Components, Packaging and Manufacturing Technology,* vol. 6, no. 1, Jan. 2016

[17] Qiong Zou, , Kaixue Ma, and Kiat Seng Yeo, "A Low Phase Noise and Wide Tuning Range Millimeter-Wave VCO Using Switchable Coupled VCO-Cores" IEEE Transactions on circuits and Systems—I: Regular Papers, vol. 62, no. 2, Feb.2015

[18] Xinyi Wang, Xin Yang, Xiao Xu, Toshihiko Yoshimasu "2.4-GHz-Band Low-Voltage LC-VCO IC with Simplified Noise Filtering in 180-nm CMOS", *IEEE 2016.*

[19] Bhattacharjee, J., D. Mukherjee, and J. Lasker, "A Monolithic CMOS VCO for Wireless LAN Applications," *Proc. IEEE 2002.Intl. Symp. Circuits Systems ISCAS 2002*, vol. 3, May 2002, p. 441.

[20] Robert Caverly , "CMOS RF IC Design Principles " Artech House , Boston London . Chapter 6.

Cell thickness optimization of dual junction InGaP/GaAs solar cell against temperature variation

Shingmila Hungyo
Dept. of Electronics and
Communication Engineering
National Institute of Technology
Mizoram-796012 , India
shingmilahungyo@gmail.com

Khomdram Jolson Singh
Dept. of Electronics and
Communication Engineering
Manipur Institute of Technology
Imphal-795004 , India
jolly4u2@rediffmail.com

Rudra Sankar Dhar
Dept. of Electronics and
Communication Engineering
National Institute of Technology
Mizoram-796012 , India
rudra.ece@nitmz.ac.in

Abstract—A numerical modelling of highly efficient InGaP/GaAs dual-junction solar cell using advanced TCAD tool Silvaco ATLAS gives a theoretical conversion efficiency up to 31.08% at 1 sun under AM1.5 illumination. Within a temperature range from 0 to 150°C, the critical performance parameter such as J_{sc}, V_{oc} and Photogeneration rates of this optimised cell were extracted. It is found that with the increase in the cell temperature, the sub cells quantum efficiencies increase slightly and due to the energy gap narrowing effect with increase in cell temperature, the red-shift phenomena of absorption limit for all sub cells are observed. It was also found that with increased in temperature, V_{oc} decreases which are due to the increase in reverse saturation current, thereby leading to the decrease in the overall FF and efficiency of the solar cell. This work therefore calculates the optimum thickness of the modelled solar cell so that maximum efficiency is gained even if the temperature is elevated.

Keywords—Multi-junction solar cell; Temperature coefficient; SILVACO TCAD

I. INTRODUCTION

PV technology is old but yet lots of R&D is needed to attain highly efficient solar cell both for space and terrestrial applications. Various optimizations are also needed to be performed in modelling of solar cell so that the performance of the solar cell does not deteriorate with different physical conditions like temperature. Earlier studies have pointed that the performance of solar cell grades with the rise in temperature [1, 2]. DJ GaInP/GaAs solar cell fabricated using metal organic chemical vapor deposition (MOCVD) technique attained 25% efficiency [3], GaInP/GaAs/Ge triple junction solar cell shows maximum efficiency of 32.4% near 372 suns [4] as the temperature rises the general solar cell open-circuit voltage of a Ge single-junction cell reduced to almost 0V temperatures over 120°C [5]. Thus, the operating efficiency and performance of photovoltaic cells heavily depends on temperature. We know that some of the primary source of energy at far flung area such as remote snow land or desert or in deep space satellites is only photovoltaic cells. However, the problem is that these photovoltaic cells are required to work under extreme temperature conditions. The main challenges in this work are therefore to thoroughly investigate the performance of the cell under this extreme temperature variation and based on this analysis, we have to design the multi-junction cell in order to improve its efficiency. It is well known that with the increase in cell's operating temperature, there is a small increase in Isc in the range of tens of micro amps/°C-cm^2 and significant decrease in V_{oc} in the range about 2mV/°C. The changes in V_{oc} is known to contribute to the majority of the changes is efficiency [6].

II. EFFECTS OF TEMPERATURE ON SOLAR CELL

The effects of temperature on a solar cell cannot be neglected, as for concentrated solar cell and solar cell installed in probe satellite used for studying hot planets closed to Sun like Mercury which needs to operate at high temperature exceeding 300°C [6]. An increase in operating temperature of solar cell's result in slight increase in short circuit current on the other hand it significantly decreased the open circuit voltage. We know that the variation of V_{oc} with temperature is the main cause affecting the cell performance. The balance between direct, indirect, and Auger recombination rates of the carriers and the photogeneration of electron-hole pairs and the change in the band gap with temperature results in the change in this voltage [7]. In most of the materials, with increase in temperature, band gap decreases thereby reducing V_{oc}. The reduction in V_{oc} results directly in decreasing the cell's efficiency. The following equation relates the semiconductor's band gap energy as a function of temperature [8]

$$E_g(T) = E_g(0) - \frac{\alpha T^2}{T + \beta}$$

Where Eg(0) is the band gap energy at zero temperature, T is the temperature, and alpha and beta are the coefficients for band gap temperature dependence for a material. Although the maximum effect on cell performance is due to the variation in band gap with temperature, the decrease in overall voltage is however from electron-hole recombination rate.

Green uses the three types of recombination in an in-depth physics-based model to show the relationship between changing V_{oc} and temperature [7]. A simplified result of his derivation showing the relationship of V_{oc} to temperature was

$$\frac{dV_{oc}}{dT} = \frac{\left(\frac{(E_{go})}{q} - V_{oc} + \frac{kT}{q}\left(\gamma \frac{fd\xi}{\xi df}\right)\right)}{T}$$

$$\xi = npexp\left(\frac{-E_g}{kT}\right)/n_{ie}^2$$

Eg is the band gap appropriate of the recombination process of interest, f is a general function in the limiting cases used in Green's paper,

$$E_{go} = E_g - T\left(\frac{dE_g}{dT}\right),$$

Where T is the temperature and kT/q is the thermal voltage. Green proves the relation of temperature sensitivity of V_{oc} with the electron-hole product in recombination that occurs throughout the cell leading to 80-90 % of the device sensitivity with temperature [7].

III. DEVICE MODELLING

A. Structure modelling

Modelling and simulation of solar cells and analysing their performance is one of the key important role in designing and development of realistic cell. To build solar cell with the desired output, the processing of the device may be repeated numerous times and testing it to determine its performance is very expensive and time consuming. A modern Technology Computer Aided Design (TCAD) software program ATLAS from Silicon Valley Company (Silvaco) can be used effectively to model the design and

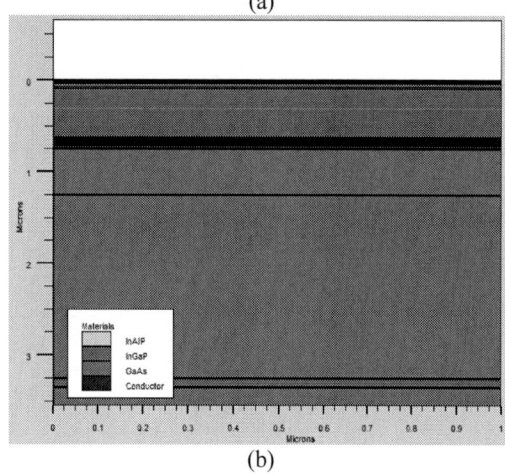

(a)

(b)

Fig.1 (a)Schematic and (b)Modelled structure of DJ of InGaP/GaAs solar cell [10]

simulate the behaviour, extract the performance output and analyse the result. The modelled structure of InGaP/GaAs solar cell is given in Fig.1. In order to facilitate the tunnelling of electrons with minimum loss, a tunnel junction and a tunnel emitter is inserted between the two junction of the top and bottom layer.

B. Physics models

In order to develop the accurate model and simulate effectively, suitable device physics including carrier generation and recombination model particularly for solar cell is compulsory. The Shockley-Read-Hall recombination (SRH) is modelled as follows

$$R_{SRH} = \frac{pn - n_{ie}^2}{TAUP0\left[n + n_{ie}exp\left(\frac{ETRAP}{kT_L}\right)\right] + TAUN0\left[n + n_{ie}exp\left(-\frac{ETRAP}{kT_L}\right)\right]}$$

Where ETRAP is the difference between the trap energy level and the intrinsic Fermi level, TL is the lattice temperature in degrees Kelvin and TAUN0 and TAUP0 are the electron and hole lifetimes. This model is activated by using the SRH parameter of the MODELS statement. The electron and hole lifetime parameters, TAUN0 and TAUP0, are user-definable in the MATERIAL statement.

The effects of temperature can be used in the Bowing model by specifying the EG1ALPH, EG1BETA, EG2ALPH and EG2BETA parameters of the MATERIAL statement. These parameters are used in Equations 3-53 through 3-55 to calculate the bandgap

$$E_{g1} = EG1300 + EG1ALPH\left[\frac{300^2}{300 + EG1BETA} - \frac{T_L^2}{T_L + EG1BETA}\right]$$

$$E_{g2} = EG2300 + EG2ALPH\left[\frac{300^2}{300 + EG2BETA} - \frac{T_L^2}{T_L + EG2BETA}\right]$$

$$E_g = E_{g2} * x + E_{g1}(1 - x) - EG12BOW * x * (1 - x)$$

The effects of temperature can be used in the Bowing model by specifying the EG1ALPH, EG1BETA, EG2ALPH and EG2BETA parameters of the MATERIAL statement. These parameters are used to calculate the bandgap

The various parameter used in the modelling of the device is presented in Table I, this parameter have been obtained from various literature review [10].

TABLE I. VARIOUS MATERIAL PARAMETER

Parameter	GaAs	InGaP	AlInP
Band gap Eg[eV] @300°K	1.42	1.9	2.4
Lattice constant α [Å]	5.65	5.65	5.65
Permittivity es/eo	13.1	11.6	11.7
Affinity [eV]	4.07	4.16	4.2

IV. RESULTS AND DISCUSSION

Short circuit current, open circuit voltage, fill factors etc. are important parameters for consideration of performance of a solar cell, this parameters are thoroughly studied and for different thickness of the solar cell also carried out, it was observed that some of the parameter improves by varying the thickness of various layer, but after certain point the performance either deteriorate or remains

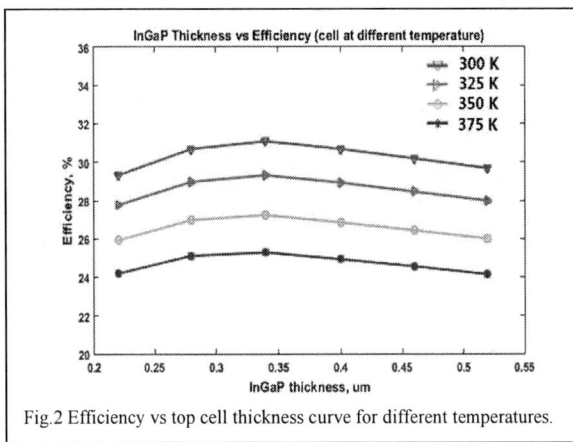

Fig.2 Efficiency vs top cell thickness curve for different temperatures.

Fig.2

A. Temperature and optimized thickness

The modelled structure have been simulated for different temperature ranging from 300K to 375K for different thickness of each of InAsP and GaAs, it was observed that maximum power was efficiency was obtained 31.085 % was obtained for InAsP and GaAs thickness of 0.34 μm and 2.1 μm respectively. Fig.2 shows the plot of efficiency of the modelled solar cell with different thickness of InGaP top layer, for different temperature. Also the detailed simulation results for various thickness of the top and bottom layer at different temperature is shown in Table II.

TABLE II.   SIMULATION RESULTS AT DIFFERENT TEMPERATURE

Temperature (°K)	InGaP base thickness (um)	GaAs base thickness (um)	Max. power (mW)	Efficiency (%)
300	0.22	1.3	39.65892388	29.31184
	0.28	2.1	41.51190148	30.68138
	0.34	**2.1**	**42.05842737**	**31.08531**
	0.4	2.9	41.51101768	30.68072
	0.46	3.7	40.83436732	30.18061
	0.52	3.7	40.14097529	29.66813
325	0.22	1.3	37.56438997	27.76377677
	0.28	2.1	39.19973794	28.97245968
	0.34	**2.1**	**39.66902577**	**29.31930951**
	0.4	2.9	39.13769412	28.92660319
	0.46	3.7	38.49350299	28.45048262
	0.52	3.7	37.85812379	27.98087494
350	0.22	1.3	35.06466115	25.91623145
	0.28	1.3	36.49009365	26.96976619
	0.34	**2.1**	**36.84558054**	**27.23250594**
	0.4	2.9	36.3107759	26.83723274
	0.46	2.9	35.75063575	26.42323411
	0.52	3.7	35.16289268	25.98883421
375	0.22	1.3	32.73228743	24.19237799
	0.28	1.3	33.97532835	25.11110743
	0.34	**2.1**	**34.21425425**	**25.28769716**
	0.4	2.9	33.72567945	24.92659235
	0.46	2.9	33.21896504	24.55208059
	0.52	3.7	32.66534961	24.14290437

B. Short circuit current and Open circuit voltage

Short circuit current is the maximum current that can flow, whereas the maximum voltage in open circuit is called the open circuit voltage. It was found that as the temperature rises, it also raises I_{sc} but at the same time V_{oc} decreases drastically, hence degrading the overall performance of the solar cell. The decrease in V_{oc} is due to the increase in reverse saturation current, the open circuit voltage is given by [9].

$$V_{oc} = |HOMO_D| - |LUMO_A| + n_{voc}\frac{kT}{q}\ln\left(\frac{\kappa}{N_A N_D}\alpha P_0\right) + \Delta_{low(F_{c,D})} + \Delta_{low(F_{c,A})}$$

Where $HOMO_D$ and $LUMO_A$ are HOMO donor and LUMO acceptor respectively, Fig. 3 shows the IV characteristics of the optimised solar cell structure.

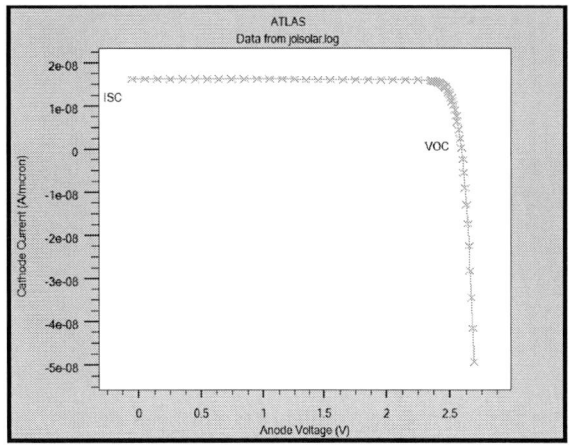

Fig. 3 Current voltage curve of the optimised InGaP/GaAs DJ solar cell.

C. Electron Current Density and Potential Developed

Electron current density and potential developed in the modelled structure is shown in Fig. 4, it can be seen that maximum potential is developed in the tunnel junction and the density of electron current is maximum at the bottom layer along the buffer and the substrate region of the device.

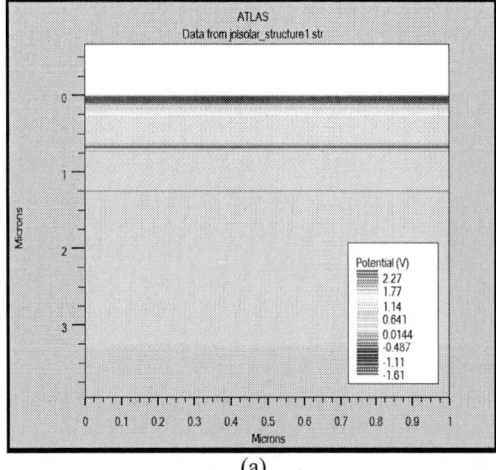

(a)

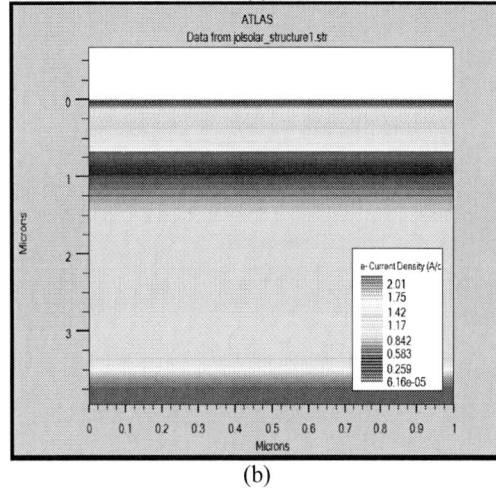

(b)

Fig.4 (a) Potential and (b) e- Current density in InGaP/GaAs DJ solar cell.

D. Photogeneration Rate and Optical Intensity

The modelled optimized device is illuminated to AM1.5 solar spectrum to simulate the Terrestrial sunlight, the LUMINIOUS 3D of SILVACO take into account the sunlight simulation. The photogeneration rate and the optical intensity in the model device is shown in Fig. 5. A maximum photogeneration rate of 23.6 (e- h+ pairs) per (s cm3) was obtained which is better than all the previously un-optimised structure. Also it can be seen from Figure 6(b) that the optical intensity of the device decreases in the bottom layer

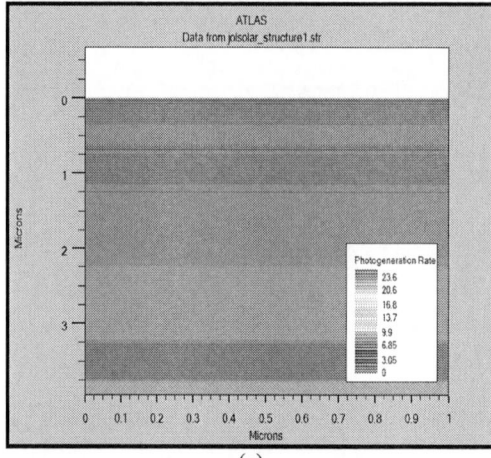

(a)

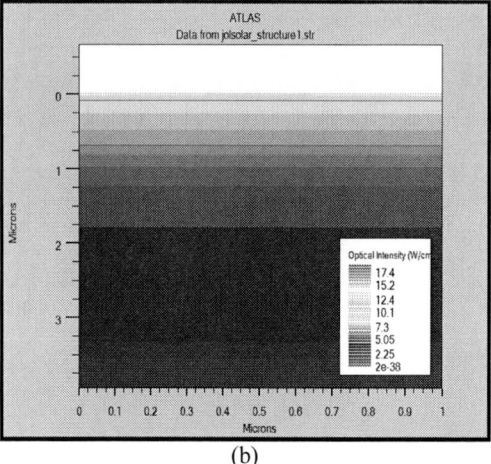

(b)

Fig. 5 (a)Photogeneration rate and (b)optical intensity in the optimised InGaP/GaAs DJ solar cell.

of the device and hence photogeneration rate in also low in that region.

V. CONCLUSION

Performance optimization of various layer of DJ InGaP/GaAs solar cell was performed over a different range of temperature and theoretical conversion efficiency upto 31.08% at 1 sun under AM1.5 illumination was obtained modelled using advanced TCAD tool Silvaco ATLAS. It was also observed that with increased in temperature V_{oc} decreases which is due to the increase in reverse saturation current, thus this leads to the decrease in the overall FF and hence efficiency of the solar cell. This work calculated the optimum thickness of the modelled solar cell so that maximum efficiency is gained even if the temperature is elevated. This data of different optimum thickness of DJ InGaP/GaAs solar cell obtained can be used for fabrication of optimised solar cell and would be helpful in the further R&D in temperature dependence of solar cell domain.

REFERENCES

[1] Wysocki, Joseph J., and Paul Rappaport. "Effect of temperature on photovoltaic solar energy conversion." Journal of Applied Physics 31, no. 3 (1960): 571-578.

[2] Singh, Priyanka, S. N. Singh, M. Lal, and M. Husain. "Temperature dependence of I–V characteristics and performance parameters of silicon solar cell." Solar Energy Materials and Solar Cells 92, no. 12 (2008): 1611-1616.

[3] Liu, L., Chen, N., Bai, Y., Cui, M., Zhang, H., Gao, F., Yin, Z. and Zhang, X., 2009. Quantum efficiency and temperature coefficients of GaInP/GaAs dual-junction solar cell. Science in China Series E: Technological Sciences, 52(5), pp.1176-1180.

[4] Cotal, H. L., et al. "Triple-junction solar cell efficiencies above 32%: the promise and challenges of their application in high-conceniration-ratio PV systems." Photovoltaic Specialists Conference, 2000. Conference Record of the Twenty-Eighth IEEE. IEEE, 2000.

[5] Nishioka, Kensuke, et al. "Evaluation of temperature characteristics of high-efficiency InGaP/InGaAs/Ge triple-junction solar cells under concentration." Solar energy materials and solar cells 85.3 (2005): 429-436.

[6] Landis, Geoffrey A., Danielle Merritt, Ryne P. Raffaelle, and David Scheiman. "High-temperature solar cell development." (2005).

[7] Green, Martin A. "General temperature dependence of solar cell performance and implications for device modelling." Progress in Photovoltaics: Research and Applications 11.5 (2003): 333-340.

[8] Sze, Simon M., and Kwok K. Ng. Physics of semiconductor devices. John wiley & sons, 2006.

[9] Cheyns, D., et al. "Analytical model for the open-circuit voltage and its associated resistance in organic planar heterojunction solar cells." Physical Review B 77.16 (2008): 165332.

[10] Singh, K.J., Sarkar, S.K.: Highly efficient ARC less InGaP/GaAs DJ solar cell numerical modeling using optimized InAlGaP BSF layers.Opt. Quantum Electron. 43, 1–21 (2012)

2019 Devices for Integrated Circuit (DevIC), 23-24 March, 2019, Kalyani, India

Metal Grain Granularity Induced Variability in Gate-All-Around Si-Nanowire Transistors at 1nm Technology Node

T. P. Dash[1*], S. Dey[2], J. Jena[3], S. Das[4], E. Mohapatra[5] and C. K. Maiti[6]

Department of Electronics and Communication Engineering,

Siksha 'O' Anusandhan (Deemed to be University), Bhubaneswar, Odisha, India-751030

[1*]taradash@soa.ac.in; [2] supravadey@soa.ac.in; [3]jhansiranijena@soa.ac.in; [4]sanghamitradas@ soa.ac.in; [5]eleenamohapatra@gmail.com; [6]ckmaiti@soa.ac.in.

Abstract – **As predicted, 5nm technology is not going to be ready for production until 2025 and it will be some sort of FinFET (possibly gate-all-around silicon nanowire or similar type of devices). It is time to search for advanced device structures such as nanowires. In this work, TCAD simulations are performed for the first time to evaluate the potential of 1nm gate length cylindrical Si channel nanowire field effect transistors (NW-FET) at extreme scaling limits. Effects of metal grain granularity (MGG) of the gate-all-around (GAA) NW-FET device have been studied to understand variability of the performance metrics such as, the threshold voltage, on-current, off-current, sub-threshold slope and drain induced barrier lowering. It is shown that the gate-all-around NW-FETs have the potential to replace FinFETs in future technology nodes because of their better channel electrostatic control.**

Keywords –*Scaling, 1N technology node, Nanowire FETs, technology computer aided design (TCAD), density-gradient (DG) model, metal grain granularity (MGG), Variability.*

I. INTRODUCTION

Silicon nanowires (SiNW) have received serious attention as biological [1] and chemical [2] sensors, and as thermoelectric [3, 4], photovoltaic [5], and nanoelectronic devices [6, 7]. For all of these applications, performance improvements are necessary in nanowire devices. Today's emphasis on low-power and low-voltage operation in industry and academe requires an understanding of the nanodevices at ultimate scaling. The obvious thing is to try and scale the devices further. As the integrated circuit technology scaling continues, the stacked horizontal nanowire FET has emerged as a potential successor to the FinFET. The nanowire field-effect transistors (NWFETs) are promised a great future in the field of micro-electronics because of the cylindrical shape. Unfortunately, stacking horizontal nanowires on top of each other puts a hard limit to the oxide thickness which is needed for thick oxide I/O transistors. IRDS considers nanowire as the ideal transistor [8]. Furthermore, the more and more severe metal grain granularity (MGG) and random dopant fluctuation (RDD) variation in scaled devices can cause intolerable threshold voltage variation [9, 10]. In this work, we present a systematic investigation on the effects of metal grain granularity fluctuations on the nanowire device performance at extreme gate length scaling of 1nm technology node.

II. DEVICE STRUCTURE FOR SIMULATION

Nanowire FETs can be implemented in a lateral or a vertical configuration. However, the diminishing cross-section to obtain the better short-channel effect immunity is traded off against the drain current enhancement. To obtain larger drive current for a given layout (nanowire diameter) with a high immunity against the short-channel effects, the vertical integration of nanowire is effective. In classical model such as Drift-Diffusion (DD) transport model, quantum effect has not been considered. But scaling down the device in nanometer range, quantum effect play a vital role in device characteristics. So these effects has to be considered for the simulation. Generally in classical theory the position of the particle can be determined precisely but in quantum mechanics the position of particle has been found out in terms of probability. In order to describe the position of electron, probability density function has been used and the behavior of an electron inside the crystal has been represented by wave function. In order to do the analysis, the cylindrical nanowire is cut perpendicularly to the cylindrical axis. For the structure shown in Fig. 1, the Poisson and Schrodinger equations have been solved using MINIMOS-NT [11] to include the quantum effects. These equations have been solved to obtain the electric potential distribution and energy levels in the device. The detail of device simulated is shown in Table-I.

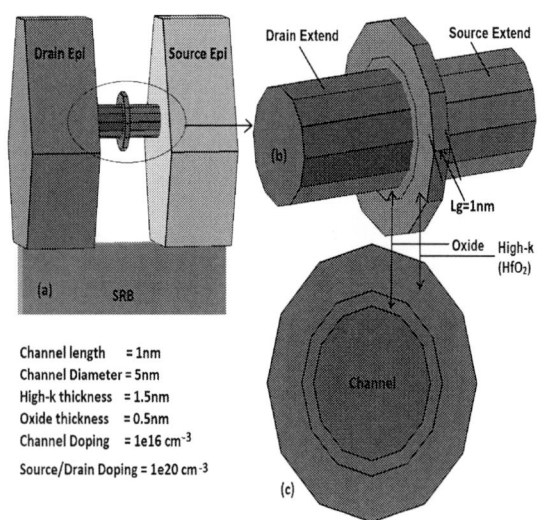

Fig. 1. Nanowire device schematic with a gate length of 1nm.

978-1-5386-6723-1/19 $31.00 © 2019 IEEE   286

TABLE I. DEVICE GEOMETRY AND OTHER PARAMETERS USED IN SIMULATION.

Parameters	Value	Units
Number of wire(s)	1	-
Nanowire Diameter	5	nm
Channel Length	1	nm
High-k (HfO$_2$) Thickness	1.5	nm
Oxide (SiO$_2$) Thickness	0.5	nm
Epi Length	14	nm
Epi Top Width	8	nm
Epi Middle Width	25	nm
Epi Bottom Width	14	nm
Substrate Height	350	nm
Channel Doping	1e16	cm^{-3}
S/D Doping	1e20	cm^{-3}

III. RESULTS AND DISCUSSION

Metal grain granularity (MGG) due to grain orientation leads to variability in effective work function of high-k metal gate in nanoscale devices. To model the variability in nanoscale devices, the spatial variation of work function needs to be accounted for. The atomistic details of the interface physics of the high-k/metal gate interface, is a critical component of work-function variability. We have performed atomistic simulations to reveal the impact of metal grain crystallographic orientation on the gate work function. In simulation, we use the Poisson-Voronoi model to simulate the polycrystalline grain structure of materials. For the present study of variability, we have generated two different work function patterns with one average grain sizes of 1nm. The work function variation in the metal gate has been shown in Fig. 2. Table II shows work function (WF) and its distribution probability with crystallographic orientation.

TABLE II. METAL GATE GRAIN WORK FUNCTION (WF) DISTRIBUTION

Orientation	Probability	WF[eV]	
<200>	50%	4.45	▬
<111>	50%	4.35	▬

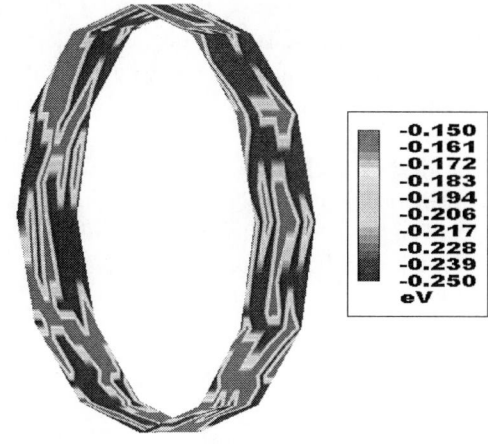

Fig. 2. Work function variation in the metal gate with 1nm grain size, has been taken for the analysis of MGG. The color bar on the right side represents the work function difference between metal and substrate in eV.

The potential distribution due to metal grain granularity has been presented in Fig. 3. Two cases are shown as with and without MGG introduction to simulation environment. Fig. 3 illustrates clearly the potential gets affected due to introduction metal grain granularity. The potential distribution is quite uniform in the outer periphery when MGG is not considered whereas it is quite non-uniform when MGG is introduced. This can also be viewed from cross-sectional view; where we observe the penetration of electric field towards the channel varies for both the cases. It ultimately results in changing the current density in the channel which has been shown in Fig. 4.

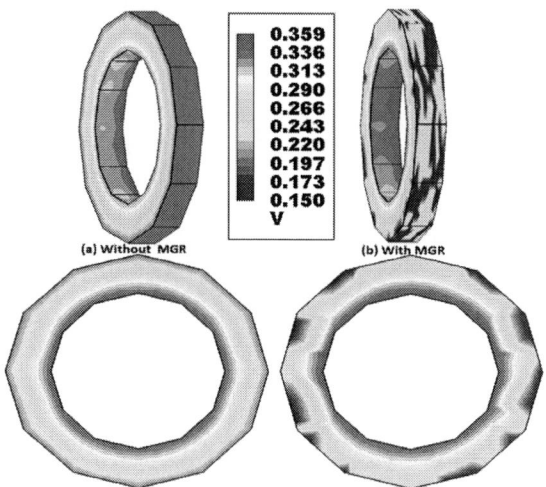

Fig. 3. The potential variation in high-k with 1nm grain size has been shown for the analysis of MGG. The color bar on the middle represents the potential variation (a) for without MGG and (b) with MGG effect consideration.

Fig. 4 shows the electron current density in the channel for with/without MGG simulation environment. Due to higher penetration of electric field intensity in with introduction of MGG condition, higher current density is observed. As we have used density gradient (DG) model with quantum corrections, the magnitude and distribution of electron density distribution changes in the wire. So the overall electron density get decreased as a result of which more gate voltage is required to achieve the same inversion charge compared to the classical case. Fig. 5 shows the transfer characteristics including metal grain granularity with average grain size of 1nm with 100 different configurations (red lines) along with the blue line condition as conventional without MGG.

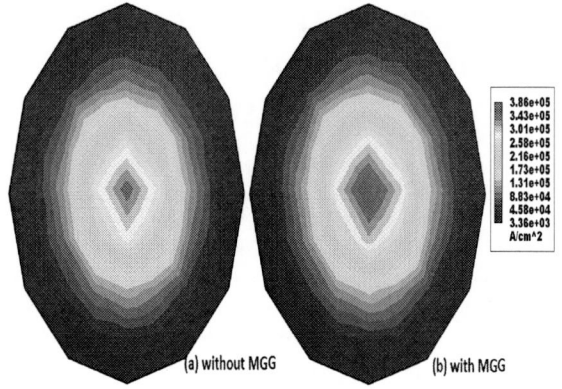

Fig. 4. Current Density in the channel (a) without MGG effect and (b) with MGG effect.

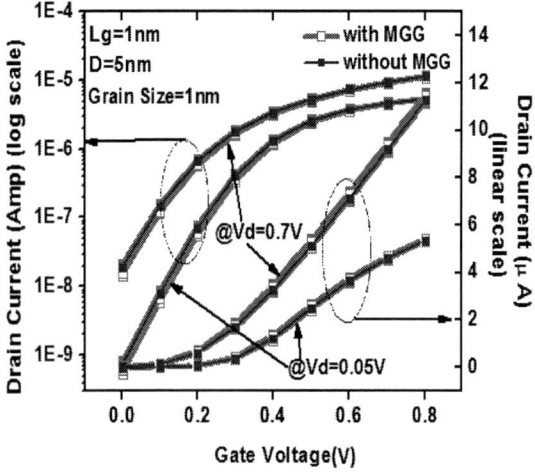

Fig. 5. Full electrical transfer characteristics simulation of 1nm gate length nanowire, including metal grain granularity with average grain size of 1nm with 100 different probability configurations, Different metal grain orientation has been introduced which causes the so called metal grain granularity. Blue line represents the normal configuration case when MGG effect is not included.

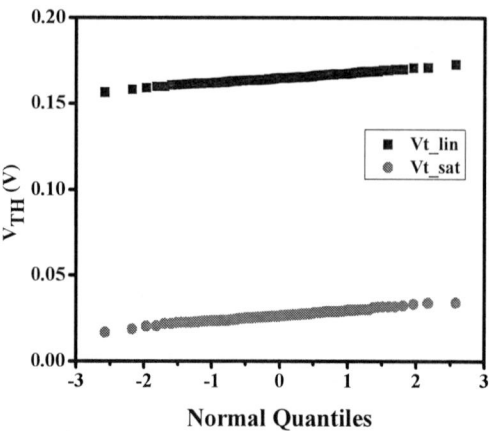

Fig. 6. The Q-Q test on V_T distribution MGG in both linear and saturation region. The distribution of V_{TH} in 1nm nanowire deviates marginally from the normal distribution on the upper tail and lower tail in both linear and saturation region. Vt_lin and Vt_sat correspond to 50mV and 800mV of drain bias, respectively.

The threshold voltage, V_{TH} is an important parameter in advance MOSFETs design. A stable, steady and well defined threshold voltage is necessary for the analog and digital circuits (minimum variation in V_{TH}). Threshold voltage should be kept within acceptable tolerance in order to get a reliable integrated circuit. However, in real nanoscale devices, V_{TH} fluctuations are induced by different sources of variability. The fluctuation of V_{TH} in the linear and saturation region are shown (normal QQ plots) in Fig. 6 due to MGG as a source of variability. For further understanding of the effect of MGG on V_{TH}, the distribution of V_{TH} for 100 different configurations of simulations are shown in Fig. 7. The mean value of V_{TH} is found to be 164.85mV and the standard deviation is 2.87mV.

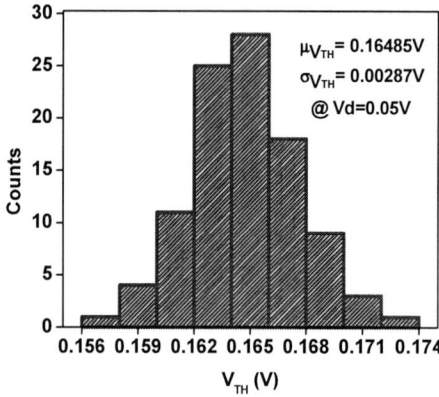

Fig. 7. Histogram plots of V_{TH} for 100 different configurations following MGG as source of statistical variability. Mean (μ) and standard deviation (σ) are shown.

The effects of MGG on the ON- and OFF-state currents have been studied and are shown in Figs. 8 and 9. The Ioff varies from 532pA to 806pA for 100 different configurations. Fig. 8 shows the distribution of Ioff for these simulation configurations. The mean value is found to be 655.028 pA and the standard deviation is 47.536pA. Similarly, the ON state current ranges from 5.3237µA to 5.4346µA. The mean and standard deviation for Ion are observed to be 5.376 µA and 0.0197µA, respectively. From these results it can be concluded that the effect of MGG on Ioff is more severe than on Ion.

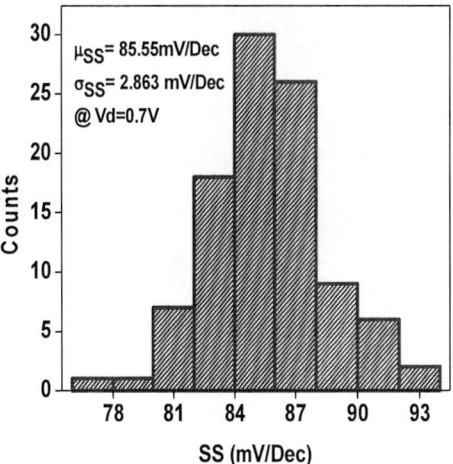

Fig. 10. Histogram plots of subthreshold slope for 100 different configurations following MGG as source of statistical variability.

The impact of MGG on SS (subthreshold slope) has been shown in Fig. 10. The subthreshold slope has been extracted for a drain bias of 0.7V from Fig. 5. It can be clearly observed that the mean value of subthreshold slope is 85.55mV/dec and the standard deviation is 2.863mV/dec. The slight degradation in SS is observed when DG corrections are included in simulations. When DG correction are included, the shape and magnitude of electron density distribution changes in the wire. So the overall electron density get decreased as a result of which more gate voltage is required to achieve the same inversion charge compared to the classical case and a V_{TH} shift is observed. Also the charge density shifts away from the interface so the control of gate on the channel reduces leading to SS degradation.

The impact of MGG on DIBL has also been investigated and is shown in Fig. 11. DIBL can be defined as ratio of absolute change of threshold voltages at linear and saturation drain bias to the change of drain bias. The mean value of DIBL is found to be 184.45mV/V and the standard deviation (SD) is 86.377mV/V, where the change of threshold voltage is in mV unit and change of drain voltage is in volt unit range.

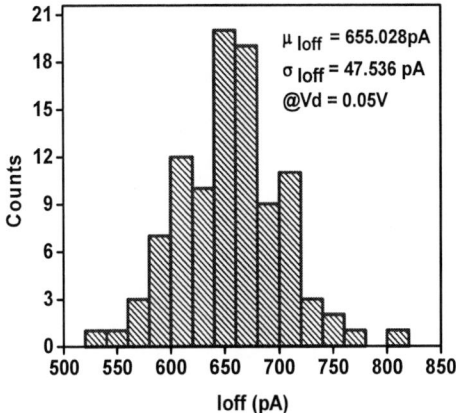

Fig. 8. Histogram plots of Ioff for 100 different configurations following MGG as source of statistical variability.

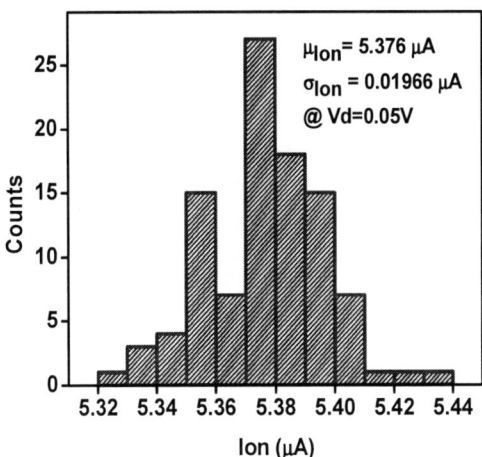

Fig. 9. Histogram plots of Ion for 100 different configurations following MGG as source of statistical variability.

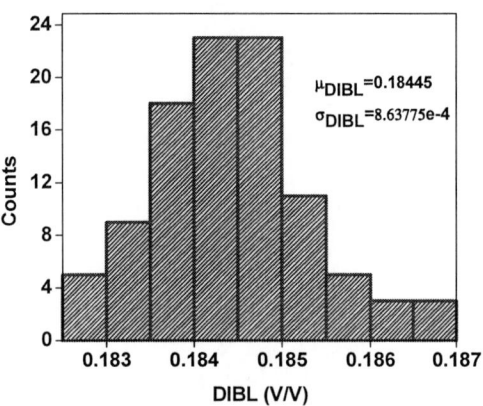

Fig. 11. Histogram plots of DIBL for 100 different configurations following MGG as source of statistical variability.

IV. CONCLUSIONS

Si-NW field-effect transistors (FETs) provide the most important class of devices for investigation and optimization of NW electronic properties, regardless of the final intended application. As predicted, 5nm FinFET technology is not going to be ready for production until 2025 and it will be some sort of FinFET (possibly gate-all-around silicon nanowire or similar type of device), it is time to search for advanced device structures such as multiple stacked nanowires (or perhaps vertical nanowires). Stacking vertically NW channels could be the ultimate of conventional CMOS scaling). Effects of metal grain granularity on the performance metrics such as, the threshold voltage, on-current, off-current, sub-threshold slope and drain induced barrier lowering of gate-all-around nanowire transistors have been studied to understand variability of 1nm gate length at extreme scaling limits.

REFERENCES

[1] Y. Cui, Q. Q. Wei, H. K. Park, and C. M. Lieber, "Nanowire nanosensors for highly sensitive and selective detection of biological and chemical species," Science, vol. 293, pp. 1289-1292, 2001.

[2] M. C. McAlpine, H. Ahmad, D. W. Wang, and J. R. Heath, "Highly ordered nanowire arrays on plastic substrates for ultrasensitive flexible chemical sensors," Nature Materials, vol. 6, pp. 379-384, 2007.

[3] A. I. Boukai, Y. Bunimovich, J. Tahir-Kheli, J. K. Yu, W. A. Goddard, and J. R. Heath, "Silicon nanowires as efficient thermoelectric materials," Nature, vol. 451, pp. 168-171, 2008.

[4] J. K. Yu, S. Mitrovic, D. Tham, J. Varghese, and J. R. Heath, "Reduction of thermal conductivity in phononic nanomesh structures," Nature Nanotechnology, vol. 5, pp. 718-721, 2010.

[5] B. Z. Tian, X. L. Zheng, T. J. Kempa, Y. Fang, N. F. Yu, G. H. Yu, J. L. Huang, and C. M. Lieber, "Coaxial silicon nanowires as solar cells and nanoelectronic power sources," Nature, vol. 449, pp. 885-U8, 2007.

[6] N. Singh, K. D. Buddharaju, S. K. Manhas, A. Agarwal, S. C. Rustagi, G. Q. Lo, N. Balasubramanian, and D. L. Kwong, "Si, SiGe nanowire devices by top-down technology and their applications," IEEE Transactions on Electron Devices, vol. 55, pp. 3107-3118, 2008.

[7] H.-H. Hsu, T.-W. Liu, L. Chan, C.-D. Lin, T.-Y. Huang, and H.-C. Lin, "Fabrication and characterization of multiple-gated poly-Si nanowire thin-film transistors and impacts of multiple-gate structures on device fluctuations," IEEE Transactions on Electron Devices, vol. 55, pp. 3063-3069, 2008.

[8] IRDS, Executive Summary. Rapport technique, International Roadmap for Devices and Systems, 2017.

[9] International technology roadmap for semiconductors: process integration, devices, and structures, 2015 edition.

[10] K. Takeuchi, T. Tatsumi, and A. Furukawa, "Channel engineering for the reduction of random-dopant-placement-induced threshold voltage fluctuation," in International Electron Devices Meeting, Technical Digest, 1997, pp. 841-844.

[11] MINIMOS-NT User's manual, 2017.

SPICE Parameter Extraction of Tri-Gate FinFETs-
An Integrated Approach

T. P. Dash[1*], S. Das[2], S. Dey[3], E. Mohapatra[4], J. Jena[5] and C. K. Maiti[6]

Department of Electronics and Communication Engineering,

Siksha 'O' Anusandhan (Deemed to be University), Bhubaneswar, Odisha, India-751030

[1*]taradash@soa.ac.in; [2]sanghamitradas@ soa.ac.in; [3]supravadey@soa.ac.in; [4]eleenamohapatra@gmail.com; [5]jhansiranijena@soa.ac.in; [6]ckmaiti@soa.ac.in.

Abstract — **The FinFET transistor is now considered the most probable successor of the bulk MOSFET transistor in the global race for miniaturization in the field of micro- and nanoelectronics. The development of integrated circuits using FinFETs is made possible only by the use of their compact models. These models must predict precisely the electrical behavior of these devices advanced technologies. In this work, we show an integrated approach for SPICE parameter extraction explicitly for nanoscale FinFETs, which is validated by comparisons with simulation results. We discuss in detail the platform necessary for the development of the model and automated SPICE parameter extraction. The predictive capability of TCAD to estimate the SPICE model parameters from process-based on the physical variations of process parameters has been examined.**

Keywords— SPICE parameter, Variability, FinFET, technology computer aided design (TCAD).

I. INTRODUCTION

The evolution of microelectronics and silicon CMOS technology goes hand-in-hand towards miniaturization. FinFET technology is intended to extend the CMOS technology up to a gate length of 3 nm [1]. Also, the industry is moving away from single gate transistors to multi-gate transistors for better control of the channel. Choosing a low-doped channel is preferable to avoid the doping fluctuations that exist for Bulk current MOSFETs, and allows an improvement of the mobility. Several compact models of FinFET and Double-gate MOSFETs have been published [2]. However, there is still a lack of analytical compact models for devices. Use of technology computer aided design (TCAD) is now essential before a device can reach production. Prediction of statistical circuit performance variation is becoming critical for 22nm technology node and below. The integrated TCAD approach used in this work is shown in Figure 1.

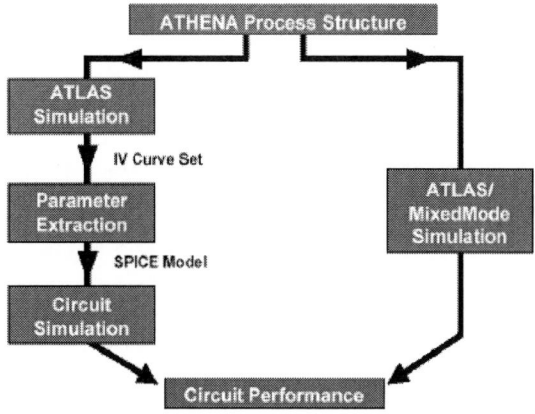

Fig. 1. Flow diagram for integrated TCAD-based device design viz., process to device simulation and SPICE model parameter extraction essential for circuit design (Source: Silvaco).

However, as the technology moves toward the technology nodes projected by the ITRS [3], the prototypes of the scaled-down transistors must be physically tested to prove that the electrical characterizations are consistent and reliable. This will insure that it follows the analytical model and proves the validity of the simulations as the device is scaled down. Once the transistors are proved to have consistent electrical characteristics, a simulation model needs to be developed. The models are developed by extracting parameters of the device with a semiconductor characterization system. Once the parameters are extracted, a simulation model can be developed so that designers can design components and systems with them. With the industry shifting towards double gated and other multi-gated transistors, a technique for parameter extraction needs to be developed to allow for better modeling of double and multi-gate transistors. Our main focus is on SPICE parameters extraction for DC modeling which will need, namely the I_d vs. V_g and I_d vs. V_d characteristics.

II. TCAD OF BULK-SI FINFET DEVICE

The TCAD flow begins with simulating a layout in process simulation to create structure representing a semiconductor device. The structure is prepared for device simulation by re-meshing and adding contacts. After device simulation is complete, the electrical response is captured from device simulation. Figure 2 shows the process-simulated FinFET 3D device structure of the modeled FinFET in this work.

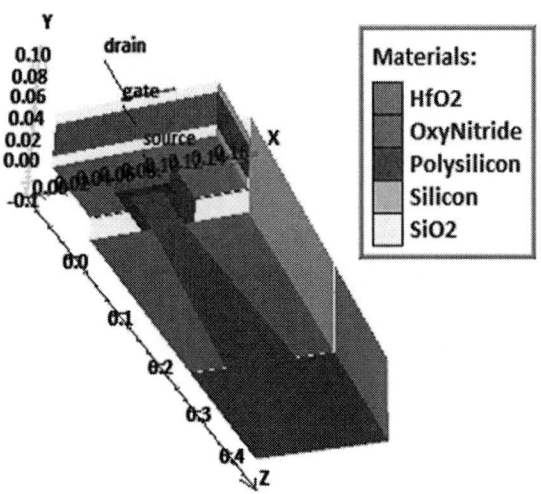

Fig. 2. Process simulated FinFET structure used for device simulation and SPICE parameter extraction.

III. SIMULATION ENVIRONMENT

While device simulators are reasonably accurate for a given transport model used, they are very slow for large-scale circuit simulation. Compact models specifically for FinFETs [4] serve as crucial link between process technology and circuit simulation, by leveraging inputs from TCAD simulation. SPICE models used in circuit design are traditionally extracted from measurements taken on working transistors fabricated in a process technology. A typical flow diagram (procedure) SPICE parameter extraction using TCAD is shown in Figure 3.

Software platform for simulation and parameter extraction are described below. The objective of parameter extraction is to find a set of parameter values for the model, valid for a given technology node. Readily available and industry-standard standalone parameter extraction software programs are Celestry's BSIMPro and SimuCAD's UTMOST III or UTMOST IV. The inputs and outputs of UTMOST IV are shown in Figure 3. Utmost IV Optimization Module [5] provides an easy to use, database-driven environment for the generation of accurate, high quality SPICE models and macro-models for analog, mixed-signal and RF applications.

The values of SPICE parameters of the model may be obtained by software like ICCAP and UTMOST which need a built-in implementation of compact model. In order to extract parameters from a device, three things are needed and will be discussed briefly below. First, a semiconductor characterization system is needed as well as a probe station. Second, the right software is needed to be able to set up different tests on the devices. Most semiconductor device analyzers come with software already installed by the original equipment manufacturer. Finally, in order to characterize a device, a methodology for parameter extraction must be followed so that device models can be designed to simulate correctly. Software used by the equipment for parameter extraction is the equipment interface. The software can convert measured and simulated data to SPICE model parameters for compact modeling. The semiconductor characterization systems in the previous section come with their own extraction software. In order to successfully design chips with higher integration and higher transmission speed, the work of modeling engineers to develop accurate device models up to tens of Gigahertz becomes more and more challenging. An absolute prerequisite for achieving this goal are accurate measurements.

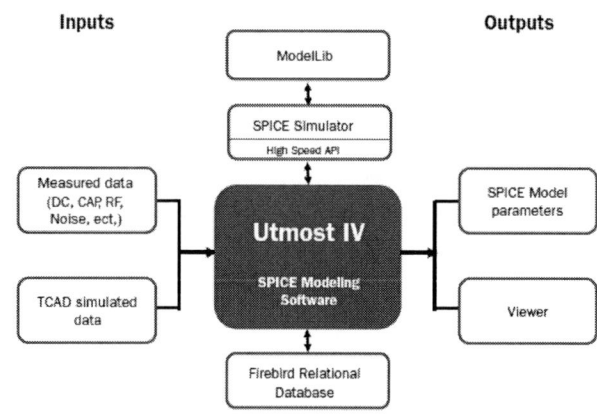

Fig. 3. UTMOST IV Optimization Module used for SPICE parameter extraction (Source: Silvaco).

The extraction is carried out through a software platform described in Figure 3. By the use of script programs, the UTMOST platform allows an automated execution of 3D simulations, and the execution of a procedure for extraction of parameters. The extraction of SPICE parameters is carried out from the data from experimental measurements of transistor electrical quantities, or from data of simulations of electrical quantities, if we have no measurement data. The measurements or simulations are typically current-voltage (I-V) and capacitance-voltage (C-V) characteristics. However, the use of a software allowing to simulate in 3D transistor is necessary to study the physical phenomena to be modeled that takes place in the device.

UTMOST can accept data from process and device simulation and direct measurements on devices. UTMOST also support the simulation of DC, transient, capacitance and s-parameter characteristics. UTMOST supports commercial device models, user-defined device models and macro models. Direct communication with the models in most popular commercial circuit simulators is also supported. The flow diagram used in UTMOST is shown in Figure 3.

IV. RESULTS AND DISCUSSIONS

The n-channel FinFET considered here has been virtually fabricated using ATHENA process simulator. During process simulation, extract statements are used to extract important process parameters. In order to obtain SPICE model extraction, those extracted process parameter are to be used in UTMOST, in proper scale, e.g., the gate oxide thickness in meter. The process-simulated FinFET structure is interfaced with device simulator ATLAS. The first ATLAS run generates I_d-V_g characteristics (see Figure 4) for two V_d = 0.05V and 0.1 V, respectively. Though two drain biases yield two transfer characteristics curve, both the characteristics are to be saved in a single log file. Then, the second ATLAS run simulates I_d-V_d characteristics at 4 different gate voltages for V_d up to 1V, respectively. All the four characteristics are saved to a single log file similar to first run.

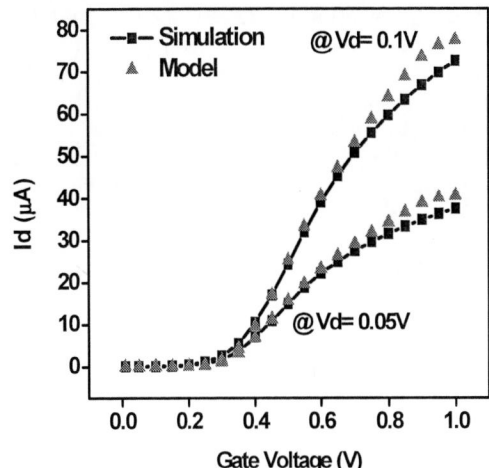

Fig. 5. Drain Current vs. Gate Voltage comparison. Model: red; simulations: black.

Figure 5 shows the comparison of the I_d-V_g characteristics for the FinFET. The slope below threshold is correctly predicted for different channel lengths as shown by the good agreement with simulations in the region of weak inversion.

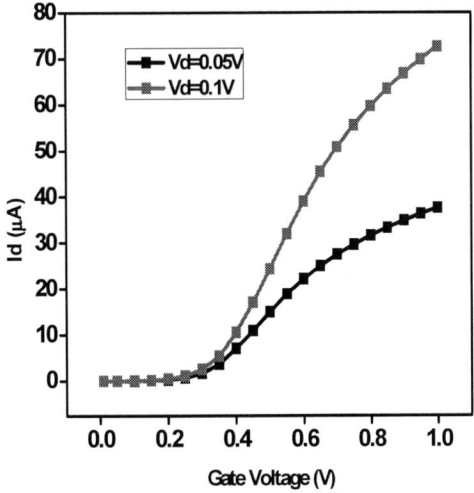

Fig. 4. Drain Current vs. Gate Voltage for the FinFET for V_d =0.05V and 0.1V.

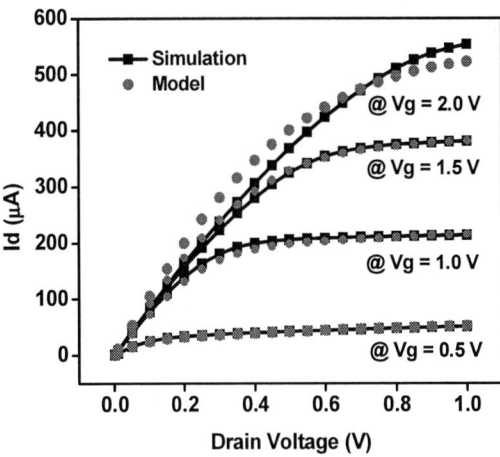

Fig. 6. Drain Current vs. Drain Voltage comparison. Model: red; simulations: black.

Once device simulation is over, the final stage is to run UTMOST in order to extract the SPICE model parameters. The critical and important details of the device structure such as gate length, oxide thickness etc., are to be transferred to UTMOST environment using the extracted results obtained from process simulator ATHENA as discussed earlier. The set of log files (I_d-V_g) and (I_d-V_d) obtained from device simulator ATLAS are loaded. Then the UTMOST tool can fit SPICE models to complete the given sets of log files. It can generate and store the complete set of UTMOST parameters to a file which can be used to see the parameters of interest (Figure 7). While the number of parameters used by device modeling software was less than ten in the 1960s, today, some programs use hundreds of different parameters necessary for precise model development. Also, modeling programs have different parameter extraction methodologies. Some of SPICE parameters (partial list) extracted are shown in Figure 7.

```
.MODEL nmos NMOS {
+LEVEL   = 14        VERSION  = 3        BINUNIT = 1
+PARAMCHK = 1        MOBMOD   = 1        RDSMOD  = 0
+IGCMOD  = 0         IGBMOD   = 0        CAPMOD  = 2
+RGATEMOD = 0        RBODYMOD = 0        TRNQSMOD = 0
+ACNQSMOD = 0        FNOIMOD  = 1        TNOIMOD = 0
+DIOMOD  = 1         PERMOD   = 1        GEOMOD  = 0
+EPSROX  = 3.9       TOXE     = 1.5e-09  DTOX    = 0
+XJ      = 1.5e-07   NDEP     = 1.7e+17  NGATE   = 0
+NSD     = 1e+20     RSH      = 0        RSHG    = 0.1
+VTH0    = -0.0322836  VFB    = -1       PHIN    = 0
+K1      = 0.5       K2       = 0        K3      = 80
+K3B     = 0         W0       = 2.5e-06  LPE0    = 1.74e-07
+LPEB    = 0         DVT0     = 2.2      DVT1    = 0.53
+DVT2    = -0.032    DVTP0    = 0        DVTP1   = 0
+DVT0W   = 0         DVT1W    = 0        DVT2W   = -0.032
+U0      = 0.0956499  UA      = 5.89289e-09  UB  = -8.28174e-18
+UC      = -4.6e-11  EU       = 1.67     VSAT    = 221595
+A0      = 13.9248   AGS      = 10.5127  B0      = 0
```

Fig. 7. UTMOST extracted BSIM3 SPICE parameters for the bulk-Si FinFET (partial list).

The results presented here show how worst-case SPICE model parameters can be derived by the simulation of variations in FinFETs fabricated in the CMOS manufacturing process. The whole process can be automated as an experiment in the Virtual Wafer Fabrication (VWF). Statistical fluctuations inherent in any IC manufacturing process cause variations in device and circuit performance. Product yield and manufacturing problems requires costly redesign cycles. SPAYN can be used to determine the underlying statistical nature of IC manufacturing processes. Resulting variations in circuit performance caused by process fluctuations can then be predicted early in the design cycle.

V. CONCLUSIONS

As modern circuits are usually very complex, the performance of such circuits is difficult to predict without accurate device models. The focus of this work is on device compact modelling methodologies and SPICE parameter extraction of bulk-Si FinFET. An N-Well CMOS involving a bulk-Si FinFET was process/device simulated and BSIM3 model parameters were extracted using UTMOST. In this work, we presented an integrated approach for SPICE parameter extraction essential for compact model development for circuit designs.

REFERENCES

[1] C. K. Maiti, S. Chattopadhyay, and L. K. Bera, *Strained-Si Heterostructure Field-Effect Devices*, CRC Press (Taylor & Francis), USA, 2007.

[2] M. V. Dunga, C.-H. Lin, X. Xi, D. D. Lu, A. M. Niknejad, and C. Hu, "Modeling advanced FET Technology in a compact model", IEEE Trans. Electron Dev. Vol. 53, pp. 1971-1978, 2006.

[3] ITRS 2015. http://www.itrs.net.

[4] Y. S. Chauhan, D. D. Lu, V. Sriramkumar, S. Khandelwal, J. Duarte, N. Payvadosi, A. Niknejad, and C. Hu, "FinFET Modeling for IC Simulation and Design", Elsevier 2015.

[5] UTMOST User's manual, Silvaco, 2017.

TCAD Modelling of 30nm Strained-Si/SiGe/Si Channel MOSFET

Lalthanpuii Khiangte, Rudra Sankar Dhar
Department of Electronics & Communication Engineering
National Institute of Technology, Mizoram
Aizawl, India
rdhar@uwaterloo.ca

Abstract—Technical-Computer-Aided-Design (TCAD) tools are the bridging element between electronics design and manufacturing. Design of novel devices using TCAD initially and analyzing all the physics prior to direct manufacturing reduces the cost and increase the optimization in device operation. Modeling of the scaled down tri-layered strained-Si/strained-SiGe/strained-Si channel based heterostructure on insulator (HOI) Metal-Oxide-Semiconductor-Field-Effect-Transistor (MOSFET) device developed using Sentaurus TCAD is illustrated in this paper. A complete modeling of the device physics that encountered for the reduced dimension device remains the crux of this paper. As a result, an enhanced drive current pertaining to the scaled HOI MOSFET is achieved, thus ensuring the further scalability of the said HOI MOSFET device for improved performance.

Keywords—TCAD, HOI, MOSFET, strained Silicon, strained Silicon-Germanium, tri-layered channel.

I. INTRODUCTION

Strained silicon technology is no new concept for the improvement of device performance as also recently observed by Khiangte et al. [1]. Therefore, with its wide range of applicability in various ways and its effects on the device operation made the research area of strained Si technology more attractive for many scholars worldwide till date. Two types of results occur as strain is applied in the channel region of the HOI MOSFET, these are: band energy level shifting and the electronic states within the structure splitting [2]. Therefore, the twofold valleys in the energy level shifts down in contrast to the fourfold valleys at the conduction band edge due to the strained channel region, which instigate splitting of the energy bands thereby increases the occupancy of electrons in twofold valleys. This results into developing a twofold degenerate energy band, congregating in enhancement of electron mobility within the channel, hence on reducing phonon scattering it suppresses intervalley transition of electrons from lower to upper valley in the nano regime of the MOSFET channel [3]–[7].

International Technology Roadmap for Semiconductors (ITRS), predicts the 40% development costs reduction in 2016 because of TCAD which confirmed the significance of TCAD tools in the global manufacturing industries [8]–[10]. This has given tremendous growth in design, development and understanding of the device physics prior to manufacturing. Thus, all the possible device physics for the design structure must be described by the inclusion of appropriate rigorous models.

The development of the three layered s-Si/s-SiGe/s-Si channel MOSFET was looked into [11], [12], that left with a question on its scalability for further miniaturization and enhancement of drive current.

II. TRI-LAYER CHANNEL MOSFET

A. Device Structure

Unlike any other strained heterostructure MOSFET, a three layered channel: s-Si/s-SiGe/s-Si was developed by Khiangte et al. [11] with an indicative enhancement of drive current over the conventional strained Si channel MOSFET device. The device design under investigation is shown in Fig. 1. As depicted, the channel is engineered by three layers: (i) Upper strained Si (ii) strained SiGe and (iii) Lower strained Si. The device dimensions and parameters used are tabulated in Table 1. Both the strained Si layers and the strained SiGe layers are of 1.5nm and 3nm thick, respectively.

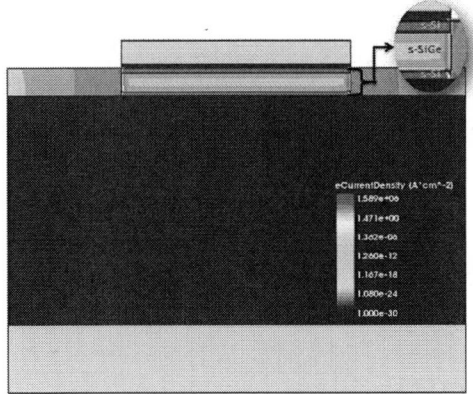

Fig. 1. Schematic of the S-Si/s-SiGe/s-Si Channel HOI MOSFET.

TABLE I.   DEVICE PARAMETERS

Parameters	Value
Strained Si (t_{si})	1.5nm, 2nm
Strained SiGe (t_{sige})	3nm, 6nm
Ge mole fraction	0.4
Gate-oxide	2nm
Source/Drain doping (N_D)	10^{20} cm^{-3}
Channel doping	10^{17} cm^{-3}
Drain Bias	50mV

B. Simulation Approach

The device structure and concept described earlier have been design and simulated using Synopsis TCAD tool [13], [14]. To model such structure with inclusion of the possible physics completely is a handfull task i.e different models are required for the band structure, mobility, quantization, effective mass etc.

Starting with the strain-induced change of the silicon band structure, several valleys in the conduction and valence bands are needed to be consider for computing the correct dependency of the carrier concentration on the quasi–Fermi level. So, a Multivalley Band structure model have been included along with the Modified Local-Density Approximation (MLDA) model that calculates the confined carrier distributions that occur near semiconductor–insulator interfaces [13]. This model consider a dependency of the quantization effect on interface orientation and stress. Also the effective mass changes for both carriers are incorporated. Additional 1D Schrodinger Model for carrier confinement in the thin s-Si layer was incorporated using non-local meshing which encountered the threshold-voltage roll-off cause due to inducement of strained in the silicon layers.

.

III. RESULTS AND DISSCUSSION

With the reduction of channel length to 30nm, various dimensions and parameters are precisely scaled and corrected values are induced for performance enhancemnet of the device and also minimizes different short channel effects. The initial consideration in this novel device fall upon the selection of s-SiGe thickness when both the s-Si layers are at a constant thickness value (s-Si = 1.5nm). Thus, the effect of s-SiGe thickness on device performance have been studied for five device as shown in Fig. 2.

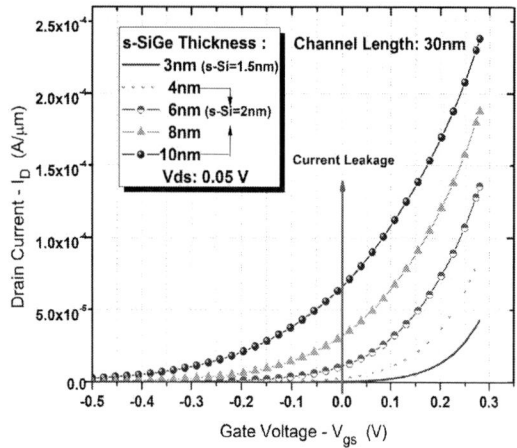

Fig. 2. HOI MOSFET with variation in s-SiGe thickness (Lg=30nm)

As a result, increase in s-SiGe thickness lead to higher off-current in MOSFET, hence many fold higher current leakage is observed. Based upon these results s-SiGe thickness of 3nm in combination with s-Si thickness of

1.5nm have been developed here. The calculated leakage current (I_{off}) for both 30nm and 50nm channel length device are illustrated as shown in Fig.3, which are in the acceptable range of leakage as suggested by C.Hu [15]. Simulated result of device characteristics are depicted in Fig.4, comparison with 50nm channel length of the same device structure [12] clearly indicate the increase quasi-balistic carrier transport in the shorter channel region.

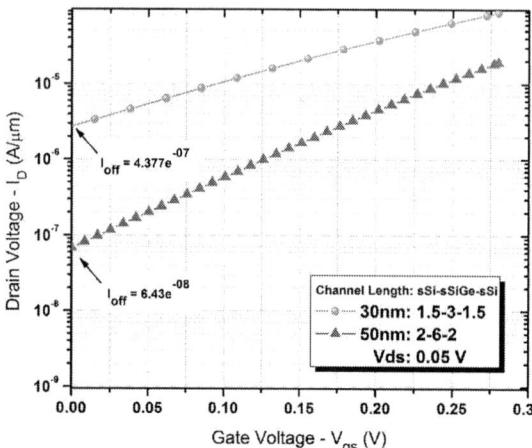

Fig. 3. Current-Voltage characteristics of 30nm channel length HOI

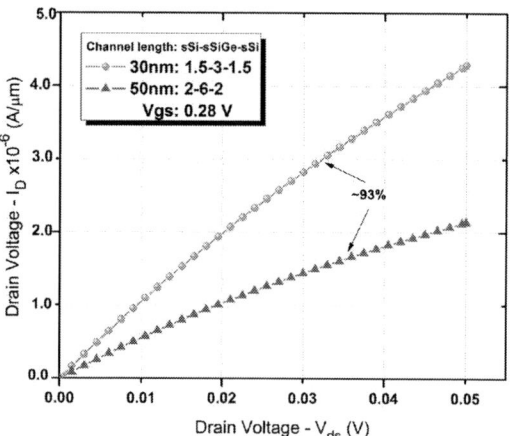

Fig. 4. Output characteristics of Tri-layered channel HOI MOSFET (Lg=30nm)

With the reduction of channel length scaled down to 30nm, the electric field in the channel region increased as is explicitly compared and presented in Fig. 5. Thus, an increase in the drift velocity of carriers is observed, which results in enrichment of drive current by ~93%.

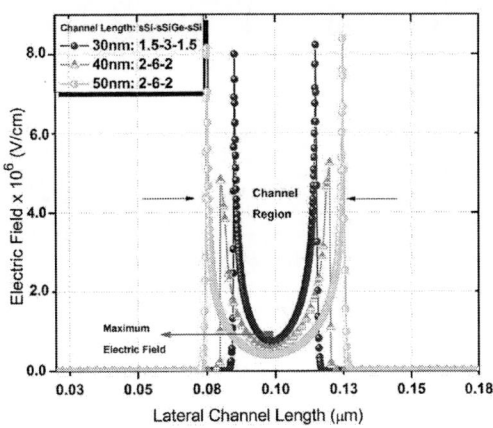

Fig. 5. Electric field along the lateral channel region of Tri-layered HOI MOSFET

IV. CONCLUSION AND OUTLOOK

Different modelling parameters for strained induced energy band structure, mobility, effective mass and quantization were incorporated in the HOI MOSFET for the scaled dimension of 30nm channel length using Sentaurus TCAD tool. An enhanced drive current with acceptable current leakage was modeled and developed as a result of increased drift velocity due to increase in electric field in the channel region.

ACKNOWLEDGMENT

The authors acknowledge NIT Mizoram and especially to the SMDP-C2SD project for providing the required amenities such as providing the workstation for simulation of the research work.

REFERENCES

[1]   L. Khiangte and R. S. Dhar, "Development of Tri-Layered s-Si/s-SiGe/s-Si Channel Heterostructure-on-Insulator MOSFET for Enhanced Drive Current," *Phys. status solidi*, p.

1800034, 2018.

[2]   N. Kharche, M. Prada, T. B. Boykin, and G. Klimeck, "Valley splitting in strained silicon quantum wells modeled with 2 miscuts, step disorder, and alloy disorder," *Appl. Phys. Lett.*, vol. 90, no. 9, p. 92109, 2007.

[3]   M. L. Lee, E. A. Fitzgerald, M. T. Bulsara, M. T. Currie, and A. Lochtefeld, "Strained Si, SiGe, and Ge channels for high-mobility metal-oxide-semiconductor field-effect transistors," *J. Appl. Phys.*, vol. 97, no. 1, p. 1, 2005.

[4]   S. E. Thompson, G. Sun, Y. S. Choi, and T. Nishida, "Uniaxial-process-induced strained-Si: Extending the CMOS roadmap," *IEEE Trans. Electron Devices*, vol. 53, no. 5, pp. 1010–1020, 2006.

[5]   T. K. Maiti, S. S. Mahato, and C. K. Maiti, "Modeling of Strain-Engineered Nanoscale MOSFETs," in *4th Int. Conf. Nanotechnol. Health Care Appl.(NateHCA-07)*, *Mumbai, India, pp. D41–D45*, 2007.

[6]   M. Willander, M. Y. Yousif, and O. Nur, "Nanostructure Effect in Si-MOSFETs," CHALMERS UNIV OF TECHNOLOGY GOETEBORG (SWEDEN) DEPT OF PHYSICS, 2001.

[7]   M. J. Kumar and T. V. Singh, "Quantum confinement effects in strained silicon mosfets," *Int. J. Nanosci.*, vol. 7, no. 02n03, pp. 81–84, 2008.

[8]   D. Z. Pan, B. Yu, and J.-R. Gao, "Design for manufacturing with emerging nanolithography," *IEEE Trans. Comput. Des. Integr. Circuits Syst.*, vol. 32, no. 10, pp. 1453–1472, 2013.

[9]   R. Minixhofer, "TCAD as an integral part of the semiconductor manufacturing environment," in *Simulation of Semiconductor Processes and Devices, 2006 International Conference on*, 2006, pp. 9–16.

[10]   I. Lysenko, D. Zykov, S. Ishutkin, and R. Meshcheryakov, "The use of TCAD in technology simulation for increasing the efficiency of semiconductor manufacturing," in *AIP Conference Proceedings*, 2016, vol. 1772, no. 1, p. 60012.

[11]   L. Khiangte and R. S. Dhar, "Development of double strained Si channel for heterostructure on insulator MOSFET," in *Man and Machine Interfacing (MAMI), 2017 2nd International Conference on*, 2017, pp. 1–3.

[12]   L. Khiangte and R. S. Dhar, "Double Strained Si Channel Heterostructure on Insulator MOSFET in sub-100nm regime."

[13]   T. Sentaurus, "Sdevice User Guide, ver," *G-2012.06, Synopsys*, 2012.

[14]   T. Sentaurus, "User Manual, Synopsys," *Inc., Mt. View, CA, Version F-2011.09*, 2010.

[15]   C. Hu, "MOSFETs in ICs–scaling, leakage, and other topics," *Mod. Semicond. Devices Integr. Circuits. Prentice Hall, New York*, 2009.

Novel Self-Pipelining Strategy for Efficient Multiplication

Rahul Pal
Department of ECE
, Bengal College of Engineering
Durgapur, India
rahulmail08@gmail.com

Jayanta Ghosh
Department of ECE
National Institute of Technology
Patna, India
jghosh@nitp.ac.in

Aloke Saha
Department of ECE
Dr. B. C. Roy Engineering College
Durgapur, India
saha81@gmail.com

Abstract— In this work a novel self-pipelining strategy with proper synchronization among subsequent stages for low-power high-speed digital multiplication is explored. The true and complementary clock inputs are alternatively applied to each subsequent self-latching stage to achieve proposed self-pipelining operation. A 4-b×4-b self-pipelined Wallace-tree multiplier based on aforesaid idea has been designed first. Next, the 4-b×4-b multiplier thus designed has been exploited in order to design proposed self-pipelined 8-b × 8-b multiplier with decomposition logic. All the designs and optimization are performed on TSMC 0.18μm CMOS technology based on BSIM3 device parameters with 1.8V supply rail and at 25°C temperature using S-Edit of Tanner EDA V.13. The performance of designed multiplier has been evaluated with T-Spice simulation using W-Edit. A benchmarking comparison with respect to latest competitive design establishes superiority of proposed idea.

Keywords—Decomposition Logic, Register-Pipelining, Self-Latching, Self-Pipelining, Tree Multiplier

I. INTRODUCTION

High-speed and low-power are the two parallel design aspirations for present digital VLSI designer [1-3]. Multiplier plays as a major bottleneck to improve speed-power performance of most computation intensive digital applications. A number of researchers are working to enhance speed-power performance for parallel multiplication and that can be evidenced in [4-15]. Conventional Register-pipelining [12] can offer improved throughput performance of digital multiplier at the expense of increased power, area and overall latency [6]. Also the uncontrolled (generated) clock-skew at different point may also lead to system malfunctioning. Designing clock-driver is another difficult job for register pipelined system [6-7].

Wave-pipelining [6-7, 16] in other hand can eliminate the problems associated with conventional register-pipelining by removing intermediate latches/registers from the circuit. In wave-pipelined system the maximum throughput depends not on the critical path delay but on the delay difference between longest and shortest path delay [16]. With this idea the area, power and latency overhead of register pipelining can be eliminated. More about wave-pipelining can be found in [6-7, 16]. However, in order to achieve speed-power efficient wave-pipelined system the time/delay equalization among all the data paths from input to output is extremely important. The data dependent delay [6-7] at leaf-cell level has a major role to achieve effective time-equalization. Also the PVT (Process

Voltage Temperature) effect may lead to wave-pipelined system unreliable [16]. Practically the efficiency of wave-pipelined system depends largely on the designer skill.

This work proposes new self-pipelining strategy that can improve speed-power performance and reliability as compared to register-pipelining and wave-pipelining for digital circuits/systems design respectively. Proposed self-pipelined structure has been constructed based on self-latching technique as presented in [15]. The complete system has been divided into self-pipelined stages and each stage of the system has been enabled and disabled using clock input in subsequent interval making novel pipelining strategy. At first a 4-b×4-b multiplier has been designed based on proposed idea. The leaf cells are designed using conventional static CMOS logic. Next, the 4-b×4-b multiplier is exploited to design 8-b×8-b self-pipelined multiplier using decomposition logic [13]. Performance of designed 8-b×8-b multiplier has been compared with most recent competitive design to find effectiveness of proposed strategy. All the simulations are performed on TSMC 0.18μm CMOS technology based on BSIM3 device model with 1.8V supply rail and at 25°C temperature using Tanner EDA V.13.

The remaining part of the manuscript has been organised as follows: Section-II explains the proposed self-pipelining strategy with multiplier structure. The design and simulation result of designed 4-b×4-b and 8-b×8-b multiplier is presented in Section-III. The paper is concluded in Section-IV.

II. PROPOSED SELF-PIPELINING STRATEGY

The proposed idea to design self-pipelined structure is explained in this section. The signal flow diagram of proposed idea is depicted in Fig.1. As shown in Fig.1, each stage of the proposed self-pipelining structure works as a self-latching stage. The idea of self-latching has been borrowed from [15]. The self-latching operation of each stage is controlled by two input signal namely "EN" and "ENB". "ENB" is the complement of "EN". When "EN" input is high (and "ENB" is low) the corresponding self-latching stage will be active and generate output as per functionality. For "EN" input low (and "ENB" input high) the corresponding self-latching stage will hold (or latch) its previous value. In proposed technique the true and complementary clock-input has been applied to Terminal "EN" and "ENB" alternatively in each subsequent self-latching stage. As a result each subsequent self-latching stage will be in active-mode and latching-mode periodically.

The throughput performance depends on the maximum delay among all the self-latching stages. The 2-input binary AND gate with self-latching technique is presented in Fig.2.

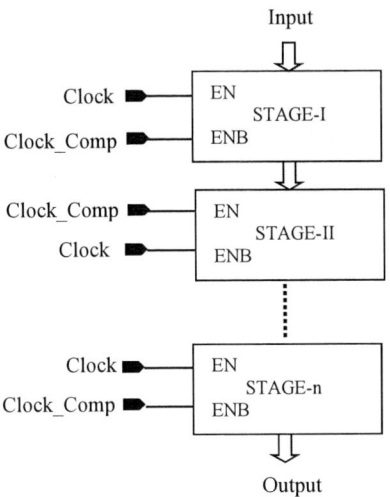

Fig.1 Proposed Self-Pipelining Strategy

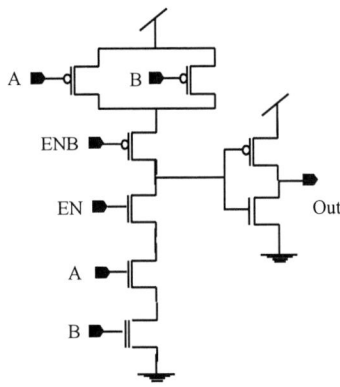

Fig.2 Self-latched 2-input AND

As shown in Fig.2, when "EN" input is high and the corresponding "ENB" input is low the AND circuit will be operative. For the reverse case ("EN" is low and "ENB" is high) the circuit will hold (or larch) its previous value. In proposed design the true and complementary clock input has been applied alternatively to "EN" and "ENB" input for each subsequent stage. For clock input high each alternative stage will be operative and other alternative stages will be in self-latching mode. For clock input low the reverse action will take place. The same self-latching strategy as shown for AND gate in Fig.2 has been applied to other logic cells also.

The proposed self-pipelined 4-b×4-b Tree multiplier architecture is shown in Fig.3. Here in the figure the 2-input Self-latching AND gate is denoted by its usual symbol. The self-latching Half-Adder has been denoted by white box and the grey box in the figure denotes self-latching Full-Adder. The true clock input has been applied to the black input pins at the left hand side of Fig.3 whereas white pins take the complementary clock input. In proposed work this 4-b×4-b multiplier has been used to generate self-pipelined 8-b×8-b multiplier using decomposition logic [13]. The architectural representation of 8-b×8-b self-pipelined multiplier is not shown for the sake of brevity. However the schematic design and simulation results are presented in section-III.

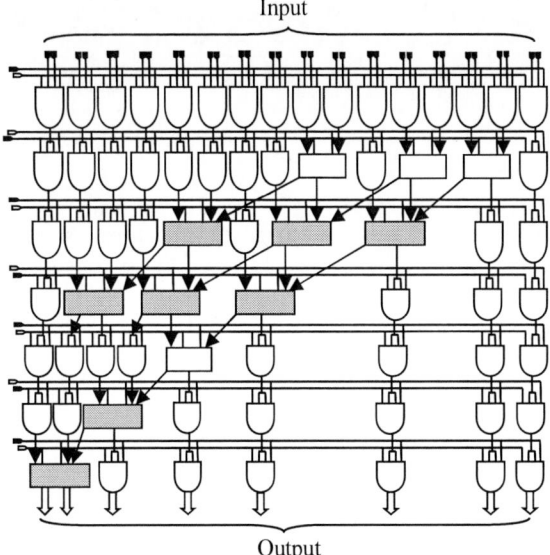

Fig.3 Proposed 4-b×4-b Self-pipelined Tree Multiplier

III. DESIGN AND EVALUATION

The proposed self-pipelined multiplier has been designed and optimized on TSMC 0.18µm CMOS technology based on BSIM3 device model with 1.8V supply rail and at 25°C temperature using S-Edit of Tanner EDA V.13. The front-end schematic design of Self-latching 2-input "AND" gate is shown in Fig.4. The transient response from T-Spice simulation using W-Edit of Tanner EDA V.13 is presented in Fig.5.

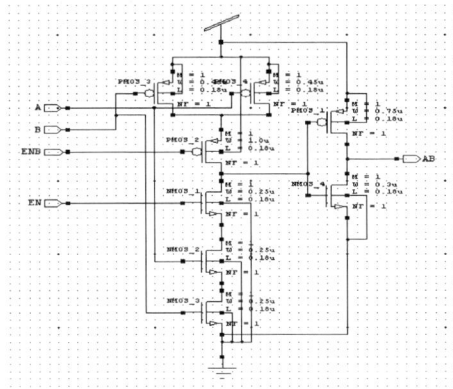

Fig.4 Schematic Design of Latch-AND

2019 Devices for Integrated Circuit (DevIC), 23-24 March, 2019, Kalyani, India

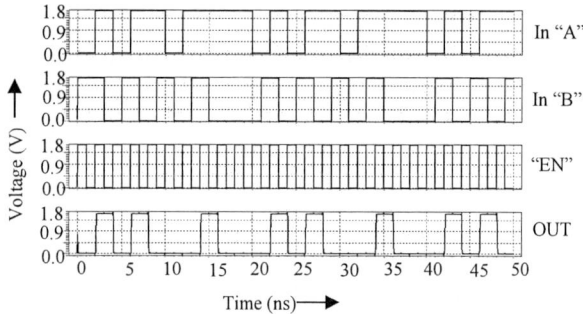

Fig.5 Transient response of Latch-AND

The circuit operation of designed self-latching 2-input AND gate has been validated from transient response presented in Fig.5. As shown in Fig.5, the AND circuit is operative for "EN" input high and holds its previous value when "EN" input is low. "ENB" input is the complement of "EN" input and is given in the figure. The schematic design of self-latching Full-Adder and its T-Spice transient response is shown in Fig.6 and Fig.7 respectively.

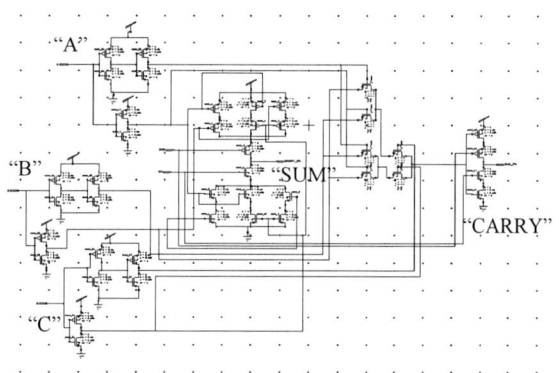

Fig.6 Proposed self-latching Full-Adder Schematic

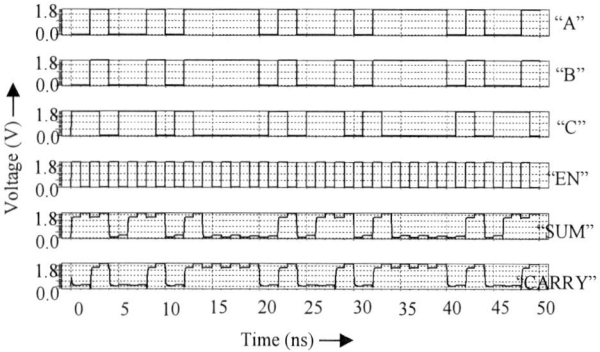

Fig.7 Transient I/O response of self-latching Full-Adder

As shown in Fig.7 the three inputs to the self-latching Full-Adder are "A", "B" and "C". The enable input is denoted by

"EN". The output of Full-Adder is "SUM" and "CARRY". The transient response (Fig.7) validates the functionality of proposed self-latching Full-Adder. The aforesaid idea has been exploited to design other self-latching leaf-cells for the proposed self-pipelining multiplier. The schematic design of 4-b×4-b self-pipelined multiplier is shown in Fig.8.

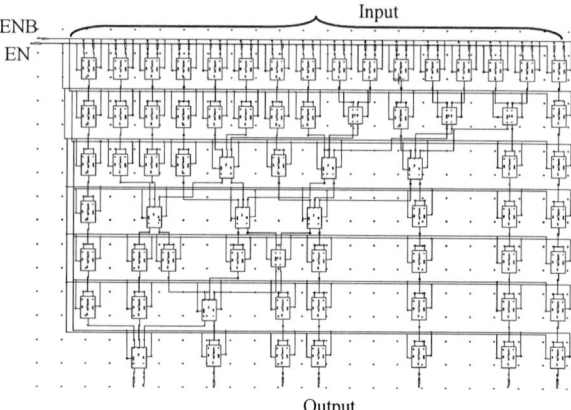

Fig.8 Proposed 4-b×4-b Self-pipelined multiplier Schematic

Next, the decomposition logic as explored in [13] has been applied to design proposed 8-b×8-b self-pipelined multiplier using 4-b×4-b multiplier and the front-end schematic of the designed multiplier is shown in Fig.9.

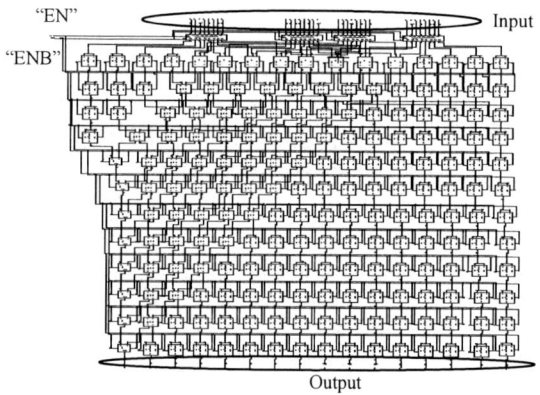

Fig.9 Proposed 8-b×8-b Self-Pipelined Multiplier

In order to evaluate performance of proposed strategy the 8-b×8-b self-pipelined multiplier has been compared with most recent competitive design proposed by J.-F. Lin in 2017 [9]. Since it is unfair to compare simulated result with measured result from fabricated chip (As presented in [9]), the multiplier proposed in [9] has been redesigned and optimized based on BSIM3 device model with 1.8V supply rail at 25°C temperature on TSMC 0.18μm CMOS technology. The comparison has been done with 10MHz input data rate with equal operating condition. As per T-Spice simulation results the proposed self-pipelined 8-b×8-b multiplier offers 35.78% reduced average power with 18.23%

978-1-5386-6723-1/19 $31.00 © 2019 IEEE
300

less delay. As per analysis the proposed self-pipelining strategy can be an effective designer choice for high-speed low-power binary multiplier.

IV. CONCLUSION

In this paper a new self-pipelining strategy to improve speed-power performance of binary multiplier is proposed. Self-latching leaf cells are used to generate proposed structure. Each successive self-latching stage in self-pipelined multiplier has been enabled and disabled alternatively using clock input. With this idea the problem associated with register-pipelining and wave-pipelining can be reduced. Initially a 4-b×4-b multiplier has been designed based on proposed technique. Next, the decomposition logic has been used to design 8-b×8-b self-pipelined multiplier using 4-b×4-b multiplier. The speed-power performance of designed 8-b×8-b self-pipelined multiplier has been compared with most recent competitive design to establish superiority of proposed idea. All the design and optimization has been performed on TSMC 0.18μm CMOS technology based on BSIM3 device parameter with 1.8V supply rail and at 25°C temperature using Tanner EDA V.13.

REFERENCES

[1] Aloke Saha, Sushil Kumar, Debajit Das and Mrinmoy Chakraborty, "LPHS Logic Evaluation on TSMC 0.18μm CMOS Technology," *Int. J. of High Speed Electronics & Systems (IJHSES)*, World Scientific, vol. 26, no. 4, December 2017.

[2] V. Jamshidi and M. Fazeli, "Design of ultra low power current mode logic gates using magnetic cells," *International Journal of Electronics and Communication (AEU)*, vol. 83, pp. 270-279, January 2018.

[3] Roshni Gupta, Rockey Gupta and Susheel Sharma, "Design of high speed and low power 4-bit comparator using FGMOS," *International Journal of Electronics and Communication (AEU)*, vol. 76, pp. 125-131, June 2017.

[4] Aloke Saha, Rahul Pal, Akhilesh G. Naik and Dipankar Pal, "Novel CMOS Multi-bit Counter for Speed-Power Optimization in Multiplier Design," *Int. J. of Electronics & Communication (IJEC)*, Elsevier, vol. 95, pp. 189-198, August 2018.

[5] Z. Zhang and Y. He, "A low error energy-efficient fixed-width booth multiplier with sign-digit-based conditional probability estimation," *IEEE Transactions on Circuits and Systems II: Express Briefs*, Volume: 65, Issue: 2, pp. 236-240, Feb. 2018.

[6] AlokeSaha, Dipankar Pal and Mahesh Chandra, "Low power 6-GHz wave-pipelined 8b×8b multiplier," *IET Circuits, Devices & Systems*, vol. 7, no. 3, pp. 124-140, May 2013.

[7] AlokeSaha, Dipankar Pal and Mahesh Chandra, "Benchmarking of DPL Based 8b×8b Novel Wave-Pipelined Multiplier," *Int. J. of Electronics Letters(IJEL)*, Taylor & Francis, vol.5, no.1, pp.115-128, January 2017.

[8] S. V. Gomes, P. Sasipriya and V. S. K. Bhaaskaran, "A low power multiplier using a 24-transistor latch-adder," *Indian Journal of Science and Technology*, vol. 8 (19), pp. 1-5, August 2015.

[9] J.-F. Lin, "Low power latch-adder based multiplier design," *Journal of Semiconductor Technology and Science*, vol. 17, no. 6, pp. 806-814, December 2017.

[10] L. Ing-Chao, C. Yu-Hung and Y. Yi-Ming, "Aging-Aware Reliable Multiplier Design With Adaptive Hold Logic", *IEEE Transactions on Very Large Scale Integration (VLSI) Systems*, Vol. 23, no. 3, pp. 544 – 556, April 2014.

[11] W. I-Chyn, P. Chien-Chang and L. Feng-Yu, "Reliable Low-Power Multiplier Design Using Fixed-Width Replica Redundancy Block", *IEEE Transactions on Very Large Scale Integration (VLSI) Systems*, Vol. 23, no. 1 , pp. 78 – 87, 13 February 2014.

[12] A. Saha, D. Pal, Mahesh Chandra and M.K. Goswami, "Novel high speed MCML 8-bit by 8-bit multiplier," *IEEE International Conference on Devices & Communications (ICDeCom-11)*, BIT Mesra, India, pp. 1-5, 24-25 February 2011.

[13] P. Ramanathan, P. T. Vanathi and S. Agarwal, "High speed multiplier design using decomposition logic", *Serbian Journal of Electrical Engineering*, vol. 6, no. 1, pp. 33-42, May 2009.

[14] S. Asif and Y. Kong, "Design of an algorithmic Wallace multiplier using high speed counters", *Tenth International Conference on Computer Engineering & Systems (ICCES)*, 2015, 23-24 December 2015.

[15] T.-Y. Kuo and J.-S. Wang, "A low-voltage latch-adder based tree multiplier," IEEE International Symposium on Circuits and Systems (ISCAS 2008), pp. 804-807, Seattle, WA, USA, 18-21 May 2008.

[16] W. P. Burleson, M. Ciesielski, F. Klass and W. Liu, "Wave-pipelining: a tutorial and research survey," *IEEE Transactions on Very Large Scale Integration (VLSI) Systems*, vol. 6, no. 3, pp. 464-474, September 1998.

978-1-5386-6723-1/19 $31.00 © 2019 IEEE

Smart Power Theft Detection System

[1]Nitin K Mucheli, [1]Umakanta Nanda, [2*]D Nayak, [2*]P K Rout, [2]S K Swain, [2]S K Das, [2*]S M Biswal

Department of Electronics and Communication Engineering
*Department of Electronics and Instrumentation Engineering
[1]Vellore Institute of Technology, Amaravati
[2]Silicon Institute of Technology, Bhubaneswar

Email: [1]nitin.18bev7013@vitap.ac.in, [1]uk_nanda@yahoo.co.in, [2*]nayak.debasish84@gmail.com,
[2*]prakashrout05@gmail.com, [2*]sudhansu.mohan@silicon.ac.in, [2]sswain@silicon.ac.in, [2]satish.das@silicon.ac.in

Abstract—Power theft is normally done by two methods that is bypassing or hooking. So to detect it, a system (current measuring and comparing) is proposed in which the household distribution of current is done indirectly from the electric pole to an intermediate distributor box and then to the individual houses. The current is measured periodically in the distributor box and is posted to the server database for each house using GSM/GPRS module. Similarly, for each house electric meter is designed which can measure the value of the current and post the same to the server database periodically using GSM/GPRS module.

At the time of the installation of the electric meter the details of the users are stored in the database through a user friendly mobile application including the address, latitude, longitude using mobile GPS and the photograph of the user's house/area. Upon successful comparison between the current values from distributor box and electric meter in the server if we get a marginal difference between the currents then the theft is detected. Finally, the details of the user are shared with the authorized mobile application including the address and photograph of the area. The latitude and longitude are also used to show the area of theft in Google maps. And hence the required steps are taken. The same process is used for hooking but on the individual electric poles.

Keywords— Power consumption, Energy meter, Arduino UNO

I. INTRODUCTION

Theft of electricity is the criminal practice of stealing electrical power. According to a study the world loses US$89.3 billion annually to electricity theft. The highest losses were in India ($16.2 billion), followed by Brazil ($10.5 billion) and Russia ($5.1 billion) [1]. Nationally, total transmission and distribution losses approach 23% and some states' losses exceed 50% [2]. A huge amount of power is required for the Integrated steel plants To meet this requirement it is calculated that approximately a capacity of 25,000 MW power plant needs to be installed apart from the captive power & blowing stations [3]. By the help of our prototype we can detect the theft and possibly save the maximum electricity which can be effectively used in the steel plants.

II. OBJECTIVE AND BRIEF WORKFLOW

The power distribution sector has been played with distribution losses (overall 30%) coupled with theft of electricity. To minimize the theft of power by bypassing or by hooking we are proposing a web based MOBILE APPLICATION which will notify about the theft of electricity that is happening in a particular area.
The main objectives of this work are,

➢ To detect the power theft automatically without engaging any man powers by developing a cost effective and efficient system.

➢ To develop a web based mobile app for the authorized officials of electricity board to keep track of all the thefts, area of thefts and the direction to reach the area under theft.

➢ To maintain the record of the total number of electric units consumed by users in the server database periodically and make online bill payment system.

➢ To develop a global website that would maintain the analytics of the thefts and the probable area under theft using multi-color graphs and pictorial representations which would make the theft analysis easier and can also predict the thefts that may happen in future.

To create this solution, work has to be done both for the hardware and software. The hardware includes the customized electric meter and the distributor box and the software includes the development of both website and mobile application. For the hardware part to be developed first, study of basics components like digital electric meter, GSM module, ACS712 module, microcontroller, ADC and their workings has to be carried out. Then the brief circuit diagrams of both the distributor box and the electric meter have to be designed (shown in Fig. 1). After that the hardware design and the PCB design of the above components should be considered. Now the PCB designs of all the components need to be integrated to a single PCB design in accordance to our customization or circuit diagram. According to our customization the PCB design of the distributor box will consist of a GSM module and more than one ACS712 module which should to be connected to a microcontroller. The ACS712 module should have a series connection with live wire. The PCB design of the customized electric meter will consist the circuits of the commercial electric meter along with the ACS712 module at the input side and a GSM module such that all of them share the same microcontroller. The microcontroller will be programmed according to the desired functioning of the customized electric meter. The analog to digital conversion circuit should also be integrated in the PCB design in order to give supply to the GSM module and microcontroller from the available AC current. The above PCB designs can be obtained using PCB design software and it must be simulated using simulation tool for any possible defects. After the successful simulation, the PCB design will be fabricated on the PCB board. The above work will be carried out under a manufacturer/vendor/lab which gives training, facilitates PCB design, fabricating, licensing and

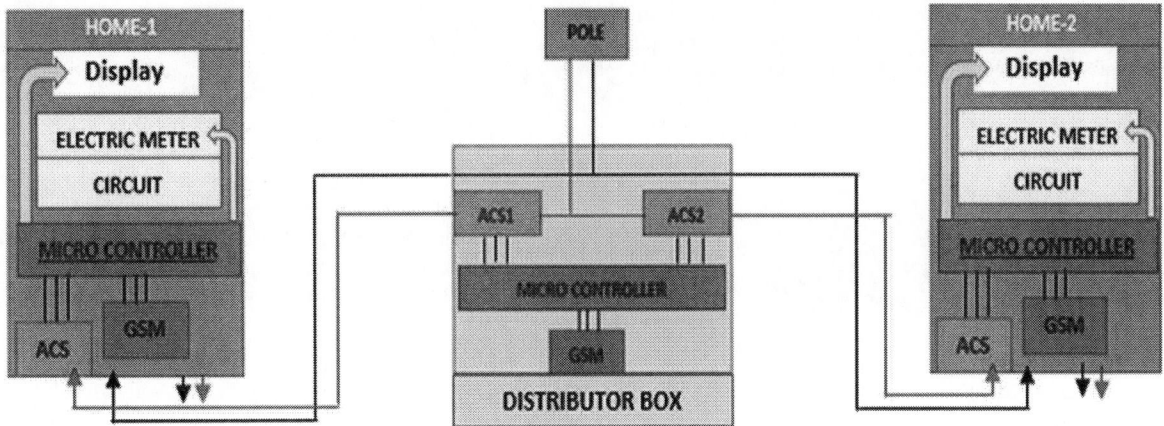

Fig. 1. Architecture of the proposed system

manufacturing related to electric meter. For the software part a global website will be developed using web development for all the users for payment system and analytics of theft and a web based mobile app will be developed using android development for the authorized officials who are in charge of detecting theft. Server side code and the database plays a crucial role in the whole system. The mobile app will be developed in such a way that the server of the app would be able to receive the values of currents from the GSM module periodically and compare them and display the area of theft along with the details of the user under theft and the directions in Google maps. Hence software part can be developed using open source software.

The novelty of the system is outlined below.

➤ This solution works on the principle of division of current in parallel path. So it is more efficient.

➤ This solution shows the exact location of area under theft along with its photograph.

➤ The analytic study of the thefts can predict the thefts that may take place in future.

➤ It can also allow the supervisor to cut out the power of a place under power theft

The solution uses a GSM sim800 module that can transmit the value of current measured to the server through GPRS technology.

III. DETAILED SYSTEM DESIGN

Current division rule has been used to detect the power theft. According to this rule, a parallel circuit acts as a current divider as the current divides in all the branches in a parallel circuit, and the voltage remains the same across them. The current division rule determines the current across the circuit impedance.

We connect a current sensing module ACS712 in both the distribution box and the meter installed in the house. The current from ACS712 is fed into a micro-controller (Arduino Uno). The data of either sides is then sent to the authorized database after the arduino is interfaced with the GSM-GPS module.

Hence the data are compared, if the difference is more than the provided threshold, POWER THEFT is detected.

The hardware and software required for our solution is:

➤ Arduino Uno

➤ Current Sensing Module (ACS 712)

➤ GSM (SIM 800)

➤ GPS Module

➤ Database (using My SQL)

A. Arduino UNO

Arduino [4] is an open-source electronics prototyping platform based on flexible, easy-to use hardware and software. As Arduino is open source, the CAD and PCB design are freely available. There are several different arduino boards are available on the market (both original and cloned) such as Arduino UNO, Arduino Nano, Arduino Mini and Arduino Mega.

Fig. 2. Arduino UNO R3

For our work specifically we used this board (Fig. 2) due to the below specifications.
➤ 6 analog input ports
➤ Power Input connector.
➤ 14 digital I/O ports (of which 6 PWM)
➤ Standard USB for data and power and programming.
➤ Female headers.
➤ 1 hardware serial port (UART)
➤ Most popular board. Ideal for starters.

B. Current sensing Module (ACS 712)

The ACS712 Current Sensor as shown in Fig. 3 offered on the internet are designed to be easily used with micro controllers like the Arduino. These sensors are based on the Allegro ACS712ELC chip. These current sensors are offered with full scale values of 5A, 20A and 30A.

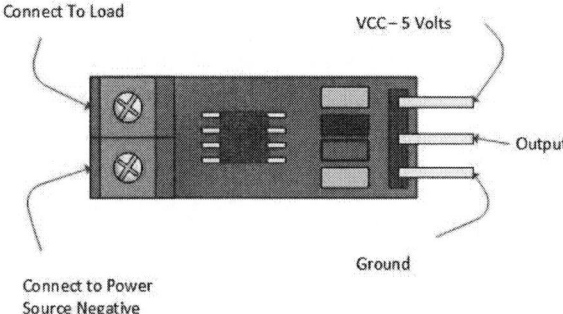

Fig. 3. Pin configuration of ACS 712

C. GSM Module

Fig. 4 GSM Module (SIM 800)

SIM800 is a quad-band GSM/GPRS module that works on frequencies GSM/GPRS module that works on frequencies GSM 850MHz, EGSM 900MHz, DCS1800MHz and PCS 1900MHz. SIM800 features GPRS multi-slot class12/class 10(optional) and supports the GPRS coding schemes CS-1,CS-2, CS-3 and CS-4; With a tiny configuration of 24*24*3mm, sim800 can meet almost all the space requirements in users' applications, such as M2M, smart phone, PDA and other devices.

D. GPS Module

To get the information of the device's geographical position, a GPS navigation device (Fig. 5) is needed that is capable of receiving information from satellites. Using suitable software, the device may display the position on a map, and it may offer directions. SIM800 has 68 SMT pads, and provides all hardware interfaces between the module and customers' boards.

So to detect this we have to make the following arrangements: Firstly we will use a GPS module to store the latitude and longitude of every pole and house on the 1st day of installation of the meter.

Connection and arrangements at the distributor box
- ➢ The live wires from the POLES should be fed into a distributor box block wise.
- ➢ The distributor box has the capability of distributing the power among the houses of particular locality.

Accordingly, subsequent distributor box will be setup for a cluster of houses.
- ➢ Hence the AC current is measured in the distributor box separately for each house using the ACS712 module and the magnitude of this current is fed into the microcontroller.
- ➢ A server side database is maintained and the measured value of current is transmitted with the help of GSM/GPRS module and is updated into the database table containing the user-id at a regular desired period (referring timestamps).

Fig. 5 GPS module

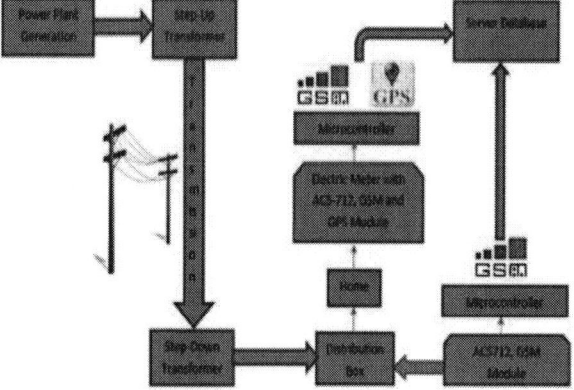

Fig.6. Block level demonstration of the system

Connection and arrangement at the electric meter installed in each house
- ➢ The main objective of this connection is to measure the total current actually entering the electric meter. So a fixed connection is made at the inlet terminal of the electric meter in such a manner that tampering of this connection is made void with the help of laser sensors and microcontroller.
- ➢ The respective connection too consists the ACS712 module to measure the AC current and fed into the microcontroller.
- ➢ Hence now the same measured current is transmitted with the help of GSM/GPRS module and is updated in the same database table for same user-id that was maintained for the distributor box at regular desired period (referring timestamps).

Google map API/ Google street view

- At this stage the theft table of database has the exact location of the theft.
- Hence the location from the database is transmitted to the satellite through satellite communication.
- And now the satellite is programmed to take the image of the location of theft accessed from the GPS of that place.

Mobile Application

- Mobile Application has been made for authorized people to get access to the area of theft and take suitable actions.
- This application has a direct access to the database of theft table.
- The images taken by the satellite are directly sent to this application with the location and the electric meter's unique id.
- The user of the mobile application gets pinged when the theft is detected.

Comparison among this and previous related work has been carried out in Table I. It can be seen that in our work both the use of mobile app and location detection has been adopted which is a clear advantage over the other systems.

Table I Comparison of this work with existing systems

Work done	Location detection	Use of mobile App
[5]	No	No
[6]	Yes	No
[7]	Yes	No
[8]	No	No
This work	Yes	Yes

IV. CONCLUSION

This method reduces the heavy power and revenue losses that occur due to power theft by customers. By this design it can be concluded that power theft can be effectively curbed by detecting where the power theft occurs and informing the authorities. The proposed system will be hidden in electric meters in such a way that as soon as the difference between current crosses the threshold value, an automatic message and email will be sent to the concerned authority along with its location and image of that particular area.

REFERENCES

[1] "Controlling electricity theft and improving revenue", World Bank report on reforming the power sector, 2010.

[2] Annual report of power and energy division of planning commission, government of India, New Delhi, 2011-12.

[3] "All India Electricity Statistics", Central Electricity Authority, Ministry of Power, Government of India, New Delhi, 2011-12.

[4] www.arduino.cc, March 31, 2017

[5] R. M. Mutupe, S. O. Osuri, M. J. Lencwe and S. P. Daniel Chowdhury, "Electricity theft detection system with RF communication between distribution and customer usage," *IEEE PES PowerAfrica*, Accra, 2017, pp. 566-572.

[6] M. Saad, M. F. Tariq, A. Nawaz and M. Y. Jamal, "Theft detection based GSM prepaid electricity system," *IEEE International Conference on Control Science and Systems Engineering (ICCSSE)*, Beijing, 2017, pp. 435-438.

[7] R. E. Ogu and G. A. Chukwudebe, "Development of a cost-effective electricity theft detection and prevention system based on IoT technology," *IEEE International Conference on Electro-Technology for National Development (NIGERCON)*, Owerri, 2017, pp. 756-760.

[8] A. S. Metering, S. Visalatchi and K. K. Sandeep, "Smart energy metering and power theft control using arduino & GSM," *2nd International Conference for Convergence in Technology (I2CT)*, Mumbai, 2017, pp. 858-961.

2019 Devices for Integrated Circuit (DevIC), 23-24 March, 2019, Kalyani, India

Analysis of Adiabatic flip-flops for Ultra Low Power Applications

S.Samanta
ECE Department, Neotia Institute of
Technology, Management & Science
West Bengal, India

R.Mahapatra
ECE Department, National Institute of
Technology, Durgapur
West Bengal, India

A.K.Mal
ECE Department ,National Institute of
Technology, Durgapur
West Bengal, India

Abstract— **In this paper we have presented adiabatic flip flops which are used for clocking in digital systems. The clocking scheme using energy recovery technique has already appeared as a successful and promising scheme for limiting power dissipation in ultra low power digital systems. Adiabatic flip flops are the key elements for this type of energy efficient adiabatic clocking scheme. The flip-flops are working in adiabatic principle. Here in this work we have done the simulation and analyze the performance of two basic types of energy recovery flip flops. These are single ended conditional capturing flip-flop and differential conditional capturing flip flop. Both the flip-flops are utilizing energy recovery scheme. For better comparison results we have also used clock gating scheme along with energy recovery technique. Using cadence 180nm technology the simulations are obtained.**

Keywords— *CMOS, PUN, PDN, VLSI, dynamic power, setup time, hold time*

Introduction

In complementary metal oxide semiconductor circuit based designs, flip-flops are essential part of the clocking circuits. They are responsible for the synchronous-asynchronous behavior to the system. The basic circuit of flip flop that is used in various digital circuits determines the system power consumption. The type of flip flop used in digital circuits determines the output load of the circuits. This is also known as clock load. This clock load directly affects the switching power consumption of circuits. Therefore, it is essential to introduce one technique to minimize the power consumption of flip flops to reduce the overall system power consumption. The power specification of modern portable digital circuits is severely limited. It is very essential to improve system power performance in the flip flop networks. Timing elements like flip-flops are very important for the performance of digital systems. This is due to the extremely large set up time and hold time. These are also essential for good performance and better efficiency. Recovery or recycling of energy in a circuit is a technique for low power digital circuits [5]. Energy recovery circuits can be designed to achieve low energy dissipations. This can be achieved by restricting the current to flow across circuit. As a result there is minimum amount of voltage drop across the circuit components. This type of circuits use pulsed power supply or sinusoidal power supply. In this scope, we have applied energy recovery or adiabatic techniques to the clock network. The basic signal which is used for clock supply is the most vital signal. Generally, the nature of the signal is capacitive. The principle used in energy recovery circuits is that, they recycle the energy from the output load capacitance to the input node. This happens during each operation cycle. The power consumption of the clock networks contributes more than 60% of the total power in high performance high speed VLSI systems [6]. Hence, low

power clocking methodologies are essential for ultra low power design. Adiabatic flip-flops can use energy recovery process from the clock network. This results significant reduction of power dissipation. In single phase sinusoidal clock these low power flip-flops can operate. This clock can be generated with very high efficiency and energy recycling elements. We have implemented the clock based flip-flops in cadence digital simulator and obtained the results.

In this paper, we have discussed and compared two basic types of energy recovery flip flops. They are single ended conditional capturing energy recovery flip- flop (SCCER) and differential conditional capturing energy recovery flip flop (DCCER). We have estimated their power dissipation and timing specification constraints. We have also introduced clock gating technique. For the energy recovery clocked flip flops it is clearly seen that they reduce power consumption and propagation delay of the system.

I. ADIABATIC LOGIC

Conventional CMOS technology is very attractive and efficient method for low power circuit design. This is because of its minimum static power dissipation. Generally, conventional CMOS has two networks. One is PUN and another is PDN. Switching or dynamic power dissipation of CMOS circuits originates as a result of charging and discharging of load capacitance. This has a relationship with supply power clock and is proportional to the square of the power supply voltage of the circuit and frequency of switching clock. The circuit capacitance and frequency factors increase the power consumption of system. Therefore conventional CMOS design needs to be changed in order to satisfy the demand of low power supply. In Fig. 1 we have shown the equivalent circuit of charging adiabatic system. In adiabatic technique, output load capacitance is charged by a constant current source. In conventional CMOS structures this is done by constant voltage source. The on resistance of PUN of PMOS network is represented by resistance, R and C_o is the output capacitance [6].Constant current source resembles a voltage ramp. Now, the energy that is dissipated through adiabatic logic is given as

$$E(diss)= (C_o V_{dd}/T)^2.RT=RC_o/T.C_o V_{dd}^2 ----------------(1)$$

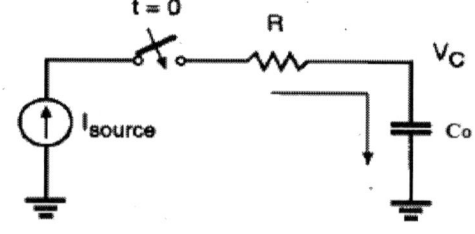

Figure 1. Equivalent circuit of adiabatic logic based circuits

978-1-5386-6723-1/19 $31.00 © 2019 IEEE

It can be noticed that the energy dissipation is directly proportional to resistance R. Therefore, the energy dissipation can be limited by decreasing the on resistance through PMOS network. By using a current source which is constant throughout a switching event, the energy can be easily transferred from power supply to load capacitor. This can be done without any energy dissipation. Again, the energy stored in the load capacitance after charging process can be sent back to the power supply voltage. This is done by simply reversing the direction of current source [7]. The recycling of energy is a very attractive feature in adiabatic logic circuits. For this process, it is also called energy recycling logic. The constant current supply must be designed in such a way so that it can be capable of retrieving the charge back to the power supply. Adiabatic circuits use a special type of power supply which is constant. This source has a special feature. Instead of using the standard one, it uses pulsed power supply. There are many important design specifications which should be taken into account in any CMOS based adiabatic circuit design. The designers follow two basic design criteria. Firstly, the implementation should result in a power efficient design using the combination of power supply and clock generator. Secondly, a transistor operating in adiabatic mode must maintain some rules. The transistor must be always in on state when there is a significant current flowing through the transistors. The transistor must be off state when there is a significant difference between the source and drain voltages of the transistor [7].

II. SCCER FLIP-FLOP

Flip flops are the examples of sequential digital circuits. Generally in digital circuits four phase transmission gate (FPTG) flip-flops are used. They are very similar to the transmission gate based flip-flops that are used conventionally. Here we have presented two new adiabatic flip-flops which are more power and area efficient. These flip-flops can recover energy from their clock input capacitance which is the basic adiabatic principle.. That means, they have an energy recovery behavior. Their storage elements and internal nodes are powered by constant power supply. The SCCER flip-flop is nothing but single-ended version of the DCCER flip-flop. The main circuit of SCCER flip-flop is shown in fig 2.There is pull up and pull down networks like MOS inviters. Pull up network contains one PMOS transistor MP1 and pull down network contains four NMOS transistors.MN1, MN2, MN3, MN4.The output QB controls the transistor MN3. It provides the conditional capturing. There is a static evaluation path in the right hand side. No conditional capturing process is required here.. Transistor MN3 is placed above MN4 in the stack. This placing of transistor reduces the sharing of charge in the circuit. That is achieved because, during the charge sharing event, the capacitance associated with transistor MN3 has already been charged. Therefore does not contribute to the charge sharing process. Due to this incident, the amount of power dissipation of flip-flops in sleep mode and in active mode are same. By the clock network, a significant amount of power is dissipated too. The basic design style separates the whole network responsible for clocking scheme, from the other components of the circuit. We use a different power supply denoted as V(clk) for the circuit responsible for clocking and it is used for measuring clock network dissipated power. Power dissipated by the clock circuit can be seen in Table1. Power dissipation can be saved by disabling the clock network during the sleep mode. The clock network can consume a large portion of power responsible for the system. The implementation of clock gating can disable the clock network from rest of the system.

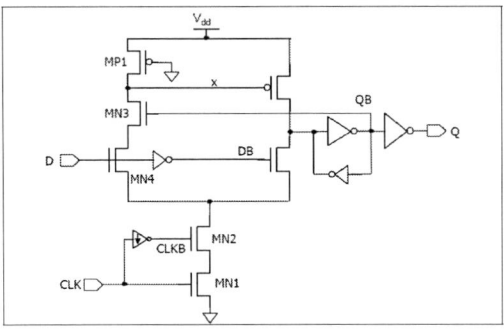

Figure 2. SCCER flip flop

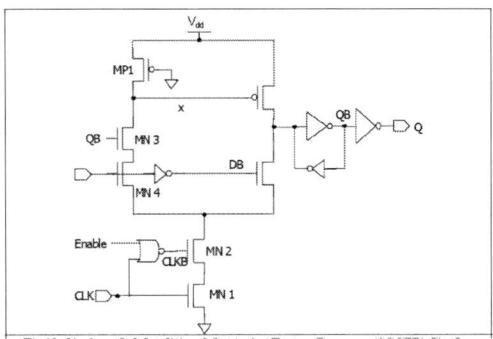

Figure 3. clock gating implemented in SCCER flip-flop

III. DCCER FLIP-FLOP

DCCER stands for Differential Conditional-Capturing Energy Recovery flip-flop. The main circuit of DCCER flip-flop contains two pull up transistors MP1 and MP2 and two pull down transistors MN1 andMN2.The pull down transistors has different names. They are also called control transistors. They provide the necessary control signal to the circuit. Like the dynamic flip-flops of sequential digital circuits, the DCCER flip-flop has their operative mode namely precharge and evaluation. In precharge phase clock is not used but small pull-up PMOS transistors MP1 and MP2 are made operative to charge the prechage node .The DCCER flip-flop uses a latch which is NAND gate based. It operates in Set/Reset latch modes. Here the conditional capturing process is obtained by using a feedback path. This feedback path is achieved through the output Q and also through QB and extended to the c transistors MN3 and MN4.These are control transistors and they help to achieve the evaluation paths too.

Generally, in energy recovery logic the clock generator circuit provides the clock continuously even when the input signal is static that means the logical circuit is in idle state.

This incident results a heavy loss of power. Clock gating is an efficient technique to minimize power dissipation in the adiabatic logic circuit. It detaches the clock generator circuit net from the logical net during the idle or non operative periods. Moreover, the clock signal and the distribution network in sequential circuit are the basic contributor to the power dissipation. Moreover, the addition of the clock signal tends to be a heavy load. All of these will add to the capacitance of the clock net.

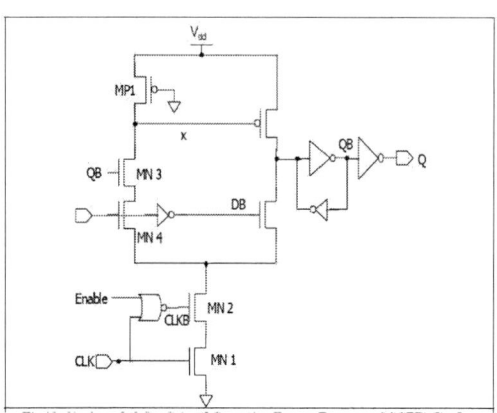

Figure 4. DCCER flip flop

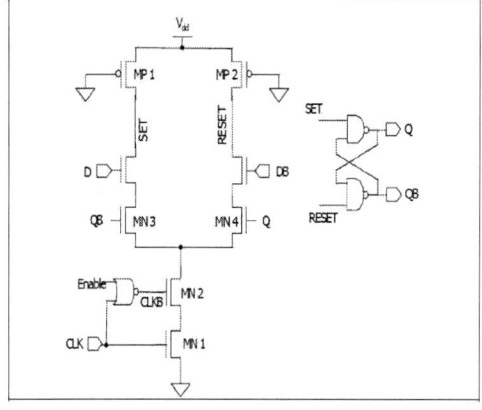

Figure 5. clock gating implemented in DCCER flip-flop

IV. SIMULATION RESULTS

The simulation is done using cadence 180 nm technology. We have calculated the power dissipated by SCCER flip-flop and DCCER flip- flop for non gating implemented circuit and gating implemented circuits. The power is estimated in different switching activities namely 0%, 50% and 100%. We have also calculated the setup time and hold time for the various energy recovery flip-flops.

TABLE I.   SET UP TIME CALCULATION

Set up time	
SCCER	*DCCER*
41ps	143ps

TABLE II.   HOLD TIME CALCULATION

Hold time	
SCCER	*DCCER*
63ps	128ps

TABLE III.   POWER DISSIPATION OF SCCER FLIP-FLOP

Switching activity	Power dissipation	
	P(Vdd)	*P(clk)*
0	4.6µw	12.2 µw
50	45.1 µw	11.1 µw
100	99.0 µw	9.1 µw

TABLE IV.   POWER DISSIPATION OF SCCER FLIP-FLOP IN SLEEP MODE

Switching activity	Power dissipation	
	P(Vdd)	*P(clk)*
0	7.7 µw	3.2 µw
50	9.8 µw	3.1 µw
100	13.5 µw	3.0 µw

TABLE V.   POWER DISSIPATION OF SCCER FLIP-FLOP IN ACTIVE MODE

Switching activity	Power dissipation	
	P(Vdd)	*P(clk)*
0	1.6 µw	11.2 µw
50	48.1 µw	10.1 µw
100	93.9 µw	8.1 µw

TABLE VI.   POWER DISSIPATION IN DCCER FLIP-FLOP

Switching activity	Power dissipation	
	P(Vdd)	*P(clk)*
0	12 µw	12.8 µw
50	53 µw	11.1 µw
100	120 µw	10 µw

TABLE VII. POWER DISSIPATION IN DCCER FLIP-FLOP IN SLEEP MODE

Switching activity	Power dissipation	
	$P(V_{dd})$	$P(clk)$
0	12 μw	13.1 μw
50	49.1 μw	10.1 μw
100	99.8 μw	8.99 μw

TABLE VIII. POWER DISSIPATION IN DCCER FLIP-FLOP IN ACTIVE MODE

Switching activity	Power dissipation	
	$P(V_{dd})$	$P(clk)$
0	12.1 μw	13.1 μw
50	52.4 μw	10.9 μw
100	110.8 μw	9.7 μw

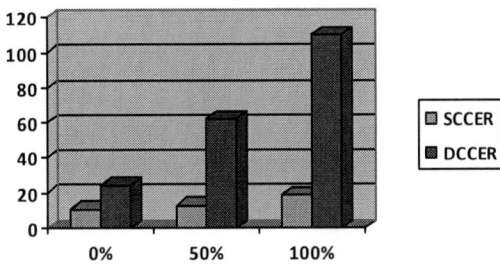

Figure 6 . Comparison of SCCER and DCCER flip-flop in sleep mode

Figure 6 shows the comparison of power dissipation with various switching activities. The switching activities range between 0% to 100% in sleep mode. In figure 7, we have shown the comparison of power dissipation with variation of switching activities in active mode. Here also we have varied the switching activities from 0% to 100%.

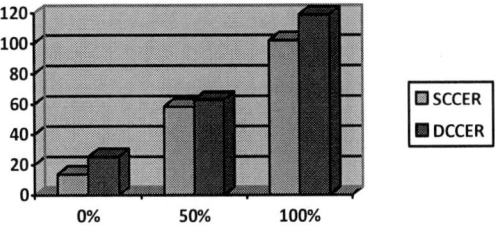

Figure 7. Comparison of SCCER and DCCER flip-flop in active mode

V. CONCLUSION

Table I shows the power dissipation of SCCER flip-flop without use of gating. For various switching activities power dissipated by the clock and power dissipated by the rest of the flip-flop have been estimated. Table II shows power dissipation of SCCER flip-flop in sleep mode. Table III shows power dissipation of SCCER flip-flop in active mode. It is clearly seen that power dissipated by the clock and power dissipation of the rest of the circuit is less in case of sleep mode in SCCER and DCCER flip-flops In sleep mode SCCER flip-flop has better performance than DCEER flip-flops. In active mode SCCER also has better performance than DCCER flip-flop. The clock gating scheme is implemented without any clock overload.This SCCER flip-flops can be used in adiabatic clocking in digital systems as they have good power and delay characteristics.

REFERENCES

[1] L.Ding,et al."A dual rail edge triggered latch" IEEE international symposium on circuits and systems,pp645- 648.May2001.

[2] G.Dickinson and J.S.Denkar "Adiabatic dynamic logic" IEEE Journal of Solid State Circuits, Vol 30,No 03,pp 311-315,March 1995

[3] Q. Wu, M. Pedram, and Xunwei Wu, "Clock-gating and its application to low power design of sequential circuits," IEEE Transactions on Circuits and Systems I, vol. 47, no. 3, pp. 415–420, Mar 2000.

[4] B. S. Kong, et al., "Conditional-capture flip-flop for statistical power reduction," IEEE Journal of Solid State Circuits, vol. 36, pp. 1263 –1271, Aug. 2001.

[5] S. L. Hurst, "Multiple-valued logic. Its status and its future," IEEE Trans. Comput., vol. C-33, pp. 1160–1179, Dec. 1984

[6]H. Partovi, et al., "Flow-through latch and edge triggered flip-flop hybrid elements," IEEE International Solid-State Circuits Conference, pp. 138 -139, Feb 1996.

[7]Y. Ye and K. Roy, "Reversible and quasi-static adiabatic logic," in European Conf. Circuit Theory and Design, 1997, pp. 912–917.

[8] Michael P. Frank, "Common mistakes in adiabatic logic design and how to avoid them," Proceedings of the International Conference on Embedded Systems and Applications, held in Las Vegas, Nevada on June 23-26, 2003, pp. 216-222, CSREA Press.

[9]Anantha P. Chandrakasan, Robert W. Brodersen, "Low Power Digital CMOS Design", Kulwer Academic Publishers, 2002.

[10]Priyanka Ojha, Charu Rana, "Design of Low Power Sequential Circuit by using Adiabatic Techniques", I.J. Intelligent Systems and Applications, 08, 45-50, July 2015.

[11]Samaneh Babayan-Mashhadi and Reza Lotfi, "Analysis and Design of a Low-Voltage Low-Power Double-Tail Comparator", IEEE Transactions on Very Large Scale Integration Systems, Vol. 22, No.2, pp. 314-316, 2014.

[12]A.P. Chandrakasan, S. Sheng, and R. W.Brodersen, "Low Power CMOS Digital Design", IEEE Journal of Solid-state Circuits, Vol. 27, No. 4, pp. 473-484, 1999.

[13].S. Samanta "Adiabatic Computing: A Contemporary Review" International conference on computers and devices for communication (CODEC) ,Dec. 2009, pp. 1-4.

[14]W. Zhang, D. Zhou, X. Hu and J.Hu, "The implementations of adiabatic flip-flops and sequential circuits with power-gating schemes," IEEE 51st Midwest Symp on. Circuits and Systems, MWSCAS 2008, Knoxville, Tennessee, pp. 767 – 770, 2008.

[15]Y. Zhang, L. Okamura, and T. Yoshihara,"An energy efficiency 4-bit multiplier with two-phase nonoverlap clock driven charge recovery logic," Electronics, IEICE Transactions on, vol. E94-C, no.4, pp. 605–612, April 2011

[16]M. P. Frank. Throwing computing into reverse. IEEE Spectrum September , 2017.

[17]M. Morrison and N. Ranganathan. Synthesis of dual-rail adiabatic logic for low power security applications. IEEE Trans. on CAD of Integrated Circuits and Systems, 33(7):975–988, 2014.

[18]William C.Athas,Lars "J."Svensson, Jeffrey G.koller, "Low- Power Digital Systems Based on Adiabatic-Switching Principles", IEEE Transcation on VLSI systems, Volume 2,Issue 4,December 2013.

2019 Devices for Integrated Circuit (DevIC), 23-24 March, 2019, Kalyani, India

A Low Power LNA using Current Reused Technique for UWB Application

Dhananjaya Tripathy[1,a], Debasish Nayak[2,a], Sudhansu Mohan Biswal[3,a], Sanjit Kumar Swain[4,b], Biswajit Baral[5,b], Satish Kumar Das[6,b]

[a]Department of Electronics and Instrumentation Engineering, Silicon Institute of Technology, Bhubaneswar, India
[b]Department of Electronics and communication Engineering, Silicon Institute of Technology, Bhubaneswar, India
[1]dhananjayavssut@gmail.com, [2]nayak.debasish84@gmail.com, [3]sudhansu.mohan@silicon.ac.in,
[4]sswain@silicon.ac.in,[5]biswaiit@silicon.ac.in,[6]satish.das@siicon.ac.in

Abstract— **Here in this paper the design of a low power low noise amplifier (LNA) is presented which works for very wideband of frequency known as UWB signals. This method uses current reused technique which helps in reducing the power consumption while maintaining the same conversion gain and NF. No inductor is used in this circuit which reduces the complexity. Here a power gain of 12.6 dB, a NF of 2.1 dB at 3.5 GHz and 8.5 GHz is achieved, while consuming very less power nearly 7.6mW. It is clear from the observations that this technique solves the major problem of power consumption that was present in the previously existing techniques.**

Keywords— *LNA, UWB, current reused, resister feedback, noise cancelling, NF*

I. INTRODUCTION

Ultra-wide band(UWB) also referred as ultra band which can be helpful for communication at a very low energy level using a major portion of the radio spectrum for short range application. It is extended over a large range of frequency i.e. up to 10.6 GHz. It has several applications like collection of data from the target sensor, radar imaging. It is also used to track the systems. UWB is superior to all other frequency bands due to zero interfere with narrowband. Here the carrier signal and the message signal lies in the same bandwidth. To spread the message over a larger frequency band UWB is the best option. It is different from the general radio transmissions by the fact that the general systems send the message by changing the phase, frequency and power of a sinusoidal signal where as UWB transmissions send message by creating radio energy at fixed time intervals and covering a high bandwidth.

During the initial stage, ultra-band low noise amplifier was designed to provide a large range of bandwidth. By the use of the passive filter technique the LNA design was done [1, 2]. This circuit operates for a large bandwidth. On the contrary due to the presence of an inductor, area of the chip was evolved as a larger constraint. Apart from the area problem the noise figure was also high. The noise was reduced by the introduction of a feedback technology where a resistance is connected as a feedback path as explained in [3]. Here the noise performance was improved. But on the other hand, the bandwidth is reduced along with little conversion gain and higher power consumption. To extend the bandwidth a dual-feedback topology is introduced which is explained in [4].Here the bandwidth is extended by a huge factor. The circuit also provided moderate conversion gain. But there was

a large degradation in the noise performance . A stable wideband common-gate (CG) LNA architecture is described in [5] which achieves better NF and gain. The major problem in this case is that the bandwidth is again reduced and power consumed is high. The conversion gain which was less in resistive shunt feedback technology [3] was improved including the cancellation of noise technique which is described in [6]. As consequence, the bandwidth of the LNA was reduced in a huge manner. Also the power consumption is increased. In most of the designs there were two problems mainly. Firstly due to the presence of inductors the area requirement was more. Secondly gain has to be sacrificed for better bandwidth and vice-versa. To suppress this problem, a new technique was suggested in [7]. Here no inductor was used. But the power consumption by the design is large and needed to be improved. Here we have introduced a current reuse noise cancelling gain enhanced technique to improve the noise and reduce the power.

In Section II a brief discussion on conventional noise cancelling technique is addressed. Section III explains the current reused resistive feedback LNA and the MOSFET level design of the proposed LNA. In Section III the observations are presented and Section IV focuses on the conclusions of the proposed circuit.

II. CONVENTIONAL NOISE CANCELLING TECHNIQUE

In this technique the noises were created with opposite phase and travel through two different paths and finally got added at the output. As the noises were having opposite phase so they will cancel each other. By the use of this technique the noise performance was good. The conventional noise cancelling LNA is shown in Fig.1.

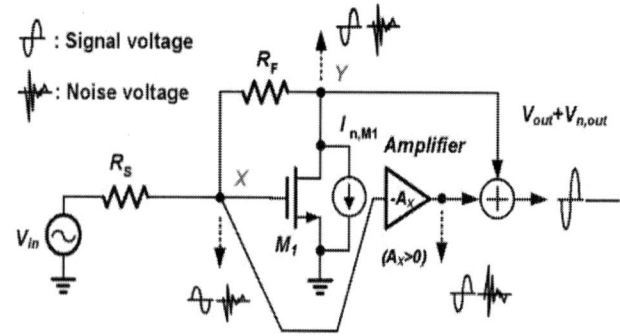

Fig 1. LNA using noise cancelling technique

978-1-5386-6723-1/19 $31.00 © 2019 IEEE

Here the MOSFET M_1 is used as the core of LNA, R_F is connected between the gate and drain of M_1 as a feedback resistor, and a amplifier A_X which will amplify the voltage at the terminal X. To analyze the effect of noise on the circuit, the noise due to MOSFET M_1 is modeled by a current source $I_{n,M1}$ from the terminal Y and ground. Due to the noise current noise voltage with same phase is generated at two terminals X and Y, which is denoted as $V_{n,x}$ and $V_{n,y}$. Whereas, V_x and V_y, the signal voltages, are having opposite-phase polarities.

Hence, the noise voltage at the output can be written as,

$$V_{n,out} = I_{n,M1}(R_S + R_F - R_S A_x) \qquad (1)$$

To cancel the noise the output noise voltage should be zero. Applying this condition the feed forward gain becomes

$$A_X = 1 + \frac{R_F}{R_S} \qquad (2)$$

If this condition is maintained during the design then the total noise in the MOSFET M_1 is reduced to zero. Now the net voltage gain A_V of the circuit for the signal voltage can be denoted as,

$$A_V = \frac{V_{out}}{V_X} = (1 - g_{M1}R_F) - A_X$$

and the total voltage gain $(A_V)_{ZN}$ without any noise can be expressed as,

$$(A_V)_{ZN} = -R_F(g_{M1} + \frac{1}{R_S}) \qquad (3)$$

The cancellation is independent of the input matching condition. Now A_V is proportional to R_F. But large R_F accounts for bandwidth degradation and thermal noise, which is a major disadvantage. To solve the problem the noise cancelling gain enhanced technique is generated.

III. CURRENT REUSED RESISTER FEEDBACK GAIN ENHANCED NOISE CANCELLING LNA

A. Current Reused Resister Feedback Gain Enhanced Noise Cancelling LNA

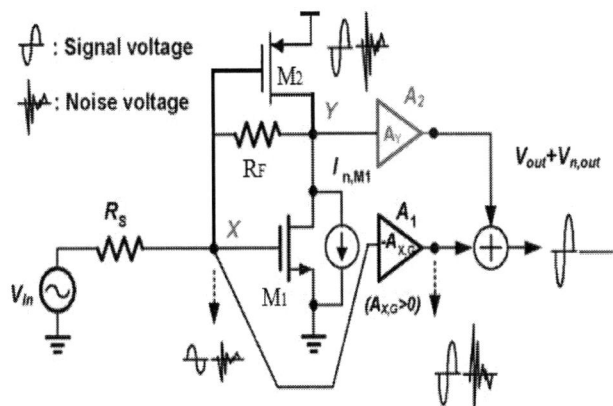

Fig 2. LNA using current reuse gain enhanced noise cancelling technique

The current reused resister feedback gain enhanced noise cancelling LNA is shown in Fig.2. Here another amplifier of gain A_y is introduced between the drain of M_1 and output of amplifier A_X and a PMOS is connected at the drain of M_1. So total noise current due to M_1 and M_2 becomes double of $I_{n,M1}$(M_1 and M_2 are of same size).

In this case the noise voltage at the output is presented as

$$V_{n,out} = I_{n,M1}((R_S + R_F)A_y - R_S A_{x,G}) \qquad (4)$$

Hence the condition for cancelling the noise becomes

$$A_{X,G} = A_Y \left(1 + \frac{R_F}{R_S}\right) \qquad (5)$$

Now the voltage at node Y is,

$$V_y = V_x(1 - (g_{m1} + g_{m2})R_F) \qquad (6)$$

Then

$$V_{out} = V_x\{(1 - (g_{m1} + g_{m2})R_F)A_y - A_{x,G}\}$$

Hence, the overall gain of the amplifier is found to be

$$A_{V,G} = \frac{V_{out}}{V_x} = (1 - (g_{m1} + g_{m2})R_F)A_y - A_{x,G} \qquad (7)$$

Putting equation(5) in equation (7)

$$(A_{V,G})_{ZN} = -A_y R_F((g_{m1} + g_{m2}) + \frac{1}{R_S}) \qquad (8)$$

if same size of M_1 and M_2 will be taken,

$$(A_{V,G})_{ZN} = -A_y R_F(2g_{m1} + \frac{1}{R_S}) \qquad (9)$$

So smaller size of M_1 and M_2 can be selected to have moderate gain which is required for a LNA. As the size of M_1 and M_2 is reduced , power consumption by each MOSFET is reduced. Again the two MOSFETs are connected in series so the NMOS will be activated by the current flowing from PMOS. Hence no extra power is consumed by M_1.

Here the zero noise voltage gain will be increased by the amplification factor A_y .So R_F value can be reduced. Therefore the increase in gain can be used to reduce the power consumption which is a major advantage.

B. MOSFET level design of the proposed LNA

The MOSFET level implementation of the LNA using current reused technique is shown in Fig.3 where the feedback resistor R_F is connected between drain and gate of M_1 and M_2 and is used to sense the noise and signal of MOSFET M_1 . M_2 is to enhance the current through M_1 . The combination of M_4 and M_5 forms the common-source amplifier A_1 with a amplifying factor of $A_{x,G}$ and the transistor M_3 connected in common gate configuration forms the amplifier A_2 with a voltage gain of A_y. R_G is the load resistor.

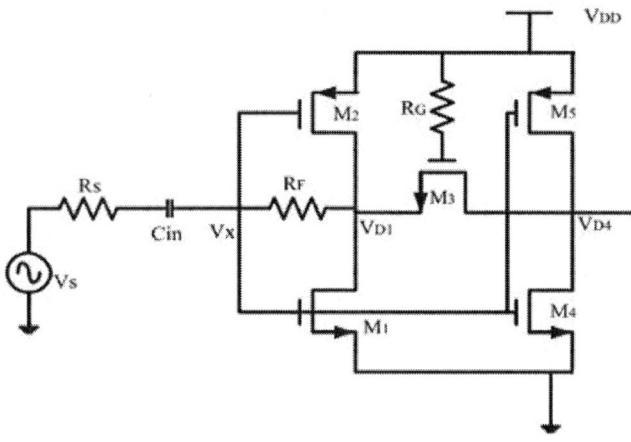

Fig 3. Circuit diagram of current reuse gain enhanced noise cancelling technique

Here amplification factor of both the amplifiers are given as

$$A_x = \frac{V_{D4}}{V_x} = (g_{m4} + g_{m5})R_2 \qquad (10)$$

$$A_y = \frac{V_{D4}}{V_{D1}} = \frac{R_2}{R_1} \qquad (11)$$

Where R_1 and R_2 are the drain resistances of the MOSFET M_1 and M_4 respectively and are expressed as

$$R_1 = \frac{1}{g_{m1}} \| r_{01} \| \frac{1}{g_{m2}} \| r_{02} \qquad (12)$$

$$R_2 = \frac{1}{g_{m4}} \| r_{04} \| \frac{1}{g_{m5}} \| r_{05} \| g_{m3} r_{03} R_1 \qquad (13)$$

Hence, $\frac{A_x}{A_y} = (g_{m4} + g_{m5})R_1$

Hence using(5) the noise cancelling condition of the LNA becomes

$$(g_{m4} + g_{m5}) \left(\frac{1}{g_{m1}} \| r_{01} \| \frac{1}{g_{m2}} \| r_{02} \right) = (1 + \frac{R_F}{R_S}) \qquad (14)$$

From equation(14) it is clear that R_F is proportional to g_{m4} and g_{m5}. Also previously it is already discussed that R_F in this technique is small compared to the conventional technique. So M_4 and M_5 size is very small helping in consuming less power.

now according to equation(9) the gain at noise cancelling stage is calculated as

$$(A_{V,G})_{ZN} =$$

$$-R_F \left(\frac{\frac{1}{g_{m4}} \| r_{04} \| \frac{1}{g_{m5}} \| r_{05} \| g_{m3} r_{03} R_1}{\frac{1}{g_{m1}} \| r_{01} \| \frac{1}{g_{m2}} \| r_{02}} \right) (2g_{m1} + \frac{1}{R_S}) \qquad (15)$$

Equation (15) shows that the gain at noise cancelling condition is proportional to R_F, g_{m1} and A_y. As here moderate gain is required so R_F and g_{m1} can be reduced in size compared to the conventional technique and the reduction in gain due to small size will be compensated by the increment in gain due to the presence of A_y. So size of M_1 and M_2 can be reduced, thereby reducing the power consumption while maintaining the same gain.

III. EXPERIMENTAL RESULTS

Here for the design and simulation of the circuit UMC180-nm CMOS process technology is used. Here the characteristic graph of NF and conversion gain is simulated using CADENCE Tool. The schematic level design of the circuit is shown in Fig. 4. In Fig. 5 the test bench diagram is shown which is required to simulate the circuit. Fig. 6 explains that the noise figure is 2.1 dB-2.4 dB for the total UWB range. The conversion gain is found to be 12.6dB at 10 GHz from Fig. 7 .

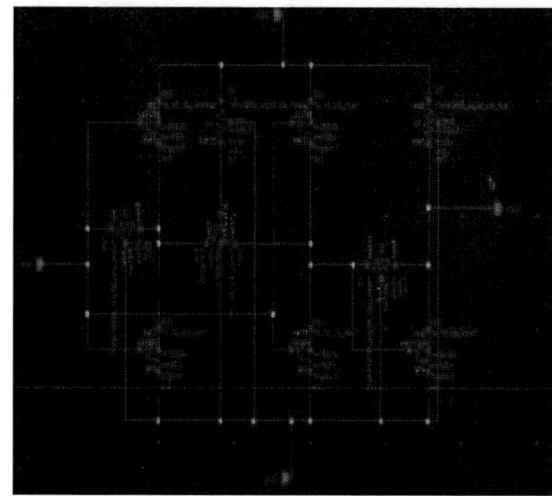

Fig 4. Schematic of the LNA

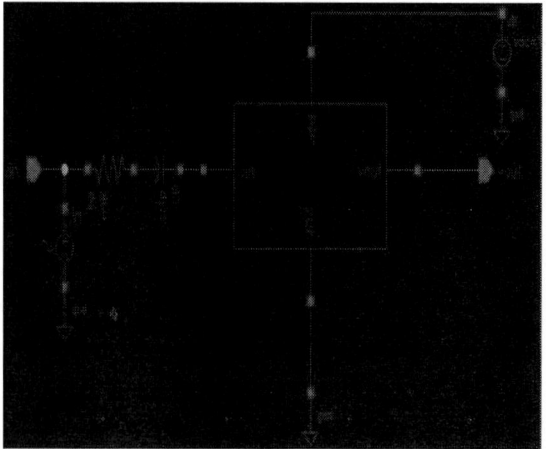

Fig 5.Test bench of the LNA

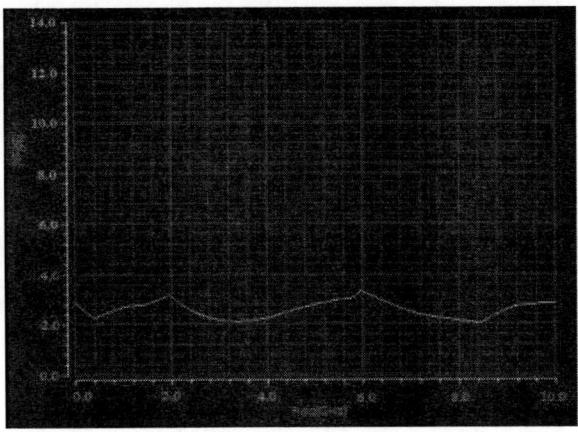

Fig 6. NF simulation of the LNA

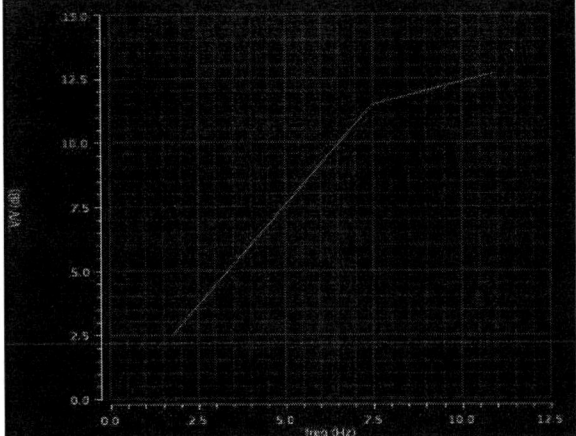

Fig 7. Conversion gain simulation of the LNA

In Table II the copmprision between the parameters of the existing techniques and the proposed technique is shown. The proposed LNA consumes 7.6mW power which is very less compared to 12.6mW[2,5],14.4mW[6],13.7mW[7]. It provides a moderate conversion gain of 12.6dB which is better than 10.9dB[2],9.8dB[3],10.7dB[4] and 10.5dB[7].It operates in the ultra band of frequency with a noise figure of 2.1dB which is lesser than 3.5[2],4.3dB[4]2.86dB[5] and (2.6-3.3) dB[7].

IV. CONCLUSION

In this paper the current reused technique is introduced to modify the conventional noise cancellation technique to design a low power LNA with moderate gain and better noise performance. It operates in the ultra band frequency range. This LNA provides a moderate conversion gain of 12.6 dB, a noise figure of 2.1 dB at 3.5 GHz and 8.5 GHz. Along with moderate gain and good NF it consumes very less power i.e. 7.6 mW. In comparision with the previous techniques it is a better one in terms of power consumption while maintaining nearly same conversion gain and NF. So, the current reused technique can be implemented to get better performance in the ultra band range of frequency and for multi-purpose applications.

REFERENCES

[1] S. Roy, J. R. Foerster, V. S. Somayazulu, D. G. Leeper,"Ultrawideband Radio Design: The Promise of High-Speed, Short-Range Wireless Connectivity", Proceedings of the IEEE, vol. 92, no. 2, pp. 295-311, february 2004.

[2] Mei-Fen Chang-Ching Wu, Wen-Shen Chou, Wuen, Kuei-Ann Wen, "A low power CMOS low noise amplifier for ultra-wideband wireless applications", IEEE Int. Symp. Circuits Syst. 5, 5063–5066, 2005.

[3] C.W. Kim, M.S. Kang, P. T. Anh, H.T. Kim, and S.G. Lee," An Ultra-Wideband CMOS Low Noise Amplifier for 3–5-GHz UWB System", IEEE JOURNAL OF SOLID-STATE CIRCUITS, VOL. 40, NO. 2, FEBRUARY 2005.

[4] M. Okushima, J. Borremans, D. Linten, G. Groeseneken, "A DC-to-22 GHz 8.4 mW compact dual-feedback wideband LNA," in Proc. IEEE Radio Frequency Integrated Circuits Symp., pp. 295–298, Jun. 2009.

[5] J. Kim, S. Hoyos, and J. Silva-Martinez, "Wideband common-gate CMOS LNA employing dual negative feedback with simultaneous noise,gain and bandwidth optimization," IEEE Trans. Microwave Theory.Tech., vol. 58, no.9, pp. 2340-2351, Sep. 2010.

[6] Yueh-Hua Yu, Yong-Sian Yang, Yi-Jan Emery Chen," A Compact Wideband CMOS Low Noise Amplifier With Gain Flatness Enhancement", IEEE JOURNAL OF SOLID-STATE CIRCUITS, VOL. 45, NO. 3, MARCH 2010.

[7] Ke-Hou Chen, Shen-Iuan Liu, "Inductorless Wideband CMOS Low-Noise Amplifier Using Noise-Canceling Technique" IEEE TRANSACTIONS ON CIRCUITS AND SYSTEMS-I: regular papers, vol. 59, no. 2, pp. 305-314 february 2012

TABLE I. COMPARISON TABLE

Ref.	CMOS process	Freq. (GHz)	Conversion Gain (dB) (Power)	NF (dB)	Inductor	Power (mW)
2	180nm	2.6-9.2	10.9	3.5	Yes	7.1
3	180nm	2-4.6	9.8	2.3	Yes	12.6
4	90nm	22.1	10.7	4.3	Yes	8.4
5	180nm	1.05-3.05	16.9	2.86	Yes	12.6
6	180nm	.04-1.2	16.4	2.1-3.4	Yes	14.4
7	65nm	10	10.5	2.7-3.3	No	13.7
This Work	180nm	3.1-10.6	12.6	2.1	No	7.6

A Novel Driver less SRAM with Indirect Read for Low Energy Consumption and Read Noise Elimination

Debasish Nayak[1,a], Umakanta Nanda[2,c], Prakash Kumar Rout[3,a], Sudhansu Mohan Biswal[4,a], Dhananjaya Tripthy[5,a]
Sanjit Kumar Swain[6,b], Biswajit Baral[7,b], Satish Kumar Das[8,b]

[a]Dept. Electronics and Instrumentation Engineering, Silicon Institute of Technology, Bhubaneswar, India
[b]Dept. Electronics and Communication Engineering, Silicon Institute of Technology, Bhubaneswar, India
[c]Dept. Electronics and Communication Engineering, Vellore Institute of Technology, Amaravati, India

[1]nayak.debasish84@gmail.com, [2]uk_nanda@yahoo.co.in, [3]prakashrout05@gmail.com, [4]sudhansu.mohan@silicon.ac.in,
[5]dhananjayavssut@gmail.com, [6]sswain@silicon.ac.in, [7]biswajit@silicon.ac.in [8]satish.das@silicon.ac.in

Abstract— The modern electronics gadget has influenced tremendously every aspects of life. The demand to add more and more functionality has forced to increase the performance of the processor. To ensure a robust data supply to the processor a high performance, stable and low power SRAM is also of utmost necessity. An indirect read SRAM cell is proposed here which eliminates the read noise insertion to increase the data stability. It also consumes 41% less energy compared to the conventional SRAM cell. The SRAM cell is designed to be written single ended using only one write access transistor. The cell reduces the energy consumption by reducing the short circuit current and also reducing the number of leakage path. The cell also has a high write speed since the storage data node is a floating node and not connected to the ground.

Index Terms— **Short-circuit energy, Low energy, Dynamic energy, indirect-read, single ended write**

I. INTRODUCTION

The continuous demand from the users to use more and more sophisticated gadgets has skyrocketed the integration of more and more logic to the hand held devices such as mobile phones, tablets etc. [1-5]. It demands the MOS transistor size also to be decreased gradually. As the transistor size decreases, though the above requirement gets satisfied but the system gets more vulnerable to the presence of noise. Hence proper care should be taken to increase the noise withstanding capacity.

In all the modern digital system the processor is feed high speed data through a static random access memory (SRAM) since the lower-end data processing elements such as DRAM cannot match the speed of the processor. As SRAM block is a repetition of large number of single SRAM bit cell hence the performance enhancement of SRAM bit cell affects the performance of the overall SRAM block. Moreover the SRAM block occupies almost 70% of the modern SOC. Hence performance enhancement of the SRAM block significantly increases the performance of the complete SoC. As SRAM

deal with data handling its stability is of utmost necessity since a single error in a data bit may lead to a catastrophic condition.

We have proposed a 6T-SRAM cell which follows indirect read technique that helps in read noise isolation. It also reduces the energy consumption of the SRAM cell in various ways.

The remaining of the paper is elaborated as given. The problem with the conventional SRAM cell is mentioned in section II. The proposed KP-SRAM cell is described in Section III and its performance analysis is carried out in Section IV. Finally the conclusion of the work is mentioned in section V.

II. PROBLEMS IN CONVENTIONAL SRAM

Conventionally the SRAM cell is two cross-coupled inverters in combination with two access transistors. During data write operation the data to be written is loaded to the bit line (BL) and the complement of the dada is loaded to the bit line bar (BLB). Then the access transistor are activated which over writes the desired data to the internal nodes of the SRAM cell.

The destructive nature of the read operation of conventional SRAM cell is its major drawback which tries unwantedly to destroy the stored data during read operation. During read task, both bit line and bit line bar are pre-charged to V_{dd} and then the access transistor pair is activated. The internal node which contains '0' drags the corresponding pre-charged bit line to '0'. The differential voltage between the BL-BLB pair is sensed by the sense amplifier and the data is read out to the data latch. During the read process as the '0'-node drags the pre-charged bit line, so the '0'-node itself is also raised little bit due to the voltage division among the access transistor and the pull-down transistor. This erroneous increase in the '0'-node voltage should be less than that of the switching voltage of the other inverter present in the SRAM to avoid erroneous data switching.

The pull-down transistor can be chosen with more width to reduce the effect of noise insertion on the data stability. But doing so will reduce the ability to write the SRAM cell. Hence the indirect read technique [6-8] has become an alternative among the design.

III. THE PROPOSED KP-SRAM CELL

The SRAM cell proposed here focuses on an alternative indirect read technique which eliminates the destructive read problem. It also uses a driverless data storing technique to save energy consumption and area.

The proposed cell consists of six transistors as shown in Fig. 1. It uses the keeper cell concept to store the data. Fig. 1 shows that the storage element of the proposed SRAM cell consists of PM1 and INV-1. To write a particular data, first the data is loaded on the WBL line and then the NM1 transistor is activated. The new data enters to the 'Q' node. If '1' is stored at Q-node then output of INV-1 will be '0' and hence PM1 will be on. So the data '1' will be stored at 'Q'. Any small decrease in 'Q' node cannot change the state of INV-1 and its output will remain as '0'. To flip the stored data the 'Q' node should be pulled down very fast such that the output of INV-1 will be changed to '1' and the PM1 will be switched off. This is possible only if the data '0' is loaded fast to WBL and then the access transistor is NM1 enabled. If the 'Q' node stores '0' then the output of the INV-1 becomes '1'. Hence the PM1 will be switch off. So no current flows from V_{dd} to the 'Q' node and the 'Q' node continues to store '1'.

The proposed cell also uses an indirect read technique for reading the stored data. The read bit line bar (RBLB) is first pre-charged and then NM3 transistor is activated by activating the RWL line. The data stored at 'Q' node drives the NM2 transistor. Depending on the data at the 'Q' node the NM2 transistor gets switched on or off. So depending on the data at 'Q' node the RBLB may or may not discharge. The discharging of RBLB indicates '1' at 'Q' node and if the RBLB does not discharge it indicates the data stored at 'Q' node is '0'.

As the 'Q' node is connected to the gate of NM2 transistor, the pre-charged RBLB cannot erroneously raise the Q-node's voltage level unlike the conventional 6T-SRAM. So the 'Q' node is not disturbed at all during read operation. This leads to a robust read operation of the proposed SRAM cell.

IV. PERFORMANCE ANALYSIS

Proposed SRAM is simulated in General purpose PDK. We have used 90 nm library for design of the cell. It is simulated in ADE Specter of Cadence. Simulation is conducted at 27^0C with a supply of 0.8V. The output of the transient analysis is shown in Fig. 2. It can be found from the Fig. 2 that '0' is written on the 'Q' node at 1ns. At 2ns data is fetched. Before 2 ns the RBLB line is pre-charged by lowering the PC signal. Since the 'Q' node contain '0' so the RBLB line remains at high level during read operation at 2ns. Similarly at 3ns the data '1' is written to the 'Q' node and at 4ns the data is read. Since the 'Q' node contains '1' so the RBLB line is discharged to '0' during the read operation taken place at 4ns.

The power utilized in the proposed cell during different operations such as writing 0, reading 0, writing 1 and reading 1, are compared to those of a 6T-cell. The energy utilizations of the proposed and conventional 6T-cell are shown in Fig. 3 and 4 respectively. The static energy utilization of the cell is the energy utilized through the idle interval between the consecutive dynamic operations. The energy consumption is also estimated and shown in Table. 1.

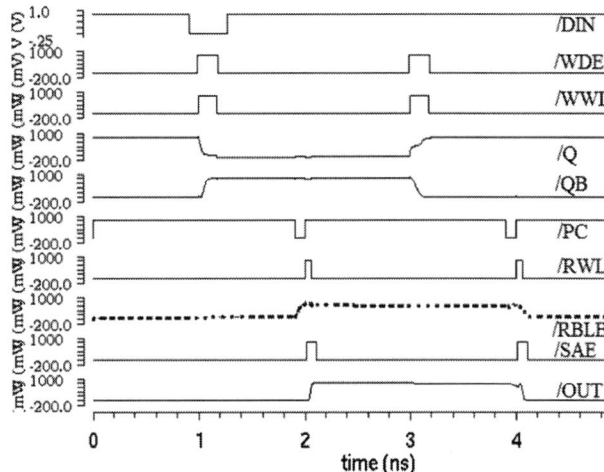

Fig. 2. Transient response of KP-SRAM cell accessed at 1 GHz.

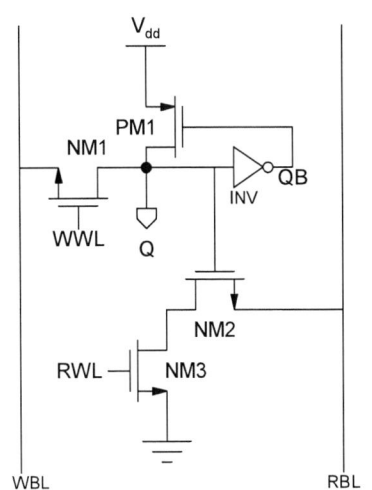

Fig. 1. Cell topology of proposed KP-SRAM.

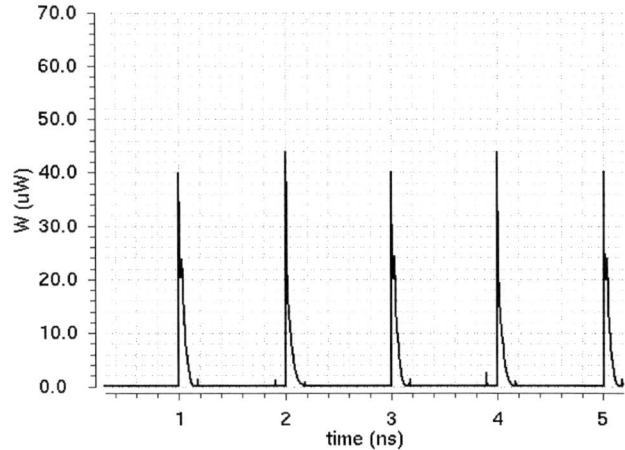

Fig. 3. Energy utilization of KP-SRAM cell during 1GHz access

TABLE I
COMPARISON OF ENERGY CONSUMPTION OF PROPOSED AND CONVENTIONAL SRAM CELLS AT 1GHz ACCESS

Task ↘ SRAM ↓ Cells	ENERGY UTILIZATION (aJ) DURING DYNAMIC OPERATIONS				ENERGY IN DYNAMIC OPERATION (aJ) = A+B+C+D = (E)	ENERGY IN STATIC OPERATION (aJ) (F)	TOTAL ENERGY (aJ) =E+F	ENERGY SAVING W.R.T. 6T-SRAM CELL (%)
	0-WRITING (A)	0-READING (B)	1-WRITING (C)	1-READING (D)				
KP-SRAM	1109	551.9	949	273.9	2883.8	25	2908.8	41.6
6T-SRAM	1654	716.2	1917	672.2	4959.4	29	4988.4	-

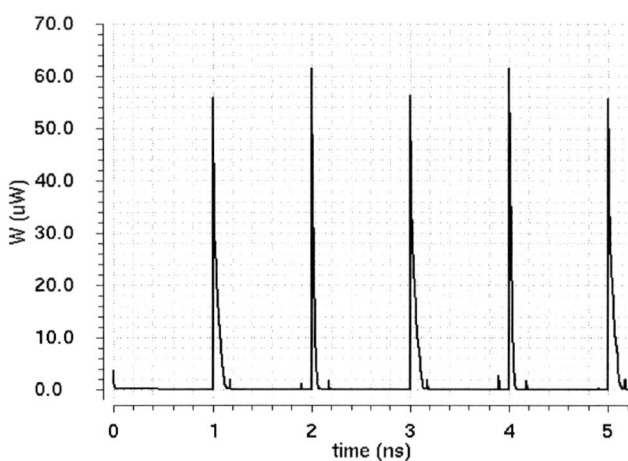

Fig. 4. Energy utilization of conventional SRAM cell during 1GHz access

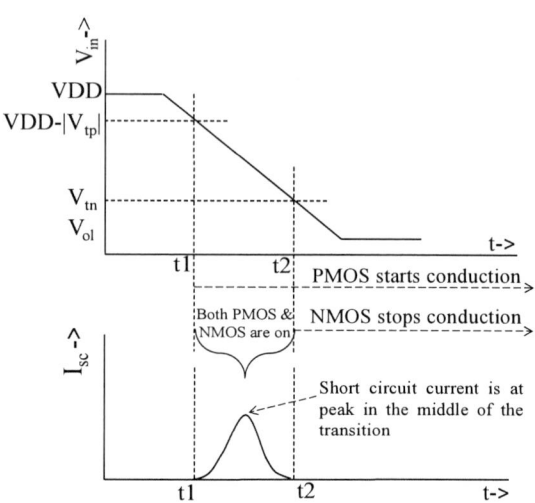

Fig. 5. Representation of short circuit loss

Table I shows that, the total energy utilization of the SRAM is dominated by dynamic energy utilization of the cell. The dynamic energy utilization of 6T-cell is 4959aJ whereas that of proposed cell stands at 2888aJ which is 41% lesser. The reduction in energy consumption is attributed to the following reasons.

During the dynamic switching of SRAM cell a major amount of energy is consumed as short [9-13] circuit path energy loss. The concept of short circuit energy consumption is shown in Fig. 5. As shown in the Fig. 5, in an inverter the pull-up PMOS is switched on when its input voltage fall down to V_{dd}-$|V_{tp}|$. But the pull down NMOS stops conduction when its input voltage drops to V_{tn}. Hence for the interval when the input voltage lies between V_{dd}-$|V_{tp}|$ and V_{tn}, both the pull-up PMOS and pull-down NMOS are in on state. Hence significant amount of current may flow unwantedly from V_{dd} to ground. This component of energy consumption is reduced in the proposed cell by reducing the number of short [14-17] circuit path.

The conventional 6T-SRAM cell consists of two inverter pairs whereas the proposed SRAM cell consists of a single inverter. So during the dynamic operation such as write operation, the 6T-SRAM cell has two short circuit paths from V_{dd} to ground. On the other hand the proposed cell has only one short circuit path. This helps in reduction of short circuit

current flow in the propose cell hence short circuit energy consumption.

In a 6T-SRAM, internal nodes are also connected to the ground terminal through the driver NMOS. But in proposed cell the 'Q' terminals floats free without being connecting to the ground. So for the 6T-SRAM the write driver has to apply more force to pull the 0-node to 1-state. But in proposed SRAM cell as the 'Q' node is floating with very less effort it can be pulled up.

The static energy consumption [18-20] of the cell is also less which can be marked from the Table I. When the 'QB' node stores '0' it has only one leakage path from V_{dd} to 'QB'. Similarly if the 'QB' node stores '1' it gets only one leakage path from 'QB' to ground. In contrast if in a conventional SRAM cell the 'QB' node stores '0' it has two leakage paths. One is from V_{dd} to 'QB' and another is from 'QB' to RBLB. Similarly if the 'QB' node stores '1' then there are also two leakage paths. One is from 'QB' to ground and another is from 'QB' to RBLB. The presence of more number of leakage paths increases the amount of leakage current and hence the leakage energy consumption.

All the above mentioned issues are well handled in the proposed KP-SRAM, which increases the cell performance along with the reduction of energy consumption and increment in cell stability

V. CONCLUSION

A novel SRAM cell architecture is proposed which uses a driverless node to store the data. It uses one end write technique to write into the cell and it also use an indirect read method to fetch the data from the cell. The indirect read technique makes the cell read disturbance free. The driverless node increases the easiness of writing and utilizes less energy during write. The presence of only one inverter in the cell reduces the short circuit power loss of the circuit. The overall power consumption of the cell is reduced by 41% compared to the conventional cell.

REFERENCE

1. A. Goel, R.K. Sharma and A.K. Gupta, "Process variations aware area efficient negative bit-line voltage scheme for improving write ability of SRAM in nanometer technologies", *IET Circuits, Devices & Systems*, vol. 6, no. 1, pp. 45-51, Jan 2012

2. D. Nayak, D.P. Acharya, P.K. Rout, U. Nanda, A high stable 8T-SRAM with bit interleaving capability for minimization of soft error rate, Microelectronics Journal, 73, (2018) 43-51

3. D. Nayak, D.P. Acharya, P.K. Rout, U. Nanda, "A novel charge recycle read write assist technique for energy efficient and fast 20 nm 8T-SRAM array," *Solid-State Electronics*, vol. 148, pp. 43–50, Oct. 2018

4. P. K. Rout, D. P. Acharya, G. Panda and D. Nayak, "Process corner variation aware design of low power current starved VCO power," in *proc. of the IEEE International Conference on Electronics and Communication Systems (ICECS)*, pp. 1-5, Feb. 2014

5. S.N. Panda, S. Padhi, V. Phanindra, U. Nanda, S.K. Pattnaik, D. Nayak, "Design and implementation of SRAM macro unit," in *proc. of the IEEE Conference on* Trends in Electronics and Informatics *(ICEI)*, pp. 119-123, May. 2017

6. S. Mukhopadhyay, R.M. Rao, Kim Jae-Joon and Chuang Ching-Te, "SRAM Write-Ability Improvement With Transient Negative Bit-Line Voltage", *IEEE Transactions on Very Large Scale Integration (VLSI) Systems*, vol. 19, no. 1, pp. 24-32, Jan 2011

7. J. M. Rabaey, A. Chandrakasan, and B. Nikolic, "Digital Integrated Circuits". Pearson 2003, pp. 658-659

8. D. Nayak, D.P. Acharya, K. Mahapatra, A Read Disturbance Free Differential Read SRAM Cell for Low Power and Reliable Cache in Embedded Processor, AEU - International Journal of Electronics and Communications, 74, (2017) 192-197

9. M. M. Khellah, A. Keshavarzi, D. Somasekhar, T. Karnik, and V. De, "Read and write circuit assist techniques for improving $V_{cc\min}$ of dense 6T SRAM cell," in *Proceeding of International Conference on Integration Circuit Design Technol., Jun. 2008, pp. 185–189*

10. D. Nayak, D.P. Acharya, P.K. Rout and K.K. Mahapatra, "Design of Low Leakage and High Writeable Proposed SRAM cell Structure", in *proc. of the IEEE International Conference on Electronics and Communication Systems (ICECS)*, pp. 1-5, Feb 2014

11. D. Nayak, D.P. Acharya, K.K. Mahapatra, "An improved energy efficient SRAM cell for access over a wide frequency range," *Solid-State Electronics*, vol. 126, pp. 14–22, Sept. 2016

12. S.K. Gupta, J.P. Kulkarni and K. Roy , "Tri-Mode Independent Gate FinFET-Based SRAM With Pass-Gate Feedback: Technology–Circuit Co-Design for Enhanced Cell Stability", *IEEE Transaction on Electron Devices*, vol. 60, no. 11, pp. 3696-3704, Nov 2013

13. D. Nayak, D.P. Acharya, K.K. Mahapatra, "Power efficient design of a novel SRAM cell with higher write ability," in *proceedings of IEEE India Conference (INDICON)*, pp. 1-6, Dec. 2015

14. E. Grossar, M. Stucchi, K.Maex and W. Dehaene, "Read Stability and Write-Ability Analysis of SRAM Cells for Nanometer Technologies", *IEEE Journal of Solid-State Circuits*, vol. 41, no. 11, pp. 2577–2588 Nov 2006

15. D. Nayak, D.P. Acharya, K.K. Mahapatra, "Current starving the SRAM Cell: a strategy to improve cell stability and power". Circuits System and Signal Processing, vol. 36, pp. 3047–3070, Dec. 2016

16. P.K. Rout, D. Nayak, and D.P. Acharya "A novel low power 3T inverter, " in *proc. of the IEEE Int. Conference on Advanced Electronic Systems (ICAES)*, 2013 , vol., no., pp. 221-224, Sep 2013

17. E. Seevinck, F. J. List and J. Lohstroh, "Static-noise margin analysis of MOS SRAM cells", *IEEE Journal of Solid-State Circuits*, vol. 22, No. 5, pp. 748-754, Oct 1987

18. Y. Morita, H. Fujiwara, H. Noguchi, Y. Iguchi, K. Nii , H. Kawaguchi and M. Yoshimoto, "An area-conscious low-voltage-oriented 8T-SRAM design under DVS environment", in *Proceeding of Symposium on VLSI Circuits*, pp. 256 -257, Jun 2007.

19. V. Naveen and A. Chandrakasan, "A 65nm 8T Sub-Vt SRAM employing sense-amplifier redundancy," in *Digest of Technical Papers IEEE International Solid-State Circuits Conference, ISSCC 2007*, pp. 328–606, 2007,

20. M. Samson, and M.B. Srinivas, "Analyzing N-Curve Metrics for Sub-Threshold 65nm CMOS SRM," in proceedings of *8th IEEE Conference on Nanotechnology*, 2008, pp.25-28, 18-21 Aug 2008

Scholastic Approach towards Economic Digital Mileage Meter with GPS

Anirban Ghosal
Asistance Prof., ECE
JIS College of Engineering
Kalyani, Nodia,WB, India
anirban.ghosal@jiscollege.ac.in

Subham Ghosh
B.Tech, ECE
JIS College of Engineering
Kalyani, Nodia,WB, India
sghosh.network@gamil.com

Ayan Saha
B.Tech, ECE
JIS College of Engineering
Kalyani, Nodia,WB, India
ayansaha114@gamil.com

Nilanjan Bhattacharjee
B.Tech, ECE
JIS College of Engineering
Kalyani, Nodia,WB, India
nilanjan.bhattacharjee77@gmail.com

Srijan Kr. Bhar
B.Tech, ECE
JIS College of Engineering
Kalyani, Nodia,WB, India
srijanbhar95@gmail.com

Indranath Sarkar
Associate Prof., ECE
JIS College of Engineering
Kalyani, Nodia,WB, India
indranath.sarkar@jiscollege.ac.in

Abstract— Fuel economy is an important factor while buying an automobile. For automobile buyers, it has an impact on the running expenses of the vehicle. By definition fuel economy is the number of kilometers of distance that can be covered per liter or gallon worth of fuel. Automotive Research Association of India conducts fuel economy tests based on the Indian Driving Cycle. However, in practically driving conditions, one can expect to get -10% to -20% (less) of the ARAI certified values. This paper aims at designing a mileage meter along with a vehicle tracking provision. The users will have a better option to examine the vehicle's performance before and after purchasing it can also be attached in an existing vehicle.

Keywords—Automobile, Mileage Meter, Vehicle Tracking, Fuel, Driving Cycle

I. INTRODUCTION

Mileage meter is a device which generally determines the distance covers by fuel consumed or the distance your engine is able to drive the vehicle per litre of fuel that it consumes [1]. It describes the efficiency of the engine. Till date, mileage meter is available only in very expensive vehicle. The development of vehicle tracking device becomes more important for vehicle owners to ensure theft prevention and quick identification towards recovery process in situations where a vehicle is missing driven by an unauthorized person. An odometer or mileage meter is an instrument used for measuring the distance travelled by a vehicle, such as a bicycle or car and it also monitors the average and real time fuel consumption. AnIndian Driving Cycle has been laboratory test, conducted on a rolling road that simulates typical Indian driving environment over 10 kilometres [12]. The device may be electronic, mechanical, or a combination of the two. With the advancement of technology human activities have been facilitated by the use of digital based tools, which includes a digital mileage meter. This device not only measures the distance covered by the vehicle per litre fuel but also helps in periodic inspection of vehicle fuel volume by using GPS module, pressure sensor and atmega328p [8]. Regular fuel volume inspection helps in timely replacement of fuel and also prevents wastage which makes the process economical.

Improvement in technology in the recent years has influenced people to know more about the accurate value of the vehicle's mileage and fuel volume rather than the approximate value of the same [2]. This also leads to the development of digital speedometers providing exact speed with some insignificant delay [3]. This paper describes of finding the exact level of fuel that is available inside a tank using the pressure sensor and displays it immediately in the LCD display unit [4]. Generally in vehicles, the fuel gauge consists of into the sensing unit and the indicator parts. A fuel gauge is an instrument used to indicate the level of fuel contained in a boiler [5-6] which are used in two wheeler vehicle and four wheeler vehicles; on the other hand, these may also be used for any tank including underground storage tanks. The GPS module tracks the position of the vehicle and records its mileage.In addition, the vehicle fuel is unable to substitute habitually and for this the engine will quickly heated and possibly engine can out of order since friction. It would be very disadvantageous and not to beneficial for vehicles users and companies [11]. Due to high number of vehicles owned each year, owners and companies are concerned to agenda of periodic vehicle scrutiny. Therefore, it is needed a system that can automatically perform scrutiny the mileage and provide experimental data to owners about the condition of the vehicle.Another designed prototype consists of an early warning system in digital odometer prototype as a timer for intervallic vehicle fuel substitute by uses two types of reduced instruction set computers microcontroller to retrieving the data with two infrared transceivers for data sender and recipient [1].

Study with this proposed project prototype aims to measure and indicate the accurate mileage of a vehicle that can receive data from fuel tank. Analyse the mileage performance by testing several parameters namely the numeral bytes of data transmit, data retrieval covering distance, the ensuing reply instant to accuracy of data generated, range of the activate GPS area with receiving latitudes and longitudes.

978-1-5386-6723-1/19 $31.00 © 2019 IEEE

II. PROPOSED PROTOTYPE AND SCHEMATIC APPROACH TOWARDS SYSTEM MODIFICATIONS

This proposed project prototype is based on how much distance is travelled by a vehicle using 1 liter of fuel. Distance can be measure by various ways. In this proposed prototype, determine the distance through GPS module more accurate. Due to GPS system it has been possible to receiving vehicle latitude and longitude by tracing. Fuel consumed is measured as the difference between the net fuel capacity and the fuel present in current whiles the vehicle in motion with using of fuel. Mileage is very essential in our daily life because fuel is getting costlier day by day and we have to be aware of about how much fuel is getting consumed.

The initiative understanding for mileage meter is demonstrated in Fig.1 and heart of the whole circuit atmega328p microcontroller (MCU) has been used to develop this project prototype. The Rx pin & TX pin of NEO 6m GPS module is connected with the digital pin 3 and digital pin 4 of MCU respectively. Here digital pin3 & digital pin 4 of MCU are used as TX & RX respectively. Next to connect the MPX 5010dp with MCU for volume measurement where the data out pin (PIN1) of the pressure sensor has connected to the A0 pin of the MCU board. After powering the pressure sensor the sensor senses the pressure of liquid with that of the pressure of air and gives an output. To determine the output given by the pressure sensor that needs to use analog read keyword for proper fetching of the data. A 16x2 LCD is used to display the whole process. An LCD display has 16 pins and connected with a 10k potentiometer with 1, 2 & 3 pins which controls the contrast value of LCD display. The pin 4 & pin 6 are connected with digital pin 13 & digital pin 12 of MCU for REGISTER SELECT & READ WRITE respectively while the pin5 of LCD has connected to the ground for keeping LOW so that writing in register have done. The DB4, DB5, DB6, DB7 pins of LCD has connected with the digital pin 11, 10, 9, 8 of MCU respectively.

The scheme of proposed prototype is kind of complicated mechanism obtained. In Fig.2 the circuitry approach is demonstrated. This GPS module has been programmed in such a manner such that at an interval of two hours this module automatically sends a message to a given SIM card which will show the latitude and longitude of the vehicle and will also indicate the whether the vehicle is in motion or stationary.

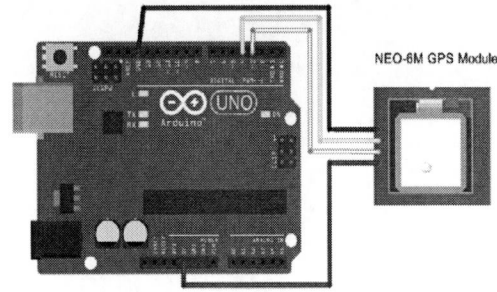

Fig. 2. Schematic circuit representation for distance calculation

NEO-6M GPS [9] module is to simplify with accuracy for distance calculation. With used of this module's GPS, it is possible to identify the vehicle latitude and longitude immediately and also distance can be measured by fundamental application. The NEO-6M GPS is connected with atmega328p microcontroller board through digital pin 3 and pin 4 along with GND and 5V power supply connection.

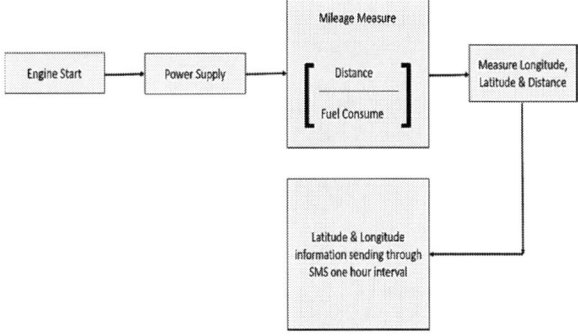

Fig. 3. Block Diagram representation of impact working principal of proposed design

The Successful testing of the prototype is shown below in the Fig.4. (a) & (b) where complete project prototype is demonstrated in Fig.4. (a) & (b) is demonstrated the measuring volume of the fuel on the LCD screen to calculated mileage.

(a)

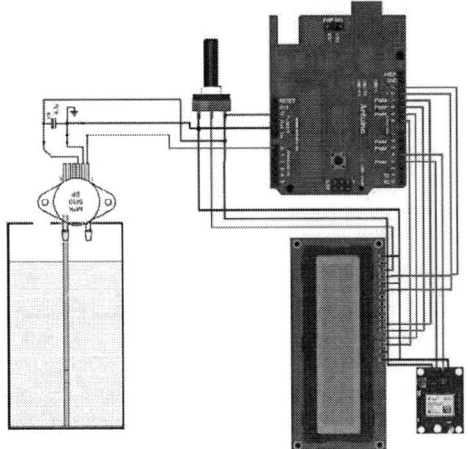

Fig. 1. Schematic circuit representation of mileage meter

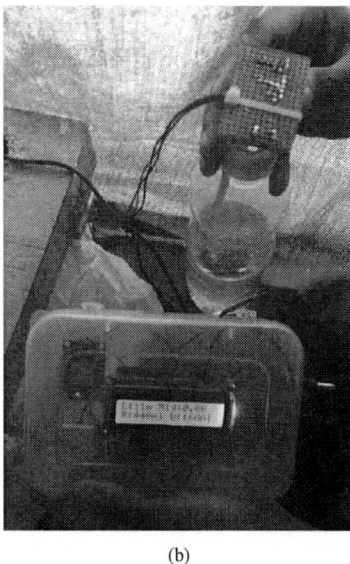

(b)

Fig. 4. (a) & (b) Successful prototype testing by digital mileage meter

The proposed prototype in this project discussed is helps to understand and calculate easily. To process all data two line LCD screen has been used [7]. Partially difficulties to access exact mileage. The testing result of initiative work is in approximately 70% success rate. But according to future aspect, proper modification and encouragement is highly required for perfect result. The working principal of this project prototype is explaining by flowchart in the Fig. 5.

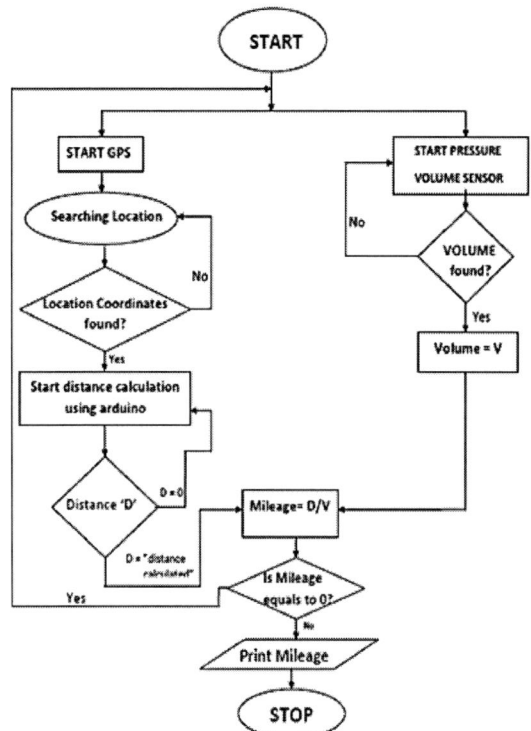

Fig. 5.   Working flowchart of the proposed prototype

III. Limitations

This proposed project prototype calculates distance by measuring the difference of the latitude longitude points of the initial and final positions. Therefore, this device would indicate the mileage as zero if the initial and final positions of the vehicle are same. This device is more appropriate for measuring mileage when the vehicle travels in a straight line rather than a curved path.

IV. Further Development Prospects

In future this proposed project can be made more accurate by using read switches to calculate the number of turns of the vehicle's wheels and measure the distance covered and mileage accordingly. This system would also make the odometer able to measure the vehicle's mileage even if it travels along a curved path. In that case the GPS module can be used for tracking the vehicle's location making the device more precise.

V. Conclution

Now a day's restricted fuel consumption is highly needful because of its high price. Fuel consumption varies with the vehicles and its manufacturing companies. As there was no such device for instantly measuringthe actual consumption of fuel digitally but, there was amanual way that can measure the fuel consumption of the vehicles of a specified company. This method was not a proper one rather it was a hectic one. In this proposed prototype to implement such a device that measure the actual fuel consumption of the vehicle. The proposed economic digital mileage meter system prototype had been tested. This project prototype proved the result completely and wasacceptable.

References

[1] Asdi Galvani, Andrian Rakhmatsyah and Giva Andriana Mutiara, "Prototype of Microcontroller-Based Odometer Reading for Early Warning in the Vehicle Lubricants Replacement", 3rd International Conference on Information and Communication Technology (ICoICT), IEEE, pp. 172-177, 2015.

[2] Arun Kumar Vala, "Microcontroller-based Speedometer-Cum-Odometer" electronics for you www. efymag.com, pp.60, November 2008.

[3] "UAF 2115 Speedometer and Mileage indicator" Edition 6251-435-1DS ITT Semiconductors, Jan. 15, 1997

[4] Rashmi R and Mrs.Rukmini Durgale, "The Novel of Embedded Based Digital Fuel Gauge," International Conference on Computing and Control Engineering (ICCCE 2012), 12 & 13 April, 2012.

[5] Jaimon Chacko Varghese and Binesh Ellupurayil Balachandran, "Low Cost Intelligent Real Time Fuel Mileage Indicator for Motorbikes", (IJITEE) ISSN: 2278-3075, Volume-2, Issue-5, April 2013.

[6] http://en.wikipedia.org/wiki/Odometer Tanggal akses 21 Maret 2011

[7] Altera Corp, "Creating Low-Cost Intelligent Display Modules With an FPGA and Embedded Processor," v 1.0, September 2008

[8] http://www.atmel.com/Images/Atmel-2486-8-bit-AVR-microcontroller-ATmega8_L_datasheet.pdf.

[9] https://www.u-blox.com/sites/default/files/products/documents/NEO-6_DataSheet_(GPS.G6-HW-09005).pdf

[10] Daniel R. McGlynn, "Vehicle Usage Monitoring And RecordingSystem", US Patent 4072850, February 1978

[11] Khoswanto. H. Thiang and Kuntoro, J. "Odometer Digital untuk Kendaraan dengan Mikrokontroler" MCS51, 2004

[12] Mr.Vishal Parekh and Dr.Varshaben Shah, "Measurement and Analysis of Indian Road Drive Cycles for Efficient and Economic Design of HEV Component", World Electric Vehicle Journal Vol. 7 - ISSN 2032-6653 –2015

[13] Michael D Murphy, "Integrated taximeter/GPS position tracking system", US5917434A, June 1995.

[14] David Bell and Philip Norton, "Combining time and gps locations to trigger message alerts", US20100141514A1, Dec 2008

[15] Xiaoguang Yu, "System and method for tracking a path of a vehicle", US20090254270A1, April 2008

2019 Devices for Integrated Circuit (DevIC), 23-24 March, 2019, Kalyani, India

Comparative Study of High K in Silicon Nano Tube FET for Switching Applications

Avtar Singh
Department of ETC
C.V.Raman College of
Engineering,Bhubhneshwar,Orrisa
avtar.ju@gmail.com

Chandan Kumar Pandey
Department of EE
NIT SILCHAR,ASSAM
chandankumarpandey@gmail.com

Saurabh chaudhury
Department of EE
NIT SILCHAR,ASSAM
Saurabh1971@gmail.com

Chandan kumar Sarkar
Department of ETC
Jadavpur
University,Kolkata
phyhod@gmail.com

Abstract—**In this work we have studied the impact of variation of k dielectric constant on Silicon Nano Tube FET for low power and high speed applications. The Silicon Nano tubular structure offers better immunity towards short channel effects (SCE's) because of the better control of channel region due to the double gate all around. By cause of gate engineered structure high K value structures possess high value of electron velocity as compare to low k dielectric structure, which helps in improving the efficiency of carrier transport. In this work we have considered a Silicon Di-oxide(SiO₂), Silicon Nitride(Si₃N₄), Hafnium Oxide(HfO₂), Hafnium Silicate (HfSiO₄),Tin oxide (SnO₂) and Titanium Oxide (TO₂) as a gate dielectric. It has been found that when the high k is replaced with SiO₂ then the switching performance of the device is enhanced which makes it suitable for the SOC applications. From the analysis it has been found that HFO2 in SINTFET will be a superior alternative for future tubular FET devices**

Keywords—High K.Silicon nano Tube FET, I_{on} / I_{Off} ratio ,tubular structure

I. INTRODUCTION

Due to the extensive scaling of the FET devices, the gate dielectric becoming very thin due to which the direct gate tunneling increases and which regularly degrades the performance of the devices. However, using SiO_2 as gate dielectric exhibits many advantages such as (i) SiO_2 can be made from silicon via thermal oxidation (ii) it is an excellent insulator (iii) it is an amorphous and has few electronics defects[1]. But the major problem with SiO_2 is that it is very easy to tunnel across it when it is very thin. This led the researchers to search for the alternative of the SiO_2 with a physically thicker layer of a high k dielectric constant[2]. Researchers are looking for the following characteristics in a new dielectric used for gate dielectric in FET devices: (i) k must be high so that can be used economically for a good number of scaling nodes.(ii) dielectric constant must be in very close to SiO_2 which helps in the thermodynamically stability of it.(iii) Band offsets must be over than 1 ev which helps in the reduction of the leakage current (iv) must have few electrically active defects and should be kinetically stable[1][3]. The high-k dielectric gate material in place of SiO_2 is the reasonable alternative for this particular bottleneck. Gusev et al. identified four basic problems for successful incorporation of high k dielectric these are (i) possible to keep the scaling to curtailed EOTs (ii) reduce the threshold instabilities caused by the large defect densities (3) confine the loss of mobility in the channel while using high k (4) metal gate is required to control the threshold voltage[1][4].

In 2015 Intel has started the manufacturing of the chips using second generation high K/metal gate stack and demonstrated the incorporation of high k dielectrics in the FINFET based chips[1], [5].

Many new structures are invented and employed ,which are much better than conventional MOSFET in terms of high k better immune to short channel effects, gate leakage current is minimum, higher driving capability, better subthreshold slope , and lower DIBL values. Further for the better controlling over the channel nanowire FET (also known as Gate All Around (GAA)) comes into the existence[6]. It is one of the benchmark configuration and showed the excellent properties in sub 10 nm technology node devices. It has provides eventual electrostatic controllability over the channel as all around channel is circumscribe by the gate electrode[7]. Withal, the need for better device structure continued and turn into the invention of silicon tube based FET devices as Silicon nanotube field effect transistor.(SINTFET)[8]–[11].

In this work first time we are propose the analysis of implantation of high k in Silicon nanotube FET (SINTFET) device structure. The high k dielectric material such as Si3N4 HFO2,SiHSO4,SnO2 and TiO2 are utilized for simulation analysis and they give much better results in reducing the leakage current with high k than SiO2.This paper is organized as follows. Section II explain the device description and simulation methodology. Section III presents the simulation results and analysis and in section IV conclusion of the paper is given.

II. DEVICE DESCRIPTION

Figure 1 (a) shows the inner view of the SINTFET device structure with high k implanted gate dielectric whereas figure 1 (b) gives the vertical cross-sectional view of the device. The Inner and outer gate of the device is made up of metal and it is located in the inner and outer side of the tubular structure. Inner gate is wrapped by the gate oxide. The drain and source regions are surrounded all across the inner gate oxide. Outer gate is surrounded by the outer channel, metal is deposited all over the channel for the better controlling of the device. Source and Drain regions are doped at 2 x10²⁰ cm⁻³ while the channel region is lowly doped (10¹⁷ cm⁻³) which helps in reducing the fluctuations by cause of random dopants. High k is used to reduce the gate leakage current exists due to the reduction in gate oxide thickness. Thickness of the gate contact is taken small so that the limit can be applied on the miller capacitances values and delay which exists due to the gate overlapped is decreased. The physical parameter used for the design of tubular structure are listed in table 1.We have consider the channel length of 14 nm and an effective oxide thickness is 1 nm. All the cuts are made in to the center of the channel so

978-1-5386-6723-1/19 $31.00 © 2019 IEEE

Table 1

Parameter	Value
Gate Length (L_g)	14nm
Source/Drain length	10 nm
Tube Diameter (Outer)	18 nm
Tube Diameter (Inner)	9 nm
Oxide thickness	1 nm
Channel doping	10^{17} cm^{-3}
Doping Concentration in	2×10^{20} cm^{-3} (P
Dielectric Material	Dielectric value
Air	1
Silicon Di-oxide(SiO$_2$)	3.9
Silicon Nitride(Si$_3$N$_4$)	7.6
Hafnium Oxide(HfO$_2$)	23
Hafnium Silicate (HfSiO$_4$)	11
Tin oxide (SnO$_2$)	9.8
Titanium Oxide (TiO$_2$)	25

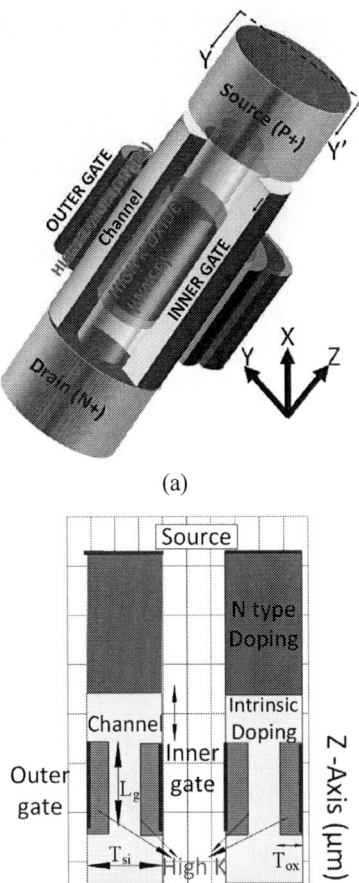

(a)

(b)

Fig.1. Simulated Silicon Nano Tube FET structure with High K (a) Inner view (b) Vertical Crossectional view.

that effect of both the gates are reflected in the analysis graphs. Description of the fabrication steps are given in patent filed by Daniel Tekleib [12]. Formation of high K will be done by using the epitaxial growth process[13].

In this work we have considered seven different dielectric constants i.e., Air , Silicon Di-oxide(SiO$_2$), Silicon Nitride(Si$_3$N$_4$), Hafnium Oxide(HfO$_2$), Hafnium Silicate (HfSiO$_4$),Tin oxide (SnO$_2$), Titanium Oxide (TO$_2$).Nano dimensions at high k dielectric is predicted to have better properties than traditional Silicon di oxide, due to their large surface area.[3]. Table 2 shows the dielectric constant values used in this work.

The incorporation of high k gate oxide in FET Devices increases the gate capacitance which results in to the decrement of the leakage current. Titanium oxide has highest dielectric constant but the band gap of it is very low which increases the leakage current.

III. RESULT DISCUSSION

The proposed structure is simulated using the 3D ATLAS device simulator of SILVACO TCAD[14]. The electric field charge dependent mobility model with concentration dependent is employed in the simulated structure in this paper. For numerical calculation newton trap method is used.

Figure 2 shows the value of threshold voltage of different dielectrics at two drain voltages i.e., 0.8V for saturation and 0.05V for linear. It is observed that the threshold voltage is increases when the value of dielectric constant is increases. This is due to the fact that as the value of dielectric constant k increases, the control over the channel increases and threshold voltage will be improved[15].

Fig 3 shows the ON- current and OFF-current on Y axis and the different gate dielectrics on X-axis. Basically it shows the effect of varying gate dielectric material on I_{on} and I_{off}. It is observed as the gate dielectric constant value increases the ON-current increases. This is due to the fact that as the value of dielectric constant k increases, the channel resistance increases and hence the ON-current slightly increases.

2019 Devices for Integrated Circuit (DevIC), 23-24 March, 2019, Kalyani, India

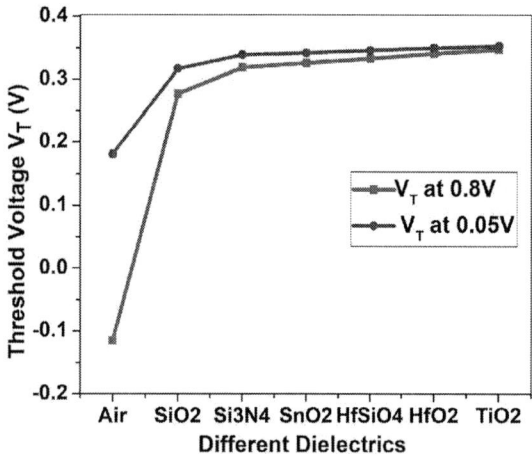

Fig. 2 Variation of threshold voltage for different dielectric constant at drain voltages 0.8V and 0.05V.

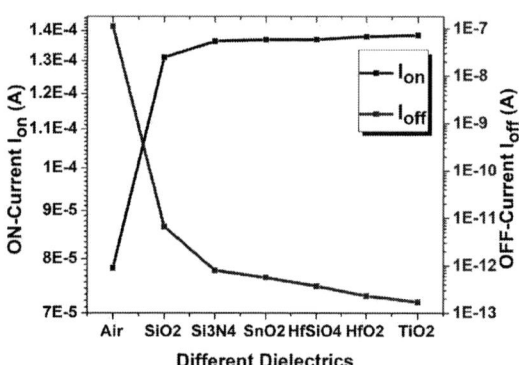

Fig. 3 Variation of ON-Current and OFF-Current for different dielectric constant

OFF- current is also plotted in the fig. 3 and it is seen that SiO_2 has high off current and it reduces with increasing k value, due to the barrier potential of the charge carrier in the channel increases with increase in dielectric constant. So it can be concluded that the leakage current is reduced by using the high k dielectric upto a great extent.

The I_{on}/I_{off} ratio against different dielectric constant values are shown in fig. 4. I_{on}/I_{off} ratio for silicon is 1.99 x 10^7, while for high-k as HfO_2 it is 5.95 x 10^8 and for TiO_2 is about 8.0 x 10^8, which is comparatively 10 times higher than the SINTFET with SiO_2 as dielectric. If the I_{on}/I_{off} ratio of the device is higher then the device has better switching which is the basic requirements for the devices which used for the low power applications. Since the Nano tube FET with HfO_2 has a large ratio, the switching of the device between ON and OFF state is spontaneous. Although TiO_2 has better I_{on}/I_{off} ration but due to the lower bandgap the leakage is more as compare to HfO2 based SINTFET.

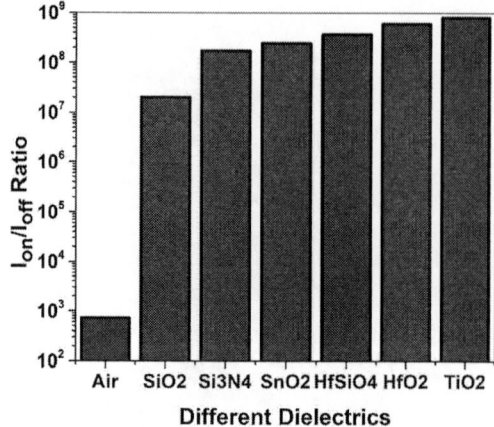

Fig. 4 I_{on}/I_{off} ratio at different dielectric constants

IV. CONCLUSION

In this work the impact of high-k gate dielectrics is analyzed for the gate engineered Silicon nano tube FET for various high –k dielectrics. Between SiO_2 and other different high k materials as gate dielectric comparative analysis of performance is carried out.. As the dielectric constant of the gate oxide increases the I_{on}/I_{off} ratio of the device is also increases which makes it suitable for the switching applications. The SINFET with SiO_2 as dielectric has on current of 0.131 µA and threshold voltage of 0.317 V while in the case of HfO_2, the on state current is .138 µA and the threshold voltage is 0.333. The incorporation of high k in silicon nano tube structure makes the device as good choice for semiconductor applications. So it can be said that HfO_2 is found to be suitable candidate to replace the conventional SiO_2 as it has moderate bandgap of 6ev which is favorable for the switching applications.

References

[1] J. Robertson and R. M. Wallace, "High-K materials and metal gates for CMOS applications," *Mater. Sci. Eng. R Reports*, vol. 88, pp. 1–41, 2015.

[2] K. Boucart and A. M. Ionescu, "Length scaling of the Double Gate Tunnel FET with a high-K gate dielectric," vol. 51, pp. 1500–1507, 2007.

[3] J. Charles Pravin, D. Nirmal, P. Prajoon, and J. Ajayan, "Implementation of nanoscale circuits using dual metal gate engineered nanowire MOSFET with high-k dielectrics for low power applications," *Phys. E Low-Dimensional Syst. Nanostructures*, vol. 83, pp. 95–100, 2016.

[4] K. E. Moselund *et al.*, "Comparison of VLS grown Si NW tunnel FETs with different gate stacks," *ESSDERC 2009 - Proc. 39th Eur. Solid-State Device Res. Conf.*, pp. 448–451, 2009.

[5] F. Karimi, M. Fathipour, H. Ghanatian, and V. Fathipour, "Improvement of Short Channel Effects in Cylindrical Strained Silicon Nanowire Transistor," *Eng. Technol.*, pp. 467–470, 2010.

[6] X. Chen and C. M. Tan, "Modeling and analysis of gate-all-around silicon nanowire FET,"

978-1-5386-6723-1/19 $31.00 © 2019 IEEE

Microelectron. Reliab., vol. 54, no. 6–7, pp. 1103–1108, 2014.

[7] A. Veloso *et al.*, "Gate-all-around NWFETs vs. triple-gate FinFETs: Junctionless vs. extensionless and conventional junction devices with controlled EWF modulation for multi-VT CMOS," *Dig. Tech. Pap. - Symp. VLSI Technol.*, vol. 2015–Augus, pp. T138–T139, 2015.

[8] D. Tekleab, "Device performance of silicon nanotube field effect transistor," *IEEE Electron Device Lett.*, vol. 35, no. 5, pp. 506–508, 2014.

[9] H. M. Fahad, C. E. Smith, J. P. Rojas, and M. M. Hussain, "Silicon nanotube field effect transistor with core-shell gate stacks for enhanced high-performance operation and area scaling benefits," *Nano Lett.*, vol. 11, no. 10, pp. 4393–4399, 2011.

[10] G. D. Jayakumar and R. Srinivasan, "SET analysis of silicon nanotube FET," *J. Comput. Electron.*, vol. 16, no. 2, pp. 307–315, 2017.

[11] Avtar Singh, Chandan Kumar Pandey, Saurabh Chaudhury, Chandan Kumar Sarkar, Effect of strain in silicon nanotube FET devices for low power applications, Eur. Phys. J. Appl. Phys. Vol, No (2019)

[12] D.Tekleab,H.H.Tran,J.W.Sleight,D. Chidambarrao " SILICON NANOTUBE MOSFET", US 8,866,266 B2 (2014) .

[13] J. Oh *et al.*, "Controlled threshold voltage of high-mobility Ge pMOSFETs with high-k/metal gate on epitaxial Ge films on Si substrates," *Int. Symp. VLSI Technol. Syst. Appl. Proc.*, pp. 40–41, 2008.

[14] Silvaco, "Silvaco ATLAS manual Device Simulation Software," pp. 2-42-2–45, 2006.

[15] A. Singh, S. Adak, H. Pardeshi, A. Sarkar, and C. K. Sarkar, "Comparative assesment of ground plane and strained based FDSOI MOSFET," *Inf. MIDEM*, vol. 19, no. 2, pp. 73–79, 2015.

High Frequency Performance of AlGaN/GaN HEMTs Fabricated on SiC Substrates

E. Mohapatra[1*], S. Das[2], T. P. Dash[3], S. Dey[4], J.Jena[5] and C. K. Maiti[6]

Department of Electronics and Communication Engineering,
Siksha 'O' Anusandhan (Deemed to be University),
Bhubaneswar, Odisha, India--751030

[1*]eleenamohapatra@soa.ac.in; [2]sanghamitradas@ soa.ac.in; [3]taradash@soa.ac.in; [4]supravadey@soa.ac.in; [5]jhansiranijena@soa.ac.in; [6]ckmaiti@soa.ac.in.

Abstract–**Power transistors based on gallium nitride (GaN) enable power electronic switches to operate at much higher switching frequencies compared to those based on silicon (Si). In this work, using TCAD simulations, we show that GaN-based high electron mobility transistors (HEMTs) can be optimized to have effectively reduced undesirable parasitic capacitances to greatly improve both the high transconductance and current gain cutoff frequency simultaneously. We report a new generation of high performance AlGaN/GaN HEMTs grown on high resistivity SiC substrates. We map out to evaluate small signal and large signal device performances against technological parameters such as the gate length, field plate length and the source-drain contact separation. The device with a gate length of 0.25μm and field plate length of 0.3μm exhibits a maximum dc drain current density of 3.66 A/mm at V_{GS}=3V with an extrinsic transconductance of 233.6 mS/mm and an extrinsic current gain cut-off frequency (f_t) of 78.9 GHz.**

Keywords- HEMT, AlGaN/GaN, SiC, Field plate, drift-diffusion model, hydrodynamic model, TCAD

I. INTRODUCTION

Power devices are gaining more popularity now a days with the extensive use of electronic devices in every field as these have a major influence on the system cost and efficiency. Importance of TCAD tools is also well known for the device early-stage design to perform performance evaluation by using computer-based models to compensate for the expensive prototyping and large signal experimental characterization of millimeter and sub-millimeter wave devices. Combination of both the high transconductance and current gain cutoff frequency(f_t) is desirable for power semiconductor devices which is a major part of the power electronic systems.

GaN/SiC technology is preferred for industrial use as the SiC substrates are more cost effective. The GaN/SiC HEMT technology can completely replace the presently used substrate/GaN epi-wafer technology due to high thermal conductivity (+30%) of the semi-insulating SiC substrate along with its low cost. GaN has a hexagonal structure named "wurtzite" and has a bandgap energy of 3.4 eV. The energy bandgap, breakdown field and electron mobility in GaN is very high which makes it extremely suitable to be used in power electronic applications[1]. The switching frequencies of GaN HEMTs can be very high in the range of megahertz and the device shows high power conversion efficiency [2].

The device performance can be improved even without any change in the semiconductor material properties, however, with dedicated new device structures and fabrication methods. For GaN-based HEMTs, among one of the techniques to improve performance is the introduction of field plate (FP) in the device. This has resulted in 2-to-4 times performance enhancement in RF GaN based devices. Currently, FP technique is the most widely used approach and even employed in commercial of AlGaN/GaN HEMTs for electric field (E-field) modulation and breakdown voltage improvements [3]. The field plate and its extension are favorable to decrease the electric field intensity at the gate edge and hence reduces the electron trapping probability which results in the reduction of low frequency noise [4].

In this paper, we have the four sections including the introduction part in section I. The device structure details and simulation environment have been discussed in section II. The model and material parameters have also been incorporated in 2D device simulator MINIMOS-NT [5] included in section II. Section III includes result and discussion part where the device characteristics have been presented in details. The effect of dimensions of field plate on the small signal gain and cutoff frequency is also presented systematically.

II. SIMULATION ENVRIONMENT

The device structures considered for simulation were epitaxially grown using MOCVD on SiC substrates. The heterostructure consists of a highly resistive c-plane GaN buffer, 50 nm thick AlGaN barrier over the GaN buffer and on the top a 5 nm thin GaN cap layer. Between the different layers of the HEMT, a sheet charge forms, that is assumed to be -4x10^14 cm^-2, -2.5x10^14 cm^-2 and 9.4x10^14 cm^-2 for cap and spacer charge, barrier and cap charge and semiconductor and barrier charge, respectively. The 2DEG formed at the heterostructure interface resulted in a sheet carrier concentration of 9.4x10^14 cm^-2 and a mobility of 1600 cm^2/V.s. Ohmic contacts were formed after epitaxial growth.

The nitride-assisted T-gate is defined by e-beam lithography with gate length of 0.25μm. The schematic of AlGaN/GaN HEMT without and with field plate are shown in Figs. 1(a) and (b), respectively. The net doping concentrations in different regions of AlGaN/GaN HEMT is shown in Fig.1(c).

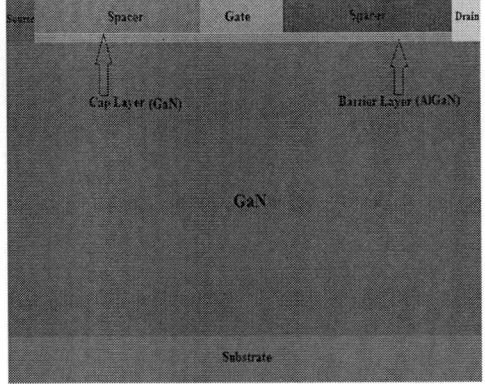

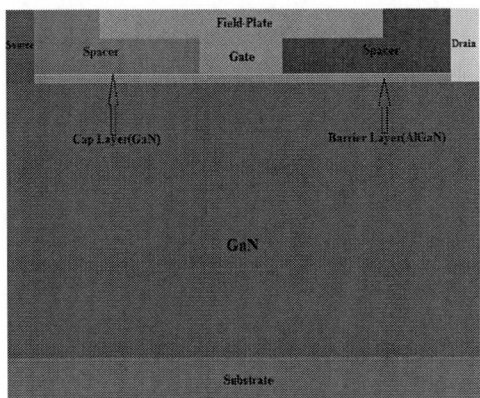

Fig. 1. Schematic of AlGaN/GaN HEMT (a) without field plate, (b) with field plate, and (c) AlGaN/GaN HEMT with net doping concentration.

TABLE I. PARAMETERS FOR ALGAN/GAN HEMTS USED IN SIMULATION.

Name of the parameters	Value
Gate Length (L_g)	250 nm
Field Plate Length	0.3μm
Cap and Spacer charge	-4x1014 cm^{-2}
Barrier and Cap charge	-2.5x1014 cm^{-2}
Semiconductor and Barrier charge	9.4x1014 cm^{-2}
Barrier Layer	50nm
Cap Layer	5nm

Here 2D simulations have been performed with the device simulation tool MINIMOS-NT [5] for the heterostructure device analysis with field plates [6]. The transport models should be chosen in such a way that it should be computationally efficient with highest precision. The hydrodynamic model has been implemented in this work as the fact that the drift-diffusion (DD) model is not suitable in terms of accuracy for submicron GaN devices [7]. The hole transport has been neglected because the AlGaN/GaN HEMTs are unipolar devices and a constant hole temperature of 300K is maintained.

The basic semiconductor equations for simulation of HEMT is given below. The electron and hole current densities are given by the expressions: [7]

$$J_n = q \cdot \mu_n \cdot n \cdot \left(\text{grad} \left(\frac{\varepsilon_C}{q} - \psi \right) + \frac{k_B}{q} \cdot \frac{N_{C,0}}{n} \cdot \text{rad} \left(\frac{n \cdot T_L}{N_{C,0}} \right) \right) \quad (3)$$

$$J_p = q \cdot \mu_p \cdot p \cdot \left(\text{grad} \left(\frac{\varepsilon_V}{q} - \psi \right) - \frac{k_B}{q} \cdot \frac{N_{V,0}}{p} \cdot \text{rad} \left(\frac{p \cdot T_L}{N_{V,0}} \right) \right) \quad (4)$$

Where J_n is the electron current density,

J_p is the hole current density,

μ_n and μ_p are the carrier mobilities,

ε_C and ε_V are the position dependent band edge energies,

$N_{C,0}$ and $N_{V,0}$ are the effective density of states, and

T_L is the lattice temperature.

For the electron and hole mobilities, μ_n and μ_p respectively, the model considers a field dependence for the DD model and a carrier temperature dependence for the HD model. The energy transport equations for the HD model are given as:

$$div\, S_n = grad \left(\frac{\varepsilon_C}{q} - \psi \right) \cdot J_n - \frac{3 \cdot k_B}{2} \cdot \left(\frac{\partial (n \cdot T_n)}{\partial t} + R \cdot T_n + n \cdot \frac{T_n - T_L}{\tau_{\epsilon,n}} \right) \quad (5)$$

$$div\, S_p = grad \left(\frac{\varepsilon_V}{q} - \psi \right) \cdot J_p - \frac{3 \cdot k_B}{2} \cdot \left(\frac{\partial (p \cdot T_p)}{\partial t} + R \cdot T_p + p \cdot \frac{T_p - T_L}{\tau_{\epsilon,p}} \right) \quad (6)$$

Where S_n and S_p are the energy fluxes.

$\tau_{\epsilon,n}$ and $\tau_{\epsilon,p}$ are the energy relaxation times

$$S_n = -k_n \cdot grad\, T_n - \frac{5}{2} \cdot \frac{k_B T_n}{q} \cdot J_n \quad (7)$$

Where k_n is the thermal conductivity

$$k_n = \left(\frac{5}{2} + c_n \right) \frac{k_B^2}{q} \cdot T_n \cdot \mu_n \cdot n \quad (8)$$

The DD mobility is modelled by

$$\mu_v^{LIF}(F_v) = \frac{\mu_v^{LI}}{\left(1 + \left(\frac{\mu_v^{LI} \cdot F_v}{v_v^{sat}} \right)^{\beta_v} \right)^{1/\beta_v}} , \quad v = n, p \quad (9)$$

Where F_v represents the driving force for electrons and holes

μ_v^{LI} is the zero-field mobility

v_v^{sat} is the saturation velocity

The HD mobility is modelled carrier temperature dependent

$$\mu_v^{LIT} = \frac{\mu_v^{LI}}{1+\alpha_v \cdot (T_v - T_L)} \qquad (10)$$

Where $\alpha_v = \frac{3 \cdot k_B \cdot \mu_v^{LI}}{2 \cdot q \cdot \tau_\epsilon \cdot (v_v^{sat})^2}$ (11)

And τ_ϵ is the energy relaxation times

v_v^{sat} is the saturation velocity

III. RESULTS AND DISCUSSION

The transfer characteristics and the small signal analysis has been studied.

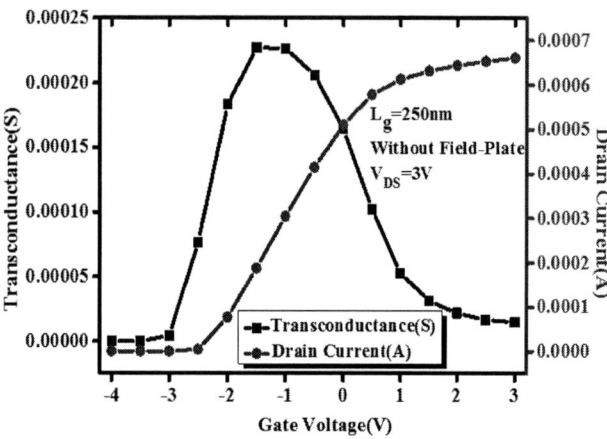

Fig. 2. I_D-V_G and G_m-V_G plots of GaN HEMT (gate length 250nm) without field plate using DD model.

The DC transfer characteristics at V_{DS}=3V is shown in Fig. 2. The I_D-V_G and transconductance (G_m) characteristics without field plate length is shown.

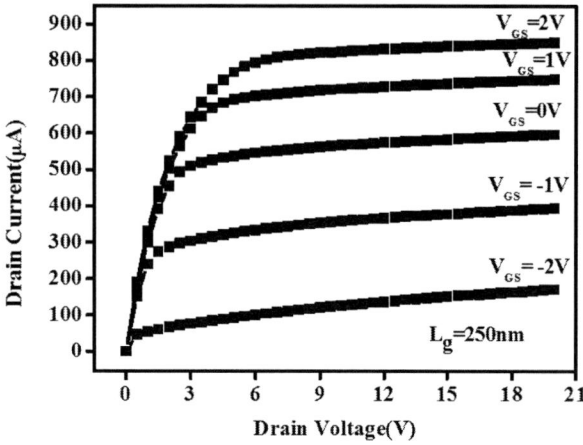

Fig. 3. I_D-V_D plot of GaN HEMT using DD model at V_{GS}= -2 to 2V without field plate.

In Fig.3 the drain current increases linearly until it reaches the saturation region. The gate bias is swept from −2 to 2V in a

step of 1V. The device exhibits maximum drain current(without field plate) at V_{GS} = 2 V and the drain bias is swept from 0 to 20V in a step of 5V.

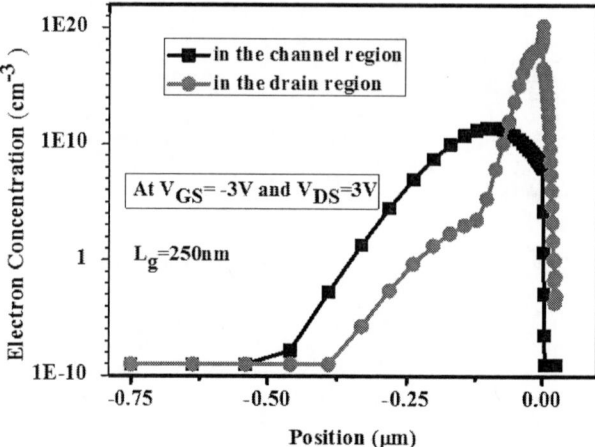

Fig. 4. Electron concentration of GaN HEMT in the off-state. The red (filled circle) color represents the electron concentration in the drain region and the black color (filled square) represent the electron concentration in the channel region.

The electron concentration of the device in off-state shown in Fig. 4. The electron concentration of the HEMT in the channel and drain region at V_{GS}=-3V and V_{DS}=3V is shown in Fig.4. The cutline is taken from the channel and drain region.

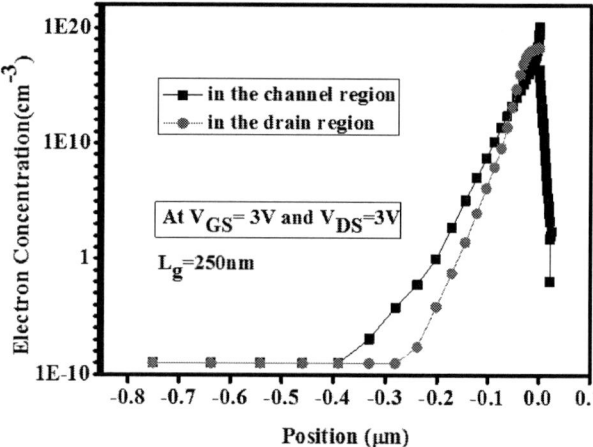

Fig 5. Electron concentration of GaN HEMT in the on-state. The red (filled circle) color represents the electron concentration in the drain region and the black (filled square) color represent the electron concentration in the channel region.

The electron concentration of the device in on-state shown in Fig.5.The electron concentration of the HEMT in the channel and drain region without field plate at V_{GS}=3V and V_{DS}=3V is shown in Fig. 5. The cutline is taken from the channel and the drain region. The most important is to estimate the effect of field plate structure to the transfer characteristics, the device current gain and cut-off frequency.

2019 Devices for Integrated Circuit (DevIC), 23-24 March, 2019, Kalyani, India

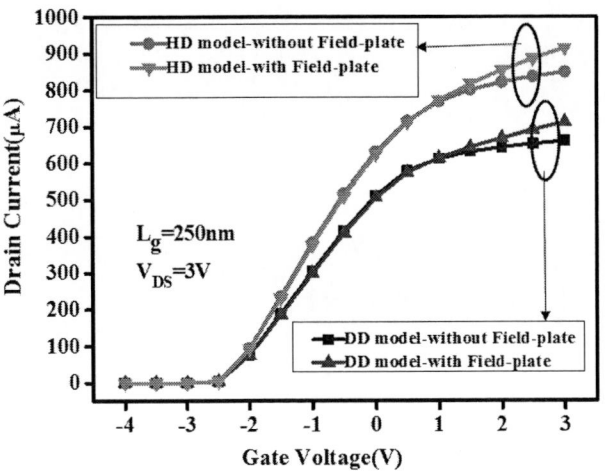

Fig 6. I_D-V_G plot of GaN HEMT at gate length 250nm with and without field plate using DD and HD model at V_{DS}=3V.

The measured transfer characteristics for V_{DS}=3V and L_g=250nm is shown in Fig. 6 along with four simulated I_D-V_G plots i.e., DD model without and with field plate and HD model without and with field plate. Here the field plate length is L_{FP}=0.3μm. The HD model simulated I_D-V_G plots shows an significant increase in the drain current compared to DD model without field plate and with field plate. With the increase in the gate voltage, the number of carrier concentration in the 2DEG is increases, and as a result of which the drain current increases. The drain current is linearly controlled by the gate voltage. There is a significant improvement in the I–V characteristics by applying the field plate technique.

The RF performance of the device is evaluated by small signal AC analysis. The current gain $|h_{21}|$ for the 0.25μm device is shown in Fig. 7. The gain decreases at higher cut-off frequency. That is due to the gate capacitance which causes a frequency dependent gate current. The device with a field plate has larger gate area due to which the gate current increases.

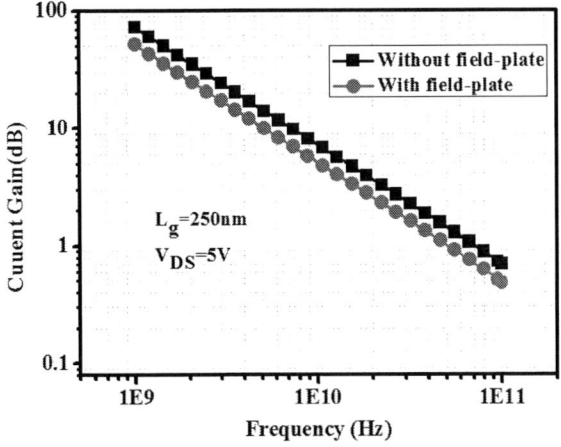

Fig 7. Current gain with and without field plate at V_{DS}=5V. The red color (filled circle) represents the current gain with field plate and the black color (filled square) represents the current gain without field plate.

The device gain decreases when field plate structure is added. The added parasitic capacitance between the field plate and channel results in an increased device gate drain capacitance (C_{GD}). The increase in C_{GD}, affects both current gain and the cut-off frequency in reducing these values. As the feedback capacitance C_{GD} increases and the extrinsic cut-off frequency i.e. f_t depends on C_{GD} in Eqn. (12) without Miller effect correction. The corresponding f_t expression is given by

$$f_t = g_m/[2\pi(C_{GS} + C_{GD})] \qquad (12)$$

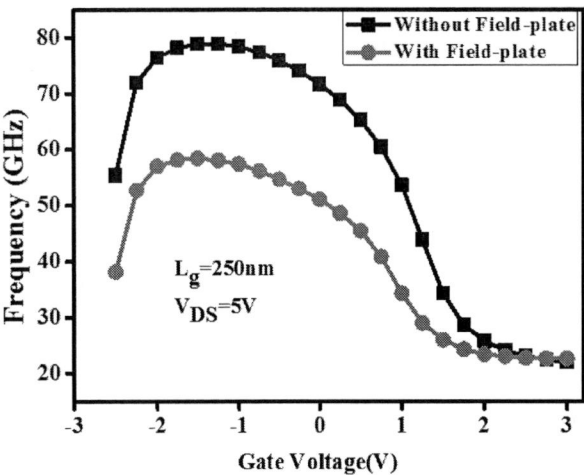

Fig 8. Cutoff frequency with and without field plate at V_{DS}=5V. The red color (filled circle) represents the cutoff frequency with field plate and the black color (filled square) represents the cutoff frequency without field plate.

The cut-off frequency is lower for the transistor with field plate shown in Fig. 8. The slightly higher cut-off frequency for gate voltage above 1.5V is due to the higher drain current compared to the one without field plate. Here the frequency characteristics are calculated without any interconnect structures or parasitic components of the measurement equipment. This causes a higher cut-off frequency than measured. The device current gain and cut-off frequency parameters for the added field plate geometry have been extracted at a V_{GS} of -3V to 3V and at a V_{DS} of 5V. The devices with field plate shows lower f_t values compared to that of the device without field plate. The formation of the gate capacitance between the field plate terminal and the device channel leads to the addition of parasitic capacitance between gate and channel of the device due to which the device current gain cut-off frequency f_t decreases. The s-parameter data have been simulated to extract f_t.

IV. CONCLUSION

In this work, we have presented the results on the performance of 0.25μm gate length AlGaN/GaN HEMTs with and without field plate length of 0.3μm. The gate connected field plates are very effective and by this technique devices exhibited significant improvements. By adding field plate geometry, the abasement in f_t values has been observed. Although present device operation is limited due to arise of excess capacitance from the field plate. At higher frequencies, the operation can be attained by reducing gate length and/or optimizing field plate geometry with slant field plates [8]. From the simulation results, it is shown that this technology has the ability to minimize both the DC and RF dispersion and parasitic capacitance. The present design has shown superior DC/RF device performance feasibility of GaN-based HEMTs for high power RF circuit applications.

REFERENCES

[1] U. K. Mishra, P. Parikh, and Y.-F. Wu, "AlGaN/GaN HEMTs—An overview of device operation and applications," Proc. IEEE, vol. 90, pp. 1022–1031, 2002.

[2] Y. Wu, M. Jacob-Mitos, M. L. Moore, and S. Heikman, "A 97.8% efficient GaN HEMT boost converter with 300-W output power at 1 MHz," IEEE Electron Device Lett., vol. 29, no. 8, pp. 824–826, Aug. 2008.

[3] Y. Ando, Y. Okamoto, H. Miyamoto, T. Nakayama, T. Inoue, and M. Kuzuhara, "10-W/mm AlGaN-GaN HFET with a field modulating plate," IEEE Electron Device Lett., vol. 24, pp. 289–291, 2003.

[4] H. Chiu, C. Yang, H. Wang, F. Huang, H. Kao, and F. Chien, "Characteristics of AlGaN/GaN HEMTs With Various Field-Plate and Gate-to-Drain Extensions," in IEEE Transactions on Electron Devices, vol. 60, pp. 3877-3882, 2013.

[5] MINIMOS-NT, Users' manual, 2017.

[6] S. Vitanov, V. Palankovski, S. Murad, T. Roedle, R. Quay, and S. Selberherr, "Predictive Simulation of AlGaN/GaN HEMTs", in Proc. Compound Semiconductor Integrated Circuit Symposium 2007, CSIC 2007, IEEE, pp. 1-4, 2007.

[7] S. Vitanov, V. Palankovski, S. Maroldt, R. Quay, S. Murad, T. Rödle, and S. Selberherr, "Physics-Based Modeling of GaN HEMTs," in IEEE Transactions on Electron Devices, vol. 59, pp. 685-693, 2012.

[8] Y. Pei, Z. Chen, D. Brown, S. Keller, S. P. Denbaars, and U. K. Mishra, "Deep-Submicrometer AlGaN/GaN HEMTs With Slant Field Plates," in IEEE Electron Device Letters, vol. 30, pp. 328-330, 2009.

Noise analysis of Dual Halo Dual Dielectric Triple Material Surrounding Gate MOSFET for RF applications

Prashant Kumar
Dept. of Electronics & Communication Engineering
J.C. Bose University of Science &
Technology YMCA, Faridabad
Faridabad, India-121006
pk.vlsi@gmail.com

Neeraj Gupta
Dept. of Electronics & Communication Engineering
Amity School of Engineering & Technology, Amity University
Haryana
Gurugram, India-122413
neerajsingla007@gmail.com

Rashmi Gupta
Dept. of Computer Science & Engineering
Amity School of Engineering & Technology, Amity University
Haryana
Gurugram, India-122413
goyal.rashmi18@gmail.com

Amit Sharma
Dept. of Mathematics
Amity School of Applied Sciences, Amity University Haryana
Gurugram, India-122413
dba.amitsharma@gmail.com

Abstract— **In this paper, the noise performance of Dual Halo Dual Dielectric Triple Material Surrounding Gate (DH-DD-TM-SG) MOSFET has been investigated. The assessment of noise performance has been carried out in terms of noise figure, noise conductance and optimum impedance. These noise metrics reveal notable cutback in noise by virtue of dual dielectric and dual halo implants in DH-DD-TM-SG MOSFET in contrast to conventional Triple Material Surrounding Gate (TM-SG) MOSFET. It is scrutinized that noise figure is shrunk by 35.7% and optimum impedance is raised by 14.49% in DH-DD-TM-SG MOSFET than its counterpart, making it a suitable device for designing of low-noise amplifiers.**

Keywords— Halo implant, Noise figure, TM-SG, Noise Conductance, Optimum impedance

I. INTRODUCTION

The future technology nodes require alleviated performance of devices in terms of high speed and low power consumption to impend the growth in semiconductor industry [1]. In the present scenario low-noise RF amplifiers operating at *THz* frequencies are in huge demands. But the contemporary MOS devices are not suitable [2]. Performance degradation is observed along with reduction in dimension of conventional MOSFET. The obnoxious effects such as drain induced barrier lowering, threshold voltage roll-off and many others short channel effects will persist with continuous downscaling in device length and deteriorate the performance of MOS devices [3-4]. The possible solution for SCEs is multigate MOSFETs. The surrounding gate MOSFET is the leading device to overcome SCEs problem in large extent because it boost up the influence of gate over the channel [5].

The DH-DD-TM-SG MOSFETs are more stunning and popular device as compared to single-gate (SG), double-gate (DG) and FinFET devices due to higher on-current [6-7]. Hence, after incorporating the advantage of CGAA MOSFET with triple material and dual dielectric further enhance the performance of the device. The cylindrical gate-all-around (CGAA) MOSFET exhibits higher cut-off frequency than conventional MOSFET [8]. Neha Gupta et al. [8] described the silicon nanowire MOSFET with Gate electrode work function engineering method. Sonam Rewari et al. [2] studied the noise immunity of nanotube junctioless MOSFET. It shows significant improvement in noise immunity as compared to conventional junctioless MOSFET. The noise figure of merit (FOM) of DH-DD-TM-SG MOSFET has been carried out in terms of simulation by using ATLAS device simulation software and compared with conventional TM-SG MOSFET [9].

II. DEVICE STRUCTURE

A 3-D view of cylindrical DH-DD-TM-SG MOSFET is shown in Fig. 1. The schematic view of DH-DD-TM-CGAA MOSFET structure is depicting in Fig. 2. It is conspicuous that the gate terminal contained three metals M_1, M_2 and M_3 with different work function. The triple metal has been constructed by employing Molybdenum (Mo) acting as gate material as its work function can be changed by varying N_2 implant [10]. The thicknesses of the inner and outer oxide layers are $t_{Sio2}=1nm$ and $t_{HfO2}=4nm$ respectively. The lengths of channel region L_1 and L_5 are halo doped with N_{dh} while the remaining parts are doped with N_{ak}, assuming that N_{dh} concentration is more than N_{ak} concentration [11]. The simulations are performed using different models: Drift-diffusion, SRH, CONMOB and Boltzmann statics have been used. Newton method has been utilized for numerical solution.

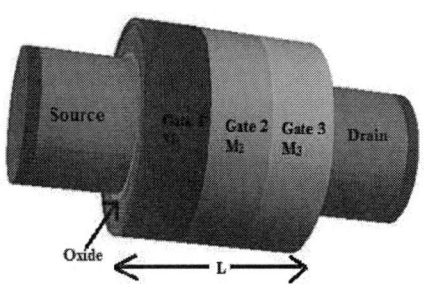

Fig.1. 3-D View of cylindrical DH-DD-TM-SG MOSFET

2019 Devices for Integrated Circuit (DevIC), 23-24 March, 2019, Kalyani, India

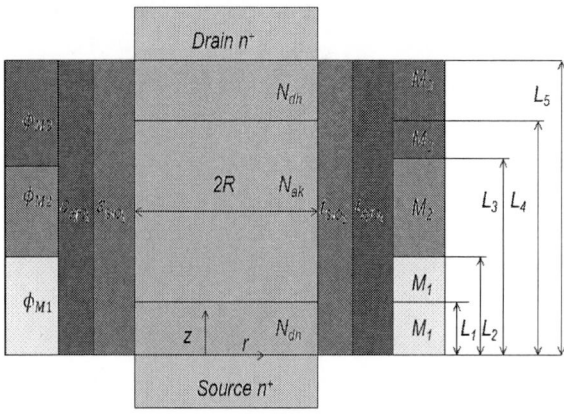

Fig.2. Schematic View of cylindrical DH-DD-TM-SG MOSFET

III. RESULTS & DISCUSSION

The noise of MOSFET has been specified in terms of noise figure, noise conductance and optimum impedance. The noise figure is an estimation of noise which is innate in the MOSFET even at the zero bias. The magnitude of noise figure should be minimum for better RF performance. The Fig. 3 shows the variation in noise figure with frequency for TM-SG and DH-DD-TM-SG MOSFET devices at channel length $L=30$ nm. Table 1 and 2 shows the variation in NF_{min}, N_C and Z_o with frequency for DH-DD-TM-SG and TM-SG MOSFET. It is seen that the magnitude of noise figure at frequency 100 *THz* for TM-SG and DH-DD-TM-SG MOSFET are 12.6 *dB* and 8.1 *dB* respectively. So, DH-DD-TM-SG MOSFET exhibits lower noise figure as compared to its counterpart. This is mainly due to the fact that DH-DD-TM-SG MOSFET reveals higher g_m as compared to TM-SG. There is inverse relationship between g_m and the noise figure [12]. So present device is less prone to radio frequency noise abasement and thus makes it suitable for LNA amplifiers.

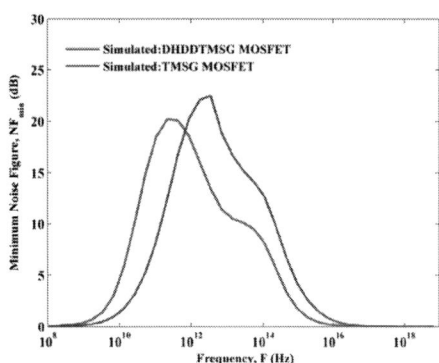

Fig.3. NF_{min} at different frequency for DH-DD-TM-SG and TM-SG MOSFETs

The power spectral density of noise generators is related to noise conductance of a MOSFET. The magnitude of noise conductance must be minimized for better RF CMOS applications. Fig. 4 shows the drift in noise conductance with frequency for TM-SG and DH-DD-TM-SG MOSFET device at $L = 30$ nm. It is seen that the magnitude of noise conductance at frequency 100 *THz* for TM-SG and DH-DD-TM-SG MOSFET are 2.36 Siemens and 0.7 Siemens respectively.

TABLE I: NF_{MIN}, N_C AND Z_O AT DIFFERENT FREQUENCY FOR DH-DD-TM-SG MOSFET

Frequency (Hz)	NF_{min} (dB)	N_C (Siemens)	Z_o (Ohms)
10^8	0.0412	6.26E-07	2.58E+06
10^{10}	5.8058	1.01E-02	2.02E+04
10^{12}	16.021	6.76E-01	1.78E+02
10^{14}	8.1107	7.04E-01	4.82E+00
10^{16}	0.1211	6.89E-01	5.39E-02
10^{18}	0.0009	6.90E-01	4.22E-04

So, DH-DD-TM-SG MOSFET exhibits lower noise figure as compared to its counterpart. This is mainly due to the fact that DH-DD-TM-SG MOSFET reveals higher g_m as compared to TM-SG. There is inverse relationship between g_m and the noise figure [12]. So present device is less prone to radio frequency noise abasement and thus makes it suitable for LNA amplifiers.

TABLE II: NF_{MIN}, N_C AND Z_O AT DIFFERENT FREQUENCY FOR TM-SG MOSFET

Frequency (Hz)	NF_{min} (dB)	N_C (Siemens)	Z_o (Ohms)
10^8	0.016	3.50E-08	2.06E+06
10^{10}	4.77	5.73E-04	1.61E+04
10^{12}	22.1	1.70E+00	1.33E+02
10^{14}	12.6	2.36E+00	4.12E+00
10^{16}	0.446	2.20E+00	5.30E-02
10^{18}	0.003	2.21E+00	4.14E-04

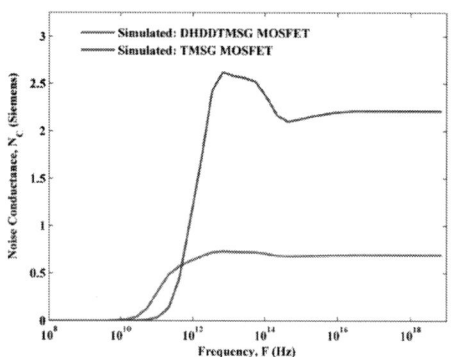

Fig.4. N_C at different frequency for DH-DD-TM-SG and TM-SG MOSFETs

Fig. 5 shows the magnitude of impedance Z_o which is the combination of real and imaginary impedances with variation in frequency. The presence of dual dielectric ensures that there is no current flowing from gate to channel. The scaling of MOSFET also reduces the thickness of the

978-1-5386-6723-1/19 $31.00 © 2019 IEEE

oxide layer to meet the future device requirements. The ideal operation of MOSFET requires high input impedance.

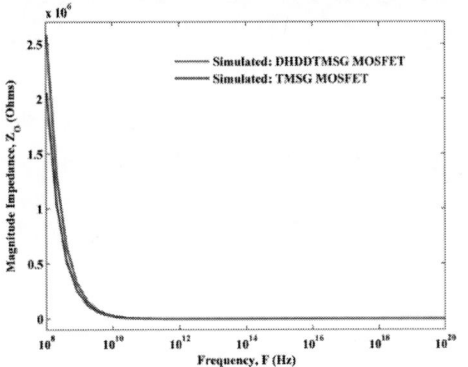

Fig.5. Z_O at different frequency for DH-DD-TM-SG and TM-SG MOSFETs

As evident from the Fig. 5 that DH-DD-TM-SG MOSFET has high impedance as compared to TM-SG MOSFET. At frequency of 100 *THz*, the input impedance of DH-DD-TM-SG MOSFET is 14.52 % higher than that of TM-SG MOSFET. So this large impedance make device suitable for RF applications.

IV. CONCLUSION

In this work, a comprehensive simulation based study of noise behavior of DH-DD-TM-SG MOSFET and TM-SG MOSFET has been carried out in terms of noise figure, noise conductance and optimum impedance at *THz* frequency. Thus noise immunity of DH-DD-TM-SG MOSFET has been improved owing to higher transconductance. The optimum value of gate metal work function has to be selected for enhancing the performance of device. Hence the low noise figure and high optimum impedance of DH-DD-TM-SG MOSFET is opening the new window for RF applications.

REFERENCES

[1] S. Rewari, S. Haldar, V. Nath, S.S. Deswal and R.S. Gupta, "Numerical modeling of Subthreshold region of junctionless double surrounding gate MOSFET (JLDSG)", Superlattices and Microstructures, 2016, 90, pp. 8-19

[2] S. Rewari, S. Haldar, V. Nath, S.S. Deswal and R.S. Gupta, "Improved analog and AC performance with increased noise immunity using nanotube junctionless field effect transistor", Applied Physics A , 2016, 122, pp. 1-10

[3] A. Sharama, A. Jain, Y. Pratap and R.S. Gupta,"Effect of high K and vacuum dielectrics as gate stack on junctionless cylindrical surrounding gate MOSFET", Solid State Electronics, 2016, 123, pp. 26-32

[4] R. Hosseini and N. Teimuorzadeh, "Simulation Study of Circuit Performance of GAA Silicon Nanowire Transistor and DG MOSFET", Physical Review & Research International, 2013, 3, (4), pp. 568-576

[5] J.H.K. Verma, S. Haldar, R.S. Gupta and M. Gupta, "Modeling and simulation of subthreshold behaviour of cylindrical surrounding double gate MOSFET for enhanced electrostatic integrity", Superlattices and Microstructures, 2015, 88, pp. 354-364

[6] J.P Colinge., "Multi-gate SOI MOSFETs", Solid State Electron, 2004, 48, (6) , pp. 897-905

[7] A. Sarkar, S. De, A. Dey, and C.K. Sarkar, "Analog and RF performance investigation of cylindrical surrounding-gate MOSFET with an analytical pseudo-2D model", Journal of Computational Electronics, 2012, 11, (2), pp. 182–195

[8] N. Gupta, and R. Chaujar, "Influence of gate metal engineering on small-signal and noise behaviour of silicon nanowire MOSFET for low-noise amplifiers", Applied Physics A ,2016, 122, pp. 1-9

[9] ATLAS user's manual: 'Device Simulation Software' (Silvaco Int., Santa Clara, CA, USA, 2013)

[10] R. Lin, Q. Lu, P. Ranade, T.J. King and C. Hu, "An adjustable work function technology using Mo gate for CMOS devices", IEEE Electron Device Letters, 2002, 23, pp. 49-51

[11] N.B. Balamurugan, K. Sankaranarayanan, P. Amutha and M.F. John, "An Analytical Modeling of Threshold and Subthreshold Swing on Dual Material Surrounding Gate Nanoscale MOSFETs for high speed Wireless Communication", Journal of Semiconductor Technology and Science, 2008, 8, (3), pp. 221-226

[12] S. Cho, K. R.Kim, B.G. Park and I.M. Kang, "RF performance and small signal parameter extraction of Junctionless Nanowire MOSFETs", IEEE Transaction on Electron Devices, 2011, 58, (5), pp. 1388–1396

2019 Devices for Integrated Circuit (DevIC), 23-24 March, 2019, Kalyani, India

Optimization of Electrical Parameters for the Gate Stack Double Gate (GSDG) MOSFET using Simplex-PSO Algorithm

Dibyendu Chowdhury
Department of ECE
Haldia Institute of Technology
Haldia, India
dibyendu.chow@gmail.com

Bishnu Prasad De
School of Electronics Engineering
KIIT University
Bhubaneswar, India
bishnu.ece@gmail.com

Kanchan Baran Maji
Department of ECE
National Institute of Technology
Durgapur, India
kbmaji@gmail.com

Sumalya Ghosh
Department of ECE
National Institute of Technology
Durgapur, India
tosg.nitd@gmail.com

Rajib Kar
Department of ECE
National Institute of Technology
Durgapur, India
rajibkarece@gmail.com

Durbadal Mandal
Department of ECE
National Institute of Technology
Durgapur, India
durbadal.bittu@gmail.com

Abstract— **In this article, the electrical parameters of the Gate Stack Double Gate (GSDG) MOSFET are optimized utilizing the Simplex-Particle Swarm Optimization (Simplex-PSO) algorithm. The electrical parameters like the OFF-state current, transconductance and subthreshold swing have been considered to formulate the overall Cost Function (CF). The overall CF is achieved by the weighted sum approach method. The results attained from the Simplex-PSO are formed to be better than the previous literature.**

Keywords— GSDG MOSFET, Transconductance, OFF-state current, Subthreshold Swing, Optimization, PSO, Simplex-PSO

I. INTRODUCTION

Double Gate MOSFET is the first member of the multi-gate transistor which provides better immunity against the short channel effects (SCE). Djeffal *et al.* [1] have designed a 2-D subthreshold model for graded channel gate stack DG MOSFET. Djeffal *et al.* [2] have also proposed the multi-objective genetic algorithm (MOGA) to optimize the electrical parameters [3-4] of GSDG MOSFET as shown in Fig. 1. Jin *et al.* [5] have developed the analytical model of the surface potential, surface electric field, and threshold voltage for dual material gate (DMG) double gate strained-Si MOSFET. Sarkar *et al.* [6] have studied the electrical performance with variation in the number of metals in the gate region for double gate structure MOSFET for analogue/RF applications. Bendib *et al.* [7] have optimized the device parameters of the GSDG MOSFET and investigated the performance of the ring oscillator and the inverter gate [8]. Swing factor can be improved by varying the gate work function [9] for the Double Gate MOSFET.

In this work, the structural dimensions and electrical performance parameters of GSDG MOSFET are optimized by Simplex-PSO algorithm [10-11]. The article is arranged as follows: Electrical parameters of GSDG MOSFET are discussed, and the overall CF is formulated in Section II. PSO and Simplex-PSO algorithms are described in Section III. Simulation results are given in Section IV. Section V concludes the article.

II. ELECTRICAL PARAMETERS FOR GSDG MOSFET

In this Section, the compact model of subthreshold and saturation parameters for 2-D short channel GSDG MOSFET are discussed.

ΔS is the subthreshold swing degradation coefficient [2-3] and is given as

$$\Delta S = \frac{\sinh\left((x_{\min} - L)/\lambda\right) - \sinh\left(x_{\min}/\lambda\right)}{\sinh\left(L/\lambda\right)} \tag{1}$$

where $x_{\min}$ is the position of minimum surface potential, L is the width of the channel, and $\lambda = \sqrt{\varepsilon_{si} t_{oxeff} t_{si} / 2\varepsilon_1}$

The subthreshold current is given as

$$I_{sub} = 2\frac{V_t}{E_s} K\left(e^{\psi_{\min}/V_t} - e^{\psi_{s\min}/V_t}\right) \tag{2}$$

where V_t represents the thermal voltage; E_s is the constant electric field; K is a constant; $\psi_{\min}$ represents the minimum potential; the minimum surface potential [1] is presented as $\psi_{s\min}$.

The OFF-state current [2-3] is estimated by

$$I_{OFF} = I_{sub}\big|_{V_{gs}=0} \tag{3}$$

The trans-conductance in saturation is represented as

$$g_m = \frac{\partial I_{ds}}{\partial V_{gs}}\bigg|_{V_{ds}} = \mu_{eff} C_{oxeff} \frac{W}{L}\left[2\times\left(V_g^* - V_0\right) - \frac{8rKT}{q} e^{\frac{q\left(V_{gs}-V_0-V_{ds}\right)}{kT}}\right] \tag{4}$$

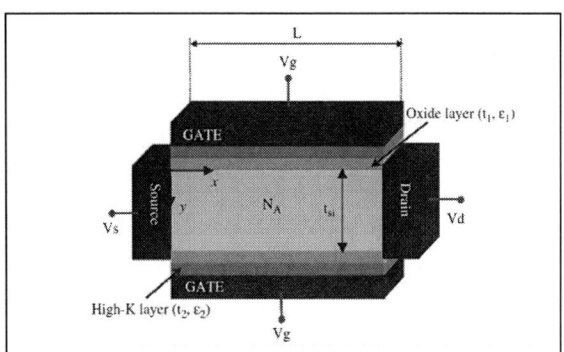

Fig.1. A cross-sectional view of the investigated GSGD MOSFET device.

978-1-5386-6723-1/19 $31.00 © 2019 IEEE

where μ_{eff} represents the effective mobility; C_{oxeff} represents the effective capacitance oxide; W is the channel width; V_g^* is the effective voltages at the front and bottom gates; V_0 is closed to the threshold voltage [2];

$$r = \left(\frac{\varepsilon_{si} \, t_{oxeff}}{\varepsilon_{ox} \, t_{si}} \right)$$ is a structural parameter.

The multi-objective optimization problem can be formulated as:

Maximization of g_m, and

Minimization of I_{OFF} and ΔS.

The overall CF is achieved by the 'weighted sum approach method' and is given as [2]

$$CF = w_1 \left(1/g_m \right) + w_2 I_{oFF} + w_3 |\Delta S| \qquad (5)$$

where w_1, w_2 and w_3 are the weight functions. Here, w_1, w_2 and w_3 are assigned equal values as 1/3. The input variable vector is represented as

$$X = \left(t_{si}, t_1, t_2, \varepsilon_2, L, \phi_m, N_A, V_{ds}, V_g \right).$$

III. EVOLUTIONARY TECHNIQUES EMPLOYED

PSO is a well-known algorithm and is described in various literature [12-13]. The description of the Simplex-PSO algorithms is given in [10-11]. The parameters of the Simplex-PSO algorithm are given in Table I.

IV. SIMULATION RESULTS AND DISCUSSIONS

In this article, electrical performances of the GSDG MOSFET are optimized through geometrical synthesis. The overall CF is individually optimized through Simplex-PSO algorithm, implemented in MATLAB 7.5. The optimized

dimensions $\left(t_{si}, t_1, t_2, L \right)$ and electrical parameters $\left(\varepsilon_2, V_{ds}, V_g, \phi_{MS}, N_A \right)$ are obtained from the Simplex-PSO algorithm. For the authentication purpose, TCAD is used to design the structure of GSDG MOSFET for electrical simulations. The Simplex-PSO algorithm is run over 50 times to achieve the best set of design parameters shown in Table II. TCAD simulation results achieved from the Simplex-PSO algorithm based optimal design of GSDG MOSFET are presented in Figs. 2-3, respectively. Simplex-PSO based design results in $t_{si} = 5nm$, $t_1 = 0.30nm$, $t_2 = 0.30nm$, $L = 15nm$, $\varepsilon_2 = 25$, $\phi_m = 4.63$ eV, $N_A = 1 \times 10^{17}$ cm^{-3}, $V_{ds} = 4.95V$, $V_{gs} = 4.70V$, $I_{OFF} = 9.05 \times 10^{-10}$A/$\mu$m, $g_m = 1.92$S/μm, S = 62.38 mV/dec, $\Delta S = 0.03815$, CF = 0.1863. Simplex-PSO requires 25 seconds to compute 5,000 parameters set with Intel Core i5-2430M CPU@3.00 GHz. Simplex-PSO yields better-optimized results compared to MOGA [2] in terms of the electrical parameters like ΔS, I_{OFF}, and g_m as shown in Table II.

TABLE I: CONTROL PARAMETERS OF SIMPLEX-PSO FOR GSDG MOSFET DESIGN

Parameters	Values [10-11]
Size of Population	50
The dimension of the optimization problem	9
Iteration cycle	100
c_0	0.8
c_1	0.6
c_2	0.08

TABLE II: OPTIMAL GSDG MOSFET PARAMETERS

Symbol	Quantity	MOGA [2]				Simplex-PSO
		Case 1	Case 2	Case 3	Case 4	
t_{si} (nm)	Silicon thickness	42.3146	48.5800	5.0035	5.1256	5
L (nm)	Channel length	91.3081	103.3133	204.9485	280.6259	15
V_{gs}(V)	Gate voltage	4.7490	4.7662	4.1372	4.1223	4.70
V_{ds}(V)	Drain source voltage	4.9999	4.9999	4.9999	4.9999	4.95
t_1 (nm)	Thickness of the SiO2	1.0004	1.0004	1.0004	1.1119	0.3
t_2 (nm)	Thickness of the high-k layer	0.5005	0.5005	0.5128	0.5891	0.3
ϕ_m(eV)	Work function	4.3492	4.9117	4.5273	5.1662	4.63
N_A (cm^{-3})	Channel doping	9.8413×10^{16}	9.9999×10^{16}	9.9994×10^{16}	9.9994×10^{16}	1×10^{17}
$\varepsilon 2$	Permittivity of the high-k layer	39.9978	33.9980	39.9980	38.0380	25
I_{OFF} (A/μm)	OFF-state current	0.0068	7.1384×10^{-26}	4.0311×10^{-22}	5.1577×10^{-30}	9.05×10^{-10}
g_m(S/μm)	Transconductance	1.4782	1.3730	1.1993	1.1113	1.92
S (mV/dec)	Subthreshold swing	-	-	-	-	62.38
ΔS	Subthreshold swing degradation coefficient	0.0075	0.0054	2.8346×10^{-16}	1.01016×10^{-20}	-0.03815
CF	Cost Function	0.2268	0.2446	0.2779	0.2999	0.1863
t (s)	Execution time	180	-	-	-	25

978-1-5386-6723-1/19 $31.00 © 2019 IEEE

2019 Devices for Integrated Circuit (DevIC), 23-24 March, 2019, Kalyani, India

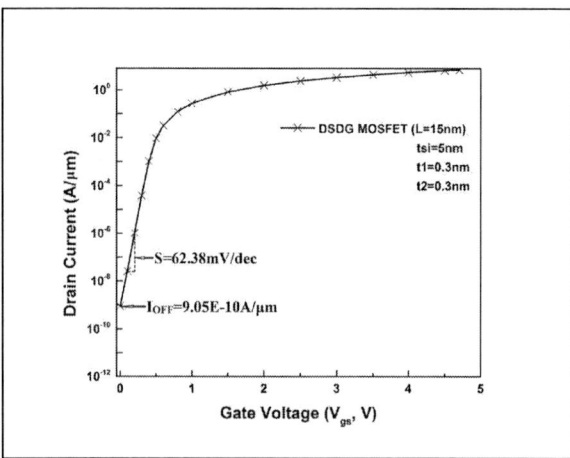

Fig.2. Drain current in OFF state.

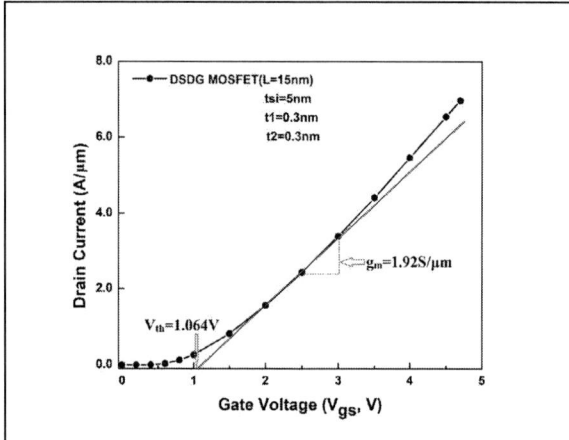

Fig. 3. The plot of drain current (I_D) versus gate voltage (V_{gs}).

V. CONCLUSION

In this article, Simplex-PSO algorithm is applied for the optimization of the electrical performance parameters of GSDG MOSFET. Three objective functions are taken for optimization: (1) minimization of ΔS, (2) minimization of I_{OFF}, (3) maximization of g_m. Simplex-PSO has efficiently calculated the optimal dimensions of the GSDG MOSFET and resulted in an improved electrical performance parameters with respect to the previous literature.

REFERENCES

[1] F. Djeffal, M. Meguellati, and A. Benhaya, "A two-dimensional analytical analysis of subthreshold behaviour to study the scaling capability of nanoscale graded channel gate stack DG MOSFETs", Physica E, Vol.41, pp. 1872–1877, 2009.

[2] F. Djeffal and T. Bendib, "Multi-objective genetic algorithms based approach to optimize the electrical performances of the gate stack double gate (GSDG) MOSFET", Microelectronics Journal, Vol.42, pp. 661–666, 2011.

[3] T. Bendib, F. Djeffal, and D. Arar, "Subthreshold behaviour optimization of nanoscale Graded Channel Gate Stack Double Gate (GCGSDG) MOSFET using multi-objective genetic algorithms" Journal of Computational Electronics, Vol.10, issue 1-2, pp. 210–215, 2011.

[4] T. Bendib and F. Djeffal, "Electrical Performance Optimization of Nanoscale Double-Gate MOSFETs Using Multiobjective Genetic Algorithms", IEEE Transactions on Electron Devices, Vol.58, No.11, November 2011.

[5] L. Jin, L. Hongxia, Y. Bo, C. Lei, and L. Bin, "A two-dimensional analytical model of fully depleted asymmetrical dual material gate double-gate strained-Si MOSFETs", Journal of Semiconductors, Vol.32, No.4, April 2011.

[6] A. Sarkar, A. K. Das, S. De, and C. K. Sarkar, "Effect of gate engineering in double-gate MOSFETs for analog/RF applications", Microelectronics Journal, Vol.43, pp. 873–882, 2012.

[7] T. Bendib, F. Djeffal, T. Bentrcia, D. Arar, and N. Lakhdar, "Multi-Objective Genetic Algorithms Based Approach to Optimize the Small Signal Parameters of Gate Stack Double Gate MOSFET", Proceedings of the World Congress on Engineering 2012, pp. 1-3, Vol. 2, 2012, London, U.K.

[8] D. Arar, F. Djeffal1, M. Meguellati1 and M. Chahdi, "An optimized GSDG MOSFET design for nanoscale circuit applications", 5th International Conference on Modeling, Simulation and Applied Optimization (ICMSAO), 2013, pp. 789-792, Hammamet, Tunisia.

[9] T. Bentrcia, F. Djeffal, and M. Meguellati, "Analytical Investigation of Swing Factor for Nanoscale Double Gate MOSFET Including Gatework-function Effect", Proceedings of the World Congress on Engineering 2015, Vol.1, pp. 426 - 429, 2015, London, U.K.

[10] B. P. De, R. Kar, D. Mandal and S. P. Ghoshal, "Optimal Selection of Components Value for Analog Active Filter Design Using Simplex Particle Swarm Optimization", International Journal of Machine Learning and Cybernetics, vol. 6, no. 4, pp. 621-636, 2015.

[11] B.P.De, K.B. Maji, B.Bag, S. Tripathi, R. Kar, D. Mandal and S.P. Ghoshal, "Optimal Design of Low Voltage, Two-stage CMOS Op-amp Using Evolutionary Techniques", Lecture Notes in Electrical Engineering, 470, pp.303-315.

[12] J. Kennedy and R. Eberhart, Particle swarm optimization, Proc. IEEE international conference on neural networks (1995), pp.1942–1948.

[13] B. P. De, R. Kar, D. Mandal and S. P. Ghoshal, Design of Symmetric Switching CMOS Inverter using PSOCFIWA, Proc. IEEE Int. Conf. Communication and Signal Processing (ICCSP'14), pp. 1818-1824.

978-1-5386-6723-1/19 $31.00 © 2019 IEEE

Analysis of Interface Trap Charges on Dielectric Pocket SOI-TFET

Chandan K. Pandey*[1], Avtar Singh[2], Saurabh Chaudhury[1]

[1]Department of Electrical Engineering, NIT Silchar
[2]C. V. Raman College of Engineering, Bhubaneswar
*chandankumarpandey@gmail.com

Abstract— **In this brief, the reliability of SOI-TFET with dielectric pocket (DP SOI-TFET) has been investigated in the presence of fixed trap charges at the interface between dielectric pocket (DP) and drain region. During numerical simulation, both types of trap charges like donor (i.e. positive) and acceptor (i.e. negative) have been considered to analyse the impact on the performance of SOI-TFET having DP in drain region used for reduction of ambipolar conduction. We have compared the device performances such as ambipolar conduction, and OFF-state current of conventional SOI-TFET with both low and high-k DP SOI-TFET in the presence of interface trap charges (ITCs). It has been found that SOI-TFET is more immune to interface trap charges when high-k material is used as DP compared to low-k. Since, high-k provides even more reduction in ambipolar conduction as compared to low-k, it can be preferred to be used a dielectric pocket in SOI-TFET.**

Keywords— **Ambipolarity, dielectric pocket, high and low-k dielectric, interface trap charges, Tunnel FETs.**

I. INTRODUCTION

Among all the new devices explored, a Tunnel FET (TFET) is considered as the best possible alternative to MOSFETs [1]. The band-to-band tunneling (BTBT) of charge carriers causing conduction allows Tunnel FETs to overwhelm the limitations of MOSFETs such as short channel effects during downscaling of device dimensions, and restriction of minimum inverse subthreshold slope to 60 mv/decade [2]. However, TFETs are still not able to replace the MOSFETs in CMOS technology due to its two major roadblocks like ambipolar conduction, and relatively low ON-state current. To amplify current during ON state, some techniques based on double-gate structure, gate dielectric with high-k, lateral heterostructures and gate-source overlapping have been suggested by the researchers in past [3]-[7]. Similarly, to remove another major hurdle i.e. ambipolar conduction in TFETs, a lot of innovative ideas such as hetero-dielectric gate structure, overlap and underlap of gate with drain, hetero-material gate, and graded channel configuration have been demonstrated by the researchers [8]-[12]. But, most of these techniques are found to degrade the device performances like reduced ON-state current, increased parasitic capacitances and increased fabrication complexity along with reduction in ambipolarity. To resolve these limitations faced by aforesaid techniques, the authors have proposed and investigated a novel structure of SOI-TFET with dielectric pocket (DP SOI-TFET) in their previous work to remove ambipolar conduction along with improved HF performances [13]-[14]. However, device reliability is also an important aspect of any semiconductor device which needs to be taken care properly. In general, during fabrication of the devices, interface trap charges (ITCs) are generated at the interface between semiconductor and oxide layer due to damaged induced by stress and radiation [15]-[17]. Since, the presence of ITCs are found to degrade the reliability and life span of the devices, its impact on device performances needs to be investigated in detail. In this brief, the authors have investigated the influence of ITCs on device performances of DP SOI-TFET in the presence of both donor and acceptor trap charges. Furthermore, we have compared the device performances of the conventional SOI-TFET with low-k and high-k DP SOI-TFET in presence of ITCs.

Unlike electrons and holes, donor and acceptor traps lie between valance and conduction band and these trap charges are created mainly due to unsaturated fourth band found at the interface. Based on electron occupancy, a donor interface trap can behave like positive ion if it is free from electron or like a neutral atom if it has not donated its electron yet. Similarly, an acceptor interface trap can behave as a negative ion if it has accepted an electron or as a neutral atom if it is free from electron [18]-[19]. ITCs have been found to degrade the peak value of lateral electric filed at tunneling interface on which conduction current in Tunnel TFETs mainly relies. Since, ITCs are present on interface between DP and drain (i.e. Silicon) at channel-drain tunneling junction only, ON-current is very likely to be unaffected with the presence of ITCs. But, ambipolar conduction which is produced by the movement of charge carriers at output tunneling interface will be definitely affected with the presence of ITCs. So, in this paper, we have focused our study mainly on the ambipolar behaviour of DP SOI-TFETs when ITCs are present at the interface between DP and drain.

II. DEVICE PARAMETERS AND SIMULATION SETUP

The 2-D schematic views of the conventional and DP SOI-TFET are shown in **Fig. 1(a)** and **(b)**, respectively. As compared to conventional SOI-TFET, a dielectric pocket is deposited on the drain region of SOI-TFET, as shown in **Fig. 1(b)**. The details of design parameters are taken from our previous work reported in [12] and listed in **Table I**.

For this work, simulations are performed with the help of Synopsys sentaurus simulator [20]. For accurate estimation of BTBT in the tunneling region, a nonlocal dynamic model is preferred over the local BTBT model. Bandgap narrowing model is also enabled to account for the effect of heavily doped

978-1-5386-6723-1/19 $31.00 © 2019 IEEE

2019 Devices for Integrated Circuit (DevIC), 23-24 March, 2019, Kalyani, India

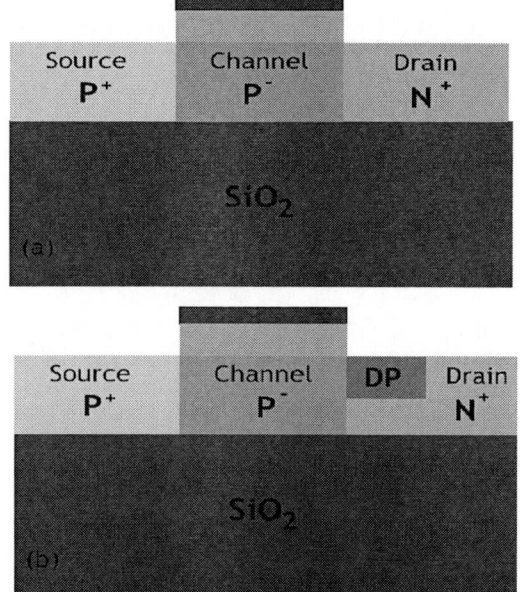

Fig. 1. (a) Conventional, and (b) Dielectric Pocket SOI-TFET.

TABLE I
LIST OF PARAMETERS USED FOR THE SIMULATION

Parameters	SOI TFET	DP SOI TFET
Length of Channel	50 nm	50 nm
Work Function of Gate	4.2 eV	4.2 eV
Length of Drain and Source	100 nm	100 nm
Source Doping (N_A)	1×10^{20} cm^{-3}	1×10^{20} cm^{-3}
Channel Doping (N_A)	1×10^{16} cm^{-3}	1×10^{16} cm^{-3}
Drain Doping (N_D)	5×10^{18} cm^{-3}	5×10^{18} cm^{-3}
Gate Oxide Thickness (t_{ox}) SiO$_2$	3 nm	3 nm
SOI Thickness (t_{si})	10 nm	10 nm
Oxide Thickness (t_{SiO2}) of SOI	30 nm	30 nm
Length of Dielectric Pocket	-	30 nm
Thickness of DP	-	5 nm
Dielectric Constant of HfO$_2$	-	25
Dielectric Constant of SiO$_2$		3.9

source and drain regions. Auger recombination along with SRH recombination are also included to estimate the subthreshold leakage current accurately. Since, a small change in the performance parameters is important to analyze the reliability of devices in presence of ITCs, the density of donor and acceptor traps are taken between 10^{11} to 10^{13} cm^{-2} suggested in [21] which has considered both radiation harm and hot carrier damage.

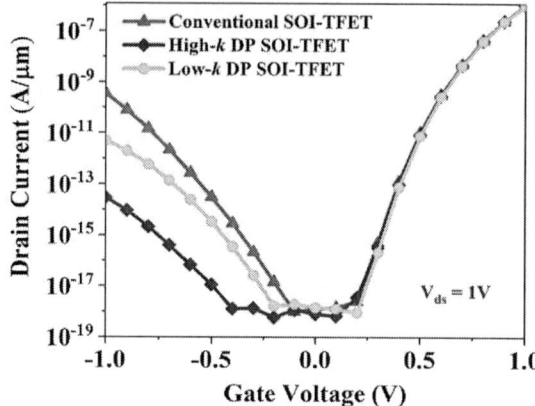

Fig. 2. $I_{ds} - V_{gs}$ plot of conventional and DP SOI-TFET with low-k and high-k dielectric.

III. RESULTS AND ANALYSIS

Fig. 2 compares the $I_{ds} - V_{gs}$ plot of the conventional SOI-TFET with low-k and high-k DP SOI-TFET. It can be observed that high-k DP SOI-TFET exhibits the lower value of ambipolar conduction compared to other two devices. Now, to analyse the reliability of DP SOI-TFET in presence of ITCs at the interface between DP and drain region, we have considered donor and acceptor trap charges individually as well as together. The density of traps and position of energy level for both the trap charges have been taken from [18] which can lead to the maximum impact on the device performances to consider the worst case scenario. Trap densities of acceptor (i.e. negative) and donor (i.e. positive) have been taken as 1×10^{13} and 5×10^{12} cm^{-2}, respectively. The energy level for negative and positive interface traps have been fixed at 0.6 eV below from conduction band and 0.2 eV above from valance band, respectively. **Fig. 3 (a)** and **(b)** compare the impact of ITCs on $I_{ds} - V_{gs}$ plot of DP SOI-TFET with low-k and high-k dielectric material. It can be observed from the figure that the ambipolar conduction is found to be reduced further for both the devices in the presence of only negative interface traps. However, the OFF-state current is also increased in the presence of only negative interface traps but this increment is not as large as observed in the presence of only positive (i.e. donor) interface traps for both the devices. Furthermore, at low negative gate bias, the presence of only donor traps is found to enhance the ambipolar conduction in both the devices, thus degrading the performance of DP SOI-TFET. When both negative and positive interface traps are considered to be present together then it can be observed that for an increment of negative gate bias, ambipolarity is initially dominated by the donor traps, thus leading to more ambipolar conduction, and later since, acceptor traps start to dominate over donor traps, ambipolar conduction is found to decrease at higher negative gate bias.

To understand the behaviour of ambipolar conduction in the presence of ITCs in detail, **Fig. 4(a)** and **(b)** exhibits the energy band profile of both the devices biased at $V_{gs} = -1V$ and $V_{ds} = 1V$. As observed from the figure, tunneling

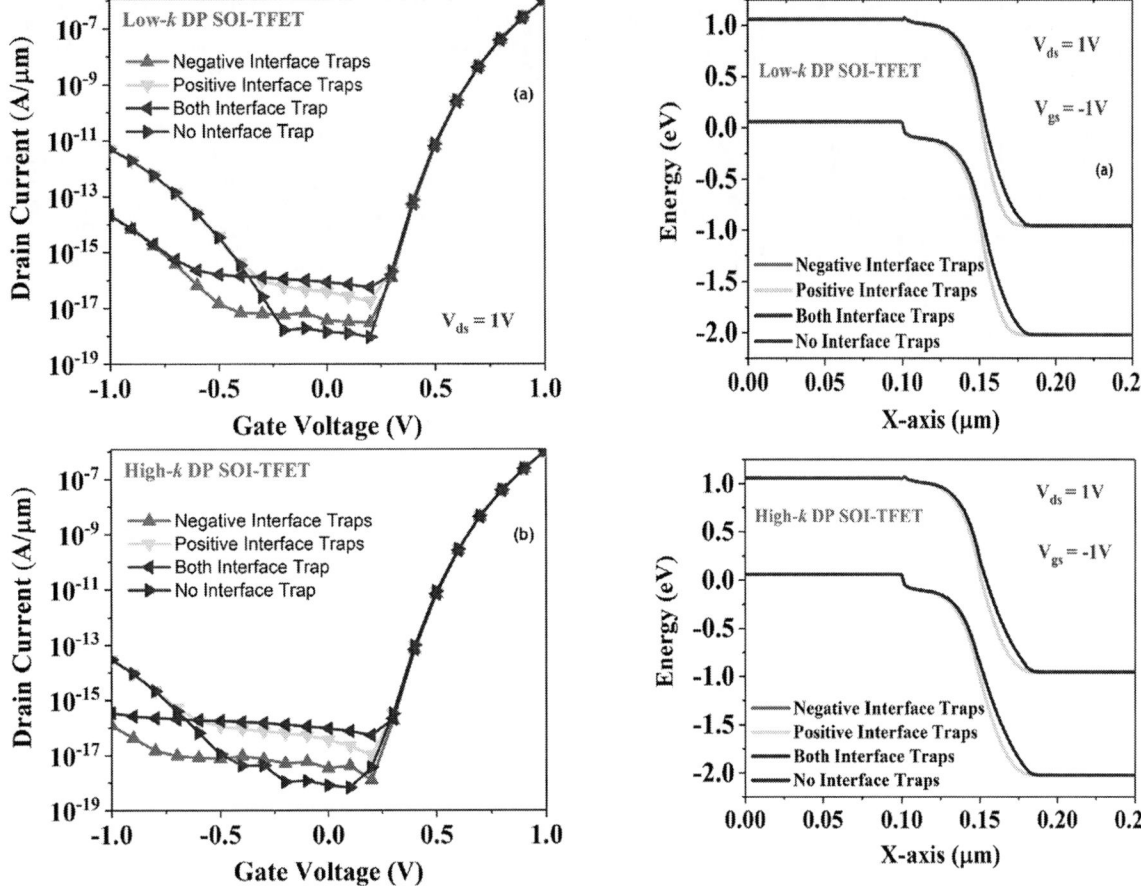

Fig. 3. Transfer characteristics of low-*k* and high-*k* DP SOI-TFET in the presence of ITCs.

Fig.4. Energy band profile of low-*k* and high-*k* DP SOI-TFET in the presence of ITCs.

width at the interface between channel and drain is increased for both the devices when only acceptor traps are considered at the interface while it is approximately found to be unchanged in the presence of only donor traps. The increment in tunneling width further causes the reduction in ambipolar conduction for the devices with negative interface traps. The main reason behind this is that the acceptor traps drag the electrons from conduction band and get negatively ionized. This, eventually, results to an enhancement in depletion region underneath the interface. However, due to increment in electron concentration in drain region, the ambipolar conduction is found to be increased in presence of donor traps which is later dominated by the electron concentration generated by BTBT at higher negative gate bias.

For a comprehensive analysis of the effect of interface traps, **Fig. 5** shows a comparison of the lateral electric field in both the devices biased at ambipolar state. We can observe that the lateral electric filed is widely distributed over drain region in the presence of only acceptor traps as well as both the traps together at higher negative gate bias, thus reducing the peak of

filed at output tunneling interface. It is mainly attributed due to the increment in depletion width inside drain region which causes more potential drop across this particular region. The decrement in peak value of this electric field at tunneling interface leads to further reduction of ambipolar conduction in DP SOI-TFET.

IV. CONCLUSIONS

In this paper, we have investigated the impact of interface trap charges on the device performances of DP SOI-TFET for low and high-*k* dielectric materials. For detail analysis, all possible scenarios of ITCs such as only donor traps, only acceptor traps and both the traps together have been considered to be present at interface between DP and drain region. Through 2-D numerical simulations, it has been found that the presence of only acceptor traps reduces the ambipolar conduction in DP SOI-TFET for both low and high-*k* dielectric material. However, the only donor traps have been found to enhance the ambipolar conduction in DP SOI-TFET for both low and high-*k* dielectric material, thus degrading the device performances.

2019 Devices for Integrated Circuit (DevIC), 23-24 March, 2019, Kalyani, India

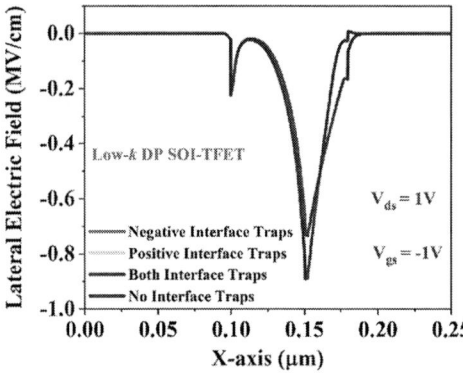

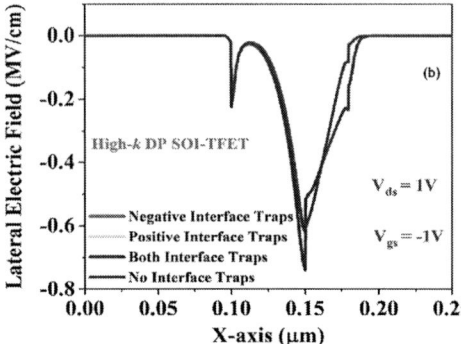

Fig. 5. Lateral electric field distribution for low-k and high-k DP SOI-TFET in presence of ITCs.

We have demonstrated that the widely spread lateral electric field over drain region caused by reduction in electron concentration at output tunneling interface actually leads to further reduction of ambipolar conduction in the presence of acceptor traps compared to its counterpart donor traps. Furthermore, it has also been found that transfer characteristics of DP SOI-TFET in presence of both the traps follow the same pattern as being obtained in presence of only acceptor traps at higher negative gate bias. Since, high-k DP provides more reduction in ambipolar conduction than low-*k* even in the presence of interface traps, it can be preferred for better device performance of SOI-TFET.

REFERENCES

[1] V. Nagavarapu, R. Jhaveri, and J. C. S. Woo, "The tunnel source (PNPN) n-MOSFET: A novel high performance transistor," *IEEE Trans. Electron Devices*, vol. 55, no. 4, pp. 1013–1019, Apr. 2008.

[2] P. F. Wang et al., "Complementary tunneling transistor for low power application," *Solid State Electron.*, vol. 48, no. 12, pp. 2281–2286, Dec. 2004.

[3] K. Boucart and A. M. Ionescu, "Double-gate tunnel FET with high-K gate dielectric," *IEEE Trans. Electron Devices*, vol. 54, no. 7, pp. 1725–1733, Jul. 2007.

[4] S. Cho et al., "Design optimization of a type-I heterojunction tunneling Field-effect transistor (I-HTFET) for high performance logic technology," *J. Semicond. Technol. Sci.*, vol. 11, no. 3, pp. 182–189, Sep. 2011.

[5] R. Wang et al., "Investigations on line-edge roughness (LER) and line width roughness (LWR) in nanoscale CMOS technology: Part II—

Experimental results and impacts on device variability," *IEEE Trans. Electron Devices*, vol. 60, no. 11, pp. 3676–3682, Nov. 2013.

[6] T. Krishnamohan, D. Kim, S. Raghunathan, and K. Saraswat, "Double gate strained-Ge heterostructure tunneling FET (TFET) with record high drive currents and << 60mV/dec subthreshold slope," *in IEDM Tech. Dig.*, Dec. 2008, pp. 1–3.

[7] W. Cao, C. J. Yao, G. F. Jiao, D. Huang, H. Y. Yu, and M. F. Li, "Improvement in reliability of tunneling field-effect transistor with p-n-i-n structure," *IEEE Trans. Electron Devices*, vol. 58, no. 7, pp. 2122–2126, Jul. 2011.

[8] W. Y. Choi and W. Lee, "Hetero-gate-dielectric tunneling field effect transistors," *IEEE Trans. Electron Devices*, vol. 57, no. 9, pp. 2317–2319, Sep. 2010.

[9] C. K. Pandey, and S. Chaudhury, "Dual-Metal Graded-Channel Double-Gate Tunnel FETs for Reduction of Ambipolar Conduction," *in 1st IEEE Intern. Conf. on Elect. Devices (EDKCON-2018)*, Kolkata, Nov. 24-25.

[10] S. Saurabh and M. J. Kumar, "Novel attributes of a dual material gate nanoscale tunnel field-effect transistor," *IEEE Trans. Electron Devices*, vol. 58, no. 2, pp. 404–410, Feb. 2011.

[11] D. B. Abdi and M. J. Kumar, "Controlling ambipolar current in tunneling FETs using overlapping gate-on-drain," *IEEE J. Electron Devices Soc.*, vol. 2, no. 6, pp. 187–190, Nov. 2014.

[12] A. S. Verhulst, W. G. Vandenberghe, K. Maex, and G. Groeseneken, "Tunnel field-effect transistor without gate-drain overlap," *Appl. Phys. Lett.*, vol. 91, no. 5, p. 053102, Jul. 2007.

[13] C. K. Pandey, D. Dash, and S. Chaudhury, "Impact of Dielectric Pocket on Analog and High-Frequency Performances of Cylindrical Gate-All-Around Tunnel FETs," *ECS J Solid State Sc.*, vol 7, no. 5, pp. 59-66, May 2018.

[14] C. K. Pandey, D. Dash, and S. Chaudhury, "Approach to suppress ambipolar conduction in Tunnel FET using dielectric pocket," *Micro & Nano Lett.*, vol 14, no. 1, pp.86-90, Jan. 2019.

[15] M. G. Pala, D. Esseni, and F. Conzatti, "Impact of interface traps on the IV curves of InAs tunnel-FETs and MOSFETs: A full quantum study," *in Proc. IEEE IEDM*, Dec. 2012, pp. 1–4.

[16] M.-L. Fan, V. P.-H. Hu, Y.-N. Chen, P. Su, and C.-T. Chuang, "Analysis of single-trap-induced random telegraph noise and its interaction with work function variation for tunnel FET," *IEEE Trans. Electron Devices*, vol. 60, no. 6, pp. 2038–2044, Jun. 2013.

[17] G. F. Jiao et al., "New degradation mechanisms and reliability performance in tunneling field effect transistors," *in Proc. IEEE IEDM*, Dec. 2009, pp. 1–4.

[18] Y. Qiu, R. Wang, Q. Huang, and R. Huang, "A comparative study on the impacts of interface traps on tunneling FET and MOSFET," *IEEE Trans. Electron Devices*, vol. 61, no. 5, pp. 1284–1291, May 2014.

[19] J. Madan and R. Chaujar, "Interfacial charge analysis of heterogeneous gate dielectric-gate all around-tunnel FET for improved device reliability," *IEEE Trans. Device Mater. Rel.*, vol. 16, no. 2, pp. 227–234, Jun. 2016.

[20] Sentaurus Device User Guide. Version M-2016.12, Synopsys, Inc., Dec 2016.

[21] S. Shabde, A. Bhattacharyya, R. S. Kao, and R. S. Muller, "Analysis of MOSFET degradation due to hot-electron stress in terms of interface state and fixed-charge generation," *Solid State Electron.*, vol. 31, no. 11, pp. 1603–1610, Nov. 1988.

978-1-5386-6723-1/19 $31.00 © 2019 IEEE

Ubiquitous and Emerging Concepts of Sensors

Manisha Sharma
Computer Science and Engineering
NIT, Jamshedpur
Jamshedpur, India
2018pgcscs07@nitjsr.ac.in

Monu Bhagat
Computer Science and Engineering
NIT, Jamshedpur
Jamshedpur, India
2018rscs002@nitjsr.ac.in

Dilip Kumar
Computer Science and Engineering
NIT, Jamshedpur
Jamshedpur, India
dilip.cse@nitjsr.ac.in

Abstract—Ubiquitous computing and sensor networks persuading the technological bars so firmly that its success is elevating in exponential manner. Along with all the prons that this technology is providing to our society in day to day life, industries, medical system, military services , etc there are still some barriers in achieving the performance to the fullest extent due to some resistances like battery life, maintenance, security and others which are listed and described in this paper. This paper will be carried out starting with the basic introduction of Ubiquitous computing and then focusing mainly on wireless sensor network technology, then its features along with the software as well as hardware constraints along with the countermeasures. The main challenge being the security assurance of the network that is briefly described in this section with layer wise attacks and actions to recover. In section 3, applications of wireless sensor network technology are mentioned that been categorized depending on the types of sensor networks. The last section contains the trending standards and technologies of wireless sensor networks which includes IEEE 802.15.4, ZigBee, Bluetooth , etc along with their classifications of characteristics that are bulleted on the table of comparison. At last , concluding the subject with a summary on all technologies ,applications and leaving future scope of betterment in the present day's technology and deployment of the newer ones.

Index Terms—WSN, ZigBee, BLE, UWB, ANT, Bluetooth, IEEE802.15.4

I. INTRODUCTION

Ubiquitous Sensor networks have become the most prevailing technology in recent times. The pervasiveness of the mobile devices and the Internet of Things has created various new openings of the technology and has increased the convenience of human lives. Ubiquitous Computing purposed with wireless intercommunicating sensors which are capable of sending/receiving or both from the environment.[13] The wireless sensor networks consists of many autonomous sensing self-powered nodes which themselves are implemented with processors, a radio interface, transducers, sensors, memory and power supply which are capable of gathering and processing the information or detect some specific events in wireless fashion.[5] When the objects get connected to the internet, they become smarter . Such objects has an exclusive quality i.e., they "known".[8] A transducer is a device which is basically used to convert energies from one domain to another is used in wireless sensor technology as the output of transducers is essential for sensors to process the data. A sensor can be

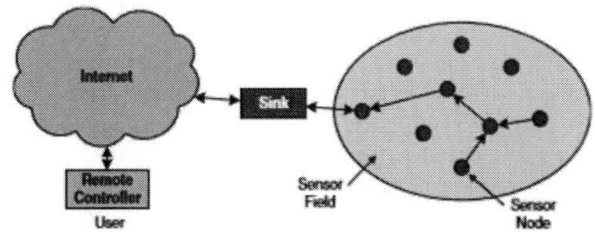

Figure 1. Wireless Sensor Network Structure

capable for observing transposition, steaminess, illumination, etc. Fixed storage is used into it to accumulate demands and raw data. [10]The power source is usually a chargeable battery. A wireless sensor network system is constructed on modularity design. The developed model has two types of nodes that are sensor nodes and coordinator nodes.[18] Sensor nodes plays as an end gadget that senses the physical parameters of surroundings for instance temperature, humidity or pressure as well as the coordinator node is commanding of collecting all the information from the sensor nodes and ultimately sending them to the end user. In the long run, ubiquitous computer and sensor networks can percolate through all the sphere of lives such as environment monitoring, military welfare, healthcare, smart cities, intelligent vehicles, building and home automation and many more countless applications that can be developed in coming future of technology are discussed briefly in section 2. [[6][17]

II. FEATURES OF WIRELESS SENSOR NETWORKS: CHALLENGES AND REQUIREMENTS

The main features of WSNs networks as discussed earlier can be scalability: the number of nodes that can be added to the network without hampering the on going process , self-organization, self-healing, energy efficiency, low complexity and low costs. The overall description of wireless sensor network bring various preferences over conventional wireless ad-hoc network inclusive of quick positioning, dynamic network, and quick managing potential. Though sensor nodes can be implemented in an ad-hoc or a preplanned technique as well. An ad-hoc implementation is best suited for huge bare areas where network of large quantity of nodes can be implied and left neglected to execute the desirable actions and announcing outcomes.[2] Network preservation such as handling coupling and identifying abortions is rigid in such a network due to huge quantity of nodes. Whereas, pre-planned implications is best-suited for finite reporting where less number of nodes are implemented at specific positions along with the edges of low maintenance and management cost of the network. However there are few drawbacks of sensor networks due to there hardware techniques, communication protocols and application designs. Therefore wireless sensor network technology must address this drawbacks to realize the various future applications.[15] Sensor nodes have a limitation of including restricted energy, restricted memory and data processing potential. The finite power lifetime of the sensors in the network inflict lifelong constraint on the network. Generally when a node is in transmit mode, the transceiver evacuate huge amount of energy from battery power than the sensors or memory chip. This problem of limited source of power could be overcome by utilizing them productively. So power effective execution is mandatory to amplify the network lifespan by applying energy effective protocols compact in essence energy-sensible of routing in network layer, power-cutting technique in MAC layer, etc.[14] Effective way to maximize the limited memory is by reducing the redundancy of data and considering the memory consuming problems like routing table, security ,etc. Dynamic structures of nodes in the network and tough environment condition may cause the collapsing of sensor nodes and downfall in the performance. The communication in wireless sensor networks is unre-liable because of fallacy vulnerable wireless medium with inflated bit error rates and various connection volume. Thereby a sensor net-work must be dependable to work perfectly and be conditional on the application demands the processed data should be transferred to the source reliably. Wireless sensor networks are endangered to node failure because of various reasons like node may fall out of power or damaged or communication among nodes might be intruded forever. This demands wireless sensor network to be vigorous to node collapses. In wireless sensor networks, the fault forbearance can be enhanced by imposing high extent of redundancy by incrementing the quantity of vertex in the network. To mark the data redundancy so that only the required information reaches the destination and the communication operating cost can be reduced by using data fusion and localized processing. The another major constraint of sensor technology is its preservation. The only required configuration of maintaining the sensor network is the absolute or limited improvement of the cipher in the vertex over the wireless medium. Every sensor node should be modernized, and the constraints on the magnitude of current nodes should be identical with the case of wired programming. The part of the codes that are consistently executing in the nodes need to assure adapted support and should have updating procedures in it.[13] As the network topology might be dynamic i.e., it might alter due to breakdown of nodes, versatility or large scale implementation therefore a wireless sensor network should be self-organized. In inclusion, many current vertex can be included to the system for

may be to replace the collapsed nodes so the network must be self constructive. It would be quite extortionate to provide distinctive address to each sensor node in the network when there are many nodes implemented in the network. therefore worldwide recognition of the sensor nodes might conduct to huge overhead. Besides because of limited storage and computational power ,it is not advisable to rely on single node's content. Thus the wireless sensor networks are demanded to use data centric paradigm which focuses on the data initiated by the sensors.

III. CHALLENGES IN SECURITY BASED ON LAYERS [12]

A. PHYSICAL LAYER

Jamming or Interference is the most familiar attack in the physical layer of wireless network. Jamming hampers the radio frequencies that is utilized by the vertex of the system. Then an intruder successively transfers over the wireless system denying the basic MAC protocol. In inclusion jamming can cause enormous power utilization at a node by introducing irrelevant data. The receiver's nodes will thus absorb power by accepting those packets. The counteraction for this attack can be channel hopping or blacklisting. Tampering is another attack on this layer. In this, nodes are endangered to interfering or any harm. The counter measure for this attack can be either protection or altering the key.

B. DATA LINK LAYER

There are numerous strikes in the link layer too. An intruder may advertently break the transmission agreement, and often dispatch messages to create crash. This type of collisions would need the retransmission of any packet influenced by collision. By the help of this method it is quite feasible for a rival to ingest a sensor node's reserves by forcing oversupply re-transmissions
. Few threats as well as there counteractions are as mentioned below in the table.

Table I
THE THREATS THAT CAN ATTACK THE DATA LINK LAYER ARE MENTIONED BELOW IN THE TABLE ALONG WITH THERE COUNTERACTIONS

THREATS	COUNTERACTIONS
Collision	CRC and time Diversity
Exhaustion	Protection of network ID
Spoofing	Use different paths to send the packets
Sybil	Regularly changing of key
De-synchronization	Using various adjacent for time synchronization
Eavesdropping	Keys protects DLPDU from eavesdropper
Traffic Analysis	Sending of dummy packets and monitoring Frequently

C. NETWORK LAYER

There can be two groups of attacks in the network layer: Passive attacks and Active attacks. Passive entries into the network without authentication does not break the functioning of the system but the antagonist to explore data, spying on the medium across the network without changing the data. It is quite tough to detect passive attacks due to the fact that this attack do not alter the working of the system. However an active attack can insert or change the data thereby restricting the working of the system where both data packets and routing control packets kept by communication. The rival can strike routing packets causing purposeless routing table at the source.

Table II
THE THREATS THAT CAN ATTACK THE NETWORK LAYER ARE
MENTIONED BELOW IN THE TABLE ALONG WITH THERE
COUNTERACTIONS

THREATS	COUNTERACTIONS
Eavesdropping	Session key protect NPDU from eavesdropper.
DoS	Protection of network particularly data link network ID
Selective forwarding	Regular network monitoring using source routing
Wormhole	Physical monitoring of Field devices and regular monitoring of network using source routing
Sybil	Resetting of device and changing of session key
Traffic Analysis	Sending of dummy packets frequently and regular monitoring of wireless sensor network.

D. TRANSPORT LAYER

Moreover the transport layer is endangered to strikes, as in the instance of flooding. Flooding is as easy such as transferring numerous link requests to a vulnerable node. In this case, operator must be designated to manage the link requests. Finally a node's resource will be worn-out thus providing the node meaningless.

E. APPLICATION LAYER

With the evolution of automation and the detectors being agile, microscopic and economical ,masses of sensor nodes are being implemented in diverse operations. The diversification of the applications of wireless sensor networks is virtually unlimited from environmental monitoring, health care, transport logistic, mobility, military welfare, personal identification to positioning and tracking and many more. In army ,detector nodes can be helpful to detect antagonist. In the instance of natural disaster , detectors can detect such calamity beforehand like hurricane, earthquake, forest fires, etc.[5] In medical healthcare system, sensor nodes targets to enhance

the current medical-aid and controlling helps especially for the constantly ill. There are many other benefits that are attained as for instance through remote monitoring i.e., the recognition of crisis status for at endangered clients/sufferers will become easier and the people with diverse grades of illness will be enabled to live a secure and independent live. The applications of wireless sensor network can be categorized on the basis of the different types of wireless sensor network. With the evolution of technology and the sensors being smarter, smaller and cheaper ,billions of sensor nodes are being implemented in diverse applications. The diversification of the applications of wireless sensor networks is virtually unlimited from environmental monitoring, health care, transport logistic, mobility, military welfare, personal identification to positioning and tracking and many more. In military ,sensor nodes can be used to detect or track of antagonist. In case of natural disaster , sensor nodes can detect such calamity beforehand like hurricane, earthquake, forest fires, etc.[5] In medical healthcare system, sensor nodes targets to enhance the existing healthcare and monitoring services especially for the chronically ill. There are many other benefits that are attained as for instance through remote monitoring i.e., the recognition of emergency conditions for at risk patients will become easier and the people with different levels of illness or disabilities will be able to have a more secure and independent lives.

The applications of wireless sensor network can be categorised on the basis of the different types of wireless sensor network.

IV. TYPES OF WIRELESS SENSOR NETWORK

A. TERRESTIIAL WIRELESS SENSOR NETWORK

This network consists a huge number of vertex nearly around thousands which are dispersed over the land in ad-hoc type of network(dynamic topology). As already discussed the power-cell is finite and generally non chargeable, this kind of network can be implemented with a alternative capacity such as solar cells. Frequently used applications of terrestrial wireless sensor networks are ecological detecting and controlling and exterior research. In military welfare, the most prevailing application is PinPtr. It is a reverse-shooter system evolved to identify and discover enemy. The ad-hoc audible detector system observes the muffle explosion and the audible blow signal that generates from the noise of firing. Sensor nodes direct their estimations to the infrastructure to estimate the gunman's position. The appearance time of the audible occurrence at various detector vertex are used to regulate the orbit of the shot and evaluate the position of the rival.[16]

B. UNDERGROUND WIRELESS SENSOR NETWORK

This network existed with detector vertices that are implemented in caves, collieries or buried to check underground situations. With the purpose of communicating or retrieving data from the subway detector vertices to the base station, extra detector nodes are implemented over the land. This kind of network is more expensive than the terrestrial kind of networks because they demand suitable devices to assure authentic transmission over land, cliffs and aqua. Furthermore it is even more difficult to maintain the battery power hidden causing it more essential to plan it more efficiently communication protocol for prolonged existence. The main applications of this network are farming controlling, geography supervision, undercover controlling of earth, aqua or fossil and army controlling.[16]

C. UNDERWATER WIRELESS SENSOR NETWORKS

This network consists of sensors implemented beneath water, for instance, into the ocean environment. Since this implementation is quite costly, therefore just scarce nodes are equipped and independent submerged machine are utilized to collect information from them. Submerged radio transmission uses audible signals that brings out many disputes for instance as limited capacity, huge diffusion stall, large delay and motion waning issues. Therefore these vertices must be implied in such a way that they are self-configuring and accommodate to utmost situations of aqua ambience. Here too vertices are implemented with finite supply of energy which irreplaceable which ultimately requires power effi-cient submarine transmission and interlinking methods. Utilizations of this network includes impurity controlling, vibrational controlling, device supervising and submarine droids.[16]

D. MULTI-MEDIA WIRELESS SENSOR NETWORKS

It comprised of economically charge detector vertices furnished with webcam and megaphones, implemented in a

pre-planned way to ensure the scope. These tools are efficient work and global unify figures. One of the best

of accumulating, equipping and fetching mass media's outlets.

capacity need, huge power consumption,

quality of service (QoS) equipment, information process- structure must have very low effective radiated power

compacting methods. Multi-media wireless sensor

networks improves the surviving wireless sensor utilization

such as tracing and controlling.[16]

E. MOBILE WIRELESS SENSOR NETWORKS
This network comprised of portable detector nodes that can be pervasive i.e., can be used anywhere and anytime and can cooperate with the ambience. Motile vertices can be rearranged and arrange selves in the mesh inclusion energetic forwarding method must ,thus, be engaged unlike immovable forwarding in. Portable wireless numerous problems such as disposition, agility guidance, positioning with agility sheering and governing of nodes, supporting sufficient detecting area, reducing power utilization mesh accordance and data division. Most basic appli-cations of this network type are managing, army.[11]

V. TECHNOLOGIES

A perception of future ubiquitous computing is that small microprocessors and sensors will be consolidated with everyday objects in order to make them intelligent. Any object when connected to internet becomes smart. Smart objects can traverse their surrounding, transmit with other smart objects, as well as can interact with humans too. The advancement of wireless sensor network relies on a huge extent of technologies, such as hardware, system software and network communication. One of the tough constraints for wireless detector networks is that the application requirements on network differs from one application to another. The outcome shows that it is not elementary to define comprehensive requirements on correctness and sampling rate, for sensor nodes and net- works. Fig. Outlines the technologies surrounding wireless sensor networks. This section describes numerous technologies supporting wireless sensor network and brief of characteristics of all the technologies is described in table 1.

A. IEEE 802.15.4 WIRELESS TECHNOLOGY

This technology is a short-range communication system proposed to provide application with moderate throughput and latency demands in wireless personal area network (WPAN). The primary features of IEEE 802.15.4 wireless technology are low complexity, low expenses, low computation, low energy utilization, low data rate transmission to be held by low cost either static or dynamic devices. The basic applications of this technology is the execution of wireless sensor networks.[16]

B. ULTRAWIDE BANDWIDTH TECHNOLOGY
This is a prevailing technology with distinct traits that trending ultra-wide bandwidth techniques, mainly for

wireless.

They should deal with the numerous constraints such as large sensor network implementations is Impulse Radio- UWB. In inclusion to the poor energy spooky width, trading

Ultra wideband features are applicable to short-distance Applications such as personal computer peripherals.[9]

The other important application is ultra-wideband

pulse Doppler radars have also used to detect essential signals been of human body, such as heart beat and respiration signals as well as humangai analysis and fall detection. It requires less power

and a high-resolution range profile.[1] to manage to identify, calculate and transmit

C. ZIGBEE AND TREE BASED TOPOLOGY

ZigBee technology wireless sensor network is . acceptable for low data rate, low-cost technology used in embedded application. ZigBee explains the application in motion, controlling. This network and technology is being developed advanced by the ZigBee foremost benefaction is giving grid Alliance and its utilizations. Basically to implement ZigBee, inclusive devices such as ZigBee coordinator, ZigBee router are demanded in inclusion to the ZigBee terminating tool.[3] Usual ZigBee nodes requires an 802.15.4/IP gateway to interface with an IP network. Therefore ZigBee is better for Wireless sensor network utilizations that do not demand to interact with the IP devices. However, the trending ZigBee IP description offers an IPv6-founded radio grid interlinking resolution. It enhances IEEE 802.15.4 by including system and safety sheets and an utilization scheme, contributing a scalable architecture with end-to- end IPv6 networking.[7]

D. BLUETOOTH TECHNOLOGY

This is a brief-scope of transmitting method considered to substitute the cables(wires) in wireless personal area network. The essential characteristics of Bluetooth wireless technology are robustness, brief energy consumption and low price. The IEEE project 802.15.1 has obtained a wireless- personal- area network standard established on the Bluetooth v1.1 groundwork stipulation. The Bluetooth Radio Frequency(physical layer) controls the unauthenticated ISM band, for many countries around 2:4 GHz in 79 frequency channels spacing of 1 MHz in the ISM band are accessible. RF performance utilizes a Gaussian shaped, binary frequency shift keying(GFSK) modulation to reduce transceiver complexity, and a forward error correction(FEC) coding technique. The bit rate is 1Mbps.[2]The vertices that are connected in this technology are systematically structured in piconet, governed by chief vertices and consisting an utmost of seven lively thralls. The tangible medium is breakdown into period units known as slots with period component of 625 seconds. Bluetooth Technology contributes the outcome of duplexing transmission by using time-division duplex scheme.[16]

F. Z-WAVE TECHNOLOGY

This is a methodology generated by the Danish company Zensys. It utilizes lower-voltage radio frequency for low-power remote control applications. This is a radio communication treaty basically used for domestic mechanization. It is a network grid topology. This technology has been standardized by the Z-wave Alliance. The main edge of this technology is its implementation in sub 1GHz band. Mesh topologies can be initiated, nevertheless the resolving methods utilized enables a limit of 232 nodes in the system. Practicable data rates are between 9.6 kbps and 40kbps. The receivers from Zensys permits 100 meters of outdoor range.[4]

G. ANT TECHNOLOGY

It is a patented methodology introducing a radio communication compact heap idea for extreme-brief energy interlinking utilizations. It is depicted to execute by using nominal price, low voltage microprocessors and receivers working in the 2.4 GHz ISM band. The ANT wireless sensor technology convention has been devised for intelligibility and effectiveness, deriving in an extreme-short energy ingestion, expanding battery life-span, a limited load on structure assets, uncomplicated structure schemes and reduced execution expenses. This technology also captivates reduced intermission, the potential to swap data rate in front of energy intake, and maintenance for announcing, outbreak and recognized proceedings for maximum net data rate of 20 kbps. In this technology, numerous topologies can inveterate such as peer-to-peer, star, tree ,or other types of mesh network. Even ANT vertices are effective for acting as thralls or chief internally a network and trading off responsibility at any time. This is an effective protocol for pragmatic networks as due to this inherit potential to support ad-hoc correlation of various vertices. [16][4]

E. BLUETOOTH LOW ENERGY TECHNOLOGY

Bluetooth is a radio methodology for brief-reach and low cost tools are considered to substitute the cords in wireless personal area networks. It operates in 2.45 GHz ISM band and utilizes recurrence hopping to control hindrance and waning. Bluetooth can wrap at transmission scope of 10-100 meters and enables data rate up to 3 Mbps. It standardized as IEEE 802.15.1, but the paradigm is no more preserved. Recently , Bluetooth is governed by the Bluetooth Special Interest Group, which accepted Bluetooth Core Specification version 5.1 in 2019. It introduced Bluetooth Low Energy (BLE) methodology that enables cost-effective Bluetooth smart devices to work for small time period, -power cells. Efficient marts for BLE- founded tools includes medical-aid, athletic and aptness, safety, and house amusement. BLE functions in the same 2.45 GHz ISM band as typical Bluetooth, but utilizes various sets of slots. Unlike of Bluetooth's 1 MHz wide 79 channels, BLE has 2 MHz wide 40 channels. It enables 1 Mbps data rates with 200 m range and has two applications substitutes : single-mode BLE tools upholds only new BLE links and dual mode BLE tools upholds both typical Bluetooth links.[4]

VI. CONCLUSION

Detector meshes presents infinite constraints, however their adaptability and their wide scope of implementations are inducing curiosity to investigate section in addition to from production houses. Detector networks and the ubiquitous technology has the capability of activating the subsequent rising in computer science. In spite of the hardware as well as software challenges like the advancement of low energy transmitting equipment, brief-voltage microprocessors, etc still if we consider the design devices which are more effective and efficient then the boons of this technology is too fruitful to be disregarded. Sensor networks provide endless opportunities as well as brings out daunting constraints. If misunderstood and misconfigured, the sensors poses to risk our data, privacy and safety, but if understood and constructed perfectly, it can enhance communications, lifestyle and delivery of services and many more. Normalization is a vital part for accomplishment of wireless sensor network marts. We have represented diverse principles and techniques accessible for wireless sensor network. For low data rate utilizations, IEEE 802.15.4 turns out to be the best easy to access new technique available, while Bluetooth low energy (BLE) can be captivating

Table III
CHARACTERISTICS OF TECHNOLOGIES

Characteristics/technology	IEEE 802.15.4(ZigBee)	UWB(IEEE 802.15.4a)	Bluetooth	Z-Wave	ANT	BLE
Frequency	2.4 GHz	3.1-10.6 GHz	2.4 GHz	Sub-1 GHz	2.4GHz	2.4 GHz
Maximum Data Rate	250 kbps	110 Mbps	3 Mbps	40 kbps	1 Mbps	1 Mbps
Range(meters)	100	10	10-100	30		200
Battery Life	Day-years	Multi-years	Depends on device	Multi-years	Years	Months-years
Network Topology	Star, P2P, Mesh		P2P	Mesh	Star,P2P, mesh	P2P
Power Consumption	Low	Low	Low	Low	Ultra-low	Ultra Low
Applications	Smart-meter, Smart grid Devices	Smart-meter, Smart grid Devices	Consumer Electronics	Home Automation	Health rate monitor	Health, fitness, smart devices

For implementations requiring raised up data rates. The option of

Techniques that should be utilized must be founded on the objective appliances

as every WSN application has various demands on the transmitting arrangements. The advancement of current techniques

is pushing wireless sensor networks and ubiquitous computing into a completely new area of applications.

REFERENCES

[1] Hande Alemdar and Cem Ersoy. "Wireless sensor networks for healthcare: A survey". In: *Computer networks* 54.15 (2010), pp. 2688–2710.

[2] Giuseppe Anastasi et al. "Energy conservation in wireless sensor networks: A survey". In: *Ad hoc networks* 7.3 (2009), pp. 537–568.

[3] Paolo Baronti et al. "Wireless sensor networks: A survey on the state of the art and the 802.15. 4 and ZigBee standards". In: *Computer communications* 30.7 (2007), pp. 1655–1695.

[4] Chiara Buratti et al. "An overview on wireless sensor networks technology and evolution". In: *Sensors* 9.9 (2009), pp. 6869–6896.

[5] Toly Chen and Horng-Ren Tsai. "Ubiquitous manufacturing: Current practices, challenges, and opportunities". In: *Robotics and Computer-Integrated Manufacturing* 45 (2017), pp. 126–132.

[6] Mohamed Anis Dhuieb, Florent Laroche, and Alain Bernard. "Context-awareness: a key enabler for ubiquitous access to manufacturing knowledge". In: *Procedia CIRP* 41 (2016), pp. 484–489.

[7] Helmy Fitriawan et al. "ZigBee based wireless sensor networks and performance analysis in various environments". In: *Quality in Research (QiR): International Symposium on Electrical and Computer Engineering, 2017 15th International Conference on.* IEEE. 2017, pp. 272–275.

[8] Michael Friedewald and Oliver Raabe. "Ubiquitous computing: An overview of technology impacts". In: *Telematics and Informatics* 28.2 (2011), pp. 55–65.

[9] Dayan Adionel Guimarães and Geraldo Gil Ramundo Gomes. "Introduction to Ultra-Wideband Impulse Radio". In: *Revista Telecomunicacoes* 14.1 (2012), pp. 49–61.

[10] Frank L Lewis. "Wireless sensor networks". In: *Smart environments: technologies, protocols, and applications* (2004), pp. 11–46.

[11] Zhenjiang Li et al. "Ubiquitous data collection for mobile users in wireless sensor networks". In: *INFOCOM, 2011 Proceedings IEEE.* IEEE. 2011, pp. 2246–2254.

[12] Hero Modares, Rosli Salleh, and Amirhossein Moravejosharieh. "Overview of security issues in wireless sensor networks". In: *2011 Third International Conference on Computational Intelligence, Modelling & Simulation.* IEEE. 2011, pp. 308–311.

[13] Alfredo J Perez, Sherali Zeadally, and Nafaa Jabeur. "Investigating security for ubiquitous sensor networks". In: *Procedia Computer Science* 109 (2017), pp. 737–744.

[14] Daniele Puccinelli and Martin Haenggi. "Wireless sensor networks: applications and challenges of ubiquitous sensing". In: *IEEE Circuits and systems magazine* 5.3 (2005), pp. 19–31.

[15] Vijay Raghunathan, Saurabh Ganeriwal, and Mani Srivastava. "Emerging techniques for long lived wireless sensor networks". In: *IEEE Communications Magazine* 44.4 (2006), pp. 108–114.

[16] Priyanka Rawat et al. "Wireless sensor networks: a survey on recent developments and potential synergies". In: *The Journal of supercomputing* 68.1 (2014), pp. 1–48.

[17] Eiko Yoneki and Jean Bacon. "A survey of Wireless Sensor Network technologies". In: *UCAM-CL-TR-646* (2005).

[18] Sergey Y Yurish. "Universal capacitive sensors and transducers interface". In: *Procedia Chemistry* 1.1 (2009), pp. 441–444.

Smart Wireless Distribution for Micro Grid System

Sayan Paramanik
Department of Electrical Engineering
Saroj Mohan Institute of Technology
Kolkata,India
sayanparamanik@gmail.com

Krishna Sarker
Department of Electrical Engineering
Saroj Mohan Institute of Technology
Kolkata,India
krishna80sarker@gmail.com

Biswajit Mahanty
Department of Electronics &
Communication Engineering
Saroj Mohan Institute of Technology
Kolkata,India
mahantybiswajit@gmail.com

Avijit Chakraborty
Department of Electrical Engineering
Saroj Mohan Institute of Technology
Kolkata,India
a.chakt@gmail.com

Debashis Chatterjee
Department of Electrical Engineering
Jadavpur University
Kolkata, India
debashisju@yahoo.com

S.K Goswami
Department of Electrical Engineering
Jadavpur University
Kolkata, India
skgoswami_ju@yahoo.co.in

Abstract— **Electric power is essential in modern life. Due to limitation of the present grid, wired household distribution system facing unpredictable technical challenges. Recently, micro grid & wireless power transfer (WPT) offers a brand new energy acquisition for increasing the reliability and decreasing power losses as well as overall cost of the system & power security. This present paper elaborates an overview of micro grid system & focused on WPT with mathematical calculation & hardware implementation. The proposed transmitter & receiver have a sensitivity of -91 dbm and bandwidth of the signal is 320 KHz to -320 KHz and device range is 7 meter. The proposed system has two phases for power transfer first is key exchange between the transmitter to the receiver and the second is encrypting the energy to allow for secure energy transmission. This paper also emphasizes on the latest technologies, and economical aspects.**

Keywords— PV Array, Wind, Magnetic Resonance, Wi-Fi module (ESP8266), Smart Meter.

I. INTRODUCTION

For the speedy growth of technology and consumption of smart appliances, the power demand is increasing at a regular ratio all over the world. In 2017 the overall electric power consumption in India, 1,066,268 GWh and earlier in 2007 it was 525,672 GWh [1]. The main reason behind this large deviation is the use of increasing electronics and electrical appliances. Due to unpredictable increment/ decrement of load, conductor or cable will affect the per unit cost of electricity. In this paper our emphasis on the distribution of micro grid power safely. Major issue of wired power distribution system, is wired losses, electromagnetic interference between conductors and non-linearity of the system. Sometimes wire system creates troublesome & embarrassment for us [2]. With increasing the power demand, the generation and power loss also be increase so, it will affect the environment (for non-renewable energy) and per unit cost. Therefore reducing losses, the saved electrical power can fulfil more energy demand and minimize the cost. As a significant technique, wireless power transfer (WPT) is able to realize the energy savings during a wireless method [3-4]. The Wireless transmission of electricity at first coined by the great inventor Nikola Tesla on the basis of using ground and atmosphere as an electrical conductor [5-7]. His dream was to develop a wireless distribution system but, due to lack of fund and technology at that time he was not able to complete the task. This is an underdeveloped technology from the time of Nikola Tesla. Research also be there this topic. This apparently enchanted manner will change our modern consumption patterns of the energy in numerous applications, like moveable electronic devices, implanted medical devices, integrated circuits, solar-powered satellites, electric vehicles (EVs), unmanned aerial vehicles (UAVs) and so forth. suggests that in its exceptional characteristics of flexibility, position-free and movableness, the WPT technique has been taken as a perfect technical resolution for energizing electric-driven devices inside some specific regions within the close to future, particularly for sensible home applications. The reason why WPT technologies are therefore crucial is concerning to two basic issues of powered devices that limit their universalization short battery life and the high initial cost. Taking an example of EVs [8], though several automobile manufacturers claim that their product will run over 120 KM per charge, once taking under consideration the vary anxiety, most EV drivers solely dare to run regarding 100 KM. Alternatively, with considerable amount of increasing batteries placed in EVs, the practice range can be extended beyond 400 KM however the corresponding initial price becomes excessive for the people. Rather than expecting the breakthrough of energy storage technology, a replacement energization manner, specifically the WPT technique, is attracting increasing attention to bypass the current technical restricted access of batteries. By utilizing the WPT technique, powered devices will harness wireless power from magnetic force field in air then charge their batteries cordlessly even within the moving state. This novel charging technology can solve their issues regards short battery life, battery storage or high initial price due to installation of an oversized range of batteries. Due to revolutionise the mode of electricity transfer, research works also going on this topic. Researchers proposed a coupled Magnetic Resonance, through which they transmit 60 watts power with 40% efficiency within the distance of 2m [9]. Design of weakly coupled planar WPT system is varying from sending end to receiving end with 80% efficiency for very shortest distance [10]. In some cases of WPT using magnetic resonant coupling [11, 12], the distance between load and source is not so large. For large distance the coil size is big. Since a false customer can make a dummy coil having reassemble of the main coil. So power transmission is not secure. All the stated literature survey [8-19] have some limitation that are not secured & bidirectional power & data transmission. The proposed Smart Secure WPT is the concept of the smart grid application. Our proposed system is capable to do bidirectional transfer of electrical power & data from a power source to a consuming device or vice versa, without using any conductors or physical connection with specific identity of the source as well as destination device. Also we transmit the frequency modulated electric power through magnetic coupled resonance circuit, which is key

978-1-5386-6723-1/19 $31.00 © 2019 IEEE

protected and no unauthorised person can use the electricity. The receiver circuit is pconnected with a unique Id & Password based smart meter. After login on smart meter we can get the electricity otherwise not. So there is almost rare chance of electricity theft. Also this system controls the load of receiver side and able to send corresponding data to the server.

II. Existing & Proposed Techniques

A. Existing Techniques:

Fig. 1. Wired distribution system

At present the black electricity distributed through wires as showed in Fig. 1. This system are facing some problems like Pollution, interconnecting wires are inconvenient, losses, electro-magnetic interference effect, di-electric losses & bulky the system. Fossil fuels are used to produce more than 90% electricity throughout the world & consequently emitting the greenhouse gas like CO_2, NO_2, SO_2 etc & causes climate hazards and also health issue. Due to rapid industrialisation & population growth more power is required to transfer through the transmission line. To increase demand & after a certain extend it endangers the system stability. For wired distribution the operation & controlling technique is not so reliable & advanced that it can't provide bi directional information flow. From wired distribution anyone can tap the electricity by hooking. Sometimes this type of system is dangerous for us. Ex:- for electric geyser if power cord contacted with water then there is a chance of electric shocked . Sometimes power cables are damaged by rats at home and faced power interruption. Also there is an unpredictable chance of fire hazards.

B. Proposed Techniques:

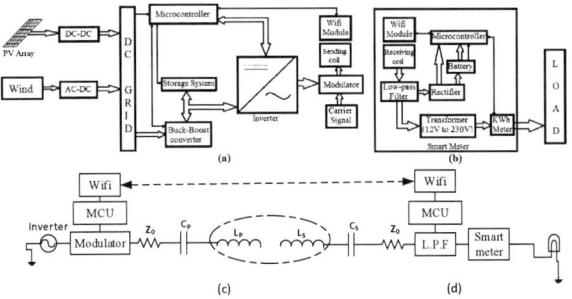

Fig. 2. Single line power diagram (SLPD) of the proposed system (a) Power generation (b) Power consumption (c) Transmitter side (d) Receiver Side

To overcome these problems we design hardware & software based smart secure wireless distribution system, which is unique id and password based anti-theft system using smart technologies like resonant coupling, Wi-Fi,

smart meter & MCU etc. in the methods of WPT the technology consists of near field & far field techniques. The far field technique is used a microwave as the carrier. The near field techniques is used for Inductive Power Transfer [20,21], compensation network [22], Magnetic Resonant coupling [23], Capacitive coupling power transfer [24]. In Fig.2 shows the SLPD of the proposed system. There is two parts for hardware development (a),(b) Power generation & consumption and (c),(d) design of transmitter side & receiver side. In our presented technology we used the wireless power transforms using Huetal's scheme [25] of encryption using nominal frequency. It provides electricity using user authentication. The key is the nominal frequency. In this method not only electricity, we also able to transfer the data from consumer to distributor or vice versa. Here at sending side C_P & L_P and receiving end C_D & L_D makes the tank circuit. Fig.3. shows the circuit design of the proposed system. Here we used Huetal's scheme for secure energy encryption. Using these techniques we generate a session key using MCU and matching the transmitter & receiver circuit through Wi-Fi. After synchronizing successfully it can send the power.

1. Generation &Transmitter side:

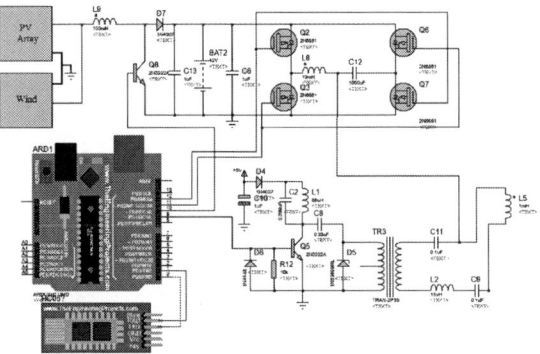

Fig. 3(a). Circuit diagram of the generation & transmitter side

In the fig. 3(a) shows the circuit design of generation & transmitter part. Here we used renewable energy to generate ecofriendly green power. Our proposed methodology described a hybrid power generation techniques to manage the power interruption. The power is fed to a DC micro grid system to run the auxiliary DC equipment. The transmitter side fed via inverter. In the transmitter section we used the frequency modulation techniques to inject high frequency with the line frequency 50Hz. The total system is an microcontroller (MCU ATMEGA328-pu) based smart system which have 2 parts of operation. In the 1st part using the MCU we generate the high frequency sine wave and merge with power line. The transformer is used to provide an electrical isolation between frequency modulator side and power line. Then the power line is fed to the inductive coil L_P to transmit the energy. After modulation the supply voltage of transmitter side is shown in (1)

$$V_S = V_M COS\left[2\pi f_C t - (M * COS(2\pi f t)]\right]$$ (1)

Where,

V_S=supply voltage, V_M= peak value of the voltage, f_C=carrier frequency, M= modulation index of the frequency modulated wave, f= supply frequency.

In the other part we use the Wi-Fi (Esp8266) technology to synchronize the receiver with transmitter. The power will be delivered when nominal frequency of both side will be same i.e, energy encryption. The frequency is inversely proportional with the capacitor so to achieve the energy encryption we have to adjust the value of capacitance and the output frequency is the key of secure connection. To avoid manual adjustment the MCU will generate the session key and send via Wi-Fi module using user authentication to deliver the power. The value of capacitor is computed using the (2)

$$\omega_t = \frac{1}{L_P C_P}$$

$$\mu\omega_0 = \frac{1}{L_P C_P}$$

$$C_P = \frac{1}{\mu\omega_0 L_P} \qquad (2)$$

Where,

ω_t= transmitting frequency, ω_0=Nominal frequency, $\mu\omega_0$=switching frequency,

L_P= transmitter coil inductance, C_P= Capacitance

At the time of power transmission the receiver unit tries to connect with transmitter side with particular user ID & password through Wi-Fi with a particular IP address. If the server connection is acceptable then sever send the nominal frequency to the receiver. This secure power cannot use without permission of the server. So at the time of apply for electricity user get an Id & Password for login with their smart meter or smart Receiver from receiver side.

2. Receiver & consumer side:

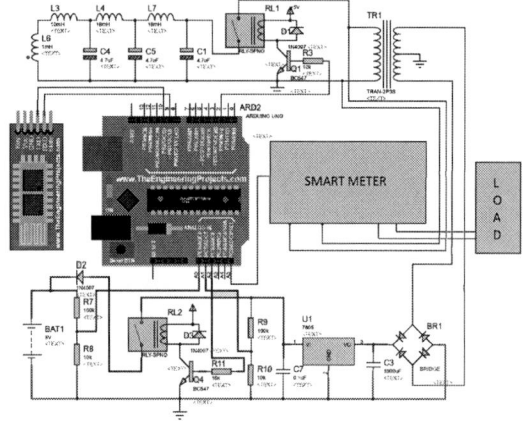

Fig. 3(b). Circuit diagram of the receiver & consumer Side

On the other hand Fig. 3(b) shows the receiver & consumer Side of the proposed smart system. The MCU and Wi-Fi module initially excited by battery. The Wi-Fi module

acknowledges the server through IP address, when the network is available. Then the smart meter wants Id & Password to connect with the transmitter network. After connection it will receives the nominal frequency key to decrypt the power. We can also connect manually by adjusting the value of tank capacitor for given frequency. After decryption, power is received through the receiver coil LD. Then the power goes through the low pass filter to get the 50Hz sine wave and eliminate the high frequency components. Hence the load is connected through smart meter for smart control. Due to smarter application, MCU also checks the permitted load capacity if it exhibits then the load is not enabled and corresponding data send to the server using Wi-Fi. Fig. 4 demonstrates the synchronizing technique of Receiving & sending coil.

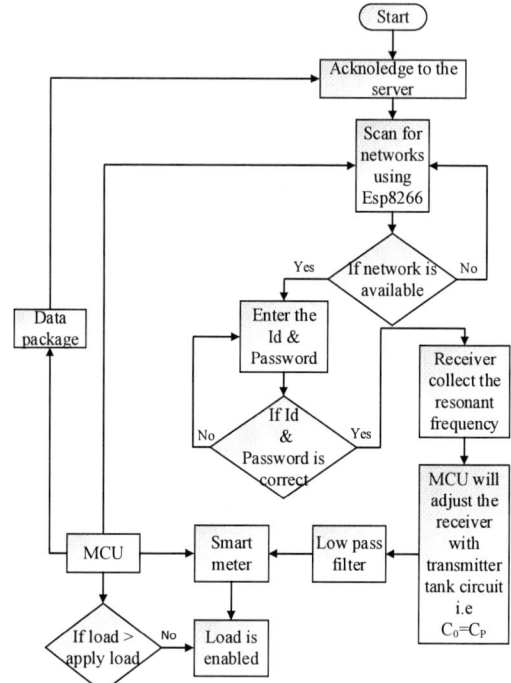

Fig. 4. Flow chart of the Power Transfer Scheme

$$C_D = \frac{1}{\mu\omega_0 L_D} \qquad (3)$$

Where,
L_D= receiver coil inductance, C_D= Capacitance
So for power transfer $C_D = C_P$

III. EQUIVALENT CIRCUIT PARAMETER CALCULATION FOR POWER TRANSFER:

The Fig. 5 shows the evaluation of the overall circuit parameter of the proposed system for magnetically coupled.

Where, V_S=supply voltage at source side, I_S=supply current at source side, Z_0=impedance of the coil, C_P=variable tank capacitor at transmitter side, L_P=transmitting coil inductance at transmitter side. M_{PS}, M_{SR}, M_{RD} =mutual inductances between 2 coils, K_{PR}, K_{PS}, K_{PD}, K_{SR}, K_{SD}, K_{RD}=cross coupling coefficients, C_S & C_R= coupling capacitors of sending & receiving side, C_D= variable tank capacitor at receiver side, Q_S=quality factor, Z_L=load impedance, I_R= current at receiving side, I_2 & I_3 =magnetizing current

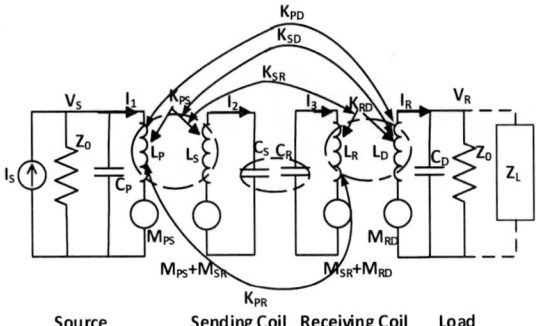

Fig. 5. Schematic diagram of circuit parameter of the proposed system

Apply KCL at source:-

$$I_S(t) = \frac{V_S(t)}{Z_0} + C_P \frac{dV_S(t)}{dt} + I_1(t)$$

Taking Laplace transform on both sides

$$I_S(S) = \frac{V_S(S)}{Z_0} + SC_P V_S(S) + I_1(S) \qquad (4)$$

Apply KCL at sending coil:-

$$0 = L_S \frac{dI_2(t)}{dt} + M_{PS} \frac{dI_1(t)}{dt} + M_{SR} \frac{dI_3(t)}{dt} + \frac{1}{C_S} \int_0^t I_2(t)$$

Taking Laplace transform on both sides

$$0 = SL_S I_2(S) + SM_{PS} I_1(S) + SM_{SR} I_3(S) + \frac{I_2(S)}{SC_S} \quad (5)$$

Apply KCL at receiving coil:-

$$0 = L_R \frac{dI_3(t)}{dt} + M_{SR} \frac{dI_2(t)}{dt} + M_{RD} \frac{dI_4(t)}{dt} + \frac{1}{C_R} \int_0^t I_3(t)$$

Taking Laplace transform on both sides

$$0 = SL_R I_3(S) + SM_{SR} I_2(S) + SM_{RD} I_R(S) + \frac{I_3(S)}{SC_R} \quad (6)$$

Apply KCL at load:-

$$0 = \frac{V_R(t)}{Z_0} + C_D \frac{dV_R(t)}{dt} + I_R(t)$$

Taking Laplace transform on both sides

$$0 = \frac{V_R(S)}{Z_0} + SC_D V_R(S) + I_R(S) \qquad (7)$$

Apply KVL at sending & load

$$V_S(S) = SL_P I_1(S) + SM_{PS} I_2(S) \qquad (8)$$

$$V_R(S) = SL_D I_R(S) + SM_{RD} I_3(S) \qquad (9)$$

Let, the core is lossless
Then the mutual impedance induced at the RECEIVING coil by the LOAD is shown in (10)

$$Z_{RM} = \frac{\omega^2 K_{RD}^2 L_R L_D}{\dfrac{1}{1/Z_0 + J\omega C_D} + J\omega L_D} \qquad (10)$$

Then the mutual impedance induced at the SENDING coil by the RECEIVING coil is shown in (11)

$$Z_{SM} = \frac{\omega^2 K_{SR}^2 L_S L_R}{Z_{RM} + 1/J\omega C_R + J\omega L_R} \qquad (11)$$

Then the mutual impedance induced at the POWER coil by the SENDING, RECEIVING & LOAD is shown in (12)

$$Z_{PM} = \frac{\omega^2 K_{PS}^2 L_P L_S}{Z_{SM} + 1/J\omega C_S + J\omega L_S}$$

$$Z_{PM} = \frac{\omega^2 K_{PS}^2 L_P L_S}{\dfrac{\omega^2 K_{SR}^2 L_S L_R}{\dfrac{\omega^2 K_{RD}^2 L_R L_D}{\dfrac{1}{1/Z_0 + J\omega C_D} + J\omega L_D} + 1/J\omega C_R + J\omega L_R} + 1/J\omega C_S + J\omega L_S}$$

$$(12)$$

Wireless energy transfer system use nearest resonance frequency to transmit the power, if the sending coils & receiving coils are identical then (13)

$$\omega_0^2 L_S C_S = \omega_0^2 L_R C_R = 1 \qquad (13)$$

The ABCD parameter of the line is:

$$\begin{bmatrix} V_S \\ V_R \end{bmatrix} = \begin{bmatrix} A & B \\ C & D \end{bmatrix} \begin{bmatrix} I_S \\ I_R \end{bmatrix}$$

Where,

$$A = \frac{V_S(S)}{V_R(S)} = \frac{SL_P I_1(S) + SM_{PS} I_2(S)}{SL_D I_R(S) + SM_{RD} I_3(S)}$$

$$C = \frac{V_R(S)}{I_S(S)} = \frac{SL_D I_R(S) + SM_{RD} I_3(S)}{\dfrac{V_S(S)}{Z_0} + SC_P V_S(S) + I_1(S)} \Omega$$

$$B = \frac{V_S(S)}{I_R(S)} = \frac{SL_P I_1(S) + SM_{PS} I_2(S)}{\dfrac{L_R}{L} I_3(S)} \Omega$$

$$D = \frac{I_S(S)}{I_R(S)} = \frac{\dfrac{V_S(S)}{Z_0} + SC_P V_S(S) + I_1(S)}{\dfrac{L_R}{L} I_3(S)}$$

The power at load side will be:

$$P = \frac{\omega_0 M_{PD}^2 Q_S I_R^2}{L_D} \qquad (14)$$

$$Z_L = \frac{V_R}{I_R}$$

$$V_R = \frac{M_{PD} I_S R_L}{L_D}$$

$$M_{PD} = \frac{V_R L_P}{I_S Z_L}$$

$$K_{PD} = \frac{1}{Q_s}\sqrt{1 - \frac{1}{4Q_s^2}}$$

Put the value of M_{PD} in (14)

$$P = \frac{\omega_0 (V_R L_P)^2 Q_S I_R^2}{(I_S Z_L)^2 L_D}$$

The power consumption is shown in (14). If the V_R L_P, I_S, Z_L, Q_S are constant then

$$P = K_1 \frac{\omega_0 I_R^2}{L_D} \tag{15}$$

Where, K_1 is the constant,

$$K_1 = \frac{(V_R L_P)^2 Q_S}{(I_S Z_L)^2}$$

Also L_D is directly proportional to I_R. Then we can considered from (15) that P α $\omega_0 I_R$.
The coil be designed as:

$$L_D = \frac{Z_L}{\omega_0 Q_s} \tag{16}$$

$$L_P = \frac{M_{PD}^2}{K_{PD} L_D}$$

$$L_P = \frac{M_{PD}^2 \omega_0 Q_s}{K_{PD} Z_L} \tag{17}$$

Depending upon load the coil diameter calculation are shown in (16, 17). From (17) we conclude that the diameter of sending coil is inversely proportional with load of receiver side.
The copper loss of the device is in (18)

$$P_{cu} = \left(-\frac{V_R(t)}{Z_0} - C_D \frac{dV_R(t)}{dt}\right)^2 Z_L \tag{18}$$

The device efficiency is shown in (19):

$$\eta = \frac{\dfrac{\omega_0 M_{PD}^2 Q_s I_R^2}{L_D}}{\dfrac{\omega_0 M_{PD}^2 Q_s I_R^2}{L_D} - \left(-\dfrac{V_R(t)}{Z_0} - C_D \dfrac{dV_R(t)}{dt}\right)^2 Z_L} \tag{19}$$

IV. DESIGN OF FILTER:

The inverter output of micro grid system contains harmonics, to eliminate those harmonics & increase the performance of the overall system, the low pass filter design as follows.
Here,

$$\Omega_P = 80\pi$$

$$\Omega_S = 220\pi$$

Let,

$$0.9 \le |H(J\Omega)| \le 1 \quad \text{For} \quad 0 \le \Omega \le 80\pi$$
$$0 \le |H(J\Omega)| \le 0.1 \qquad 220\pi \le \Omega$$

$$\frac{1}{\sqrt{1+\varepsilon^2}} = 0.9 \qquad \frac{1}{\sqrt{1+\lambda^2}} = 0.1$$

$$\varepsilon^2 = 0.2345 \quad \text{and} \qquad \lambda^2 = 99$$

$$\varepsilon = 0.484 \qquad \qquad \lambda = 9.94$$

For order of the filter defined as:

$$\frac{\log\left|\dfrac{\lambda}{\varepsilon}\right|}{\log\dfrac{\Omega_S}{\Omega_P}} \le N$$

$$\text{or,} \frac{\log\left|\dfrac{9.94}{0.484}\right|}{\log\dfrac{220\pi}{80\pi}} \le N$$

$$\text{or,} \, 2.98 \le N$$
$$\text{or,} \, N = 3$$

From Butterworth polynomial for N=3 the transfer function is shown in (20):-

$$G(S) = \frac{1}{(S+1)(S^2+S+1)} \tag{20}$$

So, cut-off frequency

$$\Omega_C = \frac{\Omega_P}{\varepsilon^{1/N}} = \frac{80\pi}{(0.484)^{1/3}} = 101.89\pi$$

$$F_C = \frac{101.89\pi}{2\pi} = 50.94 Hz$$

The transfer function G(S) for Ω_C obtained by substituting $S = \dfrac{S}{101.89\pi}$ in (20)

$$G(S) = \frac{1}{\left(\dfrac{S}{101.89\pi}+1\right)\left(\dfrac{S^2}{(101.89\pi)^2}+\dfrac{S}{101.89\pi}+1\right)}$$

$$G(S) = \frac{(101.89\pi)^3}{(S+101.89\pi)\left(S^2+S101.89\pi+(101.89\pi)^2\right)} \tag{21}$$

The transfer function of the filter is shown in (21)

V. EXPERIMENTAL RESULT

In our Wi-Fi module Data transfer rate (ESP8266) = 7 mb/s. & the hardware power transfer system having a range of 7 metre. The Fig. 6. (a) Shows the voltage output with respect to panel temperature and irradiance for maximum power with a certain temperature. Due to environment changes solar panel cannot produce constant power in every day of same year as shown in Fig. 6(b),(c).

2019 Devices for Integrated Circuit (DevIC), 23-24 March, 2019, Kalyani, India

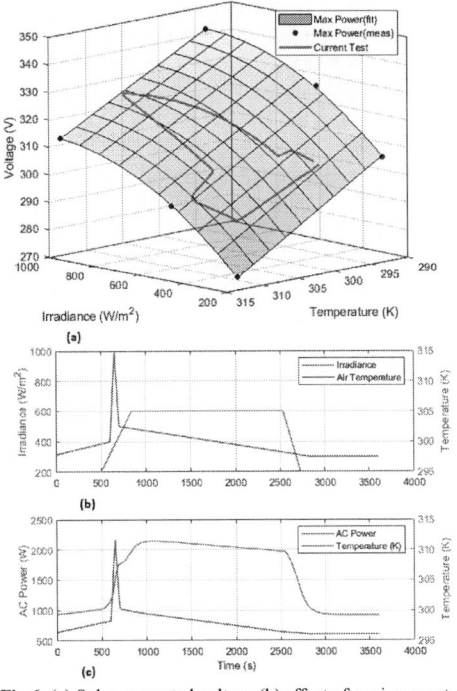

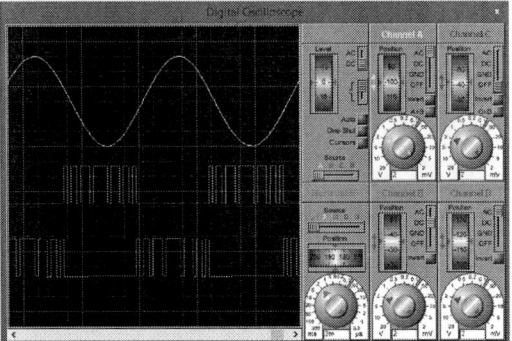

Fig.9. Power Dissipation of mosfets

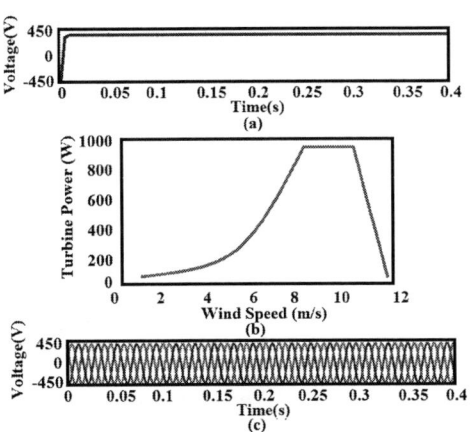

Fig.6. (a) Solar generated voltage (b) effect of environment on the panel (c) power output of panel

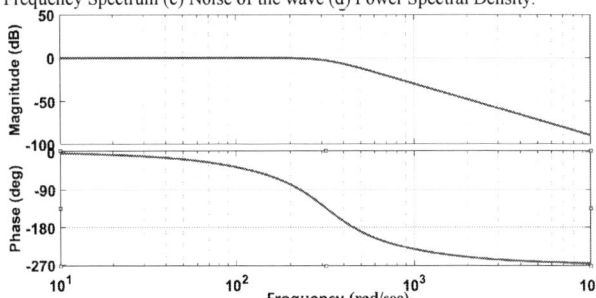

Fig.10. Gate pulse & output of the inverter

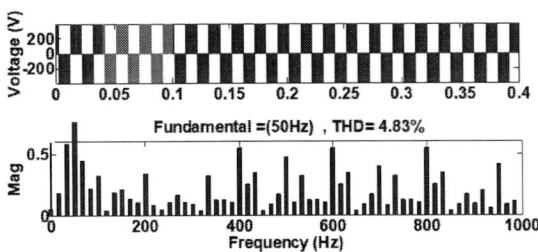

Fig.7. Simulation result of (a) PV Array (b) Wind Power Output (c) inverter output after harmonic elimination & filtering

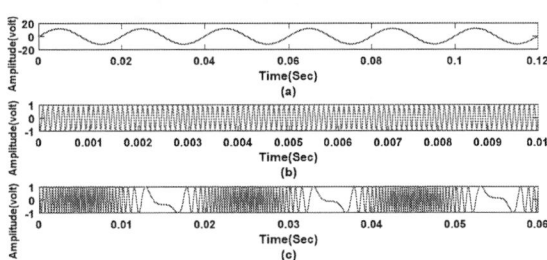

Fig.11. Simulation result of the proposed system (a) Input signal (b) Carrier wave(c) Frequency modulated

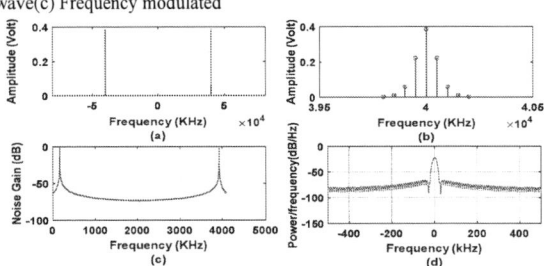

Fig.12. Simulation result of the proposed system (a) Carrier range (b) Frequency Spectrum (c) Noise of the wave (d) Power Spectral Density.

Fig.8. Matlab simulation of inverter

Fig.13. Bode plot stability of the filter

978-1-5386-6723-1/19 $31.00 © 2019 IEEE   353

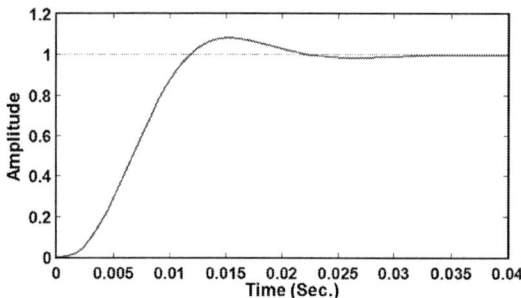

Fig.14. Step response of the filter

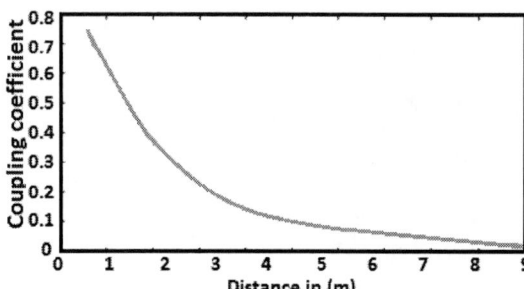

Fig.15. Coupling coefficient with variation of vertical distance

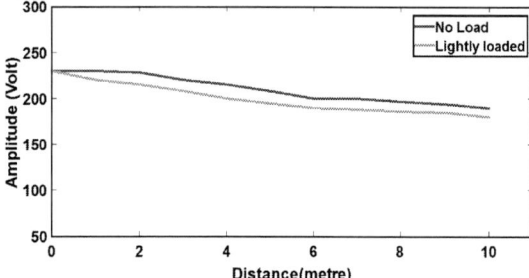

Fig.16. Voltage with variation of vertical distance

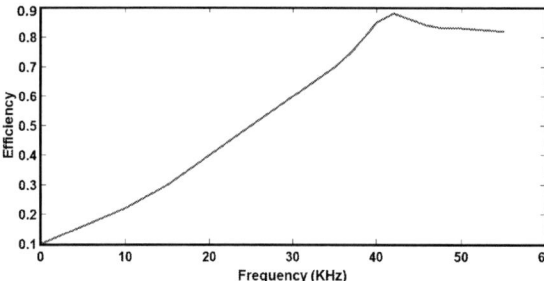

Fig.17. Relation between efficiency & Frequency

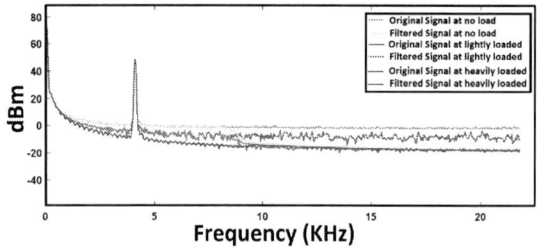

Fig.18. Matlab simulation for power transfer

Fig. 7(a) shows the Simulation result of the PV system. The plot between wind turbine powers Vs speed variation as

showed in Fig. 7(b). Fig. 7(c) shows the inverter output after harmonic elimination & filtering. Fig. 8. Shows the matlab simulation of inverter with FFT output where fundamental is in 50 Hz.Fig. 9 is displaying the power dissipation in the form of switching losses for power electronic switching. And The DSO output of hardware as shown in Fig. 10, representing the output voltage waveform of phase B of the 3 phase VSI having 180 degree control Fig. 11 shows the matlab simulation of the transmitter unit. In Fig. 11(a) shows the 50 Hz electromagnetic wave which needs to be sent. In Fig. 11(c) displays the frequency modulated output signal with carrier frequency 40 KHz, & sampling frequency 100 KHz. The total process is controlled by the MCU. Fig. 12(a) indicates the carrier range and Fig. 12(b) shows the spectrum of the modulated wave from which bandwidth is 320 KHz to -320 KHz. Fig. 12(c) indicate the noise level. Fig. 12(d) shows Welch's power spectral density. That indicates that how much power can send using this technique for a particular frequency range. Fig. 13 shows the frequency response of the designed filter using bode diagram. Its shows the stability of the filter. Fig. 14 also the step response of the filter ensuring the stability in the time domain. Fig. 15 shows the variation of coupling coefficient between the sending end and receiving end coil with respect to distance. Fig. 16 shows the receiving end voltage variation with distances.at no load & lightly loaded condition. From the mathematical calculation of efficiency (19) we get the efficiency versus frequency plot for different frequencies as showed in Fig. 17. In Fig. 18 shows matlab simulation of power transfer using different types of load at the receiving sides with filter & without filtered output. The use of power electronics devices can improve the efficiency of electric power transform.

VI. Conclusion

We developed an innovative hybrid model that can implement Smart Wireless Distribution of Micro Grid System through Wi-Fi & magnetic coupled circuit. This system solves the problems with uni-directional information flow, growing energy demand, energy wastage, reliability and security of the system. Due to smart meter the consumer cannot connect the excess load. The simulation results are strongly agreed with the theoretical results. This system will be deployed in large scale to provide better service & safety in the distribution system. There are numerous advantages with the proposed system, compare to existing systems. The system will reduce the per unit cost of energy paid by consumers and rid the landscape of conductors. The electrical energy now be distributed anywhere. As to carrier modulation losses are very small so, efficiency is high. With the help of this pioneering system, it is expected that, high electrical power can be transmitted. To the increasing device range, we can increase the area of power delivered as well as consumer of power.

VII. Economic aspects:

In terms of economic aspects, this system is more economical. This will eliminate the wire cost & space of the system. Also this model is an anti-theft model so there is no chance of electrical hooking. The system also decreases the distribution losses. Owing to this, unit cost will be reduced, and every consumer will be benefited.

REFERENCES

[1] Electricity Demand in India,
https://en.wikipedia.org/wiki/Electricity_sector_in_India#Demand

[2] V. Saltas, F. Vallianatos, D. Triantis, Dielectric properties of nonswelling bentonite: The effect of temperature and water saturation, Journal of Non-Crystalline Solids, Volume 354, Issues 52-54, 15 December 2008, Pages 5533-5541.

[3] C. T. Rim and C. Mi, Wireless Power Transfer for Electric Vehicles and Mobile Devices. Wiley-IEEE Press, 2017

[4] S. Y. R. Hui, W. Zhong, and C. K. Lee, "A critical review of recent progress in mid-range wireless power transfer," IEEE Trans. Power Electron., vol. 29, no. 9, pp. 4500-4511, 2014.

[5] History of wireless system,
https://en.wikipedia.org/wiki/World_Wireless_System

[6] N. Tesla, Experiments with Alternate Currents of Very High Frequency and Their Application to Methods of Artificial Illumination. Wildside Press, 2006.

[7] Nikola Tesla's idea for Wireless transmission
http://www.teslasociety.com/tesla_tower.htm

[8] K. T. Chau. Electric Vehicle Machines and Drives – Design, Analysis and Application. Wiley-IEEE Press, 2015.

[9] A.Kurs, A.Karalis, R.Moffatt, J.D.Joannopoulos, P.Fisher, M. Soljacic, "Wireless power transfer via strongly coupled magnetic resonances," Science, vol. 317, 2007, pp. 83-86.

[10] Z.N.Low, R.A.Chinga, R.Tseng, J.Lin, "Design and test of a high-power high-efficiency loosely coupled planar wireless power transfer system," IEEE Transaction on Industrial Electronics, vol. 56, No. 5, 2009, pp. 1801–1812.

[11] S.Cheon, Y.H.Kim, S.Y.Kang, M.L.Lee, J.M.Lee, T.Zyung, "Circuit-Model-Based Analysis of a Wireless Energy-Transfer System Via Coupled Magnetic Resonances," IEEE Trans.On Industrial Electronics,vol. 58, 2010, pp. 2906 – 2914.

[12] S.K Rahman, O.Ahmed, Md. S.Islam, A. H. M. R. Awal, Md. S.Islam, "Design and construction of wireless power transfer system using magnetic resonant coupling," American journal of Electromagnetics & Applications, vol. 2, 2014, pp.11-15 .

[13] S.Thumaty, A.Pham, H.V.Wyk, "Development of a low-IF Receiver and a Fixed Wireless Utility Network," IEEE International Conference on Microwave symposium Digest, USA 2002.

[14] R.G.C.Fuentes, M.Y.Naderiy, S.Basagniy, K.R.Chowdhury, A.C. Aparicio, E.Alarco´n, "An All-digital Receiver for Low Power," Low Bit-rate Applications using Simultaneous Wireless Information and Power Transmission, Canada 2016.

[15] P.Nintanavongsa, U.Muncuk, D.Lewis, K.R.Chowdhury, "Design optimization and implementation for RF energy harvesting circuits," IEEE Journal on Emerging and Selected Topics in Circuits and Systems, vol. 2, No. 1, 2012, pp. 24-33.

[16] Y.Liu, G.Liu, P.M.Asbeck, "High-Order Modulation Transmission Through Frequency Quadrupler Using Digital Predistortion," IEEE Transaction on Microwave Theory and Techniques, vol. 64, 2016 pp.1896 – 1910.

[17] M.Shidujaman, H.Samani, Md.Arif, "Wireless Power Transmission Trends," IEEE International Conference on Informatics, Electronics & Vision, Bangladesh 2014.

[18] Y.Park, Y.Kim, H.Ahn, Y.Lim, "The Study of the Controlling Resonant Wireless Power Transfer Using Bluetooth Communication," International Conference on Computational Science and Its Applications, Korea, 2014, pp.779-789.

[19] A. Ranaweera, T. P. Duong, B. S. Lee, and J. K. Lee, "Experimental investigation of 3D metamaterial for mid-range wireless power transfer," in Proc. 2014 IEEE Wireless Power Transfer Conference, Jeju, South Korea, 2014, pp. 92-95.

[20] G. A. Covic and J. T. Boys, "Inductive power transfer," Proc. IEEE, vol. 101, no. 6, pp. 1276-1289, 2013.

[21] I. Mayordomo, T. Drager, P. Spies, J. Bernhard, and A. Pflaum, "An overview of technical challenges and advances of inductive wireless power transmission," Proc. IEEE, vol. 101, no. 6, pp. 1302-1311, 2013.

[22] Z. Zhang, "Chapter 11 Energy Cryptography for Wireless Charging of Electric Vehicles" in Energy Systems for Electric and Hybrid Vehicles. The Institution of Engineering and Technology, 2016, pp. 319-417

[23] J. Zhang, X. Yuan, C. Wang, and Y. He, "Comparative analysis of two-coil and three-coil structures for wireless power transfer," IEEE Trans. Power Electron., vol. 32, no. 1, pp. 341-352, 2017.

[24] C. Liu, A. P. Hu, and N. C. Nair, "Modelling and analysis of a capacitively coupled contactless power transfer system," IET Power Electronics, vol. 4, no. 7, pp. 808-815, 2011.

[25] H. Xiong, Q. Wu, and Z. Chen, "Toward pairing-free certificateless authenticated key exchanges," in Proceedings of the 14th International Conference on Information Security, ser. ISC'11. Berlin, Heidelberg: Springer-Verlag, 2011, pp. 79–94.

2019 Devices for Integrated Circuit (DevIC), 23-24 March, 2019, Kalyani, India

Smart Grid Power Quality Improvement Using Modified UPQC

Sayan Paramanik
Department of Electrical Engineering
Saroj Mohan Institute of Technology
Kolkata,India
sayanparamanik@gmail.com

Krishna Sarker
Department of Electrical Engineering
Saroj Mohan Institute of Technology
Kolkata, India
krishna80sarker@gmail.com

Debashis Chatterjee
Department of Electrical Engineering
Jadavpur University
Kolkata, India
debashisju@yahoo.com

S.K Goswami
Department of Electrical Engineering
Jadavpur University
Kolkata, India
skgoswami_ju@yahoo.co.in

Abstract— The Smart Grid (SG) system typically deals with different issues involving security and Power Quality (PQ) improvement. With frequent usage of power electronic devices and nonlinear load, harmonics are inserted into the system. The well-known Flexible AC Transmission System (FACTS) devices like Unified Power Quality Conditioners (UPQC) are usually employed to resolve the issues related to voltage sag, swell, flicker, PQ, and neutral current reduction of distribution systems. An UPQC itself inserts harmonics into the system that affects the system stability for sensitive loads. This paper describes biogeography based optimization (BBO) with harmonics elimination techniques for modified UPQC connected with SG. Lower order harmonics are eliminated by proper selection of switching angles and at the same time the higher order harmonics are suppressed by injecting same order harmonics with equal magnitude but opposite in phase from the other converter. The excitation of Modified UPQC converters are obtained from PV (Photo-Voltaic) panel. The firing angles of series-shunt converter are obtained in real-time from the already stored angles in the microcontroller memory.

Keywords—UPQC, PV panel, biogeography based optimization; Quasi-sine (Q-S) wave switching, Power Quality.

I. INTRODUCTION

To overcome the problems of present grid technology, SG [1] with distributed generators (DGs) are introduced with smarter technologies. In last few decades for the massive use of power electronics devices and growth of non-linear loads injects non-sinusoidal component into the grid, thus generates the PQ degradation [2]. The voltage fluctuation [3] or system harmonics make a serious issue in power system. Numerous no of devices[4] such that Active Low Pass Filter [5], STATCOM [6], SVC [7], capacitor bank [8] etc are connected to the grid for the solution of different issues like reactive power, PQ management, voltage sag, swell, flicker, harmonics, and negative sequence current. With development of research methodology UPQC [9] is efficiently able to solve almost all the issues. For the sensitive load and manufacturing industry, maintaining PQ is the most important criteria for better process output otherwise manufacturing defects takes place. UPQC itself a power electronics switching devices with integration of series converter and shunt converter that injects harmonics or distortion in the grid which effect the system stability.

This current paper a mathematically modified switching technique for UPQC with harmonic optimization using BBO [10] algorithm is developed. Using this technique, an UPQC can effectively optimize any harmonic order and PQ issue of the SG.

II. PROPOSED TECHNIQUES

The UPQC is the grouping of a DVR and STATCOM which perform series, shunt compensating and phase shifting at the same time as in Fig.1. This is capable to control the PQ with leading and lagging reactive and real power flow through a particular route with enhance the system stability. The salient options of UPQC device are its multiple management functions, like voltage management, transient stability improvement & damping oscillation. Voltage sag & swell compensation is important for secure system operation. This is a multivariable versatile AC transmission systems controller. In this work UPQC is excited by PV array with DC link Capacitor. In the SG using PMU and flow meter, measures the phase and related data, from this data of analysis centralized controller determine the harmonics amplitude and phase. At fault condition UPQC mitigate the fault with simultaneous or individual operation of series-shunt converters. At the presence of harmonics this two converter works simultaneously and optimized the harmonics injected by the UPQC, the centralized controller calculate the switching angle for same magnitude with opposite phase harmonics and inject into the system. These two component eliminate each other and make the grid approximately harmonics free with mitigate the active and reactive power problem. The calculation for different harmonics is discussed below.

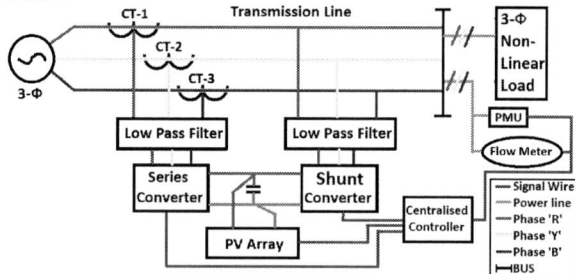

Fig.1. Schematic diagram of UPQC connected with SG

III. PROPOSED HARMONICS OPTIMIZATION TECHNIQUES

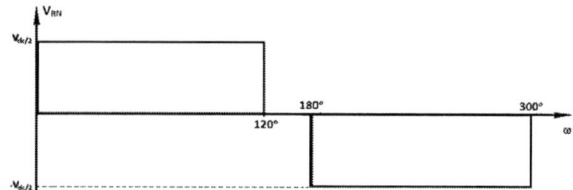

Fig.2. Quasi-sine waveforms for 120° switching

978-1-5386-6723-1/19 $31.00 © 2019 IEEE   356

For Q-S wave switching with 120° mode of operation the phase voltage V_{Rn} is shown in Fig. 2. The Fourier series of V_{Rn} is expressed in (1),

$$V_{Rn} = \frac{A_0}{2} + \sum_{n=1}^{\infty} a_n \cos(n\omega t) + b_n \sin(n\omega t) \qquad (1)$$

The waveform of Fig.2 is the odd quarter wave symmetry so even harmonics are eliminated from the system.

Here, DC part $A_0 = 0$, $a_n = 0 \ \forall$ even n

And,

$$b_n = \frac{4}{\pi} \int_0^{\pi/2} \frac{V_{dc}}{2} \sin(n\omega t) d(\omega t) \quad \forall \text{ odd n} \qquad (2)$$

$$b_n = \frac{4}{\pi} \int_{\pi/6}^{\pi/2} \frac{V_{dc}}{2} \sin(n\omega t) d(\omega t)$$

$$b_n = -\frac{2V_{dc}}{n\pi} \left[\cos(n\omega t) \Big|_{\pi/6}^{\pi/2} \right]$$

$$b_n = -\frac{2V_{dc}}{n\pi} \left[\cos\left(\frac{n\pi}{2}\right) - \cos\left(\frac{n\pi}{6}\right) \right]$$

$$\text{or } b_n = \frac{2V_{dc}}{n\pi} \left[\cos\left(\frac{n\pi}{6}\right) - \cos\left(\frac{n\pi}{2}\right) \right]$$

$$\text{or } b_n = \frac{2V_{dc}}{n\pi} \left[2.\sin\left(\frac{n\pi}{3}\right) \sin\left(\frac{n\pi}{6}\right) \right]$$

$$\text{or } b_n = \frac{4V_{dc}}{n\pi} \sin\left(\frac{n\pi}{3}\right) \sin\left(\frac{n\pi}{6}\right) \qquad (3)$$

So, the harmonics amplitude (C_n) and phase angle Φ_n expressed as (4), (5)

$$C_n = \sqrt{b_n^2}$$

$$\text{or } C_n = \frac{4V_{dc}}{n\pi} \sin\left(\frac{n\pi}{3}\right) \sin\left(\frac{n\pi}{6}\right) \qquad (4)$$

$$\varphi_n = \tan^{-1}\left(\frac{a_n}{b_n}\right)$$

$$\varphi_n = 0 \qquad (5)$$

From (4) and (5) the 3n harmonics are zero so only n=1, 5,7,11, 13, 17, 19, 23... i,e $n = 6k \pm 1$ are present into the system.

The Proposed Switching:

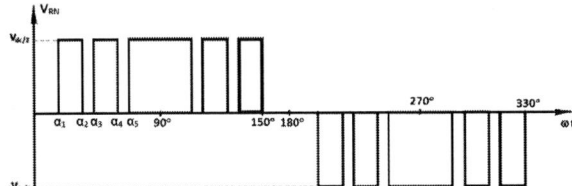

Fig.3. 3-Φ six step Q-S waveforms for 120° conduction with five different switching angles

3-Φ six-step Q-S waveforms of 120° conduction with five different switching angles is displayed in Fig. 3. In 3-Φ system first intersection point is at π/6. For phase 'A', V_{Rn} is the respective line voltage.

$$A_0 = 0, \ a_n = 0 \ \forall \text{ even n}$$

$$b_n = \frac{1}{\pi} \int_0^{2\pi} V_{Rn} \sin(n\omega t) d(\omega t)$$

$$b_n = -\frac{2V_{dc}}{n\pi} \sum_{k=1}^{m} (-1)^k \cos(n\gamma_k) \qquad (6)$$

For 5 switching let m=5 and calculate upto n=13

$$\begin{bmatrix} \cos\gamma_1 - \cos\gamma_2 + \cos\gamma_3 - \cos\gamma_4 + \cos\gamma_5 = M \\ \cos 5\gamma_1 - \cos 5\gamma_2 + \cos 5\gamma_3 - \cos 5\gamma_4 + \cos 5\gamma_5 = \varepsilon_5 \\ \cos 7\gamma_1 - \cos 7\gamma_2 + \cos 7\gamma_3 - \cos 7\gamma_4 + \cos 7\gamma_5 = \varepsilon_7 \\ \cos 11\gamma_1 - \cos 11\gamma_2 + \cos 11\gamma_3 - \cos 11\gamma_4 + \cos 11\gamma_5 = \varepsilon_{11} \\ \cos 13\gamma_1 - \cos 13\gamma_2 + \cos 13\gamma_3 - \cos 13\gamma_4 + \cos 13\gamma_5 = \varepsilon_{13} \end{bmatrix} \quad (7)$$

The $n = 6k \pm 1$ harmonics would be eliminated if that coefficient of b_n i,e $b_5 = b_7 = b_{11} = b_{13} = 0$. These can be solved using iterative method. The soft computed switching angles of converters reduces the %THD of the SG.

Conventional the output voltage $V_R(t)$ expressed in (8)

$$V_R(t) = \sum_{n=1}^{\infty} a_n \cos(n\gamma_n) + b_n \sin(n\gamma_n) \qquad (8)$$

As a result of quarter wave symmetry, even magnitude of harmonics are eliminated and for odd magnitude of harmonics switching angle range is

$$0 < \gamma_1 < \gamma_2 < \gamma_3 \ldots\ldots < \gamma_m < \pi/2$$

Eq. 6 can be obtained any harmonics and Eq. 7 calculate the suitable switching angles. The THD percentage of output phase can figured using (9) where $n = 6k \pm 1$ and k=1,2,3,4…….

$$\%THD = \left[\frac{1}{b_1^2} \sum_{n=5}^{\infty} (b_n)^2 \right]^{1/2} \times 100 \qquad (9)$$

The impartial function $F(\gamma)$ optimized the harmonics by optimizing (10).

$$F(\gamma) = F(\gamma_1, \gamma_2, \gamma_3, \ldots, \gamma_m) \qquad (10)$$

$$0 < \gamma_1 < \gamma_2 < \gamma_3 \ldots\ldots < \gamma_m < \pi/2 \qquad (11)$$

$$b_1 = M, \ b_5 \le \varepsilon_5, \ b_7 \le \varepsilon_7 \text{ and } b_n \le \varepsilon_n \qquad (12)$$

Where, b_1 is the fundamental amplitude, $\varepsilon_5, \varepsilon_7 \ldots \varepsilon_n$ are the individual harmonics permissible amplitude.

Switching angles remain arbitrarily generated to satisfy (11). Best solution within this from γ_1 to γ_m for a quarter wave are taken as the best angle. In each iteration, new search points are generated. This searching points and the best solutions is found by using BBO. Consider the Ps is the possibility that habitat contains S species. Ps changes with time t to T+ΔT as follows:

$$P_s(T+\Delta T)=P_s(T)(1-\lambda_s\Delta T-\mu_s\Delta T)+P_{s-1}\lambda_{s-1}\Delta T+P_{s+1}\mu_{s+1}\Delta T \quad (13)$$

Where λ_s is immigration and μ_s is emigration rates having S species in habitat. ΔT is the possibility of other immigration and emigration can ignored. Taking the ΔT tents to 0, of Eq. (13). Designed every habitat, bring up-to-date the possibility of species computation using Eq. (14) and compute every habitat suitability index (HSI). HSI also satisfying the constraints of suitability index variable (SIV).

$$\begin{bmatrix} \dot{P}_s=-(\lambda_s+\mu_s)P_s+P_{s+1}\mu_{s+1}, \text{for } S=0 \\ \dot{P}_s=-(\lambda_s+\mu_s)P_s+P_{s-1}\lambda_{s-1}+P_{s+1}\mu_{s+1}, \text{for } 1\le S\le S_{max}-1 \\ \dot{P}_s=-(\lambda_s+\mu_s)P_s+P_{s-1}\lambda_{s-1}+P_{s+1}\mu_{s+1}, \text{for } S=S_{max} \end{bmatrix} \quad (14)$$

The (14) can be written in single matrix (15)

$$\dot{P}=AP \quad (15)$$

From [11]

$$\mu_k=\frac{EK}{n} \text{ And } \lambda_k=I\left(1-\frac{k}{n}\right) \quad (16)$$

Where, λ_k is emigration and μ_k is immigration rate with k no of species. The maximum rate of possible immigration rate is I and the emigration rate is E. Now, for special case E = I, we have,

$$\lambda_k+\mu_k=E \quad (17)$$

The steady state value of the number of each species is given by (18)

$$P(\infty)=\frac{V}{\sum_{i=1}^{n+1}V_i} \quad (18)$$

Where, V is the eigenvector.
m is the mutation rate which inversely proportional with the solution Possibility and is given in (19)

$$m(s)=m_{max}\left(\frac{1-P_s}{P_{max}}\right) \quad (19)$$

Where, m_{max} is user defined maximum mutation rate. This (19) makes HSI solutions.
Application of BBO based SHE-PWM problem will be précised using (20). This nonlinear equations can be represented as

$$F_j(\gamma_1,\gamma_2,\gamma_3,...,\gamma_m)=0, where\ j=1,2,3,..,m \quad (20)$$

The (20) is obtained by equating (6). HSI specifies the superiority of solution set. For the considered problem, HSI represents with THD by the solution set (20).

$$F(\gamma)=0 \quad (21)$$

$$F=\left[F_1,F_2,F_3,...,F_m\right]^T \quad (22)$$

$$\gamma=\left[\gamma_1,\gamma_2,\gamma_3,...,\gamma_m\right]^T \quad (23)$$

Eq. (21) can solved using BBO technique, where the nonlinear equations give an approximate solution. The steps are as follows.
The switching angle matrix,

$$\gamma^j=\left[\gamma_1^j,\gamma_2^j\gamma_3^j,...\gamma_m^j\right]^T \quad (24)$$

The nonlinear system matrix,

$$F^j=\begin{bmatrix} \cos\gamma_1^j-\cos\gamma_2^j+\cos\gamma_3^j-\cos\gamma_4^j+\cos\gamma_5^j \\ \cos5\gamma_1^j-\cos5\gamma_2^j+\cos5\gamma_3^j-\cos5\gamma_4^j+\cos5\gamma_5^j \\ \cos7\gamma_1^j-\cos7\gamma_2^j+\cos7\gamma_3^j-\cos7\gamma_4^j+\cos7\gamma_5^j \\ \cos11\gamma_1^j-\cos11\gamma_2^j+\cos11\gamma_3^j-\cos11\gamma_4^j+\cos11\gamma_5^j \\ \cos13\gamma_1^j-\cos13\gamma_2^j+\cos13\gamma_3^j-\cos13\gamma_4^j+13\cos\gamma_5^j \end{bmatrix} \quad (25)$$

$$\left(\frac{dF^j}{d\alpha}\right)=\begin{bmatrix} -\sin\gamma_1^j+\sin\gamma_2^j-\sin\gamma_3^j+\sin\gamma_4^j-\sin\gamma_5^j \\ -\sin5\gamma_1^j+\sin5\gamma_2^j-\sin5\gamma_3^j+\sin5\gamma_4^j-\sin5\gamma_5^j \\ -\sin7\gamma_1^j+\sin7\gamma_2^j-\sin7\gamma_3^j+\sin7\gamma_4^j-\sin7\gamma_5^j \\ -\sin11\gamma_1^j+\sin11\gamma_2^j-\sin11\gamma_3^j+\sin11\gamma_4^j-\sin11\gamma_5^j \\ -\sin13\gamma_1^j+\sin13\gamma_2^j-\sin13\gamma_3^j+\sin13\gamma_4^j-\sin13\gamma_5^j \end{bmatrix} \quad (26)$$

For example let a harmonics matrix of coefficient b_n (27)

$$b_n=\begin{bmatrix} M \\ 0 \\ 0 \\ 0 \\ 0 \end{bmatrix} where, n=5 \quad (27)$$

Step: 1
Find a set of value for switching angles γ. BBO representation is in (28)

$$\begin{bmatrix} \gamma^0 \\ \gamma^1 \\ . \\ . \\ . \\ \gamma^4 \end{bmatrix}=\begin{bmatrix} \gamma_1^0,\gamma_2^0,\gamma_3^0,\gamma_4^0,......,\gamma_m^0 \\ \gamma_1^1,\gamma_2^1,\gamma_3^1,\gamma_4^1,......,\gamma_m^1 \\ . \\ . \\ . \\ \gamma_1^4,\gamma_2^4,\gamma_3^4,\gamma_4^4,......,\gamma_m^4 \end{bmatrix} \quad (28)$$

Step: 2
The values of

$$\begin{bmatrix} F(\gamma^0) \\ F(\gamma^1) \\ . \\ . \\ . \\ F(\gamma^4) \end{bmatrix}=\begin{bmatrix} F^0 \\ F^1 \\ . \\ . \\ . \\ F^4 \end{bmatrix} \quad (29)$$

Step: 3

$$F^0+\left[\frac{\partial F^0}{\partial\gamma}\right]d\gamma^0=0 \quad (28)$$

Determine $d\gamma^0=\left[d\gamma_1^0,d\gamma_2^0,d\gamma_3^0,d\gamma_4^0,......,d\gamma_m^0\right]^T$ at γ^0 then solve (28) for $d\gamma^0$

Step: 4
Repeat step 1 to step 3

$$\gamma^{j+1}=\gamma^j+d\gamma^j \quad (29)$$

The above iterative method is repeated continual (21) and fulfilled the desired accuracy. If previous method converges, then it gives a solution of (21). Otherwise make a new initial guess. The best solution must satisfy (11). The total process is illustrated in Flow chart Fig.4.

2019 Devices for Integrated Circuit (DevIC), 23-24 March, 2019, Kalyani, India

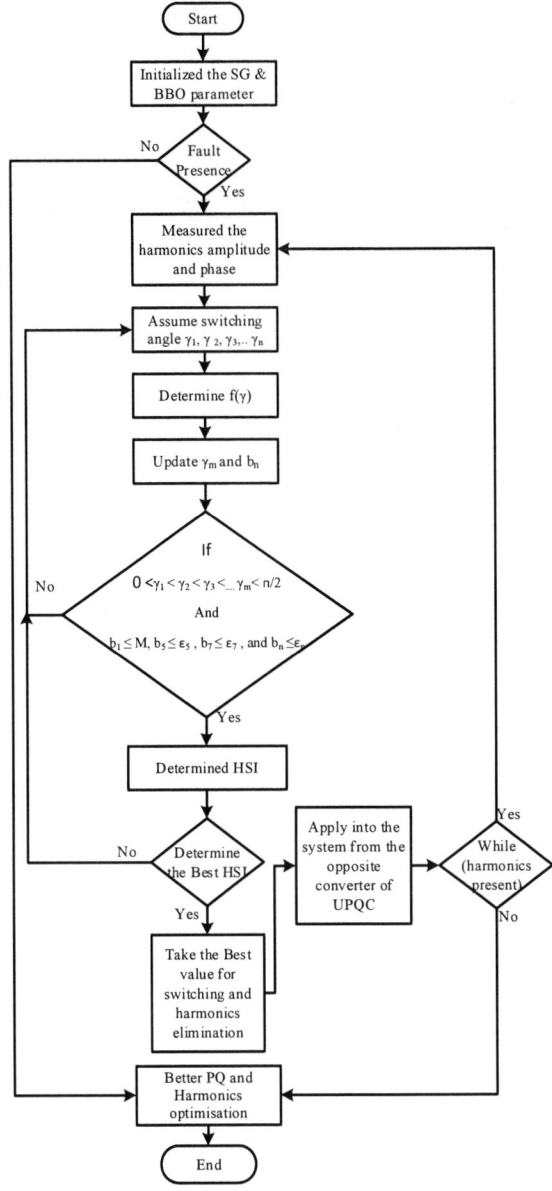

Fig.4. Flow chart of the total system operation connected with SG

5	0	23.0699
7	0	30.3626
11	0	45.6847
13	0	48.6366

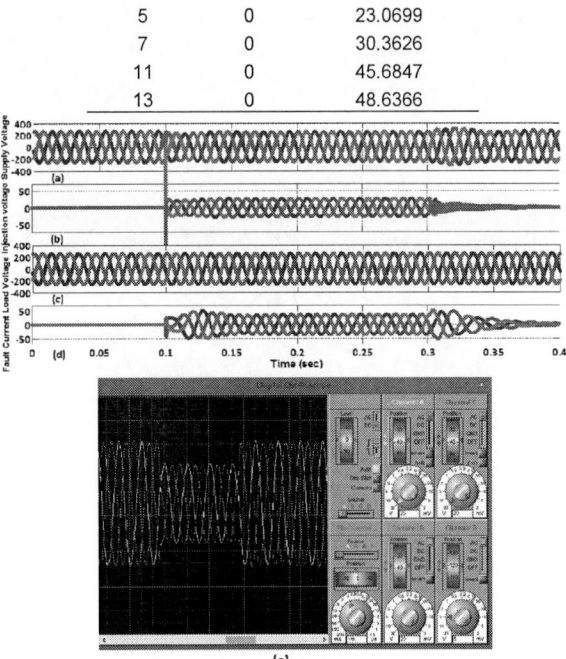

Fig.5. Simulation and DSO output for (a) nominal voltage of SG (b) inserted series voltage (c) Load voltage (d) Fault current (e) DSO output at sag condition

Fig.5 shows the simulation and Digital Storage Oscilloscope (DSO) output at sag condition with duration is 0.1 sec. to 0.3 sec. When the series converter set at M= 0.67, the calculated switching values are 28.0295, 33.2034, 44.5603, 51.9774, and 58.1643. For the calculated switching value upto 13th order harmonics are completely eliminated with the higher order harmonics are suppressed. Similarly, at M=0.73 and 0.87 the 120° optimised switching angles are shown in Table. 1.

In Fig. 6 illustrated the DSO output of the Low-High step up converter when the DC linkage voltage is not its desired value.

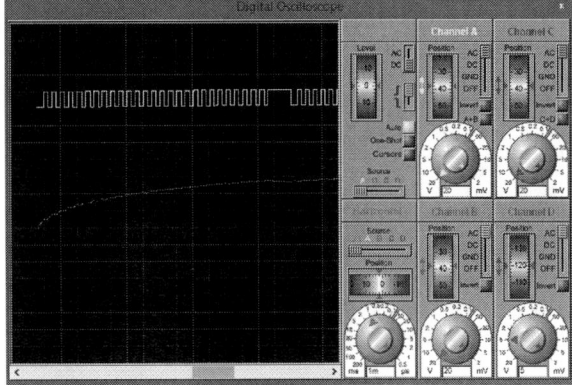

Fig.6. DSO output of Low-High step up converter

In Fig.7. demonstrated the overall hardware setup of the proposed system with gate pulses and FFT of SG.

Fig.8 shows the matlab simulation output for reactive and active power flow for series and shunt converter applying modified UPQC technique under the condition of voltage sag and swell.

IV. RESULTS AND DISCUSSION

Table 1. Switching angle at 120° mode with different M

Harmonic order	M	Switching angles (Deg.)
1	0.67	28.0295
5	0	33.2034
7	0	44.5603
11	0	51.9774
13	0	58.1643
1	0.73	22.6782
5	0	28.1435
7	0	37.9299
11	0	47.2386
13	0	54.6010
1	0.87	15.2133

978-1-5386-6723-1/19 $31.00 © 2019 IEEE

2019 Devices for Integrated Circuit (DevIC), 23-24 March, 2019, Kalyani, India

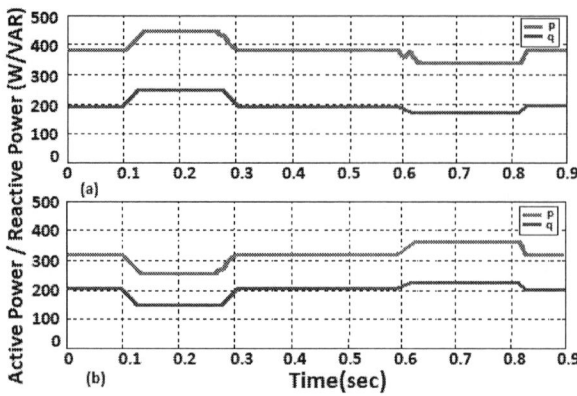

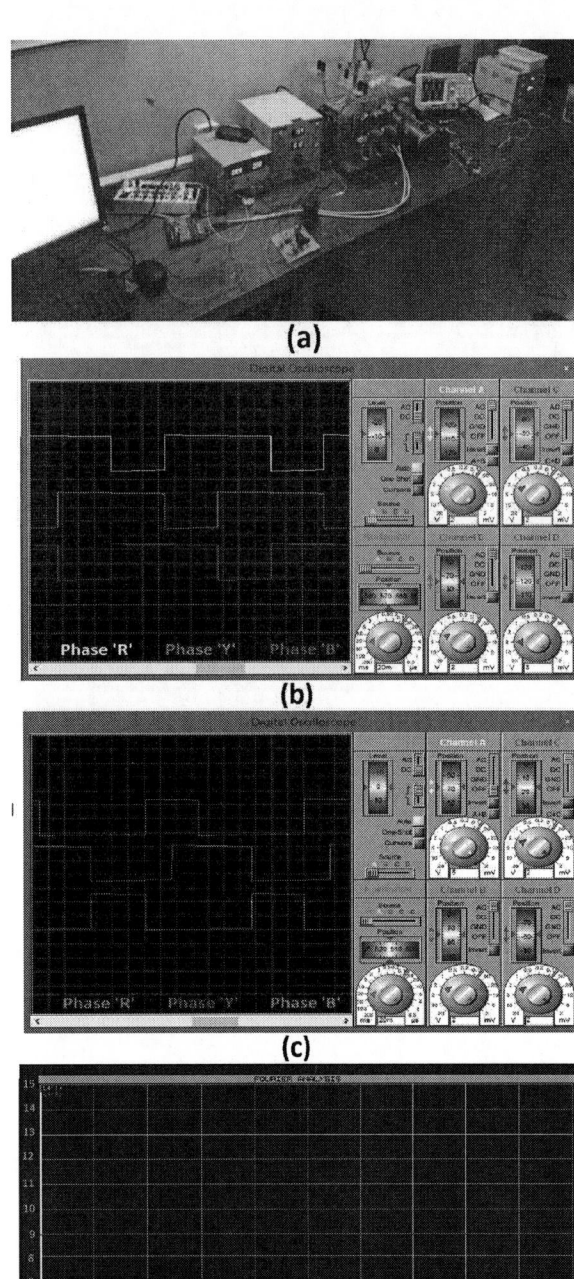

Fig.8. Matlab simulation output of Active and Reactive power at non linear R-L-C load (a) series converter (b) shunt converter.

V. CONCLUSION

The suggested technique effectively eliminates the tergated lower order harmonics at different modulation indeices by proper selection of switching angles and same time the higher order harmonics are suppressed. Also this innovative technique solve the smart grid PQ issues like sag, swell, flicker, active , reactive power and increase the overall grid performance. The proposed technique can be efficiently implemented for online application through commercially available "Raspberry pi 3", advanced microcontroller.

REFERENCES

[1] Smart Grid
https://en.wikipedia.org/wiki/Smart_grid

[2] Y.-J. Shin ; E.J. Powers ; M. Grady ; A. Arapostathis, "Power quality indices for transient disturbances", IEEE Transactions on Power Delivery, Vol. 21 , no. 1, pp. 253 – 261, 2006.

[3] H. Akagi, "New Trends in Active Filter for power conditioning", IEEE Trans. Ind. Application, vol. 32, no. 6, pp. 1312-1322, 1996.

[4] L.G Gyugyi, "Power electronics in electric utilities, static VAR compensators", Proc. IEEE, 876(4), pp. 483-494, 1988.

[5] S. Devassy ; B. Singh, "Control of solar energy integrated active power filter in weak grid system", 7th International Conference on Power Systems (ICPS), India, 2017.

[6] R. K. Varma ; R. Salehi, "SSR Mitigation With a New Control of PV Solar Farm as STATCOM (PV-STATCOM)", IEEE Transactions on Sustainable Energy, Vol. 8 , no. 4, pp. 1473 – 1483, 2017.

[7] L. Wang ; K.W. Lao ; C. S. Lam ; M.C. Wong, "Delta-connected static var compensator (SVC) based hybrid active power filter (SVC-HAPF) and its control method", 43rd Annual Conference of the IEEE Industrial Electronics Society, China, 2017.

[8] S. Samineni ; C. Labuschagne ; J. Pope, "Principles of shunt capacitor bank application and protection", 63rd Annual Conference for Protective Relay Engineers, U.S.A , 2010.

[9] S. L. Nikam ; K. K. Sandeep "Analysis of modified three-phase four-wire UPQC design", Third International Conference on Science Technology Engineering & Management, India,2017.

[10] R. Brindha ; R. Kavitha, "Harmonic optimization in seven level inverter employing hybrid BBO/MADS algorithm", International Conference on Innovations in Information, Embedded and Communication Systems (ICIIECS), India, 2015.

[11] S. Dan, "Biogeography based optimizations", IEEE Trans Evol Comput, vol. 12, no. 6, pp. 702-713, 2008.

Fig.7. DSO output for (a) hardware setup of the proposed system (b) gate pulse for upper group of thyristor (c) gate pulse for lower group of thyristor (d) FFT output of SG upto 100KHz

978-1-5386-6723-1/19 $31.00 © 2019 IEEE   360

2019 Devices for Integrated Circuit (DevIC), 23-24 March, 2019, Kalyani, India

Survey of Smart Grid Network Using Drone & PTZ Camera

Sayan Paramanik
Department of Electrical Engineering
Saroj Mohan Institute of Technology
Kolkata,India
sayanparamanik@gmail.com

Partha Sarathi Sarkar
Department of Electrical Engineering
Saroj Mohan Institute of Technology
Kolkata,India
parthasarkar048@gmail.com

Koustav Kumar Mondol
Department of Elctronics and
Communication Engineering
Saroj Mohan Institute of Technology
Kolkata,India
koustav_mondal@yahoo.in

Avijit Chakraborty
Department of Electrical Engineering
Saroj Mohan Institute of Technology
Kolkata, India
a.chakt@gmail.com

Sajib Chakraborty
System Engineer
Saroj Mohan Institute of Technology
Kolkata,India
sajib02@yahoo.co.in

Krishna Sarker
Department of Electrical Engineering
Saroj Mohan Institute of Technology
Kolkata,India
krishna80sarker@gmail.com

Abstract— Currently Drones become a brand new issue for Grid monitoring & security. As extremely innovative Drones area unit offered at the high value. Drones are facing various analysis & physical threats. The threats that may be caused by the attacks of drone has been enhanced. Recent object detection are dramatically improved in accuracy by victimization convolution neural networks. Sometimes GPS information will become defective underneath conduits, within subways, or close to high voltage (HV) power lines, that may result in flight inaccuracies. To optimize that, we ARE emerging a drone for whole grid structure checkups. Drones essentially use Global Positioning System (GPS) for independent control & monitoring, through image processing. This article defines a way for examine smart grid victimizing drones & employing a Pan-Tilt-Zoom (PTZ) camera. With the ambition of emerging associate self-directed flight & secure power lines for better Grid Performance & sort fault clearance with decreasing human life risk.

Keywords— Smart Grid, Drone, PTZ camera, GPS, SIFT algorithm.

I. INTRODUCTION

Maintenance of Power line for Smart Grid (SG) utility corporations is expensive, risky and time-consuming. Therefore, power distributors have been in searching for detecting faults on power lines employing different methods. Nowadays Drones are extensively used in numerous applications such as public surveys, Transport, Medical, photography, commercial enterprise, and defence security purpose. For these purposes, drones are needed to be controlled either by remote or smartphone or they must be fitted with self-controlling capability to approximate the self-position measurement using GPS, barometer and gyroscope. However, the operation of the drones are being affected underneath conduits, within passageways, or close to HV power lines [1], that might result in flight inaccuracies.

In recent years, advanced image processing models, employing deep convolutional neural network (CNN) architectures [2] are effectively used to monitor the drone Activities. Accuracy of recent object detection is on top of human level in some case studies, and this technology will alter and progress grid surveillance systems. More than a few quick image detection algorithms, like YOLO [3], are recommended, and recently, Google launched object detection API [4], which includes 5 deep CNN models to facilitate the object detection. The researchers compare six

state-of-the-arts convolutional object detectors, like YOLO [3] and 5 models provided by Google's object detection API-SSD with Mobile-Net [5], SSD with origin V2 [6], quicker R-CNN [7] with Resnet 101, R-FCN [8] with Resnet 101 [9], and quicker R-CNN with basis Resnet [10].

To compensate the power line issue, the authors develop a drone for Automatic inspection of grid with power lines infrastructure using PTZ camera that essentially custom barometer, gyroscope, GPS and image processing for self-control, along with estimate self-position. Once GPS is being disturbed, underneath things like those mentioned above, the GPS information can't be further properly utilized, so that independent flight will become very critical. As such the GPS direction-finding process must having the ability to search a new advanced mechanism which can automatically handle to self-estimate the proper position. For this perspective, a new method is being introduced in which ground images beneath the drone are set as base parameters to directly estimate the self-position. Specifically, the scheme includes perceiving feature points from ground spitting image structure, approximating PTZ location and angle, furthermore as planning. With this work, we tend to discuss with this method using Hardware Model as "ground spitting image gyro". The proposed smart technique also includes a feature through which it can send auto adjusted data along with all the information following the SG monitoring to the operator.

II. PROPOSED TECHNIQUES

A. Proposed Techniques:

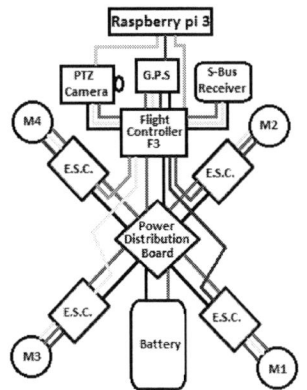

Fig. 1. Block Diagram of Drone

978-1-5386-6723-1/19 $31.00 © 2019 IEEE   361

The proposed drone is being featured with autonomous inspection and maintenance capability of the SG network. Some special considerations are required during inspection of power lines which are [11]: land-based mobile mapping data, optical aerial images, optical satellite images, unmanned aerial vehicle (UAV), images airborne laser scanner (ALS) data, synthetic aperture radar (SAR) images, and thermal images. This statistics has the ability to diagnose the health of power lines which may be rigorously affected by the earthquake & cyclones. Due to lower spatial & chronological perseverance of Satellite broadcasting images, nowadays aerial images are used. Fig. 1 depicts the block diagram of the proposed system.

1. Real-time Hardware implementation:

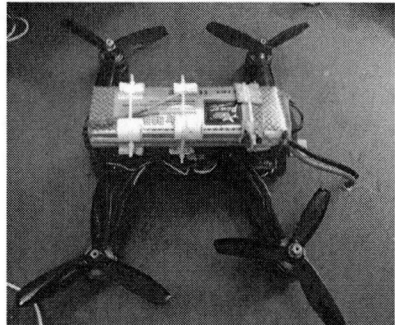

Fig. 2. Hardware setup of the drone

Fig. 2 demonstrate the hardware setup of proposed smart drone. The drone is based on Radio Frequency controlled 10 channel S-bus Receiver using 2.4.GHz frequency range. The frame of the drone is made by carbon fiber which gives mechanical support and perform as a heat exchanger.

a. Frame Type:

Depending on the size of the motor, propeller, Weight carrying capacity, the frame with following specifications are chosen. The smart drone incorporates 250mm frame having weight 145gm. The frame have a rubber damper stand to decrease the mechanical vibration at the time of flight.

TABLE 1: Propeller & Motor Selection

Frame Size (mm)	Prop. Size (inch)	Stator Size (KV)
0-150	3"	1300
150-210	4"	1800
210-250	5"	1800-2300
250-350	6"	2300
350-450	10"	2300
450>	11"	2600

b. Propeller Size:

Depending on frame size the authors select 5045 carbon fibre propeller with specifications such as 3 blades with 5 inch diameter of each and with 5 mm mounting hole at center. The weight of each propeller is 6.1 gm., so that the total weight is (6.1*4)=24.4 gm.

c. Motor stator Size:

Table 1 also presents the stator specification of the selected brushless DC (BLDC) motor.

Each BLDC motor has the specifications such as RPM/KV: 2300KV, Width: 27.9mm, Height: 31.5mm, mount hole: M3, Shaft diameter: M5, Weight: 28 gm.

The stator width and rotor height are judiciously chosen to meet the requirement for greater power and torque.

Total weight is =(28*4)=112gm.

d. ESC (Electronics speed controller) Selection:

ESC's are selected from the motor current ratings. We are used 35A, 4 in 1 multishort programmable ESC for our hardware implementation.

e. Thrust Calculation:

$$T = \left[\left(\eta * P \right)^2 * 2 * \pi * R^2 * \rho \right]^{0.3333} \qquad (1)$$

The motor thrust calculation is shown in (1)

Where,

T= Thrust in Newton

η= propeller hover efficiency =0.7 - 0.8 for low pitch props

P=shaft power = V*I

π=3.14159

R=propeller Radius in meters

P= Air density =1.22 kg/m^3

f. Battery selection:

Maximum Safe Current Draw (mA) = Battery Capacity (mAh) * Rate of Discharge Rating.

We used 5500mah LIPO battery.

2. Software platform

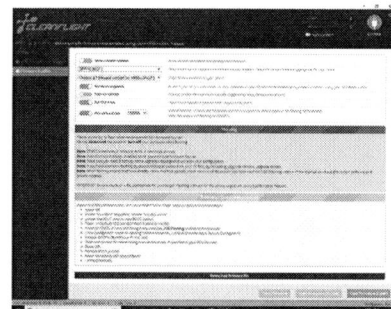

Fig. 3 Firmware update

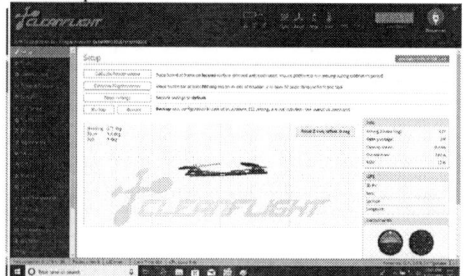

Fig. 4. Setup of Setup tab

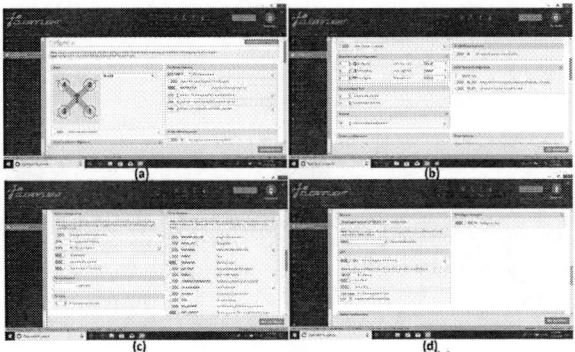

Fig. 5. Setup of Configurator Tab

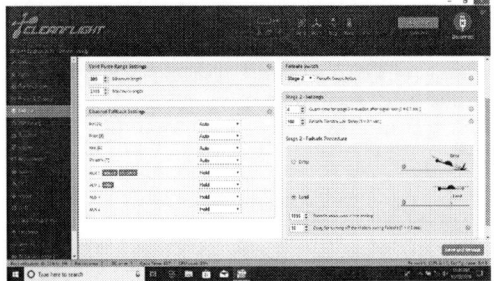

Fig. 6. Setup of Failsafe Tab

For software configuration we used "CLEANFLIGHT" open source software.

Fig.3 shows the software update procedure in a F3 flight controller which we used for our practical implementation.

Fig. 4-6 shows the software configuration of different tabs. In Fig.4 shows the calibration of Accelerometer, Magnetometer, gyro & barometer.

Here MSP port at UART 1 should always on at a braudrate of 115200. The configuration tab shown in Fig. 5, for GPS we used Indian satellite GAGAN (GPS Aided GEO Augmented Navigation). Fig.6 shows the Failsafe operation for perfect landing or return to home at the time of receiver signal lost or autonomous flight.

3. Image Processing

Threats of drone attacks & inspection are increasing because the drone technology has been quickly developed.

Drone invasion should be known at the first part of attack, and the arms ought to be enforced at the low price, as a result of drone is incredibly low cost compared to usual aerial police investigation systems. During this section, we have a tendency to introduce a drone following system employing a PTZ camera to detect fault in power line and track unauthorized drones. Lowe solves the image rotation technique which we used for our proposed system to affine alterations, intensity, and view modification in equivalent topographies using SIFT algorithm. It takes four simple phases. 1st is to estimate a scale area extreme exploitation with the distinction of Gaussian. 2nd is a key purpose localization wherever the key point aspirants are localized and distinguished by eliminating the small distinction points. 3rd is a key purpose direction assignment supported native image gradient and finally 4th is a descriptor generator to intention the native descriptor of image with key purpose, supported orientation [12].

The following system uses PTZ camera, and the object detection algorithmic rule is employed to spot whether or not Associate in nursing image from the PTZ camera contains a drone or faults. With detection of faults & other unauthorized drone, our system generates pan-tilt-zoom action and sends it to the PTZ camera to trace the drone or the faulty section. If those are known, our system obtains current horizontal and vertical Field of read from the camera, and then calculates the angle distinction between the drone and also the center purpose of the camera. The subsequent is formulas of relative degrees of pan and tilt standing. Zoom actions area unit calculated to stay the dimensions of a drone to the predefined desired portion. Portion of a particular box is that the space divided by the image size (1920 × 1080). We tend to calculate new zoom setting of PTZ camera because the current zoom increased by the quantitative relation between the specified portion and therefore the current portion. For example, if the present portion of a drone is a pair of of the image, and our desired portion of a drone is four-dimensional, then we zoom in 2X. The subsequent formula for pan (2), tilt (3) angle & zoom (4) describes however our system calculates the new pan, tilt, zoom setting.

$$\angle pan = \left(0.5 - \frac{Position\_along\_axis:x}{image\_width}\right)*HFV \quad (2)$$

$$\angle tilt = \left(0.5 - \frac{Position\_along\_axis:y}{image\_height}\right)*VFV \quad (3)$$

$$zoom = \left(\frac{desired\_portion}{current\_portion}\right)*current\_zoom \quad (4)$$

III. EXPERIMENTAL RESULT

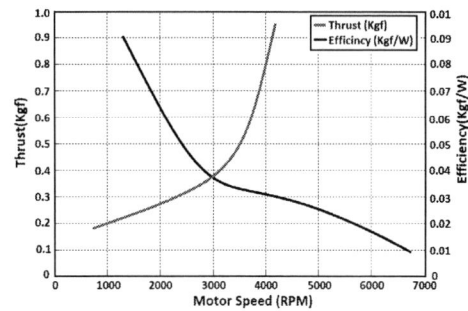

Fig. 7. Propeller Thrust & efficiency plot

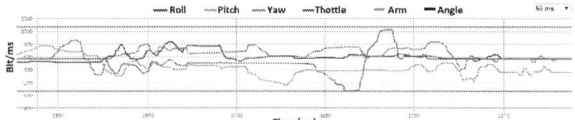

Fig. 8. Receiver Control signals

Fig. 7 shows the variations of Propeller Thrust & efficiency with respect to motor speed following a Matlab Simulink simulation for adjusting the stability. Fig. 8 shows the different control signals at the time of manually operated drone flight. The PID output for different Roll, Pitch, and Angle is shown in Fig. 9.

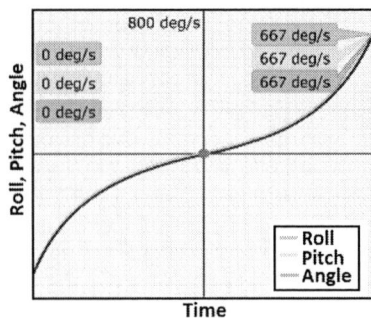

Fig. 9. PID tuning

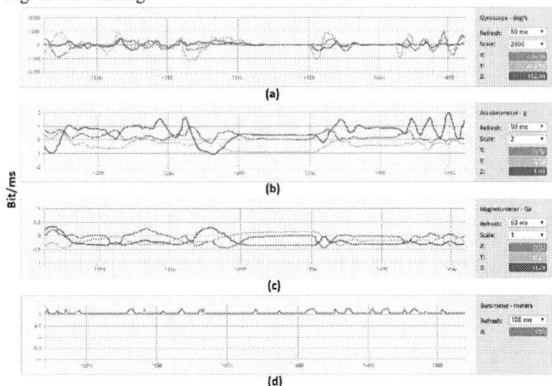

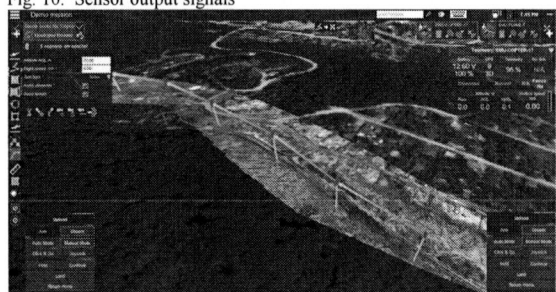

Fig. 10. Sensor output signals

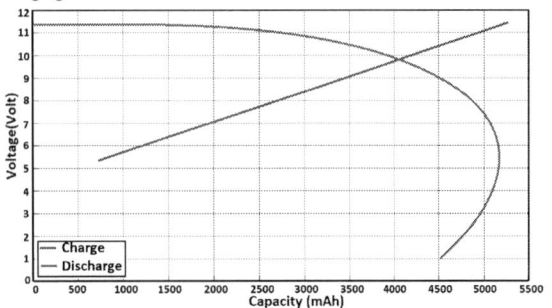

Fig. 11. View of Power lines based on PTZ camera & GPS with thermal imaging

Fig. 12. Lipo Battery Charging & Discharging

For autonomous flight, sensors are adjusted according to the longitude & latitude as depicted in Fig. 10. Due to high electromagnetic interference & electro static coupling near High voltage (H.V) power lines, the Gyroscope, magnetometer, Accelerometer & Barometer are not working properly. This Sensors disturbances are shown in Fig. 10 between the times span of 1280ms to 1350ms. To overcome this problem of autonomous flight near H.V line we used ground image processing using PTZ camera as shown in

Fig. 11. Lipo batteries have a capability of fast charging & discharging as shown in Fig. 12.

IV. CONCLUSION

In the current paper issues regarding power line maintenance and safety are briefly discussed and a hardware drone is designed and developed for Automatic inspection of grid with power lines infrastructure using PTZ camera that essentially incorporates accelerometer, magnetometer, gyroscope, barometer and GPS for autonomous control, along with self-position evaluation through image dispensation. Once GPS data is disturbed by external issues as mentioned earlier. GPS information can no longer be properly used and thus modified GPS can be feasible by the proposed technique through the smart drone. Nevertheless the smart drone is also capable to inspect the condition of power line and it also monitors and collect data from the the power line and send to the operator successfully. The proposed system also reduces the human risk for HV line checking.

REFERENCES

[1] M.A. King, "The GPS Contribution to the Error Budget of Surface Elevations Derived From Airborne LIDAR", IEEE Transactions on Geoscience and Remote Sensing, vol. 47 , no. 3, pp. 874-883, March 2009.

[2] R. C. Çalık, M. F. Demirci, "In embedded systems image classification with convolutional Neural Network", in Proceedings of the IEEE conference on 26th Signal Processing and Communications Applications , 2018, Turkey.

[3] J. Redmon, S. Divvala, R. Girshick, and A. Farhadi, "You only look once: Unified, real-time object detection", in Proceedings of the IEEE Conference on Computer Vision and Pattern Recognition, pp. 779–788, 2016.

[4] J. Huang, V. Rathod, C. Sun, M. Zhu, A. Korattikara, A. Fathi, I. Fischer, Z. Wojna, Y. Song, S. Guadarrama, K.Murphy, "Speed/accuracy trade-offs for modern convolutional object detectors," arXivpreprint arXiv:1611.10012, 2017.

[5] A. G. Howard, M. Zhu, B. Chen, D. Kalenichenko,W. Wang, T. Weyand, M. Andreetto, and H. Adam, "Mobilenets: Efficient convolutional neural networks for mobile vision applications," arXivpreprint arXiv:1704.04861, 2017.

[6] S. Ioffe and C. Szegedy, "Batch normalization: Accelerating deep network training by reducing internal covariate shift", in International Conference on Machine Learning, pp. 448–456, 2015.

[7] S. Ren, K. He, R. Girshick, and J. Sun, "Faster rcnn: Towards real-time object detection with region proposal networks", in Advances in neural information processing systems, pp. 91–99, 2015.

[8] Y. Li, K. He, J. Sun et al., "R-fcn: Object detection via region-based fully convolutional networks", in Advances in Neural Information Processing Systems, pp. 379–387, 2016.

[9] K. He, X. Zhang, S. Ren, and J. Sun, "Deep residual learning for image recognition," in Proceedings of IEEE conference on computer vision and pattern recognition, pp. 770–778, 2016.

[10] C. Szegedy, S. Ioffe, V. Vanhoucke, and A. A. Alemi, "Inception-v4, inception-resnet and the impact of residual connections on learning", in AAAI, pp. 4278–4284, 2017.

[11] L. Matikainen, M. Lehtomäki, E. Ahokas, J. Hyyppä, M. Karjalainen, A. Jaakkola, A. Kukko, T. Heinonen, "Remote sensing methods for power line corridor surveys", ISPRS, vol. 119, pp. 10–31, 2016.

[12] D. G. Lowe, "Distinctive Image Features from Scale-Invariant Keypoints," International Journal of Computer Vision, vol.50, No. 2, 2004, pp.91-110.

Effect of High-K Spacer on the Performance of Gate-Stack Uniformly doped DG-MOSFET

[1]Satish K Das, [1]Sanjit K Swain, [2]Sudhansu M Biswal, [2]Debasish Nayak, [3]Umakanta Nanda,[1]Biswajit Baral, [2]Dhananjaya Tripathy
[1]*Department of Electronics and Communication Engineering, Silicon Institute of Technology, Bhubaneswar*
[2]*Department of Electronics and Instrumentation Engineering, Silicon Institute of Technology, Bhubaneswar*
[3]*School of Electronics Engineering, Vellore Institute of Technology, Amaravati*
Email: satish.das@silicon.co.in

Abstract— In this work, we have analyzed the novelty of the Gate Stack Double Gate (DG) MOSFET with respect to different spacer variations in order to reduce the short channel effect challenges and simultaneously increasing the device performance. Silicon is used as the channel material along with the gate stacked technology for studying the analog performance and Radio Frequency (RF) performance of the device. For gate stacking, two types of oxides are used- one denoting low-K i.e SiO_2 and the other as high-K i.e- HfO_2. Spacers with various permittivities were used to understand their effects on the performance of the device. The simulation result shows that the use of spacer material affected both the analog and RF behavior of the device significantly. The computer aided design (TCAD) simulations have been carried by SILVACO International.

Keywords—DG MOSFET, Spacer layer, Gate Stack, Analog Applications, RF Behaviour.

I. INTRODUCTION

The use of metal oxide semiconductor field effect transistors (MOSFETs) has increased rapidly in the recent years, as they are highly efficient in analog and RF applications. However, as the technology changes, the demand for more efficiency comes into play which is essential for applications involving low power and high frequencies[1, 2]. So, as to maximize the potential of the MOSFETs, the process of downscaling is used. But downscaling of the device comes with a price of short channel effects (SCEs) that decreases the performance of the device [3-5].

To overcome short channel effect challenges faced by MOSFETs, certain modifications are required to be made in the design of MOSFETs like asymmetric engineering of channel [8-10], device engineering technique[11-13], and gate-stacking technique [14, 15].Conventional MOSFETs have increased leakage current when they are downscaled beyond certain limit, leading to an unnecessary power consumption [7]. The channel control is dependent on both V_{gs} and V_{ds} unlike conventional MOSFETs as the device dimensions as lowered. An answer to these problems is double gate (DG) MOSFETs which provides a better controllability of the channel [6]. In this paper, the proposed device is a DG MOSFET with SiO_2 as the low -K oxide and high-K oxide as HfO_2 which results in stack-minimization of trap density at interface and also improving performance of the device [16-19].Recent developments in the field of gate stack engineering are used to overcome the restrictions imposed by DG- MOSFETs such as further reduction of DIBL effect, lowering the leakage current and an improvement in on current to off current ratio, resulting in better switching characteristics [7]. High-K spacers with different permittivity were used which reduces the unacceptable bulk accumulation of fixed charges and lowers the mobility rate of carriers at interfaces thereby providing better analog /RF performances. However, all the previous work that are done mainly based upon either unstacked gate DG MOSFETs or with high-K spacer based DG-MOSFETs [20-26,]. So, here we have used gate stack engineering along with implementation of high-K spacer on DG-MOSFETs to study their effects on analog and RF performance, on which no work is ever done till date.

In this paper we have tried to analyse the performance of gate stacked DG MOSFET for different spacer layer materials by taking simulations of the following analog and RF parameters: Drain current, Transconductance factor (g_m), Transconductance Generation Factor (TGF), Gate to Source Capacitance (Cgs), Drain to Gate Capacitance(Cdg) and Cut-off frequency (f_T) with respect to gate to source voltage(Vgs).

II. DEVICE STRUCTURE AND SIMULATION

Fig. 1 depicts the 2-D cross sectional view of Double Gate MOSFET with high-K spacer. Various parameters are considered according to the International Roadmap for Semiconductor Technology (ITRS) standards [23]. The proposed device has gate length (L_g) of 32nm.The source/drain lengths are of 10nm, channel thickness (t_{si}) of 16nm. The low-K oxide thickness is 1nm and high-K oxide thickness is 1nm. Low- K oxide is compsed of SiO_2 whereas high- K oxide is taken to be HfO_2. The device regions, source and drain have an uniform n- type doping and a concentration of $10^{18}/cm^3$. The doping concentration of channel is $10^{15}/cm^3$. The structure is symmetric with respect to the double layered gate and stacked oxide dielectric of SiO_2 and HfO_2. Spacers are composed of high-K material having different permittivity such as K=3.9,7.5 and 25 for the simulation pyrpose.

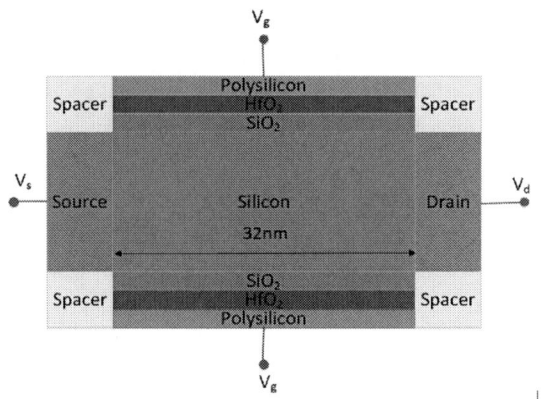

Figure 1. Cross sectional 2-D view of Gate-Stacked DG-MOSFET with High-K Spacer.

For the simulations, two-dimensional drift diffusion numerical simulation model have been considered. The drift–diffusion model makes simulations faster and gives high accuracy levels. Similarly considerations of the ionized impurity scattering and temperature effects, Aurura model is incorporated.The ksn=0 numerical method is used in the models for calculations where as

the temperature is set at 300 K for simulation. The electric field dependent mobility is used in the simulation. Mesh density is chosen carefully to accelerate the computational efficiency and for getting accurate simulation. All these simulations are done using computer aided technology (TCAD) Silvaco International [24].

III. RESULTS AND DISCUSSION

High-K oxide materials effects on various analog parameters of the proposed Gate Stack DG-MOSFET are discussed in this section. Various important analog parameters are studied by taking into the account of various high-K spacer materials of the proposed device such as Drain Current vs Gate Voltage(I_D vs V_{gs}), Transconductance vs Gate Voltage(g_m vs V_{gs}) and Transconductance Generation Factor vs Gate Voltage (g_m/I_d vs V_{gs}). For all these simulations V_{ds} is kept at 0.5V.

Fig. 2 depicts the drain current (I_d) variations with Gate to Source Voltage (V_{gs}) for different types of high-K spacer materials. This results show a clear distinction that for permittivity K=25 the current is highest at a given gate voltage among all other high-K materials taken as the spacers. This can be implied from the fact that if the permittivity of the material is increased, then capacitance is increased which in return increases charge as (Q=CV). As the charge increases current also increases as per equation (I=Qt). This indicates in better switching characteristics.

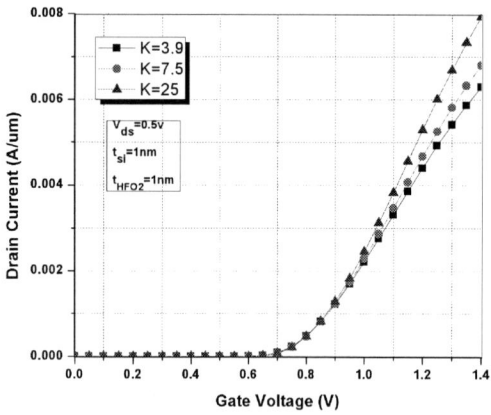

Figure 2. Drain current as a function of gate to source voltage for oxides having different permittivity as the Spacer Material .

Fig. 3 shows the variation of the Transconductance vs Gate Voltage (g_m vs V_{ga}) for various high-K materials as the spacer. It is observed that the material with highest permittivity is having a higher transconductance as compared to material with lower permittivity preferred as the spacer material. However, this can also be figure out that from the fact that for a given small range of gate voltage, there is a higher change in drain current with spacers having higher permittvities than that of the lower ones, implying high transconductance. Transconductance is also an key parameter to be used to determine the gain-bandwidth product of the device. This gain is one of the important parameter if one considers the amplification process done by MOSFETs. Higher the transconductance of the device better is the gain. Thus the device will have a better ability to amplify the signals. So the gate stacked DG MOSFETs with high K spacers having high permittivity can be considered as a better amplifier. In Fig. 4, it is notice that the variation of transconductance generation factor (g_m/i_d) with respect to the gate-source voltage of the proposed device. It is can be noted from the graph that the material with higher permittivity has higher transconductance generation factor. This is due to the fact that the device with higher permittivity based spacer will have higher transconductance. Even though the same device will have higher drain current, the relative difference is much smaller in comparison with the transconductance of (say) any two corresponding device having variations in the composition of the spacer material.The transconductance generation factor indicates the degree of effectiveness of the on achievement of a certain value of the transconductance by the drain current. It shows that how effectively the device can operate in low voltages. Thus is essential for the realization of the circuits operating for low power supply.

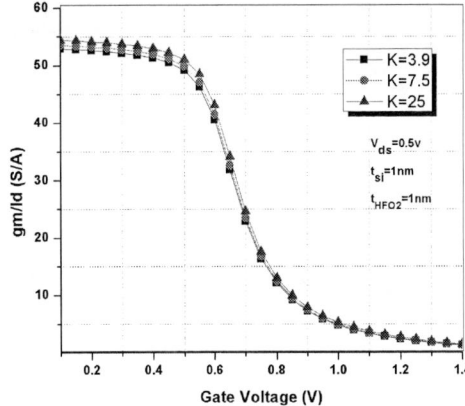

Figure 4. Transconductance generation factor as a function of gate to source voltage for oxides having different permittivity as the Spacer Material for Gate-Stacked DG-MOSFET with High-K Spacer.

For the analysis of proposed device RF performance we have to study various intrinsic capacitances. Various important parameters such as:- Gate- Source Capacitance vs Gate-Source Voltage (C_{gs} vs V_{gs}), Gate-Drain Capacitance vs Gate-Source Voltage (C_{gd} vs V_{gs}) and Cut- off frequency vs Gate-Source Voltage (f_T vs V_{gs}) are analyzed after the simulation of the proposed device.

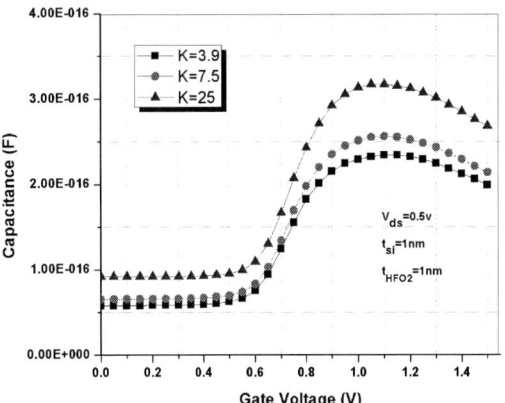

Figure 5. Gate-Source Capacitance as a function of gate to source voltage for oxides having different permittivity as the Spacer Material for Gate-Stacked DG-MOSFET with High-K Spacer.

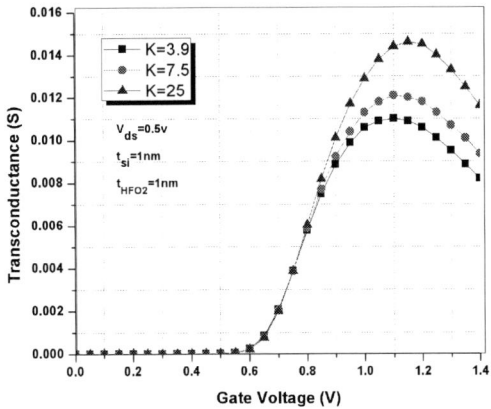

Figure 3. Transconductance as a function of gate to source voltage for oxides having different permittivity as the Spacer Material for Gate-Stacked DG-MOSFET with High-K Spacer.

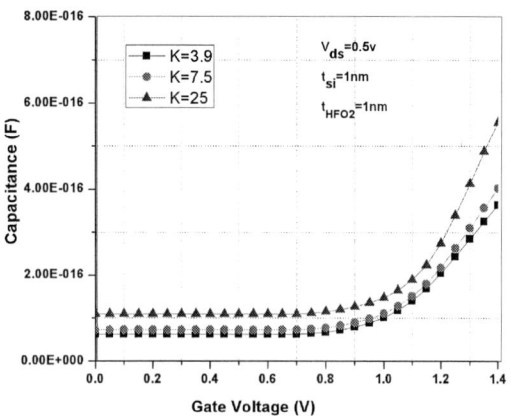

Figure 6. Gate-Drain Capacitance as a function of gate to source voltage for oxides having different permittivity as the Spacer Material for Gate-Stacked DG-MOSFET with High-K Spacer.

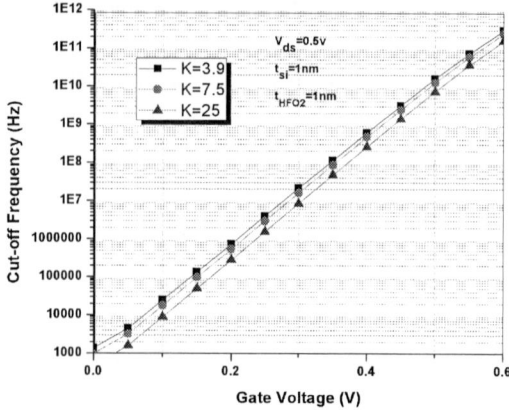

Figure 7. Cut-off Frequency as a function of gate voltage for oxides having different permittivity as the Spacer Material for Gate-Stacked DG-MOSFET with High-K Spacer.

Fig. 5 and 6 indicate the variation of the Gate- Source Capacitance vs Gate-Source Voltage (C_{gs} vs V_{gs}) and Gate- Drain Capacitance vs Gate-Source Voltage (C_{gd} vs V_{gs}) respectively. In cut-off mode of operation, inversion doesn't take place so both C_{gs} and C_{gd}=0 as there is no channel. However, as the mode of operation changes to linear, both the capacitance becomes non zero and increases gradually with increase in V_{gs}. As evident from the graphs, the capacitance of the spacer material having highest permittivity (K=25) is higher than other materials for the proposed device for the same V_{gs}. It is due to the fact that the permittivity is directly proportional to capacitance.It is depicted by the equation ($C=\varepsilon A/d$ where $\varepsilon=\varepsilon_o\varepsilon_r$) .Here 'A' and 'd' are constant for the proposed device, only variation being permittivity.

Fig. 7 shows the variation of the Cut-off frequency with Gate-Source Voltage (f_T vs V_{gs}). Cut-off Frequency is the frequency where the current gain is unity. It is given by:- $f_T=g_m/(2*\pi*C_{gg})$,where, C_{gg}= total gate capacitance given by $C_{gg}=C_{gs}+C_{gd}$. For the proposed device it can be seen that the spacer with lowest permittivity is having the highest cut-off frequency. This due to the fact that the permittivity is directly proportional to the capacitance which in turn is inversely related to the frequency. Even though the g_m play a role, as per the equation, its effect is not that much as compared to that of the capacitance. Thus, low permittivity based high-K spacer based gate stacked DG-MOSFETs is suitable for the circuit operating in high frequencies.

IV. CONCLUSION

The performance for gate stack DG-MOSFET as a function of different spacer layer is thoroughly investigated and presented in this paper. From the result many relevant information's has been figured out. It is observed that the analog parameters shows better result for the spacer material having highest permittivity (K=25) as compared to lower value of spacer material. Similarly the RF parameters i.e. cut-off frequency behaves oppositely with respect to spacer material i.e. for lower value of K it shows higher result. This study reveals that depending upon the applications this device can be carefully used for different value of spacer materials. Hence selection of high-K spacer material has a major role for this device in applications where low power analog and RF performances matters.

VI. ACKNOWLEDGEMENT

The authors are grateful to Silicon Institute of Technology, Patia, Bhubaneswar 751024, for rendering its valuable support while carrying out this research work.

REFERENCES

[1] P. H. Woerlee, M. J. Knitel, et al, "RFCMOS perfomances trends"IEEE Trans. Electron Devices, vol. 48, pp 1776-1782, 2001.

[2] V. Kilchytska, A. neve, L. Vaneaillie, et al, "Influence of device engineering on analog and RF perfomance of SOI MOSFETs" IEEE Trans. Electron Devices, vol. 50, pp 577-588, 2003.

[3] K. K. Young, "Short channel effect in fully depleted SOI MOSFETs" IEEE Transactions on Electron Devices,Vol. 36 ,pp.399-402,1989

[4] D. J. Frank et al, "Device scaling limits of Si MOSFETs and their application" Proc. IEEE , vol.89, pp 259-288, 2001 .

[5] S.K.Swain et al., "Effect of Channel Thickness and Doping Concentration on Sub-Threshold Performance of Graded Channel and Gate Stack DG MOSFETs" JOLPE, vol.11, pp 366-372 , 2015.

[6] F. Balestra, S. Cristoloveanu, M. Benachir, et al, "Double gate silicon-on-insulator with volume inversion" IEEE Electron Device Letters ,Vol.9,pp. 410-412 ,1987.

[7] K. P. Pradhan, S. K.Mohapatra.et al, "Impact of high-K gate dielectric on analog and RF perfomance of nanoscale DG-MOSFET"Microelectronics journal,vol.45 pp 144-151 ,2014

[8] F. Djeffal, Z. Ghoggali, Z. Dibi, et .al, "Analytical analsis of nanoscale multiple gate MOSFETs including effects of hot-carrier induced interface charges" Microelectronics Reliability ,Vol. 49,pp. 377-381,2009.

[9] D. Nirmal ,P.Vijaya Kumar, et al, "A review of Nanoscale Channel and gate Engineered FINFETs for VLSI Mixed Applications Using Zirconium-di-Oxide Dielectrics" JEST Review, Vol.7,pp. 119-124, 2014.

[10] N. Mohankumar, B.Samal, C.K. Sarkar, "Influence of channel and gate engineering on the analog and RF perfomance of DG MOSFETs" IEEE Transactions on Electron Devices,Vol. 57 ,pp.820-826,2010

[11] C, L. chen, J. M. Knecht, J. Kedzierski, "Channel Engineering for SOI-MOSFETs for RF applications" Proc. IEEE Int.SOIConf, 5232-5233, 2009.

[12] Y. Taur, D. A. Buchanan, W. Chen, et al, "Dimensional effects in MOSFETs…" Proc. IEEE (1997) ,Vol.85, pp.486-504.

[13] R. K. Sharma, M. Gupta, R. S. Gupta, "TCAD Assessment of Device Design Technologies for Enhanced Performance of Nanoscale DG MOSFET" IEEE Transactions on Electron Devices,Vol. 58 ,pp.2936-2943,2011

[14] C. Hu, "Gate oxide scaling limit and projection"IEDM Tech. Dig (1996), 319-322

[15] E. M. Vogel, K. Z. Ahmed, B. Hornung, et al, "Model tunnel currents for high dielectric constant dielectrics" IEEE Transactions on Electron Devices ,Vol.45,1350-1355, 1998.

[16] B. Cheng, M. Cao, R. Rao, A. Inani, et al. "The impact of high-K dielectrics and metal gate electrodes on sub-100nm MOSFETs" IEEE Trans. Electron Devices, Vol. 46, pp. 1537-11543, 1999

[17] A. Inani, B. Cheng, RV. Rao, J. Woo, "Gate stack architecture analysis and channel engineering in deep sub-micron MOSFETs" Jpn J Appl Phys, Vol. 38, pp.2266-2271, 1999

[18] M. Saxena, S. Halder, M. Gupta, RS. Gupta, "Modelling and simulation of asymmetric gate stack (ASYMGAS)-MOSFET" Solid State Electronics, Vol. 47 ,pp.2134-2134,2003

[19] E. Faith, A. Behnam, P. Hashemi, et al. "The influence of the stacked and the double material gate structures on the short channel effects in SOI MOSFETs" IEICE Trans Electron, Vol. 88, pp. 1122-1126, 2005

[20] F. Djeffal, M. Meguellati, A. Benhaya, "A two-dimensional analytical analysis of subthreshold behavior to study the scaling capability of nanoscale graded channel gate stack DG MOSFETs" Physica E,Vol.41,1872-1877, 2009.

[21] M. A. Pavanello, J. A. Martino, V. Dessard, et al, "An asymmetric channel SOI nMOSFET for reducing parasitic effects and improving output charcterstics" Electrochem. Solid-State Lett,Vol. 3,pp. 50-52,2000.

[22] A. Kranti, TM. Chung, D. Flandre, JP. Raskin, "Laterally asymmetric channel engineering in fully depleted double gate SOI MOSFETs for high perfomance analog applications" Solid State Electronics,Vol. 48 ,pp.947-959,2004

[23] Semiconductor Industry Association, International Technology Roadmap for Semiconductors, (2011)SIA San Jose.

[24] Synopsys. TCAD Sentaurus device user's manual V-2008.-09.

[25] Sanjit Kumar Swain, Arka Dutta, Sarosij Adak, et al, "Influence of Channel length and High-K oxide Thickness on Subthreshold Analog/RF Performance of Graded Channel and Gate stack DG- MOSFETs." Microelectronics Reliability (Elsevier) Volume 61, ,Pages 24-29, 2016.

[26] Sarosij Adak, Sanjit Kumar Swain, Arka Dutta et al, "Influence of Channel length and High-K oxide Thickness on Subthreshold DC Performance of Graded Channel and Gate stack DG-MOSFETs NANO: Brief Reports and Reviews (World Scientific Publishing Company) 17 2016. Volume   11,   Issue   09,   pp   1650101,   2016.

2019 Devices for Integrated Circuit (DevIC), 23-24 March, 2019, Kalyani, India

Bandwidth Increment of Piezoelectric Energy Harvester using Multi-beam Structure

Sourav Naval
*Department of Electronics and
Communication Engineering
Birla Institute of Technology, Mesra,
Ranchi, Jharkhand, India
sourav.naval@gmail.com*

Prasun Kumar Sinha
*Department of Electronics and
Communication Engineering
Birla Institute of Technology, Mesra
Ranchi, Jharkhand, India
prasunks98@gmail.com*

Nikhil Kumar Das
*Department of Electronics and
Communication Engineering
Birla Institute of Technology, Mesra
Ranchi, Jharkhand, India
nkd112119@gmail.com*

Ashutosh Anand
*Department of Electronics and
Communication Engineering
Birla Institute of Technology, Mesra,
Ranchi, Jharkhand, India
ashunitjsr@gmail.com*

Sudip Kundu
IEEE Member
*Department of Electronics and
Communication Engineering
Birla Institute of Technology, Mesra,
Ranchi, Jharkhand, India
kundu.sudip@gmail.com*

Abstract— **Piezoelectric energy harvesters have high power density and are simpler to fabricate as compared to other low power energy harvesters. There are certain issues which need to be addressed while designing a piezoelectric energy harvester. These are the requirements of maintaining a high output voltage, low resonant frequency, small size and wide bandwidth of operation. Achieving a wide bandwidth is one of the most prominent issues. It is because most of the vibrations occur over a range of frequencies. So, the challenge is to design an energy harvester which generates high output voltage over a wide range of frequencies. In this paper, a Microelectromechanical system (MEMS) based multi-beam energy harvester has been proposed. This energy harvester has been designed using two single cantilever beams and the top electrodes of both the beams are connected by a metal layer. The peak output voltage of the proposed structure is 18 V at 142 Hz. The multi-beam structure generates an output voltage of more than or equal to 5 V for a bandwidth of 15 Hz which is 1.5 times wider as compared to that of a single beam energy harvester.**

Keywords—Piezoelectric Energy Harvesters, MEMS, Wide Bandwidth, Multi-beam cantilever

I. INTRODUCTION

Energy harvesting involves extracting electrical energy from different energy sources in the environment. The energy harvesting may be of different types depending upon the source used for extraction of energy. The source may be thermal energy of a hot body, kinetic energy of a moving body, vibration energy of an oscillating body or some other form of energy. The most commonly used energy harvesters generate energy from the vibrations in the environment. The process of obtaining electrical energy from the vibrations in the surrounding environment is called vibration energy harvesting [1][2]. The principle of piezoelectricity is utilized to generate the electrical energy, i.e. when stress is applied along a piezoelectric material, an electric field is generated across it. These piezoelectric (PZ) energy harvesters are implemented using Microelectromechanical systems (MEMS) technology [3]-[5] to reduce their physical size. The small size of the energy harvesters makes them suitable for

diverse applications [6]. These energy harvesters play an important role as alternative powering mechanisms for modern age equipment, replacing the use of conventional batteries. These are particularly useful when the equipment is in an inaccessible area, say a wireless sensor node [7] or an implanted device [8], where it is difficult to replace batteries. In such cases, these energy harvesters extend the lifetime of the equipment.

A rectangular cantilever beam structure is most commonly used as an energy harvester [9]. The reason is that it is simple to design and fabricate. The cantilever energy harvester structure consists of a piezoelectric layer deposited over a substrate layer. There are metal electrodes above and below the piezoelectric layer. The structure of a PZ cantilever beam has been shown in Fig. 1. One end of the beam is fixed, and a rigid proof mass is attached on the other end of the beam which is commonly referred to as free end. The proof mass reduces the resonant frequency of the structure. The low resonant frequency of such PZ energy harvester is necessary as most of the vibrations occurring in real life are of low frequency.

The rectangular cantilever beam structure doesn't generate a considerable voltage over a wide range of frequencies. So, various researchers tried designing and implementing different structures in pursuit of achieving a wide bandwidth for energy harvester. One of the structures suggested the use of several unconnected cantilever beams, each giving maximum voltage at its resonant frequency [10]. This suffered from a drawback of requiring external hardware for

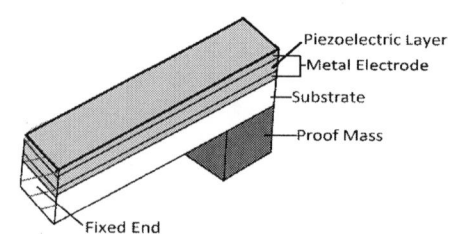

Fig. 1. Schematic Diagram illustrating rectangular cantilever beam

978-1-5386-6723-1/19 $31.00 © 2019 IEEE

selecting the beam with the maximum output voltage among the multiple beams. Another structure demonstrated an array of piezoelectric beams connected through springs [11] which led to a certain increment in bandwidth. But, the use of common springs causes power loss, which can only be avoided using low loss springs for interconnection. The concept of parallel connection of single beam structures to increase the bandwidth has been suggested by researchers [12]. But no profound analysis was available regarding these parallel multibeam structures.

In this paper, we aim to design and implement a parallel two-beam structure to widen the operating bandwidth of the energy harvester structure. The paper has been organized as follows. Section II describes the basic theory behind a piezoelectric beam and what interaction occurs between the two piezoelectric beams in the proposed two-beam structure. In Section III we describe the design of the proposed two-beam structure followed by its analysis using Finite Element Method in Section IV. Finally, the conclusion is presented in Section V.

II. Theory

When a piezoelectric beam vibrates, a stress is generated along the beam which leads to the generation of an electric field across the beam. To understand how this occurs, we assume a piezoelectric beam vibrating about a horizontal plane. When the beam vibrates along one direction, say downwards, the lower edge of the beam elongates, and the upper edge of the beam gets compressed. As a result, charges of opposite polarities are induced on the two electrodes on either side of the piezoelectric layer. So, an electric field is developed across the piezoelectric layer. A similar phenomenon occurs when the beam vibrates along the other direction, and an electric field develops in opposite direction. In this manner, a vibrating piezoelectric beam generates an output voltage.

The fundamental equations [13] to find the developed mechanical strain (S) and the electric displacement (D) in the piezoelectric materials under given mechanical stress (T), and electric field (E), are given in (1) and (2), respectively.

$$S = s^E . T + d_t . E \qquad (1)$$
$$D = d . T + \varepsilon^T . E \qquad (2)$$

Here, s^E is strain under a zero or constant electric field and ε^T is the dielectric permittivity under zero or constant stress. d and d_t are piezoelectric coefficient and reverse piezoelectric coefficient respectively. These equations govern the behavior of the piezoelectric beams.

We need to understand how the output voltage of the proposed two-beam structure at different frequencies (also the bandwidth) depends upon the vibration characteristics of both the individual beams. For this, we consider two beams- Beam 1 and Beam 2. The top electrodes of both the beams are connected by a metal layer. We assume that the vibration of Beam 1 generates a voltage $V(t)$. This generated voltage appears as an electric beam $E = \frac{V(t)}{t_p}$ across Beam 2, where t_p is the thickness of the piezoelectric layer in Beam 2. As it is

evident from (1) and (2), this electric field affects the generated strain S and electric displacement D in Beam 2. So, the vibration of Beam 1 affects the vibration of Beam 2 and similarly, the vibration of Beam 2 affects the vibration of Beam 1. The output of the two-beam structure hence depends on the vibration and output of both the beams. The output voltage of the two-beam structure at any frequency hence depends on the voltage generated by the single beams and their relative phase. If the displacements of both the beams are in the same phase, the resultant output voltage would be greater than the output voltage of the individual beams. Similarly, if the displacements of both the beams are in the opposite phase, the resultant output voltage would be less than that of the output voltage of the individual beams.

Since we have discussed the basic theory related to the single beam structures and two-beam structure, we will now discuss the design of the single beam structures and the proposed two-beam structure.

III. Structure of Energy Harvester

A. Single Beam Structures

We designed a single rectangular cantilever beam with a length of 5000 μm, a width of 1000 μm and a total thickness of 17.26 μm. The Silicon layer had a thickness of 14.53 μm and Zinc Oxide (ZnO) piezoelectric layer had a thickness of 2.73 μm. Here, Zinc Oxide was chosen as the piezoelectric material because it is biocompatible and can be used for designing an implanted device. A solid proof mass having a cubic structure with each side measuring 1 mm was attached to the free end of the cantilever beam. We designated this as Beam 1. The resonant frequency of this beam was observed to be 143 Hz. The constructed beam has been shown in Fig. 2.

The substrate layer and proof mass were made with Silicon having density 2320 kg/m³ and Young's Modulus 160 GPa. While the piezoelectric layer was made with Zinc Oxide having density 5680 kg/m³ and Young's Modulus 6.83 GPa. Then we designed another single beam with a length of 5100 μm. Other dimensions and material properties were kept same as that of the Beam 1. The resonant frequency of this beam was observed to be 138 Hz and we designated this as Beam 2.

B. Proposed Two-Beam Structure

The designed two-beam structure consists of Beam 1 and Beam 2. In this proposed structure, the top electrodes of

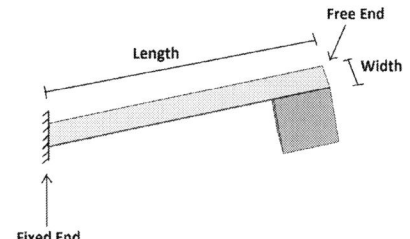

Fig. 2. Diagram of designed single cantilever beam structure

TABLE I. OUTPUT VOLTAGE AND POWER GENERATED BY BEAM 1 AT DIFFERENT FREQUENCIES

Frequency (Hertz)	Peak Voltage (Volts)	Power Generated (microwatt)
133	1.8	0.19
138	2.8	0.48
141	12.9	9
143	14.2	7.9
144	11.2	7.3
146	4.7	1.3
148	2.8	0.48
153	1.8	0.19

TABLE II. OUTPUT VOLTAGE AND POWER GENERATED BY BEAM 2 AT DIFFERENT FREQUENCIES

Frequency (Hertz)	Peak Voltage (Volts)	Power Generated (microwatt)
128	1.8	0.19
134	5.4	1.6
136	11.6	6
138	14	7.1
140	11.6	6
143	4.4	0.85
148	1.8	0.19

both the beams are connected by a metal layer (in this case aluminum) and the bottom electrodes are considered at ground potential. The constructed two-beam structure has been shown in Fig. 3.

Next, we will present the simulation of the designed single and two-beam structures and the observed results.

IV. RESULTS AND DISCUSSION

This section presents the Finite Element Method (FEM) analysis [14] of the designed structures. In this paper, we have performed the simulations on the COMSOL Multiphysics platform, which is a commonly used platform for FEM analysis. We have performed the transient analysis of the beams by applying a sinusoidal acceleration of magnitude 1g (g is acceleration due to gravity) and varying the frequency of acceleration.

The observed values of Peak Voltage and Power generated on simulating Beam 1 at different frequencies have been recorded in the TABLE I. We observe that Beam 1 has a peak output voltage of 14.2 V at its resonant frequency, i.e. 143 Hz. Here we will consider the bandwidth of operation to be frequencies at which the output voltage is more than or equal to 5 V. In case of Beam 1, the bandwidth is 10 Hz as shown in Fig. 4. The power generated by the single and two-beam structures has been shown in Fig. 5.

Similarly, we performed the transient analysis of the single beam: Beam 2, and recorded the observed Peak Voltage and Power generated at different frequencies in TABLE II.

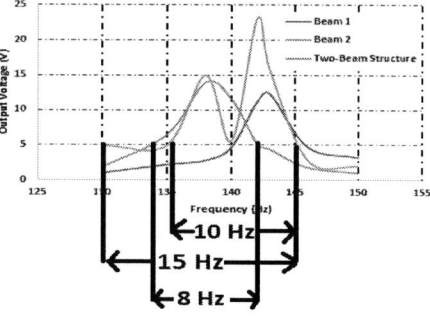

Fig. 4. Output voltage of single beam and two-beam structures at different frequencies

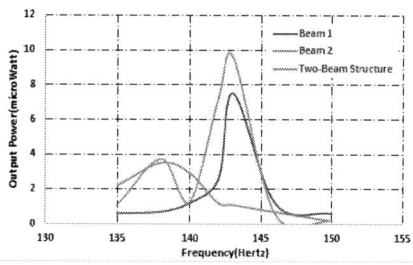

Fig. 5. Power Generated by single beam and two-beam structures at different frequencies

TABLE III. OUTPUT VOLTAGE AND POWER GENERATED BY TWO-BEAM STRUCTURE AT DIFFERENT FREQUENCIES

Frequency (Hertz)	Peak Voltage (Volts)	Power Generated (microwatt)
130	5.1	1.15
135	4.4	1
138	15	9.5
140	5.6	1.3
142	18	4.8
143	15.8	9.6
146	2.9	0.56
150	2	0.2

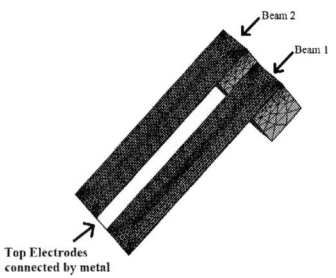

Fig. 3: Diagram of the designed two-beam structure

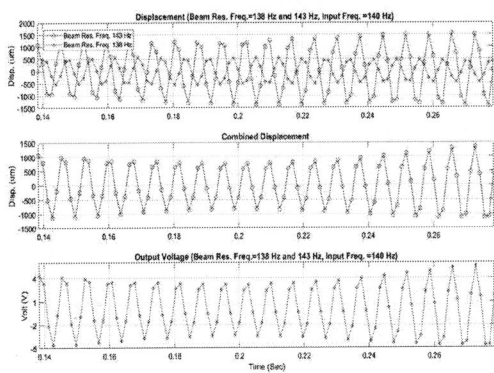

Fig. 6: Plot of displacement and output voltage for coupled two-beam structure vibrating at 140 Hz

Beam 2 shows a peak output voltage of 14 V at its resonant frequency and has a bandwidth of 8 Hz, which can be observed from Fig. 4.

Finally, we perform the transient analysis of the proposed two-beam structure and observe the peak voltage and power generated at different frequencies. This is recorded in TABLE III.

This structure generates a peak output voltage of 18 V at 142 Hz and has a bandwidth of 15 Hz. The bandwidth is significantly higher than both of the single beam structures. From Fig. 4, we observe an interesting phenomenon. The output voltage of the two-beam structure falls at 140 Hz while both the single beam structures have high output voltage at that frequency. This can only be explained by considering the relative phase of displacement of both the beams, as already discussed in Section II. We need to observe the displacement and the output voltage for the single beams and the equivalent two beam structure vibrating at 140 Hz. The result is shown in Fig. 6.

It may be clearly observed that the displacements of both beams are out of phase with one another. As a result, the net displacement of the two-beam structure reduces and hence the output voltage at that frequency is very low.

V. CONCLUSION

In this paper, a multi-beam energy harvester is designed to improve the bandwidth of operation of the piezoelectric energy harvester. Two single beam structures and a two-beam structure was designed and simulated. The proposed two-beam structure shows a significant bandwidth improvement as compared to the single beam structures. It gives a bandwidth of 15 Hz as compared to 10 Hz and 8 Hz bandwidth obtained in case of single beams Beam 1 and Beam 2 respectively. Thus, the proposed two-beam structure

has about 1.5 times wider bandwidth as compared to that of single beam structures.

ACKNOWLEDGMENT

We, all the authors, would like to express our special thanks to Science and Engineering Research Board (SERB), Govt. of India for funding to execute this research through project ref. no. ECR/2016/001279. This research was not possible without the help and support of BIT Mesra, Ranchi and SERB, Govt. of India.

REFERENCES

[1] H. Takise, T. Takahashi, M. Suzuki and S. Aoyagi, "Fabrication of piezoelectric vibration energy harvester using coatablePolyVinylideneDiFluoride and its characterisation," in *IET Micro & Nano Letters*, vol. 12, no. 8, pp. 569-574, 8 2017.
[2] Heung Soo Kim, Joo-Hyong Kim and Jaehwan Kim. A Review of Piezoelectric Energy Harvesting Based on Vibration. INTERNATIONAL JOURNAL OF PRECISION ENGINEERING AND MANUFACTURING Vol. 12, No. 6
[3] Y. Tan, Y. Dong and X. Wang, "Review of MEMS Electromagnetic Vibration Energy Harvester," in Journal of Microelectromechanical Systems, vol. 26, no. 1, pp. 1-16, Feb. 2017.
[4] Yu, H.; Zhou, J.; Deng, L.; Wen, Z. A Vibration-Based MEMS Piezoelectric Energy Harvester and Power Conditioning Circuit. *Sensors* **2014**, *14*, 3323-3341
[5] L. Deng, Z. Wen, X. Zhao, C. Yuan, G. Luo and J. Mo, "High Voltage Output MEMS Vibration Energy Harvester in d_{31} Mode with PZT Thin Film," in *Journal of Microelectromechanical Systems*, vol. 23, no. 4, pp. 855-861, Aug. 2014.
[6] Chaudhuri, Dipta & Kundu, Sudip & Chattoraj, Neela. (2018). Design and analysis of MEMS based piezoelectric energy harvester for machine monitoring application. Microsystem Technologies. 10.1007/s00542-018-4156-z.
[7] Chaudhuri, Dipta & Kundu, Sudip. (2017). MEMS piezoelectric energy harvester to power wireless sensor nodes for machine monitoring application. 584-588. 10.1109/DEVIC.2017.8074018.
[8] D. Chaudhuri, S. Kundu and N. Chattoraj, "Harvesting energy with zinc oxide bio-compatible piezoelectric material for powering cochlear implants," *2017 Innovations in Power and Advanced Computing Technologies (i-PACT)*, Vellore, 2017, pp. 1-5. doi: 10.1109/IPACT.2017.8245043
[9]Mohamed, Ramizi & Sarker, Mahidur & Mohamed, Azah. (2016). An optimization of rectangular shape piezoelectric energy harvesting cantilever beam for micro devices. International Journal of Applied Electromagnetics and Mechanics. 50. 537-548. 10.3233/JAE-150129.
[10] Jens Twiefel, Marcus Neubauer, Jörg Wallaschek, "Bandwidth Improvement for Vibration Energy Harvesting devices" AMA Conferences 2013 - SENSOR 2013, OPTO 2013, IRS 2 2013
[11] V. Meruane and K. Pichara, "A Broadband Vibration-Based Energy Harvester Using an Array of Piezoelectric Beams Connected by Springs," Shock and Vibration, vol. 2016, Article ID 9614842, 13 pages, 2016.
[12] Almuatasim Alomari, Ashok Batra, "Experimental and Modelling Study of a Piezoelectric Energy Harvester Unimorph Cantilever Arrays," in Sensors & Transducers, Vol. 192, Issue 9, September 2015, pp. 37-43
[13] Cao, J., Ling, M., Inman, D. J., & Lin, J. (2016). Generalized constitutive equations for piezo-actuated compliant mechanism. Smart Materials and Structures, 25(9), [095005].
[14] Hamrit F., Necib B., Driss Z. Analysis of mechanical structures using beam finite element method. Intern. J. of Mechanics and Applications, 2015, vol. 5, no. 1, pp. 23-30.

2019 Devices for Integrated Circuit (DevIC), 23-24 March, 2019, Kalyani, India

An approach towards the development of refreshable Braille Computer Display Unit

Moumita Ghosh
Electronics & Communication Engineering
JIS College of Engineering
Kalyani, India
moumita.ghosh90@gamil.com

Subham Ghosh
Electronics & Communication Engineering
JIS College of Engineering
Kalyani, India
sghosh.network@gamil.com

Shivam Sarkar
Electronics & Communication Engineering
JIS College of Engineering
Kalyani, India
shivamsarkar1998@gmail.com

Kaustav Saha
Electronics & Communication Engineering
JIS College of Engineering
Kalyani, India
kaustav.in.saha@gmail.com

Anirudha Ray
Electronics & Communication Engineering
JIS College of Engineering
Kalyani, India
aray.uit@gmail.com

Biswarup Neogi
Electronics & Communication Engineering
JIS College of Engineering
Kalyani, India
biswarupneogi@gmail.com

Abstract— **Modern time is the age of digitalization. People are accessing information with the advancement of digital technology. Blinds have very limited accessible resources to access this digitalized information. In this article, a novel proposal about working and development of an electronic refreshable braille display unit to build a getaway to the digital world for the blind people. An economic adoption of a tactile display having an array of six small independent vibrator motors are arranged in an array of a 3×2 matrix. The system would be enabled to initially capture information from computer screen then with suitable image processing, characters recognition and stimulation of target motor/s would enable enhanced readability of the texts.**

Keywords— *Refreshable Braille, Electronic Braille, Tactile Display, Braille computer*

I. INTRODUCTION

Statics shows around the world about 253 million people live with vision impairments of which 36 million are blind. The vast majority live in low-income situations [1], [2]. These blinds are blessed with their incredible senses. In their day to day, activities rely on auditory or tactile feedback. In 1821 Louis Braille implemented tactile, offers braille for reading and writing [3], [4], [5]. Braille is a tactile language consisting of six raised dots of varying arrangements that represent characters. Fig. 1 shows braille cell, single Braille character comprises an array of 3×2 dots [6], [7].

Fig. 1. Braille Cell

Modern technologies have improved the development of lifestyles for blind people. The rise of the digital age, an electronic refreshable braille display capable of converting

digital text from documents, computer screens etc. into physical braille [8], [9], [10].

In the year 1966, Linvill et al. proposed a concept by representing the operation of impressed on a set of photocells coupled one-to-one to piezoelectric reeds making the tactile image processing pins vibrate and facsimile the printed material presented tactually [11]. Kang et al. have proposed an enhanced compact device having set of vertically portable stimulation pins positioned on a closed rack with opening at one side allowing stimulation outward to sense by touch [12]. Ezekiel Kruglick, introducing a hierarchical perspective. The devised haptic communication system has been accepted as well as patented in the year 2012 [13]. Cheng Xu et al. introduce a system, touching fingertips on screens to interpret 2-D information supporting communication among blind peoples [14]. After a detail analysis on the existing tactile display, a concept presented by of Jacksonet et al. providing tactile feedback with a touch surface layer in form of set of pixels having aperture for emitting fluid along with a valve for opening and closing the aperture to inject the fluid at a frequency for recognition by the sensors on the human finger skin [15]. Paolo Motto Ros et al. introduced design of a flexible dynamic tactile display device an array of 8X8 matrix of pins maintaining standards of 0.7 mm pin gap along with an excellent refresh rate by utilizing piezoelectric technology, providing reduction in arbitrary patterns and contingent user needs, rendering of non-character information as well as reconfigurable rendering [16]. XinXie et al. introduced MEMS enabled, high resolution tactile display of 28-element prototype, fabricated with each element consists of piezoelectric extensional actuator along with micro fabricated scissor converting the in plane actuation into high amplitude, out-of-plane vibration being sensed by fingers, having displacement of 3×10 mm exceeds 10 µm providing force exceeding 45 mN [17]. Russomanno et al. developed a novel Electronic Braille display, having a single character consisting of an array of 2 columns and 3 rows to represent. High acceptability rates are reported from experiments by experienced Braille readers. The importance of sliding contact between the fingertip and braille reading surface is reported to be more demanding and error prone when compared to conventional Braille displays [18]. Akhtar et al. introduced an economic as well as low power consuming method of

978-1-5386-6723-1/19 $31.00 © 2019 IEEE

actuation for braille refreshable display by lead-screw actuation. No continuous power supply is required to retain the pins in the position once they have been actuated [19]. An innovative, low-cost Refreshable Braille Display (RBD) developed by Leonardis et al. A single actuated cursor that refreshes Braille cells composed of pins. Piezo-actuated pins allow for implementation in matrices [20].

The proposed system interprets English Language from computer screen to Braille Language. Vibrating Braille dots substantiated more efficient as well as economical than the traditional raised dots of Braille system.

II. PROPOSED REFRESHABLE BRAILLE DISPLAY UNIT

In the proposed Refreshable Braille display unit any language is translated into braille code. Fig 2 shows repetition of proposed system. Fig 3 shows block diagram of proposed system.

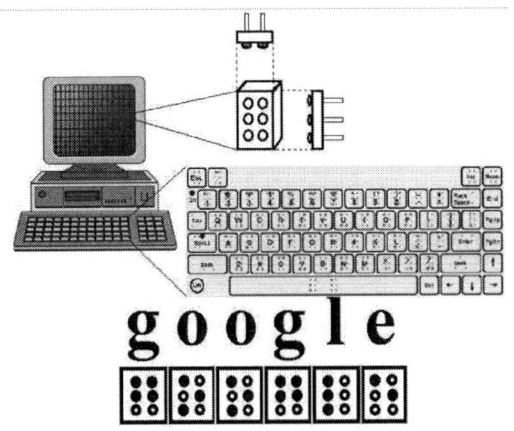

Fig. 2. Representation of The Proposed System

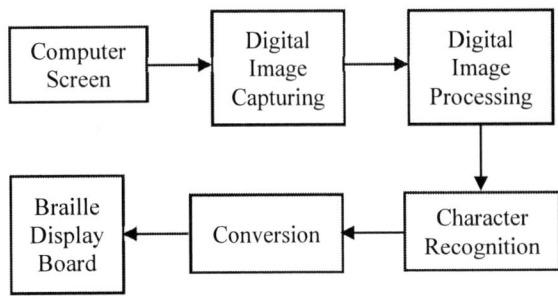

Fig. 3. Block Diagram of The Proposed System

Technologies Used Behind Development of Braille display board:

A. Digital Image Capturing

Image can be capturing several ways thought all have one common thing; the visual information requires to be converted into an electronic signal. We recognize images by the light that reflects off or passes through an object. The electronic information can be stored as an analog / digital signal. The image capturing process can be divided into four steps. These are: Capture, Store, Edit & Display.

(India) for carrying out Research & Development work in this subject. Project ID: RDUG2018028.

Fig. 4. Capturing Image from webpage

B. Digital Image Processing

After digital image capturing, to process the image for extracting the data containing the characters and its position co-ordinates, we used Optical Character Recognition Technique [21]. Optical character recognition (OCR) is the mechanical/electronic transformation of images of typewritten or printed text into machine-encoded text. Improve the chances of successful recognition, OCR software often "pre-processes" images. Techniques includes:

- De-skew – proper alignment of the document during scanned may be required to shift a few degrees clockwise / counter-clockwise to make lines of text flawlessly vertical/horizontal.
- DE speckle– remove spots as well as smoothing edges [22].
- Binarization – Convert an image into greyscale; it is essential since utmost recognition algorithms operation only on binary images since it shows to be more straightforward. The effectiveness of this process influences to improvement in the quality of the character recognition along with careful make decisions in the choice of the binarization for a given type of input image type.

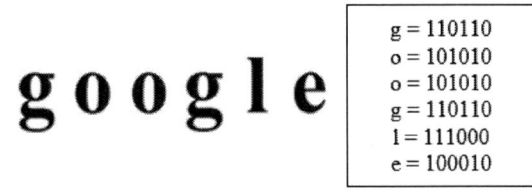

Fig. 5. Binary representation of "google"

- Word and Line detection – According to character shapes, establishes a baseline as well as separates character and words.
- Script recognition – It is becoming essential for the identification of the multilingual document before the OCR can process the specific script.

C. Character Recognition

The OCR algorithm can be classified into two basic types. Matrix matching comprises an image to save on a pixel-by-pixel basis; also recognized as "pattern recognition" / "image correlation".
It relies on the input precisely separated from the rest of the image as well as on the stored graph at the same scale in a similar font. This procedure works best with text. Feature

extraction decomposes graphs into "peculiarities" like lines, line direction, line intersections and closed loops. The traditional technique of feature disclosure in computer vision are suitable to this type of OCR, regularly observed in "intelligent" handwriting recognition and indeed most modern OCR software k-nearest neighbors classification algorithm is used to distinguish image features among stored graph to determine the nearest match.

Software like Cuneiform and Tesseract utilize a two-pass path to character recognition. The second pass is recognized as "adaptive recognition", it employs identification of the letter shapes along with high reliance on the first pass to recognize better the leftover letters on the second pass. This is beneficial for distorted fonts or low-quality scans.

D. Conversion

OCR efficiency can be enhanced if the output is restrained by a thesaurus – a listing of words that are permitted to establish a sentence. This method can be questionable if the document contains words, not in proper nouns. Tesseract applies its vocabulary to check the character segmentation level, for improved precision. More advanced OCR operations can maintain the initial layout of the page. "Near-neighbor analysis" can gain use of occurring frequencies to better errors, by noting that certain words are often seen together.

In modern times, the OCR technology providers started to squeeze OCR systems to deal with precise input adequately. Ahead of an application-specific lexicon, excellent execution can be possessed by leading into account business rules, common expression or rich information contained in color images. This approach is called "Application-Oriented OCR" or "Customized OCR" .

E. Refreshable Braille Display Board

The system intrinsically follows an embedded technology driven approach, where the hardware used is effectively and efficiently controlled by the fed software shows in Fig 7. Refreshable braille display system. The combination and smooth functioning of the system allow the tactile display to work making the blind to be able to read the content of computer by touching the vibrating pins of braille blocks designed on tactile display board. Fig 6, shows the proposed conversion of "Google page". Each Cell of the board consists of 6 vibrating pinheads, placed 2.5 mm apart from each other raised dots and the individual cells respond according to the input signals and symbolize the specific braille character. Each cell displays braille character, as per vocabulary from justified word as well as sentences. This system converts digital documents into vibro-tactile feedback output. Pins are vibrated to represent the Braille code represent in Table 1.

Table 1: Output values of pins are given below [10]:

Pin Numbers	Vibrates for English Characters
1	A,B,C,D,E,F,G,H,K,L,M,N,O,P,Q,R,U,V,X,Y,Z
2	B,F,G,H,I,J,L,P,Q,R,S,T,V,W
3	K,L,M,N,O,P,Q,R,S,T,U,V,X,Y,Z
4	C,D,F,G,I,J,M,N,P,Q,S,T,W,Y
5	D,E,G,H,J,N,O,Q,R,T,W,Y,Z
6	U,V,W,X,Y,Z

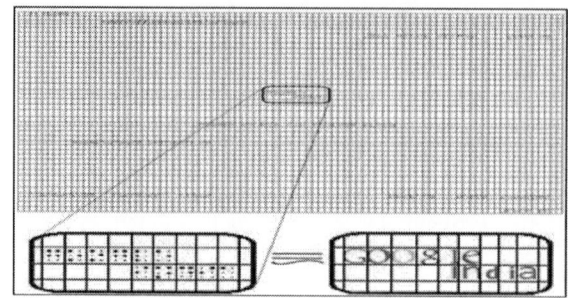

Fig. 6. Braille Representation of Google Home Page

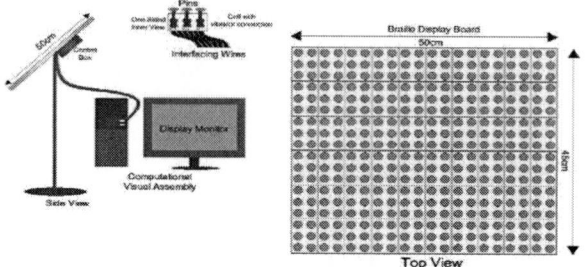

Fig. 7. Proposed Refreshable Braille Display Board

CONCLUSIONS

The article discusses about the methodology to be adopted for development of the braille display board. The board will be able to bridge the gap between the information accessibility for the blinds and will enable them to become more conversant with the pool of knowledge documented as soft data. The future scope of work remains to develop the hardware of the display board and to formulate the software part in order to have a smooth running of the system.

ACKNOWLEDGMENT

The author/s thankfully acknowledge(s) the financial support provided by The Institution of Engineers (India) for carrying out Research & Development work in this subject. Project ID: RDUG2018028.

REFERENCES

[1] R.R. Bourne, S.R. Flaxman, T. Braithwaite, M. V. Cicinelli, A. Das, J.B. Jonas, J. Keeffe et al., "Magnitude, temporal trends, and projections of the global prevalence of blindness and distance and near vision impairment: a systematic review and meta-analysis," The Lancet Global Health 5, no. 9, 2017, e888-e897.

[2] M. Goldschmidt, "Orientation and Mobility Training to People with Visual Impairments," Mobility of Visually Impaired People, 2018, pp. 237–261.

[3] E. Rex, A. Koenig, R. Baker, "Foundations of braille literacy," American Foundation for the Blind, 1994.

[4] C.M. Mellor, "Louis Braille: A touch of genius," National Braille Press, 2006.

[5] J. Jiménez, J. Olea, J. Torres, I. Alonso, D. Harder, K. Fischer, "Biography of Louis Braille and Invention of the Braille Alphabet," Surv. Ophthalmol, 54, 2009, 142–149.

[6] K.A. Toussaint, J.H.Tiger, "Teaching early braille literacy skills within a stimulus equivalence paradigm to children with degenerative visual impairments," Journal of Applied Behavior Analysis, 43(2), 2010, pp.181-194.

2019 Devices for Integrated Circuit (DevIC), 23-24 March, 2019, Kalyani, India

[7] B. Landau, L.R. Gleitman, B. Landau, "Language and experience: Evidence from the blind child," Harvard University Press, Vol. 8. 2009.

[8] A. Khasnobish, S. Datta, D. Sardar, D.N. Tibarewala, A. Konar, "Interfacing robotic tactile sensation with human vibrotactile perception for digit recognition," Robotics and Autonomous Systems, 71, 2015, 166-179.

[9] A. Ray, S. Ghosh, B. Neogi, "An Overview on Tactile Display, Haptic Investigation towards Beneficial for Blind Person," International Journal of Scientific & Engineering Research, Volume 6, Issue 5, 2015, 88-95.

[10] M. Ghosh, D. Banerjee, M. Roy, A. Ray, S. Ghosh, B. Neogi, "Arranged Braille Computer Display Board (ABCD): A Simulative Scholastic Approach towards Analytic Circuitry," International Journal of Computer Applications, 2017, 15-20.

[11] J.G. Linvill, J.C. Bliss, "A direct translation reading aid for the blind," Proceedings of the IEEE, 54(1), 1966, 40-51.

[12] S.C. Kang et al. "Tactile Display Apparatus and Method Thereof," Patent no: US8248217, Patented on: Aug 21, 2012.

[13] E. Kruglick, "Tactile Display Control," Patent no : US8289291 , Patented on: Oct 16, 2012.

[14] Cheng Xu, Ali Israr, Ivan Poupyrev, Olivier Bau, C. Harrison, "Tactile display for the visually impaired using TeslaTouch," Proceeding CHI EA '11 CHI '11 Extended Abstracts on Human Factors in Computing Systems , 2011, Pages 317-322.

[15] J. Warren, P. Mei, "Tactile display using distributed fluid ejection," U.S. Patent Application 13/259,236, filed November 8, 2012.

[16] P.M. Ros,V. Dante, L. Mesin, E. Petetti, P.D. Giudice, E. Pasero, "A newdynamic tactile display for reconfigurable braille: implementation and test," Front Neuroeng, 7:6, 2014.

[17] X. Xie, Y. Zaitsev, L.F. Velásquez-García, S.J. Teller, C. Livermore, "Scalable, MEMS-enabled, vibrational tactile actuators for high resolution tactile displays," Journal of Micromechanics and Microengineering, Volume 24 , 2014.

[18] A. Russomanno, S. O'Modhrain, R.B. Gillespie, M.W. Rodger, "Refreshing refreshable braille displays," IEEE transactions on haptics 8, no. 3 , 2015, 287-297.

[19] S. A. Akhtar, D. Prasad, "Braille refreshable display using lead screw actuation," 4th IEEE Uttar Pradesh Section International Conference on Electrical, Computer and Electronics (UPCON) , 2017, pp. 5-8.

[20] D. Leonardis, L. Claudio, "Braille cursor: An innovative and affordable refreshable braille display designed for inclusion," In International Conference on Applied Human Factors and Ergonomics, 2018, pp. 302-311.

[21] A.M. Sabu, A.S. Das, "A Survey on various Optical Character Recognition Techniques," Conference on Emerging Devices and Smart Systems (ICEDSS), 2018, pp. 152-155.

[22] M.Y. Akpinar, E. Emeklıgıl, S. Arslan, "Extracting table data from images using optical character recognition text," 26th Signal Processing and Communications Applications Conference (SIU), 2018, pp. 1-4.

978-1-5386-6723-1/19 $31.00 © 2019 IEEE

2019 Devices for Integrated Circuit (DevIC), 23-24 March, 2019, Kalyani, India

An Improved Edge Detection Method based on Median Filter

Preeti Topno*,Govind Murmu
Department of Electronics Engineering
Indian Institute of Technology (Indian School of Mines)
Dhanbad, Jharkhand,India-826004
*Corresponding Author Email: topno_preeti@yahoo.com

Abstract—To preserve the edges from corrupted noise, median filter is a popular choice for filtering algorithms. The salt and pepper noise is one of the main problems present while capturing an image. To solve this problem the edges were detected, using different edge detection operators which include Roberts, Sobel, Prewitt and Canny, prior to processing by a median filter. Several simulations were carried out over a number of digital images.The peak signal to noise ratio (PSNR) is higher using Prewitt operator than Sobel operator. Similarly, results also confirm the better performance of Canny edge detection algorithm over others for edge detection however computational cost is high in it compared to other methods.

Index Terms—Sobel edge detector, Prewitt edge detector, Canny edge detection algorithm, median filter.

I. Introduction

Image processing is done to get the better pictorial information from a digital image. One of the image processing methods is image segmentation where an image is subdivided into smaller images with some unique properties [1]. Again reducing the large amount of data while processing an image is a necessary requirement now a day. A very important part of a digital image is edges. Edges are defined as sudden change of intensities, which occurs on the boundaries between two regions of an image [2]. Due to high frequency nature of noise, detection of an edge in noisy digital images is very difficult and a challenging task.

Several researchers have worked in this area to solve this problem. A survey was done on edge detection using various computed approaches based on different techniques like Fuzzy based approach, wavelet approach etc [3], [4]. Girish et al. had compared Robert, Prewitt and Sobel operators based on real time applications on Field Programmable Gate Array (FPGA) [5]. The author proposed an adaptive algorithm to solve the problems of traditional Canny edge detection algorithm [6]. Anchal et al. used a hybrid method which combines Sobel and Canny edge detectors [7]. This approach has been implemented to get better smoothening in the output image without loosing much of detailed information. Ganesan P et al. had done a comparative study of different edge methods using various applications [8]. Dong et al. proposed an algorithm based on wavelet transform to detect the edges of underwater images [9]. Naijian et al. designed an edge detection algorithm

based on machine vision for the cutting edge process in rough laminated wood [10].

Median filter is a non-linear filter which helps in reducing the impulse noise and has the capability to keep the properties of edge. It also provides less blurring when compared to other linear smoothing filters [11], [12] such as averaging filter. In this paper an attempt has been made to use median filter for reducing the noise in edges once they are detected by different operators. The paper is organized as follows: In section II, overview of basic detection and the different classical methods of edge detection is presented briefly followed by Section III where details of the methodology used is discussed. Section IV presents the experiments and results obtained, which are discussed finally in the conclusions in section V.

II. Overview of Basic Detection

To detect discontinuities, the common way is to run a mask through the image. There are two basic detection-Point detection and line detection. The principle of point detection is to detect gray level points which is different from the background and also located in the homogeneous area. The Laplacian mask is a high pass(sharpening) filter which highlights the details of an image or to enhance the detail that has been blurred.

To detect abrupt meaningful changes of intensity, it is classified into first order and second order derivative operator. The rapid change in intensities is identified using one of the two general criteria:

1) The magnitude of first derivative of intensity is compared with a threshold value. If it is greater than threshold then the respective place is noted down.
2) All the places where second derivative of the intensity has a zero crossing are noted down.

The operators used in the first derivative are Roberts, Prewitt, Sobel. The second order derivative searches for zero crossing and the operators are LOG, Canny.

Some of the popular edge detection techniques are:

A. Roberts operator

The Roberts operator [1] is a first order derivative operator which uses a 2X2 mask to find the gradient. Approximation to the gradient is

$$\nabla f \approx |z_9 - z_5| + |z_8 - z_6|$$

978-1-5386-6723-1/19 $31.00 © 2019 IEEE   378

where $G_x = z_9 - z_5$ and $G_y = z_8 - z_6$.

Being a 2×2 operator, hardware implementations are simple with high speed operations. This operator has high accuracy for position but it is not used much due to even nature. Then it is not able to detect some parts of an image which is another problem with it.

B. Prewitt operator

In Prewitt operator [1], a 3X3 mask is used to find the first derivatives. In digital images the pixel values on left and right column of reference pixel is subtracted for x-direction. Similarly for y-direction gradient the top and bottom rows of reference pixels are subtracted. The partial derivatives using 3X3 masks of size are given by

$G_x=|(z_7 + z_8 + z_9)| - |(z_1 + z_2 + z_3)|$
and $G_y=|(z_3 + z_6 + z_9)| - |(z_1 + z_4 + z_7)|$

C. Sobel operator

As the Prewitt operator produces noisy results so some changes are done in the above two equations by giving extra weight to the centre reference coefficient. Mathematically,

$G_x= z_7 + 2z_8 + z_9 - z_1 + 2z_2 + z_3$
and $G_y = z_3 + 2z_6 + z_9 - z_1 + 2z_4 + z_7$

By taking a weight of 2 for the centre location makes the image smooth. Hence by using these two partial derivatives, gradient is given by

$$\nabla f = [G_x{}^2 + G_y{}^2]^{0.5}$$
$$= [(z_7 + 2z_8 + z_9 - z_1 + 2z_2 + z_3)^2$$
$$+ (z_3 + 2z_6 + z_9 - z_1 + 2z_4 + z_7)^2]^{0.5}$$

where z's are intensities.

D. Canny Edge Detection algorithm

This is the widely used edge detector found in literature. First the image is made smooth using a Gaussian filter with a specified standard deviation to reduce noise. The derivatives are computed using any one of the three detectors i.e., Sobel, Prewitt or Robert's Edge Detector. Next the local gradient and edge direction are calculated at each point. Then the non-maximum suppression is followed where pixels are tracked along the top of the ridges. If any pixel is not on the top of the ridge then such pixel values are set to zero. This makes a thin line for edges. The ridge pixels are then compared with two threshold values, T_1 and T_2 with $T_1 < T_2$, known as hysteresis thresholding.The ridge pixels which are greater than T_2 are known as strong edge pixels and the ridge pixels with values in between T_1 and T_2 are weak edge pixels.Then finally the weak edges which are 8-connected to the strong edges,are linked.

III. IMPROVED EDGE DETECTION USING MEDIAN FILTER

The edges found using above methods may have noise present in them. Many a times the presence of noise may produce a false edge. Thus in this paper an attempt has been made to increase the reliability of edge detection by removal of noise. There are many non-linear methods to suppress impulse noise like min, max and median filters [1]. Among all of these filters, median filter is useful to suppress noise, such as salt and pepper noise which may create a problem in edged detected so far. The proposed method also brings an improvement in the Prewitt edge detector.

To implement the idea a digital image is first corrupted by a known noise such as salt and pepper noise. The edges of this noisy image is then detected using known operators. Obviously the noise will disturb the detected edges, to improve the situation median filter is used over noisy images. In this way a recovery of edges with a combination of median filter should produce much better result which is presented in next section.

IV. EXPERIMENTS AND RESULTS

Though digital image is a low frequency spectrum but the edge and noise are both high frequency signals. In this section, we have shown the results obtained by using our propounded method to the original images.

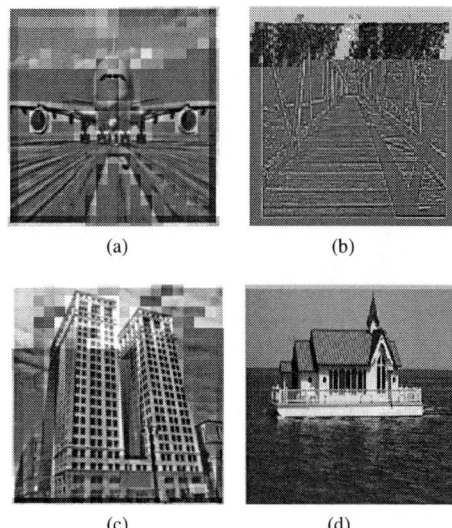

(a)   (b)

(c)   (d)

Fig. 1: Original images (a) Aero plane (b) Bridge (c) Dime building (d) Houseboat

A. The results using Sobel operator

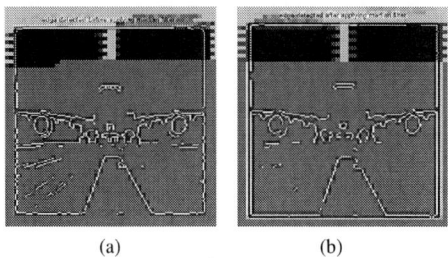

(a)   (b)

Fig. 2: Edge detected using Sobel operator (a)Before applying median filter (b)After applying median filter

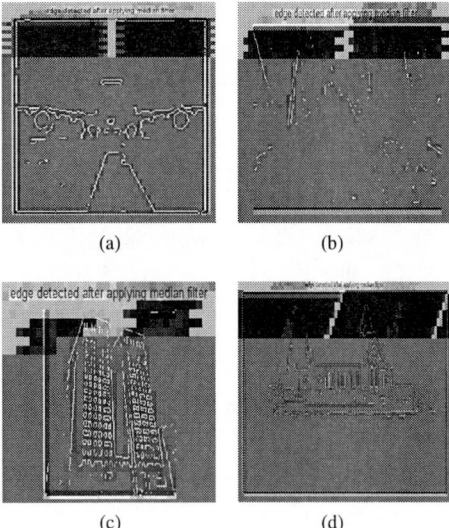

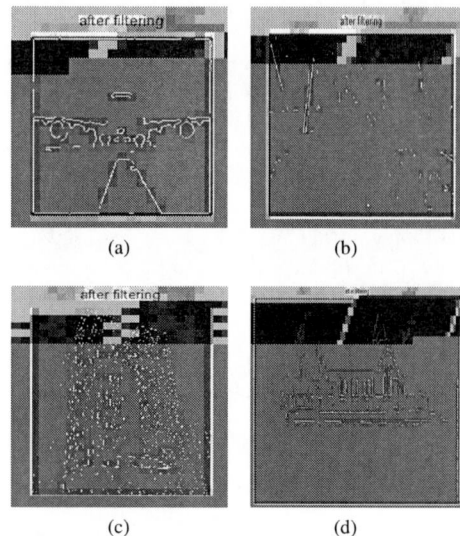

Fig. 3: Edge detected using Sobel operator (a) Aeroplane (b)Bridge (c) Dime building (d) House boat

Fig. 5: Edge detected using Prewitt operator (a) Aeroplane (b) Bridge (c) Dime building (d) House boat

These original images were downloaded from internet and used for academic and research purpose only. PSNR and MSE are used as a parameter to judge the quality of a recovered image. The PSNR is calculated using the following formula [11]:

$$PSNR = 10 \, log \left(\frac{a_{max}^2}{MSE} \right) dB \quad (1)$$

where

$$a_{max} = \sum_{i=1}^{M} \sum_{j=1}^{N} [f(i,j)]$$

and

$$MSE = \frac{1}{M \times N} \sum_{i=1}^{M} \sum_{j=1}^{N} [f_{out}(i,j) - f(i,j)]^2$$

C. The results showing Canny edge detector

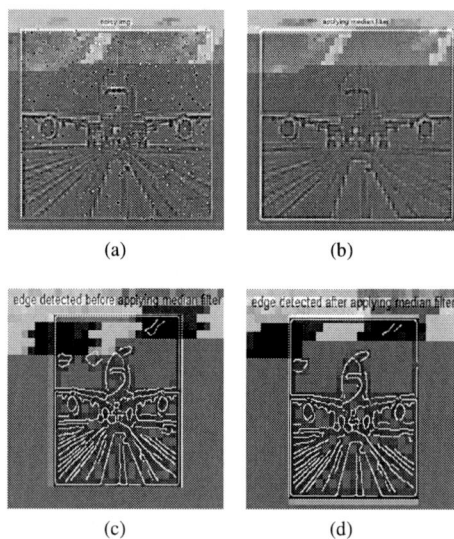

Fig. 6: Using Canny edge detector (a)Noisy image (b) Applying median filter (c) Edge detected before applying median filter (d) Edge detected after applying median filter

B. The results showing Prewitt operator

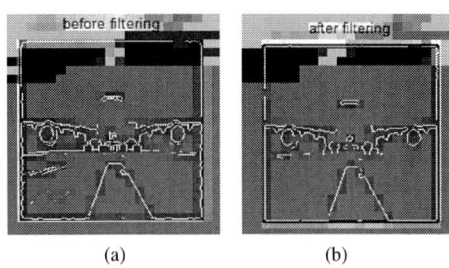

Fig. 4: Edge detected using Prewitt operator (a) Before applying median filter (b) After applying median filter

where MSE is mean squared error, $f(i,j)$ is the original image and $f_{out}(i,j)$ is the processed image.

To simulate the algorithms four digital images are downloaded. They are shown in Fig.1(a-e). The edges are then

detected by using Sobel operator. They are presented in Fig.2. As the edge detection is applied with and without median filter so the better detection is possible only with the use of median fiilter.

Similarly, Fig.3 presents the edge detected image using Prewitt operator. The edges are detected before and after applying median filter. Fig.4,5 show the edges of different images after passing through the median filter using both Sobel and Prewitt operator respectively.

978-1-5386-6723-1/19 $31.00 © 2019 IEEE

Fig.6 shows the result of Canny edge detection algorithm. The edges are more clearer than Sobel operator as it is less sensitive to noise.

Fig.6(a) shows the noisy image where the noise is present in terms of black and white dots. These black and white dots are known as salt and pepper noise. Then the image is passed through median filter as shown in Fig.6(b). So when median filter is operated over an input image, it filters out the noise and thus the images are visually better than the original noisy images. Fig.6(c) presents the edge detected image before applying median filter. Again Fig.6(d) presents the edge detected image after applying median filter. The edge detection methods were implemented on MATLAB®.

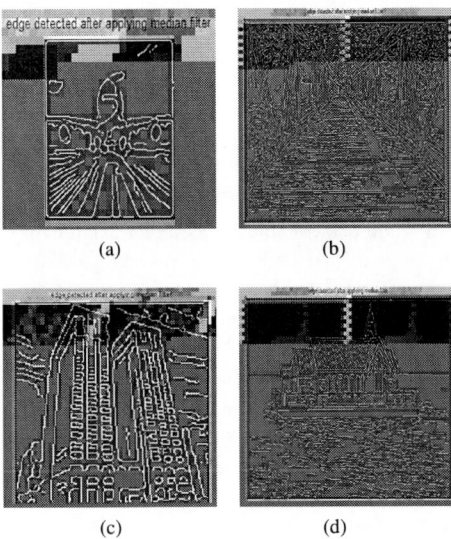

(a)   (b)

(c)   (d)

Fig. 7: Edge detected image using Canny edge detector (a)Aeroplane (b) Bridge (c) Dime building (d) house boat

The Peak Signal to Noise Ratio(PSNR) values found using the proposed algorithm are listed in the Table I. The Canny edge detection algorithm shows the highest PSNR value. This recommends that median filter can be used to get more better results.

TABLE I: Peak Signal to Noise Ratio (PSNR) in dB

Images Names	DIFFERENT EDGE DETECTION ALGORITHMS		
	Prewitt	Sobel	Canny
Aeroplane	11.54	11.52	18.85
Bridge	12.2680	11.9822	11.4180
Boy and horse	10.7104	10.2617	12.4141
Dime building	3.2299	3.1992	12.7553
Houseboat	14.5851	14.5327	11.8734

V. CONCLUSIONS

When median filter is not used, there is noise in all the edge detected outputs and the edges detected are not continuous. Classical edge detections are sensitive to noise , so noise reduction method is necessary. Sobel edge detector is accurate in edge detection but it is sensitive to noise which gets removed

by median filter. Prewitt operator is more sensitive to noise but after processing by median filter, the amount of noise present is reduced significantly. It also gives high PSNR than Sobel edge detector. Similarly Canny edge detection algorithm gives the far best results in noisy environment but computationally it is complex and has high cost. For future work, other methods can also be compared with the current proposal. The adaptive method can also be used in combination of the current work for further improvement in noise reduction.

ACKNOWLEDGEMENT

Authors would like to acknowledge IIT(ISM), Dhanbad for providing partial financial support for attending the conference.

REFERENCES

[1] Rafael C. Gonzalez,Richard E. Woods,*Digital Image Processing*, 3rd ed, Pearson Education Publication, 2009.

[2] Raman Maini, Himanshu Aggarwal,"Study and Comparison of Various Image Edge Detection Techniques", *International Journal of Image Processing(IJIP)*,vol3,Issue 1,Feb.2009.

[3] Lakshmi S. and V. Sankaranarayanan,"A Study of edge detection techniques for segmentation computing approaches",*IJCA Special Issue on Computer Aided Soft Computing Techniques for Imaging and Biomedical Applications(CASCT)*, pp.35-40, 2010.

[4] Muthukrishnan R. and Radha M.,"Edge detection techniques for image segmentation",*International Journal of Computer Science and Information Technology(IJCSIT)-2011)*,vol3.no.(6), pp.259, Dec. 2011.

[5] Girish N. Chaple, R. D. Daruwala, Manoj S. Gofane,"Comparisons of Robert, Prewitt, Sobel operator based edge detection methods for real time uses on FPGA",*International Conference on Technologies for Sustainable Development(ICTSD-2015)*, Mumbai, India, Feb. 2015.

[6] Liying Yuan, Xue Xu,"Adaptive Image Edge Detection Algorithm Based on Canny Operator", in *Advanced Information Technology and Sensor Application (AITS), 4th International Conference on*,Harbin, China, pp.28-31, 2015.

[7] Anchal Kalra and Roshan Lal Chokar,"A hybrid approach using Sobel and Canny operator for digital image edge detection, in *Micro Electronics and Telecommunication Engineering(ICMETE), IEEE International Conference on*,Ghaziabad, India, pp. 305-310, 2016.

[8] Ganesan P, G.Sajiv,"A comprehensive study of edge detection for image processing applications",*International Conference on Innovations in information, Embedded and Communication Systems (ICIIECS)*,Coimbatore, India, Mar 17, pp.1-6, 2017.

[9] Dong Xiaoheng,Li Minghang, Miao Jiashu,"Edge Detection Operator for Underwater Target Image", in *2018 IEEE 3rd International Conference on Image, Vision and Computing (ICIVC)*,Chongqing, China, pp. 91-95, Jun 27, 2018.

[10] Naijian CHEN,Xiuhua MEN,Xiangdong HAN,"Edge Detection Based on Machine Vision Applying to Laminated Wood Edge Cutting Process", in *13th IEEE Conference on Industrial Electronics and Applications (ICIEA)*,Wuhan, China, pp.449-454, May 31, 2018.

[11] Youlian Zhu, Cheng Huang,"An Improved Median Filtering Algorithm for Image Noise Reduction",*International Conference on Solid State Devices and Material Science, Physics Procedia*,Macao, vol. 25, pp.609-616, 2012.

[12] Kumar Bittu, "Spectral subtraction using modified cascaded median based noise estimation for speech enhancement." in *Proceedings of the Sixth International Conference on Computer and Communication Technology*, MNIT Allahabad, India, pp. 214-218, 2015.

Improvement of Output Power in Piezoelectric Energy Harvester under Magnetic Influence

Ashutosh Anand
Dept. of Electronics and Communication Engineering
Birla Institute of technology, Mesra
Ranchi, India
ashunitjsr@gmail.com

Sudip Kundu
Member, IEEE
Dept. of Electronics and Communication Engineering
Birla Institute of technology, Mesra
Ranchi, India
kundu.sudip@gmail.com

Abstract— **In this paper, a method is proposed to improve the performance of the piezoelectric energy harvester by introducing the two magnets in the overall structure. One of the magnets is attached to the proof mass and the other one is used to create a magnetic field in the vicinity of the proof mass. The generated magnetic field develop a repulsive force between the magnet which improves the performance of the harvester in terms of the output voltage and output power. The proposed structure generates a peak voltage of 21.2 V at the resonant frequency of 95.5 Hz at an excitation acceleration of 1g. The peak output power of the proposed structure is 3.16 μW which is 7.11 % more than the simple rectangular structure of the same dimension.**

Keywords— *Piezoelectric, Vibration, Energy harvester, Magnet*

I. INTRODUCTION

The size and power requirement of the electronic devices reduces drastically due to the advancement in VLSI technology. Thus, energy harvesting became the active research area with great scope in areas such as in biomedical devices, wireless communication, military equipment and consumers electronic devices in the past few years. Vibration-based energy harvester (EH) is the best suited to micro-power devices where solar and thermal based energy harvester not feasible and also not easily available. This EH converts the available vibration energy from the ambient environment into useful electrical energy that can be used to power low power electronics devices [1]. Piezoelectric [2][3], electromagnetic [4] and electrostatic [5][7] are the different approaches to harvest energy from the ambient vibration. The piezoelectric energy harvester is favored over electrostatics and electromagnetic because it does not require any external voltage source and easy to integrate into a MEMS environment [2]. The cantilever structure is mostly preferred for the vibration-based EH as the average strain produced in the cantilever structure for a given input force is more than other structure such as cymbal type structure, stack type structure, shell type structure [6]. The rectangular cantilever structure is easy to fabricate and implement.

The output voltage of the piezoelectric cantilever structure is maximum at its resonant frequency. However, with a slight deviation in the input vibration frequency, the output voltage falls rapidly. Some researchers try to overcome by introducing the non-linearity in the structure [8].

Non-linearity can be introduced in the piezoelectric EH by

attaching the magnetic tip mass at the free end [9]. The relative orientation, alignment, and distance of the magnet produces the nonlinear effect. Challa *et.al.* [9] designed 34 mm energy harvester and uses the magnetic force to tune the resonant frequency of EH by changing the distance between the magnets. Zhu *et.al.* [10] designed a cantilever beam of length 13 mm and width of 5 mm and uses the attractive force of axially oriented permanent magnets to tune the resonant frequency of cantilever beam of electromagnetic micro-generator. Wei-Jiun Su *et.al.* [11] designed a cantilever of length 98 mm and uses the magnets on dual cantilever beam to working bandwidth of the EH. Similarly, D. Guo *et.al.* [12] designed the array of the cantilever of length 12 cm with magnetic tip mass to improve bandwidth and performance of EH. It may be noted that in all the structures [9]-[12] used magnetic force and improve the performance of the PZ cantilever structure having a larger dimension, but as per authors knowledge, there is no such result which shows the use of magnetic force in micro-cantilever structure having a length less than 5 mm.

This paper introduces two magnets in the rectangular structure. One magnet is attached to the proof mass. The second magnet is in the vicinity of first to create a magnetic field, which improves the performance of EH without changing the fabricated size.

The paper has been organized as follows. Section II discusses the design and analysis of the structure. Section III talked about the magnetic effect on the stress of the piezoelectric layer. Section IV discusses the electrical output of the piezoelectric EH. In this section comparisons of the outputs between the proposed structure and the commonly used rectangular structure have been presented. Finally, the conclusion has been drawn in section V.

II. DESIGN AND ANALYSIS

The EH consists of the silicon-based cantilever beam with integrated Silicon (Si) proof mass along with permanent magnet at the free side of the cantilever beam as shown in figure 1. The proof mass consists of Si and permanent magnet (PM1) to reduce the resonant frequency [1]. The second magnet (PM2) is fixed at some distance from the first magnet to produce the repulsive magnetic effect. The permanent magnet behaves as a nonlinear spring to produce non-linearity. The Zinc Oxide (ZnO) is the piezoelectric material used in the cantilever to generate electrical energy. The dimension of different materials and their property are listed in Table 1 and Table 2 respectively. The magnetic field intensity of the magnet is 5 mT and the distance between the magnets (D) is 500 μm. The structure is built in COMSOL Multiphysics software which is a commonly used software for finite element analysis.

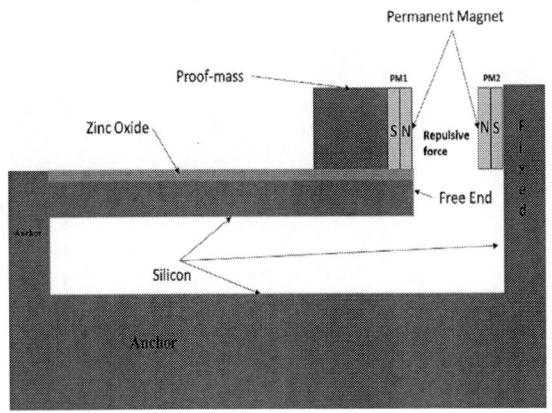

Figure 1. Block diagram of piezoelectric energy harvester with a magnetic tip mass

Table 1. The geometry of different materials used in the cantilever

	Si	ZnO	Proof mass		PM2
			Si	PM1	
Length (μm)	5000	5000	800	200	200
Width (μm)	1000	1000	1000	1000	1000
Thickness (μm)	12	2	850	850	850

Table2 Property of the materials used in the cantilever beam

Property	Material		
	Si	ZnO	Magnet
Young's Modulus	170 GPa	210 GPa	0.1 GPa
Poisson ratio	0.29	0.33	0.29
Density	2329 [kg/m³]	5680 [kg/m³]	7500 [kg/m³]

The piezoelectric EH with magnetic tip mass aligned with the fixed magnet can be mechanically modeled as the spring mass damper configuration with additional magnetic force as shown in figure 2. The equivalent electrical system is shown in figure 3. The magnetic force in the Piezoelectric EH produces nonlinear restoring force. The general electromechanical coupling equation of Piezoelectric EH can be modified to be used in the magnetic environment using newtons law and Kirchhoff law as shown in equation 1 [9].

$$M_{eq}\ddot{x}(t) + d_{eq}\dot{x}(t) + k_{eq}x - \theta v + F_m = -M_{eq}\ddot{y}(t) \quad (1)$$

$$C_p\dot{v}(t) + \frac{v}{R_l} + \theta\dot{x} = 0 \quad (2)$$

Where M_{eq}, $d_{eq,}$ and k_{eq} is the equivalent mass, damping coefficient and spring constant of piezoelectric EH. θ is the equivalent electromechanical coupling term and v is the voltage across the resistor. C_p is the equivalent capacitance of the piezoelectric layer in EH. F_m is the magnetic force acting between the magnetic tip mass and fixed magnet at the zero-deflection position, is given by equation 3.

$$F_m = -\frac{3\mu_0 m_1 m_2}{2\pi D^4} \quad (3)$$

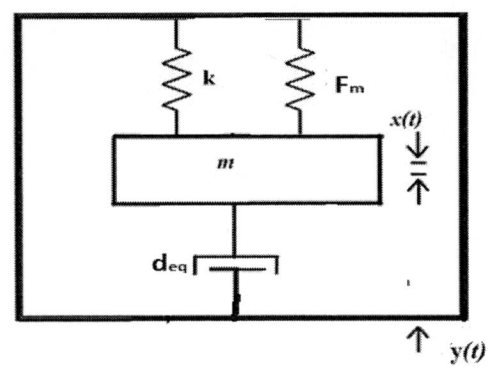

Figure 2. A mechanical model of energy harvester with magnetic force

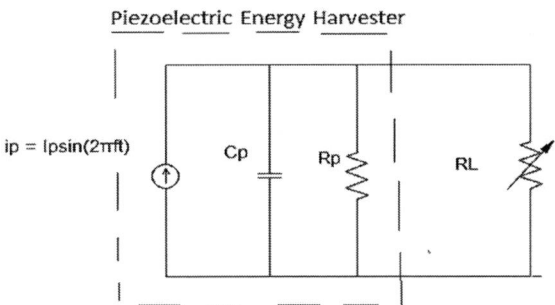

Figure 3. Equivalent electrical model of piezoelectric energy harvester

Where m_1 and m_2 are the magnetic dipoles, μ_0 is the permeability of the medium (air) and D is the distance between the magnetic dipole. The magnetic dipole can be represented as in equation 4.

$$m = \frac{2BV}{\mu_0} \quad (4)$$

Where V is the volume of the magnet which depends on the geometry of the magnets.

Whenever there is a deflection $x(t)$ in the cantilever beam, the magnetic force is modified as shown in equation 5.

$$F_m = -\frac{3\mu_0 m_1 m_2}{2\pi(D + x(t))^4} \quad (5)$$

If dipole moment of both the magnets is positive or negative then there is a repulsive magnetic force and if the dipole moment of one of the magnets is positive whereas the dipole moment of the other magnet is negative then the magnetic force is attractive. In this paper, the repulsive force is used.

III. MAGNETIC EFFECT ON THE STRESS OF PIEZOELECTRIC PATCH OF THE CANTILEVER

The output voltage *(V)* and output power *(P)* of the piezoelectric EH depend on the average stress developed in the cantilever beam [1], as shown in equation 6 and 7.

$$V = S \times g_{31} \times t_p = \frac{d_{31} \times S \times t_p}{\epsilon_0 \epsilon_r} \qquad (6)$$

$$P = \frac{V^2 R_l}{(R_l + Rp)^2} \qquad (7)$$

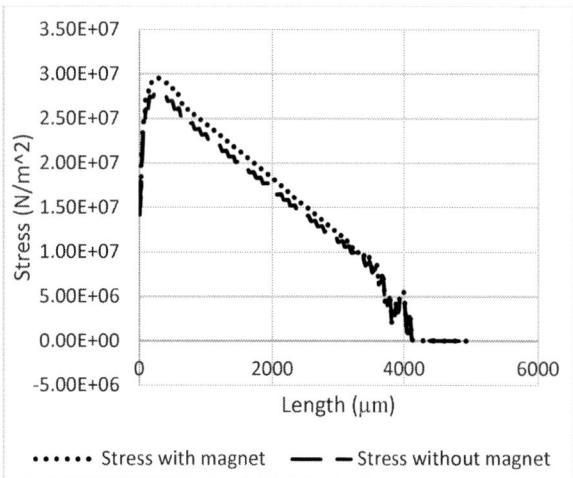

Figure 4. Variation Stress in ZnO with the length of the cantilever

Where S is stress applied in the cantilever g_{31} is the piezoelectric voltage coefficient d_{31} is the piezoelectric charge coefficient and t_p is the thickness of the piezoelectric layer. So the structure with a higher stress level gives more voltage when all the other parameter remains the same. Figure 4 compares the stress development in the piezoelectric layer of a cantilever with and without the magnetic effect. The graph shows that the larger stress developed in the piezoelectric layer when the magnetic effect is applied than EH without magnetic effect.

IV. ELECTRICAL OUTPUT OF STRUCTURES

A. Voltage output of structures

A time domain analysis is carried out on the rectangular structure with and without magnetic effect in COMSOL Multiphysics to obtain the ac peak voltage. A sinusoidal acceleration of 1g is applied to all the structure at the resonant frequency of 95.5 Hz. Figure 5 shows the voltage output of all the structure at the resonant frequency.

B. Power output of the structure

Piezoelectric EH generates maximum electrical output power when the resonant frequency matches with the ambient vibration frequency and the external load matches with the internal resistance of the harvester [1].

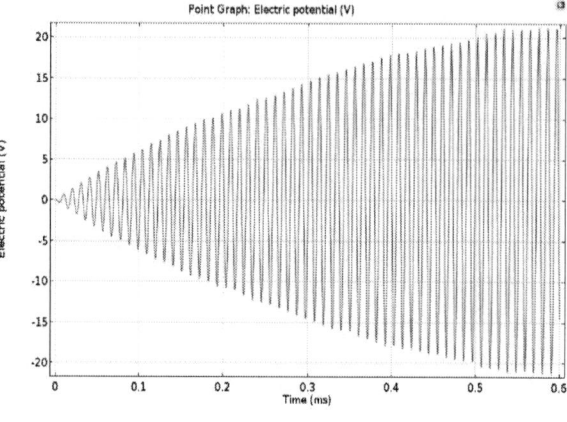

(a)

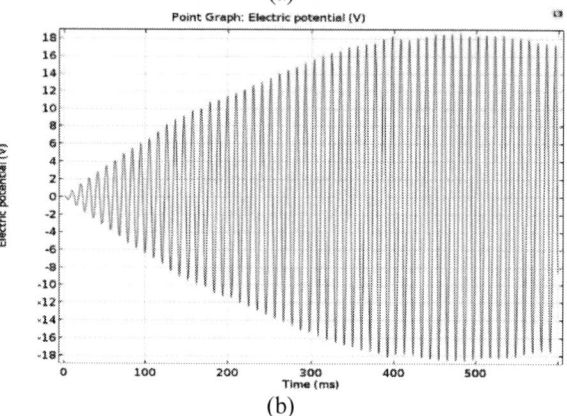

(b)

Figure 5. Voltage of (a) Rectangular with magnet (b) Rectangular without magnet

The optimum load resistance of the EH is 4.3 MΩ. This optimum load resistance is used to calculate the maximum output power. Figure 6 shows the variation of the power with the frequency. The maximum power generated by the rectangular structure with magnetic effect is more than the structure without the magnetic effect. Table 3 shows the maximum value of all the output parameter of the structures.

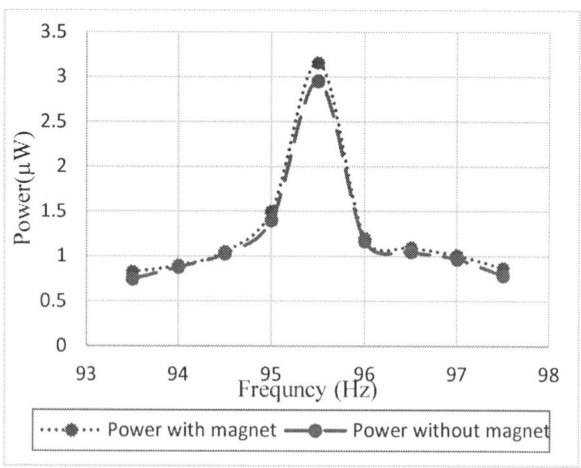

Figure 6. Power vs Frequency

Table 3 Comparison of Parameters of Structures

Parameter (Maximum value)	Rectangular with magnetic effect	Rectangular without magnetic effect.
Resonant frequency (Hz)	95.5	95.5
Voltage (V)	21.2	18.4
Power (µW)	3.16	2.95
Stress (N/m²)	2.94e+7	2.77e+7

There is an increase in the stress development in the cantilever beam with the increase in the magnetic intensity of the permanent magnet. Figure 7 shows the variation of power with the magnetic field intensity. Figure 8 shows the variation of stress developed in the cantilever beam with magnetic field intensity. The graph 7 shows that with an increase in magnetic field intensity, there is an increase in the output power of the EH. However, figure 8 shows that stress level increases tremendously with an increase in magnetic field intensity and there is a chance that cantilever can break with an increase in the magnetic intensity of the magnet. So, the strength of the magnetic field should be optimum.

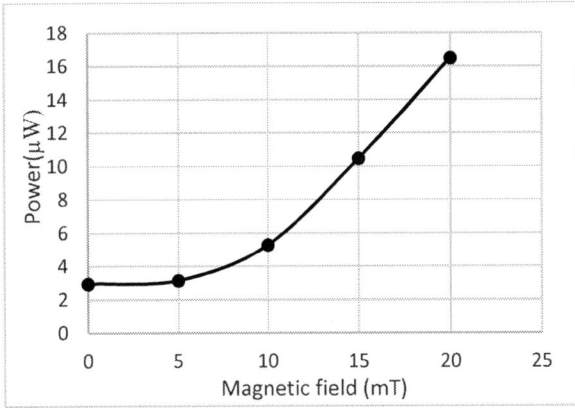

Figure 7 Power vs magnetic field

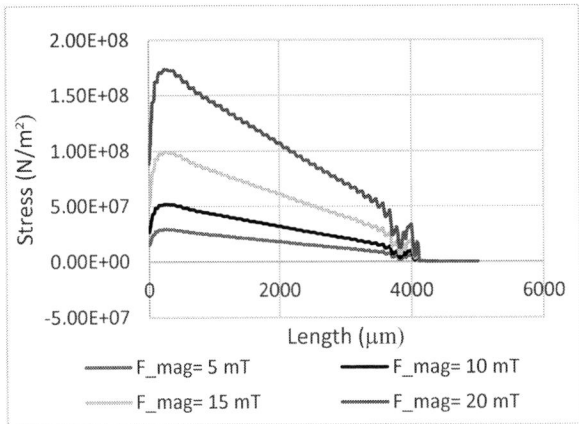

Figure 8. Stress developed in cantilever

V. CONCLUSION:

In this paper, piezoelectric EH with a rectangular structure with magnetic tip mass is designed to work in the magnetic environment. The system is designed to enhance the performance of the harvester. The structural and stress analysis is carried out using material property. The structures are simulated and compared with all the output parameter of the energy harvester. After simulation and analyzation, it is found that the rectangular structure with magnetic tip mass in the magnetic environment has more output power than the normal rectangular structure of the same dimension. The resonant frequency of the proposed structure is 95.5 Hz and generates the peak output voltage of 21.2 V and peak output power of 3.16 µW which is 7.11 % more than the simple rectangular structure of the same dimension.

ACKNOWLEDGMENT

We, all the authors, would like to express our special thanks to Science and Engineering Research Board (SERB), Govt. of India for funding to execute this research through project ref. no. ECR/2016/001279. This research was not possible without the help and support of BIT Mesra, Ranchi and SERB, Govt. of India.

REFERENCES

[1] D. Chaudhuri, S. Kundu, and N. Chattoraj, "Design and analysis of MEMS based piezoelectric energy harvester for machine monitoring application," Microsyst. Technol., https://doi.org/10.1007/s00542-018-4156-z, 2018.

[2] S. Roundy, P. K. Wright, and J. Rabaey, "A study of low level vibrations as a power source for wireless sensor nodes," Comput. Commun., vol. 26, no. 11, pp. 1131–1144, 2003.

[3] D. Chaudhuri and S. Kundu, "MEMS piezoelectric energy harvester to power wireless sensor nodes for machine monitoring application," Proc. 2nd Int. Conf. 2017 Devices Integr. Circuit, DevIC 2017, pp. 584–588, 2017.

[4] P. Glynne-Jones, M. J. Tudor, S. P. Beeby, and N. M. White, "An electromagnetic, vibration-powered generator for intelligent sensor systems," Sensors Actuators, A Phys., vol. 110, no. 1–3, pp. 344–349, 2004.

[5] P. D. Mitcheson, E. M. Yeatman, G. K. Rao, A. S. Holmes, and T. C. Green, "Energy harvesting from human and machine motion for wireless electronic devices," Proc. IEEE, vol. 96, no. 9, pp. 1457–1486, 2008.

[6] H. S. Kim, J. H. Kim, and J. Kim, "A review of piezoelectric energy harvesting based on vibration," Int. J. Precis. Eng. Manuf., vol. 12, no. 6, pp. 1129–1141, 2011.

[7] P. D. Mitcheson, P. Miao, B. H. Stark, E. M. Yeatman, A. S. Holmes, and T. C. Green, "MEMS electrostatic micropower generator for low frequency operation," Sensors Actuators, A Phys., vol. 115, no. 2–3 SPEC. ISS., pp. 523–529, 2004.

[8] M.F. Daqaq, R. Masana, A. Erturk, D.D. Quinn, "On the role of nonlinearities in vibratory energy harvesting: a critical review and discussion", Appl. Mech. Rev. 66 (4) (2014) 040801.

[9] V. R. Challa, M. G. Prasad, Y. Shi, and F. T. Fisher, "A vibration energy harvesting device with bidirectional resonance frequency tunability,"
Smart Materials and Structures, vol. 17, p. 015035, 2008.

[10] Zhu D, Roberts S, Tudor MJ, et al. "Design and experi- mental characterization of a tunable vibration-based electromagnetic micro-generator". Sensors and Actuators A 158: 284–293 (2010)

[11] W. Su, J. Zu, and Y. Zhu, "Design and development of a broadband magnet-induced dual-cantilever piezoelectric energy harvester," vol. 25, no. 4, pp. 430–442, 2014.

[12] D. Guo, "Piezoelectric Energy Harvester Array with Magnetic Tip Mass," pp. 1–6, 2016. IMECE 2015-51044

Analysis of Four-Stage Charge Pump Circuit for UHF RFID Tag Design

Savio Jay Sengupta
Dept.- E.T.C.E
Jadavpur University
Kolkata, India
saviojsengupta@gmail.com

Dipanjan Sen
Dept.- E.T.C.E
Jadavpur University
Kolkata, India
sendipanjan10@gmail.com

Subhashis Roy
Dept.- E.T.C.E
Jadavpur University
Kolkata, India
subhashisaec@gmail.com

Sudhabindu Ray
Dept.- E.T.C.E
Jadavpur University
Kolkata, India
sudhabin@yahoo.com

Subir Kumar Sarkar
Dept.- E.T.C.E
Jadavpur University
Kolkata, India
su_sircir@yahoo.co.in

Abstract—**This work presents the analysis of a charge pump circuit for UHF RFID tag design using 22nm CMOS technology. A charge pump circuit is basically a DC to DC converter. This circuit can be used in many systems for its high performance and low power consumption. Hence, this voltage charge pump circuit can be used in low voltage applications such as in RFID tag's EEPROM. This charge pump circuit has been used as a part of the power supply unit of a fully integrated RFID transponder IC. This modified circuit can generate a stable output voltage with low power dissipation and higher gain for RFID applications. Measured output of this circuit at 433MHz frequency with 1pF of pumping capacitor value, is 3.48V which is more than the value of the industry standardized voltage 3.3V. The extensive simulations are done by using T-spice simulator.**

Keywords—**RFID, Charge pump, DC to DC converter, EEPROM, T-spice, 22nm CMOS technology**

I. INTRODUCTION

RFID or Radio Frequency Identification and Detection is a well-known and emerging technology which has been used in many systems throughout the last decade. RFID is basically a wireless communication technology which is highly reliable and cost efficient. The RFID tag, also called transponder, is attached with the object which is needed to be tracked or whose value is required for a certain work. Each tag contains a unique id which identifies unique objects. To read the tags, Reader is used. The reader communicates with the tag and reads the tag ID in order to identify the object. The communication between tags and reader can be done in various frequency ranges. Each range has a unique name. From 125KHz to 134.2KHz and 140KHz to 148.5KHz frequency band is called low frequency (LF) range, from 13.55MHz to 13.567MHz frequency band is called high frequency (HF) range and 433MHz frequency band and from 858MHz to 930MHz frequency band is called ultra-high frequency range (UHF). The RFID tag contains a microchip. The tag memory may consist of different memory elements depending on the device functionality. The memory elements may be ROM or RAM or non-volatile memory such as EEPROM or flash and data buffers.

The commonly used tag memory is embedded NVM or non-volatile memory. Besides RFID tags, NVM can be used in FPGA systems, Silicon on Chip (SoC), microcontroller and so on. But previously it was a problem to use NVM in RFID tags due to various reasons. The fabrication process of NVM (EEPROM and flash memory) requires multi polysilicon and oxidation steps for thin film silicon dioxide (SiO_2). Hence high cost was incurred and the reliability was also low. Again, the time required for the fabrication is more as compared to CMOS fabrication technology. But researchers [4-6] overcame these huddles and developed CMOS logic based NVM which results in low power dissipation and lower cost.

Charge pump circuits generate a high output voltage with respect to lower supply voltage (V_{DD}). Hence the advantage is that no external voltage regulator is needed which ultimately avoids the need for an external rail. These circuits are very useful to Flash memory or EEPROM, power management chips and DC to DC converters. The charge pump circuit is basically a capacitor-based circuit. Here the voltage is pumped-up stage by stage. The amount of charge pumped-up depends upon the voltage gain of each stage. Dickson charge pump circuits are widely used. The two important components of this circuit are diode connected NMOS transistor and the pumping capacitors. Owing to threshold voltage (V_{TH}) and the body effect of NMOS transistors, the circuit should sustain a high voltage loss which in turn results in low voltage gain per stage. To improve the pumping efficiency, this circuit is modified by Jongshinet et al. [7].

The paper is organized in the following sections: In section II the architecture and working principle of the circuit is described. The section III includes the simulated results, section IV includes the advantages of the proposed circuit and section V contains the conclusion part.

II. ARCHITECHTURE AND WORKING PRINCIPLE

Charge pump with compatible time-sharing system (CTSS) had already been designed by many researchers. In this type of circuits an auxiliary pass transistor is required to turn off the switches completely during the desired time period. But in low voltage environment it is difficult for the CTSS to turn on the charge pump circuits. In order to remove the two limiting factors, i.e. threshold voltage and the effect of parasitic capacitance, CTSS is used in individual wells. In this work, pass transistors (both N and P types) are used to dynamically control the inputs to the CTSS. To turn on the CTSS more easily, a four-stage clock scheme is used. To further improve the charge, two clock signals C_1 and C_2 are generated in such a way that they are out of phase with each other, but with the same amplitude, i.e. supply voltage (V_{DD}).

The working principle of the charge pump circuit is explained as follows. When C_1 is high and C_2 is low, then between node 'a' and node 'b' the voltage is V_{ab} but at node 'c' the voltage is about $2\Delta V$. So,

$$2\Delta V > V_{THP} \text{ and } 2\Delta V > V_{THN}(V_{ab}) \qquad (1)$$

So, both the MOSFETs P2 and N_7 are now turned on due the voltage at node 'c'. Here, V_{THP} is the threshold voltage of PMOS transistor and V_{THN} is the threshold voltage for NMOS transistor.

Now when C_1 is low and C_2 is high then at node 'a' the voltage is say V_a and the voltages between node 'b' and 'c' are now above $2\Delta V$. So,

$$2\Delta V > V_{THN}(V_a) \qquad (2)$$

During this period as N_7 is now become free from the node 'c', hence N_{12} is turned on but on the other hand P_2 is turned off. The above two equations, i.e. the equation (1) and (2) are necessary for the successful operation of the charge pump circuit. Fig. 1, shows the circuit diagram of the charge pump circuit.

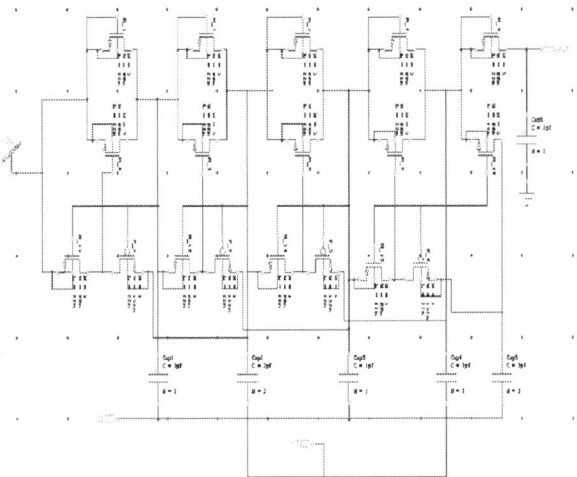

Fig 1. Circuit diagram of four stage charge pump

According Fig 1, Cap1, Cap2, Cap3, Cap4, Cap5 are the pumping capacitors and Cap6 is the smoothing capacitor. The drain node of each well is connected. Hence when clock C_1 transit from high to low, charges are injected in the well from CTSS gate during that small period of transition.

III. SIMULATED RESULT

The output of the charge pump circuit has been verified by T-spice simulator. Voltage amplitudes of two clocks C_1 and C_2 are taken similar to the voltage amplitude of the supply voltage (V_{DD}). For simulation purpose the parameters are taken as follows: pumping capacitors of value 1pF, smoothing capacitor of value 1pF, clock frequency is varied from 6.8MHz to 433MHz (Dual to UHF range) and input voltage is also varied from 0.6V to 1V. In our work 22nm CMOS technology has been used to obtain the charge pump circuit. Simulated results are given below in figures:

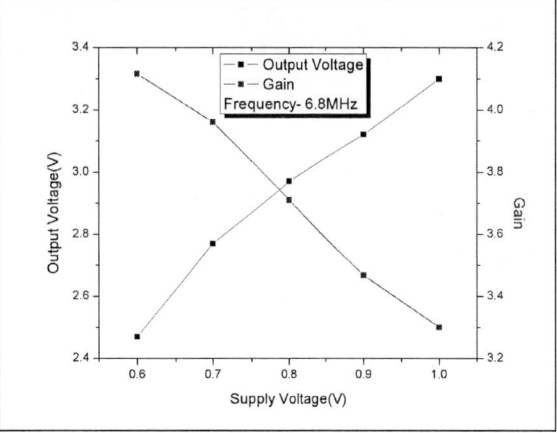

Fig 2. Output voltage and Gain versus Supply Voltage for 6.8MHz

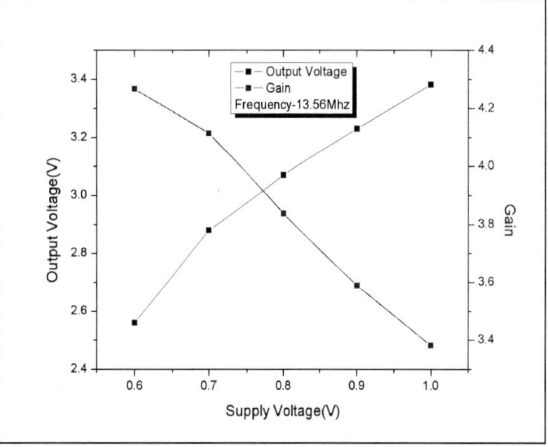

Fig 3. Output voltage and Gain versus Supply Voltage for 13.56MHz

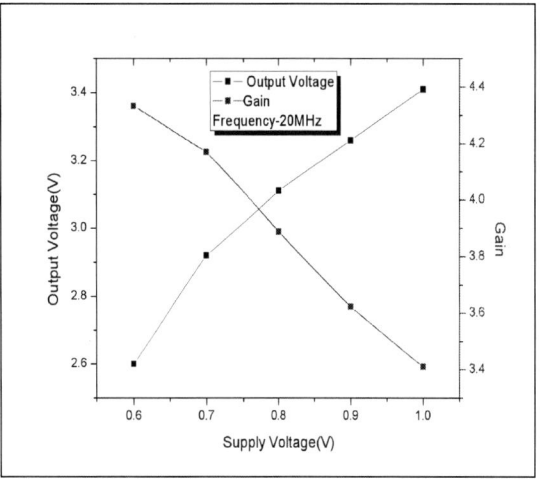

Fig 4. Output voltage and Gain versus Supply Voltage for 20MHz

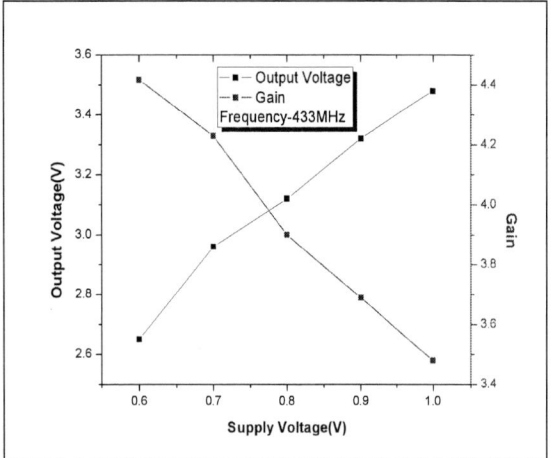

Fig 5. Output voltage and Gain versus Supply Voltage for 433MHz

Figure 2, 3, 4 and 5 shows the variation of output voltage and voltage gain with respect to supply voltage (V_{DD}). As it can be seen from the above figures, the output voltages are increasing with increase in supply voltage. But for voltage gain the characteristics is opposite to that of the output voltage, i.e. voltage gain is more for lower supply voltages and it decreases with increase in supply voltage. For frequency value 6.8MHz the maximum voltage gain is 4.12 whereas the minimum voltage gain is 3.3. For 13.56MHz frequency value, the maximum voltage gain is 4.27, whereas the minimum voltage gain is 3.382. For 20MHz frequency value, the maximum voltage gain is 4.33 and the minimum voltage gain is 3.41. For frequency value 433MHz the maximum voltage gain is 4.42 and the minimum voltage gain is 3.48. The values of voltage gain corresponding to supply voltages are provided in the following table:

Frequency (MHz)	Corresponding Voltage (Volts)	Output Voltage (Volts)	Voltage Gain
6.8	0.6	2.47	4.12
	1.0	3.3	3.3
13.56	0.6	2.56	4.27
	1.0	3.82	3.382
20	0.6	2.6	4.33
	1.0	3.41	3.41
433	0.6	2.65	4.42
	1.0	3.48	3.48

Table 1. Values of voltage gain with respect to supply voltages

It is also very clear from the simulated results and Table 1 that for all the four frequency bands and for supply voltage (V_{DD}) equals to 1volt, the output voltages are greater than equal to (>=) 3.3 volts which is one of the standardized voltages in industry. Also, it can be seen that the voltage gain is increasing with increase in frequency range, thereby exhibiting better performance at the higher frequency range.

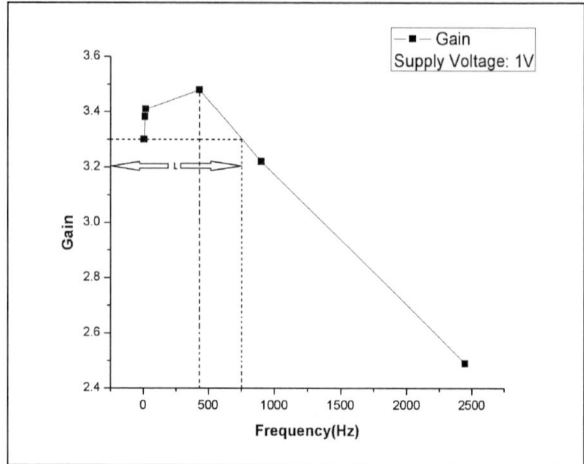

Fig 6. Gain versus Frequency

Figure 6 gives gain versus frequency plot. Now voltage gain is given by,

$$Gain = G = \frac{V_{out}}{V_{in}} \qquad (3)$$

Where,

$$V_{out} = Output\ Voltage$$

$$V_{in} = Supply\ Voltage$$

Now supply voltage (V_{DD}) is taken as 1volt for the purpose of simulation. Hence, Gain now becomes:

$$Gain = G = \frac{V_{out}}{1V} = V_{out} \qquad (4)$$

Hence, Figure 6 is the output voltage versus frequency plot. So, according to Figure 6, it can be seen that the output voltages for 1volt supply voltage, remains at 3.3 volts level or higher for up to 750-760MHz frequency. 433Mhz is in the UHF range. Hence it is evident that the proposed charge pump circuit can operate perfectly at 433Mhz. Region L in the Figure 6 defines the frequency region up to which this charge pump circuit can work properly i.e. will give stable output.

IV. ADVANTAGES OF THE PROPOSED CIRCUIT

It is well known that there are many works that have been carried out in the last decade in the field of RFID tag design and analysis. But the CMOS technology used is of 130nm, 180nm etc. for various charge pump circuits. In this proposed work, a four-stage charge pump circuit has been analyzed by using 22nm CMOS technology to obtain better circuit behavior. Advantages of this proposed work are listed below:

- 22nm CMOS technology consumes lesser chip area.

- Lower input voltage or supply voltage (V_{DD}) is required.

- By using only 1V as supply voltage, better output gain is achieved.

- The output voltage for 1V supply voltage is more than or equal to 3.3 volts which is the industrially standardized voltage.

- This charge pump circuit can operate perfectly up to 750-760MHz frequency, i.e. in UHF region.

The only drawback observed is that beyond 750-760 MHz frequency, the output voltage drops below 3.3V. This drawback will be addressed in the future works.

V. CONCLUSION

In case of RFID tag design, a low voltage charge pump circuit is one of the fundamental building blocks. In this proposed work, the analysis of the four-stage charge pump circuit has been performed by applying an input voltage from 0.6 to 1volt and as a result the corresponding output voltage is more than 3.3 volts which is a standardized voltage used in industry. By using the 22nm CMOS technology, the chip area can be minimized as nowadays the total usage of the chip area is a major concern for VLSI circuit design. The maximum gain measured is 4.42 at 433MHz frequency for a supply voltage of .6 but at the same time the voltage gain at 433MHz at 1volt supply voltage is 3.48. The proposed circuit works almost accurately at higher frequency ranges such as HF, UHF. As in modern era, RFID systems often use UHF tags and readers, therefore getting a good output gain is an important factor. This analysis shows better performance in terms of voltage gain with respect to frequency & input voltage compared to the existing works. For example, the gain of existing work is 3.30 at 20MHz clock frequency for an input voltage of 1.8V, whereas in this case the voltage gain is 3.41 at 20MHz clock frequency for an input voltage of 1V. Hence, this proposed charge pump circuit gives better performance for low voltage applications in analyzing EEPROM circuit for RFID tag design while generating the standardized high output voltage.

ACKNOWLEDGMENT

Authors of this work would like to thank the financial support obtained from the DST-WB project. Ref. No: 287(Sanc.)/ST/P/S&T/6G-43/2017.

REFERENCES

[1] Joyashree Bag, Subhashis Roy, P. K. Dutta and Subir Kumar Sarkar, "Design of a DPSK Modem Using CORDIC Algorithm and Its FPGA Implementation", IETE Journal of Research, Volume 60, Issue 5,pp-355-363, 2014.

[2] Joyashree Bag and Subir Kumar Sarkar, "Development and VLSI implementation of a data security scheme for RFID system using programmable cellular automata" International Journal Radio Frequency Identification Technology and Applications, Vol. 4, No. 2, Pp: 197-211, 2013.

[3] Joyashree Bag, Rajanna K.M and Subir Kumar Sarkar, "Design and FPGA Implementation of a Zig-bee enabled processor for RFID reader suitable for power efficient home/office automation", European Journal of Scientific Research, Vol. 97, No. 4, pp: 592-603,2013.

[4] K. Ohsaki, N. Asamoto and S. Takagaki, A single poly EEPROM cell structure for use in standard CMOS processes, IEEE Journal of Solid-State Circuits, vol. 29, no. 3, pp. 311-316, Mar. 1994.

[5] D.X. Zhao, N.Y.W. Xu, L.W. Yang and J.Y. Wang, Low-power, Singlepoly, Non-volatile Memory for Passive RFID Tags, Chinese Journal of Semiconductors, Vol. 29, pp. 99-104, 2008.

[6] B. Wang, H. Nguyen, Y. Ma and R. Paulsen, Highly Reliable 90-nm Logic Multitime Programmable NVM Cells Using Novel Work-FunctionEngineered Tunneling Devices, IEEE Transactions on Electron Devices, vol.54, no. 9, pp. 2526-2530, Sept. 2007.

[7] J. Raszka, M. Advani, V. Tiwari, L. Varisco, N.D. Hacobian, A. Mittal, M. Han, A. Shirdel and A. Shubat, Embedded flash memory for security applications in a 0.13um CMOS logic process, in Proceedings of the International Solid-State Circuits Conference (ISSCC), IEEE, 15-19 Feb. 2004.

Analytical modelling and simulation of pseudo-resistive circuit techniques for biomedical applications

Kulbhushan Sharma, Anisha Pathania, and Rajnish Sharma[*]

VLSI Centre of Excellence,
Chitkara University Institute of Engineering and Technology,
Chitkara University, Punjab, India
kulbhushan.sharma@chitkara.edu.in, anisha.pathania@chitkara.edu.in,
[*]rajnish.sharma@chitkarauniversity.edu.in

Abstract—**Pseudo-resistors (PRs) implemented on-chip have become extremely popular than passive off-chip linear resistors as they serve an effective way to emulate high value of Incremental Resistance (IR). The nonlinear V-I characteristics of PRs and their high sensitivity toward Process Voltage and Temperature (PVT) variations degrades the performance of analog circuits used in biomedical applications. In this work, different PR techniques have been modeled in subthreshold region for calculating IR expression and IR range associated with them. All the simulations and statistical analysis have been performed on 0.18 µm standard CMOS process using BSIM3V3 MOS models. It was found that statistical results and linearity improves when the bulk terminal of the MOS device is either connected to drain or a bias voltage is applied to it. Among all the PR techniques reported in this paper, Scaled Gate Linearization (SGL) emulates IR of the order of 1 TΩ with highest linearity for source to drain voltage (V_{SD}) between −1 to 0.3 V. For SGL, the nominal value of IR ($R_{SD,N}$), and mean (µ), are 999.99 GΩ and 1 TΩ respectively with negligibly small difference between these values. Moreover, the values of standard deviation (σ) and Coefficient of variance (CV = σ/µ) for SGL are too low, i.e. 1.364 MΩ, and 1.36×10^{-6} respectively. The PR techniques reported in this work are anticipated to be used in different analog circuits for numerous biomedical applications.**

Keywords— *Biomedical applications, Incremental Resistance, linearity, linearization techniques, pseudo-resistor.*

I. INTRODUCTION

Over the past several years, MOS devices capable of mimicking the behavior of linear resistors frequently referred as Pseudo-Resistors (PRs) have been used for various analog circuits applications like transconductance amplifiers [1], transimpedance amplifiers [2] voltage gain amplifiers [3], continuous time filters [4], current conveyors [5] and data converters [6] etc. Majority of the researchers have used these circuits for biomedical applications such as neural signal recording [7-9], On-Chip Bio-Impedance Spectroscopy [5] and Scanning Ion-Conductance Microscopy [2] etc. In last two decades, a large number of publications have come up which have used non-tunable [1], tunable [8, 9], Floating Gate (FG) [10], balanced [11] PR techniques etc. in neural recording applications.

In all the aforementioned applications, the values of resistance provided by these PR techniques generally ranges from several MΩ to few TΩ. These PR techniques face major challenges like limited linearity, sensitivity towards PVT, and common-mode voltage shifting. The gate shot noise component induced at gate of MOS devices and substrate leakage currents are responsible for limited linearity in IR values emulated by these PR design techniques [12]. Further, the excessive sensitivity towards PVT leads to asymmetrical V-R characteristics curves [13] and common-mode voltage shifting causes severe resistance drop mainly near quiescent point (0V) [9].

Non tunable PR techniques [1, 14] offer fixed value of IR, whereas the IR values can be altered by using tunable PR techniques. Tunable pseudo-resistors can be realized using voltage controlled and current controlled techniques [8, 9]. Former relies on variable low gate voltage and later on controlled low current magnitude. Non-conventional techniques like Common Mode Linearization (CML) [10], SGL [6], FG [15], Quasi-Floating-Gate [16] have been found useful in overcoming the problem of nonlinearity in PRs. Though these techniques improve linearity yet the range of emulated resistance is of the order of few MΩ only. This value of resistance is not suitable for some of the analog circuits such as transconductance amplifiers [1,8], transimpedance amplifiers [2] voltage gain amplifiers [3] used in applications like Electroencephalography (EEG), Electrocorticography (ECoG), and electrophysiology etc. On a close analysis, we find the reason for such low values of resistance is attributed to the device sizing, biasing and modeling in triode region of operation. Wang *et al.* [11] presented Gate-Balanced PR techniques modeled in the subthreshold region which are capable of emulating resistance of the order ~GΩ, but the linearity in the resistance is confined for a narrow voltage swing of ±277 mV with 1% variation.

Review of recent literature conveys that bootstrapping [17], robustness enhancement [18] and leakage compensation schemes [12] have been worked upon by the researchers to encounter the effect of PVT variations and substrate leakage currents. These techniques show better performance but simultaneously increase area and power overheads.

In order to benefit the researchers for designing analog circuits in the aforesaid applications, this paper presents analytical modeling and simulation of non-tunable, tunable and linearization PR design techniques in subthreshold region. Among all the PR techniques reported in this paper, SGL emulates IR of the order of 1 TΩ with highest linearity for $-1 \leq V_{SD} \leq 0.3$ V. The values of $R_{SD,N}$, and µ, are 999.99 GΩ, and 1 TΩ respectively for SGL with negligibly small difference between these values. Moreover, the values of σ and CV for SGL are too low, i.e. 1.364 MΩ, and 1.36×10^{-6} respectively.

978-1-5386-6723-1/19 $31.00 © 2019 IEEE

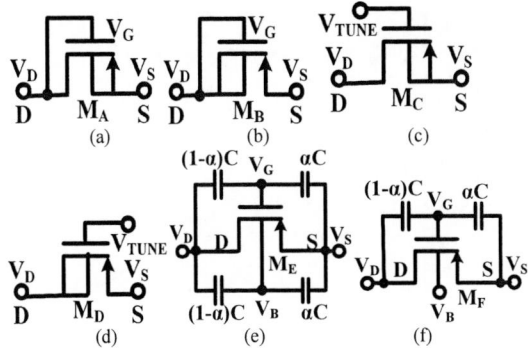

Fig. 1. Technique for implementing PRs (a) NTPR-I (b) NTPR-II (c) TPR-I (d) TPR-II (e) CML (f) SGL

All the simulations have been carried out in 0.18 μm standard CMOS process using BSIM3V3 PMOS models. Paper has been divided into four different sections. Section II specifically focusses on design methodology involved in implementing PR techniques. All simulation results are presented in Section III. While the Section IV brings forth the important conclusions drawn from the reported work.

II. DESIGN METHODOLOGY

A Tunable PR technique with enhanced linearity, low noise and independent of PVT variations is in demanding trend mainly for filters and amplifiers used in biomedical applications. In all these application, researchers have shown keen interest in determining the IR of the PRs based on different design techniques [3, 12, 13]. The IR of a PMOS device depends upon its source to drain current (I_{SD}). The I_{SD} of a diode connected PMOS device operated in subthreshold region can be expressed by [19]:

$$I_{SD} = I_{DO} \times \left(e^{\frac{-V_{BS}}{U_T}} - e^{\frac{-V_{BD}}{U_T}} \right) \times e^{\frac{V_{BG}-V_{TO}}{nU_T}} \quad (1)$$

Where $U_T = kt/q$ is the thermal voltage and I_{DO} is residual channel current which can be given by:

$$I_{D0} = 2n\mu_p C_{ox} \frac{W}{L} U_T^2 e^{\frac{-V_{TO}}{nU_T}} \quad (2)$$

The voltages V_{BG}, V_{BS}, V_{BD}, and V_{TO} in (1) are designated as voltage between bulk and gate terminals, voltage between bulk and source terminals, voltage between bulk and drain terminals and threshold voltage of PMOS device respectively. While the parameters n, μ_p, W, and L are designated as subthreshold slope factor, mobility of charge carriers, width and length of the PMOS device respectively. The IR of a MOS device can be given by [20]:

$$R_{SD} = \left(\frac{1}{\partial I_D / \partial V_{DS}} \right) \quad (3)$$

Fig. 1 shows schematics of different techniques for PRs [6, 8, 9, 10, 14, 20]. The device M_A in Non-tunable PR-I (NTPR-I) technique shown in Fig. 1(a) has its bulk terminal tied to its source and gate terminal tied to its drain. In this PR technique, the $V_{BS} = 0$, and $V_{DG} = 0$, therefore, I_{SD} can be expressed as:

$$I_{SD} = I_{DO} \times \left(1 - e^{\frac{-V_{SD}}{U_T}} \right) \times e^{\frac{V_{SD}-V_{TO}}{nU_T}} \quad (4)$$

The IR for NTPR-I technique can be given by:

$$R_{SD} = \frac{nU_T \left(1 - e^{-V_{SD}/U_T} \right)}{I_{SD}(n-1)e^{-V_{SD}/U_T} + 1} \quad (5)$$

The minimum and maximum values of IR possible with this PR technique are 3.05 Ω and 20 G Ω respectively.

The Non-tunable PR-II (NTPR-II) technique shown in Fig. 1(b) is similar to that shown in Fig. 1(a) except that the bulk terminal of MOS device M_B is tied to its drain terminal. In this technique, the $V_{BD} = 0$, $V_{DG} = 0$, thus, I_{SD} can be given by:

$$I_{SD} = I_{DO} \times \left(e^{\frac{V_{SD}}{U_T}} - 1 \right) \times e^{\frac{-V_{TO}}{nU_T}} \quad (6)$$

The IR of this NTPR-II technique can be expressed as:

$$R_{SD} = \frac{nU_T \left(e^{V_{SD}/U_T} - 1 \right)}{I_{SD}(n-1)e^{V_{SD}/U_T} + 1} \quad (7)$$

From (7) the minimum and maximum values of IR can be calculated as 2.98 Ω and 1 TΩ respectively.

Fig. 1(c) shows schematic of Tunable PR-I (TPR-I) technique in which the gate terminal of device M_C is made tunable, while, its bulk and source terminals are tied to each other. For, TPR-I the $V_{BS} = 0$ and gate voltage, $V_G = V_{TUNE}$, thus I_{SD} can be expressed as:

$$I_{SD} = I_{DO} \times \left(1 - e^{\frac{-V_{SD}}{U_T}} \right) \times e^{\frac{V_{SD}-V_{GD}-V_{TO}}{nU_T}} \quad (8)$$

The IR of TPR-I technique can be given by:

$$R_{SD} = \frac{nU_T \left(1 - e^{-V_{SD}/U_T} \right)}{I_{SD}(n-1)e^{-V_{SD}/U_T} + 1} \quad (9)$$

The minimum and maximum values from (9) can be calculated as 2.93 Ω and 100 G Ω.

The schematic of Tunable PR-II (TPR-II) technique shown in Fig. 1(d) is similar to that shown in Fig. 1(c) except that the bulk terminal of MOS device M_B is tied to its drain terminal. For, TPR-II the $V_{BS} = 0$, and $V_G = V_{TUNE}$, thus, I_{SD} can be given by:

$$I_{SD} = I_{DO} \times \left(e^{\frac{V_{SD}}{U_T}} - 1 \right) \times e^{\frac{V_{GD}-V_{TO}}{nU_T}} \quad (10)$$

The IR for TPR-II technique can be given by:

$$R_{SD} = \frac{nU_T \left(e^{V_{SD}/U_T} - 1 \right)}{I_{SD}(n-1)e^{V_{SD}/U_T} + 1} \quad (11)$$

The minimum and maximum values of IR possible with TPR-II technique are 3.59 Ω and 1 TΩ respectively.

Unlike, NTPR and TPR the CML PR technique illustrated in Fig. 1 (e) rely on scaling of capacitors (C=1pF) of MOS device M_E connected between gate-source and bulk-source terminals by (α) and capacitors connected between gate-drain and bulk-drain terminals by $(1 - α)$. In present piece of work, we have chosen the value of α equal to 0.5. The V_G and V_B of M_E are equal, and can be given by $V_G = (V_S+V_D)/2$. With CML PR design approach the I_{SD} and IR of M_E can be expressed by (12) and (13) as follows:

$$I_{SD} = I_{DO} \times \left(1 - e^{\frac{-V_{SD}}{U_T}} \right) \times e^{\frac{-V_{BD}+V_{SD}-V_{TO}}{nU_T}} \quad (12)$$

$$R_{SD} = \frac{nU_T \left(1 - e^{-V_{SD}/U_T} \right)}{I_{SD}(1+(n-1)e^{-V_{SD}/U_T})} \quad (13)$$

CML technique has minimum and maximum values of IR of 7.49 KΩ and 0.5 T Ω respectively.

Fig. 1(f) illustrates schematic of SGL PR technique. SGL technique is similar to CML technique shown in Fig. 1(e) except that the bulk voltage (V_B) is applied directly to the

2019 Devices for Integrated Circuit (DevIC), 23-24 March, 2019, Kalyani, India

TABLE I. IR Expression, Max. and Min. values of IR for PR techniques shown in Fig 1.

PR technique	IR Expression	Min. value of IR (Ω)	Max. value of IR (Ω)
NTPR-I (Fig. 1 (a))	$R_{SD} = \dfrac{nU_T\left(1-e^{-V_{SD}/U_T}\right)}{I_{SD}(n-1)e^{-V_{SD}/U_T}+1}$	3.05	20 G
NTPR-II (Fig. 1 (b))	$R_{SD} = \dfrac{nU_T\left(e^{V_{SD}/U_T}-1\right)}{I_{SD}(n-1)e^{V_{SD}/U_T}+1}$	2.98	1 T
TPR-I (Fig. 1 (c))	$R_{SD} = \dfrac{nU_T\left(1-e^{-V_{SD}/U_T}\right)}{I_{SD}(n-1)e^{-V_{SD}/U_T}+1}$	2.93	100 G
TPR-II (Fig. 1 (d))	$R_{SD} = \dfrac{nU_T\left(e^{V_{SD}/U_T}-1\right)}{I_{SD}(n-1)e^{V_{SD}/U_T}+1}$	3.59	1 T
CML (Fig. 1 (e))	$R_{SD} = \dfrac{nU_T\left(1-e^{-V_{SD}/U_T}\right)}{I_{SD}(1+(n-1)e^{-V_{SD}/U_T})}$	7.49 K	0.5 T
SGL (Fig. 1 (f))	$R_{SD} = \dfrac{nU_T\left(1-e^{-V_{SD}/U_T}\right)}{I_{SD}(1+(n-1)e^{-V_{SD}/U_T})}$	16.6 K	1.5 T

bulk terminal without making use of scaled capacitors (C=1pF). The value of V_B for M_E is 1.8 V. While the V_G of M_F can be given by $V_G=(V_S+V_D)/2$. With SGL PR approach the I_{SD} and IR of M_F can be expressed by (14) and (15) as follows:

$$I_{SD} = I_{DO} \times \left(1-e^{\frac{-V_{SD}}{U_T}}\right) \times e^{\frac{V_{BG}-V_{BD}+V_{SD}-V_{TO}}{nU_T}} \qquad (14)$$

$$R_{SD} = \frac{nU_T\left(1-e^{-V_{SD}/U_T}\right)}{I_{SD}(1+(n-1)e^{-V_{SD}/U_T})} \qquad (15)$$

SGL technique has minimum and maximum values of IR of 16.6 KΩ and 1.5 T Ω respectively.

For fair comparison, the W/ L ratio of all the MOS devices used in different PR techniques shown in Fig. 1. are taken as 0.6 μm /0.18 μm. Table I shows IR expression, maximum and minimum values emulated by the PRs shown in Fig. 1. Although the IR expressions for NTPR-I and TPR-I appears to be similar, but these expressions differ in values of I_{SD}. Same is applicable for NTPR-II and TPR-II and for CML and SGL. The S and D terminal of PR techniques shown in Fig.1 have been biased at $V_S=A\sin\omega t$ and $V_D=0V$ respectively such that voltage ($\Delta V=V_S-V_D$) is developed across them. The amplitude (A) of the a. c. signal (V_S) has been swept from -1V to 1V at low frequency of 30 Hz. Similar methodology for simulating PR design techniques has been adopted by [8, 9] too. The 0.18 μm standard CMOS process and BSIM3V3 PMOS models have been used for obtaining results of PR techniques depicted in Fig. 1 which have been discussed in Section III.

III. SIMULATION RESULTS

Fig. 2 (a) and (b) illustrates R_{SD} versus V_{SD} curve for non-tunable PR techniques shown in Fig.1 (a) and (b). It can be clearly observed from Fig. 2 (a) that the IR plot shown for NTPR-I increases as negative values of V_{SD} are decreased and decreases as positive values of V_{SD} are increased. While the plot illustrated in Fig. 2 (b) for NTPR-II construes that IR is constant for negative values of V_{SD} and decreases as positive values of V_{SD} are increased. In both these PRs problem of common mode voltage shifting exists which causes severe resistance drop mainly near quiescent point.

Fig. 3 (a) and (b) illustrates R_{SD} versus V_{SD} curve for tunable PR techniques shown in Fig.1 (c) and (d) at $V_{Tune} = -0.1, 0.1,$ and 0.2 V. The IR of tunable PR techniques can be altered by applying different tunable voltages. It can be inferred from Fig. 3 (a) that the IR of TPR-I increases as

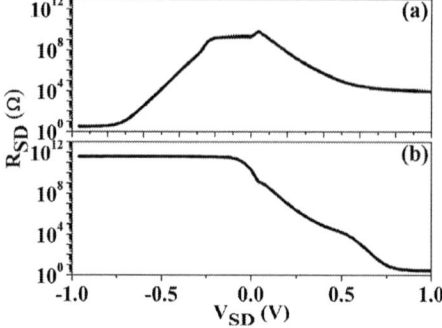

Fig. 2. R_{SD} versus V_{SD} for (a) NTPR-I and (b) NTPR-II techniques shown in Fig. 1 (a) and (b) respectively, at normal process corner, and at 27 ^{0}C.

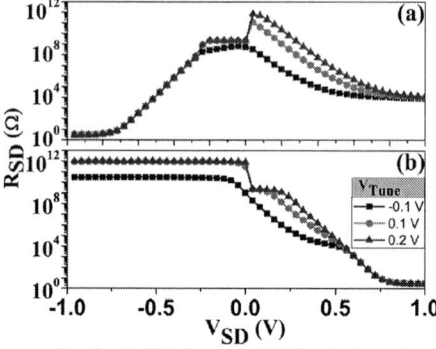

Fig. 3. R_{SD} versus V_{SD} for (a) TPR-I and (b) TPR-II techniques shown in Fig. 1 (c) and (d) respectively, at normal process corner, and at 27 ^{0}C.

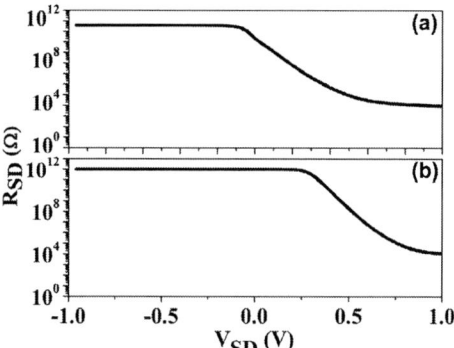

Fig. 4. R_{SD} versus V_{SD} for (a) CML and (b) SGL techniques shown in Fig. 1 (e) and (f) respectively, at normal process corner, and at 27 ^{0}C.

negative values of V_{SD} are decreased and decreases as positive values of V_{SD} are increased. While the plot shown in Fig. 3 (b) for TPR-II infers that IR is constant for negative values of V_{SD} and decreases as positive values of V_{SD} are increased. In these PRs also common mode voltage shifting exists. Nature of plots observed for TPR is in agreement with the NTPR counterparts. Same can be easily understood by having a close look at schematics shown in Fig. 1 for both types of PR techniques on which this work has been carried out. In NTPR-II and TPR-II the IR values are approximately constant for negative values of V_{SD} when compared to NTPR-I and TPR-I. The reason for such V-R characteristics is attributed to reversal of current direction when V_{SD} is positive which leads to device behavior similar to NTPR-I and TPR-I.

Fig. 4 (a) and (b) illustrates R_{SD} versus V_{SD} curve for CML and SGL techniques construed in Fig.1 (e) and (f). It is evident from the curve shown in Fig. 4 (a) that IR is constant for negative values of V_{SD} and decreases as positive values of V_{SD} are increased. The problem of common mode

978-1-5386-6723-1/19 $31.00 © 2019 IEEE

voltage shifting still exists in CML. For SGL the curve shown in Fig. 4 (b) clearly indicates IR is constant not only for negative values of V_{SD} but also for some of the positive values of V_{SD}. The problem of common mode voltage shifting near quiescent point is completely resolved using SGL technique. Moreover, this technique emulates IR of the order of 1 TΩ for $-1 \leq V_{SD} \leq 0.3$ V which is quite higher and linear than other PR techniques. Though in CML and SGL V_G tend to linearize I_{SD} but the reason for better performance of SGL over CML relies on the fact that the magnitude of V_B is greater for SGL than CML. As higher value of V_B is applied, leakage current decreases and threshold voltage requirements increase [21].

Table II depicts the statistcal results obtained by Monte Carlo Simulation (MCS) with 100 runs for PR techniques shown in Fig. 1. It can be observed from Table II that $R_{SD,N}$, and μ are 6.668 GΩ and 12.41 GΩ for NTPR-I and 960.3 GΩ, and 404.3 GΩ for NTPR-II respectively. The σ and CV for NTPR-I are 22.06 GΩ and 1.8 and for NTPR-II are 224.3 GΩ and 0.55 respectively. There exists a significant difference between $R_{SD,N}$ and μ for NTPR-I. The CV for NTPR-II is better than NTPR-I. Further, the TPR-II has CV of 6.5×10^{-3} which is far better than CV of 1.31 for TPR-I. The difference in values of $R_{SD,N}$ and μ for TPR-II is very small. For, CML the difference between $R_{SD,N}$ and μ is 10 GΩ with CV of 0.55 which is quite high. In case of SGL the values of $R_{SD,N}$, μ, σ and CV are 999.99 GΩ, 1 TΩ, 1.364 MΩ, and 1.36×10^{-6} respectively. There exisists negligible difference between $R_{SD,N}$ and μ values for SGL. Moreover, SGL has highest linearity and lowest value of CV among all PR design techniques shown in Fig. 1.

IV. CONCLUSION

This research effort presents an elegant approach to model and simulate NTPR-I, NTPR-II, TPR-I, TPR-II, CML and SGL techniques for implementing PRs in subthreshold region. It was found that statistical results and linearity improves when the bulk terminal of the MOS device is connected to drain as in case of NTPR-II and TPR-II. However, most significant improvement in linearity and statistical results can be observed by biasing the bulk terrminal and providing components of source and drain voltages at gate through scaled gate-source and drain-source capacitances as in case of SGL technique. These PR techniques are anticipated to be used in different analog circuits used in biomedical applications like EEG, ECG, ECoG, On-Chip Bio-Impedance Spectroscopy and Scanning Ion-Conductance Microscopy etc.

REFERENCES

[1] R. Harrison and C. Charles, "A Low-Power Low-Noise CMOS Amplifier for Neural Recording Application," IEEE J. Solid-State Circ., vol. 38, no. 36, pp. 958-965, June 2003.

[2] Denis Djekic, Georg Fantner, Jan Behrends, Klaus Lips, Maurits Ortmanns and Jens Anders, "A transimpedance amplifier using a widely tunable PVT-independent pseudo-resistor for high-performance current sensing applications," in proc. 43rd IEEE Europ. Solid-state Circ. Conf., Beijing, China, pp. 79 – 82, 2017.

[3] P. Mohseni and K. Najafi, "A low power fully integrated bandpass operational amplifier for biomedical neural recording applications," in Proc. IEEE EMBS/BMES Conf., Houstan, USA, pp. 2111–2112, 2002.

[4] N. Neshatvar, H. A. Nashash and L. Albasha, "Design of Low Frequency Analog Low Pass Filter Using Tunable Pseudo-resistors", in Proc. Mid. Eas. Conf. Biomed. Eng. (MECBME), Doha, Qatar, pp. 17-20, Feb. 2014.

[5] W-Y. Chung, A. A. Silverio, and V. F.S. Tsai, "A wideband current source for system-on-chip bio-impedance spectroscopy using a CCII

drive and pseudo-resistor feedback", in Proc Int. Sym. on Bioelect. Bioinf. (ISBB), Beijing, China, pp. 107 – 110 Oct. 2015.

[6] E. Ozalevli, H. Dinc, H.-J. Lo and P. E. Hasler, "Design of a binary-weighted resistor DAC using tunable linearized floating-gate CMOS resistors," in Proc. IEEE Custom Integr. Circ. Conf., San Jose, USA pp. 149–152, Sept. 2006.

[7] A. Bagheri, M. T. Salam, J. L. P. Velazquez and R. Genov, "Low-Frequency Noise and Offset Rejection in DC-Coupled Neural Amplifiers: A Review and Digitally-Assisted Design Tutorial," IEEE journal of Biomedical Circuits and System, vol. 11, no. 1, pp. 161–176, 2017.

[8] X. Zou, X. Xu, L.Yao and Y. Lian, "A 1-V 450-nW Fully Integrated Programmable Biomedical Sensor Interface Chip," IEEE J. Solid-State Circ., vol. 44, no. 4, pp. 1067-1077, Apr. 2009.

[9] H. Dehsorkh, N. Ravanshad, R. Lofti and A. Sodagar, "Analysis and Design of Tunable Amplifiers for Implantable Neural Recording Applications," IEEE J. Emer. Selec. Topi. Circ. Syst., vol. 1, no. 4, pp. 546-556, Dec. 2011.

[10] E. Ozalevli and P. E. Hasler, "Tunable highly linear floating-gate CMOS resistor using common-mode linearization techniques," IEEE Trans. Circuits Syst. I, Reg. Papers, vol. 35, no. 5, pp. 999–1010 May 2008.

[11] T. Wang, L. Liu and S.Peng, "A power- efficient highly linear reconfigurable biopotential sensing amplifier using gate-balanced pseudoresistors," IEEE Trans. Circ. and Syst.-II Express Briefs, vol.62, no.2, Feb. 2015.

[12] S. Wang, C. M. Lopez, M. Ballini, and N. V. Helleputte, "Leakage compensation scheme for ultra-high-resistance pseudo-resistors in neural amplifiers," Electron. Lett., vol. 54, pp. 270–272 2018.

[13] K.Yao, C. Gong, S. Yang and M. Shiue, "Design of a Neural Recording Amplifier with Tunable Pseudo Resistors," IEEE Int. SOC Conf. (SOCC), Taipei, Taiwan, pp. 376-379 Sep. 2011.

[14] T. Delbruck and C. A. Mead, "Analog VLSI adaptive, logarithmic widely namic-range photoreceptor," in Proc. IEEE Int. Symp. Circ. Syst., pp. 339–342, June 1994.

[15] T.-Y. Wang, M.-R. Lai, C. Twigg, and S.-Y. Peng, "A fully reconfigurable low-noise biopotential sensing amplifier with 1.96 noise efficiency factor," IEEE Trans. Biomed. Circ. Syst., vol. 8, no. 3, pp. 411–422 June 2014.

[16] A.Torralba, C.Lujan-Martinez, R.G.Carvajal, J.Galan, M. Pennisi, J.Ramirez-Angulo and A. Lopez-Martin, "Tunable Linear MOS Resistors Using Quasi-Floating-Gate Techniques," IEEE Trans. Circ. and Syst.-II Expr. Briefs, vol.56, no. 1, pp. 41-45, Jan. 2009.

[17] D. M. Caetano, D. Oliveira, J. Silva, T. Rabuske, J. Fernandes, "Tunable kΩ to GΩ Pseudo-Resistor with Bootstrapping Technique" in Proc. IEEE 60th Int. Mid. Sympo. Circ. Syst. (MWSCAS) Boston, MA, USA, pp. 6-9, Aug. 2017.

[18] R. Puddu, C. Carboni, L. Bisoni, G. Barabino, D. Pani, L. Raffo and M. Barbaro, "A precision pseudo resistor bias scheme for the design of very large time constant filters," IEEE Trans. Circ. and Syst.-II Expr. Briefs, vol. 64, no 7, pp. 762-766, 2017.

[19] C. C. Enz and E. A. Vittoz, "Charge-based MOS Transistor Modeling," Wiley, 2006.

[20] A. Tajalli and Y. Leblebici, "Extreme low-power mixed signal ic design: Subthreshold source-coupled circuits," New York: Springer, 2010.

[21] K. Roy, S. Mukhopadhyay and H. Mahmoodi-Meimand, "Leakage current mechanisms and leakage reduction techniques in deep-submicrometer CMOS circuits" Proc. of IEEE, vol. 91. no. 2, pp. 305–327, 2003.

TABLE II. PERFORMANCE COMPARISON OF STATISTICAL RESULTS OBTAINED FOR 100 MCS RUNS AT PROCESS AND CORNER SIMULATIONS FOR PR TECHNIQUES SHOWN IN FIG. 1.

PR technique	$R_{SD,N}$ (Ω)	μ (Ω)	σ (Ω)	CV (σ/μ)
NTPR-I (Fig. 1 (a))	6.668 G	12.41 G	22.06 G	1.8
NTPR-II (Fig. 1 (b))	960.3G	404.3G	224.3G	0.55
TPR-I (Fig. 1 (c))	56.08 M	88.47 M	115.8 M	1.31
TPR-II (Fig. 1 (d))	999.9G	993.6G	6.532G	6.5×10^{-3}
CML (Fig. 1 (e))	393.03 G	404.3 G	224.3 G	0.55
SGL (Fig. 1 (f))	999.99 G	1 T	1.364 M	1.36×10^{-6}

2019 Devices for Integrated Circuit (DevIC), 23-24 March, 2019, Kalyani, India

Design of an Integrator-Differentiator Block For a Transimpedance Amplifier Using 0.18µm Technology

Chander Partap Singh, Anisha Pathania, Kulbhushan Sharma, Jaya Madan, Rajnish Sharma[*]

VLSI Centre of Excellence,
Chitkara University Institute of Engineering and Technology,
Chitkara University, Punjab, India
chander.singh@chitkara.edu.in, anisha.pathania@chitkara.edu.in, kulbhushan.sharma@chitkara.edu.in,
jaya.madan@chitkara.edu.in, [*]rajnish.sharma@chitkarauniversity.edu.in

Abstract—Transimpedance amplifier (TIA) has become an integral part of front-end electronics required for current sensing applications. In this paper, an integrator-differentiator block as an integral part of TIA has been reported using 0.18 µm technology in standard CMOS N-well process. A tunable pseudo-resistor has been deployed in the proposed TIA architecture to obtain a variable gain and bandwidth of interest. The reported work also discusses the problem of saturation and clock feed-through present in the integrator block. The simulated gain and noise plots for the integrator-differentiator blocks are also presented in this work. Research effort put forth in this way in implementing the TIA may be helpful for efficient current recording and detection.

Keywords— Transimpedance amplifier; tunable pseudo-resistor; variable gain TIA

I. INTRODUCTION

A TIA frequently referred to as a current-to-voltage converter is a key block for innumerable sensing applications such as optical fiber communication [1,2], temperature sensing [3], microwave communication [4] and biomedical applications [5-8]. Review of the literature pertaining to TIA conveys that many researchers have worked upon different techniques such as continuous time approach [9,10], discrete time approach [11,12], regulated cascode scheme [1], half-shared TIA [13] and integrator-differentiator approach [14-16]. There is a requirement to implement this block in a way so as to minimize the integrator saturation and also optimize noise-bandwidth tradeoff [6].

Among all the aforementioned approaches, the integrator-differentiator technique with DC removal block proposed by Sani *et al.* [17] proves to be most useful in circumventing the concerned issues. However, this technique has also been found to be lacking with challenges of high noise and power consumption.

Conventional off-chip resistor to be deployed as a part of integrator-differentiator approach can be replaced with a high value (>MΩ) MOS based pseudo-resistor operating in weak inversion [18]. There has to be an optimized tradeoff between bandwidth and noise as at higher bandwidths the noise increases which can deteriorate the quality of the output signal obtained after amplification [19]. To sense extremely small values of current, although flicker and thermal noise are considered to be negligible [7] but the overall performance of the circuit capable of sensing the small value of current remains crucial.

In an effort to circumvent the aforementioned challenges of high power and noise consumption, this work introduces a modified integrator-differentiator approach for sensing small values of current. Proposed circuit has been designed and simulated in Cadence analog design environment using 0.18 µm CMOS technology. The circuit block has been optimized for low power and variable bandwidth requirement of the

small current sensing applications. Moreover, to attain variable gain of the TIA a Tunable pseudo-resistor (TPR) is used as a feedback component of the differentiator.

This paper is organized into three sections as follows: Section II providing the detail of the proposed Integrator-Differentiator block follows the current section of Introduction. The important conclusions drawn from the present piece of work are presented in Section III.

II. DESIGN METHODOLOGY OF PROPOSED INTEGRATOR-DIFFERENTIATOR BLOCK

Fig. 1 shows that an integral part of TIA comprises of a capacitive feedback integrator and a resistive feedback differentiator. The integrator is fed with the current source (I_{IN}). Further, the output of this integrator is then provided to the next stage through a coupling capacitor (C_{in}) in order to obtain a flat gain. The operational amplifiers; AMP1 and AMP2, have been used for integrator and differentiator stages respectively. A TPR as in [20] has been connected in the feedback path of the differentiator which accounts for variable bandwidth. Conventionally, the output voltage (V_{OUT}) of a TIA with the parallel combination of feedback capacitor (C_F) and resistor (R_F) is given by [6]:

$$V_{OUT} = \frac{R_F}{1 + j2\pi f R_F C_F} I_{IN}. \qquad (1)$$

Where I_{IN} is incoming current at the input of proposed Low Power TIA.
Similarly, the overall gain (A_V) of the integrator-differentiator TIA architecture shown in Fig. 1 can be expressed theoretically by (2) as follows:

$$A_v = V_{OUT} \Big/ I_{IN} = \frac{C_{IN}}{C_F} \times R_F. \qquad (2)$$

The bandwidth (BW) of the overall architecture of the TIA can be given by (3) as follows:

$$BW(-3dB) = \frac{g_{mo}}{A_v . C_L}. \qquad (3)$$

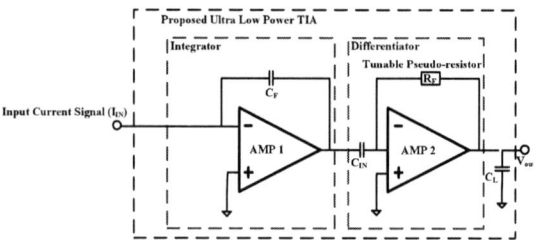

Fig. 1. Block diagram of proposed integrator-differentiator block.

978-1-5386-6723-1/19 $31.00 © 2019 IEEE   394

Where g_{mo} is the overall transconductance of TIA and C_L is the load capacitance.

The overall noise power spectral density (PSD) of the TIA ($\overline{i_{IN}^2}$) can be modeled through (4) as [6]:

$$\overline{i_{IN}^2} = \overline{i_D^2} + \overline{i_N^2}. \tag{4}$$

Where $\overline{i_D^2}$ is the noise PSD of the equivalent nanopore circuit comprising of flicker and thermal noise components, while $\overline{i_N^2}$ is the noise PSD of the TIA electronic interface. The contribution of $\overline{i_D^2}$ to the allover noise PSD of TIA is negligible as compared to the dominant noise contributor $\overline{i_N^2}$. More specifically, noise contributed by the CMOS opamp ($\overline{e_{n-op}^2}$) in a conventional TIA electronic interface can be modeled by (5):

$$\overline{e_{n-op}^2} = \frac{16kT}{g_{mi}} + \frac{2K_F}{C_{OX}WL} \times \frac{1}{f}. \tag{5}$$

Parameters k, T and g_{mi} are Boltzmann constant, temperature coefficient and transconductance of input differential pair respectively. W, L, C_{OX}, and K_F are the width, length, oxide capacitance of input differential pair and flicker noise coefficient respectively. Thus, the noise PSD of the TIA electronic interface ($\overline{i_N^2}$) becomes [6]:

$$\overline{i_N^2} = \frac{4kT}{R_F} + \overline{e_{n-op}^2}\left[\frac{1}{R_F^2} + (2\pi f)^2\left(C_F + C_{IN}\right)^2\right]. \tag{6}$$

It is clear from (6) that noise power will reduce for higher values of R_F and smaller values of total capacitance ($C_F + C_{IN}$) connected at the input node. The noise component of opamp i.e., $\overline{e_{n-op}^2}$ gets multiplied by f^2, and thus, at the low frequencies, the noise PSD rises due to 1/f noise. However, the differentiation of noise spectrum performed by the differentiator in the reported integrator-differentiator TIA scheme significantly reduces the impact of noise at low frequencies. Moreover, the first pole of the differentiator stage allows fixing the maximum bandwidth which removes the noise bandwidth tradeoff.

In present piece of work, an a.c current source of 150 pA_{pk-pk} with one of the frequency as 200 kHz and the other at a frequency of 2 MHz, and a DC current of 2 nA is used to simulate approximate behavior of the small signal model. The integrator design and the problem of integrator saturation have been individually addressed in sub-section A while the design of the differentiator has been discussed individually in sub-section B.

A. Integrator architecture

Fig. 2(a) shows the schematic of the architecture of the integrator used in the TIA while Fig. 2(b) illustrates the op-amp used to design integrator. As evident from Fig. 2(a), feedback capacitor (C_F) has been added to the feedback path of the op-amp circuit which plays an important role in determining the overall gain of the TIA. The op-amp (AMP1) shown in Fig. 2(b) used in the integrator architecture comprises of a differential pair (M1 and M2), two current mirror circuits (M3, and M4; M5, M6, and M7), and an output cascode stage (M7 and M8). The biasing current (I_{dc}) is kept 3 µA which has been mirrored from M5 to M6 and M7. The drain current of 1.5 µA flows through the devices M1 and M2. The reason for choosing the value of I_{dc} = 3 µA relies on the fact that this value meets the optimized low power and low

noise tradeoff. The devices used in the input differential pair (M1 and M2) have been sized with larger gate area in order to reduce the impact of flicker noise. Increasing the *W/L* ratio of the input differential pair pushes them into a weak-inversion region where, fortunately, the device exhibits very high g_m/I_D and reduces thermal noise. In the sub-threshold region, the transconductance (g_m) of a MOS device can be given by:

$$g_m = q\frac{I_{DS}}{nkT}. \tag{7}$$

In weak inversion region drain current (i_{DS})

$$i_{DS} = i_{DS0}\frac{W}{L}e^{\frac{V_{GS}}{nV_t}(1-e^{\frac{V_{DS}}{V_t}})}. \tag{8}$$

$$g_m = \frac{I_{DS}}{nV_T}. \tag{9}$$

Using g_m/I_D methodology [21] the calculated *W/L* values, current efficiency (g_m/I_D), and operating region of devices used in AMP1 of integrator have been tabulated in Table I. Besides this, a mention of the component values used in the integrator architecture has also been construed in Table I.

The major problem faced by the pioneer researchers in implementing the TIA for current sensing application has been integrator saturation [22] which occurs due to the fast switching in the baseline current (I_{base}) and the presence of DC components in it which can also be seen from Fig. 3. This leads to degradation of the frequency response of the TIA. Also, the feedback capacitor (C_F) needs regular switching in order to avoid clock feed-through.

As the stability of the overall circuit is affected due to the aforementioned problems an additional DC removal circuitry need to be employed which is going to be a subject matter for our future research work. Fig. 3 shows the gain and phase versus frequency plot of the integrator architecture shown in Fig. 2. Fig. 4 shows the plot between input-referred voltage noise versus frequency plot of the integrator architecture. It is worth noting here that the simulated gain of integrator stage was found to be 31 dB over a bandwidth ranging from 100 Hz to 100 kHz with a phase margin of 63°.

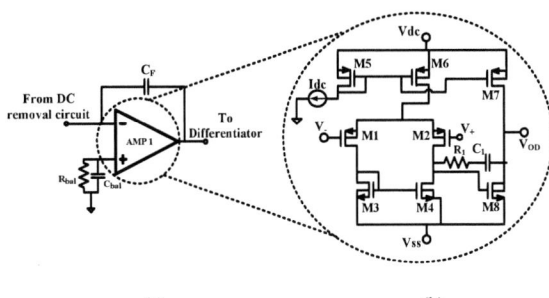

(a)   (b)

Fig. 2. (a) Schematic of integrator and (b) Op-amp used in integrator architecture (AMP1).

TABLE I. ASPECT RATIOS OF DEVICES AND COMPONENT VALUES USED IN DESIGNING INTEGRATOR ARCHITECTURE (AMP1).

Devices	W/L(µm)	g_m(µS)	g_m/I_D (V^{-1})	Region of Operation
M1,M2	200/0.5	14	28	Sub-threshold
M3,M4	0.4/5	4.5	9	Saturation
M5,M6, M7	4/8	7	7	Saturation
M8	3.2/8	13.5	13	Saturation
R_1= 1 GΩ, C_1= 1 pF, C_F= 20 pF, C_L=260fF, V_{dd}= 1.8 V, I_{dc}=3 µA.				

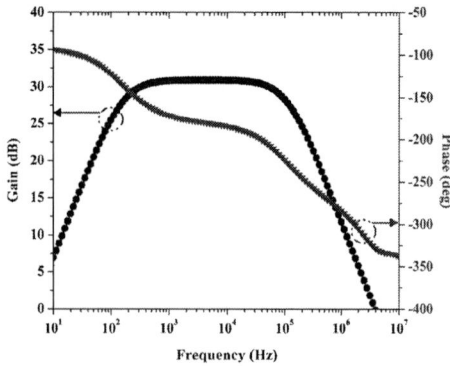

Fig. 3. Simulated transfer function of the integrator architecture shown in Fig. 2. Midband gain is 31 dB over a bandwidth ranging from 100 Hz to 100 kHz.

The achieved results in comparison with [17] show high gain and lower noise values owing to its simple circuitry and use of active devices instead of kxres resistors.

The integrated input-referred voltage noise of the integrator architecture over 100 Hz to 100 kHz frequency spectrum is found to be 0.16 µV/√Hz. The simulated integrator architecture consumes a power of 10.8 µW.

B. Differentiator architecture

The schematic of differentiator architecture and its corresponding op-amp (AMP2) topology deployed in the TIA is shown in Figs. 5(a) and (b) respectively. The schematic comprises of an input coupling capacitor (C_{IN}) at the inverting terminal, a TPR (R_F) in the feedback path of the differentiator and AMP2. The purpose of employing C_{IN} is to prevent the DC mismatch between the integrator and the differentiator which otherwise may deteriorate the strength of the signal of interest. Moreover, C_{IN} is responsible for improving the overall stability of the TIA. For low voltages across the TPR, its incremental resistance (r_{INC}) is high, $dV/dI > 10^6$ Ω. The TPR used in the feedback path of the differentiator accounts for the tunable gain and bandwidth of the TIA. The TPR is based on two series connected PMOS devices (shown in Fig. 5) whose gate terminals have been biased by a tunable voltage source (V_{TUNE}). For achieving variable gain and bandwidth of the allover TIA architecture, the V_{TUNE} is set to -80 mV (~1 MΩ) and –220 mV (~8 MΩ). Fig. 5(b) shows the op-amp (AMP2) circuit used in differentiator architecture which consist of a differential pair (M9 and M10), two current mirror circuits (M11 and M12; M13, M14, and M16), a compensation capacitor (C_3) and output cascode stage (M15 and M16). The circuit has been provided with a biasing current (I_{dc}) of 10 µA which has been mirrored from M13 to M14 and M16. The drain current of 5 µA flows through the devices M9 and M10.

Denoting the $g_{m9,10}$, $g_{m11,12}$ and $g_{m13,14}$ as transconductances of the differential pair (M$_9$, M$_{10}$), current mirror 1 (M$_{11}$, M$_{12}$) and current mirror 2 (M$_{13}$, M$_{14}$) respectively, then by sizing $g_{m9,10} >> g_{m11,12}$, $g_{m13,14}$ noise contribution of the devices can be minimized. The $g_{m11,12}$ and $g_{m13,14}$ cannot be reduced randomly beyond an extent as it poses a severe danger of instability attributed to shifting of poles near the dominant pole. Therefore, there has to be a tradeoff between area-noise and noise-stability. To ensure stability, the frequency of the poles generated by these current mirrors has to be greater than the dominant pole frequency due to C_L in the overall TIA design. Similar to AMP1 the devices M9 and M10 of AMP2 have also been operated in the

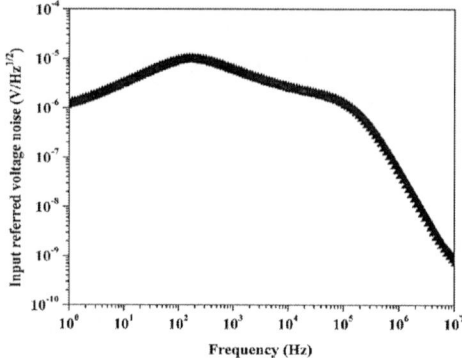

Fig. 4. Simulated input-referred noise versus frequency plot for the integrator architecture shown in Fig. 2. The integrated input-referred voltage noise of the integrator architecture over 100 Hz to 100 kHz frequency spectrum is found to be 0.16 µV/√Hz.

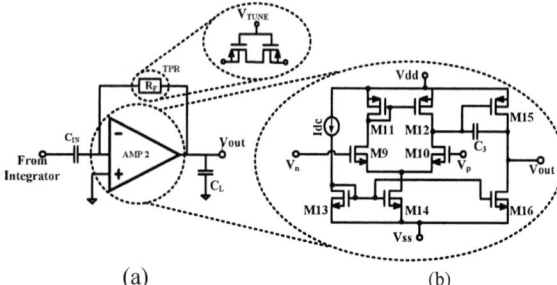

(a)   (b)

Fig. 5. (a) Schematic of differentiator and (b) Op-amp used in differentiator architecture (AMP2).

sub-threshold region for the reasons mentioned earlier in sub-section A. Table 2 tabulates the W/L values, current efficiency (g_m/I_D), and operating regions of the devices calculated by using gm/I$_D$ methodology [21]. Also, the values of the components employed in the differentiator architecture and its corresponding op-amp circuit have been charted in Table II.

It is worth mentioning here that in order to understand the behavior of differentiator architecture (shown in Fig. 5) it has been simulated for transfer function and noise. Fig. 6 shows the gain and phase versus frequency plot of the differentiator architecture and Fig. 7 shows the plot between input-referred voltage noise versus frequency of the differentiator. The simulated gain of the differentiator stage was found to be 90.8 dB over a −3 dB bandwidth of 60 kHz at $V_{TUNE} = −80$ mV, while no significant change was observed in gain and bandwidth at $V_{TUNE} = −220$ mV. However, when implemented in series with the integrator block, the overall block provides variable gain and bandwidth values for different values of V_{TUNE}. The phase margin was set to 63° for the op-amp (AMP2) used in differentiator while optimizing bandwidth-stability and noise-stability. The input-referred voltage noise of the differentiator architecture integrated over 100 mHz to 60 kHz frequency spectrum is

TABLE II.   ASPECT RATIOS OF DEVICES AND COMPONENT VALUES USED IN DESIGNING DIFFERENTIATOR ARCHITECTURE (AMP2).

Devices	W/L(µm)	g_m(µS)	g_m/I_D (V^{-1})	Region of Operation
M9,M10	8/1	92	18	Sub-threshold
M11,M12	4/2	30.5	6	Saturation
M13,M14	2/2	76	8	Saturation
M15	30.47/2	85	14	Saturation
C_{IN}= 50 nF, C_3= 1 fF, V_{dd}= 1.8 V, I_{dc}= 10 µA.				

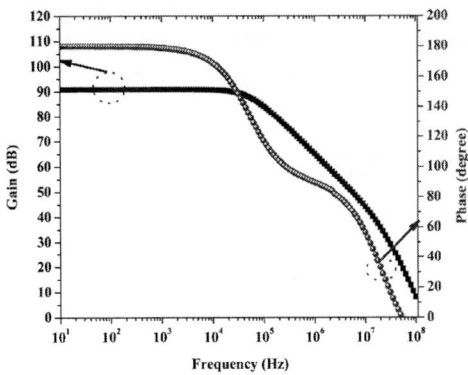

Fig. 6. Simulated transfer function of the differentiator architecture shown in Fig. 5. Midband gain is 90.83 dB over a bandwidth of 60 kHz at $V_{TUNE} = -80$ mV.

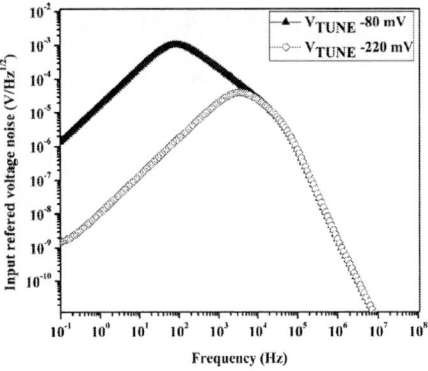

Fig. 7. Simulated input-referred noise versus frequency plot for the differentiator architecture shown in Fig. 5. The integrated input-referred voltage noise of the differentiator architecture over 100 mHz to 60 kHz frequency spectrum is found to be 0.12 µV/√Hz at both $V_{TUNE} = -80$ mV and -220 mV.

found to be 0.12 µV/√Hz at both $V_{TUNE} = -80$ mV and -220 mV. The simulated differentiator architecture consumes a power of 46.5 µW.

The differentiator architecture presented by G. Ferrari *et al.* [14] suggests the use of an additional feedback circuitry to achieve better stability but the use of passive components for achieving the desired results increases the area of the device and circuit complexity as well as contributes to noise which is undesirable. Investigation of the same and finding a resolve to that will be the focus of our future research work.

III. CONCLUSION AND FUTURE SCOPE

A low power integrator-differentiator block as an integral part of TIA for small current sensing applications is presented in this paper where an integrator is cascaded with a differentiator stage. Furthermore, the differentiator architecture is employed with the TPR in its feedback path in order to obtain variable gain and bandwidth requirements. The presented integrator-differentiator block best suits the small current sensing applications owing to the variable gain and low power requirements. The paper also discusses the saturation and clock feed-through problem present in the integrator block. Thus, to overcome these challenges research can be done to introduce a DC removal circuitry before the integrator stage and a buffer stage following an offset correction circuit can be added for further analysis of small valued currents.

REFERENCES

[1] S. M. Park and C. Toumazou, "A packaged low-noise high-speed regulated cascode transimpedance amplifier using a 0.6 nm N-well CMOS technology," in Proc. 26th Eur. Solid-State Circ., pp. 431–434, 2000.

[2] J. Lambrecht, H. Ramon, B. Moeneclaey, J. Verbist, M. Verplaetse, M. Vanhoecke, P. Ossieur, P. D. Heyn, J. V. Campenhout, J. Bauwelinck and X. Yin, "90 Gb/s NRZ optical receiver in silicon using a fully-differential transimpedance amplifier," J. Ligh. Tech., pp. 1-1, 2019.

[3] L. Toygur, A. C. Patil, J. Guo, X. Yu and S. L. Garverick, "A 300°C, SOI transimpedance amplifier with application to capacitive temperature sensing," in Proc. IEEE Sensors, pp. 3–6, 2012.

[4] L. Bary, M. Borgarino, R. Plana, T. Parra, S. J. Kovacic, H. Lafontaine and J. Graffeuil, "Transimpedance amplifier-based full low-frequency noise characterization setup for Si/SiGe HBTs," IEEE Trans. Elect. Dev., vol. 48, pp. 767–773, 2001.

[5] M. Tavakoli, L. Turicchia and R. Sarpeshkar, "An ultra-low-power pulse oximeter implemented," IEEE Trans. Biomed. Circ. Syst., vol. 4, pp. 27–38, 2010.

[6] M. Crescentini, M. Carminati, M. Tartagni and M. Bennati, "Noise limits of CMOS current interfaces for biosensors: A review," IEEE Trans. Biomed. Circ. Syst., vol. 8, pp. 278–292, 2014.

[7] J. Kim, R. Maitra, K. D. Pedrotti and W. B. Dunbar, "A patch-clamp ASIC for nanopore-based DNA analysis," IEEE Trans. Biomed. Circ.

Syst., vol. 7, pp. 285–295, 2013.

[8] G. Ferrari, F. Gozzini, A. Molari and M. Sampietro, "Transimpedance amplifier for high sensitivity current measurements on nanodevices," IEEE J. Solid-State Circ., vol. 44, pp. 1609–1616, 2009.

[9] F. Gozzini, G. Ferrari and M. Sampietro, "Linear transconductor with rail-to-rail input swing for very large time constant applications," Electron. Lett., vol. 42, 2006, pp. 1069–1070.

[10] P. Weerakoon, E. Culurciello, K. G. Klemic and F. J. Sigworth, "An integrated patch-clamp potentiostat with electrode compensation" in IEEE Trans. Biomed. Circ. Syst., pp. 117–125, 2014.

[11] A. Sunitha, G. D. Kevin, L. Manfred and M. Bradley, "Design of a CMOS potentiostat circuit for electrochemical detector arrays," IEEE Trans. Circ. Syst. I Regul. Pap., vol. 54, pp. 736–744, 2007.

[12] J. Hamed and M. G. Roman, "Chopper-stabilized bidirectional current acquisition circuits for electrochemical amperometric biosensors," IEEE Trans. Circ. Syst. I Regul. Pap., vol. 60, pp. 1149–1157, 2013.

[13] G. Mulberry, K. A. White and B. N. Kim, "Analysis of simple half-shared transimpedance amplifier for picoampere biosensor measurements," IEEE Trans. Biomed. Circ. and Syst., pp. 1-1, 2019.

[14] G. Ferrari, F. Gozzini and M. Sampietro, "A current-sensitive front-end amplifier for nano-biosensors with a 2MHz BW," IEEE Int. Solid-State Circ. Conf. Digest of Tech. Pap., pp. 164–165, 2007.

[15] Rosenstein, J. K. Wanunu, M. Merchant, C. A. Drndic, M. Shepard and L. Kenneth, "Integrated nanopore sensing platform with sub-microsecond temporal resolution," Nat. Methods, vol. 9, pp. 487–492, 2012.

[16] D. Djekic, G. Fantner, J. Behrends, K. Lips, M. Ortmanns and J. Anders, "A transimpedance amplifier using a widely tunable PVT-independent pseudo-resistor for high-performance current sensing applications," IEEE Eur. Solid-State Circ. Conf., pp. 79–82, 2017.

[17] M. Taherzadeh-Sani, S. M. H. Hussaini, H. Rezaee-Dehsorkh, F. Nabki and M. Sawan, "A 170-dB Ω CMOS TIA with 52-pA input-referred noise and 1-MHz bandwidth for very low current sensing," IEEE Trans. Very Lar. Scale Integr. Syst., vol. 25, pp. 1756–1766, 2017.

[18] K. Sharma and R. Sharma, "Highly consistent bulk driven quasi floating gate (BDQFG) PMOS pseudo-resistor design and implementation in 0.18 micron meter technology," IEEE Reg. 10 Annu. Int. Conf. Proc., pp. 488–493, 2017.

[19] E. Rebecca, A. Balan, B. Machielse, D. Niedzwiecki, J. Lin, P. Ong, K. L. Shepard and M. Drndi, "Improving signal-to-noise performance for DNA translocation in solid-state nanopores at MHz bandwidths," Nano Lett., vol. 14, pp. 7215–7220, 2014.

[20] H. Rezaee-Dehsorkh, N. Ravanshad, R. Lotfi, K. Mafinezhad and A. M. Sodagar, "Analysis and design of tunable amplifiers for implantable neural recording applications," IEEE J. Emerg. Sel. Top. Circ. Syst., vol. 1, pp. 546–556, 2011.

[21] F. P. Cortes, E. Fabris, and S. Bampi, "Analysis and design of amplifiers and comparators in CMOS 0.35 µm technology," Microelectron. Reliab., vol. 44, no. 4, pp. 657–664, 2004.

[22] K. Venta, G. Shemer, M. Puster, J. A. R-Manzo, A. Balan, J. K. Rosenstein, K. Shepard and M. Drndic, "Differentiation of short, single-stranded DNA homopolymers in solid-state nanopores," ACS Nano, vol. 7, pp. 4629–4636, 2013.

Impact Analysis of Dual Material Double Gate Oxide-Stack Junction-Less MOSFET in RFID Memory Cell Realisation

Dipanjan Sen
Dept. of E.T.C.E
Jadavpur University
Kolkata, India
sendipanjan10@gmail.com

Savio Jay Sengupta
Dept. of E.T.C.E
Jadavpur University
Kolkata, India
saviojsengupta@gmail.com

Subhashis Roy
Dept. of E.T.C.E
Jadavpur University
Kolkata, India
subhasisaec@gmail.com

Sudhabindu Ray
Dept. of E.T.C.E
Jadavpur University
Kolkata, India
sudhabin@yahoo.com

Subir Kumar Sarkar
Dept. of E.T.C.E
Jadavpur University
Kolkata, India
subirsarkar@ieee.org

Abstract— **In this proposed article, a realization of RFID memory cell has been performed using Dual Material Double Gate Stack-Oxide Junction-Less MOSFET for high speed and low power application [1] in Sub-threshold regime. SNM, Power and Delay of the Memory Cell or SRAM circuit in different operating modes have been analyzed in depth. Dual Material Double Gate Oxide-Stack Junction-Less MOSFET (DMDGS-JLT) shows promising I_{ON}/I_{OFF} ratio, less subthreshold swing and less Drain Induced Barrier Lowering or DIBL, in comparison with Double Gate Junction-Less MOSFET. So, proposed SRAM cell would be efficacious to offer less power dissipation and higher speed and a better Static Noise Margin. The impact of DMDGS-JLT in realizing RFID memory cell or SRAM has been studied in sub-threshold regime for ultra-low power tag design. Extensive simulations are performed using SILVACO ATLAS platform to validate the analyzed models. Besides, an optimum supply voltage range has been chosen to get an ultra-low power and higher speed of operation. DMDGS-JLT can be an alternative for ultra-low power Passive-RFID tag design, which results into greater time-span of the battery.**

Keywords—Dual Material, Gate-Stack Oxide, Double Gate Junction-less MOSFET, RFID Memory Cell , Passive RFID Tag, Power, Delay, SNM.

I. INTRODUCTION

In recent years, the world of VLSI has benefited immensely from the device downsizing. But the miniaturization of semiconductor device (conventional MOSFET) shows many adverse effects like- Short Channel Effect, Quantum Effects etc. As CMOS scaling is approaching its limits, Double-Gate MOSFET is becoming a very promising solution for the fabrication of high performance devices for low-power applications. Besides Low power applications are very much efficacious in portable applications. Conventional CMOS MOSFET does not offer better performances as it is not a better option for poor variability in sub-threshold or near threshold regime. Besides, due to short channel effects of conventional MOSFET voltage swing degrades significantly and does not offer reasonable noise margin. To overcome the issues due to

SCE, a non-conventional Dual Material Double-Gate Oxide-Stack Junction-Less MOSFET [1-2] can be used in realising the Digital and Analog circuit performances. The problems related to sharp source and drain junctions for sub-nano devices due to the challenging situation with doping concentration and thermal profile can be managed by using Junction-Less devices. Junction-Less [2] provides good control of the gate, reduces the issues related to the diffusion of impurities and sharp formation of doping concentration. It exhibits better On current, Sub-threshold swing than the conventional MOS devices. DMDGS-JLT [3-4] exhibits promising On/Off current ratio, and the larger value of On-Current results into promising characteristics in terms of switching speed. Therefore, Dual Material Double Gate Oxide-Stack Junction-Less MOSFET [4-5] can be used as an emerging device to analyse Memory-Cells in RFID systems [6] for ultra-low power and faster operations.

The recent aspect of constant miniaturization of device dimensions, the thickness of the oxide thickness of the gate should also be scaled down gradually and the effective oxide thickness of the device also decreases. The decrement of the T_{OX} results into direct tunnelling of charge carriers through the dielectric layer and the reliability of the device has been compromised. One of the solutions in this case is High-K dielectrics, which can replace SiO_2 in gate oxide layer. Hafnium Oxide (HfO_2) provides a high dielectric constant of the range 22 to 25, a band gap of 5.6eV and can be a better solution for low power sub-nano devices like MOSFET for its higher dielectric constant K, good thermal stability. But, still it could not overcome the Short-Channel Effects such as Gate leakage current. In this proposed article, the SRAM cell has been studied in depth to analyse the power consumption of the cell and to achieve a minimum amount of delay by using DMDGS-JLT. Static Noise Margin of the SRAM cell has also been shown to validate the stability of the memory cell.

The paper has been divided into few sections: Section I consists of the introduction. Section II shows the description of the device and Section III shows the brief description SRAM circuit in RFID [6] . Section IV presents the simulation results and discussions. Section V shows the conclusion part of the work.

978-1-5386-6723-1/19 $31.00 © 2019 IEEE

II. DEVICE DESCRIPTION

In this paper, a Dual Material Double Gate Oxide Stack Junction-Less MOSFET [7] has been designed by considering a gate length of 50nm. Here, the effective oxide thickness (t_{OXE}), silicon thickness (t_{si}), Work-Function of Gate Metal and channel doping concentration (N_D)are chosen selectively. Channel doping is 10^{17} cm^{-3}, silicon thickness is 10nm and effective oxide thickness is 2 nm (t_{HfO2}=1nm & t_{SiO2}=1nm) and the gate oxide-stack technique has been adopted. Doping concentration of the source and drain are chosen to be 10^{20} cm^{-3}. Here, the cross-sectional view of Dual Material Double Gate Oxide Stack Junction-Less MOSFET is shown.

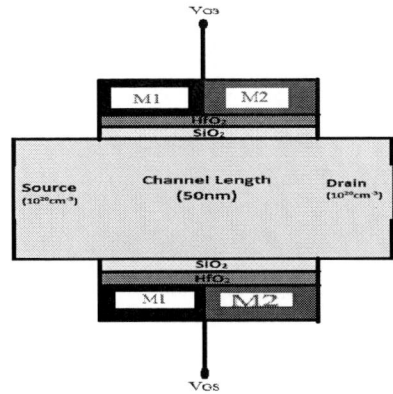

Fig.1: TCAD simulated cross-sectional form of N-type Dual Material Double Gate Oxide Stack Junction-Less MOSFET

Parameter	Values
T_{OXE} or Effective oxide thickness (HfO$_2$ & SiO$_2$)	2nm
T_{SI} or silicon thickness	10nm
Source/Drain Doping concentration	10^{20} cm^{-3}
Channel Length	50nm
Channel Doping	10^{17} cm^{-3}
Permittivity T_{OX1} (SiO$_2$)	3.9
Permittivity T_{OX2} (HfO$_2$)	24
Work-Function of M1	4.8
Work-Function of M2	4.6

Table 1: Parameter Values

Table.1 shows the process parameter values used in designing the DMDGS-JLT device shown in Fig.1.

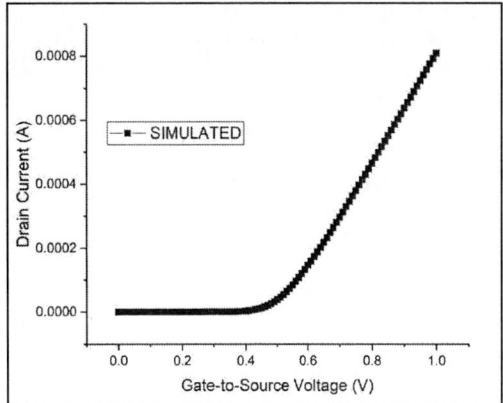

Fig.2 I_D VS V_{GS} characteristic curve of Dual Material Double Gate Oxide Stack Junction-Less MOSFET at V_{DS}=0.1V

The I_D VS V_{GS} characteristics curve of the device have been shown in Fig.2, while the drain voltage or V_{DS} is 0.1V and gate voltage is 0V (Sub-threshold region). The device threshold voltage with the mentioned process parameters is 362mV and the Sub-threshold swing is 49mV/dec. The device (DMDGS-JLT) shows a higher ON-OFF current ratio (10^{10}), which is desirable more often. The addition of High-K gate oxide stack is quite advantageous in terms of reducing junction leakage current, and the use of SiO2 as an intermediate layer between High-k gate dielectric and silicon channel helps in minimizing the scattering effect and significant improvement can be achieved in case of carrier mobility [7]. Besides, the use of work function engineering technique such as dual material gate results into a designed device with a higher immunity towards DIBL. Therefore, the DMDGS-JLT [7] can be an alternative to the conventional MOS devices to reduce SCEs and can be considered in terms of analysing circuits for low voltage, low power and high speed applications. In case of realizing a 6T SRAM cell for RFID systems, DMDGS-JLT can be considered as an alternative for ultra-low power circuit analysis.

Besides, all the software simulations are done in SILVACO ATLAS [8]. The device physical models are also mentioned, such as Shockley-Read-Hall (SRH) recombination model [9] and Auger model [10] are used for recombination. FLDMOB or Field dependant mobility model [10] is used to consider the velocity saturation under high electric field. Band-Gap Narrowing model [11] has also been used. CONMOB [11] is used for mobility degradation.

III. PROPOSED SRAM CIRCUIT & USE IN RFID

The SRAM [12] performance is analyzed by considering a non-conventional MOS device structure (DMDGS-JLT) to mitigate the Short-Channel Effects in a circuit operation. Also, the purpose of analyzing a SRAM [12] using the above mentioned device is to get a better SNM, ultra-low power dissipation and faster operation by minimizing the leakage current of the device and improving the threshold voltage roll-off for RFID applications. In RFID Systems, contactless devices for more advanced applications require a higher amount of flexibility, more storage capacity, faster operation from the embedded memory. SRAM exhibits [13] very less delay, which is required in case of high-end RFID applications. In the near and sub-threshold regime, the power dissipation will be very less and delay of the circuit will be optimized keeping in mind that in sub-threshold regime propagation delay increases. Fig. 3. Shows the system architecture of digital RFID circuits.

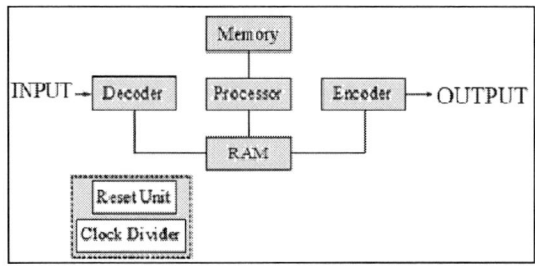

Fig. 3. System architecture of Digital RFID Circuits

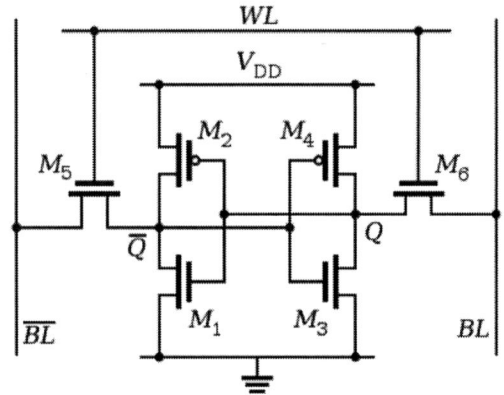

Fig.3.1 Schematic representation of Dual Material Double Gate Oxide Stack Junction-Less MOSFET based SRAM

Fig.3.1 shows the schematic form of the 6T-SRAM, which uses bi-stable latching mechanism to store a bit. In this work, a six transistor CMOS SRAM memory cell has been studied by using Dual Material Double Gate Oxide Stack Junction-Less MOSFET. The 6T-SRAM [13] contains two cross-coupled static CMOS inverters, connected back to back and also two N-type Dual Material Double Gate Oxide Stack Junction-Less MOSFET pass transistors are connected with them in Fig.3.1. The circuit shows low-power dissipation, less leakage current and better SNM [13-14]. Also, two complementary bit lines and a word line is connected with the mentioned pass transistors. The operations of the different MOSFETs depend on the value of the bit line and word line, therefore the mode of operation of the SRAM can be managed with the value of BL (bit line) and WL (word line). Hence, there are mainly three modes, such as- HOLD, READ and WRITE [14].

A. HOLD MODE

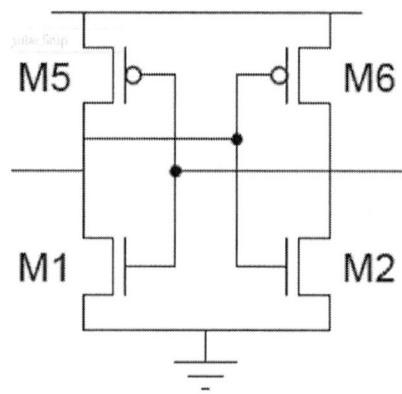

Fig.3.2. Schematic form of HOLD Mode of 6T SRAM

Fig.3.2 shows the schematic representation of the HOLD Mode of SRAM. In this mode of operation, the word line remains inactive, since both the pass-transistors are disconnected. In this case, the memory cell stores the information used in past as long as the power supply is switched-on. The cross-coupling mechanism supply data to the inverters connected back to back.

B. READ MODE

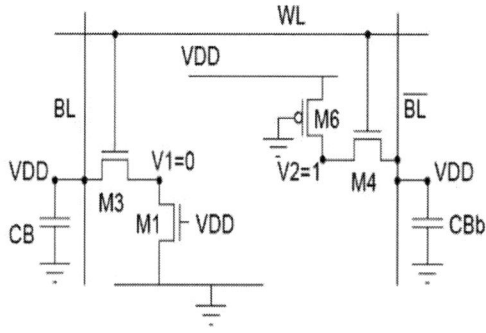

Fig.3.3. Schematic form of READ Mode of 6T SRAM

Fig.3.3 shows the schematic form of READ operation of the SRAM cell where both the lines are connected to V_{DD}, which is an unstable state for the memory cell because the values of the two bit lines should be in a complimentary form to each other. Due to this almost unstable state, the cell pulls down one of the inverters and it creates a delay. The M3 and M5, both the transistors are active. M4 goes into saturation mode and M2 becomes zero. M6 and M1 stays inactive. When the pass-transistors are turned on the load-capacitance CB_b will not change much as no current flows through M4. Through M3 and M1, a very low current passes and also the voltage drops across CB and also M2 remains off in READ mode [14-15].

C. WRITE MODE

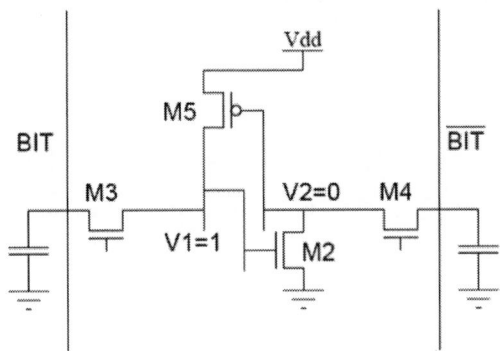

Fig.3.4. Schematic form WRITE Mode of 6T SRAM

Fig.3.4 shows the write mode of operation where two different cases exist, one is the write 'HIGH' and another is the write 'LOW'. Here, M6 and M1 are at inactive state whereas M2 and M5 are in active state. V1 and V2 are the node voltages with values V_{DD} and 0. In case of writing 'HIGH' V1 and V2 becomes zero and V_{DD} and M2 remains inactive.

IV. SIMULATION RESULT & DISCUSSION

In this paper, we have paid attention on calculating the SNM, Power dissipation, Propagation delay and Power-Delay Product for different state of operation. Dual Material Double Gate Oxide Stack Junction-Less MOSFET is used to simulate the circuit in terms of power dissipation and delay of READ and WRITE mode and it compares the result with the published artciles [14-15]. Static noise margin of cell has been measured by plotting Voltage Transfer Characteristics (VTC) of the MOS transistors. By plotting both the VTCs of the inverters, the generated graphical representation is the Butterfly Curve shown in Fig.4. The best fitted square in the butterfly curve will give the SNM value of the cell in the form of diagonal of the square.

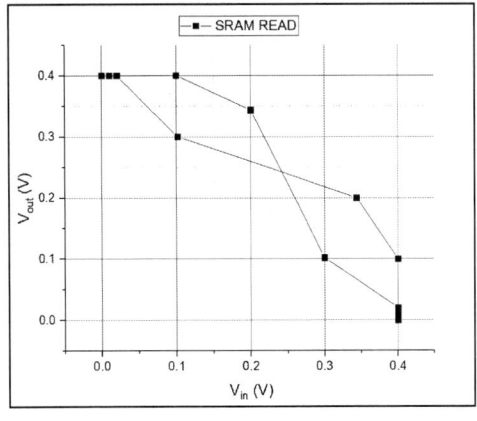

Fig.4. Butterfly Characteristics curve of 6T-SRAM cell in READ mode

A. Performance analysis of 6T SRAM in HOLD State

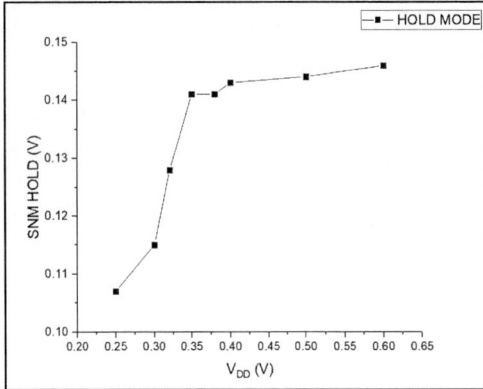

Fig. 5(a). Static Noise Margin in HOLD state of SRAM cell against supply voltage variation using DMDGS-JLT

Fig. 5(a) shows the SNM curve of HOLD state by varying the supply voltage from 250mV to 600mV. The SNM of a memory cell reduces as V_{DD} decreases. In this case, at V_{DD}=600mV the static noise margin is the maximum for HOLD state. A large value of noise margin is desirable to analyse the stability of the SRAM cell in its different mode of operation. Here, the work has been done in near and sub-threshold regime for low voltage application. By increasing the supply voltage, higher values of SNM [15] can be achieved. But in case of Sub-threshold operation, the change in the value of SNM is not so significant as the supply voltage is below the threshold of the device.

B. Performance analysis of 6T SRAM in READ State

In a SRAM cell, the READ and WRITE mode of operations are the most important one as per as the Power dissipation, Propagation delay and SNM analyses are concerned. Low-Power, faster speed of operation and large noise margin is required to design a SRAM cell for memory applications [15]. Delay of the memory cell is a major concern here to propose a SRAM cell made of Dual Material Double Gate Oxide-Stack Junction-Less MOSFETs, keeping in mind the trend of emerging devices in the field of VLSI design.

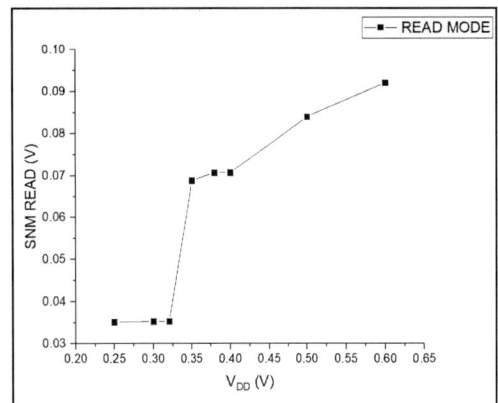

Fig. 6(a). Static Noise Margin in READ state of SRAM cell against supply voltage variation using DMDGS-JLT

2019 Devices for Integrated Circuit (DevIC), 23-24 March, 2019, Kalyani, India

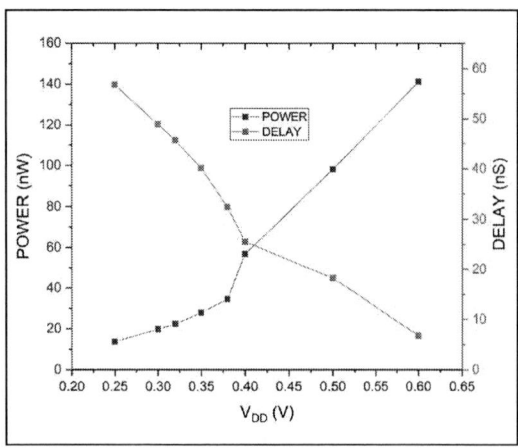

Fig.6 (b). Power & Delay plots of READ operation against V_{DD} variation using DMDGS-JLT

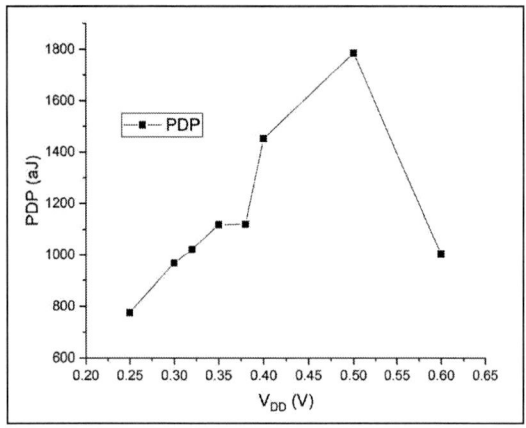

Fig. 6(c). POWER-DELAY PRODUCT plot of READ operation against V_{DD} variation using DMDGS-JLT

Fig. 6(a), 6(b) & 6(c) shows the SNM, Power dissipation, Propagation delay and Power-Delay Product of the SRAM in READ mode. From Fig. 6(a), it can be concluded that the SNM value rises as the supply voltage increases but it is less compared to the SNM of the HOLD state. It can be said that, in case of short-channel devices the stability of the SRAM memory cell falls, keeping in mind the amount of immunity to external noise. From Fig. 6(b). it can be seen that the change in power dissipation is very insignificant in the subthreshold region, which is from 0.25V to 0.35V and it exhibits low-power. But the power rapidly increases after 0.35V and it is maximum at 0.6V. Also, the propagation delay decreases with increasing supply voltage and it is minimum at 0.6V. Therefore, the average power dissipation reduces to nano-watt range with supply voltage variation in near and sub-threshold regime. Fig. 6(c). shows the PDP curve with supply voltage (V_{DD}) variation. From the PDP analysis it can be said that, the minimum value of PDP is at 0.25V. So, the optimization of the supply voltage can be performed with the help of Power-Delay Product analysis.

C. Performance analysis of 6T SRAM in WRITE State

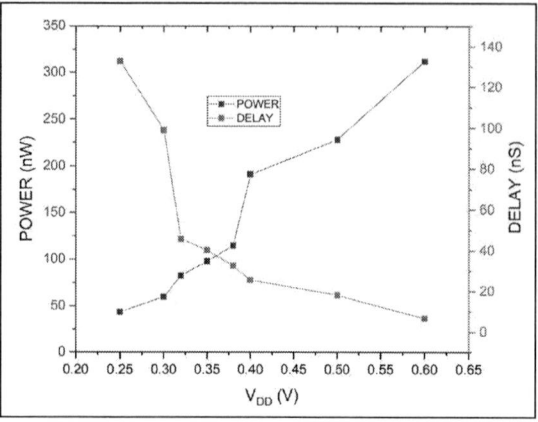

Fig. 7(a). Power & Delay plots of WRITE operation against V_{DD} variation using DMDGS-JLT

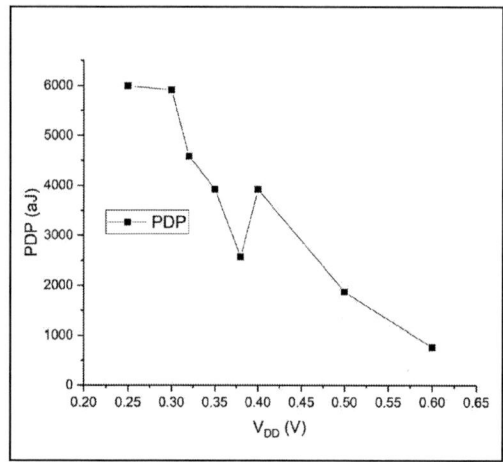

Fig. 7(b). POWER-DELAY PRODUCT plot of WRITE operation against V_{DD} variation using DMDGS-JLT

Fig. 7(a) & 7(b) shows the Power dissipation, Propagation delay and PDP of the SRAM in WRITE mode. Fig. 7(a) shows power dissipation and propagation delay curves. It can be said that the device with High-K gate oxide dissipates less power than the device which contains SiO_2 and a device with gate oxide-stack further improves the device electrical characteristics. Due to the improvement of the V_T roll-off, thus the current improves, which results into less leakage power dissipation. In case of write mode, the mechanism of bit storage makes a SRAM to work as a latch [14-15]. Small amount of leakage current of the CMOS inverters contribute towards the increase of leakage power and hence the average power dissipation becomes more than the other modes .Therefore, the average power dissipation reduces to nano-watt range with supply voltage variation in near and sub-threshold regime. Fig. 7(a). shows that at V_{DD}=0.25V, the power dissipation is minimum and from 0.25V to 0.38V, the change in power is not significant. So, for low-power RFID SRAM design, the sub-threshold regime can be considerable. Also, the propagation delay is maximum at 0.25V. Therefore, it is difficult to predict the

978-1-5386-6723-1/19 $31.00 © 2019 IEEE   402

optimum supply voltage from Fig. 7(a), thus PDP analysis should be performed to get the optimum supply voltage. Fig. 7(b). shows the PDP plot, which says that at V_{DD}=0.6V, PDP is the minimum [15].

V. CONCLUSION

In this work, the issues related to down scaling of the MOS devices have been discussed. One of the solutions to SCEs, that is Dual Material Double Gate Oxide-Stack Junction-Less MOSFET has been used to achieve better On/Off ratio, improved threshold voltage roll-off etc. D.C performance of a 6T-SRAM cell has been studied by Dual Material Double Gate Oxide-Stack Junction-Less MOSFET in Near and Sub-threshold regime for ultra-low power and high-speed applications. Also, the SNM is measured to analyse the stability of the memory cell in different state of operation. By analysing the circuit in Near and Sub-threshold regime, a very low power dissipation can be achieved for low voltage applications. Also, the propagation delay being one of the most crucial parameters can be optimised by analysing the memory cell in Near and Sub-threshold regime. The SRAM cell has been realised in depth for various RFID system designs and applications. Therefore, the Dual Material Double Gate Oxide-Stack Junction-Less MOSFET can be an alternative option in the field of CMOS circuit analysis and in the SRAM realisation for Digital RFID circuit design.

ACKNOWLEDGEMENT

Authors of this paper would like to thank the financial support obtained from the DST-WB project. Ref. No: 287 (Sanc.)/ST/P/S&T/6G-43/2017.Also, the authors would like to thankfully acknowledge the technical discussion with Dr. Manash Chanda, MSIT, Kolkata.

REFERENCES

[1] M. Chanda, S. De, and C. K. Sarkar, "Modeling of characteristic parameters for nano-scale junctionless double gate MOSFET considering quantum mechanical effect," Journal of Computational Electronics, vol. 14, no. 1, pp. 262–269, 2014.

[2] F. Jazaeri, L. Barbut, A. Koukab, and J.-M. Sallese, "Analytical model for ultra-thin body junctionless symmetric double gate MOSFETs in subthreshold regime," Solid-State Electron., vol. 82, pp. 103–110, Apr. 2013.

[3] G. Dhiman, R. Pourush, and P. K. Ghosh, "Performance Analysis of High-κ Material Gate Stack Based Nanoscale Junction Less Double Gate MOSFET," Materials Focus, vol. 7, no. 2, pp. 259–267, 2018.

[4] Pritha Banerjee and Subir Kumar Sarkar, "Modelling and Analysis of a Front High-k gate stack Dual-Material Tri-gate Schottky Barrier Silicon-on-Insulator MOSFET with a dual-material bottom gate", DOI: 10.1007/s12633-018-9940-y, Silicon, Springer, August 2018.

[5] Pritha Banerjee and Subir Kumar Sarkar, "3D Analytical Modeling of Dual Material Triple Gate Silicon-on Nothing MOSFET", IEEE Transactions on Electron Devices, Vol.64, Issue 2, pp-368-375, February 2017.

[6] Joyashree Bag and Subir Kumar Sarkar, "Development and VLSI implementation of a data security scheme for RFID system using programmable cellular automata" International Journal Radio Frequency Identification Technology and Applications, Vol. 4, No. 2, Pp: 197-211, 2013.

[7] Pritha Banerjee and Subir Kumar Sarkar, "3-D analytical modeling of high-k gate stack dual-material tri-gate strained silicon-on-nothing MOSFET with dual-material bottom gate for suppressing short channel effects", in Journal of Computational Electronics, Springer, Vol.16.No.3,pp-631-639, May 2017

[8] SILVACO International 2000, ATLAS: 2-D Device Simulation Software.

[9] N.D. Arora, J.R. Hauser, D.J. Roulston, Electron and hole mobilities in silicon as a function of concentration and temperature, IEEE Trans. Electron Devices 29 (2) (Feb 1982) 292-295.

[10] J.G. Fossum, D.S. Lee, A physical model for the dependence of carrier lifetime on doping density in nondegenerate silicon, Solid StateElectron. 25 (1982) 741-747.

[11] S. Saha, MOSFET test structures for two-dimensional device simulation, Solid State Electron. 38 (1) (Jan. 1995) 69-73.
2013.

[12] S. Y. Wu et al., "Demonstration of a sub-0.03 um2 high density 6-T SRAM with scaled bulk FinFETs for mobile SOC applications beyond 10nm node," 2016 IEEE Symposium on VLSI Technology, Honolulu, HI, 2016, pp. 1-2.

[13] C. Pan and A. Naeemi, "A Non-volatile Fast-Read Two-Transistor SRAM Based on Spintronic Devices," in IEEE Journal on Exploratory Solid-State Computational Devices and Circuits, vol. 3, pp. 93-100, Dec. 2017.

[14] Ritu Jain, Swapnil Jain," Leakage Power Estimation & Minimization In a 6T SRAM Cell Using Dual Vth, Dual Tox, & Stacking Techniques, IJECT Vol. 4, Issue Spl - 4, April - June 2013, at 1,2Dept. of ECE, Poornima Institute of Engineering & Technology, Jaipur, Rajasthan, India

[15] C. Manoj, V. Ramgopal Rao, "Impact of high-k gate dielectrics on the device and circuit performance of nanoscale FinFETs", IEEE Electron Device Lett., vol. 28, no. 4, pp. 295-297, Apr. 2007.

2019 Devices for Integrated Circuit (DevIC), 23-24 March, 2019, Kalyani, India

Design Approach for Artificial Human Ankle Movement

Susmita Das
Electronics and Instrumentation Engg.
Narula Institute of Technology
Kolkata, India
susmitad2k17@gmail.com

Dalia Nandi(Das)
Electronics and Communication Engg.
Indian Institute of Information
Technology Kalyani
Kalyani, India
daliadas311@gmail.com

Biswarup Neogi
Electronics and Communication Engg.
JIS College of Engg.
Kalyani, India
biswarupneogi@gmail.com

Abstract— In the present research work, the analysis and design aspect of artificial ankle for facility creation of transtibial amputated person is carried out. Artificial system development has achieved the most powerful impact to reproduce the presentation and functionality of amputated body parts or body organs. Ergonomics and human factors related research work is taken in this paper for the development of physically challenged people. These systems are initiated to energize the damaged limbs, in order to make the total living system operative in conjunction with the concerned artificial limb. A living body maintains the overall control of various parts of the system through sensory organs. The artificial limb is required to be properly attached to the concerned system with direct connection to the limb of the amputee patient. The artificial limb which normally needs to be controlled by motor movement is related with the electromyography signal. The EMG (Electro Myography) in this context is driving the limb through the motor providing the appropriate output signal which finally performs the required movement or operation of ankle joint.

Keywords— *Ankle joint, Biomechanics, EMG, Motor movement.*

I. INTRODUCTION

In the study of human biomechanics, it is often desirable to involve the performance assessment of a task. The artificial and mechanically operated human ankle movement [1] with proper functionality features needs the support of biomechanics. In this paper, the design scheme for plantar and dorsi flexion of natural ankle movement [2] has been attempted. The specified movements are involved to replicate the joint movements using mechanical motors with biomedical approach. The initiative towards the artificial ankle design with natural and synchronized gait style is the focal point of this work. This research will depict the idea of prosthetic model and simulation based work for bio signals of lower limb of human beings. Rehabilitation [5] procedure gives the systematic application of engineering methodologies and scientific principles to achieve the requirements and overcome the struggles faced by an amputee. Designing of a robotic mechanism with full integration of human neuromuscular system is a tough job to be implemented. It is a great challenge to replicate all types of motions from lower limb and utilize them in a resourceful way [6]. Robotics and biomechanics both have a great effect to combine their techniques to make the perfect biomechatronic mechanism.

The basic anatomy [3] of ankle joint consists of bones, muscles, ligaments, blood vessels, joints, tendons & nerves as shown in Fig 1. The upper portion of the human foot is called dorsal surface and the lower surface of the human foot is called plantar surface. The bonny area present on the lateral face of the human ankle is known as Lateral Malleolus and on the medial side of the ankle is called Medial Malleolus. The ankle joint of human being is structured by the connection of three joints such as Talus, Tibia and Fibula as shown in Fig 2.a. The ankle movement is the result of different ligaments & many soft tissues. Two most important soft tissues for ankle joint are Ligaments & Tendons as presented in Fig 2.c. The ligament attaches a bone with other bones and tendon attaches bones to muscle. The most essential tendon for activities such as walking, running and jumping is Archilles tendon as shown in Fig 2.d. that attaches two calf muscles to Calcaneus as shown in Fig 2.b. The normal movements of ankle is made to be replicated by the artificial system. The model development includes the electromyography (EMG) signal and programmed actuator or servo motor function to be implemented at the ankle joint [7] of artificial system.

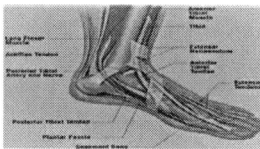

Fig 1. Basic anatomy of ankle joint of Human being

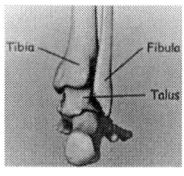

Fig 2.a. Bones of ankle joint

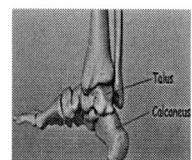

Fig 2.b. Bones of heel

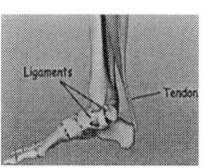

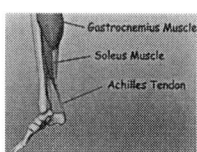

978-1-5386-6723-1/19 $31.00 © 2019 IEEE   404

Fig 2.c. Ligaments and Tendons   **Fig 2.d. Connecting Tendons**

II. METHODOLOGY

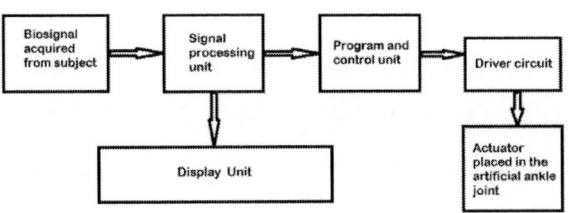

Fig 3. Block Diagram of the System development of Artificial Ankle joint

In the above mentioned block diagram as shown in Fig 3., the overview of the work-flow has been described. Initially, the bio signal is acquired from the ankle joint of a human being. The output signal is processed with noise filtering and amplification. The angular movement control of ankle joint for plantar and dorsi flexion are implemented through programming. The justified outcome of the system is presented using servo motor movement at the ankle joint of the artificial system.

The biomechanical characteristics [11] of human ankle shows some features which are planned to be functioned in the artificial ankle design stage. Important factors for design implementation [13] are given below:

1) The developed model must give the structural strength to be able to lift the amputee's lower limb weight.

2) The weight of the designed system should be compatible with the missing part of human ankle.

3) The artificial materials must have the capability to withstand sufficient shock with comfortable movement and for any mechanical damage prevention related to the mechanism for locomotion.

4) The ankle must be featured with lateral flexibility in the frontal plane (inversion and eversion) for better adaptability of natural movement.

A. Description of the Gait cycle of human walking: The ultimate objective of the prototype development is to implement the gait cycle of human walking. Human gait cycle [8] comprises of two phases and these are stance phase & swing phase. The stance phase begins with the heel strike and continues until reaches to the toe-off position. The stance phase covers approximately 60% of the gait cycle. The rest portion of gait cycle (38-40%) is covered in swing phase. The stance phase can be characterized by the three sub-phases as shown in Fig 4.

a) Plantar flexion gives 12% of human gait cycle starting from heel strike to foot flat.

b) Dorsal flexion delivers 38% of human gait cycle starting from foot flat to maximum position of dorsiflexion.

c) Powered plantar flexion completes 10-12% of human gait cycle initiating from maximum dorsiflexion to toe-off position.

A swing phase is considered as the "non-weight bearing" phase. As the big toe of one limb leaves the ground and ends just before the heel strike of the same limb. Swing phase consists of three stages such as a) Initial swing is the stage of

acceleration, b) Mid-swing, and c) Terminal swing is the stage of deceleration.

a) Initial swing is the initial third position of the swing phase from 60-73% of the gait cycle.

b) Mid swing is the middle third position of the swing phase from 73-87% of the gait cycle.

c) Deceleration is the final third position of the swing phase from 78-100% of the gait cycle.

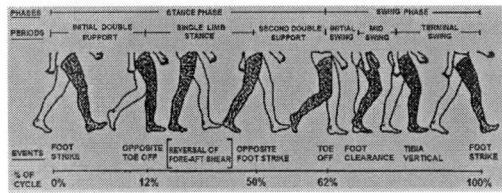

Fig 4. Pictorial presentation of Human gait during normal walking [4]

B. The angular movements of ankle: The angular range of joint rotation for the model is based upon the normal human ankle range of motion during walking as shown in Table 1 where normal walking speed is 1.4 meter/second. The maximum angular movement of 7 deg. in plantar flexion occurs at the end of stance phase, when foot is lifted off the ground. The maximum angular movement of 16 deg. in dorsiflexion takes place at the terminal controlled dorsiflexion as shown in Table1.

Table 1: Output angular changes of ankle during walking [4]

SL. No.	Walking types	\multicolumn{4}{c}{Angular changes of Ankle}
		Heel Strike (degree)	Control plantar flexion (degree)	Max. dorsiflexion (degree)	Toe-off (degree)
01	Slow	+2	-7	+16	+3
02	Normal (1.4m/s)	+1	-7	+15	-4
03	Fast	+5	-5	+14	-7

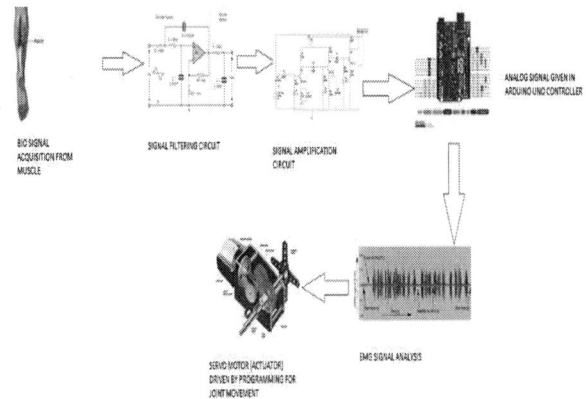

Fig 5. Schematic Diagram of Human Ankle movement

In the above schematic diagram as shown in the Fig 5, the hardware design concept of the artificial ankle motion is presented. In this model, EMG signals [10] are collected from the muscles connected to the ankle joint. The signal conditioning is done using filter and amplification circuit. The processed output is fed to the analog input pin of the Arduino interfacing board. According to the programmed [9] output, the servo motor moves or changes the direction placed at the ankle joint [12] of the prototype.

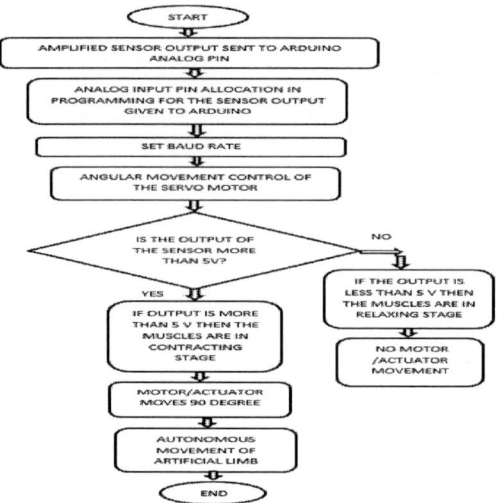

Fig 6. Work Flowchart of the motor movement of the design

III. RESULTS AND DISCUSSIONS

Two types of ankle movement are considered in this paper and these are Plantar flexion and Dorsi flexion. The output voltage values are obtained from the specified location as shown in Table 2 and the pictorial presentation of the experimental work is given in Fig 7. The observed voltages are fed to the Arduino interfacing board as the input of the developed system. According to the user-specified instruction, the output motor movement occurs.

Table 2: Output voltage values of Plantar flexion and Dorsi flexion

SL. No.	Output voltage values of Plantar flexion and Dorsi flexion	
	Plantar Flexion (millivolt)	Dorsi Flexion (millivolt)
01	9	5
02	11	5
03	12	6
04	10	3
05	10	9
06	8	3
07	9	3
08	7	4
09	15	3

SL. No.	Output voltage values of Plantar flexion and Dorsi flexion	
	Plantar Flexion (millivolt)	Dorsi Flexion (millivolt)
10	13	6
11	10	4
12	11	4
13	13	3
14	19	7
15	16	3
16	14	3
17	16	8
18	17	4
19	21	5
20	12	4
21	15	5
22	21	3
23	24	4
24	27	5
25	20	6
26	23	4
27	20	8
28	21	9
29	33	8
30	35	3

The above observations say about the output voltage variations for two types of flexion.

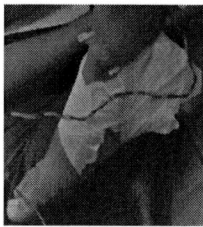

Fig 7. Pictorial representation of the data acquisition from human ankle

The end-effector movement of the developed system of lower extremity model is the final achievement of the work.

CONCLUSIONS

Traditional approach of artificial ankle joints provide mechanically supported structure to transfer forces. This makes the development of the model flexible and helps to eliminate weight to allow additional applications such as the active variation of the length of the leg within a gait cycle. The artificial limb requires electro-myographic signal to control the motor movement at ankle joint. The EMG signal driven motors with appropriate output signal performs the required operation of ankle of lower limb. An attempt shall be considered towards the development of highly efficient limb for lower amputee patient with trained walking style. This research shall lighten the path as a reference for future development in prosthetic lower extremity solution.

ACKNOWLEDGEMENT

We are grateful for the help of IIIT (Indian Institute of Information Technology) Kalyani to complete the research work.

REFERENCES

[1] O. FLIEGEL, M.D. and S.G. FEUER, M.D. Brooklyn, "Historical Development of Lower-Extremity Prostheses".

[2] "Prosthetic Management: Overview, Methods and Materials", O&P Virtual Library.

[3] Randale Sechrest, "Ankle anatomy animated tutorial".

[4] Habib Masum, Subhasis Bhaumik, Ranjit Ray, "Conceptual Design of a Powered Ankle-Foot Prosthesis for Walking with Inversion and Eversion", 2nd International Conference on Innovations in Automation and Mechatronics Engineering, ICIAME 2014, Procedia Technology 14 (2014) 228 – 235, Elsevier.

[5] Kim M. Norton, "A Brief History of Prosthetics".

[6] Keith E. Gordon, Daniel P. Ferris, "Learning to walk with a robotic ankle exoskeleton", Journal of Biomechanics 40 (2007) 2636–2644, Elsevier.

[7] Catherine R. Kinnaird and Daniel P. Ferris, "Medial Gastrocnemius Myoelectric Control of a Robotic Ankle Exoskeleton", IEEE Transactions on Neural Systems and Rehabilitation Engineering, vol. 17, no. 1, February 2009.

[8] Jan F. Veneman, Rik Kruidhof, Edsko E. G. Hekman, Ralf Ekkelenkamp, Edwin H. F. Van Asseldonk, and Herman van der Kooij "Design and Evaluation of the LOPES Exoskeleton Robot for Interactive Gait Rehabilitation", Article in IEEE Transactions on Neural Systems and Rehabilitation Engineering, October2007, DOI:10.1109/TNSRE.2007.903919 · Source: IEEE Xplore.

[9] R. Jiménez-Fabián, O. Verlinden, "Review of control algorithms for robotic ankle systems in lower-limb orthoses, prostheses, and exoskeletons", Medical Engineering & Physics 34(2012) 397-408.

[10] Jeffrey R. Koller, Daniel A. Jacobs, Daniel P. Ferris and C. David Remy, "Learning to walk with an adaptive gain proportional myoelectric controller for a robotic ankle exoskeleton", Koller et al. Journal of NeuroEngineering and Rehabilitation (2015) 12:97 DOI 10.1186/s12984-015-0086-5.

[11] Pierre Cherelle, Victor Grosu, Louis Flynn, Karen Junius, Marta Moltedo, Bram Vanderborght, Dirk Lefeber, "The Ankle Mimicking Prosthetic Foot 3—Locking mechanisms, actuator design, control and experiments with an amputee", Robotics & Multibody Mechanics, Department of Mechanical Engineering, Vrije Universiteit Brussel, Belgium Flanders Make, Belgium, Robotics and Autonomous Systems 91(2017) 327-336.

[12] Pei-Chun Kao, Cara L. Lewis, Daniel P. Ferris, "Invariant ankle moment patterns when walking with and without a robotic ankle exoskeleton", School of Kinesiology, University of Michigan, Ann Arbor, MI 48109-2214, USA, Journal of Biomechanics 43 (2010) 203– 209.

[13] Gregory S. Sawicki and Daniel P. Ferris, Human Neuromechanics Laboratory, University of Michigan-Ann Arbor, Ann Arbor, MI 48109, USA, "Mechanics and energetics of incline walking with robotic ankle exoskeletons", The Journal of Experimental Biology 212, 32-41, Published by The Company of Biologists 2009, Accepted 24 October 2008, doi:10.1242/jeb.017277.

2019 Devices for Integrated Circuit (DevIC), 23-24 March, 2019, Kalyani, India

Improvement of the Gain Accuracy of the Instrumentation Amplifier Using a Very High Gain Operational Amplifier

Maitraiyee Konar
Dept. of
Electronics & Communication Engg.
Birla Institute of Technology, Mesra
Ranchi, India
Email: maitraiyeekonar@gmail.com

Rashmi Sahu
Dept. of
Electronics & Communication Engg.
Birla Institute of Technology, Mesra
Ranchi, India
Email: pneumono02@gmail.com

Sudip Kundu
Member, IEEE
Dept. of
Electronics & Communication Engg.
Birla Institute of Technology, Mesra
Ranchi, India
Email: kundu.sudip@gmail.com

Abstract—This paper evaluates the performance of three operational amplifier (Op-Amp) based instrumentation amplifier (INA) topology using two variants of Op-Amp – one with a very high gain with respect to the other Op-Amp topology. The gain accuracy and common mode rejection ratio (CMRR) of the INA is improved with the use of the very high gain Op-Amp designed using a folded cascode amplifier. The CMRR of the INA using folded cascade Op-Amp is 114 dB which is 43 dB higher to that of INA using differential amplifier.

Keywords—*instrumentation amplifier, operational amplifier, differential amplifier, folded cascode amplifier, differential mode gain, common mode gain, CMRR*

I. INTRODUCTION

Nowadays, low power consumption without significant performance loss is a major concern in designing an electronic system. Small size with low power consumption is now the most demanding trait of any electronic device, especially for real time applications like biomedical signal acquisition systems [4], [12]. Biomedical signals are very low amplitude – low frequency signals produced by different parts of human body which are analyzed and processed for a better medical diagnosis. The devices used to retrieve such weak signals must consume less power so that the heat radiated by the device is negligible which otherwise may be harmful for the human body.

The output from the transducers and sensors are generally of very weak strength. Moreover, the output impedance of the transducers and sensors are very large. So, to perform processing on these weak signals, firstly the amplification of the signals is required which requires amplifiers with high gain and large input impedance. To eliminate the influence of common mode interferences at the sensor-amplifier interface, the CMRR of the amplifier should be extremely high. A simple differential amplifier is capable of reducing the strength of the common mode signals while enhancing the

desired differential signals. But the effect of the common mode signals on our signal of interest cannot be diminished by the differential amplifier alone. An INA is an amplifier which has both – reasonably high gain and most importantly very high CMRR, which makes it highly suitable for the sensing of low strength signals.

A three Op-Amps – based INA architecture [1], [4], [8], [9], [10] as shown in Fig 1, is the most typical INA design. It consists of two non-inverting amplifiers at the input side of the INA and a difference amplifier, also known as subtractor, forms the output stage of INA. The gain of this INA configuration is specified using the feedback resistors R_1, R_G, R_2 and R_3. The matching of the resistors together with their large value is the key to the high CMRR of this INA architecture.

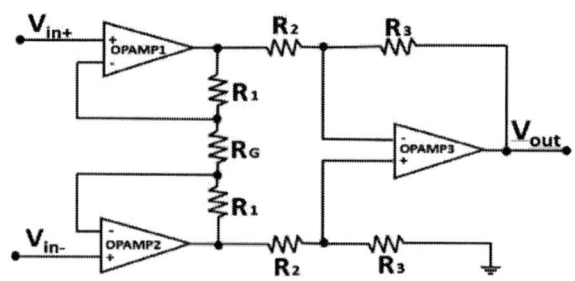

Fig 1. Three Op-Amps-Based Instrumentation Amplifier

The gain accuracy is highly dependent on the gain of the Op-Amps being used to design the INA. In this paper, three similar Op-Amps are used for implementing the INA topology, in 0.18 μm CMOS technology. Using two different Op-Amp architecture two INAs have been proposed. One Op-Amp is based on a typical differential amplifier [1], [2], [5], [6], [7] and the other is based on a folded cascode amplifier [4], [8], [10], [11]. The paper evaluates the

978-1-5386-6723-1/19 $31.00 © 2019 IEEE   408

performance of both the designed INAs. Both the INAs have been optimized with power consumption below 10 µW.

This paper is divided into the following sub-sections. Section II briefs about the three Op-Amps-based INA topology and discusses the design of this INA topology using differential amplifier for its Op-Amps. In Section III, we suggest an improvement in the gain of the Op-Amp using a folded cascode amplifier. This section also highlights the advantage of having a very high gain Op-Amp for this architecture of INA. The paper ends with the simulation results of both the proposed INAs and conclusions.

II. THREE OP-AMPS-BASED INA USING DIFFERENTIAL AND FOLDED CASCODE OP-AMP

The most conventional INA design is a three Op-Amp INA topology. The three-Op-Amp INA [1], [4], [8], [9] is basically an improvised version of difference amplifier characterized by large input impedance and very high CMRR, as shown in Fig 1. Its high input impedance is due to the two non-inverting amplifiers (OP-AMP1 and OP-AMP2) which constitutes the input stage preceding the difference amplifier (OP-AMP3) of the three-Op-Amp INA. The input stage is responsible for providing a differential mode gain of $1+2R_1/R_G$. The common mode signals are simply buffered to the inputs of the difference amplifier by the input stage with a unity gain. It is the difference amplifier which makes the INA highly insensitive to common mode signals by offering extremely small common mode gain. The differential gain obtained from the difference amplifier is determined using R_2 and R_3 as R_3/R_2. Thus, the overall gain of the INA is expressed as:

$$G = (1 + 2\,R_1/R_G) \times (R_3/R_2) \qquad (1)$$

This INA topology is highly dependent on the matching of the resistors for its performance. So, to have a control over the gain of the amplifier, the value of R_G can be tweaked accordingly without the worry of introducing any mismatch in the INA topology.

A. Design of INA using Differential Amplifier-Based Op-Amp

Fig 2 shows the differential amplifier-based Op-Amp [2], [5], [6], [7] topology. The gain offered by the inherent differential amplifier is:

$$A_{diff} = g_{m1} \times (r_{ds1}||r_{ds2}) \qquad (2)$$

where, g_{m1} is the transconductance of $M_{1A,1B}$,

r_{ds1}, r_{ds2} are the output impedance of $M_{1A,1B}$ and $M_{2A,2B}$ respectively as shown in fig 2.

From the above gain expression, it is clear that the values of r_{ds1} and r_{ds2} are very high for large gains. But, since the Op-Amp have low output impedance, a common drain amplifier is cascaded to the differential amplifier to decrease the output impedance. The common drain amplifier does not do any addition to the gain of the Op-Amp as the gain due to the common drain amplifier approximates to unity [2], [5], [6].

The gain of this Op-Amp configuration cannot go above few hundreds. The simulated DC gain obtained for this Op-Amp is 47.57 dB with a power consumption of 8 µW. Since the gain of the Op-Amp is not very high, accurate determination of the gain of the INA designed using this Op-Amp topology is very difficult.

The common mode response of the INA depends upon how good is the difference amplifier. The common mode gain of the INA can be reduced significantly by limiting the gain from the difference amplifier, i.e. by keeping the ratio ofR_3/R_2 equal to or slightly greater than 1. Also, the values of resistors $R_2$and R_3 should be very high for better suppression of the common mode signals. Thus, we can say, the differential mode gain of the INA now depends on the ratio of R_1 and R_G. Large values of R_1 and R_G ensures the gain from the input stage is closer to $1+2R_1/R_G$.

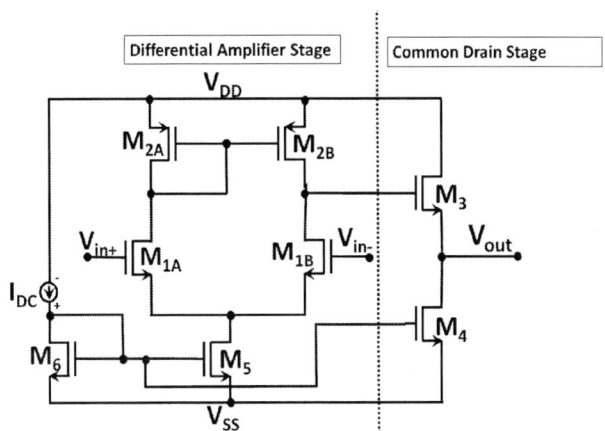

Fig 2: Differential amplifier-based Op-Amp topology

B. Design of INA using folded cascode Op-Amp

In this paper, we are presenting a very high gain Op-Amp for improving the performance of a three-Op-Amp INA topology. The configuration of the high gain Op-Amp is depicted in Fig 3, which shows a folded cascode amplifier followed by a common drain stage [4], [8], [10], [11]. A folded cascode amplifier is like a common source amplifier but with very large output impedance. The large output impedance of the folded cascode amplifier is because of the presence of common gate stage in the amplifier. The common drain stage at the output of the folded cascode amplifier makes sure that same signal is reproduced at the output of the common drain stage as at the output of the folded cascode amplifier but with low output impedance. Therefore, the gain of this folded cascode-based Op-Amp can be approximated to:

$$A_{fc} = g_{m2}\{[g_{m6}r_{ds6}(r_{ds3}||r_{ds2})]||[g_{m5}r_{ds5}r_{ds4}]\} \qquad (3)$$

where, g_{m2}, g_{m5}, g_{m6} are the transconductance of $M_{2A,2B}$, $M_{5A,5B}$, $M_{6A,6B}$ respectively.

r_{ds2}, r_{ds3}, r_{ds4}, r_{ds5} are the output impedance of $M_{2A,2B}$, $M_{3A,3B}$, $M_{4A,4B}$, $M_{5A,5B}$ respectively as shown in fig 3.

We have designed a folded cascode-based Op-Amp with a power consumption of 9 µW which is nearly same as that of the designed differential amplifier-based Op-Amp and analysed its performance. The DC gain of the designed folded cascode-based Op-Amp is 83.59 dB. But its bandwidth is quite smaller than that of the differential amplifier-based Op-Amp, which can be observed from the Table 1. This result is because, when gain is very much high obtaining a larger bandwidth is a difficult task for low power amplification.

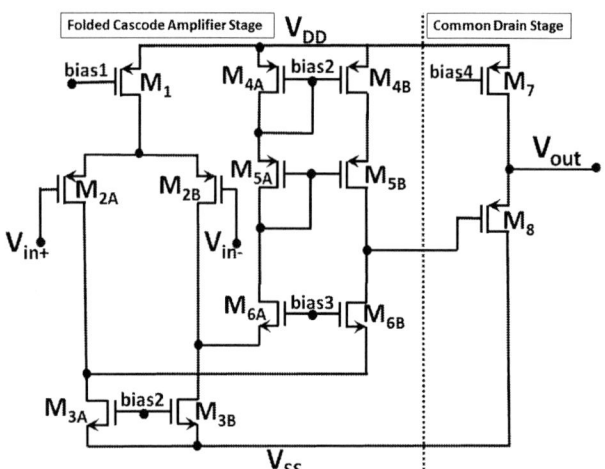

Fig 3: Folded cascode-based Op-Amp topology

The error in determining the gain of the INA designed using this high gain Op-Amp is appreciably lower than that of the differential amplifier-based Op-Amp topology as the designed folded cascode amplifier provides 100 times larger gain than that by the differential amplifier. This facilitates in keeping the difference amplifier stage of the INA in unity gain configuration with complete dependence on the non-

inverting amplifiers for the differential mode gain of the INA.

Large feedback resistors are also required by this topology, so that performance of the INA does not deteriorate. Having both R_2 and R_3 as 1 MΩ resistors enabled us to have a common mode gain of the INA to be -73.69 dB with this Op-Amp topology. The gain provided by the INA is therefore made under the control of R_1 and R_G. The values of R_1 and R_G should be high enough so as to not to spoil the performance of the INA.

III. SIMULATION RESULTS

Both the Op-Amps, discussed in previous sections were realized in 0.18 µm CMOS technology. Positive and negative voltages of magnitude 0.9 V were used to operate the 2 Op-Amps.

The differential and common mode response of the differential amplifier-based Op-Amp and the folded cascode Op-Amp is recorded in Table 1. It is clear from Table 1 that the differential mode gain, Ad of the differential amplifier-based Op-Amp is quite below 60 dB. While the differential mode gain of the designed folded cascode Op-Amp is 83.59 dB, which is very much higher than the Op-Amp designed from differential amplifier.

Table 1: Performance record of the two Op-Amp designs

Op-Amp Type	Differential mode gain, Ad (dB)	Common mode Gain, Ac (dB)	CMRR (dB)	3 dB Bandwidth (Hz)
Differential Amplifier Based	47.57	-42.82	90.4	458.8 K
Folded Cascode based	83.59	-44.34	127	13.78 K

The performance analysis of the designed INAs using the differential amplifier-based Op-Amp and the folded cascode Op-Amp is shown in Table 2. The values of the feedback resistors R_1, R_G, R_2 and R_3 were kept same as mentioned in Table 2 for both the INAs. Therefore, the expected value of the differential mode gain of both the INAs was about 43 dB. However, the performance analysis of the INAs showed that the differential mode gain of the INA designed with differential amplifier-based Op-Amp is only 20.13 dB while that of the INA designed with folded cascode Op-Amp is 40.49 dB. This shows that the gain accuracy of the INA is very much dependent on the gain of the Op-Amp. Since the gain of the Op-Amp implemented with differential amplifier is not much high, the gain accuracy of the INA designed with this Op-Amp topology is very poor. The inaccuracy in the gain of the three Op-Amps INA architecture improves significantly when implemented with the folded cascode Op-Amp having larger gain.

Table 2: Performance Summary of the two INAs

INA	Op-Amp Type	R_1 (Ω)	R_G (Ω)	R_2 (Ω)	R_3 (Ω)	Differential mode gain, Ad (dB)	3 dB Bandwidth (Hz)	Common mode Gain, Ac (dB)	CMRR (dB)
INA 1	Differential Amplifier Based	10 M	140 K	800 K	1M	20.13	8.12 M	-51.42	71.55
INA 2	Folded Cascode Based	10 M	140 K	1M	1M	40.49	1 M	-73.7	114.2

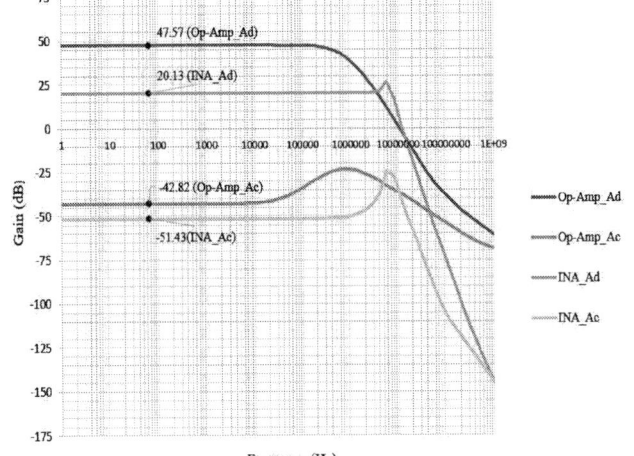

Fig 4: Graph of differential amplifier-based Op-Amp and of the INA designed using this Op-Amp

Using the folded cascode-based amplifier, not only the gain accuracy improved but the capability of the INA to reject the common mode signals boosted up. The common mode gain went as low as -73.7 dB which resulted in CMRR of the INA to be 114.2 dB. Although the bandwidth of this INA is not as high as the INA implemented using the differential amplifier-based Op-Amp, but it is sufficiently high enough for the amplification of weak signals of low frequency signals, such as biomedical signals.

Fig 4 depicts the plots of differential mode gain and common mode gain of the differential amplifier-based Op-Amp and of the INA implemented with this Op-Amp. Similarly, the plots for differential mode gain and common mode gain of the folded cascode Op-Amp and that of the INA implemented using it is shown in Fig 5.

Now to observe the process variation, 1% mismatch is introduced in the feedback resistors of the INA. Its effect on the performance of the INA is recorded in Table 3 (considering the values of the feedback resistors as given in Table 2). The CMRR of the INA using the folded cascode-based Op-Amp was still higher than the INA derived from the Op-Amp based on differential amplifier.

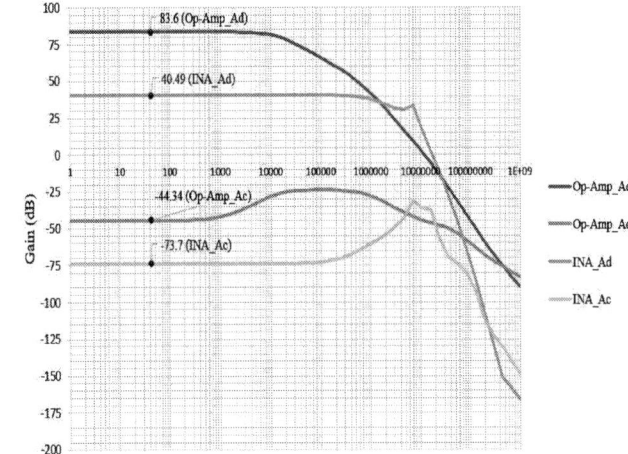

Type of Op-Amp	Mismatch in resistor	Ad (dB)	Ac (dB)	CMRR (dB)
Differential Amplifier Based	R3	20.11	-56.95	77.06
	R2	20.15	-43.60	63.75
	R1	20.13	-51.42	71.55
Folded cascode Based	R3	40.47	-46.40	86.87
	R2	40.51	-45.82	86.33
	R1	40.48	-73.7	114.2

Fig 5: Graph of folded cascode-based Op-Amp and of the INA designed using this Op-Amp

Table 3: 1% mismatch in external feedback resistor

IV. CONCLUSION

A very high CMRR INA of three-Op-Amps based INA topology has been implemented in a 0.18 μm CMOS technology. The INA operates using 1.8 V, consumes power around 27 μW. The INA has a superior common mode response for a large frequency range. The error in the gain accuracy is very low. The power consumption of the designed INA is comparably very low than some of the state-of-the-art three Op-Amps based INAs, which can be seen in Table 4.

The small size, the small supply voltage together with low power consumption makes the INA using the folded-cascode Op-Amp highly attractive option for low power applications.

Table 4: Comparison table of some of the state-of-the-art INAs

PARAMETER	[8]	[9]	[10]	[13]	This work
Technology (um)	0.35 µm CMOS	0.5 µm CMOS	0.5 µm CMOS	0.18 µm CMOS	0.18 µm CMOS
INA type	3-Op-Amps topology	3-Op-Amps topology	3-Op-Amps topology	3-Op-Amps topology	3-Op-Amps topology
Op-Amp type	Folded cascode	Different-al Amplifier	Different-al Amplifier	Folded cascode	Folded cascode
Supply(V)	1.5	-	-	-	1.8
Gain(dB)	81	45	19.9	67.7	40.48
CMRR (dB)	125	75	>110	92	114.2
Common mode gain (dB)	-44	-30	-	-24.3	-73.69
Ad 3 dB bandwidth (Hz)	-	1.1 M	20 K	1.1 M	1 M
Power Dissipation (µW)	43	~280	-	263	27

V. Acknowledgement

We, all the authors, would like to express our special thanks to Science and Engineering Research Board (SERB), Govt. of India for funding to execute this research through project ref. no. ECR/2016/001279. This research was not possible without the help and support of BIT Mesra, Ranchi and SERB, Govt. of India.

VI. References

[1] M.A. Smither, D.R. Pugh, L.M. Woolard, "C.M.R.R. analysis of the 3-op-amp instrumentation amplifier", IEEE Journals & Magazines, Electronics Letters, Volume: 13, Issue: 20.

[2] STOUT, D. F., and KAUFMAN, M, "Handbook of operational amplifier circuit design" (McGraw-Hill, New York, 1976), section 9-2.

[3] Textbook, SMITH, J. I., "Modern operational circuit design" (Wiley, New York, 1971), pp. 124-126.

[4] Hwang-Cherng Chow, Chang Gung, Jia-Yu Wang, "High CMRR instrumentation amplifier for biomedical applications", 12-15 Feb, IEEE International conference.

[5] Textbook, Ramakant A. Gayakwad, "Op-Amps and Linear Integrated Circuits"

[6] Johan Huijsing, a textbook on "Operational Amplifiers".

[7] Siddharth Malhotra, Abhinav Mishra, Rakesh BR, Anu Gupta, "Frequency Compensation in Two-Stage Operational Amplifiers for achieving High 3-dB Bandwidth", 2013 IEEE Asia Pacific Conference on Postgraduate Research in Microelectronics and Electronics.

[8] A. Harb and M. Sawan, "New Low Power low voltage high CMRR CMOS instrumentation amplifier".

[9] Manish Goswami, Smriti Khanna, "DC suppressed high gain active CMOS instrumentation amplifier for biomedical application", 2011 IEEE International Conference on Emerging Trends in Electrical and Computer Technology.

[10] Chih-Jen Yen, Wen-Yaw Chung and Mely Chen Chi, "Micro-Power Low Offset Instrumentation Amplifier IC Design For Bio-Medical System Applications", IEEE Transactions On Circuits And Systems:Regular Papers ,Vol.51,No.4, pp 691-699 April 2004.

[11] Zhang Kun+, Wu Di and Liu Zhangfa, "A High-performance Folded Cascode Amplifier", 2011 International Conference on Computer and Automation Engineering (ICCAE 2011)

[12] Textbook, Metin Akay, "Biomedical Signal Processing"

[13] Akshay Goel, Gurmohan Singh, "Novel High Gain Low Noise CMOS Instrumentation Amplifier for Biomedical Applications", 2013 International Conference on Machine Intelligence Research and Advancement.

Trade-off Characteristics of Hysteresis Comparator used in Noisy Systems

Aashita Raj

Dept. of Electronics and Communication Engineering

Birla Institute of Technology, Mesra

Ranchi, India
Email:aashita20896@gmail.com

Sai Yaswanth Divvela

Dept. of Electronics and Communication Engineering

Birla Institute of Technology, Mesra

Ranchi, India
Email:dsygnt11@gmail.com

Geetanjali Singh

Dept. of Electronics and Communication Engineering

Birla Institute of Technology, Mesra

Ranchi, India
Email:geetanjali.rs@gmail.com

Sudip Kundu
Member, IEEE

Dept. of Electronics and Communication Engineering Birla Institute of Technology, Mesra

Ranchi, India
Email: kundu.sudip@gmail.com

Abstract— **The output from digital input buffer is not stable when input noise is present near the threshold voltage of inverter. Hysteresis comparator with two threshold voltages gives stable output with noisy input signal. This paper presents the trade-off characteristics between the hysteresis voltage and the power consumption of a hysteresis comparator. An optimal design point is proposed with a hysteresis voltage up to ±180 mV with a power consumption of 0.713 μW.**

Keywords— Input pads, comparator, Hysteresis, Threshold

I. INTRODUCTION

The low magnitude input signal is very sensitive to the system noise such as noise introduced by sensors. When these input signals along with noise are processed, the output response has ringing effects. It is mandatory to remove this noise in the input stage before applying it for further processing.

Conventional input protection circuits of input pads use input buffer to remove the noise of the incoming digital signal coming from the external world and regenerate the digital signal for further processing, as shown in Fig. 1(a). Present approach uses hysteresis comparator to apply the input signal to the next stage.

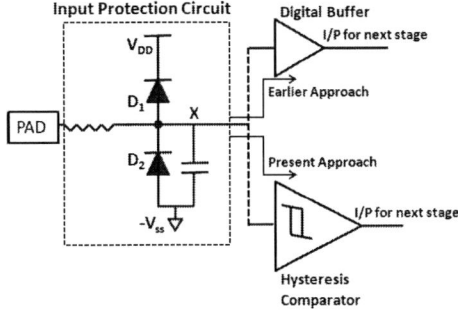

Figure 1(a): The Input protection circuit

Digital buffer is comprised of inverter chain. Inverter has a single threshold voltage V_{TH}, which switches the output when input reaches V_{TH}. When noise is attached with input at the level of V_{TH}, inverter switches the output in response to noise. This is because the input increases or decreases around V_{TH} due to noise. This is shown in figure 1(b). It is clear from figure that the output is suffered from voltage hikes, when input signal is directly applied through input buffer.

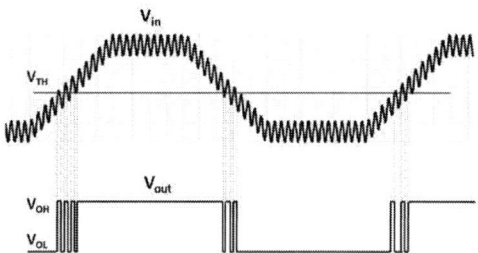

Figure 1(b): Output of Digital buffer

Since the input varies with the noise, if threshold voltage V_{TH} also varied with the input variation, the output will not change in response to noise. This creates a requirement of double threshold voltage in the input buffer.

Hysteresis comparator is a potential solution for this. Figure 1(c) shows the output voltage with hysteresis comparator. When the hysteresis comparator is used, the circuit has two threshold voltage levels, upper threshold V_{UT} and lower threshold V_{LT}. The output changes from low to high when V_{in} reaches V_{UT}. Similarly the output changes from high to low when V_{in} reaches V_{LT}. If V_{in} decreases due to noise below V_{UT}, output does not change from high to low, because the lower threshold voltage is not reached. This results in controlled output even in the noisy environment [3].

2019 Devices for Integrated Circuit (DevIC), 23-24 March, 2019, Kalyani, India

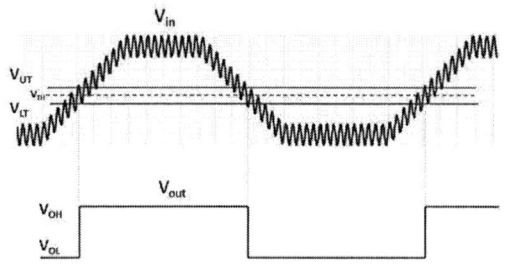

Figure 1(c): Output of Hysteresis Comparator

Figure 2 represents hysteresis curve. As depicted in figure, there are two threshold voltages for changing the output, upper threshold voltage V_{UT} and lower threshold voltage V_{LT}. As V_{in} increases, the output changes only when V_{in} reaches V_{UT}. Similarly during decrease in V_{in}, output changes only when V_{in} falls down to V_{LT}.

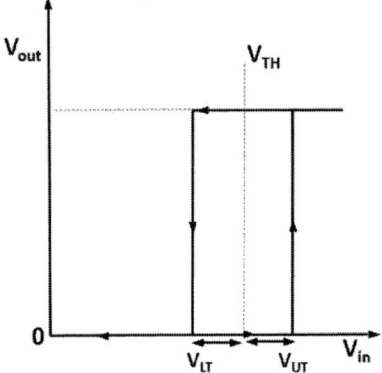

Figure 2: Hysteresis Curve with Variation of the output voltage with the input

Hysteresis comparator can be used at the end of the IC to get noise free input to the next stage. In noisy systems, hysteresis comparator can be used between two circuit blocks to suppress noise [7].

Apart from digital input pads, hysteresis comparator can be used in analog circuit when there is a requirement of comparator to change the output after comparing the input voltage with reference voltage. As described in figure 3, normal analog comparator generates spikes when the input signal is affected by noise near reference voltage V_{ref}. Hysteresis comparator gives the stable output due to upper and lower threshold voltages V_{UT} and V_{LT}, which can be set above and below the reference voltage V_{ref}.

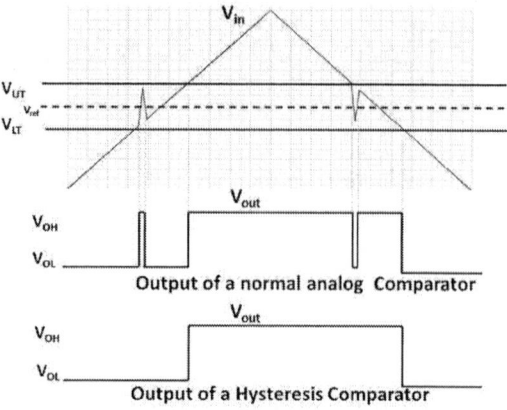

Figure 3: Output comparison of normal analog comparator and hysteresis comparator

In Section II describes the circuit design and design parameters of hysteresis comparator. Section III shows the optimal design parameters for the circuit. Section IV represents the trade-off characteristics of hysteresis comparator. Section V represents the conclusion.

II.   DESIGN OF HYSTERESIS COMPARATOR

Hysteresis comparator is designed by introducing a Positive feedback circuit in the comparator circuit [6]. Figure 3 describes the circuit design of hysteresis comparator. This is comprised of an nMOS differential pair with pMOS active load, a common source amplifier followed by an inverter. To introduce hysteresis, positive feedback is added by two cross coupled pMOS P_{2A} and P_{2B}. A current mirror is used to provide the bias [2], [3].

The input Signal V_{in} is applied as gate input V_g of N_{1A} and the reference threshold voltage V_{ref} is applied as gate input V_g of N_{1B} of differential pair. When the input signal V_{in} increases with noise, V_{ds} of N_{1A} decreases. This reduces V_{gs} of P_{2A} and results in the decreased drain current of P_{2A} and N_{1B}. Because of this V_{ds} of N_2 increases, this is fed back to the drain of N_{1A} via P_{2A}. Since this is a positive feedback path for N_{1A}, it further decreases the V_{ds} of N_{1A}. This regenerative decrease in the V_{ds} of N_{1A} keeps the output unchanged. This way the increase in input does not result in increase in the output.

978-1-5386-6723-1/19 $31.00 © 2019 IEEE   414

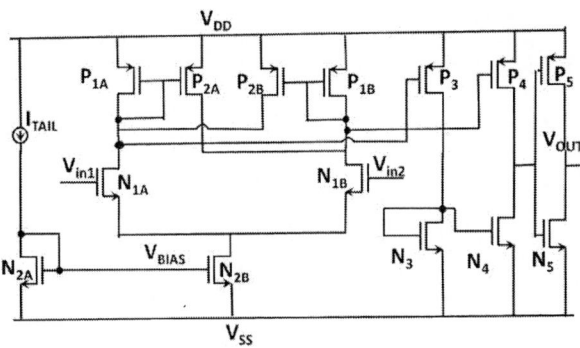

Figure 4: Circuit Diagram of the hysteresis comparator

The input noise level, which can be removed by the circuit, depends on the width of P_{2A} and P_{2B}. Hysteresis voltage increases with the increase in width of P_{2A} and P_{2B}. This is because g_m ratio of P_1 and P_2 controls the amount of current flowing through P_1 and P_2 and thus the amount of positive feedback. This protects the output from changing for a higher increase or decrease in the input signal.

The bias current I_{TAIL} is another parameter, which directly affects the noise removal capability of the circuit. Increase in I_{TAIL} increases the upper and lower threshold voltages.

Power consumption is very important to be analyzed for circuit parameter optimization. A static current passes through P_3. Increase in bias current results in significant current flow through P_3 and thus high power consumption in the circuit. We propose the optimum size of P_3 to provide high hysteresis voltage keeping low power consumption.

III. OPTIMAL DESIGN OF HYSTERESIS COMPARATOR

The hysteresis comparator is implemented using 0.18 µm CMOS technology. Before finding the optimal design, at first we design the hysteresis comparator considering the width of the P_1, P_2, P_3, N_1, N_2, N_3 MOSFET as 1 µm, 20 µm, 15 µm, 5 µm, 10 µm and 5 µm, respectively. The length of all the MOSFETs is considered at the lowest value based on the limit of the selected technology, i.e. 180 nm. The I_{TAIL} was kept at 1 µA. This design point we are referring as the nominal design point. At the nominal design point the achieved hysteresis voltages are +73.12 mV and -72.54 mV for the V_{HYS+} and V_{HYS-}, respectively.

I_{TAIL} is varied from 1 µA to 10 µA and it is observed that V_{HYS} increases with the increase in I_{TAIL}. The result is shown in table I.

TABLE I: HYSTERESIS VOLTAGE FOR DIFFERENT VALUES OF I_{TAIL}

I_{TAIL} (µA)	V_{HYS-} (mV)	V_{HYS+} (mV)
1	-72.54	73.12
2	-77.31	77.91
3	-80.94	81.62
4	-83.21	84.03
5	-85.48	86.21
6	-87.29	87.96
7	-88.88	89.93
8	-90.24	91.17
9	-91.60	93.64
10	-92.74	92.55

A. Optimal design to increase hysteresis

To achieve noise suppression capability up to 10% of maximum output voltage swing, we optimize the circuit to get hysteresis voltage $V_{HYS} \pm 180$ mV.

To improve the V_{HYS}, width ratio of P1 and P2 is varied. W_1 and W_2 depict the width of P_1 and P_2 respectively. The circuit is simulated with a bias current of 1µA and the different values of hysteresis voltage corresponding to the different values of W_1/W_2 are shown in Table II. It is clear from the table that the hysteresis voltage increases with the increase in width of P_2 MOSFETs which provide the positive feedback.

TABLE II: HYSTERESIS VOLTAGE FOR DIFFERENT VALUES OF W_1/W_2

W_1/W_2 (µm/µm)	Ratio of W_1/W_2	V_{HYS-} (mV)	V_{HYS+} (mV)
1/20	0.05	-72.78	73.09
2/20	0.10	-66.81	66.93
4/20	0.20	-50.70	51.83
6/20	0.30	-40.31	40.49
8/20	0.40	-31.49	31.70
10/20	0.50	-24.31	24.62
12/20	0.60	12.24	-2.27
14/20	0.70	19.30	-4.13
16/20	0.80	25.28	-5.70
18/20	0.90	30.31	-7.11
20/20	1.00	34.92	-8.24

To increase V_{HYS}, W_1 is decreased to minimum value of 400 nm. This reduces the ratio W_1/W_2 and improves V_{HYS}. The length L of P_1 is increased to further improve the V_{HYS}.

B. Optimal design to reduce power consumption

As depicted from table 1, hysteresis increases with increase of I_{TAIL}. Increasing I_{TAIL} results in the increase in power consumption. To reduce the power consumption we reduce the W/L ratio of P_3 from to 1 µ/1 µ. The circuit is simulated for I_{TAIL} of 1 µA and it is observed that power consumption is reduced.

The design parameter of the optimal design is fixed for the V_{HYS} ±180 mV with minimum power consumption at I_{TAIL} =1 µA. The proposed W/L values are presented in table III.

TABLE III: OPTIMUM VALUES OF DESIGN PARAMETERS

Transistors	Width (µm)	Length (µm)
P_1	0.4	10
P_2	20	0.18
P_3	1	1
P_4	60	0.5
P_5	60	0.5
N_1	5	0.18
N_2	10	0.18
N_3	15	0.5
N_4	15	0.5
N_5	15	0.5

IV. TRADE-OFF CHARACTERISTICS OF HYSTERESIS COMPARATOR

With the proposed optimal design parameters, the circuit is simulated for different values of I_{TAIL} to get the minimum power consumption. The result is presented in table V and plotted in the graph of figure IV.

It is clear from graph that there is a trade-off between the hysteresis voltage and the power consumption. The proposed optimal design point is for the V_{HYS} of 180mV. At this point the power consumption obtained is 0.713 µW.

TABLE IV: TRADE-OFF BETWEEN HYSTERESIS VOLTAGE AND POWER CONSUMPTION FOR DIFFERENT VALUES OF I_{TAIL}

I_{TAIL} (µA)	V_{HYS-} (mV)	V_{HYS+} (mV)	Power Consumption (µW)
0.01	-171.2	175.8	0.352
0.02	-180.4	180.5	0.713
0.03	-182.6	184.5	1.073
0.05	-187.8	188.5	1.794
0.10	-193.7	194.4	3.500
0.20	-199.8	200.6	6.600
0.40	-206.3	207.1	11.66
0.60	-210.2	211.6	14.41
0.80	-213.5	214.7	15.87
1.00	-215.8	217.2	16.89

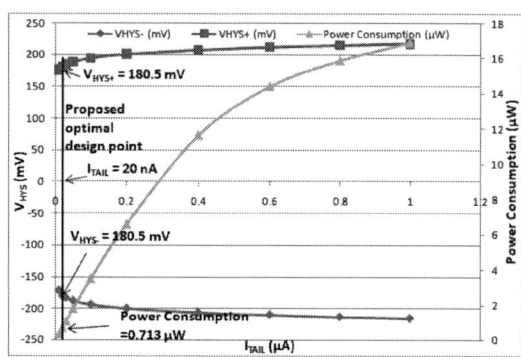

Figure 5: Trade-off between hysteresis Voltage and power consumption for different values of I_{TAIL}

TABLE V: PERFORMANCE COMPARISON WITH REPORTED HYSTERESIS COMPARATORS

	Supply Voltage (V)	Process (µm)	Power Consumption (µW)	V_{HYS-} (mV)	V_{HYS+} (mV)
This Work (proposed optimal)	1.8	0.18	0.71	-180.4	180.5
This Work (compared with [8])	1.8	0.18	0.24	-169.01	170
This work (compared with [9])	1.8	0.18	0.60	-178.10	178.6
[5]	1.8	0.18	41.58	-2.5	2.5
[8]	0.8	0.18	0.24	-30	30
[9]	3.0	0.18	0.60	-127	127

The table V compares the performance of the proposed hysteresis comparator with the other reported implementations of hysteresis comparators. It is clear from the table that, for the power consumption of 0.24 μW, proposed circuit provides a V_{HYS} of 170 mV against the V_{HYS} of 30 mV reported in [8]. Similarly for the power consumption of 0.6 μW, a V_{HYS} of 178.6 mV is obtained which is much higher than the V_{HYS} of 127 mV which is reported in [9].

V. CONCLUSION

In this paper, the use of hysteresis comparator to improve the noise robustness, for digital input pads is proposed. Based on 0.18 μm technology, a nominal design was implemented. An optimal design for circuit parameters is proposed to achieve a hysteresis voltage of ± 180 mV with a bias current of 20 nA. The power consumption of 0.713μW is obtained with this design. Meanwhile, the trends of trade-off were examined to obtain the optimal design depending on the requirement.

VI. ACKNOWLEDGMENT

We, all the authors, would like to express our special thanks to Science and Engineering Research Board (SERB), Govt. of India for funding to execute this research through project ref. no. ECR/2016/001279. This research was not possible without the help and support of BIT Mesra, Ranchi and SERB, Govt. of India.

VII. REFERENCES

[1] Mikkel C. W. Høyerby, Michael A.E Andersen: "A small-signal model of the hysteretic comparator in linear-carrier self-oscillating switch-mode controllers", Jan 2006

[2] Y. Li, A. L. Mansano, Y. Yuan, D. Zhao and W. A. Serdijn, "An ECG Recording Front-End With Continuous-Time Level-Crossing Sampling," in *IEEE Transactions on Biomedical Circuits and Systems*, vol. 8, no. 5, pp. 626-635, Oct. 2014.Takumi Saito, Satoshi Komatsu: "A Low - Voltage Hysteresis Comparator for Low Power Applications."

[3] Textbook, Neil H.E Weste. "Principles of CMOS VLSI Design"

[4] Satyabrata Nanda, Avipsa S. Nanda, G.L.K.Moganti: "A Novel Design of a High Speed Hysteresis Based Comparator in 90 nm CMOS Technology", *ICIP* 2015, pp. 388-391

[5] JIN Shuo-Wei, Jing-Jiao Li, Zhen-Ni Li"A hysteresis comparator for level-crossing ADC" *Chinese Control And Decision Conference* May 2017, pp. 7753-7757

[6] Paul M. Furth, Yen-Chun Tsen, Vishnu B. Kulkarni, Thilak K., Poriyani House Raju. On the design of low-power CMOS comparators with programmable hysteresis[C]. *53rd IEEE International Midwest Symposium on Circuits and Systems*, 2010: 1077-1080.

[7] D. Kwon and G. A. Rincon-Mora, "A rectifier-free piezoelectric energy harvester circuit," *2009 IEEE International Symposium on Circuits and Systems*, Taipei, 2009, pp. 1085-1088.

[8] Takumi Saito, Satoshi Komatsu: *"A Low – Voltage Hysteresis Comparator for Low Power Applications."* 2017 24th IEEE International Conference on Electronics, Circuits and Systems (ICECS)*, Batumi, 2017, pp. 427-430.*

[9] X. Qian and T. H. Teo, "A low-power comparator with programmable hysteresis level for blood pressure peak detection," *TENCON 2009 - 2009 IEEE Region 10 Conference*, Singapore, 2009, pp. 1-4.

2019 Devices for Integrated Circuit (DevIC), 23-24 March, 2019, Kalyani, India

An Impact of Current &Voltage Harmonics in the Power Quality Issues

Tapas Halder
Department of Electrical Engineering
Kalyani Government Engineering
College
Kalyani, India
Email : tapas_haldar@yahoo.com

Abstract—**The paper presents that the current and voltage harmonics not only contributes the power loss, uncertainty and malfunctions but also sustains the bad quality of the power in the power distributions generated by the linear and non-linear loads. The signatures, bad effects and impacts of the harmonics are decidedly highlighted to forecast the performance analysis, operation and control of a power distribution system. A simple but robust computational technique is invited here to optimize the various performance parameters, allegory of the power quality indices and excellent profile of the sine waves in the economic load dispatch carried out by experimental results for the reliable electricity or power distribution systems.**

Keywords—Harmonics, K-factor, THD, Order of harmonics, Effects of the harmonics

I. INTRODUCTION

The non linear loads in the power distributions have not symmetrical current and voltage (i.e. i-v) behaviours which contribute both even voltage and current harmonics due to presence of the circulating currents in the power system components. The contributions of the even harmonics are more detrimental than odd harmonics for the power quality issues and irregular heating effects in the system [1].

A waveform, whatever its profile may be analyzed by individual component is said to be a fundamental frequency bearing a number of frequency. As time goes on, the uses switched mode power supply widens to mitigate undesirable signals in addition to harmonics in the power system [2]

The letdown of the power devices in the information and communication sector are widely dependent huge harmonics and power loss so the good quality and un-interrupted power are fruitful to operate of the power sector. In order to meet up the demand of power, the pure sine wave in shape and size is essential at every nodal point of the power sector [3].

Owing to bulk commercial and industrial installations, the power devices of the consumers are affluent in harmonics generations. The current deformations across the loads are pronounced for the non-linear loads even pure sine wave supply at the consumer side by virtue of which it is liable for the poor power factor with trivial focus in the articles [4].

When different consumers use same power line to draw power, the voltage fluctuations are taken place by harmonic current injection of the one customers may affect the other consumers in comparisons with paper [5].

Substantial grounds of total harmonic distortion (THD) are caused by the non-limear loads derived by the adjustable

Sponsored by my beloved Father, Late Jagabandhu Halder

speed drive (ASD), fluorescence lamps, diode rectifiers, computer and data processing units, electric arc furnaces and etc, with respect to following the articles [6].

When the non-linear loads proliferate into the power systems, voltage deformations are injected to consume power from the source to loads because of variable loop impedance in the electrical energy systems in consideration with the some papers [7]-[8].

This paper predicts and point outs the major issues of the harmonics and with some special activities for the practical engineers whatever these potentials of this paper might be suggested to be treated with harmonics and enlighten a sound survey why these legends are necessary to be precautions.

II. DEFINATION AND EFFECTS HARMONICS AND IN POWER DISTRIBUTIONS

The harmonics are basically integral multiples of the fundamental frequency when these include deformed waveforms together with voltage and current waveforms.

The effects of the harmonics is usually caused by the non-linear devices with non-linear loads at which current is not directly proportional to applied voltage. Most of the case, the odd harmonics are not found in the power system. The major parts of the non-linearity are caused by the shunt and non-linear elements.

The power utilities usually supply a pure sine wave with third and fifth harmonic in the power distributions as shown in the Fig. 1.

The Fig. 1(a) represents a pure sine wave as a green curvature which has no disturbances in the power supply. It may be also represented by a following equation over a complete time period (T) as:

$$f(\omega t)=\sin(\omega t) \qquad (1)$$

The Fig. 1(b) represents a pure sine wave with third harmonic waveform (i.e. dotted red curve) which includes disturbances and distorted the sine wave but when supply is balanced, it produces zero sequence current and voltage for the load. It may be represented by a following equation over a complete time period (T) as:

$$f(\omega t)=\sin(3\omega t) \qquad (2)$$

The Fig. 1(c) forecasts a pure sine wave with fifth harmonic waveform (i.e. dotted black curve) which includes disturbances and distorted the power supply and it invites poor power factor in the sine wave of the power distribution. It also produces huge harmonic current and voltage in the power system. It may be represented by a following usual

978-1-5386-6723-1/19 $31.00 © 2019 IEEE

equation over a complete time period (T) with respect to pure sine wave as:

$$f(\omega t) = \sin(5\omega t) \qquad (3)$$

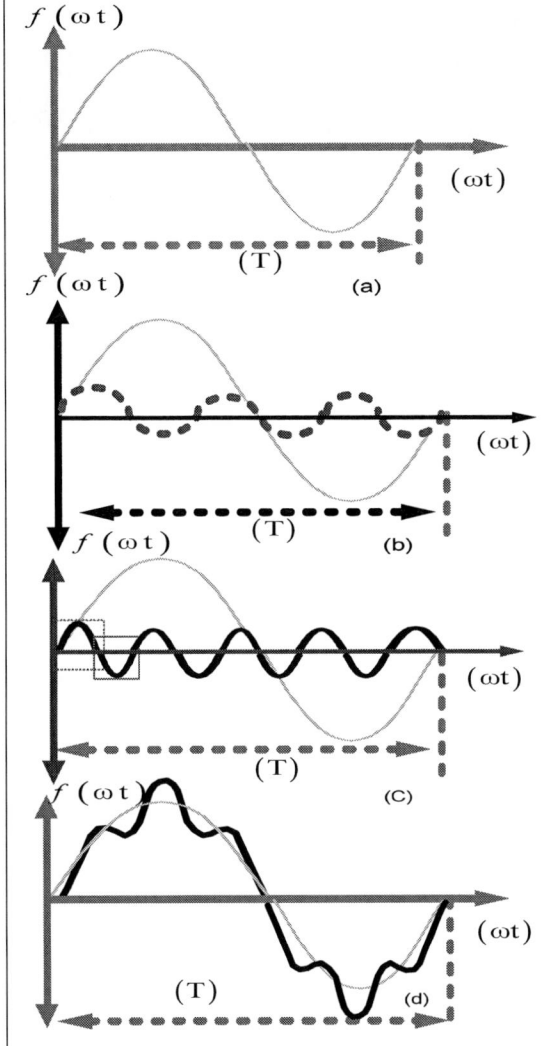

Fig. 1. Pure Sine wave with harmonics

The Fig. 1(d) forecasts a distorted sine wave with fifth harmonic waveform (i.e. bold black distorted curve) which includes also disturbances and distorted the sine wave of the power supply; it produces harmonic current and voltage for the load. It may be represented by a following equation over a time period (T) as a resultant of the pure sine and fifth harmonic wave forms in the power supply as:

$$f(\omega t) = \sin(\omega t) + \frac{\sin(5\omega t)}{5} \qquad (5)$$

III. PHASOR DIAGRAM OF THE HARMONICS

In a phasor diagram of the power distribution system with voltage (v) and current (i) waveforms with magnitudes and phase angles apart are shown in the Fig. 2 at which the balanced three phase system and sequence refer to the order. The phasors move past with respect to reference axis as a

counter clockwise direction classically known as positive phase sequence in the power system.

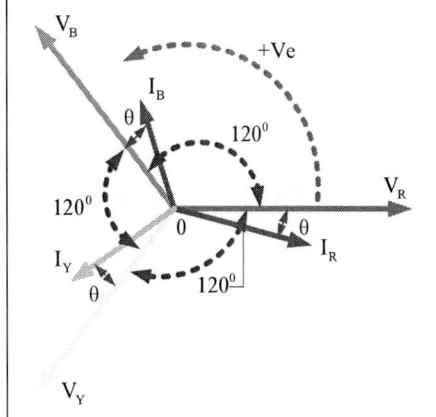

Fig. 2. Three phase voltage & current waveforms with positive (+Ve) phase sequence in power system

In order to analyze the effects of the phase angles, the fundamental phasor with positive (+Ve)) phase sequence and fifth harmonic phasor currents with negative (-Ve sequence are clearly drawn in the Fig. 3 (a) and Fig. 3(b) respectively with respect to a reference frame to forecast the performance of the power system.

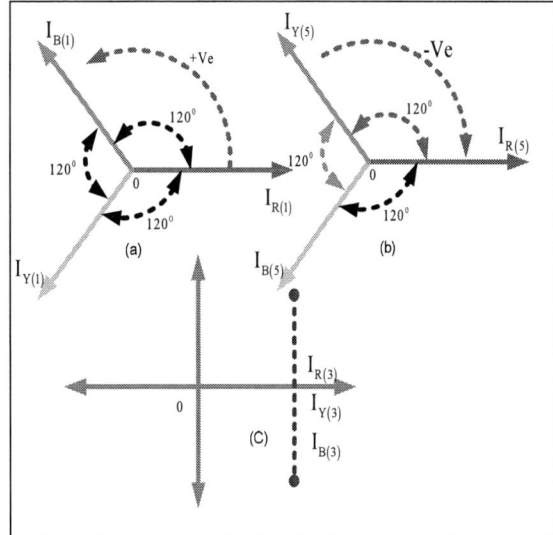

Fig. 3. Phasor diagram of the fundamental, fifth and third harmonic phasor

Further, the schematic phasor diagram is drawn in the Fig. 3 (c) to envisage to the effect of the third harmonic in the power system.

The phase wise fundament current components with the fundamental frequency may be written by the Fig. 3(a) as:

$$\left. \begin{aligned} i_{R(1)} &= I_{R(1)} \times \sin(\omega t) \\ i_{Y(1)} &= I_{Y(1)} \times \sin\left(\omega t - \frac{2\pi}{3}\right) \\ i_{B(1)} &= I_{B(1)} \times \sin\left(\omega t - \frac{4\pi}{3}\right) \end{aligned} \right\} \qquad (6)$$

If the load is balance for the three phase system, the equation (6) will be written by the flowing condition as:

$$I_{R(1)} = I_{Y(1)} = I_{B(1)} \qquad (7)$$

Now, the combination of the equations (6) & (7) yields as algebraic addition phase wise fundamental current written as:

$$i_{R(1)} + i_{Y(1)} + i_{B(1)}$$

$$= I_{R(1)} \times \sin(\omega t) + I_{Y(1)} \times \sin\left(\omega t - \frac{2\pi}{3}\right) + I_{B(1)} \times \sin\left(\omega t - \frac{4\pi}{3}\right) \quad (8)$$

$$= 0$$

The negative phase angles of the Fig. 3(a) implies that the fundamental current $\left(i_{Y(1)}\right)$ and current $\left(i_{B(1)}\right)$ follow the current $\left(i_{R(1)}\right)$ accordingly. Further, It ensures that the expression of the third currents have no phase angles. The zero phase displacement angles of the third harmonic currents correspond to zero sequence currents forming the symmetrical component in nature in the power system.

In this regard, the expression third (3rd) harmonic current may be written by the following equation as:

$$\left. \begin{aligned} i_{R(1)} &= I_{R(1)} \times \sin(3\omega t) \\ i_{Y(1)} &= I_{Y(1)} \times \sin\left\{3\omega t - \left(3 \times \frac{2\pi}{3}\right)\right\} = I_{Y(1)} \times \sin(3\omega t) \\ i_{B(1)} &= I_{B(1)} \times \sin\left\{3\omega t - \left(3 \times \frac{4\pi}{3}\right)\right\} = I_{B(1)} \times \sin(3\omega t) \end{aligned} \right\} \quad (9)$$

The equation (9) satisfies the phasor diagram as shown in the Fig. 3(c) accordingly.

The fifth harmonic currents in the power lines are shown in the Fig. 3(b) from which it may be written by the following equation as:

$$\left. \begin{aligned} i_{R(1)} &= I_{R(1)} \times \sin(5\omega t) \\ i_{Y(1)} &= I_{Y(1)} \times \sin\left\{5\omega t - \left(5 \times \frac{2\pi}{3}\right)\right\} = I_{Y(1)} \times \sin\left(5\omega t - \frac{10\pi}{3}\right) \\ i_{B(1)} &= I_{B(1)} \times \sin\left\{5\omega t - \left(5 \times \frac{4\pi}{3}\right)\right\} = I_{B(1)} \times \sin\left(5\omega t - \frac{20\pi}{3}\right) \end{aligned} \right\} \quad (10)$$

Now, the equation (10) may be simplified as written by:

$$\left. \begin{aligned} i_{R(1)} &= I_{R(1)} \times \sin(5\omega t) \\ i_{Y(1)} &= I_{Y(1)} \times \sin\left\{5\omega t - \left(5 \times \frac{2\pi}{3}\right)\right\} = I_{Y(1)} \times \sin\left(5\omega t - \frac{4\pi}{3}\right) \\ i_{B(1)} &= I_{B(1)} \times \sin\left\{5\omega t - \left(5 \times \frac{4\pi}{3}\right)\right\} = I_{B(1)} \times \sin\left(5\omega t - \frac{2\pi}{3}\right) \end{aligned} \right\} \quad (11)$$

Thus, the phase angle rotation $\left(A_R\right)$ in the radian value for the different types of the harmonics orders (h) are shown as per phase wise rotation $\left(P_R\right)$ tabulated in the Table I to envisage the rotational nature of the sine waves in the power distribution system.

Table I.

h	R-Phase	Y-Phase	B-Phase	P_R
3	$\sin \times 3(\omega t - 0)$ $= \sin(3\omega t)$ $(A_R) = 0$	$\sin \times 3\left(\omega t - \frac{2\pi}{3}\right)$ $= \sin(3\omega t)$ $(A_R) = 0$	$\sin \times 3\left(\omega t - \frac{4\pi}{3}\right)$ $= \sin(3\omega t)$ $(A_R) = 0$	0
5	$\sin(5\omega t - 0)$ $= \sin(5\omega t)$ $(A_R) = 0$	$\sin \times 5\left(\omega t - \frac{2\pi}{3}\right)$ $= \sin\left(5\omega t - \frac{2\pi}{3}\right)$ $(A_R) = -\frac{2\pi}{3}$	$\sin \times 5\left(\omega t - \frac{4\pi}{3}\right)$ $= \sin\left(5\omega t - \frac{4\pi}{3}\right)$ $(A_R) = -\frac{4\pi}{3}$	B Y R
7	$\sin(7\omega t - 0)$ $= \sin(7\omega t)$ $(A_R) = +0$	$\sin \times 7\left(\omega t - \frac{2\pi}{3}\right)$ $= \sin\left(7\omega t + \frac{2\pi}{3}\right)$ $(A_R) = +\frac{2\pi}{3}$	$\sin \times 7\left(\omega t - \frac{4\pi}{3}\right)$ $= \sin\left(7\omega t + \frac{4\pi}{3}\right)$ $(A_R) = +\frac{4\pi}{3}$	R Y B
9	$\sin(9\omega t - 0)$ $= \sin(9\omega t)$ $(A_R) = 0$	$\sin \times 9\left(\omega t - \frac{2\pi}{3}\right)$ $= \sin(9\omega t)$ $(A_R) = 0$	$\sin \times 9\left(\omega t - \frac{4\pi}{3}\right)$ $= \sin(9\omega t)$ $(A_R) = 0$	0

The equation (11) represents that the fifth harmonic currents rotate clockwise but it exists in opposite phase of the fundamental component of the currents in the system.

The derived equations from the equation (6) to (11) are used to determine the sequence wise order of the harmonics in the linear or non-linear loads in the power distribution as shown in the Table II as generic analysis and prediction.

Table II.

Harmonic Order (h)	Generalized form of the harmonics	Nature of the phase sequence
$\begin{pmatrix} 1,4,7,10,13,16,19 \\ ----\,----\,\infty \end{pmatrix}$	$\sum\limits_{h=0}^{\infty}(3h+1)$	Positive (+Ve)
$\begin{pmatrix} 2,5,8,11,14,17,20 \\ ----\,----\,\infty \end{pmatrix}$	$\sum\limits_{h=0}^{\infty}(3h+2)$	Negative (-Ve)
$\begin{pmatrix} 3,6,9,12,15,18,21 \\ ----\,----\,\infty \end{pmatrix}$	$\sum\limits_{h=0}^{\infty}(3h)$	Zero

IV. Effects of The Harmonics on The Transformer Operation

The undesirable effects and issues of the harmonic currents integrate the voltage and cooling problems of the distribution transformers in the power supply experienced by the colossal core power losses and core saturation.

The uses of the linear loads append harmonics in the power system to degenerate the power quality and the anomalous heating effects during the operation of the power systems abusing the power electronic controller by virtue of which the distorted waveforms appear across the consumer device or loads.

There are two standard schemes used for the basic estimation of the De-Rating factor $\left(D_{Rf}\right)$ adopted by the Computer Business Equipment Manufacture's Association (CBEMA) and the IEEE standards in terms of voltage and current harmonics impact on the transformer operations.

A. Computation of the De-Rating factor $\left(D_{Rf}\right)$ using the CBEMA manner

The crest factor $\left(C_f\right)$ is written by the following basic definition (i.e. the ratio of peak current $\left(I_P\right)$ and rms current $\left(I_{rms}\right)$) so it may be written as:

$$C_f = \left(I_P / I_{rms}\right) \qquad (12)$$

So, the De-Rating factor $\left(D_{Rf}\right)$ of the transformer related with the operation may be written as:

$$D_{Rf} = \left(\sqrt{2}/C_f\right) \qquad (13)$$

When the value of the crest factor $\left(C_f\right)$ of the pure sine wave is equal to $\left(\sqrt{2}\right)$, the value of the De-Rating factor $\left(D_{Rf}\right)$ yields as:

$$D_{Rf} = \left(\sqrt{2}/\sqrt{2}\right) = 1 \qquad (14)$$

Thus, 100 KVA transformers are required to be de-rated by 70 KVA when the De-Rating factor $\left(D_{Rf}\right)$ is equal to $\left(D_{Rf} = 2\right)$ to avoid overheating.

This method is unpopular because the crest value of the peak current is not true with various waveforms and it correspond same height of total harmonic distortion (THD).

B. Computation of the K- factor $\left(K\right)$ as harmonic issues

Total harmonic distortion $\left(THD_i\right)$ of the current may be defined by the ratio of the rms value of the total harmonic current $\left(\sum_{h=1}^{\infty}\left(I_{h(rms)}\right)\right)$ and rms value of the fundamental current $\left(I_{1(rms)}\right)$ as:

$$THD_i = \left(\sum_{h=1}^{\infty}\left(I_{h(rms)}/I_{1(rms)}\right)\right) \qquad (15)$$

Where, h is order of harmonic the fundamental frequency 50 Hz.

The K- factor $\left(K\right)$ of the power transformer may be defined by the following term as:

$$K = \left(\sum_{h=1}^{\infty}\left(\left(I_{h(rms)}^2 \times h^2\right)/I_{1(rms)}^2\right)\right) \qquad (16)$$

Now, total harmonic current $\left(I_{h(T)}\right)$ may be expressed by the sum of the individual harmonic current generated by the non-linear load as:

$$I_{h(T)} = \left(\sum_{h=1}^{\infty}\left(I_{h(rms)}\right)\right) \qquad (17)$$

Now, the pu value of the K- factor $\left(K\right)$ may be expressed by the ratio of the total harmonic current $\left(I_{h(T)}\right)$ considering the order of the harmonic $\left(h\right)$ and rms current $\left(I_{(rms)}\right)$ may be expressed as:

$$I_{h(pu)} = \left(I_{h(T)}/I_{rms}\right) \qquad (18)$$

Now, pu value of The K- factor $\left(K\right)$ may be expressed accordingly as:

$$K = \left(\sum_{h=1}^{\infty}\left(\left(I_{h(pu)}^2 \times h^2\right)/I_{h(pu)}^2\right)\right) \qquad (19)$$

The eddy current power loss of the $\left(K\right)$ of the transformer is dependent on the square of the frequency so the value of the K- factor comprises an useful index for the heating effects and temperature rises due to presence of the different harmonics. Further, it contributes a quantitative means of commutating the De-Rating factor $\left(D_{Rf}\right)$ of the transformers as a function of the non-linearity.

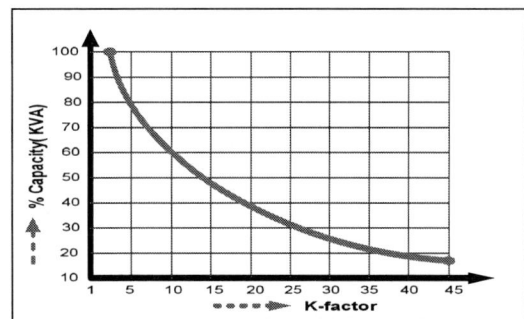

Fig. 4. Capacity (in KVA) versus K-factor of a transformer

In order to envisage a better representation of De-Rating factor $\left(D_{Rf}\right)$ of a transformer, a highly non-linear schematic curve is shown in the Fig. 4 to guesstimate the overall performance of the distribution systems.

A computational method is carried out to determine the de-rating factor of a transformer in terms of the order of the harmonic (h), core $\left(P_{(CL)}\right)$ and copper loss $\left(P_{(cu)}\right)$.

Total $\left(P_{LL}\right)$ power of a transformer may be written by hysteresis power loss, eddy current power loss $\left(P_{(ED)}\right)$ and copper loss $\left(P_{(cu)}\right)$ as:

$$P_{LL} = \left(P_{(He)} + P_{(ED)} + P_{(cu)}\right) \qquad (20)$$

The hysteresis power loss $\left(P_{(He)}\right)$ of the power transformer may be written by the peak flux density $\left(B_m\right)$, operating frequency $\left(f_s\right)$ thickness of lamination of the core materials $\left(t_{\mu m}\right)$ and a proportionality $\left(K_{He}\right)$ constant depending on the type of the core magnetic specimen as:

$$P_{(He)} = K_{He} \times \left(B_m^{\chi}\right) \times f_s \times \left(t_{\mu m}\right) \qquad (21)$$

Where, (χ) is the Strezman constant depending on the magnetic material of the core of a usual transformer? Sometimes, it may be defined by the hysteresis and eddy current power loss computation in the range of the following numerical value as:

$$1 \leq \chi \leq 2 \qquad (22)$$

Similarly, the eddy current power loss $\left(P_{(ED)}\right)$ of the power transformer may be written by the peak flux density $\left(B_m\right)$, operating frequency $\left(f_s\right)$ & thickness of the lamination of the core materials $\left(t_{\mu m}\right)$ and a proportionality constant $\left(K_{Ed}\right)$ depending on the type of the core magnetic specimen as:

$$P_{(ED)} = K_{Ed} \times \left(B_m^2\right) \times f_s^2 \times \left(t_{\mu m}\right) \qquad (23)$$

Further, total cupper loss $\left(P_{(cu)}\right)$ of the transformer may be written by rms load current $\left(I_{(rms)}\right)$ and equivalent resistance $\left(R_{(Eq)}\right)$ as:

$$P_{(cu)} = I_{(rms)}^2 \times R_{Eq} \qquad (24)$$

Total core power loss may be re-written from the equation (11) & (12) as:

$$P_{CL} = \left(P_{(ED)} + P_{(He)}\right) \qquad (25)$$

Now, from the equation (9), (13) & (14) may be re-written as:

$$P_{LL} = \left(P_{CL} + I_{(rms)}^2 \times R_{Eq}\right) \qquad (26)$$

Again, the eddy current power loss $\left(P_{(ED)}\right)$ may be written by the base eddy power loss $\left(P_{(ED)B}\right)$, numbered of harmonics (h) and rms current $\left(I_{rms}\right)$ for the determination of the K- factor (K) of a transformer.

$$P_{(ED)} = P_{(ED)B} \sum_{h=1}^{\infty} \left[\left(I_h / I_{rms}\right)\right]^2 \times h^2 \qquad (27)$$

Further, the pu representation of $\left(P_{(ED)}\right)$ may be written by the following ratio as:

$$P_{(ED)pu} = \frac{P_{(ED)}}{P_{(ED)B}} = \sum_{h=1}^{\infty} \left[\left(I_h / I_{rms}\right)\right]^2 \times h^2 \qquad (28)$$

Again, from the equation (28), it may be written by base value as:

$$P_{LL(B)} = P_{CL(B)} \times \sum_{h=1}^{\infty} \left(I_h^2 \times h^2\right) \qquad (29)$$

Further, it should be expressed by the equation (26) in pu value as:

$$P_{LL(pu)} = \frac{P_{LL(B)}}{P_{LL}} = \frac{P_{CL(B)}}{P_{CL(R)}} \times \sum_{h=1}^{\infty} \left(I_h^2 \times h^2\right) \qquad (30)$$

Further, the equation (30) may be modified by the pu value to compute the performance parameters of K-rated transformers which withstands the non- sinusoidal load from no load to full load current rating with k-factor up to the K-

rating of the transformer and total power loss in the power systems so the equation is re-written as:

$$P_{LL(pu)} = \frac{P_{LL(B)}}{P_{LL}} = P_{CL(pu)} \times \sum_{h=1}^{\infty} \left(I_h^2 \times h^2\right) \qquad (31)$$

We have also a following relation as:

$$P_{LL(pu)} = \sum_{h=1}^{\infty} \left(I_{h(pu)}^2\right) + P_{CL(pu)} \sum_{h=1}^{\infty} \left(I_h^2 \times h^2\right) \qquad (32)$$

The acceptable maximum current $\left(I_P\right)$ is given by an expression of equation (32) using the K- factors (K) of a transformer as a loss gesture:

$$I_{P(pu)} = \sqrt{\frac{\left(1 + \left(P_{CL(pu)}\right)\right)}{\left(1 + \left(K \times P_{CL(pu)}\right)\right)}} \qquad 33)$$

V. RESULTS

In order to determine K- factor (K) and de-rating factor of a transformer following case has been considered in consideration 50 kW load, $P_{CL(pu)} = 12\%$ and transformer capacity 100KVA with the following data as shown in the Table III used for yielding the computational result.

Table III.

(h)	(h^2)	(f_s) Hz	$\left(I_{h(pu)}\right)$	$\left(I_{h(pu)}^2\right)$	$\left(I_{h(pu)}^2 \times h^2\right)$
1	1	50	1.00	1.00	1
3	9	150	0.67	0.4489	4.0401
5	25	250	0.41	0.1681	4.2025
7	49	350	0.14	0.0196	0.9604
9	81	450	0.046	0.002116	0.171396
11	121	550	0.054	0.002916	0.352836
13	169	650	0.024	0.000576	0.097344
17	289	850	0.021	0.000441	0.127449
19	361	950	0.001	0.0361	13.0321
21	441	1050	0.007	0.000049	0.021609
23	529	1150	0.008	0.000064	0.033856
25	625	1250	0.003	0.000009	0.005625
				$\sum_{h=1}^{25} 1.642$	$\sum_{h=1}^{25} 11.02$

Now, the Table III is used for working out the K-factor (K) using the equation (8) as case of the result.

$$K = \left(\sum_{h=1}^{\infty} \left(\left(I_{h(pu)}^2 \times h^2\right) / I_{h(pu)}^2\right)\right) = \frac{\sum_{h=1}^{25} 11.02}{\sum_{h=1}^{25} 1.642} = 6.07$$

In order to find out De-Rating factor $\left(D_{Rf}\right)$ of the transformer using the equation (22), the value $P_{CL(pu)}$ is achieved by the manufactures' measurement. Usually, its value is equal to $\left(P_{CL(pu)} = 12\%\right)$ as a rated condition and permissible temperature rise of the transformer in the power distribution system as:

$$I_{P(pu)} = \sqrt{\frac{1 + \left(P_{CL(pu)}\right)}{1 + \left(K \times P_{CL(pu)}\right)}} = \sqrt{\frac{1 + 0.12}{1 + (6.07 \times 0.12)}} = 0.648$$

The de-rated allowable loading of the 100 kVA transformer may be computed by KVA as spurious effects of the harmonics as contribution of malfunctions:

$$\left(I_{P(pu)} \times KVA\right) = \left(0.648 \times 100\right) \approx 64.8$$

The computational results of the voltage harmonics are included in the Table IV to demonstrate the height of total harmonic distortion $\left(THD_v\right)$ against rms fundamental voltage (i.e. $V_1 = 230$ V).

Table IV.

Harmonic order (h)	Harmonic Voltage (V_h)	$(V_h)^2$	THD_v $= \sum_{h=2}^{9}\sqrt{v_h^2}\Big/v_1$
3	7	49	
5	4	16	$\approx 3.8\%$
7	3	9	
9	1.5	2.25	

A survey report of the bulk consumers' load are collected showing in the Table V with the fundamental current is equal to 500 A along with the harmonic currents.

Table V.

Harmonic order (h)	Harmonic current (A) (I_h)	$\sum_{h=1}^{7}\sqrt{\left(I_h\right)^2}$	$\dfrac{I_h}{\left(\sum_{h=1}^{7}\sqrt{\left(I_h\right)^2}\right)}$
1	500		0.8964
3	190	≈ 557.8	0.3406
5	130		0.2331
7	95		0.1613

In order to determine the K- factor of a transformer, the above Table III is used. Now, K- Factor (K) using the Table III is shown as computational results:

$$K - Factor = \left.\begin{array}{l}\left(0.8964 \times 1\right)^2 + \left(0.3406 \times 3\right)^2 \\ + \left(0.2331 \times 5\right)^2 + \left(0.1613 \times 7\right)^2\end{array}\right\} \approx 4.48$$

Further, an experiment is carried out to find out the order of the harmonics using the following experimental data as results showing in the Table VI.

Table VI.

$i(\omega t)$ (A)	1.0	1.5	1.9	1.7	1.5	1.2	1.00
(ωt) (Rad)	0	$\dfrac{\pi}{3}$	$\dfrac{2\pi}{3}$	π	$\dfrac{4\pi}{3}$	$\dfrac{5\pi}{3}$	2π

In order to find out the order of the harmonics, the above Table IV is taken into account with the following equation (34) to judge the power quality as:

$$i(\omega t) = I_m \times \sin\left(h\omega t + \theta\right) \tag{34}$$

In order to work out the value of the angle (θ), the equation (34) is used for the computation of the phase difference (i.e. (θ)) between voltage and current in the linear or the non-linear loads in the power distribution as:

$$\left.\begin{array}{l}\left(h\omega t - \theta\right) = Sin^{-1}\left(\dfrac{i(\omega t)}{I_m}\right)\Bigg|_{\omega t = 0} \Rightarrow \left(0 + \theta\right) = Sin^{-1}\left(\dfrac{1.00}{1.9}\right) \\ \Rightarrow \theta = Sin^{-1}\left(\dfrac{1.00}{1.9}\right) \approx 0.554 \text{ (Rad)}\end{array}\right\}$$

Table VII.

$i(\omega t)$ (A)	(ωt) (Rad)	$(h) = \left\{Sin^{-1}\left(\dfrac{i(\omega t)}{1.9}\right) + 0.554\right\}\Big/(\omega t)$	(hf_s) Hz
1.0	0	$\propto$	$\propto$
1.4	$\pi/3$	1.03	51.5
1.9	$2\pi/3$	1.01	50.5
1.7	π	0.528	26.4
1.5	$4\pi/3$	0.349	17.45
1.3	$5\pi/3$	0.249	12.45
1.0	2π	0.176	8.80

Now, the extended computations are incorporated with the fundamental frequency (f_s) in addition to other sub-harmonic and intermediate frequency (hf_s) as shown in the Table VII using the simplified equation (34) and maximum value of current (1.9 A) taken from the Table VI.

VI. CONCLUSIONS

Harmonics plays galloping and pilot role in the power quality issues in the power distributions. The voltage and current harmonics have huge negative impact on the transformer and power reactor owing to additional copper and core power losses. The De-Rating factor and K-rated transformers are established the various factor and power losses in the power system. The address of distorted sine wave includes wider validation of the experimental and computational results with sub harmonics and intermediate frequencies. Surveying the loads and making preparatory performance evaluations, validations and case study for the designers and practical applications.

REFERENCES

[1] W Balow, P Sain, A Chatterjee,"An Improved Technique of Wind Power Harvesting With a Doubly Fed Induction Generator (DFIG)", 2018, 8th IEEE Power India International Conference, 2018, pp. 1-6.

[2] W Balow, T Halder,"A Selective Harmonic Elimination (SHE) Technique For the Multi-Leveled Inverters", IEEE EDKCON 2018.

[3] T Halder, "An Impact of the Voltage & Current Ripples in the Power Stages of the Boost Converter", IEEE EDKCON 2018.

[4] I Biswas, B Das, W Balow, T Halder, "Probability of the Pure Sine Wave in the Power Quality Issues", 2018 8th IEEE Power India International Conference, Dec. 2018, pp. 1-6.

[5] IEEE Task Force on Modeling and Simulation "Modeling and Simulation of the propagation of harmonies in electric power networks, Part II: Sample systems and examples", IEEE Trans. on Power Delivery, Vol. 11, No. 1, Jan. 1996, pp. 466-474.

[6] N.K. Madora and A. Kusko, "Computer-Aided Design and Analysis of Power-Harmonic Filters" IEEE Trans. on Industry Applications, Vol. 36, No. 2, March /April 2000, pp. 604-613.

[7] Arrillaga J., Watson N. R., Chen S., Power system quality assessment, John Wiley & Sons, Ltd, Chichester, 2000.

[8] Chapman D., Harmonics – causes and effects. Leonardo Power Quality Application Guide –Part 3.1, 2001.

2019 Devices for Integrated Circuit (DevIC), 23-24 March, 2019, Kalyani, India

A Low Power Biopotential Amplifier based on Bulk Driven Quasi Floating Gate Technique

Preeti Sharma[1], Kulbhushan Sharma[1], H. S. Jatana[2], and Rajnish Sharma[1,*]

[1]*VLSI Centre of Excellence*
[1]*Chitkara University Institute of Engineering and Technology,*
Chitkara University, Punjab, India
[2]*Semi-conductor Laboratory (SCL), Mohali, Punjab, India*
*rajnish.sharma@chitkarauniversity.edu.in

Abstract—Design of low-power and low-noise Biopotential Amplifier (BPA) plays crucial role in the success of high end medical diagnosis systems. However, most of these researched BPAs face a major challenge of consuming large amount of power and also exhibit high values of Noise Efficiency Factor (NEF). Here we report the design of a BPA using Bulk-Driven Quasi-Floating Gate (BDQFG) technique which consumes only low-power (0.657 μW) and also exhibits NEF of 2.06. Circuit design and simulation have been performed in Cadence Analog Design Environment using standard 0.18 μm technology. Besides promising results on power and noise, design of the BPA using BDQFG technique has also been fine-tuned to achieve mid-band gain of 38.3 dB (from 2.9 Hz to -3dB frequency of 735.5 Hz) and phase margin of 80.6 °.

Keywords—*Bulk-driven, Biopotential amplifier, Operational transconductance amplifier, Pseudo-resistor, Quasi-floating gate.*

I. INTRODUCTION

Biomedical signals like Electroencephalogram (EEG), Electrocorticogram (ECoG), Electromyogram (EMG) and Electrocardiogram (ECG) are known to possess small values of amplitude typically in the range of 1 μV to 5 mV with frequency values between 0. 01 Hz to 2 kHz [1-3]. The frequency band occupied is 0.5 Hz − 100 Hz for EEG, 0.5 Hz − 200 Hz for ECoG, 6 Hz − 500 Hz for EMG and 0.01 Hz − 250 Hz for ECG [3]. There is always a requirement to amplify such small values of amplitude of signals while retaining their originality [3]. These amplified signals are used in biomedical systems to ensure correct and timely diagnosis of many chronic diseases like Paralysis, Arrhythmias, Respiratory disorders, Body fluid overload, Epilepsy and Parkinson etc. [4, 5]. A lot of researchers have proposed various closed loop amplifier architectures like Chopper Stabilized (CHS) [6], Capacitively Coupled-Capacitive Feedback (CC-CF) [7] and Directly Coupled-Active Feedback (DC-AF) [8] which have been found to display their own set of strengths and limitations. While CC-CF architecture finds its strength in providing high values of gain and Common Mode Rejection Ratio (CMRR) [9], DC-AF and CHS architectures establish their advantage in terms of DC offset rejection [10,11].

However, most of the researched and reported architectures seem to suffer from a major challenge of drawing huge amount of power along with high values of Noise Efficiency Factor (NEF). A detailed literature survey establishes the fact that the choice of Operational Transconductance Amplifier (OTA) topology majorily contributes to noise and power in the overall architecture of any BPA. Mainly research work has been focussed on Current mirror [12,13], Miller [14], Folded-cascode [15], Telescopic [16], Differential self-bias [17], Floating-gate [18] and Complementary Metal Oxide Semiconductor

(CMOS) inverter-based [19] OTA topologies to meet the requirements set for design of a close to ideal BPA. Similarly, the choice of right Pseudo-Resistor (PR) not only helps in accurate stabilization of lower cut-off frequency (f_L) of a neural amplifier but also helps in saving on overall chip area.

In this paper, a BPA comprising of an OTA and PR based on Bulk-Driven Quasi-Floating Gate (BDQFG) technique is presented. Obtained simulation results reveal the high value of the gain of the order of 38.3 dB, low values of NEF close to 2.06 and least amount of power consumption equal to 0.657 μW only for the proposed technique.

Paper has been divided in four different sections other than providing the list of references at the end. Section II providing all the details about the design and simulation methodology adopted for this research effort follows the present section of Introduction. Results in the form of simulation plots ably supported by the justification for the same are presented in "Results and Discussions", section III. Paper culminates in providing the conclusion of the research effort in section IV.

II. DESIGN AND SIMULATION METHODOLOGY

Circuit shown in Fig. 1 shows generic architecture of a BPA typically highlighting the presence of an OTA block, PRs (R_F) and Capacitances (C_1, C_2 and C_L). Primarily C_1 and C_2 play a major role in controlling the mid-band gain (A_1) owing to the basic relationship ($A_1 = C_1/C_2$), band-limiting load capacitance (C_L) helps in fine tuning the overall −3dB bandwidth (f_{-3dB}) which is approximated by g_m/A_1C_L, where g_m is the transconductance of the OTA. C_2 and R_F connected in the feedback path of OTA are responsible for determining the f_L of the BPA as given by equation 1.

$$f_L = 1/2\pi R_F C_2 \qquad (1)$$

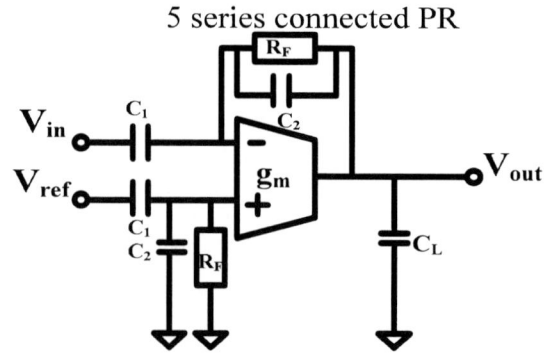

Fig. 1. Schematic of generic biopotential amplifier architecture.

978-1-5386-6723-1/19 $31.00 © 2019 IEEE   424

2019 Devices for Integrated Circuit (DevIC), 23-24 March, 2019, Kalyani, India

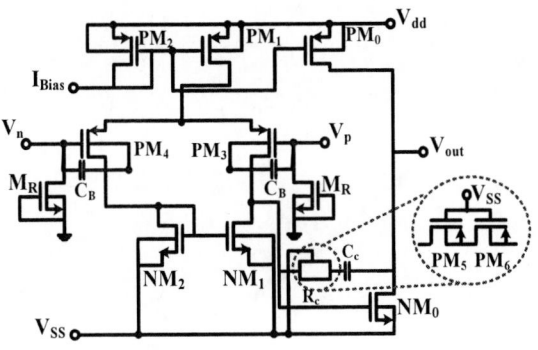

Fig. 2. BDQFG OTA topology.

Another set of R_F and C_2 connected at the non-inverting terminal of OTA help in sorting out unnecessary DC components present in the input signal [20]. Five BDQFG pseudo-resistors capable of providing 1 TΩ of resistance each have been connected in series so as to obtain overall resulting $R_F = 5$ TΩ. Choosing such high value of R_F helps in keeping the C_2 values on a much smaller side of the order of 30 fF only. Use of this kind of approach provides a huge advantage in terms of saving a large amount of area at the time of final realization of the circuit on the chip. The circuit diagram of the OTA and pseudo-resistor used in Fig. 1 have been shown in Fig. 2 and Fig. 3 respectively.

Active components used in the differential amplifier stage of OTA have been biased using BDQFG approach, which provide higher value of g_m along with low power consumption and noise in comparison to Bulk-driven, Floating-gate and Quasi-floating gate techniques [21]. Khateb et al. [22] and Kumngern et al. [23] have earlier used BDQFG technique to design a rectifier and current-conveyor respectively. Working of Miller OTA used in the present piece of work has been satisfactorily explained by Cortes et al. [24]. The input signals V_p and V_n are applied to the quasi-floating gate terminal of PM_3 and PM_4 respectively, through coupling capacitor (C_B). To provide biasing to these transistors, a very large value of resistance obtained from diode-connected transistor M_R is used. Simultaneously, the input signals are also connected to the bulk terminal of PM_3 and PM_4 through C_B making the MOS bulk-driven. R_C has been added in the circuit used in [25] so as to obtain better stability in terms of obtaining higher values of phase margin. Pseudo-resistors have been designed as per the details provided in our earlier reported work [26]. Size of the active devices have been optimized as per the details given in Table

TABLE I. ASPECT RATIOS, OPERATING REGION OF DEVICES AND COMPONENT VALUES IN BPA

Devices	W/L (μm/μm)	I_D (A)	g_m (S)	g_m/I_D (V⁻¹)	Region of operation
PM_0	1/1	479.57 n	5.95 μ	12.41	Saturation
PM_1	1/1	436.3 n	5.25 μ	12.03	Saturation
PM_2	1/1	499.9 n	6.30 μ	12.60	Saturation
PM_3, PM_4	50/27	168.1 n	3.3 μ	19.9	Weak inversion
PM_5	0.6/0.18	1.308 f	11.46 f	8.76	Triode
PM_6	0.6/0.18	1.23 f	10.8 f	8.76	Triode
NM_0	1/6.66	479.65 n	5.81 μ	12.13	Saturation
NM_1	0.5/9.09	168.15 n	2.02 μ	12.06	Saturation
NM_2	0.5/9.09	168.1 n	2.02 μ	12.13	Saturation

Vdd = 500 mV, Cc = 3.5 pF, C_1 = 3 pF, C_2 = 30 fF, C_B = 5 pF
C_L = 1 pF, R_F = 5 TΩ, Vss = −500 mV, I_{Bias} = 500 nA, R_C ≈ GΩ

I to achieve the desired set of parameters like small values of absorbed power and high values of gain. All the simulation results have been carried out in Cadence Analog Design Environment using standard 0.18 μm technology. BSIM3V3 models of transistors have been used for the implementation of different circuits presented in this work.

III. RESULTS AND DISCUSSIONS

Fig. 4 shows the plots of incremental resistance (R_{PR}) and current (I_{PR}) on y_1 and y_2 axes respectively, versus swept voltage (ΔV_{XY}) between five series connected BDQFGPRs shown in Fig. 3. As can be clearly observed from Fig. 4 that there exists a linear relationship between I_{PR} and ΔV_{XY} and value of the same can be calculated as 0.2 pA. The value of the R_{PR} over ΔV_{XY} as calculated from equation 2, is found to be 5 TΩ and has been marked as almost a straight line in Fig. 4.

$$R_{PR} = \frac{1}{\delta I_{PR}/\delta V_{XY}} \qquad (2)$$

This observed value of R_{PR} is in conformity with theoretical value of 5 TΩ resulting as an outcome of five series connected BDQFGPRs of 1 TΩ each. It is worth mentioning here that while performing simulation of five series connected BDQFGPRs shown in Fig. 3 at 50 Hz, the node V_Y has been kept at 50 μV while the node V_X has been swept from −1 V to 1 V. The reason for selection of these values lies in the fact that nodes V_Y and V_X are analogous to input and output

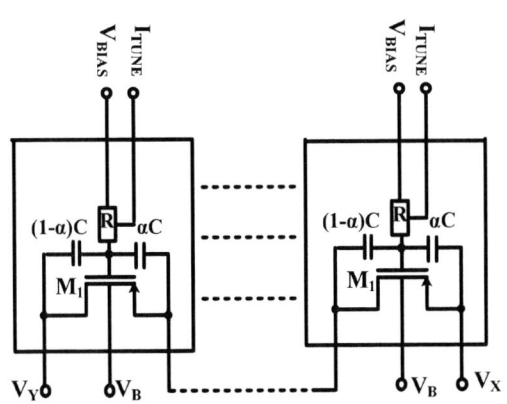

Fig. 3. Schematic of five series connected BDQFG Pseudo-resistor.

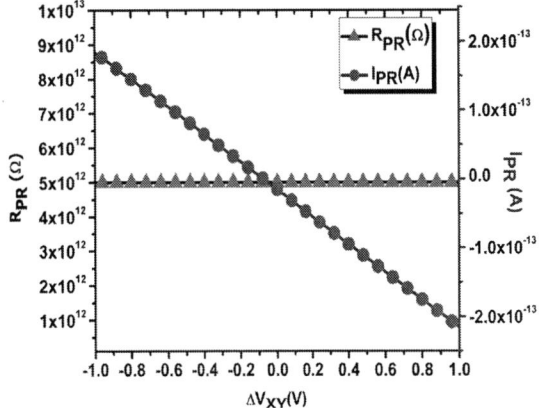

Fig. 4. I_{PR}, R_{PR} versus V_{XY} for 5 series connected BDQFGPRs.

978-1-5386-6723-1/19 $31.00 © 2019 IEEE   425

terminals of BPA shown in Fig. 1 under section II. Similar kind of simulation strategy for obtaining V-I and V-R curves for pseudo-resistors has been adopted by Kassiri *et al.* [27] and Ozalevli *et al.* [28]. These interpretations convey that such a high magnitude of incremental resistance of the order of 5 TΩ obtained from the combination of five series connected BDQFGPRs meets the requirement of low f_L of the BPA. This PR topology was used in feedback path of BPA as indicated by R_F in Fig. 1. A weak signal of amplitude 50 μV with frequency of 50 Hz was applied at the input of this BPA. With this design methodology, the simulated transfer function for the circuit is shown in Fig. 5. The results show that overall BPA mid-band gain is 38.3 dB from f_L of 2.9 Hz to $f_{(-3dB)}$ of 735.5 Hz. For a typical process corner at room temperature (27° C), the phase margin of 80.6 ° is obtained with a gain crossover frequency of 27.01 kHz. Total power consumed by the neural amplifier is equal to 0.657 μW for 1.31 μA supply current from a ± 0.5 V voltage supply. Fig. 6 shows simulated V/Hz$^{1/2}$ noise versus frequency spectrum. The total input-referred noise for neural amplifier integrated over bandwidth of 2.9 Hz to 735.5 Hz is evaluated to be 1.27 μV$_{rms}$ while a NEF of 2.06 is obtained by following equation 3.

$$NEF = V_{ni,rms}\sqrt{\frac{2}{\pi V_T}\left(\frac{I_{Bias}}{4kT}\right)\frac{1}{BW}} \qquad (3)$$

Where, I_{Bias} is the bias current, k is Boltzmann's constant (1.38 x 10^{-23} JK^{-1}), T is the temperature in Kelvins (310 K), V_T is the threshold voltage, BW is the bandwidth of the OTA and $V_{ni,rms}$ is the rms value of the input-referred noise.

Table II shows the comparison of the performance of BPA design presented in this paper along with work of other researchers [29-32]. It is evident from the Table II that the biopotential amplifier reported by [29] is based on current-mirror topology which achieved gain of 40 dB from 2.2 Hz − 30 Hz. The supply voltage applied by architecture is ±2.5 V with supply current of 0.128 μA and the power consumption of 4.58 μW. The integrated input-referred noise is 2.4 μV$_{rms}$ with the quite high value of NEF of 6.0. Liu *et al.* [30] designed a BPA based on current-reuse OTA topology configured to record local field potentials (LFP). The amplifier yielded the gain of 34 dB over bandwidth of 0.200 Hz − 430 Hz. The power consumption of the architecture reported by the authors is 0.8 μW from a 1 V supply voltage

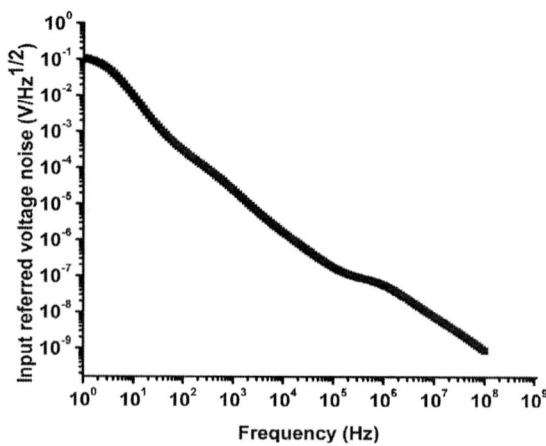

Fig. 6. Input referred voltage noise (V/Hz$^{1/2}$) versus frequency (Hz). The Integrated input-referred noise from 2.9 Hz to 735.5 Hz is 1.27 μVrms.

at 0.8 μA supply current. The reported architecture yielded integrated input-referred noise of 5.71 μV$_{rms}$ with NEF of 2.59. Gallegos *et al.* [31] demonstrated folded-cascode OTA topology which yielded the gain of 40.85 dB over 0.684 Hz − 140 Hz. The power consumed by this amplifier is 0.655 μW from a 1.8 V power supply with the supply current of 0.364 μA. The integrated input-referred noise is 4.09 μV$_{rms}$ with the high NEF of 8.05. Further, J. Zhang *et al.* [32] proposed Two-Stage Continuous-Time Common Mode Feedback (CT-CMFB). The gain corresponding to this amplifier is 39.8 dB over 0.2 Hz − 200 Hz. The supply voltage used by architecture is 2 V with supply current of 0.16 μA. The integrated input-referred noise is 2.05 μV$_{rms}$ with the high value of NEF of 2.26. Although Harrison *et al.*, Gallegos *et al.* and J. Zhang *et al.* proposed BPA with the high value of gain as compared to

TABLE II. COMPARATIVE ANALYSIS OF PARAMETERS OF DIFFERENT NEURAL AMPLIFIERS

Reference	[29]	[30]	[31]	[32]	This work
Year	2003	2012	2014	2018	2018
Technology (μm)	0.18	0.18	0.18	0.35	0.18
OTA Topology	Current mirror	Current-reuse	Folded-cascode	2- stage CT-CMFB	Miller
Supply Voltage (V)	±2.5	1	1.8	2	±0.5
Supply Current (μA)	0.128	0.8	0.364	0.16	1.31
Bias Current (nA)	32	160	56	---	500
Input-Referred Noise (μVrms)	2.4	5.71	4.09	2.05	1.27
Gain (dB)	40	34	40.85	39.8	38.3
Power Consumption (μW)	4.58	0.8	0.655	----	0.657
f_L (Hz)	2.2	0.200	0.684	0.2	2.9
f_{-3dB} (Hz)	30	430	140	200	735.5
NEF	6.0	2.59	8.05	2.26	2.06
Applications	EEG	LFP	EEG	ECG	EEG, ECOG and EMG

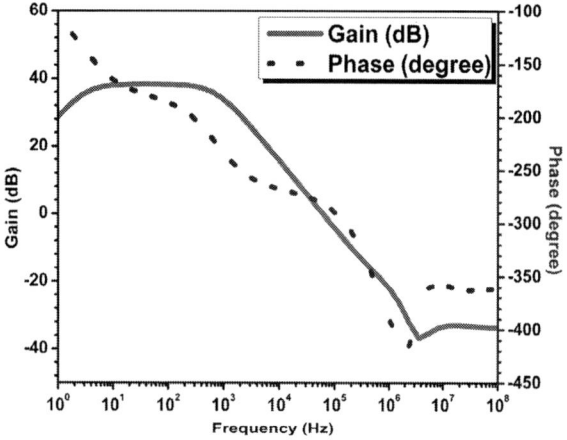

Fig. 5. Phase and gain plot for proposed neural amplifier: mid-band gain is 38.3 dB over 2.9 Hz to −3dB frequency of 735.5 Hz.

the one presented in this work, but the proposed amplifier has achieved a great combination of low-power (0.657 µW), low-NEF (2.06) and low input-referred noise (1.27 µV$_{rms}$) over 2.9 Hz − 735.5 Hz. The reason for low-values of power consumption is based on the fact that BDQFG technique decreases V_T which further reduces the biasing requirement [33]. Moreover, this technique also offers low-voltage operation of BDQFG OTA which helps in providing low values of power operation [33]. The reported OTA architectures in [29-32] used the quite high operating voltage of the value of ±2.5 V, 1.8 V, 1 V and 2 V while this proposed BDQFG OTA uses the low value of operating voltage of the order of ±0.5 V to achieve a remarkable trade-off between power and noise, the biasing current, I_{Bias} has been raised upto 500 nA in comparison to the values used in [29-31] as shown in Table II. The smaller value of NEF is attributed to the fact that the increased value of I_{Bias} and g_m of the proposed OTA helps in reduction of noise.

IV. CONCLUSION

Design of a biopotential amplifier drawing least amount of power equal to 0.657 µW along with other parameters like gain of 38.3 dB, NEF of 2.06 and input-referred noise of 1.27 µV$_{rms}$ has been demonstrated. Comparison of the results obtained as an outcome of this work along with the ones reported in literature seem to be promising for use of the proposed design of BPA in medical diagnostic equipments. Reason for the favourable results has been dedicated to use of Bulk-Driven Quasi-Floating Gate approach in OTA as well as PRs used in the design.

ACKNOWLEDGEMENT

The authors acknowledge the support received from Dr. Jaya Madan and Dr. Rahul Pandey for their valuable discussions and comments.

REFERENCES

[1] W. M. Chen, W. C. Yang, T. Y. Tsai, H. Chiueh, and C. Y. Wu, "The design of CMOS general-purpose analog front-end circuit with tunable gain and bandwidth for biopotential signal recording systems," in Proc. Annu. Int. Conf. IEEE Eng. Med. Biol. Soc. EMBS, pp. 4784–4787, 2011.

[2] J. Holleman, "Design considerations for neural amplifiers," in Proc. Annu. Int. Conf. IEEE Eng. Med. Biol. Soc. EMBS, pp. 6331–6334, 2016.

[3] C. L. Hsu, M. H. Ho, Y. K. Wu, and T. H. Chen, "Design of low-frequency low-pass filters for biomedical applications," IEEE Asia-Pacific Conf. Circ. Syst. Proceedings, APCCAS, pp. 690–695, 2006.

[4] J. Xu, M. Konijnenburg, H. Ha, R. Van Wegberg, S. Song, and D.B. Almazan, "A 36 µW, 1.1 mm² reconfigurable analog front-end for cardiovascular and respiratory signals recording," IEEE Trans. Biomed. Circ. and Syst., no. 99, pp. 1–10, 2018.

[5] C. Y. Wu, C. H. Cheng, and Z. X. Chen, "A 16-channel cmos chopper-stabilized analog front-end ECoG acquisition circuit for a closed-loop epileptic seizure control system," IEEE Trans. Biomed. Circ. and Syst., vol. 12, no. 3, pp. 543-553, 2018.

[6] H. Chandrakumar and D. Markovic, "A high dynamic-range neural recording chopper amplifier for simultaneous neural recording and stimulation," IEEE J. Solid-State Circ., vol. 52, no. 3, pp. 645–656, 2017.

[7] P. Mohseni and K. Najafi, "A fully integrated neural recording amplifier with dc input stabilization," IEEE Trans. Biomed. Eng., vol. 51, no. 5, pp. 832-83, 2004.

[8] B. Gosselin, A. E. Ayoub, J. F. Roy, M. Sawan, A. Chaudhuri and D. Guitton, "A mixed-signal multichip neural recording interface with bandwidth reduction," IEEE Biomed. Circ. and Syst. Conf., pp. 129-141, 2009.

[9] K. A. Ng and Y. P. Xu, "A compact, low input capacitance neural recording amplifier," IEEE Trans. Biomed. Circ. and Syst., vol. 7, no. 5, pp. 610–620, 2013.

[10] K. A. Ng and P. K. Chan, "A CMOS analog front-end IC for portable EEG/ECG monitoring application," IEEE Trans. Circ. and Syst., vol. 52, no. 11, pp. 2335–2347, 2005.

[11] T. Denison, K. Consoer, W. Santa, A. T. Avestruz, J. Cooley, and A. Kelly, "A 2 µw, 100 nV/Hz$^{1/2}$ chopper-stabilized instrumentation amplifier for chronic measurement of neural field potentials," IEEE J. Solid-State Circ., vol. 42, no. 12, pp. 2934–2945, 2007.

[12] M. Mollazadeh, S. Member, and K. Murari, "Micropower CMOS integrated low-noise amplification, filtering and digitization of multimodal neuropotentials," IEEE Trans. Biomed. Eng., vol. 3, no. 1, pp. 1–10, 2009.

[13] S. Farshchi, A. Pesterev, P. Nuyujukian, E. Guenterberg, I. Mody, and J. W. Judy, "Embedded neural recording with TinyOS-based wireless-enabled processor modules," IEEE Trans. Neural Syst. Rehabil. Eng., vol. 18, no. 2, pp. 134–141, 2010.

[14] L. H. Ferreira, T. C. Pimenta, and R. L. Moreno, "An ultra-low-voltage and ultra-low-power CMOS Miller OTA with Rail-to-Rail Input/Output swing," IEEE Trans. Circ. and Syst., vol. 54, no. 10, pp. 843–847, 2007.

[15] W. Wattanapanitch, M. Fee, and R. Sarpeshkar, "An energy efficient micropower neural recording amplifier," IEEE Trans. Biomed. Circ. and Syst., vol. 1, no. 2, pp. 136–147, 2007.

[16] F. Shahrokhi, K. Abdelhalim, D. Serletis, P. L. Carlen, and R. Genov, "The 128-channel fully differential digital integrated neural recording and stimulation interface," IEEE Trans. Biomed. Circ. and Syst., vol. 4, no. 3, pp. 149–161, 2010.

[17] J. Y. Lee and S. N. Hwang, "A high-gain boost converter using voltage-stacking cell," Trans. Korean Inst. Electr. Eng., vol. 57, no. 6, pp. 982–984, 2008.

[18] T. Y. Wang, L. H. Liu, and S. Y. Peng, "A power-efficient highly linear reconfigurable biopotential sensing amplifier using gate-balanced pseudo-resistors," IEEE Trans. Circ. and Syst., vol. 62, no. 2, pp. 199–203, 2015.

[19] K. A. Ng and Y. P. Xu, "A low-power, high CMRR neural amplifier system employing CMOS Inverter-Based OTAs with CMFB through supply rails," IEEE J. Solid-State Circ., vol. 51, no. 3, pp. 724–737, 2016.

[20] A. Wash, "Front-end amplifier and RC filter design for a precision SAR analog-to-digital converter," Analog Dial., vol. 46, no. 4, pp. 1–5, 2012.

[21] F. Khateb, S. Dabbous, and S. Vlassis, "A survey of non-conventional techniques for low-voltage, low-power analog circuit design," Radioeng., vol. 22, no. 2, pp. 415–427, 2013.

[22] F. Khateb, S. Vlassis, M. Kumngern, C. Psychalinos, T. Kulej R. Vrba, L. Fujcik, "1 V Rectifier based on bulk-driven quasi-floating-gate differential difference amplifiers," Circ., Syst. and Signal Process., vol. 34, no. 7, pp. 2077–2089, 2015.

[23] M. Kumngern and F. Khateb, "0.5 V fully differential current conveyor using bulk-driven quasi-floating-gate technique," IET Circ., Dev. and Syst., vol. 10, no. 1, pp. 78–86, 2016.

[24] F. P. Cortes, E. Fabris, and S. Bampi, "Analysis and design of amplifiers and comparators in CMOS 0.35 µm technology," Microelectron. Reliab., vol. 44, no. 4, pp. 657–664, 2004.

[25] R. Gupta and S. Jain, "A Miller compensated gain & phase enhanced two-stage differential OTA's using positive feedback at differential stage," Int. J. Technol. Enhanc. Emerg. Eng. Res., vol. 3, no. 04, pp. 148–154, 2015.

[26] K. Sharma and R. Sharma, "Highly consistent bulk driven quasi floating gate (BDQFG) PMOS pseudo-resistor design and implementation in 0.18 micron meter technology," IEEE Reg. 10 Annu. Int. Conf. Proc., pp. 488–493, 2017.

[27] H. Kassiri, K. Abedelhalim, and R. Genov, "Low-distortion Super-GΩ subthreshold-MOS resistor for CMOS neural amplifiers," IEEE Biomed. Circ. and Syst. Conf., pp. 270-273, 2013.

[28] E. Özalevli and P. E. Hasler, "Tunable highly linear floating-gate CMOS resistor using common-mode linearization technique," IEEE Trans. Circ. Syst. I Regul. Pap., vol. 55, no. 4, pp. 999–1010, 2008.

[29] R. R. Harrison and C. Charles, "A low-power low-noise cmos amplifier for neural recording applications," IEEE J. Solid-State Circ., vol. 38, no. 6, pp. 958–965, 2003.

[30] L. Liu, X. Zou, W. L. Goh, R. Ramamoorthy, G. Dawe, and M. Je, "800 nW 43 nV/Hz$^{1/2}$ neural recording amplifier with enhanced noise efficiency factor," Electron. Lett., vol. 48, no. 9, pp. 479, 2012.

[31] S. A. Gallegos and H. F. Huq, "A 129 nW neural amplifier and g$_m$-C filter for EEG using g$_m$/i$_d$ methodology and a current reference without resistance," IEEE 57th International Midwest Symp. Circ. and Syst., pp. 876–880, 2014.

[32] J. Zhang, H. Zhang, Q. Sun, and R. Zhang, "A low-noise and low-power amplifier with current-reused OTA for ECG recordings," IEEE Trans. Biomed Circ. and Syst., vol. 12, no. 3, pp. 700–708, 2018.

[33] F. Khateb, "Bulk-driven floating gate and bulk-driven quasi-floating gate techniques for low-voltage, low-power analog circuits design," Int. J. Electron. Commun., vol. 68, no. 1, pp. 64–72, 2014.

Synthesis of PPG Waveform Using PSPICE and Simulink Model

Ankita Mukherjea
Electronics & Instrumentation
Engineering Department
Techno Main, Saltlake
Kolkata, India

amukherjea310@gmail.com

Parshati Chaudhury
Electronics & Instrumentation
Engineering Department
Techno Main, Saltlake
Kolkata, India

chaudhuryparshati@gmail.com

Alvin Karkun
Electronics & Instrumentation
Engineering Department
Techno Main, Saltlake
Kolkata, India

alvin.karkun@gmail.com

Soumalya Ghosh
Electronics & Instrumentation
Engineering Department
Techno Main, Saltlake
Kolkata, India

soumalya468@gmail.com

Subhajit Bhowmick
Electronics & Instrumentation
Engineering Department
Techno Main, Saltlake
Kolkata, India

subhajitsan@yahoo.co.in

Abstract— Volumetric changes in the microvascular tissue can be studied by using an optically obtained plethysmogram; a photoplethysmogram (PPG).This has gained much intrigue in the modelling of PPG signal using appropriate Synthesis techniques. The obtained circuit is used to reconstruct PPG waveform. The current work is a proposal of a method for the development of an equivalent circuit to simulate the PPG waveform. Based on the amount of energy, the PPG signal was segregated into 'complex' and 'plain' zones. To construct that, the individual waves were modelled using Fourier analysis method by MATLAB Curve Fitting Tool Box. Then, after generating the Fourier series coefficients from MATLAB, we obtained the sine and cosine components of the data along with the DC component. The sine and cosine components along with the DC component are added via an adder circuit, followed by an inverter, which generates the final PPG waveform using PSPICE. The building blocks of the Simulink model have been developed using sine and cos function generator blocks, and adder blocks thereby generating an appropriate waveform. The errors between the actual normal PPG waveform and reconstructed PPG waveform using PSPICE for one cycle have been computed, which results in deviations within acceptable ranges of deviation.

Keywords—PPG Wave, Fourier Model, MATLAB Curve Fitting GUI Toolbox, PSIPCE, MATLAB Simulink

I. INTRODUCTION

Photoplethysmography (PPG) is a non-complex, inexpensive and non-intrusive method, measuring changes in blood volume. In this technique, a infrared light emitting source and a photodetector are used, which measures the intensity of transmitted or reflected light which is dependent on whether the mode is transmission or refection. The PPG waveform comprises an AC part and a DC part. The AC is a result of absorption of light by the blood flowing via arteries, while the DC is a result of the absorption of light by blood flowing through veins[1]. In the recent years, PPG has gained much intrigue in the biomedical research community due to its user-friendly and non- invasive nature. In the last ten years PPG has been used for cardiovascular monitoring and also in the monitoring of blood pressure.

Early signs of atherosclerosis, can be assessed using PPG [2]. Blood pressure, which provides considerable amount of information about the general health of a person, can also be measured following a non-invasive technique using PPG [3-4]. Sensors which detect blood pressure using the PPG recording techniques are becoming very common for pulmonary and cardiac function assessment in health monitoring[5-8]. In continuous monitoring, different disorders related to health, including study of vital signs of the patient can be decoded by a medical expert, which reduces the physical examination and frequency of patient visit to the hospital [9- 10]. At the same time, mobile wireless sensors make the patient more comfortable as it allows him to lead a normal life [11].This paper aims to mathematically model and implement a PPG synthesizer that is capable of producing PPG morphology. In the future stages of this work, this system will be made real-time by using a PPG sensing module which will be used directly on a test subject. The data hence obtained will be compressed, transmitted and recompressed at the receiving end for effective analysis.

II. METHODOLOGY

A PPG dataset was obtained from a diagnostic database under Physionet with a sampling frequency of 250Hz[12]. After obtaining the whole dataset in .txt format, MATLAB programming was used on it to obtain the full PPG wave. Selecting Fourier type model under function fitting, in the Curve Fitting Toolbox window, the coefficients were determined[13-14]. The plethysmograph signal was studied using principal component analysis with the help of Fourier series that represents a function as the sum of simple sinusoidal waves and decomposes it into a sum of infinite set of simple oscillating function. By decomposing the signal using this method, regeneration has been made possible, thus, preserving in the process, the functional relationships between the components. The relative contributions of each specific component to the signal were also found.

The frequency domain representation of this wave can be expressed by:

$$y(t) = C_0 + \sum_{n=1}^{n} \alpha_n \sin(n\omega_0 t) + \beta_n \cos(n\omega_0 t) \quad (1)$$

Where y(t) = The instantaneous value of PPG signal.

C_0 = Average value of PPG DC component.

$\omega_0 = \dfrac{2\Pi}{T}$ [ω_0 is angular frequency of fundamental component].

T is the time period of PPG interval wave.

$$\alpha_n = \frac{2}{T} \int_0^T c(t) \sin(n\omega_0 t) dt \quad (2)$$

$$\beta_n = \frac{2}{T} \int_0^T c(t) \cos(n\omega_0 t) dt \quad (3)$$

Assuming $\alpha_n = C_n \cos\theta_n$ and $\beta_n = C_n \sin\theta_n$, the above equation (1) may be written as

$$x(t) = C_0 + \sum_{n=1}^{n} C_n \sin(n\omega_0 t + \theta_n) \quad (4)$$

Where $C_n = [\alpha_n^2 + \beta_n^2]^{\frac{1}{2}} \quad (5)$

$\theta_n = \tan^{-1}\left[\dfrac{\beta_n}{\alpha_n}\right] \quad (6)$

Upon generation of all the sixth order Fourier co efficient, the PPG Synthesizer circuit was developed using PSPICE and later, using Simulink Toolbox in MATLAB. The output from the Synthesizer was superimposed on the available dataset and accordingly, the performance indices were calculated.

III. SYNTHESIS OF PPG WAVEFORM USING PSPICE

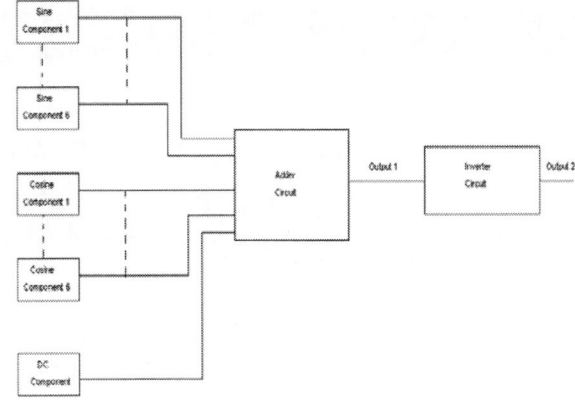

Fig.1 Block Diagram of a PPG Synthesizer Circuit

The circuit made here consists of three broad Sections:

1. **The Voltage sources**: It consists of the 6 Cosine components and 6 Sine components along with the DC component. The frequency of the AC sources is calculated at first by using the Fourier series and the amplitudes are taken from the coefficients of the MATLAB.
2. **The Adder circuit**: It consists of an IC741 which is used as an adder circuit. The different AC sources are combined together along with the DC component following the principle of superposition theorem.
3. **The Inverter Circuit**: Another IC741 is used as an inverting op-amp which is used to get the original PPG signal which gets inverted in the adder circuit.

Using the 6th order Fourier series equation:

x(t) = C_0 + (α_1 cosω_0 t) +(β_1 sinω_0t) + (α_2 cosω_0t) + (β_2 sinω_0t) +(α_3 cosω_0t) +(β_3 sinω_0t) +(α_4 cosω_0t) +(β_4 sinω_0t) +(α_5 cosω_0t) +(β_5 sinω_0t) +(α_6 cosω_0t) +(β_6 sinω_0t)

There are twelve voltage controlled switches, which are used to construct the PPG wave[15].

TABLE I

C_0	ω_0	nth value of coefficient		
		n	α_n	β_n
0.7799	8.193	1	-0.3797	0.4191
		2	-0.3478	0.002181
		3	-0.04357	-0.0851
		4	-0.02877	-0.04811
		5	0.01195	- 0.02345
		6	0.004962	0.002356

According to Table I, the controlling voltage of the switch is given. The VAMPL of sine source is given by α_n coefficient and β_n coefficient of 6th order Fourier series.

The FREQ is calculated from ω_0 ($\omega_0 = 2\pi f$, or $f = \dfrac{\omega_0}{2\pi}$) of Table I.

Along with the twelve voltage sources (six sine and six cosine) a DC Source is added along with it which is the C_0 of Table-1. After passing the adder output via an inverting amplifier, we get the desired PPG Waveform output (Fig.2), in which $A_1 \ldots A_6 = \alpha_1 \ldots \alpha_6$ and $B_1 \ldots B_6 = \beta_1 \ldots \beta_6$.

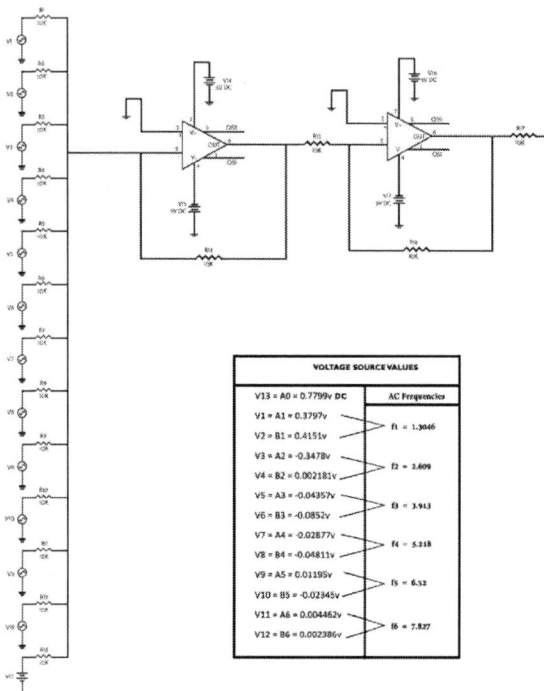

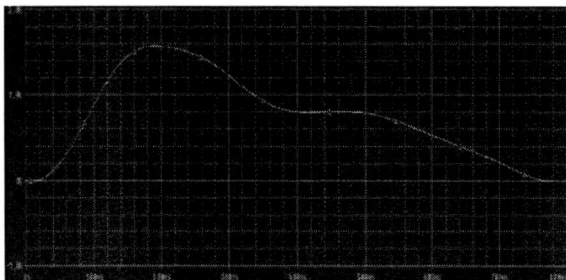

FIG. 2 PPG WAVE SYNTHESIZER CIRCUIT

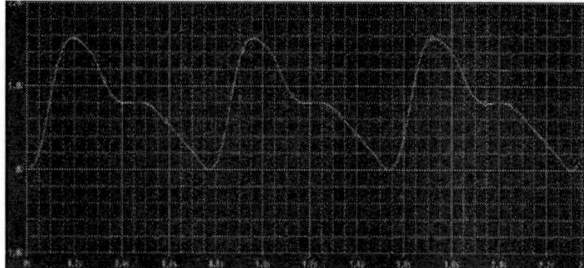

Fig.3(b) Reconstructed PPG waveform obtained from PPG Wave Synthesizer Circuit for 3 Cycles

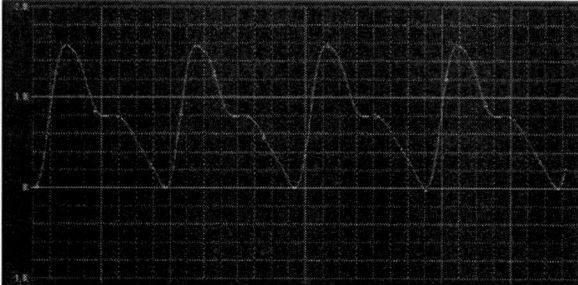

Fig. 3(c) Reconstructed PPG waveform obtained from PPG Wave Synthesizer Circuit for 4Cycles

From the PPG waveforms obtained using PSPICE, we find that the time period of 1 cycle, 3 cycles and 4 cycles are 0.8seconds, 2.4seconds and 3.2seconds respectively (Fig.3(a), Fig.3(b), Fig.3(c)).

IV. SIMULATION USING SIMULINK MODEL

A typical PPG data was obtained from MIMIC II database under Physiobank ATM having sampling frequency of 250 Hz. The modelling method truncated one beat from the entire sample. The beat was fragmented in two sections. The two sections of the beat were systole and diastole represented as AC and DC components respectively. The coefficients were thus obtained by MATLAB "CURVE FITTING GUI-TOOL BOX", selecting sixth Fourier type model.

The Simulink Circuit Diagram is shown below(Fig.4), which includes sine and cosine generator blocks for systole and a DC voltage for diastole, MATLAB function blocks for systole and diastole segment waveforms according to Sixth order Fourier model, Adder block and Inverter Block. The systole and diastole waveforms are generated by Simulink Model[16] which is superimposed to produce one cycle of PPG Waveform (Fig. 3(a)).

Fig.3(a) Reconstructed PPG waveform obtained from PPG Wave Synthesizer Circuit for 1 Cycle

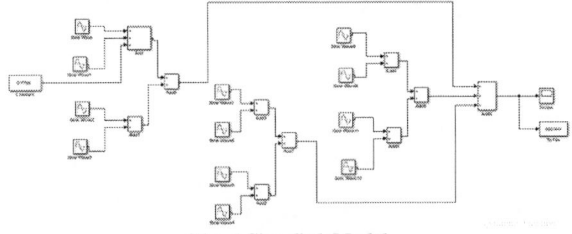

Fig. 4 Simulink Model

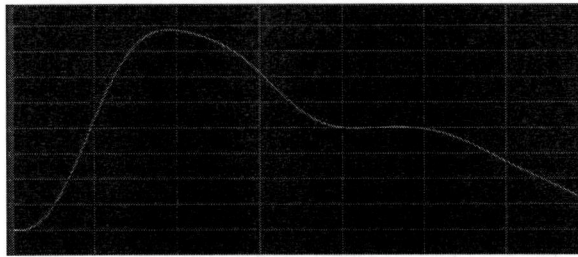

Fig. 5 Reconstructed PPG waveform obtained from Simulink Model

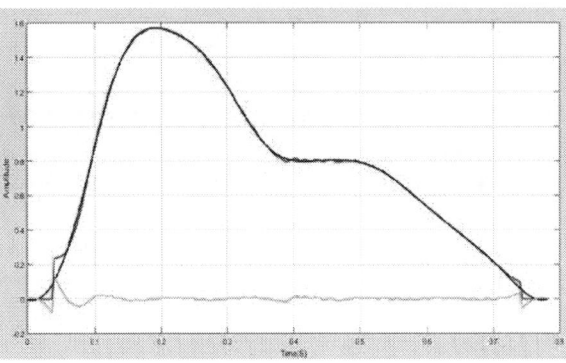

Fig. 6 Comparison between actual and reconstructed PPG waveform

V. RESULT SET

The performance of reconstructed PPG waveform are denoted as :i) % RMS difference (PRD), ii) Normalized % RMS difference (PRDN), iii) Max error (E_{max}) iv) RMS error (RMSE) and SNR. These can be expressed as:

$$PRD = 100 \times \sqrt{\frac{\sum_{n=1}^{N}(x[n] - \hat{x}[n])^2}{\sum_{n=1}^{N}(x[n])^2}}$$

$$PRDN = 100 \times \sqrt{\frac{\sum_{n=1}^{N}(x[n] - \hat{x}[n])^2}{\sum_{n=1}^{N}(x[n] - \bar{x})^2}}$$

$$E\max = \max(x[n] - \hat{x}[n])$$

$$RMSE = \sqrt{\frac{\sum_{n=1}^{N}(x[n] - \hat{x}[n])^2}{N}}$$

Signal to noise ratio denoted by SNR.

Where N = total number of samples in waveform dataset
$x[n]$ = Actual value of n-th sample
$\hat{x}[n]$ = Corresponding reconstructed value of n-th sample

TABLE II

PRD	PRDN	Emax	RMSE	SNR
4.5429e-04	0.1752	0.1277	0.0048	0.0016

VI. DISCUSSION

The performance of the designed electronic circuit is simulated using PSPICE circuit simulator tool. The PPG wave of a test subject is generated by a function generator circuit and a sixth order harmonic oscillator circuit. The parameter of harmonic oscillator has been determined by Fourier modeling method using MATLAB Curve Fitting Toolbox. On increasing the order greater than six the accuracy increases and thereby the number of oscillators also increase, thus increasing the complexity of the circuit. As per current findings, a waveform of sixth order Fourier model has been generated. For future scope, the above methods and mathematical models for real time data acquisition implementation and a sensor based PPG detection module for hardware implementation.

REFERENCES

[1] H.T. Wu, C.H. Lee, T.C. Wu and A.B. Liu, "Endothelium Function Assessment with Radial Pulse Wave Signals", Proc. 31st Annual International Conference of the IEEE EMBS, Minneapolis, Minnesota, USA, September 2-6, 2009, pp.3035-3038.

[2] K.W. Chan, K. Hung and Y. T. Zhang, "Noninvasive and cuffless measurement of blood pressure for telemedicine", Proc. 23rd Annual International Conference of the IEEE EMBS, Istanbul, Turkey, October 25-28, 2001, vol. 4, pp.3592-3593.

[3] I.C. Jeong, J.I. Ko, S.O. Hwang and H.R. Yoon, "A New Method to estimate Arterial Blood Pressure using Photoplethysmographic

Signal", Proc. Annual International Conference of the IEEE EMBS, New York City, USA, August 30-September 3, 2006, pp.4667-4670.

[4] Reyes, H. Nazeran, M. Franco and E. Haltiwanger, "Wireless Photoplethysmographic Device for Heart Rate Variability Signal Analysis", 34th Annual International Conference of the IEEE EMBS, San Diego, California USA, August 28-September 1, 2012, pp. 2092-2095.

[5] B.H. Yang, S. Rhee and H.H. Asada, "A Twenty-Four Hour Tele-Nursing System Using a Ring Sensor", Proc. IEEE International Conference on Robotics & Automation, Leuven, Belgium, May 16-20, 1998, pp.387-392.

[6] K. Davoudi, M. Shayegannia, B. Kaminska, "Vital Signs Monitoring using a New Flexible Polymer Integrated PPG Sensor", Proc. Computing in Cardiolology, September 22-25, 2013, vol. 40, pp. 265-268.

[7] E.M. Lee, J.Y. Shin, J.H. Hong, E.J. Cha and T.S. Lee, "Glass-type Wireless PPG Measuring System", Proc. 32nd Annual International Conference of the IEEE EMBS, Buenos Aires, Argentina, August 31 - September 4, 2010, pp.1433-1436.

[8] N. J. B.N. Mazlan and K.I. Wong, "A Wireless ECG Sensor and a Low- Complexity Screening Algorithm for Obstructive Sleep Apnea Detection", Proc. 2012 IEEE EMBS International Conference on Biomedical Engineering and Sciences, Langkawi, December 17-19, 2012, pp.279-283.

[9] G. Angiusand L. Raffo, "Cardiovascular Disease and Sleep Apnoea: a Wearable Device for PPG Acquisition and Research Aims", Proc. Computing in Cardiolology, September 9-12, 2012, vol. 39, pp. 513-516.

[10] A.C. Sobusiak, A. Hulewicz, M. Boltrukiewicz and D. Prokop, " Utilization of Human-Computer Interaction in Wireless Transmission of Arterial Pulse Waveforms", Proc. IEEE International ConferenceonVirtual Environments, Human-Computer Interfaces and Measurement Systems, Giardini Naxos, Italy, July 18-20, 2005, pp. 116-120.

[11] O. Postolache, J. Freire, P.S. Girão and J. M. D. Pereira, "Smart Sensor Architecture for Vital Signs and Motor Activity Monitoring of Wheelchair' Users", Proc. 2012 6th International Conference on Sensing Technology (ICST), December 18-21, 2012, pp.167-172.

[12] www.physionet.org

[13] S. C. Bera and R. Sarkar, "Fourier Analysis of Normal ECG Signal to Find it Maximum Harmonic Content by Signal Reconstruction", Journal of Sensors and Transducers, Vol. 123, pp. 106-107, Issue. 12, December 2010.

[14] S.C. Bera, R. Sarkar and Nirupama Mondal "A Review Work on Reconstruction of ECG Wave from Fourier Harmonic Components", IEEE Region 10 Colloquium and Third ICIIS, Kharagpur, India, Paper No: 400,2008.

[15] S.Bhowmick, P.K.Kundu, G.Sarkar, "Design and Simulation of Equivalent Circuit of Heart using PSPICE", Devices for Integrated Circuit, 2017

[16] S.Bhowmick, P.K.Kundu, G.Sarkar, "Synthesis of ECG Waveform using Simulink Model", International Conference on Intelligent Control Power and Instrumentation(ICICPI), 2016

A Hypothetical Analysis to Study the Variations of Complex Dielectric Permittivity for Detection of Various Stages of Cancer of a Biological Target using Microwave Tomography

Deborsi Basu
M.Tech in Department of Electronics and Communication Engineering, IEEE student member,
Kalyani Government Engineering College, Kalyani, Nadia
West Bengal, India
deborsibasu2015@gmail.com

Kabita Purkait
Associate professor in Department of Electronics and Communication Engineering,
Kalyani Government Engineering College, Kalyani, Nadia
West Bengal, India
kp_lectrpk@yahoo.co.in

Abstract - In recent days, different kinds of Cancerous tumor detection in human body using Microwave Tomography technique (MWT) is picking up huge interest of research among modern day scientists. As all other existing methods like ultrasound, mammography, magnetic resonance imaging (MRI), X- ray imaging etc. are coming up with certain difficulties to patients so Microwave imaging technique is a suitable substitute for that. In this paper, a hypothetical assumption based on the variations of complex dielectric permittivity of the water content of cells with different stages of cancer has been implemented. The complex dielectric permittivities are reconstructed with the help of suitable reconstruction techniques and images have been formed accordingly. Analysing the reconstructed values, a comparative study has been done between a normal cell and different kinds of affected cells. The new approach of this reconstruction algorithm shows a very much efficient way of discriminating all possible kinds of dielectric perturbations for various stages of cancer inside human body. Through this analysis it is shown that the improved version of reconstruction algorithm works well on the hypothetical assumption and the result found is pretty satisfactory and the margin of error is also less, based on the result it is proposed that microwave imaging technique is one of the best methods for detection of the presence cancer cell inside human body.

Keywords - Microwave Tomography Technique (MWT), Cancerous tumor, Complex dielectric permittivity, Reconstruction algorithm, Dielectric perturbations.

I. INTRODUCTION

Microwave imaging of human body using Tomography has been of interest for a number of years. Tomography is a technique where the sectioning or sectoring of a material, like any live or dead objects have been done by the penetration of a wave ranging in the suitable frequency range. It can also be applied in the various other technologies of biological as well as industrial applications like geophysics, material science, astrophysics, radiology, archaeology and others.

Reconstructing the internal distribution of an object based on the parameters of its physical architecture and represents the equivalent structure through color gradation gray scale model is done using topographic imaging. It provides multipoint measurements based on the material property and provide important information for controlling and optimization of the reconstructed images. The idea of microwave imaging technique was first brought in to limelight in 1970s [1] [2]. The main concept of MTI (Microwave Tomography Imaging) was to develop the image of an object that is kept inside an experimental setup. Based on the scattering of electromagnetic fields that was generated from a suitable source the reconstructed image has been formed. Some more developments had been seen in early 1980s [3] [4]. Bolomey and Jofre [5] analyzed about different challenges and obstructions in its implementation and suggested some suitable tomographic imaging techniques and algorithmic systems. Gradually significant developments have been shown in various tomography instruments based on X-ray technique, ultrasound, magnetic resonance imaging (MRI) techniques, c ray ,electromagnetic induction, resistance and capacitance based on electrical property and microwaves [6] [7] [8]. So, within these mentioned techniques, X-ray and MRI have been extensively used in the medical sector. Though they produce high resolution images with large contrasts in density, but they are bulky, costly and can damage body organs. Other techniques like c-ray imaging, positron emissions contain radioactive emissions that are equally dangerous for human cells. Therefore all these techniques are not appropriate for biomedical applications.

In recent days the main focus of microwave tomography has been put on biomedical applications [9] but no significant amount of work has been done so far [10]. Gradually it is picking up the scale of rising applications. So, some new applications with suitable algorithms in microwave tomographic imaging have been depicted through this work, considering its increasing demands in the field of biomedical image processing.

In this work the reconstruction algorithms have been applied with a new concept. The modified Simultaneous Iterative Reconstruction Algorithms (SIRT) based on Moment method [11] [12] which was used to detect the variations of dielectric perturbations for a particular type of cell region has been explored more into a modified Exact reconstruction algorithms [13] [14]. Now this modified form of algorithms have been used here with newly defined concept. The variations of cell complex dielectric constants with different stages of cancer has been reconstructed iteratively with the above mentioned algorithms and later compared with the initial assuming values to check the level of errors. All kinds of perturbations have been examined through this analysis and in every case it is found that the newly defined approach works very well with minimum errors.

II. DIELECTRIC PROPERTIES OF HUMAN TISSUES

The human-body can be considered as a composite volume conductor which comprises tissues with different electrical properties. Electric conduction within biological tissues occurs through movement of mobile ions. Such conduction is related to the ionic content and ionic mobility of each particular tissue type. Functional activity, as well as pathological conditions, results in structural changes in the tissue. These changes are accompanied by characteristic temporal and spectral electric behaviour. Knowledge regarding the spatial dielectric properties distribution can provide useful information regarding the functional and pathological condition of the tissues [15].

In microwave imaging, based on the differences of complex permittivity the dielectric properties of the material have been reconstructed. That is defined by the equation:

$$\varepsilon^* = \varepsilon' - j\varepsilon''$$

where ε' is the relative permittivity that demonstrates the effects of polarization of charged particles in the tissue and ε'' shows the out-of-phase losses because of the displacement currents generated due to the application of electromagnetic field.

III. DIELCTRIC PROPERTY OF CANCER CELL

Proper understanding of the different pathological conditions inside human body can be analysed easily using the dielectric properties of respective organ tissues. Electrical properties of those tissues are equally essential for this purpose. Information about tissue structure and composition can be obtained by measuring these dielectric properties.

A. Capacitance relaxation phenomenon of cancer cell

Capacitors are composed of two conducting sheets separated by a thin layer of insulating material known as a dielectric. Human body cell is very much analogical with the structure of a biological capacitor, as the cell membrane behaves like an insulating material and the charged particles that store both of the sides of the membrane behave like conducting metal plate. Capacitors lead to the phenomenon of Dielectric relaxation due to momentary delay (or lag) in the dielectric constant of a material. Dielectric relaxation results in charge storage capacity reduction under dynamic random access memory operating conditions [16] [19].

The electrical conductivity and permittivity of the materials inside a human tissue change due to any condition

of illness or affection of diseases. Cancer affected human body cells demonstrate higher permittivity related to the resistance to the formation of an electrical field. As a result Cancer cells resonate differently from normal cells. The initial stage of cancer affected correlation corresponds to Capacitance relaxation from 0.1pu to 0.5 pu, the second stage correlation corresponds to Capacitance relaxation from 0.1pu to 0.7 pu and similarly the last stage of cancer affected correlation corresponds to Capacitance relaxation from 0.1pu to 0.9 pu [17]. Similarly it had been shown that cell conductivity and relaxation frequency can be used as specific electrical signature for identification of cancer cells and differentiation of the pathological stages of malignant cells. The electrical conductivity and permittivity of cancerous cells has been found to be greater than those of normal cells [18]. The cancerous cells demonstrate greater permittivity and larger surface area for which there is large increase in capacitance components due to the Capacitance Relaxation phenomenon.

B. Assumed complex dielectric variations for different stages of cancer

From the above analysis it is clear that the increment in dielectrics of the water content in a cancer affected cell happens with the up gradation of cancerous stages. Here some assumed complex dielectric variations have been considered for different stages of cancer. The deviation of the complex dielectric in the cancer affected cells with respect to the corresponding cancer stages is drafted in a tabular form below.

TABLE 1 - DIELECTRIC VARIATIONS FOR DIFFERENT STAGES OF CANCER

Error Level – Perturbation of Dielectric in percentage with respect to the normal one (%)	Corresponding stage of cancer
0-4	Normal healthy cell
5-10	Stage 0 – Initial stage
11-20	Stage 1 - Intermediate stage
21-30	Stage 2- Semi-Advanced stage
31-40	Stage 3- Advanced Stage
41-50	Stage 4 – Critical stage

IV. EXPERIMENTAL MODEL FORMATION

 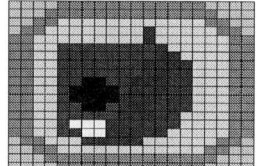

Fig 1.a   Fig 1.b

Fig 1.a *The approximated cross section of human body model (Normal condition)*

Fig 1.b *The approximated cross section of human body model (Affected condition in Liver region)*

The above model is constructed based on a realistic section of a Human body with some specific portions mentioned. The total model is of dimension 20 cm * 20 cm (Length = 20 cm, Breadth = 20cm). There are total 400 numbers of cells. The grid view depicts the individual cell specifically. The dimensions of all cells are same that is 1cm x 1cm.

A. Section Description in of the Model:

Region Specification	Name of the Region	No of cell present	Complex Permittivity
	Fat Region	60	25-j5
	Muscle Region	56	50-j23
	Muscle Type Region	123	60-j18
	Liver Region	60	46-j10
	Stomach Region	57	60-j18
	Pancreas Region	5	65-j30
	Matching medium	40	76-j40

B. Positioning of the transmitters and receivers with the model

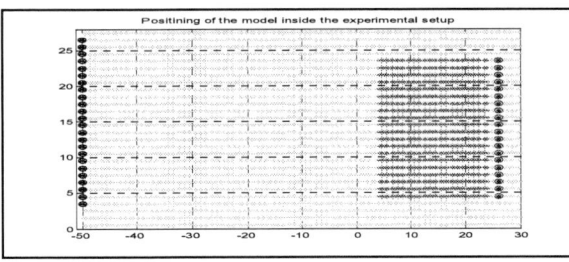

Fig 2: *Positioning of the model inside the experimental setup*

The transmitters are placed approximately 50cm apart from the model along with the matching medium. The number of transmitting antennas can be varied as per the requirement. The transmitters are placed along positive Y axis. Total 24 different numbers of positions are shown in Black Pointed region where the 24 transmitters are placed. Based on the requirements, the transmitting antenna arrangement can be moved to scan the model in various directions. A 15x15 quarter wave Dipole Array antennas with a separation of half wavelength, are placed as transmitter.

The receivers are placed in the Matching medium after the end of the Model. As per the 2-D model diagram the receivers are also placed through the positive Y axis vertically. Total 20 different numbers of positions are shown in Blue Pointed region where the 20 numbers of receivers are placed.

Single half-wave dipole antenna is considered as a receiver. So, 20 such antennas are placed in those marked positions. In the receiver locations electric fields are simulated as experimental data using Exact algorithm. The entire set up is immersed inside the matching medium (saline water in this case). Here the light blue '*' marked region is the matching medium. The model and the receiver both are placed in the near field region of the transmitting antenna.

V. AN EXACT ALGORITHM FOR IMAGE RECONSTRUCTION

Geometrical ray tracing is successful in some cases (Geo-tomography), but it will fail in case of medical tomography using microwaves. The resolution required for medical images will be of the order of millimeter which is only a fraction of the propagating wave in medium. Therefore, the received field will be determined not only by the cells along the ray path but also other cells contained within the beamwidth of the transmitter radiation pattern.

The electromagnetic fields induced in an inhomogeneous biological medium are obtained by applying the method of moment solution developed by Richmond [11]. According to the Richmond, if an inhomogeneous dielectric cylinder of any cross-section with its axis placed along z-axis is illuminated by electromagnetic wave, then only z-component of electric field will exist and permittivity of the medium is expressed as a function of transverse coordinates (i.e $\varepsilon = \varepsilon(x,y)$).The scattered field E^S is the difference between the total field (E) incident by the source in presence of dielectric cylinder and incident field is [E^i] in absence of dielectric cylinder .

$$E^S = E - E^i \ldots\ldots\ldots (1)$$

The dielectric cylinder is divided into large number of small cells so that the electric field is essentially constant over each cell. In matrix notation, the system of linear equations of electric fields can be presented as

$$[C] . [E] = [E^i] \quad\ldots\ldots\ldots(2)$$

where [E^i] is the incident field matrix at different pixels in vacuum. [E] is the total internal field in different cells in homogeneous medium and [C] is the co-efficient matrix corresponding to assumed base line homogeneous medium as described below.

Now, the system of linear equations can be represented in a compact form as shown below:

$$\sum_{n=1}^{N} C_{mn} E_n = E_m^i \quad ; \quad m= 1,2, \ldots\ldots N \quad \ldots\ldots\ldots(3)$$

where $C_{mn}= 1+ (\varepsilon_m -1)(j/2) [\pi k a_m H_1^{(2)}(k a_m) -2j]$ if n = m and
$C_{mn}=(j \pi k a_n /2)(\varepsilon_n -1) [J_1(k a_n)H_0^{(2)}(k \rho_{mn})]$, if n $\neq$ m

a_n and a_m are the radius of equivalent circular cell which has the same cross sectional area as cell n and m respectively. $H_1^{(2)}$, $H_0^{(2)}$ is the Hankel function of type two, order 1 and 0 respectively. J_1 is a Bessel Function.

These N equations are solved to determine the total electric field at the centre of each cell. For tomographic imaging a number of transmitter positions are selected and electric fields at different receiver locations corresponding to each transmitter positions are measured.

The scattered field of the dielectric cylinder can be calculated as the summation of the contributions from each one of n number of cells of the cylinder. The fields received at different receivers can be processed to produce tomographic images by the method of perturbation by Datta & Bandyopadhyay [12] [15] [20].

The permittivity values of the cells are perturbed due to the presence of inhomogeneous biological targets simultaneously by small amounts of $\Delta\varepsilon_i$ (i = 1, 2 ,...., n) and the corresponding changes in the internal fields are ΔE_i s then

$$[C^1] . [E + \Delta E] = [E^i] \quad ....\ (4)$$

Where $[C^1]$ is the co-efficient matrix corresponding to inhomogeneous medium. According to Exact Algorithm [14] [21] the change in electric field in different cells are obtained by subtracting equation (2) from equation (4) & is given by

$$\Delta E_i = - x_i E_i^1 + \sum_{j=1}^{N} x_j E_j^1 \frac{M_{ji}(0)}{\Delta(0)} \quad ....\ (5)$$

where E_i^1 ($= E_i + \Delta E_i$) is the modified field in the ith cell under perturbed condition. $\Delta(0)$ and $M_{ji}(0)$ are the determinant and cofactor of (j , i)th element of unperturbed coefficient matrix $[C]$ respectively.

If $E_{Rml}(k)$ denotes the scattered field at the lth receiver location for the kth beam in the numerical model and $E_{Rol}(k)$ denotes the calculated scattered field intensity at the same receiver location for the same beam with an assumed uniform baseline permittivity for the object.

Therefore, the total change in the scattered field at the l th receiver position corresponding to k th beam is

$$E_{Rml}(k) - E_{Rol}(k) = \sum_{i=1}^{n} x_i \sum_{j=1}^{n} E_j^1 \frac{M_{ji}(0)}{\Delta(0)} \quad ....(6)$$

The values of x_is i.e, the requisite fractional changes in permittivity values for different cells from the assumed initial permittivity for the medium can be determined from the above equation.

A. Algorithm based on Method I:

In this project one new approach has been considered on Exact Algorithm where normal standard values of complex permittivity of each cell is taken as the trial initial values for starting reconstruction. This is done because the main objective is to detect the affected zone at early stage.

The required values of x_i^1 from the normal value are obtained by the equn.

$$x_i^1 = \frac{\varepsilon_{i,cal} - \varepsilon_{i,nor}}{\varepsilon_{i,nor} - 1} \quad ....(7)$$

where $\varepsilon_{i,cal}$ is the complex permittivity of ith cell calculated from x_i obtained from eqn (6) using $E_i^1 = E_{i,nor}$ ($E_{i,nor}$ is the total field at ith cell for normal condition of complex permittivity). Internal fields are modified each time by the equn (5).

B. Algorithm based on Method II:

After successful detection of the affected region using Method I, in this method reconstruction has been done for the affected region only assuming other organs are in their normal states using equation (6).

While analyzing the complex dielectrics using this reconstruction algorithm with Method II an important fact is observed that the accuracy in the result is very high. This is achieved because the reconstruction is applied to the affected region only leaving the other unaffected regions.

VI. PROCESSING OF THE MODEL USING THE ALGORITHM

Some well defined models have been taken under consideration, processed through this algorithm and suitable reconstructed outputs have been formed. Models with both normal and diseased conditions have been made. Here the inputs and the reconstructed outputs have been shown with some colour gradation system as shown in fig. 4.

In the reconstructed output some colour variations based on the reconstructed result has been shown. The intensity of colour variation depends on how much the complex permittivity varies from its normal value.

A. Colour Gradation based Reconstructed Image Analysis:

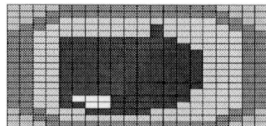 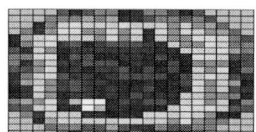

Fig: 3.1 (a)   Fig 3.1(b)

Fig 3.1(a): *Normal healthy Model with standard complex dielectrics as input,*

Fig 3.1(b): *Reconstructed image of normal model using Method I as output*

2019 Devices for Integrated Circuit (DevIC), 23-24 March, 2019, Kalyani, India

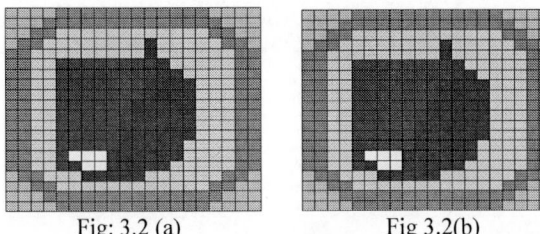

Fig: 3.2 (a)   Fig 3.2(b)

Fig.3.2 (a): *Normal healthy Model with standard complex dielectrics as input,*
Fig.3.2 (b): *Reconstructed image of normal model using Method II as output*

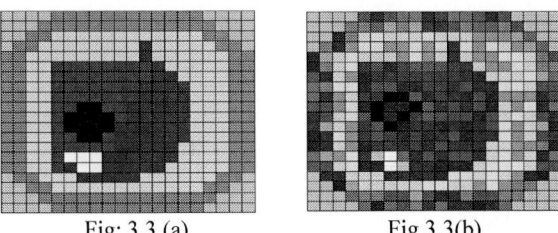

Fig: 3.3 (a)   Fig 3.3(b)

Fig 3.3(a)*: Diseased model with small (10%) perturbations as input*
Fig 3.3(b): *Reconstructed diseased model using Method I with small perturbations in dielectrics as output*

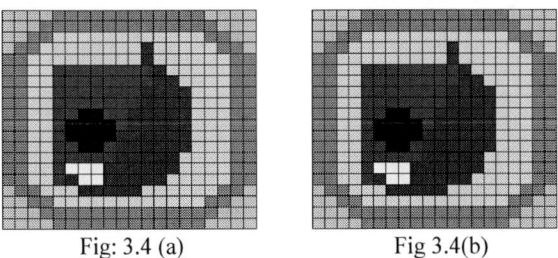

Fig: 3.4 (a)   Fig 3.4(b)

Fig 3.4(a): *Diseased model with small perturbations as input*
Fig 3.4(b): *Reconstructed diseased model using Method II with small perturbations in Dielectrics as output*

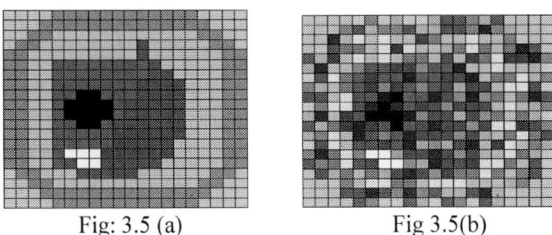

Fig: 3.5 (a)   Fig 3.5(b)

Fig 3.5(a): *Diseased model with large (50%) perturbations as input*
Fig 3.5(b): *Reconstructed diseased model using Method I with large perturbations in Dielectrics as output*

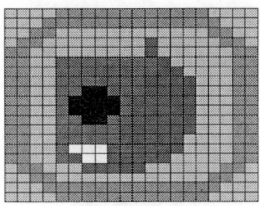

Fig: 3.6 (a)   Fig 3.6(b)

Fig 3.6(a): *Diseased model with large perturbations as input*
Fig3.6 (b): *Reconstructed diseased model using Method II with large perturbations in Dielectrics as output.*

B. *Relation between Color Gradation and Complex Dielectrics:*

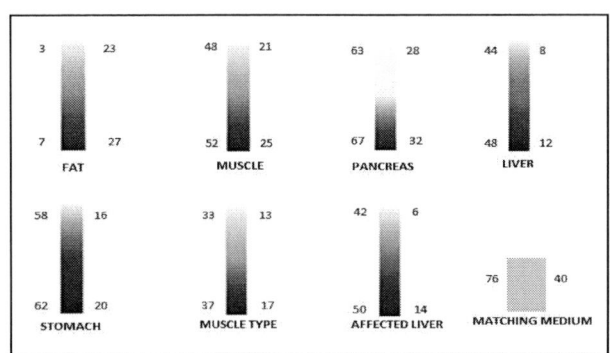

Fig.4: *Color gradation based on varying complex dielectric in different regions*

In the above normal and diseased models and their corresponding reconstructed images, the used color variations follow a particular gradation system is explained below. Each cell with a particular complex dielectric has been depicted with a particular color. The significance of the colors for a standard normal model and standard diseased model is same as explained in section IV-A. The real and imaginary parts of complex permittivity are shown using same color scale.

Every color gradation scale has two limits, the upper most limit shows the comparatively light color while the lower limit shows dark color. The light color signifies that the complex dielectric value in that region is less with respect to the standard normal value and similarly the dark color signifies higher complex dielectric value compared to the standard one. The region of matching medium (saline water) remains unchanged because no reconstruction algorithm has been applied in those regions. The gradation scale has been shown in the above figure 4.

VII.   A COMPARATIVE ANALYSIS WITH POSITIVE PERTURBATION OF COMPLEX DIELECTRIC PERMITTIVITY IN CANCEROUS CELLS

In this section the reconstructed complex dielectric permittivity values of different cells of organs in the semi human sized model have been drafted in tabular fashion.

978-1-5386-6723-1/19 $31.00 © 2019 IEEE   437

The errors found in the 5th column have been calculated using the following error calculation formula mentioned below.

Error (%) =

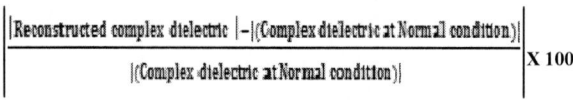

$$\frac{|\text{Reconstructed complex dielectric}| - |(\text{Complex dielectric at Normal condition})|}{|(\text{Complex dielectric at Normal condition})|} \times 100$$

Here the error signifies the deviation of complex dielectric constant value of a cancer affected cell with respect to dielectric constant of standard normal cell.

Table 2.1 -Reconstructed Dielectrics and corresponding errors for Diseased Model **(10% perturbation)** Analysis using **Method I**

Different organs of the Model	Complex Dielectric at Normal condition	Complex Dielectric at Diseased condition	Reconstructed Complex Dielectric with the proposed algorithm	Error found in the reconstructed values (%)
FAT	25-j5	25-j5	24.88968-j4.53486	0.7674
MUSCLE	50-j23	50-j23	49.86380-j22.75934	0.4071
MUSCLE TYPE	35-j15	35-j15	34.76872-j14.88708	0.6750
PANCREAS	65-j30	65-j30	64.35892-j29.88705	0.8789
LIVER	46-j10	**50.6-j11**	50.34289-j10.90528	**9.4235**
STOMACH	60-j18	60-j18	59.61635-j17.76854	0.6926
WATER	76-j40	76-j40	76-j40	0.0000

Table 2.2 -Reconstructed Dielectrics and corresponding errors for Diseased Model **(20% perturbation)** Analysis using **Method I**

Different organs of the Model	Complex Dielectric at Normal condition	Complex Dielectric at Diseased condition	Reconstructed Complex Dielectric with the proposed algorithm	Error found in the reconstructed values (%)
FAT	25-j5	25-j5	25.65748-j3.81731	1.7446
MUSCLE	50-j23	50-j23	50.47817-j22.59886	0.4899
MUSCLE TYPE	35-j15	35-j15	34.52681-j14.73704	1.4141
PANCREAS	65-j30	65-j30	64.24155-j28.23105	2.0209
LIVER	46-j10	**55.2-j12**	54.68054-j11.81957	**18.8403**
STOMACH	60-j18	60-j18	59.00829-j17.63408	1.6841
WATER	76-j40	76-j40	76-j40	0.0000

Table 2.3 -Reconstructed Dielectrics and corresponding errors for Diseased Model **(30% perturbation)** Analysis using **Method I**

Different organs of the Model	Complex Dielectric at Normal condition	Complex Dielectric at Diseased condition	Reconstructed Complex Dielectric with the proposed algorithm	Error found in the reconstructed values (%)
FAT	25-j5	25-j5	24.56575-j3.76821	2.5182
MUSCLE	50-j23	50-j23	49.67232-j22.29786	1.1599
MUSCLE TYPE	35-j15	35-j15	34.35701-j14.68418	1.8787
PANCREAS	65-j30	65-j30	64.21354-j28.23145	2.0580
LIVER	46-j10	**59.8-j13**	58.96101-j12.71510	**28.1300**
STOMACH	60-j18	60-j18	58.52692-j17.43180	2.5128
WATER	76-j40	76-j40	76-j40	0.0000

Table 2.4 -Reconstructed Dielectrics and corresponding errors for Diseased Model **(40% perturbation)** Analysis using **Method I**

Different organs of the Model	Complex Dielectric at Normal condition	Complex Dielectric at Diseased condition	Reconstructed Complex Dielectric with the proposed algorithm	Error found in the reconstructed values (%)
FAT	25-j5	25-j5	24.37191-j3.46320	3.4452
MUSCLE	50-j23	50-j23	49.45338-j22.09855	1.5809
MUSCLE TYPE	35-j15	35-j15	34.15519-j14.60405	2.4487
PANCREAS	65-j30	65-j30	63.21354-j28.23145	3.4058
LIVER	46-j10	**64.4-j14**	63.33817-j13.67475	**37.6491**
STOMACH	60-j18	60-j18	58.05680-j17.20888	3.3336
WATER	76-j40	76-j40	76-j40	0.0000

Table 2.5 -Reconstructed Dielectrics and corresponding errors for Diseased Model **(50% perturbation)** Analysis using **Method I**

Different organs of the Model	Complex Dielectric at Normal condition	Complex Dielectric at Diseased condition	Reconstructed Complex Dielectric with the proposed algorithm	Error found in the reconstructed values (%)
FAT	25-j5	25-j5	24.16331-j3.20967	4.3912
MUSCLE	50-j23	50-j23	51.82199-j21.92881	2.2427
MUSCLE TYPE	35-j15	35-j15	33.94184-j14.54840	3.0213
PANCREAS	65-j30	65-j30	64.32560-j28.21546	1.9183
LIVER	46-j10	**69-j15**	67.57407-j14.57219	**46.8471**
STOMACH	60-j18	60-j18	57.98211-j17.87874	3.1382
WATER	76-j40	76-j40	76-j40	0.0000

Table 2.6 -Reconstructed Dielectrics and corresponding errors for Diseased Model (**Mixed perturbation**) Analysis using **Method I**

Different organs of the Model	Complex Dielectric at Normal condition	Complex Dielectric at Diseased condition	Reconstructed Complex Dielectric with the proposed algorithm	Error found in the reconstructed values (%)
FAT	25-j5	25-j5	23.85823-j6.77213	2.7234
MUSCLE	50-j23	50-j23	53.43676-j21.95444	4.9687
MUSCLE TYPE	35-j15	35-j15	33.24918-j15.15035	4.0459
PANCREAS	65-j30	65-j30	63.74996-j30.48024	1.2951
LIVER	46-j10	50.6-j11	49.84553-j11.46012	8.6480
		55.2-j12	53.24825-j12.44329	16.1625
		59.8-j13	57.89548-j19.29742	26.4225
		64.4-j14	62.84843-j14.48419	37.0083
		69-j15	67.33414-j15.63046	46.8409
STOMACH	60-j18	60-j18	56.71768-j18.68981	4.6680
WATER	76-j40	76-j40	76-j40	0.0000

Table 2.7 - Reconstructed Dielectrics and corresponding errors for Diseased Model (**Mixed perturbation**) Analysis using **Method II**

Different organs of the Model	Complex Dielectric at Normal condition	Complex Dielectric at Diseased condition	Reconstructed Complex Dielectric with the proposed algorithm	Error found in the reconstructed values (%)
LIVER	46-j10	50.6-j11	50.60004-j11.00003	10.0000
		55.2-j12	55.19999-j11.99999	19.9999
		59.8-j13	59.79998-j13.00002	29.9999
		64.4-j14	64.40000-j14.00003	40.0000
		69-j15	69.00000-j15.00001	50.0000

Table 2.8 – Comparison between the errors found using **Method I** and **Method II** in the affected Liver region (Mixed Perturbation).

Different organs of the Model	Perturbations in complex dielectrics(%) at diseased organ w.r.t normal value.(error)	Error found in the reconstructed values (%) Using Method I	Error found in the reconstructed values (%) Using Method II
LIVER	10	8.6480	10.0000
	20	16.1625	19.9999
	30	26.4225	29.9999
	40	37.0083	40.0000
	50	46.8409	50.0000

From the above table (Table 2.8) it is shown that the corresponding errors have been found accurately for all the cases using **Method II**. For example, for an input perturbation of **30%** in the affected region cells Method I in Table 2.3 is giving an error of **26.6%** on an average, but in **Method II** the error is found exactly to be **30%** that is same with respect to the initial error.

So in this section the efficiency of this algorithm has been checked with some suitable cases. This Exact Algorithm with new approach is found to be very suitable in all cases of perturbations. Segregation of a normal model with a diseased one, segregation between various kinds of affected cells for different stages of cancer, differentiated error calculations have been performed nicely with the help of this algorithm. These observations will be very much helpful for medical diagnosis of cancer in future. These results also validate the new hypothetical concept which has been assumed previously.

VIII. CONCLUSION

Detection and treatment of a fatal disease like cancer remains a crucial issue for the scientists, researchers and doctors throughout the globe. In maximum cases the disease can't be cured due to late detection of its presence in human body. So any such algorithm or process in this field which have, such kind of potential to detect cancer in early stages will be a remarkable one. The cases can be defended and sometimes can be cured totally if it has been detected within correct time.

Keeping this critical issue in mind, a theoretical approach having the ability to detect small deviations of complex dielectric permittivity of a cell from its normal value that may lead to cancer has been developed. Depending upon different stages of cancer the level of perturbations varies which in turn varies the complex dielectric permittivity from a small to large extent. The Exact reconstruction algorithm is sufficient enough to approximate low to high perturbations of cell dielectric. The reconstructed images are found to be pretty satisfactory and giving the desirable output as well. All types of possible conditions for perturbations of dielectric have been observed and the result in every analysis produces very good results. Method II of the algorithm gives almost 100 percent accurate result for identification of the diseased region inside a model.

A hypothetical analysis has been made in different stages of cancerous cell detection. This assumption is extremely helpful in medical diagnosis and it restricts any misleading of treatment. The validity of this hypothesis can be understood if some experiment of medical science can provide the real life complex dielectric values at different stages of cancer affected cells. But now due to lack of such kind of practical data only a theoretical approach can be made with the present algorithm.

All the programs related to field calculations, reconstructions, model formation and complex dielectric calculations etc. have been done using FORTRAN programming and C programming with Force 2.0 and CODEBLOCS softwares respectively. All the graphical models, plots, graphs have been simulated using MATLAB programming. Colour gradation scale for model analysis has been developed using RGB percentage scaling method.

IX.   FUTURE WORKS

We will further improve this work, if real life practical diagnostic values of complex dielectric permittivity can be accumulated to compare with the output results and also study the effect of noise in the proposed algorithm for reconstruction of real life cancerous affected cases.

In future we will also like to make further extension of the used algorithms with necessary modifications and apply it onto some bigger human sized models with smaller cell size having other kinds of cell perturbations in negative direction also to make it appropriate for real life problems and will try to optimize the reconstructed images to produce more accurate outputs.

X.   REFERENCES

[1]   A. M. Cormack, " Reconstruction of densities from their projections, with applications in radiological physics ", Phys. Med. Biol., vol. 18, no. 2, pp. 195– 207, 1973.

[2]   G. N. Hounsfield, " Computerized transverse axial scanning (tomography)— part 1. Description of the system ", Br. J. Radiol., vol. 46, pp. 1016–1022, 1973.

[3]   L. E. Larsen and J. H. Jacobi, " Mcrowave Scattering Parameter Imaging on an Isolated Canine Kidney ", Medical Physics, vol. 6, pp. 394–403, 1979.

[4]   J.-C. Bolomey, A. Izadnegahdar, L. Jofre, C. Pichot, and G. Peronnet, " Microwave diffraction tomography for biomedical applications ", IEEE Trans. Microw. Theory Techn., vol. 30, no. 11, pp. 1998–2000, 1982.

[5]   J.-C. Bolomey and L. Jofre, " Three decades of active microwave imaging achievements, difficulties and future challenges ", in Proc. 2010 IEEE Int. Conf. Wireless Information Technology and Systems (ICWITS), pp. 1–4. doi: 10.1109/ ICWITS.2010.5611904.

[6]   M. S. Beck, " Process tomography: A European innovation and its applications ", Meas. Sci. Technol., vol. 7, no. 3, pp. 215–224, 1996.

[7]   Z. Wu, " Developing a microwave tomographic system for multiphase flow imaging: Advances and challenges ", Trans. Inst. Meas. Control, vol. 19, Sept. 2014. doi: 10.1177/0142331214546523.

[8]   Z. Wu and H. Wang, "Microwave Tomography for Industrial Process Imaging" , IEEE Antennas & Propagation Magazine, october 2017, 1045-9243/17©2017IEEE

[9]   L. E. Larsen and J. H. Jacobi, " Medical Applications of Microwave Imaging ", Piscataway, NJ: IEEE Press, 1986.

[10]   C. Pichot, " Inversion algorithms and measurement systems for microwave tomography of buried objects ", in Proc. 16th IEEE Instrumentation and Measurement Technology Conf., 1999, pp. 1570–1575.

[11]   J. H. Richmond, " Scattering by a Dielectric Cylinder of Arbitrary Cross Section Shape ", IEEE, Trans Antenna Propagation, 1965, pp. 334-341

[12]   A. N. Datta & B. Bandyopadhyay, " An Improved SIRT-Style Reconstruction Algorithm for Microwave Tomography ", IEEE Transactions on Biomedical Engineering, VOL. BME-32, NO. 9, September 1985,pp.719-723

[13]   S. MANDAL and K. PURKAIT, " A Modified Exact Reconstruction Algorithm for Microwave Tomography for Detection of Disease in Human Body ", ITJS, Vol – 18,F11, pp-82-92 January, 2011.

[14]   K.PURKAIT , " Development of a suitable microwave imaging system for the biomedical applications and formulation of relevant reconstruction algorithms for Microwave Tomography", Institute of Radiophysics and Electronics, University of Calcutta, July 2003

[15]   A N Dattta & B Bondopadhyay, " A complete iterative algorithm for the first order reconstruction of tomographic image by microwaves ", Innov Tech Biol Med, vol. 8, No. 4, pp.- 409, 1987.

[16]   H. P. Schwan, " Electrical properties of tissues and cell suspensions, in Advances in Biological and Medical Physics ", Academic Press, New York, pp. 147–209.

[17]   Eilon D. Kirson, Zoya Gurvich, Rosa Schneiderman, Erez Dekel, Aviran Itzhaki, Yoram Wasserman,Rachel Schatzberger and Yoram Palti (2004): " Disruption of Cancer Cell Replication by Alternating Electric Fields ", Cancer Research, 3288–3295.

[18]   G Qiao, W Duan, C Chatwin, A Sinclair and W Wang (2010): " Electrical properties of breast cancer cells from impedance measurement of cell suspensions ", Journal of Physics: Conference Series 224 012081

[19]   Gupta, A and G.U.Kharat, " Modeling of Dielectric Properties of Cancer Cell and Evaluation of Cancer   Stages : A Review ", International Journal of Recent Scientific Research Vol. 5, Issue, 2, pp.443-448, February, 2014.

[20]   A. N. Datta & B. Bandyopadhyay, " Nonlinear Extension to a Moment Method Iterative Reconstruction Algorithm for Microwave Tomography ", Proceedings of The IEEE, Vol. 74, No. 4, April 1966

[21]   K. Purkait , S. Mondal , S. N. Bose , R. Roy , S Halder , K. Adhikary, " An Approach to Study the range of perturbations in Exact Algorithm on Microwave Tomography of Biological targets with different BeamWidth ", 978-1-4799-4445-3/15/$31.00 ©2015 IEEE.

2019 Devices for Integrated Circuit (DevIC), 23-24 March, 2019, Kalyani, India

Radio Frequency Identification based Goal Line Technology for Quick Decision Making in a Football Match

Samrat Ghosh[1], Suvam Sasmal[2], Saptarshi Bhui[3], Sandipan Dutta[4], Swarnadeep Mukherjee[5], Arko Majumder[6] and Biswarup Ganguly[7]

Electrical Engineering Department
Meghnad Saha Institute of Technology
Kolkata, India

[1]iamsamratghosh@gmail.com, [2]sasmalsuvam@gmail.com, [3]saptarshibhui@gmail.com, [4]sandipandutta96@gmail.com,
[5]swarnadeepm01@gmail.com, [6]arkoscorpio5@gmail.com, [7]bganguly@msit.edu.in

Abstract—**With the growing popularity of football globally, the accuracy of each and every events and technology involved in the game is a concern. Critical situations arise when the referee cannot discriminate a goal or no goal by fine margins due to human visual limitations. Nowadays Video-Assistant Referee (VAR) and other technologies perform accurate decision-making can be implemented in a live match. But the process consumes a lot of time which reduces the fast pace of the game and can cause unnecessary distractions. This study aims to design an automatic goal-line detection system with the help of radio frequency identification (RFID) – Arduino interfacing. RFID incorporates the use of radio waves to extract the information stored in a tag attached to an object. The proposed system uses RFID tags which are fitted on the inside surface of a football. The information embedded in the tags is read by RFID readers, placed behind the goalpost. This technology does not require any additional programming and camera analysis in decision-making, thus making the system faster to help the referees in quick decision-making and maintaining the pace of the game.**

Keywords— radio frequency identification, goal-line technology, decision making, football, Arduino.

I. INTRODUCTION

In Football, the introduction and implementation of modern-day technologies came lately as compared to other popular sports. In the early 2000s various incidents in top-flight football matches at big stages of World Football created varying opinions and controversies. These incidents pointed towards the introduction of goal-line technology. However, the topic was a matter of debate for a decade. But following some errors at the 2010 FIFA World cup-especially in the Germany-England match, the Federation International Football Association (FIFA) announced that it would open the decision about implementing goal-line technology in football matches [1].

Goal-line technology can be defined as the technology by which we can determine whether a goal has been scored or not. It is completely done by electronic measures. In detail, it is a method where we can identify whether a ball has crossed the goal-line or not. The objective of this goal-line technology (GLT) is to help the referees so that they can take the correct decision while awarding a goal to the scoring side.

Similar technologies for decision-making were already evolved in other sports. In Cricket, with the help of hawk-eye [2] and hot-spot technology [3], decisions of out or not-out have been taken by third umpires. Hawk-eye system is

also used in lawn-tennis by the chair-umpires [4]. However it must be mentioned that hot-spot technology took time to be implemented on a good scale mainly due to its cost of implementation. On a similar note, due to its expense, GLT is only used currently in few top European domestic leagues as well as at major FIFA & UEFA championships.

Researches have already been done to implement goal-line technology in football matches. Spagnolo *et al.* [5] have proposed an algorithm where candidate ball regions have been analyzed for detection of the ball. Here, the image sequences, that are formed, record the goal event via camera calibration. Fraunhofer IIS [6] has proposed a radio-based system GoalRef™ which determines whether the whole ball has passed over the goal line employing magnetic fields. One field is created in the goal area whereas the other field is formed surrounding the ball whenever it is moving towards the goal. D'Orazio *et al.* [7] have proposed an automated visual system for detection of goal in real time during soccer matches. This system checks image sequences where false interpretations regarding perspective errors, high velocity of the events, occlusions can be prevented.

Ekin *et al.* [8] have presented a framework for goal detection in football using cinematic feature containing higher-level algorithms. Cruciani [9] has proposed goal detection system consisting both mobile and magneto-acoustic- double resonance induced- ground-based equipment. Ancona *et al.* [10] have proposed an image processing tool for detection of goals via single camera during football matches.

In all these above-mentioned system GLT is implemented either via a camera-based detection system or by using expensive electronic equipment. In this paper an efficient and cheap system is designed with the help of Arduino-RFID interface. By implementing this technique, the time-consuming camera based analysis for decision-making is not required. By eliminating human intervention in decision making, this system is fast and more efficient than other primitive approaches and here lies the novelty of the proposed work.

Section II describes the entire work introducing circuit model with its specifications. Experimental results with schematic diagram and program outputs are given in section III. Finally conclusion is drawn in Section IV.

II. DETAILED DESCRIPTION OF THE WORK

The design of our proposed circuit model is shown in Fig. 1. The proposed model consists of a microcontroller [11-13],

978-1-5386-6723-1/19 $31.00 © 2019 IEEE

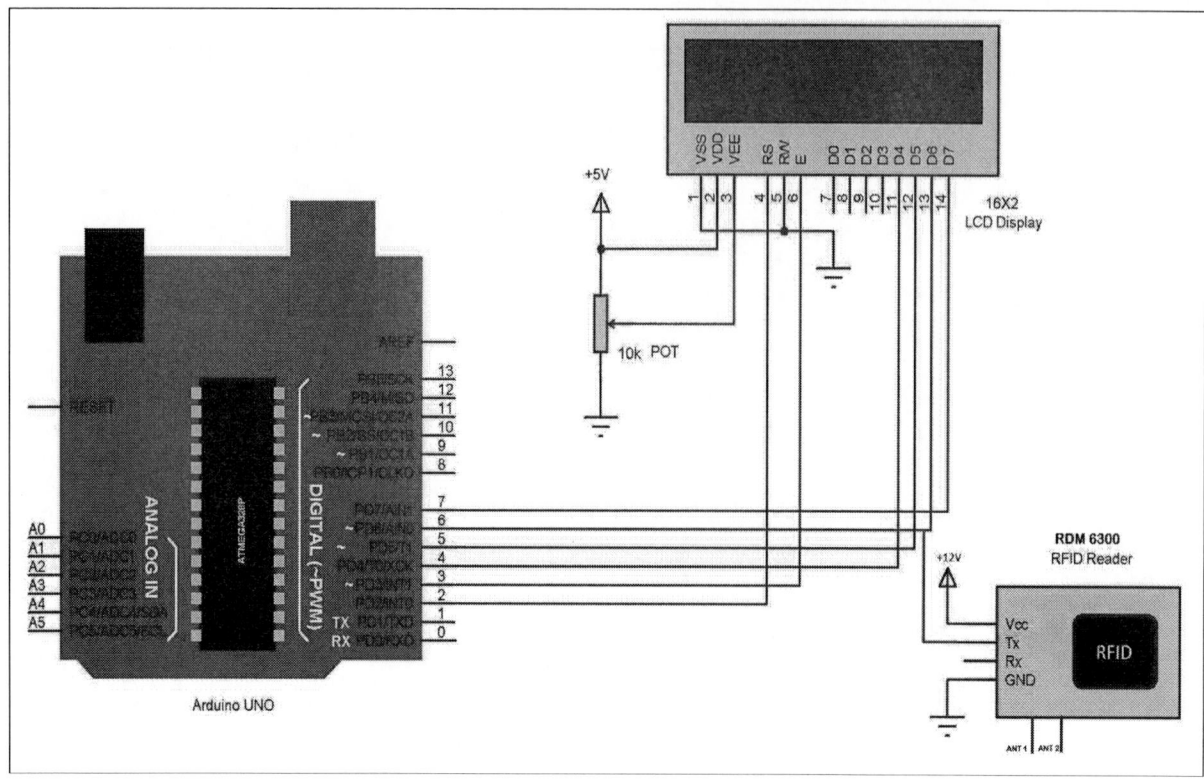

Fig. 1. Circuit model of RFID based goal-line technology

RFID reader [14-15], RFID passive tags [16] and an LCD display. The working principle of this proposed system mainly depends upon the interaction of RFID tags and RFID reader. In a RFID tag, we basically store information in the form of binary bits. This information could be any personal or medical information or any other type of information which can be stored digitally.

reader receives the feedback signal and delivers another signal to the LCD board that a goal has been scored. This process of detection is spontaneous and doesn't require any preset or reset operation. The block diagram of the proposed scheme is shown in Fig. 2.

TABLE I.   COMPONENTS OF CIRCUIT MODEL OF RFID BASED GLT

Sl. No.	Name of the component	Specifications
1.	Microcontroller	Arduino Uno
2.	RFID Reader	RDM 6300, 125 kHz
3.	RFID passive tags	125 Khz
4.	LCD display	16cm × 2cm

RFID reader basically consists of a microcontroller, RF signal generator & a receiver/signal detector. RF signal generator generates radio waves of a certain frequency and it is transmitted within a particular range via antennas. Whenever the RFID tag comes within the range of RFID reader emitted RF signal, the information stored in the tag is transferred to the reader via a feedback signal.

In our proposed system, a radio-frequency passive tag is installed inside the ball and a RFID reader consisting an antenna is placed in the rear goalpost embraced in the goal-frame. RFID reader is supplied power via Arduino. The antenna emits 125 kHz radio frequency signal continuously. When the ball crosses the goal-line *i.e.* comes within the range of antenna, it sends a feedback signal to the reader.

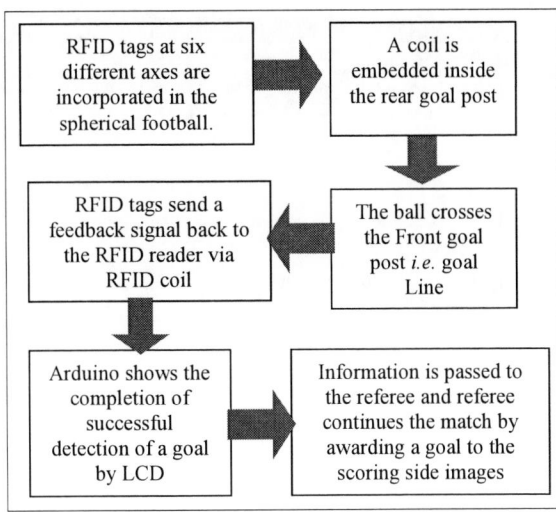

Fig. 2. Block diagram of the proposed scheme

III. EXPERIMENTAL RESULTS

The experimental arrangement is shown in Fig. 3. By

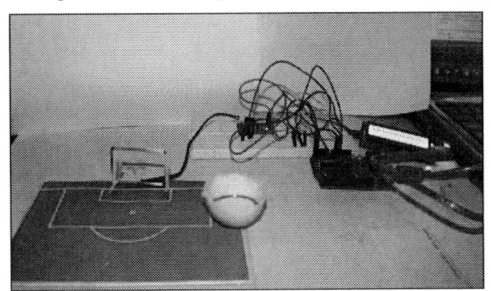

Fig. 3. Experimental arrangement for RFID based GLT

performing several stages we can succesfully implement the above-mentioned technology in a real time football match:

A. Insertion of passive RFID tags inside the ball

Passive RFID tags were inserted inside the ball so that RFID reader can easily detect the ball when the ball crosses the goalline. By using passive tags we can make the system even more cheaper and compact.

B. Placing a RFID reader coil at the rear goal-post

A RFID reader coil attached with the RFID reader is placed at the rearpost to track the ball. By varying the inductance and number of turns of the coil, the size of the coil can be differed and required protection is given to the coil so that it didn't get affected by real-time conditions.

C. Switching On the RFID reader

The RFID reader is required to switch on during its operation. It can be done either via Arduino-RFID interface or battery source. The later one is done in this proposed work so that we can cross check the identification of tags in serial monitor.

D. Setting up the indication system

At last we have to set up some indication system to referees. It can be done with the help of LCD display or a buzzer or a smart watch.

Now if and only if the ball crosses the goal-line RFID reader recognise it via RFID coil and indicate it to the referee and now he/she can award the goal to the scoring side. Total time required for this entire process is 3-4 seconds so that it couldn't hamer the first pace of the game.

The program on Arduino-Uno for RFID-Arduino interface is shown in Fig. 5 and its corresponding output ensuring the detection of a RFID passive tag is reflected in Fig. 6. The experiment is executed in a i3-4005U 4 GB RAM with 2 GB NVIDIA graphics card.

2019 Devices for Integrated Circuit (DevIC), 23-24 March, 2019, Kalyani, India

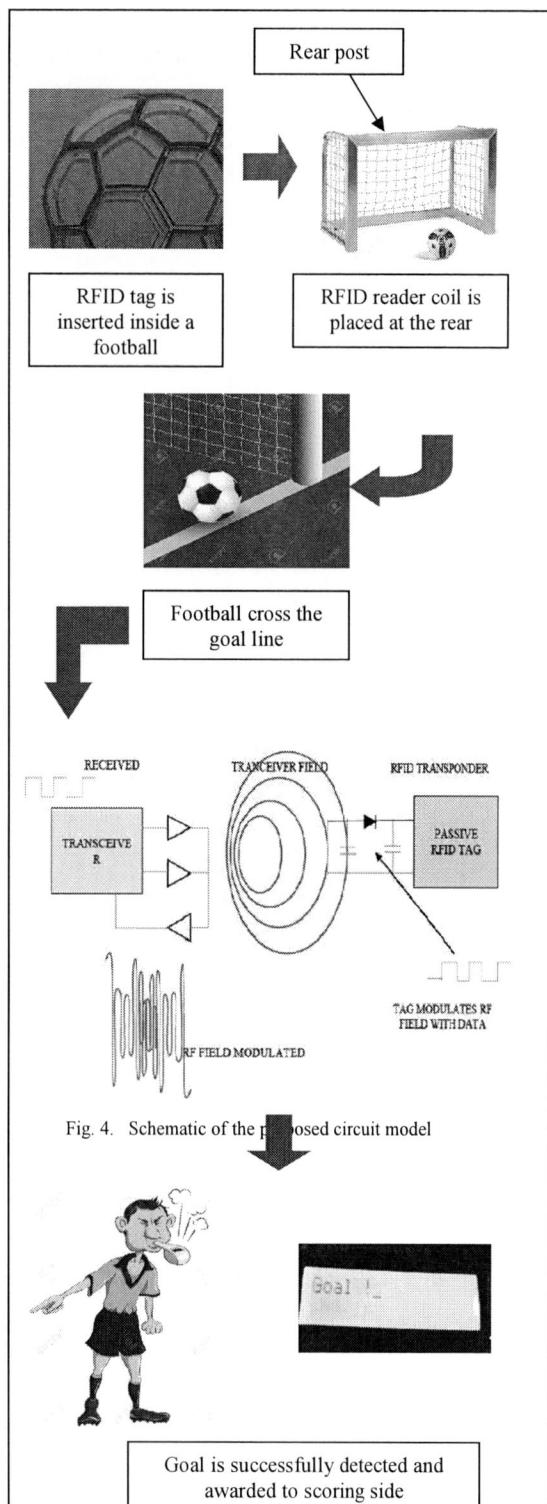

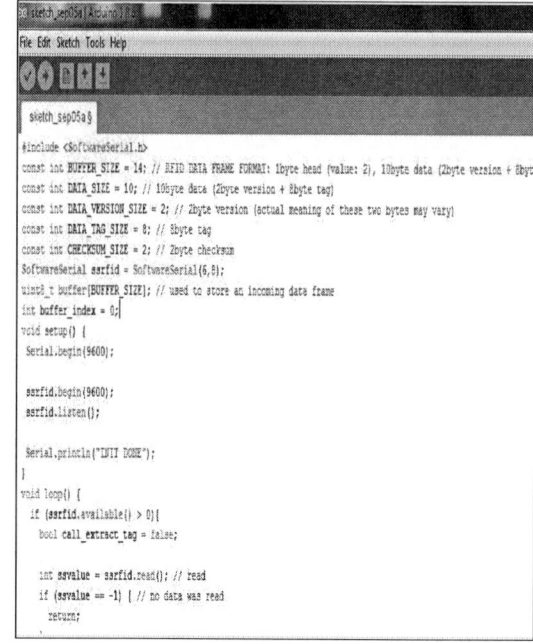

Fig. 5. Arduino Uno program for RFID interface

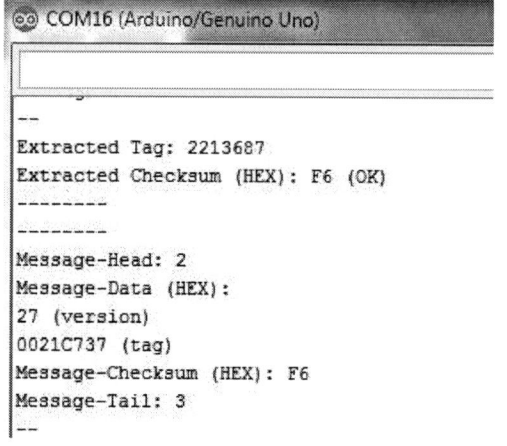

Fig. 6. Arduino Uno programming output verifying the detection of a RFID tag

IV. CONCLUSION

This paper reported a goal-line technology system for quick and correct decision-making in goal-detection using RFID. We have taken utmost care to eliminate any safety-related issue of RFID. This RFID technique is reliable so that we can easily authenticate and authorize it in RFID tagged embedded balls and RFID antenna coil assembled goalpost. This project is reliable and cheap and gives an easy way to detect goals in a live football match.

This system will definitely help both the ends *i.e.*, players and match officials to maintain the pace of the game. By implementing this technique, the time-consuming camera based analysis for decision-making will may not be required henceforth. By eliminating human intervention in decision making, this system is fast and more efficient than

978-1-5386-6723-1/19 $31.00 © 2019 IEEE

other primitive approaches. So, it is clear that by the proposed RFID based Goal-line technology system is not only better than manual camera-based system but also could be the superior technology for automatic detection of goals by far. In future, by implementing the RFID technology in order to track the movement of the players along with the ball, we can evaluate the performance of a player in real time cheaply and undoubtedly it will enrich the level of the game from the grass-root level.

REFERENCES

[1] Pires, Marcelo, and Vítor Santos. "Assessing the Impact of Internet of Everything Technologies in Football." In *World Conference on Information Systems and Technologies*, pp. 375-388. Springer, Cham, 2018.

[2] Arora, Udit, Sohit Verma, Sarthak Sahni, and Tushar Sharma. "Cricket umpire assistance and ball tracking system using a single smartphone camera." *PeerJ Preprints* 5 (2017): e3402v1.

[3] Devi, K. Nirmala. "Hotspot Detection and Analysis in Chat Environment." *Asian Journal of Research in Social Sciences and Humanities* 6, no. 7 (2016): 1000-1007.

[4] Mecheri, Sami, François Rioult, Bruno Mantel, François Kauffmann, and Nicolas Benguigui. "The serve impact in tennis: First large☐scale study of big Hawk☐Eye data." *Statistical Analysis and Data Mining: The ASA Data Science Journal* 9, no. 5 (2016): 310-325.

[5] Spagnolo, Paolo, Marco Leo, Pier Mazzeo, Massimiliano Nitti, Ettore Stella, and Arcangelo Distante. "Non-invasive soccer goal line technology: A real case study." In *Proceedings of the IEEE Conference on Computer Vision and Pattern Recognition Workshops*, pp. 1011-1018. 2013.

[6] GoalRef ™ –Goal Detection System, 2018, Available online:https://www.iis.fraunhofer.de/en/ff/lv/net/proj/goalref.html.

[7] D'Orazio, Tiziana, Marco Leo, Paolo Spagnolo, Massimiliano Nitti, Nicola Mosca, and Arcangelo Distante. "A visual system for real time detection of goal events during soccer matches." *Computer Vision and Image Understanding* 113, no. 5 (2009): 622-632.

[8] Ekin, Ahmet, A. Murat Tekalp, and Rajiv Mehrotra. "Automatic soccer video analysis and summarization." *IEEE Transactions on Image processing* 12, no. 7 (2003): 796-807.

[9] Cruciani, Gabriele. "Goal detection equipment for football." U.S. Patent Application 10/527,873, filed June 22, 2006.

[10] Ancona, Nicola, Grazia Cicirelli, Ettore Stella, and Arcangelo Distante. "Ball detection in static images with Support Vector Machines for classification." *Image and Vision Computing* 21, no. 8 (2003): 675-692.

[11] Badamasi, Yusuf Abdullahi. "The working principle of an Arduino." In *Electronics, computer and computation (icecco), 2014 11th international conference on*, pp. 1-4. IEEE, 2014.

[12] Barma, Bodhisatwa, Samrat Ghosh, Abhrodip Chaudhury, and Biswarup Ganguly. "Microcontroller Based Robotic Arm Development for Library Management System." *arXiv preprint arXiv:1812.11316* (2018).

[13] Bhattacharya, Tamal, and Prasanta Sarkar. "ARDUINO UNO based packed U cell inverter for photovoltaic application." In *2018 International Symposium on Devices, Circuits and Systems (ISDCS)*, pp. 1-6. IEEE, 2018.

[14] Farswan, Akanksha, Anil Kumar Gautam, Binod Kumar Kanaujia, and Karumudi Rambabu. "Design of Koch fractal circularly polarized antenna for handheld UHF RFID reader applications." *IEEE Transactions on antennas and propagation* 64, no. 2 (2016): 771-775.

[15] Hanwate, Aboli, and Poonam Thakare. "SMART TROLLEY USING RFID." *International Journal of Research In Science and Engineering e-ISSN* (2015): 2394-8299.

[16] Ferdous, Raquib Md, Ahmed Wasif Reza, and Muhammad Faisal Siddiqui. "Renewable energy harvesting for wireless sensors using passive RFID tag technology: A review." *Renewable and Sustainable Energy Reviews* 58 (2016): 1114-1128.

2019 Devices for Integrated Circuit (DevIC), 23-24 March, 2019, Kalyani, India

Performance Comparison of CMOS and MEMS based Thermal Energy Harvesters using Finite Element Analysis

Indrajit Sil, Sagar Mukherjee[a], Kalyan Biswas[b],

[a,b] ECE Department, MCKV Institute of Engineering, Liluah, WB, India
sil.indrajit25@gmail.com, sagarju87@gmail.com, bkalyan.ece@gmail.com

Abstract— In this paper, thermal energy harvesters (TEH) or Thermoelectric Power Generators (TPG) are designed to harvest electrical energy from heat. These Thermoelectric Power Generators are compatible to CMOS and MEMS fabrication technology. Detailed analysis of various models of thermoelectric power generators using FEA software are studied to achieve enhanced performance. Comparison has been made for both CMOS and MEMS based TPGs. From the analysis, it is observed that MEMS based TPG model produces 43.76% more output voltage than CMOS based TPG when the temperature difference across hot and cold junction is 5K. Analysis reveals that 91.23% increase in output power is also achieved with MEMS based TPG model. The design and simulation results provides a very good overview of the power generation capability of the TEG, which may be useful in future design of improved thermal energy harvesters.

Keywords—MEMS, CMOS, Thermal analysis, Energy Harvester, Finite element analysis

I. INTRODUCTION

Recent technology of portable & wireless devices includes electrochemical batteries as their power source. The necessity of periodic replacement and disposal of batteries are necessary because of their limited life period. In this regard, research on energy harvesters has gained consideration in recent years because it converts ambient energy into usable electrical output. Unused ambient energy coming from light, thermal, wind, and vibration sources may be converted to usable power. In recent years, several energy harvesting approaches have been proposed using solar, thermoelectric, electromagnetic, piezoelectric and capacitive schemes at different scales [1-4]. The thermal energy harvesters are one of the technologies which provide solution to these problems. The thermal energy harvesters work based on the principle of Seebeck effect [5-8]. It helps to harvest waste and unused heat energy from the ambient where temperature difference is found to convert into useful output energy. It does not require any external power source for its operation so it has a very long lifetime. This output energy of TPGs can be used to power low power requirement devices like rechargeable batteries, capacitors, mobile phones etc.

CMOS technology has advanced in the last decade [9-14]. Large value of Seebeck effect is found in Poly Silicon [15] which is compatible with CMOS thin film technology. With its help, the thin films of poly-Si are deposited and patterned to create an array of thermo-couples (thermopiles), in a small area. Output energy efficiency is improved [15] by making trenches below cold junctions to equal thermal resistance with surrounding air. Researchers already have demonstrated devices based on thermoelectric technologies. In a recently published work J. Xie et al[16] designed, modeled, fabricated and characterized a CMOS compatible thermal energy harvester to convert discarded heat into a usable electrical power amounting few microwatts. Detailed chemical process for fabrication of TEGs was demonstrated by A. Odia et al [17]. These works establishes the capability of CMOS based technologies.

This paper presents a detailed simulation and structural design of both CMOS and MEMS based micro TPG which is able to harvest unused heat energy which can be used to power low power devices. The thermocouples are prepared of n and p type material which are embedded between heat sink and silicon with use of thin film technology of CMOS and MEMS. The thermopiles are organized in such a way that they are parallel to direction of heat flow and in series with current flow. The heat sink is made of copper which can withstand high temperature and dissipate heat quickly which increases the temperature difference between hot and cold junction. The thickness of SiO_2 is made smaller about 0.2μm so that it does not absorb too much of heat. The length of thermocouples is decreased to accommodate more thermocouple to achieve higher output voltage and power. This paper is organized as follows. After brief introduction in Section I, Section II deals with basics theory and mathematical expression associated with TPGs. Section II shows the simulation and result of both CMOS and MEMS based TPG and their comparison is provided in section IV. Section V gives the conclusion of the findings obtained in this work.

II. THEORETICAL BACKGROUND OF THERMOELECTRIC POWER GENERATOR

The thermoelectric effect used by this thermoelectric power generator is the Seebeck effect i.e. if a pair of wire made of different material is connected together, direct current (DC) flows through the wire when there is a temperature difference across the two junctions of the wires.

The output voltage (V_o) of thermal energy harvester as given by Seebeck effect is written as

$$V_0 = m\alpha\Delta T_G \qquad (1)$$

Where m denotes the number of thermocouples, α is the relative Seebeck coefficient, ΔT_G is the temperature difference between the junctions.

The output power may be written by

$$P_O = V_o I = \frac{m^2\alpha^2\Delta T^2{}_G}{4R_L} \qquad (2)$$

Where V_O is the output voltage, I denotes the current flowing and R_L is the load resistance.

978-1-5386-6723-1/19 $31.00 © 2019 IEEE

The efficiency of TPG can be found from voltage and power factors [16]

$$\Phi_V = \frac{V_O}{A_K \Delta T_G} \qquad (3)$$

$$\Phi_P = \frac{P_O}{A_K \Delta T^2{}_G} \qquad (4)$$

Where Φ_V is the voltage factor, Φ_P is the power factor and A_K is the surface area.

III. Length Simulation and Analysis

A 3-D model is built in ANSYS Multiphysics software for thermal finite-element analysis (FEA). The detailed CMOS and MEMS based structures is discussed in this section. Planar thermopiles for energy conversion are arranged on top of the substrate due to the limited thickness of the thermoelectric material. Geometrical parameters and materials properties used are tabulated in respective sub-section. Effect of different parameters such as length of thermoelectric legs, no of thermopiles present, materials used etc. are analysed and summarized in this section.

A. CMOS based Thermoelectric Power Generator

The model of a CMOS based thermal energy harvester is shown in Fig 1. Silicon is the base material as it is most versatile component of CMOS technology. SiO_2 is formed on the top of silicon. An insulator, Si_3N_4 (Silicon Nitride) is deposited on top of SiO_2 to prevent leakage of current from the contacts and thermocouple legs to silicon and SiO_2.

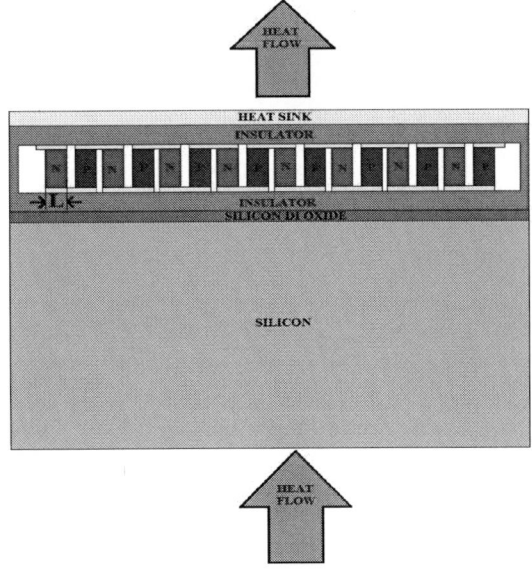

Fig. 1. Schematic diagram of Thermal energy Harvester

Thermopiles of length 'L' (shown in Fig. 1) are deposited on top of Si_3N_4, 1μm apart. Thermocouple legs are formed at a gap of 1.8μm from silicon edge to lessen the heat loss due to heat absorption from edge of silicon. Between the thermopiles, the cupper contacts are used to enable the flow of current in one direction. Another Si_3N_4 layer is deposited on top of thermocouples for insulation Copper made heat sink material is used on top of Si_3N_4. Heat flows from bottom to upward direction in a longitudinal path. Finite element

analysis is used to simulate output current obtained from the device. Relevant properties of different materials used in simulation for CMOS based TPG is given in Table I. Device parameters used in simulation are given in table II.

Table I. Properties of Materials used in CMOS based Thermal energy harvester Simulation

Materials	Thermal Conductivity (W/m-k)	Resistivity (Ohm-m)	Seebeck Coefficient
Silicon [8]	148	3.2E+03	NA
SiO_2[12]	1.5	1E+15	NA
Si_3N_4	27	1E+21	NA
Copper	400	1.68E-08	NA
n-type Poly Silicon [12]	29.7	8.9E-06	-110E-6
p-type Poly Silicon [12]	28.4	13.7e-06	130E-6

Table II. Device parameters used in the model

Materials	Length (μm)	Thickness (μm)	Width (μm)
Silicon	100	45	5
SiO_2	100	0.2	5
Si_3N_4	100	0.4	5
Thermocouple Legs	16,10 and 5	5	5
Contacts	33,21 and 11	0.4	5
Heat Sink	100	0.4	5

3-D model of CMOS based TPGs are made in ANSYS software. 5K temperature difference is applied between the hot and cold junctions of the device. The simulation result of heat distribution contour in the model is shown in Fig. 2. It can be seen that the heat distribution is vertical. From Fig. 3. It is observed that the current flow direction from hot to cold junction for the temperature difference.

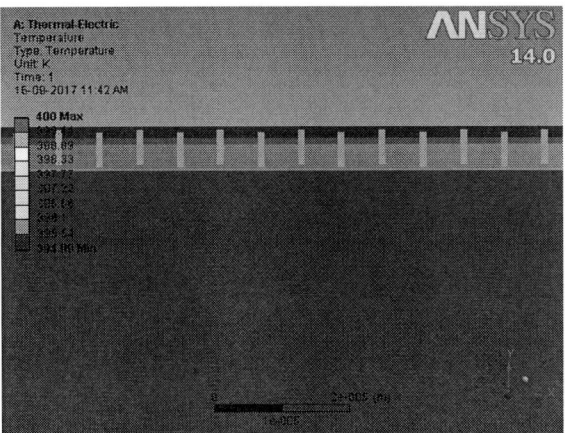

Fig. 2: Simulated temperature contour of the TEG for ΔT_G=5K

Fig. 3. Current Flow in TPG legs due to temperature difference

Shortening each leg of n and p type poly Si more number of thermopiles can be deposited in the same chip area. More thermopiles mean improved output voltage and power which is observed from Fig 4 and 5 respectively. According to Fig. 4 and 5, it is observed that the TPG with thermocouple leg of L=5μm, the output is higher than TPG with leg of L= 16 and L=10μm under same conditions. An improvement of 8.64% in output voltage and 18.6% increase in output power (at ΔTG=5K) is realized when more thermopiles are built-in within same chip area by shortening each thermoelectric legs by 5μm. But with increase in number of thermocouples in a given area by shortening each thermoelectric leg, the internal resistance of thermocouples increases. Resistance increases by 58.3% with increase of thermocouples shortening each thermoelectric legs by 5μm. So, an optimization is necessary in this regard.

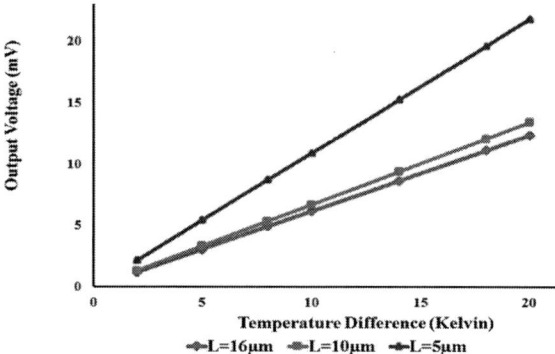

Fig. 4. Temperature Difference vs Output Voltage for CMOS TPG for different thermocouple legs length

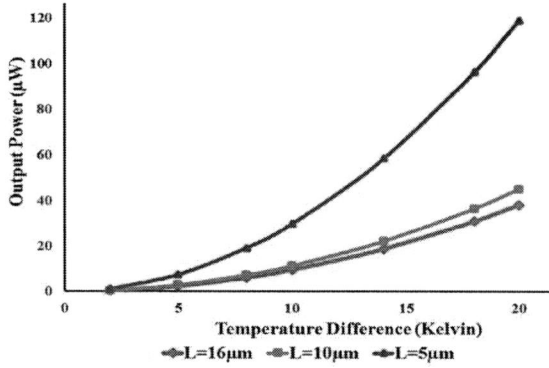

Fig. 5. Variation of output power with Temperature Difference for CMOS TPG for different thermocouple legs length.

B. MEMS based Thermoelectric Power Generator

A model of a MEMS based thermoelectric power generator is shown in Fig. 6. All the parameters used in MEMS based TPG are same used in CMOS based TPG with two exceptions. First due to advanced thin film technology of MEMS, use of SiO_2 can be ignored. Secondly n and p type thermocouples are made of materials $Bi_2Se_{0.5}Te_{2.5}$ and $Bi_{0.5}Sb_{1.5}Te_3$ respectively which are MEMS compatible. Properties of materials used in ANSYS simulation for MEMS based TPG are tabulated in Table III.

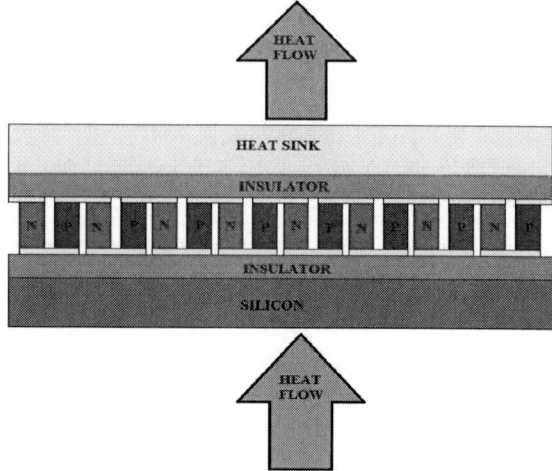

Fig. 6. MEMS based Thermal Energy Harvester

Table III. Relevant material properties used in MEMS based Thermal Energy Harvester Simulation

Materials	Thermal Conductivity (W/m-k)	Resistivity (Ohm-m)	Seebeck Coefficient
Silicon	148	3.2E+03	NA
Si_3N_4	27	1E+21	NA
Copper	400	1.68E-08	NA
n-type ($Bi_2Se_{0.5}Te_{2.5}$) [24]	1.1	1.49E-05	-228E-6
p-type ($Bi_{0.5}Sb_{1.5}Te_3$) [24]	1.1	9E-06	188E-6
Silicon	148	3.2E+03	NA

Fig. 7 and 8 shows variation of output voltage and power respectively with change in temperature difference. Same observations are made as CMOS based TPG i.e. For TPG with smaller leg (L=5μm), the output parameters readings are more than TPG with longer leg (L= 16 and L=10μm) under same conditions. An improvement of 33.36% in output voltage and 77.86% increase in output power (at ΔTG=5K) is gained when more thermopiles are incorporated in same area of chip by shortening each thermoelectric legs by 5μm.

2019 Devices for Integrated Circuit (DevIC), 23-24 March, 2019, Kalyani, India

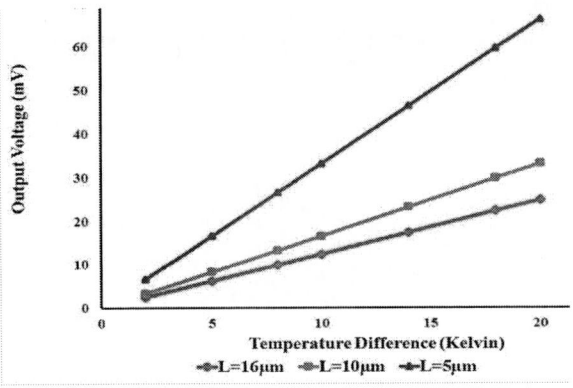

Fig. 7. Variation of Output Voltage with Temperature Difference for different thermocouple legs length in MEMS TPG

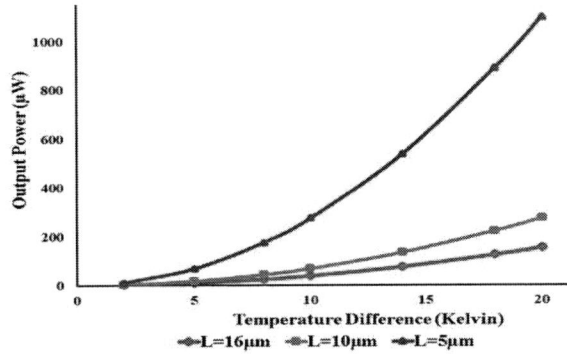

Fig .8. Variation of Output Power with Temperature Difference for different thermocouple legs length in MEMS TPG

IV. PERFORMANCE COMPARISON

In this section both CMOS and MEMS based TPG are compared with each. A comparison of CMOS based TPG & MEMS based TPG, both with L=5μm is done here. Fig. 9 and 10 shows change in Output Voltage and power with variation of temperature difference for both CMOS and MEMS based TPG. It is observed that keeping each thermoelectric legs to 5μm, the output voltage and power of MEMS based TPG demonstrates better performance than CMOS based TPG. At low temperature difference, the variation of outputs between CMOS and MEMS based devices are lower. Increasing the temperature difference of both the devices, it can be seen that variation of output readings between CMOS and MEMS based TPG increases. From the analysis, 43.76% increase of output voltage is obtained using MEMS technology where as 91.23% increase in output power is achieved. Fig. 11 and 12 shows the voltage and power factor of this two TPG models. In graph, blue colour bar represents CMOS based device and the orange colour bar represents MEMS based device. It can be observed that MEMS based TPG model gives better output voltage and power than CMOS based TPG.

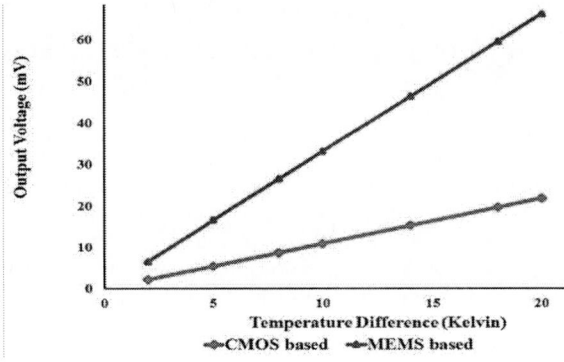

Fig. 9. Temperature Difference vs Output Voltage comparison of CMOS and MEMS based TPG (Same thermoelectric leg length L=5μm)

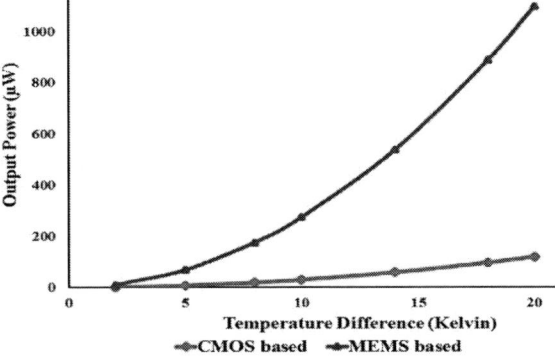

Fig. 10. Temperature Difference vs Output Power comparison of CMOS and MEMS based TPG (Same thermoelectric leg length L=5μm)

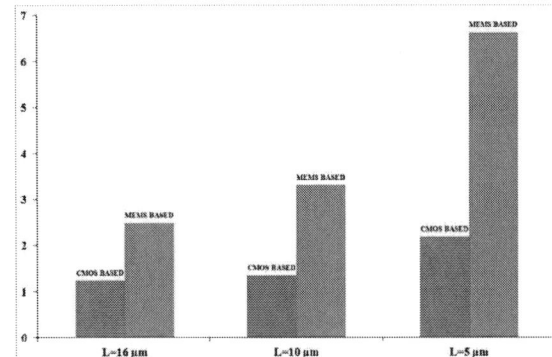

Fig. 11. Comparison of Voltage factor of CMOS and MEMS based TPG

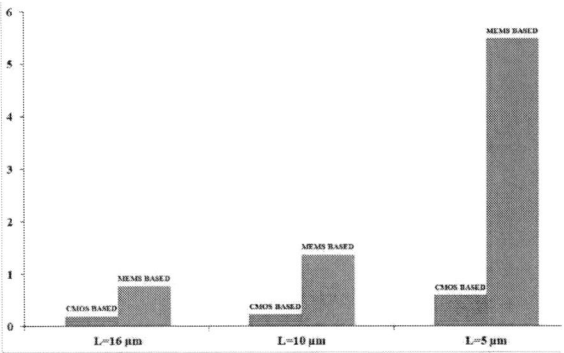

Fig. 12. Comparison of Power factor of CMOS and MEMS based TPG

978-1-5386-6723-1/19 $31.00 © 2019 IEEE

V. CONCLUSION

The Modelling, simulation and analysis of both CMOS and MEMS based TPG is presented in this paper. Properties of thermocouple legs for each technology were changed to get improved output. For CMOS based TPG with thermocouples of length L=5μm each in a chip area of 500μm^2, Output Voltage (V_o) of 5.45mV and Output Power (P_o) of 7.4μW is obtained with a temperature difference of 5K which gives Voltage Factor (Φ_v) of 2.183Vm^{-2}k^{-1} and Power Factor (Φ_p) of 0.0592mWm^{-2}k^{-2}. For MEMS based TPG using same parameters, Output Voltage of 16.5 mV and Output Power of 68.5μW is obtained, which gives Voltage Factor of 6.62Vm-2k-1 and Power Factor of 5.4 mWm^{-2}k^{-2}. The high outputs of MEMS based TPG in comparison of CMOS based TPG is because of MEMS new thin film technologies. Thermopiles made of n and p type materials of $Bi_2Se_{0.5}Te_{2.5}$ and $Bi_{0.5}Sb_{1.5}Te_3$ respectively are compatible with MEMS technology and incompatible with CMOS technology. Their Seebeck Coefficient values are 98.3% more than n and p type materials used in CMOS based TPG. Due to this reason MEMS is able to provide more improved output than CMOS based TPG. Whatever the technology choice CMOS or MEMS, the TPGs are able to perform fruitfully to harvest wasted heat energy which provides temperature differences in the ambient. These findings are useful for the choice of fabrication technology based on device structures and its applications.

REFERENCES

[1] Priya S, Inman D, (Eds.), Energy Harvesting Technologies (Springer, NY, 2009.)

[2] Leonov V.: 'Thermoelectric Energy Harvesting of Human Body Heat for Wearable Sensors', IEEE Sensors Journal, vol. 13, no. 6, pp. 2284-2291, June 2013.

[3] Beeby SP, Tudor MJ , White NM,: 'Energy harvesting vibration sources for microsystems applications', Measurement Science and Technology, 2006

[4] Bader S. Oelmann B.: 'A concept for remotely reconfigurable solar energy harvesting testbeds', IEEE SENSORS, Glasgow, October 30-November 2, 2017.

[5] Ashraf M., Masoumi N.: 'A Thermal Energy Harvesting Power Supply with an Internal Startup Circuit for Pacemakers', IEEE Transactions on Very Large Scale Integration (VLSI) Systems, Vol. 24, No.1, January 2016.

[6] Matsubara K.: 'Development of a high efficient thermoelectric stack for a waste exhaust heat recovery of vehicles', Proc. of the 21st International Conference on Thermoelectronics, August 25-29th, Portland, OR, 418–423 (2002).

[7] Yang S. M, Lee T, Cong M.: 'Design and verification of a thermoelectric energy harvester with stacked polysilicon thermocouples by CMOS process', Sensors and Actuators A: Physical, vol. 157, pp. 258-266, 2, 2010.

[8] Wojciechowski, K.T., Schmidt, M., Zybala, R. et al.: 'Comparison of Waste Heat Recovery from the Exhaust of a Spark Ignition and a Diesel Engine', Journal of Electronic Materials (2010) 39: 2034.

[9] Fleming J., Ng W., & Ghamaty S.: 'Thermo electric based power system for unmanned-air-vehicle/microair-vehicle applications', Journal of Aircraft, vol.41, 674–676 (2004).

[10] D. Guyomar, G. Sebald, E. Lefeuvre, & A. Khodayari.: 'Toward heat energy harvesting using pyroelectric material', Journal of Intelligent Material Systems & Structures, vol.20 265-271 (2009).

[11] Ziouche K., Yuan Z, Lejeune P, Lasri T, Leclercq D, and Bougrioua Z.: 'Silicon-Based Monolithic Planar Micro Thermoelectric Generator Using Bonding Technology', Journal of Microelectromechanical Systems, Volume: 26, Issue: 1, Feb. 2017.

[12] Xie J., Lee C. , Wang M.F., Liu Y, Feng H.: 'Characterization of heavily doped polysilicon films for CMOS-MEMS thermoelectric power generators', Journal of Micromechanics and Microengineering, Volume 19, Number 12, 2009

[13] Wan Q, Teh Y.K, Gao Y, Philip K. T. Mok.: 'Analysis and Design of a Thermoelectric Energy Harvesting System with Reconfigurable Array of Thermoelectric Generators for IoT Applications', IEEE Trans on Circuits and Systems I, Volume: 64, Issue: 9, Sept. 2017.

[14] Rozgic D, and Markovic D.: 'A Miniaturized 0.78-mW/cm^2 Autonomous Thermoelectric Energy-Harvesting Platform for Biomedical Sensors', IEEE Transactions on Biomedical Circuits and Systems, Volume: 11, Issue: 4, Aug. 2017

[15] Wang Z., Leonov V., Fiorini P., Hoof C. V.: 'Realization of a poly-SiGe based micromachined thermopile', Eurosensors conference, Dresden, Germany, 7 - 10 September, 2008, pp. 1420–1423.

[16] Xie J., Lee C., Feng H.: 'Design, Fabrication, & Characterization of CMOS MEMS-Based Thermoelectric Power Generators', Journal of Microelectromechanical Systems, Vol. 19, No. 2, April 2010.

[17] Odia A, Lin L.F, Douglas, J. Paul et al.: 'Modelling and "Experimental Verification of a Ge/SiGe Thermoelectric Generator', Microelectronics and Electronics (PRIME), 2015 11th Conference on 29 June-2 July 2015.

High PSNR based Image Fusion by Weighted Average Brovery Transform Method

Nidhi Taxak
M. TECH. Scholar
RCEW
Jaipur, Rajasthan, India

Sachin Singhal
Associate Professor
RCEW
Jaipur, Rajasthan, India

Abstract: - Image Fusion is a method, in which two relevant Image get combine and generate a new Image. The generated image has excellent clarity as compared to the previous input image. Image Fusion Technique is improving the performance of the images and increase the application of Image Fusion. In the Base paper, they present Image Fusion for Two-Dimensional Multiresolution 2-D image. The applications of the Image fusion is using various fields like multi - Focus Images, CT, Multi-Sensor Satellite image and MR of the Human Brain. In this Paper, working for improve PSNR(Peak Signal to Noise Ratio) and Reduce to MSE (Mean Square Error). For improve the performance of Image fusion using Weighted Average Brovery Transform. In the base paper, PSNR and MSE are comparing by use PCA, DWT, DWT-PCA, DCT-PCA, DWT-DCT-PCA methods. Proposed Weighted Average Brovery Transform method is showing better results as compare to base paper results.

Keyword :-Image Fusion, Wavelet Transform, Fused Images, Wavelet based Fusion, Multi-Resolution, Multi-sensor image fusion, multi-resolution SVD, Image Fusion Performance Evaluation Metrics.

I. INTRODUCTION

Image Fusion refers to the integration of number of images into a single image. This is done without making any compromise with the features of the original images. Images are generally fused for serving two purposes. The first one is to capture the policies of same articles or act (such as multi-modal, center and sensor images). Secondly, Image Fusion allows one to acquire pictures from each instrument modality. These can be exemplified in case of multi-center imaging where a particular image may contain a single or more questions in-center, whereas distinct images can exhibit various items in the act to be in center. When images are captured from a distance, some of them contain rich geometric determination whereas others can show up vague data. Image Fusion is of prime importance in biomedical imaging, where a single technique is not sufficient to provide the whole information about the cerebrum structure. Commonly used imaging techniques are reverberation imaging (MRI) and the registered tomography (CT). The former captures the delicate tissues present in the brain while the latter takes images of the tough tissues and bones. Therefore, a single picture is certainly not adequate. Combination of both can be quite beneficial in accurate detection of diseases.

Image Fusion is advantageous in several fields such as remote identification, medical imaging, PC vision, microscopic imaging and many more. In PC vision, relatable data of constituent images is combined in a single picture by following multi-sensor image combination method, and the resulting image is comparatively more informative [1, 2]. In case of

remote discovering applications, space borne sensors are helpful in making several calculations of image combinations. These sensors are becoming more accessible with time, favoring the image fusion. Sometimes, the images are need to be handled with rich ghostly and spatial determination in a solitary image. The difficulty here is that the available hardware is not able to provide such information. The data sources have to reconcile in order to combine the pictures into a single image. Moreover, modification in the bizarre multispectral data can also be there by the combination strategies.

Image Fusion is also beneficial while obtaining information rich images from the satellites. Satellite imaging provides two types of images, panchromatic and multispectral. The transmission of former is done with high determination while that of latter occurs with coarser determination (which is lower by about two to four times). To forward high information data, both the pictures are congregated at the recipient station.

The combination of images follows a step-wise procedure. In the first stage, enrollment is carried out where the original pictures are cut down into a specific course framework. It is required because a perfect fusion takes place only when the constituent pictures contain symmetrical basic items in terms of area, size or else. In the next step, combination of the images is carried out to form a final unique fused picture. It is very important here to choose accurate amounts of elements from the original images.

There are several ways in which the integration of the pictures is done. They vary from very simple technique for averaging of pixels to complex strategies such as wavelet change combination and central segment examination. The documentation of different ways of image fusion is done depending upon how the combination is processed, like whether the images are transferred into other space, or they are merged in the spatial area, or the changes of the images are intertwined.

The picture combination can be carried out using several techniques. The fundamental technique is the high pass sifting technique while all others are dependent upon the Laplacian pyramid, Discrete Wavelet Transform and uniform balanced channel bank.

A. Need for Image Fusion

Multi sensor data combination is a mechanism that now demands comprehensive formal responses for different applications. Areas like remote detection requests images with not only rich ghostly data but also rich spatial data in a solitary image. Such requirements cannot be fulfilled by the current technology say instruments, which are not able to provide the desired data because of either the outline or observational limits. The data integration is the one easily feasible answer to this.

B. Standard Image Fusion Method

Different techniques of image combination can be broadly grouped into two categories namely, change space combination and spatial area combination. The spatial area combination group involves the combination techniques such as vital part investigation (PCA), averaging, IHS based strategies, Brovey strategy, and high pass separating based method. The high pass separating based method is quite essential in which the careful interpretation from MS pictures is instilled with high repetition subtle elements. Although the spatial area combination methodologies are good to work with, there is an obstruction that comes across. These techniques result in the spatial contortion in the fused image which is leads to problems during the next stages of picture handling. However, there is a solution for it. With the use of recurrence area approaches over the image fusion, the spatial contortion can be avoided. The multi determination analysis is found to be quite beneficial for carrying out investigating of remote spotting images. There are number of combination techniques as mentioned above, however, a highly useful device for the combination of pictures is the discrete wavelet change. In comparison to other spatial combination strategies, these methods are found to be great in executing in bizarre and spatial nature of the combined image.

The pre-registration of to be used images is done firstly in the beginning of image fusion. If the registration is not accurate, it can lead to big faults during the process. Highly popular ways of image fusion are:

- Wavelet transform image fusion
- High pass filtering technique
- PCA based image fusion
- Pair-wise spatial frequency matching
- IHS transform based image fusion
- PCA based image fusion

There are several fields where the combination of pictures in remote discovering can be applied. One of the highest important field is the multi-determination image combination. Following are the types of images present in satellite symbolism:

- Multispectral images – These are the pictures that are obtained optically in multiple spectral or wavelength interim. Although each image belongs to the single physical territory, it scales of different ghostly bands.
- Panchromatic images – These refer to the pictures which are combined in the wide visual wavelength run but are extremely distinct.

The multispectral images are low determination in nature (30m pixel) and are captured by the LANDSAT TM satellite. Whereas, the panchromatic pictures are of high determination (10m pixel). Such images are given by the SPOT PAN satellite. Both the multispectral and panchromatic pictures are joined in the picture combination procedure to finally provide a solitary multispectral image with high determination. The different integration approaches for the image combination are dependent on Red-Green-Blue (RGB) to Intensity-Hue-Saturation (IHS) change. The combination of satellite images follows the following steps:

1. The multispectral image of low determination is transformed to make them equal in size to the panchromatic image.
2. The Red-Green-Blue assemblies of the multispectral image are changed into the into IHS segments.
3. The panchromatic image is changed as for the multispectral picture by carrying out their histogram coordinating. The intensity segment of the latter is set as reference in this process.
4. The power segment is substituted by the panchromatic image. The opposite of this is then done to achieve high determination multispectral image.

C. Image Fusion in Biomedical field

The therapeutic diagnosis and further treatment of detected diseases has been benefitted by the picture combination [3]. In this domain, various images collected during the diagnosis of the patient are considered and then integrated so as to provide more information about the illness. It is possible to form final entangled image from the pictures belonging to the both the same or different modalities [4]. In case the modalities are distinct, the data from all of them is consolidated [5]. Some examples are positron discharge tomography (PET), attractive reverberation picture (MRI), single photon emanation registered tomography (SPECT), and processed tomography (CT). Such images are found to be highly needful in the radiation and radiology oncology. For example, cerebrum tumors can be examined by the MRI images while the contrasts in the thickness of the tissues can be known by CT images.

The radiologists must consider the information from several image designs to reach at accurate results. It is very important to get precise outcomes for accurate diagnosis and further the treatment of the tumor. Therefore, the Image Fusion has advanced the medical diagnosis and treatment of life threatening diseases. It is now possible for the radiation oncologists to complete favorable position of force adjusted radiation treatment (IMRT).

II. LITERATURE REVIEW

Li et al. [1] proposed a multisensory image combination with the aid of wavelet transform technique. Here, the final blended picture is produced from the multimodal images by applying the cascaded sequence of forward and reverse wavelet transform. There is one more way in which wavelet transform is utilized. It is maximum selection scheme where the highest magnitude wavelet coefficient is picked in every sub-band. A normalized correlation has also been employed between two pictures by Burt and Kolczynski. Small-area sub-bands are used to find the subsequent coefficient for reconstruction by calculating weighted average of the two pictures in question. Another fusion strategy based on wavelet has been suggested by Zu Shu-long [7] where gradient criteria is used. Hill et al. [6] used the Dual Tree Complex Wavelet Transform (DT-CWT) to fuse the images. This technique is selective in direction and is shift invariant. Authors have come up with a wavelet transform based algorithm for the image fusion. The algorithm involves the decomposition of constituent pictures into the sub-images in which the same levels have similar resolution while the different levels have distinct resolution. During the fusion, sub-images of more frequency are employed under the comparative smoothness criterion and joint gradient. In the final stage, the recreation of these sub-images is done into the resultant solitary image with higher information. The method suggested here is used in case of multi-modal, multi-focus and remote detection pictures. Registration is the foremost thing to do during image fusion no matter which technique is used.

III. HYBRID ALGORITHMS - IMPLEMENTATION AND FLOW CHARTS

Figure 1. shows the process flow diagram of different methods and techniques. It depicts the work flow during implementation of the methods. Red image is achieved in the initial stage and in the beginning, images are pre-processed by examination of the colour frames. In case the frames are 3 (N=3), the RGB gets transformed into grey colour. The image is then resized to 256 x 256. Further, a particular hybrid algorithm is chosen depending upon which method to be used (such as DCT, PCA, DCT-PCA, DWT-DCT-PCA, DWT and DWT-PCA), and MIF processing is carried out. The final fused image is then constructed. Finally, the study of the performance is done with the help of quality matrices like Entropy, PSNR and MSE.

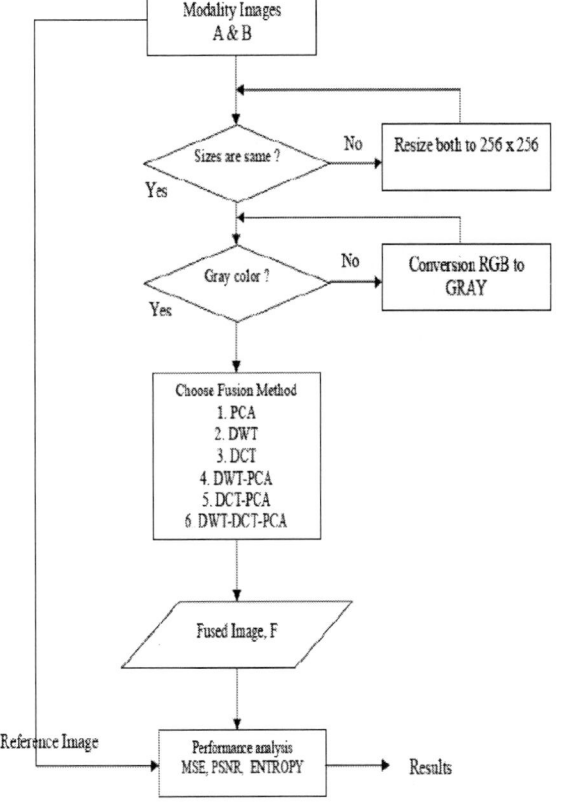

Fig 1: The complete flow diagram of MIF System

A. The Hybrid algorithm: DWT-PCA.

As shown in the process flow figure, the DWT and PCA methods have been combined to give DWT PCA – a hybrid algorithm [8]. The resulting hybrid algorithm processes as follows:

➢ In the beginning, the size of the constituent pictures is made to be equal followed by mapping of Red-Green-Blue (RGB) to grey colour. DWT method is then implemented as described in earlier section. In the hybrid algorithm, fusion rule is modified here. PCA algorithm is applied to combine the disintegrated coefficients (A1, A2, A3, A4 and B1, B2, B3, B4) of two original images (M1

and M2). The implementation of PCA algorithm is also explained elsewhere here. This step gives the fused coefficients.
➢ In the next stage, DWT is applied in inverse to reconstruct the fused picture F.
➢ In the end, the performance analysis is done by comparing the resulting fused image F with the reference image M1.

B. The hybrid algorithm: DCT-PCA.

DCT-PCA hybrid algorithm is the result of the integration of two algorithms, FDCT and PCA. The new algorithm works as follows:

➢ The first step is similar to that of DWT PCA algorithm.
➢ FDCT wrapping algorithm is then implemented to give the curvelet coefficients C1{1, i }{1,j} and C2{1, i}{1, j} [5].
➢ For fusion, PCA algorithm is applied later that provides fused Curvelet-PCA coefficient.
➢ Next, the final fused image is obtained upon application of inverse FDCT wrapping.
➢ In the final step, the performance analysis, both qualitative and quantitative is done by comparing final image with the reference image.

C. The hybrid algorithm: DWT-DCT-PCA

The three algorithms, DWT, FDCT and PCA are integrated to give a new hybrid algorithm DWT-DCT-PCA. The resulting algorithm contains merits of all the three algorithms without any problem created by their cons. This hybrid algorithm works as below:

➢ The initial step is again similar to other hybrid algorithms discussed above. The medical images with distinct modality (images M1 and M2) are registered and mapping of RGB is done to grey colours. The images are then resized to 256 x 256 dimensions.
➢ Then, DWT algorithm is implemented and group of four disintegrated frequency band coefficients is obtained for every image.
➢ FDCT wrapping algorithm is applied to each picture's decomposed coefficients. It gives respective curvelet descriptor coefficients.
➢ The decomposed coefficients are then joined selectively to give fused PCA-CT-DWT coefficients. It is done by applying PCA fusion rule.
➢ Regeneration is then done by applying inverse FDCT wrapping to respective coefficients of frequency bands discussed above.
➢ Then, the final fused image F is obtained by applying inverse DWT algorithm.
➢ Finally, the performance of the hybrid algorithm is checked by making comparisons between fused image F and reference image M1.

2019 Devices for Integrated Circuit (DevIC), 23-24 March, 2019, Kalyani, India

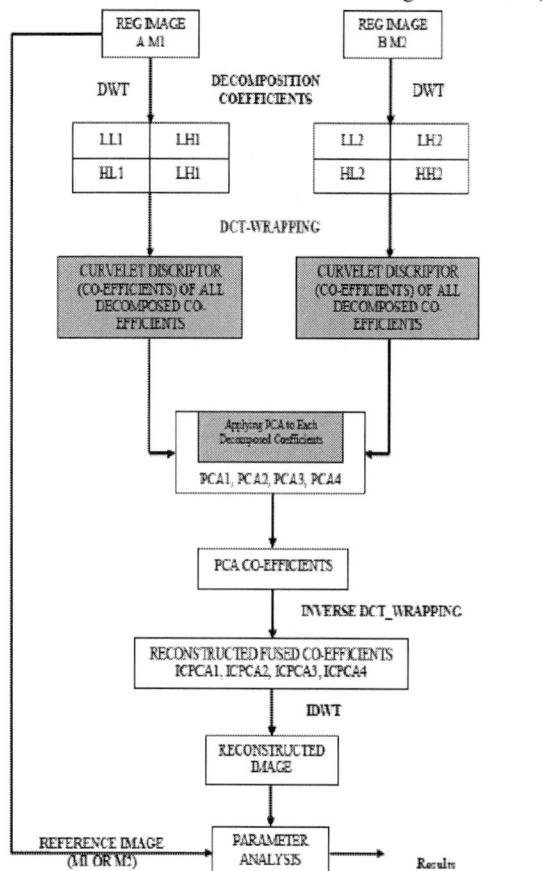

Fig 2: The process flow diagram of hybrid algorithm, DWT-DCT-PCA.

IV. PROPOSED METHODOLOGY

The Brovey transform (BT) technique was developed by Bob Brovey. This technique combines the multispectral and panchromatic images as depicted by the mathematical formulas used in this transform. In the formulas, some ratio of the Pan image is multiplied to an MS image and then division is done by the addition of MS images. Following equations dictates the fused R, G, and B images :

$$R_{new} = \frac{R}{(R+G+B)} \times PAN$$

$$G_{new} = \frac{G}{(R+G+B)} \times PAN$$

$$B_{new} = \frac{B}{(R+G+B)} \times PAN$$

The Brovey transform technique have been employed by number of researchers for the purpose of combining RGB images with a high resolution image (Pan). The BT technique also have some cons like it is limited to three bands and there occurs noteworthy radiometric distortion upon using the multiplicative techniques. Moreover, the parameters are not easy to be used, they require particular adaptation which can only performed by experienced analysts. Hence, the BT technique is not much user-friendly.

In BT, the distortion of the colors can take place in case of constituent pictures of different spectral range or of long term temporal changes.

WA(Weighted Average + BT(Brovey Transform) Algorithm

1. Initialize the weight w which can vary from 0.1 to 0.9 .
2. Calculate the weight average MX_WA using weight with respect to two input images I1 and I2.
3. Estimate the low resolution IL and high resolution IH based on the mean of the image as threshold condition.
4. Calculate the min and max resolution based on the relational operators between IL and IH as IL<IH and IL>IH.
5. Apply the summation between the min, max, IL and IH
6. Apply the final sum between the above and weight average MX_WA to get the final WA+BT fusion Image.

V. RESULTS

Image fusion processes is implemented in matlab . In the first experiment a pair of multi-focus image was taken. In figure 3 one is CT image and one is PET image. By Image fusion method, need to merge both the images.

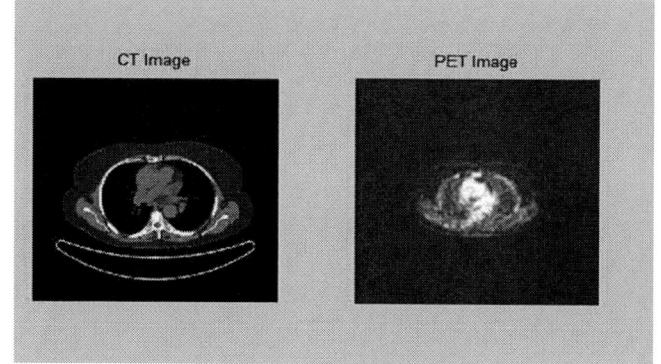

Fig 3:- Images for Image Fusion

Figure 4 is showing the output images of Image Fusion by using PCA, DWT, DCT, DWT+PCA, DCT+PCA, DWT+DCT+PCA algorithms. In this take two image and make the fusion process by the different methods/algorithms.

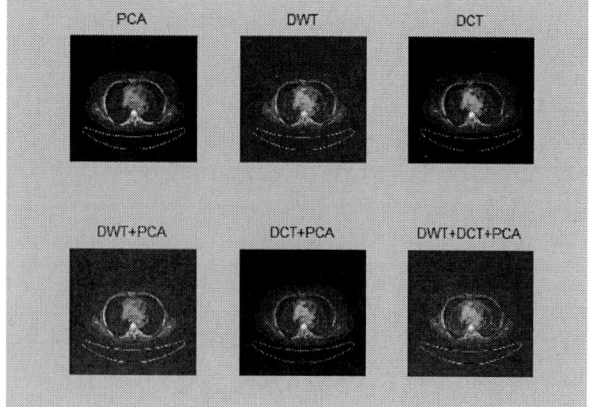

Fig 4:- Output result of Image Fusion by different method

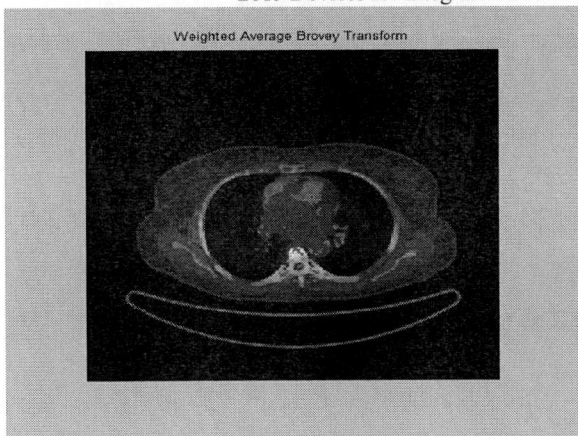

Fig 5:- Output Fusion image by Weighted Average Brovey Transform

Figure 5 is showing the output image of Fused Image by proposed method which is Weighted Average Brovey Transform.

For comparison the results of Existing methods and Proposed methods calculating PSNR and MSE. Table 5.1 is showing the values of PSNR and MSE for PCA, DWT, DCT, DWT+PCA, DCT+PCA, DWT+DCT+PCA and WA+BT. From the table 1, it is showing that PSNR is high for WA+BT method as compare to PCA, DWT, DCT, DWT+PCA, DCT+PCA, DWT+DCT+PCA while is MSE is low for WA+BT method as compare to PCA, DWT, DCT, DWT+PCA, DCT+PCA, DWT+DCT+PCA methods.

Methods	MSE	PSNR
PCA	551.5	56.84
DWT	536.7	56.96
DCT	300.7	59.47
DWT+PCA	287.3	59.67
DCT+PCA	155.5	62.34
DWT+DCT+PCA	155.3	62.34
WA+BT	146.5	62.59

Table 1 :- Comparison Table

VI. CONCLUSION & FUTURE SCOPE

A. Conclusion

In this paper, A novel image fusion technique based on WA(Weighted Average + BT(Brovey Transform) Algorithm has been presented and evaluated. The performance of this algorithm is compared with image fusion technique by PCA, DWT, DCT, DWT+PCA, DCT+PCA, DWT+DCT+PCA. It is concluded that image fusion by WA(Weighted Average + BT(Brovey Transform) Algorithm perform better as compare to PCA, DWT, DCT, DWT+PCA, DCT+PCA, DWT+DCT+PCA in terms of PSNR , MSE , CR.

B. Future Work

In future, we can further improve the performance of the PSNR and MSE by improve MSVD algorithm. In the future, we can also increase some parameters also for show the results comparison.

References

[1] Stephane G. Mallat, "*A Theory for Multi-resolution Signal Decomposition:* The Wavelet Representation", IEEE Transaction on Pattern Analysis andd Machine Intelligence. Vol. II No. 7, pages 671-693, July 1989

[2] A. Colligon, D. Vandermeulen, P. Seutens, G. Marchal, "*Registration of 3D multimodality medical imaging using surfaces and point landmarks*", *Pattern Recognition* Lett. 15, pp.461-467, (1994).

[3] S. Banerjee, D.P. Mukherjee, D. Dutta Majumdar, "*Point landmarks for the registration of CT and MR images*", Pattern Recognition Lett. 16, 1033- 1042, (1995)

[4] H. Li, B.S. Manjunath, S.K. Mitra, "*Multisensor image fusion using the wavelet transform*", GMIP: *Graphical Models Image Process.* 57(3) pp.235-245 (1995)

[5] A.A. Goshtasby, J.L. Moigne, "*Image registration Guest Editor's introduction*", Pattern Recognition 32, 1-2, (1999).

[6]PAUL Hill, Nishan Canagarajah and Dave Bull, "*Image Fusion using Complex Wavelets*", BVMC 2002

[7] Zhu Shu-long, "*Image Fusion Using Wavelet Transform*", *Symposium on Geospatial Theory, Process and Applications*, Ottawa 2002.

[8] Paul Hill, Nishan Canagarajah and Dave Bull, "*Image Fusion using Complex Wavelets*", *BVMC 2002.*

[9] Pavithra C, Dr. S. Bhargavi ," Fusion Of Two Images Based On Wavelet Transform ", International Journal Of Innovative Research In Science, Engineering And Technology Vol. 2, Issue 5, May 2013.

[10] Hari Om Shanker Mishra and Smriti Bhatnagar," MRI and CT Image Fusion Based on Wavelet Transform ", International Journal of Information and Computation Technology. ISSN 0974-2239 Volume 4, Number 1 (2014).

[11] Jitendra Gangwar, Anil Kumar, A.K.Jaiswal," Image Fusion of PET and CT Images based on Wavelet Transform ", International Journal of Computer Applications (0975 – 8887) Volume 121 – No.4, July 2015.

[12] Kusum Rani1, Reecha Sharma," Study of Image Fusion using Discrete wavelet and Multiwavelet Transform ", International Journal of Innovative Research in Computer and Communication Engineering Vol. 1, Issue 4, June 2013.

978-1-5386-6723-1/19 $31.00 © 2019 IEEE

2019 Devices for Integrated Circuit (DevIC), 23-24 March, 2019, Kalyani, India

A Novel Model for Analyzing Current-Voltage Characterization of SiC MOSFET

Sabuj Sarkar
Dept. of Electrical and Electronic Engineering
Khulna University of Engineering & Technology
Khulna-9203, Bangladesh
sabujeeekuet@gmail.com

Saikat Adhikary
Dept. of Physics
Adamas University
Kolkata-700126, India
spgrar@live.com

Md. Mostafizur Rahman
Dept. of Electronics and Communication Engineering
Khulna University of Engineering & Technology
Khulna-9203, Bangladesh
mostafiz@kuet.ac.bd

Abstract—**This paper mainly proposes a novel current-voltage characteristic model of silicon carbide metal-oxide semiconductor field effect transistor (SiC MOSFET) for achieving significant amount of drain current performance. First, drain current characteristics is performed for increasing gate-source voltage with variable mobility and channel length circumstances. Then I-V characteristics of SiC MOSFET are compared and evaluated for different operating states i.e. cut-off, linear and saturation. Later, drain current is characterized with varying the trans-conductance. Finally the drain current is simulated for the proposed novel method and it is with that of existing method. From the simulated performance, it is obvious that the performance in terms of drain current increases significantly for the novel model than that of the existing model.**

Keywords— Silicon carbide (SiC), MOSFET, trans-conductance, I-V Characteristics, drain current.

I.   INTRODUCTION

Recent advancements of SiC MOSFET play a pivotal role towards emerging generation semiconductor devices. By exploiting short-channel effects (SCEs) principle, an exclusive model of double gate shell-doping junction less (JL) MOSFET is explained in [1]. In-Ga-As quantum-well MOSFETs scaling study is carried out in [2] by considering excess off-state current. The distortion of MOSFET is studied in paper [3] by employing the charge-based EKV model. Channel length extraction of small gate length MOSFET is investigated in [4] by evaluating the values of split current-voltage. An advanced SiC power MOSFET behavioral model in presence of non linear junction capacitances is proposed in [5]. By using two newer measurement methods, SiC MOSFET characteristics as well as reproduction of their exact transient incident is studied in [6]. By exploiting a newer experimental method, performance study of 4H silicon carbide MOSFET is carried out in [7]. With the aim of minimizing the crosstalk, a novel gate low turn-off gate impedance driver is described in [8]. High voltage-high current characteristics of power MOSFET at active state is explained in [9] by using a new method. Novel 3-kV 4H-SiC reverse blocking of MOSFETs have been expressed in [10] in presence of high-voltage bidirectional switching applications. Modeling and experimental study of the current-trans-conductance dependent of the ALD1106 and ALD1107 array is described in [11]. An exclusive model is proposed in [12] in order to evaluate short-channel performance and scaling length of efficient-induced-conducting-path of multi-gate MOSFETs. A three-dimensional analytical model of triple material tri-gate silicon-on-nothing MOSFET is represented in [13].

Current-voltage characteristics of a double-gate MOSFET model under the saturation criterion are studied in [14]. Small-signal modeling as well as parameter collection criterion for an entire MOSFET is proposed in [15]. Instead of uniformly doped drift region of silicon carbide super-junction MOSFET, vertical variable doping profile is exploited in [16] for achieving a better tradeoff between breakdown voltage and specific ON-resistance ($R_{sp,on}$).

Presently, though a considerable amount performance in terms of drain current is drawn from the existing model of SiC MOSFET but that is below the desired level to some extent. The present paper proposes a novel model in order to improve the performance of SiC MOSFET further than the existing method to a significant level by analyzing the drain current characteristics.

II.   MOSFET CONSTRUCTION DIAGRAM

MOSFET is the abbreviation of metal oxide semiconductor field effect transistor. Figure 1 represents the construction diagram of a MOSFET in which source, gate, drain and body form four terminals in terms of four metallic contacts. The substrate i.e. body terminal of a MOSFET must be always kept at one of the extreme voltage level in the circuit, either the most positive for the P-MOS or the most negative for the N-MOS. Usually the body is coupled with the source of the MOSFET which converts it a three terminal device identical to field effect transistor.

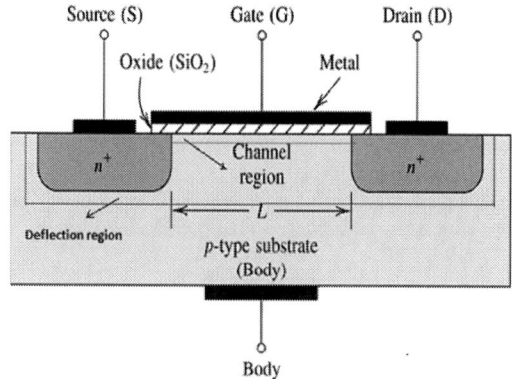

Fig. 1. Construction Diagram of MOSFET

Drain and source terminals are doped with heavily depending on the channel character i.e. p-type or n-type. The oxide layer is connected with the gate terminal while the intrinsic semiconductor is attached to the substrate i.e. body terminal. Depending on the flow of charge carriers, the channel width varies and device operates. From the source

978-1-5386-6723-1/19 $31.00 © 2019 IEEE

terminal the charge carriers pass through and leave via the drain terminal. Mostly an extremely thin layer of metal oxide insulates the gate terminal from the channel and the voltage across the drain and source controls the channel width. The capacitance exists in the MOFET forms the MOS capacitor. The source terminal is connected with body terminal that makes voltage across source to body terminal is zero. The voltage across the source and drain terminals control current flow between the source and drain. That is why it is a voltage controlled device.

Depending on the channel characteristics, MOSFETs are basically two type i.e. p-type or n-type. The current flow occurs due to the electron flow for an n-type MOSFET and due to the holes flow for a p-type MOSFET. Enhancement mode and depletion mode are the other two types of MOSFETs. For an enhancement mode when a gate-source voltage is applied the MOSFET device is switched on while for depletion mode when a gate-source voltage is applied the MOSFET device is switched off.

III. MOSFET FUNCTION THEORY

The primary function of the MOSFET is to control the current flow between source and drain by applying voltage across the source and drain. Basically silicon dioxide acts as the dielectric material that forms the capacitor of MOSFET and operates the transistor device. Under the oxide layer, a semiconductor layer exists that extends towards the source to drain end. It can be converted to p-channel and n-channel MOSFET respectively depending on the application of positive or negative gate voltage. Figure 2 explains the operating principle of n–channel MOSFET.

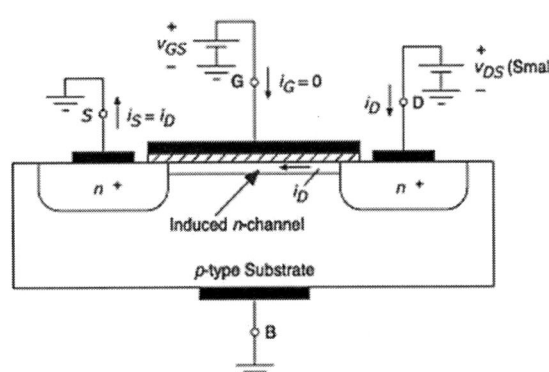

Fig. 2. Operation Diagram of MOSFET

When applied gate-source voltage is positive, holes are accumulated below the oxide portion undergo a repellent force and they are pushed at the substrate portion. The encircled negative charges are crowded in the depletion layer and they are linked with the acceptor atoms. The electron enriched source and drain is also attracted by the positive voltage and forms electron enhanced channel. Explosion of drain-source voltage causes current to flow through the source to drain. Electron density is controlled by the gate voltage. On the contrary, when gate-source voltage is negative, encircled positive charges are crowded in the depletion layer and they are linked with the donor atoms. The hole enriched source and drain is also attracted by the negative voltage and forms hole enhanced channel. Source to drain current flow is controlled by the gate-source voltage. The conduction begins only when gate-source voltage

reaches a pre-defined level i.e. threshold voltage. In linear condition, MOSFET drain current relationship is written as:

$$I_d = \mu_n * C_{ox} * \left(\frac{W}{L} \right) * \left[(V_{gs} - V_{th})V_{ds} - \frac{1}{2}V_{ds}^2 \right] \quad (1)$$

Where, I_d= drain current, μ_n= mobility of electrons, C_{ox}= oxide capacitance, W= width of the MOSFET, L= length of the channel, V_{ds}= voltage between drain to source, V_{gs}= gate source voltage, V_{th}= threshold voltage.

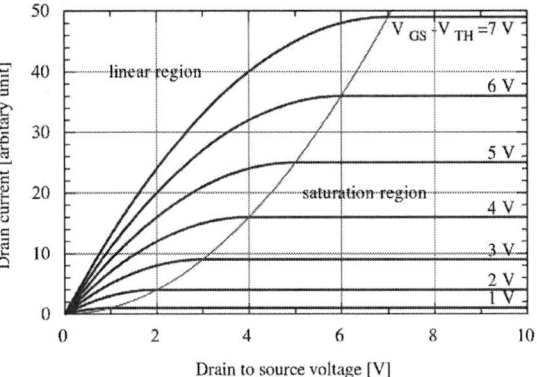

Fig. 3. Operating Mode of MOSFET

Figure 3 describes the operating mode of MOSFET in terms of I_d $vs.$ V_{ds} characteristics. Cut off, linear and saturation states are the three criterions of operation for MOSFET. When $V_{gs} < V_{th}$ and $I_d = 0$ i.e. no conductive channel is present, the device is in cut off state. When $V_{gs} < V_{th}$ and $V_{ds} < V_{ds,sat}$, the device is in the linear state of operation. When pinchoff is reached, further V_{ds} increases only increase I_d due to the formation of the higher field region. When $V_{gs} < V_{th}$ and $V_{ds} > V_{ds, sat}$, the device belongs to saturation state.

Positive-Channel MOSFET:

Basically positive-channel MOSFET is a four terminal device that contains hole enriched channel across drain and source.

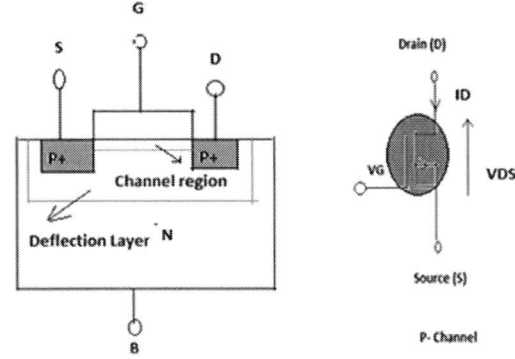

Fig. 4. P-Channel-Deflection-mode

Figure 4 describes the operation of p-channel-deflection-mode MOSFET. Highly enriched holes are kept in drain and source and substrate is filled with electrons. Bound positive charges crowded in the depletion region. Initially the enriched positive charged holes cause current to flow. Source to drain current flow is controlled by the gate-source voltage.

Negative-Channel MOSFET:

The negative-channel MOSFET is a four terminal device that contains electrons enriched channel across drain and source. Figure 5 describes the operation of negative-channel-enhanced-mode MOSFET. Highly enriched electrons are kept in drain and source and substrate is filled with holes. Bound negative charges crowded in the depletion region. Initially the enriched negative charged electrons cause current to flow. Source to drain current flow is controlled by the gate-source voltage.

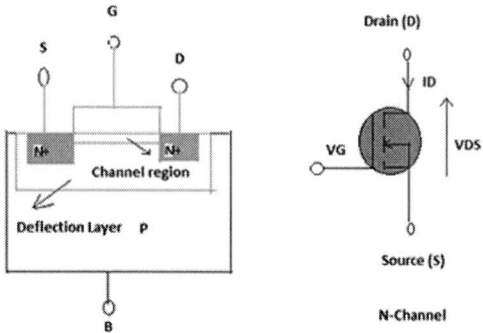

Fig. 5. N-Channel-Enhanced-mode

IV. EQUIVALENT CIRCUIT OF MOSFET

MOSFET is a three terminal device consisting drain, gate and source. The equivalent circuit diagram of a MOSFET is described by the Figure 6. Gate terminal is followed by a gate resistance. Two separate effects from gate-source and gate-drain are connected in parallel with resistance from the channel. This two effects has its individual resistance as well as capacitance from two sides. The it is connected in series with the gate and drain terminal resistance. Resistance from the source is also connected in the equivalent circuit. Thus the MOSFET device has a MOS capacitor that controls the transistor.

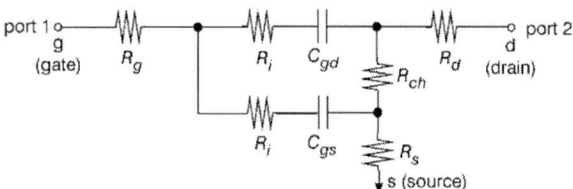

Fig. 6. MOSFET Equivalent Circuit

V. MATHEMATICAL ANALYSIS OF MOSFET FOR EXISTING AND NOVEL MODEL

MOSFET Gate-source voltage, V_{gs} controls the drain current flow. A MOSFET with its gate and drain connected together always operates in the constant current state, its i_d-V_{gs} relationship is expressed as:

$$\sqrt{I_d} = \sqrt{K} V_{gs} - \sqrt{K} V_{rs} \tag{2}$$

where the threshold voltage V_{th} depends on the source-body potential v_{sb} as

$$V_{th} = V_{th0} \pm \gamma \left[\sqrt{|v_{sb}| + 2|\emptyset_f|} \right] - \sqrt{2|\emptyset_f|} \tag{3}$$

The equation related to saturation state of MOSFET following the square law is expressed as:

$$I_d = (1/2) * \mu_n * C_{ox} * (W/L) * (V_{gs} - V_{th})^2 \tag{4}$$

Thus MOSFET operates in saturation state creates a constant current with varying gate-source voltage. That is it always behaves like a current source. Transconductance is the measure of how well the MOSFET device is able to convert the voltage into output current. The leading equations are:

$$g_m = \mu_n * C_{ox} * (W/L) * (V_{gs} - V_{th}) \tag{5}$$

$$g_m = (2 * \mu_n * C_{ox} * W/L * I_{ds})^{0.5} \tag{6}$$

$$g_m = (2 * I_{Ds})/(V_{gs} - V_{th}) \tag{7}$$

Where, g_m is the trans-conductance of the device, μ_n is the mobility of carriers, C_{ox} is the capacitance of oxide layer, W is the depletion width, L is the length of the channel, $(V_{gs} - V_{th})$ = overdrive voltage. Thus output current I_d in linear state is given by:

$$I_d = (0.5*\mu_n * C_{ox} * W/L) *[2*(V_{gs} - V_{th}) * V_{ds} - V_{ds}^2] \tag{8}$$

In the saturation state, the current equation for I_d is given by:

$$I_d = (0.5*\mu_n * C_{ox} * W/L) * [V_{gs} - V_{th}]^2 \tag{9}$$

While mobility, μ and channel length, L are varied simultaneously, according to the double gate (DG) model i.e. novel model, equation 8, 9 becomes

$$I_d = (0.5* \mu_n * C_{ox} * W/L) * [2* (2 * V_{gs} - V_{th}) * V_{ds} - V_{ds}^2] \tag{10}$$

$$I_d = (0.5*\mu_n * C_{ox} * W/L) * [2 * V_{gs} - V_{th}]^2 \tag{11}$$

For the novel model, SiC MOSFET performs improved drain current characteristics than the existing model single gate. For this case, the gate voltage will be doubled and the novel model provides greater channel performance.

VI. SIMULATED RESULTS AND PERFORMANCE ANALYSIS

The simulation performances are studied and analyzed in cooperation with Matlab simulink software. Figure 7 and 8 shows the the drain current versus gate-source voltage graphical representations for variable mobility and channel length circumstances. As observed from the two output graph that with changes in the mobility as well as the channel length, the drain current shows a variation with increase in the gate source voltage. As the mobility of carriers increases, the drain current increases. This is because, increase of mobility signifies that there are less collision and electrons can move through the channel freely and hence generate more current. With the decrease in channel length, the current increases. Because a shorter channel has more concentration of carriers and more carrier concentration will generate more current.

flow between source and drain can exist. So in linear state of operation, this device is controlled by the applied voltage.

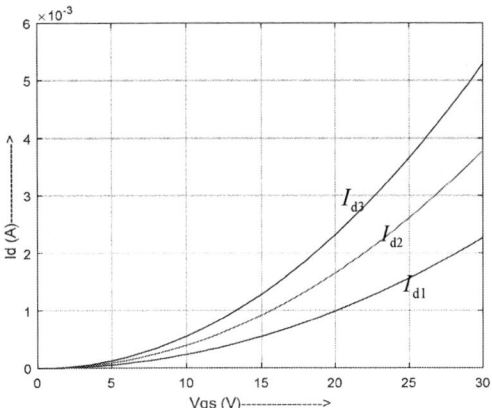

Fig. 7. V_{gs} v/s I_d characteristics when mobility is varied

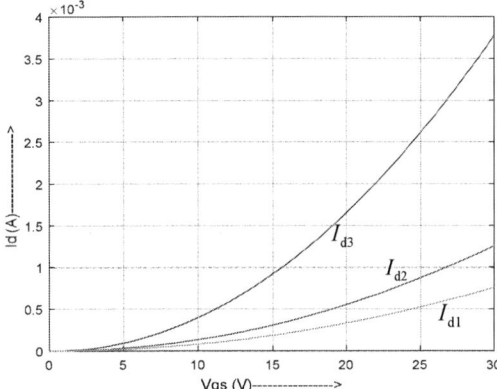

Fig. 8. V_{gs} v/s I_d characteristics when channel length is varied

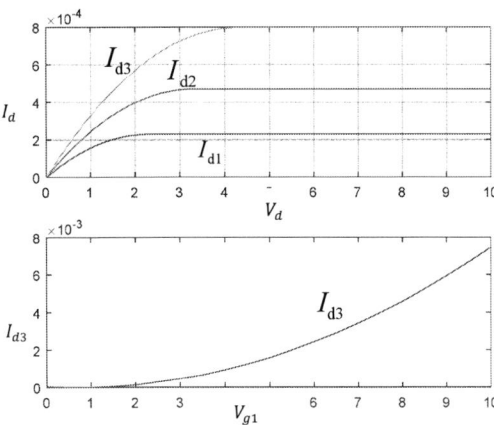

Fig. 9. I-V Characteristics of SiC MOSFET at Triode State

The graph for both the linear state and the saturation state has been plotted in Figure 9 and 10 respectively. When the gate voltage, V_{gs} is in the range $0<V_{gs}<V_{th}$, the depletion layer exists between source and drain and gate-source is below the threshold level. Thus no carrier flow occurs and device is laterally off. As the gate-source voltage increases and reaches at a state that is greater than the threshold value i.e. $V_{gs} > V_{th}$, conduction channel is formed, then a current

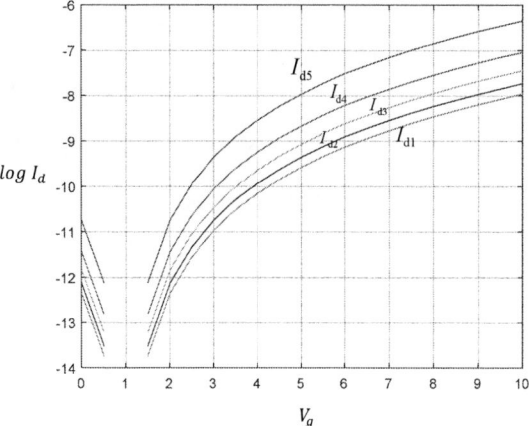

Fig. 10. I-V Characteristics of SiC MOSFET at Saturation State

Figure 10 depicts that beyond the pinch-off point and adherent to the drain portion a depletion layer is formed. As the gate voltage increases the depletion layer extends toward the source. MOSFET operates in this mode is termed as the saturation state of operation.

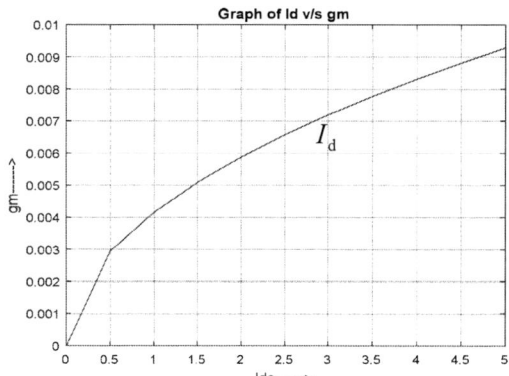

Fig. 11. g_m v/s I_d Characteristics of MOSFET

The variation of the trans-conductance with the drain current is described by figure 11 where the trans-conductance is proportional to the square root of drain current.

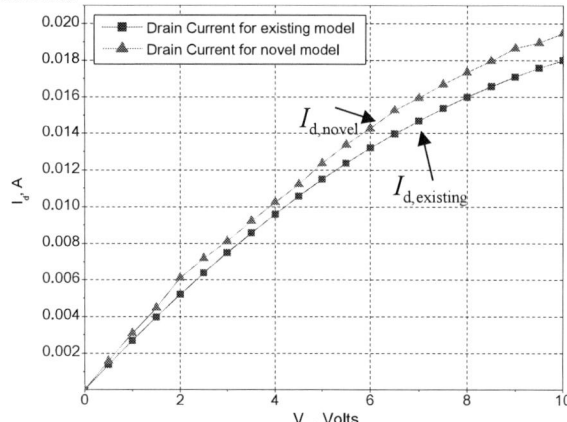

Fig. 12. Drain Current Performance of Novel Model with Existing Model for SiC MOSFET

978-1-5386-6723-1/19 $31.00 © 2019 IEEE   459

Figure 12 shows the drain current versus drain-source voltage representations of SiC MOSFET for existing as well as novel model. From the plot, it seen that for increasing drain-source voltage the drain current increases for both of the existing as well as novel. But for the novel model, the increment of drain current is significant than that of existing model.

VII. CONCLUSIONS

The simulation studies of current-voltage characteristics for SiC MOSFET are studied for variable mobility, channel length as well as different channel state circumstances. Later the current-voltage performances are evaluated for variable trans-conductance. Finally, the drain current characteristics of SiC MOSFET for novel as well as existing model are simulated. From the simulated results, it is obvious that the performance of SiC MOSFET for the novel model improved significantly than that of the existing model for variable drain-source voltage.

REFERENCES

[1] N. Jaiswal and A. Kranti, "Modeling Short-Channel Effects in Core–Shell Junctionless MOSFET," *IEEE Transactions on Electron Devices*, vol. 66, no. 1, pp. 292-299, Jan. 2019.

[2] J. Lin, X. Zhao, I. M. Clavero, D. A. Antoniadis and J. A. del Alamo, "A Scaling Study of Excess OFF-State Current in InGaAs Quantum-Well MOSFETs," *IEEE Transactions on Electron Devices*, vol. 66, no. 3, pp. 1208-1212, March 2019.

[3] F. Chicco, A. Pezzotta and C. C. Enz, "Charge-Based Distortion Analysis of Nanoscale MOSFETs," *IEEE Transactions on Circuits and Systems I: Regular Papers*, vol. 66, no. 2, pp. 453-462, Feb. 2019.

[4] S. Severi, G. Curatola, C. Kerner and K. De Meyer, "Accurate channel length extraction by split C-V measurements on short-channel MOSFETs," *IEEE Electron Device Letters*, vol. 27, no. 7, pp. 615-618, July 2006.

[5] Z. Duan, T. Fan, X. Wen and D. Zhang, "Improved SiC Power MOSFET Model Considering Nonlinear Junction Capacitances," *IEEE Transactions on Power Electronics*, vol. 33, no. 3, pp. 2509-2517, March 2018.

[6] H. Sakairi, T. Yanagi, H. Otake, N. Kuroda and H. Tanigawa, "Measurement Methodology for Accurate Modeling of SiC MOSFET Switching Behavior Over Wide Voltage and Current Ranges," *IEEE Transactions on Power Electronics*, vol. 33, no. 9, pp. 7314-7325, Sept. 2018.

[7] Martin Hauck, Johannes Lehmeyer, Gregor Pobegen, Heiko B. Weber and Michael Krieger, "An adapted method for analyzing 4H silicon carbide metal-oxide-semiconductor field-effect transistors," *Communication Physics*, vol. 2, no. 5, pp. 1-6, 2019.

[8] Y. Li, M. Liang, J. Chen, T. Q. Zheng and H. Guo, "A Low Gate Turn-OFF Impedance Driver for Suppressing Crosstalk of SiC MOSFET Based on Different Discrete Packages," *IEEE Journal of Emerging and Selected Topics in Power Electronics*, vol. 7, no. 1, pp. 353-365, March 2019.

[9] C. Salcines, A. Kruglov and I. Kallfass, "A Novel Characterization Technique to Extract High Voltage-High Current IV Characteristics of Power MOSFETs from Dynamic Measurements," *IEEE 6th Workshop on Wide Bandgap Power Devices and Applications (WiPDA)*, Atlanta, GA, 2018, pp. 1-6.

[10] S. Mori et al., "High-Temperature Characteristics of 3-kV 4H-SiC Reverse Blocking MOSFET for High-Performance Bidirectional Switch," *IEEE Transactions on Electron Devices*, vol. 64, no. 10, pp. 4167-4174, Oct. 2017.

[11] M. Cheng and V. Prodanov, "Experimental Study and Modeling of the gm-I Dependence of Long-Channel MOSFETs," *IEEE 61st International Midwest Symposium on Circuits and Systems (MWSCAS)*, Windsor, ON, Canada, 2018, pp. 570-573.

[12] J. J. Liou and T. Chiang, "A Novel Effective-Conducting-Path-Induced Scaling Length Model and Its Application for Assessing Short-Channel Performance of Multiple-Gate MOSFETs," *IEEE Transactions on Electron Devices*, vol. 65, no. 10, pp. 4535-4541, Oct. 2018.

[13] P. Banerjee, P. Saha and S. K. Sarkar, "Analytical modelling and performance analysis of gate engineered TG silicon-on-nothing metal–oxide–semiconductor field-effect transistor," *IET Circuits, Devices & Systems*, vol. 12, no. 5, pp. 557-562, 9 2018.

[14] Y. Taur and H. Lin, "Modeling of DG MOSFET I-V Characteristics in the Saturation Region," *IEEE Transactions on Electron Devices*, vol. 65, no. 5, pp. 1714-1720, May 2018.

[15] Y. Cao, W. Zhang, J. Fu, Q. Wang, L. Liu and A. Guo, "A Complete Small-Signal MOSFET Model and Parameter Extraction Technique for Millimeter Wave Applications," *IEEE Journal of the Electron Devices Society*, pp. 1-1, February 2019. (Early Access)

[16] P. Vudumula and S. Kotamraju, "Design and Optimization of SiC Super-Junction MOSFET Using Vertical Variation Doping Profile," *IEEE Transactions on Electron Devices*, vol. 66, no. 3, pp. 1402-1408, March 2019.

978-1-5386-6723-1/19 $31.00 © 2019 IEEE

Work-function modulated hetero gate charge plasma TFET to enhance the device performance

Sasmita Sahoo, Sidhartha Dash

Dept. of Electronics & Communication Engg.,
Siksha 'O' Anusandhan deemed to be University,
Bhubaneswar, India.
sasmitasahoo@soa.ac.in; sidharthadash@soa.ac.in

Guru Prasad Mishra

Dept. of Electronics & Telecommunication Engg.,
NIT Raipur, Raipur, India.
gpscmishra.etc@nitrr.ac.in

Abstract— **In this paper a simulated device configuration is introduced for doping less charge plasma tunnel FET (CP-TFET) to improving the drain current and ambipolar nature. To achieve these improvements, we have proposed drain electrode work-function modulation engineering along with hetero dielectric in CP-TFET. The use of varied work-function modulation technique in drain region significantly reduces the ambipolar current by modifying the tunneling barrier width at the proximity of drain junction. However, the combined effect of hetero dielectric with drain electrode work-function modulation results in improved ON-current (I_{ON}) and reduced ambipolar-current (I_{AMB}). The paper presents a comparative DC analysis of drain work-function modulated hetero gate charge plasma TFET (DWM-HCP-TFET) with that of conventional charge plasma TFET (CP-TFET). Analysis is done to prove the superiority of DWM-HCP-TFET over the conventional model.**

Keywords—charge plasma TFET, work-function modulation, hetero-gate dielectric engineering, ambipolar current

I. INTRODUCTION

The semiconductor industry was fully controlled by MOSFET in the last few decades. Insightful statement of Moore on device scaling opens up the path for miniaturizing the transistor further. This reduction in device dimension improves its performance in terms of minimizing the die area, cost and high-frequency parameters [1-4]. Along with all these advantages, MOSFET is still associated with some serious drawbacks such as rise in leakage current, subthreshold slope (SS) and short cannel effects (SCEs) [5-7]. The unacceptable exponential increase in leakage current gives rise to power crises in electronic circuits. In order to overcome the issues, a latest device named Tunneling Field Effect Transistor (TFET) is coming in the picture as a better substitution of conventional MOSFET [8-9]. It provides extremely low OFF-state current (I_{OFF}) and lower SS as a result of its unique current conduction technique [9-11]. Thus, TFET have the ability to minimize the power requirement further by down-scaling the operating voltage [12-14]. However, the effect of ambipolar current (I_{AMB}), lower drain current (I_{ON}) and random dopant fluctuations (RDFs) are still the major issues related to TFET technology.

To overcome the mentioned issues, for the past few years researches are focusing on junction less transistors with high body doping concentration greater than 10^{18}/cm³ [15,16]. This device presents some serious issues related to lower drain current for mobility degradation, increase in parasitic capacitance and requirement of high work-function metal to turn-off the device [17]. So a new structure evolved with intrinsic body known as doping less TFET, whose source and drain regions are formed by charge plasma concept. That uses a specific work-function metal contact over source and drain region for the induction of p⁺ and n⁺ regions. This will automatically ease the way of fabricating the device by eliminating high temperature doping and annealing process [18-19]. Still, doping less TFET is suffering from low drain current, because of the presence of large tunneling barrier at source-channel interface [19, 20], which can overcome by introducing hetero-gate dielectric engineering. Hetero-gate oxide at both the upper and lower gate improves the drain current by minimizing the tunneling barrier width at source junction without degrading OFF-state current [21]. Ambipolarity is another important characteristic of the device, which restricts its application in digital circuit design. Researchers had proposed dual metal gate [22], double metal drain contact [23] to adjust the overall electric field for minimizing the ambipolar current. In continuation with that, tri-material gate contact is also introduced in which work-function varies throughout the gate contact region [24]. This has already been fabricated by T. L. Li et al. in 2005 using co-sputtering technique, where work function adjustment from 3.93 to 4.93 eV is done linearly by using Hf_xMo_{1-x} binary alloys [25]. In 2013, K. M. Choi analyzed the work-function variation (WFV) effect on TFETs [26]. The approach of work-function modulation technique is adopted here for better results.

Here, we present a simulated drain work-function modulated doping less TFET with hetero-dielectric (DWM-HCP-TFET) and has compared the results with CP-TFET. The structure comprises a linearly modulated work-function variation from the drain-channel interface towards the drain end to generate a high potential barrier. This will help minimizing the ambipolar conduction and the use of hetero-dielectric gate improves the device current up to a specific range. The drain current and ambipolar conduction of the model is analyzed by its energy band diagram, electric field and BTBT generation rate. Improved DC parameters with reduced fabrication steps make this device a qualified candidate to be used as low power devices.

II. STRUCTUAL AND SIMULATION DETAILS

Fig. 1(a) and 1(b) shows the schematic representation of CP-TFET with drain work-function modulated hetero dielectric CP-TFET (DWM-HCP-TFET). The design and simulation of the device is processed by Sentaurus device simulator [27]. The 2D simulator uses nonlocal band-to-band tunneling model to calculate the tunneling generation rate. Drift-diffusion model and Hurkx trap-assisted tunneling model are also considered as carrier transport models in the simulation process. In order to find the carrier life time at tunneling region, Shockley-Read-Hall (SRH) recombination model is used in the simulation.

The selected source and drain work-function metals are 5.93 eV and 3.90 eV respectively for CP-TFET. On the other hand the proposed structure (DWM-HCP-TFET) is having a drain electrode of metal whose work-function linearly varies from 4.40 eV to 3.90 eV as shown in Fig. 2. Intrinsic silicon body doping is taken 1×10^{16} atoms/cm^3. Here the source, channel and drain lengths are considered as 50 nm each. The silicon body thickness (t_{si}) and gate oxide thickness (t_{ox}) are chosen as 10 nm and 1 nm respectively. Source-gate electrode (L_{SC}) and drain-gate electrode (L_{CD}) separation are considered 2 nm and 15 nm in order to match with the depletion region width at specified regions for a conventional double gate TFET. Hetero-dielectric (HfO$_2$ and SiO$_2$) gate materials of 75 nm length are applied on top and bottom gate for better results. The device with both the hetero-dielectric and drain work-function modulation engineering provides enhanced drain current and suppressed ambipolar current as compared to CP-TFET.

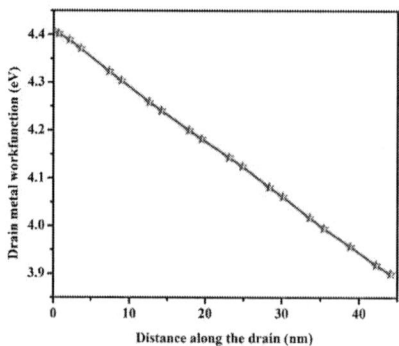

Fig. 2. The work-function variation along the drain region for the proposed device

III. ANALYSIS OF SIMULATED RESULTS

The current comparison graph of CP-TFET and DWM-HCP-TFET are shown in Fig. 3. By visualizing the graph it is understood that the proposed structure exhibit improved drain current and reduced ambipolarity. This rise in drain current is due to the introduction of hetero-dielectric gate (HfO$_2$ at source-channel interface and SiO$_2$ at drain-channel interface), which improves the carrier velocity at the tunneling junction. The bidirectional nature (ambipolar) of the device is further minimized by applying work-function modulation engineering at drain electrode. Varied work-function at drain end reduces the electric field profile.

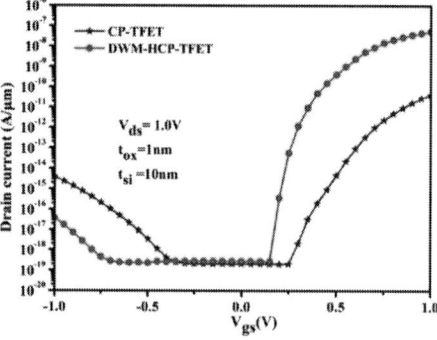

Fig. 3. Drain current analysis with different V_{gs} for CP-TFET and DWM-HCP-TFET

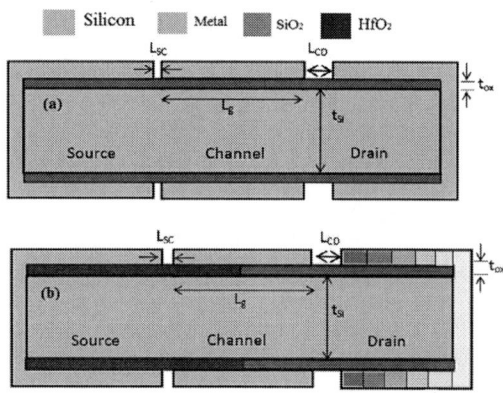

Fig. 1. Schematic representation of (a) conventional CP-TFET (b) DWM-HCP-TFET

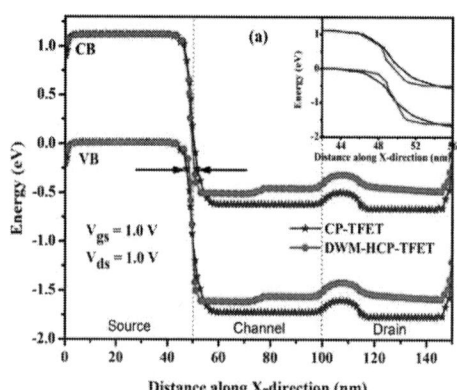

978-1-5386-6723-1/19 $31.00 © 2019 IEEE   462

2019 Devices for Integrated Circuit (DevIC), 23-24 March, 2019, Kalyani, India

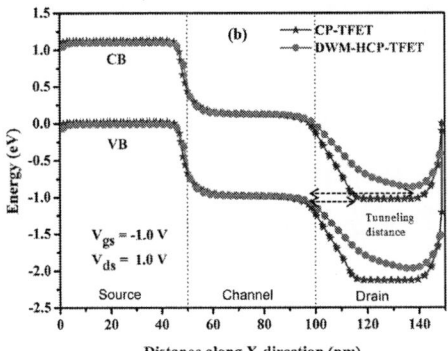

Fig. 4. Energy band analysis for CP-TFET and DWM-HCP-TFET in both the (a) ON and (b) ambipolar state

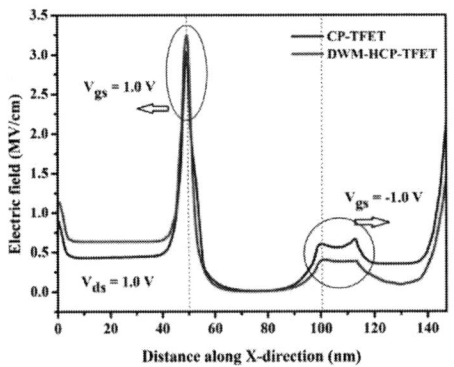

Fig. 5. Electric field comparison of DWM-HCP-TFET with CP-TFET in both the ON and ambipolar state

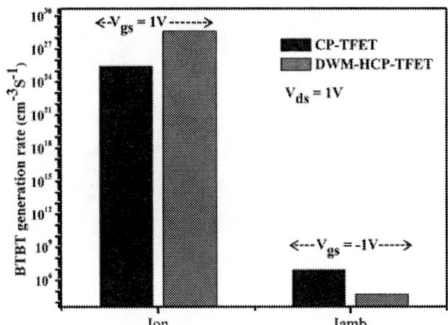

Fig. 6. Band to Band tunneling (BTBT) generation rate comparison in both the ON and ambipolar state

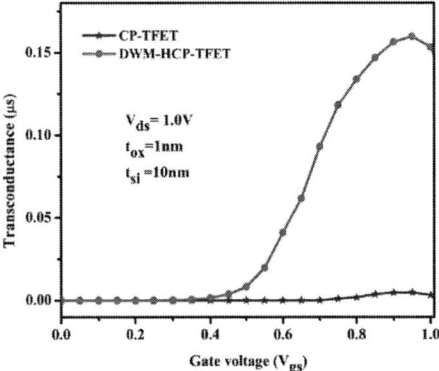

Fig. 7. Change in transconductance w.r.t. V_{gs} variation for CP-TFET and DWM-HCP-TFET

Junction abruptness is the key factor of getting a better device characteristic in case of doping less TFET. Fig. 4 describes the energy band diagram corresponds to ON and ambipolar-state of the device. In Fig 4(a), it is observed that the proposed structure with hetero-gate dielectric (HfO_2 near source region) is having a large band banding in the ON-state, which allow a maximum amount of electrons to easily pass through the junction and hence provides better drain current characteristics. On the other hand, the amount of band banding in the ambipolar-state depends on both the hetero-dielectric and work-function modulation engineering. The techniques lead to an increased tunneling barrier length at the drain-channel interface as shown in Fig. 4(b), which in turn increases the tunneling band gap and reduces the ambipolar current of the proposed structure as compared to CP-TFET. The effect of these techniques also lowers the electric field at the drain region as shown in Fig. 5. This provides a resistive path for the charge carriers (holes) to tunnel towards channel from drain end, and guides to the suppression of ambipolar current. On the contrary, the electric field of DWM-HCP-TFET is higher at the source-channel junction, because of the application of hetero-dielectric material in the vicinity of the tunneling junction. As a result of this, higher band to band tunneling generation rate is achieved for the device and this enhances the ON-current by 2 decade in comparison to CP-TFET.

In charge plasma TFET, high density regions (n^+ and p^+) are formed by applying specific work-function metal contact. This creates plasma of charge carriers that take part in tunneling process. Band to band generation rate helps to determine the tunneling probability and magnitude of tunneling volume at the interface and estimate the drain current in both the ON and ambipolar-state. Fig. 6 shows the maximum BTBT generation rate comparison of the suggested device with CP-TFET in ON and ambipolar-state (V_{gs}=1V & V_{gs}= -1V). Linearly modulated work-function of metal at drain electrode increases the tunneling barrier width at the drain region and exhibits reduced BTBT generation rate as compared to conventional charge plasma structure. Higher bandgap variation at the drain interface decreases tunneling efficiency by reducing the BTBT generation rate of DWM-HCP-TFET. It is to be noted that the device under study has a better BTBT rate due to hetero-dielectric applied at top and bottom gate, thus enhances tunneling probability and thus, ON-current improves.

The frequency analysis of the device can be better explained by its transconductance (g_m) characteristics. Converting supply voltage to output current is a significant characteristics obtained by transconductance. The comparison of transconductance (g_m) with the variation of gate to source voltage (V_{gs}) for CP-TFET and DWM-HCP-TFET is shown in Fig.7. DWM-HCP-

978-1-5386-6723-1/19 $31.00 © 2019 IEEE   463

TFET shows a higher transconductance of 0.1597 μS at V_{gs}= 0.95V. Higher transconductance value at lower gate to source voltage is achieved by the introduction of hetero-gate engineering at the proximity of source-channel junction. This helps the device to operate at lower supply voltage and shows a better performance in analog circuit design.

IV. CONCLUSION

Reduced fabrication complexity and lower RDF makes doping less TFET production more cost effective and reliable. The presence of work-function modulated drain along with hetero-dielectric engineering in DWM-HCP-TFET makes this device more capable of reducing ambipolar current and enhancing drain current as compared to conventional CP-TFET. The DC performance evaluation exhibits 2-decade improvement in drain current and 3-decade reduction of ambipolar current without affecting its OFF-current characteristics. Improved ON-current and reduced ambipolarity makes DWM-HCP-TFET device a competitor in future low power circuit applications.

REFERENCES

[1]   G. E. Moore, "Cramming more components onto integrated circuits," Proc. IEEE, vol. 86, no. 1, pp. 82–85, Jan. 1998.

[2]   P. H. Woerlee et al., "RF-CMOS performance trends," IEEE Trans. Electron Devices, vol. 48, no. 8, pp. 1776–1782, Aug. 2001.

[3]   T. Ghani et al., "Scaling challenges and device design requirements for high performance sub-50 nm gate length planar CMOS transistors," in Proc. Symp. VLSI Technol., 2000, pp. 174–175.

[4]   J. D. Meindl, "Low power microelectronics: Retrospect and prospect,"Proc. IEEE, vol. 83, no. 4, pp. 619–635, Apr. 1995.

[5]   S. Bangsaruntip, G. M. Cohen, A. Majumdar, and J. W. Sleight,"Universality of short-channel effects in undoped-body silicon nanowire MOSFETs," IEEE Electron Device Lett., vol. 31, no. 9, pp. 903–905,Jul. 2010.

[6]   M. Kumar, M. A. Hussain, and S. K. Paul, "Performance of a two input nand gate using subthreshold leakage control techniques," J. Electron Devices, vol. 14, pp. 1161–1169, Jun. 2012.

[7]   P. S. Peercy, "The drive to miniaturization," Nature, vol. 406, no. 6799, pp. 1023–1026, 2000.

[8]   A. M. Ionescu, and H. Riel, "Tunnel field effect transistors as energy efficient electronics switches," Nature, vol. 479, no. 7373, pp. 329–337, Nov. 2011.

[9]   W.Y. Choi, B.-G. Park, J. D. Lee, et al., "Tunneling field-effect transistor (TFETs) with subthreshold swing (SS) less than 60 mV/dec," IEEE Electron Device Lett., vol. 28, no. 8, pp. 743–745, Aug. 2007.

[10] U. E. Avci, D. H. Morris, and I. A. Young, "Tunnel field-effect transistors: Prospects and challenges," IEEE J. Electron Devices Soc., vol. 3, no. 3, pp. 88–95, May 2015.

[11] P. K. Dubey, Nanoelectronics Devices, Circuits and System. Elsevier: United States, 2019, pp. 3-25.

[12] S. Sahoo, S. Dash, and G. P. Mishra, "An Accurate Drain Current Model for Symmetric Dual Gate Tunnel FET Using Effective Tunneling Length," Nanoscience & Nanotechnology-Asia, vol. 9, no. 1, pp. 85-91, Jan. 2019.

[13] S. Cristoloveanu, J. Wan, and A. Zaslavsky, "A Review of Sharp-Switching Devices for Ultra-Low Power Applications," Journal of the Electron Devices Soc., vol. 4, no. 5, pp. 215-226, Sep. 2016.

[14] C. Anghel, Hraziia, A. Gupta, A. Amara, and A. Vladimirescu, "30-nm Tunnel FET With Improved Performance and Reduced Ambipolar Current," IEEE Trans. Electron Devices, vol. 58, no. 6, pp. 1649–1654, June. 2011.

[15] B. Ghosh and M. W. Akram, "Junctionless tunnel field effect transistor," IEEE Electron Device Lett., vol. 34, no. 5, pp. 584–586, May 2013.

[16] S. Tirkey, D. Sharma, B. Ram, and D. S. Yadav, "Introduction of a metal strip in oxide region of junctionless tunnel field-effect transistor to improve DC and RF performance," J. Comput. Electron., vol. 16, no. 3, pp. 714–720, Sep. 2017.

[17] J. -P. Colinge et al., "Nanowire transistors without junctions," Nature Nanotechnol., vol. 5, no. 3, pp. 225-229, Mar. 2010.

[18] B. R. Raad, S. Tirkey, D. Sharma, and P. Kondekar, "A new design approach of dopingless tunnel FET for enhancement of device characteristics," IEEE Trans. Electron Devices, vol. 64, no. 4, pp. 1830–1836, Apr. 2017.

[19] M. J. Kumar and S. Janardhanan, "Doping-less tunnel field effect transistor: Design and investigation," IEEE Trans. Electron Devices, vol. 60, no. 10, pp. 3285–3290, Oct. 2013.

[20] B. V. Chandan, K. Nigam, D. Sharma, and V. A. Tikkiwal, "A novel methodology to suppress ambipolarity and improve the electronic charcteristics of polarity-based electrically doped tunnel FET," Applied Physics A, vol. 125 , pp. 81-1-7, Jan. 2019.

[21] W. Y. Choi, and W. Lee, "Hetero-gate-dielectric tunneling field effect transistors," IEEE Trans. Electron. Devices, vol. 57, no. 9, pp. 2317–2319, Sep. 2010.

[22] B. R. Raad, D. Sharma, P. Kondekar, K. Nigam, and D. S. Yadav, "Drain work function engineered doping-less charge plasma TFET for ambipolar suppression and RF performance improvement: A proposal, design, and investigation," IEEE Trans. Electron Devices, vol. 63, no. 10, pp. 3950–3957, Oct. 2016.

[23] D. S. Yadav, A. Verma, D. Sharma, S. Tirki, and B. R. Raad, "Comparative investigation of hetero gate dielectric and drain engineered charge plasma TFET for improved DC and RF performance," Superlattices and Microstruct., vol. 111, pp. 123–133, 2017.

[24] K. Nigam, S. Pandey, P. N. Kondekar, D. Sharma, and P. K. Parte, "A Barrier Controlled Charge Plasma-Based TFET With Gate Engineering for Ambipolar Suppression and RF/Linearity Performance Improvement," IEEE Trans. Electron Devices, vol. 64, no. 6, pp. 2751–2757, June. 2017.

[25] T. L. Li, C.H. Hu, W. L. Ho, H. C. H. Wang, C.Y. Chang, "Continuous and Precise Work-Function Adjustment for Integratable Dual Metal Gate CMOS Technology Using Hf–Mo Binary Alloys," IEEE Tran. Electron Devices. vol. 52, pp. 1172-1179, May 2005.

[26] K. M. Choi, and W. Y. Choi, "Work-Function Variation Effects Of Tunneling Field-Effect Transistors (TFETs)," IEEE Electron Device Lett., vol. 34, no. 8, pp. 942–944, Aug 2013.

[27] Sentaurus Device User Guide, Synopsys, Inc., Mountain View, USA, 2014.

2019 Devices for Integrated Circuit (DevIC), 23-24 March, 2019, Kalyani, India

Design of Mems Based Piezoelectric Energy Harvester for Pacemaker

Ashutosh Anand
Dept. of Electronics and Communication Engineering
Birla Institute of Technology Mesra
Ranchi, India
ashunitjsr@gmail.com

Sudip Kundu
Member, IEEE
Dept. of Electronics and Communication Engineering
Birla Institute of Technology Mesra
Ranchi, India
kundu.sudip@gmail.com

Abstract— **In this paper, a spiral-shaped piezoelectric cantilever structure is proposed to harvest the energy from the vibration of the heartbeat. The proposed cantilever structure is 14 µm thin with the outer dimension of 5 mm x 5 mm. The spiral structure reduces the stiffness of the cantilever beam which reduces the resonant frequency. A proof mass is introduced in the spiral structure to further reduce the resonant frequency of the cantilever beam. Due to biocompatibility nature, Zinc oxide (ZnO) has been used as the piezoelectric material for the design. The proposed structure has a resonant frequency of 45.8 Hz which is suitable for harvesting energy from the human heartbeat. The harvester generates peak output open circuit voltage of 3.2 V and peak output power of 3.5 µW at the resonant frequency of 45.8 Hz for a sinusoidal acceleration of 1g.**

Keywords—Piezoelectric energy harvester, spiral structure, Biocompatibility, Pacemaker.

I. INTRODUCTION

The advancement in the VLSI technology drastically reduced the size and power requirement of the electronic devices. These devices are mostly operated with batteries and it is difficult to miniaturized the traditional battery without affecting the lifespan of the device. The low power devices powered by the battery has the limited life span and need to recharge or replace periodically. The integrated energy harvester can enhance the life span of such low power devices. Thus, energy harvesting became an active research area in the field of biomedical devices, wireless communication, military equipment and consumers electronic devices. Vibration based energy harvester (EH) is the best suited to micro power devices. The process of obtaining electrical energy from vibrations of various unused environmental sources is called vibration energy harvesting [1][2]. The sources of the vibration are generally automobiles, machines, movement of human beings and animals. These sources can produce small amount of usable power. The focus on energy harvesting from the movement of human body recently gained popularity [3].

The piezoelectric energy harvester (PZEH) is favored over other vibration based mechanism, as it does not require any external voltage source and can easily integrated in the Micro Electro Mechanical System (MEMS) environment [4]. The piezoelectric energy harvester works efficiently in the vicinity

of the resonant frequency. The most probable application for the PZEH is the replacement or recharging of the batteries of the implanted medical devices such as pacemaker. The Ni-Cd battery used in the pacemaker needs to replace periodically as they have the limited life span [5]. Many researchers tried to overcome this by extracting electrical energy from the human body to power the pacemaker.

Goto *et.al.* [6] proposed the feasibility of using automatic power generating system of the quartz watch to power the pacemaker. The system attached to the dog's heart and successfully generated 13 µJ of energy per heartbeat, which shows that automatic power generation system can be used to power pacemaker. Tashiro *et.al.* [7] present the electrostatic harvesting system which generates 36 µW power from the movement of canine heart. This system generates enough power to operate cardiac pacemaker. The honeycomb-type variable capacitor used in the harvester has the length of 5 m, width 30 mm and thickness are 5 µm. So, the proposed design is impossible to place in the thoracic cavity of an animal. Zurbuchen *et.al.* [8] designed energy harvesting mechanism derived from the working of swiss wrist watch and combine with pacemaker performed in vivo study on the pig's heart for 30 minutes. The outer diameter and the thickness of the prototype device is 27 mm and 8.3 mm, which makes it impractical for the pacemaker. Ansari *et.al.* [9] proposed the fan folded structure to achieve low frequency. It is shown that proposed structure can produce sufficient energy to power pacemaker. However, the use of PZT as piezoelectric material makes it harmful for human use and 18.4g tip mass is too heavy for the pacemaker.

Kanai *et.al.* [10] shows that the most of the high amplitude frequency are less than 50 Hz as shown in figure 1. So, the target resonant frequency of PZEH in this paper is below 50 Hz. The challenge of this research is to design the energy harvester suitable for pacemaker within very small dimension (within 6 x 6 mm^2) so that the harvester can be easily integrated with modern pacemakers [11]. In this paper, we are designing a spiral shaped PZEH to power the battery of a pacemaker from the vibration of the heartbeat.

The paper has been organized as follows. Section II discusses about the material selection of the PZEH. The design

978-1-5386-6723-1/19 $31.00 © 2019 IEEE 465

and analysis of the structure are discussed in section III. Section IV discusses about the mechanical output of the PZEH. Section V discusses about the electrical output of the piezoelectric EH. In this section, comparisons are carried out between the proposed structure and the output of some of the available literature. Finally, the conclusion has been drawn in section VI.

.

II. MATERIAL SELECTION OF PZEH

Lead Zirconate Titanate (PZT) is the most widely used piezoelectric material in the field of PZEH. Since PZT is not suitable for the implanted medical devices as it contains Lead which is harmful for humans and the environment. The properties of some of the biocompatible piezoelectric materials are investigated and compared in this paper. The materials like Zinc oxide (ZnO), Aluminium Nitrite (AlN), Barium Titante (BaTiO₃) are some of the biocompatible materials.

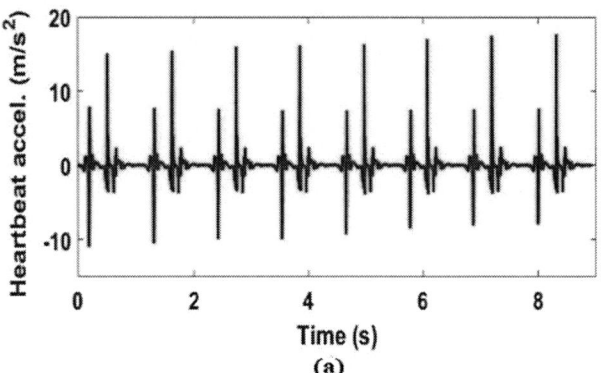

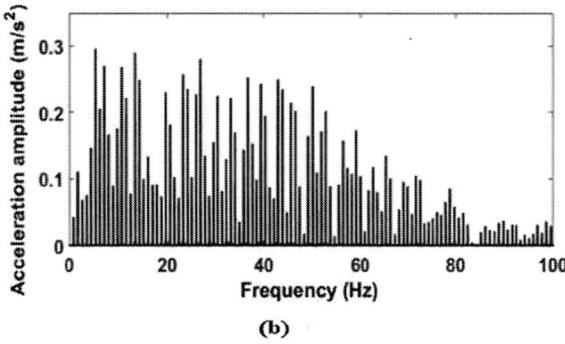

Figure 1. (a) Normal heartbeat vibrations in the time domain [9]. (b) Fourier transform of a normal heartbeat [9].

The property of the all the above material is shown in Table 1. The piezoelectric material used in the implanted bio-medical devices should be biocompatible.

The charge generated in the piezoelectric material on the application of force depends upon the charge coefficient as shown in equation 1

$$q = d_{31}F \qquad (1)$$

Where, q is the charge, F is the force applies and d_{31} is the charge coefficient.

Table1. Piezoelectric Properties of PZT-5H, AlN, BaTiO₃ and ZnO

Material	PZT-5H [1][2]	AlN [1][2]	BaTiO₃ [14] [15]	ZnO [1][2]
Charge Coefficient (d_{31}) (C/N)	741	3.84	149	11.34
Dielectric Constant (ε_r)	3400	10.256	1200	12.64
Voltage coefficient (g_{31}) (Vm/N)	0.217	0.374	0.124	0.897

The voltage generated by the piezoelectric material is given by equation 2.

$$V = \frac{q}{C} = \frac{d_{31}Ft_{pz}}{\varepsilon_0\varepsilon_r A} = g_{31}St_{pz} \qquad (2)$$

Where, ε_0 is the permittivity of air, ε_r is the relative permittivity of material, A is the area of the cross section and t_{pz} is the thickness of the Piezoelectric material, S is the stress developed and g_{31} is the voltage coefficient of the material.

The charge coefficient of PZT-5H is highest among all the above material, but voltage coefficient is highest in ZnO. So, among all these materials Zinc oxide will gives the best output voltage for the same dimension.

In this paper, zinc oxide is selected as the piezoelectric material in energy harvester because of its biocompatibility and high voltage coefficient.

III. DESIGN AND ANALYSIS

In this paper, EH is designed with silicon (Si) based rectangular spiral cantilever with the Si proof mass at the center of the structure as shown in figure 2. The challenge is to design the structure within the dimension of 6 x 6 mm² and achieve the resonant frequency below 50 Hz. The structure length and width are 5 mm respectively. The width of each arm of the spiral is 0.5 mm with a gap of 0.5 mm between each arm. The width of the arm connected to the proof mass is 1 mm. The piezoelectric material ZnO is distributed throughout the spiral structure. The thickness of the Si beam is 12 μm and thickness of ZnO is 2 μm. The dimension of the proof mass is 1x1x0.85 mm³. The structure is built and simulated in COMSOL Multiphysics software which is common finite element analysis software.

The different property of the material [1] [2] used in this paper is given in Table 2.

Table 2 Property of PZEH material

Materials	Si	ZnO
Young's Modulus (GPa)	170	210
Poisson Ratio	0.29	0.33
Density (kg/m³)	2329	5680

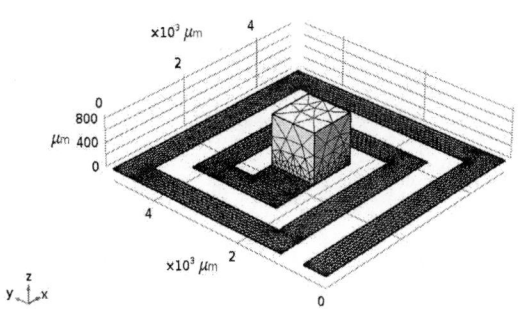

Figure 2. Meshed spiral cantilever structure

IV. MECHANICAL OUTPUT OF STRUCTURE

The classical beam theory defines the deflection of the cantilever beam at any point A in the transverse direction as given in equation 3 [16].

$$\delta_x = \frac{Fl_a^2(3l-l_a)}{6EI} \tag{3}$$

Where, F is the applied force, l is the length of the cantilever, l_a is the length of point A from fixed point, E is the young's modulus and I is the moment of inertia of the beam.

In the case of MEMS cantilever, the internal stress developed in the cantilever beam dominates over gravitation force. The Stoney's equation can be used to calculate approximate deflection of the beam as shown in the equation 4 [16].

$$\delta_x = \frac{3Sl^2(1-v)}{Et_{pz}^2} \tag{4}$$

Where, v is the poison ratio. The peak displacement of the spiral cantilever structure at different frequency is shown in figure 3. The peak displacement is 9 mm at the resonant frequency.

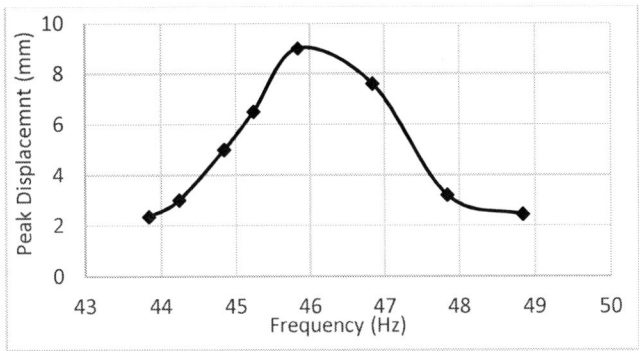

Figure 3. Peak Displacement of the spiral structure with frequency.

The output power generated in the cantilever structure depends on the average strain of the cantilever as shown by Yu Jia [16].

$$Q = d_{31} \in_{avg} Ew_{pz}l_{pz} \tag{5}$$

$$P = \frac{\omega t_{pz}Q^2}{\varepsilon_0 \varepsilon_r w_{pz}l_{pz}} \tag{6}$$

Where, Q is the charge generated in the piezoelectric layer of cantilever, d_{31} is the piezoelectric strain constant in 31 mode, $\in_{avg}$ is the average strain developed in piezoelectric layer of cantilever, w_{pz}, l_{pz} is the width and length of the piezoelectric layer, $\varepsilon_0 \ and \ \varepsilon_r$ are the permittivity of air and the dielectric constant of the piezoelectric material respectively. The figure 4 shows the strain distribution in the spiral cantilever.

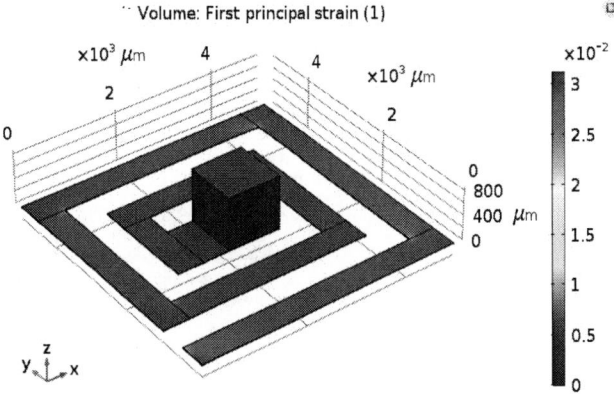

Figure 4. Strain distribution in spiral cantilever at resonant frequency.

V. ELECTRICAL OUTPUT OF STRUCTURES

A. Voltage Output of Simulated Structures

The PZEH gives the maximum output voltage and output power at the resonant frequency [1] which is given by equation 7.

$$\omega = \sqrt{\frac{K}{m}} \tag{7}$$

Where, K is the stiffness of the cantilever structure and m is the mass of the cantilever structure.

The generated voltage by the PZEH is shown in equation 2, which depends upon the applied stress, the voltage coefficient and the thickness of the piezoelectric material.

The time dependent analysis of the structure is done in COMSOL Multiphysics to obtain the time dependent output voltage. A structure is excited with the sinusoidal acceleration of 1g at a resonant frequency of 45.8 Hz. Figure 5 shows the time dependent output voltage of spiral structure. The peak output open circuit voltage generated by the spiral structure at the resonant frequency is 3.2 V.

978-1-5386-6723-1/19 $31.00 © 2019 IEEE   467

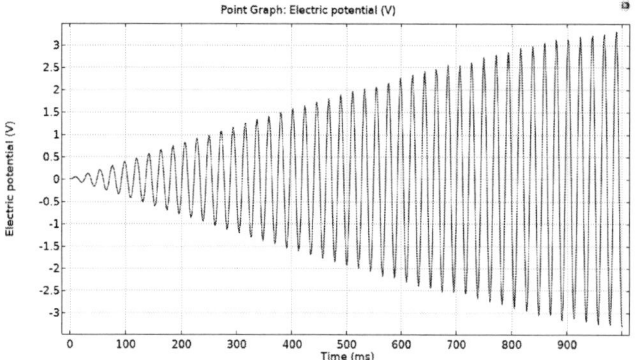

Figure 5. Time Dependent Output Voltage

Figure 6 shows the variation of the output voltage with the frequency.

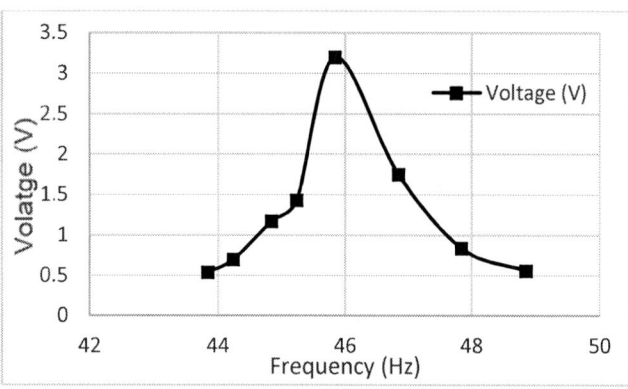

Figure 6. Variation of peak Output Voltage with Frequency

B. Power Output of Simulated Structures

The PZEH gives the maximum output power when the resonant frequency matches with excitation frequency and load resistance and internal resistance of the energy harvester are equal. The electrical equivalent of the PZEH as shown in figure 7. The voltage at the load resistance is given by equation 8.

$$V_{out} = \frac{V_{pz}}{R_{pz}+R_L} R_L \qquad (8)$$

Where, V_{out} is the output voltage across R_L, V_{pz} is the open circuit voltage of the PZEH, R_{pz} and R_L are internal and external load resistance respectively. The output power is given as in equation 9.

$$P = \frac{(V_{out})^2}{(R_{pz}+R_L)^2} R_L \qquad (9)$$

The maximum power is obtained from the PZEH when R_{PZ} and R_L are same. The equation 9 is modified as in equation 10.

$$P = \frac{(V_{out})^2}{4R_{pz}} \qquad (10)$$

The figure 9 gives the variation of output power with the load resistance. The optimum load obtained from the figure 8 is 4.3 MΩ.

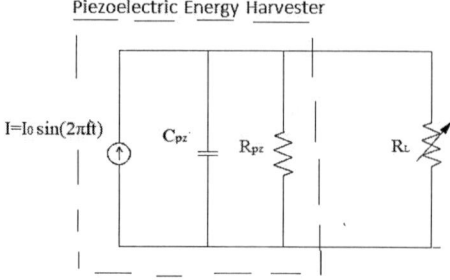

Figure 7. Electrical Equivalent of Piezoelectric Energy Harvester

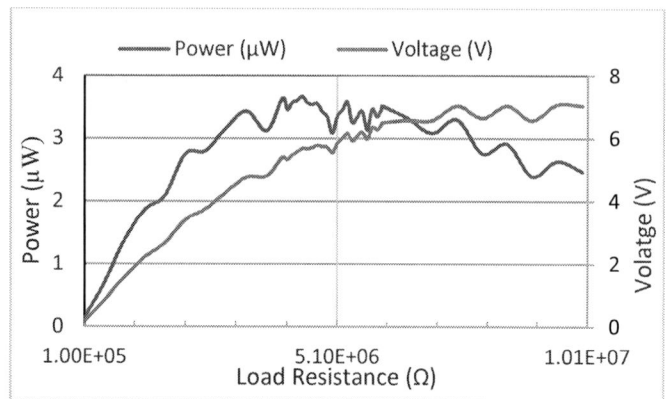

Figure 8. Variation of Output Power and Output Voltage with Load Resistance

The figure 9 gives the variation of output power with the frequency. The peak output Power of the spiral structure is 3.5 µW at the resonant frequency of 45.8 Hz. The table 3 compiles all the output parameter of the spiral structure and compares with the existing literature.

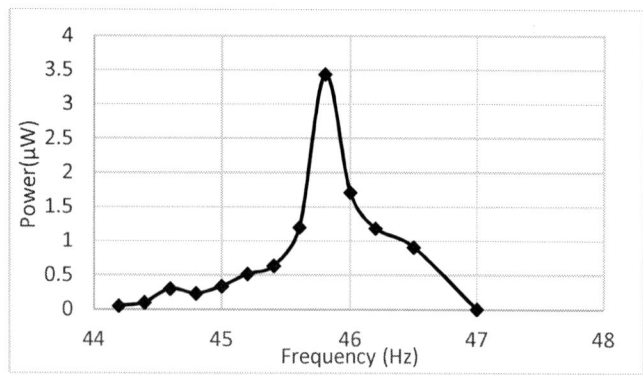

Figure 9. Variation of Output Power with Frequency

Table 3 Comparison of PZEH parameters for pacemaker with available literature.

Sno	Geomerty & Material used	Resonant frequency (Hz)	Electrical output	Remarks	Ref.
1	Fan folded structure 1x1x1 cm3, PSI-5A 4E piezo sheets	170 Hz	Power 2.12 µW	Very high frequency	[13]
2	Fan folded geometry (2 × 0.5 × 1 cm3), PZT	15.79 Hz	Power 16.25 µW	PZT is used and 18.4g tip mass is too heavy for the energy harvester.	[9]
3	Rectangular, 40x5x0.38 mm3, PZT	15 Hz	Voltage 1.5-4 V, Power 6 -18 µW	PZT is used and Impratctical geometrical dimension for pacemaker	[12]
4	Spiral 5x5x0.85 mm3, ZnO	45.8 Hz	Voltage 3.2 V Power 3.5 µW	The dimension is within the permitted size of morden pacemaker	This work

VI. CONCLUSION

In this paper, spiral shaped piezoelectric energy harvester is designed and simulated for the pacemaker. The system is designed to recharge the battery of the pacemaker so that the longevity of the pacemaker can be increased. The size of the harvester is within the permitted size of the modern available pacemaker. The proposed structure of piezoelectric energy harvester generates 3.2 V peak output open circuit voltage at the resonant frequency of 45.8 Hz at an excitation acceleration of 1g. The peak output power generated by the structure is 3.5 µW at the resonant frequency.

ACKNOWLEDGMENT

We, all the authors, would like to express our special thanks to Science and Engineering Research Board (SERB), Govt. of India for funding to execute this research through project ref. no. ECR/2016/001279. This research was not possible without the help and support of BIT Mesra, Ranchi and SERB, Govt. of India.

REFERENCES

[1] D. Chaudhuri, S. Kundu, and N. Chattoraj, "Design and analysis of MEMS based piezoelectric energy harvester for machine monitoring application," Microsyst. Technol., vol. 0123456789, 2018.

[2] D. Chaudhuri and S. Kundu, "MEMS piezoelectric energy harvester to power wireless sensor nodes for machine monitoring application," Proc. 2nd Int. Conf. 2017 Devices Integr. Circuit, DevIC 2017, pp. 584–588, 2017.

[3] Starner, T., & Paradiso, J. A. (2004). Human generated power for mobile electronics. Low power electronics design, 45, 1-35.

[4] S. Roundy, P. K. Wright, and J. Rabaey, "A study of low level vibrations as a power source for wireless sensor nodes," Comput. Commun., vol. 26, no. 11, pp. 1131–1144, 2003.

[5] V. S. Mallela, V. Ilankumaran, and N. S. Rao, "Technical Series Trends in Cardiac Pacemaker Batteries," vol. 4, no. 4, pp. 201–212.

[6] Goto H, Sugiura T, Harada Y, Kazui T. Feasibility of using the automatic generating system for quartz watches as a leadless pacemaker power source. Med Biol Eng Comput. 1999 May;37(3):377-80. PubMed PMID: 10505390

[7] Tashiro, R., Kabei, N., Katayama, K., Tsuboi, E., & Tsuchiya, K. (2002). Development of an electrostatic generator for a cardiac pacemaker that harnesses the ventricular wall motion. Journal of Artificial Organs, 5(4), 239–245. doi:10.1007/s100470200045

[8] Zurbuchen, A., Haeberlin, A., Bereuter, L., Wagner, J., Pfenniger, A., Omari, S., ... Vogel, R. (2017). The Swiss approach for a heartbeat-driven lead- and batteryless pacemaker. Heart Rhythm, 14(2), 294–299. doi:10.1016/j.hrthm.2016.10.016

[9] M. H. Ansari and M. A. Karami, "Experimental investigation of fanfolded piezoelectric energy harvesters for powering pacemakers," Smart Materials and Structures, vol. 26, no. 6, 2017.

[10] H. Kanai, M. Sato, Y. Koiwa, and N. Chubachi, "Transcutaneous measurement and spectrum analysis of heart wall vibrations," IEEE Transactions on Ultrasonics, Ferroelectrics, and Frequency Control, vol. 43, no. 5, pp. 791–810, Sep. 1996.

[11] Chew, D. S., & Kuriachan, V. "Leadless cardiac pacemakers". Current Opinion in Cardiology, 33(1), 7–13 (2018). doi:10.1097/hco.0000000000000468

[12] L. Rufer, M. Colin, and S. Basrour, "Application driven design , fabrication and characterization of piezoelectric energy scavenger for cardiac pacemakers," pp. 340–343, 2013.

[13] M. H. Ansari and M. A. Karami, "Smasis2015-9071 Energy Harvesting From Heartbeat Using Piezoelectric Beams With Fan - Folded Configuration and Added Tip Mass," pp. 1–8, 2015.

[14] S. Priya and D. J. Inman, Energy harvesting technologies. 2009.

[15] J. Gao, D. Xue, W. Liu, C. Zhou, and X. Ren, "Recent Progress on BaTiO3-Based Piezoelectric Ceramics for Actuator Applications," Actuators, vol. 6, no. 3, p. 24, 2017.

[16] Y. Jia and A. A. Seshia, "Power Optimization by Mass Tuning for Vibration Energy Harvesting," Journal Of Microelectromechanical Systems vol. 25, no. 1, pp. 108–117, 2016.

2019 Devices for Integrated Circuit (DevIC), 23-24 March, 2019, Kalyani, India

Comparative computational study of LD-HfO$_2$ and TiO$_2$ as layered dilectrics in RRAM

Deependra Chettri
(Dept. of Physics)
(SMIT, SMU)
Majhitar, India
deependrathapa28@gmail.com

Abinash Thapa
(Dept.of E&C Engg.)
(SMIT, SMU)
Majhitar, India
thapa.abinash006@gmail.com

Smita Rai
(Dept. of Physics)
(SMIT, SMU)
Majhitar, India
raismita415@gmail.com

Pronita Chettri
(Dept. of Physics)
(SMIT, SMU)
Majhitar, India
pronita2107@yahoo.co.in

Chandan Kumar Sarkar
(Dept. of Physics)
(IIEST)
Shibpur, India
phyhod@yahoo.co.in

Bikash Sharma
(Dept. of E&C Engg.)
(SMIT, SMU)
Majhitar, India
ju.bikash@gmail.com

Abstract— The development of Resistive Ram (RRAM) has been greatly promising and has emerged as highly reliable non-volatile memory. The biasing of the device creates the filament formation that is due to the moving charge and shallow localized charge. However, the conduction mechanism of resistive switching changes considerably depending on the type of the material used in the dielectric layer and the electrode. Amongst the important mechanism initiating the filament formation in the device, we have mentioned about Space Charge Limited Conduction (SCLC) and F-N tunneling in our work particularly considering the development of conductive filament (CF) with the interest in the I-V characteristics of the device. The RRAM device has huge potential and many advantages over other non-volatile technology i.e. high switching speed, high device density and low power consumption. In this paper we have considered LD (low-dimensional) forms of HfO$_2$ and TiO$_2$ dielectrics for applications in RRAM (Resistive Random Access Memory) devices. The retention of memory device is determined by its saturation curve, HfO$_2$ showed slower saturation at -18mA providing higher retention factor than. The faster decrease in curve for HfO$_2$ based RRAM at -19mA, promoting low power consumption of memory device. Therefore we have obtained that the retention and power consumption of HfO$_2$ is better than the TiO$_2$.

Keywords— SCLC, CF, DFT, F-N Tunneling

RRAM is the evolving non-volatile memory device. The device comprises of the insulated dielectric material between two conducting metals. The metal oxide element used in our survey is TiO$_2$ and HfO$_2$ studying their deviation in the behaviour. The materials used above are available in the form of nanocrystals and nanodots having high surface area and magnetic properties [1]. The HfO$_2$ has significantly higher charge capacity and lower impedance that makes it a suitable material for the neutral based application [2]. The device upon a suitable biasing gives the current flow through the conduction filament which is created in the metallic oxide layer due to the motion of moving charge and the shallow localized charge [3]. During the deposition of the charge on the interface material will oxidize due to the process of the REDOX reaction. As the oxygen states will be changed due to the loss of electron from the metallic surface that is the increase in oxidation state and as well the gain of electron. Therefore it results in decrease in the oxygen state, contributing the switching behaviour in the device [4]. The current density in a particular region takes place through various mechanisms including vacancy creation or metal

defect migration [4]. Once the filament is formed it may be reset by applying the opposite polarity [5]. This effect takes place due to the SCLC mechanism and tunneling mechanism in the oxide layer is given by the F-N tunneling that explains the filament conduction quantum mechanically [5]. The dielectric can be made to conduct through a filament or conductive path after application of sufficiently high voltage. The conductive path can arise due to the migration of the oxide ion to the interstitial sites of the oxide layer and the tunneling of the electrons through the interface. This movement of the carriers creates the filament into the device. The material system for resistive memory plays an important role in the device operation such as mobility of the charge carriers and the permittivity of the device. High resistive state (HRS) is the phase at which the current flow is minimum and the SET process is the state at which the current flows due to conductive filament path growth. This is attributed to the quantum tunneling effect at the interface as the F-N tunneling [6].

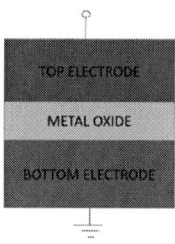

Fig1.The schematic of Resistive-RAM (RRAM) [8].

HfO$_2$ based RRAM shows prevailing performance [7] and exceptional compatibleness with existing CMOS technology at back-end line. It forms a successive partial oxidation of a conducting filament (CF) at set and rest operation [8].

L. Achsah et. al. observed the dual-oxide TiO$_2$ based RRAM with switching activity. Therefore it provides the possibilities of studying multilayer based TiO$_2$ structure in RRAM. The various material used in RRAM are the transition metal oxides, perovskite, organic materials, which are recently been studied and used in RRAM. Among it, the transition metal oxides such as HfO$_2$ and TiO$_2$, which is a high-K materials [9]. However, research on RRAM still faces great challenges, such as wide resistance distribution and the non-uniformity of V-SET and V-RESET voltages, which are related to the local filament formation and rupture in filament-type resistive switching memories [10].

978-1-5386-6723-1/19 $31.00 © 2019 IEEE   470

The resistive state of RRAM can be represented as 'write', 'read', 'erase' operation [11-12]. At first the TiO₂ based next generation non-volatile memory was developed by the Choi [13] using atomic layer deposition process. Many literature on the metal oxides like ZrO_2, $SiTiO_3$, TiO_2, NiO, Gd_2O_3, HfO_2, Silicon Sub oxides, CeO_2, polymer, organic molecules, and perovskite have been studied [14]. All the above material used in RRAM has their own advantages and drawbacks. However, it has been reflected from recent study that TiO₂ based RRAM outperforms other oxide based ones owing to its unique advantages such as high-k dielectric value exhibits both unipolar and bipolar RS phenomenon and ease of deposition over others materials [15].

L. Achsah et. al. Pt/TiO₂/TiO₂-x/Pt based RRAM device switching performance is improved by apply voltage of 1.2V [15]. Thus it provides stable performance at high frequency add the important advantage of applications high frequency RRAM device [15].

I. METHODOLOGY

The Density functional Theory (DFT) calculation is performed for HfO₂ and TiO₂ based RRAM device. The simulation are based on quantum ESPRESSO, consist of in-built DFT, plane wave and pseudopotential. The Band structure has been determined using Quantum ESPRESSO for HfO₂ and TiO₂ based RRAM.

In HfO₂ multilayer simulation tetragonal structure is considered, along with it the pseudopotential is linked with cut-off wave function 45.578Ry and cut-off charge 264.438Ry. The six layers of HfO₂ are implemented for multilayer case having thickness of 2.672nm (26.726□).

Similarly in case of TiO₂ multilayer simulation orthorhombic structure is confirmed better. The pseudopotential is linked with cut-off wave function 25.00Ry and cut-off charge 225.00Ry. The six layers of TiO₂ are implemented for multilayer case having thickness of 2.629nm (26.296□).

To study the characteristic behavior of the device we have chosen two different oxide material layers with same density, molar mass and constituent percent. The effective current density has been calculated by the space –charge limited conduction expression as,

$$J_{SCLC} = \frac{9}{8}\epsilon_i \mu\theta\frac{v^2}{d^3} \qquad (1)$$

Where, ε is permittivity, θ is the ratio of free and shallow trapped charge (1), μ is the mobility and d is the thickness of the insulated metal oxide.

The F-N tunneling explains the charge distribution that is based on the potential barrier change between the electrode and the resistance change materials. In this phenomenon the tunneling of the charges takes place through the interface through which the transportation of the charges and the interaction between two states of matter takes place. Therefore we have calculated the band structure of both the oxide that is required for the determination of the barrier height that is needed for the tunneling equation as,

$$J_{FN} = \frac{q^2E^2}{8\Pi h\phi_B}\exp(-\frac{8\pi d\sqrt{2dm^*}}{3Eh}\phi_B^{\frac{3}{2}}) \qquad (2)$$

Where, E is the electric field, h is the Planck's constant, m^* is the effective mass and ϕ_B is the barrier height.

The effective mass and barrier height of composite material is calculated using Maxwell Garnett Theory [16-17]. This parameters are implemented in eq. (1) and (2) to obtain effective current density.

II. RESULT & DISCUSSION

Quantum ESPRESSSO is the simulation tool for calculation of band gap from band structure. Therefore we have calculated the bandgap for both the oxide layers (HfO₂ and TiO₂). Quantum ESPRESSO works on density functional theory (DFT), plain wave and pseudopotential.

The DFT calculation as shown in fig. 2 (a) and (b) was performed after the construction of HfO₂ and TiO₂ structure using Quantum ESPRESSO. The atomic structure of HfO₂ and TiO₂ is formed by six layer combinations and further its lattice parameter and cell volume is determined as shown in Table1. In Table1 we present the atomistic properties of HfO₂ and TiO₂ i.e. lattice parameter, lattice type and cell volume in Angstrom. The properties attribute the crystalline structure of HfO₂ and TiO₂ crystal.

Table1. The lattice parameter, lattice type and cell volume of the HfO₂ and TiO₂ crystal

Materials	Lattice type	Lattice parameters (□)			Cell volume (□³)
		a	B	C	
HfO₂ (multilayer)	Tetragonal	5.142	5.195	5.326	140.262
TiO₂ (multilayer)	Orthorhombic	3.803	3.803	5.566	70.479

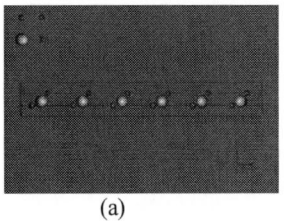
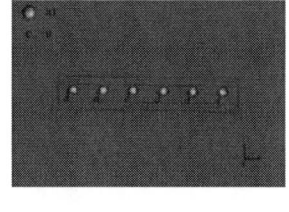

(a)

(b)

Fig.2.The Multilayer structure of (a) TiO₂ and (b) HfO₂

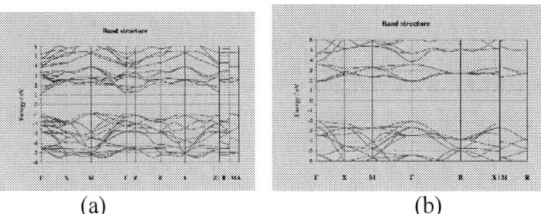

(a)   (b)

Fig.3.The Band structure of (a) TiO₂ and (b) HfO₂

The observation made from fig.3 (a) and (b) determines the band structure of TiO₂ and HfO₂. The direct band gap is observed for both HfO₂ and TiO₂ is found to be 3.8eV and 2.2 eV at Γ point respectively. Clean and undistorted band structure is observed for both the multilayer HfO₂ and TiO₂ which provides its prominent use in RRAM memory device. The band structure anvil the band gap and effective mass of electron, which is an important electronic properties for a memory device.

The calculation of electric properties for HfO_2 and TiO_2 is such as band gap, effective mass is determined from fig.3 (a) and (b).

Table2. The calculation of electric properties (band gap and effective mass of electron) obtain from band structure

Sl. No.	Materials	Number of composed atom	Band Gap (eV)	Effective mass of electron	Thickness (nm)
1.	HfO_2 (multilayer)	6	3.8	2.25	2.672
2.	TiO_2 (multilayer)	6	2.2	9	2.629

The value obtained is equated in eq. (1) and eq. (2). The equations helps in graphically represent the I-V characteristic curve for HfO_2 and TiO_2 based RRAM. As the I-V characteristic curve represent the retention, power consumption and stability of memory device.

I-V characteristic graph

The characteristic I-V Graph represents the behavior of device and also its mechanism i.e. for HfO_2 and TiO_2 based RRAM. MATLAB tool is used for simulation and graphical representation of SET and RESET in memory device. The eq. (1) and (2) are plotted with the parameter obtained from Table2 using quantum ESPRESSO. Thus the I-V characteristic graph is generated i.e. shown in fig.4 (a), (b), fig.5 (a), (b) and fig.6. (a), (b).

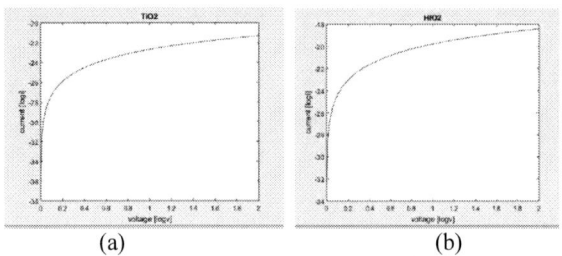

(a)   (b)

Fig.4.The SET phase of (a) TiO_2 and (b) HfO_2 based RRAM

Fig.4. (a), (b) show the SET phenomenon in which the current increases with the gradual increase in the voltage. In the fig 4 (a), (b) we observe the saturation in TiO_2 based RRAM is faster than the HfO_2. As faster saturation determine the lower retention of the memory device in case of TiO_2 (fig.4 (a)). In comparison the HfO_2 based RRAM fig.4 (b) shows slower saturation which determines better retention of the memory device. Hence, HfO_2 (low dimensional form) can be preferred over TiO_2.

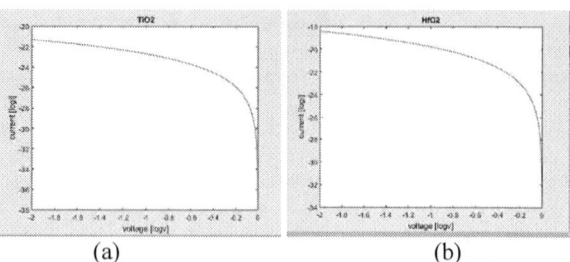

(a)   (b)

Fig.5.The RESET phase of (a) TiO_2 and (b) HfO_2 based RRAM

Fig.5. (a), (b) the RESET phase is high resistive phase where the conduction of current is not possible as the filament gets ruptured .In the fig 5(a), (b) we observe filament rupturing in the TiO_2 and HfO_2. The reverse voltage is applied to TiO_2 based RRAM, where the current curve decrease from -21microampere. Whereas in reverse voltage applied to HfO_2 based RRAM, the current curve decrease as -19.5microampere. Thus the above current graph depict the power consumption of RRAM, the HfO2 based RRAM (-19.5mA) determine the low power consumption of memory device in comparison to TiO_2 (-21mA).

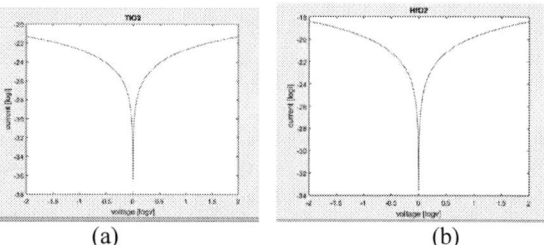

(a)   (b)

Fig.6.The SET-RESET phase of (a) TiO_2 and (b) HfO_2 based RRAM

The SET-RESET process is shown in fig.6. (a), (b) for both HfO_2 and TiO_2 based RRAM. The SET-RESET process is comparatively better for HfO_2 based device then compare to TiO_2. The SET and RESET graph indicate the reliability of memory device. In case of HfO_2 fig .6 (b) the stability factors is higher as the set and reset state is faster in comparison to TiO_2 fig.6 (a) based RRAM.

III. CONCLUSION

We have calculated the band structure of TiO_2 and HfO_2 and parallel the band gap of both the materials were determined. The band structure of HfO_2 was found better in comparison to the TiO_2 i.e. band gap of HfO_2 is 3.8eV. The band structure determined the electric property (i.e. band gap and effective mass of electron) of the materials.

On taking account of the I-V characteristics we have found out that the efficiency of the TiO_2 is less as compared to the HfO_2 because the time taken for the rupture of filament by HfO_2 is longer than the TiO_2. From the observations we conclude the retention of memory device considering two oxide viz. HfO_2 and TiO_2 based memory. The HfO_2 showed better retention as saturation curve is faster than TiO_2 based RRAM. We also predict the power consumption of memory device, the HfO_2 showed less power consumption in reverse bias curve than the TiO_2. As the SET and RESET graphs shows the reliability of HfO_2 is better than the TiO_2 which is due to the dielectric constant of the HfO_2. The dielectric constant of HfO_2 is greater as compared to the TiO_2 which makes HfO_2 retain more charge than TiO_2. Thus the overall performance i.e. retention, power consumption and stability of HfO_2 based resistive-RAM is better than TiO_2.

REFERENCES

[1] Journal of Alloys and compounds, A.V .Powell, J.G .Gore, P.D. Battle,1993.

[2] Magnetic properties of Iridium oxide, Nobuyuki Taira, Makoto Wakeshima and Yukio   Hinatsu ,Journal of physics: Condensed Matter,Volume 13,Number 23.

[3] Ee Wah Lim and Razalli Ismail,received: 6August 2015/Published 9 September 2015.

[4] Redox- based Switching memories by Rainer Waser, Regina Dittman,Georgia Staikor and Kristoff Szot.

[5] J.G.Simmons: Conduction in the thin dielectric films, Phys D:Appl.phy,Volume (4).

[6] Metal oxide based memeory by monte carlo technique.

[7] Gilmer DC, et al. Superior filament formation control in HfO2 based RRAM forhigh-performance low-power operation of 1 lA to 20lA at +/-1 V. In:Proceedings of the international symposium on VLSI-TSA, 23–25 April 2012.p. 1–2.

[8] Bersuker G et al. Metal oxide RRAM switching mechanism based on conductivefilament properties. J Appl Phys Dec. 2011;110:124518.

[9] Puglisi FM et al. An empirical model for RRAM resistance in Low-and highresistancestates. IEEE Electron Device Lett 2013;34(3):387–9.

[10] Veksler D, et al. Methodology for the statistical evaluation of the effect ofrandom telegraph noise (RTN) on RRAM characteristics. In: Proceedings of IEEEInternational Electron Devices Meeting (IEDM), 10–13 Dec. 2012. p. 961–964.

[11] Fang Z et al. Low-frequency noise in oxide-based TiN/HfOx/Pt resistive randomaccess memory cells. IEEE Trans Electron Devices 2012;59(3):850–3.

[12] Ambrogio S et al. Statistical fluctuations in HfOx resistive-switching memory:Part II—random telegraph noise. IEEE Trans Electron Devices2014;61(8):2920–7.

[13] Puglisi FM, Pavan P. Factorial Hidden Markov Model analysis of randomtelegraph noise in resistive random access memories. ECTI Trans Electr EngElectron Commun 2014;12(1):24–9.

[14] Veksler D, et al. Random telegraph noise (RTN) in scaled RRAM devices'',Proceedings of IEEE International Reliability Physics Symposium (IRPS), 14–18April 2013. p. MY.10.1–MY.10.4.

[15] Francesco Maria Puglisi, Paolo Pavan, Luca Larcher, Andrea Padovani, "Statistical analysis of random telegraph noise in HfO2-based RRAMdevices in LRS" , Article history:Available online 2 June 2015.

[16] Sharma, B., Mukhopadhyay, A., Sengupta, A., Rahaman, H., Sarkar, C. K., "Analysis of tunnelling currents in multilayer black phosphorous and MoS2 non-volatile flash memory cells", J Comput Electron. DOI: 10.1007/s10825-015-0750-9.

[17] Sengupta, A., Sarkar, C. K. and Requejo, F. G., "Comparative study of CNT, silicon nanowire and fullerene embedded multilayer high-k gate dielectric MOS memory devices", J. Phys. D: Appl. Phys., vol. 44, pp. 405101, September 2011.

Diode & Neutral Point Clamped Five-Level Inverter For the Power Quality Issues

Rabindranath Das Adhikary
Electrical Engineering
Kalyani Government Engineering
College
Kalyani, India
E-mail: rabiadhikary77@gmail.com

Writwik Balow
Electrical Engineering
Ideal Institute of Engineering
(College)
Kalyani, India
writwik.balow@gmail.com

Tapas Halder
Electrical Engineering
Kalyani Government Engineering
College
Kalyani, India
E-mail: tapas_haldar@yahoo.com

Abstract—**The five-level inverter is attractive for medium power applications with reduced voltage and current stress by virtue of which lower rating of semiconductors for the medium and high voltage applications. This inverter is productive for the reduced total harmonic distortions (THD) to accomplish the good quality of the power in addition to lower size of the filter components as cost effective solution for the aggressive power market. The inverter also make certain to avoid the complications of the static and dynamic voltage sharing when the power semiconductors are connected in series with a multi-leveled-inverter topology.**

Keywords—*Five-Level-inverter, Performance Analysis, THD computations & Performance Evaluation*

I. INTRODUCTION

In the year of 1975, the concepts and wider applications are introduced in the domain of the medium and high power applications of a neutral point diode clamped power.

The conceptions of the multilevel inverter are distantly parallel to the common inverter topology for the dc to ac power conversions. It is not only used for the adjustable speed drive (ASD) but also used for the grid integration and major performance improvements in the renewable power sector in comparisons with the articles [1]-[13].

The uses of the multi-level-inverters have betterment for the harmonics eliminations and power quality improvements with reduced the sudden change of the stresses (i.e. dv/dt), across the power semiconductors so cumulative switching power losses are decreased to increase the circuit efficiency and electromagnetic compatibility of the multi-level-inverter for the practical applications but the articles [14]-[31] are hardly discussed to ascertain the objective of this paper.

The reliability and stability are highly restricted by the bandwidth of the control loop and complexity of the control strategy in power circuit operations of the multi-level-inverter at which a quite number of the power devices are used to achieve the tight load and line regulations for the industrial applications irrespective of the failure rate and life time of the high speed of the power semiconductors in series of the multi-level-inverter topology those are the trivial contributions and ideas in these papers [32]-[45].

A number of semiconductors are used in the Multi-level-inverters to smoothen the load voltage at the optimal cost due to revolution of the VLSI and USI technology so the manufacturing cost of the multi-level-inverter are reduced upto marks rather it is now a cost effective solutions for

lower rating power devices for the medium and high power applications. So, the demand of the multi-level-inverter is topology upgraded by leaves and bounds in the industrial applications so many items of these papers [46]-[56] are almost ineffective in comparisons with this paper.

In this paper, a diode & neutral point clamped multi-level inverter (D&NPCMI) is highly proposed to analyze the short discussion, study and harmonics analysis as standard doorsteps and performances. So, the dealings, treatments and open loop control algorithms are invited here to regulate the load voltage with the optimal value the power quality index (i.e.THD) computed by the MATLAB software simulations.

II. OPERATIONS OF THE DIODE & NEUTRAL POINT CLAMPED MULTI-LEVEL INVERTER (D&NPCMI)

An usual dc bus is purely divided by an even number to make shape number of the voltage levels a diode & neutral point clamped multi-level inverter (D&NPCMI) at which the bulky power capacitor is needed in series with a neutral point (NP) in the midpoint of the two equal magnitude power capacitors (C_{DC}) in a common leg of a line so the requirements of the clamping diodes (N_D) is equal to number of the power semiconductor (N_S) pairs forming the five-level-inverter as typical count:

$$N_D = (N_S - 1) \tag{1}$$

Single phase with a leg of five level neutral point clamping (NPC) inverter is shown in the Fig. 1. In addition to this figure Fig. 1, two additional equivalent power circuits are added to assemble three phase five-level-inverter (i.e. (D&NPCMI) architecture.

In order to abridge the modelling, performance analysis and study is invited so, the performance is limited by one leg or single phase of multilevel inverter (i.e. (D&NPCMI) architecture.

Three phase five-level-inverter (i.e. (D&NPCMI) power topology is the powerful and simple orientations with the active and passive components. Substantial examinations are carried out to improve the performance so the potential architecture of the power circuit (i.e. (D&NPCMI) of this paper popularizes for the versatile applications.

The source voltage clamping by the power diode, a (D&NPCMI) topology is usual by a one leg configurations. The voltage clamping diodes are important to clamping the

---

Financial support sponsored by the Authors

978-1-5386-6723-1/19 $31.00 © 2019 IEEE

voltage and productive operations of the inverter as shown in the Fig. 1. So it would be adequate to accomplish a concise depictions on the mode of the operations of a three phase circuits because the operation of others two phases are really similar to achieve the objective of the paper.

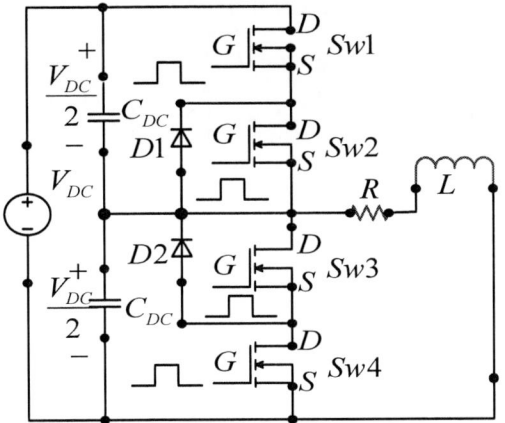

Fig. 1. Leg wise operations of the Five-Level-Inverter

In order to ascertain the common performance of the (D&NPCMI) topology, the one leg configuration is taken into account the following assumptions.

✓ Ideal supply voltage of (D&NPCMI) topology is taken for the simplified operations.

✓ All the active and passive components are operated in the ideal conditions so that forward voltage drops and reverse recovery effects of the active devices have the negligible effects through the operations.

✓ The load is considered as a highly inductive load though trivial values of the equivalent series resistance have an impact in circuit operations.

III. MODE WISE POWER CIRCUIT OPERATIONS

A. Operational mode I

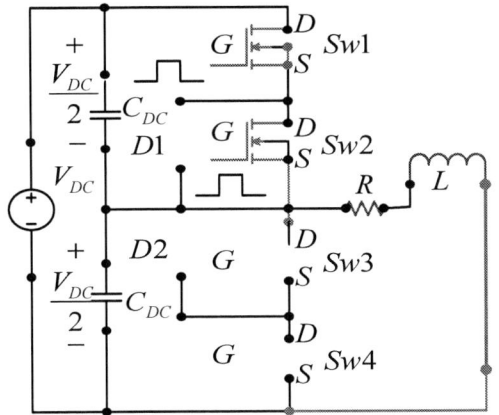

Fig. 2. Operational mode I of the (D&NPCMI) topology

The switching actions of the power circuit of the (D&NPCMI) topology is shown in the Fig. 1 are used for yielding the load voltage (V_0). So the switching actions are energised by the SPWM control strategy.

In this mode of operation of the power circuit of the (D&NPCMI) topology, the following objective is taken into picture as shown in the Fig. 2 showing the direction of the current flow and voltage (V_0) across the load $(R-L)$ as written as:

$$V_0 = +V_{DC} \qquad (2)$$

B. Operational mode II

In this mode of operation, the power semiconductor $(Sw2)$ is turned on by the gate pulse as shown in Fig. 3 when the sinusoidal pulse width modulation (SPWM) is applied by control strategy by virtue of which the clamping diode $(D2)$ get forward biased and the voltage across the capacitor (C_{DC}) is charged by the half of the supply voltage (V_{DC}) through the load $(R-L)$ so the load voltage (V_0) may be expressed as;

$$V_0 = +V_{DC}\big/2 \qquad (3)$$

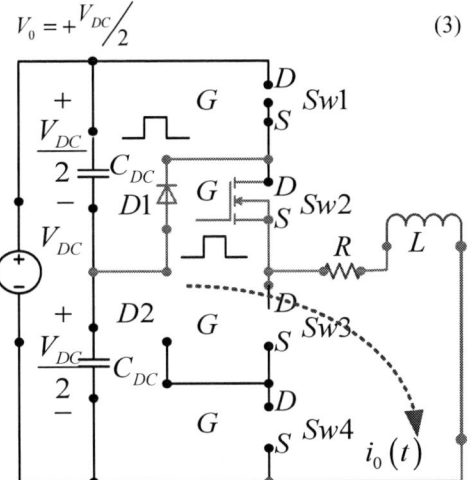

Fig. 3. Operational mode II of the (D&NPCMI) topology

C. Operational mode III

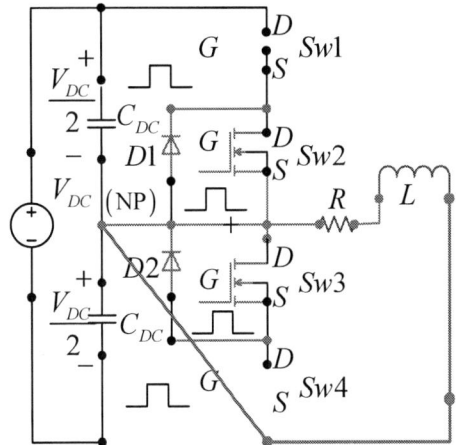

Fig. 4. Clamping mode III of Operations of the (D&NPCMI) topology

In this operational mode III of the circuit operation, load $(R-L)$ components be dynamically short circuited by the neutral point (NP) by the two controllable power semiconductors $(Sw2)$ and $(Sw3)$) along with two clamping power diodes $(D1)$ and $(D2)$ as shown in the Fig. 4. This mode of operation is usually treated as clamping action of the (D&NPCMI) topology. So the load voltage (V_0) across the element $(R-L)$ may be expressed by:

$$V_0 = 0 \qquad (4)$$

D. Operational mode IV

In this mode IV, the operation of the power circuit with a power semiconductor $(Sw3)$ is turned on by the SPWM control algorithm generated by the control actions as shown in the Fig. 5.

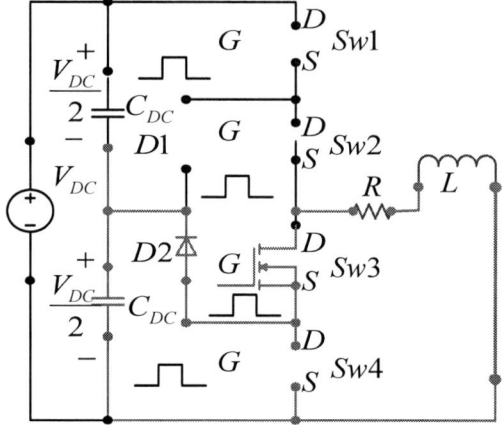

Fig. 5. Operational mode IV of the (D&NPCMI) topology

In this Fig. 5, the clamping diode $(D2)$ is being forward bias condition to switch on by the control actions. Further, in this mode, the input capacitor (C_{DC}) is charged through the control switch $(Sw3)$, power diode $(D2)$ and load voltage (V_0) current (I_0) as:

$$I_0 = \frac{V_0}{\left(\sqrt{R^2 + (L\omega)^2} \right)} \qquad (5)$$

Similarly, the load voltage (V_0) may be expressed by the following equation as:

$$V_0 = \frac{-V_{DC}}{2} \qquad (6)$$

In this mode of the (D&NPCMI) topology operation, anti-parallel power diodes $(D1)$ along with two power semiconductors $(Sw1)\&(Sw2)$ are turned on simultaneously to flow the load current (I_0) across the $(R-L)$ component as shown in the Fig. 6.

Similarly, the load voltage (V_0) may be expressed by circuit operation of mode V as shown in the Fig. 6 to find out the expression of the load voltage (V_0) as:

$$V_0 = -V_{DC} \qquad (7)$$

E. Operational mode V

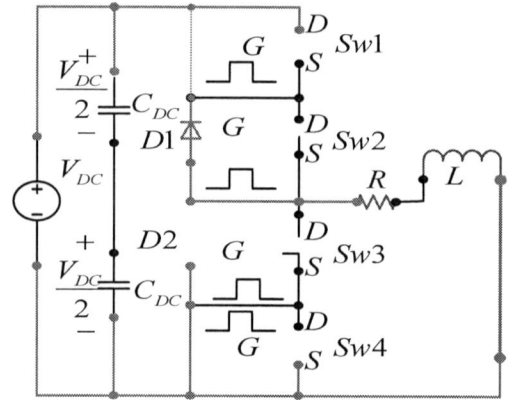

Fig. 6. Operational mode of the (D&NPCMI) topology

IV. THE SIMPLE CONTROL ACTIONS AND ALGORITHMS

The schematic diagram as shown in the Fig. 7 is highly proposed in this section to generate the sinusoidal pulse width modulation (SPWM) to get triggered the power switches as shown in the Fig. 1 to accomplish the five-level-inverter.

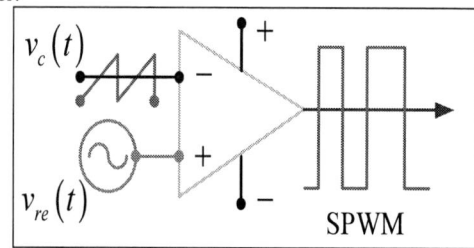

Fig. 7. Schematic SPWM generation for power devices

The pure sine wave is taken as a reference signal $(v_{re}(t))$ whereas triangular wave is the carrier wave. Now, the carrier wave $(v_c(t))$ is compared with the reference sine wave to generate the gate pulse (i.e. SPWM) by virtue of these switching actions, the load voltage (V_0) is found across the $(R-L)$ passive power component.

V. HARMONIC ANALYSIS OF THE LOAD VOLTAGE

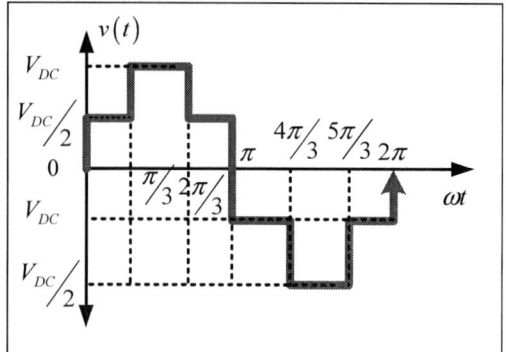

Fig. 8. Voltage waveform across R-L load

In order to determine the Fourier co-efficient, the following distinctive waveforms across the $(R-L)$ passive power component are considered as shown in the Fig. 8.

The generic Fourier series may be written by the following equation as:

$$x(t)=A_0+\sum_{n=1}^{\infty}A_nCon(n\omega t)+\sum_{n=1}^{\infty}B_n sin(n\omega t) \quad (8)$$

The average value of wave as shown in the Fig. 8 over the complete time period (T) is being zero and it has half wave symmetry so the following Fourier co-efficient are being zero as:

$$\left.\begin{array}{c}A_0\\A_n\end{array}\right\}=0\ \&\ \omega=\frac{2\pi}{T}=\frac{2\pi}{2\pi}=1 \quad (9)$$

So, the combination of the above equation (8) & (9) yields to determine the Fourier co-efficient (B_n) as:

$$B_n=\frac{2}{T}\int_0^T x(t)\times sin(n\omega t)\,dt \quad (10)$$

The equation (10) is reduced by the features of the curve as shown in the Fig. 8 and limit of the integration as:

$$B_n=\frac{2}{2\pi}\int_0^{2\pi}x(t)\times sin(nt)\,dt=$$

$$\frac{1}{\pi}\left[\int_0^{\frac{\pi}{3}}\frac{V_{DC}}{2}sin(nt)\,dt+\int_{\frac{\pi}{3}}^{\frac{2\pi}{3}}V_{DC}sin(nt)\,dt+\int_{\frac{2\pi}{3}}^{\pi}\frac{V_{DC}}{2}sin(nt)\,dt\right] \quad (11)$$

$$-\frac{1}{\pi}\left[\int_{\pi}^{\frac{4\pi}{3}}\frac{V_{DC}}{2}sin(nt)\,dt+\int_{\frac{4\pi}{3}}^{\frac{5\pi}{3}}V_{DC}sin(nt)\,dt+\int_{\frac{5\pi}{3}}^{2\pi}\frac{V_{DC}}{2}sin(nt)\,dt\right]$$

The simplification form of the equation (11) may be written by another equation as:

$$B_n=\left[\frac{V_{DC}}{n\pi}\right]\times$$

$$\left[\begin{array}{c}\frac{1}{2}\times\left(1-cos\frac{n\pi}{3}\right)+\left(cos\frac{n\pi}{3}-cos\frac{2n\pi}{3}\right)+\\\frac{1}{2}\times\left(cos\frac{2n\pi}{3}-cos\,n\pi\right)+\left(cos\frac{5n\pi}{3}-cos\frac{4n\pi}{3}\right)\\\frac{1}{2}\times\left(cos\frac{4n\pi}{3}-cos(n\pi)\right)+\frac{1}{2}\times\left(1-cos\frac{5n\pi}{3}\right)\end{array}\right] \quad (12)$$

Imposing the following conditions of the five-level-inverter (i.e. D&NPCMI) topology it may be written by:

$$\left.\begin{array}{c}V_{DC}=100\\\frac{V_{DC}}{2}=50\end{array}\right\} \quad (13)$$

Now, the combination of the equation (12) & (13) yields as simplified form:

$$B_n=\frac{50}{n\pi}\times\left[\begin{array}{c}2+cos\left(\frac{n\pi}{3}\right)-cos\left(\frac{2n\pi}{3}\right)-\\2cos(n\pi)-cos\left(\frac{4n\pi}{3}\right)+cos\left(\frac{5n\pi}{3}\right)\end{array}\right] \quad (14)$$

VI. THD COMPUTATIONS OF THE D&NPCMI TOPOLOGY

The total harmonic distortion (THD) as well as the effects of the harmonics affects the power quality, so theoretical computations are useful to find out some power quality indexes as followed. So the value of the $\left(B_{1(max)}\right)$ may be computed from the equation (14) as:

$$B_{1(max)}=\left\{\frac{6\times 50}{\pi}\right\}\approx 95.5 \quad (15)$$

The rms load voltage $\left(V_{o(rms)}\right)$ may be determined from the Fig. 8 as:

$$V_{o(rms)}=\frac{1}{\sqrt{\pi}}\times\sqrt{\left[\left(\frac{V_{DC}}{2}\right)^2\times\frac{\pi}{3}+\left(V_{DC}\right)^2\times\frac{\pi}{3}+\left(\frac{V_{DC}}{2}\right)^2\times\frac{\pi}{3}\right]}$$

$$=\frac{V_{DC}}{\sqrt{2}} \quad (16)$$

So the rms value of the $\left(B_{1(rms)}\right)$ may be computed from the equation (14) as:

$$B_{1(rms)}=\frac{B_{1(max)}}{\sqrt{2}}=\frac{6\times 50}{\pi\sqrt{2}}\approx 67.5 \quad (17)$$

THD expression may be written by the rms load voltage $\left(V_{o(rms)}\right)$ and fundamental component of the rms load voltage $\left(B_{1(rms)}\right)$ as:

$$THD=\sqrt{\frac{\left(V_{o(rms)}\right)^2-\left(B_{1(rms)}\right)^2}{\left(B_{1(rms)}\right)^2}}$$

$$=\sqrt{\left(V_{DC}/\sqrt{2}\right)^2-\left(B_{1(rms)}\right)^2\Big/\left(B_{1(rms)}\right)^2} \quad (18)$$

Now, putting the relevant value as computed by the equation (16) & (17) yields as:

$$THD=\sqrt{\left(V_{DC}/\sqrt{2}\right)^2-\left(B_{1(rms)}\right)^2\Big/\left(B_{1(rms)}\right)^2}$$

$$=\sqrt{\left[(70.71)^2-(67.5)^2\right]\Big/(67.5)^2}\approx 0.3108 \quad (19)$$

So, %THD = 31.08%

VII. RESULTS

In order to predict the THD value of three phase five-level-inverter (i.e. (D&NPCMI) power topology without uses of the filter, the simulation value of % THD is observed as 21.84% using the peak 12V & 50 Hz sinusoidal reference signal and the peak carrier wave voltage 5V with 4 KHz as shown in the Fig. 9.

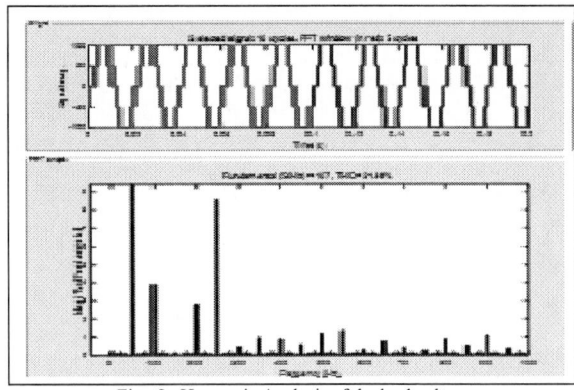

Fig. 9. Harmonic Analysis of the load voltage

Further, In order to predict the THD value of three phase five-level-inverter (i.e. (D&NPCMI) power topology without uses of the filter, the simulation value of % THD is observed as 22.87% using the peak 12V & 50 Hz sinusoidal reference signal and the peak carrier wave voltage 6V with 4 KHz as shown in the Fig. 10.

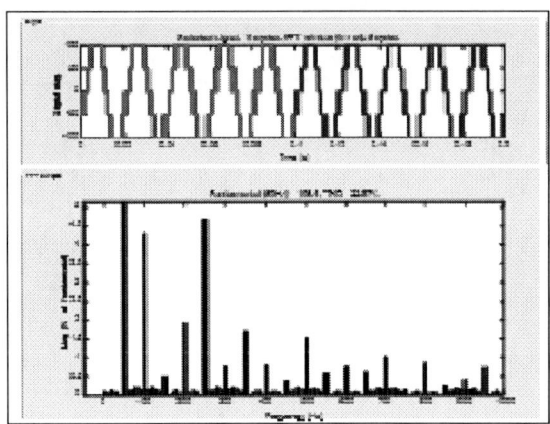

Fig. 10. Harmonic Analysis of the load voltage

Further, In order to predict the THD value of three phase five-level-inverter (i.e. (D&NPCMI) power topology without uses of the filter, the simulation value of % THD is observed as 34.87% using the peak 10V & 50 Hz sinusoidal reference signal and the peak carrier wave voltage 10V with 4 KHz as shown in the Fig. 11.

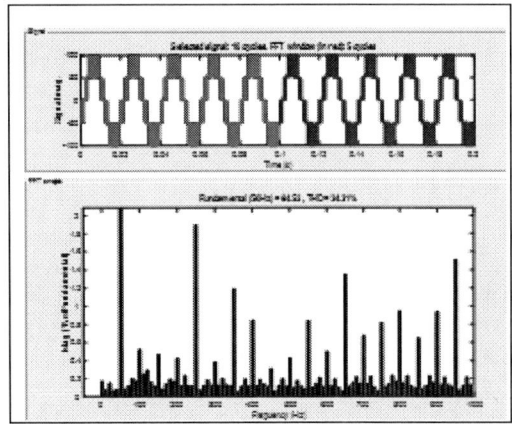

Fig. 11. Harmonic Analysis of the load voltage

VIII. CONCLUSIONS

The five-level-NPC with twelve power semiconductors is simulated in Matlab platform, the optimal value of the simulation value of % THD is observed about 21.84% using the peak 12V & 50 Hz sinusoidal reference signal and the peak carrier wave voltage 5V with 4 KHz as shown in the Fig. 9 as comparison with theoretical value (i.e. 31.08%).

Further, using the peak 12V & 50 Hz sinusoidal reference signal and the peak carrier wave voltage 6V with 4 KHz as a controller action which is observed THD, 22.87% value and it is shown in the Fig. 10 simulated by the MATLAB software.

It is found from the simulation results that the peak 12V & 50 Hz sinusoidal reference signal and the peak carrier wave voltage 5V to 6V with 4 KHz for this controller action which is observed 21.84% THD as best harmonic voltage elimination with respect to theoretical and other the peak voltage & 50 Hz sinusoidal reference signal and the peak carrier wave voltage with 4 KHz. So, it is cost effective solution of the power quality improvements and THD reductions without uses of the filter components.

REFERENCES

[1] S Jhampati, & T Halder, "Health is Wealth - A Straight Forward Diagnosis of ECG Signal of Human Heart", International Journal of Innovative Research in Science, Engineering and Technology (IJIRSET) (An ISO 3297: 2007 Certified Organization) Vol. 3, Issue 11, November 2014. ISSN: 2319-8753.

[2] T. Halder, ´A Reliability Prediction of the Flyback SMPS" India Conference (INDICON), 2015 Annual IEEE, pp. 1-6.

[3] H. K. Al-Hadidi, A. M. Gole, and D. A. Jacobson, "A Novel Configuration for a Cascade Inverter-Based Dynamic Voltage Restorer With Reduced Energy Storage Requirements," Power Delivery, IEEE Transactions on, vol. 23, 2008, pp. 881-888.

[4] T. Jimichi, H. Fujita, and H. Akagi, "Design and Experimentation of a Dynamic Voltage Restorer Capable of Significantly Reducing an Energy-Storage Element," Industry Applications, IEEE Transactions on, vol. 44, pp. 817-825, 2008.

[5] T. Halder, "A Comparative Study of the Hard & Soft Switching of the Flyback Converters",2014 6th IEEE Power India International Conference (PIICON), 2014 , pp.1-6.

[6] T. Halder, "A Flyback Converter Topology Selection Criterion for the Practical Engineer" 2016 7th India International Conference on Power Electronics (IICPE), 2016, pp.1-6.

[7] T. Halder, "A Topology Selection: An Isolated Flyback Converter"2016 IEEE 7th Power India International Conference (PIICON), 2016, pp.1-6.

[8] W Balow, P Sain, A Chatterjee,"An Improved Technique of Wind Power Harvesting With a Doubly Fed Induction Generator (DFIG)", 2018 8th IEEE Power India International Conference, 2018, pp. 1-6.

[9] T Halder, "A Reliability Prediction of The Flyback SMPS" 2015 Annual IEEE India Conference (INDICON), 2015, pp.1-6.

[10] W Balow, T Halder,"A Selective Harmonic Elimination (SHE) Technique For the Multi-Leveled Inverters", IEEE EDKCON 2018,

[11] T Halder, "An Impact of the Voltage & Current Ripples in the Power Stages of the Boost Converter", IEEE EDKCON 2018, pp. 1-6

[12] I Biswas, B Das, W Balow, T Halder, "Probability of the Pure Sine Wave in the Power Quality Issues", 2018 8th IEEE Power India International Conference, 2018, pp. 1-6.

[13] T Halder, "An Effect of the Electromagnetic Compatibility (EMC) in the Flyback Converters" 8th IEEE India International Conference on Power Electronics (IICPE-2018),2018, pp. 1-6.

[14] T Halder, "An Improved Power Loss Modeling of the MOSFET Using the Flyback SMPS" , 8th IEEE India International Conference on Power Electronics (IICPE-2018), 2018, pp. 1-6.

[15] T Halder, "An Improved Soft Switched Boost Power Converter Suitable for Power Factor Correction", 8th IEEE India International Conference on Power Electronics (IICPE-2018), 2018, pp. 1-6.

[16] T. Halder, "A Smart Grid" ,2014 6th IEEE Power India International Conference (PIICON), 2014, pp. 1-6.

[17] T Halder, "Power Loss Modeling of the Semiconductors Using the Flyback Converters"8th IEEE India International Conference on Power Electronics (IICPE-2018), 2018, pp. 1-6.

[18] S Koley, T Halder, " A Smart Modeling & Simulation of the Single Ended Primary Inductor Converter (SEPIC)", 8th IEEE India International Conference on Power Electronics (IICPE-2018), 2018, pp.1-6

[19] A Chatterjee, P Sain, T Halder, "A simple modeling & working with wind power production" IEEE 2018 IEEMA Engineer Infinite Conference (eTechNxT), 2018, pp. 1-6.

[20] MDT Hoque, AK Sinha, T Halder, "An improved strategy of energy conversion and management using PSO algorithm", Computer, Communication and Electrical Technology: Proceedings of the 2017.

[21] T. Halder, "A Robust Modeling of the Switching Personality of the MOSFET for the Inductive Load", 2017 14th IEEE India Council International Conference (INDICON), 2017, pp. 1-6.

[22] D Halder, P Biswas, T Halder, P.K. Das, "An Improved Aerodynamic Modeling of the Wind Turbine for the Wind Power Harvesting", 2017 14th IEEE India Council International Conference (INDICON), 2017, pp.1-6.

[23] P Biswas, D Halder, T. Halder, "A Zero Voltage Switching (ZVS) Boost Converter Suitable for Power Factor Correction", 2017 14th IEEE India Council International Conference (INDICON), 2017, pp. 1-6.

[24] T Halder, "An improved performance of the soft switching buck converter", 2016 2nd International Conference on Control, Instrumentation, Energy, 2016. pp.1-5.

[25] T. Halder, "The Cross Border Power Trading for Humanity & Friendship", 2016 IEEE 7th Power India International Conference (PIICON), 2016, pp. 1-6.

[26] T. Halder, "Selection of Switching Skills of the Power MOSFET in the Static Converter", 2017 14th IEEE India Council International Conference (INDICON), 2017, pp.1-6.

[27] T Hoque, AK Sinha, T. Halder, "A Smart Strategy of Automatic Generation & Control (AGC)", www.taylorandfrancis.com CRC Press 2016.

[28] T. Hoque, AK Sinha, T Halder, "An Improved Energy Conversion & Manegement Technique Using PSO Algorithm", ACCET 2016, pp. 1-5.

[29] T. Halder, "A Smart Solar Power Cultivation Using the Flyback Converter & Multi-Level Inverter", 2016 7th India International Conference on Power Electronics (IICPE), 2016, pp. 1-6.

[30] K Biswas., S Lahahiri., S Das., T. Halder, "A Trading of Carbon Credits & Green House Gas (GHG) Certificates Within Legal Frameworks"National Conference on RENEWABLE ENERGY, NCRE-2016, pp. 13-13.

[31] K Biswas., S Lahahiri., S Das., T. Halder, "A Maximum Power Point Tracker For The Solar Energy Harvesting ", National Conference on Renewable Energy, NCRE-2016, 12-12.

[32] A.K. Sinha, T Hoque, D Sau, T. Halder, "A Smart and Green Charging Station Using the Buck Converters for the Electrical Vehicle (EV)" Advances in Industrial Engineering and Management 5 (I), 2016, pp. 74-79.

[33] T. Halder, "An Improved Performance of the Soft Switching Buck Converter" Control, Instrumentation Energy & Communication (CIEC16), 2016, pp. 1-6.

[34] S. Jhampati, T. Halder, "Health is Wealth - A Straight Forward Diagnosis of ECG Signal of Human Heart", International Journal of Innovative Research in Science, Engineering and Technology (IJIRSET) (An ISO 3297: 2007 Certified Organization) Vol. 3, Issue 11, November 2014. ISSN: 2319-8753.

[35] J. Biswas & T. Halder, "Electric Shock and Design Optimizations of the Electrical Earthing Systems", ", International Journal of Innovative Research in Science, Engineering and Technology (IJIRSET) (An ISO 3297: 2007 Certified Organization) Vol. 3, Issue 11, November 2014. ISSN: 2319-8753.

[36] T. Halder, "Reliability of the Flyback Converter "ICCS-2013, 2013, pp. 275-280.

[37] Chapman D., Harmonics – causes and effects. Leonardo Power Quality Application Guide –Part 3.1, 2001.

[38] M. A. Saqib and S. A. R. Kashif," Artificial Neural Network Based Space Vector PWM for a Five-Level Diode-Clamped Inverter," AUPEC, 2010..

[39] T.Halder, "Design Optimization Techniques for the Low Power CMOS VLSI" American Scientific Publishers, vol. 5 issue I, pp. 68-73.

[40] L. Yun Wei, F. Blaabjerg, D. M. Vilathgamuwa, and L. Poh Chiang, "Design and Comparison of High Performance Stationary-Frame Controllers for DVR Implementation," Power Electronics, IEEE Transactions on, vol. 22, pp. 602-612, 2007.

[41] C. Meyer, R. W. De Doncker, L. Yun Wei, and F. Blaabjerg, "Optimized Control Strategy for a Medium-Voltage DVR,Theoretical Investigations and Experimental Results," Power Electronics, IEEE Transactions on, vol. 23, pp. 2746-2754, 2008.

[42] S.S. Saha, B. Majumdar, T. Halder, SK Biswas, "A Comparison between MOSFET and IGBT used in ZVS/ZCS phase shift PWM converter"PEITISCON-2005, 2005, pp. 204-209.

[43] P Biswas, D Halder, T Halder, PK Das," An Improved Aerodynamic Modeling of the Wind Turbine for the Wind Power Harvesting", IEEE INDICON 2017, pp. 1-6.

[44] T. Halder, "A New Simple Hybrid camp circuit for Generic Power Converters and Protection scheme" ENERGY SYSTEM PLANNING IMPLEMENTATION AND OPERATION, -ESPIO2011, organized by IEEE Calcutta Section, 2011, pp. 70-73.

[45] Md. T Hoque, A. K Sinha, T Halder, "An Improved Energy Conversion & Manegement Technique Using PSO Algorithm " ACCET, 2016, pp. 1-5.

[46] S.S. Saha, B. Majumdar, T. Halder, SK Bisaws, A MOSFET-IGBT Hybrid converter with improved efficiency using ZV-ZCS", NPEC-2005, IIT Kharagpur, India, 2005, pp. 370-373.

[47] IEEE Task Force on Modeling and Simulation, ″Modeling and Simulation of the propagation of harmonies in electric power networks, Part I: Concepts, models, and simulation techniques″, IEEE Trans. on Power Delivery, Vol. 11, No. 1, Jan. 1996, pp. 452-465.

[48] T. Halder, "A Simple Mechanics of an Electric Train Movement", RAME-2013, Proceedings of TEQIP Phase-II, at Government College of Engineering & Textile Technology Berhampur, Murshidabad, West Bengal, India, 2013, pp. 155-158.

[49] Arrillaga J., Watson N. R., Chen S., Power system quality assessment, John Wiley & Sons, Ltd, Chichester, 2000.

[50] Jose Rodriguez, Leopold G. Franquaelo, Samir Kouro Jose I. Leon, Ramon C. Portillo, Ma Angeles Martin Prats and Marcelo A. Perez," Multilevel converters: An Enabling Technology for High Power Applications", IEEE Preceding ,vol.97,no.11,Nov. 2009.

[51] N.K. Madora and A. Kusko, ″Computer-Aided Design and Analysis of Power-Harmonic Filters″ IEEE Trans. on Industry Applications, Vol. 36, No. 2, March/April 2000, pp.604-613.

[52] IEEE Task Force on Modeling and Simulation ″Modeling and Simulation of the propagation of harmonies in electric power networks, Part II: Sample systems and examples″, IEEE Trans. on Power Delivery, Vol. 11, No. 1, Jan. 1996, pp. 466-474.

[53] T. Halder, "A New Simple Hybrid camp circuit for Generic Power Converters and Protection scheme", ENERGY SYSTEM PLANNING IMPLEMENTATION AND OPERATION, -ESPIO2011, organized by IEEE Calcutta Section, 2011, pp. 70-73.

[54] P Biswas, D Halder, T Halder, "A Zero Voltage Switching (ZVS) Boost Converter Suitable for Power Factor Correction" IEEE INDICON 2017, pp. 1-6.

[55] PK Das., T. Halder, "LOAD FORECASTING OF LOAD AND ENERGY REQUIREMENTS", Kalyani Government Engineering College,Reasons-2011, ISSN 2277-1654 10, 2011.

[56] T Hoque, AK Sinha, T Halder, "A Smart Strategy of Automatic Generation & Control (AGC), www.taylorandfrancis.com CRC Press 2016 .

2019 Devices for Integrated Circuit (DevIC), 23-24 March, 2019, Kalyani, India

Semiconducting Behaviour of Metal Thin Film composed of Green Synthesized Silver Nanoparticles

Kaushik Roy
Department of Electronics &
Telecommunication Engineering
Jadavpur University
Kolkata, India
dr.kaushikroy@yahoo.com

Chandan K. Sarkar
Department of Electronics &
Telecommunication Engineering
Jadavpur University
Kolkata, India
phyhod@yahoo.co.in

Chandan K. Ghosh
School of Materials Science &
Nanotechnology
Jadavpur University
Kolkata, India
chandu_ju@yahoo.co.in

Abstract—Here, we demonstrated an easy, clean and eco-benign synthetic protocol for silver nanoparticle production from silver nitrate salt using Indian bay leaf (*Cinnamomum tamala*) extract. Production of silver nanoparticle was monitored at periodic intervals through ultra-violet visible spectroscopy (UV-Vis). X-ray powder diffraction (XRD) and high resolution electron microscopy (HRTEM) respectively indicated the crystal phases and shape of prepared nanoparticles. Infra-red spectroscopy (FTIR) of the nanoparticles revealed the role of functional organic molecules involved in surface capping and stabilization of the particles. Thin film of green synthesized silver nanoparticles was deposited carefully and then connected to regulated DC voltage to realize the conducting behaviour. The obtained result signified the semi-conductive behavior of silver thin film at ambient temperature.

Keywords—Silver nanoparticles, thin film, XRD, HRTEM, electrical conductivity

I. INTRODUCTION

Research articles on conductive behavior of semiconducting thin films are abundantly available in literature. But conductive behaviour of metallic thin films is comparatively less investigated. Probably, B. Roy and his team were first to fabricate conducting films composed of metallic nanoparticles of average diameter 5-10 nm on ceramic substrate by reduction process and resistivity of the film was estimated between 80K and 300K temperature [1]. The obtained I-V characteristics of thin film exhibits sharp rise in the value of current following threshold value of applied voltage. Another group investigated the conducting behavior of thin film made of chemically produced nano-silver [2]. The silver ions were reduced by glucose and polyvinylpyrrolidone (PVP) in the reacting medium and after production of nanoparticles, the ultra-clean quartz substrate was dip coated into the medium and dried to achieve uniform thin film. I-V characteristic of this thin film was recorded between temperature 100 and 300 K and the result

denotes a drop of resistivity with increasing temperature, which is indicative of semiconducting nature of the film.

B. Roy published another article where anomalous change in the resistance in nanocrystalline metallic system has been reported [3]. Metallic chains of nanoparticles having diameters 4 to 5 nm were grown within glass ceramics. I-V curve of this structure exhibits anomalous change in resistivity with the variation of ambient temperature. Metallic behaviour with large variation in resistivity was primarily observed at room temperature but with the increase of temperature, semiconducting behavior was noticed. This variation of resistivity was ascribed to thermal induced stress that ruptures metallic nanoparticles [4]. The particles were expected to undergo quantum size-effect and display semiconductor-like behaviour.

These cited articles demonstrated conductive behavior of thin films which are composed of physically or chemically produced nanoparticles but biologically fabricated silver nanoparticles have never been exposed to dc voltage [5,6]. Electrical conductivity of metallic films prepared using green synthesized gold or silver nanoparticles can also be investigated to get an insight of their conductivity at nanoscale and for comparative analysis with the commercial thin films available in the market [7]. On the other hand, Indian bay leaves (*Cinnamomum tamala*) are traditional Indian spice used to prepare Indian dishes and ayurvedic medicines as well [8]. Leaves of *Cinnamomum tamala* are potent source of numerous organic molecules e.g. eugenol, terpene alcohol, ascaryophyllene etc. These organic compounds may act as strong reducing and stabilizing agents for producing nanoparticles in reacting medium [9]. Here, we report clean and green production of silver nanoparticles featuring Indian bay leaf extract, then preparation of thin film by standard drop casting process and finally study of the conductive behavior of the film while connected to the dc regulated voltage.

978-1-5386-6723-1/19 $31.00 © 2019 IEEE

II. EXPERIMENTAL METHODS

A. Green synthesis of silver nanoparticles

Dried bay leaves (Figure 1) were purchased from market, authenticated and then cleaned before use. Then 50 g dried leaves was crushed, boiled in de-ionized (DI) water at temperature 60°C for 10-12 minutes and filtered to get pure bay leaf extract. Analytical grade silver nitrate was procured from Sigma Aldrich, India and its 20 mM stock solution was prepared for further use as precursor. The leaf extract was then added drop wise to silver nitrate solution (in equal volume) for reducing silver cations leading to formation of silver nanoparticles. This reacting solution was kept at 30°C for observation and after 60 min, its color turned to dark yellow from colorless suggesting the production of nano-silver. After a couple of hours of observation, the rate of reaction gradually saturated. The colloidal particles were filtered out from medium by ultracentrifugation at the rate of 6000 rpm for 10 minutes. When the centrifugation was over, the obtained clear soup was discarded and the pellet formed inside the centrifuge tube was carefully scraped and dried overnight. The dried precipitate was then crushed to prepare powdered sample of silver nanoparticles for further morphology and property analysis.

Fig. 1. Dried bay leaves (*Cinnamomum tamala*)

B. Characterization of bio-fabricated Ag nanoparticles

Production of Ag nanoparticles was examined by scanning the reacting medium under Perkin Elmer UV-Vis spectrophotometer. X-ray powder diffraction (Rigaku Ultima-III) was performed to record diffraction curve of these green synthesized nanoparticles (λ=0.154 nm). FTIR spectrometer (Shimadzu, Japan) was used to obtain IR-spectrum of the bio-fabricated nanoparticles. During preparation of grid for high resolution electron microscopy, powdered sample of nano-silver was suspended and later sonicated in de-ionized water (keeping standard concentration of 50μg/ml) for half an hour in a probe sonicator. Two drops of suspension of

colloidal Ag particles were spread on coated grid of copper mesh and vacuum-dried. Then the grid was inspected under high resolution TEM (JEOL-2010, Japan). This suspension of colloidal particles was used to prepare thin film later for conductivity assessment.

C. Study of conductivity of thin film

The nano-suspension of bioengineered particles was employed for preparation of homogeneous thin film for conductivity study. Two drops of the suspension were spread on cleaned silicon wafer and further dried overnight to obtain uniform thin film made of silver nanoparticles. Silver paste of analytical grade was procured and used to create two metallic contacts (at a distance ~ 1 mm) on deposited thin film. The prepared film was then exposed to applied regulated voltage (dc) through metal contacts along with ammeter in series connection. The regulated voltage was gradually increased from 0V with a regular increment of 0.1V. The observed values of current were in the order of μA and the respective voltage values were tabulated to draw and analyse I-V characteristic curve of this two dimensional thin metallic film.

III. EXPERIMENTAL RESULTS

A. Biosynthesis of nano-silver

After 30 min of reaction, the colour of reacting solution turned dark yellow from colorless [Figure 2 (inset)] and eventually became yellowish brown after 2 hrs. This colour change confirmed the production of colloidal nano-silver in reacting medium. Absorption spectrum of the reacting mixture consists of a distinct absorbance peak at wavelength of 450 nm (Figure 2) which confirmed the formation of silver nanoparticles as characteristic plasmonic peak of silver lies near 450 nm wavelength [10].

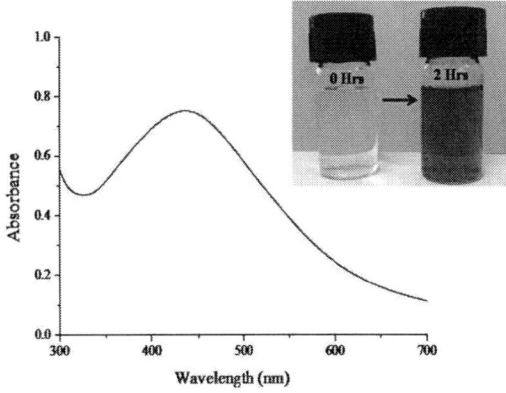

Fig.2. Recorded UV-Visible spectrum of mixture after 2 hrs

When the leaf extract interacts with silver ions, the constituents like eugenol ($C_{10}H_{12}O_2$) and alcohols present in leaf extract probably reduced the metallic ions and capped the particle surface during incubation [11]. Possible chemical reactions would be:

$$AgNO_3 \rightarrow Ag^+ + NO_3^-$$

$$C_{10}H_{12}O_2 + Ag^+ \rightarrow Ag^0 \text{ (nanoparticles)}$$

$$C_{10}H_{17}OH + Ag^+ \rightarrow Ag^0 \text{ (nanoparticles)}$$

B. Characterization of silver nanoparticles

XRD pattern of powdered nanoparticles (shown in Figure 3) consists of two distinct peaks at $2\theta = 32°$ and $38°$ that respectively correspond to (200) and (111) planes of metal silver (JCPDS card no.: 04-0783) [12].

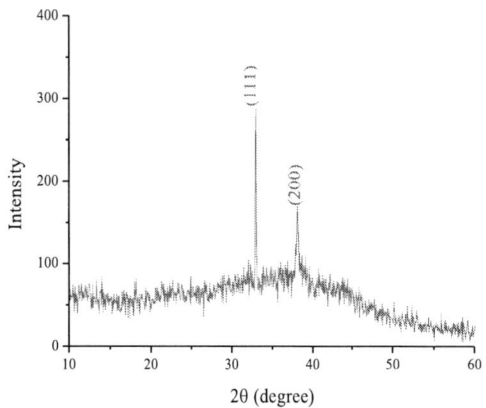

Fig.3. XRD pattern of silver nanoparticles

The electron microscopic images indicate that these green synthesized nanoparticles have spherical shape with polycrystalline nature and particle diameter ranges between 5-10 nm (Figure 4).

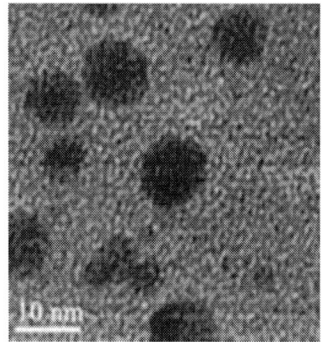

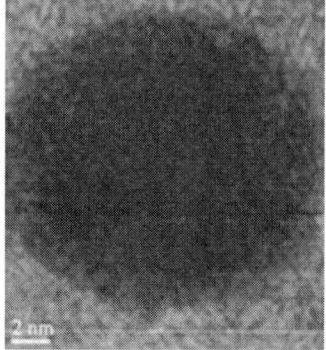

Fig. 4. Electron microscopic images of Ag nanoparticles

The IR spectrum of the particles in absorbance mode (Figure 5) exhibits five distinct peaks which may correspond to various chemical bondings. The bands noticed at 1094 and 1610 cm^{-1} can be attributed to vibrations due to stretching of C-N and C=O bonds respectively, present in amines [13]. One band observed at 2978 cm^{-1} may indicate stretching of C-H bonds of aldehydes while another band at 3290 cm^{-1} possibly signifies O-H stretching of aromatic compounds like eugenol, terpene alcohols etc. [14]. A small peak near 607 cm^{-1} denotes mark of alkane compounds on particle surface as well. IR spectroscopy result may conclude that these functional biomolecules such as phenol, eugenol, terpene alcohols and amines found in the extract of bay leaves might play the key role of stabilizing and reducing agents in time of interaction with silver cations and finally stabilized surface of the nanoparticles in liquid medium.

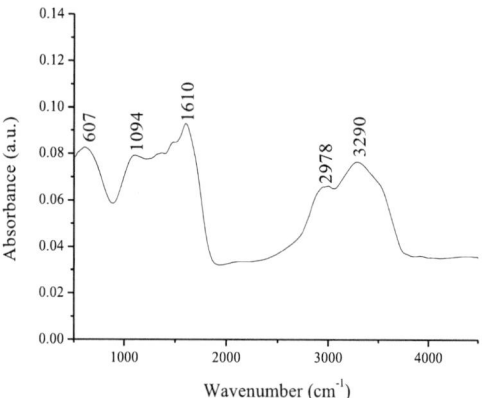

Fig. 5. IR pattern of biosynthesized nano-silver

C. Analysis of I-V curve of thin film

The I-V characteristic curve (shown in Figure 6) showed less conductive semiconducting behaviour of thin film unlike bulk metals. The value of current was observed to increase when dc voltage crossed a threshold value ($\approx 0.5V$). The reason behind this less conductive nature of the thin film has been discussed in the following section.

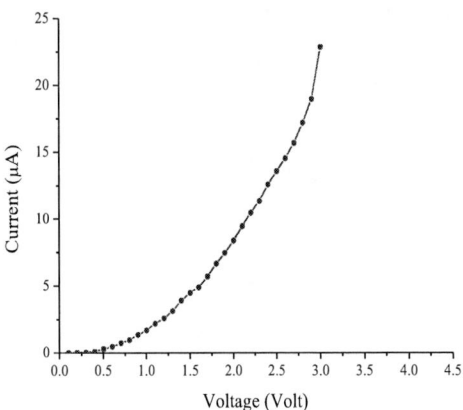

Fig.6. I-V characteristic curve of thin film

In case of bulk metals, the electrons move in all directions as there is no restriction in the movement of electrons through the materials [15]. But thin film is a layer structure of metals ranging from a few nanometres to several micrometers in thickness. So, the movement of electrons in a thin film is restricted in two dimensions only unlike three dimensional bulk structures [16]. In addition, the thin film is made of a layer of nanoparticles with different crystalline grains arranged in different orientations [17]. Electrical conductivity depends on mobility of the charge carriers like electrons in materials. The electrons can be scattered at the interfaces between different grains, called grain boundaries during movement [18]. Hence the movement of electrons will reduce at grain boundaries and eventually the electrical conductivity of film will be decreased. This explains why the thin film of metal nanoparticles here exhibits lesser conductive behavior than the bulk metals.

IV. CONCLUSION

In this report, an easy and eco-benign production of Ag nanoparticles featuring Indian bay leaf extract has been demonstrated. The metal nanoparticles were characterized using analytical tools like XRD, HRTEM and FTIR. Thin film of biogenic silver nanoparticles was prepared by its deposition on clean silicon wafer through drop coating for further study of electrical conductivity. The prepared thin film was found to exhibit less conductive nature for conduction of electricity unlike bulk metals. Semiconducting property of this film may be utilized in different circuit and device applications where semiconducting behavior is highly desired for circuit operations.

References

[1] B. Roy and D. Chakravorty, "Electrical conductance of silver nanoparticles grown in glass-ceramic," J. Phys. Conden. Matt.,vol. 2(47), Article no. 9323, 1990.

[2] M.Khan, S. Kumar, M. Ahamed, S. Alrokayan and M.S. AlSalhi, "Self-focusing and self-trapping of optical beams upon photopolymerization," Nanoscale Res. Lett.,vol. 6(1), pp. 434-442, 2011.

[3] B. Roy and D. Chakravorty, "Anomalous resistance change in nanocrystalline metallic systems," Solid State Commun., vol. 87(1), pp. 71-75, 1993.

[4] I. Astefanoaei, I. Dumitru and Al Stancu, "Size-dependent thermal stresses in the core-shell nanoparticles," Chin. Phys. B, 22(12), 128102, 2013.

[5] D. Langley, G. Giusti, C. Mayousse, C. Celle, D. Bellet and J.P. Simonato, "Flexible transparent conductive materials based on silver nanowire networks: a review," Nanotechnology,vol. 24(45), pp. 452001, 2013.

[6] A. Kim, Y. Won, K. Woo, S. Jeong and J. Moon, "All-solution-processed Indium-free transparent composite electrodes based on Ag nanowire and metal oxide for thin-film solar cells," Adv. Function. Mater.,vol. 24(17), pp. 2462-2471, 2014.

[7] K. Roy, C.K. Ghosh and C.K. Sarkar, "Degradation of toxic textile dyes and detection of hazardous Hg^{2+} by low-cost bioengineered copper nanoparticles synthesized using *Impatiens balsamina* leaf extract," Mater. Res. Bull., vol. 94, pp. 257-262, 2017.

[8] S. Kumar, N. Vasudeva and S. Sharma, "Pharmacological and pharmacognostical aspects of *Cinnamomum tamala* Nees & Eberm," J. Pharm. Res. 5(1), pp. 580-584, 2012.

[9] U. Chakraborty and H. Das, "Antidiabetic and antioxidant activities of *Cinnamomum tamala* leaf extracts in stz-treated diabetic rats,"Global J. Biotech. Biochem., vol. 5(1), pp. 12-18, 2010.

[10] K. Roy, C.K. Sarkar and C.K. Ghosh, "Fast colourimetric detection of H$_2$O$_2$ by biogenic silver nanoparticles synthesised using *Benincasa hispida* fruit extract," Nanotech. Rev.,vol. 5(2), pp. 251-258, 2016.

[11] M. Goudarzi, N. Mir, M. Mousavi-Kamazani, S. Bagheri and M. Salavati-Niasari, "Biosynthesis and characterization of silver nanoparticles prepared from two novel natural precursors by facile thermal decomposition methods," Sci. Rep. 6(Article no. 32539),doi:10.1038/srep32539, 2016.

[12] K. Anandalakshmi, J. Venugobal and V. Ramasamy, "Characterization of silver nanoparticles by green synthesis method using *Pedalium murex* leaf extract and their antibacterial activity," Appl. Nanosci. 6(3), pp. 399-408, 2016.

[13] P. Hamm, M. Lim and R.M. Hochstrasser, "Structure of the amide I band of peptides measured by femtosecond nonlinear-infrared spectroscopy," J. Phys. Chem. B, vol. 102(31), pp. 6123-6138, 1998.

[14] K. Roy and C.K. Ghosh, "Biological synthesis of metallic nanoparticles: A green alternative," Nanotechnology: synthesis to applications, Chapter-7, ISBN: 9781138032743, CRC Press, pp. 153-168, 2017.

[15] G.J. Snyderand E.S. Toberer, "Complex thermoelectric materials," NatureMater., vol. 7, pp. 105-114, 2008.

[16] S. Saxena, B. Kumar, B.Kaushik and Y.S. Negi, "Effect of contact thickness on electrical properties of organic thin film transistors,"Int. Conf. Signal Proces. Comm. India,doi: 10.1109/ICSPCom.2013.6719818, 2013.

[17] K. Roy and C.K. Ghosh, "Measurement of electrical conductivity of thin film composed of green synthesized copper nanoparticles," Intern. Conf. Electron. Dev. Comput. Tech., Doi:10.1109/EDCT.2018.8405073, pp. 1-4, 2018.

[18] R. Henriquez, S. Cancino, A. Espinosa, M. Flores, T. Hoffmann, G. Kremer, J.G. Lisoni, L. Moraga, R. Morales, S. Oyarzun, M. Antonio Suarez, A. Zúñiga and R.C. Munoz, "Electron grain boundary scattering and the resistivity of nanometric metallic structures," Phys. Rev. B, 82, 113409, 2010.

Performance Enhancement of Non-uniformly Doped Junctionless Transistors by Gate and Dielectric engineering

Muktasha Maji
School of VLSI Design and Embedded Systems
National Institute of Technology
Kurukshetra, India
muktasha_31711211@nitkkr.ac.in

Gaurav Saini
Department of Electronics and Communication Engineering
National Institute of Technology
Kurukshetra, India
gauravsaini@nitkkr.ac.in

Abstract — In this paper, the use of laterally graded doping and hetero gate high dielectric with high-k spacers which were positioned on both sides of the gate have been proposed to improve the performance of Junctionless Transistors (JLT). Further recessed gate structure is also used to compare its performance with conventional JLTs. 2-D TCAD simulations have been used to observe that the Drain-Induced Barrier Lowering (DIBL) and Sub Threshold Swing (SS) were reduced by 26% and 61% respectively. Further the current ratio improved by 10^8 times in the final structure. Our analysis focuses on the ability of the proposed design for a reduced leakage current leading to higher current ratio and also lower short channel effects (SCEs) like SS and DIBL.

Keywords — Junctionless transistor, Field Effect Transistor, Enhancement mode, graded doping, non-uniform doping, recessed gate, high-k dielectric, hetero-gate oxide, high-k spacer, subthreshold swing, DIBL

I. Introduction

THE evolution of CMOS technology into the sub 10 nm has led to physical limits because heavy doping in the ultra-shallow junctions is a challenge to the semiconductor industry. This can be explained on the basis of the diffusion laws and also the statistical nature of the doping atoms distribution in the semiconductor. Colinge *et al.* [1]-[2] proposed and further fabricated a new design called junctionless transistor (JLT) and these appear to have become a promising candidate for future technologies. Conventionally they do not have any junctions and also have no doping gradient. The main advantage of these transistors relies on improved and easier fabrication process. Also, this new design is less susceptible to short channel effects (SCEs). The key to the functioning of JLT is a thin and narrow channel to allow for full depletion of carriers in the OFF-state and heavy doping for a decent current flow in the ON-state. It is basically a variable resistor controlled by a gate electrode. It has similar electrical characteristics to those of conventional MOSFETs but the physics behind it being different. The JLTs have a major disadvantage of relatively low drain current. This can affect the device switching behavior and hence the power dissipation. In this paper, an effort has been made to improve the OFF-state current in the device without significantly affecting its ON-state current, subthreshold swing and DIBL.

Aiming to address these challenges in the junctionless design, various researchers worldwide have proposed different architectures based on several different techniques such as using analytical doping[3]-[10], recessed gate structures[11]-[12], dual material gate[13], gate stack[13] and source/drain extensions[14]. Extensive literatures based on analytical doping are available. Mondal *et al.* [3] in 2013 demonstrated that the vertical nonuniform doping in JLTs suppresses SCEs and improves the ON to OFF current ratio. In the same year, Cheng *et al.* [4] discussed feasibility of fabrication of the graded doped JLT. Bal *et al.* [5], in 2014 proposed linearly graded JLTs for better current ratio. Bahniman *et al.* [6] in the same year proposed the idea of differentially graded JLTs. Singh *et al.* [7] in 2016, gave the idea of the Gaussian-like doping in the channel of double gate (DG) JLTs. Ferhati *et al.* [9] in 2017 suggested that the use of Gaussian-like distribution helps in reducing the self-heating effects and hence improves the reliability of the device. Vandana *et al.* [10] in 2018 gave an empirical model for non-uniformly doped symmetric DG-JLTs. Thus it can be concluded that by having a non-uniform doping profile in JLT, substantial improvement can be achieved in the device performance. It is already known that a thin channel in JLT helps in depleting the channel faster, leading to improved device performance. However achievement of thin channel becomes difficult due to fabrication process limitations. Thus by recessing the area under the gate channel, a thinner channel has been achieved. Lou *et al.* [11] in 2017 has studied the performance of different recessed gate structures in JLTs. The gate-over structure was shown to have a superior current drive capacity and less leakage current. Wong *et al.* [12] in 2018 applied the recessed gate structure to JL nanowire to improve the ON current.

In this paper, a comparative study of laterally graded doping in the JLTs and the effect of recessed gate on them has been presented. Further the effects of dielectric on uniform as well as laterally graded structures have also been studied. The work discussed in this paper is presented as follows. Section II includes the simulation setup and the device structure. In Section III, the simulation results and discussions have been shown. In Section IV the conclusion is presented.

978-1-5386-6723-1/19 $31.00 © 2019 IEEE

II. SIMULATION SETUP AND DEVICE STRUCTURE

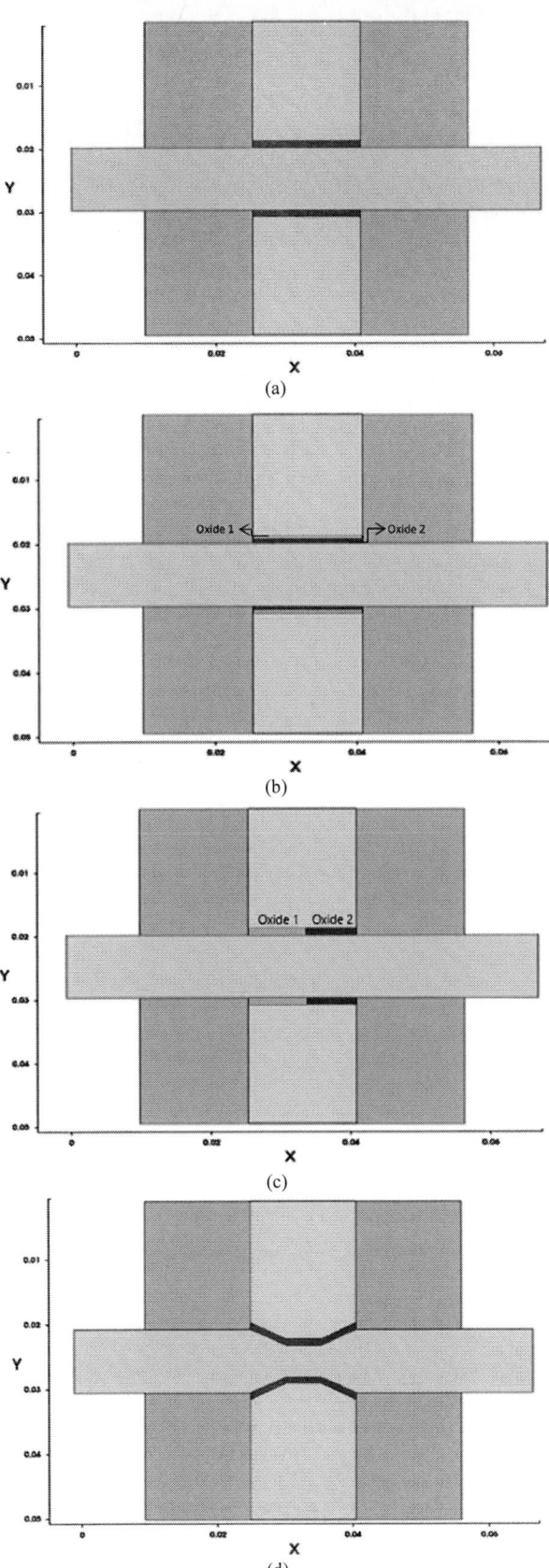

Fig. 1. Structure of (a) conventional JLT (b) horizontally stacked dielectrics (c) vertically stacked dielectrics (d) recessed gate JLT

TABLE I
DEVICE SPECIFICATIONS

Symbol	Quantity	Specification
L_{CH}	Channel Length	15 nm
T_{CH}	Channel Thickness	10 nm
T_{RCH}	Recessed Gate Channel Thickness	5 nm
L_{SP}	Spacer Length	15 nm
T_{SP}	Spacer Thickness	20 nm
$L_{S/D}$	Source/Drain Length	10 nm
$T_{S/D}$	Source/Drain Thickness	10 nm
L_{OX}	Oxide Length	15 nm
T_{OX}	Effective Oxide Thickness(EOT)	1 nm
L_G	Gate Length	15 nm
T_G	Metal Gate Thickness	19 nm
SP	Spacer Material	Si_3N_4
k_{SiO2}	Dielectric constant of SiO_2	3.9
k_{HfO2}	Dielectric constant of HfO_2	22
k_{TiO2}	Dielectric constant of TiO_2	40
k_{SP}	Dielectric constant of Si_3N_4	7.5
N_D	Constant doping concentration	10^{19} cm^{-3}
N_{DV}	Variable doping concentration	10^{20} cm^{-3} to 10^{19} cm^{-3}

In this paper all comparisons have been done at the same threshold voltage (V_{TH}) of 198 mV. Constant current method have been used for the calculation of V_{TH}. Metal work function of 5eV is used in all simulations. Dielectrics used are SiO_2, HfO_2 and TiO_2. In the hetero-gate structure (See Fig. 1. (c)), equal length of dielectrics have been used. Such structures have two gate-oxide materials with different oxide permittivity. High-k dielectric is kept near the source region and lower-k dielectric near the drain region, thus vertically stacked. In the horizontally stacked structures (See Fig. 1. (b)), thickness of SiO_2 is kept 0.6nm to maintain an effective oxide thickness (EOT) of 1nm [15]. SiO_2 is kept in contact with the channel to minimize surface roughness. In the case of lateral doping gradient, a peak phosphorus concentration of 10^{20} cm^{-3} is kept at source side and concentration is decreased till channel is reached. After that the phosphorus concentration is kept constant at 10^{19} cm^{-3} till drain. High-k spacers of Si_3N_4 are used on both sides of gate. This enhances the channel-gate fringing electric field and further confines it in the channel region. It also helps in reducing the peak of the electric fields in channel and thereby improving the subthreshold swing. In recessed gate structure (See Fig. 1. (d)), the channel length has been reduced to 5nm. Among the recessed gate structures, gate over structure has been considered because of its enhanced performance [11]. The device parameters used for simulation are mentioned in Table I.

TCAD [16] has been used for the simulations. It is widely used for research and production purposes in the semiconductor industry. It simulates device behavior based on different fundamental physical models. The models used by us are Drift Diffusion model for carrier transport, Old slot bottom model for Bandgap Narrowing, Fermi Dirac statistics model, doping dependent mobility degradation model, Shockley–Read–Hall recombination model and Auger recombination model.

III. SIMULATION RESULTS AND DISCUSSION

Simulations have been done to study the effects of variable doping concentration, recessed gate (specifically gate over structure), oxide stacks (both vertically and horizontally stacked) and different dielectrics on JLTs. The parameters studied for the different structures are ION/IOFF ratio, DIBL and SS and they have been tabulated in Table II. The electron density concentration and I_D - V_{GS} curves for all the structures have been plotted to justify our results in Table II.

drastically using the variable doping structure along with an improvement in the subthreshold swing (See Fig. 2). However, the variable doping affects the DIBL value (See Fig. 2). This increases from 27.95mV/V to 42.21mV/V. This was rectified using the oxide stack structure and finally a DIBL of 23.11mV/V was obtained. The ON to OFF current ratio simultaneously increased from 2.02×10^2 to 2.81×10^9. Improvement in subthreshold slope was from 108.178mV/dec to 67.77mV/dec.

TABLE II
SIMULATION RESULTS

Structure	Ion	Ioff	$\dfrac{I_{ON}}{I_{OFF}}$	SS (mV/dec)	DIBL(mV/V)
Constant doping with SiO₂ as dielectric	1.192×10^{-3}	5.892×10^{-6}	2.02×10^2	108.178	27.95
Variable doping with SiO₂ as dielectric	1.282×10^{-3}	2.004×10^{-8}	6.39×10^4	74.473	42.21
Variable doping with HfO₂ as dielectric	5.235×10^{-4}	2.451×10^{-11}	2.14×10^7	68.14	44.8
Variable doping with horizontally stacked SiO₂ and HfO₂	8.259×10^{-4}	2.91×10^{-12}	2.84×10^8	68.428	30.58
Variable doping with horizontally stacked SiO₂ and TiO₂	8.339×10^{-4}	5.187×10^{-12}	1.61×10^8	69.181	32.93
Variable doping with vertically stacked HfO₂ and SiO₂	6.095×10^{-3}	2.166×10^{-12}	2.81×10^9	67.773	23.11
Variable doping with vertically stacked TiO₂ and SiO₂	6.046×10^{-3}	4.332×10^{-13}	1.4×10^{10}	67.257	19.83
Constant doping with SiO₂ as dielectric and recessed gate	7.952×10^{-5}	1.021×10^{-10}	7.79×10^5	67.838	16
Variable doping with SiO₂ as dielectric and recessed gate	1.365×10^{-4}	1.449×10^{-12}	9.42×10^7	66.783	19.89
Variable doping with HfO₂ as dielectric and recessed gate	8.542×10^{-4}	1.043×10^{-12}	8.19×10^8	68.123	28.67
Variable doping with horizontally stacked SiO₂ and HfO₂ and recessed gate	1.078×10^{-3}	8.741×10^{-13}	1.23×10^9	66.841	23.07
Variable doping with horizontally stacked SiO₂ and TiO₂ and recessed gate	1.086×10^{-3}	6.353×10^{-13}	1.7×10^9	66.755	22.95
Variable doping with vertically stacked HfO₂ and SiO₂ and recessed gate	1.163×10^{-3}	1.165×10^{-13}	9.98×10^9	66.585	10.2
Variable doping with vertically stacked TiO₂ and SiO₂ and recessed gate	1.22×10^{-3}	2.149×10^{-14}	5.68×10^{10}	66.305	7.3

A. Effect of doping and dielectric oxide stack

For the study of effect of doping and dielectrics oxide stack, both horizontal and vertical, the following four different structures have been considered:
1) Constant doping with SiO₂ as dielectric
2) Variable doping with SiO₂ as dielectric
3) Variable doping with horizontally stacked SiO₂ and HfO₂
4) Variable doping with vertically stacked HfO₂ and SiO₂

With variable doping, it is observed that the OFF-state current reduces drastically without much effect on the ON-state current. The high electron density (See Fig. 3), as a result of high doping concentration at the source region helps to maintain the ON-state current. At the same time lower doping concentration in the channel leads to lower electron density in the channel (See Fig. 3). This facilitates in switching off the device easily leading to a lower OFF-state current. It is further observed that the electron density in the OFF-state reduces

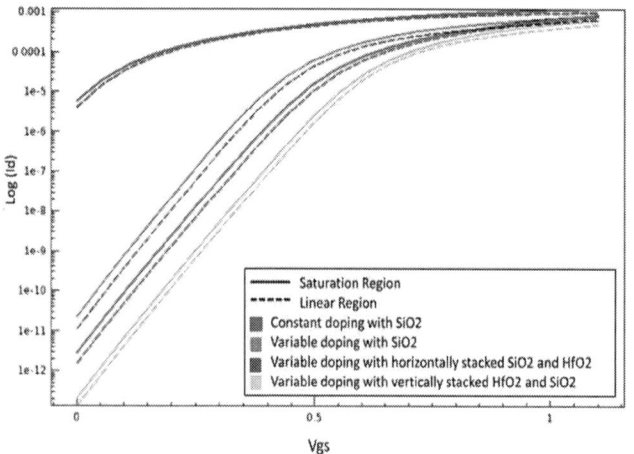

Fig. 2. Drain current vs Gate Voltage curve to study the effect of doping and oxide stack

2019 Devices for Integrated Circuit (DevIC), 23-24 March, 2019, Kalyani, India

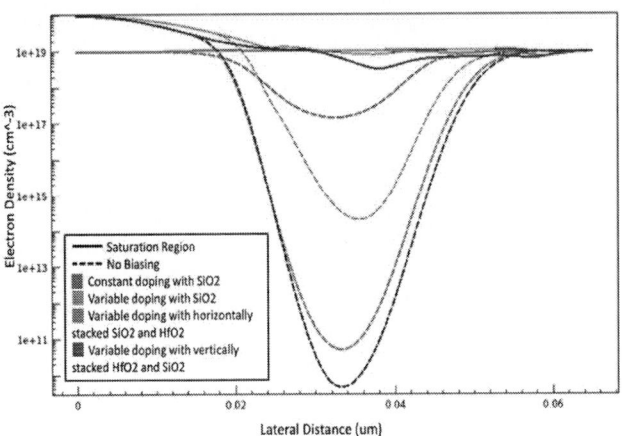

Fig. 3. Electron Density curve to study the effect of doping and oxide stack

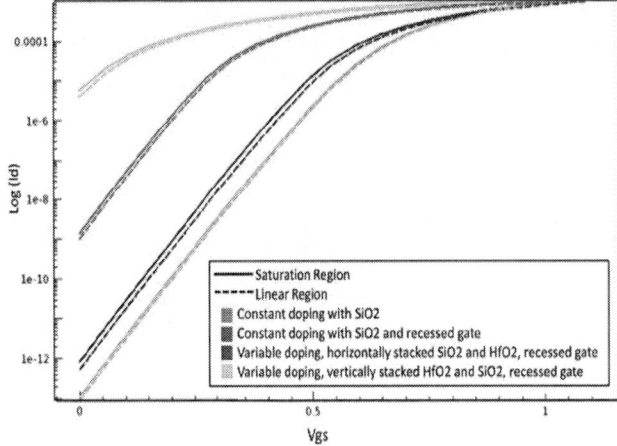

Fig. 4. Drain current vs Gate Voltage curve to study the effect of recessed gate and oxide stack

B. Effect of recessed gate and dielectric oxide stack

For this study, the following structures have been considered:
1) Constant doping with SiO_2 as dielectric
2) Constant doping with SiO_2 as dielectric and recessed gate
3) Variable doping with horizontally stacked SiO_2 and HfO_2 and recessed gate
4) Variable doping with vertically stacked HfO_2 and SiO_2 and recessed gate

Recessed gate structure reduces the OFF-state current drastically without much effect in the ON-state current. This is because of a smaller channel thickness which makes it easier to switch off the device. As a result, a higher ON to OFF current ratio is achieved. Also as evident from Fig. 5, the OFF-state electron density reduces largely without much change in the ON-state electron density, leading to a further improvement in the ON to OFF current ratio. There is an improvement in the subthreshold slope as well (See Fig. 4). However, the major advantage of using recessed gate structure over conventional structures is in terms of DIBL (See Fig. 4) which reduces from 27.95mV/V to 16mV/V. On using variable doping in this structure along with dielectric stack, there is a further improvement in the current ratio without much effect on subthreshold swing or DIBL. The final structure has current ratio of 9.98 x 10^9 with DIBL of 10.2mV/V and subthreshold swing of 66.58mV/dec. One possible method to recess the channel is to use high precision atomic layer etching.

C. Effect of dielectrics

To study the effect of dielectrics, the following four structures have been considered:
1) Variable doping and recessed gate with SiO_2 as dielectric
2) Variable doping and recessed gate with HfO_2 as dielectric
3) Variable doping and recessed gate with vertically stacked HfO_2 and SiO_2
4) Variable doping and recessed gate with vertically stacked TiO_2 and SiO_2

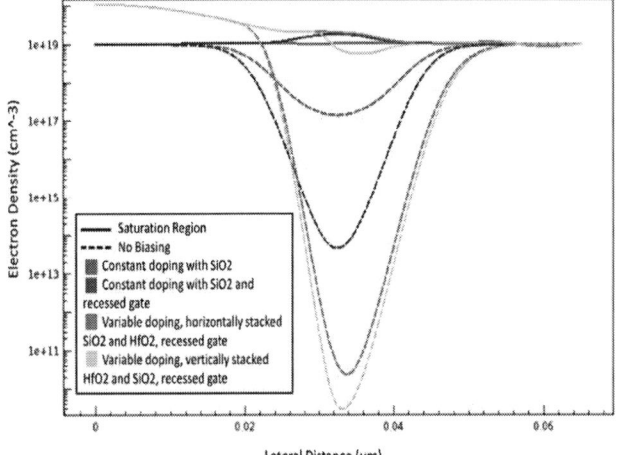

Fig. 5. Electron Density curve to study the effect of recessed gate and oxide stack

With increasing dielectric constant the OFF-state current is reduced, while the ON-state current is not much affected and remains fairly constant, thereby increasing the ON to OFF current ratio(See Fig. 6). Thus we observe that by just changing the dielectric from SiO_2 to HfO_2, there is a substantial decrease in OFF-state current. This can be explained by decreased electron concentration when HfO_2 is used (See Fig. 7). Further on using vertically stacked dielectric structure, the current ratio is further enhanced. It is observed from Fig. 7, that when vertically stacked TiO_2 and SiO_2 are used as dielectric, the electron concentration in the channel region reduces to a minimum and thus the OFF-state current reaches to a minimum level of 2.149 x 10^{-14} A/um. As observed from Fig. 6, DIBL also reduces, with a value of 7.3mV/V for vertically stacked TiO_2 and SiO_2. The subthreshold swing is not much affected (See Table II). Thus it can be inferred that with the use of higher dielectric constant material as oxide along with hetero gate structure the current driving capability is improved in JLT with lower SCEs.

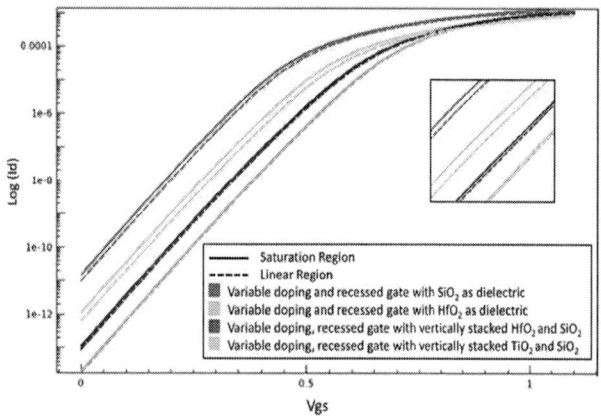

Fig. 6. Drain current vs Gate Voltage curve to study the effect dielectrics

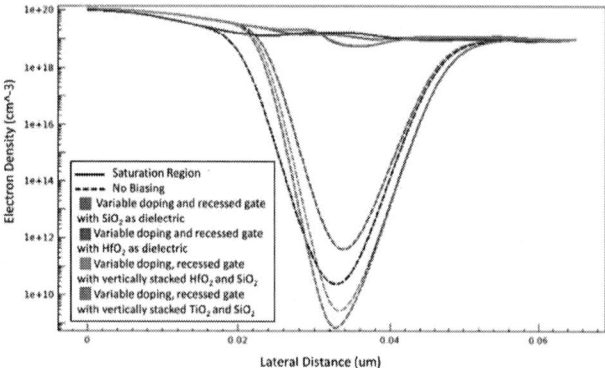

Fig. 7. Electron Density curve to study the effect of dielectrics

IV. CONCLUSION

In this paper, the electrical characteristics of various JLT structures have been investigated. Simulation results show that the distribution of doping plays an important role in controlling the electron density and thus in the OFF-state current of JLTs. When we recess the gate region allowing higher doping for lower contact and channel resistance, the ON to OFF current ratio is improved substantially. Further, the effect of dielectrics on the electrical characteristics of JLTs has been studied. Comparing with the conventional JLTs the OFF-state current was reduced from 5.892×10^{-6} A/µm to 2.149×10^{-14} A/µm. The subthreshold swing and DIBL was improved from 108.18mV/dec to 66.3mV/dec and 27.95mV/V to 7.3 mV/V respectively. Thus we have reduced the SCEs as well as increased the ON to OFF current ratio in JLTs.

REFERENCES

1. J. P. Colinge et al., "Nanowire transistors without junctions," *Nature Nanotechnology*, vol. 5–66, no. 3, pp 225–229, 2010.
2. J. P. Colinge et al., "Junctionless nanowire transistor (JNT): Properties and design guidelines," *Solid-State Electron.*, vols. 65–66, pp. 33–37, Nov./Dec. 2011.
3. P. Mondal, B. Ghosh, and P. Bal, "Planar junctionless transistor with non-uniform channel doping," *Applied. Physics. Letter.*, vol. 102, no. 13, pp. 133505-1–133505-3, 2013.
4. K. Cheng, B. B. Doris, A. Khakifirooz, P. Kulkarni, and T. H. Ning, "Method for fabricating junctionless transistor," *U.S. Patent 0078777 A1*, Mar. 28, 2013.
5. P. Bal, B. Ghosh, P. Monda, and M. W. Akram, "A laterally graded junctionless transistor," *J. Semicond.*, vol. 35, no. 3, pp. 034001–034004, 2014.
6. B. Ghosh, N. Surana, and M.W. Akram, "Differentially graded junctionless transistor," *Journal of Electron Devices*, vol. 19, 2014, pp. 1680-168
7. B. Singh, D. Gola, K. Singh, E. Goel, S. Kumar, and S. Jit, "Analytical modeling of channel potential and threshold voltage of double-gate junctionless FETs with a vertical Gaussian-like doping profile," *IEEE Trans. Electron Devices*, vol. 63, no. 6, pp. 2299–2305, Jun. 2016.
8. S. Dubey, P. K. Tiwari, and S. Jit, "A two-dimensional model for the potential distribution and threshold voltage of short-channel doublegate metal-oxide-semiconductor field-effect transistors with a vertical Gaussian-like doping profile," *J. Appl. Phys.*, vol. 108, no. 3, pp. 034518, 2010.
9. H. Ferhati, F. Douak, and F. Djeffal, "Role of non-uniform channel doping in improving the nanoscale JL DG MOSFET reliability against the self-heating effects," *Superlattices Microstruct.*, vol. 109, pp. 869–879, Sep. 2017.
10. V. Kumari, A. Kumar, M. Saxena and M. Gupta, "Empirical Model for Nonuniformly Doped Symmetric Double-Gate Junctionless Transistor," *IEEE transactions on Electron Devices*, vol. 65, no. 1, January 2018
11. H. Lou, W. Li, Y. Yang and X. Lin, "The Performance Investigation of Junctionless Transistor By Considering Different Recessed Gates," *National Science Funds of China(*NCFC) No. 61504051 and No. 61774009 and Shenzhen Key Lab Project under Grant ZDSYS 20170303140513705.
12. H. Y. Wong, N. Braga, and R. V. Mickevicius, "Enhancement-Mode Recessed Gate and Cascode Gate Junctionless Nanowire With Low-Leakage and High-Drive Current," *IEEE transactions on Electron Devices*, vol. 65, no. 9, September 2018.
13. V. Pathak, G. Saini, "A Graded Channel Dual-Material Gate Junctionless MOSFET for Analog Applications," *Procedia Computer Science*, vol. 125, pp 825-831, 2018
14. M. Gupta and A. Kranti, "Raised Source/Drain Germanium Junctionless MOSFET for Subthermal OFF-to-ON Transition," *IEEE Transactions on Electron Devices*, vol. 65, no. 6, June 2018
15. S.K. Sharma, B.Raj and M.Khosla, "Performance Enhancement of Junctionless Nanowire FET with Laterally Graded Channel Doping and High-k Spacers," *IEEE 4th Global Conference on Consumer Electronics*, ISBN: 978-1-4799-8751-1, 2015
16. TCAD Sentaurus Device User's Manual, Mountain View, CA: Synopsys, 2014.

A Low Power Hardware Implementation of Lifting based Reversible Watermarking for Medical Image

Poulami Jana
Department of ECE, Moulana
Abul Kalam Azad University of
Technology, Kolkata, W.B, India
Email:
poulami.mtech@gmail.com

Goutam Kumar Maity
Department of Physics,
Pingla Thana Mahavidyalaya,
Paschim Medinipur,W.B, India

Email: gkm20021976@gmail.com

Himadri Mandal
Department Electronics and
Communication, Calcutta Institute of
Technology, Uluberia, Howrah, India
Email:
himadrimandal2007@gmail.com

Abstract- A lifting domain reversible data hiding is presented here. A content based watermark is produced from selective (4×4) sized coefficient blocks of particular sub-band. One covert key is utilized to secure the watermark for access management which is specified by user. The secure watermark is implanted within same coefficient block (4×4). In receiver side the retrieval of original image is done by the authentic user's secure key. For real time application the 'very large scale integration' (VLSI) architecture of this proposed encoder is designed in hardware. Simulation of the encoder is done by 'field programmable gate array' (FPGA) kit. The experimentation is performed over a range of benchmark images. The results justify the supremacy of the method. The encoder module consumes only 59.20mW power when operated at 120.135MHz frequency in case of real time implementation.

Keywords- Lifting based DWT; FPGA; VHDL.

I. INTRODUCTION

With the remarkable furtherence of multimedia communication and networking domain, attention has grown in some medical fields like telemedicine, teleradiology, telesurgery, and remote patient monitoring. These applications demand the interchange of medical information among many hospitals or medical persons. Though this progress has promoted the exposure of distributed diagnosis and residence healthcare, but it introduces threats and violations on the safety and secrecy of the 'electronic patient records' (EPR). Secrecy of the patient's record sometimes becomes essential for confidentiality. Authentication is also necessary to validate the sending person along with the receiving person of the data. Digital watermarking becomes a vital approach where an extra data (watermark) is implanted into host multimedia content for copyright security and confirmation. The embedding of watermark often introduces distortion. In medical applications where the accessibility of authentic data is compulsory, permanent degradation of the original data is not agreeable. In this field the reversibility or reliability of watermarking is significant. Numerous reversible watermarking [1] algorithms are proposed by various researchers to formulate a satisfactory solution of this above stated problem. Sometimes a secret user define covert key is also used for governing the access of data with full quality. A access management algorithm depending on new 'quantization-index-modulation' (QIM) is offered by Phadikar et al. [2] to accomplish improved

security of the host and hidden data but the technique does not maintain the reversibility. Kim et al. [3] gave another idea of reversible technique using 'absolute-moments-block-truncation-coding' (AMBTC). It has also the profit of high payload capacity with admirable visual quality. RW system which uses the 'fixed point pixel' method is declared by Divecha et al. [4] which is well-suited for medical image.

Quite a lot of researchers enormous attempt to formulate better performing algorithms of lossless data hiding but remarkably only a little of them are executed in hardware. Digital arrangements in real time are necessary in significant fields like medical analysis where usually hardware based system are required. The major facility of the hardware realization is that the developers can acquire full control over different control signals, timing action and the employment of valuable hardware resources. Recently, designers provide huge attempt to coordinate real-time hardware based implementation in 'very-large-scale-integration' (VLSI). The attempts are taken for the 'data security' and 'protection of the copyright'. The hardware design of reversible watermarking may be actualized in the platforms like custom 'integrated circuit' (IC) [5], or 'field-programmable gate array' (FPGA) [6]. A custom IC based design is implemented by Mohanty et al. [5] where the operating frequency of the design is low for big image. Hardware architecture of an image compression based low power encoder for sensor camera used in wireless network is designed by Kaddachi et al. [7]. Implementation of the scheme is done in FPGA and also in ASIC. Darji et al. [8] planned hardware of one wavelet based watermarking scheme. The architecture consumes 103mW of power when operated at a frequency of 87.82MHz. Gosh et al. [9] introduce a VLSI realization of a watermarking scheme in frequency domain and the system obtained 90.131 MHz maximum frequency. Hazra et al. [10] planned an architectural design of spetial domain RW. The power consumption of the system is total 1.319watt. Das et al. [11] proposed VLSI architecture of 'reversible image watermarking' (RIW) using 'difference expansion' (DE). For a (8×8) image block 226.733 ns latency can achieve. For 150 MHz clock the throughput is 35.284 Mbps.

In this present work, we introduce a reversible scheme with its simulation results. The results establish the robustness with reversibility of the proposed technique. Also VLSI architecture of the proposed watermarking encoder is designed. Lastly the architecture is implemented on FPGA to minimize the implementation cost and time.

The prime contributions of this work are precisely given as:

• Least number of FPGA resources (registers, look-up table, and block-RAM) are adopted by appropriate design and planning of encoder with advance 'Xilinx XST' synthesis.

• Low power utilization: The power requirement to implement the RW encoder is computed by 'Xilinx X-power analyser'. Analyzed report shows that the planned encoder construction requires only 59.20mW power to process an image of size (256 × 256).

• Higher execution speed: The modelled encoder can drive at a high frequency of 120.135MHz.

The paper is further oriented as follows: The detailed algorithm is discussed within Section II. In Section III the designed VLSI based encoder architecture is described. Section IV represents the software and also the hardware simulation results respectively. Ultimately the conclusion and future scope of the work are presented in Section V.

II. PROPOSED ALGORITHM

Herein we describe the encoding technique and also the decoding technique (algorithm) of proposed RW scheme. The lifting based 'discrete wavelet transform' (DWT) encoder performs image pixels modulation. The lifting based decoder demodulates the received watermarked image to retrieve high quality host image.

A. Watermark Encoding

The block representation of the proposed encoder is revealed in Fig. 1. The procedure is summed up in Algorithm 1. Lifting based wavelet transform is performed on the original host image. Then the LH2 sub band is chosen for watermark generation and embedding. The binary watermark is produced from small blocks of size (4×4). It further secured by a covert key and also implanted within the consequent blocks. The detailed process of encoding is explained in Algorithm1.

Algorithm 1
Inputs: Host image (8 bit-gray-scale), secret key (**K**).
Step1: One 8 bit image is chosen as cover image.
Step2: Lifting based 2 levels DWT is applied on the pixels. The 4 sub bands LL2, LH2. HL2, HH2 are produced.
Step3: LH2 sub band is sub divided into a set of (4×4) sized blocks.
Step4: Calculate the row average of the coefficients using subsequent rule: $$(C_{av}) = \frac{\sum C_{ij}}{12} \quad \text{if}, C_{ij} \neq (C_{2j}, C_{4j})$$
Step5: Binary watermark (W) is generated using the subsequent rule: $$(W) = 0 \quad \text{if} C_{av} \text{ even}$$ $$= 1 \text{ if } C_{av} \text{ odd}.$$
Step6: Emigrate the generated content dependent watermark for access control applying following rule: $$(W') = W \oplus K$$ Here, K is the covert user defined key and W' is the permuted watermark.
Step7: Pick each (4×4) sub divided LH sub band and modulate. $$C'_{ij} = C_{ij} + W'$$ where $C_{ij} = (C_{2j}, C_{4j})$.
Step8: Lifting based IDWT is applied to display the access controlled watermarked image.
Outputs: 8-bit modulated or watermarked image.

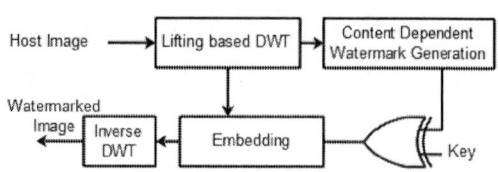

Fig. 1. Encoder block.

B. Watermark Decoding

The decoding procedure of the proposed reversible watermarking scheme is shown in Fig. 2. The Algorithm 2 describes the decoding process. The received encoded host is scrambled by 2-level lifting schemes. The LH2 sub band is taken to regenerate and extract the content dependent watermark using Algorithm2.

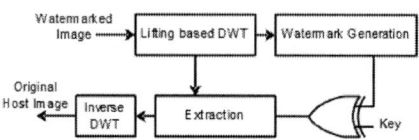

Fig. 2. Decoder block.

Algorithm 2
Inputs: Received watermarked image, secret key (**K**).
Step1: The watermarked image is received.
Step2: Perform lifting based DWT on the received image pixels.
Step3: The LH2 sub band is sub-divide into set of (4×4) sized blocks like encoder.
Step4: Calculate row-wise average of each (4×4) sized block's coefficients according to following rule $$(C_{av}) = \frac{\sum c_{ij}}{8}$$ $$where \; C_{ij} \neq (C_{2j}, C_{4j})$$
Step5: Redevelop the watermark using following rule $$(W) = 0 \text{ if } C_{av} = \text{even}$$ $$= 1 \text{ if } C_{av} = \text{odd}$$
.**Step6:** Permute the watermark with the user's covert key (used at encoder) according to the following rule to get excellent quality image. $$(W') = W \oplus K .$$
Step7: Select each (4×4) blocks of LH sub band and apply the following rule to retrieve the original coefficients. $$C_{ij} = C'_{ij} - W'$$ $$\text{where } C_{ij} = (C_{2j}, C_{4j}).$$
Step8: Apply inverse lifting based DWT to get back the original image.
Outputs: Recovered image pixels

III. VLSI ARCHITECTURE OF PROPOSED REVERSIBLE WATERMARKING ENCODER

This section we elaborate the hardware architecture of the watermark encoder as described in Algorithm 1. The hardware architecture is programmed by hardware description language (HDL). Simulation and implementation is done for Zynq series FPGA board.

A. Data Path of Encoder Module

The designed architecture of encoder circuit consist of a number of sub-circuit modules such as (a) 'Image RAM', (b) 'DWT/IDWT', (c) 'Watermark generation Embedding'. Programming of all this sub-modules are done discretely and tested in 'XILINX ISE' setting. Fig. 3 is the fundamental top-level outlook of the encoder.

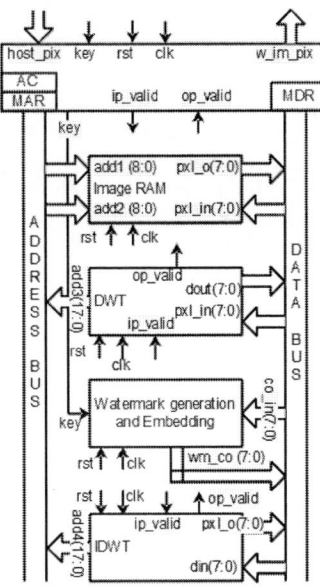

Fig. 3. Encoder data path.

B. Embedding Process:

The host image pixel values are stored in the 'Image RAM'. The whole encoding process is controlled by the control unit. It supplies proper clock and timing signal to the different units. The 'memory data register' (MDR) and 'memory addresses register' (MAR) of the 'control unit' act as temporary storage for addresses and data for the whole operation.

The stepwise embedding is described below:
- **Step1:** Load the image pixel values in Image RAM.
- **Step2:** The stored pixels are transferred to DWT module. The outputs of DWT module are lifting coefficients.
- **Step3:** The coefficients are transferred to watermark generation and embedding unit. The content dependent watermark is generated and embedded based on Algorithm1 (Step4 to Step7). The secret binary key K is used in this whole process.
- **Step4:** Modified coefficients are transferred to IDWT unit whose outputs are watermarked pixel values.

IV.PERFORMANCE EVOLUTION

In present section, the performance of the Proposed algotithm' is evaluated on a number of test images of size (256×256). The simulation of the present scheme is done by Intel i3CPU of 1.7GHz with 8 GB RAM. Circuit synthesis is done by Xilinx ISE 14.5. The encoder circuit is realized in Zynq series FPGA (XC7Z010-CLG400-3) to evaluate the system's performance in real time. Experimentations are organized on different standard images.

A. Software Simulation

The proposal's performance is experienced over a collection of test images (grayscale and medical images). Among them original 'Barbara' image is shown in Fig. 4(a). Fig. 4(b) indicates the watermarked image. Fig. 4(c) is decoded image after watermark extraction.

Fig. 4. (a) Host Image, (b) Image after Encoding, (c) Image after Decoding.

Similarly Fig. 5(a) and Fig. 5(b) indicate the original and watermarked medical image. Fig. 5(c) represents the decoded medical image after watermark extraction. The measurement of robustness of the scheme and also the visual quality of the watermarked image is evaluated by PSNR and 'mean structural similarity index measurement' (MSSIM) respectively

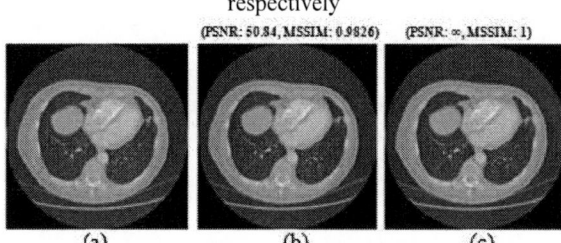

Fig. 5. (a) Host Medical Image, (b) Medical Image after Encoding, (c) Medical Image after Decoding.

Image	Before Decoding		After Decoding	
	PSNR	MSSIM	PSNR (dB)	MSSIM
Girl-face	53.60dB	0.9935	∞	1
Boat	51.65dB	0.9894	∞	1
MRI	50.95dB	0.9889	∞	1
Skull	50.67dB	0.9868	∞	1

TABLE1: PSNR & MSSIM COMPARISON OF DIFFERENT TEST IMAGES BEFORE DECODING AND AFTER DECODING

In Table1 the simulated PSNR and MSSIM results of the method for a range of gray scale and medical images are given. The good values of PSNR of the watermarked images specify the resistance power of the scheme against noise. The high values of MSSIM signify the high-quality visual characteristic of the encoded image. The PSNR values and the MSSIM values for each and every retrieved image are ∞ and 1 respectively. The previous results assure the reversibility of the method. The comparison of the reported scheme with similar methods in terms of MSSIM and PSNR values are depicted in Table2.

TABLE2: MSSIM AND PSNR COMPARISON WITH OTHER METHODS

Methods	Working Domain	PSNR	MSSIM
Maity and Maity [6]	RCM	30.41dB	0.86
Ghosh et al. [1]	Spatial	33.78dB	0.91
Proposed	DWT	53.60dB	0.9935

A. Hardware Realization

Circuit level implementation of the designed encoder architecture is done in Zynq-series FPGA. Xilinx-ISE 14.5 version synthesis tool is allocated to construct the implementation logic circuit of the watermarking encoder. The use of internal logic from FPGA board is summarized in Table3. The implementation of the encoder circuit considers (256×256) sized image blocks.

TABLE 3: DEVICE UTILIZATION SUMMARY FOR ENCODER

Target board details	Vertex 7 (XC7VX330T-FFG1157-3)
Image Size	(256×256)
Number of utilized Slice LUT	802
Number of utilized Slice Registers	421
Fully used LUTs	179
Number of used Bonded IOBs	73
BUFG	2
Number of utilized block RAM	2

TABLE4: FREQUENCY AND POWER COMPARISON

Methods	Frequency
Darji et al.[8]	87.82 MHz
Gosh et al.[9]	90.131 MHz
Proposed	120.135 MHz

Table 4 results depict that the scheme operates at a higher frequency than the other schemes reported in literature. It can achieve 36.79% higher frequency than [8] and 33.28% higher frequency than [9].

TABLE5: POWER COMPARISON

Image Size		(256×256)	
Power Consumption	StaticPower	Dynamic Power	Total Power
For Encoder Module (with no power aware design)	30.67mW	37.23mW	67.90mW
For Encoder Module (with power aware design)	30.38mW	28.82mW	59.20mW

The assessment of power expenditure is calculated using 'Xilinx X-power analyzer' (XPA). It is an interactive graphical tool. This tool can study the consumption of power for Xilinx FPGA devices. The power utilization information are given in Table5.The entire consumed power for encoder module is 67.90mW (for (256×256) sized image) with no power aware designing methodology. The simulated results depicts that when the hardware design is fabricated with power attentive or aware technique then we able to save

12.81% power during encoding of an (256×256) sized image.

V. CONCLUSION AND FUTURE SCOPE OF PRESENT WORK

In this present paper a scheme of reversible watermarking for gray scale images is presented. A watermark is generated depending on the image contents by a consumer specific covert key. The reversibility and robustness of the scheme is ensured by the simulation results. Efficiency of the designed hardware (encoder) results least employment of 'Slice LUT' (802) with low consumption of power (67.90mW) while operating at a frequency of 120.135MHz. In future plane, the effort will be given on designing the power aware FPGA based decoder architecture. The advantage of the low power architecture will be useful in today's popular electronics gazettes.

REFERENCES

[1] Ghosh, S., Das, N., Das, S., Maity, S. P., & Rahaman, H. "An adaptive feedback based reversible watermarking algorithm using difference expansion", In Proc. of the IEEE 2nd International Conference on Recent Trends in Information Systems, pp. 207-212, 2015

[2] Phadikar, A., Mandal, H., Maity, G.K., Chiu, T.L., "A new model of QIM data hiding for quality access control of digital image", In Soft-Computing and Networks Security (ICSNS) vol. 1, pp. 1-5, 2015. doi: 10.1109/ICSNS.2015.7292441

[3] Kim, C., Shin, D., Leng, L., Yang, C.N., "Lossless data hiding for absolute moment block truncation coding using histogram modification", Journal of Real-Time Image Processing, vol. 14, pp. 101-114, 2018. doi: /10.14257/ijsia.2014.8.2.31

[4] Divecha, N.H., Jani, N.N., "Reversible watermarking technique for medical images using fixed point pixel", In Fifth International Conference on Communication Systems and Network Technologies vol. 1,pp. 725-730, 2015. doi:10.1109/CSNT.2015.287

[5] Mohanty, S.P., Kougianos, E., Ranganathan, N., "VLSI architecture and chip for combined invisible robust and fragile watermarking", IET Computers and Digital Techniques, vol. 1, no. 5, pp. 600–611, September 2007.

[6] Maity, H.K., Maity, S.P., "FPGA implementation of reversible watermarking in digital images using reversible contrast mapping", Journal of Systems and Software vol. 96, pp. 93-104, 2014. doi: 10.1016/j.jss.2014.05.079

[7]Kaddachi, M.L, Soudani, A., Lecuire. V., Tork,i K., Makkaoui, L., Moureaux, J.M., "Low power hardware-based image compression solution for wireless camera sensor network", Computer Standards & Interfaces, vol. 34, pp. 14-23, 2012.doi: 10.1016/j.csi.2011.04.001

[8] Darji, A. D., Lad, T. C., Merchant, S. N., Chandorkar, A. N., "Watermarking Hardware Based on Wavelet Coefficients Quantization Method", Circuits, Systems, and Signal Processing, vol. 32, no. 6, pp. 2559-2579, December 2013.

[9]Ghosh, S., Talapatra, S., Sharma, J., Chatterjee, N., Rahaman, H., Maity, S.P, "Dual Mode VLSI Architecture for Spread Spectrum Image Watermarking using Binary Watermark", Procedia Technology, vol. 6,pp. 784-91,2012. doi: 10.1016/j.protcy.2012.10.095

[10]Hazra, S., Ghosh, S., De,S., Rahaman, H., "FPGA implementation of semi-fragile reversible watermarking by histogram bin shifting in real time", Journal of Real-Time Image Processing, vol.14, pp. 193-221, 2018. doi: 10.1007/s11554-017-0672-9

[11] Das, S., Maity, R., Maity, N.P., "VLSI-Based Pipeline Architecture for Reversible Image Watermarking by Difference Expansion with High-Level Synthesis Approach" ,Circuits Systems and Signal Processing, vol. 37, pp. 1575-93. 2018.doi: 10.1007/s00034-017-0609-3

Performance Analysis of Staggered Heterojunction based SRG TFET biosensor for health IoT application

Sudhansu Mohan Biswal,Sanjit Kumar Swain,Biswajit Baral,Debasish Nayak,Umakanta Nanda,Satish Kumar Das
DhananjaTripthy

Dept. of Electronics and Communication Engineering, Silicon Institute of Technology, Bhubaneswar, India
. Dept. of Electronics and Communication Engineering, Vellore Institute of Technology, Amaravati, India
sudhansu.mohan@silicon.ac.in

Abstract—**This paper presents the performance of SRG Tunnel FET biosensor. Here the different device parameters are deliberate to meet the requirement of the technological development. Focus is made on how TFET can be a substitute ahead of CMOS characteristics for biosensor mean and to more enhance the low power design policy to facilitate it for IOT purpose. There is a experimentally demonstration has been done for nanogap cavity A nanogap cavity region. Investigation of the staggered gap hetero junction SRG TFET was done. The results are obtained using Silvaco software.**

keywords: biosensor; TFET; SRG, Transconducctance Staggered heterojunction; Double gate TFET

INTRODUCTION

Biosensor detects biomolecules. The existence or nonexistence of biomolecules causes the change in device characteristics. The V_{TH} or drain current I_D of the device will vary with the change of biomolecules with assured permittivity close to in nanogap hollow space sections comparing to the occurrence of air molecules in the cavity sections. According to (ITRS) report [1] Tunnel FET (TFET) is a most promising device for getting a steep sub threshold swing [2]. In compare to a conventional MOSFET subthreshold swing is(>60mv/dec). The energy barrier width for BTB Tunneling determines the quantity of drain current through a Tunnel FET [3]. A double-gate (DG) Tunnel FET shows the further benefit of enhanced electrostatic control [3-4] to get better I_{ON} (on current) SS (Subthreshold Swing). TFET-based biosensor can prevail over the fundamental restriction, for example maximum sensitivity as well as imperfect exposure in biosensor, based on conventional MOSFET because of their dissimilar charge transporter booster method in the figure of band to band tunnelling.

Recently, III–V semiconductor based tunnel field effect transistors reveals considerable development in term of I_{ON} and (Subthreshold Swing) as compared to Si based tunnel field effect transistors due to its smaller effective tunneling barrier width, lower effective mass and low bandgap [5-7]. From health point of view Internet-of-Things(IoTs) have inspired to design a biosensors made up of nanoscale transistor. Therefore, in this work, we primarily focus on such as Tunnel

FET can be a new candidate for biosensor and low power circuit design approach for IoT applications.

DEVICE STRUCTURAL DESIGN AND SIMULATION

Figure 1 shows the 2-D structure of the device which has been simulated with TCAD Device simulator SILVACO ATLAS[10].

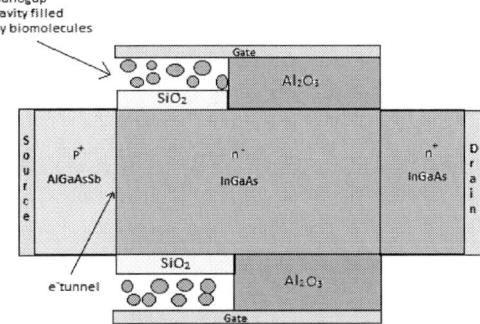

Fig. 1: 2-D structure of SRG Tunnel FET-based biosensor.

Thickness of Al_2O_3 Tox=14 nm and equivalent oxide (SiO_2) of thickness 1nm has been used as gate oxide. The Fermi level exist in between the conduction band drain and valence band of source section. The electrons are present in n- type drain and holes are present in p- type source. At zero bias condition due to heavily doped p+ and n+ region, the band becomes ally to each other. When reverse biased is applied the height of potential barrier decreases and electric field increases. This causes the electrons from the conduction band of n- type drain to transfer from the valence band of p- type source side, due to which the current increases and max current flows in this structure.

For the simulation purpose a nonlocal band-to-band tunneling (BTBT) model has been considered. And for the recombination characteristics of model Shockley Read Hall (SRH) recombination technique has been tailored. For the carrier transport means Drift Diffusion Mode Space (DDMS) approach, has been taken in account.

In Table I the detail parameters of the device has been given.

978-1-5386-6723-1/19 $31.00 © 2019 IEEE

TABLE I. **Parameters taken for the strucutre of SRG TFET**

Parameters	Value	Unit
Gate length(L_g)	30	nm
Source and the drain length	20	nm
Nano cavity region length	10	nm
SRG TFET its radius	10	nm
(EOT of t_1)	1	nm
Al_2O_3 deposit thikness	14	nm
the source doping(p^+)	2×10^{19}	cm^{-3}
the drain doping (n^+)	5×10^{19}	cm^{-3}
Channel doping (n^-)	5×10^{16}	cm^{-3}
Workfunction of gate	4.61	eV

Simulations for the SRG TFET has been perform as shown in Fig. 1, by means of the version 5.11.24.C Silvaco Atlas device simulator. For the source and drain contacts Au/Ni electrode is used respectively

RESULTS AND DISCUSSION

Here we have simulated three different permittivity values (i.e. 2.5,12,22) of different biomolecules in the region of nanogap cavity and concluded that as the permittivity value increases, surface potential increases. Therefore, k value 22 have the highest surface potential here.

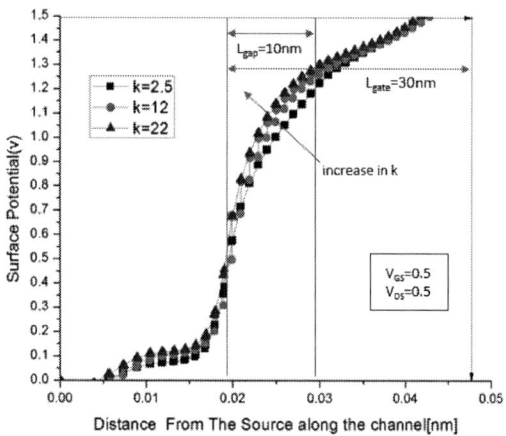

Fig. 2: Comparison of surface potential down the channel for altered permittivity of the bio molecules nearby the cavity regions

Fig. 2 shows with increase in permittivity of bio molecules, there will decrease in tunnelling barrier width because of amend in valuable capacitance of gate C_{gg}, which marks in abrupt ascend in the potential profile.

Figure 3 represents the variation of drain current w.r.t gate to source voltage (V_{GS}) for dissimilar permittivity of the bio molecules. The electrical characteristics of the device are

exaggerated by the unbiased and charged bio molecules that attach to the untie channel region. At source side if there is cavity and the dielectric constant of the bio molecules is distorted in the cavity region, on current I_{ON} will change.[10-16]

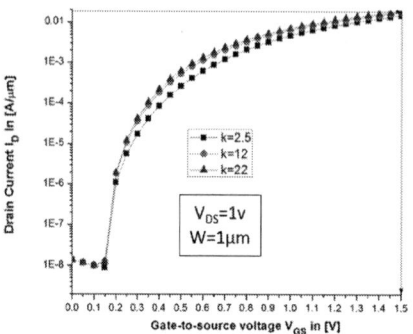

Fig. 3: Plot ofvarion of drain current w.r.t V_{GS}(gate-to-Source voltage)

TABLE II: Material parameter used for the SRG Tunnel FET

	$Al_{0.5}Ga_{0.5}As_{0.3}Sb_{0.7}$	$In_{0.8}Ga_{0.2}As$
Affinity of Electron	3.6	4.73
Light mass	0.11	0.034
Electron effective mass	0.09	0.036
Lattice constant in A^0	5.98	5.98
Bandgap in eV	1.36	0.50
Dielectric constant		14.5528

The drain current increases with increase in permittivity of bio molecules in the nanogap cavity .

The change in current noticeably indicate the chance of sensing bio molecules by detecting the change in drain current by the occurrence of biomolecules.

Trans conductance acting a key part in the intend of electronics circuit design, for example operational amplifiers (OP-AMPs) and operational trans conductance amplifiers (OTAs) for the effect on gain, effective bandwidth, noise concert and offset.

Figure 4 shows the deviation of trans conductance g_m with cgange in V_{GS} for different bio molecules. Because of exceptional sub threshold swing of TFET devices SRG TFET shows more transconductance than a conventional transistor , with increase in permittivity of bio molecules. Figure 4 shows the decrease in g_m because of decrease in drain current.

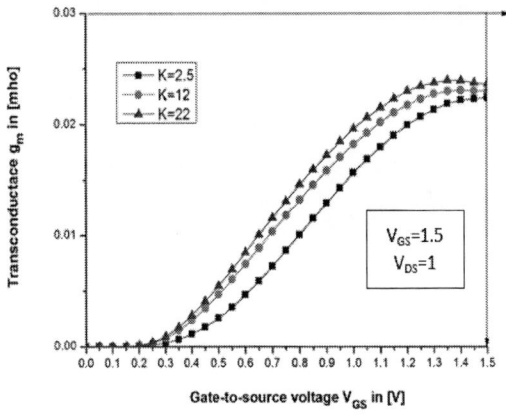

Fig-4: Plot of the comparison of the variation of trans conductance (g_m) w.r.t Gate-to-source voltage V_{GS} for different bio molecules

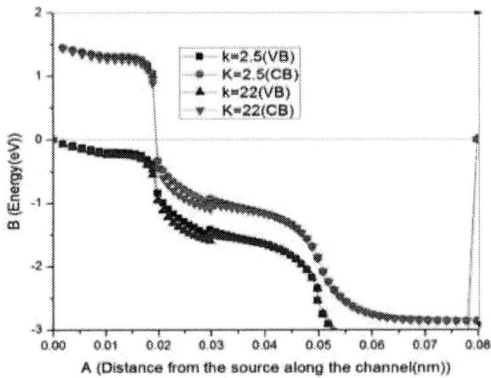

Fig. 5: Comparison of band energy as a function of the location down the channel starting the source side to drain side.

Fig 5 show deviation in energy band for different bio molecules of different permittivity . With the air molecules in cavities, the barrier width will increase between channel coduction band and valence band of source, then the tunneling of electron will decrease. Therefore, the conduction is also low. When the bio molecules permittivity k (k>1) are there in the cavities then the barrier width is reduced and the current conduction will due to the increase in electron tunneling probability[13-16]

Fig. 6 and Fig. 7 shows that the variation in threshold voltage as well as sensitivity, which is more for a hetero junction based surrounding gate biosensor. The recent report gives the idea for sensitivity [9] of a biosensor which is defined by the segregation of V_{TH} for presence of air molecules and bio molecules in the cavity region. Therefore the change of threshold voltage is most important FOMs to detect the detect the biomolecules [16].

Mathematically,

$$V_{Th} = V_{Th\ (gap = air)} - V_{Th\ (gap = filled)}$$

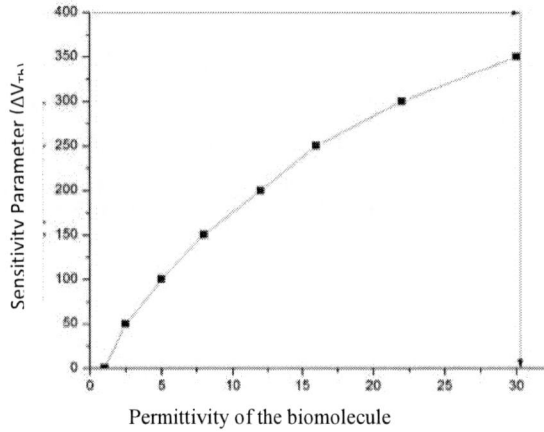

Fig. 6: Comparison of sensitivity with respect to dielectric constant of the bio molecules

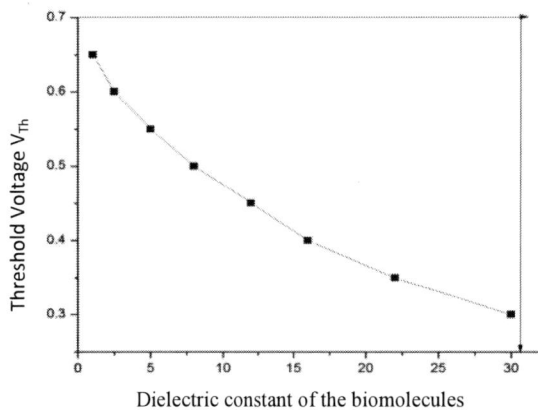

Fig. 7: Plot of Threshold Voltage as a function of permittivity of the bio molecules

CONCLUSION

In summary, an investigation on performance of detection and sensitivity of the bio molecule of the staggered hetero junction-based biosensor has been carried out. The outcomes reveals that the projected biosensor has an proficient potential candidate substitute to conventional Tunnel FET based biosensors because of high sensitivity. Therefore, biosensor based on Surrounding Gate TFET can be used for detection of low concentration bio molecules. Therefore, the projected biosensor can cover the technique for a proposition change in bio sensing applications.

REFERENCES

[1] W. Y. Choi, B.-G. Park, J. D. Lee, and T.-J. K. Liu, IEEE Electron Dev. Lett. 8, 743 (2007).

[2]   S.O. Koswatta, M.S. Lundstrom, and D.E. Nikonov, IEEE Trans. Electron Dev. 56,456 (2009)

[3]   A. S. Verhulst, B. Sor_ee, D. Leonelli, W. G. Vandenberghe, and G. Groeseneken, J. App. Phys. 107, 024518 (2010)

[4]   K. Boucart and A. M. Ionescu, IEEE Trans. Electron Dev. 54, 1725 (2007)

[5]   Ganapathi, Kartik, Youngki Yoon, and Sayeef Salahuddin. "Analysis of InAs vertical and lateral   and-to-band tunneling transistors: Leveraging vertical tunneling for improved performance." Applied Physics Letters 97, no. 3 (2010): 033504.

[6]   L. Zhang, X. Lin, J. He, M. Chan, IEEE Trans. Electron Devices 59 (2012) 3217–3223.

[7]   E.-H. Toh, G.H. Wang, G. Samudra, Y.-C. Yeo, Appl. Phys. Lett. 90 (2007) 263507-1–2635073

[8]   Manual, ATLAS User'S. "Silvaco Int." *Santa Clara, CA* 5 (2008).

[9] Narang, R.; Reddy, K.V.S.; Saxena, M.; Gupta, R.S.; Gupta, M., "A Dielectric-Modulated Tunnel-FET-Based Biosensor for Label-Free Detection: Analytical Modeling Study and Sensitivity Analysis,"Electron Devices, IEEE Transactions on, vol. 59, no.10, pp. 2809

[10] Chakraborty, Avik, and Angsuman Sarkar. "Staggered Heterojunctions-Based Nanowire Tunneling Field-Effect Transistors for Analog/Mixed-Signal System-on-Chip Applications." NANO 10, no. 02 (2015): 1550027.

[11]   Kanungo, S.; Gupta, P.S.; Rhaman, H., "Effects of Germanium mole fraction variation at the source of a dielectrically modulated Tunneling FET based biosensor," Devices, Circuits and Systems (ICDCS), 2014 2nd International Conference on , vol., no., pp.1,5, 6-8 March 2014.

[12]   Baravelli, Emanuele, Elena Gnani, Roberto Grassi, Antonio Gnudi, Susanna Reggiani, and Giorgio Baccarani. "Complementary n-and p-type TFETs on the same InAs/Al 0.05 Ga 0.95 Sb platform." In Solid-State Device Research Conference (ESSDERC), 2013 Proceedings of the European, pp. 69-72. IEEE, 2013.

[13]   B. Gu, T. J. Park, J.-H. Ahn, X.-J. Huang, S. Y. Lee, and Y.-K. Choi, "Nanogap field-effect transistor biosensors for electrical detection of avian influenza", Small, vol. 5, no. 21, pp. 2407–2412, Nov. 2009

[14]   Sarkar, Deblina, and Kaustav Banerjee. "Proposal for tunnel-field-effect-transistor as ultra-sensitive and label-free biosensors." Applied Physics Letters 100, no. 14 (2012): 143108.

[15]   Abdi, Dawit Burusie, and M. Jagadesh Kumar. "Dielectric modulated overlapping gate-on-drain tunnel-FET as a label-free biosensor." Superlattices and Microstructures 86 (2015): 198-202.

[16] Rigante, Sara, Paolo Scarbolo, Mathias Wipf, Ralph L. Stoop, Kristine Bedner, Elizabeth Buitrago, Antonios Bazigos et al. "Sensing with Advanced Computing Technology: Fin Field-Effect Transistors with High-k Gate Stack on Bulk Silicon." *ACS nano* 9, no. 5 (2015): 4872-4881.

Hardware Architecture of a Decoder for Fractal Image Compression

Hasanujjaman

Dept. of Electronics & Comm. Engg.
Kalyani Govt. Engineering College
Kalyani, West Bengal, India
hasanujjaman.diet@gmail.com

Prof. U. Biswas

Dept. of Computer Science and Engg.
University of Kalyani
Kalyani, West Bengal, India

Prof. M K Naskar

Dept. of Electronics & Tele. Comm. Engg.
Jadavpur University
Jadavpur, West Bengal, India

Abstract—**Fractal image compression is a comparatively new and less explored technique in the domain of image processing. The main problem is it's very high image compression time due to the huge number of 'range' - 'domain' comparisons it has to undergo. If efficiently utilised, fractal image compression gives the best compression ratio which is highest among other contemporary techniques. In this paper we have proposed efficient hardware of a decoder for the fractal image compression. Controlled parallelism has been incorporated to speed up the decoding process. The whole design has been simulated and synthesized using verilog HDL.**

Keywords— *Hardware Architecture, Fractal Image Compression(FIC), Decoder, Time Complexity, HDL.*

I. INTRODUCTION

Fractal image compression is based on the concept of "Partitioned Iterated Function System" or PIFS [1-3]. It simply means that certain parts of an image is similar to certain other parts of the image. So, rather than storing the data regarding both the similar parts, we can store the knowledge regarding which part of the image is similar to which other part and then reconstruct the image when needed using a sample default image. This will take up less memory space than the original image. In today's world, with the advent of high quality digital cameras, huge amount of data are produced and are needed to be transported from place to place. It either needs a highly sophisticated transmission system which can provide a large bandwidth or a faster transmission rate. Both these constraints are costly and thus arises the need to compress data. Data compression can be both lossless and lossy. Unfortunately FIC is a lossy compression technique and thus the reconstructed image will not fully correspond with the original image.

Image compression is becoming a vital issue regarding endurances of compression and processing of images or relevant data in the transferring platforms. However, the major applications of image processing are mostly associated with the capturing image along with storing the data within the cameras and transferring of data through internet usage or wireless substances. It is also assumed that the focus of the researchers is comprised of using less accomplished hardware slots, in order to bring an efficient outcome. Additionally, the processes of image compression through the mean applications of fractal image processing had already attempted by most of the associated researchers.

Likewise, a famous researcher Saad and Abdullah, in their work of 'High- speed implementation of fractal image compression in low-cost FPGA'[4] had done a comparative analysis of other important works by researchers[5-7]. In which the analysis is done based upon their bit and hardware usage, coding time and compression ratio. In which the best or better results regarding the hardware implementation of the fractal image processing had easily attempted by another researcher. During the process of encoding, the processing in the hardware sections had emerged a fast operation result. In which the compression of the image along with the required timing section for the overall process has already shown a better outcome result in accommodated factors. But the major drawback of this process is associated with the decoding process. In which the researchers have already attempted a number of ways towards proving successful results outcomes. In addition to the main context, the outcomes result regarding the timing sections has already reduced to some extent. However, on investigating the in-depth attribution of working propositions, it has been identified that the working of Vidya [5] and others on 'Architecture for Fractal image compression' has fulfilled the demands of using less hardware associated with the proposed working activities. In addition to the main context, it is also identified that the use of less image size from the other works had helped him to intercept the less usage of hardware sections. Due to the which, the time accommodations by the researcher has also reduced to some extent, which generates the major advantages for this work to get the focus for further on natural implementation. Although, the activities of controlled parallelism in which the usage of subtraction, controlled unit and comparator words simultaneously over the block range proved more benefits encoding and decoding perceptions of working acts that the researcher implemented within his paper. However, this implementation of the controlled parallelism educed more time for completion for the overall operations. Thus, the major drawback of this research work is focused upon ensuring more timing accommodation towards the working activities, rather than usage of efficient hardware sections for the image processing.

II. ENCODING PROCESS

In the encoding process a proper domain is needed to represent a particular range, each and every available range there will be represented by a domain. so we can declare a domain pool which consist of all the possible domains and their modified counterparts after applying all the isometries. so we have an objective that we are to represent each range with the help of one domain applying proper isometry. So the encoder provides 3 basic info , the domain which can closely approximate the particular range ,the isometry applied and the average gray level difference in the domain and range in terms of offset. These 3 information will be loaded into the memory at the output of the encoder.

The range pixels are approximated by, $R(i,j)=SxD(i,j)+O$. where contrast scale factor 'S' and the brightness offset factor 'O' .The larger is the domain pool, the greater chances are there for a good quality match to occur. In order to expand the domain pool, the domains are used to create it's 8 isometries namely,

1. Identity:

    $D(i,j)=D(i,j)$

2. Reflection about mid vertical axis:

    $D(i,nd-j+1)=D(i,j)$. [where nd is the pixel dimension of domain]

3. Reflection about mid horizontal axis:

    $D(nd-i+1,j)=D(i,j)$

4. Reflection about first diagonal:

    $D(j,i)=D(i,j)$

5. Reflection about second diagonal:

    $D(nd-j+1,nd-i+1)=D(i,j)$

6. 90° rotation:

    $D(j,nd-i+1)=D(i,j)$

7. 180° rotation:

    $D(nd-i+1,nd-j+1)=D(i,j)$

8. 270° rotation:

    $D(nd-j+1,i)=D(i,j)$

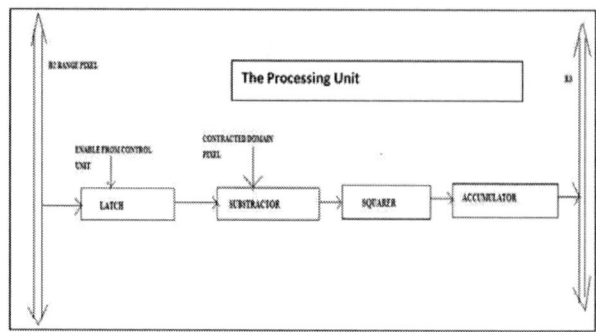

Figure 1: Processing Element.

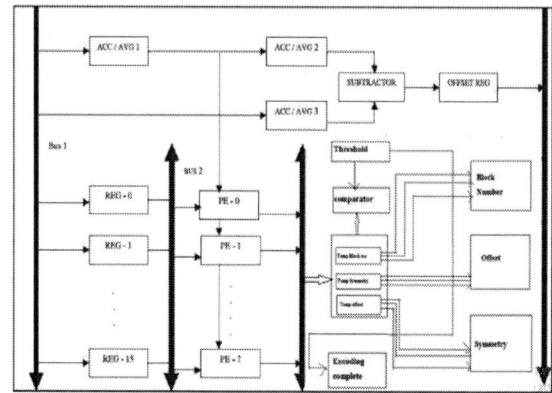

FIGURE 2: HARDWARE ARCHITECTURE FOR ENCODER

II. DECODING PROCESS

Decoding process: as we have mentioned in the encoder section clearly that at the output of the encoder we have stored 3 basic info – block no, symmentry no and offset. So at the end of the encoding process we will connect this memory to the input of the encoder. the encoder will function like the following---
1. Sense the block no from the memory and then read the entire domain starting from that address.
2. Store this contracted domain in temporary memory
3. Next the symmetry no is read from the memory which will provide the enable to the proper sequencer block e.g. Symmetry 0 will provide enable to sequencer 1 and so on.
4. The functionality of the sequencer is that it will select proper sequence of the registers to be selected to provide proper isometry ,in the architecture of the decoder we can easily see that there is 16 set of registers which are fed from from the temporary memory storing the contracted domains . as the sequencer will provide the enables to proper registers at proper timings so the symmetry will be implemented this way in order to reconfigure the required range in the image.
5. The offset is also selected from the memory and stored duely in the offset register, this offset value will be added with each of the contracted domain pixels at proper times

III. IMPLEMENTATION ISSUES

1. The bit length of the data stored in the memory at the output of the encoder,the amount of space it acquires and the method of how to read them from the memory. As the isometry and offset are of 8 bit length so eachplace of the memory has to be of 8 bit length in order to accommodate those informations in the memory. Another fact is that we have taken a 128*128 image and to store these no of pixels we require 14 address lines . but as here the memory is of 8 bit word length so we are required to load the data to such a length that it is of integer multiple of 8. Hence we are using 16 bit block no address and for that we are required to store that data in two consecutive memory locations. So we need 4 no of 8 bit locations in the memory to store the information of each range. Only 32 bits are reqired to store a 4*4 range (16*8=126 bits).

2. Now let us come to the part of reading that data from that memory. While reading we need to ensure that the data from 2 consecutive location go to the block no register and next two to the symmetry and offset register. Hence to selct them in this way we have used a 2 bit counter. it's output has been given to the enables of these 3 registers. When it will be 00 or 01 the block no register will get the enable. When it's 10 symmetry register will be selected and when it's

11 offset register wil get selected. We can also conclude that the increment operation of the address counter of the encoder output memory and the 2 bit counter will be synchronized and hence they need to be given with same clk frequency.

3. After the 2nd clock pulse the address counter of domain memory will sense the starting domain address and within the next two clock pulses it will load the entire contracted domain in the temporary memory.

4. 4 bit counter is used to select all the 16 loactions of the temporary memory,it's used to store all the 16 contracted domains as well as loading that data to the 16 set of registers.

5. The sequencer here will provide a 4 bit output in order to select proper register and to do this a 4 to 16 decoder is used here of which 16 output lines are connected to the enables of the 16 registers .

6. As we did not use any sort processing element concept in order to implement the symmetry conditions to the contracted domain, hence no need to use the squarer and hence we are saving huge amount of time as far as time complexity is concerned

Architectural Diagram of Decoder:

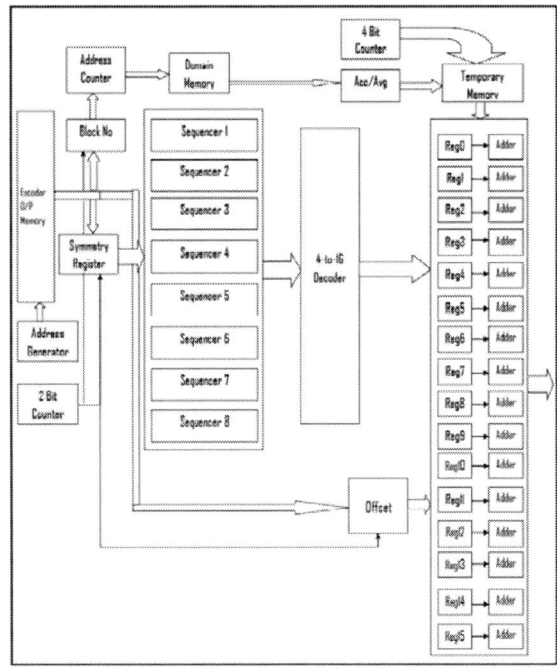

FIGURE 3: HARDWARE ARCHITECTURE OF THE PROPOSED DECODER

The Relation Of The Shift Register Output ,Counter Output And The Required Range Register Selection Combination Is Shown In The Table Below---

SHIFT REGISTER								COUNTER				SEQUENCER OUTPUT				RANGE REGISTER
S7	S6	S5	S4	S3	S2	S1	S0	C3	C2	C1	C0	O3	O2	O1	O0	SELECTED
0	0	0	0	0	0	0	1	0	0	0	0	0	0	0	0	0
0	0	0	0	0	0	1	0	0	0	0	0	0	0	1	1	3
0	0	0	0	0	1	0	0	0	0	0	0	1	1	0	0	12
0	0	0	0	1	0	0	0	0	0	0	0	0	0	0	0	0
0	0	0	1	0	0	0	0	0	0	0	0	1	1	1	1	15
0	0	1	0	0	0	0	0	0	0	0	0	0	0	1	1	3
0	1	0	0	0	0	0	0	0	0	0	0	1	1	1	1	15
1	0	0	0	0	0	0	0	0	0	0	0	1	1	0	0	12
0	0	0	0	0	0	0	1	0	0	0	1	0	0	0	1	1
0	0	0	0	0	0	1	0	0	0	0	1	0	1	1	1	7
0	0	0	0	0	1	0	0	0	0	0	1	1	1	0	1	13
0	0	0	0	1	0	0	0	0	0	0	1	0	1	0	0	4
0	0	0	1	0	0	0	0	0	0	0	1	1	0	1	1	11
0	0	1	0	0	0	0	0	0	0	0	1	0	1	1	1	7
0	1	0	0	0	0	0	0	0	0	0	1	1	1	1	0	14
1	0	0	0	0	0	0	0	0	0	0	1	1	0	0	0	8
0	0	0	0	0	0	0	1	0	0	1	0	0	0	1	0	2
0	0	0	0	0	0	1	0	0	0	1	0	1	0	0	0	8
0	0	0	0	0	1	0	0	0	0	1	0	1	1	1	0	14
0	0	0	0	1	0	0	0	0	0	1	0	1	0	0	0	8
0	0	0	1	0	0	0	0	0	0	1	0	0	1	1	1	7
0	0	1	0	0	0	0	0	0	0	1	0	1	0	1	1	11
0	1	0	0	0	0	0	0	0	0	1	0	1	1	0	1	13
1	0	0	0	0	0	0	0	0	0	1	0	0	1	0	0	4
0	0	0	0	0	0	0	1	0	0	1	1	0	0	1	1	3
0	0	0	0	0	0	1	0	0	0	1	1	0	0	0	0	0
0	0	0	0	0	1	0	0	0	0	1	1	1	1	1	1	15
0	0	0	0	1	0	0	0	0	0	1	1	0	0	1	1	3
0	0	0	1	0	0	0	0	0	0	1	1	0	1	0	1	5
0	1	0	0	0	0	0	0	0	0	1	1	1	1	0	0	12
1	0	0	0	0	0	0	0	0	0	1	1	0	0	0	0	0
0	0	0	0	0	0	0	1	0	1	0	0	0	1	0	0	4
0	0	0	0	0	0	1	0	0	1	0	0	0	1	1	1	7
0	0	0	0	0	1	0	0	0	1	0	0	1	0	0	0	8
0	0	0	0	1	0	0	0	0	1	0	0	0	0	0	1	1

FIGURE 4: OPERATION OF SEQUENCER FOR RANGE SELECTION

IV. RESULT AND DISCUSSION

As mentioned previously the objective of this work was to reduce the time more and more ,and hence we used simple sequencer logic instead of using processing elements in the decoder. In the previous literatures on this topic there was a threshold condition to decide whether the decoding is complete or not. Here in our proposed work we state that we will run the same architecture as much as 10 times repeatedly and then stop the entire process. We have seen in software that after 10th iteration for the image there is not much change that we observe in terms of improvement of psnr.

IV. REFERENCES

1. M.F. Barnsley, L.P. Hurd, Fractal Image Compression, AK Peters, Ldt., 1993.
2. Y. Fisher, Fractal image compression: Theory and application, doi:10.1142/ S0218348X94000442.
3. A.E. Jacquin, Image coding based on a fractal theory of iterated contractive image transformations, IEEE Trans. Image Process. 1 (1) (1992) 18–30.
4. A-M.H.Y. Saad, M.Z. Abdullah, "High-speed implementation of fractal image compression in low cost FPGA ," Microprocessors and Microsystems 47 (2016) 429–440 .
5. D. Vidya, R. Parthasarathy, T. Bina, N. Swaroopa, Architecture for fractal image compression, J. Syst. Archit. 46 (14) (2000) 1275–1291
6. D.J. Jackson, H. Ren, X. Wu, K.G. Ricks, A Hardware architecture for real-time image compression using a searchless fractal image coding method, J. RealTime Image Process. 1 (3) (2007) 225–237
7. K.P. Acken, M.J. Irwin, R.M. Owens, A Parallel asic architecture for efficient fractal image coding, J. VLSI Signal Process. Syst. Signal, Image Video Technol. 19(2) (1998) 97–113.

Fractal Image Compression of an Atomic Image using Quadtree Decomposition

Hasanujjaman

Dept. of Electronics & Comm. Engg.
Kalyani Govt. Engineering College
Kalyani, West Bengal, India
hasanujjaman.diet@gmail.com

Arnab Banerjee

Dept. of Electronics & Comm. Engg.
Kalyani Govt. Engineering College
Kalyani, West Bengal, India

Prof. U. Biswas

Dept. of Computer Science and Engg.
University of Kalyani
Kalyani, West Bengal, India

Prof. M K Naskar

Dept. of Electronics & Tele. Comm. Engg.
Jadavpur University
Jadavpur, West Bengal, India

Abstract—**Researchers have made several efforts towards reduction of Compression ratio and improvement of PSNR of an image compressing algorithm. However little attention has been given to reduce the time complexity of the same except for few hardware approaches. In this particular work, a mystical atomic image was created, through the process of splitting the main image into two different blocks. Atomic image was formed by strategically calculating the significant bits from even and odd portion of the original image in spatial domain. Furthermore the most popular methods were implemented on atomic image for getting a lower time complexity, as well as, increased compression ratio and acceptable PSNR. As a result of application of the proposed algorithm we obtained maximum PSNR of 30.12dB for Lena Image and maximum compression ratio of 25.96 for MRI image**

Keywords— Fractal Image Compression(FIC), Quadtree Partitioning, Atomic Image, PSNR, Compression Ratio.

I. Introduction

Image compression is a data compression for a digital image, that lead to the minimizing of the size in order to reduce the cost of storage without hampering the quality or intensity of the image. As an effect of which more number of data can be stored in the single or allocated disk space. Image compression is not only related with the concepts of compression the images, but also in Video media, were there is number of images were contributed or adjusted in slides can also be reduced by the various methods present in this processing of data. For gray value image coding algorithms[1-2] following three types of encoding schemes are incorporated by the researchers (a) Transform coding[3-5] (b) Vector quantization [6-8] and Subband decomposition [9-10]. There were variety of images present like the gray scale image, were only the intensity variance from 0 to 255 are present, colour images that main consist of three colour Red, Green and blue, binary image, and even more. The compression comprises of two types of methods for compression, lossy and lossless. These methods were mainly differentiate while the images were reconstructed after being encoded in these forms of compression. In the lossy encoding, the reconstructed image may not constitute of the same information as it has previously allocated. In lossless compression, the reconstructed image is as same as the original image but a variations is data is different from the original image in somewhat negligible rate. In this particular work the compression of the image was conducted upon the concept of splitting the image into even and odd portions in accordance of the pixel blocks then QuadTree decompose, which is efficient for the reason of small space allocation of data storage. The bunch of data will be further proceed through the Huffman coding in which the data were further reduced so that the required space will further be reduced. As the main aim of this work is to provide the much compression ratios as possible, but while decoding the main contents of image will be reformed back into the lossy form with much developed PSNR.

A. QuadTree Transforms

The QuadTree method is very much efficient for the reducing characteristics in space allocations. Researchers have shown their interest of quadtree decomposition encoding for image[11-16] some those highlighted on advantages of quadtree like structure and others on efficiency and storage. If the image is consists of 2m to 2m blocks then it can be described at m+1 level of decomposition. At each level of decomposition except last level , the node intensity is defined as follows.

$$for \; n = 0,1,2, \ldots \ldots \ldots \ldots, m$$

$$for \; x,y = 0,1,2, \ldots \ldots \ldots, 2^{m-n} - 1$$

$$p_n(x,y) = \frac{1}{4}\sum_{i=0}^{1}\sum_{j=0}^{1} p_{n-1}(2x + i, 2y + j)$$

where p_n is the pixel intensity at resolution level n.

The threshold that is used by the above equation for quadtree decomposition is as follows.

$$\bigcap_{i,j=0}^{1}(|p_i(x,y) - p_{i-1}(2x + i, 2y + j)|) \leq T_{th}$$

FIGURE 1: THE DECOMPOSITION IMAGE OF THE ORIGINAL LENA IMAGE

The process of the QuadTree is used for lossless image processing, reason of which the further partition of the original block will lead to the development of the duplicate image while decoding. Moreover, the process of the decoding lead to the development of the more like the original image if the further partitioning of the blocks can be made. This process may lead to the increment of data, but the further processing upon the data would lead to the reductions of the original image contaminants.

II. PROPOSED ALGORITHM

1. Divides the original image into even and odd part.
2. Divide the one portion using QuadTree decomposition of various thresholds, and minimum and maximum dimensions.
3. Record the values of x and y coordinates by mean value of block size from Quadtree Decompositions.
4. Record the complete encoding of fractal coding information using Huffman Coding and calculate the compression ratio.
5. For the generated bit codes, applying Huffman decoding to reconstruct the main bit stream.
6. Then decode the QuadTree Decomposition by computing the means In empty pixels.
7. For the seeking of originality, we can compute the reconstructed image into original dimensions, with blank pixels filled up with same side by allocations, and calculate the PSNR.

splitting of the image into two different blocks of even and odd in accordance of the partitioning the row and column of the intensity matrix. These process of partitioning the image into two different blocks lead to the decrease in space allocations as well as decreases the compression ratio for further process. Furthermore, the process of splitting the original content of the image into two different blocks lead to the splitting in the contrast, sharpness and the allocated margins of the original image. Although the process is causing lossy of the original information's, but the advantage of less space requirement is an achievement.

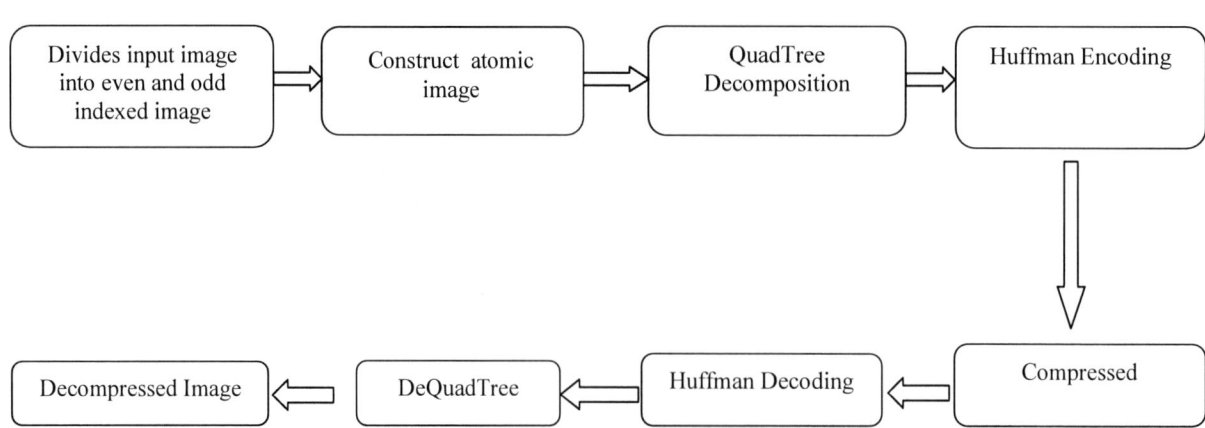

Figure 2: the block diagram for the proposed work

III. Result and Discussion

With the variance of the threshold in the full image processing, initially the time consumption is more along with the high obtainable of the PSNR is obtained. Moreover, the compression ratio at initial is obtained much less. But with the increment of the threshold the compression ratio obtained is much more than the initial stage, along with the less time for processing, with the associated supporting PSNR.

SATELLITE IMAGE FOR RURAL AREA

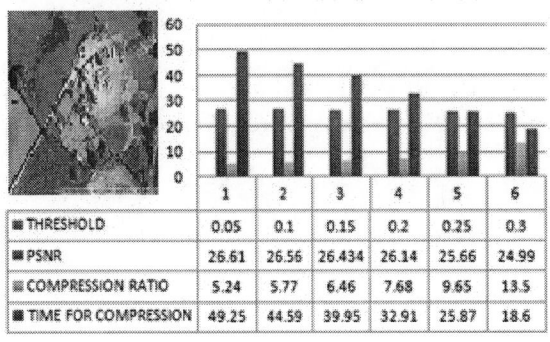

	1	2	3	4	5	6
THRESHOLD	0.05	0.1	0.15	0.2	0.25	0.3
PSNR	26.61	26.56	26.434	26.14	25.66	24.99
COMPRESSION RATIO	5.24	5.77	6.46	7.68	9.65	13.5
TIME FOR COMPRESSION	49.25	44.59	39.95	32.91	25.87	18.6

LENA

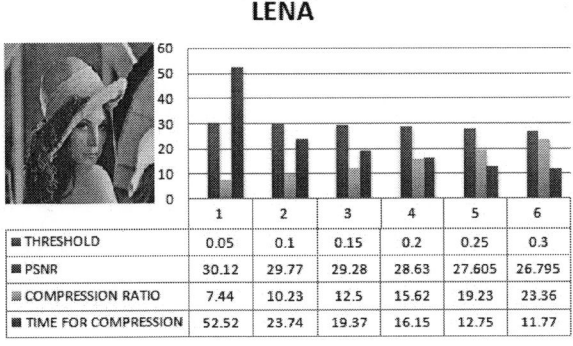

	1	2	3	4	5	6
THRESHOLD	0.05	0.1	0.15	0.2	0.25	0.3
PSNR	30.12	29.77	29.28	28.63	27.605	26.795
COMPRESSION RATIO	7.44	10.23	12.5	15.62	19.23	23.36
TIME FOR COMPRESSION	52.52	23.74	19.37	16.15	12.75	11.77

MRI SCAN IMAGE

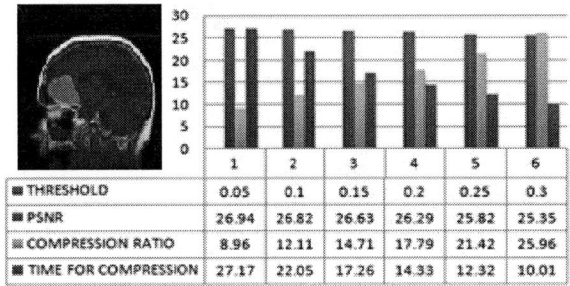

	1	2	3	4	5	6
THRESHOLD	0.05	0.1	0.15	0.2	0.25	0.3
PSNR	26.94	26.82	26.63	26.29	25.82	25.35
COMPRESSION RATIO	8.96	12.11	14.71	17.79	21.42	25.96
TIME FOR COMPRESSION	27.17	22.05	17.26	14.33	12.32	10.01

BABOON

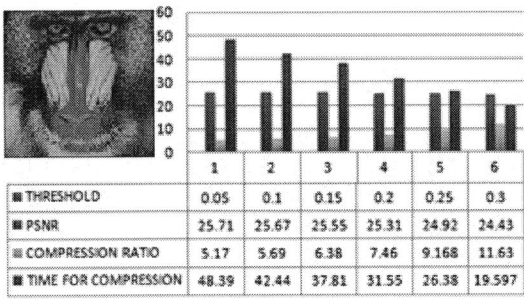

	1	2	3	4	5	6
THRESHOLD	0.05	0.1	0.15	0.2	0.25	0.3
PSNR	25.71	25.67	25.55	25.31	24.92	24.43
COMPRESSION RATIO	5.17	5.69	6.38	7.46	9.168	11.63
TIME FOR COMPRESSION	48.39	42.44	37.81	31.55	26.38	19.597

BARBARA

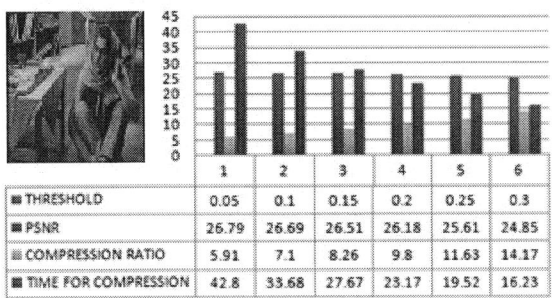

	1	2	3	4	5	6
THRESHOLD	0.05	0.1	0.15	0.2	0.25	0.3
PSNR	26.79	26.69	26.51	26.18	25.61	24.85
COMPRESSION RATIO	5.91	7.1	8.26	9.8	11.63	14.17
TIME FOR COMPRESSION	42.8	33.68	27.67	23.17	19.52	16.23

References : Will Add Later

1. A. K. Jain, "Image data compression: A review," Proc. IEEE, vol. 69, no. 3, pp. 349-389, Mar. 1981.
2. A. K. Jain, "Advances in mathematical models for image processing," Proc. IEEE, vol. 69, no. 5, pp. 502-528, May 1981.
3. W. K. Pratt, Image Transmission Techniques. New York: Academic, 1979.
4. N. Ahmed, T. Natarjan, and K. R. Rao, "Discrete cosine transform," IEEE Trans. Comput., vol. C-23, pp. 90-93, Jan. 1974.
5. A. K. Jain, "A sinusoidal family of unitary transforms," IEEE Trans. Pattern Analysis Mach. Intell., vol. PAMI-1, pp. 356-365, Oct. 1979.
6. R. M. Gray, "Vector quantization," IEEE ASSP Mag., Apr. 1984
7. R. L. Baker, "Vector quantization of digital images," Ph.D. dissertation, Stanford University, Stanford, CA, June 1984.
8. A. Gersho and B. Ramamurthi, "Image coding using vector quantization,"in Proc. Int. Conf ASSP, Apr. 1982, pp. 428-431.
9. M. Vetterli, "Multidimensional subband coding: Some theory and algorithms," Signal Processing, vol. 6, pp. 97-112, Apr. 1984.

10. J. W. Woods and S. D. O'Neil, "Subband coding of images," IEEE Trans. Acoust., Speech, Signal Processing, vol. ASSP-34, no. 5, pp. 1278-1288, Oct. 1986.

11. D. J. Vaisey and A. Gersho, "Variable block-size image coding," in Proc. IEEE Int. Conf Acoust. Speech Signal Processing (ICASSP), Apr. 1987, pp. 25.1.1-25.1.4.

12. J. Vaisey and A. Gersho, "Image compression with variable block size segmentation," IEEE Trans. Signal Processing, vol. 40, pp. 2040-2060, Aug. 1992.

13. C.-Y. Chiu and R. L. Baker, "Quad-tree product vector quantization of images," in Proc. SPIE Conf. Advances Image Compression Automat. Target Recogn., Mar. 1989, pp. 142-153, vol. 1099.

14. P. Strobach, "Tree-structured scene adaptive coder," IEEE Trans. Commun., vol. 38, pp. 477486, Apr. 1990.

15. P. Strobach, "Quadtree-structured recursive plane decomposition coding of images," IEEE Trans. Signal Processing, vol. 39, pp. 1380-1397, June 1991.

RF/Analog & Linearity performance analysis of a downscaled JL DG MOSFET on GaAs substrate for Analog/mixed signal SOC applications

Biswajit Baral[1], Sudhansu Mohan Biswal[2], Sanjit Swain[3], Satish Kumar Das[4], Debasish Nayak[5], Dhananjaya Tripathy[6]

Department of Electronics and Telecommunication Engineering
Silicon Institute of Technology, Bhubaneswar

biswajit@silicon.ac.in, sudhansu.mohan@silicon.ac.in, sswain@silicon.ac.in, dnayak@silicon.ac.in, satish.das@silicon.ac.in, dhananjayavssut@gmail.com

Abstract— **With time the design of RF and Analog application based circuits in MOSFET industry is changing and day by day it's becoming more and more difficult as device modeling has now entered the deep-subnanometer regime. Performance of junction less transistor is remarkable in digital application due to their ease of fabrication and superior SCEs. This paper highlights the DC, ANALOG, RF and LINEARITY performance of a Junction less Double Gate MOSFET (JL DG MOSFET) by downscaling the Channel length with the help of numerical TCAD device simulator (SILVACO). The figure of merits for DC, ANALOG, RF & LINEARITY parameters for example Transconductance (g_m), Gain transconductance frequency product (GTFP), output resistance (R_{out}), cut-off frequency (f_T), Gain bandwidth product (GBW), Transconductance generation factor (g_m/I_d), maximum frequency (f_{max}), Intermodulation distortion (IMD), variable intercept point (VIP) are studied and impact of downscaling in gate length have been recorded. The results conclude that down scaled junction less DG MOSFET show a great pledge to turn out to be a feasible competitor for use in SOC application.**

Keywords— **Junction less transistor, Transconductance (g_m), Gain band width product(GBW), Transconductance generation factor (g_m/I_d), Intermodulation Distortion(IMD).**

I. INTRODUCTION

Due to the additional remarkable performances in digital application, CMOS technology is now ruling over the RF Industry, which was earlier occupied by BJT and MOSFETs, and bipolar–CMOS devices [1]. With continuous downscaling of CMOS technology along with better figure of merits, now SOC application is becoming more effective and economic, However the realization of analog circuit with digital systems on the same integrated circuit in a cost efficient way and with better performance is a big challenge [2]. Now market needs a energy efficient battery operated device enormously so it's a challenging task to design a extreme low power application. But continuous downscaling of channel length leads to many challenges like increase in SCEs (short channel effects), decrease the Gate controllability over channel (DIBL), high leakage current. Considering this in recent years many advanced MOS structures have been proposed in the

nanoscale regime to replace the conventional devices so to maintain the pace of the technology. Double Gate MOS transistor is one of the advanced structure presented in the year 1980s [3,4] which has the ability to improve short channel effects by deploying a double-gate architecture in place of the traditional single-gate architecture. According to [5,6] DG-MOSFET is an essential promising structure than conventional bulk MOSFET because it offers Superior Gate controllability, Reduction in DIBL, Provide Higher Scalability, Capability to Reduced SCEs, Low channel doping, Enhanced carrier mobility. However DGMOS structure also suffers from significant SCEs for channel length below 50nm [7]. However, to realize a fanatic sharp doping profile among source and drain (n-type) region with a substrate region (p-type) was a tough task for the manufacture of multi-gate MOSFETs on the nanometer range [8]. To overcome the above stated problem junctionless field effect transistors (JLFETs), was developed. As in conventional MOSFETs, there is no pn-junctions in JLFET among the Source and Drain region and also between the substrate and channel region. Hence the JLFET can be structured in two different ways as n–n–n-type and p–p–p-type i.e. n-channel and p-channel respectively. In case of junctionless (JL) transistors the fabrication complexity is less due to its low thermal budget, riddance from the requirement of a thin abrupt junction and enhanced short-channel effect (SCEs) performance [9-10]. The JLMOSFET reduces its fabrication requirements as well as the require of higher doping carrier concentration, essential for switching from n-type to p-type can be avoided. In addition to that this device can work as good as a conventional MOSFET. Currently, few researchers made related studies on JLMOSFETs with two gate electrodes i.e. double-gate structure, where they have reported about the theoretical fundamentals to understand the device behavioral structural design investigation of the switching characteristics of the JLMOSFET with variation of drain-to-source voltages (V_{DS}) and the variation of potential at different working conditions [11]. As reported the JLMOSFET is promising device architecture for continuous scaling with minimized SECs [12]. In order to overcome SCEs and high leakage current Junction less (JL) transistor has been proposed, it also offers simplicity to design in SOC application [13]. Due to additional advantages and mitigation these shortcomings, JL-DG-MOSFET has been investigated in this paper which focuses on the DC, Analog, RF & Linearity performance of JL-DG-MOSSFET. The substrate of analyzed JL-DG-MOSFET is GaAs and Oxide layer is SiO_2[14-17]. Now Linearity property of any decide is highly important to predict their performance in mobile communication, computing & multimedia application has resulted interest in SOC application based on CMOS technologies, which is low cost. For all RF communication system Linearity analysis is the most important parameter. The inter modulation distortion of a device & it's n^{th} order harmonics need to be mitigated at the output. Various figure of merits (FOMs) such as variable intercept point2 (VIP$_2$), variable intercept point3 (VIP$_3$), IIP$_3$, 1dB-compression point, different n^{th} order transconductance co-efficient(like gm$_1$, gm$_2$, gm$_{3,...}$) prove its better performance for low-noise and ongoing future SOCs applications [18-19]. Henceforth, other parameters like work function, substrate doping, substrate potential and electric field are also studied to analyze the device performance parameters. The result obtained from this model are verified and simulated for obtaining a good result, validating the model. This paper is presented as given below: Section II deals with device structure and simulation set up. Section III contains the analog performance parameter analysis with change in gate length. section IV describes the RF performances of the device with downscaling gate length. Section V describes the Linearity performance analysis. Finally section VI, the conclusions is presented.

II. DEVICE STRUCTURE AND SIMULATION

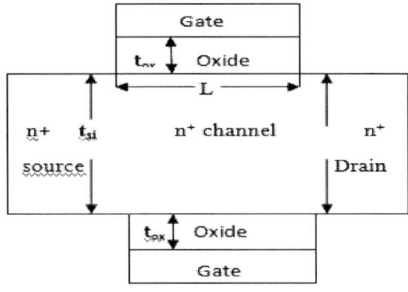

Fig. 1: 2-dimensional cross-sectional view of the DG JL FET

The structure under consideration have uniform doping type and concentration 10^{19} cm^{-3} throughout the structures. L is the channel length with an oxide thickness t$_{ox}$=0.8nm,thickness of GaAs body t$_{si}$=10nm. The work function of the p-type polysilicon gate ϕ_m=5.5eV is considered. The MOS model is selected using Newton Maxtrap method. while AuGe/Ni/Au is

used as an ohmic contacts in different terminals. The 2-dimensional device performance parameters are simulated for JL-DG-MOSFET using SILVACO TCAD SIMULATOR [20]. In this analysis Fermi Dirac carrier statistics along with Drift Diffusion (DD) model has been used to model carrier mobility transport. To model recombination characteristics Shockley-Read-hall (SRH) recombination model along with Auger Recombination model has been adapted. To solve coupled differential equation like continuity equation and current equation of electron and hole in ATLAS, Newton and Gummel's numerical iteration method are taken for consideration. Here In this analysis the Quantum Mechanical Effect (QME) has not been considered as Quantum Mechanical Effects and Balastic transport can be neglected for the devices having channel length less than 10nm.

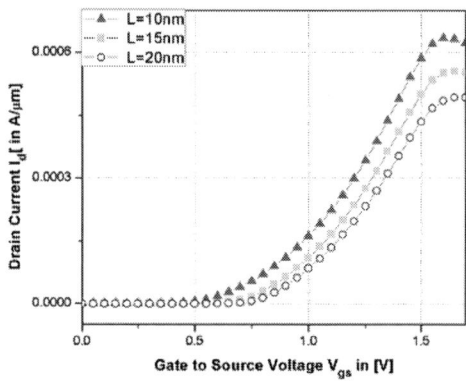

Fig. 2: Variation of I$_D$ with change in V$_{GS}$ for given channel length L=10nm, 15nm and 20nm with device parameter values V$_{DS}$=1v, t$_{si}$=10nm and t$_{OX}$=0.8nm

For different gate length the simulated characteristics of a Junctionless double Gate Hetrostuctrue MOSFET shown in Fig.2. It reveals that channel length of 10nm provides a higher drain current because of the higher gate controllability over the channel.

III. ANALOG PERFORMANCE INVESTIGATION

This portion deals with the analysis of the analog performance parameters of the Junction-less DG MOSFET.

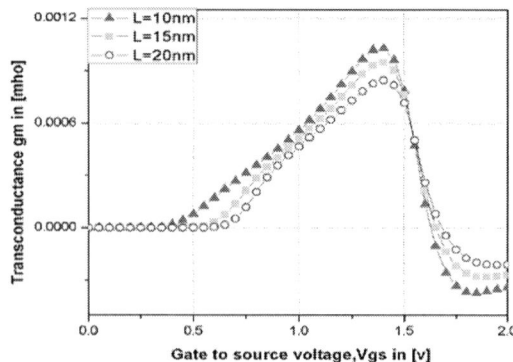

Fig. 3. Shows the change of transconductance (g$_m$) with change in V$_{GS}$ (Gate-to-source) voltage for separate gate Length.

From Fig.3 it is evident that with decrease of the channel length the transconductance decreases.

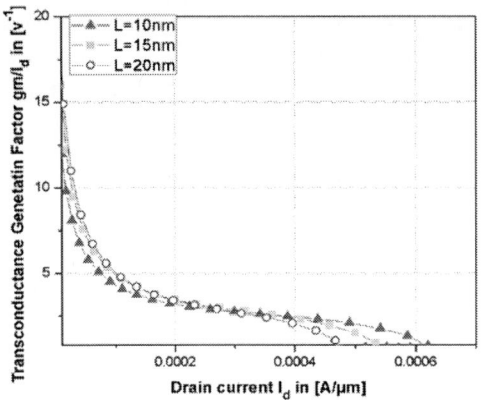

Fig.4. Variation of Transconductance Generation Factor (TGF) with change in Drain Current with device parameter values Vds=1v, t$_{si}$=10nm and t$_{ox}$=0.8nm for different Channel Length.

Fig. 4, this figure of merit shows that the measure of efficiency to convert the current into transconductance. It indicates that lower TGF reduces the ability of input device and high power dissipation. It is clear from Fig. 4 that for lower gate length the TGF value reduces.

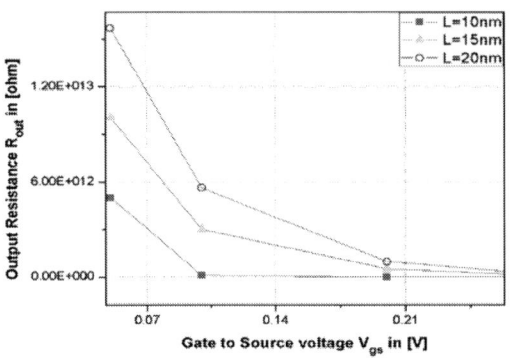

Fig. 5. Change in Output Resistance (R$_{out}$) as a function of Gate-to-source voltage V$_{GS}$.

Due to necessity of standard output requirement a high output resistance is needed. On the other hand it is clear from Fig. 5 that output resistance decreases by reducing the channel gate length because of larger penetration of gain electric field, not permitting the drain current to get its saturation value.

IV. RF PERFORMANCE INVESTIGATION

This segment shows evaluation of the Radio Frequency performance parameters of Junction-less DG MOSFET. The two major parameters for finding the device capability towards RF application is Cutoff frequency (f$_T$) and the maximum frequency of oscillation (f$_{max}$). The cut off frequency (f$_T$) is defined as the frequency where current gain is unity and the maximum frequency of oscillation(f$_{max}$) is

defined as the frequency where the power gain of the circuit is unity [21]. Standard figure of merits (FOMs) require to examine the Radio Frequency performances are,

(a) Cut-off frequency (f$_T$)

(b)Maximum frequency of oscillation (f$_{max}$)

(c) Gain Bandwidth Product (GBW)

The general representation of f$_T$, f$_{max}$, GBW are given as

$$f_T = \frac{g_m}{2\pi C_{gs}\sqrt{1 + 2\frac{C_{gd}}{C_{gs}}}} \approx \frac{g_m}{2\pi\left(C_{gd} + C_{gs}\right)} \approx \frac{g_m}{2\pi C_{gg}} \tag{1}$$

$$f_{max} \approx \frac{g_m}{2\pi C_{gs}\sqrt{4\left(R_s + R_i + R_g\right)\left(g_{ds} + g_m\frac{c_{gd}}{c_{gs}}\right)}} \tag{2}$$

$$GBW = \frac{g_m}{2\pi \times 10 \times C_{gd}} \tag{3}$$

where, Cgs , Cgg , Cgd denotess the Gate to Source, Gate to Drain capacitance and total gate capacitance respectively.

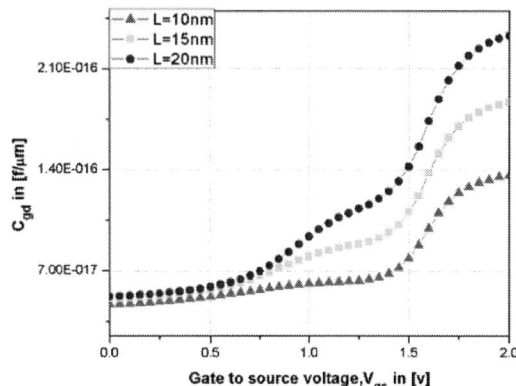

Fig. 6 plot of Gate-to-Drain Capacitance C$_{GD}$ with variation in V$_{GS}$.

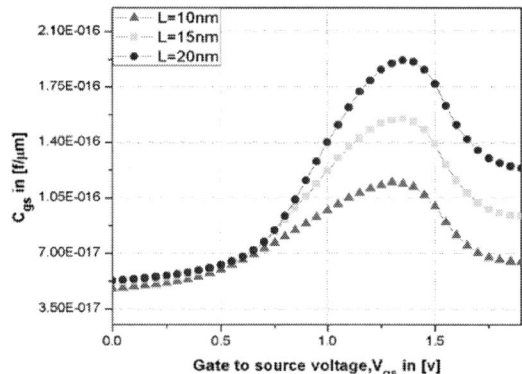

Fig. 7: Change in Gate to Source Capacitance (C$_{GS}$) with change in V$_{GS}$ for separate gate Length .

Fig. 6 and 7 shows the plot of C_{GD}, C_{GS} with change in V_{GS} for different channel length. from both the Figures it is found that that gate length decreases with decrease in C_{GD} and C_{GS} value.

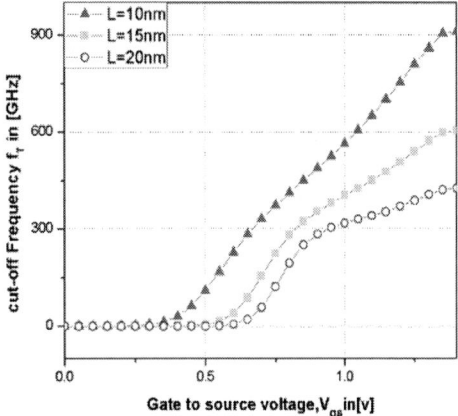

Fig. 8: Plot of Cut-off Frequency (f_T) with change in Gate to Source Voltage V_{GS} for different gate Length.

At the sub-threshold region f_T is proportional to a $1/L^2$ as tranconductance proportional to $1/L$ and C_{gd}/C_{gs} proportional to L. The cut off frequency (f_T) reduces and increases with change in I_D until it reaches a maximum value at a particular gate bias voltage. At maximum transconductance f_T is at maximum point and gate to source/drain capacitance is minimum. Higher value of L gives lower f_T value, showing a relation between bandwidth and power efficiency. By enhancing the interface charge trap density higher value of cut off frequency can be achieved. Fig. 8 shows that lower the channel length the cut of frequency increases.

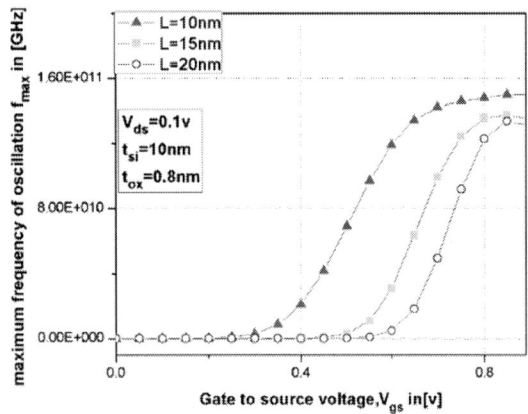

Fig. 9: Variation of maximum frequency of oscillation (f_{max}) as a function of Gate to Source Voltage V_{GS}

The change in f_{max} with respect to drain current I_D is proportional to $1/L^2$ shown in fig. 9. On the other hand, it is observed that f_{max} is also equally proportional to the parasitic resistances like (R_g, R_i, R_s).

V. LINEARITY TESTING

Linearity testing is an vital prerequisite for RF system in order to ensure minimum inter modulation distortion and higher order harmonics at the output of the device. Inter modulation Distortion because of non linearity; produce unwanted signals with different frequencies [22-26]. These unwanted (noise) signals may interfere, change or even corrupt the desired output components. So it is essential to analyze the linearity testing parameters for JL DG MOSFETs. A transistor level linearization is more suitable for power amplifiers in portable systems, which requires an analysis of the linearity behavior at the device level.

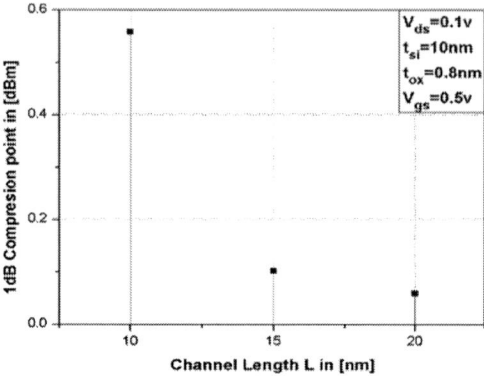

Fig. 10: plot of 1dB compression point verses Gate to Source Voltage V_{GS} for Channel Lengths L=10nm,15nm and 20nm with device parameter values V_{DS}=1v,t_{si}=10nm and t_{ox}=0.8n.

It is clear from Fig. 10 that there is an increase in 1dB compression point as channel length gradually decreases.

VI. CONCLUSION

In this study, the different performance parameters like DC, RF, Analog and linearity of a JL DGMOSFET is clearly analyzed in terms of transconductance Generation Factor (g_m/I_{DS}), cutoff frequency (f_T), maximum oscillation frequency (f_{MAX}). Comparison has been performed for different channel length and it was observed that device shows excellent RF performance for shorter channel length but the analog performance of the Device was poor. Hence, it is concluded from the above analysis that JL DG MOSFET is a strong competitor for next generation Radio Frequency System on Chip mixed signal applications with an enhanced linearity and RF performance. The above study can further extended by changing the substrate from GaAs to InGaAs with suitable oxide variation and also gives scope to further research by introducing gate stack or gate engineering concepts.

REFERENCES

[1] N. Mohankumar,Binit Syamal, Chandan Kumar Sarkar N. Mohankumar, "Influence of Channel and Gate Engineering on the Analog and RF Performance of DG MOSFETs," *IEEE TRANSACTIONS ON ELECTRON DEVICES*, vol. 57, no. 4, pp. 820-826, April 2010.

[2] P. H. Woerlee, M. J. Knitel, R. van Langevelde, D. B. M. Klaassen, L. F. Tiemeijer, A. J. Scholten, and A. T. A. Zegers-

van Duijnhoven, "RFCMOS performance trends," IEEE Trans. Electron Devices, vol. 48, no. 8, pp. 1776–1782, Aug. 2001.

[3] T. M. Chung, B. Olbrechts, U. Sodervall, S. Bengtsson, D. Flandre, and J.-P. Raskin, "Planar double-gate SOI MOS devices: Fabrication by wafer bonding over pre-patterned cavities and electrical characterization," *Solid-State Electron. Elsevier*, vol. 51, pp. 231–238, 2007.

[4] X. Lin, C. Feng, S. Zhang, W. Ho, and M. Chan, "Characterization of double gate MOSFETs fabricated by a simple method on a recrystallized silicon film," *Solid State Electron. Elsevier*, vol. 48, pp. 2315–2319, 2004

[5] Wong H-SP, Chan KK, Taur Y. Self-aligned (top and bottom) dou-ble-gate MOSFET with a 25 nm thick silicon channel. Technical Digest., International, Electron De-vices Meeting, 1997, IEDM '97, 10-10 December, 1997; 427–430. DOI: 10.1109/IEDM.1997.650416

[6] 2. Ernst T, Cristoloveanu S, Ghibaudo G, Ouisse T, Horiguchi S, Ono Y, Takahashi Y, Murase K. Ultimate thin double-gate SOI MOSFETs. IEEE Trans Electron Devices 2003; 50:830–838.

[7] 3. Reddy GV, Kumar MJ. Investiga-tion of the novel attributes of a single-halo double gate SOI MOSFET: 2D simulation study. Microelectron J 2004; 35:761–765

[8] Colinge J P, Lee C W, Afzalian A, et al. Nanowire transistors without junctions. Nat Nanotechnology, 2010, 5(3): 225

[9] Lin ZM, Lin HC, Liu KM and Huang TY. Analytical model of subthreshold current and threshold voltage for fully depleted double-gated junctionless transistor. *Japanese Journal of Applied Physics* 2012; 51(2S):02BC14.

[10] Jin X, Liu X, Wu M, Chuai R, Lee JH and Lee JH. Modelling of the nanoscale channel length effect on the subthreshold characteristics of junctionless field-effect transistors with a symmetric double-gate structure. *Journal of Physics D: Applied Physics* 2012; 45(37):375102.

[11] C.-W. Lee, A. Afzalian, N. D. Akhavan, R. Yan, I. Ferain, and J.-P. Colinge, "Junctionlessmultigate field-effect transistor," *Appl. Phys. Lett.*, vol. 94, no. 5, pp. 053511-1–053511-2, Feb. 2009.

[12] J.-P. Colinge, C.-W. Lee, A. Afzalian, N. D. Akhavan, R. Yan, I. Ferain, P. Razavi, B. O'Neill, A. Blake, M. White, A.-M. Kelleher, B. McCarthy, and R. Murphy, "Nanowire transistors without junctions," *Nature Nanotechnol.*, vol. 5, no. 3, pp. 225–229, 2010.

[13] J.P. Colinge, C.W. Lee, A. Afzalian, N. Dehdashti Akhavan, R. Yan, I. Ferain, P. Razavi, B. O'Neill, A. Blake, M. White, A.M. Kelleher, B. McCarthy, R. Murphy, " Nanowire transistors without junctions," *Nature Nanotechnology*, vol. 5, no. 3, pp. 225-229,2010

[14] Lee CW, Afzalian A, Akhavan ND, Yan R, Ferain I and Colinge JP. Junctionless multigate field-effect transistor. *Applied Physics Letters* 2009; 94(5):053511

[15] V. Kilchytska, A. Nève, L. Vancaillie, D. Levacq, S. Adriaensen,H. van Meer, K. De Meyer, C. Raynaud, M. Dehan, J.-P. Raskin, and D. Flandre, "Influence of device engineering on the analog and RF performance of SOI MOSFETs," *IEEE Trans. Electron Devices*, vol. 50, no. 3, pp. 577–588, Mar. 2003.

[16] D. J. Frank, R. H. Dennard, E. Nowak, P. M. Solomon, Y. Taur, and H.-S. P. Wong, "Device scaling limits of Si MOSFETs and their application dependencies," *Proc. IEEE*, vol. 89, no. 3, pp. 259–288, Mar. 2001.

[17] S. S. Suryagandh, M. Garg, and J. C. S. Woo, "A device design methodology for sub-100-nm SOC applications using bulk and SOI MOSFETs," *IEEE Trans. Electron Devices*, vol. 51, no. 7, pp. 1122–1128, Jul. 2004.

[18] J. Liang, H. Xiao, R. Huang, P.Wang, and Y.Wang, "Design optimization of structural parameters in double gate MOSFETs for RF applications," *Semicond. Sci. Technol.*, vol. 23, no. 5, pp. 1–8, May 2008.

[19] Yogesh Pratap, Subhasis Haldar, R.S Gupta and Mridula Gupta, " Linearity Performance Investigation of high-k Spacer based Junctionless Nanowire Transistor (JLNWT) for RFIC Design," *ISDRS,*2013.

[20] W. Ma, S. Kaya, A. Asenov, "Study of RF linearity in sub-50nm MOSFETs Using Simulations," *Journal of Computational Electronics,*vol. 2, no. 2-4, pp. 347-352,2000.

[21] ATLAS Device Simulation Software, Silvaco Int., Version 5.14.0.R

[22] J. Liang, H. Xiao, R. Huang, P.Wang, and Y.Wang, "Design optimization of structural parameters in double gate MOSFETs for RF applications," *Semicond. Sci. Technol.*, vol. 23, no. 5, pp. 1–8, May 2008.

[23] S. Kaya, W. Ma, and A. Asenov, (2003) "Design of DG-MOSFETs for High Linearity Performance"*IEEE International SOI Conference*, pp. 68-69.

[24] H. Zhang, E. Sánchez-sinencio, and L. Fellow, "Linearization techniques for CMOS low noise amplifiers: A tutorial," *IEEE Trans. Syst. Circuits*, vol. 58, no. 1, pp. 22–36, Jan. 2011.

[25] Y. Ding and R. Harjani, "A CMOS high efficiency +22 dBm linear power amplifier," in *Proc. IEEE Custom Integr. Circuit Conf.*, Oct. 2004,pp. 557–560.

[26] B. Razavi, "RF Microelectronics," *Prentice-Hall*, 1998.

Effect of High-K Spacer on the Performance of Non-Uniformly doped DG-MOSFET

[1]Sanjit K Swain, [1]Satish K Das, [2]Sudhansu M Biswal, [4]Sarosij Adak, [3]Umakanta Nanda, [1]Asmit Amlan Sahoo, [2]Debasish Nayak, [1]Biswajit Baral, [2]Dhananjaya Tripathy

[1]Department of Electronics and Communication Engineering, Silicon Institute of Technology, Bhubaneswar
[2]Department of Electronics and Instrumentation Engineering, Silicon Institute of Technology, Bhubaneswar
[3]School of Electronics Engineering, Vellore Institute of Technology, Amaravati
[4]Brainware University, Kolkata

Email: sswain@silicon.co.in

Abstract— This paper presents the performance of non-uniformed doped double gate (DG) MOSFET with different spacer variations with an aim to analysis the effects of short channel and various performance metrics. In this work we have taken silicon as the channel material with non-uniform doping for studying the analog and RF performances. Spacer's materials having different permittivities were used to understand their effect on the device performance. Based on the simulations, we can conclude that analog and Radio Frequency performance of the device shows an significant improvement with addition of spacer layer. We have used computer aided design (TCAD) simulations by SILVACO International.

Keywords—DG MOSFET, Spacer layer, Non-uniform Channel, AnalogPerformance, RF Performance.

I.   INTRODUCTION

Due to their ability of being scaled down, The MOS based devices are chosen for the analog and radio frequency applications. Down scaling is done in order to make the device more tangible for the application in small power rating, also for high frequency operation[1, 2]. However, down scaling of the device leads to the short channel effects (SCEs) which hinders the performance of the device [3-5]. To overcome them, various modifications are made in the design of MOSFETs such different channel engineering techinques[8-10], different device engineering modifications [11-13], and also gate-stacking configurations [14, 15].

It is found that when the conventional MOSFETs are downscaled, they have high leakage current, which leads to unnecessary power consumption [7]. The channel is controlled by both V_{gs} and V_{ds} when the device dimensions as lowered. A solution to these problems is double gate (DG) MOSFETs which provides a better control of the channel [6]. In this work, the device proposed is a DG MOSFET having SiO_2 as the oxide. The device follows a non uniformed doping configuration i.e it follows the high-low doping configuration across the channel. This is also called Graded channel which acts as an answer to reduce few effects of short channel device. It also has added advantage of reducing the effects of parasitic bipolar and also eliminates the effect of hot carrier phenomenon. High-K spacers having various permittivities are used for suppressing the fixed charge accumulation, high interfacial charge concentrations and also reducing the effect of mobility degradation at the interfacial region. This configurations therefore suitable for providing better analog /RF performances [16]. Various proposed device that are simulated are based upon either DG-MOSFETs with uniform doping profile or with high-K spacer based DG-MOSFETs [7,16-22,25,27,28,].Thus, we have used graded channel engineering along with high-K spacer on Double Gate -MOSFETs to study their effects on analog and RF performance, on which no work is ever done.

The following parameters: Gate Voltage vs Drain Current, Transconductance vs Gate Voltage (g_m vs V_{gs}), Transconductance Generation Factor v/s Gate

978-1-5386-6723-1/19 $31.00 © 2019 IEEE

Voltage(g_m/I_d vs V_{gs}), Gate toSource Capacitance vs Gate Voltage (C_{gs} vs V_{gs}), Drain to Gate Capacitance vs Gate Voltage (C_{gd} vsV_{gs}) and Cut-off Frequency and Gate Voltage (f_T v/s V_{gs}) are analysed by taking the simulations for the proposed non uniform channel Double Gate structure with different spacer materials .

II. DEVICE CONFIGURATION AND SIMULATION SETUP

The cross sectionsl structure of the proposed device with high-K spacer and high-low doping profile is shown in Fig. 1. All the parameters are considered as per the International Roadmap for Semiconductor Technology (ITRS) standards [23]. The device has gate length (L_g) of 32nm.The source and drain lengths are of 10nm whereas the channel thickness (t_{si}) of 16nm. The oxide thickness of SiO_2 is 1nm. The source and drain of the device are having n-type uform doping with the concentration of $10^{18}/cm^3$. The channel region is having a high-low doping profile with doping concentration of $10^{16}/cm^3$ for the N^+ region and $10^{15}/cm^3$ for the N^- region leading to the formation of the non uniformed channel. The said device is symmetrical struture with respect to the double layered gate and with oxide dielectric of SiO_2. For the simulations, spacers materials used are high-K oxides having permittivity K=3.9,7.5 and 25 respectively .

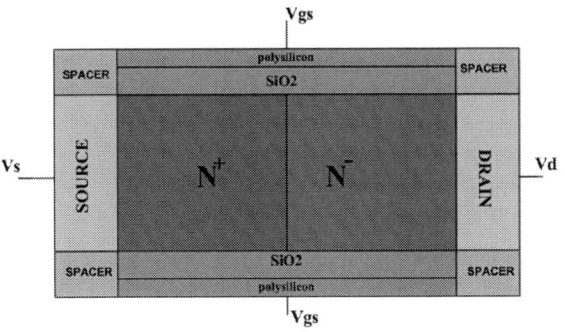

Figure 1. Cross sectional view of non uniformly doped DG-MOSFET with High-K Spacer.

The numerical simulations are considered to be two-dimensional drift diffusion model. This model results in high speed simulations with a good accuracy as it is used for carrying out the simulations of carrier transport for the semiconductors process. To consider the effects of the ionized impurity scattering and temperature effects, Aurura model is also added. The numerical method, ksn=0 is used in the models for calculations and with a temperature setting of 300K is considered for the

simulation. The electric field dependent mobility is used in the simulation model. For the accurate simulation results as well as to accelerate the computational efficiency, we have carefully taken the mesh density of the structure. All these simulations are performed using computer aided technology (TCAD) Silvaco International [24].

III. RESULTS WITH DISCUSSION

Here we have tried to explained the outcome of variation of high-K spacer layers for the proposed non uniformed (GC) DG-MOSFET and analyzed various performance parameters. Different analog parameters are studied by considering various high-K spacer materials of the proposed device. V_{ds} is kept at 0.5V for all the simulations. Parameters such as drain current with gate voltage (I_D vs V_{gs}), transconductance factor with gate voltage (g_m vs V_{gs}) and also transconductance generation factor with gate voltage (g_m/I_d vs V_{gs}) are studied.

The effect of drain current for change in gate voltage (V_{gs}) for variation of high-K spacer materials are shown in Fig.2. For permittivity K=25, it is clearly shown that the current is highest at a given gate voltage as compared to all other high-K materials taken as the spacers. It is easily inferred from the fact that if the permittivity of the material is high, then capacitance is also high whose direct impact is in rise of current. The results indicate towards the better switching capabilities of the device.

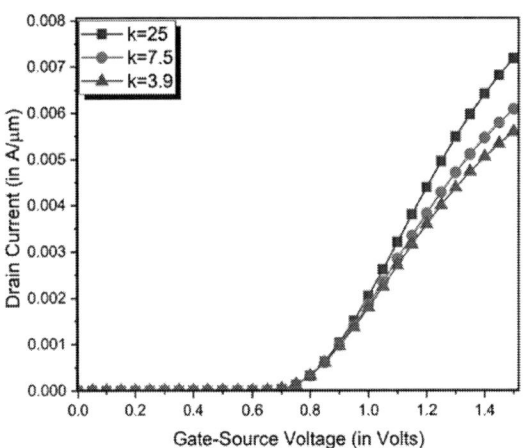

Figure 2. Change of drain current with gate voltage for oxides having different spacer material.

Fig. 3 reveals the change in Transconductance vs Gate Voltage (g_m vs V_{ga}) for different high-K materials as the spacer. The material with highest permittivity (here K=25) is having a higher transconductance as compared to material with lower permittivity preferred as the spacer

978-1-5386-6723-1/19 $31.00 © 2019 IEEE   511

material can be observed from the graph. The transconductance factor is a crucial parameter for analog and RF performance operations. It is depicted as- $(g_m = \partial I_d / \partial V_{gs})$. It can rephrase as- for a given small change of gate voltage, it will have a higher change in drain current with spacers having greater permittivity's than that of the lower ones. Transconductance is used in determining the gain. This gain is further used to determine the gain-bandwidth product of the device. The gain is a vital parameter if one takes account of the amplification process done by MOSETs. Higher the transconductance of the device better is the gain. Thus the device can amplify the signals in better way. Thus, the GC-DG MOSFETs with high-K spacers having high permittivity can be better amplifiers.

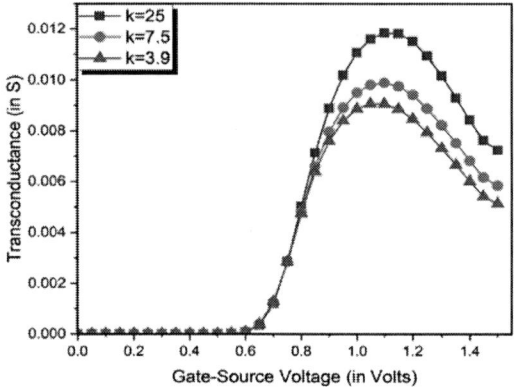

Figure 3. Change of transconductance with gate voltage for oxides having different spacer material.

Fig. 4 showcases the simulated results of the change in the value of (g_m / I_d) for different value of gate voltage for the proposed device. It is indicated from the graph that the material with higher permittivity has higher transconductance generation factor. It can be implied from the fact that the device with higher permittivity based spacer will have higher transconductance. The relative difference is larger in when one accounts for the transconductance of (say) any two corresponding device having variations in the composition of the spacer material as compared with their relative difference in the drain current. The transconductance generation factor indicates that how effectively a certain value of transconductance can be achieved by the drain current. Its consequence is the effective operation of the device in low voltages. This accounts for the realization of the circuits operating in lower supply voltages.

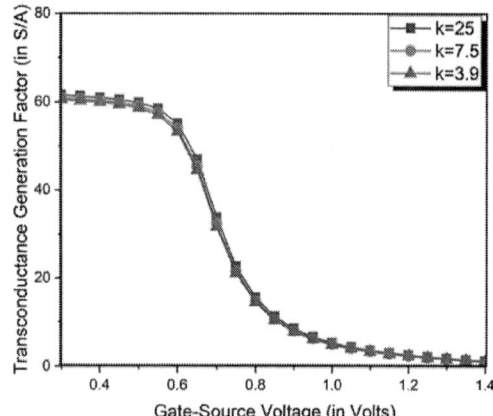

Figure 4. Change of transconductance generation factor with gate voltage for oxides having different spacer material .

In this section we have tried to explore different RF related parameters so as to study the device frequency response capability [26]. Various important parameters such as:- gate capacitance with gate voltage, gate to drain Capacitance with gate voltage and the unity gain cut- off frequency are studied during simulation of the said device. All the simulations are done by taking the $V_{DS} = 0.5$ volt.

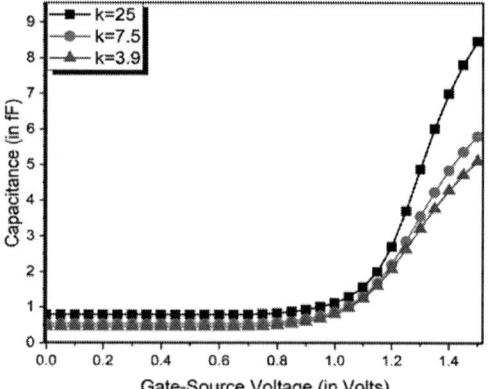

Figure 5. Change of gate to source capacitance with gate voltage for different spacer material.

The change in the Gate toSource Capacitance with respect to variation of gate voltage (C_{gs} vs V_{gs}) and similarly study of Gate to Drain Capacitance are discussed in the Fig. 5 and 6 respectively.

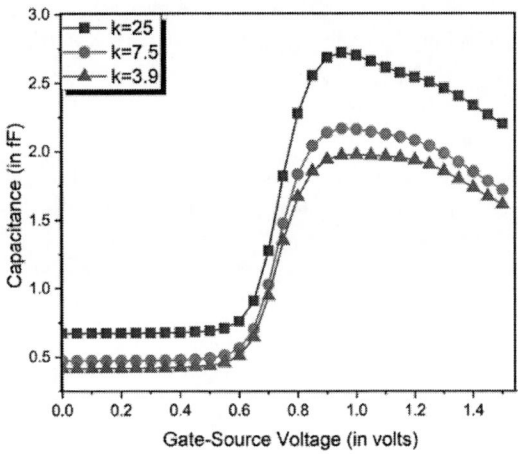

Figure6. Change of gate to drain capacitance with gate voltage for different spacer materials

Both C_{gs} and $C_{gd}=0$ in cut-off mode of operation. This is due to the fact that that inversion doesn't take place as there is no channel. But, when the mode becomes linear, both the capacitance becomes non zero and increases gradually with rise in V_{gs}. The capacitance of the spacer material having highest permittivity (K=25) is highest amongst all the materials considered as spacers for the proposed device at the same V_{gs}. The reason is depicted from the fact that the permittivity is directly proportional to capacitance. It is depicted by the equation ($C=\varepsilon A/d$ where $\varepsilon=\varepsilon_o\varepsilon_r$). Furthermore, 'A' and 'd' are constant for the proposed device, only variation being permittivity.

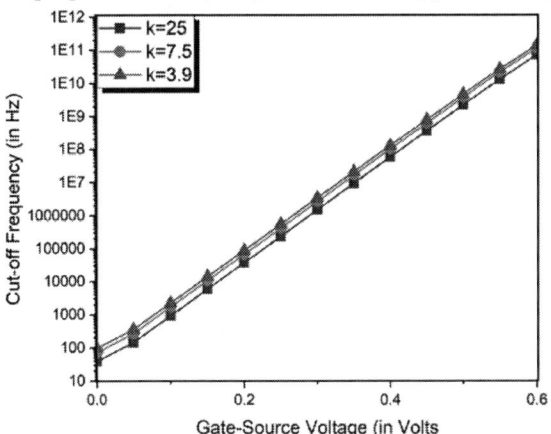

Figure 7. Change of Cut-off Frequency with gate voltage for oxides having different spacer material.

In Fig.7, The effect of gate voltage for different spacer layer to study the effect of cut-off frequency (f_T vs V_{gs}) are studied. The cut-off frequency states that the device must have current gain approximately unity at that frequency. It can be given by the equation:-

$$f_T = g_m/(2*\pi*C_{gg})$$

where, C_{gg} is the Gate Capacitance given as $C_{gg}=C_{gs}+C_{gd}$. The spacer material with lowest permittivity is having

the highest cut-off frequency is indicated by the graph. It can be inferred from the fact that the permittivity has the direct effect on the capacitance which is inversely related to the frequency. Even though there is a factor in play-g_m, its effect is lower than that of the capacitance. From the above discussion it is clear that low permittivity based High-K spacer for GC-DG-MOSFETs is suitable for the circuit operating in high frequencies.

IV. CONCLUSION

From above discussion we have studied the output of the considered device for different high-K oxide dielectric material in spacer region. Some important outcomes have been noticed which will really helpful for future analysis. It is deduce that the intrinsic gain increases with higher value of spacer material. It is also found from the simulation results that for TiO_2 the gain found to be 25.7 with Ro $1.03\times10^7\Omega$. Similarly the g_d value is found to be larger for TiO_2 i.e. 9.67×10^{-8}S .This tends to increased the early voltage V_{EA} for better analog output of the device. Similarly the RF performances also render good result with high-K spacer region. Hence these results signify that the spacer layer selection plays the key role for this promising device to shows better analog and RF output.

V. ACKNOWLEDGEMENT

The authors are really grateful to convey their regards to Silicon Institute of Technology, Bhubaneswar 751024, India for providing the require support to carry out the above research work

VI. REFERENCES

[1] P. H. Woerlee, M. J. Knitel, et al, "RFCMOS perfomances trends"IEEE Trans. Electron Devices, vol. 48, pp 1776-1782, 2001.

[2] V. Kilchytska, A. neve, L. Vaneaillie, et al, "Influence of device engineering on analog and RF perfomance of SOI MOSFETs" IEEE Trans. Electron Devices, vol. 50, pp 577-588, 2003.

[3] K. K. Young, "Short channel effect in fully depleted SOI MOSFETs" IEEE Transactions on Electron Devices,Vol. 36 ,pp.399-402,1989

[4] D. J. Frank et al, "device scaling limits of Si MOSFETs and their application" Proc. IEEE , vol.89, pp 259-288, 2001 .

[5] S.K.Swain et al., "Effect of Channel Thickness and Doping Concentration on Sub-Threshold Performance of Graded Channel and Gate Stack DG MOSFETs" JOLPE, vol.11, pp 366-372 , 2015.

[6] F. Balestra, S. Cristoloveanu, M. Benachir, et al, "Double gate silicon-on-insulator with volume inversion" IEEE Electron Device Letters ,Vol.9,pp. 410-412 ,1987.

[7] Sanjit Kumar Swain, Arka Dutta, Sarosij Adak, Sudhansu Kumar Pati, Chandan Kumar Sarkar,"Influence of Channel length and High-K oxide Thickness on Subthreshold Analog/RF Performance of Graded Channel and Gate stack DG- MOSFETs." Microelectronics Reliability (Elsevier) Volume 61, ,Pages 24-29, 2016.

[8] Sarosij Adak, Sanjit Kumar Swain, Arka Dutta,Hafizur Rahaman, Chandan Kumar Sarkar, Influence of Channel length and High-K oxide Thickness on Subthreshold DC Performance of Graded Channel and Gate stack DG-MOSFETs NANO: Brief Reports and Reviews (World Scientific Publishing Company) Accepted: 17 2016. Volume 11, Issue 09, pp 1650101, 2016.

[9] F. Djeffal, Z. Ghoggali, Z. Dibi, et .al, "Analytical analsis of nanoscale multiple gate MOSFETs including effects of hot-carrier induced interface charges" Microelectronics Reliability ,Vol. 49,pp. 377-381,2009.

[10] N. Mohankumar, B.Samal, C.K. Sarkar, "Influence of channel and gate engineering on the analog and RF perfomance of DG MOSFETs" IEEE Transactions on Electron Devices,Vol. 57 ,pp.820-826,2010

[11] C, L. chen, J. M. Knecht, J. Kedzierski, "Channel Engineering for SOI-MOSFETs for RF applications" Proc. IEEE Int.SOIConf, 5232-5233, 2009.

[12] Y. Taur, D. A. Buchanan, W. Chen, et al, "Dimensional effects in MOSFETs…" Proc. IEEE (1997) ,Vol.85, pp.486-504.

[13] R. K. Sharma, M. Gupta, R. S. Gupta, "TCAD Assessment of Device Design Technologies for Enhanced Performance of Nanoscale DG MOSFET" IEEE Transactions on Electron Devices,Vol. 58 ,pp.2936-2943,2011

[14] C. Hu, "Gate oxide scaling limit and projection"IEDM Tech. Dig (1996), 319-322

[15] E. M. Vogel, K. Z. Ahmed, B. Hornung, et al, "Model tunnel currents for high dielectric constant dielectrics" IEEE Transactions on Electron Devices ,Vol.45,1350-1355, 1998.

[16] B. Cheng, M. Cao, R. Rao, A. Inani, et al. "The impact of high-K dielectrics and metal gate electrodes on sub-100nm MOSFETs" IEEE Trans. Electron Devices, Vol. 46, pp. 1537-11543, 1999

[17] A. Inani, B. Cheng, RV. Rao, J. Woo, "Gate stack architecture analysis and channel engineering in deep sub-micron MOSFETs" Jpn J Appl Phys, Vol. 38, pp.2266-2271, 1999

[18] M. Saxena, S. Halder, M. Gupta, RS. Gupta, "Modelling and simulation of asymmetric gate stack (ASYMGAS)-MOSFET" Solid State Electronics, Vol. 47 , pp.2134-2134,2003

[19] E. Faith, A. Behnam, P. Hashemi, et al. "The influence of the stacked and the double material gate structures on the short channel effects in SOI MOSFETs" IEICE Trans Electron, Vol. 88, pp. 1122-1126, 2005

[20] F. Djeffal, M. Meguellati, A. Benhaya, "A two-dimensional analytical analysis of subthreshold behavior to study the scaling capability of nanoscale graded channel gate stack DG MOSFETs" Physica E,Vol.41,1872-1877, 2009.

[21] M. A. Pavanello, J. A. Martino, V. Dessard, et al, "An asymmetric channel SOI nMOSFET for reducing parasitic effects and improving output charcterstics" Electrochem. Solid-State Lett,Vol. 3,pp. 50-52,2000.

[22] A. Kranti, TM. Chung, D. Flandre, JP. Raskin, "Laterally asymmetric channel engineering in fully depleted double gate SOI MOSFETs for high perfomance analog applications" Solid State Electronics,Vol. 48 ,pp.947-959,2004

[23] Semiconductor Industry Association, International Technology Roadmap for Semiconductors, (2011)SIA San Jose.

[24] Silvaco International Simulator device user's manual V-2008.-09.

[25] S. K. Pati, K. Koley, A. Dutta, et al, "Study of body and oxide thickness on analog and RF performance of underlap DG-MOSFETs" MicroelectronicsReliability Vol. 54,pp. 1137-1142,2014.

[26] I. M. Kang, H. Shin, "Non-quasi-static small –signal modelling and analytical parameter extraction of SOI FinFETs" IEEE Trans. Electron Devices, Vol. 5 ,pp.205-210,2006.

[27] K. Koley, et al. "Subthreshold analog/RF perfomance of underlap DG FETs with asymmetric source/drain extensions" Microelectronics Reliability,Vol.53,pp. 520-527, 2013

[28] R. K. Sharma, M. Bucher, "Device Design Engineering for Optimum Analog/RF perfomance of Nanoscale DG MOSFET" IEEE Transactions on Nanotechnology,Vol. 58 ,pp.2936-2943,2011

All-optical Walsh-Hadamard code Generation using MZI

Supriti Samanta
Department of Physics,
Jadavpur University,
Kolkata, W.B, India.
Email:
supritisamanta28@gmail.com

Goutam Kumar Maity
Department of Physics,
Pingla Thana Mahavidyalaya,
Maligram,West Bengal,
India.
Email: gkm20021976@gmail.com

Subhadipta Mukhopadhyay
Department of Physics,
Jadavpur University,
Kolkata, W.B, India.
Email:
phy.smukherjee@gmail.com

Abstract- Code division multiple access (CDMA) or multi-carrier CDMA (MC-CDMA), Walsh-Hadamard codes are extensively used for its attribute of orthogonality and hence it leads to good cross-correlation property. In the last time, in CDMA communications, spread spectrum codes is Walsh-Hadamard codes and it is generated simply. Walsh-Hadamard codes are completely orthogonal binary user codes, that arehaving a lot of favorable applications in communications based on synchronous multicarrier. However, the optical applications of Walsh-Hadamard codes are also important for optical CDMA. To achieve this goal, all-optical Walsh-Hadamard codes generation using Mach–Zehnder interferometer (MZI) is explored in this paper.

Keywords- Semiconductor Optical Amplifier, Mach–Zehnder interferometer, Walsh-Hadamard code.

I. INTRODUCTION

In CDMA systems, the spread spectrum (SS) technique is used for transmission of information by employing spreading codes. The signal is having a bandwidth higher than the minimum requirement for sending information in SS modulation techniques. For a long term, Walsh-Hadamard codes have been working as SS codes in CDMA communications due to it is generated easily. On the other hand, O-CDMA is considered to be a promising technology because it provides high performance in local area networks along with the assignment of the capacity of network. Spreading codes with required autocorrelation and cross-correlation properties for CDMA have been also improved, like Gold, Kasami, Walsh-Hadamard codes etc. In CDMA/multicarrier CDMA (MC-CDMA), Walsh codes are used due to its orthogonal nature and good auto-correlation property [1-2].

The realm of electronics not only uses presently the design and applications of Walsh-Hadamard codes, but also is extend to optics [3]. The purpose of this work is to implementation of an all-optical walsh-Hadamard code generation using MZI. The main objective is achieved by desigining of Walsh-Hadamard code production with the minimum number of MZI switches. By taking into consideration the advantages of the nonlinear dynamics in semiconductor optical amplifiers (SOA) based MZI, Walsh-Hadamard code generation is presents here. The prime contributions of this work are precisely given as:

• Least number of MZI switches.
• Low power utilization.
• Higher execution speed.

The paper is further oriented as follows: The working principle and tree architecture of MZI are reported in Section II. The Walsh code generation using MZI based switches is also reported in Section III. Section IV shows performance evaluation and ultimately the conclusion and future scope of the work are presented in Section V.

II. WORKING PRINCIPLE AND TREE ARCHITECTURE OF MZI

This section presents the basic working principle of MZI and MZI based tree architecture. High speed (Tb/sec) operation can be achieved by this all-optical scheme [4].

I. *The working principle of MZI based optical switch*

Presently, optical processing operations are performed by interferomatic switches and are considered to be an important technology. Fig. 1 represents the schematic diagram of the SOA based MZI. It is required to divide the incoming signal that is to be switched between the arms of the interferometer such that in the non appearance of a control signal of an MZI [5-9]. In each arm of a MZI, two SOAs are inserted. In this switch, port-1 and port-2 are two input ports, and port-3 and port-4 are two output ports. Coupler C1 effectively splits the incoming signal into two arms allowing simultaneous propagation. The incoming signal which is entering port-1 at first is having a wavelength λ_2. In port-2, it suffers a set aside unlock to the insertion of data. At the same time, to the upper arm, the pulsed signal which has the wavelength λ_1 enters, all the way through coupler C2, the power of the pulsed signal is passes in the upper arm of the MZI. Next, the saturation of SOA-1occures and its index of refraction changes, and on the other hand, there occurs an unsaturated gain at SOA-2. The outcome which is also a differential phase shift occurs between the two arms of the MZI. Hence, in the port-3 (bar port), light is present, as shown in Fig. 1. When cross port is observed i.e. port-4 is considered, no light is present. On the other hand, On the other hand, both SOAs (SOA-1 and SOA-2) bear a same unsaturated gain when the control signal is absent. In this case, light is present in the port-4 (cross port) and no light is present in the port-3 (bar port), For blocking the control signal of wavelength λ_1 signal, in front of the output ports, optical filters (F) are used. Fig. 2 represents the schematic block diagram of the MZI and Table 1 represents the Truth table of Fig. 2.

978-1-5386-6723-1/19 $31.00 © 2019 IEEE

2019 Devices for Integrated Circuit (DevIC), 23-24 March, 2019, Kalyani, India

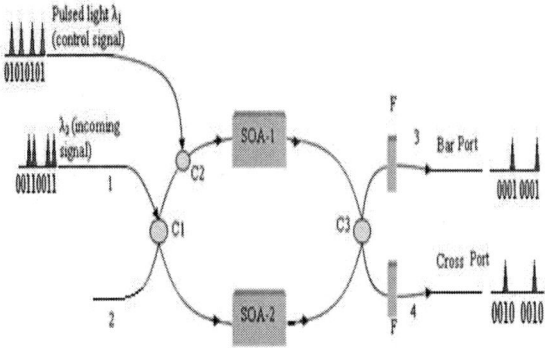

Fig. 1: MZI-based optical switch

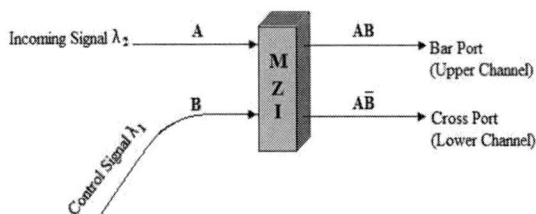

Fig. 2: Block diagram of SOA based MZI switch

Table 1: Truth Table of SOA based MZI switch

Incoming Signal	Control Signal	Bar Port	Cross Port
0	0	0	0
0	1	0	0
1	0	0	1
1	1	1	0

The intensity transmission in bar port and cross port can be represented as

$$E_3(t) = \frac{1}{4} G_1 \{ k_1 k_2 + (1-k_1)(1-k_2) R_G - 2\sqrt{k_1 k_2 (1-k_1)(1-k_2) R_G} \cos(\Delta\Phi) \} \qquad (1)$$

$$E_4(t) = \frac{1}{4} G_1 \{ k_1 (1-k_2) + (1-k_1)k_2 R_G - 2\sqrt{k_1 k_2 (1-k_1)(1-k_2) R_G} \cos(\Delta\Phi) \} \qquad (2)$$

Where k_1 and k_2 are the ratios of the couplers C1 and C2 respectively and $R_G = G_2/G_1$, G_1 and G_2 are the time-dependent gain,

$\Delta\Phi(t) = -\frac{\alpha}{2}\ln\left(\frac{G_2}{G_1}\right)$, α is the linewidth gain factor (taken 7.5 here), For straightforwardness, we take $k_1 = k_2 = 1/2$. The output signal powers at bar port and cross port are,

$$P_j(t) = P_{ip}(t).T_j(t), \qquad j = 3, 4. \qquad (3)$$

Here, the power of the incoming signal is represented by $P_{ip}(t)$. Incoming signal and control signal are concurrently presents then the control pulse saturates SOA-1 under the change in carrier density inside SOA.

II. Tree Architecture of MZIs: System Description of Demultiplexer

The Optical Tree architecture (OTA) can be designed using MZI switches. The first stage OTA is designed by single MZI and the second stage OTA is designed by three MZI's as shown in Fig. 3 (a) and Fig. 3 (b) respectively. First, I_1 and I_2 are the two control signals taking binary values 1 or 0. There are two considered states. One is light's presence which is denoted by one state and on yhe other hand another state is light's absence which is denoted by zero state. Hence we can demonstrate the two control signals I_1 and I_2 in four different wayes. Details operation are shown in Table 2 and Table 3.

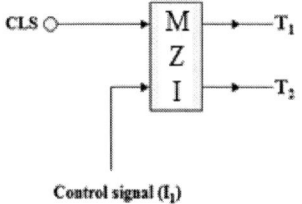

(a)

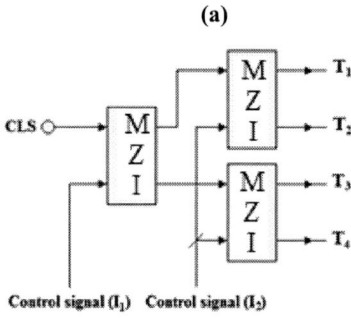

(b)

Fig. 3 (a). The basic building block of MZI, (b). MZI based optical Tree architecture

Table 2. The truth table of MZI based OTA of Fig. 3(a).

Control Signal	Output
I_1	
1	T_1
0	T_2

Table 3. The truth table of MZI based OTA of Fig. 3(b).

Control Signals		Output
I_1	I_2	
1	1	T_1
1	0	T_2
0	1	T_3
0	0	T_4

978-1-5386-6723-1/19 $31.00 © 2019 IEEE   516

III. MZI BASED WALSH CODE GENERATION

The Walsh-Hadamard codes are mutually orthogonal error correcting codes. This code is binary strings of length 'm' to a binary codeword of length 2^m .These are mutually orthogonal and it creates orthogonal spreading codes in the multicarrier CDMA system. The Walsh-Hadamard codes are generate using the following algorithm.

$$H_{M/2} = \begin{bmatrix} H_{M/2} & H_{M/2} \\ H_{M/2} & -H_{M/2} \end{bmatrix}; \text{ where } H_0 = [1].$$ I

It is more convenient to change the encoding alphabet from {1,-1} to {1,0}, We have, Hadamard matrix and inverse Hadamard matrix becomes

$$H_2 = \begin{bmatrix} 1 & 1 \\ 1 & 0 \end{bmatrix}; \qquad -H_2 = \begin{bmatrix} 0 & 0 \\ 0 & 1 \end{bmatrix}$$

Fig. 4 represents the working principle of Walsh-Hadamard code generation. Here, Input A and Input B are the two input coherent light sources and these are also represents I_1 and I_2 signals respectively. Input A and Input B enter into the MZI and then light beams are allowed to incident on (2×2) MZI matrix which is shown below.

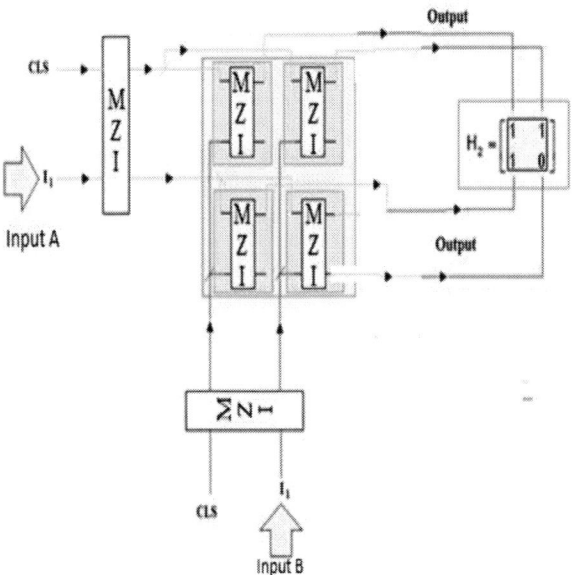

Fig. 4. All-optical Walsh-Hadamard Code (2x2) generation using two inputs.

Now, (2x2) Walsh-Hadamard Code generation with explanation are discussed as follows:
- When light at Input A and Input B are emerging, then the light comes out from C_{11} pixels of (2x2) MZI block. This represents that for A=1 and B=1, the output becomes 1which is also fulfilled in Table 4.
- When light at Input A is present but Input B is absent, then the light emerges from C_{12} pixel of (2x2) MZI block. This represents that for A=1 and B=0, the output becomes 1which is also satisfied in Table 4.
- When light at Input A is absent but light at Input B emerges, then the light comes out from C_{21} pixel of

(2x2) MZI block. This shows that for A=1 and B=1, the output becomes 1which is also satisfied in Table 4.

Table 4. The truth table of the Walsh-Hadamard Code of Fig.4.

A \ B	1	0
1	1	1
0	1	0

- When light at Input A and Input B are absent then no light emerges from C_{22} pixel of (2x2) MZI block. This implies that for A=0 and B=0, the output is 0 as indicated in Table 4.

IV.PERFORMANCE EVOLUTION

This section indicates the performance evaluation through simulation by Matlab-9. Here, Hadamard H2 (2×2) matrix output beam intensity (I ≈ 108 w.m-2) for different values of Input A and Input B:

Case-I: C_{11}, C_{12}, C_{21}, and C_{22} are four pixels in the (2x2) MZI matrix. When light at Input A and Input B are present, then the light will emerge from C_{11} pixel of (2x2) MZI matrix. Only the H_{11} pixel will get light from C_{11} pixel of (2x2) MZI matrix and it is shown in Fig. 5(a) and it indicates the output 1.

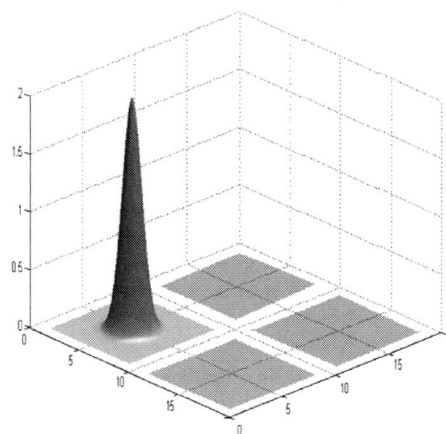

Fig. 5 (a). Walsh-Hadamard matrix: H_{11}=1, H_{12}=0, H_{21}=0, and H_{22}=0, for A= 1 and B = 1.

Case- II: When light at Input A is present but light at Input B is absent, then the light will emerge from C_{12} pixel of (2x2) MZI matrix. Finally, the H_{12} pixel will receive light from C_{12} pixel of (2x2) MZI matrix which is shown Fig. 5(b) and it indicates the output of logical state 1.

2019 Devices for Integrated Circuit (DevIC), 23-24 March, 2019, Kalyani, India

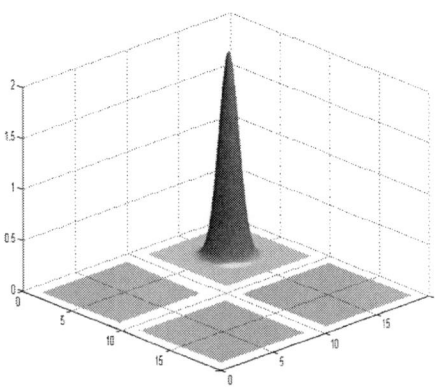

Fig. 5 (b). Walsh-Hadamard matrix: $H_{11}=0$, $H_{12}=1$, $H_{21}=0$, and $H_{22}=0$, for A =1, and B= 0.

Case-III: When light at Input A is absent but light at Input B is present, then the light will emerge from C_{21} pixel of (2x2) MZI matrix. So, only the H_{21} pixel will receive the light from C_{21} pixel of (2x2) MZI matrix as shown in Fig. 5(c) and it is the output of logical state 1.

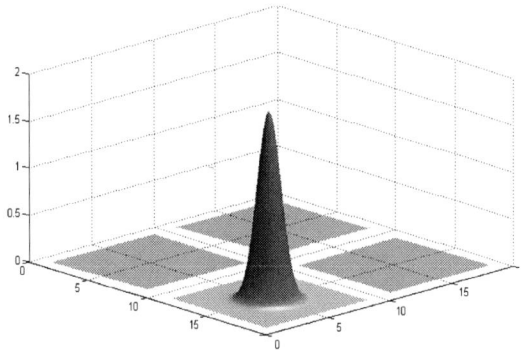

Fig. 5 (c). Walsh-Hadamard matrix: $H_{11}=0$, $H_{12}=0$, $H_{21}=1$, and $H_{22}=0$, for A= 0, and B= 1.

Case-IV: When light at Input A and Input B are absent, then the no light will emerge from C_{22} pixel (2x2) MZI matrix as shown in Fig. 4. Finally, H_{22} pixel will not receive the light from C_{22} pixel of (2x2) MZI matrix as shown in Fig. 5(c) and it is the output of logical state 0.

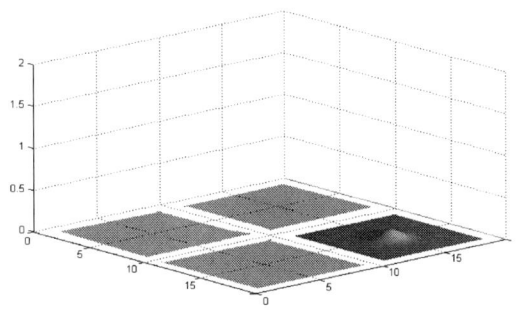

Fig. 5(d). Walsh-Hadamard matrix: $H_{11}=0$, $H_{12}=0$, $H_{21}=0$, and $H_{22}=0$, for A = 0, and B =0.

V. CONCLUSION AND FUTURE SCOPE OF WORK

This paper represents all-optical Walsh-Hadamard codes generation using Mach–Zehnder interferometer (MZI). The design of the Walsh-Hadamard code generator in all-optical domain has been discussed and demonstrated. The basic block of code for this module is the MZI. Walsh codes encodes n bit messages into 2^n bit orthogonal codewords. Walsh-Hadamard codes have been orthogonality and good auto-correlation property, so it is used extensively in CDMA/Multi-carrier CDMA (MC-CDMA). Walsh code can also be used for optical code division multiplexing (OCDM).

REFERENCES

[1] B M. Popovic, "Spreading sequences for multicarrier CDMA systems", IEEE Trans. Communication, 47, 1999.

[2] H. J. Caulfield, C. S. Vikram and A. Zavalin, "Optical Logic Redux," *Optik*, 117, 199-209, (2006).

[3] T. Chattopadhyay, G. K. Maity and J. N. Roy, "Designing of the all-optical tri-state logic system with the help of optically nonlinear material", Journal of Nonlinear Optical Physics & Materials, pp.315-328, 17(3), (2008).

[4] J. N. Roy, "Mach-Zehnder interferometer based tree architecture for all-optical logic and arithmetic operations Optik-International Journal for Light and Electron Optics 120(7), 318(2009).

[5] G. K. Maity, S. P. Maity, T. Chattopadhyay, J. N. Roy, "Mach-Zehnder Interferometer Based All-Optical Fredkin Gate," *International Conference on Trends in Optics and Photonics*, Kolkata, India, (2009).

[6] G. K. Maity, J. N. Roy, S. P. Maity, " Mach-Zehnder Interferometer based all-optical Peres Gate," *International Conference on Advances in Computing and Communications*, Kochi, India, 22-24, (2011).

[7] G. K. Maity, S. P. Maity, J. N. Roy: *MZI based Modified Trinary Number System*. International Conference on Computer, Communication, Control and Information Technology (C3IT-2012), Elsevier, (2012).

[8] Ashis Kumar Mandal, Supriti Samanta, Goutam Kumar Maity, Nabin Baran Manik, "Implementation of Reed-Muller Expansion Technique Using Mach-Zehnder Interferometer Based All-Optical Reversible Gate" *Advances in Optical Science and Engineering, (Springer), 166, 497-505, (2015).*

[9] Debarati Nath, Puja Dey, Debajit Deb, Jayanta Kumar Rakshit, Jitendra Nath Roy, "Fabrication and characterization of organic semiconductor based photodetector for optical communication"CSI transactions on ICT (Springer), 5(2), pp. 149-160, (2017).

978-1-5386-6723-1/19 $31.00 © 2019 IEEE

Author Index

A. Pandey	35	Arifuddin Sohel	124
Aashita Raj	413	Arindam Biswas	9
Aayesha Al Khadir	124	Arindam Biswas	1
Abhijit Das	11	Aritra Acharyya	1
Abhishek Kumar Singh	39	Arko Majumder	441
Abhishek Kumar Singh	35	Arnab Banerjee	501
Abinash Thapa	470	Arnab Mukhopadhyay	131
Ajit K Panda	134	Arpan Deyasi	108
Ajit Panda	61	Arpan Deyasi	252
Akash Roy	186	Arpan Deyasi	26
Akho John Richa	75	Arpan Deyasi	137
Alekhya Yalla	211	Arpan Deyasi	264
Aloke Saha	298	Arttatran Sahu	61
Alvin Karkun	428	Arun Kumar	89
Amena Najeeb	124	Ashis Kumar Das	181
Amit Kumar Sharma	331	Ashis Kumar Mal	306
Amogh Banerjee	196	Ashuntosh Anand	382
Amretashis Sengupta	131	Ashuntosh Anand	465
Angsuman Sarkar	26	Ashutosh Anand	370
Angsuman Sarkar	137	Ashutosh Kumar Dikshit	35
Angsuman Sarkar	156	Ashutosh Kumar Dikshit	39
Angsuman Sarkar	264	Asish K Mukhopadhyay	16
Angsuman Sarkar	269	Asish K Mukhopadhyay	167
Angsuman Sarkar	269	Asmit Amlan Sahoo	510
Anirban Ghosal	318	Atanu Kundu	186
Anirban Neogi	261	Avijit Chakraborty	361
Anirbit Sengupta	11	Avijit Chakraborty	348
Aniruddha Ghosal	128	Avik Chakraborty	264
Aniruddha Ghosal	151	Avtar Singh	322
Aniruddha Ghosal	151	Avtar Singh	337
Anirudh Aggarwal	221	Ayan Bhattacharya	206
Anirudha Ray	374	Ayan Saha	318
Anisha Pathania	394	B. Maji	167
Anisha Pathania	390	Babita Kumari	128
Anita Pal	9	Bansibadan Maji	16
Anita Pal	9	Bansibadan Maji	16
Ankita Mukherjea	428	Bijoy Goswami	206
Anup Dey	202	Bikash Sharma	470
Anup Kr. Bhattacharjee	9	Bikram Biswas	202
Anupam Tarafdar	261	Bishnu Prasad De	334
Anwesha Adhikary	26	Biswajit Baral	493
Archak Sadhukhan	146	Biswajit Baral	505

3rd international conference on "2019 Devices for Integrated Circuit (DevIC)"

Biswajit Baral	510	Deependra Chettri	470
Biswajit Baral	310	Deepneha Dutta	161
Biswajit Baral	314	Deobrata Kumar	141
Biswajit Baral	365	Dhananjaya Tripathy	314
Biswajit Mahanty	348	Dhananjaya Tripathy	365
Biswarup Ganguly	441	Dhananjaya Tripathy	505
Biswarup Neogi	374	Dhananjaya Tripathy	310
Biswarup Neogi	404	Dhananjaya Tripathy	510
Brahmdutta Dixit	35	Dhananjaya Tripthy	493
Buddhadev Pradhan	42	Dibyendu Chowdhury	334
C. K. Maiti	21	Dilip Kumar	341
C. K. Maiti	70	Dilip Kumar	177
Chanchal Dey	172	Dilip Kumar	141
Chandan K. Ghosh	480	Dinesh Kumar Dash	118
Chandan K. Sarkar	480	Dinesh Kumar Dash	240
Chandan Kumar Pandey	216	Dipanjan Sen	386
Chandan Kumar Pandey	322	Dipanjan Sen	398
Chandan Kumar Pandey	337	Dr. Kaushik Mazumdar	128
Chandan Kumar Pandey	245	Dr. Tapas Halder	474
Chandan Kumar Sarkar	248	Durbadal Mandal	334
Chandan Kumar Sarkar	322	E. Mohapatra	21
Chandan Kumar Sarkar	470	E. Mohapatra	70
Chander Pratap Singh	394	Eleena Mohapatra	65
Chinmay Kumar Maiti	65	Eleena Mohapatra	94
Chinmay Kumar Maiti	94	Eleena Mohapatra	99
Chinmay Kumar Maiti	99	Eleena Mohapatra	291
Chinmay Kumar Maiti	286	Eleena Mohapatra	326
Chinmay Kumar Maiti	291	Eleena Mohapatra	286
Chinmay Kumar Maiti	326	Gaurav Saini	484
Chiradeep Mukherjee	167	Geetanjali Singh	413
Dalia Nandidas	404	Goutam Kumar Maity	515
Debadipta Basak	206	Goutam Kumar Maity	489
Debarati Chakraborty	264	Govind Murmu	378
Debashis Chatterjee	356	Guru Prasad Mishra	45
Debashish Dash	216	Guru Prasad Mishra	45
Debasish Nayak	302	Guru Prasad Mishra	49
Debasish Nayak	310	Guru Prasad Mishra	53
Debasish Nayak	493	Guru Prasad Mishra	461
Debasish Nayak	314	H S Jatana	424
Debasish Nayak	505	Hafizur Rahaman	131
Debasish Nayak	510	Hasanujjaman	497
Debasish Nayak	510	Hasanujjaman	501
Deborsi Basu	433	Himadri Mandal	489
Dedasish Nayak	365	Inamul Hussain	245

3rd international conference on "2019 Devices for Integrated Circuit (DevIC)"

Indrajit Das	196	Lalthanpuii Khiangte	295
Indrajit Sil	446	M. K. Naskar	497
Indranath Sarkar	318	Maitraiyee Konar	408
Indranil Kushary	146	Maliha Naaz	124
Indronil Mazumder	256	Manisha Sharma	341
J. Jena	21	Md Golam Mohiuddin	196
J. N. Roy	1	Md. Mostafizur Rahman	456
J. R. Jena	70	Md. Tausif Mallick	11
J.S. Rana	35	Mehebub Alam	113
J.S. Rana	39	Mohinder Bassi	84
Jamuna Kanta Sing	236	Monu Bhagat	141
Jaya Madan	394	Monu Bhagat	341
Jayanta Ghosh	298	Moumita Ghosh	374
Jhansirani Jena	65	Mrinal Kanti Naskar	501
Jhansirani Jena	94	Muktasha Maji	484
Jhansirani Jena	286	Namrata Dhanda	79
Jhansirani Jena	99	Narayan Sahoo	134
Jhansirani Jena	291	Neeraj Gupta	331
Jhansirani Jena	326	Nidhi Taxak	451
Jyoti Ranjan Panda	261	Nikhil Kumar Das	370
Kabita Purkait	433	Nilanjan Bhattacharjee	318
Kajal Maji	30	Nillohit Mukhrjee	39
Kajal Maji	30	Nisarga Chand	156
Kalyan Biswas	446	Nitin Krishna Mucheli	302
Kamal	35	P. Chakrabarti	35
Kamal	39	P. Chakrabarti	39
Kanchan Baran Maji	334	P. K. Rout	302
Kaushik Roy	480	P.S.T.N. Srinivas	89
Kaustav Saha	374	Pampa Debnath	137
Kavindra Kandpal	221	Papri Chakraborty	26
Khomdram Jolson Singh	281	Parshati Chaudhury	428
Khuraijam Nelson Singh	231	Partha Sarathi Sarkar	361
Koelgeet Kaur	206	Phuntso Chotten	75
Kousik Mukherjee	58	Poulami Jana	489
Kousik Mukherjee	58	Pradip Das	146
Kousik Mukherjee	30	Prakash Kumar Rout	314
Koustav Kumar Mondol	361	Prakash Marimuthu	5
Krishna Sarker	348	Prakash Marimuthu	5
Krishna Sarker	356	Pramod Kumar Tiwari	89
Krishna Sarker	361	Pranab Kishore Dutta	231
Kulbhushan Sharma	394	Prashant Kumar	181
Kulbhushan Sharma	424	Prashant Kumar	331
Kulbhushan Sharma	390	Prashant Kumar	191
Kumar Ritu Raj Singh	177	Prasun Kumar Sinha	370

3rd international conference on "2019 Devices for Integrated Circuit (DevIC)"

Preeti Topno	378	Sangita Das	252
Pritam Gayen	146	Sangita Panda	61
Pritha Banerjee	118	Sanjit Kumar Swain	302
Pritha Banerjee	240	Sanjit Kumar Swain	310
Priyanka Saha	240	Sanjit Kumar Swain	314
Priyanka Saha	118	Sanjit Kumar Swain	365
Pronita Chettri	470	Sanjit Kumar Swain	493
Rabindranath Das Adhikary	474	Sanjit Kumar Swain	156
Rahul Pal	298	Sanjit Kumar Swain	510
Rajat Mahapatra	306	Sanjit Swain	505
Rajib Kar	334	Saptarshi Bhui	441
Rajib Majumdar	146	Saptarshi Biswas	236
Rajnish Sharma	390	Saptarshi Sil	261
Rajnish Sharma	394	Saradindu Panda	16
Rajnish Sharma	424	Saradindu Panda	167
Rajrup Mitra	186	Sarosij Adak	156
Rashmi Gupta	331	Sasmita Sahoo	461
Rashmi Sahu	408	Satish Kumar Das	310
Ratul Ghosh	26	Satish Kumar Das	314
Ritesh Prasad	177	Satish Kumar Das	505
Rohit K Singh	39	Satish Kumar Das	365
Rudra Dhar	295	Satish Kumar Das	302
Rudra Sankar Dhar	281	Satish Kumar Das	493
Rupam Goswami	221	Satish Kumar Das	510
S. Das	70	Saurabh Chaudhury	245
S. Das	21	Saurabh Chaudhury	337
S. Dey	21	Saurabh Chaudhury	216
S. Dey	70	Saurabh Chaudhury	322
S.K Goswami	356	Savio Jay Sengupta	398
Sabuj Sarkar	456	Savio Sengupta	386
Sachin Singhal		Sayan Chatterjee	275
Sagar Mukherjee	446	Sayan Hazra	226
Saikat Adhikary	456	Sayan Paramanik	348
Sajib Chakraborty	361	Sayan Paramanik	356
Samik Samanta	306	Sayan Paramanik	361
Samrat Ghosh	441	Shalini Singh	196
Sandipan Dutta	441	Sharma Preeti	424
Sangeeta K Palo	134	Shekhar Verma	84
Sanghamitra Das	99	Shekhar Verma	79
Sanghamitra Das	286	Shibaditya Chakraborty	261
Sanghamitra Das	291	Shingmila Hungyo	281
Sanghamitra Das	326	Shivam Sarkar	374
Sanghamitra Das	94	Shrabanti Das	275
Sanghamitra Dash	65	Shubhrajyoti Kundu	113

3rd international conference on "2019 Devices for Integrated Circuit (DevIC)"

Siddhartha Sankar Thakur	113		Sudip Chakraborty	1
Sidhartha Dash	461		Sudip Kundu	370
Sikha Mishra	49		Sudip Kundu	382
Sikha Mishra	53		Sudip Kundu	408
Smita Rai	470		Sudip Kundu	413
Somak Karan	172		Sudip Kundu	465
Sonali Gupta	196		Sumalya Ghosh	334
Soumalya Ghosh	428		Suman Halder	181
Soumita De	252		Suman Halder	191
Soumya Mohanty	53		Suman Lata Tripathi	79
Soumya Mohanty	49		Suman Lata Tripathi	84
Sourav Basak	104		Sumit Baneerjee	113
Sourav Naval	370		Sunipa Roy	248
Srijan Bhar	318		Suporna Bhowmick	264
Subhadipta Mukhopadhyay	515		Suprava Dey	99
Subhajit Bhowmick	428		Suprava Dey	286
Subham Banerjee	42		Suprava Dey	291
Subham Ghosh	318		Suprava Dey	326
Subham Ghosh	374		Suprava Dey	65
Subhashis Maitra	161		Suprava Dey	94
Subhashis Roy	202		Supriti Samanta	515
Subhashis Roy	386		Surbhi Singh	177
Subhashis Roy	398		Susanta Kumar Tripathy	216
Subhashis Roy	202		Susmita Das	404
Subhashree Bhol	49		Sutanni Bhowmick	206
Subhradip Mondal	146		Suvam Sasmal	441
Subhrapratim Nath	236		Swapan Das	248
Subir Kumar Sarkar	118		Swaraj Banerjee	104
Subir kumar Sarkar	202		Swarnadeep Mukherjee	441
Subir Kumar Sarkar	206		Swarnav Mukhopadhyay	108
Subir Kumar Sarkar	236		Sweta Mohanty	53
Subir Kumar Sarkar	240		T. P. Dash	21
Subir Kumar Sarkar	398		T.P. Dash	70
Subir Sarkar	386		Tapas Halder	418
Sudhabindu Ray	386		Tara Prasanna Das	65
Sudhabindu Ray	398		Tara Prasanna Dash	94
Sudhakar Das	61		Tara Prasanna Dash	326
Sudhansu Mohan Biswal	302		Tara Prasanna Dash	99
Sudhansu Mohan Biswal	314		Tara Prasanna Dash	286
Sudhansu Mohan Biswal	365		Tara Prasanna Dash	291
Sudhansu Mohan Biswal	505		Tarini C Tripathy	134
Sudhansu Mohan Biswal	510		Trinath Sahu	61
Sudhansu Mohan Biswal	510		Trinath Sahu	134
Sudhansu Mohan Biswal	310		U. Biswas	497

3rd international conference on "2019 Devices for Integrated Circuit (DevIC)"

Umakanta Nanda	211
Umakanta Nanda	302
Umakanta Nanda	314
Umakanta Nanda	365
Umakanta Nanda	493
Umakanta Nanda	510
Urmi Dey	252
Utpal Biswas	501
V.V. Kharche	35
Virendra Prasad Maurya	191
Writwik Balow	474
Yaswanth Divvela	413

IEEE
445 Hoes Lane
Piscataway, NJ 08854-4141

ISBN 978-1-5386-6723-1